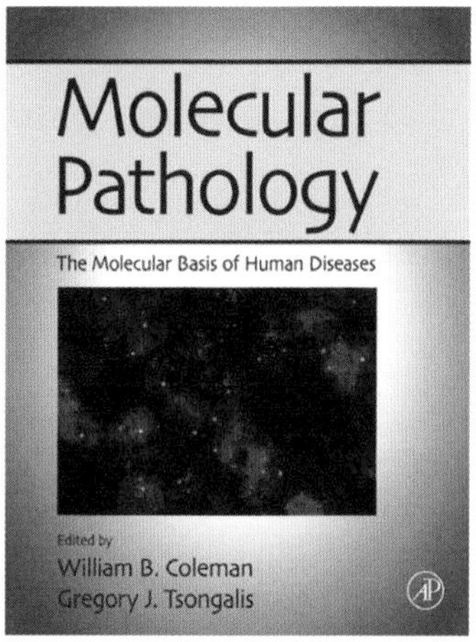

GW00994927

As a FREE special offer to
we are offering over 300 online :

EXAM MASTER!

Please be sure to take advantage of this purchase bonus.

EXAM MASTER OnLine™ Access Instructions

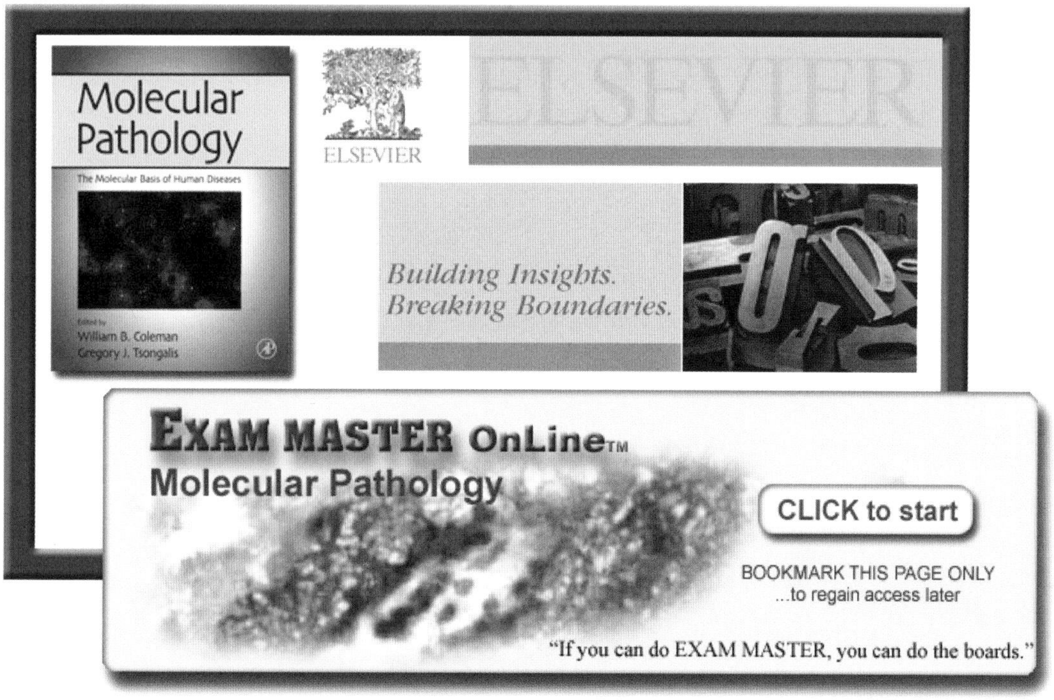

Gateway Code: http://www.exammaster2.com/wdsentry/elsevier.htm

EM OnLine Registration:

1. Go to your custom gateway: http://www.exammaster2.com/wdsentry/elsevier.htm
   - Click **"Click to Start"** and Click **"Not Registered Yet?"**
   - To continue, click **[I Accept]** on the Licensing Agreement page
2. Complete the registration form, click **[Submit Registration]**
3. Confirmation of registration and your temporary password will be e-mailed to you immediately
4. Log-in with your "User Name" and "Temporary Password"
5. The **Welcome Page** explains how to access the chapter questions

EXAM MASTER Corporation
www.exammaster.com

Academic Press is an imprint of Elsevier
30 Corporate Drive, Suite 400, Burlington, MA 01803, USA
525 B Street, Suite 1900, San Diego, California 92101-4495, USA
84 Theobald's Road, London WC1X 8RR, UK

Copyright © 2009, Elsevier Inc. All rights reserved.

No part of this publication may be reproduced or transmitted in any form or by any means, electronic or mechanical, including photocopy, recording, or any information storage and retrieval system, without permission in writing from the publisher.

Permissions may be sought directly from Elsevier's Science & Technology Rights Department in Oxford, UK: phone: (+44) 1865 843830, fax: (+44) 1865 853333, E-mail: permissions@elsevier.com. You may also complete your request online via the Elsevier homepage (http://elsevier.com), by selecting "Support & Contact" then "Copyright and Permission" and then "Obtaining Permissions."

Notice

Medicine is an ever-changing field. Standard safety precautions must be followed, but as new research and clinical experience broaden our knowledge, changes in treatment and drug therapy may become necessary or appropriate. Readers are advised to check the most current product information provided by the manufacturer of each drug to be administered to verify the recommended dose, the method and duration of administrations, and contraindications. It is the responsibility of the treating physician, relying on experience and knowledge of the patient, to determine dosages and the best treatment for each individual patient. Neither the publisher nor the authors assume any liability for any injury and/or damage to persons or property arising from this publication.

Library of Congress Cataloging-in-Publication Data
APPLICATION SUBMITTED

British Library Cataloguing-in-Publication Data
A catalogue record for this book is available from the British Library.

ISBN: 978-0-12-374419-7
For information on all Academic Press publications
visit our Web site at www.elsevierdirect.com

Printed in China
08 09 10 9 8 7 6 5 4 3 2 1

Working together to grow
libraries in developing countries

www.elsevier.com | www.bookaid.org | www.sabre.org

ELSEVIER BOOK AID International Sabre Foundation

Molecular Pathology

The Molecular Basis of Human Disease

Edited by

William B. Coleman, Ph.D.

Department of Pathology and Laboratory Medicine
UNC Lineberger Comprehensive Cancer Center
University of North Carolina School of Medicine
Chapel Hill, NC

Gregory J. Tsongalis, Ph.D.

Department of Pathology
Dartmouth Medical School
Dartmouth Hitchcock Medical Center
Norris Cotton Cancer Center
Lebanon, NH

ELSEVIER

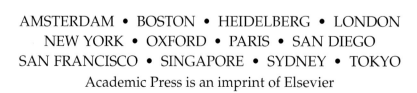

AMSTERDAM • BOSTON • HEIDELBERG • LONDON
NEW YORK • OXFORD • PARIS • SAN DIEGO
SAN FRANCISCO • SINGAPORE • SYDNEY • TOKYO
Academic Press is an imprint of Elsevier

Molecular Pathology

Dedication

The wealth of information contained in this textbook represents the culmination of innumerable small successes that emerged from the ceaseless pursuit of new knowledge by countless experimental pathologists working around the world on all aspects of human disease. Their ingenuity and hard work have dramatically advanced the field of molecular pathology over time and in particular in the last two decades. This book is a tribute to the dedication, diligence, and perseverance of the individuals who have contributed to the advancement of our understanding of the molecular basis of human disease, especially the graduate students, laboratory technicians, and postdoctoral fellows, whose efforts are so frequently taken for granted, whose accomplishments are so often unrecognized, and whose contributions are so quickly forgotten.

Molecular Pathology: The Molecular Basis of Human Disease is dedicated to the memory of two bright, resourceful, hard-working, young molecular pathologists who were taken from life at an early age: Dr. Rhonda Simper Ronan (who passed away on September 3, 2007) and Dr. Sharon Ricketts Betz (who passed away on July 24, 2008). Both Rhonda and Sharon were accomplished experimental pathologists that were on their way to promising careers in molecular pathology. They were dear friends and cherished colleagues to many people at their respective institutions. We share the sadness of their families, friends, and colleagues. This book is dedicated to their courage and example, and to the inspiration that they provide. Their memory will forever remind us that there is too much work left to be done for us to rest on our accomplishments.

We also dedicate *Molecular Pathology: The Molecular Basis of Human Disease* to the many people that have played crucial roles in our successes. We thank our many scientific colleagues, past and present, for their camaraderie, collegiality, and support. We especially thank our scientific mentors for their example of research excellence. We are truly thankful for the positive working relationships and friendships that we have with our faculty colleagues. We also thank our students for teaching us more than we may have taught them. We thank our parents for believing in higher education, for encouragement through the years, and for helping our dreams into reality. We thank our brothers and sisters, and extended families, for the many years of love, friendship, and tolerance. We thank our wives, Monty and Nancy, for their unqualified love, unselfish support of our endeavors, understanding of our work ethic, and appreciation for what we do. Lastly, we give a special thanks to our children, Tess, Sophie, Pete, and Zoe, for providing an unwavering bright spot in our lives, for their unbridled enthusiasm and boundless energy, for giving us a million reasons to take an occasional day off from work just to have fun.

William B. Coleman
Gregory J. Tsongalis

Contents

PART

IV Molecular Pathology of Human Disease

List of Contributors

Dara L. Aisner, M.D., Ph.D.
Department of Pathology and Laboratory Medicine, University of Pennsylvania School of Medicine, Hospital of the University of Pennsylvania, Philadelphia, PA, USA

M. Michael Barmada, Ph.D.
Department of Human Genetics, Graduate School of Public Health, University of Pittsburgh, Pittsburgh PA, USA

Philippe Bedard, M.D.
Translational Research Unit, Jules Bordet Institute Université, Libre de Bruxelles, Brussels, Belgium

David O. Beenhouwer, M.D.
Department of Medicine, David Geffen School of Medicine at University of California, and Division of Infectious Diseases, Veterans Affairs Greater Los Angeles Healthcare System, Los Angeles, CA

Jaideep Behari, M.D., Ph.D.
Department of Medicine, Division of Gastroenterology, Hepatology, and Nutrition, University of Pittsburgh School of Medicine, Pittsburgh, PA, USA

Maria Berdasco, M.D.
Cancer Epigenetics and Biology Program, Catalan Institute of Oncology, Barcelona, Catalonia, Spain

Carlise R. Bethel, Ph.D.
Sidney Kimmel Comprehensive Cancer Center, Department of Pathology, and the Brady Urological Research Institute, Johns Hopkins University School of Medicine, Baltimore, MD, USA

Joseph R. Biggs, Ph.D.
Departments of Pathology and Biological Sciences, University of California, San Diego, La Jolla, CA, USA

Grant C. Bullock, M.D., Ph.D.
Department of Pathology, University of Virginia Health System, Charlottesville, VA, USA

Sheldon M. Campbell, M.D., Ph.D., F.C.A.P.
Department of Laboratory Medicine, Yale University School of Medicine Pathology and Laboratory Medicine, VA Connecticut Healthcare System, West Haven, CT, USA

Wai-Yee Chan, Ph.D.
Laboratory of Clinical Genomics, National Institute of Child Health and Human Development, NIH, Bethesda, MD, and Departments of Pediatrics, Biochemistry & Molecular Biology, Georgetown University School of Medicine, Washington, D.C., USA

William B. Coleman, Ph.D.
Professor and Director of Graduate Studies, Department of Pathology and Laboratory Medicine, Curriculum in Toxicology, Program in Translational Medicine, UNC Lineberger Comprehensive Cancer Center, University of North Carolina School of Medicine, Chapel Hill, NC, USA

Angelo M. De Marzo, M.D., Ph.D.
Sidney Kimmel Comprehensive Cancer Center, Department of Pathology, and the Brady Urological Research Institute, Johns Hopkins University School of Medicine, Baltimore, MD, USA

Phuong Dinh, M.D.
Translational Research Unit, Jules Bordet Institute Université, Libre de Bruxelles, Brussels, Belgium

Vladislav Dolgachev, Ph.D.
Department of Pathology, University of Michigan Medical School, Ann Arbor, MI, USA

Virginia Espina, M.S., M.T.
George Mason University, Center for Applied Proteomics and Molecular Medicine Manassas, VA, USA

Manel Esteller, Ph.D
Cancer Epigenetics and Biology Program, Catalan Institute of Oncology, Barcelona, Catalonia, Spain

Carol Farver, M.D.
Director, Pulmonary Pathology, Vice-Chair for Education, Pathology and Laboratory Medicine Institute, Department of Anatomic Pathology, Cleveland Clinic Foundation, Cleveland, OH, USA

Claudia Fredolini, Ph.D.
George Mason University, Center for Applied Proteomics and Molecular Medicine Manassas, VA, USA

William K. Funkhouser, M.D., Ph.D.
Department of Pathology and Laboratory Medicine, University of North Carolina School of Medicine, Chapel Hill, NC, USA

Matthias E. Futschik, Ph.D.
Institute for Theoretical Biology, Charité, Humboldt-Universität, Berlin, Germany, Centre for Molecular and Structural Biomedicine, University of Algarve, Faro, Portugal

Avrum I. Gotlieb, M.D.C.M.
Department of Pathology, Toronto General Research Institute, University Health Network, Department of Laboratory Medicine and Pathobiology, University of Toronto, Toronto, Ontario, Canada

Robert Hevner, Ph.D., M.D.
Professor, Neurological Surgery/Neuropathology, University of Washington School of Medicine, USA

W. Edward Highsmith, Ph.D.
Molecular Genetics Laboratory, Department of Laboratory Medicine and Pathology, Mayo Clinic, Rochester, MN, USA

C. Dirk Keene, M.D., Ph.D.
Associate Professor, Neurological Surgery/Neuropathology, University of Washington School of Medicine, USA

Wolfgang Kemmner, Ph.D.
Department of Surgery and Surgical Oncology, Robert-Rössle-Klinik Berlin, Charité – Universitätsmedizin Berlin, Berlin, Germany

Nigel S. Key, M.D.
Department of Medicine, University of North Carolina School of Medicine, Chapel Hill, NC, USA

Hong Kee Lee, Ph.D.
Department of Pathology, Dartmouth Medical School, Dartmouth Hitchcock Medical Center and Norris Cotton Cancer Center, Lebanon, NH, USA

Joel A. Lefferts, Ph.D.
Department of Pathology, Dartmouth Medical School, Dartmouth Hitchcock Medical Center and Norris Cotton Cancer Center, Lebanon, NH, USA

John J. Lemasters, M.D., Ph.D.
Center for Cell Death, Injury and Regeneration, Departments of Pharmaceutical & Biomedical Sciences and Biochemistry & Molecular Biology, Medical University of South Carolina, Charleston, SC, USA

Markus M. Lerch, M.D
Department of Internal Medicine A, Ernst-Moritz-Arndt-Universität Greifswald, Greifswald, Germany

Lance Liotta, Ph.D.
George Mason University, Center for Applied Proteomics and Molecular Medicine, Manassas, VA, USA

Amber Chang Liu, M.Sc.
Department of Pathology, Toronto General Research Institute, University Health Network, Department of Laboratory Medicine and Pathobiology, University of Toronto, Toronto, Ontario, Canada

Karen Lu, M.D.
Department of Gynecologic Oncology, University of Texas M.D. Anderson Cancer Center, Houston, TX, USA

Nicholas W. Lukacs, Ph.D.
Professor of Pathology, Director Molecular and Cellular Pathology Graduate Program University of Michigan Medical School, Ann Arbor, MI, USA

Alice Ma, M.D.
Department of Medicine, University of North Carolina School of Medicine, Chapel Hill, NC, USA

Arlene Martin, M.D.
Department of Pediatrics, Georgetown University School of Medicine, Washington, DC, USA

Malcolm M. Martin, M.D.
Department of Pediatrics, Georgetown University School of Medicine, Washington, DC, USA

Julia Mayerle
Department of Internal Medicine A, Ernst-Moritz-Arndt-Universität Greifswald, Greifswald, Germany

John A. McGrath, M.D. FRCP
Genetic Skin Disease Group, St John's Institute of Dermatology, King's College London, Guy's Campus, London, UK

Kara A. Mensink, M.S.
Molecular Genetics Laboratory, Department of Laboratory Medicine and Pathology, Mayo Clinic, Rochester, MN, USA

Samuel Chi-ho Mok, Ph.D.
Department of Gynecologic Oncology, University of Texas, M.D. Anderson Cancer Center, Houston, TX, USA

Satdarshan (Paul) Singh Monga, M.D.
Director-Division of Experimental Pathology, Associate Professor of Pathology and Medicine, University of Pittsburgh, School of Medicine Pittsburgh, PA, USA

Thomas J. Montine, M.D., Ph.D.
Division of Neuropathology, Department of Pathology, University of Washington, Harborview Medical center, Seattle, WA, USA

Jason H. Moore, Ph.D.
Computational Genetics Laboratory, Norris-Cotton Cancer Center, Departments of Genetics and Community and Family Medicine, Dartmouth Medical School, Lebanon, NH

Karl Münger, Ph.D.
Department of Medicine, Brigham and Women's Hospital, Harvard Medical School, Boston, MA, USA

Zoltan Nagymanyoki, M.D., Ph.D.
Division of Gynecologic Oncology, Department of Obstetrics and Gynecology, Brigham and Women's Hospital, Harvard Medical School, Boston, MA, USA

William G. Nelson, M.D., Ph.D.
Sidney Kimmel Comprehensive Cancer Center, Department of Oncology, and the Brady Urological Research Institute, Johns Hopkins University School of Medicine, Baltimore, MD, USA

Margret Oethinger, M.D., Ph.D.
Department of Clinical Pathology, Clinical Microbiology Section, Cleveland Clinic, Cleveland, OH, USA

Alan LP Pang, Ph.D.
Laboratory of Clinical Genomics, National Institute of Child Health and Human Development, NIH, Bethesda, MD, USA

Emanuel Petricoin, Ph.D.
George Mason University, Center for Applied Proteomics and Molecular Medicine, Manassas, VA, USA

Ashley Rivenbark, Ph.D.
UNC Lineberger Comprehensive Cancer Center, Department of Biochemistry and Biophysics, University of North Carolina at Chapel Hill, Chapel Hill, NC, USA

C. Harker Rhodes, M.D., Ph.D.
Norris-Cotton Cancer Center, Department of Pathology, Dartmouth Medical School, Lebanon, NH

Tara Rubinas, M.D.
Department of Pathology and Laboratory Medicine, University of North Carolina School of Medicine, Chapel Hill, NC, USA

Reinhold Schafer, Ph.D.
Laboratory of Molecular Tumor Pathology, Charité – Universitätsmedizin Berlin, Berlin, Germany

Matthias Sendler
Department of Internal Medicine A, Ernst-Moritz-Arndt-Universität Greifswald, Greifswald, Germany

Antonia Sepúlveda, M.D., Ph.D.
Department of Pathology and Laboratory Medicine, University of Pennsylvania School of Medicine, Hospital of the University of Pennsylvania, Philadelphia, PA, USA

Christine Sers, Ph.D.
Laboratory of Molecular Tumor Pathology, Charité – Universitätsmedizin Berlin, Berlin, Germany

Lawrence M. Silverman, Ph.D.
Department of Pathology, University of Virginia Health System, Charlottesville, VA, USA

Natasha Snoj, M.D.
Translational Research Unit, Jules Bordet Institute Université, Libre de Bruxelles, Brussels, Belgium

Joshua A. Sonnen, Ph.D.
Department of Pathology, Neuropathology Division, University of Washington, USA

Christos Sotiriou, M.D., Ph.D.
Translational Research Unit, Jules Bordet Institute Université, Libre de Bruxelles, Brussels, Belgium

Gregory J. Tsongalis, Ph.D.
Department of Pathology, Dartmouth Medical School, Dartmouth Hitchcock Medical Center and Norris Cotton Cancer Center, Lebanon, NH, USA

Vesarat Wessagowit, M.D., Ph.D.
The Institute of Dermatology, Rajvithi Phyathai, Bangkok, Thailand

David C. Whitcomb, Ph.D.
Department of Human Genetics, Graduate School of Public Health, University of Pittsburgh,
Division of Gastroenterology, Hepatology, and Nutrition, Department of Medicine, University of Pittsburgh Medical Center, Pittsburgh, PA, USA

Kwong-kwok Wong, Ph.D.
Department of Gynecologic Oncology, University of Texas M.D. Anderson Cancer Center, Houston, TX, USA

Dani S. Zander, M.D.
Professor and Chair of Pathology, University Chair in Pathology, Penn State Milton S. Hershey Medical Center/ Penn State University College of Medicine, Department of Pathology, Hershey, PA, USA

Dong-Er Zhang, Ph.D.
Departments of Pathology and Biological Sciences, University of California, San Diego, La Jolla, CA, USA

Weidong Zhou, M.D., Ph.D.
George Mason University, Center for Applied Proteomics and Molecular Medicine, Manassas, VA, USA

Preface

Pathology is the study of disease. The field of pathology emerged from the application of the scientific method to the study of human disease. Thus, pathology as a discipline represents the complimentary intersection of medicine and basic science. Early pathologists were typically practicing physicians who described the various diseases that they treated and made observations related to factors that contributed to the development of these diseases. The description of disease evolved over time from gross observation to microscopic inspection of diseased tissues based upon the light microscope, and more recently to the ultrastructural analysis of disease with the advent of the electron microscope. As hospital-based and community-based registries of disease emerged, the ability of investigators to identify factors that cause disease and assign risk to specific types of exposures expanded to increase our knowledge of the epidemiology of disease. While descriptive pathology can be dated to the earliest written histories of medicine and the modern practice of diagnostic pathology dates back perhaps 200 years, the elucidation of mechanisms of disease and linkage of disease pathogenesis to specific causative factors emerged more recently from studies in experimental pathology. The field of experimental pathology embodies the conceptual foundation of early pathology – the application of the scientific method to the study of disease – and applies modern investigational tools of cell and molecular biology to advanced animal model systems and studies of human subjects. Whereas the molecular era of biological science began over 50 years ago, recent advances in our knowledge of molecular mechanisms of disease have propelled the field of molecular pathology. These advances were facilitated by significant improvements and new developments associated with the techniques and methodologies available to pose questions related to the molecular biology of normal and diseased states affecting cells, tissues, and organisms. Today, molecular pathology encompasses the investigation of the molecular mechanisms of disease and interfaces with translational medicine where new basic science discoveries form the basis for the development of new strategies for disease prevention, new therapeutic approaches and targeted therapies for the treatment of disease, and new diagnostic tools for disease diagnosis and prognostication.

With the remarkable pace of scientific discovery in the field of molecular pathology, basic scientists, clinical scientists, and physicians have a need for a source of information on the current state-of-the-art of our understanding of the molecular basis of human disease. More importantly, the complete and effective training of today's graduate students, medical students, postdoctoral fellows, and others, for careers related to the investigation and treatment of human disease requires textbooks that have been designed to reflect our current knowledge of the molecular mechanisms of disease pathogenesis, as well as emerging concepts related to translational medicine. Most pathology textbooks provide information related to diseases and disease processes from the perspective of description (what does it look like and what are its characteristics), risk factors, disease-causing agents, and to some extent, cellular mechanisms. However, most of these textbooks lack in-depth coverage of the molecular mechanisms of disease. The reason for this is primarily historical – most major forms of disease have been known for a long time, but the molecular basis of these diseases are not always known or have been elucidated only very recently. However, with rapid progress over time and improved understanding of the molecular basis of human disease the need emerged for new textbooks on the topic of molecular pathology, where molecular mechanisms represent the focus.

In this volume on *Molecular Pathology: The Molecular Basis of Human Disease* we have assembled a group of experts to discuss the molecular basis and mechanisms of major human diseases and disease processes, presented in the context of traditional pathology, with implications for translational molecular medicine. This volume is intended to serve as a multi-use textbook that would be appropriate as a classroom teaching tool for medical students, biomedical graduate students, allied health students, and others (such as advanced undergraduates). Further, this textbook will be valuable for pathology residents and other postdoctoral fellows that desire to advance their understanding of molecular mechanisms of disease beyond what they learned in medical/graduate school. In addition, this textbook is useful as a reference book for practicing basic scientists and physician scientists that perform disease-related basic science and translational research, who require a ready information resource on the molecular basis of various human diseases and disease states. To be sure, our understanding of the many causes and molecular mechanisms that govern the development of human diseases is far from complete. Nevertheless, the amount of information related to these molecular mechanisms has increased tremendously in recent years and areas of thematic and conceptual consensus have emerged. We have made an effort to integrate accepted principles with broader theoretical concepts in an attempt to present a current and comprehensive view of the molecular basis of human disease. We hope that *Molecular Pathology: The Molecular Basis of Human Disease* will accomplish its purpose of providing students and researchers with in-depth coverage of the molecular basis of major human diseases in the context of traditional pathology so as to stimulate new research aimed at furthering our understanding of these molecular mechanisms of human disease and advancing the theory and practice of molecular medicine.

William B. Coleman
Gregory J. Tsongalis

Foreword

Traditionally, pathologists were involved in clinical practice as physicians trained in the diagnosis of human disease using morphologic and clinical laboratory techniques. Pathologists also serve as consultants to other clinicians and as teachers for medical students, allied health students, graduate students, residents, and fellows. Recent advances in basic sciences, including cell biology, genetics, and molecular biology, have opened up a new era of research in pathology that is being applied to the practice of molecular pathology.

The recent development of powerful new tools for quantitative imaging and molecular analysis of human tissue, in combination with traditional morphological, clinical, and radiologic imaging, has enabled the practice of molecular pathology. Quantitative imagining techniques have been developed that allow multiparameter imaging of single cells in three-dimensional tissues (confocal scanning laser microscopy) or in tissue sections or cytological preparations (laser scanning cytometry). With laser capture microdissection, it is possible to precisely identify and collect individual cell populations for subsequent molecular and proteomic analyses. New microarray technologies, coupled with molecular genetics, allow comprehensive analysis of genetic and chromosomal alterations in normal and diseased cells and tissues. At the same time, new transgenic models of human disease using conditional, tissue-specific gene targeting are increasingly important in identifying early steps in the pathogenesis of disease and in assessing novel approaches for disease prevention, intervention, and therapy. Pathologists play a key role in interpretation of anatomic changes in transgenic animals and in correlating these changes with human disease.

These scientific and technological advances have developed rapidly over the past two decades leading to significant advances in understanding the molecular and genetic mechanisms of human disease. These basic research advances are now being translated into novel approaches for the prevention, diagnosis, and treatment of major human diseases including cardiovascular disease, hereditary and metabolic diseases, and cancer. This new textbook bridges traditional morphologic and clinical pathology and the emerging discipline of molecular pathology. The readers of this textbook written for medical students, graduate students in biomedical research, allied health professionals, residents, and fellows will experience the excitement of these recent scientific and technological advances in modern pathology. A unique feature of this textbook is integration of molecular and genetic pathology with traditional clinical and pathologic descriptions of human disease using a "systems biology" approach. The concept of "molecular pathogenesis" of disease is illustrated with selected diseases to demonstrate how perturbations in molecular pathways contribute to the evolution of disease from normal cells and tissues. This textbook provides a basic foundation in mechanisms of disease, molecular and genetic pathology, and the systems biology approach to disease pathogenesis integrated with morphologic manifestations of human disease. Current and future strategies in molecular diagnosis of human disease are described and placed in the context of morphological and clinical laboratory diagnosis. The impact of molecular diagnosis on treatment decisions and the practice of personalized medicine conclude this textbook with the ultimate vision of translation of basic research in molecular and genetic pathology to clinical practice. This textbook provides a valuable resource for students, biomedical researchers, and physician-scientists who will be working together in interdisciplinary research teams to achieve this vision and embark on the future practice of molecular medicine.

Agnes B. Kane, M.D., Ph.D.
Professor and Chair
Department of Pathology and Laboratory Medicine
Director, NIEHS Training Program in Environmental Pathology, The Warren Alpert Medical School of Brown University

Acknowledgments

The editors would like to acknowledge the significant contributions of a number of people to the successful production of *Molecular Pathology: The Molecular Basis of Human Disease.*

We would like to thank the individuals that contributed to the content of this volume. The remarkable coverage of the state-of-the-art in the molecular pathology of human disease would not have been possible without the hard work and diligent efforts of the 67 authors of the individual chapters. Many of these contributors are our long-time colleagues, collaborators, and friends, and they have contributed to other projects that we have directed. We appreciate their willingness to contribute once again to a project that we found worthy. We especially thank the contributors to this volume that were willing to work with us for the first time. This group also includes some of our long-time friends and colleagues, as well as some new friends. We look forward to working with all of these authors again in the future. Each of these contributors provided us with an excellent treatment of their topic and we hope that they will be proud of their individual contributions to the textbook. Furthermore, we would like to give a special thanks to our colleagues that co-authored chapters with us for this textbook. There is no substitute for an excellent co-author when you are juggling the several responsibilities of concurrently editing and contributing to a textbook. Collectively, we can all be proud of this volume as it is proof that the whole can be greater than the sum of its parts.

We would also like to thank the many people that work for *Academic Press* and *Elsevier* that made this project possible. Many of these people we have not met and do not know, but we appreciate their efforts to bring this textbook to its completed form. Special thanks goes to three key people that made significant contributions to this project on the publishing side, and proved to be exceptionally competent and capable. Ms. Mara Conner (*Academic Press*, San Diego, CA) embraced the concept of this textbook when our ideas were not yet fully developed and encouraged us to pursue this project. She was receptive to the model for this textbook that we envisioned and worked closely with us to evolve the project into its final form. We thank her for providing excellent oversight (and for displaying optimistic patience) during the construction and editing of the textbook. Ms. Megan Wickline (*Academic Press*, San Diego, CA) provided excellent support to us throughout this project. As we interacted with our contributing authors, collected and edited manuscripts, and through production of the textbook, Megan assisted us greatly by being a constant reminder of deadlines, helping us with communication with the contributors, and generally providing support for details small and large, all of which proved to be critical. Ms. Christie Jozwiak (*Elsevier*, Burlington, MA) directed the production of the textbook. She worked with us closely to ensure the integrity of the content of the textbook as it moved from the edited manuscripts into their final form. Throughout the production process, Christie gave a tremendous amount of time and energy to the smallest of details. We thank her for her direct involvement with the production and also for directing her excellent production team. It was a pleasure to work with Mara, Megan, and Christie on this project. We hope that they enjoyed it as much as we did, and we look forward to working with them again soon.

William B. Coleman
Gregory J. Tsongalis

Part

I

Essential Pathology — Mechanisms of Disease

Chapter

1

Molecular Mechanisms of Cell Death

John J. Lemasters

INTRODUCTION

A common theme in disease is death of cells. In diseases ranging from stroke to congestive heart disease to alcoholic cirrhosis of the liver, death of individual cells leads to irreversible functional loss in whole organs and ultimately mortality. For such diseases, prevention of cell death becomes a basic therapeutic goal. By contrast in neoplasia, the purpose of chemotherapy is to kill proliferating cancer cells. For either therapeutic goal, understanding the mechanisms of cell death becomes paramount.

MODES OF CELL DEATH

Although many stresses and stimuli cause cell death, the mode of cell death typically follows one of two patterns. The first is necrosis, a pathological term referring to areas of dead cells within a tissue or organ. Necrosis is typically the result of an acute and usually profound metabolic disruption, such as ischemia/reperfusion and severe toxicant-induced damage. Since necrosis as observed in tissue sections is an outcome rather than a process, the term oncosis has been introduced to describe the process leading to necrotic cell death [1,2], but the term has yet to be widely adopted in the experimental literature. Here, the terms oncosis, oncotic necrosis, and necrotic cell death will be used synonymously to refer both to the outcome of cell death and the pathogenic events precipitating cell killing.

The second pattern is programmed cell death, most commonly manifested as apoptosis, a term derived from an ancient Greek word for the falling of leaves in the autumn. In apoptosis, specific stimuli initiate execution of well-defined pathways leading to orderly resorption of individual cells with minimal leakage of cellular components into the extracellular space and little inflammation [3,4]. Whereas necrotic cell death occurs with abrupt onset after adenosine triphosphate (ATP) depletion, apoptosis may take hours to go to completion and is an ATP-requiring process without a clearly distinguished point of no return. Although apoptosis and necrosis were initially considered separate and independent phenomena, an alternate view is emerging that apoptosis and necrosis can share initiating factors and signaling pathways to become extremes on a phenotypic continuum of necrapoptosis or aponecrosis [5–7].

STRUCTURAL FEATURES OF NECROSIS AND APOPTOSIS

Oncotic Necrosis

Cellular changes leading up to onset of necrotic cell death include formation of plasma membrane protrusions called blebs, mitochondrial swelling, dilatation of cisternae of the endoplasmic reticulum (ER) and nuclear membranes, dissociation of polysomes, and cellular swelling leading to rupture with release of intracellular contents (Table 1.1, Figure 1.1). After necrotic cell death, characteristic histological features of loss of cellular architecture, vacuolization, karyolysis, and increased eosinophilia soon become evident (Figure 1.2). Cell lysis evokes an inflammatory response, attracting neutrophils and monocytes to the dead tissue to dispose of the necrotic debris by phagocytosis and defend against infection (Figure 1.3). In organs like heart and brain with little regenerative capacity, healing occurs with scar formation, namely replacement of necrotic regions with fibroblasts and collagen, as well as other connective tissue components. In organs like the liver that have robust regenerative capacity, cell proliferation can replace areas of necrosis with completely normal tissue within a few days. The healed liver tissue shows with little or no residua of the necrotic event, but if regeneration fails, collagen deposition and fibrosis will occur instead to cause cirrhosis.

Molecular Pathology © 2009, Elsevier, Inc. All Rights Reserved.

Table 1.1 Comparison of Necrosis and Apoptosis

Necrosis	Apoptosis
Accidental cell death	Controlled cell deletion
Contiguous regions of cells	Single cells separating from neighbors
Cell swelling	Cell shrinkage
Plasmalemmal blebs without organelles	Zeiotic blebs containing large organelles
Small chromatin aggregates	Nuclear condensation and lobulation
Random DNA degradation	Internucleosomal DNA degradation
Cell lysis with release of intracellular contents	Fragmentation into apoptotic bodies
Inflammation and scarring	Absence of inflammation and scarring
Mitochondrial swelling and dysfunction	Mitochondrial permeabilization
Phospholipase and protease activation	Caspase activation
ATP depletion and metabolic disruption	ATP and protein synthesis sustained
Cell death precipitated by plasma membrane rupture	Intact plasma membrane

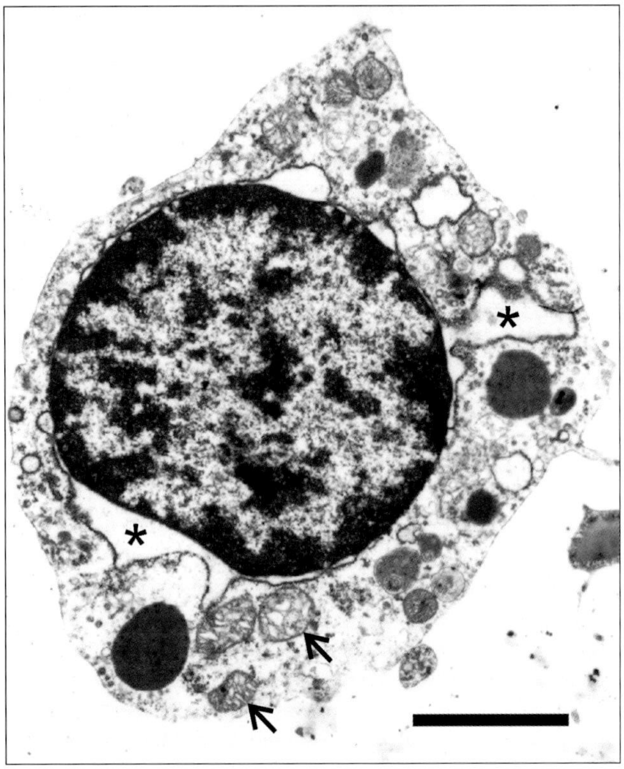

Figure 1.1 **Electron microscopy of oncotic necrosis to a rat hepatic sinusoidal endothelial cell after ischemia/reperfusion**. Note cell rounding, mitochondrial swelling (arrows), rarefaction of cytosol, dilatation of the ER and the space between the nuclear membranes (*), chromatin condensation, and discontinuities in the plasma membrane. Bar is 2 μm.

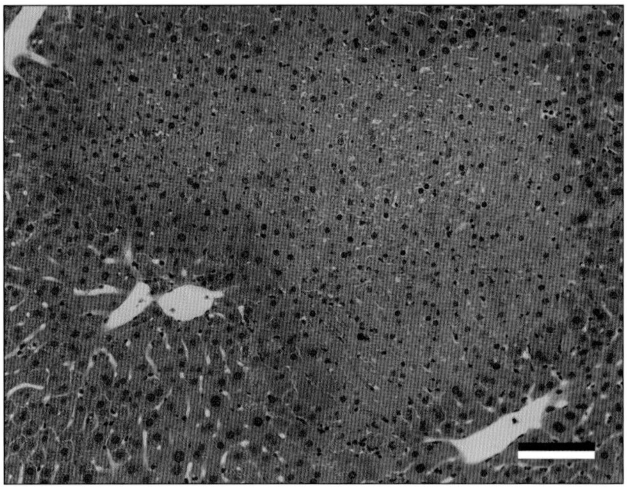

Figure 1.2 **Histology of necrosis after hepatic ischemia/reperfusion in a mouse**. Note increased eosinophilia, loss of cellular architecture, and nuclear pyknosis and karyolysis. Bar is 50 μm.

Apoptosis

Unlike necrosis, which usually represents an accidental event in response to an imposed unphysiological stress, apoptosis is a process of physiological cell deletion that has an opposite role to mitosis in the regulation of cell populations. In apoptosis, cell death occurs with little release of intracellular contents, inflammation, and scar formation. Individual cells undergoing apoptosis separate from their neighbors and shrink rather than swell. Distinctive nuclear and cytoplasmic changes also occur, including chromatin condensation, nuclear lobulation and fragmentation, formation of numerous small cell surface blebs (zeiotic blebbing), and shedding of these blebs as apoptotic bodies that are phagocytosed by adjacent cells and macrophages for lysosomal degradation (Table 1.1, Figure 1.3). Characteristic biochemical changes also occur, typically activation of a cascade of cysteine-aspartate proteases, called caspases, leakage of proapoptotic proteins like cytochrome *c* from mitochondria into the cytosol, internucleosomal deoxyribonucleic acid (DNA) degradation, degradation of poly (ADP-ribose) polymerase (PARP), and movement of phosphatidyl serine to the exterior leaflet of the plasmalemmal lipid bilayer. Thus, apoptosis manifests a very different pattern of cell death than oncotic necrosis (Table 1.1, Figure 1.3).

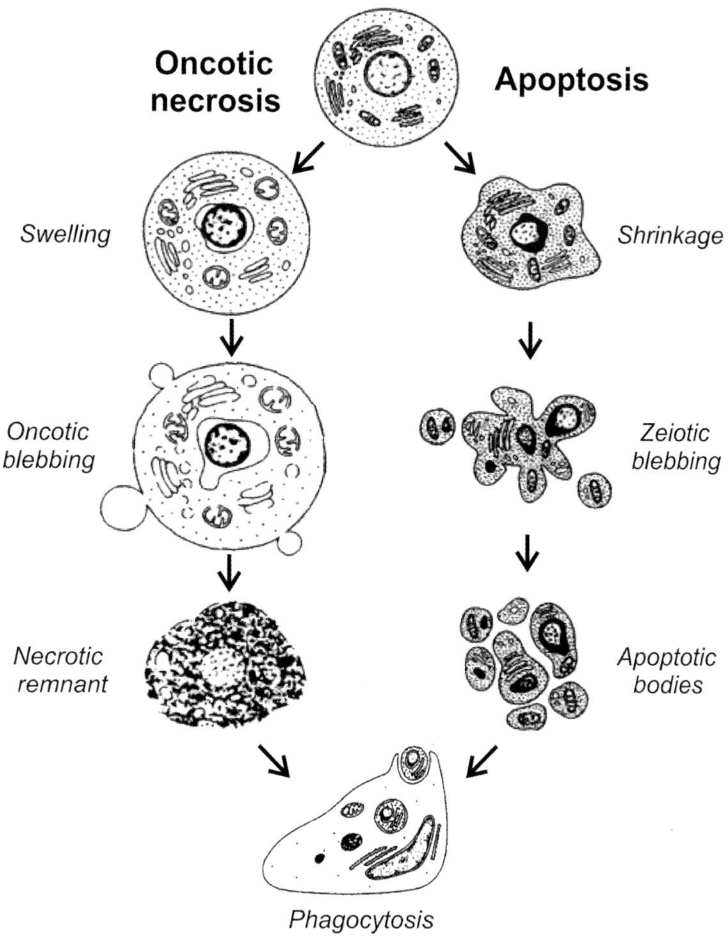

Figure 1.3 **Scheme of necrosis and apoptosis.** In oncotic necrosis, swelling leads to bleb rupture and release of intracellular constituents which attract macrophages that clear the necrotic debris by phagocytosis. In apoptosis, cells shrink and form small zeiotic blebs that are shed as membrane-bound apoptotic bodies. Apoptotic bodies are phagocytosed by macrophages and adjacent cells. Adapted with permission from [2].

CELLULAR AND MOLECULAR MECHANISMS UNDERLYING NECROTIC CELL DEATH

Metastable State Preceding Necrotic Cell Death

Cellular events culminating in necrotic cell death are somewhat variable from one cell type to another, but certain events occur regularly. As implied by the term oncosis, cellular swelling is a prominent feature of oncotic necrosis [1,8]. In many cell types, swelling of 30–50% occurs early after ATP depletion associated with formation of blebs on the cell surface (Figure 1.4) [9,10]. These blebs contain cytosol and ER but exclude larger organelles like mitochondria and lysosomes. Bleb formation is likely due to cytoskeletal alterations after ATP depletion, whereas swelling arises from disruption of cellular ion transport [11,12]. Mitochondrial swelling and dilatation of cisternae of ER and nuclear membranes accompany bleb formation (see Figure 1.1). After longer times, a metastable state develops, which is characterized by mitochondrial depolarization, lysosomal breakdown, bidirectional leakiness of the plasma membrane to organic anions (but not cations), intracellular Ca^{2+} and pH dysregulation, and accelerated bleb formation with more rapid swelling [13–16]. The metastable state lasts only a few minutes and culminates in rupture of a plasma membrane bleb (Figure 1.4) [14–16]. Bleb rupture leads to loss of metabolic intermediates such as those that reduce tetrazolium dyes, leakage of cytosolic enzymes like lactate dehydrogenase, uptake of dyes like trypan blue, and collapse of all electrical and ion gradients across the membrane. This all-or-nothing breakdown of the plasma membrane permeability barrier is long-lasting, irreversible, and incompatible with continued life of the cell.

Some work suggests that opening of a nonspecific anion channel in the plasma membrane initiates the metastable state [17,18]. Although potassium and sodium channels open early after metabolic disruption, cellular impermeability to chloride limits the

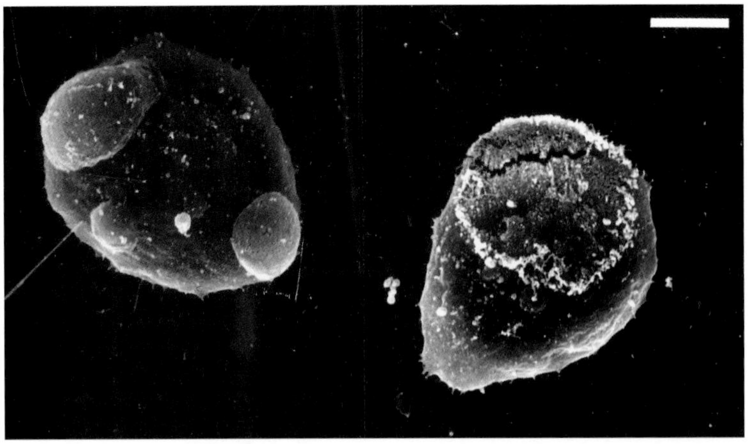

Figure 1.4 **Bleb rupture at onset of necrotic cell death**. After metabolic inhibition with cyanide and iodoacetate, inhibitors of respiration and glycolysis, respectively, a surface bleb of the cultured rat hepatocyte on the right has just burst. Note the discontinuity of the plasma membrane surface in the scanning electron micrograph. The hepatocyte on the left is also blebbed, but the plasma membrane is still intact, and viability has not yet been lost. Bar is 5 μm. Adapted with permission from [16].

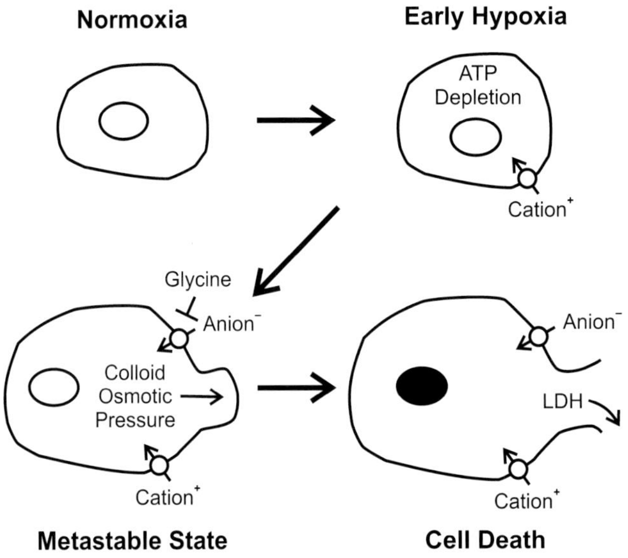

Figure 1.5 **Plasma membrane permeabilization leading to necrotic cell death**. Early after hypoxia and other metabolic stresses, ATP depletion leads to inhibition of the Na,K-ATPase and opening of monovalent cation channels causing cation gradients (Na^+ and K^+) to collapse. Swelling is limited by impermeability to anions. Later, glycine and strychnine-sensitive anion channels open to initiate anion entry and accelerate bleb formation and swelling. Swelling continues until a bleb ruptures. With abrupt and complete loss of the plasma membrane permeability barrier, viability is lost. Supravital dyes like trypan blue and propidium iodide enter the cell to stain the nucleus, and cytosolic enzymes like lactate dehydrogenase (LDH) leak out. With permission from [209].

rate of swelling. At onset of the metastable state, a relatively nonspecific chloride-conducting anion channel appears to open, permitting electroneutral uptake of electrolytes (principally sodium and chloride) and initiating rapid swelling driven by colloid osmotic (oncotic) forces (Figure 1.5). Rapid swelling continues until one of the plasma membrane blebs ruptures. Bleb rupture is the final irreversible event precipitating cell death, since removal of the instigating stress (e.g., reoxygenation of anoxic cells) leads to cell recovery prior to bleb rupture but not afterwards [15]. Glycine and the glycine receptor antagonist strychnine protect against necrotic cell killing. Protection is associated with inhibition of this anion death channel and suppression of swelling in the metastable state. Glycine protection occurs without restoration of ATP or prevention of other metabolic derangements [12,18–21]. The glycine-gated chloride channel (GlyR) appears responsible for glycine cytoprotection, since glycine is not cytoprotective in cells not expressing the GlyRα1 subunit, and since GlyRα1 confers glycine cytoprotection in such cells [22].

Mitochondrial Dysfunction and ATP Depletion

Ischemia as occurs in strokes and heart attacks is perhaps the most common cause of necrotic cell killing [8]. In ischemia, oxygen deprivation prevents ATP formation by mitochondrial oxidative phosphorylation, a process providing up to 95% of ATP utilized by highly aerobic tissues [23]. The role of mitochondrial dysfunction in necrotic killing can be assessed experimentally by the ability of glycolytic substrates to rescue cells from lethal cell injury (Figure 1.6) [24]. As an alternative source of ATP, glycolysis partially replaces ATP production lost after mitochondrial dysfunction. Maintenance of as little as 15% or 20% of normal ATP then rescues cells from necrotic death. Glucose and glycogen are prototypic glycolytic substrates that delay or prevent anoxic cell killing in most cell types. However, an important function of the liver is to maintain blood glucose levels constant, and hepatocytes do not consume glucose even during anoxia. For hepatocytes, fructose is a much better glycolytic substrate, and fructose but not glucose prevents loss of viability of hepatocytes during anoxia, respiratory inhibition, and inhibition of the mitochondrial ATP synthase [25,26].

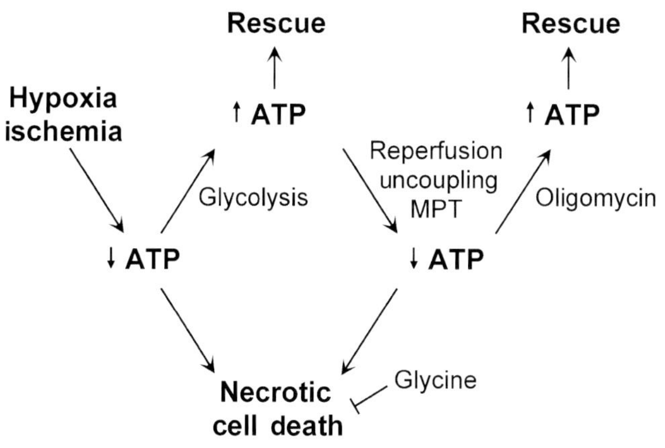

Figure 1.6 Progression of mitochondrial injury. Respiratory inhibition inhibits oxidative phosphorylation and leads to ATP depletion and necrotic cell death. Glycine blocks plasma membrane permeabilization causing necrotic cell death downstream of ATP depletion. Glycolysis restores ATP and prevents cell killing. Mitochondrial uncoupling as occurs after reperfusion due to the mitochondrial permeability transition (MPT) activates the mitochondrial ATPase to futilely hydrolyze glycolytic ATP, and protection against necrotic cell death is lost. By inhibiting the mitochondrial ATPase, oligomycin prevents ATP depletion and rescues cells from necrotic cell death if glycolytic substrate is present. With permission from [209].

In aerobic cells, exogenous glucose and fructose at high concentrations cause intracellular ATP to decrease because of ATP consumption by hexokinase and fructokinase, the first enzymes in the glycolytic metabolism of the respective two hexoses. As glucose-6-phosphate and other sugar phosphates accumulate, intracellular inorganic phosphate (Pi) also decreases [27]. In fructose-treated livers, decreased ATP has been interpreted as evidence of toxicity, but decreased Pi offsets the decline of ATP such that fructose-treated hepatocytes and livers maintain their ATP/ADP·Pi ratio (phosphorylation potential). Phosphorylation potental, rather than ATP concentration, ATP/ADP ratio or energy charge (defined as $([ATP]+{}^{1}/_{2}[ADP])/([ATP]+[ADP]+[AMP]))$, is the relevant thermodynamic variable reflecting cellular bioenergetic status [28]. Moreover, in anoxic livers and hepatocytes, glycolysis of fructose and endogenous glycogen increases ATP to protect against necrotic cell killing [25,26]. Fructose also protects against hepatocellular toxicity by oxidant chemicals, suggesting that mitochondria are also a primary target of cytotoxicity in oxidative stress [29].

Mitochondrial Uncoupling in Necrotic Cell Killing

Mitochondrial injury and dysfunction are progressive (Figure 1.6). Anoxia and inhibition with a toxicant like cyanide inhibit respiration to cause ATP depletion and ultimately necrotic cell death. Glycolysis can replace this ATP supply, although only partially in highly aerobic cells, to rescue cells from necrotic killing. In the absence of respiration, mitochondrial membrane potential ($\Delta\Psi$) is sustained by reversal of the ATP synthase reaction (mitochondrial adenosine triphosphatase, or F_1F_0-ATPase). However, when mitochondria become permeable to protons (uncoupling), then maximal stimulation of F_1F_0-ATPase occurs. Since glycolytic ATP production cannot keep pace, ATP levels fall profoundly, mitochondria depolarize, and necrotic cell death ensues. In the presence of glycolytic substrate, oligomycin, an inhibitor of the F_1F_0-ATP synthase, prevents uncoupler-induced ATP depletion and subsequent cell death. Because oligomycin does not reverse uncoupling, mitochondrial $\Delta\Psi$ is not restored [24,26]. Cytoprotection by oligomycin requires glycolytic ATP generation, since in the absence of glycolysis, oligomycin is itself toxic by inhibiting oxidative phosphorylation. Cytoprotection requiring the combination of glycolytic substrate and oligomycin indicates cytotoxicity mediated by mitochondrial uncoupling as shown, for example, for calcium ionophore toxicity and oxidative stress [29,30].

Mitochondrial Permeability Transition

Inner membrane permeability

In oxidative phosphorylation, respiration drives translocation of protons out of mitochondria to create an electrochemical proton gradient composed of a negative inside membrane potential ($\Delta\Psi$) and an alkaline inside pH gradient (ΔpH). ATP synthesis is then linked to protons returning down this electrochemical gradient through the mitochondrial ATP synthase. This chemiosmotic proton circuit requires the mitochondrial inner membrane to be impermeable to ions and charged metabolites. Thus, metabolite exchange for oxidative phosphorylation occurs via specific transporters and exchangers in the inner membrane, including the adenine nucleotide translocator (ANT), which exchanges ATP for adenosine diphosphate (ADP); the phosphate transporter; and one of several transporter systems for uptake of respiratory substrates like pyruvate and fatty acids. By contrast, the outer membrane is nonspecifically permeable to ions and hydrophilic metabolites, which move across the outer membrane through a channel called the voltage dependent anion channel (VDAC) [31,32]. Despite its name, VDAC has only weak anion selectivity and conducts freely most solutes up to a molecular mass of about 5 kDa.

Mitochondrial permeability transition pore

In the mitochondrial permeability transition (MPT), the mitochondrial inner membrane abruptly becomes nonselectively permeable to solutes of molecular weight up to about 1500 Da [33–35]. Ca^{2+}, oxidative stress, and numerous reactive chemicals induce the MPT, whereas cyclosporin A and pH less than 7 inhibit. Onset of the MPT causes mitochondrial depolarization, uncoupling, and large amplitude mitochondrial swelling driven by colloid osmotic forces. Opening of highly conductive permeability transition (PT) pores in the mitochondrial inner membrane underlies the MPT. Patch clamping shows that conductance through permeability transition pores (PT pores) is so great that opening of a single PT pore may be sufficient to cause mitochondrial depolarization and swelling [36].

The composition of PT pores is uncertain. In one model, PT pores are formed by ANT from the inner membrane, VDAC from the outer membrane, the cyclosporin A binding protein cyclophilin D (CypD) from the matrix, and possibly other proteins (Figure 1.7) (reviewed in [37,38]). Although once widely accepted, the validity of this model has been challenged by genetic knockout studies showing that the MPT still occurs in mitochondria that are deficient in ANT and VDAC [39–41]. Moreover, although CypD is responsible for pore inhibition by cyclosporin A, a cyclosporin A-insensitive MPT still occurs in CypD deficient mitochondria [34,42]. An alternative model for the PT pore is that oxidative and other stresses damage membrane proteins that then misfold and aggregate to form PT pores in association with CypD and other molecular chaperones (Figure 1.7) [43].

pH-dependent ischemia/reperfusion injury

Ischemia is an interruption of blood flow and hence oxygen supply to a tissue or organ. In ischemic tissue, anaerobic glycolysis, hydrolysis of ATP, and release of protons from acidic organelles cause tissue pH to decrease by a unit or more. The naturally occurring acidosis of ischemia actually protects against onset of necrotic cell death. Acidosis also dramatically delays cell killing from oxidant chemicals, ionophores, and alkylating agents [44–48].

Although acidosis protects against cell killing during ischemia, reoxygenation and recovery of pH after reperfusion act to precipitate necrotic cell death. In cultured cells and perfused organs, ischemia/reperfusion injury can be reproduced using anoxia at acidotic pH to simulate ischemia followed by reoxygenation at normal pH to simulate reperfusion. Reperfusion in this model causes necrotic cell killing with release of intracellular enzymes like lactate dehydrogenase and nuclear labeling with vital dyes like trypan blue and propidium iodide [12,49–56]. Much of reperfusion injury leading to necrotic cell death is attributable to recovery of pH, since reoxygenation at low pH prevents cell killing entirely, whereas restoration of normal pH without reoxygenation produces similar cell killing as restoration of pH with reoxygenation, a so-called pH paradox (Figure 1.8).

Cell death in the pH paradox is linked to intracellular pH. Ionophores like monensin that accelerate recovery of intracellular pH from the acidosis of ischemia after reperfusion also accelerate necrotic cell killing [51]. Conversely, inhibition of Na^+/H^+ exchange with dimethylamiloride or Na^+-free medium delays recovery of intracellular pH and prevents reperfusion-induced necrotic cell killing almost completely [51,52,54,55]. Cell killing in the pH paradox is linked specifically to intracellular pH and occurs independently of changes of cytosolic and extracellular free Na^+ and Ca^{2+} [44,51,52,55].

Role of the mitochondrial permeability transition in pH-dependent reperfusion injury

pH below 7 inhibits PT pores, and recovery of intracellular pH to 7 or greater after reperfusion induces the MPT, as shown directly by confocal/multiphoton microscopy [55]. During ischemia, mitochondria depolarize because of respiratory inhibition, but the mitochondrial inner membrane remains impermeant to fluorophores like calcein which can only pass through PT pores (Figure 1.8). After reperfusion at normal pH, mitochondria repolarize initially, as shown by uptake of membrane potential-indicating fluorophores (Figure 1.9). Subsequently and in parallel with recovery of intracellular pH to neutrality, the MPT occurs, leading to permeabilization of the inner membrane to calcein and mitochondrial depolarization (Figures 1.8 and 1.9). ATP depletion then follows, and necrotic cell death occurs.

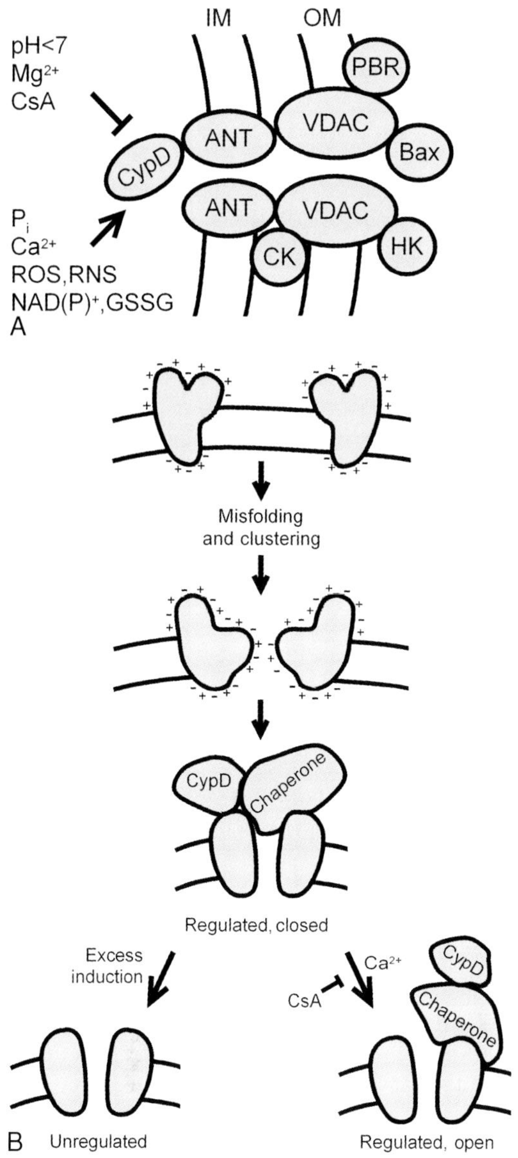

Figure 1.7 Models of mitochondrial permeability transition pores. In one model (**A**), PT pores are composed of the adenine nucleotide translocator (ANT) from the inner membrane (IM), cyclophilin D (CypD) from the matrix and the voltage-dependent anion channel (VDAC) from the outer membrane (OM). Other proteins, such as the peripheral benzodiazepine receptor (PBR), hexokinase (HK), creatine kinase (CK), and Bax may also contribute. PT pore openers include Ca^{2+}, inorganic phosphate (Pi), reactive oxygen and nitrogen species (ROS, RNS), and oxidized pyridine nucleotides ($NAD(P)^+$) and glutathione (GSSG). An alternative model (**B**) proposes that oxidative and other damage to integral inner membrane proteins leads to misfolding. These misfolded proteins aggregate at hydrophilic surfaces facing the hydrophobic bilayer to form aqueous channels. CypD and other chaperones block conductance of solutes through these nascent PT pores. High matrix $Ca2^+$ acting through CypD leads to PT pore opening, an effect blocked by cyclosporin A (CsA). As misfolded protein clusters exceed the number of chaperones to regulate them, constitutively open channels form. Such unregulated PT pores are not dependent on Ca^{2+} for opening and are not inhibited by CsA. With permission from [210].

Reperfusion at acidotic pH to prevent recovery of pH and reperfusion in the presence of PT pore blockers (e.g., cyclosporin A and its derivatives) prevents mitochondrial inner membrane permeabilization, depolarization, and cell killing (Figures 1.8 and 1.9). Notably, cyclosporin A protects when added only during the reperfusion phase, as now confirmed by decreased infarct size in patients receiving percutaneous coronary intervention (PCI) for ischemic heart disease [55–59]. Thus, the MPT is the proximate cause of pH-dependent cell killing in ischemia/reperfusion injury.

Minocycline, a semisynthetic tetracycline derivative that protects against neurodegenerative disease, trauma, and hypoxia–ischemia [60–68], also inhibits the MPT and protects against cell killing after reperfusion of livers stored by cold ischemia for transplantation [69]. However, the mechanism of MPT blockade differs from cyclosporin A. Rather than blocking through an interaction with CypD, minocycline blocks the MPT by inhibiting mitochondrial calcium uptake. This observation implies that calcium uptake into mitochondria is a prerequisite for MPT onset after reperfusion.

Oxidative stress

Reactive oxygen species (ROS) and reactive nitrogen species (RNS), including superoxide, hydrogen peroxide, hydroxyl radical, and peroxynitrite, have long been implicated in cell injury leading to necrosis (Figure 1.10). In hepatocytes, mitochondrial NAD(P)H oxidation after oxidative stress disrupts mitochondrial Ca^{2+} homeostasis to cause an increase of mitochondrial Ca^{2+}, which in turn stimulates intramitochondrial ROS formation and onset of the MPT [29,70–73]. In cardiac myocytes after reperfusion, intramitochondrial ROS formation also occurs to initiate the MPT and subsequent necrotic cell death (Figure 1.9) [56]. Notably, inhibition of the MPT with cyclosporin A does not prevent mitochondrial ROS generation after reperfusion (Figure 1.9). Although intramitochondrial Ca^{2+} may be permissive for MPT onset after reperfusion, massive Ca^{2+} overloading and hypercontracture in myocytes does not occur until after the MPT, and MPT inhibitors prevent Ca^{2+} overloading and cell death [56]. In neurons, excitotoxic stress with glutamate and N-methyl-D-aspartate (NMDA) receptor agonists also stimulates mitochondrial ROS formation, leading to the MPT and excitotoxic injury [74–77].

Iron potentiates injury in a variety of diseases and is an important catalyst for hydroxyl radical formation from superoxide and hydrogen peroxide (Figure 1.10) [78–81]. Increased intracellular chelatable iron contributes to cell death after cold ischemia/reperfusion [82,83], and addition of a membrane permeable Fe^{3+} complex causes the MPT and consequent necrotic and apoptotic cell death [84].

During oxidative stress and hypoxia/ischemia, lysosomes rupture [14,85–87], and lysosomal rupture releases chelatable (loosely bound) iron with consequent pro-oxidant cell damage [88–91]. This

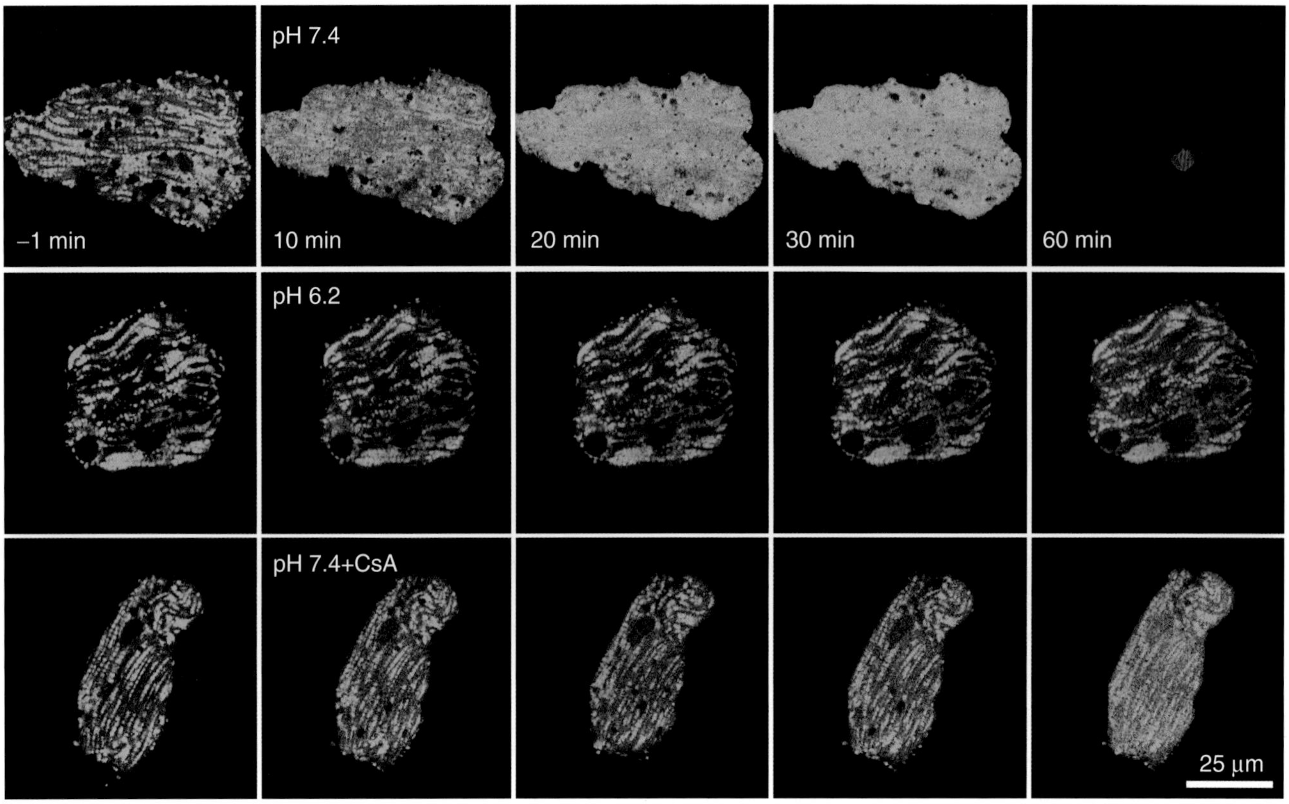

Figure 1.8 **Mitochondrial inner membrane permeabilization in adult rat cardiac myocytes after ischemia and reperfusion**. After loading mitochondria of cardiac myocytes with calcein, cells were subjected to 3 h of anoxia at pH 6.2 (ischemia) followed by reoxygenation at pH 7.4 (**A**), pH 6.2 (**B**), or pH 7.4 with 1 μM CsA (**C**). Red-fluorescing propidium iodide was present to detect loss of cell viability. Note that green calcein fluorescence was retained by mitochondria at the end of ischemia (1 min before reperfusion), indicating that PT pores had not opened. After reperfusion at pH 7.4, mitochondria progressively released calcein over 30 min. at which time calcein was nearly evenly distributed throughout cytosol. After 60 min, all cellular calcein was lost, and the nucleus stained with PI, indicating loss of viability. After reperfusion at pH 6.2 (**B**) or at pH 7.4 in the presence of CsA (**C**), calcein was retained and cell death did not occur. Thus, reperfusion at pH 7.4 induced onset of the MPT and necrotic cell death that were blocked with CsA and acidotic pH. Adapted with permission from [56].

iron is taken up into mitochondria by the mitochondrial calcium uniporter and helps catalyze mitochondrial ROS generation. Iron chelation with desferal prevents this ROS formation and decreases cell death in oxidative stress and hypoxia/ischemia [56,91,92].

Protein kinase signaling and the MPT

Reperfusion with nitric oxide suppresses MPT onset and reperfusion-induced cell killing after ischemia [93]. A signaling cascade of guanylyl cyclase, cyclic guanosine monophosphate (cGMP), and cGMP-dependent protein kinase (protein kinase G) mediates nitric oxide protection against MPT onset. In isolated mitochondria, a combination of cGMP, cytosolic extract as a source of protein kinase, and ATP blocks the Ca^{2+}-induced MPT. Thus, protein kinases act directly on mitochondria to negatively regulate MPT [93]. Protein kinase C epsilon (PKCε), together with inhibition of glycogen synthase kinase-3 beta (GSK-3ε), also lead to MPT inhibition, whereas c-Jun nuclear kinase-2 (JNK-2) promotes the MPT in reperfusion injury and acetaminophen hepatotoxicity [94–99].

Other Stress Mechanisms Inducing Necrotic Cell Death

Poly (ADP-Ribose) Polymerase

Single strand breaks induced by ultraviolet (UV) light, ionizing radiation, and ROS (particularly hydroxyl radical and peroxynitrite) activate PARP isoforms 1 and 2 (PARP-1/2). PARP assists in the repair of single-strand DNA breaks by recruiting scaffolding proteins, DNA ligases, and polymerases that mediate base excision repair [100]. With excess DNA damage, PARP attaches ADP-ribose to the strand breaks and elongates ADP-ribose polymers attached to the DNA.

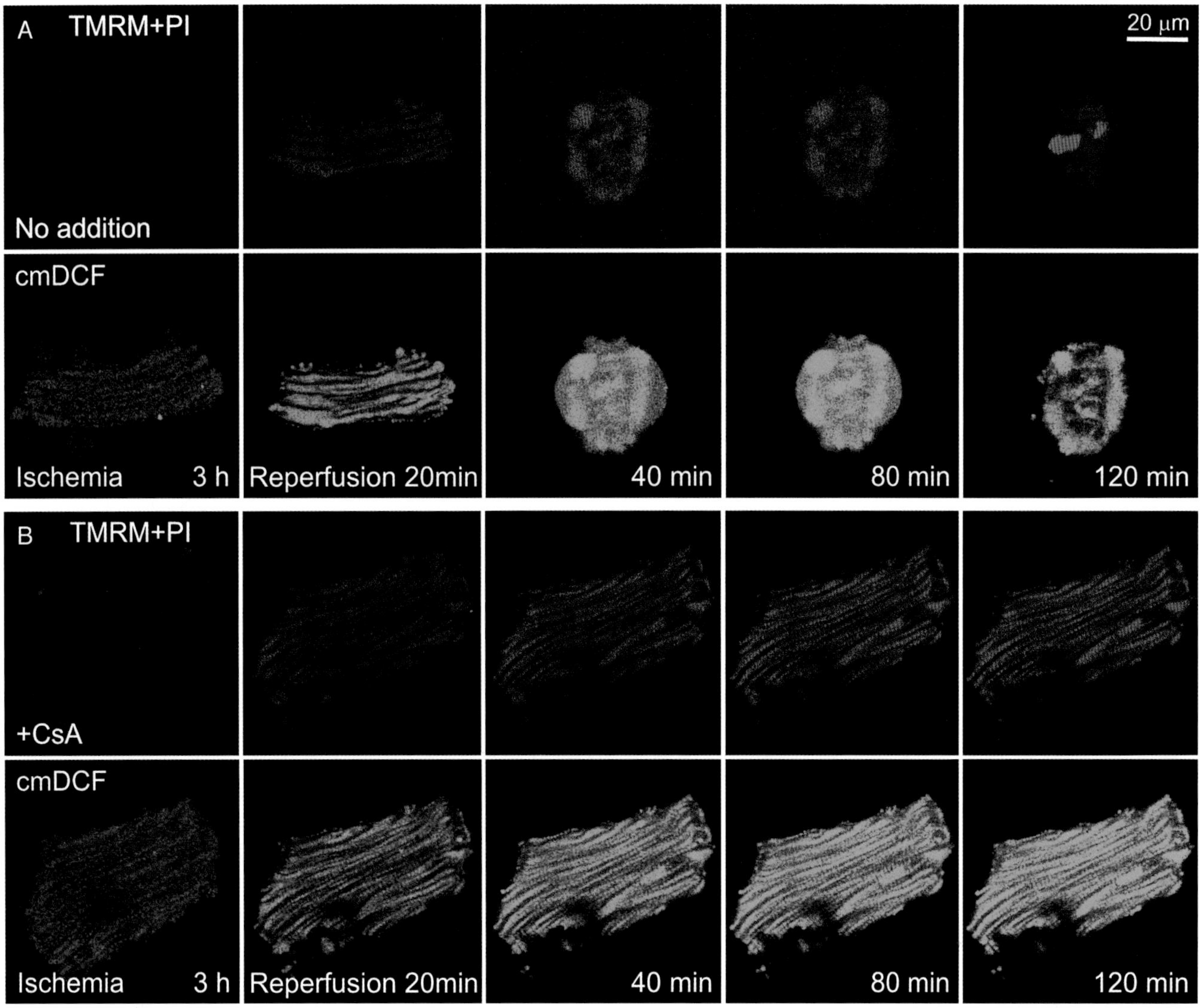

Figure 1.9 Mitochondrial ROS formation after reperfusion. Myocytes were co-loaded with red-fluorescing tetramethylrhodamine methyester (TMRM) and green-fluorescing chloromethyldichlorofluorescin (cmDCF) to monitor mitochondrial membrane potential and ROS formation, respectively. At the end of 3 h of ischemia, mitochondria were depolarized (lack of red TMRM fluorescence). After 20 min of reperfusion, mitochondria took up TMRM, indicating repolarization, and cmDCF fluorescence increased progressively inside mitochondria (**A**). Subsequently, hypercontraction and depolarization occurred after 40 min, and viability was lost within 120 min, as indicated by nuclear labeling with red-fluorescing propidium iodide. When cyclosporin A was added at reperfusion (**B**), mitochondria underwent sustained repolarization, and hypercontracture and cell death did not occur. Nonetheless, mitochondrial cmDCF fluorescence still increased. By contrast, reperfusion with antioxidants prevented ROS generation and MPT onset with subsequent cell death (data not shown). Thus, mitochondrial ROS generation induces the MPT and cell death after ischemia/reperfusion. Adapted with permission from [56].

Consumption of the oxidized form of nicotinamide adenine dinucleotide (NAD^+) in this fashion leads to NAD^+ depletion, disruption of ATP-generation by glycolysis and oxidative phosphorylation, and ATP depletion-dependent cell death [100,101]. Mice deficient in PARP-1 or PARP-2 and mice treated with PARP inhibitors are protected against such DNA damage-induced necrosis [102,103]. Glycosidases cleave long ADP-ribose polymers attached to DNA into large oligomers. Such oligomers can translocate to mitochondria to cause mitochondrial dysfunction and possibly the MPT [104].

PARP-dependent necrosis is an example of so-called programmed necrosis since PARP actively promotes a cell death-inducing pathway that otherwise would not occur. Necrotic cell death also frequently occurs when apoptosis is interrupted, as by caspase (cysteine-aspartate protease) inhibition. Such caspase independent cell death is the consequence of mitochondrial dysfunction (e.g., depolarization, loss of

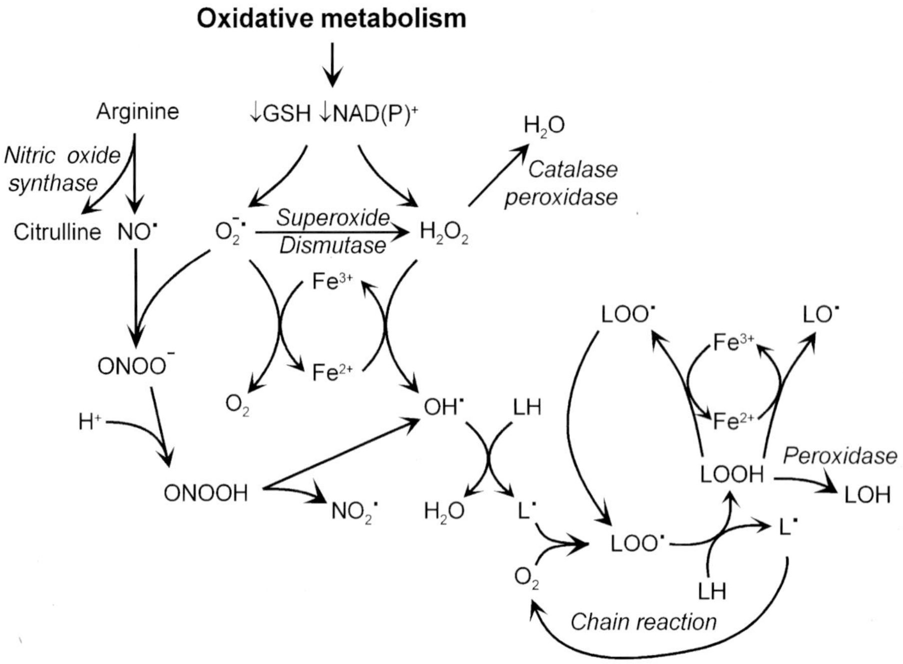

Figure 1.10 **Iron-catalyzed free radical generation.** Oxidative stress causes oxidation of GSH and NAD(P)H, important reductants in antioxidant defenses, promoting increased net formation of superoxide ($O_2^{\bullet-}$) and hydrogen peroxide (H_2O_2). Superoxide dismutase converts superoxide to hydrogen peroxide, which is further detoxified to water by catalase and peroxidases. In the iron-catalyzed Haber Weiss reaction (or Fenton reaction), superoxide reduces ferric iron (Fe^{3+}) to ferrous iron (Fe^{2+}), which reacts with hydrogen peroxide to form the highly reactive hydroxyl radical (OH•). Hydroxyl radical reacts with lipids to form alkyl radicals (L•) that initiate an oxygen-dependent chain reaction generating peroxyl radicals (LOO•) and lipid peroxides (LOOH). Iron also catalyzes a chain reaction generating alkoxyl radicals (LO•) and more peroxyl radicals. Nitric oxide synthase catalyzes formation of nitric oxide (NO•) from arginine. Nitric oxide reacts rapidly with superoxide to form unstable peroxynitrite anion ($ONOO^-$), which decomposes to nitrogen dioxide and hydroxyl radical. In addition to attacking lipids, these radicals also attack proteins and nucleic acids.

cytochrome *c* needed for respiration) or other metabolic disturbance.

Plasma membrane injury

An intact plasma membrane is essential for cell viability. Detergents and pore-forming agents like mastoparan from wasp venom defeat the barrier function of the plasma membrane and cause immediate cell death [105]. Immune-mediated cell killing can act similarly. In particular, complement mediates formation of a membrane attack complex that in conjunction with antibody lyses cells [106]. Complement component 9, an amphipathic molecule, inserts through the cell membrane, polymerizes, and forms a pore visible as a tubular channel in electron micrographs [107]. Indeed, a single membrane attack complex may be sufficient to cause swelling and lysis of an individual erythrocyte. Pores also cause calcium entry, leading to mitochondrial dysfunction and most likely onset of the MPT. The resulting bioenergetic failure is further aggravated by pore-dependent ATP leakage across the plasma membrane.

PATHWAYS TO APOPTOSIS

Roles of Apoptosis in Biology

Apoptosis is an essential event in both the normal life of organisms and in pathobiology. In development, apoptosis sculpts and remodels tissues and organs, for example, by creating clefts in limb buds to form fingers and toes [3]. Apoptosis is also responsible for reversion of hypertrophy to atrophy and immune surveillance-induced killing of preneoplastic cells and virally infected cells [108,109]. In the gastrointestinal tract especially, renewal of epithelial cells can occur every few days, and cell deletion by programmed cell death closely matches mitotic proliferation [110].

The literature on apoptosis is large and complex and continues to grow. One pattern that has emerged is that each of several organelles gives rise to signals initiating apoptotic cell killing [111–113]. Often these signals converge on mitochondria as a common pathway to apoptotic cell death. In most apoptotic signaling, activation of caspases 3 or 7 from a family of caspases (Table 1.2) begins execution of the final and committed phase of apoptotic cell death. Caspase 3/7 has many targets.

Table 1.2 **Mammalian caspases.** Caspases are evolutionarily conserved aspartate specific cysteine-dependent proteases that function in apoptotic and inflammatory signaling [212]. Initiator caspases are involved in the initiation and propagation of apoptotic signaling, whereas effector caspases act on a wide variety of proteolytic substrates to induce the final and committed phase of apoptosis. Initiator and inflammatory caspases have large prodomains containing oligomerization motifs such as the caspase recruitment domain (CARD) and the death effector domain. Effector caspases have short prodomains and are proteolytically activated by large prodomain caspases and other proteases. Proteolytic cleavage of procaspase precursors forms separate large and small subunits that assemble into active enzymes consisting of two large and two small subunits. Caspase activation occurs in multimeric complexes that typically consist of a platform protein that recruits procaspases either directly or by means of adaptors. Such caspase complexes include the apoptosome and the death-inducing signaling complex (DISC). Caspase 14 plays a role in terminal keratinocyte differentiation in cornified epithelium [213].

Initiator Caspases	Molecular Weight of Proenzyme (kDa)	Active subunits (kDa)	Prodomain	Amino Acid Target Sequence for Proteolysis
Caspase 2	51	19/12	Long with CARD	VDVAD
Caspase 8	55	18/11	Long with two DED	(L/V/D)E(T/V/I)D
Caspase 9	45	17/10	Long with CARD	(L/V/I)EHD
Caspase 10	55	17/12	Long with two DED	(I/V/L)EXD
Caspase 12	50	20/10	Long with CARD	ATAD
Effector Caspases				
Caspase 3	32	17/12	Short	DE(V/I)D
Caspase 6	34	18/11	Short	(T/V/I)E(H/V/I)D
Caspase 7	35	20/12	Short	DE(V/I)D
Inflammatory Caspases				
Caspase 1	45	20/10	Long with CARD	(W/Y/F)EHD
Caspase 4	43	20/10	Long with CARD	(W/L)EHD
Caspase 5	48	20/10	Long with CARD	(W/L/F)EHD
Caspase 11	42	20/10	Long with CARD	(V/I/P/L)EHD
Other Caspases				
Caspase 14	42	20/10	Short	(W/I)E(T/H)D

Degradation of the nuclear lamina and cytokeratins contributes to nuclear remodeling, chromatin condensation, and cell rounding [114,115]. Degradation of endonuclease inhibitors activates endonucleases to cause internucleosomal DNA cleavage. The resulting DNA fragments have lengths in multiples of 190 base pairs, the nucleosome to nucleosome repeat distance. In starch gel electrophoresis, these fragments produce a characteristic ladder pattern. DNA strand breaks can also be recognized in tissue sections by the terminal deoxynucleotidyl transferase-mediated dUTP nick-end labeling (TUNEL) assay [116]. Additionally, caspase activation leads to cell shrinkage, phosphatidyl serine externalization on the plasma membrane, and formation of numerous small surface blebs (zeiosis). Unlike necrotic blebs, these zeiotic blebs contain membranous organelles and are shed as apoptotic bodies. However, not all apoptotic changes depend on caspase 3/7 activation. For example, release of apoptosis-inducing factor (AIF) from mitochondria and its translocation to the nucleus promotes DNA degradation in a caspase 3-independent fashion [117].

Pathways leading to activation of caspase 3 and related effector caspases like caspase 7 are complex and quite variable between cells and specific apoptosis-instigating stimuli, and each major cellular structure can originate its own set of unique signals to induce apoptosis (Figure 1.11). Proapoptotic signals are often associated with specific damage or perturbation to the organelle involved. Consequently, cells choose death by apoptosis rather than life with organelle damage.

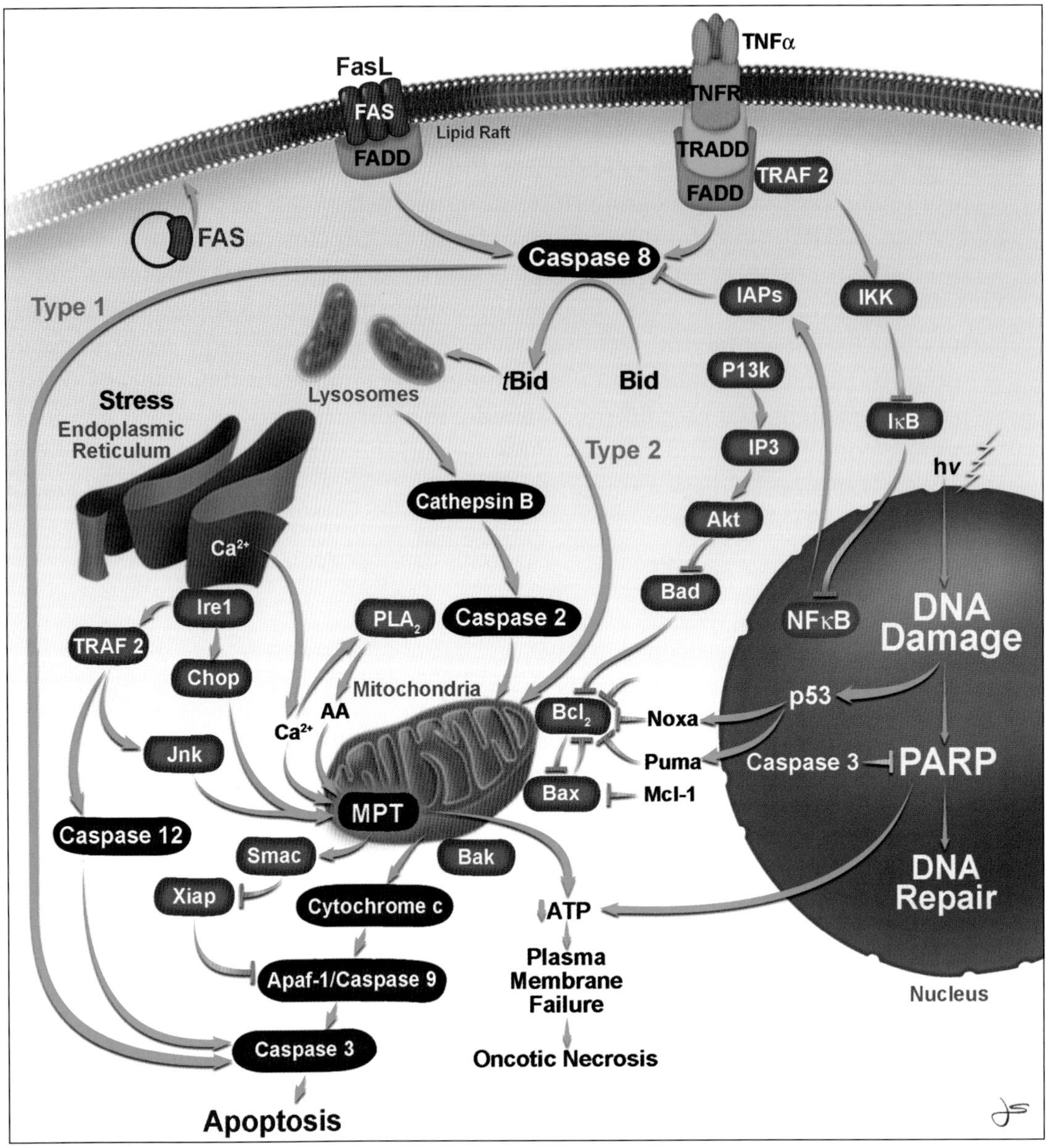

Figure 1.11 **Scheme of apoptotic signaling from organelles.** Adapted with permission from [113].

Plasma Membrane

The plasma membrane is the target of many receptor-mediated signals. In particular, various 'death ligands' (e.g., tumor necrosis factor α, or TNFα; Fas ligand; tumor necrosis factor-related apoptosis-inducing ligand, or TRAIL) acting through their corresponding receptors (TNF receptor 1, or TNFR1; Fas; death receptor 4/5, or DR4/5) initiate activation of apoptotic pathways [112,118,119]. Binding of a ligand like TNFα leads to receptor trimerization and formation of so-called Complex I through association of adapter proteins (e.g., receptor interacting protein-1, or RIP1, and TNF receptor-associated death domain protein, or TRADD). After receptor dissociation, Complex II, or death-inducing signaling complex (DISC), forms through association with Fas-associated protein with death domain (FADD) and pro-caspase 8, which are internalized. Pro-caspase 8 becomes activated and in turn proteolytically activates other downstream effectors (Figure 1.12). In Type I signaling, caspase 8 activates caspase 3 directly, whereas in Type II signaling, caspase 3 cleaves Bid (novel BH3 domain-only death agonist) to truncated Bid (*t*Bid) to activate a mitochondrial pathway to apoptosis [120]. Similar signaling occurs after association of FasL with Fas (also called CD95) and TRAIL with DR4/5.

Many events modulate death receptor signaling in the plasma membrane. For example, the extent of gene and surface expression of death receptors is an

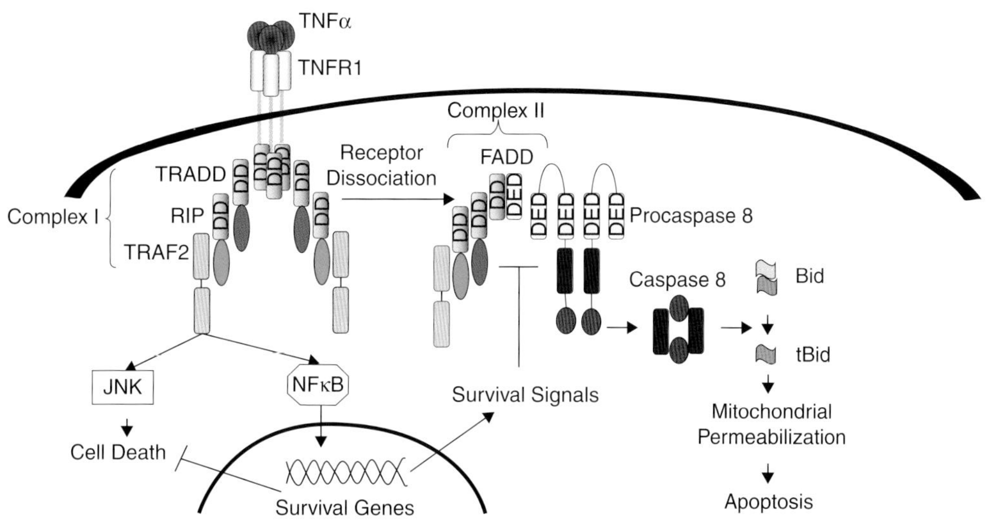

Figure 1.12 TNFα apoptotic signaling. TNFα binds to its receptor, TNFR1, and Complex I forms composed of TRADD (TNFR-associated protein with death domain), RIP (receptor-interacting protein), TRAF-2 (TNF-associated factor-2]. Complex I activates NFκB (nuclear factor kappa B), and JNK (c-jun N-terminal kinase). NFκB activates transcription of survival genes, including antiapoptotic inhibitor of apoptosis proteins (IAPs), antiapoptotic Bcl-XL, and inducible nitric oxide synthase. Complex I then undergoes ligand-dissociated internalization to form DISC Complex II. Complex II recruits FADD (Fas-associated death domain) via interactions between conserved death domains (DD) and activates procaspase 8 through interaction with death effector domains (DED). Active caspase 8 cleaves Bid to *t*Bid, which translocates to mitochondria leading to mitochondrial permeabilization, cytochrome *c* release, and apoptosis. Adapted with permission from [111].

important determinant in cellular sensitivity to death ligands. Stimuli like hydrophobic bile acids can recruit death receptors to the cell surface and sensitize cells to death-inducing stimuli [121]. Surface recruitment of death receptors may also lead to self-activation even in the absence of ligand. Death receptors localize to lipid rafts containing relatively rigid cholesterol and sphingomyelin, the latter concentrated in the outer leaflet of the bilayer. After death receptor activation, sphingomyelin hydrolysis occurs in these rafts via acid sphingomyelinase, an enzyme that may also incorporate into the plasma membrane through fusion of endomembrane vesicles. Ceramide formed from sphingolipids self-associates through hydrogen bonding to promote raft coalescence and formation of molecular platforms that cluster signal transducer components of DISC [118]. Ceramide is apoptogenic in many cells, and rearrangement of lipid rafts into larger platforms appears to be co-stimulatory factor in death receptor-mediated apoptosis [122]. Glycosphingolipids, such as ganglioside GD3, also integrate into DISCs to promote apoptosis [123]. Consistent with the importance of surface expression and sphingolipids metabolism in apoptosis, toxic hydrophobic bile acids cause a sphingomyelinase and ceramide-dependent activation of the reduced form of nicotinamide adenine dinucleotide phosphate (NADPH) oxidase generating an ROS signal activating Yes (a cellular viral sarcoma proto-oncogenic tyrosine kinase (Src) family member)-, JNK-, and epithelial growth factor receptor (EGFR)-dependent CD95 (Fas) tyrosine phosphorylation, recruitment of Fas to the cell surface, and formation of the DISC [124].

MITOCHONDRIA

Cytochrome *c* release

Mitochondria often play a central role in apoptotic signaling. In the so-called Type II pathway, DISC-activated caspase 8 proteolytically cleaves the protein Bid to a truncated fragment, *t*Bid [119,125–128]. Bid is a BH3 only domain member of the B-cell lymphoma-2 (Bcl2) family that includes both pro- and antiapoptotic proteins (Figure 1.13). *t*Bid formed after caspase 8 activation translocates to mitochondria where it interacts with either Bak (Bcl2 homologous antagonist/killer) or Bax (a conserved homolog that heterodimerizes with Bcl2), two other proapoptotic Bcl2 family members, to induce cytochrome *c* release through the outer membrane into the cytosol [129]. Cytochrome *c* in the cytosol interacts with apoptotic protease activating factor-1 (Apaf-1) and procaspase 9 to assemble haptomeric apoptosomes and an ATP (or deoxyadenosine triphosphate, or dATP)-dependent cascade of caspase 9 and caspase 3 activation [130,131]. Caspase 3, an effector caspase, in turn acts on a variety of substrates to induce the final morphological and biochemical events of apoptosis.

In many cell lines, cytochrome *c* release from the space between the mitochondrial inner and outer membranes appears to occur via formation of specific pores in the mitochondrial outer membrane. Except for the requirement for either Bak or Bax, the molecular composition and properties of cytochrome *c* release channels remain poorly understood [129,132]. In HeLa cells (a cell line derived from an atypical cervical adenocarcinoma), activated caspase 3 acts retrogradely

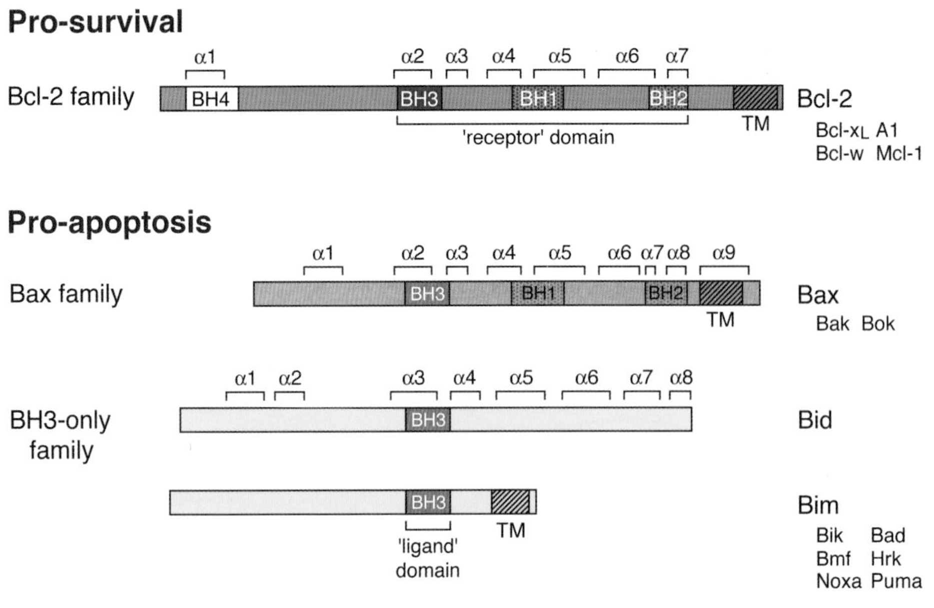

Figure 1.13 Bcl2 family proteins. BH1–4 are highly conserved domains among the Bcl2 family members. Also shown are α-helical regions. Except for A1 and BH3 only proteins, Bcl2 family members have carboxy-terminal hydrophobic domains to aid association with intracellular membranes. With permission from [211].

to proteolytically degrade a subunit of the reduced form of nicotinamide adenine dinucleotide (NADH)-ubiquinone oxidoreductase (mitochondrial respiratory Complex I) to inhibit respiration and produce mitochondrial depolarization [133,134]. By contrast in hepatocytes exposed to death ligands like TNFα, TRAIL, Fas ligand, Type II signaling leads to PT pore opening [135–138]. Cytochrome *c* release then occurs due to large amplitude mitochondrial swelling and rupture of the outer membrane [130,131,133,135, 136,138].

Because the MPT represents such a severe perturbation to mitochondria, onset of the MPT virtually assures cell death, but ensuing apoptosis or necrosis depends on other factors. If the MPT occurs rapidly and affects most mitochondria of a cell, as happens after severe oxidative stress and ischemia/reperfusion, a precipitous fall of ATP (and dATP) will occur that actually blocks apoptotic signaling by inhibiting ATP-requiring caspase 9/3 activation. With ATP depletion, oncotic necrosis ensues. However, if the MPT occurs more slowly and asynchronously among the mitochondria of a cell, or when alternative sources for ATP generation are present (e.g., glycolysis), then necrosis is prevented and caspase 9/3 becomes activated and caspase-dependent apoptosis occurs instead (Figure 1.14) [5, 30,139–142]. Indeed, because cytochrome *c* is an essential component of the mitochondrial respiratory chain, cytochrome *c* release by any mechanism causes respiratory dysfunction. Cross-talk between apoptosis and necrosis also occurs in other ways. For example, after TNFα binding to its receptor, recruitment of RIP1 to TNFR1 can activate NADPH oxidase leading to superoxide generation and sustained activation of JNK, resulting in oncotic necrosis rather than apoptosis [143].

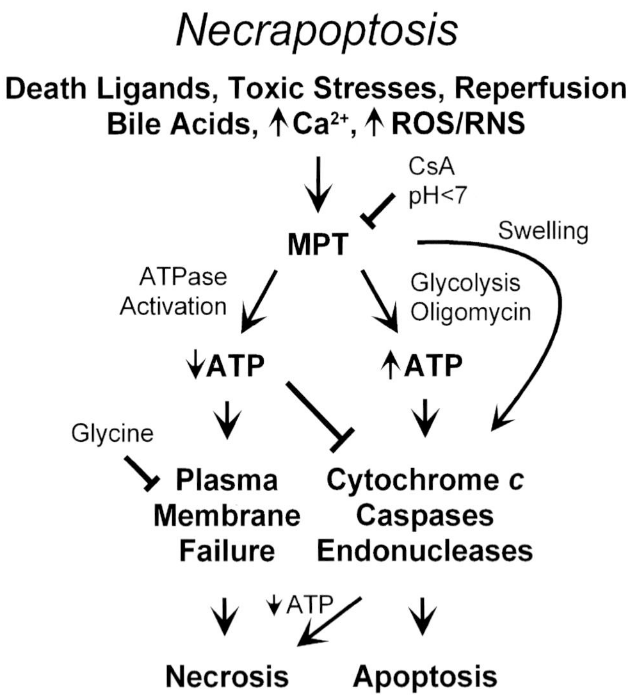

Figure 1.14 Shared pathways to apoptosis and necrosis.

Controversy remains concerning the specific roles of the MPT versus specific outer membrane permeabilization in cytochrome *c* release in apoptotic signaling. An argument against the MPT in apoptotic cell killing is that apoptosis occurs without mitochondrial swelling, but careful electron microscopy shows that mitochondrial swelling with outer membrane rupture occurs in a broad range of apoptotic models [144]. Whatever the mechanisms are, virtually all soluble mitochondrial intermembrane proteins are released along with cytochrome *c* during apoptotic signaling, including proapoptotic AIF, Smac, and endonuclease A, among others. Indeed, the outer membrane becomes permeable to all solutes up to at least two million Da molecular mass [145]. One proposal is that the cytochrome *c* release channel is a lipid ceramide barrel structure that grows progressively in size after proapoptotic signaling [146,147]. Future investigations are needed to better define the mechanisms of cytochrome *c* release. Multiple mechanisms are likely, since redundancy of apoptotic signaling exists in virtually all other aspects of apoptosis. For example, although various individual cell types are described as having Type I or Type II apoptotic signaling, in reality both pathways can co-exist, one simply producing faster signaling that the other [136]. Redundancy of signaling and multiplicity of mechanisms implies that apoptotic pathways must be individually assessed for each cell type of interest.

Regulation of the Mitochondrial Pathway to Apoptosis

Mitochondrial pathways to apoptosis vary depending on expression of procaspases, Apaf-1, and other proteins. Some terminally differentiated cells, particularly neurons, do not respond to cytochrome *c* with caspase activation and apoptosis, which may be linked to lack of Apaf-1 expression [148,149]. Antiapoptotic Bcl2 proteins, like Bcl2, Bcl extra long (Bcl-xL), and myeloid cell leukemia sequence 1 (Mcl-1), block apoptosis and are frequently overexpressed in cancer cells (Figure 1.13) [150,151]. Antiapoptotic Bcl2 family members form heterodimers with proapoptotic family members like Bax and Bak, to prevent the latter from oligomerizing into cytochrome *c* release channels. VDAC in the outer membrane is an anchoring point for Bax and other proteins regulating cytochrome *c* release [152,153].

Inhibitor of apoptosis proteins (IAPs), including X-linked inhibitor of apoptosis protein (XIAP), cellular IAP1 (c-IAP1), cellular IAP2 (c-IAP2), and survivin, oppose apoptotic signaling by inhibiting caspase activation [154,155]. Many IAPs can recruit E2 ubiquitin-conjugating enzymes and catalyse the transfer of ubiquitin onto target proteins, leading to proteosomal degradation. Some IAPs inhibit apoptotic pathways upstream of mitochondria at caspase 8, whereas others like XIAP inhibit caspase 9/3 activation downstream of mitochondrial cytochrome *c* release. Additional proteins like Smac suppress the action of IAPs, providing an "inhibitor of the inhibitor" effect promoting apoptosis. Smac is a mitochondrial intermembrane protein that is released with cytochrome *c* [154,156,157]. Smac inhibits XIAP and promotes apoptotic signaling after mitochondrial signaling. Thus, high Smac to XIAP ratios favor caspase 3 activation after cytochrome *c* release. Other proapoptotic proteins released from the mitochondrial intermembrane space during apoptotic signaling include AIF (a flavoprotein oxidoreductase that promotes DNA degradation and chromatin condensation), endonuclease G (a DNA degrading enzyme), and HtrA2/Omi (high temperature requirement A2, a serine protease that degrades IAPs) [154,158–162].

Disruption of mitochondrial function induces fragmentation of larger filamentous mitochondria into smaller more spherical structures. Such changes are also often prominent in apoptosis. Mitochondrial fission is mediated by dynamin-like protein type 1 (Drp1], a large cytosolic GTPase mechanoenzyme, and fission-1 (Fis1) in the outer membrane. Drp1 forms complexes with proapoptotic Bcl2 family members like Bax to promote cytochrome *c* release during apoptosis [163]. Mitochondrial fusion depends on optic atrophy-1 (Opa1) in the inner membrane, which is mutated in dominant optic atrophy, and mitofusin 1 and 2 (Mfn1/2), two proteins in the outer membrane [164]. Fission events in mitochondria seem to promote apoptotic signaling, since dynamin-like protein type 1 (Drp-1) overexpression promotes apoptosis, whereas Mfn1/2 overexpression retards apoptosis [164].

Antiapoptotic Survival Pathways

Ligand binding to death receptors can also activate antiapoptotic signaling to prevent activation of apoptotic death programs. Binding of the adapter protein, TNFR-associated factor 2 (TRAF2), to death receptors activates IκB kinase (IKK), which in turn phosphorylates IκB, an endogenous inhibitor of nuclear factor κB (NFκB), leading to proteosomal IκB degradation [119,165]. IκB degradation relieves inhibition of NFκB and allows NFκB to activate expression anti-apoptotic genes, including IAPs, Bcl-xL, inducible nitric oxide synthase (iNOS), and other survival factors [166,167]. Nitric oxide from iNOS produces cGMP-dependent suppression of the MPT, as well as S-nitrosation and inhibition of caspases [168,169]. As a consequence in many models, apoptosis after death receptor ligation occurs only when NFκB signaling is blocked, as after inhibition of proteosomes or protein synthesis [119,135,166]. In liver, for example, induction of hepatocellular apoptosis with TNFα requires co-treatment with galactosamine, which depletes uridine triphosphate (UTP) and thus blocks messenger ribonucleic acid (mRNA) synthesis and gene expression [170,171].

The phosphoinositide 3-kinase (PI3) kinase/protooncogene product of the viral oncogene v-akt (Akt) pathway is another source of antiapoptotic signaling [172,173]. When phosphoinositide 3-kinase (PI3 kinase) is activated by binding of insulin, insulin-like growth factor (IGF) and various other growth factors to their receptors, phosphatidylinositol trisphosphate (PIP3) is formed that activates Akt/protein kinase B, a serine/threonine protein kinase. One consequence is the phosphorylation and inactivation of Bad (heterodimeric partner for Bcl-xL), a proapoptotic Bcl2 family member, but other antiapoptotic targets of PI3 kinase/Akt signaling also exist. In cell lines, withdrawal of serum or specific growth factors typically induces apoptosis due to suppression of the PI3 kinase/Akt survival pathway.

NUCLEUS

In the so-called extrinsic pathway, death receptors initiate apoptosis by either a Type I (nonmitochondrial) or Type II (mitochondrial) caspase activation sequence. In the intrinsic pathway, by contrast, events in the nucleus activate apoptotic signaling, such as DNA damage caused by ultraviolet or ionizing (gamma) irradiation. DNA damage leads to activation of the p53 nuclear transcription factor, and expression of genes for apoptosis and/or cell-cycle arrest, especially the proapoptotic Bcl2 family members PUMA, NOXA, and Bax for apoptosis, and p21 for cell cycle arrest, especially the proapoptotic Bcl2 family members p53 upregulated modulator of apoptosis (PUMA), NOXA and Bax for apoptosis, and 21 kDa promoter (p21) for cell cycle arrest (Figure 1.11) [174,175]. PUMA, NOXA, and Bax translocate to mitochondria to induce cytochrome *c* release by similar mechanisms as discussed previously for the extrinsic pathway [176–178]. To escape p53-dependent induction of apoptosis, many tumors, especially those from the gastrointestinal tract, have loss of function mutations for p53.

DNA damage also activates PARP. With moderate activation, PARP helps mend DNA strand breaks, but with strong activation PARP exhausts cellular ATP and causes necrotic cell death. However, caspase 3 proteolytically degrades and inactivates PARP to prevent both DNA repair and this pathway to necrosis. Thus, DNA damage can lead to either necrosis or apoptosis depending on which occurs more quickly—PARP activation and ATP depletion, or caspase 3 activation and PARP degradation [179,180].

ENDOPLASMIC RETICULUM

The ER also gives rise to proapoptotic signals. Oxidative stress and other perturbations can inhibit ER calcium pumps to induce calcium release into the cytosol. Uptake of this calcium into mitochondria from the cytosol may then induce a Ca^{2+}-dependent MPT and subsequent apoptotic or necrotic cell killing (Figure 1.11) [181,182]. ER calcium release into the cytosol can also activate phospholipase A2 and the formation of arachidonic acid, another promoter of the MPT [183].

ER calcium depletion also disturbs the proper folding of newly synthesized proteins inside ER cisternae to cause ER stress and the unfolded protein response (UPR) [184–186]. Blockers of glycosylation, inhibitors of ER protein processing and secretion, various toxicants, and synthesis of mutant proteins can also cause ER stress. Calcium-binding chaperones, including glucose-regulated protein-78 (GRP78) and glucose-regulated protein-94 (GRP94), mediate detection of unfolded and misfolded proteins and promote folding of nascent unfolded proteins into a proper mature protein conformation. In the absence of unfolded/misfolded proteins, GRP78 inhibits specific sensors of ER stress, but in the presence of unfolded proteins GRP78 translocates from the sensors to the unfolded proteins to cause sensor activation by disinhibition. The main sensors of ER stress are RNA-activated protein kinase (PKR), PKR like ER kinase (PERK), type 1 ER transmembrane protein kinase (IRE1), and activating transcription factor 6 (ATF6). PKR and PERK are protein kinases whose activation leads to phosphorylation of eukaryotic initiation factor-2a (eIF-2α). Phosphorylation of eIF-2α suppresses ER protein synthesis, which decreases the rate of delivery of newly synthesized unfolded protein to the ER lumen and relieves the unfolding stress [187]. Ire1 is both a protein kinase and a riboendonuclease that initiates splicing of a preformed mRNA encoding X-box-binding protein 1 (XBP) into an active form [188]. ATF6 is another transcription factor that translocates to the Golgi after ER stress where proteases process ATF6 to an amino-terminal fragment that is taken up into the nucleus [189]. Together Ire1 and ATF6 increase gene expression of chaperones and other proteins to alleviate the unfolding stress.

A strong and persistent UPR induces Ire1- and ATF6-dependent expression of C/EBP homologous protein (CHOP) and continued activation of Ire1 to initiate apoptotic signaling (Figure 1.11). Association of TRAF2 with activated IRE1 leads to activation of caspase 12 and JNK [190,191]. Caspase 12 activates caspase 3 directly, whereas JNK and CHOP promote mitochondrial cytochrome *c* release as a pathway to caspase 3 activation. ER stress also induces Ca^{2+} release, which is taken up by mitochondria to promote mitochondrial signaling.

In addition to mitochondria, Bcl-2 family members associate with ER membranes, and proapoptotic Bcl-2 family members like Bax and Bak act to increase the size of the ER calcium store, thus increasing the proapoptotic potential of ER calcium release [192]. Chaperones require ATP to induce proper protein folding [193]. ATP depletion and ER stress may act synergistically, since perturbations that decrease ATP will augment ER stress.

LYSOSOMES

Lysosomes and the associated process of autophagy (self-digestion) are another source of proapoptotic signals. So-called autophagic cell death is characterized by an abundance of autophagic vacuoles in dying cells and is especially prominent in involuting tissues, such as post-lactation mammary gland [194]. In autophagy, isolation membranes (also called phagophores) envelop and then sequester portions of cytoplasm to form double membrane autophagosomes. Autophagosomes fuse with lysosomes and late endosomes to form autolysosomes [195,196]. The process of autophagy acts to remove and degrade cellular constituents, an appropriate action for a tissue undergoing involution. Originally considered to be random, much evidence

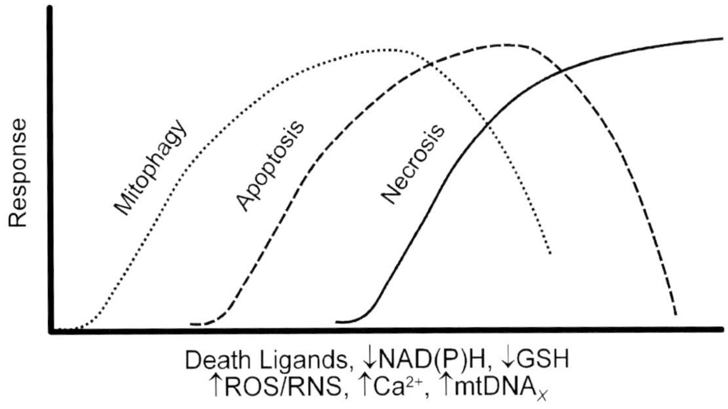

Figure 1.15 **Progression of mitophagy, apoptosis, and necrosis**. Stimuli that induce the MPT produce a graded cellular response. Low levels of stimulation induce autophagy as a repair mechanism. With more stimulation, apoptosis begins to occur in addition due to cytochrome *c* release after mitochondrial swelling. Necrosis becomes evident after even stronger stimulation as ATP becomes depleted. With highest stimulation, autophagy and apoptosis as ATP-requiring processes become inhibited, and only oncotic necrosis occurs. MPT inducers include death ligands, oxidation of NAD(P)H and GSH, formation of reactive oxygen species (ROS) and reactive nitrogen species (RNS), and mutation of mitochondrial DNA ($mtDNA_X$) which causes synthesis of abnormal mitochondrial membrane proteins.

suggests that autophagy can be selective for specific organelles, especially if they are damaged [197,198]. For example, stresses inducing the MPT seem to signal autophagy of mitochondria [199].

Whether or not autophagy promotes or prevents cell death is controversial [200–203]. In some circumstances, autophagy promotes cell death because suppression of expression of certain autophagy genes decreases apoptosis, and some have suggested that autophagic cell death as its own category of programmed cell killing. Under other conditions, autophagy protects against cell death, since disruption of autophagic processing and/or lysosomal function promotes caspase-dependent cell death [201,202,204]. When autophagic processing and lysosomal degradation are disrupted, cathepsins and other hydrolases can be released from lysosomes and autophagosomes to initiate mitochondrial permeabilization and caspase activation. Cathepsin B is released from lysosomes (or related structures such as late endosomes) during TNFα signaling to augment death receptor-mediated apoptosis and contribute to mitochondrial release of cytochrome *c* [205,206]. In addition, lysosomal extracts cleave Bid to *t*Bid, and cathepsin D, another lysosomal protease, activates Bax [207,208].

Necrapoptosis/Aponecrosis

In many and possibly most instances of apoptosis, mitochondrial permeabilization with release of cytochrome *c* and other proapoptotic factors is a final common pathway leading to a final and committed phase. At higher levels of stimulation, the same factors that induce apoptosis frequently also cause ATP depletion and a necrotic mode of cell death. Such necrotic cell killing is a consequence of mitochondrial dysfunction. Such shared pathways leading to different modes of cell death constitute necrapoptosis (or aponecrosis) [5–7]. In general, apoptosis is a better outcome for the organism since apoptosis promotes orderly resorption of dying cells, whereas necrotic cell death releases cellular constituents into the extracellular space to induce an inflammatory release that can extend tissue injury. Because of shared pathways, an admixture of necrosis and apoptosis occurs in many pathophysiological settings.

For stimuli inducing the MPT, a graded response seems to occur (Figure 1.15) [6]. When limited to a few mitochondria, the MPT stimulates mitochondrial autophagy (mitophagy) and elimination of the damaged organelles—a repair mechanism. With greater stimulation, more mitochondria undergo the MPT, and apoptosis begins to occur due to cytochrome *c* release from swollen mitochondria leading to caspase activation. Cathepsin leakage from an overstimulated autophagic apparatus likely also promotes apoptotic signaling. Indeed, autophagy is often a prominent feature in cells undergoing programmed cell death. As the majority of mitochondria undergo the MPT, oxidative phosphorylation fails and ATP plummets, which precipitates necrotic cell death while simultaneously suppressing ATP-requiring apoptotic signaling.

CONCLUDING REMARK

Apoptosis and necrosis are prominent events in pathogenesis. An understanding of cell death mechanisms forms the basis for effective interventions to either prevent cell death as a cause of disease or promote cell death in cancer chemotherapy.

ACKNOWLEDGMENTS

This work was supported, in part, by Grants DK37034, DK73336, DK70844, DK070195, and AA016011 from the National Institutes of Health.

REFERENCES

1. Majno G, Joris I. Apoptosis, oncosis, and necrosis. An overview of cell death. *Am J Pathol.* 1995;146(1):3–15.
2. Van CS, Van Den BW. Morphological and biochemical aspects of apoptosis, oncosis and necrosis. *Anat Histol Embryol.* 2002;31(4): 214–223.
3. Kerr JF, Wyllie AH, Currie AR. Apoptosis: A basic biological phenomenon with wide-ranging implications in tissue kinetics. *Br J Cancer.* 1972;26(4):239–257.
4. Searle J, Harmon BV, Bishop CJ, et al. The significance of cell death by apoptosis in hepatobiliary disease. *J Gastroenterol Hepatol.* 1987;2:77–96.
5. Lemasters JJ. V. Necrapoptosis and the mitochondrial permeability transition: Shared pathways to necrosis and apoptosis. *Am J Physiol.* 1999;276(1 Pt 1):G1–G6.
6. Lemasters JJ. Modulation of mitochondrial membrane permeability in pathogenesis, autophagy and control of metabolism. *J Gastroenterol Hepatol.* 2007;22(Suppl 1):S31–37.
7. Formigli L, Papucci L, Tani A, et al. Aponecrosis: Morphological and biochemical exploration of a syncretic process of cell death sharing apoptosis and necrosis. *J Cell Physiol.* 2000; 182(1):41–49.
8. Trump BF, Goldblatt PJ, Stowell RE. Studies of necrosis in vitro of mouse hepatic parenchymal cells. Ultrastructural alterations in endoplasmic reticulum, Golgi apparatus, plasma membrane, and lipid droplets. *Lab Invest.* 1965;14:2000–2028.
9. Lemasters JJ, Ji S, Thurman RG. Centrilobular injury following hypoxia in isolated, perfused rat liver. *Science.* 1981;213(4508): 661–663.
10. Lemasters JJ, Stemkowski CJ, Ji S, et al. Cell surface changes and enzyme release during hypoxia and reoxygenation in the isolated, perfused rat liver. *J Cell Biol.* 1983;97(3):778–786.
11. Gores GJ, Herman B, Lemasters JJ. Plasma membrane bleb formation and rupture: A common feature of hepatocellular injury. *Hepatology.* 1990;11(4):690–698.
12. Nishimura Y, Romer LH, Lemasters JJ. Mitochondrial dysfunction and cytoskeletal disruption during chemical hypoxia to cultured rat hepatic sinusoidal endothelial cells: The pH paradox and cytoprotection by glucose, acidotic pH, and glycine. *Hepatology.* 1998;27(4):1039–1049.
13. Gores GJ, Nieminen A-L, Wray BE, et al. Intracellular pH during 'chemical hypoxia' in cultured hepatocytes. *J Clin Invest.* 1989; 83:386–396.
14. Zahrebelski G, Nieminen AL, al Ghoul K, et al. Progression of subcellular changes during chemical hypoxia to cultured rat hepatocytes: A laser scanning confocal microscopic study. *Hepatology.* 1995;21(5):1361–1372.
15. Herman B, Nieminen AL, Gores GJ, et al. Irreversible injury in anoxic hepatocytes precipitated by an abrupt increase in plasma membrane permeability. *FASEB J.* 1988;2(2):146–151.
16. Nieminen AL, Gores GJ, Wray BE, et al. Calcium dependence of bleb formation and cell death in hepatocytes. *Cell Calcium.* 1988; 9(5–6):237–246.
17. Dong Z, Patel Y, Saikumar P, et al. Development of porous defects in plasma membranes of adenosine triphosphate-depleted Madin-Darby canine kidney cells and its inhibition by glycine. *Lab Invest.* 1998;78(6):657–668.
18. Nishimura Y, Lemasters JJ. Glycine blocks opening of a death channel in cultured hepatic sinusoidal endothelial cells during chemical hypoxia. *Cell Death Differ.* 2001;8(8):850–858.
19. Weinberg JM, Davis JA, Abarzua M, et al. Cytoprotective effects of glycine and glutathione against hypoxic injury to renal tubules. *J Clin Invest.* 1987;80:1446–1454.
20. Miller GW, Schnellmann RG. Cytoprotection by inhibition of chloride channels: The mechanism of action of glycine and strychnine. *Life Sci.* 1993;53(15):1211–1215.
21. Dong Z, Venkatachalam MA, Weinberg JM, et al. Protection of ATP-depleted cells by impermeant strychnine derivatives: Implications for glycine cytoprotection. *Am J Pathol.* 2001;158(3): 1021–1028.
22. Pan C, Bai X, Fan L, et al. Cytoprotection by glycine against ATP-depletion-induced injury is mediated by glycine receptor in renal cells. *Biochem J.* 2005;390(Pt 2):447–453.
23. Saraste M. Oxidative phosphorylation at the fin de siecle. *Science.* 1999;283(5407):1488–1493.
24. Nieminen AL, Dawson TL, Gores GJ, et al. Protection by acidotic pH and fructose against lethal injury to rat hepatocytes from mitochondrial inhibitors, ionophores and oxidant chemicals. *Biochem Biophys Res Commun.* 1990;167(2):600–606.
25. Anundi I, King J, Owen DA, et al. Fructose prevents hypoxic cell death in liver. *Am J Physiol.* 1987;253(3 Pt 1):G390–G396.
26. Nieminen AL, Saylor AK, Herman B, et al. ATP depletion rather than mitochondrial depolarization mediates hepatocyte killing after metabolic inhibition. *Am J Physiol.* 1994;267(1 Pt 1):C67–C74.
27. Mayes PA. Intermediatry metabolism of fructose. *Am J Clin Nutr.* 1993;58(Suppl):754S–765S.
28. Nicholls DG, Ferguson SJ. *Bioenergetics 3.* London: Academic Press; 2002.
29. Imberti R, Nieminen AL, Herman B, et al. Mitochondrial and glycolytic dysfunction in lethal injury to hepatocytes by t-butylhydroperoxide: Protection by fructose, cyclosporin A and trifluoperazine. *J Pharmacol Exp Ther.* 1993;265(1): 392–400.
30. Qian T, Herman B, Lemasters JJ. The mitochondrial permeability transition mediates both necrotic and apoptotic death of hepatocytes exposed to Br-A23187. *Toxicol Appl Pharmacol.* 1999; 154(2):117–125.
31. Colombini M. VDAC: The channel at the interface between mitochondria and the cytosol. *Mol Cell Biochem.* 2004;256–257 (1–2):107–115.
32. Rostovtseva TK, Komarov A, Bezrukov SM, Colombini M. VDAC channels differentiate between natural metabolites and synthetic molecules. *J Membr Biol.* 2002;187(2):147–156.
33. Hunter DR, Haworth RA, Southard JH. Relationship between configuration, function, and permeability in calcium-treated mitochondria. *J Biol Chem.* 1976;251:5069–5077.
34. Bernardi P, Forte M. The mitochondrial permeability transition pore. *Novartis Found Symp.* 2007;287:157–164; discussion 164–169.
35. Kim JS, He L, Lemasters JJ. Mitochondrial permeability transition: A common pathway to necrosis and apoptosis. *Biochem Biophys Res Commun.* 2003;304(3):463–470.
36. Zoratti M, Szabo I. The mitochondrial permeability transition. *Biochim Biophys Acta.* 1995;1241(2):139–176.
37. Halestrap AP, Brennerb C. The adenine nucleotide translocase: A central component of the mitochondrial permeability transition pore and key player in cell death. *Curr Med Chem.* 2003;10 (16):1507–1525.
38. Crompton M, Virji S, Doyle V, et al. The mitochondrial permeability transition pore. *Biochem Soc Symp.* 1999; 66:167–179.
39. Kokoszka JE, Waymire KG, Levy SE, et al. The ADP/ATP translocator is not essential for the mitochondrial permeability transition pore. *Nature.* 2004;427(6973): 461–465.
40. Krauskopf A, Eriksson O, Craigen WJ, et al. Properties of the permeability transition in VDAC1(-/-) mitochondria. *Biochim Biophys Acta.* 2006;1757(5–6):590–595.
41. Juhaszova M, Wang S, Zorov DB, et al. The identity and regulation of the mitochondrial permeability transition pore: Where the known meets the unknown. *Ann NY Acad Sci.* 2008;1123: 197–212.
42. Forte M, Bernardi P. Genetic dissection of the permeability transition pore. *J Bioenerg Biomembr.* 2005;37(3):121–128.
43. He L, Lemasters JJ. Regulated and unregulated mitochondrial permeability transition pores: A new paradigm of pore structure and function? *FEBS Lett.* 2002;512(1–3):1–7.
44. Gores GJ, Nieminen AL, Wray BE, et al. Intracellular pH during "chemical hypoxia" in cultured rat hepatocytes. Protection by intracellular acidosis against the onset of cell death. *J Clin Invest.* 1989;83(2):386–396.
45. Bronk SF, Gores GJ. Efflux of protons from acidic vesicles contributes to cytosolic acidification of hepatocytes during ATP depletion. *Hepatology.* 1991;14(4 Pt 1):626–633.
46. Penttila A, Trump BF. Extracellular acidosis protects Ehrlich tumor cells and rat renal cortex against anoxic injury. *Science.* 1974;185:277–278.
47. Bonventre JV, Cheung JY. Effects of metabolic acidosis on viability of cells exposed to anoxia. *Am J Physiol.* 1985;249:C149–C159.
48. Gores GJ, Nieminen AL, Fleishman KE, et al. Extracellular acidosis delays onset of cell death in ATP-depleted hepatocytes. *Am J Physiol.* 1988;255(3 Pt 1):C315–C322.

49. Currin RT, Gores GJ, Thurman RG, et al. Protection by acidotic pH against anoxic cell killing in perfused rat liver: Evidence for a pH paradox. *FASEB J.* 1991;5(2):207–210.
50. Bond JM, Herman B, Lemasters JJ. Protection by acidotic pH against anoxia/reoxygenation injury to rat neonatal cardiac myocytes. *Biochem Biophys Res Commun.* 1991;179(2):798–803.
51. Bond JM, Chacon E, Herman B, et al. Intracellular pH and Ca2+ homeostasis in the pH paradox of reperfusion injury to neonatal rat cardiac myocytes. *Am J Physiol.* 1993;265(1 Pt 1):C129–C137.
52. Harper IS, Bond JM, Chacon E, et al. Inhibition of Na+/H+ exchange preserves viability, restores mechanical function, and prevents the pH paradox in reperfusion injury to rat neonatal myocytes. *Basic Res Cardiol.* 1993; 88(5):430–442.
53. Zager RA, Schimpf BA, Gmur DJ. Physiological pH. Effects on posthypoxic proximal tubular injury. *Circ Res.* 1993;72:837–846.
54. Kaplan SH, Yang H, Gilliam DE, et al. Hypercapnic acidosis and dimethyl amiloride reduce reperfusion induced cell death in ischaemic ventricular myocardium. *Cardiovasc Res.* 1995;29(2): 231–238.
55. Qian T, Nieminen AL, Herman B, et al. Mitochondrial permeability transition in pH-dependent reperfusion injury to rat hepatocytes. *Am J Physiol.* 1997;273(6 Pt 1):C1783–C1792.
56. Kim JS, Jin Y, Lemasters JJ. Reactive oxygen species, but not Ca2+ overloading, trigger pH- and mitochondrial permeability transition-dependent death of adult rat myocytes after ischemia-reperfusion. *Am J Physiol Heart Circ Physiol.* 2006;290(5):H2024–H2034.
57. Griffiths EJ, Halestrap AP. Protection by cyclosporin A of ischemia/reperfusion-induced damage in isolated rat hearts. *J Mol Cell Cardiol.* 1993;25(12):1461–1469.
58. Griffiths EJ, Halestrap AP. Mitochondrial non-specific pores remain closed during cardiac ischaemia, but open upon reperfusion. *Biochem J.* 1995;307(Pt 1):93–98.
59. Piot C, Croisille P, Staat P, et al. Effect of cyclosporine on reperfusion injury in acute myocardial infarction. *N Engl J Med.* 2008; 359(5):473–481.
60. Chu HC, Lin YL, Sytwu HK, et al. Effects of minocycline on Fas-mediated fulminant hepatitis in mice. *Br J Pharmacol.* 2005;144 (2):275–282.
61. Friedlander RM. Mechanisms of disease: Apoptosis and caspases in neurodegenerative diseases. *N Engl J Med.* 2003;348(14): 1365–1375.
62. Gao W, Washington MK, Bentley RC, et al. Antiangiogenic agents protect liver sinusoidal lining cells from cold preservation injury in rat liver transplantation. *Gastroenterology.* 1997;113(5): 1692–1700.
63. Kelly KJ, Sutton TA, Weathered N, et al. Minocycline inhibits apoptosis and inflammation in a rat model of ischemic renal injury. *Am J Physiol Renal Physiol.* 2004;287(4):F760–F766.
64. Wang J, Wei Q, Wang CY, et al. Minocycline up-regulates Bcl-2 and protects against cell death in mitochondria. *J Biol Chem.* 2004;279(19):19948–19954.
65. Wang X, Zhu S, Drozda M, et al. Minocycline inhibits caspase-independent and -dependent mitochondrial cell death pathways in models of Huntington's disease. *Proc Natl Acad Sci USA.* 2003; 100(18):10483–10487.
66. Wells JEA, Hurlbert RJ, Fehlings MG, et al. Neuroprotection by minocycline facilitates significant recovery from spinal cord injury in mice. *Brain.* 2003;126:1628–1637.
67. Yong VW, Wells J, Giuliani F, et al. The promise of minocycline in neurology. *Lancet Neurology.* 2004;3(12):744–751.
68. Zhu S, Stavrovskaya IG, Drozda M, et al. Minocycline inhibits cytochrome c release and delays progression of amyotrophic lateral sclerosis in mice. *Nature.* 2002;417(6884):74–78.
69. Theruvath TP, Zhong Z, Pediaditakis P, et al. Minocycline and N-methyl-4-isoleucine cyclosporin (NIM811) mitigate storage/reperfusion injury after rat liver transplantation through suppression of the mitochondrial permeability transition. *Hepatology.* 2008;47(1): 236–246.
70. Imberti R, Nieminen AL, Herman B, et al. Synergism of cyclosporin A and phospholipase inhibitors in protection against lethal injury to rat hepatocytes from oxidant chemicals. *Res Commun Chem Pathol Pharmacol.* 1992;78(1):27–38.
71. Nieminen AL, Saylor AK, Tesfai SA, et al. Contribution of the mitochondrial permeability transition to lethal injury after exposure of hepatocytes to t-butylhydroperoxide. *Biochem J.* 1995;307(Pt 1):99–106.
72. Nieminen AL, Byrne AM, Herman B, et al. Mitochondrial permeability transition in hepatocytes induced by t-BuOOH: NAD(P)H and reactive oxygen species. *Am J Physiol.* 1997;272 (4 Pt 1):C1286–C1294.
73. Byrne AM, Lemasters JJ, Nieminen AL. Contribution of increased mitochondrial free Ca2+ to the mitochondrial permeability transition induced by tert-butylhydroperoxide in rat hepatocytes. *Hepatology.* 1999;29(5):1523–1531.
74. Greenwood SM, Connolly CN. Dendritic and mitochondrial changes during glutamate excitotoxicity. *Neuropharmacology.* 2007;53(8):891–898.
75. Stefanis L. Caspase-dependent and -independent neuronal death: Two distinct pathways to neuronal injury. *Neuroscientist.* 2005;11(1):50–62.
76. Dubinsky JM, Brustovetsky N, Pinelis V, et al. The mitochondrial permeability transition: The brain's point of view. *Biochem Soc Symp.* 1999;66:75–84.
77. Nieminen AL, Petrie TG, Lemasters JJ, et al. Cyclosporin A delays mitochondrial depolarization induced by N-methyl-D-aspartate in cortical neurons: Evidence of the mitochondrial permeability transition. *Neuroscience.* 1996;75(4):993–997.
78. Kehrer JP. The Haber-Weiss reaction and mechanisms of toxicity. *Toxicology.* 2000;149(1):43–50.
79. Swanson CA. Iron intake and regulation: Implications for iron deficiency and iron overload. *Alcohol.* 2003;30(2):99–102.
80. Petersen DR. Alcohol, iron-associated oxidative stress, and cancer. *Alcohol.* 2005;35(3):243–249.
81. Brewer GJ. Iron and copper toxicity in diseases of aging, particularly atherosclerosis and Alzheimer's disease. *Exp Biol Med (Maywood).* 2007;232(2):323–335.
82. Kerkweg U, Li T, de Groot H, et al. Cold-induced apoptosis of rat liver cells in University of Wisconsin solution: The central role of chelatable iron. *Hepatology.* 2002;35(3):560–567.
83. Rauen U, Kerkweg U, de GH. Iron-dependent vs. iron-independent cold-induced injury to cultured rat hepatocytes: A comparative study in physiological media and organ preservation solutions. *Cryobiology.* 2007;54(1):77–86.
84. Rauen U, Petrat F, Sustmann R, et al. Iron-induced mitochondrial permeability transition in cultured hepatocytes. *J. Hepatology.* 2004;40:607–615.
85. Wildenthal K, Decker RS. The role of lysosomes in the heart. *Adv Myocardiol.* 1980;2:349–358.
86. Ollinger K, Brunk UT. Cellular injury induced by oxidative stress is mediated through lysosomal damage. *Free Radic Biol Med.* 1995;19(5):565–574.
87. Kurz T, Terman A, Brunk UT. Autophagy, ageing and apoptosis: The role of oxidative stress and lysosomal iron. *Arch Biochem Biophys.* 2007;462(2):220–230.
88. Persson HL, Yu Z, Tirosh O, et al. Prevention of oxidant-induced cell death by lysosomotropic iron chelators. *Free Radic Biol Med.* 2003;34(10):1295–1305.
89. Yu Z, Persson HL, Eaton JW, et al. Intralysosomal iron: A major determinant of oxidant-induced cell death. *Free Radic Biol Med.* 2003;34(10):1243–1252.
90. Kurz T, Gustafsson B, Brunk UT. Intralysosomal iron chelation protects against oxidative stress-induced cellular damage. *FEBS J.* 2006;273:3106–3117.
91. Uchiyama A, Kim J-S, Kon K, et al. Translocation of iron from lysosomes into mitochondria is a key event during oxidative stress-induced hepatocellular injury. *Hepatology.* 2008;In press.
92. Gores GJ, Flarsheim CE, Dawson TL, et al. Swelling, reductive stress, and cell death during chemical hypoxia in hepatocytes. *Am J Physiol.* 1989;257(2 Pt 1):C347–C354.
93. Kim JS, Ohshima S, Pediaditakis P, et al. Nitric oxide protects rat hepatocytes against reperfusion injury mediated by the mitochondrial permeability transition. *Hepatology.* 2004;39(6): 1533–1543.
94. Costa AD, Garlid KD. Intramitochondrial signaling: Interactions among mitoKATP, PKCε, ROS, and MPT. *Am J Physiol Heart Circ Physiol.* 2008;295(2):H874–H882.
95. Gomez L, Paillard M, Thibault H, et al. Inhibition of GSK3beta by postconditioning is required to prevent opening of the

mitochondrial permeability transition pore during reperfusion. *Circulation.* 2008;117(21):2761–2768.
96. Theruvath TP, Snoddy MC, Zhong Z, et al. Mitochondrial permeability transition in liver ischemia and reperfusion: Role of c-Jun N-terminal kinase 2. *Transplantation.* 2008;85(10):1500–1504.
97. Mozaffari MS, Schaffer SW. Effect of pressure overload on cardioprotection of mitochondrial KATP channels and GSK-3beta: Interaction with the MPT pore. *Am J Hypertens.* 2008;21(5):570–575.
98. Obame FN, Plin-Mercier C, Assaly R, et al. Cardioprotective effect of morphine and a blocker of glycogen synthase kinase 3 beta, SB216763 [3–(2,4–dichlorophenyl)-4(1–methyl-1H-indol-3-yl)-1H-pyrrole-2,5-dione], via inhibition of the mitochondrial permeability transition pore. *J Pharmacol Exp Ther.* 2008;326(1):252–258.
99. Latchoumycandane C, Goh CW, Ong MM, et al. Mitochondrial protection by the JNK inhibitor leflunomide rescues mice from acetaminophen-induced liver injury. *Hepatology.* 2007;45(2):412–421.
100. Hassa PO, Haenni SS, Elser M, et al. Nuclear ADP-ribosylation reactions in mammalian cells: Where are we today and where are we going? *Microbiol Mol Biol Rev.* 2006;70(3):789–829.
101. Berger NA, Whitacre CM, Hashimoto H, et al. NAD and poly (ADP-ribose) regulation of proteins involved in response to cellular stress and DNA damage. *Biochimie.* 1995;77(5):364–367.
102. Pacher P, Szabo C. Role of poly(ADP-ribose) polymerase 1 (PARP-1) in cardiovascular diseases: The therapeutic potential of PARP inhibitors. *Cardiovasc Drug Rev.* 2007;25(3):235–260.
103. Shall S, de MG. Poly(ADP-ribose) polymerase-1: What have we learned from the deficient mouse model? *Mutat Res.* 2000;460(1):1–15.
104. Heeres JT, Hergenrother PJ. Poly(ADP-ribose) makes a date with death. *Curr Opin Chem Biol.* 2007;11(6):644–653.
105. Bernheimer AW, Rudy B. Interactions between membranes and cytolytic peptides. *Biochim Biophys Acta.* 1986;864(1):123–141.
106. Cole DS, Morgan BP. Beyond lysis: How complement influences cell fate. *Clin Sci (Lond).* 2003;104(5):455–466.
107. Podack ER, Tschopp J. Membrane attack by complement. *Mol Immunol.* 1984;21(7):589–603.
108. Kim R, Emi M, Tanabe K, et al. The role of Fas ligand and transforming growth factor beta in tumor progression: Molecular mechanisms of immune privilege via Fas-mediated apoptosis and potential targets for cancer therapy. *Cancer.* 2004;100(11):2281–2291.
109. Black DM, Behrns KE. A scientist revisits the atrophy-hypertrophy complex: Hepatic apoptosis and regeneration. *Surg Oncol Clin N Am.* 2002;11(4):849–864.
110. Stappenbeck TS, Wong MH, Saam JR, et al. Notes from some crypt watchers: Regulation of renewal in the mouse intestinal epithelium. *Curr Opin Cell Biol.* 1998;10(6):702–709.
111. Malhi H, Gores GJ, Lemasters JJ. Apoptosis and necrosis in the liver: A tale of two deaths? *Hepatology.* 2006;43(2 Suppl 1):S31–S44.
112. Malhi H, Gores GJ. Cellular and molecular mechanisms of liver injury. *Gastroenterology.* 2008;134(6):1641–1654.
113. Lemasters JJ. Dying a thousand deaths: Redundant pathways from different organelles to apoptosis and necrosis. *Gastroenterology.* 2005;129(1):351–360.
114. Oshima RG. Apoptosis and keratin intermediate filaments. *Cell Death Differ.* 2002;9(5):486–492.
115. Kramer A, Liashkovich I, Oberleithner H, et al. Apoptosis leads to a degradation of vital components of active nuclear transport and a dissociation of the nuclear lamina. *Proc Natl Acad Sci USA.* 2008;105(32):11236–11241.
116. Gavrieli Y, Sherman Y, Ben-Sasson SA. Identification of programmed cell death in situ via specific labeling of nuclear DNA fragmentation. *J Cell Biol.* 1992;119(3):493–501.
117. Lorenzo HK, Susin SA. Therapeutic potential of AIF-mediated caspase-independent programmed cell death. *Drug Resist Updat.* 2007;10(6):235–255.
118. Peter ME, Krammer PH. The CD95(APO-1/Fas) DISC and beyond. *Cell Death Differ.* 2003;10(1):26–35.
119. Ding WX, Yin XM. Dissection of the multiple mechanisms of TNF-alpha-induced apoptosis in liver injury. *J Cell Mol Med.* 2004;8(4):445–454.
120. Scaffidi C, Fulda S, Srinivasan A, et al. Two CD95 (APO-1/Fas) signaling pathways. *EMBO J.* 1998;17:1675–1687.
121. Higuchi H, Gores GJ. Bile acid regulation of hepatic physiology: IV. Bile acids and death receptors. *Am J Physiol Gastrointest Liver Physiol.* 2003;284(5):G734–G738.
122. Gulbins E, Kolesnick R. Raft ceramide in molecular medicine. *Oncogene.* 2003;22(45):7070–7077.
123. Morales A, Colell A, Mari M, et al. Glycosphingolipids and mitochondria: Role in apoptosis and disease. *Glycoconj J.* 2004;20(9):579–588.
124. Reinehr R, Haussinger D. CD95 activation in the liver: Ion fluxes and oxidative signaling. *Arch Biochem Biophys.* 2007;462(2):124–131.
125. Liu X, Kim CN, Yang J, et al. Induction of apoptotic program in cell-free extracts: Requirement for dATP and cytochrome c. *Cell.* 1996;86:147–157.
126. Luo X, Budihardjo I, Zou H, et al. Bid, a Bcl2 interacting protein, mediates cytochrome c release from mitochondria in response to activation of cell surface death receptors. *Cell.* 1998;94(4):481–490.
127. Li H, Zhu H, Xu CJ, et al. Cleavage of BID by caspase 8 mediates the mitochondrial damage in the Fas pathway of apoptosis. *Cell.* 1998;94(4):491–501.
128. Gross A, Yin XM, Wang K, et al. Caspase cleaved BID targets mitochondria and is required for cytochrome c release, while BCL-XL prevents this release but not tumor necrosis factor-R1/Fas death. *J Biol Chem.* 1999;274(2):1156–1163.
129. Wei MC, Zong WX, Cheng EH, et al. Proapoptotic BAX and BAK: A requisite gateway to mitochondrial dysfunction and death. *Science.* 2001;292(5517):727–730.
130. Green DR, Kroemer G. The pathophysiology of mitochondrial cell death. *Science.* 2004;305(5684):626–629.
131. Ferraro E, Corvaro M, Cecconi F. Physiological and pathological roles of Apaf1 and the apoptosome. *J Cell Mol Med.* 2003;7(1):21–34.
132. Ow YL, Green DR, Hao Z, et al. Cytochrome c: Functions beyond respiration. *Nat Rev Mol Cell Biol.* 2008;9(7):532–542.
133. Bossy-Wetzel E, Newmeyer DD, Green DR. Mitochondrial cytochrome c release in apoptosis occurs upstream of DEVD-specific caspase activation and independently of mitochondrial transmembrane depolarization. *EMBO J.* 1998;17(1):37–49.
134. Ricci JE, Munoz-Pinedo C, Fitzgerald P, et al. Disruption of mitochondrial function during apoptosis is mediated by caspase cleavage of the p75 subunit of complex I of the electron transport chain. *Cell.* 2004;117(6):773–786.
135. Bradham CA, Qian T, Streetz K, et al. The mitochondrial permeability transition is required for tumor necrosis factor alpha-mediated apoptosis and cytochrome c release. *Mol Cell Biol.* 1998;18(11):6353–6364.
136. Hatano E, Bradham CA, Stark A, et al. The mitochondrial permeability transition augments Fas-induced apoptosis in mouse hepatocytes. *J Biol Chem.* 2000;275(16):11814–11823.
137. Black D, Bird MA, Samson CM, et al. Primary cirrhotic hepatocytes resist TGFbeta-induced apoptosis through a ROS-dependent mechanism. *J Hepatol.* 2004;40(6):942–951.
138. Zhao Y, Ding WX, Qian T, Watkins S, Lemasters JJ, Yin XM. Bid activates multiple mitochondrial apoptotic mechanisms in primary hepatocytes after death receptor engagement. *Gastroenterology.* 2003;125(3):854–867.
139. Kim JS, Qian T, Lemasters JJ. Mitochondrial permeability transition in the switch from necrotic to apoptotic cell death in ischemic rat hepatocytes. *Gastroenterology.* 2003;124(2):494–503.
140. Jaeschke H, Lemasters JJ. Apoptosis versus oncotic necrosis in hepatic ischemia/reperfusion injury. *Gastroenterology.* 2003;125(4):1246–1257.
141. Kon K, Kim JS, Jaeschke H, et al. Mitochondrial permeability transition in acetaminophen-induced necrosis and apoptosis of cultured mouse hepatocytes. *Hepatology.* 2004;40(5):1170–1179.
142. Denecker G, Vercammen D, Declercq W, et al. Apoptotic and necrotic cell death induced by death domain receptors. *Cell Mol Life Sci.* 2001;58(3):356–370.
143. Kim YS, Morgan MJ, Choksi S, et al. TNF-induced activation of the Nox1 NADPH oxidase and its role in the induction of necrotic cell death. *Mol Cell.* 2007;26(5):675–687.

144. Sesso A, Marques MM, Monteiro MM, et al. Morphology of mitochondrial permeability transition: Morphometric volumetry in apoptotic cells. *Anat Rec A Discov Mol Cell Evol Biol.* 2004;281(2):1337–1351.
145. Kuwana T, Mackey MR, Perkins G, et al. Bid, Bax, and lipids cooperate to form supramolecular openings in the outer mitochondrial membrane. *Cell.* 2002;111(3):331–342.
146. Siskind LJ, Kolesnick RN, Colombini M. Ceramide channels increase the permeability of the mitochondrial outer membrane to small proteins. *J Biol Chem.* 2002;277(30):26796–26803.
147. Siskind LJ, Davoody A, Lewin N, et al. Enlargement and contracture of C2-ceramide channels. *Biophys J.* 2003;85(3):1560–1575.
148. Chang LK, Putcha GV, Deshmukh M, et al. Mitochondrial involvement in the point of no return in neuronal apoptosis. *Biochimie.* 2002;84(2–3):223–231.
149. Johnson CE, Huang YY, Parrish AB, et al. Differential Apaf-1 levels allow cytochrome c to induce apoptosis in brain tumors but not in normal neural tissues. *Proc Natl Acad Sci USA.* 2007;104(52):20820–20825.
150. Scorrano L, Korsmeyer SJ. Mechanisms of cytochrome c release by proapoptotic BCL-2 family members. *Biochem Biophys Res Commun.* 2003;304(3):437–444.
151. Packham G, Stevenson FK. Bodyguards and assassins: Bcl-2 family proteins and apoptosis control in chronic lymphocytic leukaemia. *Immunology.* 2005;114(4):441–449.
152. Tsujimoto Y, Shimizu S. The voltage-dependent anion channel: An essential player in apoptosis. *Biochimie.* 2002;84(2–3):187–193.
153. Adachi M, Higuchi H, Miura S, et al. Bax interacts with the voltage-dependent anion channel and mediates ethanol-induced apoptosis in rat hepatocytes. *Am J Physiol Gastrointest Liver Physiol.* 2004;287(3):G695–G705.
154. Vaux DL, Silke J. IAPs, RINGs and ubiquitylation. *Nat Rev Mol Cell Biol.* 2005;6(4):287–297.
155. Salvesen GS, Duckett CS. IAP proteins: Blocking the road to death's door. *Nat Rev Mol Cell Biol.* 2002;3(6):401–410.
156. Du C, Fang M, Li Y, et al. Smac, a mitochondrial protein that promotes cytochrome c-dependent caspase activation by eliminating IAP inhibition. *Cell.* 2000;102(1):33–42.
157. Verhagen AM, Ekert PG, Pakusch M, et al. Identification of DIABLO, a mammalian protein that promotes apoptosis by binding to and antagonizing IAP proteins. *Cell.* 2000; 102(1):43–53.
158. Cande C, Cohen I, Daugas E, et al. Apoptosis-inducing factor (AIF): A novel caspase-independent death effector released from mitochondria. *Biochimie.* 2002;84(2–3):215–222.
159. Suzuki Y, Imai Y, Nakayama H, et al. A serine protease, HtrA2, is released from the mitochondria and interacts with XIAP, inducing cell death. *Mol Cell.* 2001;8(3):613–621.
160. Parrish J, Li L, Klotz K, et al. Mitochondrial endonuclease G is important for apoptosis in C. elegans. *Nature.* 2001;412(6842): 90–94.
161. Li LY, Luo X, Wang X. Endonuclease G is an apoptotic DNase when released from mitochondria. *Nature.* 2001;412(6842):95–99.
162. Suzuki Y, Takahashi-Niki K, Akagi T, et al. Mitochondrial protease Omi/HtrA2 enhances caspase activation through multiple pathways. *Cell Death Differ.* 2004;11(2): 208–216.
163. Karbowski M, Youle RJ. Dynamics of mitochondrial morphology in healthy cells and during apoptosis. *Cell Death Differ.* 2003;10(8):870–880.
164. de Brito OM, Scorrano L. Mitofusin 2: A mitochondria-shaping protein with signaling roles beyond fusion. *Antioxid Redox Signal.* 2008;10(3):621–633.
165. Greten FR, Karin M. The IKK/NF-kappaB activation pathway—A target for prevention and treatment of cancer. *Cancer Lett.* 2004;206(2):193–199.
166. Yin XM, Ding WX. Death receptor activation-induced hepatocyte apoptosis and liver injury. *Curr Mol Med.* 2003;3(6): 491–508.
167. Hatano E, Bennett BL, Manning AM, et al. NF-kappaB stimulates inducible nitric oxide synthase to protect mouse hepatocytes from TNF-alpha- and Fas-mediated apoptosis. *Gastroenterology.* 2001;120(5):1251–1262.
168. Li J, Bombeck CA, Yang S, et al. Nitric oxide suppresses apoptosis via interrupting caspase activation and mitochondrial dysfunction in cultured hepatocytes. *J Biol Chem.* 1999;274 (24):17325–17333.
169. Kim JS, Ohshima S, Pediaditakis P, et al. Nitric oxide protects rat hepatocytes against reperfusion injury mediated by the mitochondrial permeability transition. *Hepatology.* 2004;39(6): 1533–1543.
170. Keppler D, Holstege A, Weckbecker G, et al. Potentiation of antimetabolite action by uridylate trapping. *Adv Enzyme Regul.* 1985;24:417–427.
171. Lehmann V, Freudenberg MA, Galanos C. Lethal toxicity of lipopolysaccharide and tumor necrosis factor in normal and D-galactosamine-treated mice. *J Exp Med.* 1987;165(3):657–663.
172. Downward J. PI 3-kinase, Akt and cell survival. *Semin Cell Dev Biol.* 2004;15(2):177–182.
173. Song G, Ouyang G, Bao S. The activation of Akt/PKB signaling pathway and cell survival. *J Cell Mol Med.* 2005;9(1):59–71.
174. Harris SL, Levine AJ. The p53 pathway: Positive and negative feedback loops. *Oncogene.* 2005;24(17):2899–2908.
175. Pei XH, Xiong Y. Biochemical and cellular mechanisms of mammalian CDK inhibitors: A few unresolved issues. *Oncogene.* 2005;24(17):2787–2795.
176. Miyashita T, Reed JC. Tumor suppressor p53 is a direct transcriptional activator of the human bax gene. *Cell.* 1995;80(2): 293–299.
177. Oda E, Ohki R, Murasawa H, et al. Noxa, a BH3-only member of the Bcl-2 family and candidate mediator of p53-induced apoptosis. *Science.* 2000;288(5468): 1053–1058.
178. Yu J, Zhang L, Hwang PM, et al. PUMA induces the rapid apoptosis of colorectal cancer cells. *Mol Cell.* 2001;7(3):673–682.
179. Nicoletti VG, Stella AM. Role of PARP under stress conditions: Cell death or protection? *Neurochem Res.* 2003;28(2):187–194.
180. Decker P, Muller S. Modulating poly (ADP-ribose) polymerase activity: Potential for the prevention and therapy of pathogenic situations involving DNA damage and oxidative stress. *Curr Pharm Biotechnol.* 2002;3(3):275–283.
181. Szabadkai G, Rizzuto R. Participation of endoplasmic reticulum and mitochondrial calcium handling in apoptosis: More than just neighborhood? *FEBS Lett.* 2004;567(1):111–115.
182. Orrenius S, Zhivotovsky B, Nicotera P. Regulation of cell death: The calcium-apoptosis link. *Nat Rev Mol Cell Biol.* 2003;4(7): 552–565.
183. Penzo D, Petronilli V, Angelin A, et al. Arachidonic acid released by phospholipase A(2) activation triggers Ca(2+)-dependent apoptosis through the mitochondrial pathway. *J Biol Chem.* 2004;279(24):25219–25225.
184. Liu CY, Kaufman RJ. The unfolded protein response. *J Cell Sci.* 2003;116(Pt 10):1861–1862.
185. Rao RV, Ellerby HM, Bredesen DE. Coupling endoplasmic reticulum stress to the cell death program. *Cell Death Differ.* 2004;11(4):372–380.
186. Ji C, Kaplowitz N. Hyperhomocysteinemia, endoplasmic reticulum stress, and alcoholic liver injury. *World J Gastroenterol.* 2004;10(12):1699–1708.
187. Harding HP, Novoa I, Bertolotti A, et al. Translational regulation in the cellular response to biosynthetic load on the endoplasmic reticulum. *Cold Spring Harb Symp Quant Biol.* 2001;66:499–508.
188. Urano F, Bertolotti A, Ron D. IRE1 and efferent signaling from the endoplasmic reticulum. *J Cell Sci.* 2000;113(Pt 21):3697–3702.
189. Shen J, Chen X, Hendershot L, et al. ER stress regulation of ATF6 localization by dissociation of BiP/GRP78 binding and unmasking of Golgi localization signals. *Dev Cell.* 2002;3(1):99–111.
190. Urano F, Wang X, Bertolotti A, et al. Coupling of stress in the ER to activation of JNK protein kinases by transmembrane protein kinase IRE1. *Science.* 2000;287(5453):664–666.
191. Yoneda T, Imaizumi K, Oono K, et al. Activation of caspase-12, an endoplastic reticulum (ER) resident caspase, through tumor necrosis factor receptor-associated factor 2-dependent mechanism in response to the ER stress. *J Biol Chem.* 2001;276(17): 13935–13940.
192. Scorrano L, Oakes SA, Opferman JT, et al. BAX and BAK regulation of endoplasmic reticulum Ca2+: A control point for apoptosis. *Science.* 2003;300(5616):135–139.
193. Brostrom MA, Brostrom CO. Calcium dynamics and endoplasmic reticular function in the regulation of protein synthesis:

Implications for cell growth and adaptability. *Cell Calcium.* 2003; 34(4–5):345–363.
194. Bursch W, Ellinger A, Gerner C, et al. Programmed cell death (PCD). Apoptosis, autophagic PCD, or others? *Ann N Y Acad Sci.* 2000;926:1–12.
195. Levine B, Klionsky DJ. Development by self-digestion: Molecular mechanisms and biological functions of autophagy. *Dev Cell.* 2004;6(4):463–477.
196. Klionsky DJ. Autophagy: From phenomenology to molecular understanding in less than a decade. *Nat Rev Mol Cell Biol.* 2007;8(11):931–937.
197. Kiel JA, Komduur JA, van dK I, et al. Macropexophagy in Hansenula polymorpha: Facts and views. *FEBS Lett.* 2003;549 (1–3):1–6.
198. Bellu AR, Komori M, van dK I, et al. Peroxisome biogenesis and selective degradation converge at Pex14p. *J Biol Chem.* 2001;276(48):44570–44574.
199. Elmore SP, Qian T, Grissom SF, et al. The mitochondrial permeability transition initiates autophagy in rat hepatocytes. *FASEB J.* 2001;15(12):2286–2287.
200. Tsujimoto Y, Shimizu S. Another way to die: Autophagic programmed cell death. *Cell Death Differ.* 2005;12(Suppl 2): 1528–1534.
201. Debnath J, Baehrecke EH, Kroemer G. Does autophagy contribute to cell death? *Autophagy.* 2005;1:66–74.
202. Gozuacik D, Kimchi A. Autophagy as a cell death and tumor suppressor mechanism. *Oncogene.* 2004;23(16):2891–2906.
203. Vicencio JM, Galluzzi L, Tajeddine N, et al. Senescence, apoptosis or autophagy? When a damaged cell must decide its path—A mini-review. *Gerontology.* 2008;54(2):92–99.
204. Boya P, Gonzalez-Polo RA, Casares N, et al. Inhibition of macroautophagy triggers apoptosis. *Mol Cell Biol.* 2005;25(3): 1025–1040.
205. Guicciardi ME, Deussing J, Miyoshi H, et al. Cathepsin B contributes to TNF-alpha-mediated hepatocyte apoptosis by promoting mitochondrial release of cytochrome c. *J Clin Invest.* 2000;106(9):1127–1137.
206. Foghsgaard L, Wissing D, Mauch D, et al. Cathepsin B acts as a dominant execution protease in tumor cell apoptosis induced by tumor necrosis factor. *J Cell Biol.* 2001;153(5):999–1010.
207. Stoka V, Turk B, Schendel SL, et al. Lysosomal protease pathways to apoptosis. Cleavage of bid, not pro-caspases, is the most likely route. *J Biol Chem.* 2001;276(5):3149–3157.
208. Bidere N, Lorenzo HK, Carmona S, et al. Cathepsin D triggers Bax activation, resulting in selective apoptosis-inducing factor (AIF) relocation in T lymphocytes entering the early commitment phase to apoptosis. *J Biol Chem.* 2003;278(33): 31401–31411.
209. Lemasters JJ, Qian T, He L, et al. Role of mitochondrial inner membrane permeabilization in necrotic cell death, apoptosis, and autophagy. *Antioxid Redox Signal.* 2002;4(5):769–781.
210. Kim JS, He L, Qian T, et al. Role of the mitochondrial permeability transition in apoptotic and necrotic death after ischemia/reperfusion injury to hepatocytes. *Curr Mol Med.* 2003; 3(6):527–535.
211. Cory S, Adams JM. The Bcl2 family: Regulators of the cellular life-or-death switch. *Nat Rev Cancer.* 2002;2(9):647–656.
212. Lamkanfi M, Festjens N, Declercq W, et al. Caspases in cell survival, proliferation and differentiation. *Cell Death Differ.* 2007;14(1):44–55.
213. Denecker G, Ovaere P, Vandenabeele P, et al. Caspase-14 reveals its secrets. *J Cell Biol.* 2008;180(3):451–458.

Chapter

2

Acute and Chronic Inflammation Induces Disease Pathogenesis

Vladislav Dolgachev . Nicholas W. Lukacs

INTRODUCTION

The recognition of pathogenic insults can be accomplished by a number of mechanisms that function to initiate inflammatory responses and mediate clearance of invading pathogens. This initial response when functioning optimally will lead to a minimal leukocyte accumulation and activation for the clearance of the inciting agent and have little effect on homeostatic function. However, often the inciting agent elicits a very strong inflammatory response, either due to host recognition systems or due to the agent's ability to damage host tissue. Thus, the host innate immune system mediates the damage and tissue destruction in an attempt to clear the inciting agent from the system. No matter, these initial acute responses can have long-term and even irreversible effects on tissue function. If the initial responses are not sufficient to facilitate the clearance of the foreign pathogen or material, the response shifts toward a more complex and efficient process mediated by lymphocyte populations that respond to specific residues displayed by the foreign material. Normally, these responses are coordinated and only minimally alter physiologic function of the tissue. However, in unregulated responses the initial reaction can become acutely catastrophic, leading to local or even systemic damage to the tissue or organs, resulting in degradation of normal physiologic function. Alternatively, the failure to regulate the response or clear the inciting agent could lead to chronic and progressively more pathogenic responses. Each of these potentially devastating responses has specific and often overlapping mechanisms that have been identified and lead to the damage within tissue spaces. A series of events take place during both acute and chronic inflammation that lead to the accumulation of leukocytes and damage to the local environment.

LEUKOCYTE ADHESION, MIGRATION, AND ACTIVATION

Endothelial Cell Expression of Adhesion Molecules

The initial phase of the inflammatory response is characterized by a rapid leukocyte migration into the affected tissue. Upon activation of the endothelium by inflammatory mediators, upregulation of a series of adhesion molecules is initiated that leads to the reversible binding of leukocytes to the activated endothelium. The initial adhesion is mediated by E and P selectins that facilitate slowing of leukocytes from circulatory flow by mediating rolling of the leukocytes on the activated endothelium [1–4]. The selectin-mediated interaction with the activated endothelium potentiates the likelihood of the leukocyte to be further activated by endothelial-expressed chemokines, which mediate G-protein coupled receptor (GPCR) induced activation [5–7]. If the rolling leukocytes encounter a chemokine signal and an additional set of adhesion molecules is also expressed, such as intracellular adhesion molecule-1 (ICAM-1) and vascular cell adhesion molecule-1 (VCAM-1), the leukocytes firmly adhere to the activated endothelium. The mechanism of chemokine-induced adhesion of the leukocyte is dependent on actin reorganization and a confirmational change of the β-integrins on the surface of the leukocytes [8–12]. Subsequently, the firm adhesion allows leukocytes to spread along the endothelium and to begin the process of extravasating into the inflamed tissue following chemoattractant gradients that guide the leukocyte to the site of inflammation. Each of these events has been thoroughly examined over the past several years and has resulted in a better-defined process of coordinated events that

Molecular Pathology © 2009, Elsevier, Inc. All Rights Reserved.

lead the leukocyte from the vessel lumen into the inflamed tissue.

The initial binding of the leukocytes to E and P selectins is mediated by interaction with glycosylated ligands expressed on the leukocytes, most commonly P-selectin glycoprotein ligand 1 (PSGL1) [13–15]. The ability to capture leukocytes onto the inflamed vessel wall is both dependent on the ability of selectin interactions to rapidly bind to its ligands and on the flow rate of the leukocytes in post-capillary vessels where the lower shear rate allows the productive interactions to occur. Once the leukocyte is tethered by the selectin molecules on the vessel wall, a series of binding and release events allow the leukocyte to roll along the activated endothelial cell surface. While the activation and signaling processes that mediate leukocyte release from the selectin interaction continue to be the focus of investigation, current information suggests that these processes may be dependent on key signaling molecules that are initiated upon selectin binding to its ligands. The expression of phosphoinositide 3-kinase-γ (PI3Kγ) in activated endothelium appears to be critical for securing the tethered leukocyte to the vessel wall, whereas spleen tyrosine kinase (SYK) signals downstream of PSGL1 in the bound leukocyte and regulates the rolling process [16–20].

In addition to selectin-mediated leukocyte rolling, there may be a number of instances in which specific integrins can also participate in leukocyte rolling. The CNS, intestinal track, and lung are three examples where this has been specifically observed. Early studies utilizing selectin-deficient animals demonstrated that leukocytes continue to marginate and migrate into the lung during bacterial infections despite the absence of selectin function. Using *in vitro* flow analysis, cells expressing $\alpha4\beta7$ integrins can roll on immobilized mucosal vascular addressin cell adhesion molecule-1 (MADCAM1) that is highly expressed in the intestinal track [21–23]. Likewise, very late antigen 4 (VLA4) supports monocyte and lymphocyte rolling *in vitro* and lymphocyte rolling in the CNS [24–28]. While it is not clear under what circumstances the integrin-mediated rolling occurs, it is likely functional under low shear flow conditions when an initial tethering event is not as crucial. It is most likely that the cooperation of selectin-mediated and integrin-mediated rolling and adhesion is required for leukocyte rolling.

The transition from leukocyte rolling to firm adhesion depends on several distinct events to occur in the rolling leukocyte. First, the integrin needs to be modified through a G protein-mediated signaling event enabling a conformational change that exposes the binding site for the specific adhesion molecule. Second, the density of adhesion molecule expression needs to be high enough to allow the leukocyte to spread along the activated endothelium and appropriate integrin clustering on the leukocyte surface. Finally, it appears that a phenomenon known as outside-in signaling (recently reviewed [29]) is also necessary for strengthening the adhesive interactions through several important signaling events that include FGR and HCK, two SRC-like protein tyrosine kinases (PTKs). Together, these coordinated events facilitate preparation of leukocytes for extravasation through the endothelium into the inflamed tissue.

Transendothelial cell migration of leukocytes requires that numerous potential obstacles be managed. After firmly adhering to the activated endothelium, leukocytes appear to spread and crawl along the border until they reach an endothelial cell junction that has been appropriately "opened" by the inflamed environment. While it has not been completely established, it appears that endothelial cell junctions that support transmigration of leukocytes express higher levels of adhesion molecules that allow a haptotatic gradient for the crawling cells to traverse through. This paracellular route of migration is a favored and well-supported mechanism that is optimized by tissue expressed chemoattractants for mediating the crawling into the junctional region without harming the endothelial cell border. A number of molecules have been implicated in this route of migration, but PECAM1 has been the most thoroughly studied and appears to be functionally required for the process with targeted expression at the endothelial cell junction region [30,31]. Another protein, junction adhesion molecule-A (JAM-A), has also been shown to be associated with migration of cells through the tight junctions of endothelial layer of vessels and is found on the surface of several leukocyte populations including PMNs [32,33]. It appears that PECAM1 and JAM-A are utilized in a sequential manner to allow movement through the endothelial barrier [34,35].

An alternative and recently described mode of transendothelial migration is via a transcellular route. This was initially described using *in vitro* analyses, but has also been identified in the CNS and during inflammatory responses in other tissues. These migratory pathways through the endothelial cell membrane and cytoplasmic region are associated with vesiculo-vacuolar organelles (VVOs), which have been described as membrane-associated passageways that can allow rapid leukocyte transendothelial migration into inflamed tissue compartments [36,37]. It has been suggested that these mechanisms may occur under conditions of dense adhesion molecule expression that appears to trigger the formation of the intracellular passageways in association with caveolin-1 [38]. While this mechanism of leukocyte extravasation may represent a less frequent method of migration across inflamed endothelium, under certain conditions, such as in high shear where paracellular migration could be more difficult, it may represent an important form of migration.

The final obstacle for the leukocyte to traverse prior to entering into the tissue from the vessel is the basement membrane. The model that has been proposed over the years suggests that metalloproteinases (MMPs) are activated to degrade the basement membrane extracellular matrix (ECM), enabling leukocytes to penetrate toward the site of inflammation [39–41]. While evidence *in vitro* suggests that matrix degradation is necessary and that MMPs are required, it has not been clearly identified how the basement membrane is traversed by leukocytes without substantial damage to the integrity of the vessel wall.

Chemoattractants

Over the past several years researchers have identified multiple families of chemoattractants that can participate in the extravasation of leukocytes. Perhaps the most readily accessible mediator class during inflammation is the complement system. These proteins are found in circulation or can be generated *de novo* upon cellular stimulation. Upon activation, bacterial products or immune complexes (as previously reviewed [42,43]) through the alternative or classical pathways mediate cleavage of C3 and/or C5 into C3a and C5a that can provide an immediate and effective chemoattractant to induce neutrophil and monocyte activation. The role of C3a as an anaphylactic agent illustrates the importance of this early activation event on mast cell biology. In addition, C5a stimulates neutrophil oxidative metabolism, granule discharge, and adhesiveness to vascular endothelium. Interestingly, C5a activates endothelial cells via C5aR to induce expression of P-selectin that can further increase local inflammatory events. C3a lacks these latter activities. Altogether, these functions of C3a and C5a indicate that they are potent inflammatory mediators. While these chemoattractant molecules have previously been well described, recent literature has provided additional evidence that has rekindled excitement toward targeting these factors for therapeutic intervention [44,45]. These diseases include sepsis, acute respiratory distress syndrome (ARDS), asthma, arthritis, as well as several other acute and chronic inflammatory diseases.

A second mediator system that is involved in early and immediate leukocyte migration is the leukotrienes, a class of lipid mediators that are preformed in mast cells or are quickly generated through the efficient arachadonic acid pathway induced by 5-LO [46,47]. In particular, leukotriene B4 (LTB4) has especially been implicated in the early induction of neutrophil migration, but also can generate long-term problems during inflammation. LTB4 can be rapidly synthesized by phagocytic cells (PMNs and macrophages) following stimulation with bacterial LPS or other pathogen products. Furthermore, the LTB4 receptor has been implicated in recruitment of T lymphocytes that mediate chronic inflammatory diseases, including rodent models of asthma and arthritis, as well as in transplantation rejection models. In particular, the LTB4 receptor BLT1 has been implicated in preferential recruitment of Th2 type T-lymphocytes during allergic responses [48–50]. Thus, besides their potent function as a neutrophil chemoattractant in acute inflammatory events, LTB4 and BLT1 also play important roles in chronic immune reactions. As such, they are now being considered targets for therapy in chronic immune responses. In addition to LTB4, the cysteinyl leukotrienes, LTC4, D4, and E4, also appear to have some chemotactic activity [51,52], suggesting that targeting the conversion of arachodonic acid into the leukotriene pathway may be generally beneficial.

Chemokines represent a large and well-characterized family of chemoattractants composed of over 50 polypeptide molecules that are expressed in numerous acute and chronic immune responses [53]. Given the number of individual molecules contained in the chemokine family, there is confusion of function. The promiscuous binding relationship between multiple members with a single receptor, as well as a specific receptor able to bind multiple chemokines contributes considerably to our relative lack of understanding of the biology of the family. The determination of what leukocyte populations are recruited to a particular tissue during a response is dictated by the chemokine ligands that are induced and the specific receptors that are displayed on subsets of leukocytes. This latter aspect can be best observed during acute inflammatory responses, such as in bacterial infections, when the cellular infiltrate is primarily neutrophils and it is the production of CxCR binding chemokines that mediate the process [54]. Likewise, when more insidious pathogens are present and the acute inflammatory mechanisms are not able to control the infectious process, immune cytokines, such as IFN and IL-4, tend to drive the production of chemokines that facilitate the recruitment of mononuclear cells, macrophages, and lymphocytes to the site of infection [54,55]. This allows a more sophisticated immune response to develop for clearance of the pathogen. Thus, although there are numerous chemokines being produced during any single response, the overall profile of the response may be directed for recruitment of cells that are most appropriate to deal with the particular stimuli. In addition to their ability to bind to cellular receptors, chemokines are also able to bind to glycosaminoglycans (GAGs) [55]. Unlike with complement and lipid chemoattractants, this allows chemokines to accumulate within tissue and on endothelium for long periods without being washed away or otherwise cleared through various biologic processes. This provides several advantages with respect to the maintenance of chemoattractant gradients in inflamed tissue over long periods of time, with the most intense signals maintained at the site of inflammation. Chemokines are important at the endothelial border where they mediate firm adhesion of leukocytes undergoing selectin-associated rolling to activate their β-integrins to the activated endothelial cells and subsequently direct migration of these cells to the site of inflammation. Finally, at higher concentrations (such as those found at the site of inflammation), chemokines induce leukocyte activation for effector function (for instance, degranulation). Thus, the progressive movement of leukocytes from the endothelial cell border in activated vessels through their arrival at the site of inflammation relies on the coordinated expression and interaction of chemokines and adhesion molecules.

The activation of leukocytes by chemoattractants is mediated by a common signaling cascade via G_i-protein coupled receptor pathways [56]. These pathways can be monitored by calcium mobilization and cellular migration mediated by the G_i alpha and G_i beta/gamma subunits, respectively. Signaling through these subunits results in the activation of important downstream pathways including PI3K, MAPK, and FAK. The PI3K pathway has been the most thoroughly studied after a chemotactic signal and depends on the P110 subunits α, β, γ [57]. The importance of PI3K was best demonstrated in studies using cells deficient in PI3K P110γ, as well as in localization studies that identified activated P110γ at the leading edge of migrating cells [58–60]. Once activated, PI3K appears to couple with the p85 adaptor protein that mediates activation of other pathways, including rac, rho, and Cdc42 GTPases leading to actin polymerization. These PI3K-induced responses can be actively regulated by PTEN (phosphatase and tensin homology deleted on chromosome ten protein) and involved in directional sensing [61,62]. Interestingly, while PI3K appears to be localized to the leading edge during migration, PTEN is found in the trailing edge and is excluded from the leading edge via mechanisms related to GTPase-associated amplification of PIP3 activation [63–67]. Thus, the activation of this system clearly regulates directional leukocyte movement by differential cellular localization of key signaling molecules that regulate actin and other structural proteins.

ACUTE INFLAMMATION AND DISEASE PATHOGENESIS

The initiation of a rapid innate immune response to invading pathogens is essential to inhibit the colonization of microorganisms or to sequester toxic and noxious substances. Once an infection is established, pathogenic bacteria have the capability to multiply and expand at a rate that can surpass the ability of the host to clear and destroy the bacteria. A number of mechanisms have developed to inhibit the establishment of pathogenic bacteria in tissues. The primary mechanism is activation of edema and local fluid release to flood the affected tissue, along with early activation of the complement system in response to bacterial components, resulting in cleavage of C3 and C5. These early inflammatory mediators provide a relatively effective and rapid initiation of PMN and mononuclear phagocyte infiltration to sites of infection. The recruited phagocytic cells engulf invading pathogenic bacteria and quickly activate to begin producing LTB4 as well as early response cytokines, such as IL-1 and TNF, that enhance phagocytosis and killing. The early response cytokines subsequently activate resident cell populations to produce other important mediators of inflammation, such as IL-6 and IL-8 (CxCL8), and promote cytokine cascades that lead to continued leukocyte migration and activation. These early events are critical for regulation of the intensity of the inflammatory response as well as effective containment of the pathogens and foreign substances. This multipronged approach to the activation of the inflammatory response and inhibition of the pathogen expansion is normally tightly regulated. However, in situations in which the inflammatory stimuli are intense, such as in a bolus dose of bacteria, severe trauma, or in burn victims, the acute inflammatory response can become dangerously unregulated. In these types of situations when mediators are produced in an unregulated manner, the host/patient can quickly become subjected to a systemic inflammatory response even though the initial insult may be quite localized. This is a result of mediator production (especially TNF and IL-1) that is systemically delivered to multiple organs, creating an overproduction of leukocyte chemoattractants in distal organs and inducing inflammatory cell influx. These types of responses can quickly damage target organs including liver, lung, and kidney. In this form of septic response, the overwhelming PMN recruitment and activation to multiple organs can lead to tissue damage and organ dysfunction.

The ensuing cytokine storm that develops in the affected tissues results from a cascade of cytokine and chemokine production, leading to uncontrolled leukocyte infiltration and activation that damages the tissue leading to organ dysfunction. While these events can affect any tissue of the organism, the liver and lung appear to be primary targets due to their relatively high numbers of resident macrophage populations that can quickly respond to the inflammatory cytokine signals. In the lung, the development of acute respiratory distress syndrome (ARDS) is often observed in patients experiencing a septic insult. Although early research focused on TNF and IL-1 as lead targets to combat these responses, clinical trials using specific inhibitors, along with more clinically relevant animal models, have not shown any benefit to blocking these central inflammatory mediators in acute diseases, such as sepsis. These failures are likely due to the fact that the early response cytokines are produced and cleared prior to the induction of the most severe aspects of disease. Thus, by the time these mediators were targeted in the sick patient, their detrimental inflammatory function has already been performed. More recent targets include complement receptors (C3aR and C5aR), as well as numerous chemokines and their receptors to attempt to block the recruitment and activation of the cell populations during the responses. However, it is not yet clear whether merely blocking a signal mediator or receptor system will be sufficient or consistent enough to significantly alter the outcome of severely unregulated septic responses.

In the case of viral infections, the system must deal with clearance in a different manner since the ability to recognize and phagocytize virus particles is not reasonable due to their size. Instead, the system is geared to recognize the organisms once a target cell has been infected with the virus. One of the most effective early means of blocking the spread is through the immediate production of type I interferon (IFN). This class of mediators facilitates the blockade of spread by both altering the metabolism of infected cells and by promoting the production of additional antiviral factors in uninfected cells to reduce the chance of successful viral assembly and further spread. While the antiviral effects of type I IFNs was initially identified many years ago, researchers continue to attempt to fully understand the mechanisms that are initiated by this class of mediators.

PATTERN RECOGNITION RECEPTORS AND INFLAMMATORY RESPONSES

Toll-Like Receptors

The initiation of acute inflammation and the progression of chronic disease are often fueled by infectious agents that provide strong stimuli to the host. These responses evolved to be beneficial for the rapid recognition of pathogenic motifs that are not normally present in the host during homeostatic circumstances. These pathogen recognition systems form the basis for our innate immune system, which is rapidly activated to destroy pathogens prior to their colonization. However, at the same time the overactivation of this system can contribute to significant pathology in the host. While there are now a number of diverse families of pattern recognition systems, the best-characterized family is the toll-like receptors (TLR) [68,69]. The Toll system was first discovered in *Drosophila* as a crucial part of the antifungal defense for the organism. The TLR family in mammalian species consists of transmembrane receptors that reside either on the cell surface or within the endosome and that characteristically consist of leucine rich repeats (LRR) for motif recognition and an intracytoplasmic region for signal transduction (Figure 2.1). While TLRs are most notably identified on immune/inflammatory cell populations, they can also be expressed on nonimmune structural cells and provide an activation signal for the initiation of inflammatory mediator production. Cellular activation signals are transmitted by TLRs via cytoplasmic adapter molecules that initiate a cascade of now well-defined activation pathways including NFkB, IRF3, IRF7, as well as a link to MAPK pathways [70,71]. These activation pathways provide strong stimuli that alert the host with "danger signals" that allow effective immune cell activation.

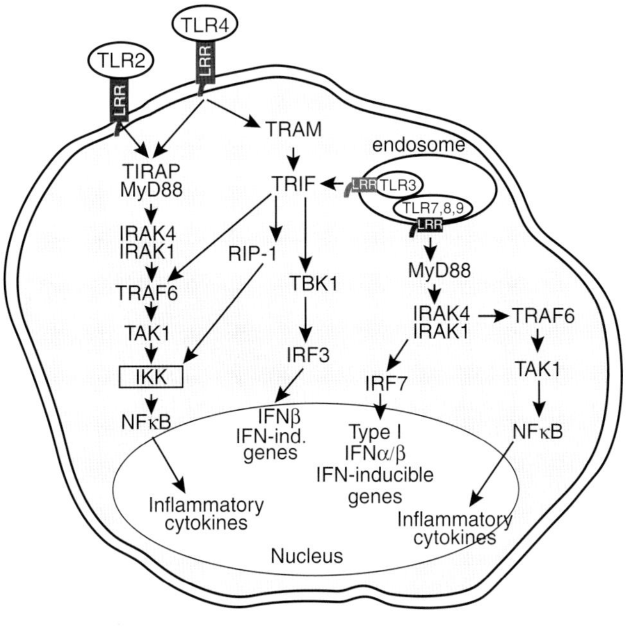

Figure 2.1 TLR activation leads to induction of diverse inflammatory, chemotactic, and activating cytokine production.

One of the first molecules in this family that was identified was toll-like receptor 4 (TLR4), which primarily recognizes lipopolysaccharide (LPS; also known as endotoxin), a component of the cell wall of gram-negative bacteria. The TLR4 activation pathway is unique among TLRs as it signals via multiple adaptor proteins, including MyD88, TRIF, and MD-2, making it the most dynamic TLR within the family [72]. It is the TLR4 pathway that is likely the most prominent during sepsis for the strong systemic activation during acute inflammatory responses. In addition to recognizing LPS, RSV F protein [73] and other complex carbohydrate and lipid molecules from pathogens, TLR4 has also been demonstrated to recognize free fatty acids from host adipose tissue and may contribute to the inflammatory syndrome observed during obesity and in type 2 diabetes [74]. This latter activation cascade may contribute to an unknown number of diseases where obesity and insulin resistance predispose individuals to more severe disease progression. This aspect will surely become a focus in future investigations related to disease severity and obesity.

Other TLR family members recognize distinct and now better-defined factors that allow the immediate activation of the innate immune system and subsequent signaling of the adaptive immune responses. TLR2 appears to have the most diverse range of molecules that are recognized directly including peptidoglycan, mycoplasm lipopeptide, a number of fungal antigens, as well as a growing number of carbohydrate residues on parasitic, fungal, and bacterial moities. In addition, TLR2 can heterodimerize with TLR1 and TLR6 to further expand its recognition capabilities. TLR5 specifically recognizes flagellin and is therefore important for recognizing both gram-negative and gram-positive bacteria. While the previously described TLRs are expressed on the cell membrane, a number of TLRs are predominantly expressed in endosomic membrane compartments of innate immune cells, including TLR3, TLR7, TLR8, and TLR9 [9]. These pathogen recognition receptors (PRR) are involved in recognition of nucleic acid motifs including dsRNA (TLR3), ssRNA (TLR7 and TLR8), and unmethylated CpG DNA (TLR9). Together, these TLRs function in the recognition of viral and bacterial pathogens that enter the cell via receptor-mediated endocytosis or that are actively phagocytized. All of these TLRs exclusively utilize the MyD88 adaptor pathway for activation, except TLR3, which uses TRIF. Thus, these pathways are important for the initiation of innate cytokines, including TNF, IL-12, and type I IFN [75,76], as outlined in Figure 2.1. However, the activation of antigen-presenting cells via the TLR pathways is also extremely important for integrating acute inflammatory events mediated by the innate immune system with acquired immunity and therefore also implicates TLR activation with chronic inflammation.

Cytoplasmic Sensors of Pathogens

While the TLR proteins have been the best characterized, it is now evident that they are not the only molecules that are important for recognition of pathogenic insults. One of the early observations that surround the TLR recognition system was the fact that they are expressed either on the surface membrane or on the endoplasmic membranes. However, if pathogens infect directly into the cytoplasm or escape endosomal degradation pathways, the host cell must have the ability to recognize and deal with the cytoplasmic insult. Cells have developed a number of additional recognition systems to specifically identify pathogen products in the cytoplasm. Similar to the TLR system, Nod-like receptors (NLR; also known as catapiller proteins) are able to recognize specific pathogen patterns that are distinct from host sequences [77–80]. Nod1 and Nod2 sense bacterial molecules produced by the synthesis and/or degradation of peptidoglycans (PGN) (Figure 2.2). Specifically, Nod1 recognizes PGNs that contain meso-diaminopimelic acid produced by gram-negative bacteria and some gram-positive bacteria, while Nod2 recognizes muramyl dipeptide that is found in nearly all PGNs. Other members of this family include NALPs (NACHT-, LRR-, and pyrin-domain-containing proteins), IPAF (ICE-protease activating factor), and NAIPs (neuronal apoptosis inhibitor proteins) [79,81]. An interesting aspect of NLRs is that they contain a conserved caspase associated receptor domain (CARD) that was initially related to proteins involved in programmed cell death or apoptosis. These proteins specifically activate cells through a complex of proteins known as the inflammasome. The central protein of the inflammasome is caspase 1, which is bound by the CARD of the NLRs. A simplified model of the NLR activation pathway suggests that binding to caspase 1 leads to processing of pro-IL-1b and pro-IL-18 to their active and released forms. Interestingly, these same activation pathways can lead to programmed cell death, possibly depending on the activation state of the responding cell and/or the intensity of the NLR signal. In addition to the inflammasome-mediated activation of proinflammatory cytokine release, NOD proteins have also been shown to induce NFkB and MAPK activation through a RICK/RIP2 signaling pathway [82]. This opens up a number of additional gene activation events via this bacterial-induced pathway.

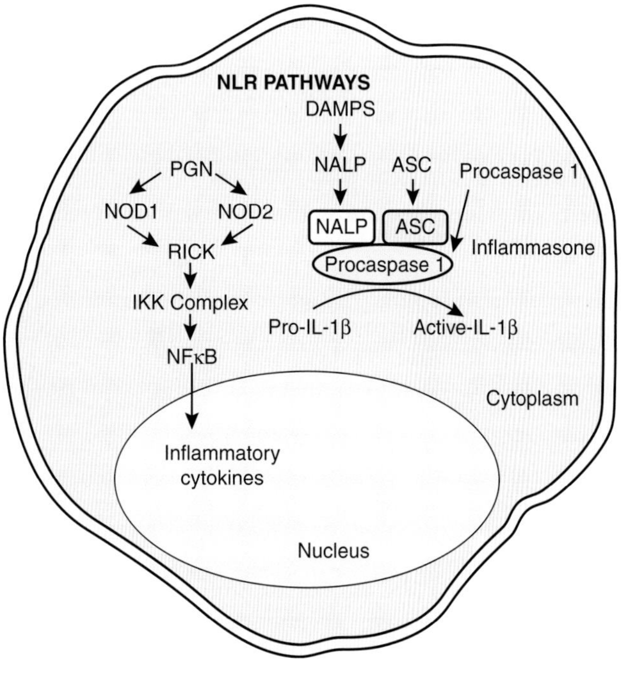

Figure 2.2 Cytoplasmic pathogen receptors allow activation of cytokines responsible for the upregulation of inflammatory processes.

Beyond the activation of cytosolic sensors by bacterial products, host organisms are also armed with a system that recognizes viral products that either escape endosomic recognition or directly infect in a cytosolic manner. Several years ago it was recognized that PKR proteins had the ability to recognize dsRNA, an intermediate step in nearly every viral infection. Interestingly, one of the apparent functions of PKR is that it phosphorylates the alpha-subunit of eukaryotic translation initiation factor 2 (eIF2), which causes inhibition of cellular and viral protein synthesis [83–85]. This latter function works in concert with type I IFN that is also a product of PKR activation for creating an antiviral environment [86]. PKR has a long history of investigation related to antiviral effects of infected cells and appears to provide an important aspect to the response against a productive virus infection.

More recent investigations have identified two helicase proteins that have the ability to recognize dsRNA, RIG-I (retinoic acid-inducible gene) and MDA5 (melanoma differentiation-associated gene) [87,88]. The activation of the protein products of either of these genes in the cytoplasm leads to an immediate activation of type I IFNs (Figure 2.3). The signaling pathway that RIG-I utilizes was initially surprising since it also utilizes its CARD to interact with a mitochondria-associated protein, MAVS [89]. This interaction leads to a scaffold involving TRAF-3, TBK-1, and IRF-3 activation [90]. While this pathway continues to be defined, it appears to be very important in several forms of viral infection. Interestingly, various viruses may differentially activate the two helicases, and this differential activation may define the type and intensity of the ensuing immune responses. Overall, the number of proteins that have the ability to recognize dsRNA not only demonstrates redundancy in the system but also indicates the importance of being able to detect this specific PAMP that is a clear sign of a productive viral infection.

Pattern Recognition and Pathologic Consequences

One of the potential pathogenic side effects of activation of cells by TLRs, NLRs, and other pathogen-sensing receptor systems is the initiation of a wide range of

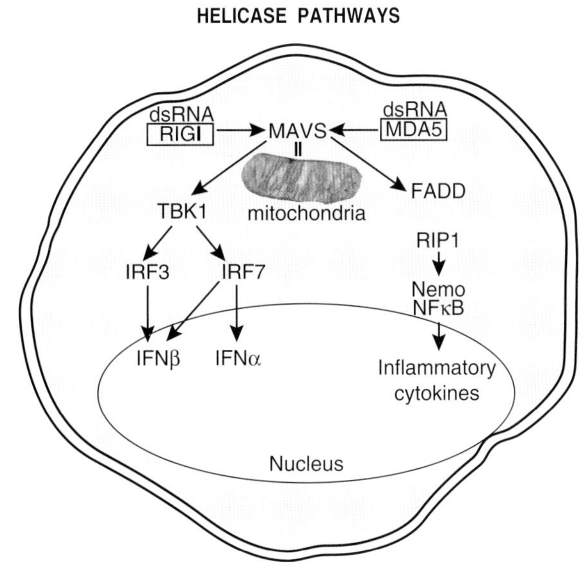

Figure 2.3 Helicase proteins induce early response and activating cytokines by recognition of dsRNA.

inflammatory molecules that, if not properly controlled, could lead to detrimental pathogenic responses. The focus for researchers and clinicians has been centered on the role that pattern recognition receptors play in the initiation of immune responses. However, these mechanisms can also swing the opposite way by promoting inflammation that causes tissue pathology. Two important activation systems that are strongly upregulated by PRRs are the type I IFN (IRF-mediated) and early response cytokine (NFkB-mediated) cascades. Both of these systems have a strong impact on the intensity and direction of the inflammatory responses through their ability to drive chemokine production. In addition to the cytokine cascades that lead to increased levels of these chemotactic molecules, the PRR systems directly regulate the chemokine expression, as many of the chemokine genes have either or both NFkB and/or IRF transcription factor binding sites in their promoter regions. Thus, a local inflammatory response can quickly become amplified and spread the inflammatory damage to neighboring, uninvolved areas. This response is greatly enhanced and perpetuated in more persistent pathogenic insults or with nonpathogenic stimuli that cannot be cleared, such is the case with silica.

Regulation of Acute Inflammatory Responses

The most successful therapy to date for controlling inflammation has been the use of steroidal compounds that nonspecifically inhibit the production of many inflammatory cytokines. The continued dependence on this strategy, although often effective, demonstrates our lack of complete understanding of the mechanisms that control inflammatory responses. A number of well-described regulators may be suited for management of acute inflammatory responses. Several anti-inflammatory mediators have been investigated, including IL-10, TGFβ, IL-1 receptor antagonist (IL-1ra), as well as IL-4 and IL-13 [91–93]. Perhaps the most attractive anti-inflammatory cytokine with broad-spectrum activity is IL-10. IL-10 is predominantly produced by macrophage populations, Th2-type lymphocytes, and B-cells, but can be produced by airway epithelial cells and by several types of tumor cell populations. The function of IL-10 appears to be important during normal physiologic events, as IL-10 gene knockout mice develop lethal inflammatory bowl dysfunction [94]. The importance of this cytokine has been demonstrated in models of endotoxemia and sepsis that neutralized IL-10 during the acute phase and led to increased lethality. In addition, administration of IL-10 to mice protects them from a lethal endotoxin challenge [95,96]. These latter observations can be attributed to the ability of IL-10 to downregulate multiple inflammatory cytokines, including TNF, IL-1, IL-6, IFN-γ, expression of adhesion molecules, and production of nitric oxide. The downregulation of the inflammatory cytokine mediators by IL-10 therapy suggest an extremely potent anti-inflammatory agent that might be used for intervention of inflammation-induced injury. The ability of IL-10 to act as a potent anti-inflammatory cytokine has been shown in other disease states, such as in transplantation responses. However, in severe acute inflammatory diseases, such as sepsis, IL-10 also appears to have a role in promoting secondary or opportunistic infections due to its inhibitory activity [97,98]. IL-10 may also have a role in promoting end-stage disease if not properly regulated. Thus, its role as a therapeutic has been questioned.

Other cytokines also function to suppress the inflammatory response. Transforming growth factor beta (TGFβ) has antiproliferative effects in macrophages and lymphocytes, downregulates cytokine production, and inhibits endotoxin-induced inflammation [99]. In TGFβ gene knockout mice, a progressive inflammatory response was observed early in neonatal development (day 14 after birth) throughout the body including heart, lung, salivary glands, and in virtually every organ at later time points. Thus, TGFβ, like IL-10, plays a critical role in homeostatic regulation of inflammatory responses within the body. However, the use of this cytokine as a therapeutic must be viewed very cautiously as TGFβ may promote fibrogenic outcomes and can skew the T-cell response toward a Treg or Th17 phenotype. In addition, the overexpression of TGFβ has demonstrated enhanced inflammatory responses during endotoxemia [100,101]. Therefore, the therapeutic potential for this cytokine is very limited, if not nonexistent. Another cytokine which has demonstrated anti-inflammatory functions is IL-4. IL-4 is a member of the Th2-type cytokine family and has the ability to downregulate the production of inflammatory cytokines from macrophages. However, like TGFβ, IL-4 has a number of alternate functions that should be viewed carefully, including upregulation of VCAM-1, IgE antibody isotype switching, Th2 cell skewing, and fibroblast activation. IL-4 has been shown to play an important role in the progression of pulmonary granulomatous responses, in allergic airway eosinophilia, and in proliferation and collagen-gene expression in pulmonary fibroblast populations [102,103]. The use of either TGFβ or IL-4 as a therapeutic anti-inflammatory agent will likely not be considered.

CHRONIC INFLAMMATION AND ACQUIRED IMMUNE RESPONSES

Perhaps one of the most difficult and important aspects of disease pathogenesis to regulate is when and how to turn off an immune/inflammatory response. Clearly, pathogen clearance is a primary focus for our immune system, and leukocyte accumulation and activation are a critical event that must be coordinated with the continued presence of pathogens. PRR activation is a critical recognition system that not only activates important cytokine and chemokine pathways for increasing leukocyte function, but also initiates critical antigen presentation cell (APC) functions to optimize lymphocyte activation for pathogen clearance. However, uncontrolled or inefficient immune responses can lead to continual inflammatory cell recruitment and tissue damage that, if persistent and unregulated, can result in organ dysfunction. There are numerous pathogen- and non-pathogen-related diseases that have been classically regarded as being caused by chronic inflammation, including rheumatoid arthritis (RA), chronic obstructive pulmonary disease (COPD), asthma, mycobacterial diseases, multiple sclerosis (MS), and viral hepatitis, among others. However, more recently additional diseases that have not been traditionally grouped with those associated with chronic inflammation have now been recognized to have defects in regulation of inflammation. These disease states include atherosclerosis, numerous obesity-related diseases, as well as various cancers. Persistent inflammatory mediators can be induced by an antigenic or pathogenic stimulus, such as in the case of allergic, bacteria or viral responses, or chronic inflammation may be driven by nonantigenic stimuli that are persistent and cannot be effectively cleared, such as in the case of silicosis. In addition to the persistence of the inflammatory response during chronic inflammatory disorders, there is also a shift in the cellular composition of the leukocyte populations that accumulate. While neutrophils and macrophages may continue to be the end-stage effector cells that mediate the damage, a significant component of the inflammatory responses now comprises lymphocyte infiltration. The presence of activated T- and B-lymphocytes likely indicates the presence of a persistent antigen that induces cell-mediated and humoral immune responses. It is critical to regulate T-lymphocytes since they are central to the activation and regulation of the acquired immune response, as well as the intensity of PMN and macrophage activation.

T-Lymphocyte Regulation of Chronic Inflammation

The nature, duration, and intensity of episodes of chronic inflammatory events are largely determined by the presence and persistence of antigen that is recognized and cleared by acquired immune responses. Thus, the regulation of T-cells is central to the outcome of the inflammatory/immune responses and is mediated through a combination of cytokine environment and transcription factor regulation (Figure 2.4). When a pathogenic insult is encountered, the most effective immune response is a cell-mediated Th1-type response, which is induced by IL-12 with Stat4 activation and characterized by IFN production along with T-bet transcription factor expression [104]. However, long term this immune response can be devastating to the host, as unregulated it can rapidly destroy local tissue and organ function. Thus, the immune response must be modulated and begin to shift to a less harmful response for the tissue, which in T-cells is regulated by IL-4 production and STAT6 activation leading to GATA3 transcription factor expression. While this shift in responses does not represent a sudden switch, rather

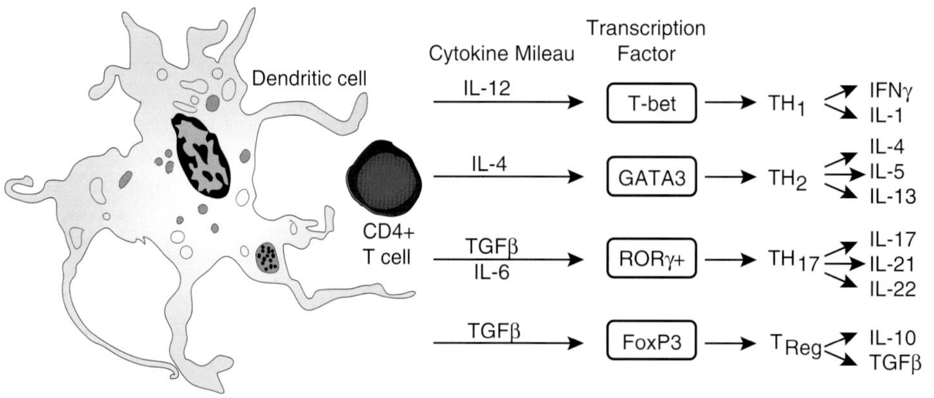

Figure 2.4 The cytokine environment during antigen presentation controls T-cell differentiation by regulation of specific transcription factor activation.

a gradual transition, the long-term consequences of chronic inflammation are often a result of a combined cytokine phenotype that leads to altered macrophage function and continual tissue damage. One of the aspects of the Th2 response that can be detrimental is the shift toward tissue remodeling designed to promote both restoration of function and host protection.

Clearly, regulation of both the Th1- and Th2-type immune responses is central to resolving inflammatory responses and limiting damage once the inciting agent has been removed. An area that has held a significant level of interest has been the differentiation of T regulatory (Treg) cells during the development of chronic responses [105–107]. This cell subset has been divided into several subpopulations, including natural Treg cells and inducible Treg cells (iTreg). Natural Treg cells develop in the thymus and are essential for control of autoimmune diseases, whereas inducible iTreg cells develop following an antigen-specific activation event and appear to function to modulate an ongoing response. In addition, Treg cells can also be subdivided based on the mechanism of inhibition that they use, such as production of IL-10 and/or TGFβ or use of CTLA-4. Some common ground has been forged in these cell populations based on the expression of Foxp3 transcription factor, although apparently not all Treg populations express this protein [108]. Support for the importance of Foxp3+ Treg cells comes from studies with mice missing this factor and in humans who have mutations in FOXP3, both of which develop multiorgan autoimmune diseases [109,110]. Thus, no matter the nomenclature, this cell population appears to be centrally important for the regulation of immune responses, and defects in this pathway may lead to chronic disease phenotypes.

More recent investigations have further enhanced our understanding of the role of T-cell subsets in chronic disease with the explosion of data that has described a newer subset of T-cells (Th17) that characteristically produce IL-17 [111]. This subset of T-cells has only recently begun to be understood during chronic inflammatory diseases. IL-17 producing cells were first described as an important component of antibacterial immunity. Subsequently, the Th17 subset has been identified as having a central role in the severity of autoimmune responses, in cancer, in transplantation immunology, as well as in infectious diseases [111,112]. Interestingly, the critical aspect of whether T-cells will differentiate into a Th17 cell depends on the expression RORγt transcription factor [113]. Similar to the Treg cells, the differentiation of these cells depends on exposure to TGFβ, but is additionally dependent on IL-6 or other STAT3 signals along with RORγt. The differentiation of Th17 cells also appears to be enhanced by IL-23, an IL-12 family cytokine that is upregulated in APC populations upon TLR signaling. Thus, in a coordinated effort to better remove an infectious pathogen, utilization of a combination of Th1 and Th17 responses appears to be the most detrimental trigger for exacerbating a chronic autoimmune disease. While it is not yet clear, the relationship between Treg and Th17 development and regulation appears to be closely controlled, perhaps dependent on whether IL-6 is present in the inflammatory environment along with TGFβ [114–116]. Thus, the differentiation of CD4+ T helper cell subsets clearly depends on the immune environment that the cells find themselves. Rather than clearly differentiated cell subsets, T-cell activation represents a continuum of activation depending on the immune environment and pathogenic cues that the T-cells receive from APC when being activated and differentiating. As outlined in Figure 2.4, T-cell activation and cytokine phenotypes depend on the distinct transcription factor expressed for the activation of the particular T-cell subset. Clearly, these subsets and the cytokines that they produce dictate the outcome of a chronic response not only based on the cascade of mediators that they induce but also by the leukocyte subsets that are used as end-stage effectors during the responses.

B-Lymphocyte and Antibody Responses

The pathogenic role of the humoral immune system has been implicated in a number of chronic disease phenotypes, including allergic responses, autoimmune diseases, arthritis, vasculitis, and any other disease where immune complexes are deposited into tissues [117–120]. Antibodies produced by B-lymphocytes are a primary goal of the acquired immune response to combat infectious organisms at mucosal surfaces, in the circulation, and within tissues of the host. To be effective, antibodies need to have the ability to bind to specific antigens on the surface of pathogen and through their Fc portion to facilitate phagocytosis by macrophages and PMNs for clearance and complement fixation for targeted killing of the microorganism by the lytic pathway. However, these features can also lead to detrimental aspects and tissue damage due to inappropriate activation of inflammatory effector cell populations. In particular, one of the more pathogenic side effects of antibody effector function is the inappropriate activation of PMNs leading to release of their granular products and destroying host tissue. This is often a problem in autoimmune diseases such as systemic lupus (SLE) and rheumatoid arthritis, where a wide array of antibodies directed against self-antigens is formed. In SLE, immune complexes form and are often deposited in the skin, driving a local inflammatory response and vasculitis. Even more serious for the host is the deposition of immune complexes in the kidney, leading to glomerulonephritis and possible severe kidney disease [118,119]. Autoantibodies directed against tissue antigens can also induce damage due to FcR cross-linking on phagocytic cells including PMN, macrophages, and NK cells. This is often a central mechanism for initiating local inflammation and damage within autoimmune responses, such as with joints of RA patients with autoantibodies directed against matrix proteins.

The induction of allergic responses in developed nations has been steadily increasing for the past three decades. Incidence of food, airborne, and industrial allergies has a significant impact on the development of chronic diseases in the skin, lung, and gut, including atopic dermatitis, inflammatory bowel disease (IBD), and asthma. The production of IgE leading to mast cell and basophil activation is the central mechanism that regulates the induction of these diseases [121]. While it continues to be controversial whether there is an increase in mast cell numbers during the development of these diseases, it is clear that their local activation by IgE is a key initiating mechanism for the induction of the response. As the antibody isotype produced by the B-cell is governed by the T-lymphocyte response and production of specific cytokines, determining mechanisms that regulate T-cells during allergic diseases has been central to research in these fields. In particular, IL-4 production from T-lymphocytes is key to isotype switching to IgE in B-cells [122–124]. More recently, pharmaceutical focus on inhibiting IgE-mediated responses has centered on directly clearing IgE from the host using an anti-IgE antibody as a therapeutic [125]. The use of this reagent appears not only to clear IgE from the system prior to its interaction with the FcεR on mast cells and basophils, but also may target the IgE-producing B-cells and kill them. Thus, this strategy may help alleviate the initial activation and chronic mast cell and basophil-mediated inflammatory responses. This treatment appears to be especially efficacious in patients with severe food allergies who are at risk for anaphylactic responses.

Exacerbation of Chronic Diseases

While much of the research in this field has centered on understanding the factors involved and defining targets for therapy of chronic disease, less research has focused on what exacerbates and/or extends the severity of these diseases. This becomes much more difficult when the mechanisms and causative agents that initiate the chronic response are not clearly identified or may be heterogeneous within the patient population. No matter the disease, it appears that a common initiating factor for the exacerbation is an infectious stimulus, bacterial or viral. In fact, in diseases such as multiple sclerosis (MS) or SLE, a number of viruses and bacteria have been implicated as the causative agents for the initiation and/or exacerbation of the responses [126–128]. The reality is that any stimulus that initiates a strong inflammatory signal has the ability to break the established maintenance and reinitiate the chronic response in individuals with an underlying disease. This is most often manifested in diseases where an antigenic response is the underlying cause of the chronic disease and the antigen is environmental or host available, such as allergic asthma or autoimmunity. The strong activation of immune cells locally within the affected tissue would provide the reactivation of a well-regulated immune response. The mediators that are upregulated, IL-12 and/or IL-23, in DC might dictate the type of effector response that is initiated, such as Th1 and/or Th17, respectively.

A common activation pathway for these responses is the use of molecules that can quickly and effectively recognize pathogens, such as the TLR family members. While these molecules are clearly expressed on immune cell populations and facilitate an effective host response, it may be their inappropriate expression on nonimmune cells, such as epithelial cells and fibroblasts, which presents the host with the most detrimental response. The expression of TLRs on nonimmune cells within chronic lesions would predispose these cells to strong infectious stimuli that initiate (or reinitiate) an inflammatory response through the activation and expression of cytokines and inflammatory mediators. While it has not been clearly established, a number of studies have indicated that TLR expression on nonimmune cells in chronic lesions is upregulated and would presumably predispose these tissues to hyperstimulation during infectious insults. This alone could cause the exacerbation of a chronic response without any other specific stimuli. The continual reactivation of tissue inflammation with infectious insults, such as in the lung and gut, could provide the mechanism for tissue damage and potentially remodeling that over time could lead to gradual but continuous organ dysfunction. The addition of other antigenic stimuli, such as an auto-antigen, would further enhance the damaging responses and lead to more severe and accelerated disease phenotypes along with end-stage disease that often accompanies tissue remodeling and ineffective repair.

TISSUE REMODELING DURING ACUTE AND CHRONIC INFLAMMATORY DISEASE

Repair of damaged tissues is a fundamental feature of biological systems and properly regulated has little harmful effect on normal organ function. Abnormal healing and repair, however, can lead to severe problems in the function of organs, and in some cases the perpetuation of the remodeling and repair can result in end-stage disease. Damage to tissues can result from various acute or chronic stimuli, including infections, autoimmune reactions, or mechanical injury. In some cases acute inflammatory reactions, such as ARDS in septic patients, can result in a rapid and devastating disorder that is complicated by significant lung fibrosis and eventual dysfunction. However, more common chronic inflammatory disorders of organ systems (including pulmonary fibrosis, systemic sclerosis, liver cirrhosis, cardiovascular disease, progressive kidney disease) and the joints (such as rheumatoid arthritis and osteoarthritis) are a major cause of morbidity and mortality and enormous burden on healthcare systems. A common feature of these diseases is the destruction and remodeling of extracellular matrix (ECM) that has a significant effect on tissue structure and function [129]. A delicate balance between deposition of ECM by myofibroblasts and ECM degradation by tissue leukocytes determines the tissue restructuring during repair processes, and proper function versus development of pathologic scarring. Chronic inflammation, tissue necrosis, and infection lead to persistent myofibroblast activation and excessive deposition of ECM (including collagen type I, collagen type III, fibronectin, elastin, proteoglycans, and lamin [130–136]), which promote formation of a permanent fibrotic scar. Most chronic fibrotic disorders have a persistent irritant that stimulates production of proteolytic enzymes, growth factors, fibrogenic cytokines, and chemokines. Together, they orchestrate excessive deposition of connective tissue and a progressive destruction of normal tissue organization and function (Figure 2.5).

A mechanism that counteracts deposition of ECM and formation of fibrotic foci is activation of matrix metalloproteinases (MMPs), which represent a class of catalytic enzymes that degrade various components of ECM. All MMPs are composed of shared molecules but have different primary structures (reviewed in [137]). MMPs are produced by cells in latent form and need to be activated by removal of a propeptide from the active site. Some MMPs are constitutive or homeostatic (MMP-2) and expressed in most cells under normal conditions. Others are inducible (MMP-9) or inflammatory. Activities of MMPs are always dependent on a balance between proteinases and natural inhibitors (TIMPs, tissue inhibitors of metalloproteinases) [137]. It may be that the upregulation of TIMPs during chronic disease can be more detrimental for developing severe fibrosis than downregulation of MMPs. Overall, the maintenance of ECM deposition may depend on the active removal and proper arrangement of the components allowing proper restructuring of the tissue and basement membrane.

Profibrogenic Cytokines and Growth Factors Involved in Fibrotic Tissue Remodeling

Alterations in the balance of cytokines can lead to pathological changes, abnormal tissue repair, and tissue fibrosis. The most well-studied cytokines involved in these processes include transforming growth factor β (TGFβ), tumor necrosis factor-α (TNFα), platelet-derived growth factor (PDGF), basic fibroblasts growth factor (BFGF), monocyte chemoattractant protein-1 (MCP-1), macrophage inflammatory protein-1-α (MIP-1-α), and interleukin-1 (IL-1), IL-13, and IL-8 [138–143].

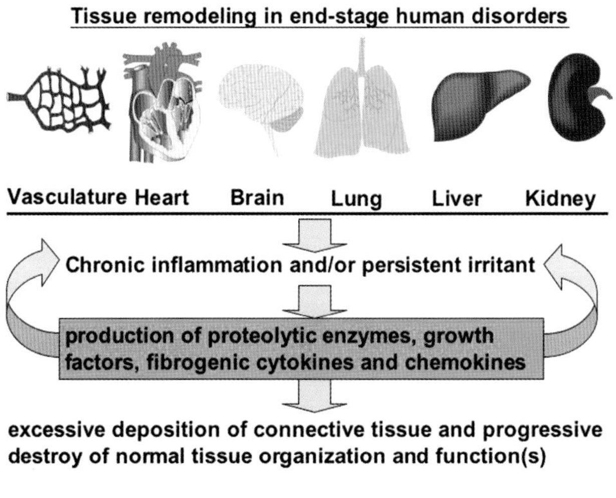

Figure 2.5 The persistent production of cytokines, fibrogenic growth factors, and proteolytic enzymes can result in tissue remodeling and eventual organ dysfunction.

TGFβ

TGFβ is one of the most well-studied profibrotic cytokines. Upregulation of TGFβ1 has been associated with pathological fibrotic processes in many organs (Table 2.1), such as chronic obstructive pulmonary disease [144], cataract formation [145], systemic sclerosis [146], renal fibrosis [133,147], heart failure [140,148–152], and many others [133,134,136, 153–162]. The TGFβ family represents a group of multifunctional cytokines that includes at least five known isoforms [133,134,136,143,153–162], three of which are expressed by mammalian cells (TGFβ1–3). TGFβ-induced effects are mainly mediated by signaling via the TGFβ receptors. TGFβ isoforms are known to induce the expression of ECM proteins (such as collagen I, collagen III, and collagen V; fibronectin; and a number of glycoproteins and proteoglycans normally associated with development) in mesenchymal cells, and to stimulate the production of protease inhibitors that prevent enzymatic breakdown of the ECM. In addition, some ECM proteins (such as fibronectin) are known to be chemoattractants for fibroblasts and are released in increased amounts by fibroblasts and epithelial cells in response to TGFβ [162]. Elevated TGFβ expression in affected organs correlates with abnormal connective tissue deposition observed during

Table 2.1 TGFβ Contributes to Fibrosis of These Diseases by Excessive Matrix Accumulation

Organ	Disease
Eye	Graves ophthalmopathy
	Conjunctival cicatrization
Lung	Pulmonary fibrosis
	Pulmonary sarcoids
Heart	Cardiac fibrosis, cardiomyopathy
Liver	Cirrhosis
	Primary biliary cirrhosis
Kidney	Glomerulosclerosis
	Interstitial fibrosis
Pancreas	Chronic or fibrosing pancreatitis
Skin	Hypertrophic scar
	Keloids
	Scleroderma
Subcutaneous tissue	Dupuytren's contracture
Endometrium	Endometriosis
Peritoneum	Sclerosing peritonitis
	Postsurgical adhesion
Retroperitoneum	Retroperitoneal fibrosis
Bone	Renal osteodystrophy
Muscle	Polymyositis, dermatomyositis
	Muscular dystrophy
	Eosinophilia-myalgia
Bone marrow	Myelofibrosis
Neuroendocrine	Carcinoid

the beginning of fibrotic diseases [143,163]. For example, fibroblasts derived from lungs of idiopathic pulmonary fibrosis patients show an enhanced synthetic activity in response to growth factors, whereas normal fibroblasts show a predominantly proliferative response [164]. Thus, while TGFβ is necessary for normal repair processes, its overexpression plays a pivotal role in deposition of ECM and end-stage disease.

TNFα

Another cytokine that has been often linked to chronic remodeling diseases is tumor necrosis factor alpha. TNFα is a well-known early response cytokine that is key in initiating responses and is rapidly expressed in response to many kinds of stress (mechanical injury, burns, irradiation, viruses, bacteria) [165,166]. However, the role of TNFα is much more complex than simply serving as a trigger of cytokine cascades. TNFα is a proinflammatory mediator that is involved in the extracellular matrix network, as shown in the healing infarct and collagen synthesis by cardiac fibroblasts [167]. The pathologic events in asthma also correlate with increased TNFα production both *in vivo* and *in vitro* in cellular isolates from asthmatic patients. The most convincing evidence that TNFα plays a role in pathophysiology of asthma was observed in normal subjects receiving inhaled recombinant TNFα [168]. These studies demonstrate a significant increase in airway hyper-responsiveness and decreases in FEV1 within those subjects receiving TNFα compared to the placebo group linked to its ability to change extracellular matrix and upregulate pathways specific for leukocyte recruitment. A direct role of TNFα in fibrogenesis was recently demonstrated using human epithelioid dermal microvascular endothelial cell cultures. Exposure of those cells to TNFα for 20 days induced permanent transformation into myofibroblasts. Similar transformation events following chronic inflammatory stimulation *in vivo* may explain one source of myofibroblasts in skin fibrogenesis [169].

Rheumatoid arthritis (RA) represents a chronic disease that highlights the importance of TNFα. TNFα-dependent cytokine cascades were identified in the *in vitro* culture of synovium from joints of patients with RA and led to studies of TNFα blockage in experimental animal models of RA. Using a collagen-induced human RA model in DBA/J mice, researchers showed that anti-TNF antibodies ameliorate arthritis and reduce joint damage [170,171]. This led to the use of TNF blockade in human RA. With the success of anti-TNF treatment in RA, this approach was tested in a number of other chronic disorders [165], including inflammatory bowel disease, asthma, and graft versus host (GVD) during bone marrow transplantation. Successful targeting of TNF has been observed in a number of chronic diseases where anti-TNF therapy has been approved for use and is now also being examined in numerous additional inflammatory, infectious, and neoplastic diseases (Table 2.2). However, anti-TNF therapy is not appropriate in all diseases, as there have been a number of failed clinical trials, including those using anti-TNF treatment for MS and congestive heart failure. In addition, a potential side effect of using anti-TNF therapy is the increased susceptibility to infectious organisms.

IL-13

While the profile of cytokines expressed during a specific response can be classified in several ways, for many chronic diseases a Th2-type immune environment accompanies a profibrotic response. In particular, IL-13 expression is often linked to the progression of end-stage disease phenotypes, both in animal models and in human diseases [102,172]. The initial studies examining IL-13 in fibrosis demonstrated that by neutralization of the IL-13 *in vivo* resulted in a significant reduction in remodeling and end-stage disease. These findings were repeated using IL-13 deficient mice and followed up by demonstrating that although IL-13 does not directly cause myofibroblast conversion, it does enhance ECM production by

Table 2.2 Targeting of TNFα in Chronic Inflammatory Disorders

Approved for Use	Not Completed Trials and Pilot Studies	Clinical Failures
- Rheumatoid arthritis	- Ulcerative colitis	- Congestive heart failure
- Juvenile rheumatoid arthritis	- Behçet syndrome	- Multiple sclerosis
- Crohn's disease	- Vasculitis (small and large vessel)	- Chronic obstructive pulmonary disease
- Psoriatic arthritis	- Glomerulonephritis	- Sjögren syndrome
- Ankylosing spondylitis	- Systemic lupus erythematosus	- Wegener granulomatosis
- Psoriasis	- Joint prosthesis loosening	
	- Hepatitis	
	- Polymyositis	
	- Systemic sclerosis	
	- Amyloidosis	
	- Sarcoidosis	
	- Ovarian cancer	
	- Steroid-resistant asthma	
	- Refractory uveitis	

myofibroblasts. Similar findings have been made in other organ systems, especially in lung and liver remodeling in infectious and allergic diseases. Convincingly, the overexpression of IL-13 in the lung, by itself, induced a significant remodeling response [173]. However, the remodeling of the lung was accompanied by a substantial eosinophil accumulation. The role of the eosinophil was crucial, as when IL-13 transgenic mice were bred with CCR3–/— mice, the fibrotic responses were nearly abrogated, correlating to an absence of eosinophil accumulation [174]. These studies further suggested that this was due to the production of TGFβ by the eosinophils. This conclusion is supported by numerous other studies that have indicated that eosinophils can facilitate remodeling via their ability to produce TGFβ. Thus, a complex cascade is likely to be necessary to initiate and maintain long-term fibrotic responses and the reason for the relatively rare occurrence of end-stage disease in most patient populations.

IL-10

One of the most surprising contributors to tissue remodeling may be IL-10, since it has been demonstrated to exert an overall suppressive effect on most cytokines, including TNF and IL-13. Thus, initially it was envisioned that IL-10 would be very useful in chronic diseases like rheumatoid arthritis, inflammatory bowel disease, and liver cirrhosis [175–177]. This was supported by studies using rIL-10 in experimental models during the induction stage prior to or during the initiation of the fibrotic responses. However, IL-10 also regulates ECM metabolism. The role for IL-10 may be most devastating in later stages of disease where it inhibits the production of MMPs that would help break down excessive ECM deposition and induce TIMPs that would promote the further blockade of ECM degradation [178]. Therefore, the use of rIL-10 in established disease might have the undesirable effect of altering and worsening the remodeling responses. In fact, using an overexpression model in the lung led to the overproduction of collagen and increased remodeling, and was at least coincident with increased IL-13 production, even without an inciting agent [179]. Thus, the role of IL-10 during fibrotic diseases likely depends on the nature and cause of the responses as well as the stage of the disease when the function of IL-10 is investigated.

REFERENCES

1. Zarbock A, Ley K. Mechanisms and consequences of neutrophil interaction with the endothelium. *Am J Pathol.* 2008;172:1–7.
2. Etzioni A. Leukocyte adhesion deficiencies: Molecular basis, clinical findings, and therapeutic options. *Adv Exp Med Biol.* 2007; 601:51–60.
3. Smith CW. Possible steps involved in the transition to stationary adhesion of rolling neutrophils: A brief review. *Microcirculation.* 2000;7:385–394.
4. McEver RP, Cummings RD. Role of PSGL-1 binding to selectins in leukocyte recruitment. *J Clin Invest.* 1997;100:S97–S103.
5. Alon R, Grabovsky V, Feigelson S. Chemokine induction of integrin adhesiveness on rolling and arrested leukocytes local signaling events or global stepwise activation? *Microcirculation.* 2003;10:297–311.
6. DiVietro JA, Smith MJ, Smith BR, et al. Immobilized IL-8 triggers progressive activation of neutrophils rolling in vitro on P-selectin and intercellular adhesion molecule-1. *J Immunol.* 2001;167: 4017–4025.
7. Luscinskas FW, Gerszten RE, Garcia-Zepeda EA, et al. C-C and C-X-C chemokines trigger firm adhesion of monocytes to vascular endothelium under flow conditions. *Ann NY Acad Sci.* 2000; 902:288–293.
8. Johnston B, Butcher EC. Chemokines in rapid leukocyte adhesion triggering and migration. *Semin Immunol.* 2002;14:83–92.
9. Laudanna C, Kim JY, Constantin G, et al. Rapid leukocyte integrin activation by chemokines. *Immunol Rev.* 2002;186:37–46.
10. Ley K. Arrest chemokines. *Microcirculation.* 2003;10:289–295.
11. Szabo MC, Butcher EC, McIntyre BW, et al. RANTES stimulation of T lymphocyte adhesion and activation: Role for LFA-1 and ICAM-3. *Eur J Immunol.* 1997;27:1061–1068.
12. Kishimoto TK, Jutila MA, Berg EL, et al. Neutrophil Mac-1 and MEL-14 adhesion proteins inversely regulated by chemotactic factors. *Science.* 1989;245:1238–1241.
13. Sperandio M. Selectins and glycosyltransferases in leukocyte rolling in vivo. *FEBS J.* 2006;273:4377–4389.
14. Furie B, Furie BC. Role of platelet P-selectin and microparticle PSGL-1 in thrombus formation. *Trends Mol Med.* 2004;10:171–178.
15. Yang J, Furie BC, Furie B. The biology of P-selectin glycoprotein ligand-1: Its role as a selectin counterreceptor in leukocyte-endothelial and leukocyte-platelet interaction. *Thromb Haemost.* 1999;81:1–7.
16. Schymeinsky J, Then C, Sindrilaru A, et al. Syk-mediated translocation of PI3Kdelta to the leading edge controls lamellipodium formation and migration of leukocytes. *PLoS ONE.* 2007; 2:e1132.
17. Bruyninckx WJ, Comerford KM, Lawrence DW, et al. Phosphoinositide 3-kinase modulation of beta(3)-integrin represents an endogenous "braking" mechanism during neutrophil transmatrix migration. *Blood.* 2001;97:3251–3258.
18. Willeke T, Schymeinsky J, Prange P, et al. A role for Syk-kinase in the control of the binding cycle of the beta2 integrins (CD11/CD18) in human polymorphonuclear neutrophils. *J Leukoc Biol.* 2003;74:260–269.
19. Mocsai A, Zhou M, Meng F, et al. Syk is required for integrin signaling in neutrophils. *Immunity.* 2002;16:547–558.
20. Siddiqui RA, English D. Phosphatidylinositol 3′-kinase-mediated calcium mobilization regulates chemotaxis in phosphatidic acid-stimulated human neutrophils. *Biochim Biophys Acta.* 2000; 1483:161–173.
21. Nagamatsu H, Tsuzuki Y, Matsuzaki K, et al. Regulation of T-lymphocyte trafficking by ICAM-1, MAdCAM-1, and CCR7 in microcirculation of appendicular and intestinal lymphoid tissues. *Microcirculation.* 2004;11:493–502.
22. Abitorabi MA, Mackay CR, Jerome EH, et al. Differential expression of homing molecules on recirculating lymphocytes from sheep gut, peripheral, and lung lymph. *J Immunol.* 1996;156: 3111–3117.
23. Hamann A, Andrew DP, Jablonski-Westrich D, et al. Role of alpha 4-integrins in lymphocyte homing to mucosal tissues in vivo. *J Immunol.* 1994;152:3282–3293.
24. Laschinger M, Engelhardt B. Interaction of alpha4-integrin with VCAM-1 is involved in adhesion of encephalitogenic T cell blasts to brain endothelium but not in their transendothelial migration in vitro. *J Neuroimmunol.* 2000;102:32–43.
25. Male D, Rahman J, Pryce G, et al. Lymphocyte migration into the CNS modelled in vitro: Roles of LFA-1, ICAM-1 and VLA-4. *Immunology.* 1994;81:366–372.
26. Meerschaert J, Furie MB. Monocytes use either CD11/CD18 or VLA-4 to migrate across human endothelium in vitro. *J Immunol.* 1994;152:1915–1926.
27. Chuluyan HE, Issekutz AC. VLA-4 integrin can mediate CD11/CD18-independent transendothelial migration of human monocytes. *J Clin Invest.* 1993;92:2768–2777.
28. Hakkert BC, Kuijpers TW, Leeuwenberg JF, et al. Neutrophil and monocyte adherence to and migration across monolayers

of cytokine-activated endothelial cells: The contribution of CD18, ELAM-1, and VLA-4. *Blood.* 1991;78:2721–2726.
29. Alon R, Dustin ML. Force as a facilitator of integrin conformational changes during leukocyte arrest on blood vessels and antigen-presenting cells. *Immunity.* 2007;26:17–27.
30. Muller WA. The role of PECAM-1 (CD31) in leukocyte emigration: Studies in vitro and in vivo. *J Leukoc Biol.* 1995;57:523–528.
31. DeLisser HM, Newman PJ, Albelda SM. Molecular and functional aspects of PECAM-1/CD31. *Immunol Today.* 1994;15:490–495.
32. Naik MU, Naik UP. Junctional adhesion molecule-A-induced endothelial cell migration on vitronectin is integrin alpha v beta 3 specific. *J Cell Sci.* 2006;119:490–499.
33. Corada M, Chimenti S, Cera MR, et al. Junctional adhesion molecule-A-deficient polymorphonuclear cells show reduced diapedesis in peritonitis and heart ischemia-reperfusion injury. *Proc Natl Acad Sci USA.* 2005;102:10634–10639.
34. Woodfin A, Reichel CA, Khandoga A, et al. JAM-A mediates neutrophil transmigration in a stimulus-specific manner in vivo: Evidence for sequential roles for JAM-A and PECAM-1 in neutrophil transmigration. *Blood.* 2007;110:1848–1856.
35. Nourshargh S, Krombach F, Dejana E. The role of JAM-A and PECAM-1 in modulating leukocyte infiltration in inflamed and ischemic tissues. *J Leukoc Biol.* 2006;80:714–718.
36. Dvorak AM, Feng D. The vesiculo-vacuolar organelle (VVO). A new endothelial cell permeability organelle. *J Histochem Cytochem.* 2001;49:419–432.
37. Dvorak AM, Kohn S, Morgan ES, et al. The vesiculo-vacuolar organelle (VVO): A distinct endothelial cell structure that provides a transcellular pathway for macromolecular extravasation. *J Leukoc Biol.* 1996;59:100–115.
38. Feng D, Nagy JA, Dvorak HF, et al. Ultrastructural studies define soluble macromolecular, particulate, and cellular transendothelial cell pathways in venules, lymphatic vessels, and tumor-associated microvessels in man and animals. *Microsc Res Tech.* 2002;57:289–326.
39. Marom B, Rahat MA, Lahat N, et al. Native and fragmented fibronectin oppositely modulate monocyte secretion of MMP-9. *J Leukoc Biol.* 2007;81:1466–1476.
40. Hu Y, Ivashkiv LB. Costimulation of chemokine receptor signaling by matrix metalloproteinase-9 mediates enhanced migration of IFN-alpha dendritic cells. *J Immunol.* 2006;176:6022–6033.
41. Schor H, Vaday GG, Lider O. Modulation of leukocyte behavior by an inflamed extracellular matrix. *Dev Immunol.* 2000;7:227–238.
42. Guo RF, Ward PA. Role of C5a in inflammatory responses. *Annu Rev Immunol.* 2005;23:821–852.
43. Guo RF, Ward PA. Mediators and regulation of neutrophil accumulation in inflammatory responses in lung: Insights from the IgG immune complex model. *Free Radic Biol Med.* 2002;33:303–310.
44. Allegretti M, Moriconi A, Beccari AR, et al. Targeting C5a: Recent advances in drug discovery. *Curr Med Chem.* 2005;12:217–236.
45. Ricklin D, Lambris JD. Complement-targeted therapeutics. *Nat Biotechnol.* 2007;25:1265–1275.
46. Osher E, Weisinger G, Limor R, et al. The 5 lipoxygenase system in the vasculature: Emerging role in health and disease. *Mol Cell Endocrinol.* 2006;252:201–206.
47. Patarroyo M, Prieto J, Rincon J, et al. Leukocyte-cell adhesion: A molecular process fundamental in leukocyte physiology. *Immunol Rev.* 1990;114:67–108.
48. Tager AM, Bromley SK, Medoff BD, et al. Leukotriene B4 receptor BLT1 mediates early effector T cell recruitment. *Nature Immunol.* 2003;4:982–990.
49. Islam SA, Thomas SY, Hess C, et al. The Leukotriene B4 lipid chemoattractant receptor BLT1 defines antigen-primed T-cells in humans. *Blood.* 2006;107:444–453.
50. Luster AD, Tager AM. T-cell trafficking in asthma: Lipid mediators grease the way. *Nature Rev Immunol.* 2004;4:711–724.
51. Fregonese L, Silvestri M, Sabatini F, et al. Cysteinyl leukotrienes induce human eosinophil locomotion and adhesion molecule expression via a CysLT1 receptor-mediated mechanism. *Clin Exp Allergy.* 2002;32:745–750.
52. Busse W, Kraft M. Cysteinyl leukotrienes in allergic inflammation: Strategic target for therapy. *Chest.* 2005;127:1312–1326.
53. Rot A, Von Andrian UH. Chemokines in innate and adaptive host defense: Basic chemokinese grammar for immune cells. *Annu Rev Immunol.* 2004;22:891–928.
54. Kobayashi Y. Neutrophil infiltration and chemokines. *Crit Rev Immunol.* 2006;26:307–316.
55. Sanchez-Sanchez N, Riol-Blanco L, Rodriguez-Fernandez JL. The multiple personalities of the chemokine receptor CCR7 in dendritic cells. *J Immunol.* 2006;176:5153–5159.
56. Lombardi MS, Kavelaars A, Heijnen CJ. Role and modulation of G protein-coupled receptor signaling in inflammatory processes. *Crit Rev Immunol.* 2002;22:141–163.
57. Sotsios Y, Ward SG. Phosphoinositide 3-kinase: A key biochemical signal for cell migration in response to chemokines. *Immunol Rev.* 2000;177:217–235.
58. Chang JD, Sukhova GK, Libby P, et al. Deletion of the phosphoinositide 3-kinase p110gamma gene attenuates murine atherosclerosis. *Proc Natl Acad Sci USA.* 2007;104:8077–8082.
59. Reif K, Okkenhaug K, Sasaki T, et al. Cutting edge: Differential roles for phosphoinositide 3-kinases, p110 gamma and p110 delta, in lymphocyte chemotaxis and homing. *J Immunol.* 2004; 173:2236–2240.
60. Naccache PH, Levasseur S, Lachance G, et al. Stimulation of human neutrophils by chemotactic factors is associated with the activation of phosphatidylinositol 3-kinase gamma. *J Biol Chem.* 2000;275:23636–23641.
61. Comer FI, Parent CA. Phosphoinositides specify polarity during epithelial organ development. *Cell.* 2007;128:239–240.
62. Devreotes P, Janetopoulos C. Eukaryotic chemotaxis: Distinctions between directional sensing and polarization. *J Biol Chem.* 2003;278:20445–20448.
63. Wu Y, Hannigan MO, Kotlyarov A, et al. A requirement of MAPKAPK2 in the uropod localization of PTEN during FMLP-induced neutrophil chemotaxis. *Biochem Biophys Res Commun.* 2004;316:666–672.
64. Iijima M, Huang YE, Luo HR, et al. Novel mechanism of PTEN regulation by its phosphatidylinositol 4,5-bisphosphate binding motif is critical for chemotaxis. *J Biol Chem.* 2004;279:16606–16613.
65. Merlot S, Firtel RA. Leading the way: Directional sensing through phosphatidylinositol 3-kinase and other signaling pathways. *J Cell Sci.* 2003;116:3471–3478.
66. Huang YE, Iijima M, Parent CA, et al. Receptor-mediated regulation of PI3Ks confines PI(3,4,5)P3 to the leading edge of chemotaxing cells. *Mol Biol Cell.* 2003;14:1913–1922.
67. Iijima M, Devreotes P. Tumor suppressor PTEN mediates sensing of chemoattractant gradients. *Cell.* 2002;109:599–610.
68. Takeda K, Kaisho T, Akira S. Toll-like receptors. *Annu Rev Immunol.* 2003;21:335–376.
69. Akira S, Hemmi H. Recognition of pathogen-associated molecular patterns by TLR family. *Immunol Lett.* 2003;85:85–95.
70. Moynagh PN. TLR signalling and activation of IRFs: Revisiting old friends from the NF-kappaB pathway. *Trends Immunol.* 2005;26:469–476.
71. O'Neill LA. Signal transduction pathways activated by the IL-1 receptor/toll-like receptor superfamily. *Curr Top Microbiol Immunol.* 2002;270:47–61.
72. Beutler B, Jiang Z, Georgel P, et al. Genetic analysis of host resistance: Toll-like receptor signaling and immunity at large. *Annu Rev Immunol.* 2006;24:353–389.
73. Kurt-Jones EA, Popova L, Kwinn L, et al. Pattern recognition receptors TLR4 and CD14 mediate response to respiratory syncytial virus. *Nature Immunol.* 2000;1:398–401.
74. Shi H, Kokoeva MV, Inouye K, et al. TLR4 links innate immunity and fatty acid-induced insulin resistance. *J Clin Invest.* 2006;116:3015–3025.
75. Colonna M. TLR pathways and IFN-regulatory factors: To each its own. *Eur J Immunol.* 2007;37:306–309.
76. Pasare C, Medzhitov R. Toll-like receptors: Linking innate and adaptive immunity. *Adv Exp Med Biol.* 2005;560:11–18.
77. Freche B, Reig N, van der Goot FG. The role of the inflammasome in cellular responses to toxins and bacterial effectors. *Semin Immunopathol.* 2007;29:249–260.
78. Lee MS, Kim YJ. Signaling pathways downstream of pattern-recognition receptors and their cross talk. *Annu Rev Biochem.* 2007; 76:447–480.
79. Kufer TA, Sansonetti PJ. Sensing of bacteria. NOD a lonely job. *Curr Opin Microbiol.* 2007;10:62–69.

80. Philpott DJ, Girardin SE. The role of toll-like receptors and Nod proteins in bacterial infection. *Mol Immunol.* 2004;41:1099–1108.
81. Kanneganti TD, Lamkanfi M, Nunez G. Intracellular NOD-like receptors in host defense and disease. *Immunity.* 2007;27: 549–559.
82. Park JH, Kim YG, Shaw M, et al. Nod1/RICK and TLR signaling regulate chemokine and antimicrobial innate immune responses in mesothelial cells. *J Immunol.* 2007;179:514–521.
83. de Haro C, Mendez R, Santoyo J. The eIF-2alpha kinases and the control of protein synthesis. *FASEB J.* 1996;10:1378–1387.
84. Wek RC. eIF-2 kinases: Regulators of general and gene-specific translation initiation. *Trends Biochem Sci.* 1994;19:491–496.
85. Jagus R, Gray MM. Proteins that interact with PKR. *Biochimie.* 1994;76:779–791.
86. Kaempfer R. RNA sensors: Novel regulators of gene expression. *EMBO Rep.* 2003;4:1043–1047.
87. Kato H, Takeuchi O, Sato S, et al. Differential roles of MDA5 and RIG-I helicases in the recognition of RNA viruses. *Nature.* 2006;441:101–105.
88. Yoneyama M, Kikuchi M, Natsukawa T, et al. The RNA helicase RIG-I has an essential function in double-stranded RNA-induced innate antiviral responses. *Nature Immunol.* 2004;5:730–737.
89. Seth RB, Sun L, Ea CK, et al. Identification and characterization of MAVS, a mitochondrial antiviral signaling protein that activates NF-kappaB and IRF 3. *Cell.* 2005;122:669–682.
90. Seth RB, Sun L, Chen ZJ. Antiviral innate immunity pathways. *Cell Res.* 2006;16:141–147.
91. Bradley JR. TNF-mediated inflammatory disease. *J Pathol.* 2008;214:149–160.
92. Prud'homme GJ. Pathobiology of transforming growth factor beta in cancer, fibrosis and immunologic disease, and therapeutic considerations. *Lab Invest.* 2007;87:1077–1091.
93. Christodoulou C, Choy EH. Joint inflammation and cytokine inhibition in rheumatoid arthritis. *Clin Exp Med.* 2006;6:13–19.
94. Davidson NJ, Leach MW, Fort MM, et al. T helper cell 1-type CD4+ T cells, but not B cells, mediate colitis in interleukin 10-deficient mice. *J Exp Med.* 1996;184:241–251.
95. Howard M, Muchamuel T, Andrade S, et al. Interleukin 10 protects mice from lethal endotoxemia. *J Exp Med.* 1993; 177: 1205–1208.
96. Gerard C, Bruyns C, Marchant A, et al. Interleukin 10 reduces the release of tumor necrosis factor and prevents lethality in experimental endotoxemia. *J Exp Med.* 1993;177:547–550.
97. Steinhauser ML, Hogaboam CM, Kunkel SL, et al. IL-10 is a major mediator of sepsis- induced impairment in lung antibacterial host defense. *J Immunol.* 1999;162:392–399.
98. Greenberger MJ, Strieter RM, Kunkel SL, et al. Neutralization of IL-10 increases survival in a murine model of Klebsiella pneumonia. *J Immunol.* 1995;155:722–729.
99. Karres I, Kremer JP, Steckholzer U, et al. Transforming growth factor-beta 1 inhibits synthesis of cytokines in endotoxin-stimulated human whole blood. *Arch Surg.* 1996;131:1310–1316.
100. Garcia-Lazaro JF, Thieringer F, Luth S, et al. Hepatic overexpression of TGF-beta1 promotes LPS-induced inflammatory cytokine secretion by liver cells and endotoxemic shock. *Immunol Lett.* 2005;101:217–222.
101. Vodovotz Y, Kopp JB, Takeguchi H, et al. Increased mortality, blunted production of nitric oxide, and increased production of TNF-alpha in endotoxemic TGF-beta1 transgenic mice. *J Leukoc Biol.* 1998;63:31–39.
102. Jakubzick C, Kunkel SL, Puri RK, et al. Therapeutic targeting of IL-4- and IL-13–responsive cells in pulmonary fibrosis. *Immunol Res.* 2004;30:339–349.
103. McGaha TL, Bona CA. Role of profibrogenic cytokines secreted by T cells in fibrotic processes in scleroderma. *Autoimmun Rev.* 2002;1:174–181.
104. Kaplan MH, Grusby MJ. Regulation of T helper cell differentiation by STAT molecules. *J Leukoc Biol.* 1998;64:2–5.
105. Suvas S, Rouse BT. Regulation of microbial immunity: The suppressor cell renaissance. *Viral Immunol.* 2005;18:411–418.
106. Maizels RM. Infections and allergy—helminths, hygiene and host immune regulation. *Curr Opin Immunol.* 2005;17:656–661.
107. Paust S, Cantor H. Regulatory T cells and autoimmune disease. *Immunol Rev.* 2005;204:195–207.
108. Chatila T. The regulatory T cell transcriptosome: E pluribus unum. *Immunity.* 2007;27:693–695.
109. Chang X, Zheng P, Liu Y. FoxP3: A genetic link between immunodeficiency and autoimmune diseases. *Autoimmun Rev.* 2006;5:399–402.
110. Bacchetta R, Passerini L, Gambineri E, et al. Defective regulatory and effector T cell functions in patients with FOXP3 mutations. *J Clin Invest.* 2006;116:1713–1722.
111. McGeachy MJ, Cua DJ. Th17 cell differentiation: The long and winding road. *Immunity.* 2008;28:445–453.
112. Weaver CT, Harrington LE, Mangan PR, et al. Th17: An effector CD4 T cell lineage with regulatory T cell ties. *Immunity.* 2006;24:677–688.
113. Ivanov II, McKenzie BS, Zhou L, et al. The orphan nuclear receptor RORgammat directs the differentiation program of proinflammatory IL-17+ T helper cells. *Cell.* 2006;126:1121–1133.
114. Ichiyama K, Yoshida H, Wakabayashi Y, et al. Foxp3 inhibits RORgamma t-mediated IL-17A mRNA transcription through direct interaction with RORgamma t. *J Biol Chem.* 2008;283: 17003–17008.
115. Mucida D, Park Y, Kim G, et al. Reciprocal TH17 and regulatory T cell differentiation mediated by retinoic acid. *Science.* 2007;317:256–260.
116. Nishihara M, Ogura H, Ueda N, et al. IL-6-gp130-STAT3 in T cells directs the development of IL-17+ Th with a minimum effect on that of Treg in the steady state. *Int Immunol.* 2007;19:695–702.
117. Ward PA, Lentsch AB. Endogenous regulation of the acute inflammatory response. *Mol Cell Biochem.* 2002;234–235: 225–228.
118. Bugatti S, Codullo V, Caporali R, et al. B cells in rheumatoid arthritis. *Autoimmun Rev.* 2007;7:137–142.
119. Nimmerjahn F, Ravetch JV. Fc-receptors as regulators of immunity. *Adv Immunol.* 2007;96:179–204.
120. Okroj M, Heinegard D, Holmdahl R, et al. Rheumatoid arthritis and the complement system. *Ann Med.* 2007;39: 517–530.
121. Kinet JP. The high-affinity IgE receptor (Fc epsilon RI): From physiology to pathology. *Annu Rev Immunol.* 1999;17:931–972.
122. Davies JM, O'Hehir RE. Immunogenetic characteristics of immunoglobulin E in allergic disease. *Clin Exp Allergy.* 2008; 38:566–578.
123. Boyce JA. Mast cells: Beyond IgE. *J Allergy Clin Immunol.* 2003; 111:24–32.
124. Bacharier LB, Geha RS. Molecular mechanisms of IgE regulation. *J Allergy Clin Immunol.* 2000;105:S547–S558.
125. Avila PC. Does anti-IgE therapy help in asthma? Efficacy and controversies. *Annu Rev Med.* 2007;58:185–203.
126. Poole BD, Scofield RH, Harley JB, et al. Epstein-Barr virus and molecular mimicry in systemic lupus erythematosus. *Autoimmunity.* 2006;39:63–70.
127. Ascherio A, Munger K. Epidemiology of multiple sclerosis: From risk factors to prevention. *Semin Neurol.* 2008;28:17–28.
128. McCoy L, Tsunoda I, Fujinami RS. Multiple sclerosis and virus induced immune responses: Autoimmunity can be primed by molecular mimicry and augmented by bystander activation. *Autoimmunity.* 2006;39:9–19.
129. Laurent GJ, Chambers RC, Hill MR, et al. Regulation of matrix turnover: Fibroblasts, forces, factors and fibrosis. *Biochem Soc Trans.* 2007;35:647–651.
130. Samuel CS, Lekgabe ED, Mookerjee I. The effects of relaxin on extracellular matrix remodeling in health and fibrotic disease. *Adv Exp Med Biol.* 2007;612:88–103.
131. Gomperts BN, Strieter RM. Fibrocytes in lung disease. *J Leukoc Biol.* 2007;82:449–456.
132. Duprez DA. Aldosterone and the vasculature: Mechanisms mediating resistant hypertension. *J Clin Hypertens.* 2007;9:13–18.
133. Sato M, Muragaki Y, Saika S, et al. Targeted disruption of TGF-beta1/Smad3 signaling protects against renal tubulointerstitial fibrosis induced by unilateral ureteral obstruction. *J Clin Invest.* 2003;112:1486–1494.
134. Jagadeesan J, Bayat A. Transforming growth factor beta (TGFbeta) and keloid disease. *Int J Surg.* 2007;5:278–285.
135. Gressner OA, Weiskirchen R, Gressner AM. Biomarkers of liver fibrosis: Clinical translation of molecular pathogenesis or based

on liver-dependent malfunction tests. *Clin Chim Acta.* 2007; 381:107–113.
136. Burke JP, Mulsow JJ, O'Keane C, et al. Fibrogenesis in Crohn's disease. *Am J Gastroenterol.* 2007;102:439–448.
137. Hu J, Van den Steen PE, Sang QX, et al. Matrix metalloproteinase inhibitors as therapy for inflammatory and vascular diseases. *Nature Rev Drug Discov.* 2007;6:480–498.
138. Breitkopf K, Haas S, Wiercinska E, et al. Anti- TGF-beta strategies for the treatment of chronic liver disease. *Alcohol Clin Exp Res.* 2005;29:121S–131S.
139. Hinz B, Phan SH, Thannickal VJ, et al. The myofibroblast: One function, multiple origins. *Am J Pathol.* 2007;170:1807–1816.
140. Leask A. TGFbeta, cardiac fibroblasts, and the fibrotic response. *Cardiovasc Res.* 2007;74:207–212.
141. Wynn TA. Common and unique mechanisms regulate fibrosis in various fibroproliferative diseases. *J Clin Invest.* 2007;117:524–529.
142. Boerrigter G, Burnett Jr JC. Nitric oxide-independent stimulation of soluble guanylate cyclase with BAY 41–2272 in cardiovascular disease. *Cardiovasc Drug Rev.* 2007;25:30–45.
143. Verrecchia F, Mauviel A. Transforming growth factor-beta and fibrosis. *World J Gastroenterol.* 2007;13:3056–3062.
144. de Boer WI, Alagappan VK, Sharma HS. Molecular mechanisms in chronic obstructive pulmonary disease: Potential targets for therapy. *Cell Biochem Biophys.* 2007;47:131–148.
145. Wederell ED, Brown H, O'Connor M, et al. Laminin-binding integrins in rat lens morphogenesis and their regulation during fibre differentiation. *Exp Eye Res.* 2005;81:326–339.
146. Verrecchia F, Mauviel A, Farge D. Transforming growth factor-beta signaling through the Smad proteins: Role in systemic sclerosis. *Autoimmun Rev.* 2006;5:563–569.
147. Negri AL. Prevention of progressive fibrosis in chronic renal diseases: Antifibrotic agents. *J Nephrol.* 2004;17:496–503.
148. Bujak M, Frangogiannis NG. The role of TGF-beta signaling in myocardial infarction and cardiac remodeling. *Cardiovasc Res.* 2007;74:184–195.
149. Dabek J, Kulach A, Monastyrska-Cup B, et al. Transforming growth factor beta and cardiovascular diseases: The other facet of the 'protective cytokine.' *Pharmacol Rep.* 2006;58:799–805.
150. Everett TH 4th, Olgin JE. Atrial fibrosis and the mechanisms of atrial fibrillation. *Heart Rhythm.* 2007;4:S24–S27.
151. Lim H, Zhu YZ. Role of transforming growth factor-beta in the progression of heart failure. *Cell Mol Life Sci.* 2006;63:2584–2596.
152. Ruiz-Ortega M, Rodriguez-Vita J, Sanchez-Lopez E, et al. TGF-beta signaling in vascular fibrosis. *Cardiovasc Res.* 2007;74: 196–206.
153. Eickelberg O, Morty RE. Transforming growth factor beta/bone morphogenic protein signaling in pulmonary arterial hypertension: Remodeling revisited. *Trends Cardiovasc Med.* 2007;17:263–269.
154. Klahr S, Morrissey J. Obstructive nephropathy and renal fibrosis: The role of bone morphogenic protein-7 and hepatocyte growth factor. *Kidney Int Suppl.* 2003;S105–S112.
155. Petersen M, Thorikay M, Deckers M, et al. Oral administration of GW788388, an inhibitor of TGF-beta type I and II receptor kinases, decreases renal fibrosis. *Kidney Int.* 2008;73:705–715.
156. Akool el S, Doller A, Muller R, et al. Nitric oxide induces TIMP-1 expression by activating the transforming growth factor beta-Smad signaling pathway. *J Biol Chem.* 2005;280:39403–39416.
157. Jones SE, Kelly DJ, Cox AJ, et al. Mast cell infiltration and chemokine expression in progressive renal disease. *Kidney Int.* 2003;64:906–913.
158. Grandaliano G, Pontrelli P, Cerullo G, et al. Protease-activated receptor-2 expression in IgA nephropathy: A potential role in the pathogenesis of interstitial fibrosis. *J Am Soc Nephrol.* 2003; 14:2072–2083.
159. Gruber BL, Marchese MJ, Kew RR. Transforming growth factor-beta 1 mediates mast cell chemotaxis. *J Immunol.* 1994; 152:5860–5867.
160. Kolodsick JE, Toews GB, Jakubzick C, et al. Protection from fluorescein isothiocyanate-induced fibrosis in IL-13-deficient, but not IL-4-deficient, mice results from impaired collagen synthesis by fibroblasts. *J Immunol.* 2004;172:4068–4076.
161. de Iongh RU, Wederell E, Lovicu FJ, et al. Transforming growth factor-beta-induced epithelial-mesenchymal transition in the lens: A model for cataract formation. *Cells Tissues Organs.* 2005;179:43–55.
162. Gabbiani G. The myofibroblast in wound healing and fibrocontractive diseases. *J Pathol.* 2003;200:500–503.
163. Diegelmann RF, Evans MC. Wound healing: An overview of acute, fibrotic and delayed healing. *Front Biosci.* 2004;9:283–289.
164. Hetzel M, Bachem M, Anders D, et al. Different effects of growth factors on proliferation and matrix production of normal and fibrotic human lung fibroblasts. *Lung.* 2005; 183:225–237.
165. Feldmann M, Pusey CD. Is there a role for TNF-alpha in anti-neutrophil cytoplasmic antibody-associated vasculitis? Lessons from other chronic inflammatory diseases. *J Am Soc Nephrol.* 2006;17:1243–1252.
166. Friedman WJ. Cytokines regulate expression of the type 1 interleukin-1 receptor in rat hippocampal neurons and glia. *Exp Neurol.* 2001;168:23–31.
167. Siwik DA, Chang DL, Colucci WS. Interleukin-1beta and tumor necrosis factor-alpha decrease collagen synthesis and increase matrix metalloproteinase activity in cardiac fibroblasts in vitro. *Circ Res.* 2000;86:1259–1265.
168. Walsh CJ, Sugerman HJ, Mullen PG, et al. Monoclonal antibody to tumor necrosis factor alpha attenuates cardiopulmonary dysfunction in porcine gram-negative sepsis. *Arch Surg.* 1992; 127:138–144.
169. Chaudhuri V, Zhou L, Karasek M. Inflammatory cytokines induce the transformation of human dermal microvascular endothelial cells into myofibroblasts: A potential role in skin fibrogenesis. *J Cutan Pathol.* 2007;34:146–153.
170. Marinova-Mutafchieva L, Williams RO, Funa K, et al. Inflammation is preceded by tumor necrosis factor- dependent infiltration of mesenchymal cells in experimental arthritis. *Arthritis Rheum.* 2002;46:507–513.
171. Williams RO, Feldmann M, Maini RN. Anti-tumor necrosis factor ameliorates joint disease in murine collagen-induced arthritis. *Proc Natl Acad Sci USA.* 1992;89:9784–9788.
172. Wynn TA. IL-13 effector functions. *Annu Rev Immunol.* 2003; 21:425–456.
173. Zhu Z, Homer RJ, Wang Z, et al. Pulmonary expression of interleukin-13 causes inflammation, mucus hypersecretion, subepithelial fibrosis, physiologic abnormalities, and eotaxin production. *J Clin Invest.* 1999;103:779–788.
174. Fulkerson PC, Fischetti CA, Rothenberg ME. Eosinophils and CCR3 regulate interleukin-13 transgene-induced pulmonary remodeling. *Am J Pathol.* 2006;169:2117–2126.
175. Papadakis KA, Targan SR. Role of cytokines in the pathogenesis of inflammatory bowel disease. *Annu Rev Med.* 2000;51:289–298.
176. Williams RO. Collagen-induced arthritis in mice: A major role for tumor necrosis factor-alpha. *Methods Mol Biol.* 2007; 361:265–284.
177. Tilg H, Kaser A, Moschen AR. How to modulate inflammatory cytokines in liver diseases. *Liver Int.* 2006;26:1029–1039.
178. Lacraz S, Nicod LP, Chicheportiche R, et al. IL-10 inhibits metalloproteinase and stimulates TIMP-1 production in human mononuclear phagocytes. *J Clin Invest.* 1995;96:2304–2310.
179. Lee CG, Homer RJ, Cohn L, et al. Transgenic overexpression of interleukin (IL)-10 in the lung causes mucus metaplasia, tissue inflammation, and airway remodeling via IL-13-dependent and independent pathways. *J Biol Chem.* 2002;277:35466–35474.

Chapter

3

Infection and Host Response

Margret D. Oethinger . Sheldon M. Campbell

MICROBES AND HOSTS—BALANCE OF POWER?

Disease is one of the major driving forces of evolution. Humans have a generation time of roughly 20 years, and even small mammals reproduce in weeks to months. In contrast, microbial generation times range from minutes to days. Thus, microbes evolve hundreds to thousands of times more rapidly than their vertebrate hosts. In this context, it is hardly remarkable that microbes have found numerous ways to exploit multicellular creatures for their own ends. Rather, it is remarkable that humans and other higher organisms have managed to survive at all.

Large multicellular creatures represent concentrated, extremely rich nutrient sources for microbes. Therefore, the survival of multicellular creatures requires that they have sufficient defenses to prevent easy invasion and consumption. Recent advances in basic immunology illustrate the breadth and depth of the adaptations that have evolved to protect multicellular organisms from microbial invasion. However, the wondrous complexity and power of mammalian host defenses serve only as a backdrop to the even more astonishing complexity of microbial strategies for evading them. In the great scheme of things, humans (and other multicellular organisms) survive only because the microbes let us live.

Because pathogens exert immense selective pressure, many aspects of host physiology have a role in preventing infection, in addition to the elements normally thought of as the immune system. Physical and chemical barriers such as skin, mucosal surfaces, and gastric acidity prevent invasion by microbes. Behavioral adaptations, such as avoidance of decayed food and slapping at insects, are most likely driven by preventing exposure to microbial threats. These aspects of host response to infection, while important, will not be further discussed here.

In most cases, the pathologies induced by microbial pathogens primarily serve to aid microbial spreading to new hosts. Thus, coughing, sneezing, and diarrhea are all mechanisms for microbial spread, and the distress caused to the host is merely incidental. We will illustrate major host response mechanisms using examples of five microbes that are exposed to, but circumvent, those responses.

THE STRUCTURE OF THE IMMUNE RESPONSE [1–3]

The response to invading microbes consists of three major arms: (i) the innate immune system, which recognizes pathogens and cellular damage; (ii) adaptive immunity, which mounts a pathogen-specific response; and (iii) effector mechanisms, directed by both innate and adaptive mechanisms, which inactivate pathogens (listed in Table 3.1). These divisions are somewhat arbitrary. In fact, nearly every cell in the body participates to a degree in all three functions; there really is no clear division between the immune system proper and the rest of the body. Of course, cells such as lymphocytes, phagocytes, and dendritic cells are much more deeply committed to defense against pathogens than most other specialized cell types (skeletal muscle cells, for instance).

Two major categories of molecules are recognized by the innate immune system: (i) microbial components and (ii) markers of tissue damage or death. Microbial molecules recognized by the innate immune system include peptidoglycan, lipopolysaccharide, or double-stranded RNA. Pathogen-Associated Molecular Patterns (PAMPs) are detected by the Pattern Recognition Receptors (PRRs) on host cells. The first-described and most important class of PRRs are the toll-like receptors (TLRs). Markers of tissue damage and death recognized by the innate immune system include tissue factor and other markers of cellular distress. Vascular damage can trigger the activation of coagulation and kinin pathways. Triggering PRRs in turn leads to a cascade of events which invoke the other two functions of the immune response. Depending on the tissue site, types of microbial structures, and category of cellular distress

Molecular Pathology © 2009, Elsevier, Inc. All Rights Reserved.

Table 3.1 Host Effector Mechanisms

Name	Properties	Effector Mechanisms
	Soluble Effectors	
Complement System	Proteolytic cascade, activated by antibody, directly by microbial components, or via PRRs.	Direct destruction of pathogens via pore-formation. Recruit inflammatory cells. Enhance phagocytosis and killing.
Coagulation System	Proteolytic cascade, activated by tissue and vascular damage.	Prevents blood loss. Bars access to bloodstream. Proinflammatory.
Kinin System	Proteolytic cascade triggered by tissue damage.	Proinflammatory. Causes pain response. Increases vascular permeability to allow increased access by plasma proteins.
Antibodies	Antigen-specific proteins produced by B-cells. Recognize a broad range of antigens.	Directly neutralize pathogens. Activate complement. Opsonize pathogens to enhance phagocytosis and killing
	Cellular Effectors	
Monocyte/ Macrophage	Have PRRs to recognize pathogens; activated by specific T-cells and chemokines.	Phagocytosis and microbial killing via multiple mechanisms. Antigen presentation.
Dendritic Cell	Ingest large amounts of extracellular fluid; migrate to lymph node to present antigen to naïve T-lymphocytes.	Antigen uptake, transport, and presentation to T-lymphocytes. Initiate adaptive immune response.
Neutrophil	Have PRRs to recognize pathogens, activated antibody and complement.	Phagocytosis and microbial killing via multiple mechanisms.
Eosinophil	Recognize antibody-coated parasites.	Killing of multicellular pathogens.
Basophil/Mast cell	Associated with IgE-mediated responses.	Release of granules containing histamine and other mediators of anaphylaxis.
NK-cell	Lymphocyte lacking antigen-specific reactivity; recognize PAMPs of intracellular pathogens, activated by chemokines and by membrane proteins of infected cells.	Induce death of infected cells via membrane pores and induced apoptosis.
B-lymphocyte	Recognize antigens presented by APCs; regulated by T-cells and chemokines.	Produce antibody.
T-lymphocyte	Recognize antigens presented by APCs; regulate major portions of both adaptive and innate immunity.	Directly kill infected cells via membrane pores and induced apoptosis. Activate macrophages. Many other functions.

recognized, cells of the adaptive immune system and a broad range of effector mechanisms are recruited. Cellular recruitment is largely mediated by chemokines (peptide messengers that modulate immune cellular responses) and nonpeptide inflammatory mediators (such as prostaglandins).

Antigens processed by phagocytic and nonphagocytic cells are presented to lymphocytes, which then mount an adaptive immune response. The adaptive immune response is embodied in (i) T-lymphocytes, which regulate immune responses, invoke powerful effector mechanisms, and participate directly in cytotoxic effector responses; and (ii) B-lymphocytes, which produce antibodies. Antibodies are both direct effectors of the immune response and mediators of innate and adaptive immunity. Antibodies directly neutralize some organisms, but also invoke and enhance further effector mechanisms by opsonizing microbes to direct their ingestion by phagocytes and by initiating complement activity.

The essence of the adaptive immune response is somatic genetic variation which produces diverse, antigen-specific molecules (antibodies and T-cell receptors). Each lymphocyte produces only a single receptor or antibody. Lymphocytes are elaborately selected to eliminate self-reactive molecules and to favor cells making receptors or antibodies to pathogens. While this process is complex, time-consuming, and wasteful (many lymphocytes are eliminated for each clone which survives), the specificity of the adaptive immune response makes it a central component of the mammalian defense system. The most powerful effector mechanisms include phagocytic cells (such as neutrophils and activated macrophages), and certain soluble factors (such as complement). These effectors are directed and controlled by the antigen-specific immune response, but have the potential to cause significant damage to host tissues.

A huge range of effector mechanisms limit and eliminate infection either by direct antimicrobial activity, or by creating physical or chemical barriers to microbial proliferation and spread. A partial list of effectors is provided in Table 3.1.

It is important to re-emphasize that the distinction between the innate immune system, adaptive immunity, and effector mechanisms is rather arbitrary, and many cellular and humoral responses involve all three. For example, the proteins of the coagulation cascade recognize tissue damage and certain microbial components, recruit inflammatory cells, and by blocking blood flow to infected tissues, limit spread of infections. T-lymphocytes carry TLRs, which play a significant role in the activation and selection of pathogen-reactive cells; express T-cell receptors, which recognize specific pathogen antigens; recruit and activate specific effector cells; and participate directly in cell-dependent cytotoxicity to eliminate cells infected with viruses and other intracellular pathogens.

REGULATION OF IMMUNITY [4,5]

Because inflammatory responses are metabolically costly and capable of causing enormous damage to tissues, the immune system is tightly regulated. Soluble effector systems, such as the complement and coagulation cascades, have soluble inhibitors which usually confine these responses to the area where the initiating stimulus occurs. Cellular effectors are activated and inhibited via both chemokines and direct signaling via adhesion molecules and direct ligand-receptor interactions with regulatory cells, mostly T-lymphocytes. Adaptive immune responses undergo elaborate screening. In most cases self-reactive cells are screened out, and cells reactive to current infectious challenges are activated and proliferate. In turn, the adaptive immune system directs the activities of the innate immune system. Antibodies activate complement and direct phagocytosis, and T-cells activate macrophages and secrete chemokines that modulate innate cellular responses.

PATHOGEN STRATEGIES

To evade the flexible and powerful system of host defenses, successful pathogens have evolved complex strategies. As examples, we have selected five pathogens: (i) the African trypanosomes (*Trypanosoma brucei* species), bloodstream-dwelling protists that evade antibody and complement via a remarkable strategy for generating antigenic diversity; (ii) *Staphylococcus aureus*, which employs a variety of strategies to evade, and in some cases to overload, innate and adaptive immune responses; (iii) *Mycobacterium tuberculosis*, an intracellular bacterium which actually proliferates inside an effector immune cell, the macrophage, by manipulating the phagosome-lysosome trafficking of its vacuole; (iv) herpes simplex virus, a complex DNA virus which successfully disrupts intracellular mechanisms of viral control; and (v) human immunodeficiency virus, a small RNA virus which turns the immune system on itself, and generates enormous molecular diversity during infection of a single host to evade and subvert the immune response over a period of years to decades. While these five organisms hardly demonstrate the incredible range of pathogen strategies of pathogenesis, they will allow for the discussion of the major aspects of immune function, in the context of meeting infectious challenges.

THE AFRICAN TRYPANOSOME AND ANTIBODY DIVERSITY: DUELING GENOMES

Blood is a tissue with many functions, including transport of nutrients and oxygen, carriage of hormones and other chemical messengers, and defense against invading microbes, among others. Many of these functions are mediated by the cellular components of the blood—red blood cells, assorted types of leukocytes, and platelets. However, the acellular portion of the blood (the plasma) contains a host of molecules involved in defense. These include the proteins involved in coagulation (which prevent spread of microbes via the blood and have powerful proinflammatory effects), a variety of regulatory cytokines, as well as dedicated antimicrobial molecules such as antibodies and complement. Complement was initially described as a component of plasma which enhanced the antibacterial powers of antibodies. Hence, this activity that *complemented* the activity of antibodies became known as complement. Complement can be activated by antibodies (the classical pathway) or by interaction with specific molecular signatures, including the alternative and lectin pathways. Once activated, the proteases of the complement system cleave targets, many of which are also proteases, continuing the cascade of activation and propagation of the response. The activated components of complement opsonize pathogens to enhance phagocytosis, attract immune cells, and serve as co-receptors to enhance adaptive responses, and directly damage some bacteria by forming a membrane attack complex.

Antibodies are a major arm of the adaptive immune response. Table 3.2 summarizes structure and major functions of the five immunoglobulin classes: IgD, IgM,

Table 3.2 Antibody Classes and Functions

Class	Location*	Structure	Function
IgD	Surface of B-cells only	2 κ or λ light chains, 2 δ heavy chains	Unknown; expressed early in differentiation along with IgM.
IgM	Plasma	2 κ or λ light chains, 2 μ heavy chains, arranged in pentamers with 1 J chain	Activates complement; first functional immunoglobulin formed in immune response.
IgG	Widely distributed in extracellular fluid	2 κ or λ light chains, 2 γ heavy chains	Complement activation, transfer to neonate via placenta, opsonization, neutralization of viruses and other pathogens.
IgA	Mucosal tissues, surfaces, and secretions	2 κ or λ light chains, 2 α heavy chains arranged in dimers with 1 J chain	Important in mucosal immunity, has opsonizing activity.
IgE	Bound to mast cells and basophils	2 κ or λ light chains, 2 ε heavy chains	Binds to and activates mast cells and basophils; important in defense versus parasites.

* All immunoglobulin classes are found on B-cells as antigen receptors.

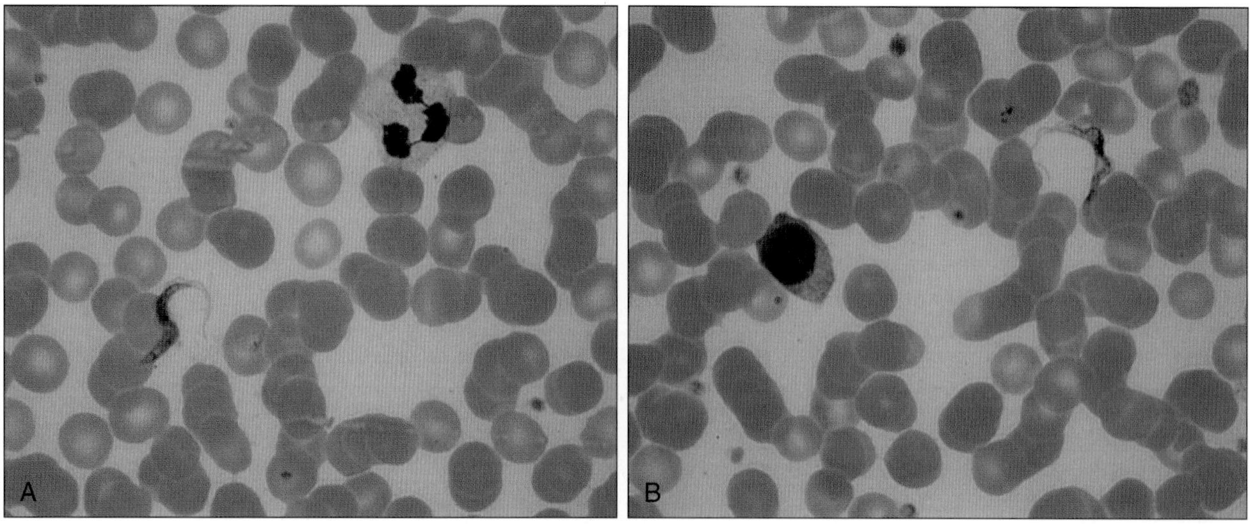

Figure 3.1 ***Trypanosoma brucei* in the blood.** The African trypanosome is typically seen in peripheral blood smears from infected patients. In the bloodstream, this extracellular parasite is exposed to complement, antibody, and white cells, but survives to produce a persistent infection (Wright-Giemsa stain, 1000x).

IgG, IgA, and IgE. Antibodies can be produced in large amounts, with extraordinary affinity and selectivity for specific pathogens. Antibodies function to activate complement, direct effector cells to the pathogen, neutralize, and sequester microbes.

Thus, the bloodstream is an extremely hostile environment for microbes. Yet the agents of African sleeping sickness, *Trypanosoma brucei (b.) rhodesiense* and *T. b. gambiae*, establish and maintain extracellular, bloodstream infections that can last weeks, months, and occasionally years. Figure 3.1 shows trypanosomes in the blood.

Generation of Antibody Diversity: Many Ways of Changing [6,7]

An antigenic stimulus is required to induce B-lymphocytes to produce antibody, but the mere presence of antigen is insufficient to induce a robust response. Co-stimulation, either by helper T-cells or by particular antigens (which induce T-cell independent antibody production via interaction with a pattern-recognition receptor) is required. Binding of complement to antigens activates co-receptors which markedly enhance the response. A complex series of stimulation, selection, and differentiation events result in maturation of the B-cell into a mature antibody-producing plasma cell.

Structurally, antibodies are divided into variable and constant regions. The constant regions determine the antibody class: IgD, IgM, IgG, IgA, or IgE. The different classes of antibodies have different effector functions and different destinations, as described in Table 3.2. The genome is not large enough to contain a separate gene for each antibody needed to respond to any antigen that the host might encounter. Thus, antibody diversity is generated by five different mechanisms (Figure 3.2): (i) the inherent diversity of the variable-region sequences, (ii) the genetic recombination of those regions into functional immunoglobulins, (iii) the combination of different heavy and light chain variable regions to form a functional antigen-binding site, (iv) junctional diversity introduced during the joining process, and (v) somatic hypermutation of the V-region in activated B-cells.

The variable regions of one heavy and one light chain combine to form the antigen-binding site of immunoglobulins (F_{ab}). During B-cell differentiation, V (variable) gene segments are recombined with J (joining) regions and C (constant) regions to form functional immunoglobulin genes. In heavy chain differentiation, a D region is also included. As B-cell differentiation proceeds, genetic rearrangements attach an antibody class-specific C region to each V-J or V-D-J region to produce genes for IgD, IgM, IgG, IgA, or IgE.

The genome contains substantial diversity of variable regions. Table 3.3 lists the numbers and combinations. In theory, the genetic diversity of the variable-region genes and their combinations can produce 1.9×10^6 different antigen-binding sites. In practice, it is likely that many of these combinations are useless, unstable, or even recognize self-antigens, so the true diversity of pathogen recognition sequences available from simple recombination of germ-line elements is much lower. During the genetic recombination events which join the variable regions, additional diversity is generated at the junctions. Two different mechanisms add nucleotides at the V-J and V-D junctions. In the first mechanism, P-nucleotides are added by a complex set of enzymatic reactions that create palindromic sequences. In the second mechanism, N-nucleotides are added randomly by terminal deoxynucleotide transferase. Another, less well-characterized mechanism may remove nucleotides. In combination, these mechanisms add diversity in a semi-random way. Finally, somatic hypermutation of the variable regions occurs during

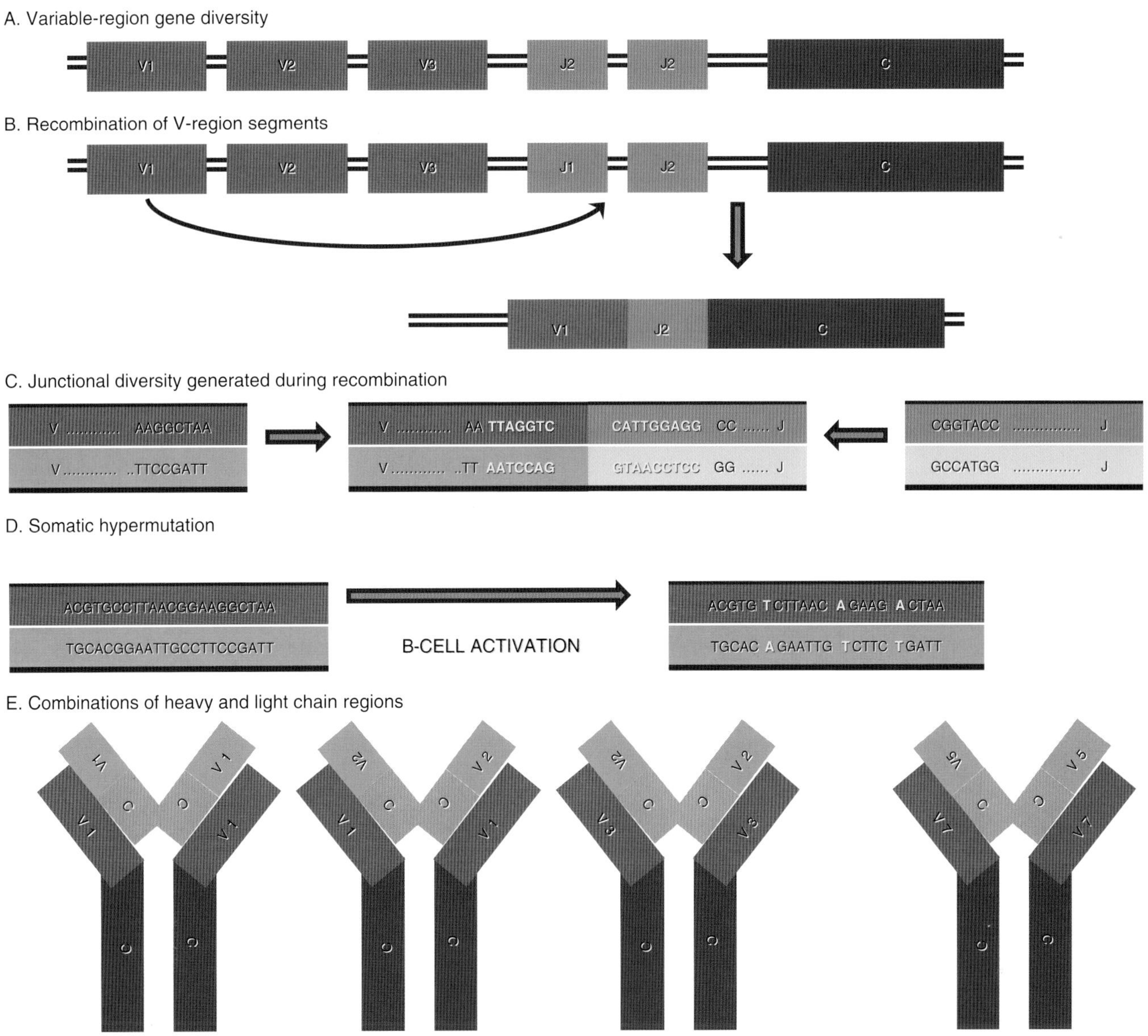

Figure 3.2 **Mechanisms of generating antibody diversity.** C regions and sequences are in shades of purple; J regions and sequences are in shades of green, and V regions and sequences are in shades of red. Altered or mutated sequences are in yellow. Designations of sequences (V1, etc.) are arbitrary and not meant to represent the actual arrangement of specific elements. (A) The inherent germline diversity of V and J regions provides some recognition diversity. (B) The combinations of V and J regions (V, D, and J in heavy-chains) provide additional diversity. (C) The V-J junctions undergo semirandom alterations during recombination, generating more variants. (D) In activated B-cells, the variable regions are hypermutated. (E) V regions of both light and heavy chains combine to form the antigen-recognition zone of the antibody. They can combine in different ways to provide still more variety of antigen recognition.

Table 3.3 Variable-region Gene Diversity

Immunoglobulin Class	Region	# of Genes
λ Light chains	V	30
120 total combinations	J	4
κ Light chains	V	40
200 total combinations	J	5
Heavy chains	V	40
6,000 total combinations	D	25
	J	6
Overall; 1.9 × 10⁶ combinations		

proliferation of activated B-cells. Point mutations are introduced into the variable regions at a very high rate. This requires transcriptional activation, but other factors must be involved to target mutations to the variable regions. During maturation, the antibody serves as the B-cell antigen receptor. The processes of stimulation, co-stimulation, and clonal deletion eliminate self-reactive cells and select for high-affinity receptors. The mechanisms for generating diversity, and for selecting high-affinity receptors combine to produce an immune response of high specificity and of increasing efficacy.

Trypanosoma brucei and Evasion of the Antibody Response: Diversity Responds to Diversity [8–10]

African trypanosomes are unicellular, flagellate parasites carried by the tsetse fly (*Glossina* species). The fly injects the infectious metacyclic form of the parasite into the host. Subsequently, the parasite invades the subcutaneous tissues, then the regional lymph nodes, and finally the bloodstream. One of the salient characteristics of the infectious trypanosome is a homogeneous glycophosphatidylinositol (GPI)-linked surface protein called the variant surface glycoprotein (VSG). VSG is proinflammatory. As an antigen, VSG is recognized by B-cells and T-cells, and an antibody response is effectively generated. Most (but not all) trypanosomes are destroyed by antibody, complement, and phagocytosis.

A subpopulation of organisms manages to change its coat protein to a new VSG that is structurally similar but antigenically distinct from the original VSG. Again, the immune system responds to the antigen by producing an antibody. Once again, most of the flagellates are destroyed, but a few produce still another variation of the coat protein and the cycle continues. In experimental trypanosome infections, hundreds of cycles have been observed. In patients, parasitemia can persist for months or even years. This ability to change surface proteins as rapidly as the host immune system can generate new antibodies is the fruit of a set of genetic mechanisms hauntingly similar to the mechanisms which generate the antibody diversity the parasite is successfully evading.

On a genomic level, the trypanosome contains hundreds of VSG genes and pseudogenes, only one of which is expressed in a particular cell at a particular time. VSGs are transcribed as parts of polycistronic mRNAs from telomeric regions of the chromosome known as expression sites (ES). There are roughly 20 ES per cell. Most of the VSG genes not part of an ES (called silent VSGs) are found at the telomeres of roughly 100 minichromosomes of 50–150 kb each. These minichromosomes most likely evolved specifically to expand the accessible VSG repertoire of the organism. Finally, a few VSG genes and large numbers of pseudogenes (which are truncated, containing frameshift or in-frame stop-codon mutations, or lacking the biochemical properties of expressed VSGs) are found in tandem-repeat clusters in subtelomeric locations. The structure of the VSG genes is depicted in Figure 3.3A.

There is only one active ES per cell, but several mechanisms can lead to expression of a new VSG, as depicted in Figure 3.3B. These mechanisms include (i) activation of a new ES (there are roughly 20 per genome) with inactivation of the original ES *in situ*, (ii) recombination of a VSG gene into the active ES via homologous recombination and telomere exchange, and (iii) segmental gene conversion of a portion or portions of VSG gene or genes into the ES VSG. Complex chimeric VSG containing elements of one or more silent VSG genes or pseudogenes may be produced. The regulation of the ES in the living trypanosome is still mysterious. Transcription of VSGs is performed by RNA polymerase I, the RNA polymerase normally associated with ribosomal RNA production, rather than by RNA polymerase II, which is typically associated with production of mRNA. The ES promoter appears to be constitutively active and unregulated. Control of gene expression is mediated through post-transcriptional RNA processing and elongation. The active ES is located in a specialized nuclear region known as the expression site body. The molecular mechanisms underlying activation and inactivation of a given ES are not yet understood, but are hypothesized to involve a competition for transcription factors.

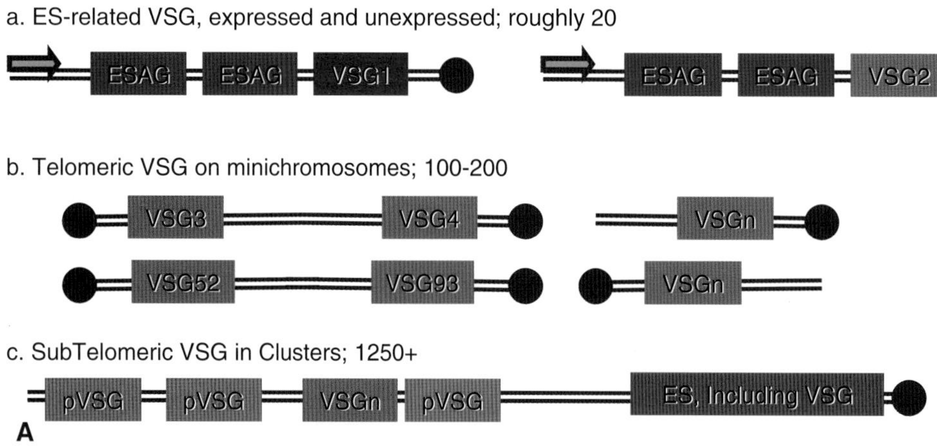

Figure 3.3A Mechanisms for generating variant surface glycoprotein diversity in trypanosomes: VSG genome structure. VSG sequences are in shades of red, others are purple. Silent VSG genes are dark red; expressed VSG genes are bright red, and VSG pseudogenes are pink. The large dots at the end of the chromosome represent telomeres. Green arrows are VSG promoters. ESAG are Expression Site Associated Genes, non-VSG genes, which are part of the polycistronic transcript driven by the VSG promoter. Designations of sequences (VSG1, etc.) are arbitrary and not meant to represent the actual arrangement of specific elements.

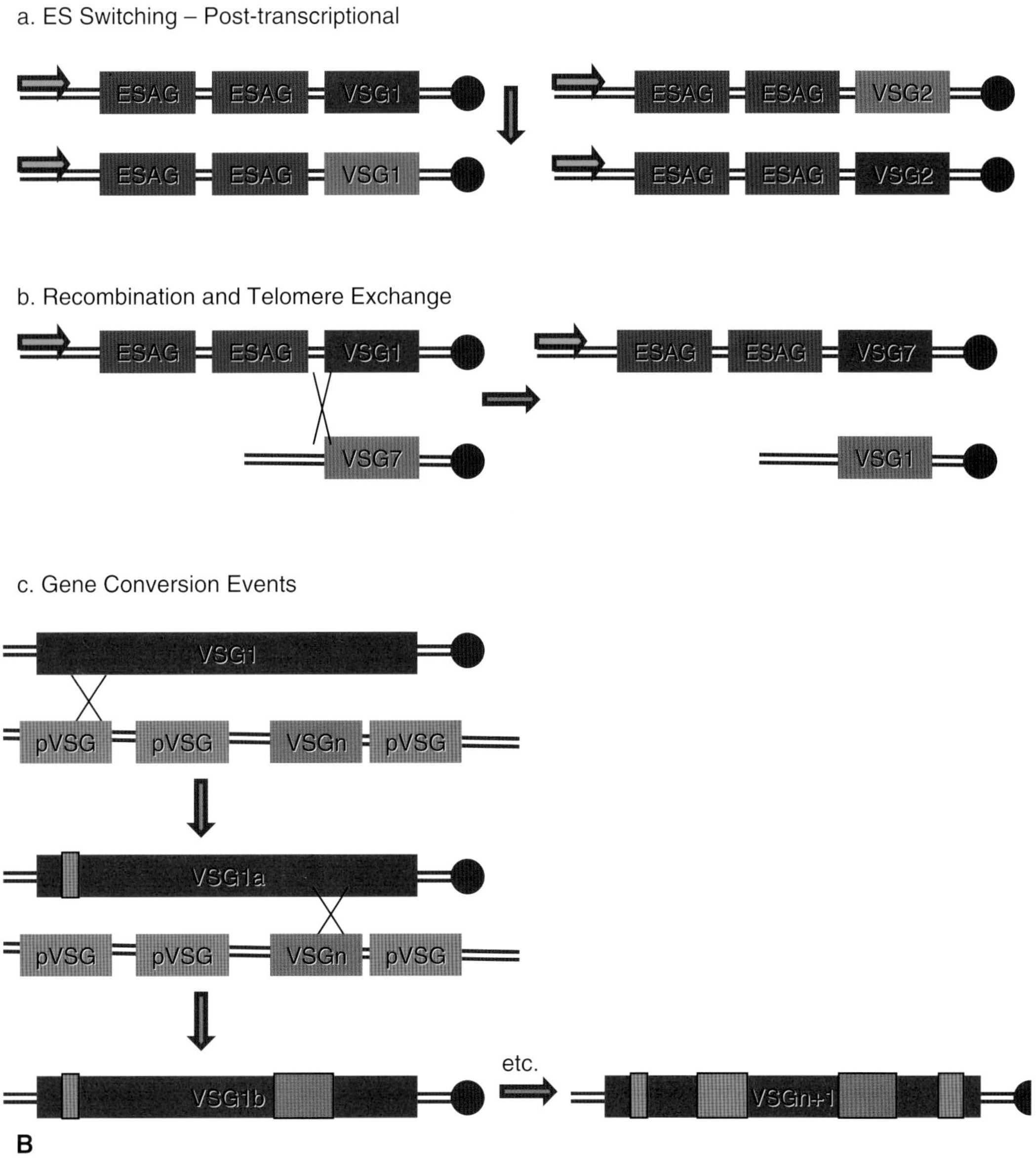

Figure 3.3B **Mechanisms for generating variant surface glycoprotein diversity in trypanosomes: Expressing new VSG.** VSG sequences are in shades of red, others are purple. Silent VSG genes are dark red; expressed VSG genes are bright red, and VSG pseudogenes are pink. The large dots at the end of the chromosome represent telomeres. Green arrows are VSG promoters. The Xs represent recombination or gene conversion events. ESAG are Expression Site Associated Genes, non-VSG genes which are part of the polycistronic transcript driven by the VSG promoter. Designations of sequences (VSG1, etc.) are arbitrary and not meant to represent the actual arrangement of specific elements. a. Post-transcriptional regulation causes different VSGs, located in alternative telomeric ESs, to be expressed. b. Recombination can switch a VSG gene from a minichromosome or other telomere to an ES. c. Gene conversion events can alter the sequence of VSGs located at ESs or elsewhere, drawing upon the sequence diversity not only of the silent VSGs but also of the VSG pseudogene pool.

If switching between VSGs were simply random, one would expect an initial wave of parasites and then a second wave containing all the possible VSG variants which would overwhelm the host. However, this does not occur. There appears to be a hierarchy of switching mechanisms, so the more probable switching events occur early in infection, and VSGs generated by less probable mechanisms occur later in the infection. After switching between ESs, recombination of VSGs into an active ES is the next most commonly observed mechanism, followed by more-complex gene conversion and recombination events. The product of these genetic mechanisms is a semiprogrammed progression of surface coats in the population of organisms infecting a single host, which allows for prolonged parasitemia, and an expanded period of time in which the organism can be picked up by an insect vector for transmission to a new host. Perhaps not coincidentally, these mechanisms also contribute to variation over historical and evolutionary time of the parasite's VSG repertoire, as new sequences are assembled from the diverse genetic repertoire of potential VSG elements.

Trypanosomes are not the only organisms to utilize antigen switching or antigenic variation to evade immune responses. *Borrelia burgdorferi*, *Plasmodium*, *Neisseria gonorrhoeae*, and other organisms have various mechanisms of changing their antigenic constituents, but the mechanisms employed by the African trypanosome are the most spectacular.

The Trypanosomacidal Serum Factor Story: Host Adaptations to Specific Pathogens and Pathogen Response [11]

Trypanosoma brucei rhodesiensae (which causes a uniformly fatal human disease) and *Trypanosoma brucei brucei* (which is nonpathogenic for humans but causes disease in cattle) are extremely closely related. However, normal human serum lyses *T. b. brucei*, while *T. b. rhodesiensae* is unharmed, although it quickly loses this serum resistance if passaged *in vitro* without serum. The substance in serum associated with trypanosomal lysis, Trypanosomal Lytic Factor (TLF), is associated with HDL, and has been identified as apolipoprotein L-I (apoL-I). This is a human-specific apolipoprotein of relatively recent evolutionary origin (probably from a gene duplication and divergence event) that exerts its lytic activity after endocytosis into the trypanosomal lysosome. The protein associated with serum resistance in *T. b. rhodesiensae* (Serum Resistance Associated protein—SRA) is coded by a highly modified VSG gene which binds to and inactivates apoL-I.

Highly prevalent pathogens which cause severe illness exert powerful selective pressure on host populations. Specific host genetic adaptations to particular pathogens are widely known. Sickle-cell trait as a defense against falciparum malaria is the best documented [12]. In their turn, microbes adapt to the novel host environments that they encounter.

STAPHYLOCOCCUS AUREUS: THE EXTRACELLULAR BATTLEGROUND

Staphyloccus aureus (S. aureus) is a Gram-positive extracellular bacterium that is part of our commensal flora, living on the mucosal surfaces of humans and other mammals (Figure 3.4). It is a versatile pathogen in both community-acquired and hospital-acquired infections that range from superficial infections of skin and soft tissue to potentially life-threatening systemic disease. The first lines of defense against *S. aureus* are the recognition molecules and effector cells of the innate immune system; but *S. aureus* engages a multitude of mechanisms to subvert the innate immune response of the host.

The Innate Immune System: Recognition of Pathogens [13]

After *S. aureus* breaches intact skin or mucosal lining, which constitutes the border by which the human body shields inside from outside, it first encounters resident

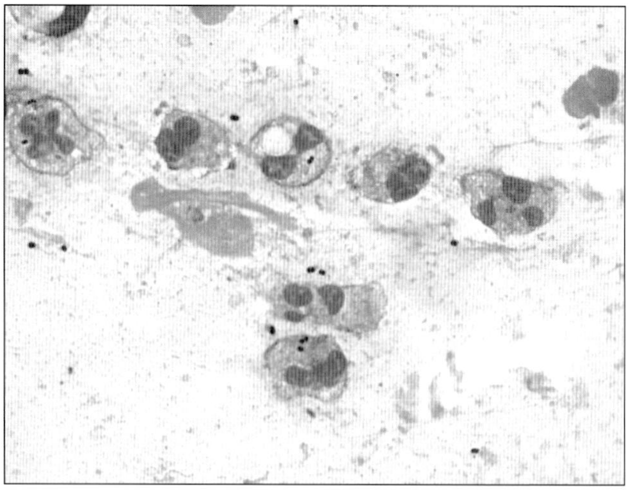

Figure 3.4 ***Staphylococcus aureus* and neutrophils.** A Gram-stained smear from a patient with *S. aureus* pneumonia. Staphylococci are located both extracellularly and within neutrophils (Gram stain, 1000x).

macrophages. These are long-lived phagocytic cells that reside in tissue and participate in both innate and adaptive immunity. The macrophage expresses several receptors that are specific for bacterial constituents, such as LPS (endotoxin) receptors, TLR (see Table 3.4), mannose receptors, complement receptor C3R, glucan receptors, and scavenger receptors. After bacteria or bacterial constituents (such as free bacterial DNA with CpG-rich oligonucleotide sequences, lipopolysaccharide, or peptidoglycan) bind to their receptors, the macrophage engulfs them. The phagosome fuses with the lysosome to form the phagolysosome, degradative enzymes and antimicrobial substances are released, and the content of the phagolysosome is digested, then the fragments presented to the adaptive immune system via MHC class II.

Activation of TLRs kicks off a common intracellular signal transduction pathway that involves an adaptor protein called MyD88 and the interleukin-1 receptor associated kinase (IRAK) complex. Ultimately, the transcription factor NFκB translocates to the nucleus of the macrophage and induces expression of inflammatory cytokines. Important cytokines that are secreted by macrophages in response to bacterial products include IL-1, IL-6, tumor necrosis factor α (TNFα), the chemokine CXCL8, and IL-12. These molecules have powerful effects and start off the local inflammatory response (Figure 3.5). A critical task of the macrophages is to recruit neutrophils to the site of infection in an attempt to keep the infection localized.

As killers of bacteria, macrophages pale in comparison with neutrophils, which are short-lived, dedicated phagocytes circulating in the blood, awaiting a call from the macrophage to enter infected tissue. They are plentifully supplied with antimicrobial substances and mechanisms, stored in their granules and elsewhere, and themselves programmed to die, on average within days after being released from the bone marrow. Neutrophils have receptors on their surfaces for inflammatory mediators derived both from the human host

Table 3.4 Recognition of Microbial Products Through Toll-like Receptors

Receptor	Ligands	Microorganisms Recognized	Notes
TLR-2 (TLR-1, -6) heterodimers	Peptidoglycan, bacterial lipoprotein and lipopeptide, porins, yeast mannan, lipoarabinomannan, glycophosphatidylinositol anchors	Gram-positive bacteria, mycobacteria, *Neisseria*, yeast, trypanosomes	Carried on macrophages
TLR-3 homodimer	Double-stranded RNA	Viral RNAs	
TLR-4 homodimer	Lipopolysaccharide (LPS)	Gram-negative bacteria	Carried on macrophages
TLR-5 homodimer	Flagellin	Gram-negative bacteria	Carried on intestinal epithelium; interacts directly with ligand
TLR-9 homodimer	DNA with unmethylated CpG motifs	Bacteria	Intracellular receptor

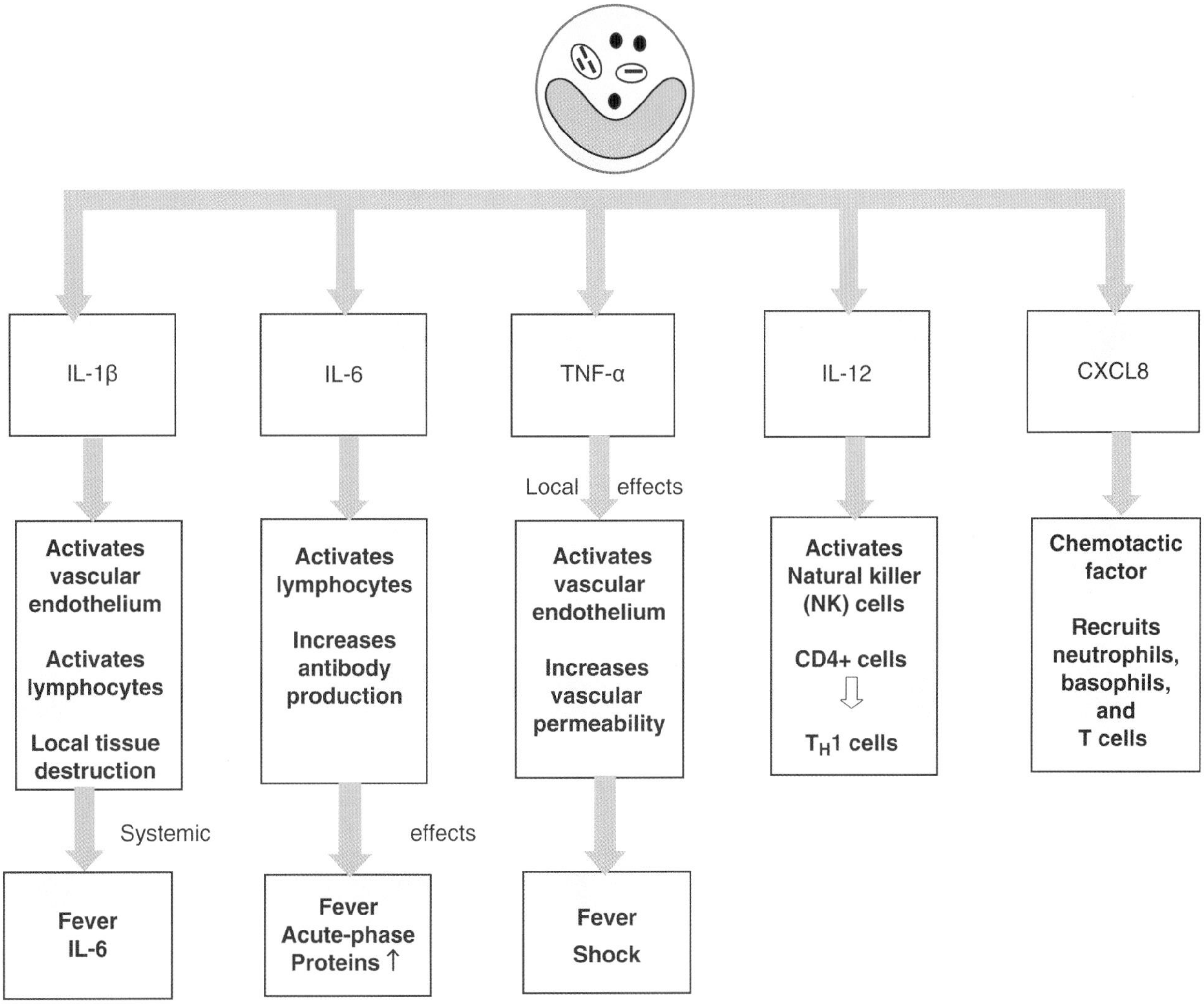

Figure 3.5 Chemokines secreted by macrophages in response to bacterial challenge. Chemokines secreted by macrophages have both local and systemic effects which mobilize defenses to infection, but may have unfortunate consequences as well.

(CXCL8 from activated macrophages, C3a and C5a during complement activation) and from the bacteria themselves (for instance, bacteria-specific formylated peptides). Neutrophils home to sites of infection along the gradient of these chemoattractants.

The complement system is also part of the innate immune system. In the presence of microorganisms, the spontaneous low-level hydrolysis of complement factor C3 to iC3 is increased and leads to activation of C3, which ultimately leads to deposition of C3b on the microbial surface. This represents the alternative pathway of complement activation, and serves two purposes. First, C3b covalently bound to microbial surfaces tags them for more efficient phagocytosis since phagocytic cells have complement receptors. Second, the accumulation of C3b on the bacterial cell surface changes the specificity of the C3 convertase of the alternative pathway, Bb-C3b, to cleave C5, which then harbors the terminal complement proteins to form the lytic membrane attack complex consisting of C5b-C9. Gram-negative bacteria, but not Gram-positive bacteria such as *S. aureus*, are successfully lysed by the pore-forming membrane attack complex [14].

Most infections are efficiently cleared by the ubiquitous and induced responses of innate immunity described. Once an infectious agent escapes innate mechanisms and spreads from the point of entry, it faces the adaptive immune response that is characterized by an extensive process in the draining lymph node in which pathogen-specific lymphocyte clones are selected, expanded, and differentiated.

S. aureus is such a successful pathogen because it expresses a multitude of virulence genes (Table 3.5) that act together to evade and subvert the three main axes of the innate immune responses: (i) recruitment and actions of inflammatory cells, (ii) antimicrobial peptides, and (iii) complement activation.

Table 3.5 Examples of Virulence Factors Responsible for Immune Evasion by *Staphylococcus aureus* (Modified after [10])

Name of Factor	Abbrev.	Function	Interference with Host Response
		Anti-inflammatory Peptides	
Chemotaxis inhibitory protein of *S. aureus*	CHIPS	Binds to C5aR and formylated protein receptor (FPR)	Blocks chemotaxis
Staphylococcal complement inhibitor	SCIN	Stabilizes C2a-C4b and Bb-C3b convertases	Inhibits complement
		Toxins	
Staphylococcal superantigen-like protein-5	SSL-5	Binds to P-selectin glycoprotein ligand-1 (PSGL-1)	Inhibits neutrophil recruitment
Staphylococcal superantigen-like protein-7	SSL-7	Binds to complement C5; binds to IgA	Inhibits complement
β-hemolysin	Hlb	Lysis of cytokine-containing cells	Cytotoxicity
γ-hemolysin	Hlg	Lysis of erythrocytes and leukocytes	Cytotoxicity
Panton-Valentine leukocidin	PVL	Stimulates and lyses neutrophils and macrophages	Change in gene expression of staphylococcal proteins; important in necrotizing pneumonia
Leukocidins D, E, M	LukD, LukE, LukM	Lysis of erythrocytes and leukocytes	Cytotoxicity
Exotoxins with superantigen activity enterotoxins	Se	Food poisoning when ingested; septic shock when systemic	Bridge MHC-II-TCR without antigen presentation; confer nonspecific T-cell activation and/or T-cell anergy; downregulate chemokine receptors
Toxic shock syndrome toxin-1	TSST-1		
		Secreted Expanded Repertoire Adhesive Molecules	
Coagulase	Coa	Activates prothrombin and binds fibrin	Antiphagocytic
Extracellular adherence protein	Eap	Binds to endothelial cell membrane molecules, binds to ICAM-1 and T-cell receptors	Blocks neutrophil and T-cell recruitment; inhibits T-cell proliferation
Extracellular fibrinogen binding protein	Efb	Binds to fibrinogen; binds to complement factors C3 and inhibits its deposition on the bacterial cell surface	Inhibits complement activation beyond C3b, thereby blocking opsonophagocytosis; binds to platelets and blocks fibrinogen-induced platelet aggregation
		Microbial Surface Components Recognizing Adhesive Matrix Molecules	
Clumping factor A	ClfA	Binds to fibrinogen	Antiphagocytic
S. aureus protein A	Spa	Binds to Fc portion of IgG and TNF receptor 1	Antiopsonic, antiphagocytic; modulates TNF signaling
		Extracellular Enzymes	
Catalase	CatA	Inactivates free hydrogen peroxide	Required for survival, persistence, and nasal colonization
Staphylokinase	Sak	Plasminogen activator	Antidefensin; cleaves IgG and complement factors
		Capsular Polysaccharides	
Capsular polysaccharide type 1, 5, and 8	CPS 1, CPS 5, CPS 8	Masks complement C3 deposition	Antiphagocytic effect

Inhibition of Inflammatory Cell Recruitment and Phagocytosis

While the macrophage and other primary defenses are sending messages about the presence of a pathogen, which recruit neutrophils and other inflammatory cells to the site of infection, *S. aureus* is busy blocking, scrambling, or subverting those messages. *S. aureus* contains an arsenal of antiadhesive and antimigratory proteins that specifically interfere with every step of host inflammatory cell recruitment. Staphylococcal chemotaxis inhibitory protein of *S. aureus* (CHIPS), present in approximately 60% of *S. aureus* strains, blocks neutrophil stimulation and chemotaxis by competing with the physiologic ligands at the complement receptor C5aR and formylated peptide receptor (FPR). Once near the site of infection, neutrophils must leave blood vessels through the vascular endothelium to reach the site of infection. This process starts with the rolling of leukocytes on activated endothelial cells by sticking to P-selectin expressed on the endothelial cell surface. Staphylococcal superantigen-like protein-5 (SSL-5) blocks this interaction. The next steps are adhesion to and transmigration through the endothelium, mediated in part by ICAM-1, the intercellular adhesion molecule-1 expressed on endothelial cells. *S. aureus* answers with production of extracellular adherence protein (Eap) that binds to ICAM-1, thereby interfering with extravasation of neutrophils at the site of infection [15,16].

Once a neutrophil manages to get close to a *S. aureus* cell, the bacterium still has means to evade phagocytosis. *S. aureus* expresses surface-associated antiopsonic proteins and a polysaccharide capsule that compromise efficient phagocytosis by neutrophils (Figure 3.6**a**). Protein A is a wall-anchored protein of *S. aureus* that binds the Fc portion of IgG and coats the surface of the bacterium with IgG molecules that are in the incorrect orientation to be recognized by the neutrophil Fc receptor (Figure 3.6**c**). Clumping factor A (ClfA) is a fibrinogen-binding protein present on the surface of *S. aureus* that binds to fibrinogen and coats the surface of the bacterial cells with fibrinogen molecules, additionally complicating the recognition process (Figure 3.6**e**).

Inactivation of Antimicrobial Mechanisms

One of the cardinal features of *S. aureus* is its ability to secrete several cytolytic toxins (hemolysins, leukocidins—see Table 3.5) that damage the membranes of host

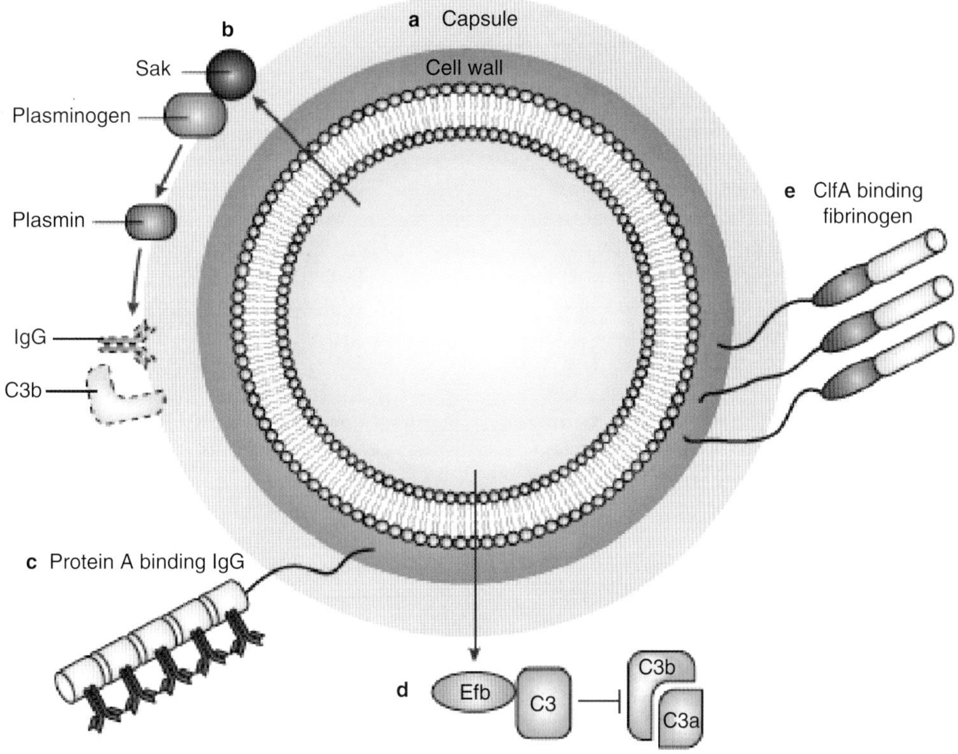

Figure 3.6 **Mechanisms by which *Stapylococcus aureus* evades opsonophagocytosis.** The figure illustrates (**a**) the capsular polysaccharide, which can compromise neutrophil access to bound complement and antibody; (**b**) the extracellular staphylokinase (Sak), which activates cell-bound plasminogen and cleaves IgG and C3b; (**c**) protein A with 5 immunoglobulin G (IgG) Fc-binding domains; (**d**) fibrinogen-binding protein (EfB), which binds complement factor C3 and blocks its deposition on the bacterial cell surface. Complement activation beyond C3b attachment is prevented, thereby inhibiting opsonization. (**e**) Clumping factor A (ClfA), which binds the γ–chain of fibrinogen. Reprinted by permission from Nature Publishing Group, Nature Reviews Microbiology, Volume 3, copyright 2005, page 952.

cells. They contribute to the development of abscesses with pus formation by direct killing of neutrophils.

If *S. aureus* is successfully engulfed by a neutrophil, its end has not yet come. It is well endowed with surface modifications and other mechanisms to help it survive in the phagosome. Transcriptional microarray analysis of mRNA from *S. aureus* following ingestion by neutrophils revealed a large number of differentially regulated genes [16]. Many known stress-response genes, including superoxide-dismutases, catalase, and the leukotoxin Hlg, were upregulated immediately after ingestion. *S. aureus* is able to interfere with endosome fusion and the release of antimicrobial substances. Two superoxide dismutase enzymes help *S. aureus* to avoid the lethal effects of oxygen free radicals that are formed during the respiratory burst of the neutrophil. Modifications to the cell wall teichoic acid and other cell wall components change the cell surface charge such that the affinity of cationic, antimicrobial defensin peptides is reduced. Staphylokinase binds defensin peptides, and the extracellular metalloprotease aureolysin cleaves and inactivates certain defensin peptides [14,15].

Inhibition of Complement Activation: You Can't Tag Me!

The prerequisite for complement activation is cleaving C3 into a soluble C3a and covalent attachment of C3b to the surface of *S. aureus*. This is either carried out by the C3 convertase C4bC2a (classical and lectin pathway) or C3bBb (alternative pathway). *S. aureus* secretes a protein called *Staphylococcus* complement inhibitor (SCIN) that stabilizes both C3 convertases and renders them less active. Similarly, the extracellular fibrinogen-binding protein Efb blocks C3 deposition on the bacterial surface (Table 3.6**d**). The effect is reduced opsonization and hence reduced phagocytosis. However, *S. aureus* not only prevents complement factor deposition, but is also capable of eliminating bound C3b and IgG through a very clever mechanism. Host plasminogen that is attached to the bacterial cell surface is activated by the *S. aureus* enzyme staphylokinase to plasmin, which then cleaves surface-bound C3b and IgG, resulting in reduced phagocytosis by neutrophils (Table 3.6**b**).

Staphylococcal Toxins and Superantigens: Turning the Inflammatory Response on the Host

One of the most serious, life-threatening infections with *S. aureus* is toxic shock syndrome. It is caused by secreted exoenzymes and exotoxins (Table 3.5). Enterotoxins cause a fairly common, short-lived, benign gastroenteritis (food poisoning) when ingested, but act as a superantigen in systemic infections. The potent immunostimulatory properties of superantigens are a direct result of their simultaneous interaction with the V_b domain of the T-cell receptor and the MHC class II molecules on the surface of an antigen-presenting cell. Superantigens derive their name from the fact that they are able to polyclonally activate a large fraction of the T-cell population (2–20%) at picomolar concentrations, compared to a normal antigen-induced T-cell response where 0.001–0.0001% of the body's T-cells are activated. They bind to the variable part of the β chain of the T-cell receptor and to MHC class II molecules present on antigen-presenting cells without the need to be presented by antigen-presenting cells and cause an immune response that is not specific to any particular epitope on the superantigen. The cross-linking of MHC II molecule and TCR induces a signaling pathway that leads to proliferation of T-cells and a massive release of cytokines. This systemic cytokine storm causes extravasation of plasma and protein, resulting in decreased blood volume and low blood pressure. Activation of the coagulation cascade leads to disseminated intravascular coagulation (DIC), which further compromises perfusion of end organs and eventually results in multi-organ failure and death. It is ironic that in toxic shock most of the deleterious effects on the host tissue are not related to actions of the bacteria, but the exaggerated host immune response [16]. After all, it is usually not in the interest of *S. aureus* to kill the host.

MYCOBACTERIUM TUBERCULOSIS AND THE MACROPHAGE

Mycobacterium tuberculosis is an extremely successful pathogen, and is one of the most important causes of worldwide morbidity and premature death. The mycobacteria are also known as the acid-fast bacilli (AFB). The staining property of acid-fastness is mediated by the thick, lipid-rich cell wall of the organisms. The unique cell wall structure of the mycobacteria is involved in management of the host cell. Spread by the aerosol route from person to person, organisms in droplet nuclei are deposited in the alveoli, where they encounter – and enter – their first immunological barrier, the alveolar macrophages.

Many pathogens utilize the intracellular compartment to evade host responses. While the intracellular lifestyle avoids many host defenses, such as complement and neutralizing antibody, other mechanisms of immunity operate intracellularly. In order to thrive intracellularly, pathogens must enter the cell. Utilization of the phagocytic pathway of entry carries the risk of fusion with lysosomes and destruction. Utilization of other pathways is more energy-intensive for the microbe, limits the potential for the parasite to exploit cellular mechanisms of transport and trafficking of endosomes, and may expose the microbe to cytoplasmic pattern-recognition receptors which will activate other defenses. After entering the cell, pathogens must survive and reproduce within whichever compartment is entered. This may require inhibition of lysosomal fusion, surviving constitutive intracellular inhibitory or killing mechanisms, transport of nutrients, and exploiting other aspects of host cell physiology. Next intracellular pathogens must limit exposure to the cell-mediated immune system. Professional phagocytes are specifically equipped to present antigen via MHC class II, but most nucleated cells can present antigen via MHC class I pathways, which are specifically designed to initiate immune responses to intracellular pathogens by stimulating cytotoxic CD8+ T-cells. Finally, intracellular pathogens must prevent destruction of the host cell and pass successfully to another.

While tubercle bacilli are quite capable of extracellular growth and proliferation, in the early stages of infection this does not occur for long, since alveolar macrophages rapidly phagocytose them. The initial interaction between *M. tb* and the macrophage takes place in the absence of adaptive immunity. The mycobacterial surface appears to contain a rich array of TLR agonists that drive uptake of the organism and recruitment of inflammatory cells to the site of infection. For most microbes, that would be the end of the line; shortly after phagocytosis phagosome-lysosome fusion occurs, regulated by an elaborate array of proteins and glycolipds that control trafficking of membrane-bound organelles, and a variety of killing mechanisms are invoked. Acid pH, reactive oxygen and nitrogen intermediates, degradative enzymes, and toxic peptides kill and digest the invading pathogens. However, *Mycobacterium tuberculosis* survives and reproduces within the macrophage. Infected macrophages, though initially incapable of killing mycobacteria, nonetheless secrete chemokines which recruit other inflammatory cells to the area [3].

Mycobacterium and Macrophage: The Pathogen Chooses Its Destiny [17,18]

The mycobacterial manipulation of the macrophage begins on uptake, and some of the mechanisms utilized by the pathogen are shown in Figure 3.7. In phagocytic cells, different receptors drive different processing of the phagosome. Thus, pathogens can control their intracellular fate by choosing the receptor that recognizes them. In the case of mycobacteria, multiple receptors may be involved, and the details of their interaction *in vivo* are not yet known. However, one of the major receptors is complement receptor type 3 (CR3), a pathway which prevents macrophage activation, in contrast with phagocytosis driven by other receptors.

Once internalized in a phagosome, several mycobacterial molecules actively interfere with phagosome-lysosome fusion, and cellular processes regulated by phosphorylation and dephosphorylation signals are controlled through modification of normal signals. Phosphatidylinositol-3-phosphate (PI3P), a molecule synthesized by host phosphatidylinositol-3-kinase (PI3K), serves to direct proteins involved in the fusion process to the endosomal membrane. *M. tuberculosis* inhibits PI3K and secretes an acid phosphatase that degrades PI3P. In addition, other secreted phosphatases interfere with endosome trafficking. A mycobacterial protein kinase, PknG, is required for survival in macrophages. Dominant-negative mutants for this protein kinase are killed secondary to lysosomal fusion. PknG is secreted from mycobacteria and enters the host cell cytosol. Its most likely role there is to phosphorylate a host protein involved in endosome trafficking, preventing fusion of the mycobacterial compartment. Mycobacterial cell wall lipoarabinomannan (LAM) also blocks phagosome-lysosome fusion through an unknown mechanism.

Phagosome-lysosome fusion also involves calcium-dependent signaling. The calcineurin inhibitors cyclosporine A and tacrolimus, which are used clinically as

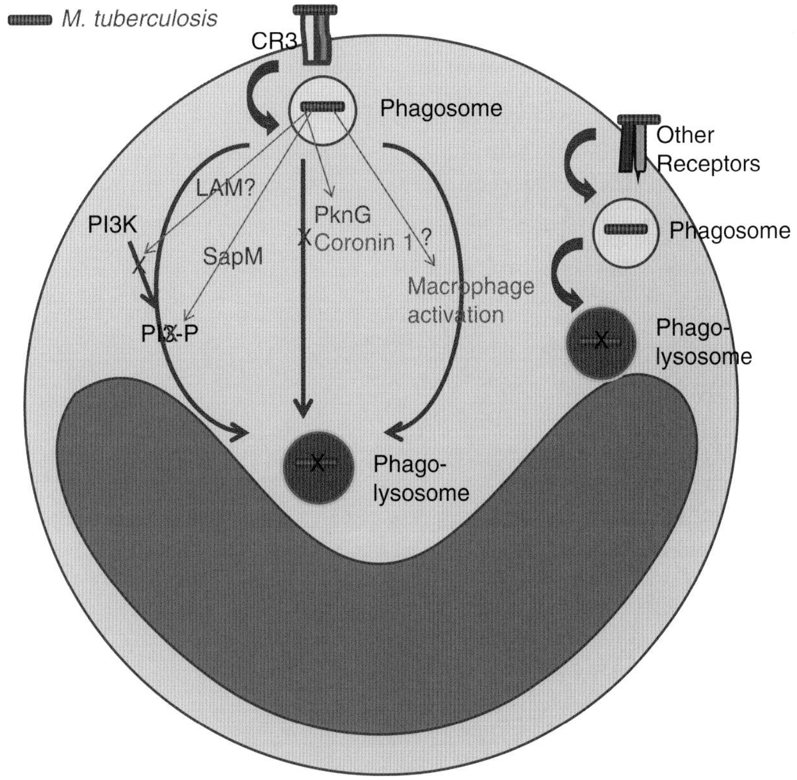

Figure 3.7 ***Mycobacterium tuberculosis* and the macrophage.** Gray arrows are host endosome processing pathways; the pink lines are mycobacterial mechanisms. Fusion of mycobacteria into mature phagolysosomes usually leads to death of the organism, so mycobacteria select their endocytotic pathway and interfere with mechanisms designed to result in phagosome-lysosome fusion.

immunosuppressants, paradoxically prevent mycobacterial proliferation in macrophages. A host protein, coronin 1, responsible for activation of calcineurin, appears to associate selectively with phagosomes containing living mycobacteria. A coronin-dependent calcium influx activates calcineurin, which blocks fusion. How mycobacteria invoke this pathway is still unknown.

Despite all these mechanisms, *M. tuberculosis* have very little ability to block phagosome-lysosome fusion in *activated* macrophages stimulated via cytokines and TLRs. Probably the most important cytokines in activating macrophages to kill mycobacteria are interferon-γ and TNFα. Both in animal models and in humans, suppression of these pathways by drugs or mutations results in vulnerability to tuberculosis. A number of *M. tuberculosis* components inhibit elements of the activation pathway. LAM and its glycosylated derivatives can modulate signaling pathways initiated by interferon-γ and by TLRs, and can block mitogen-activated-protein-kinase (MAPK) pathways within the network of activities that lead to activation. However, the macrophage is not entirely a passive host to the tubercle bacillus. The most effective response to this infection involves secretion of chemokines to recruit the adaptive immune system and other inflammatory cells, and to present mycobacterial antigen to T-cells.

The Adaptive Response to *M. tuberculosis*: Containment and the Granuloma [19]

Antigens from intracellular pathogens located in the endosomal compartment are primarily presented on class II MHC molecules and recognized by CD4 T-cells. This is not an exclusive arrangement since some antigen from the endosomal compartment is transported to the cytosol and presented to CD8 T-cells via MHC class I. Processing of antigen for class II requires acid pH and active acid proteases typically found in lysosomes; evidently not all *M. tuberculosis* succeed in inhibiting phagosome maturation and fusion. In addition to MHC class I and II, the membrane protein CD1 also binds antigen for presentation to T-cells. CD1 is specialized for presentation of hydrophobic molecules, such as the rich cell wall lipid and glycolipid components of mycobacteria, and is recognized by T-cells which express neither CD4 nor CD8.

CD4 T-cells mature into several populations of effector T-cells after extensive differentiation and selection. The best-known division is into Th_1 and Th_2 type cells. Th_2-type CD4 T-cells express cytokines which activate and induce class-switching in B-cells, resulting in an immune response centered around antibody production; opsonization; and handling of extracellular bacteria, viruses, and parasites. In contrast, the major activity of Th_1-type cells is to activate macrophages via interferon-γ, IL-2, TNFα, and other cytokines.

Differentiation of CD4 T-cells into the Th_1 and Th_2 lineages is controlled by the cytokines they encounter during the early stages of activation. Interferon-γ, secreted by macrophages and dendritic cells, induces differentiation toward the Th_1 phenotype. Since interferon-γ is a major cytokine produced by Th_1-type cells, the Th_1-type immune response is self-enforcing and tends to be stable unless other influences perturb the balance [3–5].

Activated macrophages are capable of killing ingested mycobacteria. The lesion that results from the effective immune response to *M. tuberculosis* infection is the granuloma, shown in Figure 3.8. Structurally, a tuberculous granuloma consists of a central area of necrosis, surrounded by macrophages (both activated and nonactivated), then a mixture of lymphocytes, macrophages, tissue cells, and fibroblasts. The lymphocytes are mainly Th_1-type CD4 cells, though CD8 cells are present and seem to have a role in immunity to tuberculosis.

The granuloma structure is effective in containing, but typically not in eradicating, the mycobacteria.

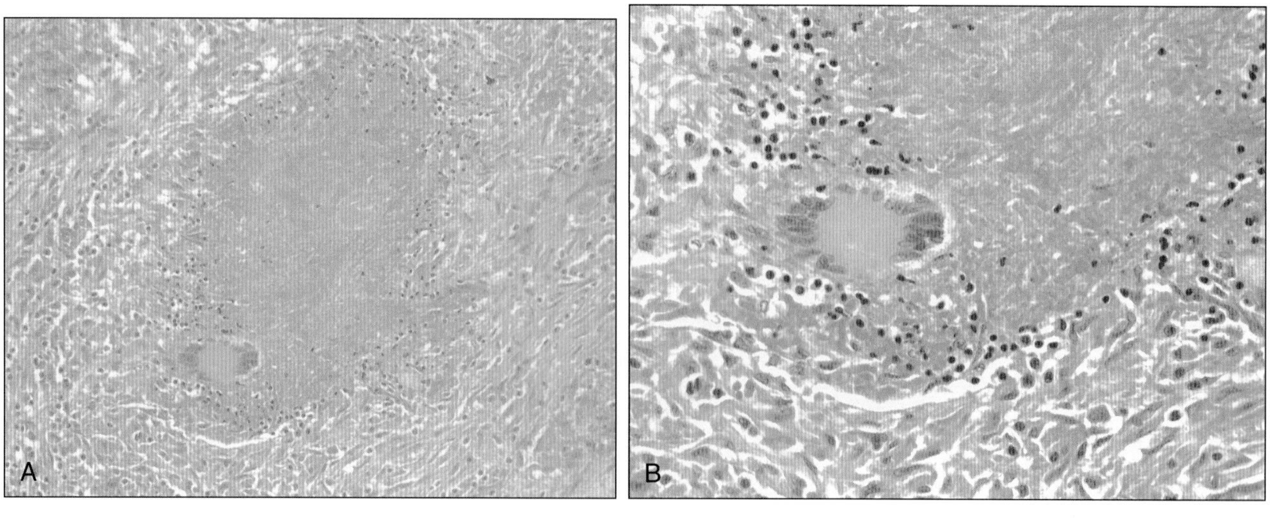

Figure 3.8 A mycobacterial granuloma. H&E stained sections of a mycobacterial granuloma. Central necrosis and an inflammatory response consisting of macrophages, lymphocytes, and fibroblasts are apparent. A large multinucleated giant cell, characteristic of the granulomatous reaction, is also present.

Persisting bacteria may remain viable for the life of the host, either within the necrotic center of the granuloma or in dynamic equilibrium within the inflammatory region of the lesion, or both. Under conditions of waning or suppressed immunity, the persisting mycobacteria can proliferate, spread, cause disease, and also escape, typically via the airborne coughed-out route, to a new host. *Mycobacterium tuberculosis* exploits the intracellular compartment to maintain itself in the host. Antibody responses to tuberculosis infection are weak, not protective, and possibly harmful. Instead, cell-mediated immune responses leading to macrophage activation, directed primarily by CD4 T-cells, lead to an at least partially protective response. A rather delicate balance of factors works in the interest of the pathogen. It is able to maintain itself for prolonged periods of time in a single host, awaiting an opportunity for transmission.

HERPES SIMPLEX VIRUS: TAKING OVER

Herpes Simplex Virus (HSV) types 1 and 2 are ubiquitous DNA viruses that cause a broad spectrum of disease, from painful oral and genital lesions to life-threatening brain and systemic infections. HSV is responsible for both acute and reactivation disease. Once infected, a person usually harbors latent HSV for life.

Defense Against Viruses: Subversion and Sacrifice [2,3,20,21]

Viruses present unique challenges to the immune system. Typically, viruses enter host cells and then utilize the host protein-synthetic and other apparatus to assemble new viral particles. Entry into cells is rarely by phagocytosis until a mature immune response produces opsonized viral particles. Instead, viruses recognize and enter host cells via pathways that place them directly into the cytoplasm. Mechanisms by which viruses damage cells are illustrated in Figure 3.9.

In response to viral infection, the immune system has mechanisms for recognizing and managing viruses. There are pattern-recognition receptors which recognize the molecular signatures of viruses. In particular, TLR-3 recognizes double-stranded RNA, TLR-7 recognizes single-stranded RNA, and TLR-9 recognizes unmethylated CpG-containing DNA. The mannose-binding protein CD 206 recognizes some viral glycoproteins, including one HSV protein. Double-stranded RNA is also recognized by the cytoplasmic proteins RIG-1 and MDA-5. Recognition of viral components by TLR-3, RIG-1, or MDA-5 causes activation of the interferon-regulatory factors IRF3 and IRF7 and the nuclear translocation of regulatory factor NFκB, which induces a range of proinflammatory mechanisms, most

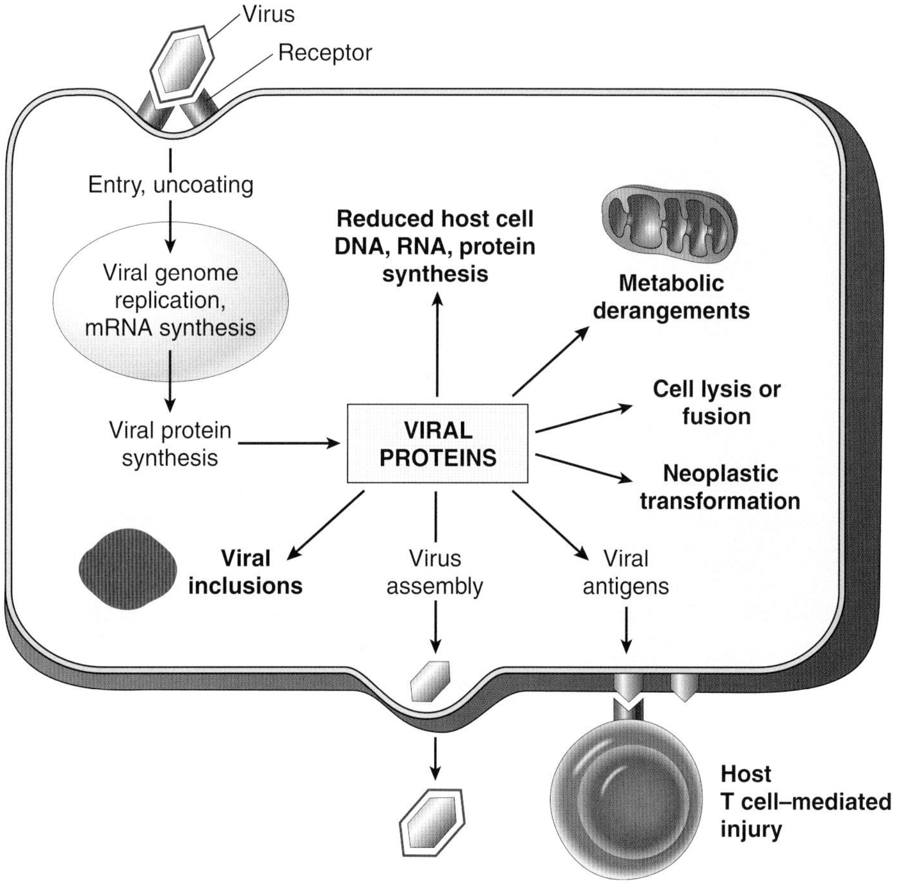

Figure 3.9 **How viruses damage cells**. Mechanisms by which viruses cause injury to cells. Reprinted by permission from Elsevier Saunders. Robbins and Cotran: Pathologic Basis of Disease, 7th edition, copyright 2004, page 357.

importantly, production of the antiviral chemokines interferon (IFN) α and β.

IFNα/β induces a series of events which form a primary defense against viral infections. Via a series of protein phosphorylation events triggered by the Janus-family kinase linked to the interferon receptor, a number of antiviral activities are induced. These include (i) 2′–5′ oligoadenylate synthetase, an enzyme that produces 2′–5′-linked polyadenylates, which activate ribonuclease L to digest viral RNAs; and (ii) dsRNA-dependent protein kinase R (PKR), which modifies eukaryotic initiation factor 2α, leading to arrest of translation. In addition to the intracellular activities of IFNα/β which target viral replication, other activities include upregulation of IFNα/β synthesis (a positive feedback loop); (ii) alteration of the proteasome, the protein degradation system of the cell, to favor production of peptides for presentation via MHC class I; (iii) upregulation of MHC class I expression and antigen presentation; and (iv) activation of macrophages, dendritic cells, and NK cells. Cytoplasmic antigens, such as most viral peptides, are exported from the cytosol to the endoplasmic reticulum. The heterodimeric proteins responsible for transport are called Transporters associated with Antigen Processing 1 and 2 (TAP1 and TAP2). TAP1 and TAP2 are induced by interferons, as are the MHC class I molecules which bind the antigen. After transport to the plasma membrane, MHC class I presents antigen to cytotoxic CD8 T-cells.

The final defense of virally infected cells is apoptosis. Cells activate a cascade of events which lead to programmed cell death through both induction by cytotoxic T-cells (with appropriate co-stimulus by dendritic cells and CD4 T-cells) and via other mechanisms triggered by viral infection. Apoptosis is an energy-requiring, deliberate, highly regulated process. Two major signaling pathways trigger apoptosis: (i) via extrinsic death receptors, such as the TNFα receptor, and (ii) via intracellular signals. In each case, a series of proteases called caspases are activated, and destroy the critical infrastructure of the cell in a systematic way.

Herpes Simplex Virus on the High Wire: A Delicate Balancing Act [22,23,24,25,26]

HSV has a large genome for a virus, consisting of ~150 kb, with at least 74 genes. An image of an HSV infection is shown in Figure 3.10. While fewer than half the genes are required for replication in cell cultures, viruses isolated from human hosts almost always have the full complement. Those accessory genes not required for growth *in vitro* are mainly involved in evading or inhibiting host responses.

A protein known as vhs (*v*iral *h*ost *s*hutoff) is an RNAse that degrades mRNA. Cellular responses to viral infection typically involve activation of response genes, and vhs globally inhibits such responses.

Because interferons play such a central role in defense against viral infection, a number of HSV proteins inhibit components of the IFNα/β response system (Table 3.6). US11 binds directly to PKR, preventing it from shutting off translation. US11 is actually trafficked intercellularly, and acts on adjacent uninfected but interferon-stimulated cells as well as the infected cell, priming them for infection. Meanwhile, ICP 34.5 activates host protein phosphorylase 1α, which dephosphorylates and reactivates initiation factor 2α should it have been inactivated by PKR. HSV protein ICP0 acts in the nucleus to inhibit IRF3 and IRF7, as well as several other interferon-response nuclear proteins. In the cytoplasm, ICP0 appears to block the activity of ribonuclease L. A latency-associated transcript (LAT) locus directly inhibits expression of interferon in neurons.

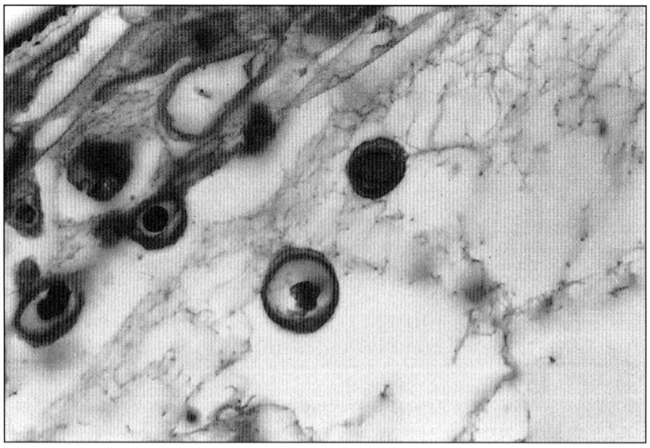

Figure 3.10 Histopathology of a herpes simplex lesion. Herpesvirus blister in mucosa. High-power view shows host cells with glassy intranuclear herpes simplex inclusion bodies. Reprinted by permission from Elsevier Saunders. Robbins and Cotran: Pathologic Basis of Disease, 7th edition, copyright 2004, page 366.

HSV also acts to reduce presentation of antigen to the adaptive immune system. ICP47 blocks entry of peptides to the ER via binding to TAP. The activities that inhibit interferon actions also inhibit the interferon-mediated increase in MHC class I expression. In addition, HSV exerts broad inhibitory activities when it invades dendritic cells, inducing downregulation of co-stimulatory surface proteins, adhesion molecules, and class I MHC. Since dendritic cells play a central role in antigen presentation and control of the adaptive response, this inhibition, mediated in part by the US3 protein kinase, most likely slows the response to HSV infection.

HSV proteins even exert control over apoptosis. Infection with HSV initially makes cells resistant to apoptosis by either the extrinsic or intrinsic pathways. A number of viral proteins seem to be involved, and the complete pathway has yet to be determined. Two essential viral genes appear to be regulatory elements that control expression of several other viral genes, and some host proteins are also involved. However, later in infection in some cell types, apoptosis is induced by HSV. The pathogen appears to create a delicate balance between inhibition of apoptosis early in infection, prior to production of virions, and induction of apoptosis late in the infective cycle.

Table 3.6 Interferon Actions and HSV Reactions

Mechanism	Effect	HSV Response
Activities That Inhibit Viral Gene Expression		
Activation of ribonuclease L	Digest viral RNAs	ICP0 inhibits ribonuclease L
ds-RNA-dependent phosphorylation of ribosomal initiation factor (PKR)	Arrest protein synthesis	Block the kinase responsible for phosphorylation; increase activity of phosphorylase which restores activity
Activities That Enhance Inflammatory Responses		
Alteration of proteasome to favor production of peptides for class I MHC	Increase presentation of antigen to adaptive immune system	Unknown
Upregulation of MHC class I and associate mechanisms	Increase presentation of antigen to adaptive immune system	Block TAP transport of antigen, which in turn limits externalization of MHC class I
Activation of antigen-presenting and effector cells	Accelerate antibody and cell-mediated immune responses; induce apoptosis in infected cells	Infection of these cell types leads to downregulation of response elements, especially in dendritic cells
Upregulation of interferon synthesis	Positive-feedback loop to limit infectability of nearby cells	vhs globally inhibits host gene expression; ICP0 blocks multiple transduction mechanisms of IFN signaling

HSV is an extremely prudent pathogen. It fails to completely inhibit the immune response, and local control of HSV infection is achieved relatively rapidly, usually with minimal lasting damage. Before this occurs, however, the virus has invaded neurons and entered a latent stage of infection, with only a small number of genes being transcribed at a low level. Neurons express low levels of MHC class I, which is further downregulated by HSV. Periodically, viral replication is turned on and viral particles are transported down the axon to its terminal near a mucosal surface, where the virus can invade epithelial cells and initiate a lesion, with more opportunities for transmission to a new host.

HIV: THE IMMUNE GUERILLA

By evolution, our immune system has developed several strategies to fight viral infections. In most chronic viral infections, both virus-specific T helper cells and cytotoxic T-lymphocytes (CTL) are required to effectively eliminate an infected cell. In turn, viruses have evolved numerous ways to evade the host immune system. Since the early 1980s, a new infectious disease of epidemic proportion has successfully emerged and spread around the globe: Acquired Immune Deficiency Syndrome (AIDS).

For more than two decades, the human immunodeficiency virus (HIV) has infected millions of people worldwide each year, mainly through mucosal transmission during unprotected sexual intercourse. In 2007, an estimated 33 million people lived with HIV globally, 2.5 million people were newly diagnosed with HIV infection, and 2.1 million patients died from AIDS. Since 1981, more than 25 million people have died from AIDS as a result of HIV infection (http://www.unaids.org/) [27].

HIV is special in that this virus not only evades the immune response, but directly attacks the very effector cells that play a pivotal role in the fight against viruses, namely T-lymphocytes, macrophages, and dendritic cells. One of the paradoxes of HIV infection is that the virus elicits a broad immune response that is not completely protective, while it causes immune dysfunction on several levels [28]. HIV infection is rarely successfully terminated by the immune system, but nearly always continues for many years and slowly progresses to AIDS and death if left untreated. Since the discovery of HIV in 1981, there has been an explosion of research aimed at deciphering the mechanism of infection, understanding why it cannot be controlled by our immune system, and at developing an effective vaccine.

Structure and Transmission of HIV—Small But Deadly

The human immunodeficiency virus is a human retrovirus belonging to the lentivirus group. Two genetically distinct forms exist, HIV-1 and HIV-2, but they cause similar syndromes and elicit the same host response. HIV is an RNA virus that utilizes reverse transcriptase (RT) and other enzymes to convert its genome from RNA into an integrated proviral DNA. Its viral core contains the major capsid protein p24, nucleocapsid proteins, two copies of viral RNA, and three viral enzymes (protease, reverse transcriptase, and integrase). The viral particle is covered by a lipid bilayer that is derived from the host cell membrane [29]. Two glycoproteins protrude from the surface: glycoproteins (gp)120 and gp41, which are critical for HIV infection of cells (Figure 3.11). In contrast with the 150 kb HSV genome, HIV has to accomplish all its tasks with a genome of only 9.8 kb.

The mode of transmission of HIV is mainly through close contact such as sexual intercourse: viral particles contained in semen enter the new host via microscopic lesions. Parenteral transmission, through blood products, sharing needles among drug users, or vertical transmission from mother to baby, are also important routes of transmission.

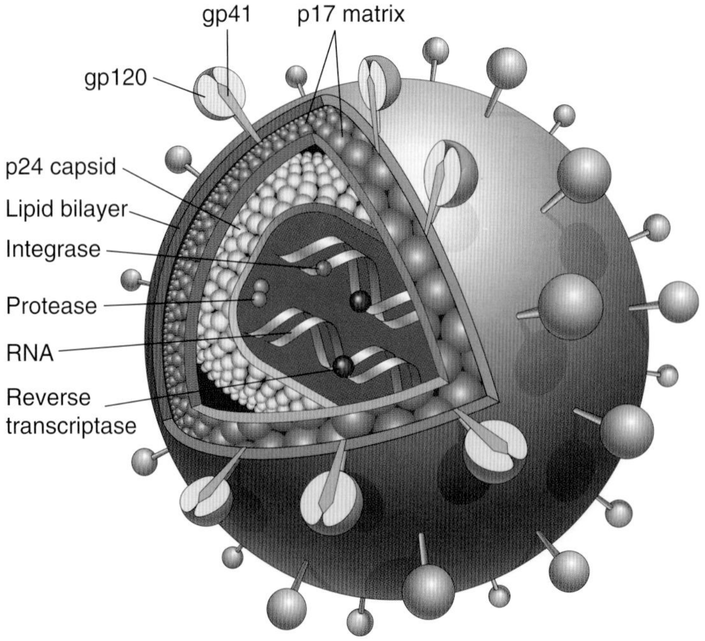

Figure 3.11 **The structure of the HIV virion.** Schematic illustration of an HIV virion. The viral particle is covered by a lipid bilayer that is derived from the host cell. Reprinted by permission from Elsevier Saunders. Robbins, and Cotran: Pathologic Basis of Disease, 7th edition, copyright 2004, page 247.

Invasion of Cells by HIV: Into the Lion's Den [21]

The high affinity receptor that is used by HIV to enter host cells is the CD4 receptor, hence the major target for HIV is lymphoid tissue—more specifically, CD4+ T lymphocytes, macrophages, and dendritic cells (Figure 3.12). The first encounter between HIV and the naïve host takes place in the mucosa and draining lymph node [30]. Dendritic cells play an important role in the infectious process. They are not only primary target cells, but also powerful professional antigen-presenting cells that are infected either directly or via capture of virus on their stellate processes. They can present antigens via MHC class I and class II molecules, stimulating both T-helper and CTL responses.

The presence of CD4 on host cells is not sufficient to mediate infection. The receptor needs to be accompanied by the presence of one of two chemokine receptors used as co-receptors: either CXCR4 or CCR5. R4 viruses utilize CXCR4 as co-receptor which is expressed on lymphocytes, but not on macrophages. Hence, these viruses are called lymphocyte-tropic or T-tropic viruses. R5 viruses utilize CCR5 as co-receptor which is expressed on monocytes/macrophages, lymphocytes, and dendritic cells. R5 viruses are called macrophage-tropic or M-tropic viruses despite the fact that they can infect several cell types. Viruses that can use both CXCR4 and CCR5 as co-receptors are called dual tropic viruses. In the early phase of HIV infection, R5 (M-tropic) viruses dominate, but over the course of the infection the tropism often changes due to mutations in the viral genome, and R4 (lymphocyte-tropic) viruses increase in numbers.

The initial step in infection of any of the CD4+ cells is the binding of gp120 to CD4 molecules, which leads to a conformational change of the viral protein which now recognizes the co-receptor CCR5 or CXCR4. This interaction then triggers conformational change of gp41, which is noncovalently bound to pg120, and fusion of the viral bilayer with the host cell membrane. The HIV genome enters the host cell and reverse transcribes its RNA genome into cDNA (proviral DNA). In quiescent host cells, HIV cDNA may remain in the cytoplasm in linear form. In dividing host cells, the cDNA enters the nucleus and is then integrated in the host genome. In the case of infected T-cell, proviral DNA may be transcribed, virions formed in the cytoplasm, and complete viral particles bud from the cell membrane. If there is extensive viral production (productive infection), the host cell dies. Alternatively, the HIV genome may remain silent, either in the cytoplasm or integrated as provirus into human chromosomes, for months or even years (latent infection). Since macrophages and dendritic cells are relatively resistant to the cytopathic effect of HIV, they are likely important reservoirs of infection.

Clinically, the patient is asymptomatic or has flu-like symptoms during this first phase of HIV infection, also called the acute HIV syndrome. Approximately 40–90% of patients develop self-limiting symptoms (sore throat, myalgias, fever, weight loss, and a rash) 3–6 weeks after infection. This phase is characterized by widespread seeding of the lymphoid tissues, loss of activated CD4+ T-cells, and the highest level of viremia at any time during infection—unfortunately with high infectivity exactly when the infection is usually undiagnosed. The initial infection is readily controlled

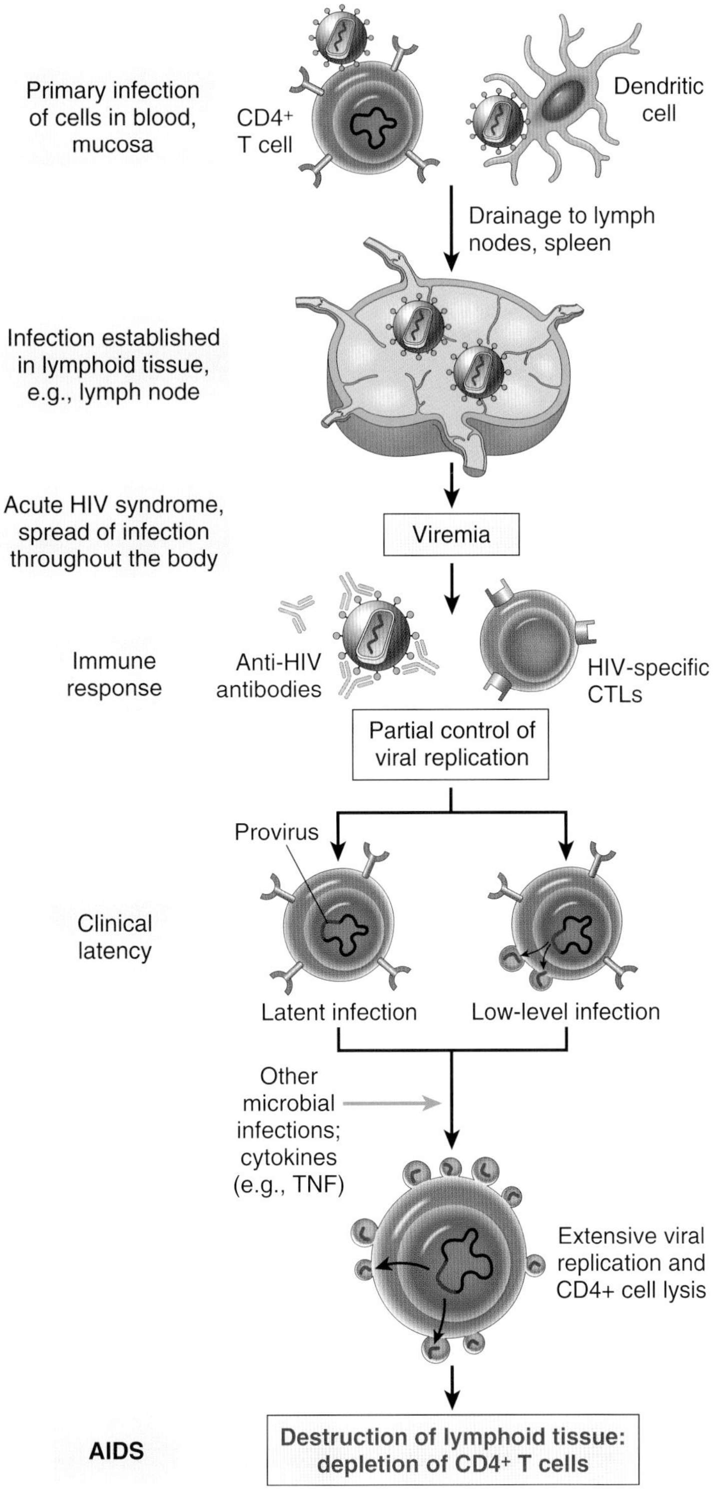

Figure 3.12 **Pathogenesis of HIV-1 infection.** Pathogenesis of HIV-1 infection. Initially, HIV-1 infects T cells and macrophages directly or is carried to these cells by Langerhans cells. Viral replication in the regional lymph nodes leads to viremia and widespread seeding of lymphoid tissue. The viremia is controlled by the host immune response, and the patient then enters a phase of clinical latency. During this phase, viral replication in both T cells and macrophages continues unabated, but there is some immune containment of virus. Ultimately, CD4+ cell numbers decline due to productive infection and other mechanisms, and the patient develops clinical symptoms of full-blown AIDS. Reprinted by permission from Elsevier Saunders. Robbins, and Cotran: Pathologic Basis of Disease, 7th edition, copyright 2004, page 248.

by the development of an HIV-specific cytotoxic T-lymphocyte (CTL) response and humoral response with antibodies raised against the envelope glycoproteins (Figure 3.12), and the patient again becomes asymptomatic. However, the antibody response is ineffective at neutralizing the virus. Thus, the role that antibodies play in controlling HIV disease is unclear. The level of viremia and the viral load in the lymphoid tissue at the end of the acute HIV syndrome define the so-called set point, which differs between individual patients and has prognostic implications. It is the result of a multifactorial process that is not yet clearly understood.

During the following clinical latency phase (also called the middle or chronic phase of HIV infection), the immune system is relatively intact. However, individual T-cells throughout the body, when activated by antigen contact or HIV itself, release intact virions and undergo apoptosis. Thus, it is misleading to talk about a latent infection in the context of HIV, since the definition of latency in the viral world implies a lack of viral replication. In the case of HIV, there is continuous HIV replication, predominantly in the lymphoid tissues, which may last for years. This means that HIV infection lacks a phase of true microbiological latency. Indeed, the latent phase of HIV infection is a dynamic competition between the actively replicating virus and the immune system.

In a twist of fate, the life cycle in latently infected T-cells comes to completion (and usually leads to cell lysis) at the very moment when the T-cell is needed most—upon activation. On the molecular level this is achieved by sharing the transcription factor NFκB. After T-cells are activated by antigen or cytokines (such as TNFα, IL-1), signal transduction results in translocation of NFκB into the nucleus and upregulation of the expression of several cytokines. Flanking regions of the HIV genome also contain similar κB sites that are triggered by the same signal transduction molecule. Thus, the physiologic response of the T-cell stimulates virus production and leads ultimately to cell lysis and death of the infected cell. In addition, CD4+ T-cell loss is caused by mechanisms other than the direct cytopathic effect of the virus. Infected T-cells are killed by CTL cells that recognize HIV antigen presented on their cell surface. Even uninfected CD4+ T-cells, so-called innocent bystanders, are killed. Chronic activation of the immune system starts them down the pathway to apoptosis (programmed cell death), in this case activation-induced.

The last phase of HIV infection is progression to full-blown AIDS. A vicious cycle of increasingly productive viremia, loss of CD4+ cells, increased susceptibility to opportunistic infections, further immune activation, and progression of cell destruction develops. The clinical picture is characterized by a breakdown of host defense, a dramatic increase in circulating virus, and clinical disease. Patients present usually with long-lasting fever, fatigue, weight loss, and diarrhea [29]. The onset of certain opportunistic infections such as invasive candidiasis, mycobacteriosis, or pneumocystosis (Figure 3.13), secondary neoplasms, or HIV-associated encephalitis marks the beginning of AIDS. Prior to the highly active antiretroviral therapy (HAART) era, AIDS was a death sentence, but treatment with several antiviral drugs has changed the fate of HIV-infected patients greatly.

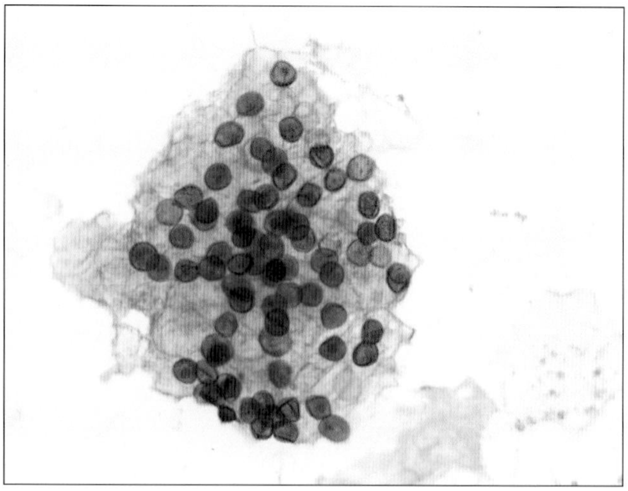

Figure 3.13 Opportunistic pathogen in AIDS. Cluster of *Pneumocystis jirovecii* cysts in bronchoalveolar lavage of an HIV-positive patient stained with toluidin blue (oil immersion, magnification 1000x).

Care has to be taken to avoid the development of drug resistance, which is due to the extreme plasticity of the HIV genome. Even early on during the infection, mutations are frequent due to error-prone replication by HIV and the structural flexibility of the viral envelope. Mutation and variation both in viral antigens and in viral physiology play an important role in the pathogenesis of HIV disease. This is also one reason why the quest for an HIV vaccine has remained elusive [31].

The devastating clinical course of AIDS and the unique pathologic features of HIV infection, with significant viremia persisting for years, demonstrate the essential role of the CD4 T-cell in adaptive immunity (Table 3.7). While most CD4 T-cells have relatively modest effector function, they are the central regulators of the immune response. CD4 T-cells are essential for maturation and development of B-cells and CD8+ cytotoxic T-cells, as well as for activation of macrophages. Patients with late-stage HIV infection become vulnerable to a host of opportunistic pathogens because they suffer from deteriorating antibody production, cell-mediated immunity, and decreased production and altered function of every kind of immune cell.

PERSPECTIVES

The five pathogens discussed in detail here do not begin to cover the breadth of microbial interactions with the host. In fact, these comparatively short discussions fail to cover the breadth of the interactions these few organisms have. Microbes, for the most part, do not enter the forbidding interior of mammalian hosts with a single, general-purpose toxin or strategy for

Table 3.7 Immune Dysfunction in AIDS (Modified after [22])

Altered Monocyte/Macrophage Functions
Decreased chemotaxis and phagocytosis
Decreased HLA class II antigen expression
Decreased antigen presentation capacity
Increased secretion of IL-1, IL-6, and TNFα
Altered T-cell Functions *in vivo*
Preferential loss of memory T-cells
Susceptibility to opportunistic infections
Susceptibility to neoplasms
Altered T-cell Functions *in vitro*
Decreased proliferate response to antigens
Decreased specific cytotoxicity
Decreased helper function for B-cell Ig synthesis
Decreased IL-2 and IFN-γ production
Polyclonal B-cell Activation
Hypergammaglobulinemia and circulating immune complexes
Inability to mount antibody response to new antigen
Refractoriness to normal B-cell activation *in vitro*

causing disease. Instead, they come equipped with a variety of very specific molecular tools, each with a particular target, which allow them to survive, replicate, and be transmitted to a new host.

It's difficult to overestimate the subtlety of the interactions of pathogens with the host on the molecular level. Pathogens rarely eradicate an immune response. Rather, because of cross-talk and redundancy on a molecular level between elements of the immune system, they attenuate, misdirect, and delay the host response. Overall, the complexity and flexibility of the interaction benefits both the host and the pathogen. The pathogen benefits because, even if it is ultimately eliminated from the host, it survives and proliferates long enough and well enough to be transmitted. The host benefits because a temperate response is less likely to result in severe collateral tissue damage. In addition, if host responses were so rigid and forceful that the pathogen was forced to kill the host or die itself, the microbes, ultimately, would win, due to their rapid evolution. The mammalian immune system, for all its extraordinary complexity and power, is a compromise between metabolic cost and efficacy, between elimination and containment of pathogens, and is able to live with what it cannot destroy.

REFERENCES

1. Murphy K, Travers P, Walport M. *Janeway's Immunobiology*. 7th ed. New York: Garland Science; 2008 [Chapter 1].
2. Murphy K, Travers P, Walport M. *Janeway's Immunobiology*, 7th ed. New York: Garland Science; 2008 [Chapter 2].
3. Murphy K, Travers P, Walport M. *Janeway's Immunobiology*. 7th ed. New York: Garland Science; 2008 [Chapter 10].
4. Murphy K, Travers P, Walport M. *Janeway's Immunobiology*. 7th ed. New York: Garland Science; 2008 [Chapter 6].
5. Murphy K, Travers P, Walport M. *Janeway's Immunobiology*. 7th ed. New York: Garland Science; 2008 [Chapter 7].
6. Murphy K, Travers P, Walport M. *Janeway's Immunobiology*. 7th ed. New York: Garland Science; 2008 [Chapter 3].
7. Murphy K, Travers P, Walport M. *Janeway's Immunobiology*. 7th ed. New York: Garland Science; 2008 [Chapter 9].
8. Taylor JE, Rudenko G. Switching trypanosome coats: What's in the wardrobe? *Trends Genetics*. 2006;22:614–620.
9. Pays E. Regulation of antigen gene expression in *Trypanosoma brucei*. *Trends Parasitology*. 2005;21:517–520.
10. Pays E, Vanhamme L, Perez-Morga D. Antigenic variation in *Trypanosoma brucei*: Facts, challenges and mysteries. *Current Opinion Microbiology*. 2004;7:369–374.
11. Vanhamme L, Pays E. The trypanosome lytic factor of human serum and the molecular basis of sleeping sickness. *International J Parasitology*. 2004;34:887–898.
12. Smith TG, Ayi K, Serghides L, et al. Innate immunity to malaria caused by *Plasmodium falciparum*. *Clinical Invest Medicine*. 2002; 25:262–272.
13. Parham P. *The Immune System*. New York: Garland Science; 2005.
14. Chavakis T, Preissner KT, Herrmann M. The anti-inflammatory activities of *Staphylococcus aureus*. *TRENDS Immunology*. 2007; 28:408–418.
15. Foster TJ. Immune evasion by staphylococci. *Nature Reviews*. 2005;3:948–958.
16. Voyich JM, Braughton KR, Sturdevant DE, et al. Insights into mechanisms used by *Staphylococcus aureus* to avoid destruction by human neutrophils. *J Immunology*. 2005;175:3907–3919.
17. Houben EN, Nguyen L, Pieters J. Interaction of pathogenic mycobacteria with the host immune system. *Current Opinion Microbiology*. 2006;9:76–85.
18. Pieters J. *Mycobacterium tuberculosis* and the macrophage: Maintaining a balance. *Cell Host Microbe*. 2008;3:399–407.
19. Russell DG. Who puts the tubercle in tuberculosis? *Nature Reviews Microbiology*. 2007;5:39–47.
20. Murphy K, Travers P, Walport M. *Janeway's Immunobiology*. 7th ed. New York: Garland Science; 2008 [Chapter 5].
21. Murphy K, Travers P, Walport M. *Janeway's Immunobiology*. 7th ed. New York: Garland Science; 2008 [Chapter 8].
22. Mori I and Nishiyama Y. Accessory genes define the relationship between the herpes simplex virus and its host. *Microbes and Infection*. 2006;8;2556–2562.
23. Mguyen ML and Blaho JA. Apoptosis during herpes simplex virus infection. *Advances in Virus Research*. 2007;69;67–97.
24. Gill N, Davies EJ, and Ashkar AA. The role of Toll-like receptor ligands/agonists in protection against genital HSV-2 infection. *American Journal of Reproductive immunology*. 2008;59;35–43.
25. Finberg RW, Knipe DM, and Kurt-Jones EA. Herpes simplex virus and Toll-like receptors. *Viral Immunology*. 2005;18;457–465.
26. Cunningham AL, Diefenbach RJ, Miranda-Saksena M, et al. The cycle of herpes simplex virus infection: virus transport and immune control. *Journal of Infectious Disease*. 2006;194;S11–18.
27. HIV-1. UNAIDS. 2007 AIDS epidemic update. 2008. http://www.unaids.org/en/ accessed July 11, 2008.
28. Johnston MI, Fauci AS. An HIV vaccine—Evolving concepts. *N Engl J Med*. 2007;356:2073–2081.
29. Abbas AK. Diseases of immunity. In: Kumar V, Abbas A, Fausto N, eds. *Robbins and Cotran Pathologic Basis of Disease*. Philadelphia: Elsevier Saunders; 2004:245–258.
30. Hladik F, Sakchalathorn P, Ballweber L, et al. Initial events in establishing vaginal entry and infection by human immunodeficiency virus type-1. *Immunity*. 2007;26:257–270.
31. Walker BD, Burton DR. Toward an AIDS vaccine. *Science*. 2008; 320:760–764

Chapter

4 Neoplasia

William B. Coleman . Tara C. Rubinas

INTRODUCTION

Cancer does not represent a single disease. Rather, cancer is a collection of myriad diseases with as many different manifestations as there are tissues and cell types in the human body, involving innumerable endogenous or exogenous carcinogenic agents, and various etiological mechanisms. What all of these disease states share in common are certain biological properties of the cells that compose the tumors, including unregulated (clonal) cell growth, impaired cellular differentiation, invasiveness, and metastatic potential. It is now recognized that cancer, in its simplest form, is a genetic disease or, more precisely, a disease of abnormal gene expression. Recent research efforts have revealed that different forms of cancer share common molecular mechanisms governing uncontrolled cellular proliferation, involving loss, mutation, or dysregulation of genes that positively and negatively regulate cell proliferation, migration, and differentiation (generally classified as proto-oncogenes and tumor suppressor genes). The molecular mechanisms associated with neoplastic transformation and tumorigenesis of specific cell types is beyond the scope of this chapter. Rather, in the discussion that follows, we will introduce basic and essential concepts related to neoplastic disease as a foundation for more detailed treatment of the molecular carcinogenesis of major cancer types provided elsewhere in this book.

In this chapter, we provide an overview of cancer statistics and epidemiology, highlighting cancer types of importance to human health in the United States and worldwide, with a brief review of risk factors for the development of cancer. Subsequently, we discuss the classification of neoplasms, focusing on the general features of benign and malignant neoplasms, with an overview of nomenclature for human neoplasms, a description of preneoplastic conditions, and consideration of special subsets of neoplastic disease (cancers of childhood, hematopoietic neoplasms, and hereditary cancers). Next, we discuss in some detail the distinguishing characteristics of benign and malignant neoplasms, with a focus on anaplasia and cellular differentiation, rate of growth, local invasiveness, and metastasis. We conclude with a discussion of the clinical aspects of neoplasia, including an overview of cancer-associated pain, cancer cachexia, paraneoplastic syndromes, and methods for grading and staging of cancer.

CANCER STATISTICS AND EPIDEMIOLOGY

Cancer Incidence

Cancer is an important public health concern in the United States and worldwide. Due to the lack of nationwide cancer registries for all countries, the exact numbers of the various forms of cancer occurring in the world populations are unknown. Nevertheless, estimations of cancer incidence and mortality are generated on an annual basis by several domestic and world organizations. Estimations of cancer incidence and mortality for the United States are provided annually by the *American Cancer Society* (ACS) and the National Cancer Institute's Surveillance, Epidemiology, and End Results (SEER) program. Global cancer statistics are provided by the *International Agency for Research on Cancer* (IARC) and the *World Health Organization* (WHO). Monitoring of long-range trends in cancer incidence and mortality among different populations is important for investigations of cancer etiology. Given the long latency for formation of a clinically detectable neoplasm (up to 20–30 years) following initiation of the carcinogenic process (exposure to carcinogenic agent), current trends in cancer incidence probably reflect exposures that occurred many years (and possibly decades) before. Thus, correlative analysis of current trends in cancer incidence with recent trends in occupational, habitual, and environmental exposures to known or suspect carcinogens can provide clues to cancer etiology. Other factors that influence cancer incidence include the size and average age of the affected population. The average age at the time of cancer diagnosis for all tumor sites is

 © 2009, Elsevier, Inc. All Rights Reserved.

approximately 67 years [1]. As a higher percentage of the population reaches age 60, the general incidence of cancer will increase proportionally. Thus, as the life expectancy of the human population increases due to reductions in other causes of premature death (due to infectious and cardiovascular diseases), the average risk of developing cancer will increase.

General Trends in Cancer Incidence

The *American Cancer Society* estimates that 1,437,180 new cases of invasive cancer were diagnosed in the United States in 2008 [2]. This number of new cancer cases reflects 745,180 male cancer cases (52%) and 692,000 female cancer cases (48%). The estimate of total new cases of invasive cancer does not include carcinoma *in situ* occurring at any site other than in the urinary bladder, and does not include basal and squamous cell carcinomas of the skin. In fact, basal and squamous cell carcinomas of the skin represent the most frequently occurring neoplasms in the United States, with an estimated occurrence of >1 million total cases in 2008 [2]. Likewise, carcinoma *in situ* represents a significant number of new cancer cases with 67,770 newly diagnosed breast carcinomas *in situ* and 54,020 new cases of melanoma carcinoma *in situ* [2].

Estimated site-specific cancer incidence for both sexes combined are shown in Figure 4.1. Cancers of the reproductive organs represent the largest group of newly diagnosed cancers in 2008 with 274,150 new cases. This group of cancers includes prostate (186,320 new cases), uterine corpus (40,100 new cases), ovary (21,650 new cases), and uterine cervix (11,070 new cases), in addition to other organs of the genital system (vulva, vagina, and other female genital organs; testis, penis, and other male genital organs). The next most frequently occurring tumors originated in the digestive tract (271,290 new cases), respiratory system (232,270 new cases), and breast (184,450 new cases). The majority of digestive system tumors involved colon (108,070 new cases), rectum (40,740 new cases), pancreas (37,680 new cases), stomach (21,500 new cases), liver and intrahepatic bile duct (21,370 new cases), and esophagus (16,470 new cases), in addition to the other digestive system organs (small intestine, gallbladder, and others). Most new cases of cancer involving the respiratory system affected the lung and bronchus (215,020 new cases), with the remaining cases affecting the larynx or other components of the respiratory system. Other sites with significant cancer burden include the urinary system (125,490 new cases), lymphomas (74,340 new cases), skin (67,720 new cases), leukemias (44,270 new cases), and the oral cavity and pharynx (35,310 new cases).

Among men, cancers of the prostate, respiratory system (lung and bronchus), and digestive system (colon

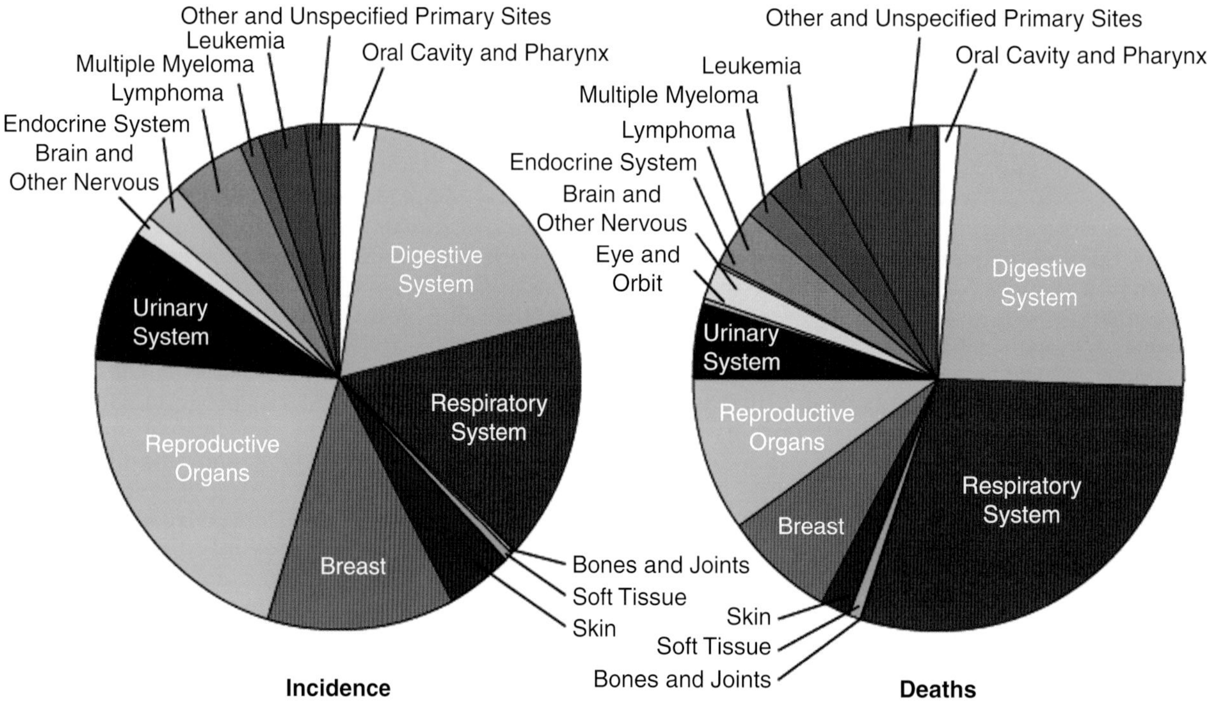

Figure 4.1 **Cancer incidence and mortality by site for both sexes (United States, 2008).** The relative contributions of the major forms of cancer to overall cancer incidence and cancer-related mortality (both sexes combined) were calculated from data provided by Jemal et al. [2]. Cancers of the reproductive organs include those affecting the prostate, uterine corpus, ovary, uterine cervix, vulva, vagina, testis, penis, and other organs of the male and female genital systems. Cancers of the digestive system include those affecting esophagus, stomach, small intestine, colon, rectum, anus, liver, gallbladder, pancreas, and other digestive organs. Cancers of the respiratory system include those affecting the lung, bronchus, larynx, and other respiratory organs.

and rectum) occur most frequently [2]. Together, these cancers account for 62% of all cancers diagnosed in men. Prostate is the leading site, accounting for 186,320 new cases and 25% of cancers diagnosed in men. Among women, cancers of the breast, respiratory system (lung and bronchus), and digestive system (colon and rectum) occur most frequently [2]. Cancers at these sites combine to account for 59% of all cancers diagnosed in women. Breast is the leading site for tumors affecting women, accounting for 182,460 new cases and 26% of all cancers diagnosed in women.

General Trends in Cancer Mortality in the United States

Mortality attributable to invasive cancers produced 565,650 cancer deaths in 2008 [2]. This reflects 294,120 male cancer deaths (52% of total) and 271,530 female cancer deaths (48% of total). Estimated numbers of cancer deaths by site for both sexes are shown in Figure 4.1. The leading cause of cancer death involves tumors of the respiratory system (166,280 deaths), the majority of which are neoplasms of the lung and bronchus (161,840 deaths). The second leading cause of cancer deaths involve tumors of the digestive system (135,130 deaths), most of which are tumors of the colorectum (49,960 deaths), pancreas (34,290 deaths), stomach (10,880 deaths), liver and intrahepatic bile duct (18,410 deaths), and esophagus (14,280 deaths). Together, tumors of the respiratory and digestive systems account for 53% of cancer deaths.

Trends in cancer mortality among men and women mirror in large part cancer incidence. Cancers of the prostate, lung and bronchus, and colorectum represent the three leading sites for cancer incidence and cancer mortality among men [2]. In a similar fashion, cancers of the breast, lung and bronchus, and colorectum represent the leading sites for cancer incidence and mortality among women [2]. While cancers of the prostate and breast represent the leading sites for new cancer diagnoses among men and women (respectively), the majority of cancer deaths in both sexes are related to cancers of the lung and bronchus. Tumors of the lung and bronchus are responsible for 31% of all cancer deaths among men and 26% of all cancer deaths among women. The age-adjusted death rate for lung cancer among men has increased dramatically over the last 60–70 years, while the death rates for other cancers (like prostate and colorectal) have remained relatively stable. The lung cancer death rate for women has increased in an equally dramatic fashion since about 1960, becoming the leading cause of female cancer death in the mid-1980s after surpassing the death rate for breast cancer [2].

Global Cancer Incidence and Mortality

The IARC GLOBOCAN project estimates that 10,862,496 new cancer cases were diagnosed worldwide in 2002 [3]. This number of new cases represents 5,801,839 male cancer cases (53%) and 5,060,657 female cancer cases (47%). Mortality attributed to cancer for the same year produced 6,723,887 deaths worldwide. This reflects 3,795,991 male cancer deaths (56%) and 2,927,896 female cancer deaths (44%). The leading sites for cancer incidence and mortality worldwide in 2002 included tumors of the lung, stomach, prostate, breast, colorectum, esophagus, and liver. Lung cancer accounted for the most new cancer cases and the most cancer deaths during this period of time, with 1,352,132 new cases and 1,178,918 deaths for both sexes combined. The leading sites for cancer incidence among males included lung (965,241 new cases), prostate (679,023 new cases), stomach (603,419 new cases), colorectum (550,465 new cases), and liver (442,119 new cases) [3]. Combined, cancers at these five sites account for 44% of all cancer cases among men. The leading causes of cancer death among men included tumors of the lung (848,132 deaths), stomach (446,052 deaths), liver (416,882 deaths), colorectum (278,446 deaths), and esophagus (261,162 deaths) [3]. Deaths from these cancers account for 59% of all male cancer deaths. The leading sites for cancer incidence among females included breast (1,151,298 new cases), cervix uteri (493,243 new cases), colorectum (472,687 new cases), lung (386,891 new cases), and stomach (330,518 new cases) [3]. Combined, cancers at these five sites account for 56% of all cancer cases among women. The leading causes of cancer death among females directly mirrors the leading causes of cancer incidence: breast (410,712 deaths), lung (330,786 deaths), cervix uteri (273,505 deaths), stomach (254,297 deaths), and colorectum (250,532 deaths) [3]. Combined, these five cancer sites account for 52% of female cancer deaths.

Risk Factors for the Development of Cancer

Risk factors for cancer can be considered anything that increases the chance that an individual will develop neoplastic disease. Individuals who have risk factors for cancer development are more likely to develop the disease at some point in their lives than the general population (lacking the same risk factors). However, having one or more risk factors does not necessarily mean that a person will develop cancer. It follows that some people with recognized risk factors for cancer will never develop the disease, while others lacking apparent risk factors for cancer will develop neoplastic disease. While certain risk factors are clearly associated with the development of neoplastic disease, making a direct linkage from a risk factor to causation of the disease remains very difficult and often impossible. Some risk factors for cancer development can be modified, while others cannot. For instance, cessation of cigarette smoking reduces the chance that an individual will develop cancer of the lung, bronchus, or other tissues of the aerodigestive tract. In contrast, a woman with an inherited mutation in the *BRCA1* gene carries an elevated lifetime risk of developing breast cancer. In general, risk factors for cancer include age, sex, family medical history, exposure to cancer-causing factors in the environment, and various lifestyle choices (such as tobacco and alcohol use, diet, and sun exposure). Different forms of neoplastic disease affecting various tissues have different risk factors. Some of the major risk factors that contribute to cancer development include (i) age, (ii) race, (iii) gender, (iv) family history, (v) infectious agents, (vi) environmental exposures, (vii) occupational exposures, and (viii) lifestyle exposures.

Age, Race, and Gender as Risk Factors for Cancer Development

Cancer is predominantly a disease of old age. Thus, advancing age is the most important risk factor for cancer development for most people [4,5]. In fact, most malignant neoplasms are diagnosed in patients over the age of 65. According to the National Cancer Institute SEER Statistics (http://seer.cancer.gov/index.html) the median age at diagnosis for cancer of the lung/bronchus is 71 years old (<2% of cases occur in people <45 years old), the median age at diagnosis for cancer of the prostate is 68 (<10% of cases occur in men <55 years old), the median age at diagnosis for cancer of the breast is 61 years old (<2% of cases occur in women <35 years old), the median age at diagnosis for cancer of the colorectum is 71 years old (<5% of cases occur in people <45 years old), the median age at diagnosis for cancer of the liver is 65 years old (<5% of cases occur in people <45 years old), the median age at diagnosis for cancer of the ovary is 63 years old (<5% of cases occur in women <35 years old), and the median age at diagnosis of melanoma is 59 years old (<10% of cases occur in people <35 years old). The age-specific incidence and death rates for cancers of the prostate, breast (female), lung (both sexes combined), and colorectum (both sexes combined) for the period of 1992–1996 are shown in Figure 4.2. The trends depicted in this figure clearly show that the majority of each of these cancer types occur in individuals of advanced age. In the case of prostate cancer, 86% of all cases occur in men over the age of 65, and 99.5% occur in men over the age of 50. Likewise, 97% of prostate cancer deaths occur in men over the age of 65 (Figure 4.2). In contrast, female breast cancer occurs much more frequently in younger individuals. Nonetheless, 63% of cases occur in women over the age of 65, and 88% occur in women over the age of 50 (Figure 4.2). A notable exception to this relationship between advanced age and cancer incidence involves some forms of leukemia and other cancers of childhood. Acute lymphocytic leukemia (ALL) occurs with a bimodal distribution, with highest incidence among individuals less than 20 years of age,

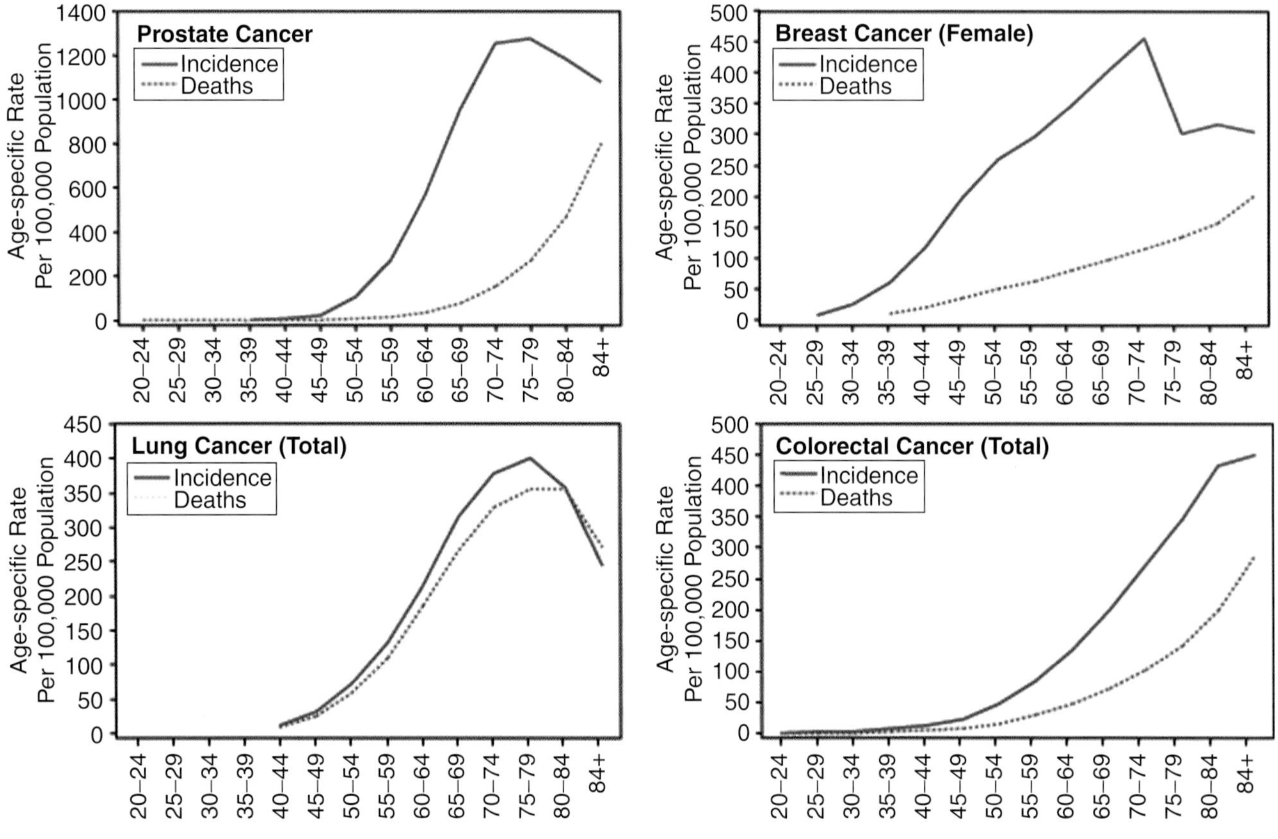

Figure 4.2 Age-specific incidence and mortality rates for selected sites, 1992–1996. The age-specific rates for breast cancer incidence and mortality are for females only. The age-specific rates for lung cancer and colorectal cancer are combined for both sexes. These data were adapted from the NCI SEER Statistics Database (http://seer.cancer.gov/index.html). Rates are per 100,000 population and are age-adjusted to the 1970 standard population of the United States.

and a second peak of increased incidence among individuals of advanced age. The majority of ALL cases are diagnosed in children, with 40% of cases diagnosed in children under the age of 15, and 45% of cases occurring in individuals under the age of 20. Despite the prevalence of this disease in childhood, a significant number of adults are affected. In fact, 32% of ALL cases are diagnosed in individuals over the age of 65 years of age. In contrast to ALL, the other major forms of leukemia demonstrate the usual pattern of age-dependence observed with solid tumors, with large numbers of cases in older segments of the population.

Cancer incidence and mortality can vary tremendously with race (and/or factors that are associated with race) and ethnicity (http://seer.cancer.gov/index.html). In the United States, African Americans and Caucasians are more likely to develop cancer than individuals of other races or ethnicity. African American men demonstrate a cancer incidence for all sites combined of approximately 664 cases per 100,000 population, and Caucasian men exhibit a cancer incidence rate of 557 cases per 100,000 population. In contrast, Native American men show the lowest cancer incidence among populations of the United States with 321 cases per 100,000 population for all sites combined. Likewise, African American women demonstrate a cancer incidence for all sites combined of approximately 397 cases per 100,000 population, and Caucasian women exhibit a cancer incidence rate of 424 cases per 100,000 population, while Native American women show the lowest cancer incidence with 282 cases per 100,000 population for all sites combined. These differences in race-related cancer incidence rates can be magnified when site-specific cancers are considered. The overall incidence of prostate cancer in the United States (for all men) is 163 cases per 100,000 men. African American men have a significantly higher incidence rate (249 cases per 100,000 men) compared to Caucasian men (157 cases per 100,000). In contrast, Native American men have a significantly lower incidence of prostate cancer (73 cases per 100,000 men) compared to African American and Caucasian men. The mechanisms that account for these differences are not known, but may be related to genetic factors or differences in various physiological factors (for instance, androgen hormone levels). Mortality due to cancer also differs among patients depending on their race or ethnicity. Similar to the cancer incidence rates, mortality due to cancer is higher among African Americans (322 and 189 per 100,000 population for men and women, respectively) and Caucasians (235 and 161 deaths per 100,000 population for men and women, respectively) than other populations, including Asian/Pacific Islanders, American Indians, and Hispanics. Factors that are known to contribute to racial differences in cancer-related mortality include (i) differences in exposures (for instance, smoking prevalence), (ii) access to regular cancer screening (for breast, cervical, and colon cancers), and (iii) timely diagnosis and treatment. For both cancer incidence and mortality, racial and ethnic variations for all sites combined differ from those for individual cancer sites.

Gender is clearly a risk factor for cancers affecting certain tissues such as breast and prostate where there are major differences between men and women. However, there are numerous other examples of cancers that appear to develop preferentially in men or women, and/or where factors related to gender increase risk. For instance, liver cancer affects men more often than women. The ratio of male to female incidence in the United States is approximately 2:1 [2], and worldwide is approximately 2.4:1 [3]. However, in high incidence countries or world regions, the male to female incidence ration can be as high as 8:1 [6,7]. This consistent observation suggests that sex hormones and/or their receptors may play a significant role in the development of primary liver tumors [8,9]. Some investigators have suggested that hepatocellular carcinomas overexpress androgen receptors [10], and that androgens are important in the promotion of abnormal liver cell proliferation [11,12]. Others have suggested that the male predominance of liver cancer is related to the tendency for men to drink and smoke more heavily than women, and are more likely to develop cirrhosis [13].

Family History as a Risk Factor for Cancer Development

Familial cancers have been described for most major organ systems, including colon, breast, ovary, and skin. These cancers are associated with genetic predisposition to development of disease. Hereditary cancers are typically characterized by (i) early age at onset (or diagnosis), (ii) neoplasms arising in first degree relatives of the index case, and (iii) in many cases, multiple or bilateral tumors. Epidemiologic evidence has consistently pointed to family history as a strong and independent predictor of breast cancer risk. Thus, women with a first-degree relative (mother or sister) that have been diagnosed with breast cancer are at elevated risk for development of the disease themselves. A substantial amount of research has led to the discovery of several breast cancer susceptibility genes, including *BRCA1*, *BRCA2*, and *p53* [14], which may account for the majority of inherited breast cancers. It has been estimated that 5%–10% of breast cancers occurring in the United States each year are related to genetic predisposition [15]. Despite the recognition of multiple genetic and environmental risk factors for development of breast cancer, approximately 50% of affected women have no identifiable risk factors other than being female and aging [16]. In many of these cases, the genetic predisposition to cancer development may be related to small but measurable risks associated with genetic variations (polymorphic variations) at multiple loci [17]. Other risk factors for development of breast cancer include, advancing age (over 50 years of age), early age at menarche, late age at menopause, first childbirth after age of 35, nulliparity, family history of breast cancer, obesity, dietary factors (such as high-fat diet), and exposure to high dose radiation to the chest before age 35 [18–22]. Many of these risk factors may interact with genetic polymorphisms and other genetic determinants to drive the development of breast cancer.

Infectious Agents as Risk Factors for Cancer Development

There are several excellent examples of infectious agents and cancer causation. These include (i) liver cancer and hepatitis virus infection, (ii) cervical cancer and human papillomavirus infection, and (iii) Epstein-Barr virus infection and Burkitt's lymphoma. Human cancers are also associated with certain bacterial and parasite infections.

Liver cancers are frequently associated with hepatitis virus infection and related conditions [23,24]. Both hepatitis B virus (HBV) and hepatitis C virus (HCV) are associated with development of hepatocellular carcinoma [25,26]. Primary liver cancers are usually associated with chronic hepatitis [27–29], and 60%–80% of hepatocellular carcinomas occurring worldwide develop in cirrhotic livers [29–31], most commonly nonalcoholic posthepatitic cirrhosis. However, hepatitis virus infection is not thought to be directly carcinogenic. Rather, HBV and/or HCV infection produces hepatocyte necrosis and regeneration (cell proliferation), which makes the liver susceptible to endogenous and exogenous carcinogens. Thus, preneoplastic nodules in the liver tend to occur in regenerative nodules of the injured liver (chronic hepatitis or cirrhosis). In certain geographic areas (such as China), large portions of the population are concurrently exposed to the hepatocarcinogen aflatoxin B_1 and HBV, which increases their relative risk for development of liver cancer [32].

Cervical cancer is the second most common cancer of women worldwide, with 493,243 new cases and 273,505 deaths in 2002 [3]. Human papillomavirus infection is strongly associated with the development of certain human cancers [33]. Infection with high-risk HPV types (in particular HPV16 and HPV18) are specifically associated with development of cervical cancer (Figure 4.3). Case-control and prospective epidemiological studies have shown that HPV infection precedes high-grade dysplasia and invasive cancer, and represents the strongest independent risk factor for the development of cervical cancer [34–36]. Molecular analyses of cervical cancers suggest that HPV plays a key role in cervical carcinogenesis since >90% of cervical cancer biopsies contain DNA sequences of high-risk HPV types. Within these biopsies, the viral DNA is present in every tumor cell and is also found in metastases derived from the same neoplasms.

Epstein-Barr virus (EBV) infection has been suggested to be mechanistically involved in the development of several forms of neoplastic disease in humans [37]. EBV is a member of the human herpesvirus subfamily *Gammaherpesviridae*. EBV is a ubiquitous virus that infects >90% of the human population. Primary infection with EBV usually occurs in childhood, and in most cases there is no apparent clinical course. However, in a subset of infected individuals, primary

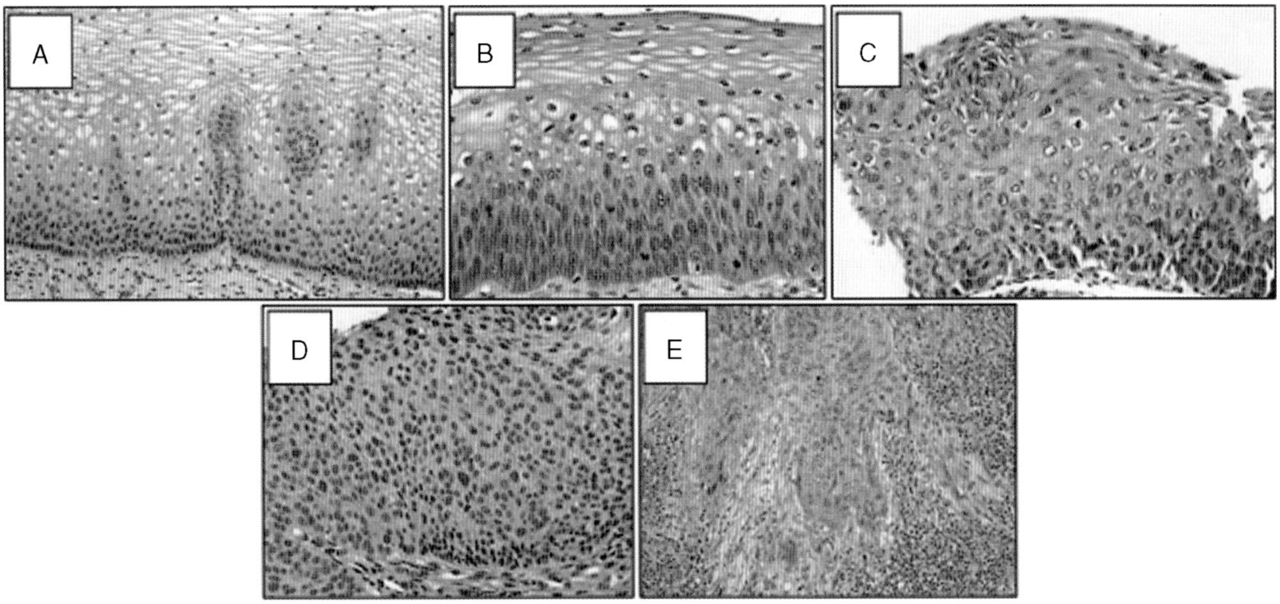

Figure 4.3 **Progression of dysplasia in the cervix. (A)** Normal squamous epithelium of the cervix. The basal (bottom) layer of cells appears dark, and there is cellular maturation as the squamous cells move to the surface. **(B)** Low-grade squamous dysplasia. The basal layer appears thicker with mild nuclear pleomorphism and mitotic figures involving the lower third of the epithelium. In this image, several HPV-infected cells, known as koilocytes, are present in the middle third of the epithelium. A koilocyte has a wrinkled nucleus with cytoplasmic clearing. **(C)** Moderate squamous dysplasia. The dysplastic squamous cells and mitotic figures involve the lower two-thirds of the epithelium. There is still some maturation toward the surface of the epithelium. **(D)** High-grade squamous dysplasia. Dysplastic squamous cells and mitotic figures involve the full thickness of the epithelium. There is no invasion into the underlying stroma. **(E)** Invasive squamous cell carcinoma. Dysplastic squamous cells invade into the underlying stroma in a haphazard fashion. There is prominent inflammation in the stroma in response to the invasive carcinoma.

EBV infection can result in infectious mononucleosis, a self-limited lymphoproliferative disease, particularly when primary infection is delayed into adolescence. It is not yet known if EBV primarily infects the oropharyngeal epithelia with secondary infection of B-cells which circulate through the mucosa-associated lymphoid tissue, or whether B-cells also represent a primary target site for EBV infection [38]. EBV causes a latent infection in lymphoid cells which persists for life in a small subpopulation of B-lymphocytes [39]. Although EBV has a very high transforming potential *in vitro*, it rarely causes tumors in humans, suggesting that EBV-infected cells are subject to continuous surveillance by cytotoxic T-cells. However, EBV infection is also closely associated with the development of lymphoid and epithelial malignancies in apparently immunocompetent hosts [40]. The human neoplasm that is classically associated with EBV infection is Burkitt's lymphoma [41,42]. Burkitt's lymphoma is often multifocal and composed of monoclonal proliferations of B-cells. On a worldwide scale, Burkitt's lymphoma is a low incidence tumor. However, its prevalence shows striking geographic variations. For instance, Burkitt's lymphoma occurs in an endemic form in equatorial Africa where it can represent the most common childhood tumor. Other cancers that are associated with EBV infection include nasopharyngeal cancer, Hodgkin disease, and certain cancers of the stomach.

Environmental and Occupational Exposures as Risk Factors for Cancer Development

Environmental exposures represent significant risk factors for certain forms of cancer. The most well studied hepatocarcinogen is a natural chemical carcinogen known as aflatoxin B_1 that is produced by the *Aspergillus flavus* mold [43–45]. This mold grows on rice or other grains (including corn) that are stored without refrigeration in hot and humid parts of the world. Ingestion of food that is contaminated with *Aspergillus flavus* mold results in exposure to potentially high levels of aflatoxin B_1. Aflatoxin B_1 is a potent, direct-acting liver carcinogen in humans, and chronic exposure leads inevitably to development of hepatocellular carcinoma.

Another naturally occurring carcinogen is the radioactive gas radon, which has been suggested to increase the risk of lung cancer development. This gas is ubiquitous in the earth's atmosphere, creating the opportunity for exposure of vast numbers of people. However, passive exposure to the background levels of radon found in domestic dwellings and other enclosures is not sufficiently high to increase lung cancer risk appreciably [46]. High-level radon exposure has been documented among miners working in uranium, iron, zinc, tin, and fluorspar mines [47,48]. These workers show an excess of lung cancer (compared to nonminers) that varies depending on the radon concentration encountered in the ambient air of the specific mine [47,48]. Radon can also contaminate drinking water. Water that comes from deep, underground wells in rock with high radium concentrations may have high levels of radon, whereas the radon levels of surface water (drawn from lakes or rivers) are usually negligible. It is generally believed that water contamination does not significantly contribute to exposure to radon.

Although once heavily studied and thought to be a major mechanism of human cancer induction, exposure to chemical carcinogens does not represent an important risk factor for most of the general population. Nevertheless, several chemicals, complex chemical mixtures, industrial processes, and/or therapeutic agents have been associated with development of malignant neoplasms in exposed human populations. These exposures may include therapeutic exposure to the radioactive compounds (such as thorium dioxide or Thorotrast for the radiological imaging of blood vessels) and occupational exposures to certain industrial chemicals (such as vinyl chloride monomer, asbestos, bis[chloromethyl] ether, or chromium). Chemical carcinogens are classified as direct agents or indirect agents. Indirect agents are chemicals that require metabolic conversion to become an ultimate carcinogen. Examples of indirect carcinogenic agents include certain polycyclic hydrocarbons (such as those found in tobacco smoke and in smoked meats and fish), and azo dyes (such as β-naphthylamine). Since indirect carcinogens require metabolic activation for their conversion into ultimate carcinogens with genotoxic activity (DNA-damaging activity), a great deal of research has focused on the enzymatic pathways that are required for carcinogen activation. Many of these carcinogen activating pathways involve cytochrome P-450-dependent monooxygenase enzymes which are encoded by highly polymorphic genes. Thus, the activity of these enzymes tends to vary among individuals depending on the specific form of the enzyme that is carried. Hence, susceptibility to a specific chemical carcinogen may depend in part on the specific form of the enzyme that is expressed, making it probable that molecular analyses to determine these genetic polymorphisms may become as important in the prediction of cancer risk in the future as consideration of exposures. Several other agents with carcinogenic potential in humans include vinyl chloride, arsenic, nickel, chromium, various insecticides, various fungicides, and polychlorinated biphenyls. These agents are encountered most frequently through occupational exposures. In addition, certain food preservatives (such as nitrites) are of significant concern as potential carcinogens. Nitrites can produce nitrosylation of amines in various foodstuffs, resulting in formation of nitrosamine compounds which are suspected to have carcinogenic potential in humans.

Lifestyle Exposures as Risk Factors for Cancer Development

Numerous risk factors for cancer development fall into the category of lifestyle exposures and generally reflect lifestyle choices [49]. These exposures include (i) consumption of tobacco products, (ii) consumption of excessive amounts of alcohol, and (iii) excessive exposure to sunlight, among others.

Several major forms of human cancer are associated with exposures to tobacco products. Cancers of the lung, mouth, larynx, bladder, kidney, cervix, esophagus, and pancreas are related to consumption of tobacco products, including cigarettes, cigars, chewing tobacco, and snuff. Cigarette smoking alone is the suggested cause for one-third of all cancer deaths. Several lines of evidence strongly link cigarette smoking to lung cancer. Smokers have a significantly increased risk (11-fold to 22-fold) for development of lung cancer compared to nonsmokers [50], and cessation of smoking decreases the risk for lung cancer compared to continued smoking [50,51]. Furthermore, heavy smokers exhibit a greater risk than light smokers, suggesting a dose-response relationship between cigarette consumption and lung cancer risk [50,51]. Numerous mutagenic and carcinogenic substances have been identified as constituents of the particulate and vapor phases of cigarette smoke, including benzo[*a*]pyrene, dibenza[*a*]anthracene, nickel, cadmium, polonium, urethane, formaldehyde, nitrogen oxides, and nitrosodiethylamine [52]. There is also evidence that smoking combined with certain environmental (or occupational) exposures results in potentiation of lung cancer risk. Urban smokers exhibit a significantly higher incidence of lung cancer than smokers from rural areas, suggesting a possible role for air pollution in development of lung cancer [53].

Excessive alcohol consumption has been associated with increased risk of certain forms of cancer [54]. For instance, a connection between alcohol consumption and increased risk of breast cancer has been established [55,56]. Likewise, alcohol consumption is associated with development of cancers of the gastrointestinal tract [57]. Given that the liver is a target for alcohol-induced damage, it is not surprising that chronic alcohol consumption is associated with an elevated risk for primary liver cancer [58–60]. However, it is important to note that whereas heavy sustained alcohol consumption is associated with risk of liver cancer, moderate consumption of alcohol is not [24]. Alcohol is not directly carcinogenic to the liver; rather it is thought that the chronic liver damage produced by sustained alcohol consumption (hepatitis and cirrhosis) may contribute secondarily to liver tumor formation [61]. For some other major cancer sites (such as lung), the role of alcohol consumption as a co-factor in cancer development is not clear [62].

Most/all skin cancer is related to unprotected exposure to strong sunlight. All of the major forms of skin cancer (basal cell carcinoma, squamous cell carcinoma, and malignant melanoma) have been linked to sunlight exposure [63]. The carcinogenic agent in sunlight that accounts for the neoplastic transformation of skin cells is ultraviolet (UV) radiation [63]. Basal cell carcinoma is a malignant neoplasm of the basal cells of the epidermis that occurs predominantly in areas of sun-damaged skin. Thus, sun bathing and sun tanning using artificial UV light sources represent significant lifestyle risk factors for development of these tumors. Basal cell carcinoma is now diagnosed in some people at very young ages (second or third decade of life), reflecting increased exposures to UV irradiation early in life. Some researchers have suggested that the increasing frequencies of skin cancer can be partially attributed to depletion of the ozone layer of the earth's atmosphere [64], which filters out (thereby reducing) some of the UV light produced by the sun. Squamous cell carcinoma is a malignant neoplasm of the keratinizing cells of the epidermis. As with basal cell carcinoma, extensive exposure to UV irradiation is the most important risk factor for development of this tumor. Likewise, development of malignant melanoma occurs most frequently in fair-skinned individuals and is associated to some extent with exposure to UV irradiation. This accounts for the observation that Caucasians develop malignant melanoma at a much higher rate than individuals of other races and ethnicity.

CLASSIFICATION OF NEOPLASTIC DISEASES

The word neoplasia is derived from the Greek words meaning "condition of new growth." The term tumor is commonly used to refer to a neoplasm. Tumor literally means "a swelling." In the early 1950s, R. A. Willis provided a description of neoplasm that we still utilize today: "A neoplasm is an abnormal mass of tissue the growth of which exceeds and is uncoordinated with that of the normal tissues and persists in the same manner after the cessation of the stimuli which evoked the change" [65]. In similar fashion, Kinzler and Vogelstein describe tumors to be the result of a disease process in which a single cell acquires the ability to proliferate abnormally, resulting in an accumulation of progeny cells [66]. Furthermore, Kinzler and Vogelstein define cancers to represent tumors that have acquired the ability to invade the surrounding normal tissues [66]. This definition highlights one of the most important distinguishing factors in the overall classification of neoplasms—the distinction between benign and malignant tumors. The division of neoplastic diseases into benign and malignant categories is extremely important, both for understanding the biology of these neoplasms and for recognizing the potential clinical challenges for treatment. At the most basic level, neoplasms are classified as benign or malignant. Further subclassification of malignant neoplasms draws distinctions to (i) cancers of childhood versus cancers that primarily affect adults, (ii) solid tumors versus hematopoietic neoplasms, and (iii) hereditary cancers versus sporadic neoplasms.

Development of neoplastic disease is a multistep process through which cells acquire increasingly abnormal proliferative and invasive behaviors. Neoplasia also represents a unique form of genetic disease, characterized by the accumulation of multiple somatic mutations in a population of cells undergoing neoplastic transformation [67,68]. Genetic and epigenetic lesions represent integral parts of the processes of neoplastic transformation, tumorigenesis, and tumor progression. Several forms of molecular alteration have been described in human cancers, including gene amplifications, deletions, insertions, rearrangements, and point mutations [68]. In many cases specific genetic lesions have been identified that are associated with neoplastic transformation and/or tumor progression in a particular tissue or cell type [67]. Epigenetic alterations (epimutations) in neoplastic disease include genome-wide hypomethylation of DNA (possibly resulting in induction of oncogene expression) [69–72], gene-specific hypermethylation events (resulting in silencing of tumor suppressor genes) [73–75], other changes in chromatin packaging [76–78], and aberrant post-transcriptional regulation of gene expression (related to abnormal microRNA expression) [79–81]. Statistical analyses of age-specific mortality rates for different forms of human cancer predict that multiple mutations or epimutations in specific target genes are required for the genesis and outgrowth of most clinically diagnosable neoplasms. In accordance with this prediction, it has been suggested that tumors grow through a process of clonal expansion driven by mutation or epimutations, where the first mutation/epimutation leads to limited expansion of progeny of a single cell, and each subsequent mutation/epimutation gives rise to a new clonal outgrowth with greater proliferative potential. The idea that carcinogenesis is a multistep process is supported by morphologic observations of the transitions between premalignant (benign) cell growth and malignant tumors. In colorectal cancer (and some other tumor systems), the transition from benign lesion to malignant neoplasm can be easily documented and occurs in discernible stages, including benign adenoma, carcinoma *in situ*, invasive carcinoma, and eventually local and distant metastasis (Figure 4.4) [82,83]. Moreover, specific genetic alterations have been shown to correlate with each of these well-defined histopathologic stages of tumor development and progression [82,83]. However, it is important to recognize that it is the accumulation of multiple genetic alterations in affected cells, and not necessarily the order in which these changes accumulate, that determines tumor formation and progression.

Both benign and malignant neoplasms are composed of (i) neoplastic cells that form the parenchyma, and (ii) the host-derived non-neoplastic stroma that is composed of connective tissue, blood vessels, and other cells, and supports the tumor parenchyma. The tumor stroma serves a critical function in support of the growth of the neoplasm by providing a blood supply for oxygen and nutrients. In nearly all cases, the parenchymal cells determine the biologic behavior (and clinical course) of the neoplasm. Further, the parenchymal cell type of the neoplasm determines how the lesion is named.

Benign Neoplasms

The classification of neoplasms into benign and malignant categories is based on a judgment of the potential clinical behavior of the tumor. This judgment is based primarily on observations of the cellular features of the neoplasm, the growth pattern of the tumor, and various clinical findings. Benign neoplasms are characterized by features that suggest a lack of aggressiveness. The most important characteristic of benign neoplasms is the absence of local invasiveness (Figure 4.5). Thus, while these neoplasms grow and expand, they do not invade locally or spread to secondary tissue sites (remain localized). Since benign neoplasms remain localized, they are often amenable to surgical removal. However, it is important to note that benign neoplasms can cause adverse effects in the patient. Problems associated with benign neoplasms depend on (i) the size of the tumor, (ii) the location of the tumor, and (iii) secondary consequences related to presence of the neoplasm. Many benign neoplasms attain large size, impinge on important structures (like nerves or blood vessels), resulting in various types of local effects. Consider a few examples. Many/most brain

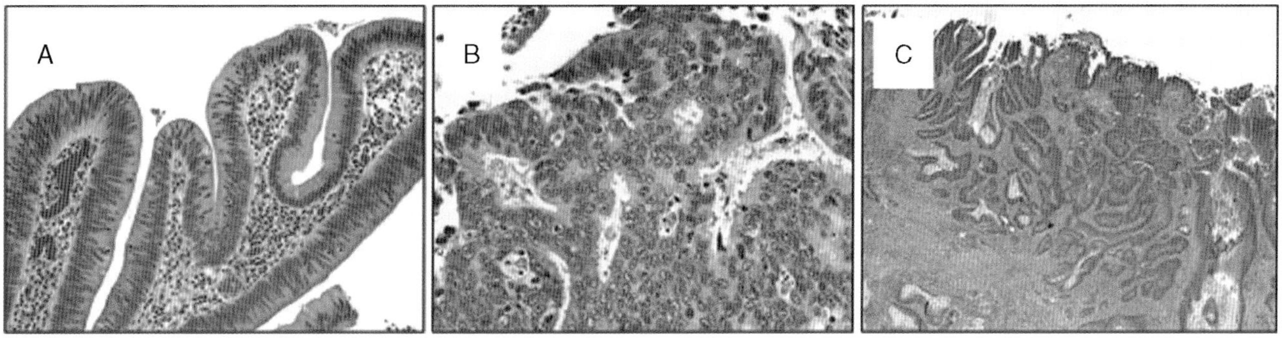

Figure 4.4 **Progression of neoplastic transformation in the colon. (A)** Low-grade glandular dysplasia. This image is of an adenomatous colon polyp, which is, by definition, polypoid low-grade dysplasia of the colonic glandular epithelium. The nuclei are hyperchromatic (dark) with pseudostratification (overlapping) and a higher nuclei to cytoplasmic ratio (the nucleus is occupying a larger percentage of cell volume than in a normal colon epithelial cell). **(B)** High-grade glandular dysplasia. Sometimes adenomatous polyps can harbor high-grade dysplasia. Features of high-grade dysplasia include both increased architectural complexity, including back-to-back glands without intervening stroma, and increased degree of nuclear pleomorphism. **(C)** Invasive adenocarcinoma. Glands, lined by dysplastic cells, are haphazardly invading through the musculars mucosa and into the underlying submucosa.

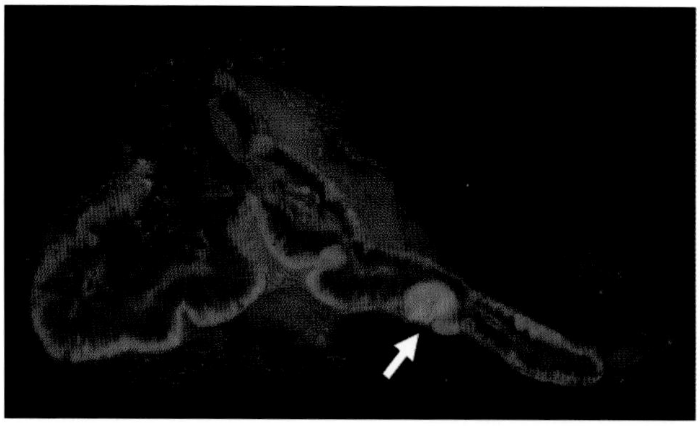

Figure 4.5 **Adrenal adenoma.** The normal cortex of the adrenal gland is yellow-gold in appearance. The white arrow points to a well-delineated, round lesion arising in the adrenal cortex. It does not appear to invade into the adjacent tissue. Courtesy of Kirsten Boland, MHS, PA(ASCP).

tumors are considered benign by virtue of the fact that they do not invade locally or produce distant metastases. However, as expanding space-filling lesions, these neoplasms can cause severe effects on the host due to the application of pressure to nearby aspects of the brain or brainstem. For instance, a benign meningioma can cause cardiac and respiratory arrest by compressing the medulla. Likewise, hemangiomas represent a benign neoplastic lesion of blood vessels which creates a blood-filled cavity. Some hemangiomas (such as those affecting the liver) can become large and frequently impinge on the capsule of the organ. Lesions of this sort are subject to rupture, producing life-threatening bleeding in the patient. As these examples illustrate, despite a lack of invasive behavior, benign neoplasms can produce adverse effects in the patient and can have life-threatening consequences depending on their size and location.

Benign neoplasms are named by attaching the suffix *-oma* to the cell type from which the tumor originates. Thus, a benign neoplasm of fibrous tissue is termed a fibroma, a benign neoplasm of cartilaginous tissue is termed a chondroma, a benign neoplasm of osteoid tissue is termed an osteoma, a benign neoplasm arising from lipocytes is termed a lipoma, a benign neoplasm arising from blood vessels is termed a hemangioma, and a benign neoplasm arising from smooth muscle cells is termed a leiomyoma. The nomenclature for benign epithelial tumors is more complex. These neoplasms are classified either on the basis of their microscopic or macroscopic pattern, or according to their cells of origin. Thus, a benign epithelial neoplasm producing glandular patterns or a tumor arising from glandular cells is termed an adenoma, a benign epithelial neoplasm growing on any surface that produces microscopic or macroscopic finger-like fronds is termed a papilloma, a benign epithelial neoplasm that projects above a mucosal surface to produce a macroscopically visible structure is termed a polyp, and cystadenoma refers to hollow cystic masses.

Malignant Neoplasms

Malignant neoplasms are collectively known as cancers. Malignant neoplasms display aggressive characteristics, can invade and destroy adjacent tissues, and spread to distant sites (metastasize). These features of invasiveness distinguish malignant from benign neoplasms (Figure 4.6). Adverse effects associated with malignant neoplasms are generally associated with tumor burden on the host once the cancer has spread throughout the body. Specific adverse effects can arise from (i) the size and location of the primary tumor, (ii) consequences of local invasion and spread from the primary site, and (iii) consequences associated with tumor colonization of tissue sites distant to the primary tumor. Most commonly, the cause of death associated with malignant neoplasms can be attributed to the metastatic spread of the tumor. Common sites of metastasis for malignant epithelial neoplasms include the lungs, liver, bone, and brain. The lung is the most common site for cancer metastasis (involving most cancer types and primary sites), which may be accomplished by hematogenous spread (through the blood), lymphatic spread, or by direct invasion. Likewise, cancers of the breast, lung, and colon are well known to spread to the liver, but cancers associated with any site in the body (including leukemias and lymphomas) can colonize the liver. Metastatic cancer found in the lungs or liver is characterized by the presence of multiple (often numerous) cancer nodules that can replace large percentages of the normal tissue, and in liver can produce marked hepatomegaly. Adverse effects associated with lung metastasis include respiratory insufficiency and/or failure. Likewise, patient death related to liver metastasis results from various manifestations of liver insufficiency and/or failure.

The nomenclature for malignant neoplasms is very similar to that for benign neoplasms. Malignant neoplasms arising in mesenchymal tissues or its derivatives

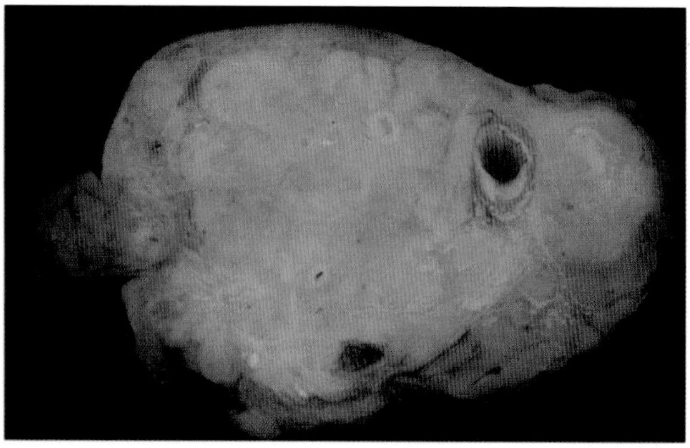

Figure 4.6 **Pancreatic adenocarcinoma.** This is a cut section of a pancreatic tumor. It is poorly delineated (meaning it is difficult to identify the exact borders of the lesion) and appears to be infiltrating into adjacent adipose tissue (patchy bright yellow areas along the bottom of the specimen). Courtesy of Kirsten Boland, MHS, PA(ASCP).

are called sarcomas. Sarcomas are designated by their histogenesis. Thus, a malignant neoplasm of fibrous tissue is termed a fibrosarcoma, a malignant neoplasm originating in cartilaginous tissue is termed a chondrosarcoma, a malignant neoplasm arising from lipocytes is termed a liposarcoma, a malignant neoplasm of osteoblasts is termed an osteosarcoma, a malignant neoplasm arising in blood vessels is termed an angiosarcoma, and a malignant neoplasm arising from smooth muscle cells is termed a leiomyosarcoma. Malignant epithelial neoplasms are called carcinomas. Carcinomas can be subclassified as adenocarcinoma and squamous cell carcinoma. Adenocarcinoma describes a malignant neoplasm in which the neoplastic cells grow in a glandular pattern. Squamous cell carcinoma describes a malignant neoplasm with a microscopic pattern that resembles stratified squamous epithelium. In each case, the nomenclature for a given tumor will specify the organ system of origin for the neoplasm (for instance, colonic adenocarcinoma or squamous cell carcinoma of the skin). In some cases, malignant neoplasms grow in an undifferentiated pattern that is inconsistent with a classification of adenocarcinoma or squamous cell carcinoma. In these cases, the neoplasm is termed a poorly differentiated carcinoma.

Mixed Cell Neoplasms

Some neoplastic cells undergo divergent differentiation during tumor formation, giving rise to tumors of mixed cell type. Examples of mixed cell tumors include that of salivary gland origin and breast fibroadenoma. These neoplasms are composed of epithelial components dispersed throughout a fibromyxoid stroma that may contain cartilage or bone. In the case of mixed tumor of salivary gland origin, all of the cellular components of the neoplasm are believed to derive from epithelial and myoepithelial cells of the salivary glands. Therefore, the appropriate designation for these neoplasms is pleomorphic adenoma to reflect the differences in cellular differentiation among cells of common origin. Breast fibroadenoma is a benign neoplasm that contains a mixture of proliferating ductal elements (adenoma) contained in a loose fibrous tissue (fibroma). In contrast to most mixed cell neoplasms where all of the cell types observed in the lesion are neoplastic, in breast fibroadenoma only the fibrous component is thought to be neoplastic and the proliferating ductal elements are normal.

In contrast to mixed cell tumors where all of the cellular components of the neoplasm are believed to derive from the same germ layer, teratomas contain recognizable mature or immature cells or tissues representative of more than one germ-cell layer, and sometimes all three. Teratomas typically originate from totipotential cells such as those found in the ovary and testis. In rare instances, teratomas originate from sequestered midline embryonic rests. The totipotent cells that give rise to teratomas have the capacity to differentiate into any cell type and can give rise to all of the tissues found in the adult body. It follows that teratomas are commonly composed of various tissue elements that can be recognized, including skin, hair, bone, cartilage, tooth structures, and others. Teratomas are further classified as benign (also referred to as mature) or malignant (also referred to as immature) (Figure 4.7). As the designation implies, mature teratomas contain well-formed tissue elements that appear normal, but arise in the abnormal context of the neoplasm. In contrast, immature teratomas are composed of abundant poorly differentiated (primitive) blast-like cells (blastema).

Confusing Terminology in Cancer Nomenclature

Some malignant neoplasms are conventionally referred to using terms that are suggestive of benign neoplasms based on the usual nomenclature for naming tumors. For example, lymphoma is a malignant neoplasm of lymphoid tissue, mesothelioma is a malignant neoplasm of the mesothelium, melanoma is a malignant neoplasm arising from melanocytes, and seminoma is a malignant neoplasm of the testicular epithelium. In each of these

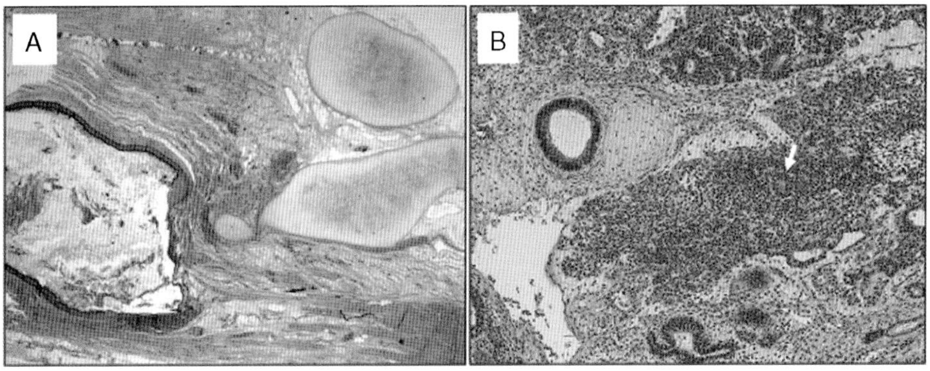

Figure 4.7 **Neoplasms of mixed cell type. (A)** Mature teratoma. This example demonstrates mature cartilage (right) and skin (left). **(B)** Immature teratoma. This type of teratoma contains areas of primitive-appearing hyperchromatic cells, known as blastema. There is focal rosette formation (white arrow).

examples that refers to a specific tumor type, the name given implies a benign neoplasm even though all of these are malignant neoplasms. Likewise, hepatocellular carcinomas are often called hepatomas. Unfortunately, these tumor designations are well established in medical terminology and are unlikely to be corrected. In addition to these tumor misnomers, several other examples of confusing terminology exist. For instance, hamartoma refers to a developmental malformation that presents as a mass lesion of disorganized tissue that is indigenous to that particular tissue site. Hamartomas are not neoplastic, even though the name implies a neoplasm. These malformations can be found in any tissue, but a common site for these lesions is the lung. Hamartomas of the lung generally contain islands of cartilage, along with connective tissue and various types of cells.

Preneoplastic Lesions

Neoplastic disease develops in patients over long periods of time and typically is preceded by development of one or more preneoplastic lesions. Well-characterized preneoplastic lesions include (i) metaplasia, (ii) hyperplasia, and (iii) dysplasia.

Metaplasia represents a reversible change in tissues characterized by substitution of one adult cell type (epithelial or mesenchymal) by another adult cell type (Figure 4.8). Metaplasia is a reactive condition, reflecting an adaptive replacement of cells that are sensitive to stress by cells that are resistant to the adverse conditions encountered by the tissue. Metaplasia is a well-recognized precursor for development of malignant neoplasms at various tissue sites [84]. Examples of metaplasia include the adaptive changes that occur in the lungs of individuals who smoke cigarettes and the adaptive changes that occur in the esophagus of individuals with reflux disease. In cigarette smokers, columnar to squamous epithelial metaplasia occurs in the respiratory tract in response to chronic irritation caused by inhalation of cigarette smoke. The ciliated columnar epithelial cells of the normal trachea and bronchi become replaced by stratified squamous epithelial cells in a pattern that may be focal or more widely distributed. Squamous metaplasia in the respiratory tract is accompanied by loss of function secondary to loss of the ciliated epithelial cells. In addition, development of squamous cell carcinoma of the lung may originate in focal areas of squamous metaplasia [85]. Barrett's esophagus is an example of squamous to columnar metaplasia in response to refluxed gastric acid. The adaptive change is from stratified squamous epithelial cells to intestinal-like columnar epithelial cells, which are resistant to the effects of the gastric acid. Barrett's esophagus is frequently the site of development of esophageal adenocarcinomas [86,87].

Hyperplasia reflects an increase in the number of cells in an organ or tissue, typically resulting in an increased volume (or size) of the affected organ or tissue (Figure 4.9). Hyperplasia can be physiologic or

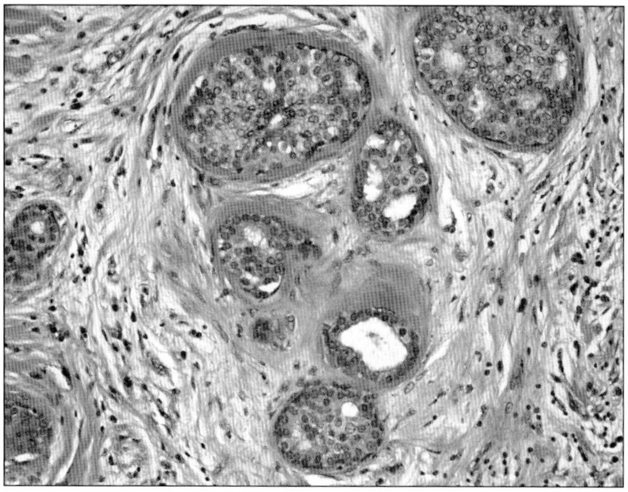

Figure 4.9 Epithelial hyperplasia of the breast. Normal mammary ducts are lined by a single layer of cuboidal epithelial cells (as in the duct second from the center bottom). The lumens of the remaining ducts in this image are filled with bland appearing-epithelium cells, representing prominent ductal epithelial hyperplasia.

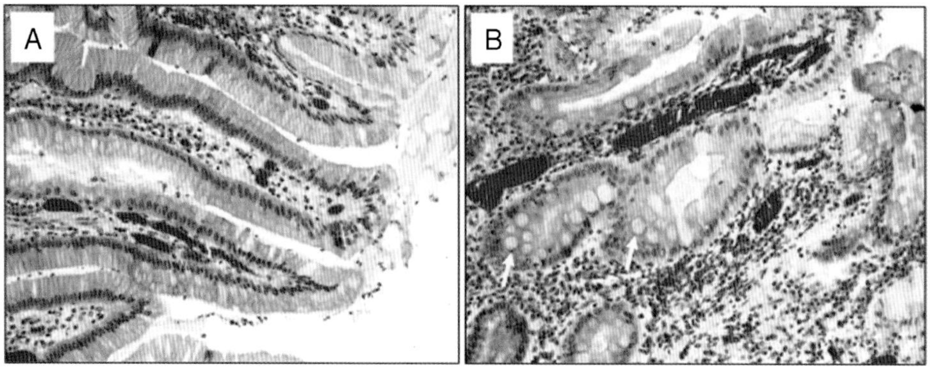

Figure 4.8 Metaplastic change in the stomach. (A) Normal glandular epithelium of the stomach. **(B)** Intestinal metaplasia. This image displays glandular epithelium of the stomach with goblet cells (white arrows). Goblet cells contain a large, round intracytoplasmic vacuole of mucin and are normally found in the small and large intestine only.

pathologic. Physiologic hyperplasia can be classified as (i) hormonal hyperplasia or (ii) compensatory hyperplasia. Most forms of pathologic hyperplasia result from excessive (abnormal) hormonal or growth factor stimulation. Benign prostatic hyperplasia (BPH) is a commonly occurring condition among older men that is typical of pathologic hyperplasia [88,89]. In BPH, abnormal stimulation of the prostate tissue by androgen hormones results in a benign proliferation resulting in hypertrophy of the prostate gland. The abnormal cell proliferation that occurs in hyperplasia is controlled to the extent that upon cessation of the stimulus (elimination of growth factor or hormone) cell proliferation will halt and the hyperplastic tissue will regress. Hyperplasia can precede neoplastic transformation, and many hyperplastic conditions are associated with elevated risk for development of cancer. For example, patients with endometrial hyperplasia (also referred to as endometrial intraepithelial neoplasia) are at increased risk for endometrial cancer [90].

Dysplasia is a proliferative lesion that is characterized by a loss in the uniformity of individual cells in a tissue and loss in the architectural orientation of the cells in a tissue. Thus, dysplasia can be simply described as a condition of disorderly but non-neoplastic cellular proliferation. Dysplastic cells show many alterations that are suggestive of their preneoplastic character, including cellular pleomorphism, hyperchromatic nuclei, high nuclear-to-cytoplasmic ratio, and increased numbers of mitotic figures (Figure 4.10). In some cases, mitotic cells appear in abnormal locations within the tissue. For instance, mitotic figures may appear outside of the basal layers of dysplastic stratified squamous epithelium. Lesions characterized by extensive dysplastic changes involving the entire thickness of the epithelium but remaining confined within the normal tissue are classified as carcinoma *in situ* (Figure 4.11). In many cases, dysplasia and/or carcinoma *in situ* are considered immediate precursors of invasive cancers. It follows that dysplastic changes are often found adjacent to invasive cancers. In the case of heavy cigarette smokers or individuals with Barrett's esophagus, development of dysplasia often portends progression to invasive cancer. However, not all dysplastic lesions will develop into a malignant neoplasm.

Cancers of Childhood

Cancer is primarily a disease of adults, and occurs much more frequently in older individuals. However, malignant neoplasms do occur in children. These neoplasms are rare but not uncommon. In 2008, an estimated 10,730 new cases of cancer and 1,490 cancer-associated deaths occurred in children under 15 years old [91]. The major childhood cancers include leukemia (33% of all childhood cancers), brain tumors and other neoplasms of the nervous system (21% of all childhood cancers), neuroblastoma (7% of all childhood cancer), Wilms' tumor (5% of all childhood cancer), Non-Hodgkin lymphoma (4% of all childhood cancer), rhabdomyosarcoma (3.5% of all childhood cancer), retinoblastoma (3% of all childhood cancer), osteosarcoma (3% of all childhood cancer), and Ewing's sarcoma (1.5% of all childhood cancer). Acute lymphocytic leukemia is the most common childhood leukemia, representing >80% of all childhood leukemias. The incidence of acute lymphocytic leukemia peaks in children 2–5 years old. In contrast, the incidence of neuroblastoma and retinoblastoma peaks earlier, in children 1–2 years old. Neuroblastoma is a malignant neoplasm of the sympathetic nervous system and is the most frequently occurring neoplasm among infants, with peak incidence among children <1 year old (Figure 4.12). In fact, 40% of neuroblastomas are diagnosed in the first 3 months of life. Retinoblastoma is a tumor that originates in the retina of the eye. This neoplasm affects children as well as adults. Retinoblastoma occurring in children is typically associated with a genetic mechanism involving mutation of the *Rb1* gene [92]. Wilms' tumor tends to occur in children <10 years old, with greatest incidence in children <5 years old. Wilms' tumor (also known as nephroblastoma) is the most commonly occurring pediatric kidney tumor. Astrocytomas represent the most frequently occurring brain tumor of children (52% of all childhood brain tumors), with ependymoma (9%), primitive

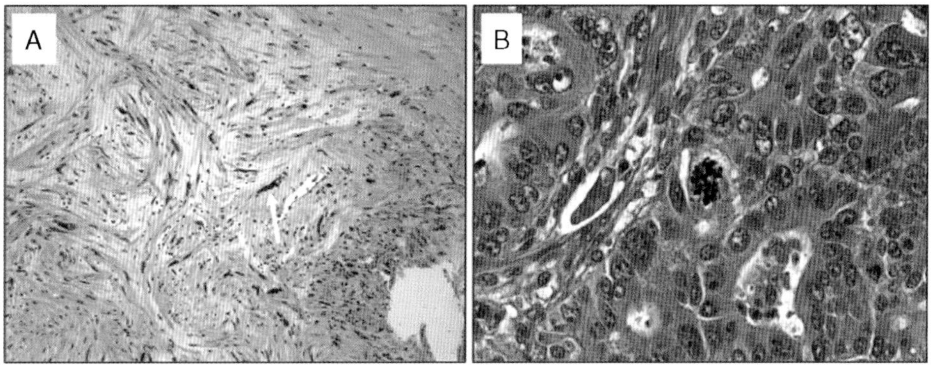

Figure 4.10 Nuclear pleomorphism and abnormal mitotic figures in neoplastic cells. (A) Pleomorphism. This is an example of leiomyosarcoma. Several of the malignant stromal cells are very large and different in shape from neighboring cells. **(B)** Mitotic figure (center). Malignant neoplasms often have an increased number of mitotic figures.

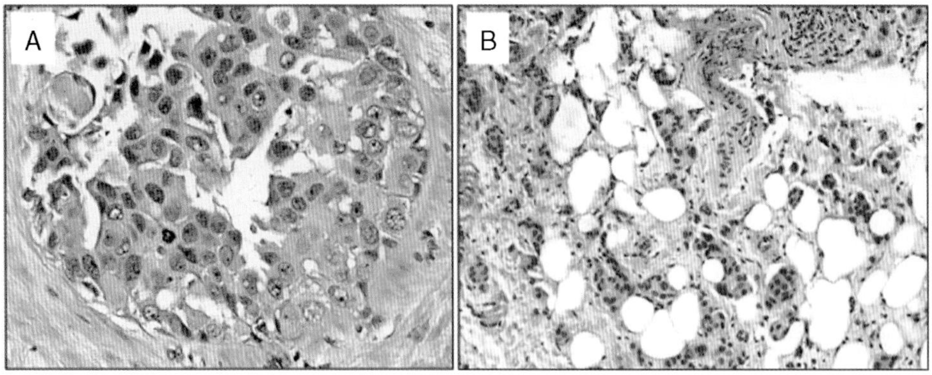

Figure 4.11 **Progression of tumorigenesis in the breast. (A)** Ductal carcinoma *in situ* (DCIS) of the breast. The lumen of this mammary duct is filled with pleomorphic cells with high nuclear to cytoplasmic ratios. Scattered mitotic figures are noted. This neoplastic process is *in situ* carcinoma because the dysplastic cells are confined to the lumen of the duct and have not invaded into the surrounding tissue. **(B)** Invasive adenocarcinoma of the breast. Glands lined by dysplastic cells haphazardly infiltrate into adipose tissue.

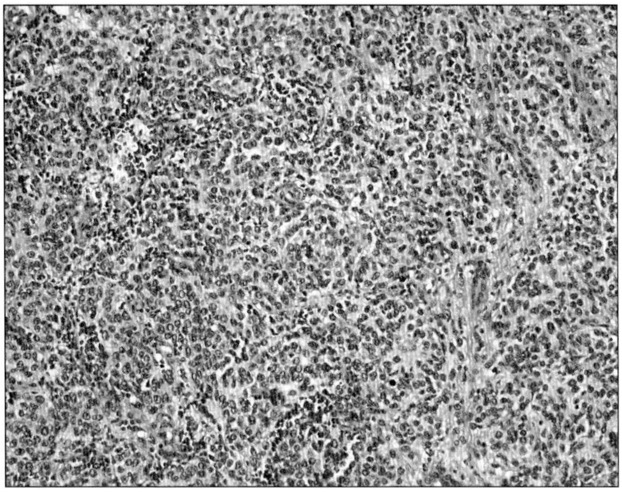

Figure 4.12 **Childhood neuroblastoma.** Neuroblastoma is considered one of the so-called small, round, blue cell tumors, as it is composed of small, hyperchromatic, monomorphic cells with scant cytoplasm.

neuroectodermal tumors (21%), and other gliomas (15%) representing most of the balance. Ependymomas and primitive neuroectodermal tumors occur most often in younger children (<5 years old), while astrocytomas are diagnosed with approximately the same frequency in children between birth and 15 years old. Rhabdomyosarcoma is the most common soft tissue sarcoma of children, typically occurring in children <10 years old. Osteosarcoma and Ewing's sarcoma are the most commonly occurring bone tumors among children. Osteosarcomas derive from primitive bone-forming mesenchymal stem cells and most often occur near the metaphyseal portions of the long bones. There is a bimodal age distribution of osteosarcoma incidence, with peaks in early adolescence and in adults >65 years old. Ewing's sarcoma is believed to be of neural crest origin and occur roughly evenly between the extremities and the central axis. Like osteosarcoma, Ewing's sarcoma is a disease primarily of childhood and young adults, with highest incidence among children in their teenage years.

Hematopoietic Neoplasms

The majority of human cancers can be classified as solid tumors (grouped as carcinomas or sarcomas). The exceptions to this classification include the malignant neoplasms of hematopoietic origin, including lymphoma, myeloma, and leukemia (Figure 4.13). In 2008, 138,530 new malignant neoplasms of hematopoietic origin will be diagnosed, representing approximately 9.5% of all new cancers [2]. These cancers include 74,340 cases of lymphoma, 19,920 cases of myeloma, and 44,270 cases of leukemia [2]. Non-

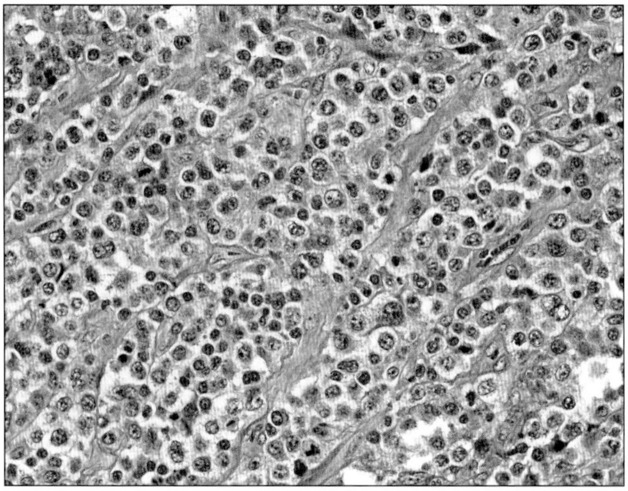

Figure 4.13 **Lymphoma.** This image of lymphoma demonstrates sheets of large cells with prominent nucleoli. Subclassification of the lymphoma depends on several additional laboratory tests.

Hodgkin's lymphoma occurs much more frequently than Hodgkin's lymphoma and represents 89% of all lymphomas. The four major forms of leukemia (acute lymphocytic leukemia, chronic lymphocytic leukemia, acute myeloid leukemia, and chronic myeloid leukemia) account for 87% of all leukemias. The two most prevalent forms of leukemia are chronic lymphocytic leukemia and acute myeloid leukemia, which combine to account for 64% of all cases of leukemia [2].

Hereditary Cancers

A number of familial cancer syndromes and hereditary cancers have been recognized and characterized. Several rare genetic disorders involving dysfunctional DNA repair pathways are associated with elevated risk for cancer development. These disorders include xeroderma pigmentosum, ataxia telangiectasia, Bloom's syndrome, and Fanconi anemia. Individuals affected by these conditions are prone to development of various malignancies when exposed to specific DNA damaging agents. Patients with xeroderma pigmentosum display hypersensitivity to ultraviolet light and increased incidence of several types of skin cancer, including basal cell carcinoma, squamous cell carcinoma, and malignant melanoma [93,94]. Patients with ataxia telangiectasia exhibit hypersensitivity to ionizing radiation and chemical agents, and are predisposed to the development of B-cell lymphoma and chronic lymphocytic leukemias, and affected women demonstrate an increased risk of developing breast cancer [95,96]. Patients with Fanconi's anemia demonstrate sensitivity to DNA cross-linking agents and are predisposed to malignancies of the hematopoietic system, particularly acute myelogenous leukemia [97]. Patients with Bloom's syndrome demonstrate an increased incidence of several forms of cancer, including leukemia, skin cancer, and breast cancer [98,99]. These patients exhibit chromosomal instability manifest as abnormally high levels of sister chromatid exchange [100–102].

Colorectal cancer is a fairly common disease worldwide, and particularly in populations from Western nations. A substantial fraction of colorectal cancers exhibit a genetic component, and several familial colorectal cancer syndromes are recognized, including familial adenomatous polyposis (FAP) and hereditary nonpolyposis colon cancer (HNPCC). Genes associated with each of these conditions have been identified and characterized. Of these familial colorectal cancer syndromes, HNPCC has been determined to be related to defective DNA repair [103]. HNPCC is characterized by the occurrence of predominantly right-sided colorectal carcinoma with an early age of onset and an increased risk for the development of certain extracolonic cancers, including cancers of the endometrium, stomach, urinary tract, and breast [103–105]. Tumors associated with HNPCC exhibit a unique form of genomic instability, which represents a unique mechanism for a genome-wide tendency for instability in short repeat sequences (microsatellites), which was originally termed the replication error phenotype [106–109]. The molecular defect responsible for microsatellite instability in HNPCC involves the genes that encode proteins required for normal mismatch repair [110,111].

Two addition familial cancer syndromes have been described that exhibit clinical features similar to that of HNPCC or FAP. The Muir-Torre syndrome is defined by the development of at least one sebaceous gland tumor and a minimum of one internal tumor, which is frequently colorectal carcinoma [112,113]. This syndrome shares several features with HNPCC syndromes Lynch I and Lynch II, including the occurrence of microsatellite instability in a subset of tumors [114,115]. This observation suggests the possible involvement of abnormal mismatch repair mechanisms in the genesis of a subset of Muir-Torre syndrome tumors. Turcot's syndrome is defined by the occurrence of a primary brain tumor and multiple colorectal adenomas [116]. The molecular basis for this syndrome has been suggested to involve mutation of the *APC* gene or mutation of a mismatch repair gene in tumors exhibiting microsatellite instability [117].

The Li-Fraumeni syndrome was initially characterized among several kindreds with excess cancer incidence [118]. Patients with Li-Fraumeni syndrome develop various types of neoplasms, including breast cancer, soft tissue sarcomas and osteosarcomas, brain tumors, leukemias, and several others [119–121]. Cancer susceptibility among individuals with Li-Fraumeni syndrome follows an autosomal dominant pattern of inheritance and is highly penetrant (90% by age 70), but many neoplasms develop early in life [121]. It is now known that Li-Fraumeni syndrome is associated with germline mutations in the *p53* tumor suppressor gene [122,123].

It is well known that approximately 5%–10% of breast cancers are related to genetic predisposition [15]. These breast cancers are typically associated with a strong family history of breast cancer development. A substantial amount of research has led to the discovery of several breast cancer susceptibility genes, including *BRCA1* [124], *BRCA2* [125,126], and *p53*, which may account for the majority of inherited breast cancers [14]. In some cases, families at increased susceptibility for breast cancer also show elevated rates of ovarian cancer. Patients that are affected by familial breast and ovarian cancer syndrome tend to have germline mutation of *BRCA1* [127–129].

Familial melanoma is associated with (i) a family history of melanoma, (ii) the presence of large numbers of common or atypical nevi, (iii) a history of primary melanoma or other (nonmelanoma) skin cancers, (iv) immunosuppression, (v) susceptibility to sunburn, or (vi) a history of blistering sunburn. Given the linkage between excess exposure to sunlight and development of skin cancers (including melanoma), it is not surprising susceptibility to sunburn would confer an increased risk for melanoma. This susceptibility is particularly pronounced in individuals with fair complexion characterized by freckling, blue eyes, red hair, and skin that burns readily in response to sunlight and fails to tan. Two highly penetrant melanoma susceptibility genes have been identified: *CDKN2A* (which encodes cyclin-dependent kinase inhibitor 2A) and *CDK4* (which encodes cyclin-dependent kinase 4) [130,131]. *CDKN2A* is found on chromosome 9p21 and *CDK4* resides on 12q13. Germline inactivating mutations of the *CDKN2A* gene are the most common cause of inherited susceptibility to melanoma, while mutations of *CDK4* occur much more rarely [131]. Nevertheless, germline mutations of *CDKN2A* are rare, and many account for a very small proportion of melanoma susceptibility among the general population. The *CDKN2A* gene encodes two important cell-cycle regulatory proteins: $p16^{INK4A}$ and $p14^{ARF}$. While other melanoma susceptibility loci have been mapped through genome-wide linkage analysis, the gene targets that contribute to familial melanoma predisposition remain undiscovered for a large proportion of recognized kindreds [132]. Ongoing research is focused on the identification of low-penetrance melanoma susceptibility genes that confer a lower melanoma risk with more frequent variations. For instance, specific variants of the *MC1R* and the *OCA2* genes have been demonstrated to confer an increase in melanoma risk [133,134].

Familial cancers affecting a number of other tissues and organs have been described or suggested. Familial cancer of the pancreas has been reported and suggested to follow an autosomal dominant inheritance pattern [135,136]. A familial form of gastric cancer has been suggested to represent approximately 10% of all gastric cancers and is associated with the *CDH1* gene on chromosome 16q22.1 [137]. A genome-wide scan of 66 high-risk prostate cancer families produced evidence of disease linkage to chromosome 1q24–25 [138,139]. Subsequently, the chromosome 1q24–25 susceptibility locus was connected to the *HPC1* gene [138].

CHARACTERISTICS OF BENIGN AND MALIGNANT NEOPLASMS

The distinction between benign and malignant neoplasms is based on observations related to cellular features of the cells composing the lesion, growth pattern of the neoplasm, and various clinical findings. Four fundamental features are particularly important when comparing and contrasting the characteristics of benign and malignant neoplasms: (i) cellular differentiation and anaplasia, (ii) rate of growth, (iii) presence of local invasion, and (iv) metastasis.

Cellular Differentiation and Anaplasia

In the evaluation of neoplasms, the extent of cellular differentiation and anaplasia of the parenchymal cells that constitute the neoplastic elements of the lesion are assessed. The extent of cellular differentiation describes the degree to which the neoplastic cells resemble their normal counterparts based on morphology and function. Benign neoplasms are typically composed of well-differentiated cells that closely resemble normal cells. Hence, a lipoma is composed of mature fat cells that contain cytoplasmic lipid vacuoles. Likewise, a chondroma is composed of mature cartilage cells that synthesize their usual cartilaginous matrix. In well-differentiated benign neoplasms, cellular proliferation rates are low and mitoses are infrequent. However, when observed, mitotic figures appear normal in these neoplasms. In contrast to benign tumors, malignant neoplasms exhibit an extremely wide range of cell differentiation. Many malignant neoplasms are very well differentiated, but it is not uncommon for these tumors to lack differentiated features and/or to appear completely undifferentiated. Malignant neoplasms of intermediate phenotype are designated as moderately differentiated.

In general, there is a direct relationship between the degree of cellular differentiation and the functional capabilities of the cells that compose the neoplasm. Thus, benign neoplasms (as well as some well-differentiated malignant neoplasms) of the endocrine glands can elaborate hormones characteristic of their origin. Likewise, well-differentiated squamous cell carcinomas (at various tissue sites) elaborate keratin (giving rise to histologically recognizable keratin pearls), and well-differentiated hepatocellular carcinomas synthesize bile salts. In contrast, many malignant neoplasms express genes and produce proteins or hormones that would not be expected from the cells of origin of the cancer. Some cancers synthesize fetal proteins that are not expressed by comparable cell types in adults. For instance, many hepatocellular carcinomas express α-fetoprotein, which is not expressed in mature hepatocytes of the adult liver [140,141]. Further, malignant neoplasms of nonendocrine origin can excrete ectopic hormones producing various paraneoplastic syndromes [142,143]. Certain lung cancers produce antidiuretic hormone (inducing hyponatremia in the patient), adrenocorticotropic hormone (resulting in Cushing syndrome), parathyroid-like hormone or calcitonin (both of which are implicated in hypercalcemia), gonadotropins (causing gynecomastia), serotonin and bradykinin (associated with carcinoid syndrome), or others.

Malignant neoplasms that are composed of undifferentiated cells are said to be anaplastic. Lack of cellular differentiation (or anaplasia) is considered a hallmark of cancer. The term anaplasia means "to form backward," which implies dedifferentiation (or loss of the structural and functional differentiation) of normal cells during tumorigenesis. However, it is now well recognized that many cancers originate from stem cells in tissues [144–146]. Thus, the lack of cellular differentiation exhibited by these neoplasms results from a failure to differentiate, rather than through a process of dedifferentiation of highly differentiated (specialized) cells. Other mechanisms that might account for loss of differentiation in malignant neoplasms include epithelial to mesenchymal transition, mesenchymal to epithelial transition, and transdifferentiation between cell differentiation states related to cellular plasticity [147–149]. In general, malignant neoplasms that are composed of anaplastic cells (which are also typically rapidly growing) are unlikely to have specialized functional activities. Anaplastic cells tend to exhibit marked nuclear and cellular pleomorphism (extreme variation in nuclear or cell size

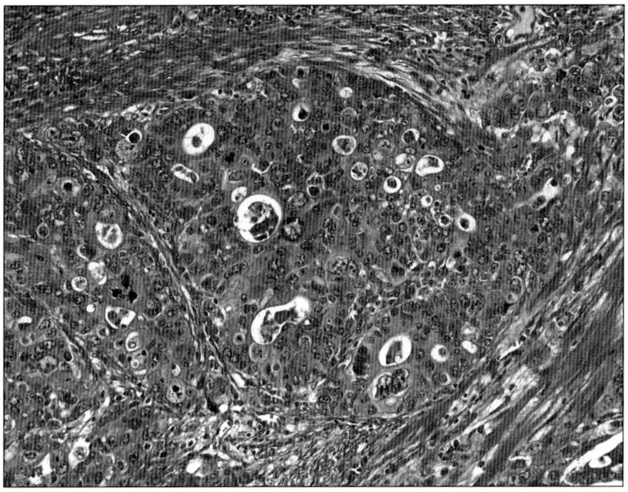

Figure 4.14 **Cellular pleomorphism.** The dysplastic cells in this adenocarcinoma demonstrate marked pleomorphism, meaning the cells vary widely in size and shape. The arrangement of the dysplastic cells is disordered, with only some residual gland formation.

and shape) (Figure 4.14). The nuclei observed in anaplastic cells are typically hyperchromatic (darkly staining) and large, resulting in altered nuclear-to-cytoplasmic ratios (which may approach 1:1 instead of 1:4 or 1:6 as observed in normal cells). Very often malignant neoplasms contain giant cells (relative to the size of neighboring cells in the neoplasm), and these cells will contain an abnormally large nucleus or will be multinucleated. The nuclear pleomorphism observed in anaplastic cells of malignant neoplasms is characterized by nuclei that are highly variable in size and shape. These nuclei exhibit chromatin that appears coarse and clumped. Furthermore, the nucleoli may be very large relative to that observed in the nuclei of normal cells, possibly reflecting the extent of transcriptional activities taking place in these highly active cells. Given that malignant neoplasms often exhibit high rates of cell proliferation, numerous mitotic figures may be seen, and these mitotic figures are often abnormal. Typically, anaplastic cells will fail to organize into recognizable tissue patterns. This lack of cellular orientation reflects loss of normal cellular polarity, as well as a failure of normal structures to form.

Rate of Growth

In general, benign neoplasms grow more slowly than malignant neoplasms. However, there are exceptions to this rule, and in some cases benign neoplasms will display an elevated growth rate. For instance, some benign neoplasms will exhibit changes in growth rate in response to hormonal stimulation or in response to alterations in blood supply. Leiomyomas are benign neoplasms of the uterus that originate in smooth muscle and are significantly influenced by circulating levels of estrogens. Thus, these neoplasms may display an elevated cell proliferation and increase rapidly in size in response to the hormonal changes seen in pregnancy. Once hormonal levels normalize (after childbirth), the lack of sufficient hormone levels to sustain high rates of cellular proliferation results in greatly diminished growth of the neoplasm. Subsequently, with the elimination of hormones related to menopause, these tumors may become fibrocalcific. Despite variations in growth rate among neoplasms and some physiological exceptions (such as leiomyoma in pregnancy), most benign neoplasms proliferate slowly over time (months to years) and increase in size slowly.

The rate of cell proliferations (and lesion growth) of malignant neoplasms generally correlates with the extent of cellular differentiation of the cells that compose the tumor. Thus, poorly differentiated neoplasms exhibit high rates of cell proliferation and tumor growth, and well-differentiated neoplasms tend to grow more slowly. Nevertheless, there is considerable variability in the relative growth rate among malignant neoplasms. There is abundant evidence suggesting that most if not all malignant neoplasms take many years (and perhaps decades) to develop and emerge clinically. Many or most malignant neoplasms grow relatively slowly and at a constant rate. However, there may be several patterns of growth among malignant neoplasms. Some of these lesions grow slowly for long periods of time before entering into a phase characterized by more rapid expansion. In this case, the rapid expansion phase probably reflects the emergence of a more aggressive subclone of neoplastic cells. In certain rare instances, the proliferation of a malignant neoplasm will diminish to a level of undetectable cellular proliferation, or the neoplasm will regress spontaneously (due to widespread necrosis of the neoplastic cells). Most malignant neoplasms progressively enlarge over time in relation to their cellular growth rate. Rapidly growing malignant neoplasms tend to contain a central area of necrosis that develops secondary to ischemia related to inadequate blood supply (Figure 4.15).

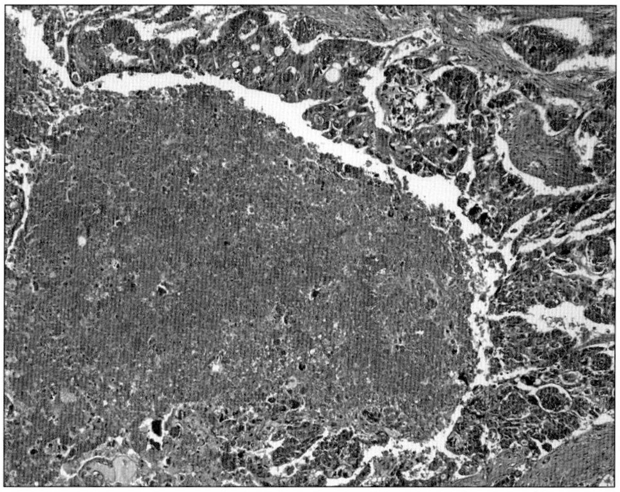

Figure 4.15 **Tumor necrosis.** Necrotic (dead tissue) debris is present within the center of the malignant glands in this adenocarcinoma.

Since the tumor blood supply is derived from normal tissues at the site of the neoplasm, the formation of new blood vessels to supply the expanding neoplasm may lag behind the proliferation of the neoplastic cells, resulting in inadequate supply of oxygen and other nutrients to the tumor mass.

Presence of Local Invasion

Benign neoplasms, by definition, remain localized at their site of origin. These neoplasms do not have the capacity to infiltrate surrounding tissues, invade locally, or metastasize to distant sites. Benign neoplasms typically have smooth borders, are sharply demarcated from the normal tissue at the tumor site, and are frequently encapsulated by a fibrous capsule that forms a barrier between the neoplastic cells and the host tissue. However, not all benign neoplasms have a capsule, and the lack of a capsule around a neoplasm does not indicate that the neoplasm is malignant. The capsule of a benign neoplasm is primarily the product of the elaboration of tumor stroma. In addition, the tumor capsule may derive in part from the fibrous debris resulting from necrotic cell death of tissue cells adjacent to the neoplasm. Benign liver adenomas represent an example of a benign neoplasm that commonly exhibits a capsule. However other benign neoplasms lack a well-developed capsule. Uterine leiomyomas do not infiltrate adjacent normal tissues and are typically discretely demarcated from the surrounding smooth muscle by a zone of compressed and attenuated normal myometrium, but these neoplasms do not elaborate a capsule. There are a few examples of benign tumors that are neither encapsulated nor discretely defined (for instance, some vascular benign neoplasms of the dermis).

The growth of malignant neoplasms is characterized by progressive infiltration, invasion, destruction, and penetration of surrounding normal (non-neoplastic) tissues (Figure 4.6 and Figure 4.11B). Malignant neoplasms form well-developed capsules. However, some slow-growing malignant neoplasms appear histologically to be encased in stroma resembling a capsule. The invasive (malignant) nature of these neoplasms is revealed by close microscopic examination, which reveals penetration of the margins of the stroma by neoplastic cells and invasion of adjacent tissue structures. The infiltrative or locally invasive nature of the growth of malignant neoplasms requires that broad margins of non-neoplastic tissue must be resected during surgical excision of a tumor to ensure complete removal of all neoplastic cells. Local invasiveness and infiltration of adjacent tissue structures represent features that are strongly suggestive of a malignant neoplasm. These features represent the most reliable predictor of malignant behavior next to the development of distant metastasis.

Metastasis

The term metastasis describes the development of secondary neoplastic lesions at distant tissue locations that are separated from the primary site of a malignant neoplasm. The clinical finding of metastasis provides a definitive classification of a primary neoplasm as malignant. Once again, benign neoplasms lack invasive behavior and do not metastasize. A significant percentage of patients with newly diagnosed malignant neoplasms exhibit clinically evident metastases at the time of diagnosis (~30%), and others have occult metastases at the time of diagnosis (~20%). Some malignant neoplasms demonstrate a propensity for metastasis. Breast cancers tend to spread to bone, lung, liver, brain, and some other sites. The size of a primary breast cancer at the time of diagnosis is directly proportional to the probability for the development of distant metastasis over time [150]. In general terms, malignant neoplasms that are anaplastic (less well differentiated) and larger are more likely to metastasize. However, there are numerous exceptions; extremely small cancers may metastasize and have poor prognosis, while some large tumors may remain localized and not produce metastases. For some malignant neoplasms, initial diagnosis of a primary tumor may occur at late stage after distant metastases have already developed. For example, osteogenic sarcomas typically metastasize to the lungs by the time the primary neoplasm is detected. However, not all malignant neoplasms exhibit a strong tendency for metastasis. For instance, basal cell carcinomas of the skin are highly invasive at the primary site of the tumor and only rarely form distant metastases. Likewise, many/most malignant neoplasms of the brain are highly invasive at their primary sites but rarely give rise to distant metastatic lesions.

Malignant neoplasms metastasize and spread to distant sites in the patient through (i) direct invasion or seeding within body cavities, (ii) spread through the lymphatic system, or (iii) hematogenous spread via the blood. Direct invasion or spreading by seeding occurs when a malignant neoplasm invades a natural body cavity. This pathway of tumor spread is characteristic of malignant neoplasms of the ovary. These cancers often widely disseminate across peritoneal surfaces, producing a significant tumor burden without invasion of the underlying parenchyma of the abdominal organs. In this example, the ovarian carcinoma cells demonstrate an ability to establish at new sites (previously uninvolved peritoneal surfaces), but lack the capacity to invade into new tissue sites. Pancreatic carcinoma tends to spread by direct invasion into the peritoneal cavity and can invade the stomach by direct extension. Likewise, lung carcinomas may invade into the plural cavity, directly invade the diaphragm, and eventually gain access to the peritoneal cavity after invasion through the diaphragm.

Malignant epithelial neoplasms (carcinomas) tend to metastasize through the lymphatic system (Figure 4.16). Hence, lymph node involvement represents an important factor in the staging of many cancers. However, given the numerous anatomic interconnections between the lymphatic and vascular systems, it is possible that some of these cancers may disseminate through either or both pathways. While enlargement of lymph nodes near a primary neoplasm may arouse suspicion

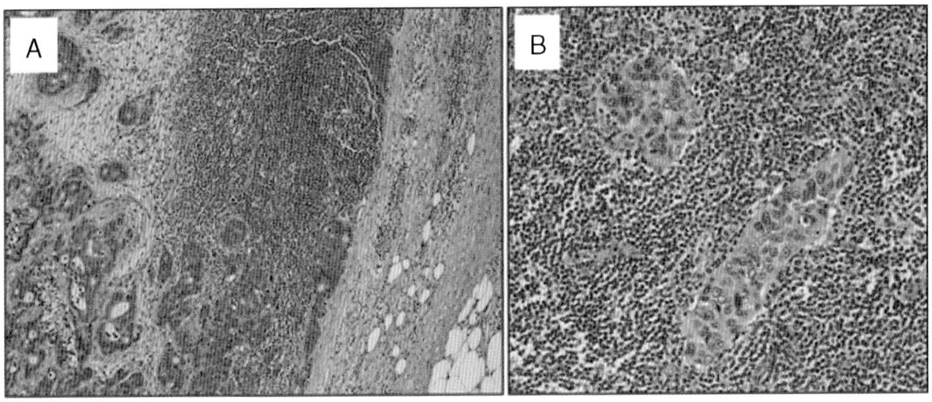

Figure 4.16 **Metastatic spread of malignant cells through the lymphatic system. (A)** Metastatic adenocarcinoma to a lymph node, low power. Residual lymph node tissue is present along the upper portion of the image. Infiltrating glands of adenocarcinoma are noted at the lower aspect of the image. **(B)** Metastatic carcinoma to a lymph node, high power. Two nests of plump malignant epithelial cells are present within the sea of smaller lymphocytes of the lymph node.

of lymph node involvement and metastatic spread of the primary malignant neoplasm, simple lymph node swelling does not accurately predict spread of neoplastic cells. Other processes can account for lymph node enlargement adjacent to malignant neoplasms. Reactive changes may occur when necrotic debris from a tumor encounters a lymph node. Patterns of lymph node seeding by metastatic tumor cells mainly depend on the site of the primary malignant neoplasm and the natural pathways of lymphatic drainage from that tissue site. Thus, lung cancers arising in the respiratory passages metastasize initially to regional bronchial lymph nodes and subsequently to the tracheobronchial and hilar nodes. In this case, the extent of spread can be inferred from which lymph nodes are positive for neoplastic cells. Breast cancer presents a slightly more complicated picture. Malignant neoplasms of the breast that arise in the upper outer quadrant of the breast will spread initially to axillary lymph nodes, while malignant tumors arising in the medial region of the breast may spread through the chest wall into the lymph nodes along the internal mammary artery. Seeding of the supraclavicular and infraclavicular lymph nodes by metastatic breast cancer cells can occur subsequent to the initial spread despite the initial route of tumor cell movement. In some cases, breast cancer cells appear to traverse lymphatic vessels without colonizing the lymph nodes that are immediately proximal to the site of the primary neoplasm, but do implant in lymph nodes that are subsequently encountered (producing metastatic lesions that are referred to as skip metastases). Given the complexity and potentially complicated pattern of lymphatic spread by metastatic breast cancer, it is not surprising that sentinel node biopsy has become a useful procedure for surgical staging of this malignant neoplasm. A sentinel lymph node is defined as the first lymph node in a regional lymphatic basin that receives lymphatic drainage from the site of a primary neoplasm. It can be identified by injection of colored dyes (or radiolabeled tracers) into the site of the primary tumor with monitoring of the movement of the dye to the downstream lymph nodes [151]. Histopathological evaluation of the sentinel lymph node provides a reliable indication of the extent of spread of a breast cancer and can be used to plan treatment [152].

Dissemination of a malignant neoplasm through the blood is referred to as hematogenous spread. Hematogenous spread of malignant mesenchymal neoplasms (sarcomas) occurs commonly, but this pathway is also readily utilized by malignant epithelial neoplasms (carcinomas). Invasion of the vascular system by neoplastic cells can involve either the arterial circulation (and associated vessels) or the venous circulation (and associated vessels). However, the arteries are penetrated less readily than are veins, making venous involvement more likely. In the process of venous invasion by a malignant neoplasm, the neoplastic cells in the blood follow the normal pattern of venous blood flow draining the primary site of the neoplasm. It is well recognized that the liver and lungs represent the most frequently involved secondary sites in hematogenous spread of malignant neoplasms. This is due to the fact that all portal venous blood drainage flows to the liver and all caval blood drainage flows to the lungs. In many cases, it is thought that metastatic tumor cells will arrest and colonize the first capillary bed that they encounter after gaining access to the venous circulation. However, numerous clinical and experimental observations combine to suggest that the anatomic location of the primary neoplasm and the natural pathways of venous blood flow from that site do not completely explain the patterns of metastatic disease observed with many forms of cancer [153–155]. Thus, the so-called seed and soil hypothesis was developed to explain the tendency of certain cancers to preferentially spread to certain tissue sites. For example, prostatic carcinoma preferentially metastasizes to bone, bronchogenic carcinomas tend to spread to the adrenals and the brain, and neuroblastomas are prone to colonize the liver and bones. In contrast, whereas skeletal muscles represent a substantial organ system that is rich in capillary beds, it is rarely a site for metastatic tumor colonization. These observations suggest that factors intrinsic to the neoplastic cells and their secondary sites of involvement determine the ability of cancer cells to efficiently colonize a given tissue.

CLINICAL ASPECTS OF NEOPLASIA

As a malignant neoplasm develops, expands, and spreads, it can produce a number of effects on the host, including fever, anorexia, weight loss and cachexia, infection, anemia, and various hormonal and neurologic symptoms. Major manifestations of tumor effects on the host include (i) discomfort (pain), (ii) cachexia, and (iii) paraneoplastic syndromes.

Cancer-Associated Pain

Pain is frequently associated with malignant neoplasms, particularly in advanced disease. This cancer-associated pain is often difficult to treat and unrelenting, producing major challenges for management and typically requiring the use of narcotic drugs. The cause of the pain can be related to destruction of tissue by the neoplasm, infection, stretching of internal organs (due to tumor involvement), pressure (from an expanding neoplasm), or obstruction (secondary to tumor impingement). Destruction of bone tissue by metastatic cancer results in characteristic bone pain. Many cancer patients have decreased immune function, resulting in infections that might not otherwise be encountered. For example, cancer patients with decreased immunity may develop infection by herpes zoster, which causes extreme nerve pain (causalgia). Pain associated with stretching of internal organs occurs with cancers of various tissues, including liver, pancreas, stomach, and other sites. The capsule of the liver is invested with blood vessels and nerves. Thus, with expansion of the tissue secondary to expansion of neoplastic growths (whether primary neoplasms or metastatic lesions), stretching of the capsule occurs, producing pain associated with perturbation of these nerves. Metastatic cancer of the liver is a major cause of hepatomegaly, and pain in the upper right quadrant is often an early symptom.

Cancer Cachexia

Cachexia refers to a progressive loss of body fat and lean body mass, accompanied by profound weakness, anorexia, and anemia. This condition affects many cancer patients with advanced disease. In fact, very often the size and extent of spread of the malignant neoplasm correlates with the severity of cachexia. However, the cachexic condition of the patient is not directly related to the nutritional needs of the neoplasm. Rather, cancer cachexia results from the action of soluble factors (such as cytokines) produced by the neoplasm. Cancer patients very often experience anorexia (loss of appetite) which can result in reduced caloric intake [156]. Nevertheless, cancer patients often display increased basal metabolic rate and expend high numbers of calories in spite of reduced food caloric intake. It follows that as caloric needs increase (or remain high) in the patient, body fat reserves and lean tissue mass are consumed to meet the energy needs of the individual. The mechanistic basis for this metabolic disturbance has not been fully elucidated. However, some evidence suggests that certain cytokines (such as TNFα) may mediate cancer-associated cachexia [157,158]. Other factors are probably involved in cancer cachexia as well, including proteolysis-inducing factor (which mediates breakdown of skeletal muscle proteins) [159] and other molecules with lipolytic action [160]. Cancer cachexia cannot be treated effectively, although strategies for management of this aspect of cancer are being investigated [161]. At the present time, the most effective treatment of cancer-associated cachexia is removal of the underlying cause (the malignant neoplasm).

Paraneoplastic Syndromes

Paraneoplastic syndromes refer to groupings of symptoms that occur in patients with malignant neoplasms that cannot be readily explained by local invasion or distant metastasis of the tumor, or the elaboration of hormones indigenous to the tissue of origin of the neoplasm. Paraneoplastic syndromes occur in 10%–15% of cancer patients. These paraneoplastic syndromes present challenges to management of the cancer patient, potentially leading to significant clinical problems that affect quality of life, or contributing to potentially lethal complications. However, the symptoms associated with paraneoplastic syndromes may also represent an early manifestation of an occult neoplasm, presenting an opportunity for cancer detection and diagnosis. There are a number of different paraneoplastic syndromes that are associated with many different tumors. Well-characterized examples of paraneoplastic syndromes include (i) hypercalcemia, (ii) Cushing syndrome, and (iii) hypercoagulability (Trousseau syndrome). Hypercalcemia refers to a condition related to elevated plasma calcium concentrations. This condition occurs in 20%–30% of patients with metastatic cancer and is the most common paraneoplastic syndrome [162]. Hypercalcemia affects multiple organ systems, can be life-threatening, and predicts a poor outcome for the patient. There are four types of hypercalcemia that have been recognized: (a) osteoclastic bone resorption due to metastasis to bone, (b) humoral hypercalcemia caused by systemic secretion of parathyroid hormone-related protein (PTHrP), (c) secretion of vitamin D, and (d) ectopic secretion of parathyroid hormone. Cushing syndrome is caused by ectopic secretion of corticotropin (hypercortisolism) and produces several major effects, including hyperglycemia, hyperkalemia, hypertension, and muscle weakness [163]. Cushing syndrome is associated with several types of malignant neoplasm, including small cell lung carcinoma, pancreatic carcinoma, and various neural tumors. Trousseau syndrome is characterized by hypercoagulopathy and produces the major effect of venous thrombosis of the deep veins in the cancer patient [164]. Small cell lung carcinoma and pancreatic carcinoma can cause Trousseau syndrome by elaborating platelet-aggregating factors and procoagulants from the tumor or its necrotic products.

Grading and Staging of Cancer

In the clinical setting, it is important to be able to predict the relative aggressiveness of a given malignant neoplasm as a guide for designing treatment. To accurately predict the clinical course of a neoplasm and the probable outcome for the patient, clinicians make observations related to the histological aggressiveness of the cells that compose the neoplasm and the apparent extent and spread of the disease. This analysis yields a clinical description of the patient that reflects a score for the grade and stage of the malignant neoplasm.

The histological grade of a malignant neoplasm is based on (i) the degree of cellular differentiation of the neoplastic cells that compose the lesion and (ii) an estimate of the growth rate of the neoplasm (based on mitotic index). In the past, it was thought that the degree of histological differentiation reflected the relative aggressiveness of a malignant neoplasm. However, it is now recognized that this notion oversimplifies the biology of neoplastic diseases. Nevertheless, the histological grade of certain cancers continues to have value for predicting clinical course. These cancers include those of the cervix, endometrium, colon, and thyroid. In general, histologic grade I refers to neoplasms that display 75%–100% differentiation, grade II reflects 50%–75% differentiation, grade III refers to 25%–50% differentiation, and grade IV describes tumors with <25% differentiation. Current methods also take into consideration the mitotic activity of the neoplastic cells, the degree of infiltration of adjacent tissues, and the amount of tumor stroma that is present.

Staging of malignant neoplasms is based on (i) the size of the primary lesion, (ii) the extent of spread to regional lymph nodes, and (iii) the presence or absence of distant metastases. Effective staging of neoplastic disease relies on clinical assessment of the patient, radiographic examination (using CT, MRI, or other technologies), and in some cases surgical exploration. The major convention for staging of malignant neoplasms is known as the TNM system. In this system, the T refers to the primary tumor (with T1, T2, T3, and T4 reflecting increasing size of the primary tumor), the N refers to regional lymph node involvement (with N0, N1, N2, and N3 reflecting progressively advancing node involvement), and M refers to the presence or absence of distant metastases (with M0 and M1 reflecting absence and presence of distant metastases, respectively). Staging guidelines vary for specific types of cancer. For instance, in primary hepatocellular carcinoma: T0 indicates no evidence of primary tumor; T1 indicates a solitary tumor, 2 cm or less in size, without vascular invasion; T2 indicates a solitary tumor, 2 cm or less in size, with vascular invasion or multiple tumors limited to one lobe (<2 cm) without vascular invasion; or a solitary tumor (>2 cm) without vascular invasion; T3 indicates a solitary tumor more than 2 cm in size with vascular invasion or multiple tumors limited to one lobe (<2 cm), with vascular invasion or multiple tumors limited to one lobe (>2 cm); and T4 indicates multiple tumors in more than one lobe of the liver, involving a major branch of the portal or hepatic vein(s) or invasion of adjacent organs; N0 indicates no regional lymph node metastasis; N1 indicates regional lymph node metastasis; M0 indicates no distant metastasis; and M1 indicates the presence of distant metastasis. In contrast to the staging of liver cancers, staging of colorectal cancer is fairly straightforward and primarily involves consideration of the depth of tumor invasion through the colon wall. Adenocarcinoma invasive into the submucosa of the colon is considered a T1 lesion, while invasion into but not through the muscularis propria is classified as T2. When the adenocarcinoma extends through the muscularis propria into the subserosa it falls under the category of T3. T4 adenocarcinomas extend through the colon wall and directly invade adjacent organs or structures (such as bladder or another portion of the gastrointestinal tract). Like the staging classification for the liver, the N (nodal) and M (metastasis) status of the patient depends on the number of lymph nodes present and presence or absence of distant metastasis, respectively.

REFERENCES

1. Hankey BF, Gloeckler Ries LA, Miller AB, et al. Overview. *Cancer Statistics Review 1973–1989.* NIH Publication Number 92-2789 ed, 1992;I.1–17.
2. Jemal A, Siegel R, Ward E, et al. Cancer statistics, 2008. *CA Cancer J Clin.* 2008;58:71–96.
3. Parkin DM, Bray F, Ferlay J, et al. Global cancer statistics, 2002. *CA Cancer J Clin.* 2005;55:74–108.
4. Newell GR, Spitz MR, Sider JG. Cancer and age. *Semin Oncol.* 1989;16:3–9.
5. Miller RA. Gerontology as oncology. Research on aging as the key to the understanding of cancer. *Cancer.* 1991;68:2496–2501.
6. Stevens RG, Merkle EJ, Lustbader ED. Age and cohort effects in primary liver cancer. *Int J Cancer.* 1984;33:453–458.
7. Simonetti RG, Camma C, Fiorello F, et al. Hepatocellular carcinoma. A worldwide problem and the major risk factors. *Dig Dis Sci.* 1991;36:962–972.
8. De Maria N, Manno M, Villa E. Sex hormones and liver cancer. *Mol Cell Endocrinol.* 2002;193:59–63.
9. Giannitrapani L, Soresi M, La Spada E, et al. Sex hormones and risk of liver tumor. *Ann NY Acad Sci.* 2006;1089:228–236.
10. Eagon PK, Francavilla A, DiLeo A, et al. Quantitation of estrogen and androgen receptors in hepatocellular carcinoma and adjacent normal human liver. *Dig Dis Sci.* 1991;36:1303–1308.
11. Ohnishi S, Murakami T, Moriyama T, et al. Androgen and estrogen receptors in hepatocellular carcinoma and in the surrounding noncancerous liver tissue. *Hepatology.* 1986;6:440–443.
12. Carr BI, Van Thiel DH. Hormonal manipulation of human hepatocellular carcinoma. A clinical investigative and therapeutic opportunity. *J Hepatol.* 1990;11:287–289.
13. Lui WY, Lin HL, Chau GY, et al. Male predominance in hepatocellular carcinoma: New insight and a possible therapeutic alternative. *Med Hypotheses.* 2000;55:348–350.
14. Gayther SA, Pharoah PD, Ponder BA. The genetics of inherited breast cancer. *J Mammary Gland Biol Neoplasia.* 1998;3:365–376.

15. Sutcliffe S, Pharoah PD, Easton DF, et al. Ovarian and breast cancer risks to women in families with two or more cases of ovarian cancer. *Int J Cancer.* 2000;87:110–117.
16. Madigan MP, Ziegler RG, Benichou J, et al. Proportion of breast cancer cases in the United States explained by well-established risk factors. *J Natl Cancer Inst.* 1995;87:1681–1685.
17. Ponder BA, Antoniou A, Dunning A, et al. Polygenic inherited predisposition to breast cancer. *Cold Spring Harb Symp Quant Biol.* 2005;70:35–41.
18. Kelsey JL, Berkowitz GS. Breast cancer epidemiology. *Cancer Res.* 1988;48:5615–5623.
19. Hsieh CC, Trichopoulos D, Katsouyanni K, et al. Age at menarche, age at menopause, height and obesity as risk factors for breast cancer: Associations and interactions in an international case-control study. *Int J Cancer.* 1990;46:796–800.
20. Kelsey JL, Gammon MD, John EM. Reproductive factors and breast cancer. *Epidemiol Rev.* 1993;15:36–47.
21. Lipworth L. Epidemiology of breast cancer. *Eur J Cancer Prev.* 1995;4:7–30.
22. Okobia MN, Bunker CH. Epidemiological risk factors for breast cancer—A review. *Niger J Clin Pract.* 2005;8:35–42.
23. Kremsdorf D, Soussan P, Paterlini-Brechot P, et al. Hepatitis B virus-related hepatocellular carcinoma: Paradigms for viral-related human carcinogenesis. *Oncogene.* 2006;25:3823–3833.
24. Yu MC, Yuan JM. Environmental factors and risk for hepatocellular carcinoma. *Gastroenterology.* 2004;127:S72–78.
25. Barazani Y, Hiatt JR, Tong MJ, et al. Chronic viral hepatitis and hepatocellular carcinoma. *World J Surg.* 2007;31:1243–1248.
26. But DY, Lai CL, Yuen MF. Natural history of hepatitis-related hepatocellular carcinoma. *World J Gastroenterol.* 2008;14:1652–1656.
27. Popper H, Thung SN, McMahon BJ, et al. Evolution of hepatocellular carcinoma associated with chronic hepatitis B virus infection in Alaskan Eskimos. *Arch Pathol Lab Med.* 1988;112:498–504.
28. Unoura M, Kaneko S, Matsushita E, et al. High-risk groups and screening strategies for early detection of hepatocellular carcinoma in patients with chronic liver disease. *Hepatogastroenterology.* 1993;40:305–310.
29. Altmann HW. Hepatic neoformations. *Pathol Res Pract.* 1994; 190:513–577.
30. Johnson PJ, Williams R. Cirrhosis and the aetiology of hepatocellular carcinoma. *J Hepatol.* 1987;4:140–147.
31. Tiribelli C, Melato M, Croce LS, et al. Prevalence of hepatocellular carcinoma and relation to cirrhosis: Comparison of two different cities of the world—Trieste, Italy, and Chiba, Japan. *Hepatology.* 1989;10:998–1002.
32. Lutwick LI. Relation between aflatoxin, hepatitis-B virus, and hepatocellular carcinoma. *Lancet.* 1979;1:755–757.
33. zur Hausen H. Papillomavirus infections—A major cause of human cancers. *Biochim Biophys Acta.* 1996;1288:F55–78.
34. Bosch FX, Munoz N, de Sanjose S, et al. Risk factors for cervical cancer in Colombia and Spain. *Int J Cancer.* 1992; 52:750–758.
35. Koutsky LA, Holmes KK, Critchlow CW, et al. A cohort study of the risk of cervical intraepithelial neoplasia grade 2 or 3 in relation to papillomavirus infection. *N Engl J Med.* 1992;327: 1272–1278.
36. Dillner J, Lehtinen M, Bjorge T, et al. Prospective seroepidemiologic study of human papillomavirus infection as a risk factor for invasive cervical cancer. *J Natl Cancer Inst.* 1997;89:1293–1299.
37. Serraino D, Piselli P, Angeletti C, et al. Infection with Epstein-Barr virus and cancer: An epidemiological review. *J Biol Regul Homeost Agents.* 2005;19: 63–70.
38. Niedobitek G, Young LS. Epstein-Barr virus persistence and virus-associated tumours. *Lancet.* 1994;343:333–335.
39. Rickinson AB. On the biology of Epstein-Barr virus persistence: A reappraisal. *Adv Exp Med Biol.* 1990;278:137–146.
40. Young LS, Rickinson AB. Epstein-Barr virus: 40 years on. *Nat Rev Cancer.* 2004;4:757–768.
41. Magrath I. The pathogenesis of Burkitt's lymphoma. *Adv Cancer Res.* 1990;55:133–270.
42. Brady G, MacArthur GJ, Farrell PJ. Epstein-Barr virus and Burkitt lymphoma. *J Clin Pathol.* 2007;60:1397–1402.
43. Abnet CC. Carcinogenic food contaminants. *Cancer Invest.* 2007; 25:189–196.
44. Reijula K, Tuomi T. Mycotoxins of aspergilli: Exposure and health effects. *Front Biosci.* 2003;8:s232–235.
45. Wogan GN, Hecht SS, Felton JS, et al. Environmental and chemical carcinogenesis. *Semin Cancer Biol.* 2004; 14:473–486.
46. Blot WJ, Xu ZY, Boice Jr JD, et al. Indoor radon and lung cancer in China. *J Natl Cancer Inst.* 1990;82:1025–1030.
47. Archer VE, Gillam JD, Wagoner JK. Respiratory disease mortality among uranium miners. *Ann NY Acad Sci.* 1976;271:280–293.
48. Harley NH, Harley JH. Potential lung cancer risk from indoor radon exposure. *CA Cancer J Clin.* 1990;40:265–275.
49. Irigaray P, Newby JA, Clapp R, et al. Lifestyle-related factors and environmental agents causing cancer: An overview. *Biomed Pharmacother.* 2007;61:640–658.
50. Shopland DR, Eyre HJ, Pechacek TF. Smoking-attributable cancer mortality in 1991: Is lung cancer now the leading cause of death among smokers in the United States? *J Natl Cancer Inst.* 1991;83:1142–1148.
51. Garfinkel L, Silverberg E. Lung cancer and smoking trends in the United States over the past 25 years. *CA Cancer J Clin.* 1991;41:137–145.
52. Rustemeier K, Stabbert R, Haussmann HJ, et al. Evaluation of the potential effects of ingredients added to cigarettes. Part 2: Chemical composition of mainstream smoke. *Food Chem Toxicol.* 2002;40:93–104.
53. Haenszel W, Loveland DB, Sirken MG. Lung-cancer mortality as related to residence and smoking histories. I. White males. *J Natl Cancer Inst.* 1962;28:947–1001.
54. Seitz HK, Becker P. Alcohol metabolism and cancer risk. *Alcohol Res Health.* 2007;30:38–41, 44–37.
55. Cargiulo T. Understanding the health impact of alcohol dependence. *Am J Health Syst Pharm.* 2007;64:S5–11.
56. Michels KB, Mohllajee AP, Roset-Bahmanyar E, et al. Diet and breast cancer: A review of the prospective observational studies. *Cancer.* 2007;109:2712–2749.
57. Seitz HK, Maurer B, Stickel F. Alcohol consumption and cancer of the gastrointestinal tract. *Dig Dis.* 2005;23:297–303.
58. Ikeda K, Saitoh S, Koida I, et al. A multivariate analysis of risk factors for hepatocellular carcinogenesis: A prospective observation of 795 patients with viral and alcoholic cirrhosis. *Hepatology.* 1993;18:47–53.
59. Nalpas B, Feitelson M, Brechot C, et al. Alcohol, hepatotropic viruses, and hepatocellular carcinoma. *Alcohol Clin Exp Res.* 1995;19:1089–1095.
60. Noda K, Yoshihara H, Suzuki K, et al. Progression of type C chronic hepatitis to liver cirrhosis and hepatocellular carcinoma—Its relationship to alcohol drinking and the age of transfusion. *Alcohol Clin Exp Res.* 1996;20:95A–100A.
61. Lieber CS. Interaction of alcohol with other drugs and nutrients. Implication for the therapy of alcoholic liver disease. *Drugs.* 1990;40(Suppl 3):23–44.
62. Bandera EV, Freudenheim JL, Vena JE. Alcohol consumption and lung cancer: A review of the epidemiologic evidence. *Cancer Epidemiol Biomarkers Prev.* 2001;10:813–821.
63. Leiter U, Garbe C. Epidemiology of melanoma and nonmelanoma skin cancer—The role of sunlight. *Adv Exp Med Biol.* 2008;624:89–103.
64. de Gruijl FR, Longstreth J, Norval M, et al. Health effects from stratospheric ozone depletion and interactions with climate change. *Photochem Photobiol Sci.* 2003;2:16–28.
65. Willis RA. *The Spread of Tumors in the Human Body.* London: Butterworths; 1952.
66. Kinzler KW, Vogelstein B. Introduction. In: Vogelstein B, Kinzler KW, eds. *The Genetic Basis of Human Cancer.* 2nd ed. New York: McGraw-Hill; 2002:3–6.
67. Bishop JM. Molecular themes in oncogenesis. *Cell.* 1991;64:235–248.
68. Lengauer C, Kinzler KW, Vogelstein B. Genetic instabilities in human cancers. *Nature.* 1998;396:643–649.
69. Feinberg AP, Vogelstein B. Hypomethylation of ras oncogenes in primary human cancers. *Biochem Biophys Res Commun.* 1983;111: 47–54.
70. Feinberg AP, Vogelstein B. Hypomethylation distinguishes genes of some human cancers from their normal counterparts. *Nature.* 1983;301:89–92.

71. Feinberg AP, Vogelstein B. Alterations in DNA methylation in human colon neoplasia. *Semin Surg Oncol.* 1987;3:149–151.
72. Goelz SE, Vogelstein B, Hamilton SR, et al. Hypomethylation of DNA from benign and malignant human colon neoplasms. *Science.* 1985;228:187–190.
73. Esteller M. Epigenetic gene silencing in cancer: The DNA hypermethylome. *Hum Mol Genet.* 2007;16(Spec No 1):R50–59.
74. Esteller M. Cancer epigenomics: DNA methylomes and histone-modification maps. *Nat Rev Genet.* 2007;8:286–298.
75. Herranz M, Esteller M. DNA methylation and histone modifications in patients with cancer: Potential prognostic and therapeutic targets. *Methods Mol Biol.* 2007;361:25–62.
76. Palii SS, Robertson KD. Epigenetic control of tumor suppression. *Crit Rev Eukaryot Gene Expr.* 2007;17:295–316.
77. Wang GG, Allis CD, Chi P. Chromatin remodeling and cancer, Part II: ATP-dependent chromatin remodeling. *Trends Mol Med.* 2007;13:373–380.
78. Wang GG, Allis CD, Chi P. Chromatin remodeling and cancer, Part I: Covalent histone modifications. *Trends Mol Med.* 2007; 13:363–372.
79. Calin GA, Croce CM. MicroRNA signatures in human cancers. *Nat Rev Cancer.* 2006;6:857–866.
80. Garzon R, Fabbri M, Cimmino A, et al. MicroRNA expression and function in cancer. *Trends Mol Med.* 2006;12:580–587.
81. Hagan JP, Croce CM. MicroRNAs in carcinogenesis. *Cytogenet Genome Res.* 2007;118:252–259.
82. Fearon ER, Vogelstein B. A genetic model for colorectal tumorigenesis. *Cell.* 1990;61:759–767.
83. Vogelstein B, Fearon ER, Hamilton SR, et al. Genetic alterations during colorectal-tumor development. *N Engl J Med.* 1988;319: 525–532.
84. Quinlan JM, Colleypriest BJ, Farrant M, et al. Epithelial metaplasia and the development of cancer. *Biochim Biophys Acta.* 2007;1776:10–21.
85. Benner SE, Lippman SM, Hong WK. Chemoprevention of lung cancer. *Chest.* 1995;107:316–321.
86. Fitzgerald RC. Molecular basis of Barrett's oesophagus and oesophageal adenocarcinoma. *Gut.* 2006;55:1810S–1820S.
87. Hornick JL, Odze RD. Neoplastic precursor lesions in Barrett's esophagus. *Gastroenterol Clin North Am.* 2007;36:775–796.
88. Tiwari A, Krishna NS, Nanda K, et al. Benign prostatic hyperplasia: An insight into current investigational medical therapies. *Expert Opin Investig Drugs.* 2005;14:1359–1372.
89. Untergasser G, Madersbacher S, Berger P. Benign prostatic hyperplasia: Age-related tissue-remodeling. *Exp Gerontol.* 2005; 40:121–128, v.
90. Kurman RJ, Kaminski PF, Norris HJ. The behavior of endometrial hyperplasia. A long-term study of "untreated" hyperplasia in 170 patients. *Cancer.* 1985;56:403–412.
91. Society AC. *Cancer Facts and Figures, 2008.* Atlanta: American Cancer Society; 2008.
92. Madhavan J, Ganesh A, Kumaramanickavel G. Retinoblastoma: From disease to discovery. *Ophthalmic Res.* 2008;40:221–226.
93. Cleaver JE, Kraemer KH. Xeroderma pigmentosum and Cockayne syndrome. In: Scriver CR, Beaudet AL, Sly WS, Valle D, eds. *The Metabolic and Molecular Bases of Inherited Disease.* 7th ed. New York: McGraw-Hill; 1995:4393–4419.
94. Bootsma D, Kraemer KH, Cleaver JE, et al. Nucleotide excision repair syndromes: Xeroderma pigmentosum, Cockayne syndrome, and trichothiodystrophy. In: Vogelstein B, Kinzler KW, eds. *The Genetic Basis of Human Cancer.* 2nd ed. New York: McGraw-Hill; 2002:211–237.
95. Gatti RA. Ataxia-telangiectasia. In: Vogelstein B, Kinzler KW, eds. *The Genetic Basis of Human Cancer.* 2nd ed. New York: McGraw-Hill, 2002:239–266.
96. Swift M, Morrell D, Massey RB, et al. Incidence of cancer in 161 families affected by ataxia-telangiectasia. *N Engl J Med.* 1991; 325:1831–1836.
97. Auerbach AD, Buchwald M, Joenje H. Fanconi anemia. In: Vogelstein B, Kinzler KW, eds. *The Genetic Basis of Human Cancer.* 2nd ed. New York: McGraw-Hill; 2002:289–306.
98. German J, Passarge E. Bloom's syndrome. XII. Report from the Registry for 1987. *Clin Genet.* 1989;35:57–69.
99. German J. Bloom's syndrome. XX. The first 100 cancers. *Cancer Genet Cytogenet.* 1997;93:100–106.
100. Langlois RG, Bigbee WL, Jensen RH, et al. Evidence for increased in vivo mutation and somatic recombination in Bloom's syndrome. *Proc Natl Acad Sci USA.* 1989;86:670–674.
101. Chaganti RS, Schonberg S, German J. A manyfold increase in sister chromatid exchanges in Bloom's syndrome lymphocytes. *Proc Natl Acad Sci USA.* 1974;71:4508–4512.
102. German J, Schonberg S, Louie E, et al. Bloom's syndrome. IV. Sister-chromatid exchanges in lymphocytes. *Am J Hum Genet.* 1977;29:248–255.
103. Boland CR. Hereditary nonpolyposis colorectal cancer (HNPCC). In: Vogelstein B, Kinzler KW, eds. *The Genetic Basis of Human Cancer.* 2nd ed. New York: McGraw-Hill; 2002: 307–322.
104. Mecklin JP, Jarvinen HJ. Tumor spectrum in cancer family syndrome (hereditary nonpolyposis colorectal cancer). *Cancer.* 1991;68:1109–1112.
105. Vasen HF, Offerhaus GJ, et al. The tumour spectrum in hereditary non-polyposis colorectal cancer: A study of 24 kindreds in the Netherlands. *Int J Cancer.* 1990;46:31–34.
106. Ionov Y, Peinado MA, Malkhosyan S, et al. Ubiquitous somatic mutations in simple repeated sequences reveal a new mechanism for colonic carcinogenesis. *Nature.* 1993;363:558–561.
107. Thibodeau SN, Bren G, Schaid D. Microsatellite instability in cancer of the proximal colon. *Science.* 1993;260:816–819.
108. Aaltonen LA, Peltomaki P, Leach FS, et al. Clues to the pathogenesis of familial colorectal cancer. *Science.* 1993;260: 812–816.
109. Peltomaki P, Lothe RA, Aaltonen LA, et al. Microsatellite instability is associated with tumors that characterize the hereditary non-polyposis colorectal carcinoma syndrome. *Cancer Res.* 1993;53:5853–5855.
110. Evans DG, Walsh S, Hill J, et al. Strategies for identifying hereditary nonpolyposis colon cancer. *Semin Oncol.* 2007;34:411–417.
111. Woerner SM, Kloor M, von Knebel Doeberitz M, et al. Microsatellite instability in the development of DNA mismatch repair deficient tumors. *Cancer Biomark.* 2006;2:69–86.
112. Ponti G, Ponz de Leon M. Muir-Torre syndrome. *Lancet Oncol.* 2005;6:980–987.
113. Schwartz RA, Torre DP. The Muir-Torre syndrome: A 25-year retrospect. *J Am Acad Dermatol.* 1995;33:90–104.
114. Honchel R, Halling KC, Schaid DJ, et al. Microsatellite instability in Muir-Torre syndrome. *Cancer Res.* 1994;54:1159–1163.
115. Honchel R, Halling KC, Thibodeau SN. Genomic instability in neoplasia. *Semin Cell Biol.* 1995;6:45–52.
116. Turcot J, Despres JP, St Pierre F. Malignant tumors of the central nervous system associated with familial polyposis of the colon: Report of two cases. *Dis Colon Rectum.* 1959;2:465–468.
117. Hamilton SR, Liu B, Parsons RE, et al. The molecular basis of Turcot's syndrome. *N Engl J Med.* 1995;332:839–847.
118. Li FP, Fraumeni Jr JF. Soft-tissue sarcomas, breast cancer, and other neoplasms. A familial syndrome? *Ann Intern Med.* 1969;71: 747–752.
119. Garber JE, Goldstein AM, Kantor AF, et al. Follow-up study of twenty-four families with Li-Fraumeni syndrome. *Cancer Res.* 1991;51:6094–6097.
120. Li FP, Fraumeni Jr JF, Mulvihill JJ, et al. A cancer family syndrome in twenty-four kindreds. *Cancer Res.* 1988;48:5358–5362.
121. Strong LC, Williams WR, Tainsky MA. The Li-Fraumeni syndrome: From clinical epidemiology to molecular genetics. *Am J Epidemiol.* 1992;135:190–199.
122. Malkin D, Li FP, Strong LC, et al. Germ line p53 mutations in a familial syndrome of breast cancer, sarcomas, and other neoplasms. *Science.* 1990;250:1233–1238.
123. Srivastava S, Zou ZQ, Pirollo K, et al. Germ-line transmission of a mutated p53 gene in a cancer-prone family with Li-Fraumeni syndrome. *Nature.* 1990;348:747–749.
124. Miki Y, Swensen J, Shattuck-Eidens D, et al. A strong candidate for the breast and ovarian cancer susceptibility gene BRCA1. *Science.* 1994;266:66–71.
125. Wooster R, Bignell G, Lancaster J, et al. Identification of the breast cancer susceptibility gene BRCA2. *Nature.* 1995;378: 789–792.
126. Wooster R, Neuhausen SL, Mangion J, et al. Localization of a breast cancer susceptibility gene, BRCA2, to chromosome 13q12–13. *Science.* 1994;265:2088–2090.

127. Claus EB, Schwartz PE. Familial ovarian cancer. Update and clinical applications. *Cancer.* 1995;76:1998–2003.
128. Easton D, Ford D, Peto J. Inherited susceptibility to breast cancer. *Cancer Surv.* 1993;18:95–113.
129. Lynch HT, Lynch J, Conway T, et al. Hereditary breast cancer and family cancer syndromes. *World J Surg.* 1994;18:21–31.
130. Fargnoli MC, Argenziano G, Zalaudek I, et al. High- and low-penetrance cutaneous melanoma susceptibility genes. *Expert Rev Anticancer Ther.* 2006;6:657–670.
131. Lomas J, Martin-Duque P, Pons M, et al. The genetics of malignant melanoma. *Front Biosci.* 2008;13:5071–5093.
132. Platz A, Ringborg U, Hansson J. Hereditary cutaneous melanoma. *Semin Cancer Biol.* 2000;10:319–326.
133. Fernandez L, Milne R, Bravo J, et al. MC1R: Three novel variants identified in a malignant melanoma association study in the Spanish population. *Carcinogenesis.* 2007;28:1659–1664.
134. Jannot AS, Meziani R, Bertrand G, et al. Allele variations in the OCA2 gene (pink-eyed-dilution locus) are associated with genetic susceptibility to melanoma. *Eur J Hum Genet.* 2005;13: 913–920.
135. Lynch HT, Fitzsimmons ML, Smyrk TC, et al. Familial pancreatic cancer: Clinicopathologic study of 18 nuclear families. *Am J Gastroenterol.* 1990;85:54–60.
136. Lynch HT, Smyrk T, Kern SE, et al. Familial pancreatic cancer: A review. *Semin Oncol.* 1996;23:251–275.
137. Guilford P, Hopkins J, Harraway J, et al. E-cadherin germline mutations in familial gastric cancer. *Nature.* 1998;392:402–405.
138. Gronberg H, Xu J, Smith JR, et al. Early age at diagnosis in families providing evidence of linkage to the hereditary prostate cancer locus (HPC1) on chromosome 1. *Cancer Res.* 1997;57:4707–4709.
139. Smith JR, Freije D, Carpten JD, et al. Major susceptibility locus for prostate cancer on chromosome 1 suggested by a genome-wide search. *Science.* 1996;274:1371–1374.
140. Pang RW, Joh JW, Johnson PJ, et al. Biology of hepatocellular carcinoma. *Ann Surg Oncol.* 2008;15:962–971.
141. Yao DF, Dong ZZ, Yao M. Specific molecular markers in hepatocellular carcinoma. *Hepatobiliary Pancreat Dis Int.* 2007;6:241–247.
142. Patel AM, Davila DG, Peters SG. Paraneoplastic syndromes associated with lung cancer. *Mayo Clin Proc.* 1993;68:278–287.
143. Marchioli CC, Graziano SL. Paraneoplastic syndromes associated with small cell lung cancer. *Chest Surg Clin N Am.* 1997; 7:65–80.
144. Charafe-Jauffret E, Monville F, Ginestier C, et al. Cancer stem cells in breast: Current opinion and future challenges. *Pathobiology.* 2008;75:75–84.
145. Das S, Srikanth M, Kessler JA. Cancer stem cells and glioma. *Nat Clin Pract Neurol.* 2008;4:427–435.
146. Zou GM. Cancer initiating cells or cancer stem cells in the gastrointestinal tract and liver. *J Cell Physiol.* 2008;217:598–604.
147. Hay ED. An overview of epithelio-mesenchymal transformation. *Acta Anat (Basel).* 1995;154:8–20.
148. Hugo H, Ackland ML, Blick T, et al. Epithelial—mesenchymal and mesenchymal—epithelial transitions in carcinoma progression. *J Cell Physiol.* 2007;213:374–383.
149. Turley EA, Veiseh M, Radisky DC, et al. Mechanisms of disease: Epithelial-mesenchymal transition—Does cellular plasticity fuel neoplastic progression? *Nat Clin Pract Oncol.* 2008;5:280–290.
150. Heimann R, Hellman S. Aging, progression, and phenotype in breast cancer. *J Clin Oncol.* 1998;16:2686–2692.
151. Singh Ranger G, Mokbel K. The evolving role of sentinel lymph node biopsy for breast cancer. *Eur J Surg Oncol.* 2003;29:423–425.
152. Shuster TD, Girshovich L, Whitney TM, et al. Multidisciplinary care for patients with breast cancer. *Surg Clin North Am.* 2000;80:505–533.
153. Amerasekera S, Turner M, Purushotham AD. Paget's "seed and soil" hypothesis revisited. *J Buon.* 2004;9:465–467.
154. Fokas E, Engenhart-Cabillic R, Daniilidis K, et al. Metastasis: The seed and soil theory gains identity. *Cancer Metastasis Rev.* 2007;26:705–715.
155. Piris A, Mihm Jr MC. Mechanisms of metastasis: Seed and soil. *Cancer Treat Res.* 2007;135:119–127.
156. Tisdale MJ. Wasting in cancer. *J Nutr.* 1999;129:243S–246S.
157. Beutler B. Cytokines and cancer cachexia. *Hosp Pract (Off Ed).* 1993;28:45–52.
158. Malik ST. Tumour necrosis factor: Roles in cancer pathophysiology. *Semin Cancer Biol.* 1992;3:27–33.
159. Tisdale MJ. Loss of skeletal muscle in cancer: Biochemical mechanisms. *Front Biosci.* 2001;6:D164–174.
160. Siddiqui R, Pandya D, Harvey K, et al. Nutrition modulation of cachexia/proteolysis. *Nutr Clin Pract.* 2006;21:155–167.
161. Bosaeus I. Nutritional support in multimodal therapy for cancer cachexia. *Support Care Cancer.* 2008;16:447–451.
162. Stewart AF. Clinical practice. Hypercalcemia associated with cancer. *N Engl J Med.* 2005;352:373–379.
163. Pivonello R, De Martino MC, De Leo M, et al. Cushing's Syndrome. *Endocrinol Metab Clin North Am.* 2008;37: 135–149, ix.
164. Carrier M, Le Gal G, Wells PS, et al. Systematic review: The Trousseau syndrome revisited: Should we screen extensively for cancer in patients with venous thromboembolism? *Ann Intern Med.* 2008;149:323–333.

Part

II

Concepts in Molecular Biology and Genetics

Chapter

5

Basic Concepts in Human Molecular Genetics

Kara A. Mensink ▪ W. Edward Highsmith Jr.

INTRODUCTION

Molecular diagnostics is the branch of laboratory medicine or clinical pathology that utilizes the techniques of molecular biology to diagnose disease, predict disease course, select treatments, and monitor the effectiveness of therapies. Molecular diagnostics is associated with virtually all clinical specialties and is a vital adjunct to several areas of clinical and laboratory medicine, but is most predominantly aligned with infectious disease, oncology, and genetics. The subject of this chapter is molecular genetics, which is concerned with the analysis of human nucleic acids as they relate to disease.

Since the completion of the first working draft of the human genome sequence in 2000 and the completion of the polished sequence in 2003, progress in molecular genetics has been swift and shows no signs of abating. Relatively few gene tests were clinically available in the late 1990s, whereas over 1,000 are available today. Further, molecular genetic testing has proven useful and robust enough to expand into population-based screening. Molecular testing serves as the final confirmatory test for several disorders included as part of expanded newborn screening programs, and in 2003, the *American Colleges of Medical Genetics and Obstetrics and Gynecology* recommended that population-based carrier screening for cystic fibrosis using molecular testing be implemented in the United States.

Molecular genetics as a discipline and as a clinical laboratory service does not exist in a vacuum. Rather, it is intimately tied to molecular and cell biology and the central paradigm of molecular biology—that genes code for proteins. Thus, it is through the analysis of genes that insight into the genesis of protein malfunction can be achieved. Such examination specifically entails an assessment of how the DNA sequence of a gene compares with its wild-type or normal sequence. Ultimately, protein malfunctions related to gene mutations lead to organ dysfunction and disease states. This chapter will review the fundamentals of molecular genetics, and is divided into five sections that review concepts intrinsic to molecular genetics. Where possible, comments on the direct clinical application of these concepts have been incorporated. The first section focuses on the molecular structure of DNA, DNA transcription, and protein translation. The second section focuses on molecular pathology, DNA replication, and DNA repair mechanisms. The third section provides a basic overview of transmission genetics. The fourth section highlights the relationship between genes, proteins, and phenotype and includes rationale for molecular genetic testing. The final section reviews allelic heterogeneity and corresponding choice of analytical methodology.

MOLECULAR STRUCTURE OF DNA, DNA TRANSCRIPTION, AND PROTEIN TRANSLATION

The human genome is composed of 3 billion base pairs of DNA. This is not present as one continuous piece of double-stranded DNA, but is distributed among 22 pairs of autosomal chromosomes and 2 sex chromosomes. The DNA is associated with a large number of proteins (histones and others) that serve regulatory functions and package the genetic material into these large chromosomal units. Chromosomes range in size from the 33.4 Mb of chromosome 22 to the 263 Mb of chromosome 1, which is the largest chromosome. Along the length of each chromosome, DNA is organized into linear domains consisting of genes (primarily nonrepetitive DNA), repetitive elements, and apparently functionless regions, much like beads on a string (Figure 5.1). Approximately half of the human genome consists of repetitive DNA, while the other half consists of nonrepetitive sequence. Nonrepetitive DNA includes regulatory sequences, intronic sequence, and protein coding (exon) sequence. Protein coding regions account for a relatively small fraction of genes within the human

Molecular Pathology © 2009, Elsevier, Inc. All Rights Reserved.

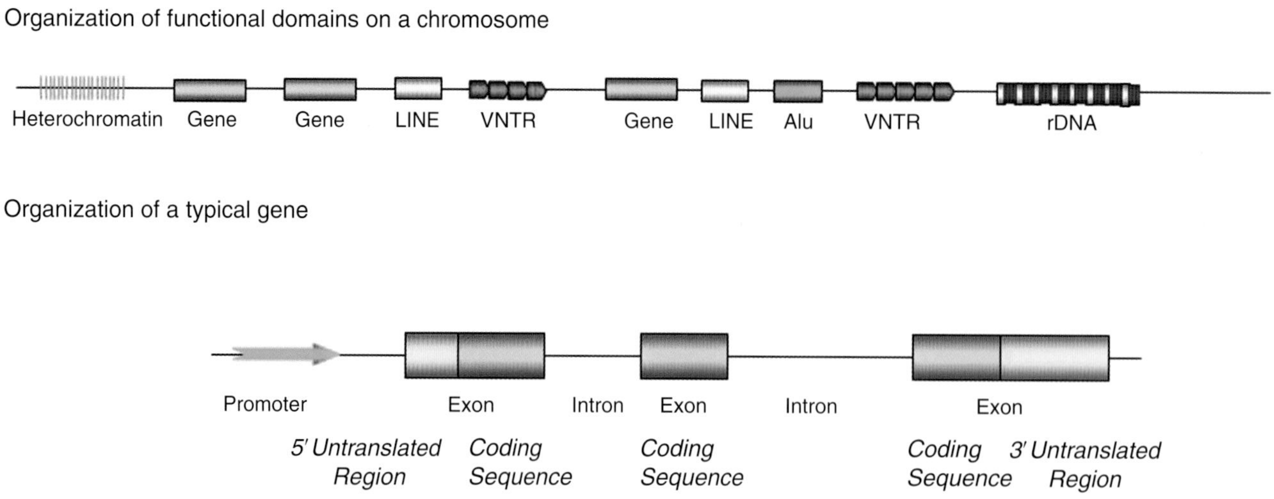

Figure 5.1 Top: The functional domains of a chromosome; Bottom: Organizational structure of a typical gene.

genome. In fact, it is estimated that only 6% of the human genome consists of protein coding, nonrepetitive DNA.

Repetitive DNA can be subdivided into several different categories or families. Generally, repetitive DNA tends to occur either in clusters of tandem repeats or as repetitive elements of various lengths dispersed throughout the genome. Clusters of tandem repeats can be localized to one or many locations. Such clusters are commonly referred to as satellite DNA. For example, α-satellite DNA is a clustered repetitive DNA sequence family that is localized to the centromeres of all human chromosomes. Repetitive sequences that are not localized to a particular area or areas of the genome are referred to as dispersed repetitive elements. The *Alu* family and LINE families are examples of dispersed repetitive elements.

Repetitive DNA is sometimes referred to as junk DNA because it does not code for an apparently active RNA transcript or functional protein. Although much remains to be discovered about the roles of so-called junk DNA, the label has been determined a misnomer. Appreciation for the role of junk DNA in protein folding and localization, DNA packaging and chromosome structure, and regulation of gene expression is increasing.

Genes are found among the nonrepetitive DNA in the genome. Genes code for specific protein chains, each with a specific function in cell physiology. A gene is composed of regulatory elements, which determine where, when, and how a gene is transcribed and coding regions, which are broken into segments, termed exons (expressed sequences). An example of a regulatory element is the promoter, which is the site where gene transcription is initiated. The exons are separated by noncoding regions of DNA called introns (intervening sequences). The average gene is about 2.7 kb (2700 bp) of DNA in length. The smallest gene, *H1A*, is located on chromosome 6 and encodes a histone protein which functions, along with several other histone proteins, to compact DNA in the cell nucleus. *H1A* is 0.5 kb (500 bp) long and has no introns. One of the largest genes, *DMD*, encodes the dystrophin protein and is located on the X chromosome (Xp21). Mutations in the *DMD* gene are associated with X-linked recessive Duchenne muscular dystrophy. *DMD* measures 2.4 Mb (2,400,000 bp), consists of 79 exons, and takes at least 16 hours to transcribe. The size of a gene may influence the molecular diagnostic laboratory's ability to design a clinical test for a particular disorder and certainly impacts the selection of the technology used to detect mutations.

Chemically, genes are composed of 2-deoxyribonucleic acid (DNA). DNA is a linear, nonbranching polymer of nucleotides. Repeating ribose and phosphate subunits form a backbone; and attached to each of the ribose moieties is a purine (adenine, guanine) or pyrimidine (thymine or cytosine) base. Following standard nomenclature for the naming of ring containing compounds, the nitrogenous bases have their various carbon and heteroatom components numbered 1–6 (for the pyrimidines) or 1–9 (for the purines) and the ribose positions are indicated by numbers 1′–5′. The bases are attached to the ribose subunits at the 1′ position of the sugar molecule. The ribose subunits are joined by phosphodiester linkages between the 5′ position of one ribose to the 3′ position of the next (Figure 5.2). Thus, the molecule is not symmetrical and there is directionality implicit in a DNA strand. There is a 5′ end of a DNA strand and a 3′ end. Two DNA strands bind together to form the familiar double helical structure of double-stranded DNA (Figure 5.3). In order for a double helix to be stable, there must be a complementary base on the opposite strand for every base on a strand of DNA. The complementary pairs of bases are adenosine and thymine (A:T) and guanine and cytosine (G:C). The two strands join in an antiparallel fashion (one strand runs 5′ to 3′ and the other 3′ to 5′). The ribose sugars form the scaffolding for the complementary nitrogenous bases connected by hydrogen bonds on the inside of the molecule. The DNA double helix is dynamic, and the weak hydrogen bonding between complementary bases allows for the DNA strands to easily denature and reassociate with themselves. In the laboratory, the process of separating (denaturing) double-stranded DNA and then allowing

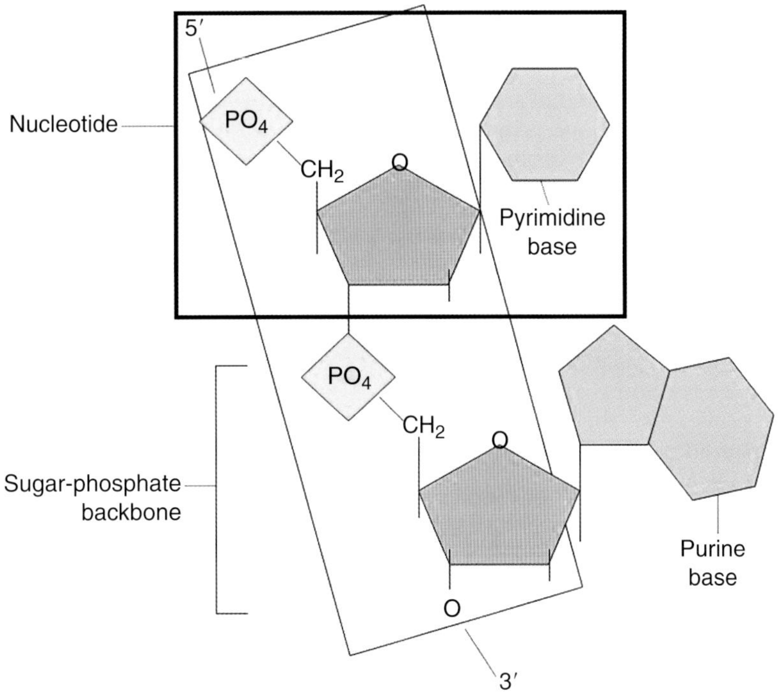

Figure 5.2 **Schematic view of nucleotide structure and how nucleotides join to form the DNA polymer.**

the complementary single strands of DNA to reassociate and return to a double-stranded configuration is called hybridization. The basis of many of the laboratory techniques central to molecular diagnostics hinge on hybridization and the remarkable specificity of a nonrepetitive sequence of bases that make up a single strand of DNA to bind to its complementary sequence and no other. *In vivo*, the denaturing and reassociation of double stranded DNA is inherent to the process of gene transcription.

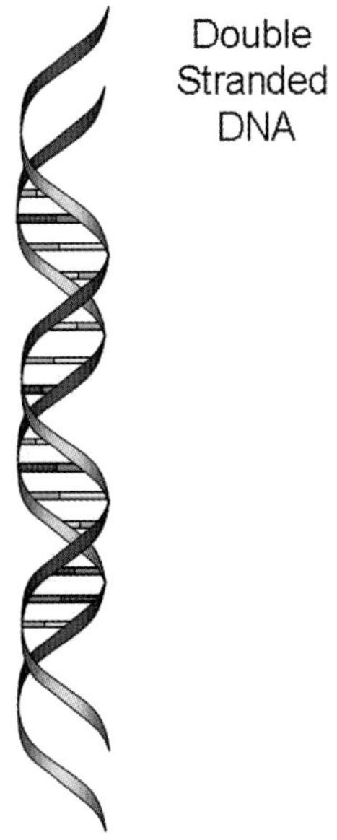

Figure 5.3 **Schematic view of the double helical structure of double stranded DNA.** Blue ribbons represent the sugar-phosphate backbone. Green/yellow and pink/lavender links represent complementary purine/pyrimidine pairs.

Transcription of DNA

Transcription is the first process in the cascade of events that lead from the genetic code contained in DNA to synthesis of a specific protein. The product of gene transcription is ribonucleic acid (RNA). The structure of RNA is similar to DNA, with three exceptions. First, the ribose sugar of RNA has two hydroxyl groups at the 2′ and 3′ carbons. Second, the base uracil (U) replaces thymine (T). And third, most RNA molecules are single rather than double stranded. There are four general types of RNA (Table 5.1). Ultimately, the specific type of RNA that results from the transcription of a structural gene is messenger RNA (mRNA). Transcription of DNA

Table 5.1 Five Types of RNA

Type of RNA	Summary
mRNA	The transcript product of a structural gene that encodes an amino acid sequence.
tRNA	Transfer RNA molecules recognize codons of mRNA and facilitate incorporation of each successive amino acid during protein synthesis.
rRNA	Integral component of the ribosomal machinery used for translating DNA transcript into protein.
Small RNA (example: snRNA)	Many small RNA molecules exist, and each has different functions in RNA modification. For example, snRNA assists with splicing intron transcripts out of precursor mRNA.
microRNA	A newly described species of RNA involved in gene regulation.

into RNA is catalyzed by RNA polymerase. RNA polymerase consists of multiple subunits that work together to recognize where the transcriptional complex should assemble, synthesize the RNA single-stranded transcript, and dissociate from the DNA template once synthesis is complete. Under the influence of the gene promoter, various transcription factors are attracted to the upstream (5′) end of the gene. The transcription factors recruit the RNA polymerase and initiate transcription of the coding region of a gene into RNA. Simultaneous reading of the DNA template (antisense strand) and elongation of the RNA product by the RNA polymerase complex proceeds in the 5′ to 3′ direction. Elongation ceases when the RNA polymerase complex recognizes the DNA terminator sequence and disassociates from the primary mRNA transcript and the double-stranded DNA. The mRNA transcript is complementary to the antisense strand and a replicate (with the exception of uracil replacing thymidine) of the sense strand (Figure 5.4A).

Once the primary RNA sequence has been synthesized, the RNA transcript requires modification for stability and translational efficiency. The primary transcript (precursor mRNA or pre-mRNA) contains both the coding (exon) and noncoding (intron) sequences, and the intron material has to be removed prior to translation and protein synthesis. Sequences flanking the exons, the donor and acceptor splice sites, recruit a series of proteins that remove the introns from the transcript and splice the exons together to form the mature mRNA (Figure 5.4B). Additional post-transcriptional modification includes the attachment of 7-methylguanosine CAP to the 5′ end of the mRNA and the addition of a polyA tail that consists of a variable number (usually 80–250) of adenine nucleotides at the 3′ end of the mRNA. Both the CAP and the polyA tail are thought to help stabilize the mRNA molecule, assist with its transport out of the nucleus into the cytoplasm, and may also help to regulate translation of mRNA into protein. Once splicing has occurred and the CAP and polyA tail have been added, RNA modification is complete and the mature mRNA transcript is exported to the cytoplasm for translation into protein (Figure 5.4B).

Protein Translation

After the mature mRNA transcript is transported to the cytoplasm, it is translated into protein by the ribosomes in the endoplasmic reticulum. The multiplicity of function performed by the various units of ribosomal machinery in concert to achieve translation of the DNA transcript (mRNA) and protein synthesis is elegantly complex. Ribosomes consist of two multiprotein subunits, each with an RNA component (rRNA) and several active centers.

Recall that mature mRNA essentially represents only the exonic or coding regions of a given gene. The base sequence within these coding regions is grouped into informational units of three bases, called codons. Each codon either codes for a specific amino acid or serves a regulatory function, such as stopping or starting protein chain synthesis. To initiate the process of translating mRNA into protein, the small ribosome subunit binds to mature mRNA at the CAP site and scans the mRNA sequence for its start codon, which is AUG. After the AUG codon is recognized, the large ribosome subunit binds a specific aminoacyl-tRNA, Met-tRNA, and the process of protein synthesis begins. An aminoacyl-tRNA (referred to as a charged tRNA) is an RNA molecule with an anticodon complementary to the mRNA codon that carries a specific amino acid. The specific amino acid that each charged tRNA carries is determined by the mRNA codon and is associated with the tRNA anticodon. As the ribosome translocates itself along mRNA in a 5′ to 3′ direction, it catalyzes the successive binding of charged tRNAs to their associated mRNA codons. The ribosome catalyzes the chemical joining of amino acids together by creating peptide bonds between the amino and carboxyl groups of each successively added amino acid (Figure 5.4C). It is this flow of genetic information (DNA transcription to RNA and RNA translation to protein) that is termed the central dogma (or paradigm) of molecular biology.

MOLECULAR PATHOLOGY AND DNA REPAIR MECHANISMS

Mutation and Genetic Variation

There is no single sequence of the human genome. Although the entire genome sequence from any given human is approximately 99.9% identical to the genome sequence of any other individual human, there are on the order of 3 million sequence variations between any two unrelated persons. It is the similarity of the genomes between individuals that defines them as human beings, and it is the differences that distinguish individuals. Although the majority of the sequence differences between individuals likely have no biological importance and do not contribute to physiological or observable differences, many clearly do have subtle effects and give rise to the remarkable diversity of the human race.

A large number of genetic variations occur at measurable frequencies in the population. Such variations are termed polymorphisms. Although often used to denote a nonpathogenic variation, the strict definition of the term polymorphism is a variation that is present at a frequency of 1% or greater in the population. The most common type of sequence variation is a difference between single nucleotides at a particular place in the genome, or locus. For example, at a certain position, one individual may have a thymine residue, whereas another may have a cytosine. This type of variation is termed a single nucleotide polymorphism (SNP). To date, over 10 million different SNPs have been characterized and are emerging as extremely useful tools for understanding genetic diversity and localizing disease genes. Another type of polymorphism involves not the substitution of one nucleotide for another, but variation in the number of copies of a string of nucleotides. One of the most common of this type of variation is the variation of the number of

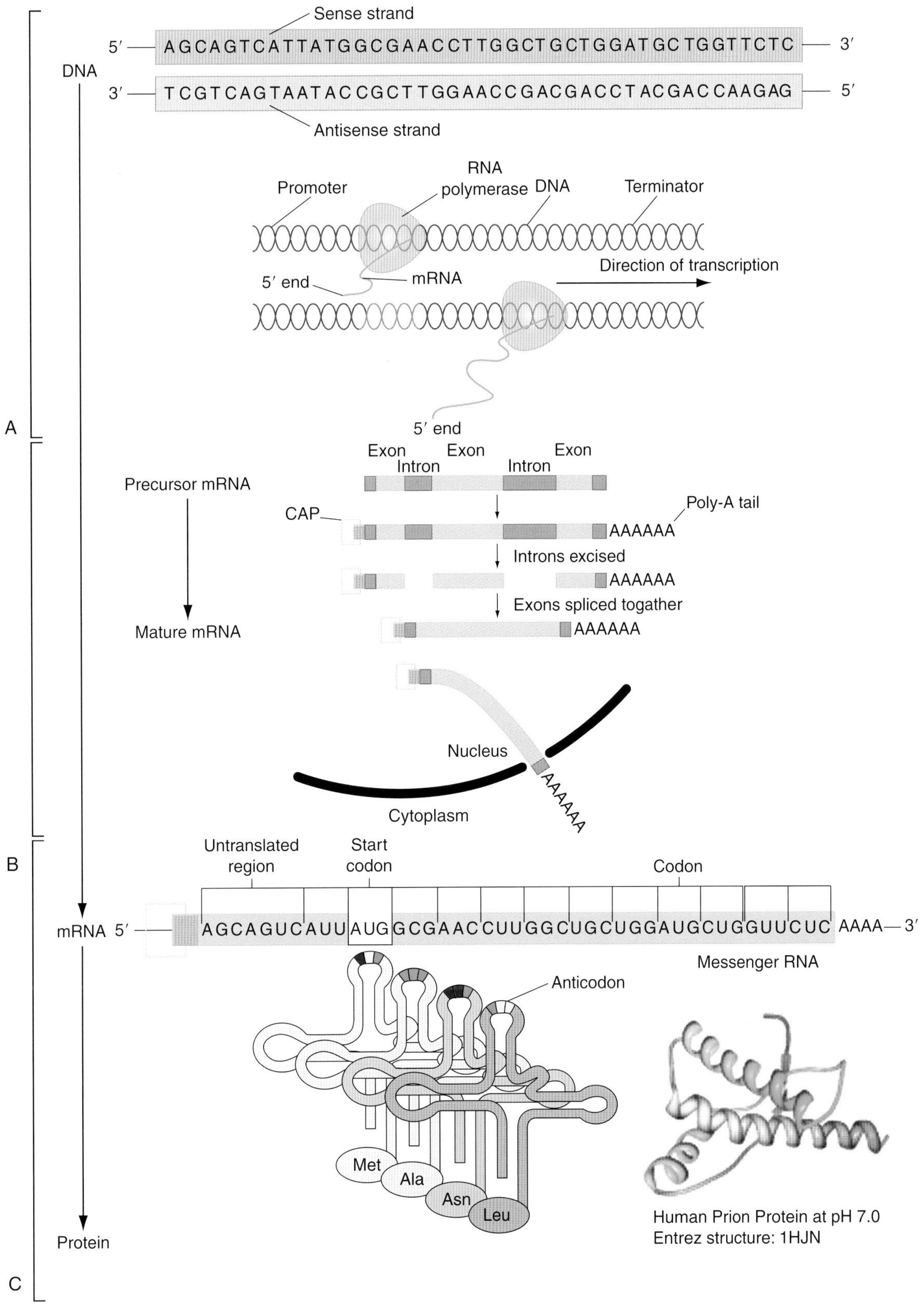

Figure 5.4 **Panel A: Top- Two strands of DNA illustrating the complementary bases that link to form the double stranded DNA molecule. Bottom- Schematic of DNA transcription with initiation at the promoter, elongation of the mRNA product as the RNA polymerase translocates in the 5′ to 3′ direction, and termination at the termination codon. Panel B: Processing of mRNA illustrating the addition of the CAP and poly-A tail, splicing out of intron sequences, and transport of the mature mRNA molecule from the nucleus to the cytoplasm. Panel C: Using the human prion protein as an example, this schematic illustrates the translation of mature mRNA to protein. Translation begins at the start codon (AUG) with protein synthesis in the 5′ to 3′ direction.** The structure of the charged tRNA molecule can be appreciated as each complementary tRNA anticodon recognizes its corresponding mRNA codon. The final result is a representation of the folded human prion protein as referenced in Entrez Structure http://www.ncbi.nim.nih.gov/sites/entrez?dp=structure (October, 2008).

copies of a repetitive sequence at a given locus. When the length of the repetitive unit is small (1 to tens of nucleotides), this type of polymorphism is termed a simple sequence repeat, or STR. When the length is longer, hundreds to thousands of nucleotides, they are termed variable number of tandem repeats, or VNTRs. STRs are a very common source of genetic differences between individuals and have been important in gene mapping studies. Currently, due to the high rate of heterozygosity, forensic laboratories utilize STR analysis extensively. Another type of polymorphism that has recently become appreciated through the use of comparative genome hybridization microarrays (CGH arrays) involves the deletions and duplications of regions of the genome. These regions can be quite large, up to several million bases in length, and may include genes. The role of these copy number variants in human variation and disease is not yet understood.

A genetic mutation is a sequence variant that has a pathogenic effect. Some mutations are relatively common in the population, and meet the 1% population frequency criteria to be formally termed polymorphic; the cystic fibrosis mutation deltaF508 in the Northern European population and the sickle cell anemia mutation in the African populations are examples. Some mutations are very rare in the population. Not infrequently, mutations are found that affect a single family. These very rare mutations are termed private mutations. Pathogenic (disease-causing) mutations often involve changes in the base sequence that composes a codon (or coding unit). However, mutations can occur in regulatory elements such as splice sites and promoter regions as well. A mutation occurring in the portion of DNA that codes for a protein can result in (i) a change of one amino acid to another, (ii) a change in an amino acid codon to one coding for a termination signal (stop), or (iii) no change in the amino acid at that position. These types of changes are termed missense, nonsense, and silent mutations, respectively. A missense change can result in no change in the function of the protein, a total loss of function, a partial loss of function, or a change of function. Partial or total loss of function usually results in a pathological state, as does a change in function. The pathogenic effect of a loss of function of a gene product can be direct, such as the loss of chloride channel function that causes cystic fibrosis, or indirect, such as the loss of function of regulators of gene expression that can result in cancers. Examples of mutations that cause changes in function include those that cause constitutive activation of a function that is normally under regulation by the cell.

It is important to note that not all losses (even complete losses) of protein function lead to an abnormal phenotype or disease. For example, the common (10% allele frequency in the Caucasian population) 32 base-pair deletion of the *CCR5* gene (CC motif chemokine receptor 5) has been associated with reduced susceptibility to infection with HIV, but even homozygotes (approximately 1% of the Caucasian population), with no CCR5 protein, do not demonstrate any observable effects on normal physiology.

Other types of mutations that can occur include deletions and insertions of nucleotides in and surrounding coding regions. These types of mutations, often abbreviated indels, can be small, from one to a few dozen bases, or large, covering large segments of chromosomes and including multiple genes. Since codons consist of a trio of bases, if a small indel occurs within the coding region of a protein which contains a number of bases that is divisible by three, the indel is said to be in-frame, as it will not shift the reading frame of the mRNA being translated into protein in the ribosome. If, on the other hand, the number of bases is not divisible by three, the indel will alter the reading frame of the protein and will typically result in the ribosome encountering a stop codon within a few dozen bases. Larger indels can involve whole exons, multiple exons, whole genes, or even multiple genes.

Abnormal expression of genes can also result from changes to the chemical structure of genes that are not a result of a change of the DNA sequence. Methylation of the nucleotide bases is a postsynthetic modification to DNA that affects the expression of genes. Abnormal patterns of DNA methylation can cause abnormal gene expression (transcription) and disease states. Repeat base sequences that have no apparent informational content regarding protein structure exist throughout the DNA. The expansion of the number of repeats in a gene has been associated with specific diseases, and this change in the gene structure is inheritable.

DNA Replication

DNA is synthesized as part of the DNA replication process that occurs during the S phase of the mitotic cell cycle and the first phase of meiosis. The DNA replication process involves multiple specialized enzymes (Table 5.2) that work together to synthesize two double-stranded daughter strands from one double-stranded parent strand. DNA polymerase synthesizes the daughter strands. The replication fork is the site at which

Table 5.2 DNA Replication Enzymes

Replication Enzymes	Function
Helicase	Breaks hydrogen bonds linking the two strands of the DNA double helix.
Topoisomerase	Mitigates the supercoiling effect that occurs in advance of the replication fork.
Single-strand binding protein (SSBP)	Acts as a retractor, preventing the single strands of the DNA double helix from rejoining.
RNA primase	Synthesizes the RNA primer that is required to initiate synthesis of the new daughter strands.
DNA polymerase	Synthesizes DNA daughter strands. Certain DNA polymerases also act as part of the DNA repair machinery.
DNA ligase	Links newly synthesized DNA fragments (Okazaki fragments).

double-stranded DNA is separated and DNA polymerase synthesizes the new daughter strands. Ahead of the replication fork is the parent double-stranded DNA; behind the replication fork are the newly synthesized daughter strands. There are no DNA polymerases that synthesize in the 3′ to 5′ direction. Thus, the DNA replication process is referred to as semidiscontinuous because, while one of each of the two new daughter strands is able to be replicated continuously in the 5′ to 3′ direction (leading strand), the other strand (lagging strand) must be copied in short 5′ to 3′ segments (Okazaki fragments) that are 100–1000 nucleotides in length. Okazaki fragments are joined together by the action of a ligase enzyme to complete the lagging strand. The fidelity of the replication is estimated to approach 99.98%. In the rare case that DNA replication incorporates an incorrect base, DNA proofreading and repair systems work to correct the error and prevent detrimental consequence.

DNA Repair

In the broadest sense, DNA repair mechanisms work to correct, or in some way mitigate, the effects of DNA replication inaccuracy and exogenous or endogenous genetic insult. Generally, when the integrity of wild-type DNA is compromised, the error is either corrected, overlooked, or programmed cell death occurs. A number of DNA repair pathways are known and can be roughly characterized into the following functional categories: (i) direct reversal, (ii) excision repair, and (iii) DNA double-strand break repair. A brief description of each follows. Although they are often studied separately, it is impossible to completely separate one from another because the various mechanisms are highly interconnected and act cooperatively as part of a large cellular arsenal with the common goal of genome integrity maintenance.

Direct Reversal of DNA Damage

Correction of DNA damage by direct reversal is a type of DNA repair that predominantly involves action by a single enzyme repair system. Consider the enzymatic photoreactivation (EPR) reaction that works to repair damage induced by ultraviolet (UV) light. The formation of pyrimidine dimers (most commonly thymidine) is one type of pathologic cellular response to excess UV exposure. When present, the bulky pyrimidine dimers impede the DNA replication and transcription process. In a relatively simple light dependent reaction, DNA photolyase acts to restore the pyrimidines to their correct monomer conformation. Direct reversal by DNA photolyase is not the only way the cell responds to UV damage. In fact, cellular response to UV damage also commonly involves one or more of the excision repair mechanisms.

Excision Repair of DNA Damage

Correction of DNA damage by excision repair involves groups of proteins that act together to excise the incorrect base(s) or nucleotide(s), replace them with the correct sequence, and ligate the corrected strand back together. There are three DNA excision repair systems: mismatch repair (MMR), base excision repair (BER), and nucleotide excision repair (NER). Generally, these excision repair mechanisms can be distinguished by considering the context within which the error occurs, whether removal involves a base(s) or nucleotide(s), and the number of bases(s)/nucleotide(s) removed. Each type of excision repair system also invokes the use of unique proteins. For example, the protein complexes that work to recognize DNA mismatch in the MMR mechanism (MSH2/MSH6 and MSH2/MSH3) are different than the protein complexes (XPC-RAD23B and UV-DDB) that recognize and invoke global genome nucleotide excision repair (GG-NER).

Mismatch Repair of DNA Damage DNA replication inaccuracy is the context within which the mismatch repair (MMR) pathway preserves genomic integrity. The primary purpose of MMR is to prevent mutations accrued during the DNA replication process from propagating and becoming the start of a mutant lineage by recognizing and excising the mismatched nucleotide, resynthesizing DNA, and ligating the broken strand back together. Germline mutations in genes coding for MMR proteins Lynch syndrome.

Base Excision Repair of DNA Damage Base excision repair (BER) involves the excision of a single base rather than the nucleotide and is most commonly used to repair damage caused by endogenous DNA insult and is especially important for cellular response to oxidative DNA damage. BER involves removing the base from the deoxyribose-phosphate chain by a specific glycosylase, endonuclease action, DNA polymerase Beta, and either DNA ligase I or DNA ligase III/XRCC1 complex.

Nucleotide Excision Repair of DNA Damage Nucleotide excision repair (NER) is predominantly invoked in response to genomic damage caused by UV exposure. NER involves the excision of an oligonucleotide, rather than a single base (BER) or single nucleotide (MMR). It is also a substantially more complex process that includes at least 30 different proteins. Two subpathways of NER, termed global genome repair NER (GG-NER) and transcription-coupled NER (TC-NER), have been recognized. Typically, GG-NER is used when errors occur in nontranscribed areas of the genome and TC-NER, as the name implies, corrects errors that occur in areas of active gene expression. Mutations in NER genes are associated with disorders such as xeroderma pigmentosa (XP) and Cockayne syndrome.

DNA Double-Strand Repair of DNA Damage

DNA double-strand repair is an important DNA repair mechanism that uses a number of proteins, many of which are similar to or the same as those used during meiotic recombination. DNA double-strand breaks (DSBs) can result from a number of exogenous and endogenous agents including ionizing radiation exposure, chemical exposure, and somatic DNA recombination or transposition events. Nonhomologous end-joining (NHEJ) and homologous recombination (HR) are the two primary DNA double-strand repair mechanisms. In addition to its role as a DNA repair mechanism, NHEJ helps to maintain B-cell and T-cell diversity and subsequently a healthy immune system, by correcting intentional breaks created during V(D)J recombination. Short homologous sequences (microhomologies) found on the single-stranded tails of the broken DNA are used to help rejoin the strands in NHEJ, whereas HR relies on homologous (or very close to homologous) sequence to repair the broken strands. HR is typically used when DNA replication is halted due to a single-strand break or another unrepaired lesion that causes collapse of the replication fork. Because it uses a homologous or near homologous template, HR is often thought to be more accurate than its NHEJ counterpart. However, both mechanisms show high accuracy, as well as imperfection.

MODES OF INHERITANCE

A detailed family history provides the foundation for genetic diagnosis and risk assessment. Visually recorded using standardized symbols and nomenclature [1], the pedigree provides the tool by which inheritance patterns are elucidated, and subsequent risk assessment is calculated. In addition to diagnosis and risk assessment, the pedigree is a powerful research tool, aiding in the discovery of new genes and helping to better understand the phenotypic expression of genes already discovered. Observations made from controlled monohybrid and dihybrid crosses of peapod plants in the 1860s formed the foundation for Gregor Mendel's landmark laws of heredity that still govern basic pedigree interpretation today. Since that time, much has been learned and the study of inheritance is now far more complex than Mendel himself may have imagined. This section of the chapter reviews modes of inheritance and factors that may influence pedigree interpretation.

Mendelian Inheritance

The concepts that two copies of a gene segregate from each other (law of segregation) and are transmitted unaltered (particulate theory of inheritance) from parents to their offspring help to explain the concepts of dominant and recessive traits. When the presence of one copy of a particular allele results in phenotypic expression of a particular trait, the trait is dominant. When two copies of a particular allele must be present for the phenotypic expression of a trait, the trait is recessive. Note that it is the phenotypic expression that is described as dominant or recessive, not the allele or gene itself. Thus, patterns of inheritance are distinguished by where the gene resides within the genome (autosome or sex chromosome) and whether or not phenotypic expression occurs in the heterozygous or homozygous state. Traditionally recognized Mendelian patterns of inheritance include autosomal dominant, autosomal recessive, X-linked recessive, X-linked dominant, and Y-linked (holandric). When each of these inheritance patterns is represented in a pedigree diagram, distinguishing features can be visually recognized (Table 5.3).

Autosomal Dominant Inheritance

Autosomal dominant inheritance is designated when no difference in phenotypic expression is observed between heterozygous and homozygous genotypes. Visually, the autosomal dominant pedigree shows multiple affected generations in a vertical pattern, an equal distribution of males and females affected, and both males and females transmit the phenotype (including males transmitting the phenotype to other males). Typically, dominant disorders occur when a mutation confers an inappropriate activity on a gene product. Examples include Huntington disease and other polyglutamine disorders where expansion of a triplet repeat with in a polyglutamine tract causes cellular toxicity, or familial amyloidosis where mutant transthyretin protein is relatively unstable and deposits as amyloid in tissues. However, some dominant disorders, such as those involved in the majority of the inherited cancer syndromes, occur with inheritance of a single copy of a gene where the mutant copy has not acquired a novel, pathogenic function, but is inactivated. Further, the great majority of cells and tissues carrying single copies of these mutant genes are functionally normal. The resolution of this apparent paradox came when, after observing familial cases of bilateral retinoblastoma (RB) and comparing those cases with sporadic (nonfamilial) cases of unilateral RB, Knudson proposed that two hits or mutational events were needed for the initiation of tumor growth. In the case of familial RB, a germline mutation in one tumor suppressor allele was postulated, and there was a much higher probability of tumor initiation because the individual was born with one hit (mutation). Therefore, somatic mutations that hit or render nonfunctional the remaining normal allele would be tumorigenic. In contrast, in a normal individual (one not carrying a mutant *RB* gene in the germline), that same somatic event would not lead to the initiation of a tumor because one functional allele would remain. Tumor initiation takes place in a normal individual only if two somatic events occur at the same locus. As tumor initiation is not observed with a single abnormal allele, it is said that tumor suppressor genes act as recessive alleles at the cellular level, but as dominant disorders at the organism level. Additional examples of dominant disorders associated with inactivating mutations in tumor suppresser genes include Lynch syndrome (inactivation of one of the mismatch repair genes *MLH1*, *MSH2*, or *MSH6*) or familial breast cancer (inactivation of *BRCA1* or *BRCA2*).

Autosomal Recessive Inheritance

Autosomal recessive inheritance is designated when phenotypic expression is observed only when both copies of a gene are inactivated or mutated. Visually,

Table 5.3 Mendelian Inheritance Patterns

Inheritance Pattern	Example Pedigree	Clinical Example
Autosomal dominant	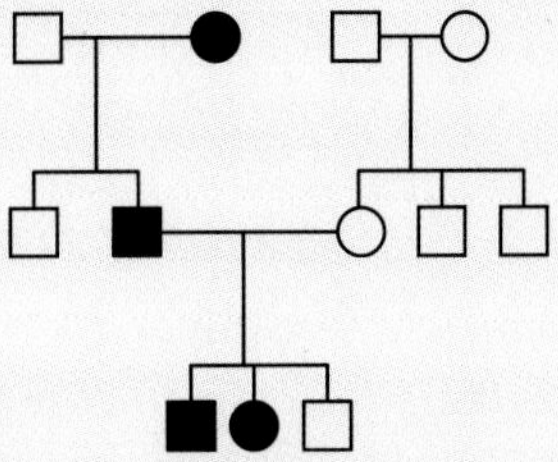	Huntington disease Myotonic dystrophy Retinoblastoma Lynch syndrome Neurofibromatosis I TTR associated-amyloidosis and many others
Autosomal recessive	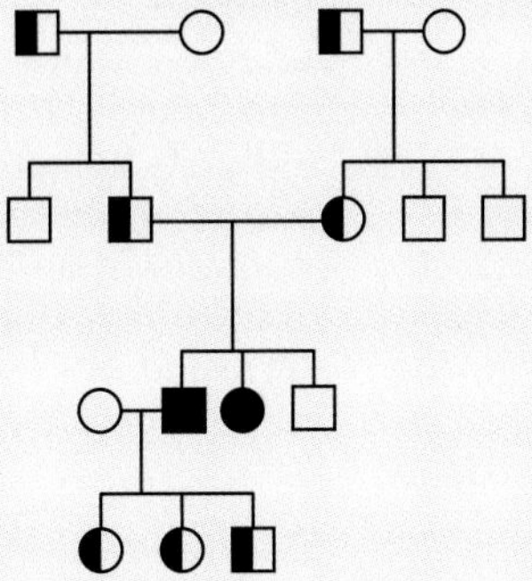	Cystic fibrosis Galactosemia Autosomal recessive (AR) deafness AR epidermolysis bullosa Tay-Sachs disease Klippel-Feil syndrome and many others
X-Linked recessive	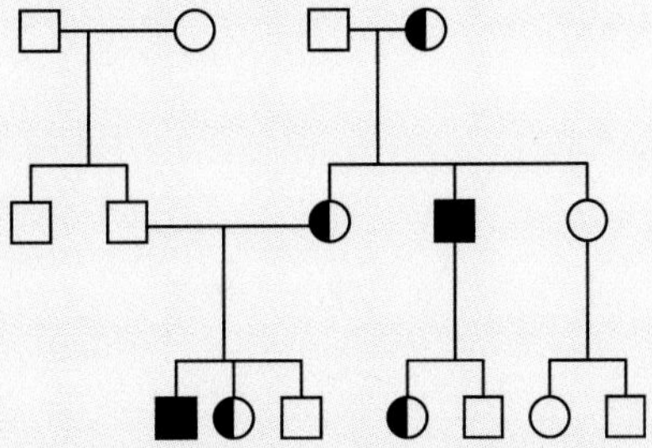	Duchenne muscular dystrophy Hemophilia A X-linked ichthyosis X-linked mental retardation Opitz syndrome Emery-Dreifuss muscular dystrophy And many others
X-Linked dominant	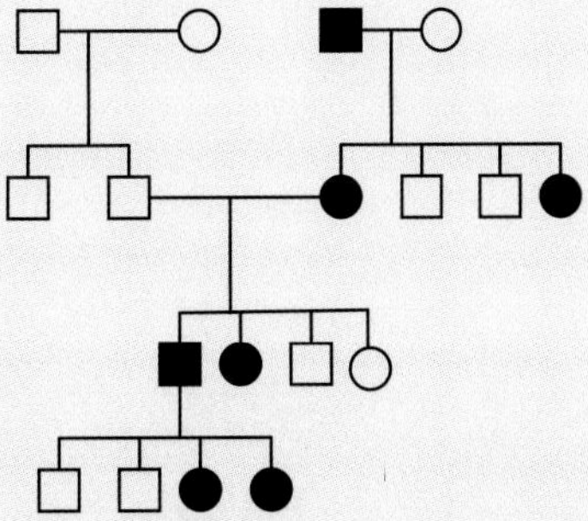	Vitamin D-resistant rickets Coffin-Lowry syndrome and others
Y-Linked (Holandric)	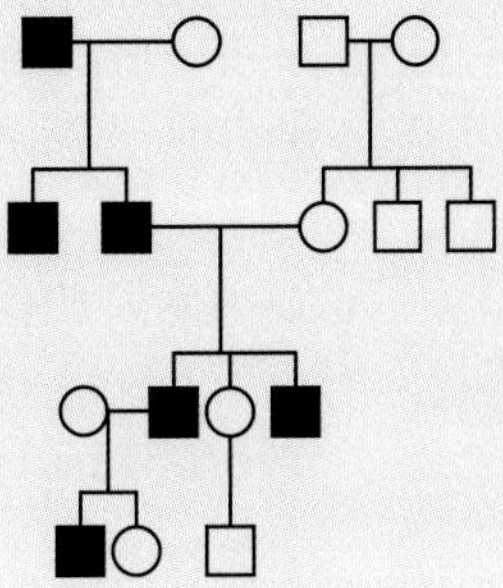	Hairy ears Y-linked deafness Very few others

the autosomal recessive pedigree typically shows a horizontal pattern where multiple affected individuals can be observed within the same sibship, and an equal number of males and females are affected. In instances of autosomal recessive inheritance, each parent of an affected individual has a heterozygous genotype composed of one copy of the mutated gene and one copy of the normal/functional gene. When a pedigree is analyzed, individuals who must be genetic carriers of the disorder in question, such as parents of an affected child, are termed obligate carriers. Other individuals in the pedigree may be at risk for being carriers. The risk to be a carrier is defined by each individual's position in the pedigree relative to affected individuals, or known carriers. For example, a sibling of an obligate carrier of sickle cell anemia has a 50% probability of being a sickle cell carrier, while a first cousin of a cystic fibrosis patient has a 25% chance of being a carrier.

Typically in recessive disorders, having only a single copy of a mutant gene is insufficient for manifestation of disease. Alternatively stated, one copy is enough for normal cellular and tissue function. Examples of recessive disorders include enzyme deficiencies, such as galactosemia (galactose-1-phosphate uridyl transferase) or phenylketonuria (phenylalanine hydroxylase), or deficiency of transport proteins, such as cystic fibrosis (CFTR). Consanguinous mating or mating between related individuals increases the risk of autosomal recessive phenotypic expression for certain genes because the proportion of shared genes among offspring is increased.

X-Linked Recessive Inheritance

X-linked recessive inheritance is designated when phenotypic expression is observed predominantly in males of unaffected, heterozygous mothers. All female offspring of affected males are obligate carriers. Visually, the pedigree typically shows a horizontal pattern of affected individuals with no instance of direct male-to-male transmission. However, males may transmit the disorder to a grandson through carrier female daughters.

It is not uncommon for X-linked recessive disorders to appear in a family such that before a certain generation the disease is not apparent, but is observed to be segregating in the family after that generation. This phenomenon is due to new mutations appearing *de novo* in an individual. This was explained by the American geneticist Haldane, and his theory is referred to as the Haldane hypothesis. If the reproductive fitness of a male affected with an X-linked recessive disorder is low or nil, then in a population one-third of all affected X chromosomes will be removed from the gene pool every generation. An example of decreased reproductive fitness among males is Duchenne muscular dystrophy. If the incidence of the disease is constant, then one-third of cases must be due to mutations arising *de novo* in a family.

No doubt the most famous family to be afflicted with an X-linked recessive condition is the House of Saxe-Coburg and Gotha, the British Royal family. Queen Victoria, apparently a carrier of a new hemophilia A mutation, had one affected son, Prince Leopold, and two daughters who were carriers of the disease. The daughter of Princess Alice, Princess Alix, was married to Tsar Nicholas II of Russia and the mother of the affected Tsarovich Alexei. The current royal family, the House of Windsor, is descended from Queen Victoria through an unaffected male, King Edward VII, so that branch of the family does not carry the hemophilia A mutation.

X-Linked Dominant Inheritance

X-linked dominant inheritance is designated when phenotypic expression is observed predominantly in females (ratio of about 2:1) and all daughters of affected males are affected and none of the sons of affected males are affected. Visually, the pedigree typically shows a vertical pattern of affected individuals, with no instance of direct male-to-male transmission. X-linked dominant conditions are substantially less common than X-linked recessive disorders. An example is X-linked, vitamin D-resistant rickets, which is caused by mutations in the *PHEX* (phosphate-regulating endopeptidase homolog, X-linked) gene located at Xp22.

X-Linked Dominant Male Lethal Inheritance

X-linked dominant male lethal inheritance is designated when phenotypic expression is observed only in females. Visually, the pedigree typically shows a vertical pattern with an increased rate of spontaneous abortion and where approximately 50% of the daughters from affected mothers are also affected. Although the great majority of cases of Rett syndrome are not familial (but sporadic), familial cases have been described and would be classified as an X-linked dominant male lethal disorder. Rett syndrome is caused by mutations in the *MeCP2* gene (methyl CpG binding protein 2). Rett syndrome is a neurodevelopmental disorder characterized by arrested development between 6 and 18 months of age, regression of acquired skills, loss of speech, stereotypical hand movements, microcephaly, seizures, and mental retardation. Affected males rarely survive to term, and the majority of affected females do not reproduce.

Y-Linked or Holandric Inheritance

Y-linked or Holandric inheritance is designated when phenotypic expression is observed only in males with a Y chromosome. Visually, the pedigree shows only male-to-male transmission. Hairy ears are an example of a Y-linked trait. Few disease states have been shown to be Y-linked. However, there is one report of a multigenerational Chinese family with Y-linked deafness.

Non-Mendelian Inheritance

Epigenetic Inheritance—Imprinting

When the phenotypic expression of a gene is essentially silenced dependent on the gender of the transmitting parent, the gene is referred to as imprinted. The phenomenon of imprinting renders the affected genes functionally haploid. In the case of imprinted genes, the functional haploid state disadvantages the imprinted gene because the gene is more susceptible to adverse effects of uniparental disomy, recessive mutations, and epigenetic (like DNA methylation-dependent gene silencing) defects. Visually, pedigrees that represent imprinting may appear similar to autosomal recessive or sporadic pedigrees and show a horizontal pattern. Imprinting disorders may also appear autosomal dominant and show a grandparental effect in the case of imprinting center mutations. Males and females are equally affected, and transmission is dependent on the gender of a parent. The two most well-known imprinting disorders are Prader-Willi and Angelman syndrome. Prader-Willi syndrome is caused by an absence of paternally contributed 15q11–13 (PWS/AS) region, whereas Angelman syndrome is caused by an absence of maternal contribution at the same locus. In the case of PWS, lack of paternally contributed genes at 15q11–13 (regardless of mechanism) results in unmethylated and overexpressed genes in this region. The same is true for Angelman syndrome. However, it is a lack of maternally contributed genes that causes the phenotype in this instance. The clinical phenotype of each disorder is distinct, but both are associated with mental retardation.

Inheritance Through Mitochondrial DNA

The inheritance of mitochondrial disease is complicated by the fact that mitochondrial disease can be either the result of mutations in nuclear DNA (nDNA) and thereby subject to the Mendelian forms of inheritance described previously or the result of mutations in organelle-specific mitochondrial DNA (mtDNA). Since the mitochondrial genome is maternally inherited, pedigrees demonstrating mitochondrial inheritance show an affected mother with all of her offspring (male and female) affected. The common phenomenon of heteroplasmy, where mtDNA mutations are present in only a portion of the mitochondria within a cell, can make laboratory analysis and clinical assessment difficult. It is estimated that only 10–25% of all mitochondrial disease is the result of maternally inherited mutations in the mitochondrial genome. Therefore, mitochondrial disease should not always be equated with mitochondrial inheritance.

Multifactorial Inheritance

Sorting out whether a particular phenotype is predominantly the result of inherited genetic variation, environmental influence, or some combination therein can be difficult. When the combined effects of both inherited and environmental factors cause disease, the disorder is said to exhibit multifactorial inheritance. Multifactorial inheritance is associated with most, if not all cases of complex, common disease (cancer, heart disease, asthma, autism, mental illness, and others). Typically, multiple loci or multiple genes are associated with the same complex disease phenotype. Such genetic heterogeneity works additively, such that the net effect of multiple mutations in multiple genes exacerbates and/or detracts from a particular clinical phenotype.

Sporadic Inheritance

Sporadic inheritance, where only one isolated case occurs within a family, is the most common pedigree pattern observed in clinical practice. Chromosomal abnormalities and new dominant mutations typically demonstrate sporadic inheritance. It is easy to imagine how autosomal recessive and X-linked recessive disorders can often appear sporadic, especially in situations where family size is small or clinical knowledge about extended family is limited. Thus, both Mendelian and non-Mendelian explanations for sporadic inheritance, each with its own recurrence risks, can apply. As a result, clinicians tend to refer to isolated cases as apparently sporadic rather than absolutely sporadic. Since noninherited disorders are associated with virtually negligible recurrence risk as compared to those exhibiting Mendelian inheritance, those associated with chromosomal abnormalities, and those associated with new dominant mutations, it is important to make every effort to distinguish apparently sporadic cases from truly sporadic ones. However, it is often not possible to make this determination and recurrence risk can be narrowed only to a broad range encompassing all possibilities.

Differences in Phenotypic Expression Can Complicate Pedigree Analysis

The occurrence of reduced penetrance, variable expressivity, anticipation, and gender influence or limitation can confound pedigree analysis. The clinical subtlety and nuance associated with each phenomenon can impact recognition of the correct inheritance pattern (usually autosomal dominant, but not always) and result in overlooked or even incorrect diagnoses. Further, accurate recurrence risk is dependent on correct diagnosis and pedigree assessment.

Genetic Penetrance

The penetrance of a genetic disorder is measured by evaluating how often a particular phenotype occurs given a particular genotype or vice versa. Some disorders show 100% penetrance, where all individuals with a particular genotype express disease, while others show reduced penetrance, such that a proportion of individuals with a particular genotype never develop any features (even mild) of the associated clinical phenotype. Thus, penetrance is the probability that any phenotypic effects resulting from a particular genotype will occur. Certain factors are known to influence the gene penetrance for specific disorders. For example, phenotypic expression of a particular phenotype may be modified by age, termed age-related penetrance. Sometimes, as age increases, penetrance increases. For example, only 25% of individuals with a specific Huntington disease genotype (41 repeats) exhibit symptoms at age 50, while 75% exhibit symptoms at age 65. Although less common, penetrance can also decrease with age. Gender-related penetrance has been observed in cases of hereditary hemochromatosis where some females with a particular HH genotype show no evidence of iron accumulation in contrast with their affected male siblings who are known to have the identical genotype. Reduced penetrance can sometimes obscure an autosomal dominant inheritance pattern because, while some family members may have affected offspring, they themselves are not affected due to reduced penetrance of the disorder.

Sex-Influenced Disorders

Sex-influenced disorders are disorders that demonstrate gender-related penetrance. When the probability of phenotypic expression is more likely given a specific gender, the disorder is said to be sex-influenced. *BRCA2*-related hereditary breast/ovarian cancer (HBOC) and *APOE4*-associated late onset familial Alzheimer disease are sex-influenced disorders. *BRCA2*-related HBOC is an autosomal dominant disorder associated most predominantly with increased risk for breast and/or ovarian cancer. Although less common than breast or ovarian cancer, *BRCA2* carriers may also be at increased risk for several other cancers including neoplasms of the skin, prostate, pancreas, larynx, esophagus, colon, stomach, gallbladder, bile duct, and hematopoietic system. In cases of HBOC caused by *BRCA2* mutation, about 6% of males as opposed to 86% of females are expected to develop breast cancer by age 70. With respect to *APOE4*-associated late onset familial Alzheimer disease, women who are heterozygous for *APOE4* alleles are at 2-fold increased risk to develop late onset familial Alzheimer disease as compared to males with the same genotype.

Sex-Limited Disorders

Sex-limited disorders refer to autosomal disorders that are nonpenetrant for a particular gender. Male limited precocious puberty is one example. Males heterozygous for mutations in the *LCGR* gene located on chromosome 2 exhibit this phenotype, but females with the same genotype do not. Very few sex-limited disorders have been documented.

Variable Expressivity

Variable expressivity refers to the difference in severity of disease among affected individuals, both between related and unrelated individuals. It is important to note that even between related individuals (with the same genotype) variable expressivity occurs. Variable expressivity is distinct from penetrance because it implies a degree of affectedness, not whether or not the individual is affected at all. The majority of inherited disease demonstrates some degree of variable expressivity. Variable expressivity can complicate pedigree analysis because individuals with subtle clinical manifestations can be mistaken for unaffected individuals. Neurofibromatosis type I is an autosomal dominant neurocutaneous disorder that affects 1/3000 individuals. There is a high degree of variable expressivity observed within family members that carry the same *NF1* mutation. Some affected individuals with the same mutation may show only a few café-au-lait macules of the skin, while others may be more severely affected with large invasive plexiform neurofibromas and hundreds of cutaneous and subcutaneous neurofibromas.

Pleiotropy

Pleiotropy refers to disorders where multiple, seemingly unrelated organ systems are affected. For example, one individual in a pedigree may exhibit cardiac arrhythmia, whereas another individual with the same disorder in either the same or different pedigree shows muscle weakness and deafness. Since the manifestations of disease are so vastly and usually inexplicably different, disorders that show a high degree of pleiotropy are often difficult to diagnose. As a group, mitochondrial disorders typically show a high degree of pleiotropy, as any organ system can be affected, to almost any degree, with any age of onset.

Anticipation

A disorder shows anticipation when an earlier age of onset or increased disease severity occurs in successive generations. Anticipation is predominantly associated with neurodegenerative trinucleotide repeat disorders (spinocerebellar ataxias, Huntington disease, myotonic dystrophy, etc). In such cases, the number of trinucleotide repeats expands through generations, and is correlated with severity of disease and age of onset. However, not all disorders that exhibit anticipation are trinucleotide repeat disorders. Dyskeratosis congenita-Scoggins type characterized by nail dystrophy, skin hyperpigmentation, and mucosal leukoplakia shows anticipation via a mechanism of progressive telomere shortening in successive generations. While the mechanism remains unclear, anticipation is observed in families with a specific *TTR* gene mutation (V30M) associated with amyloidosis.

Other Factors That Complicate Pedigree Analysis

Genetic Mosaicism

Mosaicism occurs when two or more genetically distinct cell lines are derived from a single zygote. The timing of the post-zygotic event(s) and tissues involved determine the clinical consequence and help to distinguish one type of mosaicism from another. Gonosomal mosaicism occurs early in embryonic development and is more likely to involve gonadal tissue and result in phenotypic expression. The clinical effects are often milder for mosaic individuals where only a proportion of cells carry a particular mutation, as compared to those who inherit germline mutations where all cells are affected. When mosaicism is confined to gonadal tissue, there are usually no clinical consequences to the gonadal mosaic individual. However, such individuals are at higher risk for having affected offspring. Thus, since gonosomal mosaic parents have some proportion of mutant germ cells, they can (and do) have nonmosaic, affected offspring. There is no practical way to exclude the possibility of gonadal mosaicism or effectively test for it. This can cause a dilemma with respect to providing accurate recurrence risks to families. Gonadal mosaicism has been found to be more common for certain disorders and some empiric risk estimates have been determined. For example, Duchenne muscular dystrophy, an X-linked disorder, has an empiric risk for gonadal mosaicism of 10–30%. This means that even when a mother of an affected boy tests negative for a *DMD* gene mutation, a male who inherits the same X chromosome as an affected sibling has a 10–30% chance of being affected.

Consanguinity

Consanguinity is both a social and genetic concept. Generally, it refers to marriage or a reproductive relationship between two closely related individuals. The degree of relatedness between two individuals defines the proportion of genes shared between them. The offspring of consanguineous couples are at increased risk for autosomal recessive disorders due to their increased risk for homozygosity by descent. A frequent way that consanguinity can complicate pedigree analysis is when a provider is unaware of consanguinity at the time they are evaluating the pedigree and what appears to be an autosomal dominant inheritance pattern is associated with an autosomal recessive disease phenotype.

Preferential Marriage Between Affected Individuals

Increased reproductive risk can be the result of preferential marriage between affected individuals. It is not uncommon for similarly affected individuals to attend the same school (for example, deaf high schools) or make connections at support groups (for example, *Little People of America*). And a proportion of such relationships may develop such that an affected couple may decide to start a family. Such selective mating can increase the likelihood of pseudodominance within a pedigree because the mating environment is selected such that an autosomal recessive disorder appears more frequently than expected. An increased risk for autosomal dominant disorders is also present. In cases where both reproductive partners are affected with the same autosomal dominant condition recurrence risk ranges from 66% (when homozygous dominant inheritance is not compatible with life) to 75%.

Other Considerations for Pedigree Construction and Interpretation

Pedigree construction requires equal amounts of skill, science, and art. Those obtaining them must have a strong base of medical genetic knowledge, so as to know the important questions to ask and construct the pedigree correctly. In addition, careful attention must be paid to establishing trust, navigating social relationships, educating, and communicating effectively. Family histories are deeply personal, and the psychosocial impact of the required informational gathering can be significant. Further, inaccurate information can result in misinterpretation and ultimately misdiagnosis. Whenever possible, reported diagnoses must be confirmed with medical records. The effort this requires should not be minimized, as privacy and confidentiality must be upheld for all family members throughout the process.

Clinical molecular genetics seeks to identify genetic variation and to determine whether or not the observed genetic variation has a phenotypic effect. Certainly, the latter cannot be accomplished without astute and thorough clinical evaluation and family history. Even an apparently negative family history is an important one that can guide test selection and result interpretation. In addition, the impact of pedigree analysis on genomic research is formidable. As a result of detailed pedigree assessment, numerous genes have been discovered, genotype:phenotype correlations elucidated, natural history knowledge obtained, and certainly inheritance patterns revealed.

CENTRAL DOGMA AND RATIONALE FOR GENETIC TESTING

The clinical relevance of molecular genetics is fundamentally rooted in the central paradigm of molecular biology: genes encode proteins. Genes are the blueprint for the proteins that form the macromolecules of cellular structure and function. Cells, their respective functions, and the interactions between them translate to the observable characteristics, or clinical phenotype, of an organism. Endogenous and exogenous molecular, cellular, and organismal environments also play an important role in influencing clinical phenotype. So, the expression of DNA at the molecular level coupled with environmental effects leads to more tangible morphological and physiological traits at the level of the organism. However, organisms do not exist in isolation. Each organism functions as part of a population within a larger species and external environment. A species, and the organisms within it, is subject to evolutionary forces, including natural selection, genetic drift, and gene flow. Such forces ultimately impose, overlook, propagate, or extinguish genetic variation. The dynamic relationships between genetic variation, proteins, cells, organisms, populations, and environment(s) connect genetic laboratories to clinical practice, as evaluating for genetic variation (molecular genetics, cytogenetics) and/or its biochemical consequence (biochemical genetics) provides an explanation and/or causative evidence for clinical phenotype and diagnosis.

Diagnostic and Predictive Molecular Testing

The clinical applications of molecular genetic testing can be generalized into two groups based on whether the clinical information sought is intended for diagnostic or predictive purposes. Occasionally overlap between diagnostic and predictive testing occurs. Although most commonly performed for the purpose of diagnosing a disorder in a symptomatic individual, diagnostic testing can also be informative for presymptomatic at-risk individuals. The degree of gene penetrance must be known in order for this diagnostic yet predictive testing to impart clinical value. Penetrance does not necessarily have to be 100% to be useful to the patient and/or family; for example, *BRCA1* mutations are associated with a lifetime risk of approximately 60–80% for the development of breast cancer (Table 5.4).

The molecular genetic test for Huntington disease (HD) illustrates how the same test can be used to determine diagnosis for affected individuals and to predict affected status for as yet unaffected individuals. If testing is performed on a 25-year-old asymptomatic individual known to be at 50% risk for HD and the result is consistent with a repeat expansion known to be fully penetrant by age 47, the result of predictive testing is consistent with a diagnosis of HD during the presymptomatic period. Predictive diagnostic testing can be more difficult to interpret in cases where gene penetrance is not so absolute. For example, if the same 25-year-old asymptomatic individual at 50% risk for HD was found to have an expansion mutation in the reduced penetrance range (36–39 repeats), the ultimate diagnosis is not so absolute.

The second broad group of molecular genetic tests includes those performed for the purpose of revising an already known risk. Predictive molecular testing typically employs molecular screening tests to more accurately determine the individual and familial/reproductive risks for an individual that is already a member of a high-risk population. Typically, a targeted mutation analysis method is used. For example, for Caucasian individuals of Northern European ancestry and no family history of cystic fibrosis (CF), the risk for being a heterozygous carrier is 1/25. After such an individual is screened for the 23 mutations in the *CFTR* gene that account for approximately 90% of CF alleles in the Northern European population, a negative screen result decreases that risk slightly over 10-fold, to 1/265. It is important to recognize that while many predictive molecular screening tests are focused on evaluating at-risk individuals for autosomal recessive carrier status, other subgroups of predictive screens help to distinguish germline from somatic disease or revise prognosis or risk related to complex disease based on presence or absence of disease-associated SNPs. For example, colon cancer is typically a sporadic disease that has a genetic component but does not typically follow a simple Mendelian inheritance pattern of a single-gene disorder. However, a small fraction, approximately 5% of cases, do indeed follow a Mendelian inheritance pattern and are due to inherited mutations in single genes. It is clear that identification of these families can have enormous importance for family members because individuals who are shown to carry the familial mutation can greatly benefit from enhanced monitoring and prophylactic measures. Similarly, family members that are shown not to carry the familial mutation are freed from the need for intensive monitoring and are returned to the same risk as the general population for the development of colon cancer. Unfortunately, simple pedigree analysis is seldom sufficient to identify such families because colon cancer is not a particularly rare condition, and it is not uncommon for multiple family members, who often share many environmental risk factors, to develop sporadic colon cancer. In addition, the penetrance of the disorder may not be complete, thereby making it more difficult to recognize a specific inheritance pattern. Some inherited colon cancer syndromes, such as familial adenomatous polyposis colon cancer (FAP) have a distinctive phenotype (many thousands of colonic polyps) so ascertainment of families is usually straightforward. However, other syndromes, such as Lynch syndrome (or HNPCC), which is due to mutations in the MMR pathway, cannot be distinguished from sporadic colon cancer using clinical or pathological criteria. One might investigate all of the relevant MMR genes in cases suggestive of Lynch syndrome, early age of onset (<50 years), or familial clustering, but such testing (which could involve whole gene sequencing and deletion analysis for up to three large genes) could be very expensive. Further, selection of potential cases on clinical grounds and family history often has a relatively low yield. Because tumors from Lynch syndrome patients are defective in MMR, they exhibit a type of genomic instability known as microsatellite instability. Since approximately 30% of sporadic tumors have microsatellite instability, screening tumor specimens from individuals at risk for Lynch syndrome for loss of expression for mismatch repair proteins and microsatellite stability can eliminate approximately 70% of colon cancer patients from a Lynch syndrome diagnostic algorithm and greatly reduce cost. If microsatellite instability is present in a tumor from an at-risk individual, germline testing can subsequently be performed. If microsatellite instability is not present, suspicion for an underlying germline defect is low.

Table 5.4 Disease Associated Penetrance for Common Hereditary Cancer Syndromes

Familial Cancer Syndrome	Gene	Lifetime Penetrance
Hereditary Breast/Ovarian Cancer	BRCA1	60–80%
Hereditary Breast/Ovarian Cancer	BRCA2	60–80%
Retinoblastoma	RB1	>99%
Familial Adenomatous Polyposis	APC	>99% by age 40
Lynch syndrome	MLH1, MSH2, MSH6	75% (may be slightly lower in females)

Benefits of Molecular Testing

Psychosocial benefits of a confirmed molecular genetic diagnosis may include (i) reduced anxiety associated with a known versus unknown diagnosis, (ii) reduced anxiety if the diagnosis confirmed is considered by the patient to be less severe among those being considered for patient, (iii) reduced anxiety associated with a cease in the diagnostic odyssey that many patients with rare disorders experience (multiple medical consults, procedures, and laboratory tests associated with a continued search for a diagnosis). In addition, psychosocial benefits may accrue from implementation of a more individualized and, in some cases, preventive medical management approach. For individuals undergoing presymptomatic testing, benefits may also include a sense of empowerment, regardless of their test result and a sense of relief if they test negative. Knowledge of one's risk for having children with a genetic disorder also assists with family planning, with individuals and couples being able to access genetic counseling and prenatal diagnosis.

Clinical benefits of a molecular diagnosis often include the ability for the care provider to recommend a preventive medicine and treatment plan based on the known natural history of a particular disorder. Genotype:phenotype correlations have been established for some particular mutations/disorders, such that a more individualized medical approach to care with respect to severity of disease, expected age of onset for presymptomatic cases, and increased risk for certain associated complications can be determined. So, the genotype result may give care providers and their patients information that could lead to more individualized medical management. Also, when a molecular diagnosis is confirmed, predictive testing options become available for at-risk family members. Consider an individual at 50% risk for FAP undergoing presymptomatic testing. Identification of the causative mutation for this individual allows for early intervention by screening and prophylactic colectomy, and informative presymptomatic testing option for at-risk family members. A negative test result directs implementation of a more appropriate, less aggressive screening strategy.

Risks Associated with Molecular Testing

Risks and limitations of genetic testing should always be reviewed and openly discussed with patients as part of the informed consent process prior to testing. Risks associated with molecular genetic testing are most often psychological and financial. Limitations of molecular genetic testing are usually related to confounding results, interpretive restrictions, or imperfections of the method used.

Although it can be of profound benefit, the knowledge of a molecular genetic test result can also be a risk, regardless of whether the result confirms the presence of disease. A positive test result can be devastating, and a true negative result can sometimes invoke survivor guilt. Consider two siblings who undergo presymptomatic testing where one sibling tests positive for a life-threatening disorder and the other tests negative. Both siblings will likely experience psychological repercussions of testing. Such psychological risks must be discussed before sample collection and may influence an individual's decision to undergo testing. Financial risks may include inability to obtain life insurance or certain types of health insurance should an individual test positive. However, the newly enacted Genetic Information Nondiscrimination Act (GINA) should reduce these risks. Many insurance plans do not cover the costs of genetic testing or will cover only part of the expense. Given that many molecular genetic tests are expensive due to the high costs of the technology used and the highly skilled personnel required to process and interpret the sample, personal financial cost to the patient can be substantial.

Limitations of molecular genetic testing should also be discussed with the patient as part of the informed consent process prior to sample collection. Molecular genetic testing is often misunderstood as perfectly decisive. While new technologies and detection rates are continuously improving, not all mutations are identified such that the risk for a false negative result is always a possibility. An even more common problem involves the identification of alterations whose medical or functional significance is not clear. These genetic alterations are termed variants of uncertain significance and can be especially complicated to interpret. Although rare, laboratory errors such as performing the wrong test or mislabeling samples can also occur. Disease and test-specific limitations of molecular testing are truly method, disease, and case specific, and it would be impossible to address each of them here.

Considerations for Selection of a Molecular Test

Selecting an appropriate molecular genetic test is dependent on the purpose for testing, the clinical information known, the sample(s) and testing methods available, and the clinical information sought. Molecular screening tests usually involve methods that investigate for common mutations (for example, targeted mutation detection by RFLP), whereas diagnostic testing methods are typically more comprehensive (for example, DNA sequencing). The molecular methods used for the purpose of revising a known risk can sometimes be the same, but are often different than those used for diagnostic purposes. When evaluating the method to be used, the expected detection rate for individuals that are classically affected with the disorder in question and the clinical context of the patient being tested should be considered. To maximize the informative value of presymptomatic testing, in most cases, one must know the familial mutation(s). Practically, this translates into the necessity that an affected individual should be tested before presymptomatic testing is performed on at-risk family members. Preferred testing algorithms developed by expert clinicians and laboratorians are especially useful, though ultimately each clinical situation is different and should be considered within its own unique context. An increasing number of molecular genetics laboratory directors are employing genetic counselors that act as a liaison between the ordering provider and the laboratory to serve as a resource for the identification of case-specific benefits, risks, and limitations to testing, as well as to assist with test selection, case coordination, and interpretation of results.

ALLELIC HETEROGENEITY AND CHOICE OF ANALYTICAL METHODOLOGY

The great majority of analyses performed in the clinical molecular genetics laboratory are based on the polymerase chain reaction (PCR). PCR is a technique, developed by Kary Mullis in 1984 (then at Cetus Corp.), for the rapid, *in vitro* amplification of specific DNA sequences. The rapid introduction of PCR into research and later into clinical laboratory practice has revolutionized the practice of molecular biology. In 1993, Dr. Mullis was awarded the Nobel Prize in Chemistry for his achievement.

Knowledge of the sequence of the region of DNA flanking the area of interest is required for PCR. Two synthetic oligodeoxynucleotides (primers), typically 20 to 30 bases in length, are prepared (or purchased) such that one of the primers is complementary to an area on one strand of the target DNA 5′ to the sequences to be amplified, and the other primer is complementary to the opposite strand of the target DNA, again 5′ to the region to be amplified. To perform the amplification, one places the sample DNA in a tube along with a large molar excess of the two primers, all four deoxynucleotide triphosphates (dNTPs), buffer, magnesium ion, and a thermostable DNA polymerase. Successive rounds of heating to 93–95°C to denature the DNA, cooling to 50–60°C to allow annealing of the oligonucleotides, and heating to 72°C (the temperature optimum for the DNA polymerase isolated from *Thermus aquaticus*) result in synthesis of the DNA that lies between the two primers. The amount of amplified DNA being synthesized doubles (approximately) with every temperature cycle. The amount of DNA produced is exponential with respect to cycle number. After 30 cycles of denaturation, annealing, extension, 2^{30}, or approximately 10^9, copies of the DNA sequences lying between the two primers will have been generated. In a typical experiment starting with 20–100 ng of human DNA, 30 cycles of amplification will produce enough DNA from a single copy gene to be visualized on an ethidium bromide stained gel. As each cycle takes 2–5 minutes, amplification of a specific sequence can easily be accomplished in several hours. After amplification, the DNA can be analyzed by one of several techniques, depending on the specific problem.

Specific Versus Scanning Methods

Analytical methods in molecular genetics can be grouped into two broad categories: mutation detection techniques, which are used to investigate the actual base sequence at a particular locus, and quantitative methods, in which PCR-based techniques are used to quantify specific nucleic acid sequences. Mutation detection strategies can be further grouped into specific or scanning techniques.

Specific Mutation Detection

Specific mutation detection entails straightforward, and largely routine, procedures that can be used to analyze DNA samples for previously identified mutations using an assay designed for maximum specificity. This approach targets known mutations in potentially large cohorts of patients or small panels of specific mutations in disorders characterized by one or a few common alleles. Results from these types of analyses may confirm or establish clinical diagnoses. Furthermore, in families at risk for a particular genetic disease, specific or targeted mutation detection allows for rapid screening of an entire family for the mutation identified in the proband (the first member of a family to be diagnosed with a genetic disorder), thereby permitting accurate carrier determinations that may aid reproductive decisions. Rapid testing of large numbers of patients permits an assessment of the frequency of a mutation among disease-causing alleles, thereby determining which mutations are most prevalent in different patient populations and guiding the creation of effective clinical mutation testing panels. Examples of genetic disorders that are characterized by low allelic heterogeneity and are most often investigated using specific mutation detection methods include hypercoagulable states due to Factor V Lieden or prothrombin mutations, hemochromatosis, galactosemia, and alpha-1-antitrypsin deficiency. Although cystic fibrosis has a high degree of allelic heterogeneity (over 1500 mutations identified), carrier screening is typically done with a panel of 23–100 mutations, which detect approximately 90% of mutations in the target population of Northern European Caucasians.

The specific mutation detection methods can themselves be divided into those that utilize electrophoretic- or hybridization-based methods (Table 5.5). Both types

Table 5.5 Examples of Electrophoretic- and Hybridization-based Specific Mutation Detection Methods

Electrophoretic Methods	
Restriction enzyme digestion	Typically lab developed
Allele-specific PCR	Elucigene, Tepnel
Allele-specific primer extension	ABI SNaPshot
PCR-oligonucleotide ligation	Cystic fibrosis V 3.0, Abbott/Celera
Hybridization Methods	
Allele-specific hybridization	Resequencing arrays, Affymetrix
Allele-specific primer extension	Tag-It, Luminex
Ligation-PCR	Golden Gate, Illumina

of platforms are robust, and in experienced hands yield reproducible results. Both types of systems are in widespread use in clinical and research laboratories. One criterion for choice between these general platforms is the cost incurred per sample analyzed. In the authors' experience, when the number of samples to be analyzed at one time (samples per batch) is low, electrophoretic methods are often the most cost effective to develop, validate, and implement. However, when the number of samples per batch is larger (greater than 8–12 samples), then the hybridization-based techniques, many of which can be adapted to 96 well microplate formats or real-time, are often more cost effective.

Mutation Scanning Approaches

Mutation scanning methods interrogate DNA fragments for all sequence variants present. By definition, these strategies are not predicated on specificity for specific alleles, but are designed for highly sensitive detection for all possible variants. In principle, all sequence variants present will be detected without regard to advance knowledge of their pathogenic consequences. Once evidence for a sequence variant is found, the sample must be sequenced to determine its molecular nature. The advantage of using a scanning method followed by sequencing of only positive PCR products is that the scanning methods are typically less costly to perform than DNA sequencing. Although a number of mutation scanning methods have been developed, including single-strand conformation polymorphism (SSCP), heteroduplex analysis (HA), conformation-specific gel electrophoresis (CSGE), thermal gradient gel electrophoresis (TGGE), and melt curve analysis, they have been almost completely replaced by what is considered the gold standard mutation scanning method—DNA sequencing. There are a number of disease-associated genes that have high allelic heterogeneity, or very few recurrent mutations in the population, that are typically addressed for diagnostic purposes by whole gene sequencing, including *BRCA1* and *BRCA2*, the mismatch repair genes; *MSH2*, *MLH1*, and *MSH6; CFTR* (for diagnostic, nonscreening applications); biotinidase (*BTD*); and medium chain acylCoA dehydrogenase (*ACADM*). Only when they are combined with appropriate genetic data and *in vitro* functional studies can investigators distinguish disease-causing mutations from polymorphisms without clinical consequence. In the research laboratory, mutation screening is a critical and obligatory final step toward identifying genes that underlie genetic disease. In the clinical laboratory, these methods are applied toward the detection of mutations in diseases marked by significant allelic heterogeneity. As the number of laboratories offering whole-gene sequencing assays for an increasing number of genes grows, the amount of variation in coding regions is beginning to be understood to be significantly greater than previously thought. This has the consequence that obtaining a previously unknown sequence variation in a patient sample is not uncommon. The interpretation of such results is challenging, and is not a solved problem.

Interpretation of Molecular Testing Results

Of the three types of coding region mutation caused by single nucleotide changes, two are often relatively straightforward to interpret. It is generally assumed that nonsense mutations (or indels giving rise to an in-frame stop codon) are deleterious and are likely to be associated with a disease phenotype. Similarly, silent mutations are most often assumed to be benign. Exceptions exist, of course; silent mutations occurring at the first or last bases of an exon may influence RNA splicing. In addition, silent mutations may interrupt an exonic splice enhancer, again leading to altered splicing. An example of the disruption of an exonic splice enhancer is found in spinal muscular atrophy (SMA). The great majority of *SMA* is caused by deletion of exon 7 of the telomeric copy of the survival motor neuron gene *(SNMt)*. There exists a very highly homologous gene, the centromeric copy *(SMNc)* that has only 5 nucleotide changes relative to *SNMt*. Why is the presence of this gene, which is structurally normal in almost all cases of SMA, not sufficient to prevent neuronal death even if the telomeric copies are mutated? One of the nucleotide differences between *SNMc* and *SNMt* is a C to T change in the centromeric copy. Although a silent mutation from an amino acid standpoint (both sequences code for Valine), the T allele is not recognized as an exonic splice enhancer. Thus, the *SMNc* gene transcript also lacks exon 7, and is unable to compensate for the lack of the *SNMt* gene.

The interpretation of missense changes is challenging. Many examples (affecting many different genes) exist in which missense changes are either pathogenic or benign. The distinction typically requires the examination of multiple families carrying a given missense mutation, and/or functional studies of recombinant, mutant protein. When a novel missense change is encountered in a clinical laboratory setting, these studies are not available. Thus, novel missense changes are typically referred to as variants of uncertain significance (VUS).

There are two schools of thought with respect to how VUS should be reported. One school holds that unless the laboratory can give a clean interpretation and offer documentation as to whether a given variant is known to be pathogenic or benign, the report should simply indicate that a VUS was detected. Thus, the contribution of the genetic test to the management of the patient is nil; it is as if the test were not performed (and cannot be performed). Clearly, the advantage in this approach is that one is not tempted to overinterpret the results, potentially leading to an incorrect medical decision. The disadvantage is the frustration on the part of the patient (and healthcare provider) that a rather expensive test (typically) has been performed and no useful information was obtained. The other school of thought holds that the laboratory should use all the tools available and when possible make a probabilistic statement as to the potential effect of the variant. The advantage to this approach is that the final decision as to how the result will be used in guiding patient care remains with the patient and his/her healthcare provider. The clear disadvantage is the possibility that the result provided may lead to incorrect medical management. Because of this, interpretation of VUS should be done very carefully.

A number of tools to aid in the interpretation of missense changes have been developed. As an increasing number of species have had their complete genome sequence determined, it is possible to use a variety of sequence alignment tools to compare the amino acid found at a particular location in the human gene to that found in multiple other species. The rationale for this is the notion that if an amino acid is invariant across species, it is more likely to be important for protein function, and a missense change at a highly conserved residue is more likely to be pathogenic. On the other hand, if a given amino acid position is poorly conserved, a missense change may be more likely to be benign.

Several groups have developed algorithms quantifying the probability based on sequence conservation that a given missense change is pathogenic or benign. Two of these tools, SIFT and POLYPHEN, are freely available online at http://blocks.fhcrc.org/sift/SIFT_related_seqs_submit.html and http://genetics.bwh.harvard.edu/pph/, respectively. These tools are useful but are far from perfect. For example, one study evaluated both of these online tools against sequence variations that were known to be either benign or pathogenic in two well-characterized genes, *beta-globin* and *G6PD*, as well as two others, *TNFRSF1A* (tumor necrosis factor receptor-associated periodic syndrome) and the *MEFV* gene (familial Mediterranean fever). The two programs were found to be between 70% and 80% sensitive and specific [2].

One mechanism by which sequence conservation strategies can be foiled occurs when a pathogenic change results in the substitution for an amino acid that is the normal sequence in another species. For example, the most common medium chain acyl co-A dehydrogenase *(MCAD)* deficiency mutation (accounting for approximately 70–80% of mutant alleles) is the lysine to glutamic acid change at codon 329 (K329E). Sequence alignment-based tools such as SIFT or POLYPHEN do not identify this as a pathogenic change because the corresponding codon in the mouse is normally glutamic acid.

Another widely used tool is the BLOSUM62 matrix. This matrix is derived from the BLOCKS database (http://bioinformatics.weizmann.ac.il/blocks/blocks_release.html). This database consists of alignments of peptides from many proteins from many different species. One use of the database is to understand how certain motifs (nucleotide binding clefts, leucine zippers, helix-turn-helix, and others) are conserved and frequently utilized in many different proteins. One can select groups of peptides based on the amino acid sequence alignment. For example, one may wish to study only very well conserved motifs and use only groups of peptides that are 90% identical. Or, one may wish to study more distantly related sequences and choose groups that are only 40% identical. One use of this database is to assist in making new alignments. For any given percentage of sequence homology selected, it is possible to calculate a probability that any given amino acid will be substituted for another in normal, wild-type proteins. For example, leucine to isoleucine changes are more common in nature than aspartic acid to tryptophan changes. It has been empirically determined that at a cutoff of 62% identity, the BLOSUM62 matrix is the most useful in creating new alignments by calculating the probability that an amino acid in a new protein aligns with a given position in one or more other proteins. BLOSUM62 has also been used as a tool for classifying VUS as either deleterious or benign. Amino acid changes that are frequently found are judged to be more likely to be benign, whereas those that are infrequent are more likely to be pathogenic. Note that this method relies only on the global probability of one amino acid being substituted for another and does not utilize any gene-specific information. The BLOSUM62 matrix is freely available at http://www.ncbi.nlm.nih.gov/Class/BLAST/BLOSUM62.txt.

In addition to sequence conservation and global substitution probabilities, the chemical characteristics of the amino acids have been used to characterize missense changes. A composite of three quantifiable chemical properties—composition, polarity, and molecular volume—has been defined and is termed the Grantham score. When a substitution results in a large change in the Grantham score, reflecting a large change in the chemical nature of the amino acid, the change is more likely to be deleterious. In contrast, a small change in the Grantham score indicates that the chemical nature of the residue has not changed appreciably, is less likely to alter protein function, and is more likely benign. Similar to the BLOSUM62 method, this scoring system refers to a global standard and does not take the sequence of the particular gene into account. However, it is possible to combine the Grantham score method with sequence alignment tools. In this approach, the difference in the Grantham score across the normal protein alignment is made. That is, after the construction of an alignment of normal orthologs (homologous proteins from different species), the difference between the minimum and maximum Grantham score for the residue under investigation is calculated. This number is termed the Grantham distance (the distance in Grantham space between the amino acids at the codon in question across species). Then, the difference in the Grantham score for that codon in the wild-type and mutant protein is calculated. This number is termed the Grantham variation. If the Grantham variation is larger than the Grantham distance, then the change is more likely to be pathogenic. However, if the VUS results in an amino acid substitution that gives a change in Grantham score that is smaller than that seen across species, the change is more likely to be benign. One online tool for these calculations (Align GDGV) is freely available at http://agvgd.iarc.fr/agvgd_input.php.

All of the strategies discussed in the preceding paragraphs are helpful in attempting to understand the clinical significance of VUSs, but none of them are perfect—far from it. The sensitivity and specificity for all of them seem to be in the 70–80% range. However, since they query different properties of the gene and of amino acid substitution, it is possible that, when used together, the quality of the results may be improved. One study investigated the use of all for the preceding methods as applied to five different genes, The authors found that when the results of all 4 methods agreed, the final calls had a predictive value of approximately 88%. Further, they noted that mutations at residues that were completely conserved across species had a 92–97% probability of being deleterious [3]. Another group of investigators have incorporated the results of these characterization tools and combined them with classical pedigree and linkage analysis in a comprehensive Bayesian approach. Using this method, this group has been able to classify several missense mutations in the *BRCA1* gene as either pathogenic or benign [4].

The *in-silico* characterization of missense VUS changes is still in its infancy, and much more work needs to be done in this area. As the era of whole genome sequencing rapidly approaches, the urgency of the need to characterize novel changes is increasing.

CONCLUSION

Molecular genetics utilizes the laboratory tools of molecular biology to relate changes in the structure and sequence of human genes to functional changes in protein function, and ultimately to health and disease. New technology, such as is being developed for the $1,000 genome project, promises to greatly increase the reach and scope of molecular genetics. Indeed, some subspecialties, such as biochemical and cytogenetics may ultimately merge with molecular genetics and offer the medical community a more comprehensive and integrated approach to understanding the role of our genomic variation in health and disease. However, the interpretations of results from the clinical molecular genetics laboratory will always be rooted in the fundamentals of molecular and cell biology and in the central paradigm—that genes encode proteins. It will be from these roots that modern, personalized medicine will grow.

REFERENCES

1. Bennett RL, Steinhaus KA, Uhrich SB, et al. Recommendations for standardized human pedigree nomenclature. Pedigree Standardization Task Force of the National Society of Genetic Counselors. *Am J Hum Genetics*. 1995;56:745–752.
2. Tchernitchko D, Goossens M, Wajcman H, et al. In silico prediction of the deleterious effect of a mutation: Proceed with caution in clinical genetics. *Clin Chem*. 2004;50:1974–1978.
3. Chan PA, Duraisamy S, Miller PJ, et al. Interpreting missense variants: Comparing computational methods in human disease genes CDKN2A, MLH1, MSH2, MECP2, and tyrosinase (TYR). *Hum Mutat*. 2007;28:683–693.
4. Goldgar DE, Easton DF, Deffenbaugh AM, et al. Breast Cancer Information Core (BIC) Steering Committee. Integrated evaluation of DNA sequence variants of unknown clinical significance: Application to BRCA1 and BRCA2. *Am J Hum Genetics*. 2004; 75:535–544.

Chapter

6

The Human Genome: Implications for the Understanding of Human Disease

Ashley G. Rivenbark

INTRODUCTION

Genetics is the study of cells, individuals, heredity, variation, and the population within each organism. The modern science of genetics started in the mid-19th century with the work of Gregor Mendel when he observed that organisms inherit traits in a discrete manner—later called genes [1]. In a matter of about five decades, perhaps the most far reaching endeavor that the field of genetic research has ever attempted was accomplished in the sequencing of the human genome. Applied research in genetics has produced many benefits, including the recognition of the molecular basis of human genetic disorders and cancer. For example, many genetic diseases have been discovered as a result of a single mutation or a specific chromosomal rearrangement and are now understood at the molecular level, including, but not limited to, sickle-cell anemia, hemophilia, cystic fibrosis, Duchenne muscular dystrophy, Tay-Sachs disease, Down syndrome, Li-Fraumeni syndrome, Wilm's tumor, Prader-Willi syndrome, Angelman's syndrome, and many metabolic disorders. This would not have been accomplished in a timely manner for many of these diseases if it were not for the entire sequence of the human genome produced by the Human Genome Project in 2003. This chapter will touch in summary on the structure and organization of the human genome, the overview of the Human Genome Project, and the impact it has made on the identification of disease-related genes, as well as the sources of variation in the human genome, types of genetic disease, and human malignancies we have come to understand based on the sequencing of the human genome. This chapter is not comprehensive but covers essential concepts and observations and, I hope, gives the reader an understanding that there has never been a more thrilling time to be immersed in the study of human genetics.

STRUCTURE AND ORGANIZATION OF THE HUMAN GENOME

In the 1940s, deoxyribonucleic acid (or DNA) was shown to be the genetic material of all living organisms. In 1953, James D. Watson and Francis H. C. Crick discovered the double helical structure of DNA. This discovery single-handedly revolutionized molecular biology and biological sciences. Many advances in genetics followed, and researchers established that the units of genetic information (genes) encoded information for the synthesis of enzymes/proteins. The complete set of genetic information or DNA instructions is referred to as a *genome*.

DNA Carries Genetic Information

The genetic blueprint of life occurs in the form of DNA, which is faithfully packaged within the nucleus of each cell in our body. In order to understand genetic disease, we need to first examine the structure of DNA. DNA is quite simple chemically; it is composed of phosphate, deoxyribose sugar, and four nitrogenous bases. These nucleotide bases in DNA are adenosine (A), thymine (T), cytosine (C), and guanine (G). The A is always paired to the T and the C is always paired to the G. Each triplet combination of

Molecular Pathology © 2009, Elsevier, Inc. All Rights Reserved.

these nucleotides makes up our genetic code and constitutes a code word. These words or bases are strung together to make genes, which instructs the cell how to make a specific protein. There are approximately 20,000–25,000 genes in the human genome that are transcribed to ribonucleic acid (RNA), and then translated to produce tens of thousands of proteins. DNA regions that are found and expressed in the mRNA sequence are called exons, and the DNA sequences that are not found in the final mRNA product are called introns. In addition, there are regions of intergenic DNA located between functional genes. For a certain phenotype or observable characteristic, a gene can give rise to several different combinations called alleles. An individual's genotype is made up of different alleles that arise from both parents. Alleles can be dominant or recessive. The dominant allele will result in a certain phenotype if only one copy is present, but there must be two copies of a recessive allele to result in the same phenotype. If a genotype (or pair of alleles) of an individual is two dominant alleles or two recessive alleles, the individual is said to be homozygous, and if an individual inherits two different alleles (one dominant and one recessive), then that individual is heterozygous. Ultimately, one gene can give rise to multiple transcripts that give rise to multiple proteins with different functions.

The genomes of any two people are more than 99% similar; therefore, the small fraction of the genome that varies among humans is very important. These variations of DNA are what make humans unique. However, variations in DNA can occur in the form of genetic mutations in which a base is missing or changed. This results in an aberrant protein and can lead to disease. Through studies of the genetic variation of humans, it is hoped that we can gain insight into phenotypic variation and disease susceptibility. In addition, it is thought that DNA structure plays a role in certain human genetic diseases. Certain trinucleotide (CTG and CCG) repeat sequences have been shown to be found in genes whose aberrant expression leads to disease. The severity of the disease is associated with the number of repeats; diseased individuals have greater than 50 repeats, whereas normal individuals have very few repeats.

General Structure of the Human Genome

DNA packaging, and how it gets organized in the nucleus of a cell, turns out to be a job for a class of proteins called histones. These histones can be altered by a number of chemical modifications, which have been shown to regulate the accessibility of the underlying DNA. The core nucleosome particle is composed of 147 base pairs of DNA wrapped around an octomer of four core histone proteins. These nucleosomes fold into 30 nm chromatin fibers, which are the components that make up a chromosome. The human genome is 3×10^9 base pairs or a length of about 1 meter, which compacts into a nucleus that is only 10^{-5} meters in diameter. Regulation of chromatin has profound consequences for the cell, as the ability to open and close the environment in which DNA is packaged is the primary mechanism by which the genes encoded within the DNA get expressed into proteins. The structure of chromatin is now well understood, but how chromatin is packaged into a chromosome is not. Chromosomes are clearly visible with dyes that react with DNA, which can then be visualized under a primitive light microscope. The word *chromosome* is derived from Greek and describes a colored body, which reflects the ability to visualize dense regions [2]. Dense, compact regions of chromosomes are referred to as heterochromatin consisting of mostly untranscribed and inactive DNA. Regions called euchromatin are less compact and consist of more highly transcribed genes. The genetic code that is inherited is in the DNA sequence, although the way in which the DNA is packaged into chromatin plays an important role in controlling and organizing the information that the DNA holds. When packaged into chromatin, some information is accessible and some is not, which depends on chemical modifications to the chromatin proteins (histones). Chromatin is dynamic, and the accessible regions of DNA change during human development or different disease states. This process of altering gene expression in a stable, heritable manner without changing the DNA code is referred to as epigenetics.

Chromosomal Organization of the Human Genome

Our genome contains 46 chromosomes with 22 autosomal pairs and two sex chromosomes. These chromosomes differ about 4-fold in size from chromosome 1 to chromosome 21, which is largest to smallest, respectively. Each of the 46 chromosomes in human cells contains a centromere (central region) and telomere (ends of the chromosome) composed of genes (2%), regulatory elements (1–2%), noncoding DNA (50%), which includes chromosome structural elements, replication origins, repetitive elements, and other sequences (45%) [3]. At the end of the 19th century it was accepted by numerous researchers that chromosomes formed the basis of inherited traits. There are approximately 60 trillion cells in the human body, which all originate from a single fertilized cell. The cells in the body undergo cell division, or mitosis, in which the chromosomes are condensed and genomic DNA is faithfully replicated. The nomenclature used to define the segments on a chromosome was determined by G-banding chromosomal staining, where the mitotic chromosomes were digested with trypsin and followed with Giemsa staining, which stains centromeric regions [4]. The short arm region of the chromosome (usually displayed above the centromeres) is referred to as the p arm (for instance, 17p) and the long arm region of the chromosome (displayed below the centromeres) is called the q arm (for instance, 13q) with each band having a number associated with it [4]. Chromosomal banding studies using Giemsa staining have shown that heterochromatin comprises 17%–20% of the human chromosome and consists of different families of alpha satellite DNAs and other higher order repeats [5].

Subchromosomal Organization of Human DNA

DNA features

Once the human genome was sequenced, it was recognized that our genome consists of a significant number of different repetitive DNA sequences. Several classes of repetitive DNA exist in the human genome, including Alu repeats, mammalian interspersed repeat (MIR), medium reiteration (MER), long terminal repeat (LTR), and long interspersed nucleotide elements (LINE) [6,7]. Repetitive DNA such as short interspersed nuclear fragments (SINEs), including MIRs and LINEs, are dispersed throughout the genome, whereas satellite DNAs are clustered in discrete areas (centromeres) [2]. Repetitive sequences have been recently proposed to be involved in genome compaction [2]. Genomic regions of satellite DNA are condensed throughout the cell cycle, and there is evidence that LINEs are involved in X-chromosome condensation [2,8]. Single nucleotide polymorphisms (SNPs) or a single nucleotide base change occurs by random and independent mutations, and in a high degree of variability among chromosomes.

CpG islands appear in approximately 50% of human genes and are located preferentially at the promoter region of genes, flanking the transcription start site. It has been estimated that there are around 30,000 to 45,000 total CpG islands in the human genome [9]. A CpG island is defined as a region with greater than 200 base pairs with a G+C percentage that is greater than 50% with an observed/expected CpG ratio that is greater than 0.6 [10,11]. CpG dinucleotides are sites for DNA methylation and in turn can downregulate gene expression. DNA methylation has been shown to be important during gene imprinting and tissue-specific gene expression. Gene inactivation by aberrant DNA methylation has been correlated with cancer in many different cell types. In addition, the identification of CpG islands throughout the genome can help predict promoter regions for human genes.

Gene structure

Now that the human genome has been sequenced, one of the next steps is to utilize genomic tools to obtain a picture of how DNA is targeted by transcription factors and cofactors to lead to gene expression. These proteins (transcription factors and cofactors) control whether a gene is on or off. Transcription factor binding sites are thought to contain conserved sequence motifs of 6–20 base pairs. Transcription factor binding proteins bind to *cis*-acting elements including promoters, enhancers, silencers, splicing regulators, chromosome boundary elements, insulator elements, and locus control regions to control gene expression that regulates cell development and fate [3]. The goal of the Encyclopedia of DNA Elements (ENCODE) Project is to identify and define all of these sequences in the human genome. Nuclease-hypersensitive sites are regions of DNA that interact with transcription factors in the chromatin environment *in vivo* [3]. Trans-acting factors bind chromatin at DNAse1-hypersensitive sites (DHSs), which occur at accessible chromatin regions [3]. Interestingly, CpG islands are associated with DHSs that are either constitutive or tissue-specific [3].

OVERVIEW OF THE HUMAN GENOME PROJECT

Decoding the DNA sequence that is made up of 3 billion base pairs was highly anticipated throughout the scientific and nonscientific community alike. The contribution of the Human Genome Project (HGP) to scientific research has undeniably contributed significantly toward understanding the causation of human disease, and the interaction between the environment and heritable traits defining human conditions [6]. The sequencing of the human genome was first proposed by Robert Sinsheimer (chancellor of the University of California at Santa Cruz) in 1985 [6,12]. This idea was met with some critiques from the scientific community, many thinking the idea was premature and crazy [6,12]. However, in 1988, Nobel Laureate James Watson gave the HGP a significant boost when he began to lead a National Institutes of Health (NIH) component of the project after joint funding from the NIH and the Department of Energy (DOE) [12]. In 1990, the HGP was officially initiated and was proposed to take 15 years, with a budget of $3 billion. The first 5 years of the HGP under the direction of James Watson was determined to map the genetic and physical features of the human genome [13], and his comment was "only once would I have the opportunity to let my scientific life encompass the path from the double helix to the 3 billion steps of the human genome" [14]. The managers of the HGP were Francis Collins at the NIH, Michael Morgan at The Wellcome Trust, and Aristides Patrinos at the DOE [12]. Although the United States made the largest contribution to the HGP, it was an international effort with contributions from Britain, France, Germany, Canada, China, and Japan [13]. Several species of bacteria and yeast had been completely sequenced in 1996, and this progress spurred the attempt at sequencing the human genome on to a more pilot scale [13]. Eight years into the project in 1998, the plan included a sequencing facility to be built that would help sequence the human genome in only a 3-year period, ahead of schedule [6]. The HGP agreed to release all sequences to the public and on June 26, 2000, a working draft of the human genome became available. Almost 3 years later on April 14, 2003, the HGP accomplished its ultimate goal and announced that the sequencing of the human genome was completed [15]. See Figure 6.1 for a timeline depiction highlighting the course of events leading up to the completion of the human genome. Remarkably, this announcement occurred almost

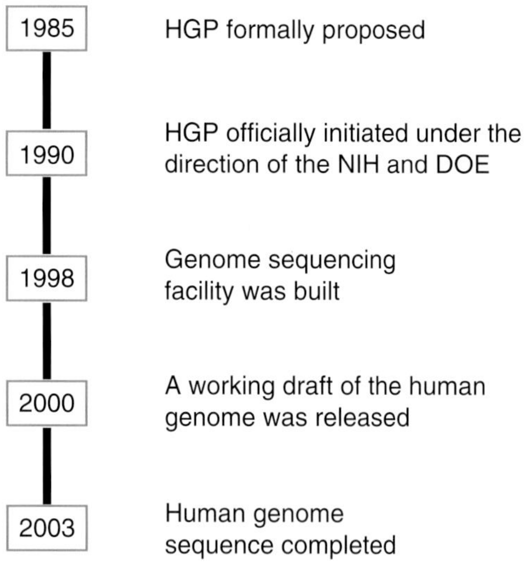

Figure 6.1 Human Genome Project timeline. The vertical timeline emphasizes the major years of the Human Genome Project (HGP) boxed in green with a summary of what occurred in that year to the right of the date.

50 years to the date of Watson and Crick's influential publication of the DNA double helix [15,16]. Therefore, genomic science has rapidly gone from the identification of the structure of DNA to the sequencing of the human genome (and many other organisms) in a span of 50 years.

The Human Genome Project's Objectives and Strategy

The primary goal of the HGP was to obtain the complete DNA sequence of the human genome by 2005 [17]. Through use of a whole genome-random shotgun method and a whole-genome assembly, along with a regional chromosome assembly, and through combination of sequence data from Celera (a private sector company that agreed to help sequence the human genome for profitable purposes [13]) and the publicly funded genome center, a 2.91 billion base pair consensus sequence was derived from the DNA of 5 individuals [6]. At first the whole-genome shotgun approach proposed in 1997 by Weber and Mayers for the sequencing of the human genome [18] was not well received [19]. However, at that time (almost 8 years into the sequencing of the human genome), only 5% of the genome sequence had been completed, and it was clear that the goal of finishing by 2005 was unattainable [6,20]. At that time PE Biosystems (now Applied Biosystems) developed a sequencer called the ABI PRISM 3700 DNA analyzer, [6] which was going to be a part of Celera [21]. Now with the ability to sequence with an automated, high-throughput capillary DNA sequencer, as well as new developments in tracking for whole-genome assembly, the chosen test case of the whole-genome assembly on a eukaryotic genome was *Drosophila melanogaster* [22]. The *Drosophila* genome, comprising 120 Mb of euchromatic DNA, was sequenced over a 1-year period [22–24].

The HGP enrolled 21 donors and collected approximately 130 ml of blood from males and females from a variety of ethnic backgrounds [6]. From the 21 donors, 5 were chosen, including two males and three females: two Caucasians, one Hispanic Mexican, one Asian Chinese, and one African American [6]. In order for the shotgun sequencing method to be fully utilized, the plasmid DNA libraries needed to be uniform in size, nonchimeric, and representative of the whole genome (rather than randomly representing the genome) [6]. Therefore, DNA from each donor was inserted in either a 2 Kb, 10 Kb, or 50 Kb plasmid library [6].

Human Genome Project Findings and Current Status

The HGP findings were extensive and exciting, as expected. Therefore, only highlighted findings related to the understanding of human disease are reported here. A complete and detailed analysis was published by Venter et al. [6]. In the wake of the human genome sequence, there was considerable acceleration in the success of the identification of genes that were important for the development of disease [13]. In 1990, fewer than 10 genes had been identified by positional cloning, but by 1997 that number grew to more than 100 genes [13].

The HGP defined 26,383 genes with confidence using a unique rule-based system called Otto [6]. Regions of sequence that were likely gene boundaries were matched up with BLAST and partitioned by Otto, and grouped into bins of related sequence that may define a gene [6]. Known genes were then matched to the corresponding cDNA and were annotated as a predicted transcript. However, the genome sequence has variations and frameshifts, and it was not always possible to predict a transcript that agrees 100%. Therefore, if a transcript matched the genome assembly for at least 50% of its length at greater than 92% identity, then the region was annotated by Otto [6]. It was predicted that an average gene in the human genome is approximately 27,874 bases [6]. These variations in the human genome are being cataloged and will provide clues for the risk and diagnosis of common genetic diseases. More than 2 million single nucleotide polymorphisms have been identified [13]. DNA arrays are now being employed to study the gene expression patterns of as many as 10,000 genes at one time [25]. This analysis will help researchers understand the gene expression patterns between normal tissues and diseased tissues.

The HGP examined the genome for regions that were gene-rich and regions that were devoid of genes. A gene-poor region was defined as a region greater than 500 Kb lacking an open reading frame. Under these conditions, about 20% of the genome contains gene-poor regions, and they were not evenly distributed throughout the genome [6]. Gene-poor chromosomes were 4, 13, 18, X, reflecting 27.5% gene-free regions of the total 492 Mb; and gene-rich chromosomes were 17, 19, 22, having only 12% gene-free regions within their 171 Mb [6]. The next few years following the sequencing of the human genome were spent closing the sequencing gaps of all the chromosomes [13]. Chromosomes 21 and 22 were completed first [26,27].

The HGP correlated CpG islands with gene start sites of computationally annotated genome transcripts and the entire human genome sequence. The HGP compared the variation of the CpG island computation with Larsen et al. [11] and used two different thresholds of CG dinucleotide likelihood, including the original ratio of 0.6 [6]. The analysis showed a strong correlation between first coding exons and CpG islands [6]. Genome-wide repeat elements were examined by the HGP. They observed that approximately 35% of the human genome was composed of different repeat elements, with chromosome 19 having 57% repeat density, the highest repeat density as well as the highest gene density [6]. Gene density and Alu repeat elements exhibit an association, whereas this was not observed with the other classes of repeat elements [6].

The human genome sequence and the variations contained within must be utilized to identify the gene target of all hereditary diseases [13]. However, this is an exciting but huge task; nevertheless, much large-scale gene duplication was identified in the HGP. These included duplications that were known to be associated with proteins involved in disease such as bleeding disorders, developmental disease, and cardiovascular conduction abnormalities [6]. The duplications were located throughout the genome. However, there were gene families that were scattered in blocks within the genome, such as the olfactory receptor family [6]. Chromosome 2 contains two very large duplications that are shared by two different chromosomes, 14 and 12. The first duplicated region is a block of 33 proteins spread in eight different regions spanning 20 Mb of 2p, and these genes are also found on chromosome 14 spanning 63 Mb [6]. The second duplication is on 2q and chromosome 12. This duplication includes two of the four known Homeotic (*Hox*) gene clusters, and the other two *Hox* gene clusters are also seen as duplications on two different chromosomes [6]. These *Hox* genes play a fundamental role in controlling embryonic development, X-inactivation, and renewal of stem cells. According to the HGP, SNPs occur frequently at about 1 per 1200–1500 bp, but only less than 1% affect protein assembly and function. These analyses were based on the potential of an SNP to impact protein function based on SNPs that are located within predicted gene coding regions [6]. Interestingly, the frequency of SNPs is highest in intronic regions, followed by intergenic regions, and then exonic regions [6].

Sequences of known proteins were compared to predicted proteins by the HGP. Analysis demonstrated that out of the predicted 26,588 proteins, 12,809 (41%) of the gene products could not be classified and were termed proteins with unknown function [6]. The remainder of the proteins were classified into broad groups based on at least two lines of evidence. Importantly, the molecular functions of the majority of the predicted proteins by the HGP are transcription factors and proteins that regulate nucleic acid metabolism [6]. Many other proteins were receptors, kinases, and hydrolases, as well as proto-oncogenes, and proteins involved in signal transduction, cell cycle regulators, and proteins that modulate kinase, G protein, and phosphatase activity [6]. Large-scale analysis for characterizing proteins (proteomics) by their structure, function, modifications, localization, and interactions are being accomplished [28] and utilized to gain understanding of their role in disease and cell differentiation [13].

A big challenge now that the human genome is sequenced is to understand how the DNA code is transcribed into biological processes that determine cell development and fate. The human genome sequence is only the first level of understanding and all functions of genes and the factors that regulate them must be defined. For example, in a disease such as diabetes mellitus, there may be up to 10 genes that result in an increased risk for the development of this very common disorder [13]. It is exciting to think that small molecule drugs are being designed to block or stimulate a certain pathway. For example, the main etiology for chronic myelogenous leukemia is a translocation between chromosomes 9 and 22, and a drug was designed to inhibit the kinase activity of the bcr-abl kinase [29], which is the protein that is produced as a result of this translocation [13]. This understanding will help explain how these underlying molecular processes when disturbed can lead to human genetic diseases and cancer. Francis S. Collins and Victor A. McKusick have predicted that "by the year 2010, it is expected that predictive genetic tests will be available for as many as a dozen common conditions, allowing individuals who wish to know this information to learn their individual susceptibilities and to take steps to reduce those risks for which interventions are or will be available" [13]. This is an exciting possibility, and as research escalates to help provide more preventative medicine and more early treatment options, we are getting closer to personalized medicine.

IMPACT OF THE HUMAN GENOME PROJECT ON THE IDENTIFICATION OF DISEASE-RELATED GENES

The identification of disease-related genes has been quite a laborious process. However, several strategies have been employed by using the information that is known about a candidate gene, including the knowledge of the protein/enzyme involved in the disease, the location of the gene within a chromosomal region, and a known animal model of the human disease in question [30]. Linkage analysis, microsatellite markers, large DNA fragment-cloning techniques, and expressed sequence tags (ESTs) are important in the identification of genes responsible for human diseases [31]. The HGP has without a doubt facilitated the strategies used for the identification of disease-related genes. The human DNA sequence has provided the template by which mutations are identified as well as cDNA, and genomic source data in order to provide numerous candidate genes for future studies [32]. Presently, there are more genes than disease phenotypes, which have helped identify many if not all single gene disorders. The challenge now is to identify genes that are involved in polygenic disorders such as diabetes, hypertension, and most cancers [30,33].

Positional Gene Cloning

Positional cloning is used to determine the location of a gene without the understanding of its function and isolating the gene starting from the knowledge of its physical location in the human genome [30,33,34]. The progress of positional cloning moved slowly at first because of its laborious nature with methods that required chromosome walking and the identification of expressed sequences [30,33]. In 1986, Stuart Orkin and colleagues first reported their success in positional cloning the X-linked gene for chronic granulomatous disease [35]. Usually, the first step in positional cloning is linkage analysis of the disorder in disease-prone families to determine chromosomal location, and then subsequent isolation and testing of genes for mutations that are segregating with the chromosomal location of this disorder [32,33]. On average there may be 20 to 50 genes in this chromosomal location, and the gene contributing to the disease can be segregated often on the basis of some presumptions of the disorder in question [32] (Figure 6.2). Positional cloning has been used to identify a number of inherited gene disorders as well as human cancers (Table 6.1).

Functional Gene Cloning

In functional cloning the protein is known and a gene is isolated based on fundamental knowledge and/or function of the protein product causing the human disease without information known about chromosomal location [30,33]. The amino acid sequence and/or the antibodies available are used to determine the gene coding sequence. A cDNA library can then be screened with a oligonucleotide probe (antibody or degenerate oligonucleotides) based on the nucleotide sequence of the gene and polymerase chain reaction (PCR) can amplify the cDNA using oligonucleotides from the amino acid sequence [30] (Figure 6.3). Functional cloning has been used to identify genes causing human diseases such as phenylketonuria and sickle cell anemia [33]. However, our fundamental understanding of human disease is lacking, and this gene discovery tool is not really available often [33].

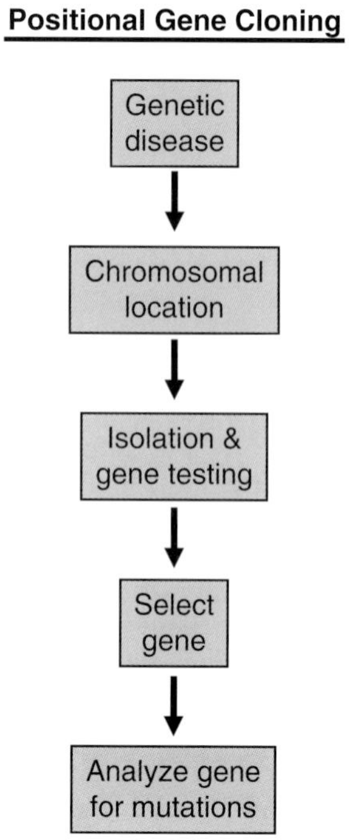

Figure 6.2 Positional gene cloning. The flow diagram highlights the steps that are used to determine the location of a disease-related gene without understanding its function. Linkage analysis is used to map the chromosomal location of the disease-causing gene.

Candidate Gene Approach

In the candidate gene approach, the cloning of a specific gene depends on having some functional information about the disease and relies on the availability of information on genes that had been previously isolated [30,33]. This may not be the best way to clone genes because an informed guess is made about the kind of protein that may be responsible for the human disorder. Missense mutations in the *p53* genes were cloned using the candidate gene approach and were shown to be the cause of Li-Fraumeni syndrome, an inherited cancer disorder [33].

Positional Candidate Gene Approach

The positional candidate approach relies on the sequence of the human genome in that disease-related genes have been mapped to the correct chromosomal location and a survey of the sequence of that region is used to identify genes that are good candidates for cloning and testing [33]. Candidate genes are analyzed by comparing the amino acid sequence of the genes to that of proteins with known functions and then studied in affected individuals in order to determine what gene(s) is responsible for the genetic disorder [31,33] (Figure 6.4). The gene responsible for Marfan syndrome, an autosomal dominant disorder of connective tissue, was mapped using the positional candidate approach. Marfan syndrome was mapped

Table 6.1 Inherited Disease-Related Genes Identified by Positional Cloning

Disease	Year
Chronic granulomatous disease	1986
Duchenne muscular dystrophy	1986
Retinoblastoma	1986
Cystic fibrosis	1989
Wilms tumor	1990
Neurofibromatosis type 1	1990
Testis determining factor	1990
Choroideremia	1990
Fragile X syndrome	1991
Familial polyposis coli	1991
Kallmann syndrome	1991
Aniridia	1991
Myotonic dystrophy	1992
Lowe syndrome	1992
Norrie syndrome	1992
Menkes disease	1993
X-linked agammaglobulinemia	1993
Glycerol kinase deficiency	1993
Adrenoleukodystrophy	1993
Neurofibromatosis	1993
Huntington disease	1993
von Hippel-Lindau disease	1993
Spinocerebellar ataxia I	1993
Lissencephaly	1993
Wilson disease	1993
Tuberous disease	1993
McLeod syndrome	1994
Polycystic kidney disease	1994
Dentatorubral pallidoluysian atrophy	1994
Fragile X "E"	1994
Achondroplasia	1994
Wiskott Aldrich syndrome	1994
Early onset breast/ovarian cancer	1994
Diastrophic dysplasia	1994
Aarskog-Scott syndrome	1994
Congenital adrenal hypoplasia	1994
Emery-Dreifuss muscular dystrophy	1994
Machado-Joseph disease	1994
Spinal muscular atrophy	1995
Chondrodysplasia punctata	1995
Limb-Girdle muscular dystrophy	1995
Ocular albinism	1995

to 15q by linkage analysis as well as the fibrillin gene [36,37]. When DNA from patients with Marfan syndrome was analyzed, mutations were found in the fibrillin gene [33,38]. Another example of genes identified by the positional candidate approach is four genes found in the mismatch repair process that had been previously implicated in hereditary nonpolyposis colon cancer [39].

SOURCES OF VARIATION IN THE HUMAN GENOME

The DNA sequence between humans is 99.9% identical. Therefore, it is important to examine the sequence variation between individuals to gain insight into phenotypic variation, as well as disease susceptibility. Single nucleotide polymorphisms (SNPs); short

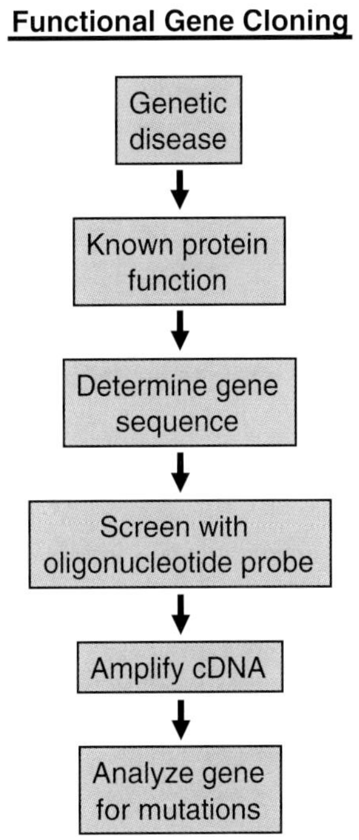

Figure 6.3 Functional gene cloning. The flow diagram highlights the steps that are used to determine the location of a disease-related gene when the function of the protein product is known. Oligonucleotide probes are used to screen a selected gene sequence.

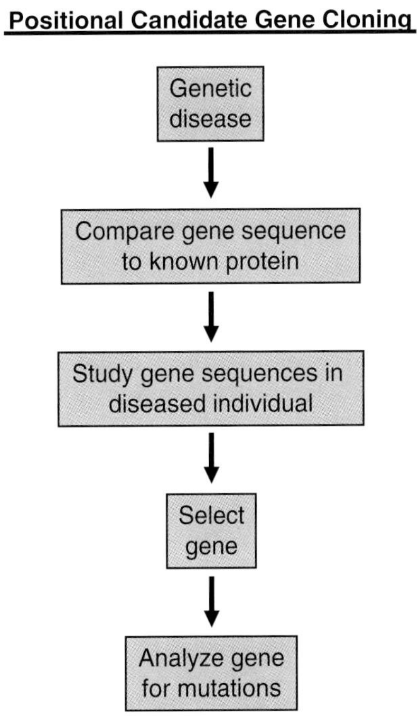

Figure 6.4 Positional candidate gene cloning. The flow diagram highlights the steps that are used to determine the location of a disease-related gene when the chromosomal location is known and that location is used to identify good candidates for testing.

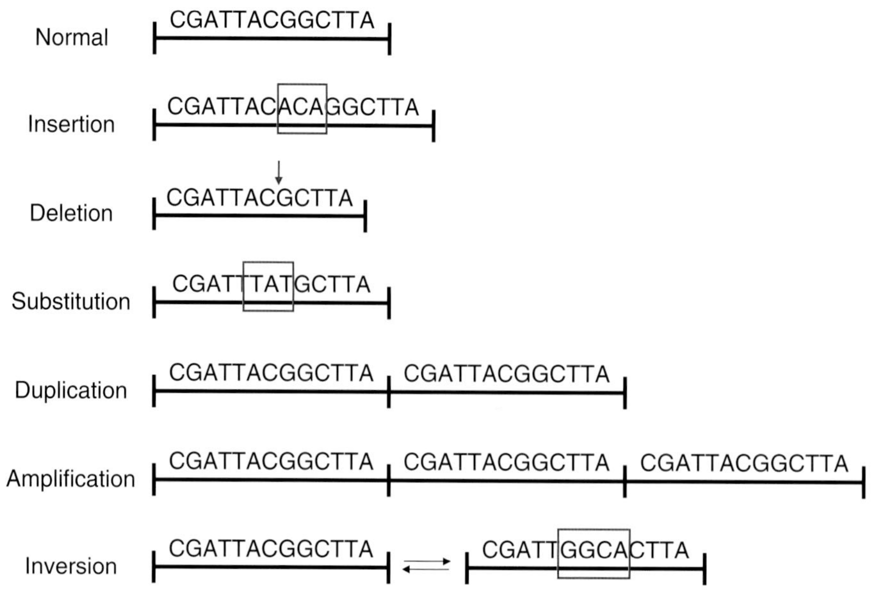

Figure 6.5 DNA rearrangements. Structural variations in the human genome result from rearrangements of DNA from gross changes to single nucleotide alterations. The letters are used to depict nucleotides in a DNA sequence and lines are used to depict a gene. Each DNA rearrangement is outlined by a red box. The red arrow represents a nucleotide deletion.

tandem repeats; micro/minisatellites; and less than 1Kb insertions, deletions, inversions, and duplications are responsible for most of the genetic variation in the human population [40] (Figure 6.5). These genome variations can give rise to diseases through a gain or loss of dosage-sensitive genes [41]. Through the sequencing of the human genome, new techniques such as genome-scanning arrays and comparative DNA sequence analysis have been developed to examine the composition of the human genome [42–46]. These technologies have been important in finding copy-number variants or segments of DNA that are 1 Kb or larger, including insertions, deletions, and duplications [40]. Genomic disorders are influenced by the genome architecture around the recombination event and share a common mechanism for genomic rearrangement, that is, nonallelic homologous recombination or ectopic homologous recombination between low-copy repeats that flank the rearranged DNA segment [41]. Inversions are created by nonallelic homologous recombination events that occur between inverted low-copy repeats, whereas a nonallelic homologous recombination by direct low-copy repeats results in a duplication or deletion [47]. In addition, nucleotide substitutions and point mutations cause alterations in protein sequence and can result in disease.

TYPES OF GENETIC DISEASES

Genetic Diseases Associated with Gene Inversions

Structural variants have been identified in the general population to be the cause of genetic disease in the offspring of parents who exhibit certain DNA inversions. In patients with *Williams-Beuren syndrome,* there is a 1.5 Mb inversion at 7q11.23 that occurs in approximately one-third of the patients' parents with a 5% frequency of this inversion in the general population [48]; this syndrome has an incidence of 1/20,000–50,000 [49]. An inversion that is 4 Mb at 15q12 is associated with *Angelman's syndrome,* and about half of the parents of these patients have this variation as well as 9% of the general population [50]; this syndrome has an incidence of 1/10,000–20,000 [41]. There are diseases in which inversions found in the patients affected have not been detected in the general population [40]. Patients with *hemophilia A* have a 400 Kb inversion in intron 22 in the *factor VIII* gene [51], and two copies are located 400 Kb telomeric in an inverted orientation; this nonallelic homologous recombination event results in inactivation of the *factor VIII* gene [52]. In addition, a smaller inversion in the emerin gene in *Emery-Dreifuss muscular dystrophy* has been identified [53]. *Hunter syndrome* is an X-linked dominant disorder. Nonallelic homologous recombination between the iduronate 2-sulphatase gene (*IDS*) and an *IDS* pseudogene generates a genomic inversion resulting in a disruption of the functional *IDS* gene occurring in approximately 13% of *Hunter syndrome* patients [54,55]. Within the Japanese population, fathers of *Soto syndrome* patients, a microdeletion syndrome, carry a 1.9 Mb inversion variant at 5q35 that predisposes their offspring to this disease [56]. Constitutional translocations in the human genome can be mediated by a polymorphic inversion at olfactory-receptor gene clusters at loci 4p16 and 8p23 which occur at frequencies of 12.5% and 26%, respectively [57]. Heterozygous carriers of these translocations exhibit no phenotypic characteristics, whereas their offspring who inherit these translocations show phenotypes from mild dysmorphic features to *Wolf-Hirschhorn syndrome* characterized by growth defects and severe mental retardation [58]. These examples signify the importance in continuing to characterize inversions within the human genome in the general population in order to examine the risk these variations have on the carriers' offspring.

Genetic Diseases Associated with Gene Deletions

Genomic disorders can be responsible for commonly occurring diseases. For instance, *α-thalassemia* affects 5%–40% of the population in Africa and 40%–80% in South Asia, and results from a homologous deletion of an approximately 4 Kb fragment that is flanked by two α-globin genes on 16q13.3. A nonallelic homologous recombination event between these two copies of the α-globin genes results in the deletion of one functional copy [41,59,60]. Red and green pigment genes are located on Xq23, and individuals who have normal color vision have one copy of the red pigment gene and one or more copies of the green pigment gene [61]. In *red-green color blindness*, which affects 4%–5% of males, deletions or fusions caused by nonallelic homologous recombination occur [62]. In patients with *incontinentia pigmenti*, an 11 Kb deletion occurs by nonallelic homologous recombination between two low-copy repeats with one in the diseased gene (*NEMO*) and one 4 Kb downstream of the gene [63,64]. *Hereditary neuropathy with liability to pressure palsy* (HNPP) is a common autosomal dominant neurological disorder that is caused by a 1.4 Mb deletion of a genomic fragment on 17p12 [65]. The gene *NF1* that encodes for *neurofibromatosis type 1* is located on 17q11.2 and a 1.5 Mb deletion encompassing this gene accounts for 5%–22% of patients with this disease [66,67]. Patients with *DiGeorge syndrome/velocardiofacial syndrome* (DGS/VCFS) can exhibit a 3 Mb deletion within a region-specific repeat unit, LCR22 that is flanked by LCR22A and D, or a 1.5 Mb deletion that is flanked by LCR22A and B located on chromosome 22q11.2 [68]. Patients with this congenital disease experience recurrent infection, heart defects, and known facial features. *Smith-Magenis syndrome* (SMS) effects 1/25,000 individuals and is caused by a 4 Mb deletion on several loci contained within chromosome 17 and depending on the loci involved determines the severity of mental retardation exhibited by the patient [69].

Genetic Diseases Associated with Gene Duplications

Charcot-Marie-Tooth disease (CMT) is an inherited autosomal dominant trait that occurs in about 1/25,000 individuals and is characterized by atrophy of the muscles in the legs, progressing over time to the hands, forearms, and feet. There are two clinical classes of CMT: type I and type II. In CMT type I (CMT1A) 75% of individuals have a duplication in one of the peripheral myelin protein 22 (*PMP22*) genes [70]. Duplication on both chromosomes at 17p12 produces a severe form of CMT1A where essentially there are four copies of the *PMP22* gene [71]. A central nervous system disorder affecting the myelin sheath covering the nerve fibers in the brain is called *Pelizaeus-Merzbacher disease*. The majority of patients with this disease have a duplication of the proteolipid protein gene (*PLP1*) which is found on Xq21–22 [72].

GENETIC DISEASES AND CANCER

The HGP and the completion of the human genome sequence have made a huge impact for the practice of medicine and molecular genetic research. The human genome sequence has helped make advances in the development of designer drugs targeting molecular pathways that disrupt diseases caused by single genes or a complex array of gene products. It is beyond the scope of this chapter to provide a comprehensive review of genetic diseases and/or the genetic causes of cancer. However, included are a few examples of how the HGP has advanced our understanding of human disease and will continue to make an impact forever.

Cystic Fibrosis (CF)

The cystic fibrosis transmembrane regulator (*CFTR*) gene is located on chromosome 7q31.2 and was the first gene to be identified by the HGP [73]. Researchers defined the *CFTR* gene by the positional cloning approach [74]. The function of the CFTR protein is to regulate chloride secretion and the inhibition of sodium absorption across the cell membrane [75,76]. Approximately 1547 mutations of the *CFTR* gene have been described, with the most common mutation as a 3-base pair DNA deletion that results in a loss of the amino acid phenylalanine at position 508 occurring in 66% of CF patients [73,76,77]. Out of these mutations only 23 have been shown directly to cause sufficient loss of *CFTR* to confer CF disease, and these mutations are seen in 85% of the diseased population [78]. Interestingly, two or more *CFTR* mutations can be located in *trans* on two separate chromosomes and this will confer CF. However, if the mutations are found in *cis* on the same chromosome, this is not associated with the disease [78]. Unfortunately, this distinction between *cis* and *trans* on chromosomes is not made in most commercial laboratories [78]. In addition, the different mutations will confer different phenotypic responses to CF with some resulting in milder forms of the disease [75]. Approximately 9.7% of genotyped individuals in the Cystic Fibrosis Foundation Patient Registry have at least one unidentified mutation, but the majority (90%) of *CFTR* mutations can be picked up by regular screening methods [78]. The discovery of the *CFTR* gene has given researchers a better understanding of the etiology, the genetic bases, and the pathobiology of CF [76].

CF is an autosomal recessive disease that occurs in approximately 1 in 3500 newborns, is the most common lethal inherited genetic disease among the

Caucasian population, and affects almost 30,000 Americans [75,76,78]. Treatment advances for patients with CF have increased the survival age from mid-teens in the 1970s, to late-twenties to thirties in the 1990s, to more than 36 years old today [78]. A CF diagnosis is based on several clinical characteristics, a familial history of CF or a positive CF newborn screening test, and a mutation in the *CFTR* gene and/or protein [79]. Newborn screening has been implemented in 40 states currently, and one hopes that by the year 2010 it will be available in all states [78]. CF is a disease that is caused by the improper regulation of the ion channels between the cell cytoplasm and the surrounding fluid, resulting in the inability of the exocrine epithelial cells to transport fluid and electrolytes in and out of the cells [80,81]. CF patients cannot effectively clear inhaled bacteria and have an abnormal accumulation of viscous, dehydrated mucus, and because of this, an excessive inflammatory response to pathogens [75,80].

Francis Collins summarized how important the HGP has been in understanding genetic diseases and used CF as an example:

> *Cystic fibrosis has become the paradigm for the study of genetic diseases and indeed, for the medicine of the future. The notion that it is possible to identify genes whose structure and function are unknown and to use that information to understand given disease and develop "designer" therapies is becoming the central paradigm of biomedical research, and cystic fibrosis is the disease that leads that charge. [82]*

Phenylketonuria (PKU)

Mutations in the phenylalanine hydroxylase gene (*PAH*) encoding the protein L-phenylalanine hydroxylase causes a mental retardation disease called phenylketonuria (PKU) [83]. The inability to hydrolyze phenylalanine to tyrosine can lead to hyperphenylalaninemia and if untreated has a toxic effect on the brain [83]. In the 1980s, the *PAH* gene was cloned and sequenced and mapped to chromosome 12, region 12q23.2 [84,85]. The PKU phenotype is not a simple disease, nor does it have a simple explanation. As with many other genetic diseases, each patient with PKU has to be treated differently [83]. With the help of the HGP, understanding PKU and its resulting phenotype has been extremely beneficial [83]. The locus for *PAH* covers 1.5 Mb of DNA with SNPs, repeat sequences, polymorphisms, and *cis* control elements embedded in the sequence, as well as harboring five other genes, and thus providing for a wide range of disease-causing mutations [83,86]. The *PAH* gene is expressed in the liver and kidney [87].

PKU caused a paradigm shift of attitudes about genetic disease by becoming one of the first disorders to show a treatment effect [83]. PKU is an autosomal recessive inherited disease, causing mental retardation; a mousy odor; light pigmentation; peculiarities of sitting, standing, and walking; as well as eczema and epilepsy [83]. The average incidence of PKU in the United States is 1 in 8000 [88]. PKU is one of the first genetic diseases to have an effective rational therapy [83]. PKU can be identified with a biochemical test in newborns and can be treated by a phenylalanine-free, tyrosine-supplemented diet, which permits normal or near-normal cognitive development [83]. In the adolescent and adult patients, it was difficult until recently to adopt the diet recommendations of PKU based on deficiencies in both organoleptic properties and nutrient content in the food [89]. Fortunately, many diet deficiencies are being overcome, and diagnosis is occurring earlier so that patients start the recommended diets sooner, and are more aggressive throughout life [83].

Breast Cancer

The majority of hereditary breast and ovarian cancers are caused by mutations in the breast cancer-predisposing gene 1 or 2 (*BRCA1* or *BRCA2*). *BRCA1* was found by candidate gene approach in 1991, and *BRCA2* was located by linkage analysis and positional cloning in 1995 using familial breast cancer pedigrees with multiple cases of breast cancer in many generations [90]. *BRCA1* is located on 17q21 encoding an 1863-amino acid polypeptide, and *BRCA2* is found on chromosome 13q12-13 encoding 3418 amino acids [90,91]. *BRCA1* has been implicated in cell-cycle regulation, chromatin remodeling, protein ubiquitylation, and both proteins are involved in DNA repair [92]. In the Ashkenazi Jewish population, there are founder mutations that occur at specific locations in *BRCA1* (185delAG and 5382insC) and *BRCA2* (617delT), but most mutations occur anywhere along the gene, including frameshift or nonsense mutations as well as deletions or duplications [93,94]. DNA-based methods have been recently employed to conduct analysis of both *BRCA1* and *BRCA2* for the presence of genomic rearrangements [95]. The prevalence of genomic rearrangements in *BRCA1* is higher than that of *BRCA2*, accounting for 8%–19% of the total mutations in *BRCA1* and 0%–11% in *BRCA2* mutations [96].

Breast cancer affects one in eight women in the United States, and a woman born in the United States has an average lifetime risk of 13% for developing breast cancer. Familial breast cancer is associated with 10%–20% of all breast cancer cases [97]. Mutations in the *BRCA1* and *BRCA2* genes in women have a 60%–80% increase of developing breast cancer [98]. In addition, women who carry a mutation in the *BRCA1* gene have a 15%–60% lifetime risk for ovarian cancer, which is a much more increased risk when compared to a mutation in the *BRCA2* gene (10%–27%) [99]. Women with *BRCA2* mutations tend to develop ovarian cancer after age 50 [100]. Women who are under the age of 50 and are *BRCA1* mutation carriers have a 57% chance of being diagnosed with breast cancer, and only a 28% chance of developing breast cancer if the *BRCA2* gene is mutated [101]. Interestingly, men who harbor a *BRCA2* mutation are estimated to have a 6% chance of being diagnosed with breast cancer [102]. *BRCA1* breast tumors are found to be more poorly differentiated, whereas *BRCA2* tumors tend to be more high-grade tumors when compared to sporadic breast tumors (nonhereditary) [103,104]. *BRCA1* breast tumors when compared to sporadic breast tumors are frequently negative for estrogen receptor, progesterone receptor, and HER-2/Neu overexpression [105]. It is recommended for women with *BRCA* mutations to begin monthly breast self-examinations at the age of 18 and clinical breast examinations and annual mammograms beginning at age 25 [106].

Nonpolyposis Colorectal Cancer (HNPCC)

Hereditary nonpolyposis colorectal cancer (HNPCC) is caused by mutations in the mismatch repair (MMR) genes *MLH1*, *MSH2*, *MSH6*, and *PMS2* [107]. A hypermutation phenotype was discovered in 1993 in families with HNPCC similar to that observed in MMR-deficient bacteria and yeasts [108]. Linkage analysis and positional cloning in HNPCC families subsequently identified *MSH2* and *MLH1* genes, and mutations in these genes account for 60%–80% of HNPCC diagnosis [109,110]. Additionally, the MMR genes *PMS2* and *MSH6* were associated with HNPCC [111,112]. A higher risk of colorectal cancer occurs in *MSH2* and *MLH1* mutation carriers as compared to *MSH6* or *PMS2* mutation carriers [113]. The *MSH2* and *MSH6* genes are located on chromosome 2p22-p21 and 2p16, respectively. *MLH1* is found on 3p21.3, and *PMS2* is located on chromosome 7p22 [114].

There are approximately 160,000 new cases of colorectal cancer diagnosed in the United States each year, with HNPCC accounting for 2%–7% of diagnosed colorectal cancer, affecting about 1 in 200 individuals [114]. The average age of HNPCC diagnosis is 44 years old [115]. HNPCC can also be called Lynch syndrome in honor of Dr. Henry T. Lynch, professor at Creighton University Medical Center [114]. HNPCC is an autosomal dominant trait and exhibits phenotypic characteristics of less than 100 colonic polyps and early onset of multiple tumors in the colon [114]. The Amsterdam criteria were established for the clinical designation of a family with HNPCC: (i) three or more relatives with colon cancer, one of them must be a first degree relative (parent, child, sibling) of the other two, (ii) at least two affected generations, (iii) one or more members of a family must develop colon cancer before the age of 50, and (iv) familial adenomatous polyposis (FAP) should be excluded from the diagnosis [115].

PERSPECTIVES

Millions of people around the world waited and watched for the completed human genome sequence to be released with the expectation that it would benefit humankind. Decades ago it was not anticipated that genomic disorders would represent such a common cause of human genetic disease [41]. Now from this perspective humans are the best model organisms that we have in order to study the human genome, disease, and its associated phenotypes [41]. Currently, there is a huge amount of sequence information that has been generated from genome sequencing projects, including vertebrates and nonvertebrates [41]. One objective of the human genome sequence is to derive medical benefit from analyzing the DNA sequence of humans. It is undeniable that genomic science will begin to unlock more of the mysteries of complex hereditary factors in heart disease, cancer, diabetes, schizophrenia, and many more [12]. Genetic tests have become available for individuals who have a strong family history or are more susceptible to a particular disorder, such as breast cancer or colorectal cancer [13]. Healthcare professionals will become practitioners of genomic medicine as more genetic information about common illnesses is available and healthy individuals want to protect themselves from illness [13]. Clinicians will have to grasp the understanding and advances of molecular genetics, and a group of physicians, nurses, and other clinicians called *The National Coalition for Health Professional Education in Genetics* has been organized to help prepare for the genomics era [116]. Within the next decade, it is exciting to think that designer drugs will be available for diabetes mellitus, hypertension, mental illness, and many other genetic disorders [13]. In addition, it is likely that all tumors could have a molecular fingerprint associated with them and the promise of individualized medicine by tailoring prescribing practices and management to that person's unique molecular profile [12,13]. Also within a decade or two, it may be possible to sequence the genome of an individual human with minimal laboratory cost (maybe $1000) [15]. If this becomes reality, we can imagine the possibilities for scientific research, clinical care, treatment options, and overall dramatically changing the face of medicine [15]. Francis Collins, one of the pioneers of the Human Genome Project, stated it best when he said, "if the past 50 years of biology is any indication of the future, the best is certainly yet to come" [12].

REFERENCES

1. Weiling F. Historical study: Johann Gregor Mendel 1822–1884. *Am J Med Genet*. 1991;40:1–25.
2. Maeshima K, Eltsov M. Packaging the genome: The structure of mitotic chromosomes. *J Biochem*. 2008;143:145–153.
3. Higgs DR, Vernimmen D, Hughes J, et al. Using genomics to study how chromatin influences gene expression. *Annu Rev Genomics Hum Genet*. 2007;8:299–325.
4. Klug WS, Cummings MR. Chromosome structure and DNA sequence organization. *Concepts of Genetics*. 6th ed. Upper Saddle River, New Jersey: Prentice Hall; 2000.
5. Francke U. Digitized and differentially shaded human chromosome ideograms for genomic applications. *Cytogenet Cell Genet*. 1994;65:206–218.
6. Venter JC, Adams MD, Myers EW, et al. The sequence of the human genome. *Science*. 2001;291:1304–1351.
7. Smit AF, Riggs AD. MIRs are classic, tRNA-derived SINEs that amplified before the mammalian radiation. *Nucleic Acids Res*. 1995;23:98–102.
8. Lyon MF. The Lyon and the LINE hypothesis. *Semin Cell Dev Biol*. 2003;14:313–318.

9. Antequera F, Bird A. Number of CpG islands and genes in human and mouse. *Proc Natl Acad Sci USA*. 1993;90:11995–11999.
10. Gardiner-Garden M, Frommer M. CpG islands in vertebrate genomes. *J Mol Biol*. 1987;196:261–282.
11. Larsen F, Gundersen G, Lopez R, et al. CpG islands as gene markers in the human genome. *Genomics*. 1992;13:1095–1107.
12. Collins FS, Morgan M, Patrinos A. The Human Genome Project: Lessons from large-scale biology. *Science*. 2003;300:286–290.
13. Collins FS, McKusick VA. Implications of the Human Genome Project for medical science. *JAMA*. 2001;285:540–544.
14. Watson JD. The human genome project: Past, present, and future. *Science*. 1990;248:44–49.
15. Guttmacher AE, Collins FS. Welcome to the genomic era. *N Engl J Med*. 2003;349:996–998.
16. Watson JD, Crick FH. Molecular structure of nucleic acids; a structure for deoxyribose nucleic acid. *Nature*. 1953;171:737–738.
17. Francomano CA, Collins FA. The human genome project: Implications for medical practice. *Today's Internist*. 1997;38:11–15.
18. Weber JL, Myers EW. Human whole-genome shotgun sequencing. *Genome Res*. 1997;7:401–409.
19. Green P. Against a whole-genome shotgun. *Genome Res*. 1997;7: 410–417.
20. Pennisi E. Funders reassure genome sequencers. *Science*. 1998; 280:1185.
21. Collins FS, Patrinos A, Jordan E, et al. New goals for the U.S. Human Genome Project: 1998–2003. *Science*. 1998;282:682–689.
22. Rubin GM, Yandell MD, Wortman JR, et al. Comparative genomics of the eukaryotes. *Science*. 2000;287:2204–2215.
23. Adams MD, Celniker SE, Holt RA, et al. The genome sequence of Drosophila melanogaster. *Science*. 2000;287:2185–2195.
24. Myers EW, Sutton GG, Delcher AL, et al. A whole-genome assembly of Drosophila. *Science*. 2000;287:2196–2204.
25. Lockhart DJ, Winzeler EA. Genomics, gene expression and DNA arrays. *Nature*. 2000;405:827–836.
26. Hattori M, Fujiyama A, Taylor TD, et al. The DNA sequence of human chromosome 21. *Nature*. 2000;405:311–319.
27. Dunham I, Shimizu N, Roe BA, et al. The DNA sequence of human chromosome 22. *Nature*. 1999;402:489–495.
28. Pandey A, Mann M. Proteomics to study genes and genomes. *Nature*. 2000;405:837–846.
29. Druker BJ, Lydon NB. Lessons learned from the development of an abl tyrosine kinase inhibitor for chronic myelogenous leukemia. *J Clin Invest*. 2000;105:3–7.
30. Ballabio A. The rise and fall of positional cloning? *Nat Genet*. 1993;3:277–279.
31. Kiyosawa H, Kawashima T, Silva D, et al. Systematic genome-wide approach to positional candidate cloning for identification of novel human disease genes. *Intern Med J*. 2004;34:79–90.
32. Nelson DL. Positional cloning reaches maturity. *Curr Opin Genet Dev*. 1995;5:298–303.
33. Collins FS. Positional cloning moves from perditional to traditional. *Nat Genet*. 1995;9:347–350.
34. Collins FS. Positional cloning: Let's not call it reverse anymore. *Nat Genet*. 1992;1:3–6.
35. Royer-Pokora B, Kunkel LM, Monaco AP, et al. Cloning the gene for an inherited human disorder—chronic granulomatous disease—on the basis of its chromosomal location. *Nature*. 1986;322:32–38.
36. Kainulainen K, Pulkkinen L, Savolainen A, et al. Location on chromosome 15 of the gene defect causing Marfan syndrome. *N Engl J Med*. 1990;323:935–939.
37. Magenis RE, Maslen CL, Smith L, et al. Localization of the fibrillin (FBN) gene to chromosome 15, band q21.1. *Genomics*. 1991; 11:346–351.
38. Dietz HC, Cutting GR, Pyeritz REGR, et al. Marfan syndrome caused by a recurrent de novo missense mutation in the fibrillin gene. *Nature*. 1991;352:337–339.
39. Leach FS, Nicolaides NC, Papadopoulos N, et al. Mutations of a mutS homolog in hereditary nonpolyposis colorectal cancer. *Cell*. 1993;75:1215–1225.
40. Feuk L, Carson AR, Scherer SW. Structural variation in the human genome. *Nat Rev Genet*. 2006;7:85–97.
41. Inoue K, Lupski JR. Molecular mechanisms for genomic disorders. *Annu Rev Genomics Hum Genet*. 2002;3:199–242.
42. Solinas-Toldo S, Lampel S, Stilgenbauer S, et al. Matrix-based comparative genomic hybridization: Biochips to screen for genomic imbalances. *Genes Chromosomes Cancer*. 1997;20:399–407.
43. Sebat J, Lakshmi B, Troge J, et al. Large-scale copy number polymorphism in the human genome. *Science*. 2004;305:525–528.
44. Iafrate AJ, Feuk L, Rivera MN, et al. Detection of large-scale variation in the human genome. *Nat Genet*. 2004;36:949–951.
45. Tuzun E, Sharp AJ, Bailey JA, et al. Fine-scale structural variation of the human genome. *Nat Genet*. 2005;37:727–732.
46. Feuk L, MacDonald JR, Tang T, et al. Discovery of human inversion polymorphisms by comparative analysis of human and chimpanzee DNA sequence assemblies. *PLoS Genet*. 2005;1: e56.
47. Valero MC, de Luis O, Cruces J, et al. Fine-scale comparative mapping of the human 7q11.23 region and the orthologous region on mouse chromosome 5G: The low-copy repeats that flank the Williams-Beuren syndrome deletion arose at breakpoint sites of an evolutionary inversion(s). *Genomics*. 2000;69: 1–13.
48. Osborne LR, Li M, Pober B, et al. A 1.5 million-base pair inversion polymorphism in families with Williams-Beuren syndrome. *Nat Genet*. 2001;29:321–325.
49. Greenberg F. Williams syndrome. *Pediatrics*. 1989;84:922–923.
50. Gimelli G, Pujana MA, Patricelli MG, et al. Genomic inversions of human chromosome 15q11-q13 in mothers of Angelman syndrome patients with class II (BP2/3) deletions. *Hum Mol Genet*. 2003;12:849–858.
51. Lakich D, Kazazian HH, Jr., Antonarakis SE, et al. Inversions disrupting the factor VIII gene are a common cause of severe haemophilia A. *Nat Genet*. 1993;5:236–241.
52. Naylor J, Brinke A, Hassock S, Green PM, Giannelli F. Characteristic mRNA abnormality found in half the patients with severe haemophilia A is due to large DNA inversions. *Hum Mol Genet*. 1993;2:1773–1778.
53. Small K, Iber J, Warren ST. Emerin deletion reveals a common X-chromosome inversion mediated by inverted repeats. *Nat Genet*. 1997;16:96–99.
54. Timms KM, Bondeson ML, Ansari-Lari MA, et al. Molecular and phenotypic variation in patients with severe Hunter syndrome. *Hum Mol Genet*. 1997;6:479–486.
55. Bondeson ML, Dahl N, Malmgren H, et al. Inversion of the IDS gene resulting from recombination with IDS-related sequences is a common cause of the Hunter syndrome. *Hum Mol Genet*. 1995;4:615–621.
56. Visser R, Shimokawa O, Harada N, et al. Identification of a 3.0-kb major recombination hotspot in patients with Sotos syndrome who carry a common 1.9-Mb microdeletion. *Am J Hum Genet*. 2005;76:52–67.
57. Giglio S, Broman KW, Matsumoto N, et al. Olfactory receptor-gene clusters, genomic-inversion polymorphisms, and common chromosome rearrangements. *Am J Hum Genet*. 2001;68: 874–883.
58. Giglio S, Calvari V, Gregato G, et al. Heterozygous submicroscopic inversions involving olfactory receptor-gene clusters mediate the recurrent t(4;8)(p16;p23) translocation. *Am J Hum Genet*. 2002;71:276–285.
59. Higgs DR, Old JM, Pressley L, et al. A novel alpha-globin gene arrangement in man. *Nature*. 1980;284: 632–635.
60. Lauer J, Shen CK, Maniatis T. The chromosomal arrangement of human alpha-like globin genes: Sequence homology and alpha-globin gene deletions. *Cell*. 1980;20:119–130.
61. Vollrath D, Nathans J, Davis RW. Tandem array of human visual pigment genes at Xq28. *Science*. 1988;240:1669–1672.
62. Nathans J, Piantanida TP, Eddy RL, et al. Molecular genetics of inherited variation in human color vision. *Science*. 1986; 232:203–210.
63. Aradhya S, Woffendin H, Jakins T, et al. A recurrent deletion in the ubiquitously expressed NEMO (IKK-gamma) gene accounts for the vast majority of incontinentia pigmenti mutations. *Hum Mol Genet*. 2001;10:2171–2179.
64. Smahi A, Courtois G, Vabres P, et al. Genomic rearrangement in NEMO impairs NF-kappa B activation and is a cause of incontinentia pigmenti. The International Incontinentia Pigmenti (IP) Consortium. *Nature*. 2000;405:466–472.

65. Chance PF, Alderson MK, Leppig KA, et al. DNA deletion associated with hereditary neuropathy with liability to pressure palsies. *Cell.* 1993;72:143–151.
66. Upadhyaya M, Ruggieri M, Maynard J, et al. Gross deletions of the neurofibromatosis type 1 (NF1) gene are predominantly of maternal origin and commonly associated with a learning disability, dysmorphic features and developmental delay. *Hum Genet.* 1998;102:591–597.
67. Valero MC, Pascual-Castroviejo I, Velasco E, et al. Identification of de novo deletions at the NF1 gene: No preferential paternal origin and phenotypic analysis of patients. *Hum Genet.* 1997;99:720–726.
68. Edelmann L, Pandita RK, Spiteri E, et al. A common molecular basis for rearrangement disorders on chromosome 22q11. *Hum Mol Genet.* 1999;8:1157–1167.
69. Chen KS, Manian P, Koeuth T, et al. Homologous recombination of a flanking repeat gene cluster is a mechanism for a common contiguous gene deletion syndrome. *Nat Genet.* 1997; 17:154–163.
70. Pasternak JJ. Molecular genetics of neurological disorders. In: Pasternak JJ, ed. *An Introduction to Human Molecular Genetics: Mechanisms of Inherited Diseases.* Bethesda, Maryland: Fitzgerald Science Press, 1999: 286–287.
71. Lupski JR, de Oca-Luna RM, Slaugenhaupt S, et al. DNA duplication associated with Charcot-Marie-Tooth disease type 1A. *Cell.* 1991;66:219–232.
72. Inoue K, Osaka H, Imaizumi K, et al. Proteolipid protein gene duplications causing Pelizaeus-Merzbacher disease: Molecular mechanism and phenotypic manifestations. *Ann Neurol.* 1999;45:624–632.
73. Tolstoi LG, Smith CL. Human Genome Project and cystic fibrosis—A symbiotic relationship. *J Am Diet Assoc.* 1999;99:1421–1427.
74. Rommens JM, Iannuzzi MC, Kerem B, et al. Identification of the cystic fibrosis gene: Chromosome walking and jumping. *Science.* 1989;245:1059–1065.
75. Davies JC, Alton EW, Bush A. Cystic fibrosis. *BMJ.* 2007;335: 1255–1259.
76. Davis PB. Cystic fibrosis: New perceptions, new strategies. *Hosp Pract (Off Ed).* 1992;27:79–83, 87–88, 93–94.
77. Kerem B, Rommens JM, Buchanan JA, et al. Identification of the cystic fibrosis gene: Genetic analysis. *Science.* 1989;245: 1073–1080.
78. Farrell PM, Rosenstein BJ, White TB, et al. Guidelines for diagnosis of cystic fibrosis in newborns through older adults: Cystic Fibrosis Foundation consensus report. *J Pediatr.* 2008;153:S4–S14.
79. Rosenstein BJ, Cutting GR. The diagnosis of cystic fibrosis: A consensus statement. Cystic Fibrosis Foundation Consensus Panel. *J Pediatr.* 1998;132:589–595.
80. Tsui LC, Durie P. Genotype and phenotype in cystic fibrosis. *Hosp Pract (Minneap).* 1997;32:115–118, 123–129.
81. Quinton PM. Cystic fibrosis: A disease in electrolyte transport. *Faseb J.* 1990;4:2709–2717.
82. Wilson JM, Wilson CB. Cystic fibrosis: Pointing to the medicine of the future. In: *Highlights (Selected Proceedings from the Tenth Annual North American Cystic Fibrosis Conference, October 24–27, 1996).* 1997: 1–2.
83. Scriver CR. The PAH gene, phenylketonuria, and a paradigm shift. *Hum Mutat.* 2007;28:831–845.
84. Woo SL, Lidsky AS, Guttler F, et al. Cloned human phenylalanine hydroxylase gene allows prenatal diagnosis and carrier detection of classical phenylketonuria. *Nature.* 1983;306:151–155.
85. Kwok SC, Ledley FD, DiLella AG, et al. Nucleotide sequence of a full-length complementary DNA clone and amino acid sequence of human phenylalanine hydroxylase. *Biochemistry.* 1985;24: 556–561.
86. Konecki DS, Wang Y, Trefz FK, et al. Structural characterization of the 5′ regions of the human phenylalanine hydroxylase gene. *Biochemistry.* 1992;31:8363–8368.
87. Wang Y, DeMayo JL, Hahn TM, et al. Tissue- and development-specific expression of the human phenylalanine hydroxylase/chloramphenicol acetyltransferase fusion gene in transgenic mice. *J Biol Chem.* 1992;267:15105–15110.
88. DiLella AG, Kwok SC, Ledley FD, et al. Molecular structure and polymorphic map of the human phenylalanine hydroxylase gene. *Biochemistry.* 1986;25:743–749.
89. Scriver CR. Mutants: Consumers with special needs. *Nutr Rev.* 1971;29:155–158.
90. Wooster R, Neuhausen SL, Mangion J, et al. Localization of a breast cancer susceptibility gene, BRCA2, to chromosome 13q12–13. *Science.* 1994;265:2088–2090.
91. Miki Y, Swensen J, Shattuck-Eidens D, et al. A strong candidate for the breast and ovarian cancer susceptibility gene BRCA1. *Science.* 1994;266:66–71.
92. Narod SA, Foulkes WD. BRCA1 and BRCA2: 1994 and beyond. *Nat Rev Cancer.* 2004;4:665–676.
93. Struewing JP, Hartge P, Wacholder S, et al. The risk of cancer associated with specific mutations of BRCA1 and BRCA2 among Ashkenazi Jews. *N Engl J Med.* 1997;336:1401–1408.
94. Nagy R, Sweet K, Eng C. Highly penetrant hereditary cancer syndromes. *Oncogene.* 2004;23:6445–6470.
95. Palma MD, Domchek SM, Stopfer J, et al. The relative contribution of point mutations and genomic rearrangements in BRCA1 and BRCA2 in high-risk breast cancer families. *Cancer Res.* 2008;68:7006–7014.
96. Woodward AM, Davis TA, Silva AG, et al. Large genomic rearrangements of both BRCA2 and BRCA1 are a feature of the inherited breast/ovarian cancer phenotype in selected families. *J Med Genet.* 2005;42:e31.
97. Madigan MP, Ziegler RG, Benichou J, et al. Proportion of breast cancer cases in the United States explained by well-established risk factors. *J Natl Cancer Inst.* 1995;87:1681–1685.
98. Ford D, Easton DF, Bishop DT, et al. Risks of cancer in BRCA1-mutation carriers. Breast Cancer Linkage Consortium. *Lancet.* 1994;343:692–695.
99. Risch HA, McLaughlin JR, Cole DE, et al. Prevalence and penetrance of germline BRCA1 and BRCA2 mutations in a population series of 649 women with ovarian cancer. *Am J Hum Genet.* 2001;68:700–710.
100. Chen S, Parmigiani G. Meta-analysis of BRCA1 and BRCA2 penetrance. *J Clin Oncol.* 2007;25:1329–1333.
101. Ford D, Easton DF, Stratton M, et al. Genetic heterogeneity and penetrance analysis of the BRCA1 and BRCA2 genes in breast cancer families. The Breast Cancer Linkage Consortium. *Am J Hum Genet.* 1998;62:676–689.
102. Liede A, Karlan BY, Narod SA. Cancer risks for male carriers of germline mutations in BRCA1 or BRCA2: A review of the literature. *J Clin Oncol.* 2004;22:735–742.
103. Lakhani SR, Jacquemier J, Sloane JP, et al. Multifactorial analysis of differences between sporadic breast cancers and cancers involving BRCA1 and BRCA2 mutations. *J Natl Cancer Inst.* 1998;90:1138–1145.
104. Chappuis PO, Nethercot V, Foulkes WD. Clinico-pathological characteristics of BRCA1- and BRCA2-related breast cancer. *Semin Surg Oncol.* 2000;18:287–295.
105. Lakhani SR, Gusterson BA, Jacquemier J, et al. The pathology of familial breast cancer: Histological features of cancers in families not attributable to mutations in BRCA1 or BRCA2. *Clin Cancer Res.* 2000;6:782–789.
106. Burke W, Daly M, Garber J, et al. Recommendations for follow-up care of individuals with an inherited predisposition to cancer. II. BRCA1 and BRCA2. Cancer Genetics Studies Consortium. *JAMA.* 1997;277:997–1003.
107. Rahman N, Scott RH. Cancer genes associated with phenotypes in monoallelic and biallelic mutation carriers: New lessons from old players. *Hum Mol Genet.* 2007;16(Spec. No. 1): R60–66.
108. Aaltonen LA, Peltomaki P, Leach FS, et al. Clues to the pathogenesis of familial colorectal cancer. *Science.* 1993;260: 812–816.
109. Fishel R, Lescoe MK, Rao MR, et al. The human mutator gene homolog MSH2 and its association with hereditary nonpolyposis colon cancer. *Cell.* 1993;75:1027–1038.
110. Syngal S, Fox EA, Eng C, et al. Sensitivity and specificity of clinical criteria for hereditary non-polyposis colorectal cancer associated mutations in MSH2 and MLH1. *J Med Genet.* 2000; 37:641–645.
111. Nicolaides NC, Papadopoulos N, Liu B, et al. Mutations of two PMS homologues in hereditary nonpolyposis colon cancer. *Nature.* 1994;371:75–80.

112. Miyaki M, Konishi M, Tanaka K, et al. Germline mutation of MSH6 as the cause of hereditary nonpolyposis colorectal cancer. *Nat Genet.* 1997;17:271–272.
113. Plaschke J, Engel C, Kruger S, et al. Lower incidence of colorectal cancer and later age of disease onset in 27 families with pathogenic MSH6 germline mutations compared with families with MLH1 or MSH2 mutations: The German Hereditary Nonpolyposis Colorectal Cancer Consortium. *J Clin Oncol.* 2004;22: 4486–4494.
114. Pasternak JJ. Molecular genetics of cancer syndromes. In: Pasternak JJ, ed. *An Introduction to Human Molecular Genetics.* Bethesda, MD: Fitzgerald Science Press; 1999: 401–404.
115. Vasen HF, Watson P, Mecklin JP, et al. New clinical criteria for hereditary nonpolyposis colorectal cancer (HNPCC, Lynch syndrome) proposed by the International Collaborative group on HNPCC. *Gastroenterology.* 1999;116:1453–1456.
116. Collins FS. Preparing health professionals for the genetic revolution. *JAMA.* 1997;278:1285–1286.

Chapter

7

The Human Transcriptome: Implications for the Understanding of Human Disease

Matthias E. Futschik . Wolfgang Kemmner . Reinhold Schäfer . Christine Sers

INTRODUCTION

The most fascinating aspect of transcriptomics is that the entire set of messenger RNA (mRNA) molecules or transcripts produced in a population of cells or in tissues can be analyzed simultaneously. The present microarray technology produces devices equivalent to the size of a stamp for gene expression profiling. Analyzing the transcriptome is a challenging task, since the mRNA content of a biological entity is heterogeneous and can vary substantially. The abundance of individual transcripts varies from a few copies to hundreds or thousands of copies per cell [1]. The kinds and copy numbers of individual transcripts expressed at a given time depend on the developmental stage, on external conditions, and environmental stimuli. Quantitative and qualitative alterations of mRNAs can be directly linked to the molecular mechanism of disease or reflect the downstream consequences of these disease processes.

This chapter outlines the methodological prerequisites for transcriptome analysis and describes typical applications in molecular cell biology and pathology. During the last three decades, technology development and experimental approaches aiming at mRNA analysis were significantly fueled by molecular cancer research. One of the main reasons for progress in this area was the availability of relevant cell lines that could be propagated indefinitely and served as reproducible sources of RNA and of sufficient quantities of normal and diseased tissues. A strong motivation lay in the demand for distinguishing as many transcripts as possible in normal and tumorigenic cells to understand cancer-specific alterations in gene expression. While early work along these lines was mostly related to pathogenesis, more recent applications deal with diagnostic issues such as tumor outcome, prognosis, and therapy response prediction.

GENE EXPRESSION PROFILING: THE SEARCH FOR CANDIDATE GENES INVOLVED IN PATHOGENESIS

To date, microarray-based expression profiling is accepted as the gold standard in transcriptome analysis. Microarray technology has gradually improved over the last decade both in academic and industrial/commercial settings to meet high technical and bioinformatic quality standards. Before microarrays were available for most researchers in sufficient quantity and quality (as well as affordable at reasonable costs), alternative techniques were instrumental in answering questions related to the quantity of transcripts expressed ubiquitously, to identifying tissue-specific expression patterns and candidate genes related to disease.

Early Gene Expression Profiling Studies

Intriguingly, the question of how many transcription units distinguish normal from tumor cells, a question that is expected to be the domain of transcriptomics using microarrays, was addressed nearly at the same time when techniques in molecular biology permitted the identification and thorough analysis of individual mRNAs. In 1977 and 1980, researchers described the northern blot technique for transferring electrophoretically separated

Molecular Pathology © 2009, Elsevier, Inc. All Rights Reserved.

RNA from an agarose gel to paper strips, the coupling of the RNA to the paper surface and the detection of specific RNA bands by hybridization with ^{32}P-labeled DNA probes followed by autoradiography for the first time [2,3]. A report published in 1980 provided evidence for the complexity of cellular transformation at the RNA level, when scientists had studied the RNA pool of chicken embryonic breast muscle cells infected with Rous sarcoma virus (RSV). The authors of the paper compared the hybridization kinetics of nuclear RNA preparations from normal and RSV-transformed cells, respectively, incubated in solution with tracer amounts of labeled single-copy chicken DNA. Based on the assumption that an average transcription unit is about 10 times larger than its corresponding mRNA, the authors concluded that the observed increase in the number of stable transcription products in transformed cells relative to normal cells was equivalent to approximately 1,000 transcription units [4]. Several years later, other scientists used a more sophisticated approach for contrasting mRNA patterns of cellular material obtained from colon tumor biopsies [5]. The researchers took advantage of the molecular cloning techniques that allowed the establishment of a set of complementary DNAs (cDNAs) obtained by reverse transcription of mRNA. The reference cDNA library of some 4,000 clones represented abundant and middle abundant RNA sequences. Replicas of the library were then hybridized to ^{32}P-labeled cDNA probes synthesized from polyadenylated RNA from small biopsies obtained from normal and neoplastic intestinal mucosa. The comparison of normal colonic mucosa with carcinomas showed expression alterations of $\sim$7% of the cloned sequences and was extrapolated to the entire, yet unknown, set of transcripts. The number of alterations was smaller between normal mucosa and benign adenomas indicating that transcriptional changes accumulate during cancer progression.

cDNA Libraries and Data Mining

Further advances in deciphering cancer-related transcripts were driven by increased efforts in cDNA cloning, and sequence analysis. Collections of cDNAs were obtained from various normal and diseased tissues, as well as from reference cell lines. The functional characterization of transcribed sequences progressed at the same time. However, due to the complexity of gene function in biological systems, functional information lagged significantly behind sequence information. The large cDNA collections deposited in expression databases often provided only partial sequence information. The corresponding cDNAs known to be expressed in various tissues or cell types analyzed were designated expressed sequence tags (or ESTs). With increasing entries into these EST catalogues, it became feasible to merge overlapping partial sequences and eventually to define full-length open-reading frames (ORFs). As a practical consequence of the global gene expression information provided by cDNA/EST databases, an approach termed the electronic northern became feasible. The electronic northern analysis facilitated prediction of expression changes between normal and diseased tissues. Extensive mining of EST databases using stringent statistical tests permitted identification of candidate genes whose altered (stimulated or reduced) expression correlated with the disease state [6].

cDNA Subtraction

The data mining approach was limited by the existing sequence information and the available gene annotations. To circumvent this bias, researchers established several elegant methods that permitted enrichment of mRNA sequences (or cDNAs) associated with special experimental conditions such as cellular stress or oncogenic transformation, or with particular cellular features such as tumorigenicity or metastatic potential. The methods established were cDNA subtraction, differential display PCR (DD), representational difference analysis, and serial analysis of gene expression (SAGE).

In general, cDNA subtraction is a method for separating cDNA molecules that distinguish related cDNA samples, for instance, prepared by reverse transcription of mRNA from normal precursor cells and related neoplastically transformed cells. The basis of subtraction is that cDNAs prepared from two different cell types to be compared are rendered single-stranded, subsequently mixed, and incubated to allow annealing of sequences common to both cell species. These sequences will hybridize, while sequences unique to one of the cells will stay single-stranded. In the classical subtraction approach, single-stranded and double-stranded cDNAs were separated by hydroxylapatite chromatography [7]. Subsequently, the unique cDNA fragments are cloned and sequenced. The major drawback of this method is that the enrichment of differentially expressed sequences usually does not exceed a factor of 100, that abundant mRNAs (cDNAs) are over-represented due to the lack of normalization, and that rare transcripts are not detected at all. These inherent disadvantages were overcome by development of a method called suppression subtractive hybridization (SSH), a PCR-based subtraction method that combines normalization and subtraction into a single procedure. Differential amplification of unique cDNA fragments is achieved by ligating different primers to each restricted cDNA originating from the cell types to be compared prior to the annealing step and the PCR. The normalization step equalizes the abundance of cDNA fragments within the target population, and the subtraction step excludes sequences that are common to the cell populations being contrasted. Using this method, the probability of recovering differentially expressed cDNAs of low abundance is largely increased (by a factor of 1,000 or more) [8]. For example, SSH was used to report on transformation target genes related to oncogenic RAS signaling on a genome-wide scale [9] (Figure 7.1). RDA is a technique that combines subtractive hybridization with PCR-mediated kinetic enrichment for the detection of differences between two complex genomes [10]. Later, the protocol was modified to look for differences in transcript expression as well [11].

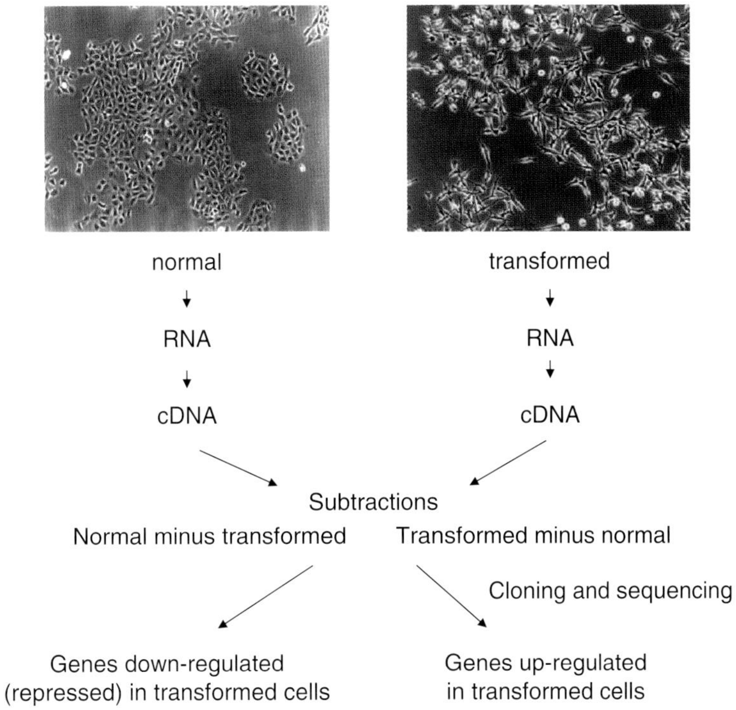

Figure 7.1 **A cDNA subtraction approach to identify genes differentially expressed upon conversion from the normal to the transformed state.** In this example, immortalized normal epithelial cells (phase contrast microscopy, magnification 100-fold) were transformed by the *KRAS* oncogene.

Differential Display PCR

Differential display PCR is a method to separate and clone individual mRNAs that are differentially expressed by means of the polymerase chain reaction. A set of oligonucleotide primers is used, one being anchored to the polyadenylated tail of a subset of mRNAs, the other being short and arbitrary in sequence to allow annealing at different sites relative to the first primer. The mRNA subpopulations defined by the primer pairs are amplified after reverse transcription and the products resolved (displayed) on DNA sequencing gels. Differential display visualizes mRNA compositions of cells by displaying subsets of mRNAs as short cDNAs. The beauty of this approach is that many samples can be run in parallel to reveal differences in mRNA composition [12]. The differentially expressed cDNA fragments can be recovered by cloning techniques. An early application of DD was the identification of genes differentially expressed in breast cancer versus mammary epithelial cells [13].

Serial Analysis of Gene Expression

While the previously discussed approaches directly aim at identifying important differences between closely related cell types, the key element of the SAGE method is to represent all transcripts in a given cell type in a quantitative manner. The basic principle of SAGE is that short nucleotide sequence tags of 10 to 14 base pairs contain sufficient information to uniquely identify transcripts. Moreover, concatenation of these short sequence tags permits an efficient analysis of transcripts serially by sequencing of multiple tags within a single cloned element [14]. More recent variants of the method are based on longer sequence tags and integrate microarray technology [15]. Two years after the initial publication of the method, Johns Hopkins University researchers for the first time reported on gene expression profiles in normal and cancer cells based on SAGE [14]. The authors confirmed previous findings on RNA abundance in cells that had been obtained with the help of Rot curves that display RNA-DNA reassociation kinetics [16]. The total number of transcripts varied from approx. 14,000 to 20,000 between cell populations. Most transcripts (86%) were expressed at fewer than 5 copies per cell; however, the bulk of the mRNA mass consisted of more abundant transcripts (more than 5 copies per cell). The relative expression levels of transcripts were determined by dividing the number of tags observed in tumor and normal tissue. Most transcripts were expressed at similar levels. However, 548 of 14,000 to 20,000 transcripts were overrepresented or underrepresented in tumor versus normal cells. The average difference in expression for these transcripts was 15-fold. About 20% of them were less than 3-fold different. The authors also addressed the issue of whether cultured cell lines, frequently used in molecular cancer research, display gene expression patterns that mimic those found in the organ microenvironment. Interestingly, 72% of transcripts expressed at reduced levels in cancer specimens were also expressed at lower levels in cell lines. Likewise, 43% of transcripts exhibiting elevated expression in cancers were also upregulated in cell lines. Useful links and SAGE databases can be found at http://www.sagenet.org/.

A procedure very similar to SAGE is used to study cellular microRNAs (miRNAs), which are short ~22-nucleotide segments of RNA that have been found to play an important role in gene regulation. Small RNAs are isolated, linkers are added to each of them, and the RNA is converted to cDNA. Afterwards the linkers containing internal restriction sites are digested with the appropriate restriction enzyme and the sticky ends are concatamerized. The concatamers are ligated into plasmid vectors and cloned, followed by sequencing. In this way, the expression levels of miRNA can be quantitatively assessed by counting the number of times they are present.

TRANSCRIPTOME ANALYSIS BASED ON MICROARRAYS: TECHNICAL PREREQUISITES

To date, microarrays are utilized by most researchers in studies related to (i) pathogenic processes at a genome-wide level, (ii) examination of drug effects, and (iii) elucidation of clinical features of disease that cannot be recognized by currently available molecular techniques or conventional histopathology and immunopathology. Microarray technology was pioneered by Pat Brown and colleagues at Stanford University. These researchers not only published the first applications of microarray to study biological questions, but also described the necessary technical devices in detail [17]. In this way, they contributed to the rapid dissemination of the technology. In parallel, microarray technology was developed at Affymetrix (Santa Clara, CA) [18]. The central element common to the various forms of these techniques is that DNA-molecules, cDNA fragments or oligonucleotides are arrayed and immobilized at defined positions on a solid support or matrix. This method extends the existing technique of membrane-based arrays that were interrogated using radioactively labeled cDNA. The probes assembled on solid supports are hybridized with complementary and fluorescent dye-labeled RNA or DNA molecules (targets) derived from biological specimens such as cells, tissues, or blood. Fluorescent dye staining intensity after hybridization obtained within the position of the probe is a measure of the abundance of the corresponding nucleotide sequence in the complex mixture of RNA/cDNA targets. The different kinds of microarrays available today are distinguished by the number, density, design, and size of oligonucleotides or cDNA probes; the manner of chip manufacturing; and the experimental protocols for target hybridization.

Typical Workflow of a Microarray Experiment Starting from Tissue Samples

Frozen tissue samples are dissected, fixed on glass slides, and stained. Histological characterization reveals the composition of the tissue, including the cell types of interest (and their frequency), the extent of necrotic areas (which contain degraded RNA), and presence of fatty tissue (from which RNA extraction is difficult). Optionally, laser capture microdissection can be used to precisely dissect the cells and tissue areas of interest. Subsequently, RNA isolation is performed, RNA yield is determined, and RNA quality is checked by electrophoresis. Isolated RNA is used for synthesis of labeled sample nucleic acid, mostly cDNA or antisense RNA (aRNA), which is quality-controlled by absorbance measurements. Most commonly, the last step is hybridization of the fluorescent dye-labeled sample nucleic acid to the probe DNA on the microarray (Figure 7.2).

Production of Microarrays

Microarrays represent a solid support (typically a glass slide or silicon surface) onto which probes are covalently linked using a chemical matrix (via epoxy-silane or amino-silane). The probes are dispensed either by contact

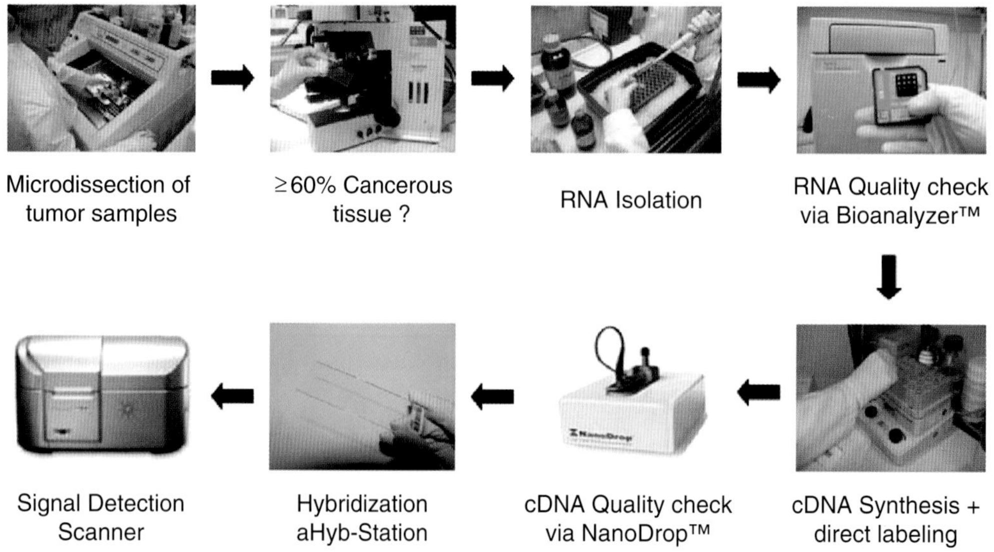

Figure 7.2 Laboratory workflow of a typical microarray experiment.

spotting or applied as micro-droplets by techniques resembling ink-jet procedures used in printing. One of the industrial suppliers (Affymetrix Inc.) produces microarrays using photolithographic methods as in silicone chip production. The procedure allows production of high-density arrays containing millions of probes covered in a partially transparent hybridization chamber. In the original fabric, the probes are short sequences of 25 nucleotides. Each potential target sequence is represented with up to 11 different complementary oligonucleotides and 11 paired mismatch probes. A mismatch probe contains a single mismatch located directly in the middle of the 25-base probe sequence. While the perfect match probe shows fluorescence only when a sample nucleic acid binds to it, the paired mismatch probe is used to detect and eliminate any false or contaminating fluorescence within that measurement. Other industrial suppliers (such as Agilent, Illumina, Milteny, and others) and academic facilities use oligonucleotides of $\geq$60 nucleotides or cDNA fragments to improve hybridization specificity. These microarrays are manufactured as open slides and are handled openly or in special hybridization chambers during the entire chip processing.

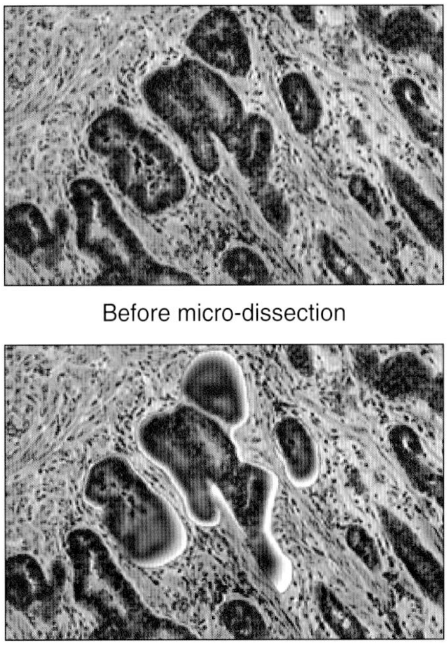

Figure 7.4 Laser microdissection of tissue samples. Cell material was removed from specific epithelial areas.

Preparation of Target RNA

During surgical removal of malignant tumors, one of the first steps is the interruption of the arterial blood supply. From this moment on, the tumor tissue is exposed to hypoxia at body temperature (Figure 7.3). The duration between artery ligation and the final removal of the tumor can vary considerably and is not subject to standardization under clinical conditions. Following tumor resection, logistical constraints may lead to further considerable delay before the tumor material is finally shock-frozen at –80°C. Thus, this lengthy process might lead to a considerable extent of target RNA degradation.

Laser Microdissection of Tumor Tissue

If the sample material is contaminated with nontumor tissue (such as stromal cells and lymphocytes) or if necrotic areas (with cellular debris) occur, data analysis will be severely hampered. Therefore, researchers often try to obtain homogeneous sample material. A convenient (although potentially laborious) approach is laser-assisted microdissection (Figure 7.4). Starting from a complex tissue architecture, areas with carcinoma cells only, stromal material, or any other area of interest, such as material located at the invasion front of the tumor, are obtained. In the example shown, colorectal carcinoma cells have been microdissected with the use of a laser beam. For each sample, 5 μm tissue sections were microdissected and RNA was extracted by a column-based procedure including DNAase digestion. Microdissection of 5×10^6 μm^2 per specimen yields about 10–20 ng RNA in about 2 hours of working time.

RNA Quality Control, Labeling, and Target Amplification

Figure 7.5 shows an electropherogram reflecting RNA quality according to the following criteria: (i) clear

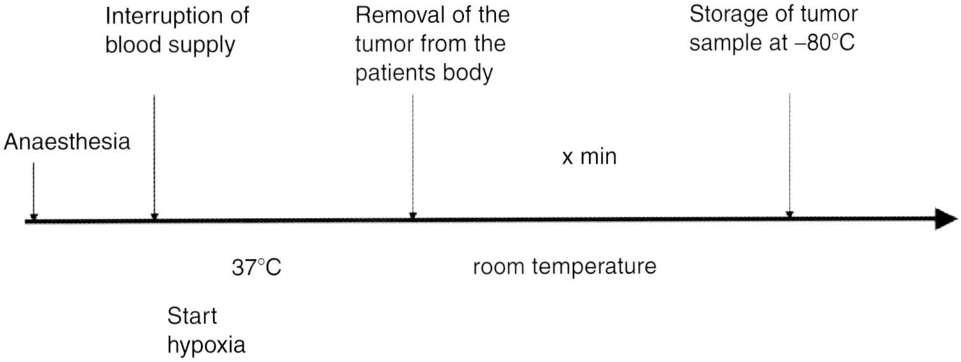

Figure 7.3 Sequential steps leading to the preservation of tumor tissue during surgery.

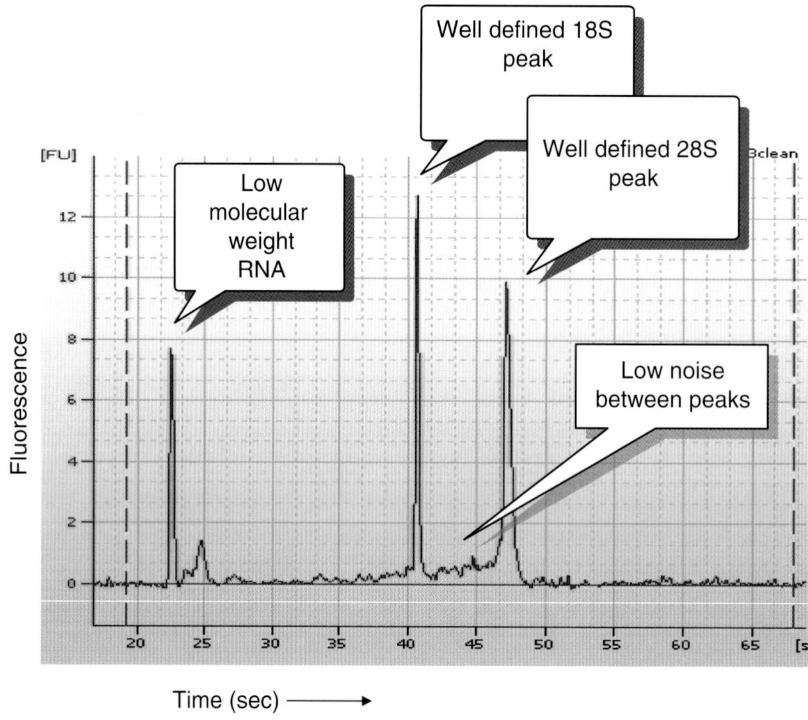

Figure 7.5 RNA quality assessment using Bioanalyser fluorescent spectroscopy.

and well-defined 18 S and 28 S peaks of ribosomal RNA, (ii) low noise between the peaks, and (iii) no or only minimal evidence for low molecular weight material. For hybridization with the probe nucleic acids on the microarray, a labeled target sample nucleic acid is needed. RNA extracted from clinical specimens is used for synthesis of labeled cDNA without amplification of the sample RNA, or for production of aRNA (Figure 7.6). The process of aRNA synthesis allows high amplification of the sample material, which is of relevance if only small amounts of sample material are available, such as after laser microdissection. Whether amplification changes the outcome of the experiment by asymmetric amplification of high-abundance and low-abundance genes is still a matter of discussion. Common amplification procedures utilize Bacteriophage T7, T3, or SP6 RNA polymerases to transcribe RNA from a DNA template (Figure 7.7). The DNA template must have an appropriate polymerase binding site (called T7 in the figure) in its sequence, upstream of the region to be transcribed. A complex of this binding sequence ~20 base pairs in length linked to an oligo-dT sequence is incorporated into the cDNA by reverse transcription of the sample RNA (first strand synthesis). The RNA is then degraded by RNase-treatment, and the second strand is fabricated by DNA-polymerase. The resulting double-stranded cDNA serves as the template for the T7-RNA-polymerase producing RNA in antisense direction (aRNA), compared to the orientation of the template RNA. The entire procedure can be repeated resulting in a 1,000-fold or higher amplification of the RNA. By including labeled nucleotides (NTP) in the *in vitro* transcription reaction, one can incorporate labels into the synthesized RNA (using biotin-labeled UTP to generate biotin-labeled RNA or any kind of UTP-bound fluorescent dye to generate fluorescently labeled RNA). Because signal intensity of red and green fluorophores might not be identical, it is mandatory to invert or swap the fluorescent dyes used for labeling. For instance, in the case of Cy3/Cy5 fluorophores, the green signal intensity is often stronger than the red one. To compensate for this, the labeling reactions are exchanged between the two targets and microarray hybridization is repeated.

Microarray Hybridization: Two-Color Experiment

Hybridization of the target nucleic acid molecule to the probe DNA on the chip is most commonly detected and quantified by fluorescence. This requires a target molecule labeled with a fluorophore such as Cy3 or Cy5. The aim is to determine the relative abundance of the target molecule within the sample solution. Spotting of the probe molecules to the surface is a critical step. In order to account for spot-to-spot variations due to differing amounts of probe molecules in the spots, two-color experiments are used and the ratio of the two fluorophores in any single spot is determined (Figure 7.8). In a typical two-color experiment, RNA is extracted from tumor tissue and neighboring normal tissue. The RNA samples are labeled with different fluorophores, for instance, tumor RNA with a red fluorophore, normal RNA with a green one. Both samples are hybridized together on the microarray. If the spot then appears in red in the

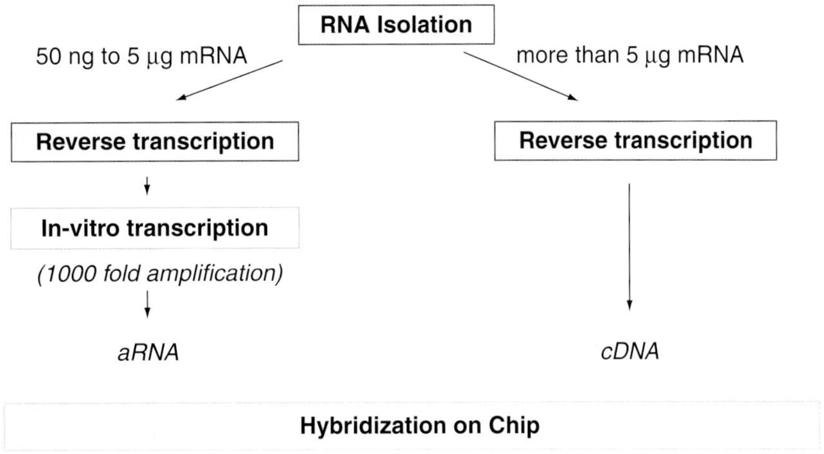

Figure 7.6 **Overview of RNA processing.**

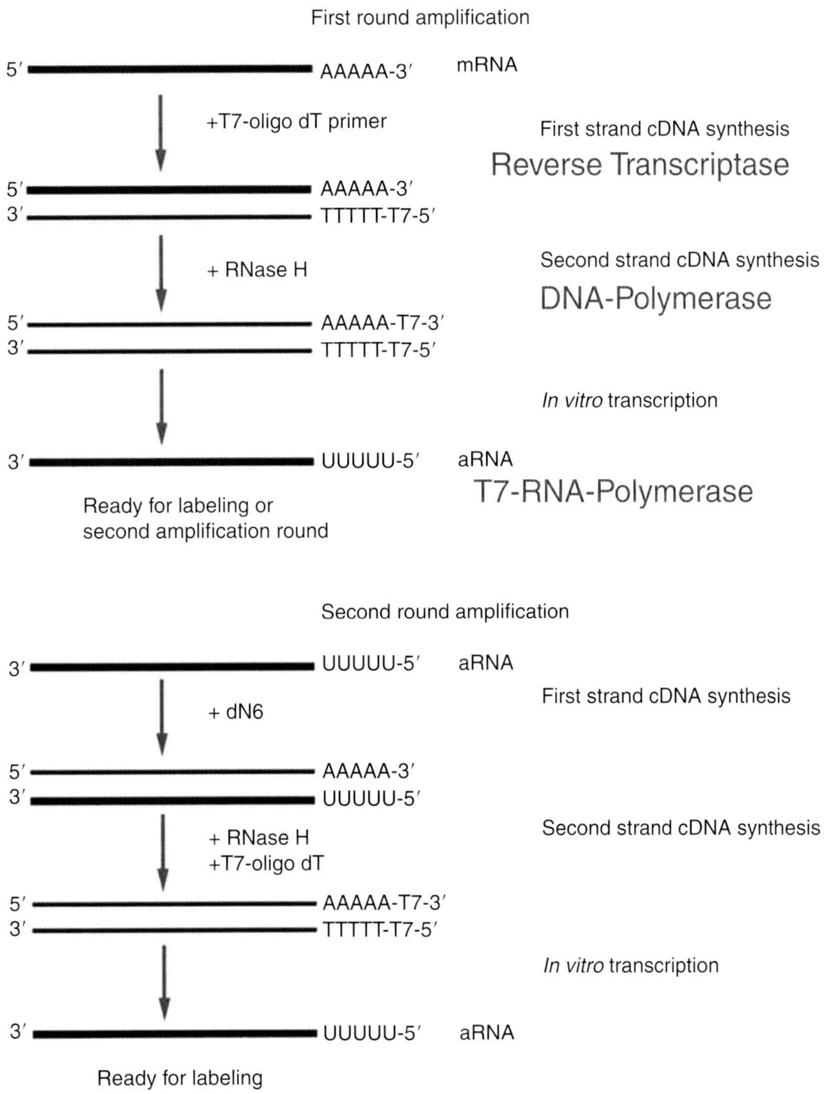

Figure 7.7 **RNA amplification (T7 *in vitro* labeling).** Antisense RNA is generated using T7 polymerase. The microarray is populated with sense oligonucleotides corresponding to the genes of interest.

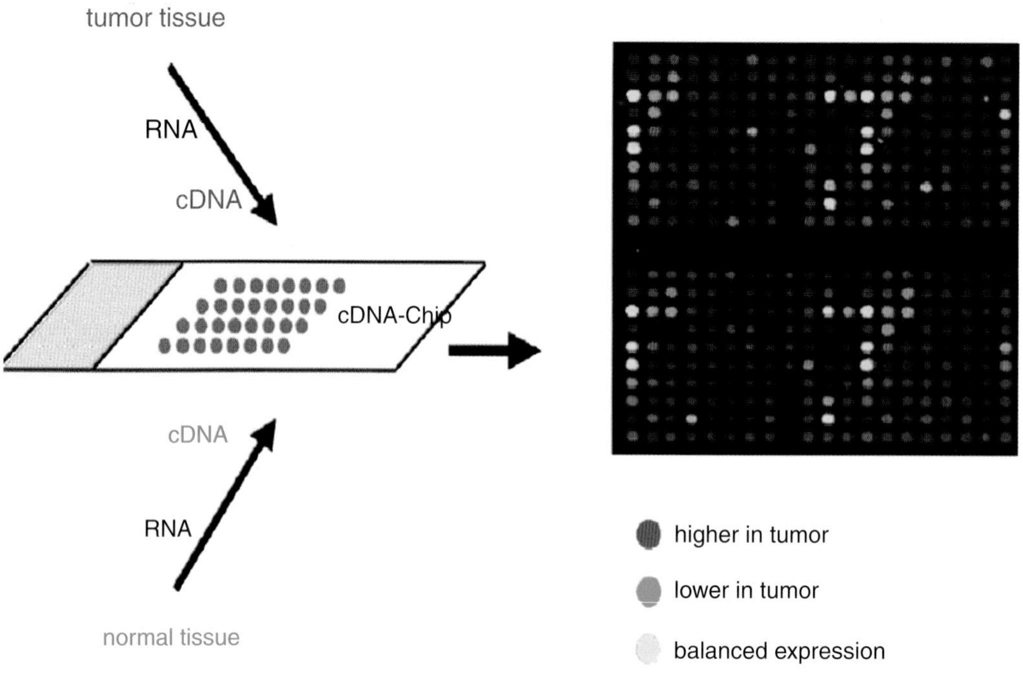

Figure 7.8 Two-color microarray experiment.

microarray scanner, this means higher expression (upregulation) of the corresponding gene in tumor tissue compared to normal tissue. If the spot looks green, higher expression of that gene in normal tissue compared to tumor tissue has been detected. If the gene is equally expressed in tumor versus normal tissue, the combination of the two fluorophores will produce a yellow color.

Image Analysis and Data Processing

Microarrays assess gene activities indirectly by measuring the fluorescence intensities of labeled target cDNA hybridized to cDNA or oligonucleotide probes on the array. There are different detection methods based on light emission (fluorescence), such as confocal laser scanning used by the scanners produced by Affymetrix, Agilent, Axon, HP, or CCD Imaging (Axon, Applied Precision). Other methods for detection of gene expression on microarrays include electrochemically based detection (Motorola) or radiolabeling (Molecular Devices, Hitachi). Most commonly, microarray scanners use laser light for excitation and matching filters and photomultiplier tubes for detection. During scanning, excitation light from the laser source hits the spots on the microarray. Fluorescent probes on the array emit Stokes-shifted light in response to the excitation light, and the emission light is collected by the photomultiplier tube. Scanning plays a pivotal role in the DNA microarray processing workflow and can profoundly affect the quality and reliability of microarray data. Typical sources of error from a microarray scanner include (i) noise in the background light, (ii) nonuniformity of the scan field, (iii) variations in laser brightness and detector gain, and (iv) spectral cross-talk between dye channels. Scanning of the hybridized microarray leads to an intensity picture displaying bright and dark spots (Fig. 7.8, right side). While high resolution scanning (5 μm–10 μm) is the standard, some scanners are capable of scanning with 2 or 3 μm resolution. Using image analysis programs, the raw fluorescence intensity signals are transformed into numerical values for gene expression. This can involve several procedures to ensure the reliability of data. For example, spots which show defects due to printing errors, scratches, and the influence of dust particles should be excluded. Additionally, spot intensities might have to be corrected for any background fluorescence due to nonspecific hybridization. Finally, the obtained measures for gene expression can be analyzed with bioinformatic and statistical methods.

MICROARRAYS: BIOINFORMATIC ANALYSIS

Finding meaningful structures and information in an ocean of numerical values obtained in microarray experiments is a formidable task and demands various approaches of data processing and analysis. In fact, microarray data analysis poses major challenges due to the sheer enormous lots of data produced. Although the type of data analysis naturally depends on the research questions posed, common steps in the analysis include (i) data preprocessing and normalization, (ii) detection of genes with significant fold changes, (iii) clustering and classification of expression profiles, and (iv) functional profiling (Figure 7.9) [19]. These steps are only partially separated. For example, the choice of preprocessing and normalization procedures can have

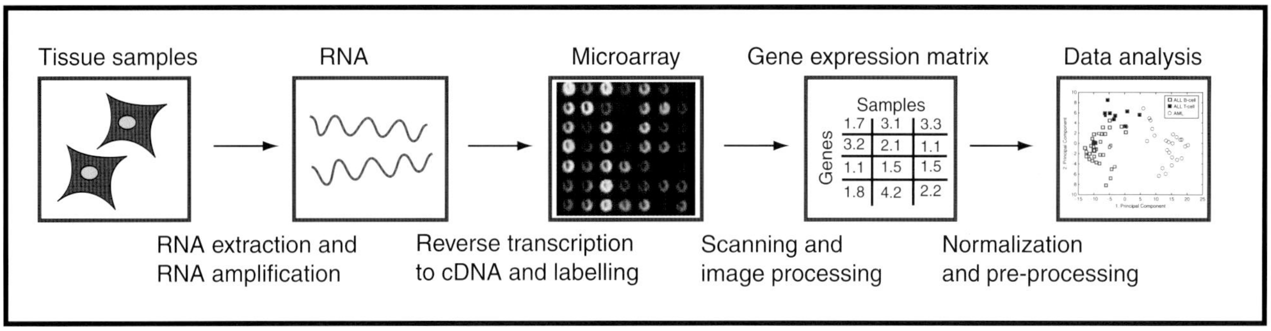

Figure 7.9 Principal steps of a microarray experiment [19]. Reproduced with permission by Elsevier.

considerable impact on the results of clustering and classification. Also, the analysis methods to use might depend on the choice of microarray technology. Here we focus the data analysis for two-color spotted microarray. Other microarray platforms might require different bioinformatic approaches, especially for data-preprocessing and normalization.

Preprocessing, Visualization, and Normalization

Preprocessing

A first preprocessing step for two-channel microarray data is commonly the logarithmic transformation of signal ratios, which offers several advantages. First, fold changes of the same order of magnitude become symmetrical around zero for upregulaton and downregulation. For example, using $\log_2$ transformation, a positive or negative fold change of two is displayed as 1 or −1, respectively. Second, the spot intensities are usually more equally distributed along the scale, which enables an easier detection of intensity bias or saturation effects (Figure 7.10). Third, the variance of intensities is more homogenous with respect to a log intensity scale compared to a linear one. A homogenous variance is often required for statistical tests.

Data Visualization

Plot representations are simple but very helpful tools to detect artifacts or other trends in microarray data. The most basic plots present the two channel intensities versus each other on linear or log scales (Figure 7.10A and Figure 7.10B). More recently, *MA*-plots have become a popular tool for displaying the logged intensity ratio (M) versus the mean logged intensities (A). Although *MA*-plots basically are only a 45° rotation with a subsequent scaling, they reveal intensity-dependent patterns more clearly than the original plot (Figure 7.10C) [20].

Normalization

Raw microarray data are often compromised by systematic errors, such as differences in detection efficiencies, dye labeling, and fluorescence yields. Such signals are corrected by normalization procedures [21]. Although normalization is only an intermediate step in the analysis, it considerably influences the final results. Depending on the experimental design and microarray techniques applied, two main normalization schemes are used: (i) between-slide normalization (to compare signal intensities between different microarrays), and (ii) within-slide normalization (for adjustment of signals of a single microarray). While between-slide normalization is commonly used for one-color chip technology, within-slide normalization is applied mainly to two-color arrays for balancing both channels. An approach referred to as simple global normalization (a within-slide procedure) assumes that the majority of assayed genes are not differentially expressed and that the total amount of transcripts remains constant. Therefore, the ratios can be linearly scaled to the same constant median value in both channels. Alternatively, a set of so-called housekeeping genes can be selected, which are thought to be equally expressed in both samples. The median of these genes can then be taken to adjust the intensity in both channels by a linear transformation, so that the intensity medians of the housekeeping genes are the same. If a dye bias is suspected, the use of an intensity-dependent normalization procedure might be justified. A widespread method is to locally regress the logged signal ratios M with respect to the logged intensities A and to subtract the regressed ratios from the raw ratios. The derived residuals of the regression provide the normalized fold changes (Figure 7.10C). Additional normalization procedures are required, if measured spot intensity ratios show a spatial bias across the array.

Detection of Differential Gene Expression

The standard task in microarray data analysis is the detection of gene expression changes. In early microarray studies, a fixed threshold for fold changes (such as two-fold) was arbitrarily defined to identify differentially expressed genes. However, the setting of fixed thresholds may yield a large number of false positives. Since the measured intensity signals usually are noisy, genes may show differential expression purely due to random signal fluctuations. Particularly, signals related

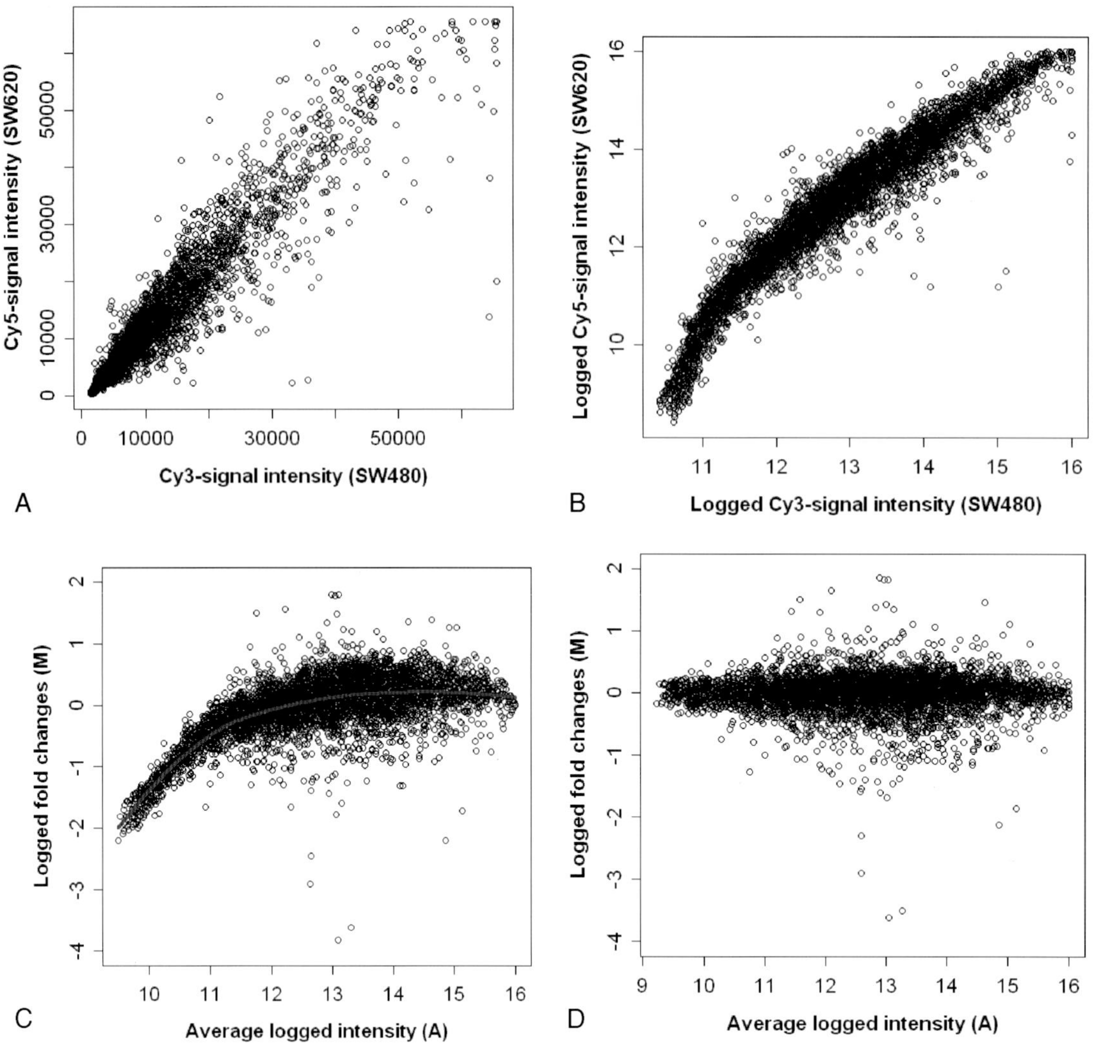

Figure 7.10 Plot representations for signal intensities of a two-color array comparing colorectal cancer cell lines derived from primary carcinoma (SW480; labeled by Cy3) and from a metastasis (SW620; labeled by Cy5) [20]. The spot intensities in both fluorescence channels are shown using linear (A) and log^2-scale (B). The use of log^2-scale reveals nonlinear behavior, reflecting a dye bias toward Cy3 for low-intensity spots. The MA-plot presents this dye bias even more clearly and also a saturation effect in the Cy5 channel for large intensities. (C) To correct the dye bias, one can perform a local regression (red line) of M (D). The obtained residuals of the local regression, i.e., normalized logged fold changes, are well balanced around zero in MA-plot.

to weakly expressed genes are affected by high background noise and therefore require selection based on a larger threshold than strongly expressed genes. To distinguish more stringently noise from meaningful changes in gene expression, statistical tests are nowadays commonly used. Such tests assess the statistical significance of changes based on a set of assumptions about the distribution of the random errors. These errors are not correlated with any experimental variable and unlike systematic errors cannot be corrected by normalization. Random errors also set a limit of detectable changes of gene expression in microarray experiments. To estimate the random errors, experimental replicates are essential. After the random error is estimated, statistical significance can be assigned to changes in gene expression in the framework of a statistical test. Replication of microarray analysis provides also a valuable index of the overall quality of the experiment. Ideally, the goal is a high degree of consistency between different replicates. For subsequent visualization of the results of statistical tests, so-called volcano plots have become a popular mean. They offer the advantage of displaying both significance and fold changes observed (Figure 7.11).

Statistical testing is based on the assessment of the validity of explicitly formulated hypotheses. In general, a null hypothesis H_0 (for instance, that a gene is not differentially expressed) and a contradictory alternative hypothesis H_a (for instance, that a gene is differentially expressed) is set up. The alternative hypothesis is supported if there is evidence against the null hypothesis. The steps in hypothesis testing are as follows: (i) setting up H_0 and H_a, (ii) use of a test statistic to compare the observed values with the values predicted by

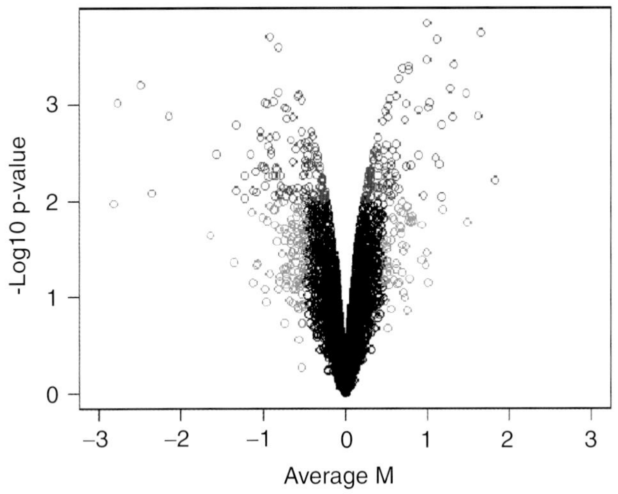

Figure 7.11 The volcano plot is a graph that shows both fold changes and statistical significance of recovered genes. The graph displays negative $\log^{10}$–transformed *p*-values against the $\log^2$-fold changes (M). Volcano plots can be used for the selection of significant genes with a minimal required fold change. Data taken from the experiment described in Figure 7.10 [20]. Genes (displayed in blue and red) having statistically significant differential expression ($p < 0.01$) lie above a horizontal line. Genes (displayed in green and red) with larger fold changes than 1.6 lie outside a pair of vertical lines. Genes which fulfill both criteria are highlighted in red.

H_0, and (iii) definition of a region for the test statistic for which H_0 is rejected in favor of H_a. The level of significance of a test is the probability that the test statistic falls in the rejection region, if H_0 is true. The incorrect rejection of H_0 is called a type I error (in contrast to type II errors where H_0 is not rejected although it is false). The probability p that H_0 is true given the observed test statistic is called the *p*-value of the test.

A variety of statistical tests has been proposed for the identification of changes in gene expression. A classical test for comparing the mean gene expression values in two biological samples is the Student's t-test. Note that this test assumes the independence and normality of the expression values. The null hypothesis is that the mean value of both samples is equal. Depending on the alternative hypothesis, two types of t-tests exist. For one-tailed t-tests, the alternative hypothesis includes the sign of the differences, whereas for the two-tailed test, positive and negative differences are treated equally. Alternatively, permutation tests are used that do not assume any particular data distribution. Permutation tests rely solely on the observed data examples and can be applied with a variety of test statistics. The basic idea of a permutation test is simple. Given labeled data, all permutations of the labels should be equally likely. Evaluating a chosen test statistic for all permutations, an empirical distribution of the test statistic can be derived. The percentage of random permutations that score higher than the actual observed case gives the significance level. However, a major restriction is that permutation tests can be computationally very intensive.

It is the nature of a microarray experiment that generally thousands of genes, if not the transcriptome as a whole, are tested for differential expression. If multiple tests are performed in parallel, the level of significance for the whole set of tests does not equal the level of significance for the single tests. For example, the probability P of rejecting a true null hypothesis in at least one of 1,000 simultaneous tests with a significance level of 0.001 is 63%. Therefore, an adjustment of the overall significance level and the *p*-values is necessary. A popular approach to circumvent the problematic interpretation of *p*-values in multiple testing is the calculation of the false discovery rate (*FDR*), which is defined as the proportion of false positives among significantly regulated genes. For instance, a *FDR* of 0.2 indicates that 20% of significant genes are likely to be false positives.

Classification

In its widest definition, classification is the assignment of a set of objects to a set of classes. In microarray data analysis, classification is commonly used to assign RNA specimens to different classes, for instance, those that distinguish types of tumors. Classification can be performed in an unsupervised and supervised manner. If class labels are not known in advance, the process is called unsupervised. If class labels exist, the process of classification is called supervised. However, note that in the context of microarray data analysis, the term classification usually refers to supervised classification, whereas unsupervised classification is generally referred to as clustering. The aim of supervised classification methods is to correctly assign new examples based on a set of examples of known classes. Thus, a classifier should generalize from known class examples to new unclassified examples. In microarray data analysis, the objects are provided as gene expression profiles and the classifiers have to identify decision boundaries between classes based on these given profiles (Figure 7.12). To achieve this goal, classifiers are optimized in a so-called (supervised) learning or training phase. After the optimization, the accuracy of the classifier can be tested using new examples of known class origin.

Challenges

Classification of tissue samples based on microarray experiments faces several major challenges. First, microarray data can contain a high level of noise. Experimental procedures such as tissue handling, RNA extraction, labeling, amplification, and hybridization can introduce additional variability in the measured expression levels. Furthermore, oligonucleotide and cDNA probes may not be specific to one gene, and several different genes may hybridize to the same probes (cross-hybridization). Second, in the typical microarray experiment thousands

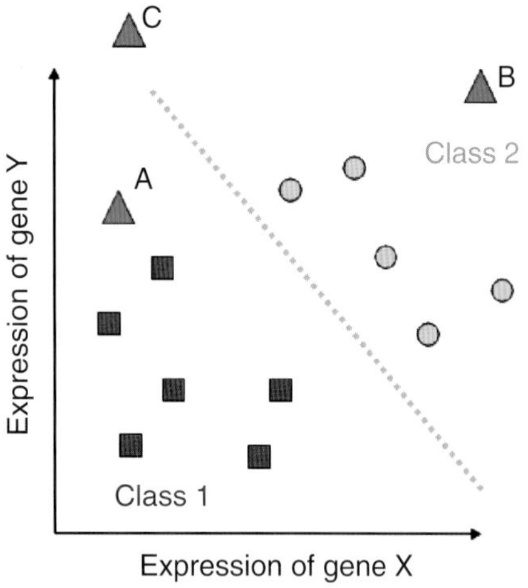

Figure 7.12 Extrapolation in classification: A classifier is trained on the sample from classes 1 and 2 based on the expression values of the two genes X and Y. The dashed line represents the border line derived by the classifier between the classes. Thus, new examples (represented by the dashed line) will be classified according to their gene expression values for X and Y. Thus, example A will be assigned to class 1, whereas example B will be assigned to class 2. The classification of C remains problematic, since it is located close to the border line and different to previously seen examples. Further tests would be advisable in this case.

of genes are monitored, while the number of RNA samples examined is usually restricted to hundreds or less. It is well known that classifiers generally perform poorly when the number of examples is small compared to the number of genes used for classification. Third, tissue samples are frequently heterogeneous in their composition. Thus, different cell types are represented in a single tissue sample used for RNA extraction. This heterogeneity can cloud the separation of the classes of interest, such as the distinction of cancer and normal tissue.

Gene Selection

Generally, large numbers of genes without changes in mRNA abundance introduce noise and may yield a poor classification performance. Gene selection aims at improving classification by excluding noninformative genes and thereby reducing the number of genes for the classifier. Genes are excluded, if they only weakly contribute to the classification or not at all. Gene selection can be incorporated in a classification system in two different ways. First, gene selection and classification can be treated separately from the classification model. Genes are selected with respect to predefined criteria such as Pearson correlation or the significance in the Student's t-test. This approach often has the advantage of being computationally inexpensive and easy to process. However, the selected genes are frequently highly correlated to each other and are likely to be redundant. Alternatively, the selection of genes is determined by the classification methods themselves in an iterative manner. This constitutes an integrated approach since an optimal set of features depends on the choice of the classifier.

Classification Methods

Numerous methods for classification have been applied to microarray data. One of the most basic methods is the k-nearest neighbor method (with k as a positive integer). The classification rule is simple: A new example is assigned to the class most common among its k-nearest neighbors. The distances of the examples are calculated based on their similarity in the expression profiles. For instance, if k is 1, then the example is simply assigned to the class of its nearest neighbor. Other currently popular classifiers are support vector machines based on statistical learning theory and belonging to the class of kernel-based methods. The basic concept of support vector machines is the transformation of input vectors into a highly dimensional feature space, where a linear separation may be possible between the positive and negative class members. In this feature space the support vector learning algorithm maximizes the margin between positive and negative class members of the training set in order to achieve a good generalization.

Cross-Validation

Biological samples included in microarray analysis generally constitute only a small fraction of a larger sample cohort of interest. However, if a classifier is optimized based on a small number of examples, it will frequently show decreased performance on new data, a phenomenon usually called overfitting. An approach to prevent overfitting is k-fold cross-validation. It splits the data into k segments of which $k - 1$ segments are used for the training and one segment for the testing of the classifier. This is repeated k times, so that every segment is used for testing. The classification error in the validation procedure is then the sum over the error in the k tests. This approach has the advantage that a large part of the data can be retained for the training of the classifier, while the validation error is evaluated using all data examples equally. In the extreme case that k equals the number of data objects, the cross-validation is also referred to as the leave-one-out or jackknife method. If different models are compared by cross-validation, the model yielding the lowest validation error is generally selected.

Visualization

Data visualization is also an important component in the assessment of class distributions. It provides a global picture of the separation of samples and helps to identify potential outliers. However, a major challenge is the accurate representation of high-dimensional microarray data, where samples are defined by the expression values of thousands of genes. In contrast, data plots are restricted to two or three dimensions. A standard method for representing high-dimensional microarray data is based on principal component analysis (PCA). The goal of this method is to find an optimal linear projection to a lower dimensional space. Practically, PCA leads a projection from the original gene expression values to an orthogonal basis of principal components. The principal components give the directions of the maximal variance in the data (Figure 7.13).

Cluster Analysis

Clustering, or unsupervised classification, has been studied for many decades in pattern recognition and related fields. Clustering methods generally aim at identifying subsets (clusters) in data sets based on the similarity between single objects. Similar objects are assigned to the same cluster, while dissimilar objects are assigned to different clusters. Cluster analysis, which can be understood as exploratory data analysis, is applied to search for patterns that may reveal relationships between individual examples. Frequently, the data structures detected by cluster analysis can give first insights into the underlying data-producing mechanisms. It is especially useful if prior knowledge is little or nonexistent since it requires minimal prior assumptions. This feature has made clustering a widely applied tool in microarray data analysis. One of the main purposes of clustering is to infer the function of novel genes by grouping them with genes of well-known functionality. This method is based on the observation that genes with similar expression patterns

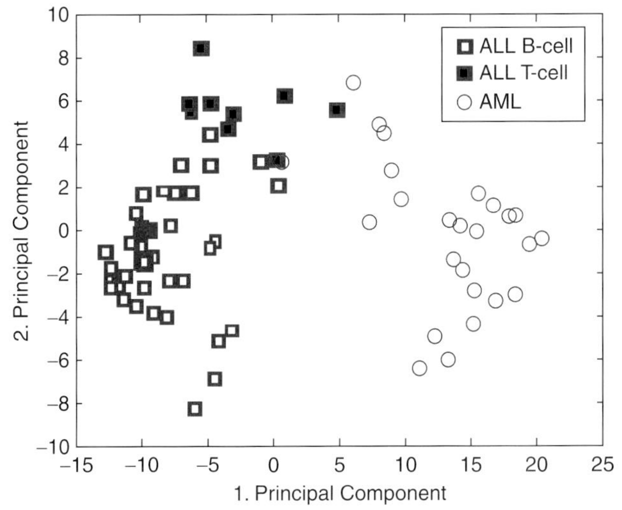

Figure 7.13 **Principal component analysis of leukemia samples based on 100 genes that have the largest squared Pearson correlation with the two classes of leukemia, ALL and AML [35].** The first two principal components include 63.3% of the total variance of the data. Most ALL and AML samples can be separated based on the first two principal components. However, note that the AML outlier makes a perfect separation difficult [20]. From: *Science* 1999; **286**: 531–537. Reprinted with permission from AAAS.

(co-expressed genes) are often functionally related and are controlled by the same regulatory mechanisms (co-regulated genes). Therefore, expression clusters are frequently enriched by genes of certain related functions. If a novel gene of unknown function falls into such a cluster associated with a certain biological function, it seems likely that this gene also plays a role in the same process. This guilt-by-association principle enables assigning possible functions to a large number of genes by clustering of co-expressed genes [22].

Clustering methods can be divided into hierarchical and partitional clustering. Hierarchical clustering creates a set of nested partitions, so that partitions on a higher hierarchical level comprise partitions on lower levels. The sequential partitioning is conventionally presented as a dendrogram. A dendrogram displays the clusters in a tree structure. The length of the branches represents the similarity between clusters. The shorter the branches, the more similar the clusters are. Usually, hierarchical clustering is performed in a stepwise agglomerative manner, starting with the single objects as singular clusters and gradually merging the clusters until all objects belong to a single cluster. To decide which clusters to merge, one calculates their similarity at every step of the clustering procedure. Clusters which show the largest similarity are subsequently joined. In microarray data analysis, both genes analyzed and biological samples can be hierarchically clustered. If these tasks are performed simultaneously, the procedure is also referred to as two-way clustering. As an alternative approach, partitional clustering splits the data in several separate clusters without the definition of a cluster hierarchy. It is commonly used to detect temporal gene expression patterns in time-series experiments. The most popular method for partitional clustering is *k*-means clustering. It starts with *k* randomly initiated cluster centers and splits the data in *k* partitions with a given integer *k* based on the distance to the nearest cluster center. By repeated recalculation of the cluster centers and partitioning of the objects, this method aims to iteratively minimize the within-cluster variation.

Before a cluster analysis is performed, it is important to standardize the expression values, as co-expressed genes frequently show similar changes in expression but may differ in the overall expression rate. Therefore, the expression values of genes are usually adjusted to have a mean value of zero and a standard deviation of one. This ensures that genes with similar changes in expression have similar standardized expression values and thus will tend to cluster together.

A crucial question is how many clusters can be retrieved from microarray data. This is generally difficult to answer for gene expression data as the detected clusters frequently are inhomogeneous and may show substructures, which can be interpreted as clusters themselves. While hierarchical clustering is able to indicate the different levels of clustering in the resulting dendrogram, partitional clustering algorithms lack the ability to indicate substructures in clusters. It is also important to note that common clustering methods always produce clusters due to the underlying algorithms. To critically assess reliability of the clusters that result from this analysis, several measures for cluster validity have been introduced. Many of them assess the quality of clusters based on criteria such as compactness and isolation. Alternatively, clusters that emerge from a given analysis can be examined based on their robustness relative to the noise in the data set. For this approach, one would artificially add noise to the data before clustering and compare the newly identified clusters with the original ones. Clusters that remain the same despite added noise are likely to be more reliable than clusters which vanish in the presence of noise.

Functional Profiling and Other Enrichment Analyses

Frequently, large numbers of differentially expressed genes are detected in microarray experiments, making the overall interpretation of the results difficult. If further research is not focused on a few candidate genes, a helpful tool for understanding the complexity of the data set is functional profiling. This approach aims at identifying biologically informative classes of genes that are likely to be affected in the experiment. The underlying framework is given by Gene Ontology (GO), a popular database providing gene annotations in a systematic manner for various species [23]. In GO, genes are assigned to a defined set of categories describing molecular functions, biological processes, and cellular compartments. The categories themselves are placed in a tree-like structure with parent-child relationships. Categories at low levels are fairly general (for instance, those related to cell death) in contrast to more specific categories at higher levels (such as those that function in the regulation of caspases). Since GO is computer-accessible, the assignment of annotations to a list of genes has become much easier and rapid. After automatic gene annotation, functional profiling is performed by determining which GO category is represented more frequently than expected in the list of differentially expressed genes. Collecting involved GO categories provides a more holistic picture than the inspection of individual genes. Nowadays, numerous software tools are available for functional profiling of microarray experiments. Besides the list of differentially expressed genes, the list of genes represented on the microarray is a necessary prerequisite for the analysis. Comparing the functional composition of both lists, one can calculate the statistical significance for enrichment of differentially expressed genes in a biological process. The user typically obtains a list of significantly enriched GO categories associated with a particular experimental condition or disease state. However, there are important caveats with respect to using GO. Results can vary considerably when using different software tools. In addition, while there is a considerable number of manually curated gene annotations in GO, the majority of human genes have been annotated solely by computational means [24].

The concept of functional profiling to examine enrichment of genes belonging to defined functional categories can be applied in a general way. Another example of enrichment analysis is the examination of the chromosomal location of differentially expressed genes. This strategy yields a first indication for potential underlying changes in the chromosomal structure, such as copy number alterations or deletions, integrating transcriptomics and genomics (Figure 7.14).

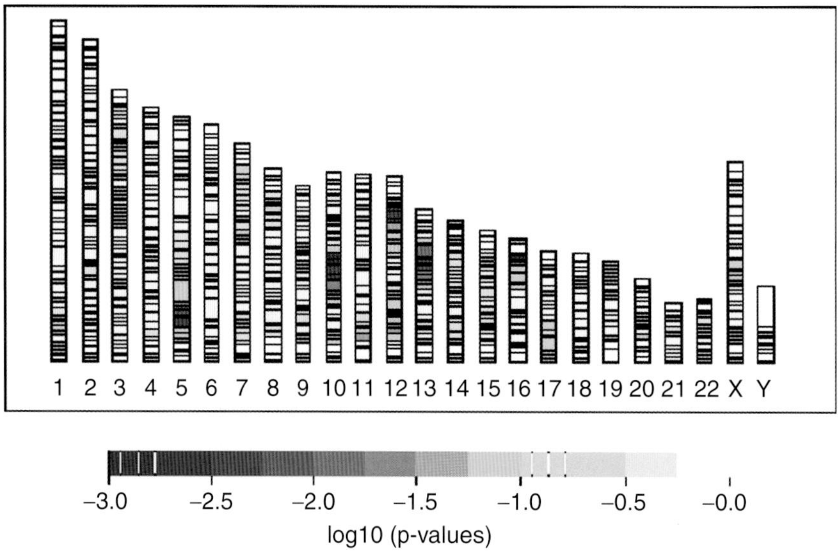

Figure 7.14 Chromosomal localization of genes exhibiting differential expression. The statistical significance for local enrichment of upregulated genes in a metastatic colorectal cancer cell line compared to a primary carcinoma line (SW480) is shown. To detect possible changes in the chromosomal structure of the two related cell lines, researchers mapped differentially expressed genes to their corresponding chromosomal loci. Subsequent enrichment analysis using a sliding window technique indicated several potential chromosomal alterations.

Microarray Databases

Microarray experiments produce massive quantities of gene expression data. Therefore, it has become good practice to deposit generated microarray data in publicly accessible databases. This practice is typically requested by journal editors prior to publication of the data. This allows independent researchers not only to scrutinize data obtained by others for their own interests, but also to validate the original analyses. In fact, the practice of sharing microarray data has allowed the community of bioinformaticians and statisticians to develop new methods and compare them with existing ones based on publicly accessible data sets. Such comparisons have been extremely valuable, since results from microarray experiments rely not only on the raw data, but to a substantial part on the applied computational methods. However, the interpretation of microarray experiments requires a common forum providing various types of information on the examined samples and experimental conditions, arrayed genes, microarray platforms, and applied computational approaches. Therefore, standards for publishing microarray data have been established. The most important one is the Minimum Information About a Microarray Experiment (MIAME) standard [25]. This standard requires deposition of raw microarray data, normalized data, sample annotation, experimental design, description of the microarray, and experimental conditions. Additionally, the development of large central microarray databases has facilitated data sharing. One of the first repositories was the Stanford Microarray Database (http://genome-www5.stanford.edu/), including a large collection of two-color array experiments. Currently, the two major public microarray databases are Gene Expression Omnibus provided by the National Center for Biotechnology Information (http://www.ncbi.nlm.nih.gov/geo/) and Array Express (http://www.ebi.ac.uk/microarray-as/ae/) provided by the European Bioinformatics Institute. Both databases follow the MIAME standard and provide several options to users for depositing their own microarray data and for accessing information from others.

MICROARRAYS: APPLICATIONS IN BASIC RESEARCH AND TRANSLATIONAL MEDICINE

An Early Example for Microarray-Based Gene Expression Profiling Aimed at Understanding Metabolism

Ten years ago, microarray analysis was still in its infancy. Most bioinformatic tools available today had not been developed. Here we present one of the early applications of microarray technology published in 1997 by Pat Brown's group at Stanford University [26] as a paradigmatic example. The Brown group used a microarray representing 6,600 yeast genes to study a process known as diauxic shift. In glucose-rich medium, yeast cells generate energy by fermentation and convert the substrate glucose to acetaldehyde, which is then reduced to ethanol by alcohol dehydrogenase. When glucose is consumed, cells switch from fermentation to respiration and utilize the produced ethanol as a carbon source to generate glycogen. To study gene activity during this process, the researchers labeled cDNA obtained from cells before reaching exponential growth phase with the red fluorescent dye Cy3 as a reference. RNA was prepared at several time points during growth phase and substrate shift, reverse transcribed into cDNA, and labeled with the green fluorescent dye Cy5. Then the Cy5-labeled cDNA (RNA) targets and the Cy3-labeled reference were hybridized to the arrays, and the relative intensities of Cy3 versus Cy5 were measured for each time point. With increasing yeast cell growth indicated by enhancement of the optical density of the cultures, the number of differentially expressed genes increased, as did the level of differential expression indicated by the intensity of red and green staining (Figure 7.15). While in sparse culture, only 0.3% of the genes were altered and the maximal difference in expression was 2.7-fold, 30% of the genes were altered at the final time point of the experiment. More than 300 genes exhibited a differential expression of more than 4-fold. This experiment confirmed that alterations of expression can be efficiently determined in a time-resolved manner by microarray analysis. It also suggested that besides the genes, whose biochemical function was well known already, a number of genes that had not been characterized, approximately 400 at the time of the analysis, could potentially play a role in the diauxic shift, growth control, and energy generation. In summary, these candidate genes were placed into a potential functional framework. This became one of the major goals of microarray experiments in subsequent microarray studies, not only in yeast but also in mammalian systems including human cells and tissues.

In the yeast microarray experiment, the Stanford researchers went one step further and asked the question, if co-expressed genes are regulated in a similar fashion. Several distinct gene clusters comprising elements that exhibit the same expression pattern of upregulation or downregulation over time were identified. When the gene promoters of the co-expressed genes were analyzed, common regulatory sequences were recovered. For example, all but one gene (*IDp2*) contained a regulatory element named CSRE—carbon source responsive element (Figure 7.16). The CSRE is required to activate transcription of the genes involved in gluconeogenesis and the glyoxylate cycle in yeast. And indeed, all of the genes found in this cluster play a role in the glyoxylate cycle (*MLS1, IDP2, ICL1*), in the conversion of acetate to acetyl-CoA (*ACR1*), and in the production of fructose-6-phosphate (*FBP1*). In summary, the basic conclusions from the yeast experiment were (i) similar function is associated with co-regulation, (ii) co-regulation provides a way to define novel functional modules, (iii) co-regulation provides a way to define potential functions for unknown genes, (iv) co-regulation is based on similar transcriptional regulatory factors, and (v) co-regulation is a basis for the identification of regulatory mechanisms.

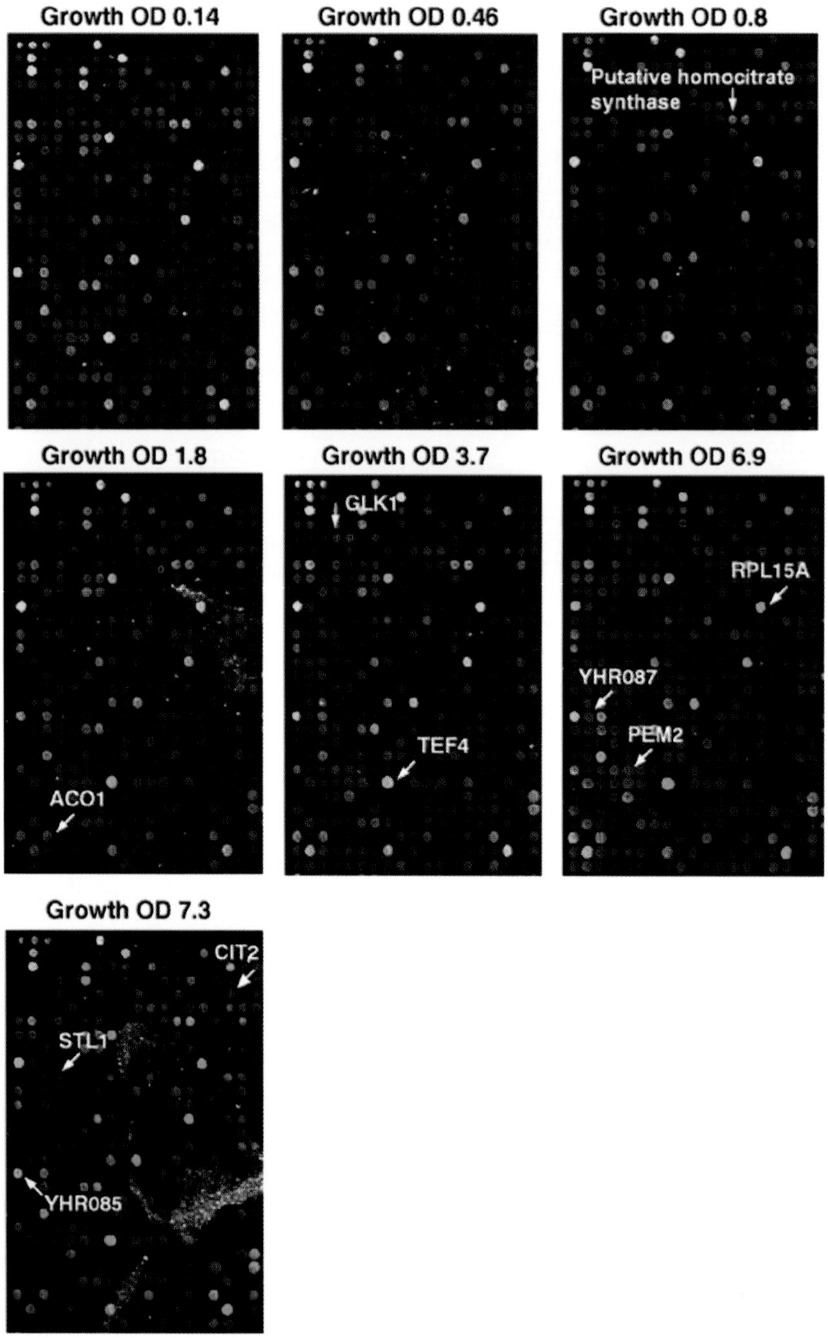

Figure 7.15 **Gene expression changes associated with increased culture density over time [26].** In each of the arrays used to analyze gene expression during the diauxic shift, red spots represent genes that were induced relative to the initial time point, and green spots represent genes that were repressed. Note that distinct sets of genes are induced and repressed in the different experiments. Cell density as measured by optical density (OD) at 600 nm was used to monitor the growth of the culture. From: *Science* 1997; **278**: 680–686. Reproduced with permission by AAAS.

Elucidating the Transcriptional Basis of the Serum Response in Human Cells

Diploid human fibroblasts, like most other cell cultures, require the presence of serum growth factors in their culture medium. Routinely, these factors are supplied by adding fetal calf serum to the culture medium. Cultured cells can be made quiescent by serum deprivation. When fetal calf serum is added to such cells, they quickly resume cell cycle progression and proliferation. This cellular reaction is called the serum response, which was chosen as another early example to demonstrate the power of microarray analysis [27]. This time the Stanford researchers obtained RNA from serum-starved cultures and prepared target cDNA labeled with Cy3. RNA from all other time points following serum stimulation was used to prepare Cy5-labeled targets. In two-color microarray experiments, the targets were hybridized to a human cDNA array representing 8,600 human sequences. About 4,000 of them were known human genes, 2,000 sequences were related to these annotated genes, while the remaining genes were ESTs without known function. Figure 7.17 shows a subset of genes ($n = 517$) whose expression changed up to 8-fold during serum stimulation.

The transcriptional response toward serum stimulation is very rapid; the earliest changes can be observed as early as after 15 minutes. The genes can be divided into several clusters, which exhibit a common regulatory scheme. Some clusters show a characteristic pattern of upregulation followed by downregulation (cluster C) or the reverse pattern (clusters E and B). Based on current knowledge in molecular cell biology and data mining for gene functions, the Stanford researchers performed a functional gene clustering. This analysis was done at a time before the Gene Ontology became available. One functional cluster of co-regulated genes comprised transcription factors including the ones known to be involved in the immediate early gene response, permitting rapid responses without the need for protein synthesis. Another cluster included phosphatases. Their functional relevance was not known at the time of the analysis. Today it is well established that the phosphatases limit signaling kinase activity, which is rapidly stimulated upon growth stimulation, by negative feedback. Not surprisingly, the researchers recovered genes encoding cell cycle regulatory proteins. Inhibitory genes were quickly downregulated, paving the way for re-entry of the serum-starved cells into the cell cycle. With a short delay, cell cycle stimulatory genes were upregulated, among them cyclin D1 and DNA topoisomerase, which is required for chromosome segregation at mitosis. A more surprising feature of the analysis was the appearance of genes with known functions in wound healing. This referred not only to genes whose products function intracellularly, but also to genes whose products play a role in remodeling clot structure and the extracellular matrix, as well as in intercellular signaling. While previous studies had aimed at elucidating intracellular events in wound healing, gene expression profiling of the serum response indicated the relevance of extracellular events during the first 24 hours in this process.

Microarray Applications in Cancer Pathogenesis and Diagnosis

A recent PubMed search revealed that the majority of microarray and gene expression profiling studies in medicine are devoted to some aspect of cancer. Cancer studies far outnumber similar studies in cardiovascular diseases, neurodegenerative diseases, infection, inflammation, and other diseases (Table 7.1). Therefore, we have chosen some prominent applications of microarrays in the field of cancer as paradigms to demonstrate the power of transcriptome analysis.

The current themes of transcriptomics in cancer analysis are related to the mechanisms of pathogenesis, cancer classification, and outcome prediction. To elucidate the mechanisms of tumorigenesis and metastasis, particularly to study the complexity of the underlying processes, researchers frequently use microarrays. Cancer classification based on microarray studies aims at identifying characteristics beyond anatomical site and histopathology. Outcome prediction tries to overcome the limitations of current diagnostic procedures by establishing gene-based criteria to indicate and predict tumor prognosis and therapy response, even for individual cancer patients. Basically, there are three types of microarray-based approaches: (i) class comparison,

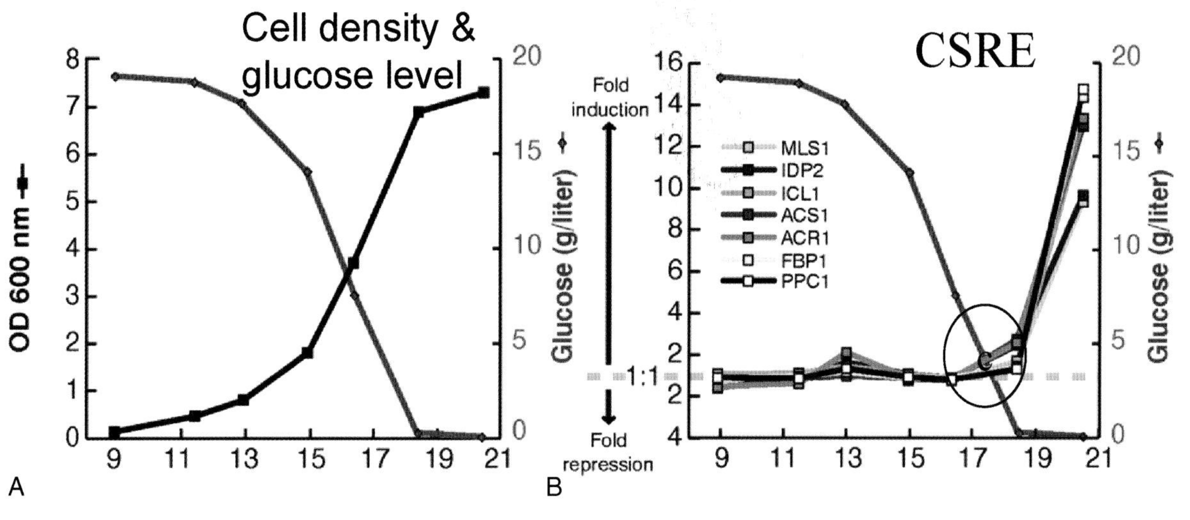

Figure 7.16 **Analysis of regulatory modules within the promoters of co-regulated genes associated with the diauxic shift [26].** (A) Growth curve of yeast cells shown as increasing optical density (black line) upon glucose consumption (red line). (B) Induction of a group of genes carrying a carbon source element (CSRE) within their promoters. The decreasing glucose level (red line) allows determination of a threshold for the onset of gene expression (grey and black lines) mediated by the CSRE. From: *Science* 1997; **278**: 680–686. Reproduced with permission by AAAS.

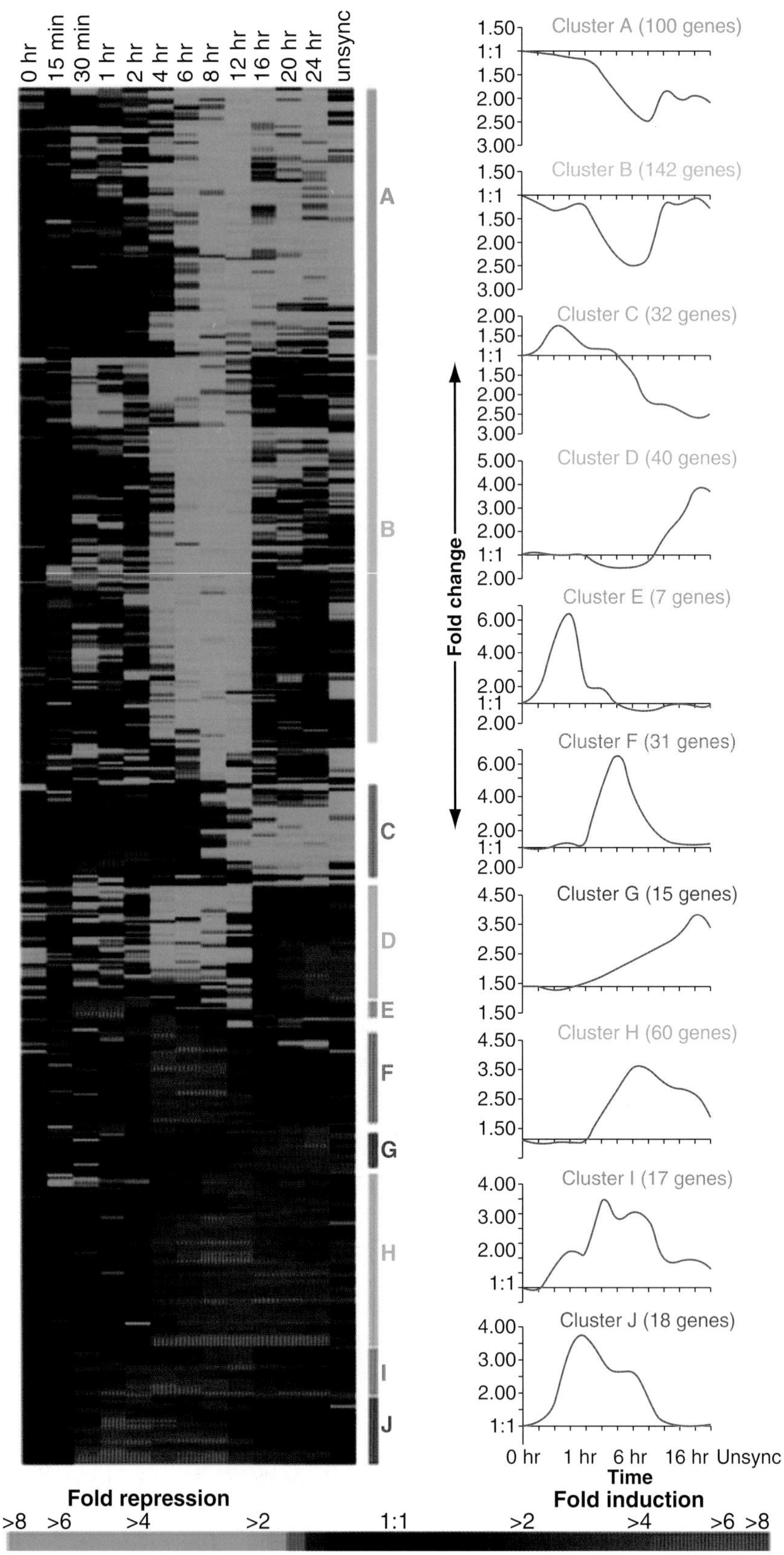

Figure 7.17 Hierarchical clustering of genes induced or repressed during serum response in human fibroblasts [27]. Ten gene clusters (A–J) harboring 517 genes, which show significant alterations in gene expression over time, are depicted. For each gene, the ratio of mRNA levels in fibroblasts at the indicated time intervals after serum stimulation compared to their level in the serum-deprived (time zero) fibroblasts is represented by a color code, according to the scale for fold-induction and fold-repression shown at the bottom. The diagram at the right of each cluster depicts the overall tendency of the gene expression pattern within this cluster. The term *unsync* denotes exponentially growing cells. From: *Science* 1999; **283**: 83–87. Reproduced with permission by AAAS.

Table 7.1 Number of Published Microarray and Gene Expression Profiling Applications in Research. Results of a PubMed Search Dated July 13, 2008, Using Single Keywords and Combinations of Two Keywords Without Limits to Publication Years.

		2nd Keyword	
1st Keyword	**None**	**Gene Expression Profiling**	**Microarray**
None	-	35,712	25,841
Pharmacology	4,244,442	8,358	5,668
Diseases	3,520,603	8,169	5,904
Cancer	*2,172,278*	*11,513*	*8,657*
Pathology	1,813,867	7,790	5,708
Cardiovascular diseases	*1,460,061*	*1,149*	*616*
Development	1,266,779	8,529	5,491
Immunology	1,094,163	3,021	2,078
Infection	*892,647*	*1,665*	*1,254*
Nutrition	404,775	528	371
Drug development	281,029	1,798	1,206
Inflammation	*280,552*	*1,318*	*1,001*
Neurodegenerative diseases	*152,735*	*457*	*277*
Toxicology	82,952	649	460

(ii) class discovery, and (iii) class prediction. Using class comparison, one tries to compare the expression profiles of two (or more) predefined classes. For example, two tissue samples, normal versus malignant cells or tissues, different developmental stages, or cells treated with drugs under different conditions. Using class discovery, one tries to identify novel subtypes within an apparently homogenous population. In this case, microarray analysis is used to identify features that cannot be distinguished by other available tools. The starting point usually is a homogenous group of specimens, in which a concealed proportion behaves aberrantly or exhibits invisible or unknown features. The problem of cancer treatment falls into this category, since patients who are stratified into treatment groups according to standard histopathological criteria often respond differently to therapy. We will see that microarray studies can help to successfully address this urgent clinical problem. Class prediction means to find a set of features that are predictive for a certain, predefined class. This is perhaps the most sophisticated type of microarray application. It is usually based on class discovery, but now the characterization of a novel class is intended. Rather, the idea is to establish a classifier. A classifier is a set of features, like genes, proteins, micro-RNAs that are surrogate markers for a certain class. This is the common approach to identify predictive gene sets or gene signatures that can predict clinical outcome or therapy response.

Identification of Hidden Subtypes Within Apparently Homogenous Cancers

The group of T. Sorlie identified 456 genes out of 8,000 genes on a microarray that discriminated between tumor subclasses in a cohort of 65 tumors from 42 breast cancer patients [28]. Gene expression patterns of breast carcinomas helped to distinguish tumor subclasses with clinical implications. Using hierarchical clustering, the researchers distinguished five distinct tumor groups characterized by their gene expression pattern: the basal epithelial cancer type, the luminal epithelial cancer types A–C, a group displaying expression of the breast cancer oncogene *ERBB2* (*HER2*), and a group without any known feature. There was yet another group showing features of normal breast epithelial cells (Figure 7.18). In the next step of the analysis, the researchers addressed the question as to whether these different groups are characterized by distinct clinical parameters. Therefore, they compared the groups by certain statistical methods, among others by univariate statistical analysis, for either overall survival or relapse-free survival monitored for up to 4 years (Figure 7.19). The patient groups that were *ERBB2*-positive or were characterized as basal epithelial breast tumors had the shortest survival times. While this information was not new for the *ERBB2*-positive tumors, the basal epithelial breast cancers belong to a novel group with an obviously bad prognosis. One characteristic of this tumor type is the high frequency of *TP53* mutations. The tumor suppressor gene *TP53*, well known as the guardian of the genome, is lost or mutated in more than 50% of all advanced human cancers, and might be responsible for the bad prognosis. There was also a difference in clinical outcome between the luminal-type breast cancers. Most strikingly, luminal A tumors exhibited a very good outcome at least within 4 years, while luminal B or luminal C tumors were intermediate. In conclusion, this study opened the door to further screen many tumors for gene signatures indicative of the clinical performance of breast cancer patients. With respect to cancer treatment, the most important issue is to find gene sets predictive for the susceptibility or resistance to therapy, particularly to chemotherapy, and to clinical outcome in the absence of other

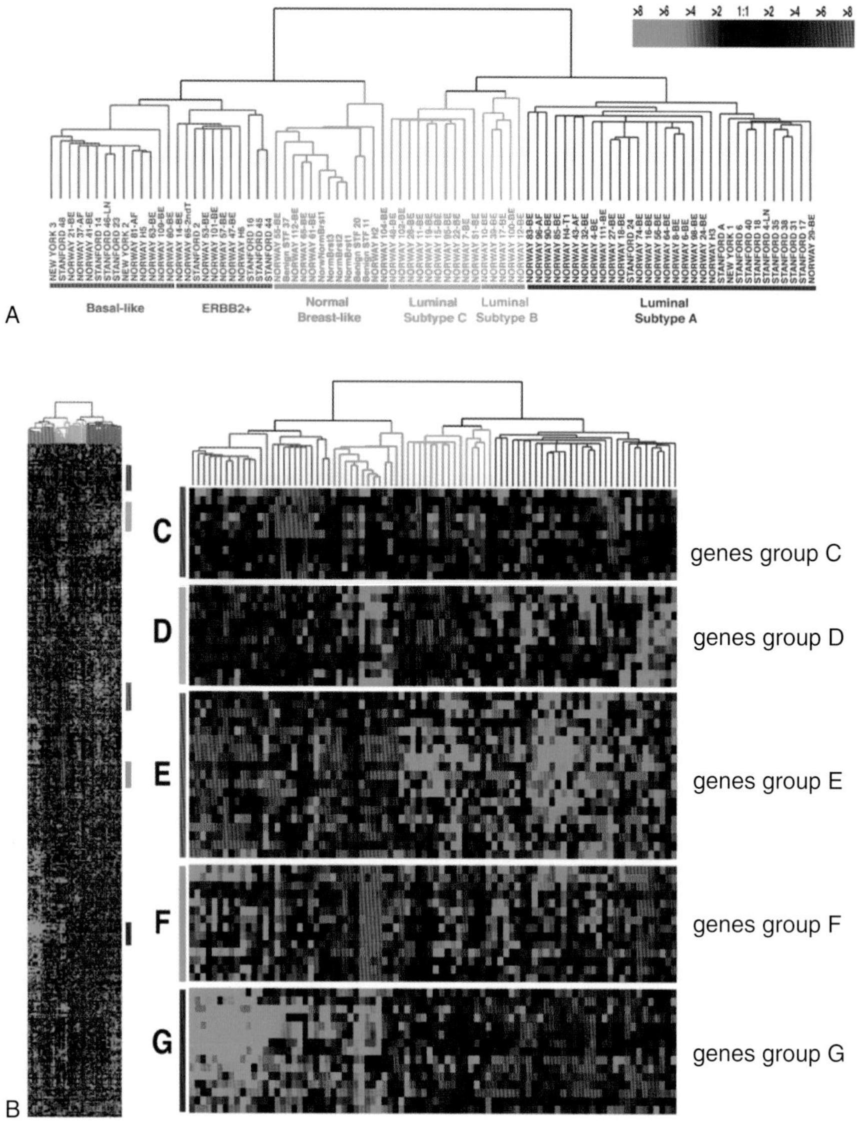

Figure 7.18 **Differential breast cancer gene expression [28].** Gene expression patterns of 85 experimental samples (78 carcinomas, 3 benign tumors, 4 normal tissues) analyzed by hierarchical clustering using a set of 476 cDNA clones. (A) Tumor specimens were divided into 6 subtypes based on their differences in gene expression: luminal subtype A, dark blue; luminal subtype B, yellow; luminal subtype C, light blue; normal breast-like, green; basal-like, red; and ERBB2+, pink. (B) The full cluster diagram obtained after two-dimensional clustering of tumors and genes. The colored bars on the right represent the characteristic gene groups named C to G and are shown enlarged in the right part of the graph: (C) ERBB2 amplification cluster, (D) novel unknown cluster, (E) basal epithelial cell-enriched cluster, (F) normal breast epithelial-like cluster, (G) luminal epithelial gene cluster containing ER (estrogen receptor). From: *Proc Natl Acad Sci USA* 2001; **98**: 10869–10874. Reproduced with permission from the National Academy of Sciences USA.

conventional indicators. So far, questions related to chemotherapy resistance and drug sensitivity have also been addressed by microarray studies, but have not been advanced to the clinical level.

Gene Expression Profiling Can Predict Clinical Outcome of Breast Cancer

Breast cancer patients with the same stage of disease exhibit markedly different treatment responses and overall outcome. However, histopathological assessment of these cancers does not have sufficient power to discriminate which patients will perform well versus those that will not. The strongest predictors for metastases (such as lymph node status and histological grade) fail to classify accurately breast tumors according to their clinical behavior. None of the signatures of breast cancer gene expression reported to date allow for patient-tailored therapy strategies. The study published by van't Veer et al. [29] in the Netherlands has pioneered gene array-based breast cancer diagnostics. The study was based on a well-characterized cohort of breast cancer patients ($n = 117$). This included 78 sporadic primary

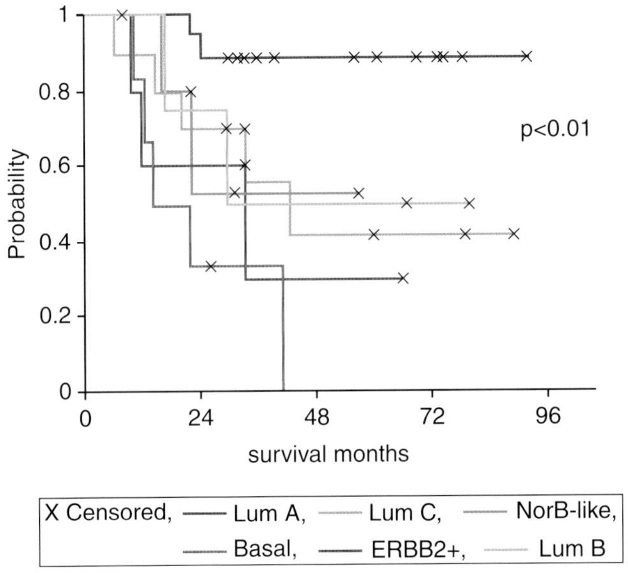

Figure 7.19 Survival analysis (Kaplan-Meier plot) of patient groups distinguished according to gene expression profiling [28]. The Y-axis shows the survival probability for each individual group; the X-axis represents the time scale according to patient follow-up data. All groups identified by gene expression profiling are shown. Luminal type A, dark blue; luminal type B, yellow; luminal type C, light blue; normal type, green; ERBB2-like type, pink; and basal type, red. Patients with ERBB2-like or basal type tumors had the shortest survival times; luminal-type A patients had the best prognosis. All others showed an intermediate probability and were not clearly distinguishable. From: *Proc Natl Acad Sci USA* 2001; **98**: 10869–10874. Reproduced with permission from the National Academy of Sciences USA.

invasive ductal and lobular breast carcinomas of less than 5 cm in size. The tumor stages were T1 or T2, nodal status N0 (without axilliary metastases), patient age <55 years at diagnosis without a history of previous malignancies. The patients received surgical treatment followed by radiotherapy, but no adjuvant chemotherapy (except for 5 patients). The follow-up period of the patient cohort was 5 years. Tissue samples contained more than 50% tumor cells by pathological inspection; estrogen receptor (ER) and progesterone receptor (PR) status were known. The cohort was supplemented by 20 hereditary tumors carrying *BRCA1/BRCA2* mutations that were of similar histology to the sporadic cancers. Target RNA/cDNA was labeled and hybridized to an oligonucleotide array representing more than 24,000 human sequences and more than 1,000 control sequences. The reference target used in this system was a pooled cRNA derived from an RNA mixture of all patients. This means that gene expression of each sample was determined relative to the pool of all samples. The hybridizations were performed in duplicate and ~5,000 genes appeared significantly regulated more than 2-fold with a *p*-value of less than 0.01.

In the first step of bioinformatic analysis, expression profiles of 98 cancer samples analyzed were clustered hierarchically according to similarities among the 5,000 genes (Figure 7.20A and Figure 7.20B). This revealed two distinct groups of tumors. In the upper group, 34% had developed distant metastasis within 5 years, while in the lower group 70% exhibited metastatic spread. There was also a clear association with ER expression, which when lacking indicates a bad prognosis. Therefore, ER-negative tumors that did not express ER and also some of the known ER targets were filtered out (Figure 7.20C). In addition, the second group of tumors expressed a B-cell and T-cell gene signature. The tumors were thus characterized by a lymphocyte infiltration and clearly separated from the ER-negative group (Figure 7.20D).

In a supervised classification procedure, the researchers from the Netherlands used the gene expression profiles obtained from the sporadic tumors only. In the first step of the classification procedure, the ~5,000 genes that were significantly regulated in more than 3 of 78 tumors were selected from the 25,000 genes represented on the array. The correlation of each gene expression profile with the clinical outcome of patients was calculated, and 231 genes were found to be significantly associated with disease progression. In the second step, the 231 informative genes were rank-ordered according to their correlation coefficient. In the third step, the number of genes in this preliminary prognosis classifier was optimized by cross-validation, particularly by the leave-one-out procedure. The final result was a signature of 70 genes, which predict the clinical outcome—distant metastasis within 5 years—with an accuracy of 83% (Figure 7.21). This means that of 78 patients, 65 were assigned to the right category, poor prognosis (Figure 7.21, cluster below the yellow line) or good prognosis (above). Five patients with poor prognosis and 8 patients with good prognosis were misclassified. Van't Veer et al. used an independent set of 19 lymph-node negative breast tumors (Figure 7.21C) to validate their classifier. This time, 2 of 19 patients were assigned to the wrong group. Thus, the classifier predictive of a short interval to distant metastases (poor prognosis signature) in patients without tumor cells in local lymph nodes at diagnosis (lymph node negative patients) showed a similar performance on this test set of tumors as compared to the training set.

Today, three gene expression-based prognostic breast cancer tests have been licensed for use. These are MammaPrint (Agendia BV, Amsterdam, the Netherlands; based on the work described above), Oncotype DX (Genomic Health, Redwood City, California), and H/I (AvariaDX, Carlsbad, California). However, a recent comparative study showed that for all tests offered, the relationship of predicted to observed risk in different patient populations and their incremental contribution over conventional predictors, optimal implementation, and relevance to patients receiving current therapies need further study [30]. A particular caveat on the currently available predictors was also provided in a paper published in 2005 [31]. The authors re-evaluated data from 8 different microarray-based studies with more than 800 tumor samples. The results suggested that the list of genes identified as predictors was highly unstable

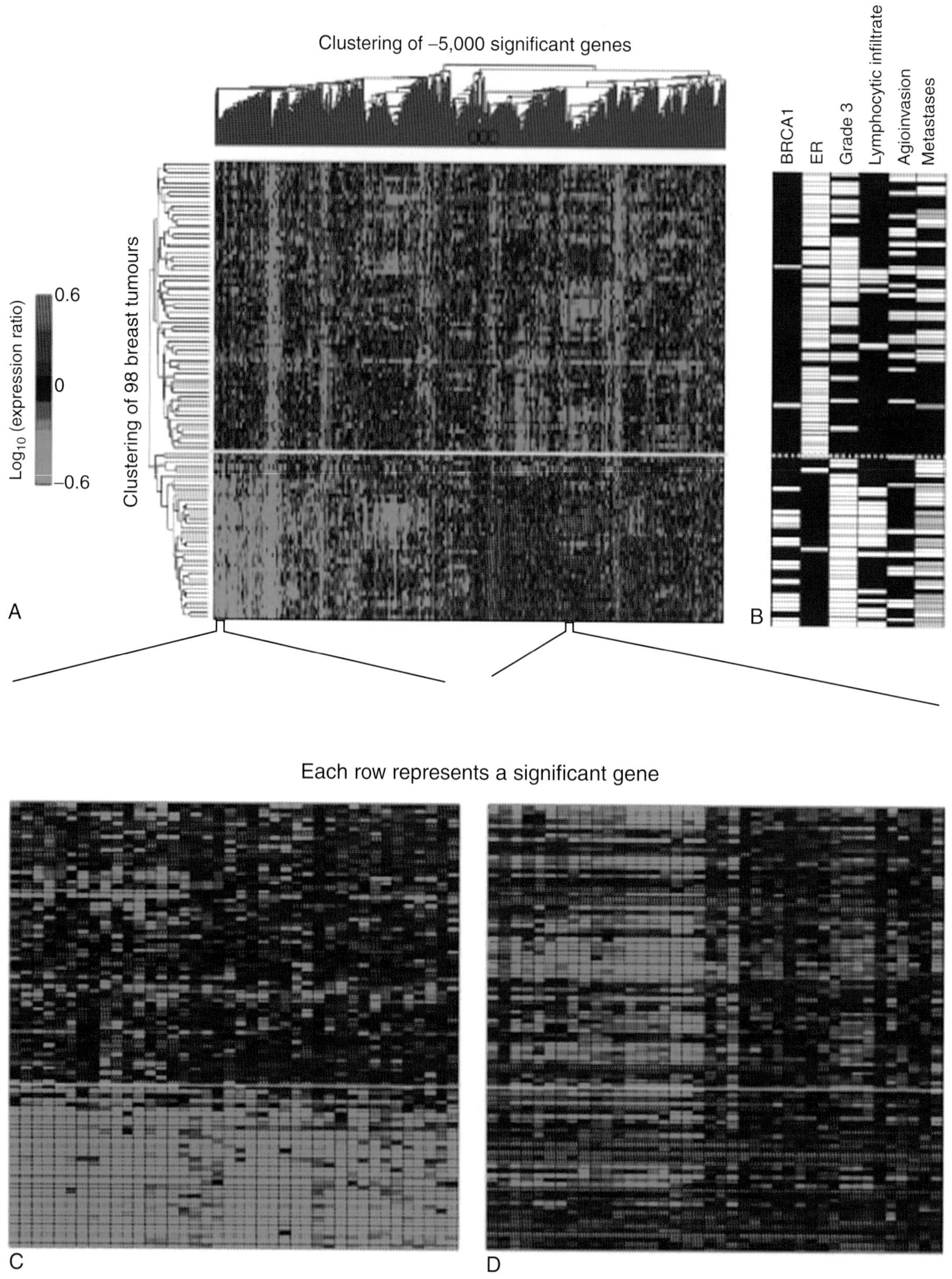

Figure 7.20 **Microarray-based prediction of breast cancer prognosis [29].** Two-dimensional clustering of 98 tumor samples based on approx. 5,000 significantly regulated genes. (A) Clustering, (B) molecular characteristics of tumors, BRCA1 mutation and estrogen receptor status (ER), grade, lymphocyte infiltration, blood vessel count, and distant metastases occurring within 5 years following diagnosis. The group above the yellow line is defined as the good prognosis group (34% of patients developed distant metastasis), the group below as the bad prognosis group (70%). (C) Expression pattern of subgroup associated with estrogen receptor expression, (D) subgroup exhibiting lymphocytic infiltration. Reprinted by permission from MacMillan Publishers Ltd: Nature 2002.

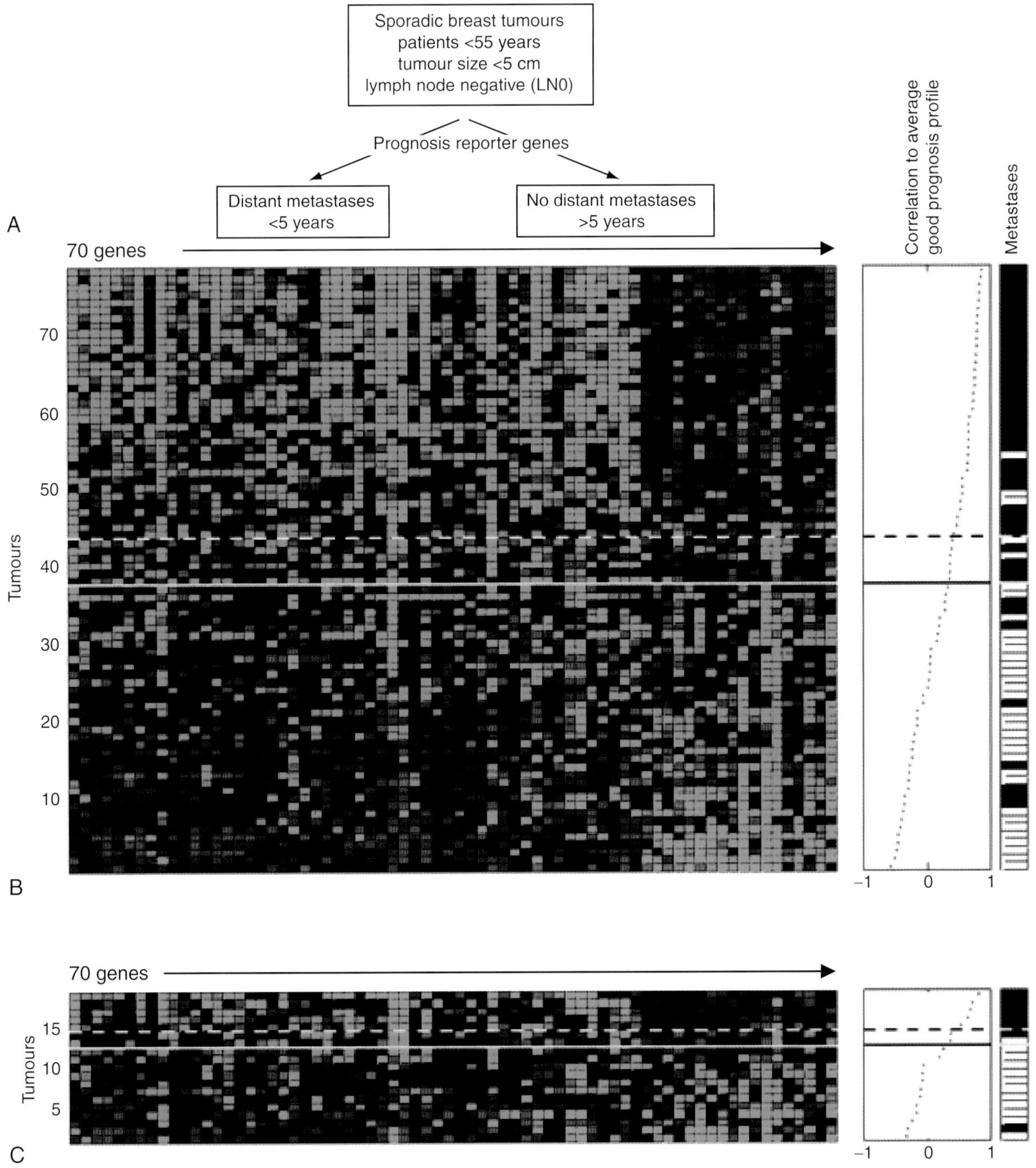

Figure 7.21 **Identification of the prognostic breast cancer gene set using a supervised approach [29].** The 231 genes identified as being most significantly correlated to disease outcome were used to recluster, as described in the text. Each row represents a tumor and each column a gene. The genes are ordered according to their correlation coefficient with the two prognostic groups. The tumors are ordered according to their correlation to the average profile of the good prognosis group. The solid line marks the prognostic classifier showing optimal accuracy; the dashed line marks the classifier showing optimized sensitivity. Patients above the dashed line have a good prognosis signature, while patients below the dashed line have a poor prognosis signature. The metastasis status for each patient is shown on the right. White bars indicate patients who developed distant metastases within 5 years after the primary diagnosis; black indicates disease-free patients. Reprinted by permission from MacMillan Publishers Ltd: Nature 2002.

and that the molecular signatures strongly depended on the selection of patients in the training set. Notably, 5 of 7 studies re-evaluated did not classify patients better than chance.

From Gene Expression Signatures to Simple Gene Predictors

In 1999, a group of scientists from highly ranked medical schools in the United States assembled a specialized microarray representing genes preferentially expressed in lymphoid cells. The so-called lympho-chip harbored more than 17,000 cDNA probes derived from libraries specific for germinal center B-cells, diffuse large B-cell lymphoma (DLBCL), follicular lymphoma, mantle cell lymphoma, chronic lymphatic leukemia (CLL), genes induced or repressed in T-cell or B-cell activation, supplemented by lymphocyte-specific genes and cancer genes [32]. The consortium interrogated these chips using targets prepared from normal cells and tumors to define signatures for the different immune cell types, under different conditions and developmental stages. Particularly, the researchers analyzed the most prevalent adult lymphomas using the lympho-chip. They identified signatures for distinct types of diffuse large B-cell lymphoma (DLBCL) exhibiting a bad prognosis, follicular lymphoma (FL) exhibiting a low proliferation rate, and for chronic lymphatic leukemia (CLL) with slow progression (>20 years). In addition, profiles were obtained from normal lymphocytes (tonsil, lymph node) as well as from several lymphoma and leukemia cell lines. Clustering analysis placed the CLL and FL profiles close to those of resting B-cells, while genes of the so-called proliferation signature were weakly expressed in these tumors. DLBCL, the highly proliferative, more aggressive disease, had higher expression levels of proliferation-associated genes. An additional signature characterized germinal center B-cells, which was clearly different from the resting blood B-cells and from the *in vitro* activated B-cells. This indicated that that germinal center B-cells represent a distinct stage of B-cells and do not simply resemble activated B-cells located in the lymph node.

When the scientists reclustered all DLBCL cases, particularly considering the genes that define the germinal center B-cells, they could clearly separate two different subclasses of DLBCL. One of them strictly showed the signature of the germinal center B-cells, while the other one was clearly distinct. These data suggested that a certain class of DLBCL was derived from germinal center B-cells and retained its differentiation signature even after malignant transformation. By investigating the genes exclusively expressed in either of the DLBCL types and reclustering, the authors defined two signatures representative of either the germinal center-type (GC-like) and what they called the activated-type DLBCL (Figure 7.22). Analysis of the clinical follow-up showed that the GC-like tumors have a much better prognosis than the activated type of DLBCL (Figure 7.23). Did the result of the microarray study provide novel information up to this stage of investigation? When the authors compared the microarray-based classification to the standard classifiers that define high and low clinical risk, there was obviously no significant classification progress (Figure 7.23B). However, when the low-risk patients initially classified conventionally are further stratified by subgrouping them into the GC and activated-type DLBCL types, the molecular classifier was superior. Subsequent functional classification of genes associated with activated-type DLBCL revealed an NFkB pathway signature that comprises several anti-apoptotic genes. The functional studies culminated in the finding that inhibition of that pathway affected growth of activated-type DLBCL, while GC-DLBCL cells were insensitive [33].

Several further microarray studies confirmed that gene signatures were associated with clinical outcome of diffuse B-cell lymphoma. However, among these studies there were disparities with regard to the number and the nature of informative genes. A recent study tried to circumvent the technical and bioinformatic issues of microarray analysis by using quantitative real-time polymerase-chain-reaction. Scientists from Miami and Stanford studied the expression of 36 genes that had previously been reported to be of predictive value among 66 lymphoma patients. The prediction of survival could be based on only 6 genes [34]. This result opens the interesting perspective that selecting informative genes that have been filtered through genome-wide microarray studies may permit the application of conventional methods in the future and may obviate microarray applications in routine clinical testing.

PERSPECTIVES

Microarrays have developed into an indispensable tool for transcriptome analysis in basic research, translational studies, and clinical investigations. In experimental pathology, gene expression activities under various conditions can be assessed at an unprecedented quantity, speed, and precision. Commercial microarray platforms exhibit a high degree of standardization allowing service laboratories and academic core facilities to offer the technology to users from industry and academia, respectively, who do not have the means to develop their own specific expertise in this field. Together with other –omics technologies, transcriptomics will be an essential component in worldwide efforts to understand normalcy and disease at the systems level. Already now, transcriptomic approaches are a standard strategy for data collection in systems biology and systems medicine.

In the clinical situation, the current instabilities of predictive gene signatures will probably be scrutinized by enforcing standard operating procedures and efforts aiming at the general standardization of diagnostic approaches, as was the case in the optimization period of microarray technology. The strong need for predictive markers in the clinic, the issue of personalized medicine, and the requirement to study the effects of old and novel drugs at the genome level are expected to

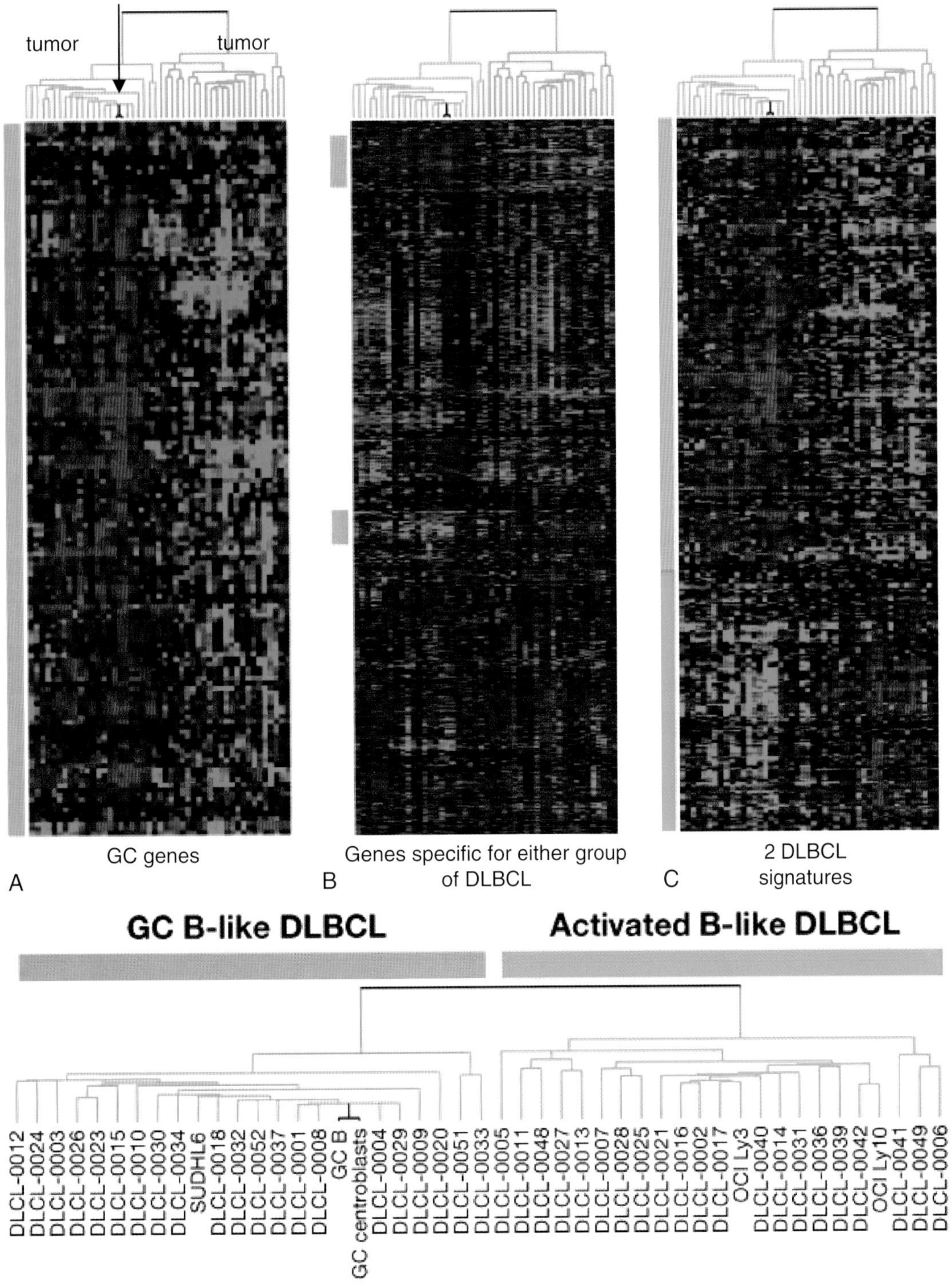

Figure 7.22 **Gene signatures representing germinal center (GC)-like Diffuse Large B-cell Lympbomas (DLBCL) and activated B cell-like Diffuse Large B-cell Lymphomas (DLBCL) [32].** (A) Genes characteristic for normal germinal center B-cells were used to cluster the tumor samples. This process defines two distinct classes of B-cell lymphomas: GC-like DLBCL and activated B-like DLBCL. (B) Genes that were selectively expressed either in GC-like DLBCL (yellow bar) or activated B-like DLBCL (blue bar) were identified in the tumor samples. (C) Result of hierarchical clustering that generated GC-like and activated B-cell-like DLBCL gene signatures. Reprinted by permission from MacMillan Publishers Ltd: Nature 2000.

increase the use of microarray technologies even further. Alternative high-throughput approaches such as proteomic profiling combined with mass spectroscopy or deep sequencing will probably not be regarded as competitive approaches. Rather, these techniques will further increase our knowledge on complex biological phenomena and pathogenic mechanisms. New types of microarrays have become available that allow analysis

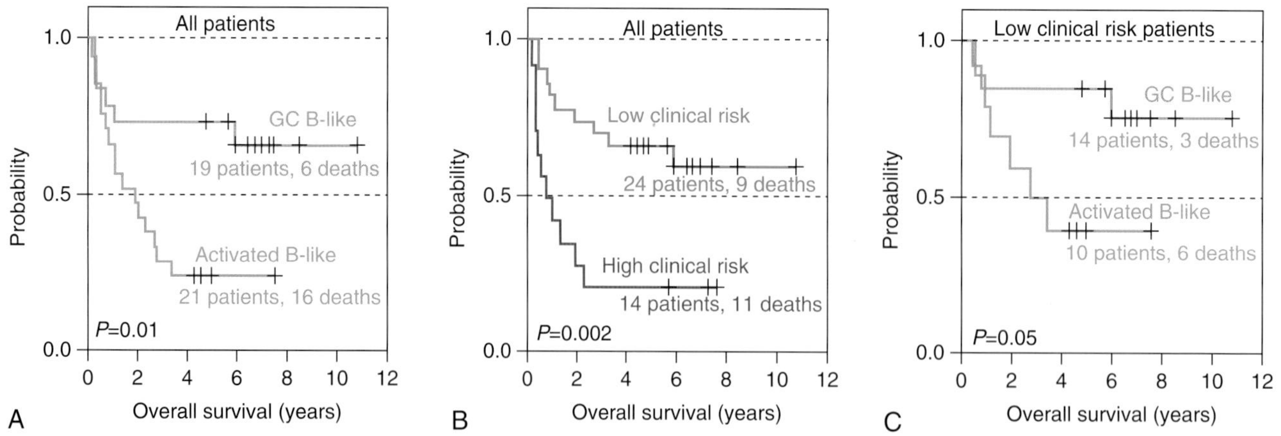

Figure 7.23 **Survival analysis of diffuse large B-cell lymphoma patients distinguishable according to gene expression profiling, conventional clinical criteria, and a combination of both sets of criteria [32].** (A) DLBCL patients grouped on the basis of gene expression profiling. The GC-like (germinal center-like) and the activated B-cell-like show clearly different survival probabilities. (B) DLBCL patients grouped according to the International Prognostic Index (IPI) form two groups with clearly different survival, independent of gene expression profiling. Low clinical risk patients (IPI score 0–2) and high clinical risk patients (IPI score 3–5) are plotted separately. (C) Low clinical risk DLBCL patients (IPI score 0–2) shown in B were grouped on the basis of their gene expression profiles and exhibited two distinct groups with different survival probabilities.Reprinted by permission from MacMillan Publishers Ltd: Nature 2000.

of alternative splicing at the level of the transcriptome, as well as to analyze the expression of microRNAs, a novel class of gene expression regulators in development, normal physiology, and disease. Rapid progress will also be made in understanding the molecular basis for the transcriptional alterations that can be assessed by microarrays, by combining chromatin immuno-precipitation (ChIP) and microarray analysis (ChIP-on-chip). Last but not least, efforts are being made to develop chip technologies that permit a truly quantitative estimation of mRNA expression. It is tempting to speculate that these novel chip technologies will gradually replace the currently available microarrays, facilitate transcriptome analysis with even higher precision, and obviate extensive validation (based on real-time PCR, immunohistochemistry, or other methods) and quantification procedures. Finally, microRNAs that have been overlooked for many years in transcriptomics have started to demonstrate their impact on gene regulation and possibly also predictive power in tumor analysis. Since the microRNAome is much smaller than the transcriptome, a current challenge is to understand the regulatory relationships between small RNAs and their mRNA targets. After all, the race is open for deciphering the protein and RNA master regulators of the transcriptome.

REFERENCES

1. Velculescu VE, Madden SL, Zhang L, et al. Analysis of human transcriptomes. *Nature Genetics.* 1999;23:387–388.
2. Alwine JC, Kemp DJ, Stark GR. Method for detection of specific RNAs in agarose gels by transfer to diazobenzyloxymethyl-paper and hybridization with DNA probes. *Proc Natl Acad Sci USA.* 1977;74:5350–5354.
3. Thomas PS. Hybridization of denatured RNA and small DNA fragments transferred to nitrocellulose. *Proc Natl Acad Sci USA.* 1980;77:5201–5205.
4. Groudine M, Weintraub H. Activation of cellular genes by avian RNA tumor viruses. *Proc Natl Acad Sci USA.* 1980;77:5351–5354.
5. Augenlicht LH, Wahrman MZ, Halsey H, et al. Expression of cloned sequences in biopsies of human colonic tissue and in colonic carcinoma cells induced to differentiate in vitro. *Cancer Res.* 1987;47:6017–6021.
6. Schmitt AO, Specht T, Beckmann G, et al. Exhaustive mining of EST libraries for genes differentially expressed in normal and tumour tissues. *Nucleic Acids Res.* 1999;27:4251–4260.
7. Fargnoli J, Holbrook NJ, Fornace AJ Jr. Low-ratio hybridization subtraction. *Anal Biochem.* 1990;187:364–373.
8. Diatchenko L, Lau YF, Campbell AP, et al. Suppression subtractive hybridization: A method for generating differentially regulated or tissue-specific cDNA probes and libraries. *Proc Natl Acad Sci USA.* 1996;93:6025–6030.
9. Zuber J, Tchernitsa OI, Hinzmann B, et al. A genome-wide survey of RAS transformation targets. *Nature Genetics.* 2000;24:144–152.
10. Lisitsyn N, Wigler M. Cloning the differences between 2 complex genomes. *Science.* 1993;259:946–951.
11. Hubank M, Schatz DG. Identifying differences in mRNA expression by representational difference analysis of cDNA. *Nucleic Acids Res.* 1994;22:5640–5648.
12. Liang P, Pardee AB. Differential display of eukaryotic messenger RNA by means of the polymerase chain reaction. *Science.* 1992;257:967–971.
13. Liang P, Averboukh L, Keyomarsi K, et al. Differential display and cloning of messenger RNAs from human breast cancer versus mammary epithelial cells. *Cancer Res.* 1992;52:6966–6968.
14. Velculescu VE, Zhang L, Vogelstein B, et al. Serial analysis of gene expression. *Science.* 1995;370:484–487.
15. Matsumura H, Reuter M, Kruger DH, et al. SuperSAGE. *Methods Mol Biol.* 2008;387:55–70
16. Bishop JO, Morton JG, Rosbash M, et al. Three abundance classes in HeLa cell messenger RNA. *Nature.* 1974;250: 199–204.
17. Schena M, Shalon D, Davis RW, et al. Quantitative monitoring of gene expression patterns with a complementary DNA microarray. *Science.* 1995;370:467–470.
18. Lipshutz RJ, Morris D, Chee M, et al. Using oligonucleotide probe arrays to access genetic diversity. *Biotechniques.* 1995;19:442–447.
19. Futschik ME, Reeve A, Kasabov N. Evolving connectionist systems for knowledge discovery from gene expression data of cancer tissue. *Artif Intell Med.* 2003;28:165–189.
20. Futschik ME. Gene expression profiling of metastatic and nonmetastatic colorectal cancer cell lines. *Genome Letters.* 2002; 1:26–34.

21. Quackenbush J. Microarray data normalization and transformation. *Nature Genetics*. 2002;32:496–501.
22. Eisen MB, Spellman PT, Brown PO, et al. Cluster analysis and display of genome-wide expression patterns. *Proc Natl Acad Sci USA*. 1998;95:14863–14868.
23. Ashburner M, Ball CA, Blake JA, et al. Gene ontology: Tool for the unification of biology. The Gene Ontology Consortium. *Nature Genetics*. 2000;25:25–29.
24. Rhee SY, Wood V, Dolinski K, et al. Use and misuse of the gene ontology annotations. *Nature Rev Genetics*. 2008;9:509–515.
25. Brazma A, Hingamp P, Quackenbush J, et al. Minimum information about a microarray experiment (MIAME) toward standards for microarray data. *Nature Genetics*. 2001;29:365–371.
26. DeRisi JL, Iyer VR, Brown PO. Exploring the metabolic and genetic control of gene expression on a genomic scale. *Science*. 1997;278:680–686.
27. Iyer VR, Eisen MB, Ross DT, et al. The transcriptional program in the response of human fibroblasts to serum. *Science*. 1999; 283:83–87.
28. Sorlie T, Perou CM, Tibshirani R, et al. Gene expression patterns of breast carcinomas distinguish tumor subclasses with clinical implications. *Proc Natl Acad Sci USA*. 2001;98:10869–10874.
29. van't Veer LJ, Dai H, van de Vijver MJ, et al. Gene expression profiling predicts clinical outcome of breast cancer. *Nature*. 2002;415:530–536.
30. Marchionni L, Wilson RF, Wolff AC, et al. Systematic review: Gene expression profiling assays in early-stage breast cancer. *Ann Intern Med*. 2008;148:358–369.
31. Michiels S, Koscielny S, Hill C. Prediction of cancer outcome with microarrays: A multiple random validation strategy. *Lancet*. 2005;365:488–492.
32. Alizadeh AA, Eisen MB, Davis RE, et al. Distinct types of diffuse large B-cell lymphoma identified by gene expression profiling. *Nature*. 2000;403:503–511.
33. Davis RE, Brown KD, Siebenlist U, et al. Constitutive nuclear factor kappaB activity is required for survival of activated B cell-like diffuse large B cell lymphoma cells. *J Exp Med*. 2001;194:1861–1874.
34. Lossos IS, Czerwinski DK, Alizadeh AA, et al. Prediction of survival in diffuse large- B-cell lymphoma based on the expression of six genes. *N Engl J Med*. 2004;350:1828–1837.
35. Golub TR, Slonim DK, Tamayo P, et al. Molecular classification of cancer: Class discovery and class prediction by gene expression monitoring. *Science*. 1999;286:531–537.

Chapter

8

The Human Epigenome: Implications for the Understanding of Human Disease

Maria Berdasco . Manel Esteller

INTRODUCTION

Epigenetic processes, defined as the heritable patterns of gene expression that do not involve changes in the sequence of the genome, and their effects on gene repression are increasingly understood to be such a way of modulating phenotype transmission and development. The patterns of DNA methylation, histone modifications, microRNAs, and several chromatin-related proteins of sick cells usually differ from those of healthy cells, highlighting the importance of epigenetic regulation in most human pathologies. The aim of the present review is to provide an overview of how epigenetic factors contribute to the development of human diseases such as abnormal imprinting-causative pathologies, cancer malignancies, as well as autoimmune, cardiovascular, and neurological disorders. These studies have provided extensive information about the mechanisms that contribute to the phenotype of human diseases, but also provided opportunities for therapy.

EPIGENETIC REGULATION OF THE GENOME

The primary goal of the Human Genome Project was to determine the sequence of the three billion base pairs that make up the human genome and to identify its approximately 25,000 genes from physical and functional standpoints. Most of the human genome (95%) had been sequenced by the end of 2003; of the remainder it is likely that the centromeres and telomeres will remain unsequenced until suitable new technology is developed. Current research efforts are directed toward analyzing the evolution of the genome, differences between individuals (polymorphisms), and environmental effects. The near-complete sequence has provided a rich source of scientific knowledge and has initiated new fields of genomic applications. However, the sequence itself does not predict how the genome is packaged in chromatin to ensure the differential expression of genes, which is essential for development and differentiation [1]. Epigenetic processes, defined as the heritable patterns of gene expression that do not involve changes in the sequence of the genome, and their effects on gene activation and inactivation are increasingly understood to be such a way of modulating phenotype transmission and development. Epigenetics links the fields of genetics and developmental biology and can explain, at least in part, the biological process by which identical genotypes establish patterns of differential gene expression that are stable through cell division.

The Human Epigenome Project

To date, small-scale studies of specific epigenetic marks have provided limited information about the regulation of genes from different pathways, for example, the hypermethylation-dependent silencing of tumor suppressor genes in cancer or the mutational inactivation of *MeCP2* in Rett patients. However, we need to develop an understanding of these processes that is based on a broader perspective. A range of matters remains to be resolved, such as the relationships between the epigenetic players (the epigenetic code) and how the environment and/or aging modulate the epigenetic marks. Much of this could be achieved

Molecular Pathology © 2009, Elsevier, Inc. All Rights Reserved.

by analyzing the epigenetic patterns on a genome-wide scale, an approach that has at last become possible thanks to recent technological advances. For example, comprehensive DNA-methylation maps (called the methylome) could be assessed by combining the methyl-DIP strategy with tilling or promoter microarray analyses [2–4]. In a similar manner, the use of the chromatin immunoprecipitation technique followed by genomic microarray hybridization (ChIP-on-chip) has begun to provide extensive maps of histone modifications [5–7]. Although the groundbreaking discoveries in the field of human disease were initially performed in cancer cells, the characterization of the error-bearing epigenomes underlying other disorders, such as neurological, cardiovascular, and immunological pathologies, has only just begun [8]. The fact that epigenetic aberrations control the function of the human genome and contribute to normal and pathological states justifies carrying out a comprehensive human epigenome project. The goal of the Human Epigenome Project is "to identify all the chemical changes and relationships among chromatin constituents that provide function to the DNA code." This "will allow a fuller understanding of normal development, aging, abnormal gene control in cancer, and other diseases as well as the role of the environment in human health" [1]. It is important to bear in mind that there is no single epigenome, but rather many different ones that are characteristic of normal and diverse human disorders, so it is essential to define the chosen starting material [1]. The information extracted from the whole-genome assays will help us understand the role of the epigenetic marks, and could have translational research benefits for diagnosis, prognostic, and therapeutic treatment [8]. Defining human epigenomes associated with human disorders will help select patients who are likely to benefit from epigenomic therapies or prevention strategies, determine their efficacy and specificity, and lead to the identification of surrogate markers and end-points of its effects. The aim of the present review is to provide an overview of how epigenetic factors contribute to the development of human diseases such as abnormal imprinting-causative pathologies, cancer malignancies, as well as autoimmune, cardiovascular, and neurological disorders with aberrant epigenetic profiles.

GENOMIC IMPRINTING

Epigenetic Regulation of Imprinted Genes

Genomic imprinting is a genetic phenomenon by which epigenetic chromosomal modifications drive differential gene expression according to the parent-of-origin. Expression is exclusively due either to the allele inherited from the mother (such as the *H19* and *CDKN1C* genes) or to that inherited from the father (such as *IGF2*). It is an inheritance process that is independent of the classical Mendelian model. Most imprinted genes, which have been identified in insects, mammals, and flowering plants, are involved in the establishment and maintenance of particular phases of development. Nucleus transplantation experiments in mouse zygotes carrying reciprocal translocations carried out in the early 1980s suggested that imprinting may be fundamental to mammalian development [9]. Assays confirmed that normal gene expression and development in mice require the contribution of both maternal and paternal alleles. However, it was not until 1991 that the first imprinted genes, insulin-like growth factor 2 (*IGF2*) and its receptor (*IGF2R*), were identified [10]. Since then, 83 imprinted genes have been identified in mice and humans, about one-third of which are imprinted in both species [11]. It has recently been predicted that 600 genes have a high probability of being imprinted in the mouse genome, and a similar genome-wide analysis predicts humans to have about half as many imprinted genes [12]. The molecular mechanisms underlying genomic imprinting are poorly understood. As imprinting is a dynamic process and the profile of imprinted genes varies during development, regulation must be epigenetic. DNA methylation has been widely described as the major mechanism involved in the control of genes subjected to imprinting. One model for this regulation is based on the cluster organization of imprinted genes. This structure within clusters allows them to share common regulatory elements, such as noncoding RNAs and differentially methylated regions (DMRs). DMRs are up to several kilobases in size, rich in CpG dinucleotides (such as CpG islands), and may contain repetitive sequences. DNA methylation of DMRs is thought to interact with histone modifications and other chromatin proteins to regulate parental allele-specific expression of imprinted genes. Furthermore, the aforementioned regulatory elements usually control the imprinting of more than one gene, giving rise to imprinting control regions (ICRs). This cluster organization, observed in 80% of imprinted genes, and the specific DNA methylation patterns associated with DMRs are two of the main characteristics of imprinted genes. Deletions or aberrations in DNA methylation of ICRs lead to loss of imprinting (LOI) and inappropriate parental gene expression [13]. Imprinted genes have diverse roles in growth and cellular proliferation, and specific patterns of genomic imprinting are established in somatic and germline cells [12,14]. Imprinting is erased in germline cells, and reprogramming involving a *de novo* methyltransferase is necessary to ensure sex-specific gene expression in the individual. Methyl groups are incorporated into most ICRs in oocytes, although only some ICRs are methylated during spermatogenesis. After fertilization, the specific methylation profiles of ICRs must be maintained during development in order to mediate the allelic expression of imprinted genes. In the primordial germline cells of the developed individual, these imprinting marks must be freshly erased by DNA demethylation to allow the subsequent establishment of new oocyte-specific and sperm-specific imprints. As a consequence, mammalian imprinting can be described as a development-dependent cycle based on germline establishment, somatic maintenance, and erasure.

Imprinted Genes and Human Genetic Diseases

Since expression of imprinted genes is monoallelic, and thereby functionally haploid, there is no protection from recessive mutations that the normal diploid genetic complement would provide [12]. For this reason, genetic and epigenetic aberrations in imprinted genes are linked to a wide range of diseases [15]. Modulation of perinatal growth and human pregnancy has played a central role in the evolution of imprinting, and many of the diseases associated with imprinted genes involve some disorders of embryogenesis. This is the case of the hydatidiform mole disorder, where all nuclear genes are inherited from the father. In most cases, this androgenesis arises when an anuclear egg is fertilized by a single sperm, after which all the chromosomes and genes are duplicated. However, fertilization by two haploid sperm (diandric diploidy) may occasionally occur. Most cases are sporadic and androgenetic, but recurrent hydatidiform mole has biparental inheritance with disrupted DNA methylation of DMRs at imprinted loci [16]. In contrast, the disorder of ovarian dermoid cysts arises from the spontaneous activation of an ovarian oocyte that leads to the duplication of the maternal genome [17]. These abnormalities suggest that normal human development is possible only when the paternal and maternal genomes are correctly transmitted. Parent-of-origin effects involved in behavioral and brain disorders have been widely reported at the prenatal and postnatal stages of development [18]. A postnatal growth retardation syndrome associated with the *MEST* gene expression is an illustrative example. The effect of introducing a targeted deletion into the coding sequence of the mouse *MEST* gene strongly depends on the paternal allele [19]. When the deletion is paternally derived, *Mest*+/– mice are viable and fertile, but mutant mice show growth retardation and high mortality. *Mest*–/+ animals, with a maternally derived deletion, show none of these effects. This suggests that the phenotypic consequences of this mutation are detected only through paternal inheritance and are the result of imprinting. There is also evidence that some imprinting effects are associated with increased susceptibility to cancer. Absence of expression of a tumor suppressor gene could be the result of LOI or uniparental disomy (UPD) in imprinted genes. Conversely, LOI or UPD of an imprinted gene that promotes cell proliferation (an oncogene) may allow gene expression to be inappropriately increased. These aberrations could have a widespread effect if the aberrant imprinting occurs in an ICR, resulting in the epigenetic dysregulation of multiple imprinted oncogenes and/or tumor suppressor genes [20]. The most common genetic disorders associated with imprinting aberrations are described next.

Prader-Willi Syndrome and Angelman Syndrome

Prader-Willi syndrome (PWS; OMIM #176270) and Angelman syndrome (AS; OMIM #105830) are very rare genetic disorders with autosomal dominant inheritance in which gene expression depends on parental origin. The clinical features of both syndromes are quite similar; they are neurological disorders with mental retardation and developmental aberrations [21]. PWS is characterized by diminished fetal activity, feeding difficulties, obesity, muscular hypotonia, mental retardation, poor physical coordination, short stature, hypogonadism, and small hands and feet, amongst other traits. AS is characterized by mental retardation, movement or balance disorder, characteristic abnormal behaviors, increased sensitivity to heat, absent or little speech, and epilepsy. The prevalence of the two syndromes is not accurately known, but is estimated to be between 1 in 12,000 and 1 in 15,000 live births, respectively. Most cases of PWS are caused by the deficiency of the paternal copies of the imprinted genes on chromosome 15 located in the 15q11–q13 region, while AS affects maternally imprinted genes in the same region. This deficiency could be due to the deletion of the 15q11–q13 region (3–4 Mb), parental uniparental disomy of chromosome 15, or imprinting defects [22]. The imprinted domain on human chromosome 15q11–q13 is regulated by an ICR that is responsible for establishing the imprinting in the gametes and for maintaining the patterns during the embryonic phases. ICR regulates differential DNA methylation and chromatin structure, and in consequence, differential gene expression affecting the two parental alleles. The ICR in 15q11–q13 appears to have a bipartite structure; one part seems to be responsible for the control of paternal expression and the other for maternal gene expression [23]. PWS is in effect a contiguous gene syndrome resulting from deficiency of the paternal copies of the imprinted *SNRF/SNRPN* gene, the necdin gene, and possibly other genes [24,25]. It has been estimated that the region could contain more than 30 genes, so PWS probably results from a stochastic partial inactivation of important genes [26]. While PWS appears to be more closely related to deficiencies caused by chromatin aberrations affecting several genes, AS is associated with mutations in single genes. For instance, the most common genetic defect leading to AS is a ~4 Mb maternal deletion in chromosomal region 15q11–13, which causes an absence of *UBE3A* expression in the maternally imprinted brain regions. Mutations in the gene encoding the ubiquitin-protein ligase E3A (*UBE3A*) have been identified in 25% of AS patients [27]. The *UBE3A* gene is present on both the maternal and paternal chromosomes, but differs in its pattern of methylation [28]. Paternal silencing of the *UBE3A* gene occurs in a brain-region-specific manner, with the maternal allele being active almost exclusively in the Purkinje cells, hippocampus, and cerebellum [29]. Another maternally expressed gene, *ATP10C* (aminophospholipid-transporting ATPase), is also located in this region and has been implicated in AS [30]. Like *UBE3A*, it exhibits imprinted, preferential maternal expression in human brain [31].

Beckwith-Wiedemann Syndrome

Beckwith-Wiedemann syndrome (BWS; OMIM #130650) is a well-characterized human disease involving imprinted genes that are epigenetically regulated, and studies of BWS patients have contributed much to the understanding of normal imprinting. BWS is a rare genetic or epigenetic overgrowth syndrome with an estimated prevalence of about 1 in 15,000 births and a high mortality rate in the newborn (about 20%). Neonatal patients are mainly characterized by exomphalos, macroglossia, and gigantism, but other symptoms may occur, such as organomegaly, adrenocortical cytomegaly, hemihypertrophy, and neonatal hypoglycemia [32]. There is also an increased risk of developing specific tumors, such as Wilms' tumor and hepatoblastoma [33]. The imprinted domain BWS-related genes are located on 11p15 and are regulated by a bipartite ICR. Two clusters of imprinting genes have been described [34]: (i) *H19/IGF2* (imprinted, maternally expressed, untranslated mRNA/insulin-like growth factor 2); and (ii) *p57KIP2* (a cyclin-dependent kinase inhibitor), *TSSC3* (a pleckstrin homology-like domain), *SLC22A1* (an organic cation transporter), *KvLQT1* (a voltage-gated potassium channel), and *LIT1* (*KCNQ1* overlapping transcript 1). Both clusters are regulated by two DMRs differentially methylated regions: DMR1, which is responsible for *H19/IGF2* control and is methylated on the paternal but not the maternal allele, and DMR2, which is located upstream of *LIT1* and is normally methylated on the maternal but not the paternal allele. BWS can appear as a consequence of two separate mechanisms [35]. First, some patients are characterized by UPD, which consists of the complete genetic replacement of the maternal allele region with a second paternal copy, and/or LOI affecting the *IGF2*-containing region [36] and/or the *LIT1* gene [37], which causes a switch in the epigenotype of the *H19/IGF2* subdomain or the *p57/ KvLQT1/LIT1* subdomain, respectively. When UPD affects *IGF2*, it yields a double dose of this autocrine factor, resulting in tissue overgrowth and increased cancer risk [38]. The LOI mechanism involves aberrant methylation of the maternal *H19* DMR [39]. Second, maternal replacements of the allele and localized abnormalities of allele-specific chromatin modification on *p57KIP2* (also known as *CDKN1C*) [40] or *LIT1* could also contribute to the BWS phenotype [41]. In conclusion, BWS is a model for the hierarchical organization of epigenetic regulation in progressively larger domains [42]. Additionally, mutations in NSD1 (*nuclear receptor-binding SET domain-containing protein 1*), the major cause of the Sotos overgrowth syndrome, have been described in BWS patients, demonstrating the role this gene plays in imprinting the chromosome 11p15 region [43].

CANCER EPIGENETICS

Cancer encompasses a fundamentally heterogeneous group of disorders affecting different biological processes, and is caused by abnormal gene/pathway function arising from specific alterations in the genome. Initially, cancer was thought to be solely a consequence of genetic changes in key tumor suppressor genes and oncogenes that regulate cell proliferation, DNA repair, cell differentiation, and other homeostatic functions. However, recent research suggests that these alterations could also be due to epigenetic disruption. The study of epigenetic mechanisms in cancer, such as DNA methylation, histone modification, nucleosome positioning and microRNA expression, has provided extensive information about the mechanisms that contribute to the neoplastic phenotype through the regulation of expression of genes critical to transformation pathways. These alterations and their involvement in tumor development are briefly reviewed in the following sections.

DNA Hypomethylation in Cancer Cells

The low level of DNA methylation in tumors compared with that in their normal-tissue counterparts was one of the first epigenetic alterations to be found in human cancer [44]. It has been estimated that 3–6% of all cytosines are methylated in normal human DNA, although cancer cell genomes are usually hypomethylated with malignant cells, featuring 20–60% less genomic 5-methylcytosine than their normal counterparts [45]. Global hypomethylation in cancer cells is generally due to decreased methylation in CpGs dispersed throughout repetitive sequences, which account for 20–30% of the human genome, as well as in the coding regions and introns of genes [45]. A chromosome-wide and large-promoter-specific study of DNA methylation in a colorectal cancer cell line using the methyl-DIP approach has revealed extensive hypomethylated genomic regions located in gene-poor areas [2]. Importantly, the degree of hypomethylation of genomic DNA increases as the lesion progresses from a benign cellular proliferation to an invasive cancer [46]. From a functional point of view, hypomethylation in cancer cells is associated with a number of adverse outcomes, including chromosome instability, activation of transposable elements, and LOI (Figure 8.1). Decreased methylation of repetitive sequences in the satellite DNA of the pericentric region of chromosomes is associated with increased chromosomal rearrangements, mitotic recombination, and aneuploidy [47,48]. Intragenomic endoparasitic DNA, such as L1 (long interspersed nuclear elements) [49] and Alu (recombinogenic sequence) repeats, are silenced in somatic cells and become reactivated in human cancer. Deregulated transposons could cause transcriptional deregulation, insertional mutations, DNA breaks, and an increased frequency of recombination, contributing to genome disorganization, expression changes, and chromosomal instability [49]. DNA methylation underlies the control of several imprinted genes, so the effect on the loss of imprinting must also be considered. Wilms' tumor, a nephroblastoma that typically occurs in children, is the best-characterized imprinting effect associated with increased susceptibility to cancer [50]. Other changes in the expression of imprinted genes caused by changes in methylation have been demonstrated in malignancies such as osteosarcoma [51], hepatocellular carcinoma [52], and bladder cancer [53]. Finally,

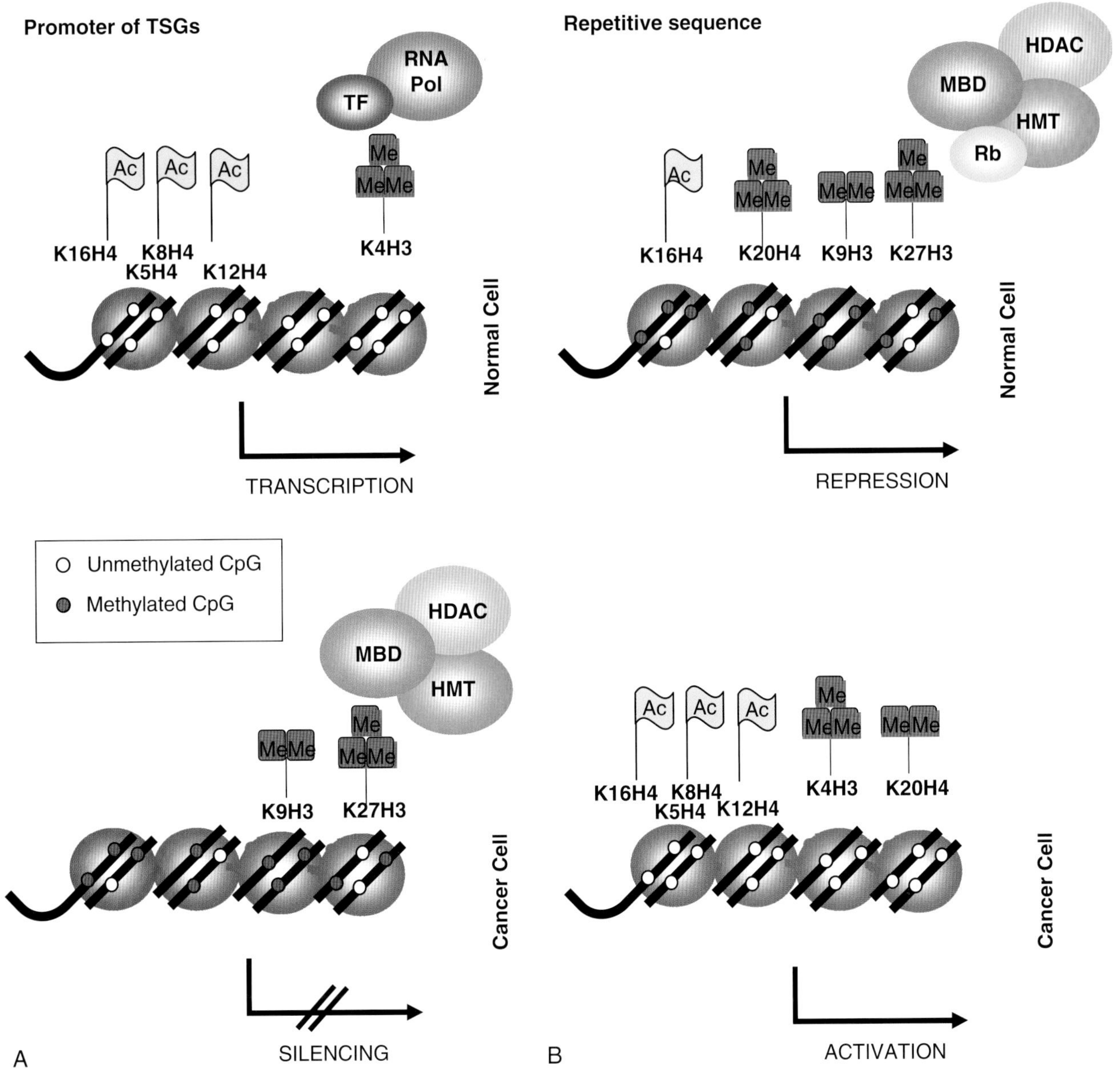

Figure 8.1 A model for the disruption of histone modifications and DNA methylation patterns in cancer cells. Nucleosome arrays are located in the context of genomic regions that include (A) promoters of tumor suppressor genes (TSGs), (B) repetitive sequences in heterochromatin regions. Nucleosomes consisting of two copies of histones H2A, H2B, H3, and H4 are represented as gray cylinders. DNA (black lines) is wrapped around each nucleosome. In normal cells, TSG promoter regions are unmethylated and enriched in histone modification marks associated with active transcription, such as acetylation of histone H4 (at lysine K5, K8, K12, and K16), or trimethylation of histone H3 (at lysine K4). Transcription machinery recognizes these active marks and transcription of TSGs is allowed. In the same cells, the repetitive genomic DNA is silenced due to the high degree of DNA methylation and histone-repressive marks: histone H4 is densely trimethylated at lysine 20, and histone H3 is dimethylated at lysine 9 and trimethylated at lysine k27. This epigenetic profile is disrupted in transformed cells. TSG promoters are silenced by the loss of the histone-active marks and gain of promoter hypermethylation. Repetitive sequences are activated by replacement of the repressive marks, leading to activation of endoparasitic sequences, genomic instability, or loss of imprinting.

DNA methylation acts a mechanism for controlling cellular differentiation, allowing the expression only of tissue-specific and housekeeping genes in somatic differentiated cells. It is possible that some tissue-specific genes became reactivated in cancer [54] in a hypomethylation-dependent manner. Activation of *PAX2*, a gene that encodes a transcription factor involved in proliferation and other important cell activities, and *let-7a-3*, an miRNA gene, have been implicated in endometrial and colon cancer [55,56].

Hypermethylation of Tumor Suppressor Genes

Aberrations in DNA methylation patterns of the CpG islands in the promoter regions of tumor suppressor genes are accepted as being a common feature of human cancer (Figure 8.1). The initial discovery of silencing was performed in the promoter of the retinoblastoma (*Rb*) tumor suppressor gene [57], but hypermethylation of genes like *VHL* (associated with von Hippel-Lindau disease), *p16INK4a, 8-11 hMLH1* (a homologue of *Escherichia coli* MutL), and *BRCA1* (breast-cancer susceptibility gene 1) has also been described [58]. The presence of CpG island promoter hypermethylation affects genes from a wide range of cellular pathways, such as cell cycle, DNA repair, toxic catabolism, cell adherence, apoptosis, and angiogenesis, among others [45], and may occur at various stages in the development of cancer. In recent years, a CpG-island hypermethylation profile of human primary tumors has emerged, which shows that the CpG island hypermethylation profiles of tumor suppressor genes are specific to the cancer type [59,60]. Each tumor type can be assigned a specific, defining DNA "hypermethylome," rather like a physiological or cytogenetic marker. These marks of epigenetic inactivation occur not only in sporadic tumors but also in inherited cancer syndromes, in which hypermethylation may be the second lesion in Knudson's two-hit model of cancer development [61]. It is expected that improvements in genome-wide epigenomic studies will increase the number of hypermethylated tumor suppressor genes in a broad spectrum of tumors. However, to date 100–400 instances of gene-specific methylation have been noted in a given tumor. Sometimes the epigenetic alteration of a tumor suppressor gene has genetic consequences, for example, when a DNA repair gene (*hMLH1, BRCA1, MGMT*, or Werner's syndrome gene) is silenced by promoter methylation and functionally blocked [58,62].

Histone Modifications of Cancer Cells

The histone modification network is very complex. Histone modifications can occur in various histone proteins (including H2B, H3, H4) and variants (such as H3.3) and affect different histone residues (including lysine, arginine, serine). Several chemical groups (methyl, acetyl, phosphate) may be added in different degrees (for example, monomethylation, dimethylation, or trimethylation). Furthermore, the significance of each modification depends on the organism, the biological process, and the chromatin-genomic region. Due to this diversity of permutations and combinations, little is known about the patterns of histone modification disruption in human tumors. Recent results have shown that the CpG promoter-hypermethylation event in tumor suppressor genes in cancer cells is associated with a particular combination of histone markers: deacetylation of histones H3 and H4, loss of H3K4 trimethylation, and gain of H3K9 methylation and H3K27 trimethylation (Figure 8.1) [63]. Increased acetylated histones H3 and H4, and H3/K4 at the *p21* and *p16* transcription start sites after genistein induction (an isoflavone found in the soybean with tumor suppressor properties), in the absence of *p21* promoter methylation, have been reported [64]. The association between DNA methylation and histone modification aberrations in cancer also occurs at the global level. In human and mouse tumors, histone H4 undergoes a loss of monoacetylated and trimethylated lysines 16 and 20, respectively, especially in the repetitive DNA sequences [46]. Subsequent studies showed that loss of trimethylation at H4K20 is involved in disrupting heterochromatic domains and may reduce the response to DNA damage of cancer cells [65]. Immunohistochemical staining of primary prostatectomy tissue samples revealed that patterns of H3 and H4 were predictors of clinical outcome independently of tumor stage, preoperative prostate-specific antigen levels, and capsule invasion in prostate cancer [66].

Epigenetic Regulation of microRNAs in Cancer

miRNA expression patterns must be tightly regulated during development in a tissue-specific manner, and play important roles in cell proliferation, apoptosis, and differentiation [67]. miRNA expression profiles differ between normal and tumor tissues, and also among tumor types [68], whereby some microRNAs are down-regulated in cancer (like tumor suppressor genes). This comparison suggests that miRNAs could be silenced by epigenetic mechanisms [69]. DNA methylation has been shown to be the regulatory mechanism in at least two microRNAs, miR-127 and miR-124a [69]. miR-127, which negatively regulates the protooncogene *BCL6* (B-Cell Lymphoma 6), is usually expressed in normal cells but is silenced by DNA methylation in cancer cells [70]. Similarly, the hypermethylation-dependent silencing of miR-124a results in an increase in expression of cyclin D-kinase 6 oncogene *(CDK6)*, and is recognized as a common feature of a wide range of tumors [71].

Aberrations in Histone-Modifier Enzymes

Aberrations in the epigenetic profiles, with respect to DNA methylation and histone modifications, could also be a consequence of genetic disruption of the epigenetic machinery. A preliminary set of genes involved in epigenetic modifications with mutations in cancer cells but not in the normal counterparts has been found[45]. A list of genes involved in epigenetic modifications that are disrupted in human cancer is presented in Table 8.1. Although alterations in the levels of DNMTs and MBD-containing proteins are commonly observed in human tumors [72,73], no genetic lesion has been described in the DNA-methylation machinery in cancer cells. The picture is different for histone modifier enzymes. In leukemia and sarcoma, chromosomal translocations that involve histone-modifier genes, such as histone acetyltransferases (HATs) [74] [such as cyclic AMP response-element-binding protein (CREB)-binding protein-monocytic leukemia zinc finger (*CBP-MOZ*)] and histone methyltransferases (HMTs) [such as mixed-lineage leukemia 1 (*MLL1*)]

Table 8.1 Disruption of Genes Involved in DNA Methylation and Histone Modifications in Cancer

Gene	Alteration	Tumor type
	Alterations affecting DNA methylation enzymes (DNMTs)	
DNMT1	Overexpression	Various
DNMT3b	Overexpression	Various
	Alterations involving Methyl-CpG-binding proteins (MBPs)	
MeCP2	Overexpression, rare mutations	Various
MBD1	Overexpression, rare mutations	Various
MBD2	Overexpression, rare mutations	Various
MBD3	Overexpression, rare mutations	Various
MBD4	Mutations in microsatellite instable tumors	Colon, stomach, endometrium
	Alterations disrupting histone acetyltransferases (HATs)	
p300	Mutations in microsatellite instable tumors	Colon, stomach, endometrium
CBP	Mutations, translocations, deletions	Colon, stomach, endometrium, lung, leukemia
pCAF	Rare mutations	Colon
MOZ	Translocations	Hematological malignancies
MORF	Translocations	Hematological malignancies, leiomyomata
	Alterations disrupting histone deacetylases (HDACs)	
HDAC1	Aberrant expression	Various
HDAC2	Aberrant expression, mutations in microsatellite instable tumors	Various
	Alterations affecting histone methyltransferases (HMTs)	
MLL1	Translocation	Hematological malignancies
MLL2	Gene amplification	Glioma, pancreas
MLL3	Deletion	Leukemia
NSD1	Translocation	Leukemia
EZH2	Gene amplification, overexpression	Various
RIZ1	Promoter CpG-island hypermethylation	Various
	Alterations affecting histone demethylases	
GASC1	Gene amplification	Squamous cell carcinoma

Adapted from [45].

[75], nuclear-receptor binding SET domain protein 1 (*NSD1*) [76], and nuclear-receptor binding SET-domain protein 3 (*NSD3*) [77], create aberrant fusion proteins [8]. In solid tumors, both HMT genes such as *EZH2* [78], mixed-lineage leukemia 2 (*MLL2*) [79], or *NSD3* [80], and a demethylase [Jumonji domain-containing protein 2C (*JMJD2C/GASC1*)] are known to be amplified [81]. Genetic aberrations also disrupt expression of histone deacetylases, such as histone deacetylase 2 (*HDAC2*), which could be affected by mutational frameshift inactivation in colon cancer [82], and chromatin remodeling proteins, such as *HLTF* (helicase-like transcription factor) [83], *BRG1* (Brahma-related gene 1) [84], and other components of the SWI/SNF family of proteins.

HUMAN DISORDERS ASSOCIATED WITH EPIGENETICS

Aberrant Epigenetic Profiles Underlying Immunological, Cardiovascular, Neurological, and Metabolic Disorders

The patterns of DNA methylation, histone modifications, microRNAs, and several chromatin-related proteins of sick cells usually differ from those of healthy cells, highlighting the importance of epigenetic regulation in most human pathologies. Most of our knowledge regarding human epigenetic diseases was first obtained from cancer cells, but nowadays there is increasing interest in understanding the role of epigenetic modifications in the etiology of human disease. This section summarizes the involvement of specific epigenetic aberrations in immunological, cardiovascular, and neurological disorders. DNA methylation is the best-characterized epigenetic modification in many pathways of immunology, and is the source of much of our knowledge about the molecular network of the immune system. Classical autoimmune disorders, such as systemic lupus erythematosus (SLE), an autoimmune disease characterized by the production of a variety of antibodies against nuclear components and which causes inflammation and injury of multiple organs, and rheumatoid arthritis, a chronic systemic autoimmune disorder which primarily causes inflammation and destruction of the joints, are characterized by massive genomic hypomethylation [85,86]. This decrease in DNA methylation levels is highly reminiscent of the global demethylation observed in the DNA of tumor cells compared with their normal tissue counterparts [8]. How this hypomethylation in T-cells induces SLE is not well understood. It has been proposed that demethylation induces overexpression of integrin adhesive receptors and leads to an autoreactive response. Identification of the full set of genes deregulated by DNA hypomethylation could help to explain these immunological disorders and will also enable the development of effective therapies to cure SLE [85]. More pathologies with epigenetic regulation of immunology have been reported, including the epigenetic silencing of the ABO histo-blood group genes [87], the silencing of human leukocyte antigen (HLA) class I antigens [88], and the melanoma antigen-encoding gene (MAGE) family. It is known that *MAGE* gene expression is epigenetically repressed by promoter CpG methylation in most cells, but *MAGE* genes may be expressed in various tumor types via CpG demethylation and can act as antigens that are recognized by cytolytic T-lymphocytes [89]. Alterations of specific genomic DNA methylation levels have been described not only in the fields of oncology and immunology but also in a wide range of biomedical and scientific fields. In neurology, for example, mutations in *MeCP2*, which encodes methyl CpG binding protein 2, cause most cases of Rett syndrome (RTT) and autism [90]. Future research needs to determine whether aberrant DNA methylation is important in the complex etiology of other frequent neurological pathologies, such as schizophrenia and Alzheimer's disease. Beyond this, DNA methylation changes are also known to be involved in cardiovascular disease, the biggest killer in Western countries [8]. For example, aberrant CpG island hypermethylation has been described in atherosclerotic lesions [91]. Since DNA methylation and histone modifications are mechanistically linked, it is likely that different changes in DNA methylation are associated with changes in histone modifications in human diseases. To date, we have been largely ignorant of how these histone modification markers are disrupted in human diseases. A preview of the patterns of histone modifications and their cellular location has only been described for cancer malignancies [45]. For other human pathologies, our knowledge of the alterations in histone modification patterns comes from the use of epigenetic drugs. Therapeutic assays have demonstrated that histone deacetylase (HDAC) inhibitors can improve deficits in synaptic plasticity, cognition, and stress-related behaviors in a wide range of neurological and psychiatric disorders, including Huntington's and Parkinson's diseases, anxiety and mood disorders, and Rubinstein-Taybi and Rett syndromes [92]. Abnormal histone modification patterns associated with specific gene expression have also been described in lupus CD4+ T-cells [93], and HDAC inhibitors are able to reverse gene expression significantly [85].

Genetic Aberrations Involving Epigenetic Genes

Genetic alterations of genes coding for enzymes that mediate chromatin structure could result in a loss of adequate regulation of chromatin compaction, and finally, the deregulation of gene transcription and inappropriate protein expression. Although the consequences in cancer malignancies have been widely described, in this section we extend the review to include other genetic diseases involving the function of several enzymes of the epigenetic machinery. The phenotype of these diseases also helps to clarify the role of various chromatin proteins in cell proliferation and differentiation. These include disorders arising from alterations in chromatin remodeling factors, alterations of the components of the DNA methylation machinery, and aberrations disturbing histone modifiers.

Syndromes of Disordered Chromatin Remodeling

Alpha-Thalassemia X-Linked Mental Retardation Syndrome Alpha-thalassemia X-linked mental retardation (ATRX; OMIM #301040) syndrome is characterized by distinctive craniofacial features, genital anomalies, and severe developmental delays with hypotonia and mental retardation. Some individuals with ATRX syndrome have an unusual mild form of hemoglobin H disease, but alpha-thalassemia is not a constant characteristic of the syndrome [94]. *ATRX* is the gene associated with ATRX syndrome and codes for a centromeric heterochromatin-binding protein from the SNF2 family of helicase/ATPases. Centromeric ATRX is required for maintaining a bipolar metaphase II spindle and for establishing correct chromosome alignment during meiosis. Its activity has been associated with epigenetic modifications, such as histone deacetylation and DNA methylation [95].

CHARGE Syndrome The acronym CHARGE (OMIN #214800) stands for Coloboma of the eye, Heart anomaly, choanal Atresia, Retardation of mental and somatic development, Genital and/or urinary abnormalities, and Ear abnormalities and/or deafness [96]. The estimated incidence of CHARGE syndrome is 1 in 8,500–12,000 births. It is an autosomal dominant condition with genotypic heterogeneity. Most cases are due to the mutation or deletion of a member of the chromodomain helicase DNA-binding domain (*CHD7*) family of ATP-dependent chromatin-remodeling enzymes [97], and in consequence, a protein involved in chromatin structure and gene expression.

Cockayne Syndrome Patients with Cockayne syndrome type B (CSB; OMIM #133540) present with failure to thrive, short stature, premature aging, neurological alterations, photosensitivity, delayed eruption of the primary teeth, congenital absence of some permanent teeth, partial macrodontia, atrophy of the alveolar process, and caries [98]. CSB is a genetic disorder with autosomal recessive inheritance that is caused by mutation of the excision-repair cross-complementing group 6 (*ERCC6*) gene. The protein, with ATP-stimulated ATPase activity, is part of the nucleotide excision repair (NER) pathway, a complex system that eliminates a broad spectrum of structural DNA lesions, including ultraviolet (UV)-induced dimers and DNA cross-links [99].

Schimke Immunoosseous Dysplasia Schimke immunoosseous dysplasia (SIOD; OMIM #242900) is an autosomal recessive disorder with the diagnostic features of growth retardation, renal failure, recurrent infections, cerebral infarcts, slowly progressive immune defects, such as T-cell immunodeficiency and skin pigmentation, that begin in childhood [100]. Mutations in a component of the SWI/SNF complex, the *SMARCAL1* gene, are responsible for this condition [101].

DNA Methylation-Associated Diseases

Immunodeficiency-Centromeric Instability-Facial Anomalies Syndrome

Immunodeficiency-centromeric instability-facial anomalies syndrome (ICF syndrome; OMIM #242860) is characterized by immunodeficiency in association with centromere instability of chromosomes 1, 9, and 16, and facial anomalies. ICF syndrome patients exhibit facial anomalies, such as hypertelorism, low-set ears, epicanthal folds, and macroglossia. It also features variable reductions in serum immunoglobulin levels that cause most ICF syndrome patients to succumb to infectious diseases before adulthood [102]. It is a rare autosomal recessive disorder that is notable for the variable severity of chromosomal abnormalities. Essentially, these involve juxtacentromeric heterochromatin formation of chromosomes 1, 9, and 16; an increased frequency of somatic recombination between the arms of these chromosomes; and a marked tendency to form multibranched configurations [103]. The observation that the cytogenetic abnormalities in the phenotype of ICF syndrome could be compared with the decondensation effect found in a normal pro-B lymphoblastoid cell line treated with the DNA methylation inhibitors 5-azacytidine or 5-azadeoxycytidine suggested the involvement of genomic DNA methylation in this disorder. In fact, 40% of ICF syndrome patients owe their condition to mutations in the *DNMT3B* gene, which encodes a *de novo* DNA-methyltransferase that acts on CpG-rich satellite DNAs [104]. In this way, ICF syndrome can be biochemically characterized by hypomethylation of CpG sites in some heterochromatic regions.

Rett Syndrome

Rett syndrome (RTT; OMIM #312750) is a severe neurodevelopmental disorder that primarily occurs in females at an estimated frequency of 1 in 15,000 births. It is characterized by arrested development between 6 and 18 months of age, regression of acquired skills, loss of speech, stereotypical movements (classically of the hands), microcephaly, seizures, and mental retardation. Rarely, RTT has also been described in male patients with somatic mosaicism or an extra X chromosome [105]. Mutations in the coding regions of the *MeCP2* (methyl CpG-binding protein 2) gene are found in approximately 70–80% of classic RTT patients [106], while the remaining cases result from large deletions of the *MeCP2* gene [107], mutations in noncoding regions of the *MeCP2* gene, and mutations in other genes [108]. An atypical Rett syndrome with infantile spasms that may be related to mutations in the cyclin-dependent kinase-like 5 (*CDKL5*) gene has also been described [109]. The *MeCP2* gene encodes a protein that preferentially binds methylated CpG dinucleotides and, in turn, mediates transcriptional repression through the recruitment of histone deacetylases, histone methyltransferases, and polycomb proteins, among others [110]. In addition, cells of RTT patients may have an abnormal secondary chromatin structure, with putative broad-ranging effects on expression of genes that otherwise are not mutated [111]. It has been proposed that *MeCP2* disruption in RTT patients causes aberrant upregulation of the genes that should normally be silenced in the nervous system, like the *BDNF* gene or imprinted genes [112]. The discovery that *MeCP2* was the gene responsible for RTT syndrome allows it to be used for early diagnosis and prenatal detection. In addition, the finding that epigenetic regulation has a role in the pathogenesis of RTT has suggested opportunities for therapy.

Human Malignancies Associated with Alterations in Histone Modifications

Coffin-Lowry Syndrome

The symptoms of Coffin-Lowry syndrome (CLS; OMIM #303600) are usually more severe in males than in females since it is an X-linked condition. Males with CLS syndrome usually exhibit severe-to-profound mental retardation and delayed development, but in female patients this may be entirely absent, or mild-to-profound [113]. Other features, such as skeletal malformations, distinctive facial features, microcephaly, and soft hands with short-tapered fingers, may be observed. CLS has an estimated incidence of 1 in 40,000–50,000 births. It is caused by mutations in the *RSK2* (ribosomal protein S6 kinase) gene [114]. RSKs are involved in activating the mitogen-activated kinase cascade and in stimulating cell proliferation (transition between the G0 and G1 phases of the cell cycle), and differentiation. RSK2 is responsible for the phosphorylation of the transcription factor CREB (cyclic AMP-responsive-binding protein) and histone H3 as a mechanism for responding to mitogenic stimulation by epidermal growth factor (EGF) [115]. Analysis of mRNA and protein expression of RSK2 on lymphocytes extracted directly from blood of patients revealed a mutation at a putative phosphorylation site that would be critical for RSK2 activity. Both assays are rapid and practical techniques for the diagnosis of CLS [116].

Rubinstein-Taybi Syndrome

Rubinstein-Taybi syndrome (RSTS; OMIM #180849) is an autosomal dominant disorder characterized by mental retardation, broad thumbs and toes, distinctive facial abnormalities, and short stature. RSTS patients have an increased risk of developing tumors, especially congenital glaucoma. The incidence has been estimated to be in 1 in 100,000 births [117]. In almost all cases, the syndrome is caused by mutation of the gene encoding the transcriptional coactivator CREB-binding protein (CREBBP) [118]. Mutations in the *EP300* gene are responsible for a small percentage of RSTS cases [119]. Both CREBBP and EP300 are histone acetyltransferases. A direct connection between loss of acetyltransferase activity and RSTS has been described [120], which indicates that the disorder is caused by aberrant chromatin regulation.

Sotos Syndrome

Sotos syndrome (OMIM #117550) is a well-known overgrowth syndrome with autosomal dominant inheritance characterized by excessive growth during childhood, advanced bone age, macrocephaly, characteristic facial gestalt, and various degrees of developmental delay [121]. It may also be associated with variable minor features, including cardiac and renal anomalies, seizures, and scoliosis. The risk of neoplasm formation in Sotos patients has been estimated to be about 2–3% [122], which is greater than in the general population. The most frequent tumor types are leukemia and lymphoma, but others, such as neural crest tumors, small-cell lung cancer, and sacrococcygeal teratomas have been reported [123]. Sotos syndrome is caused by haploinsufficiency of the nuclear receptor-binding SET domain-containing protein 1 gene (*NSD1*). Several mutations cause the loss of *NSD1* function in 90% of patients with Sotos syndrome [121], such as truncating mutations, missense mutations, 5q35 macrodeletions and microdeletions encompassing the *NSD1* gene or the t(5;11)(q35;p15.5) translocation, which fuses *NSD1* to nucleoporin-98 (*NUP98*). *NSD1* contains a SET domain that confers intrinsic histone methyltransferase activity that is specific to lysine 36 of histone H3 (H3K36) and lysine 20 of histone H4 (H4K20), epigenetic marks with ascribed roles in chromatin structure and gene function.

ENVIRONMENT AND THE EPIGENOME

Gene expression strongly depends on environmental signals—the external and endogenous stimuli—that make up the cellular context. Since epigenetic states can potentially be reverted, these gene-environment interactions could be integrated by epigenetic modifications of the genome [124]. Disease susceptibility also could be modified by environmental factors that alter the profile of DNA methylation and histone modifications in normal cells during prenatal and postnatal life. Alterations in the epigenome, and subsequent differential gene expression, could be directly affected by diet, xenobiotic chemicals, behavior, and exogenous stimuli, such as doses of radiation [12]. Aberrant epigenetic patterns could be a consequence of changes in the metabolism of the groups of donor chemicals, such as methyl or acetyl groups. Diets that are deficient in folate and methionine, which are necessary for normal biosynthesis of S-adenosylmethionine (SAM), the methyl donor for methylcytosine, lead to DNA hypomethylation and aberrant imprinting of insulin-like growth factor 2 (*IGF2*) [125]. Mutations in methylenetetrahydrofolate reductase (*MTHFR-677T* allele), which is critical for directing the folate pool to remethylate homocysteine to methionine, reduce the levels of DNA methylation responsible for the pathogenesis of diseases related to neural tube defects [126], spina bifida [127], and colorectal cancer [128]. Associations are also known between deficiency of methionine synthase, a folate-dependent and vitamin B(12)-dependent enzyme, and megaloblastic anemia, hemolytic uremic syndrome, and pulmonary hypertension, among others [129]. Exposure to metals, such as arsenic [130], cadmium [131], lead [132], nickel [133], and chromium [134], is linked to changes in the expression of epigenetically controlled genes via interactions with DNA-methylation-associated enzymes, histone acetyltransferase, and histone deacetylase enzymes. Endocrine-active chemicals, like estrogenic and anti-androgenic toxins, decrease male fertility by manipulating DNA methylation levels [135]. Among the behavioral factors, maternal grooming is an influential environmental factor affecting the methylation state of the glucocortocoid receptor in the hippocampus of rat pups [136]. Reproductive factors like assisted reproductive technology act as an environmental modulator of the epigenome, and are associated with the hypomethylation noted in Beckwith-Wiedemann and Angelman syndromes [137]. Studies performed with genetically identical individuals are an excellent method for understanding the environment-epigenome interaction [138,139]. In this way, the epigenome of human monozygotic twins could also vary because of differences in environmental factors. Recently, it was noted that although twins are epigenetically indistinguishable during the early years of life, older monozygous twins exhibit marked differences in their overall content and genomic distribution of 5-methylcytosine DNA and histone acetylation, which affects their gene-expression portrait [139].

REFERENCES

1. Jones PA, Martienssen R. A blueprint for a Human Epigenome Project: the AACR Human Epigenome Workshop. *Cancer Res.* 200; 6:11241–11246.
2. Weber M, Davies JJ, Wittig D, et al. Chromosome-wide and promoter-specific analyses identify sites of differential DNA methylation in normal and transformed human cells. *Nat Genet.* 2005;37:853–862.
3. Zhang X, Yazaki J, Sundaresan A, et al. Genome-wide high-resolution mapping and functional analysis of DNA methylation in arabidopsis. *Cell.* 2006;126:1189–1201.
4. Jacinto FV, Ballestar E, Esteller M. Methyl-DNA immunoprecipitation (MeDIP): Hunting down the DNA methylome. *Biotechniques.* 2008;44:35, 37, 39 passim.
5. Kurdistani SK, Tavazoie S, Grunstein M. Mapping global histone acetylation patterns to gene expression. *Cell.* 2004;117:721–733.
6. Bernstein BE, Kamal M, Lindblad-Toh K, et al. Genomic maps and comparative analysis of histone modifications in human and mouse. *Cell.* 2005;120:169–181.
7. Azuara V, Perry P, Sauer S, et al. Chromatin signatures of pluripotent cell lines. *Nat Cell Biol.* 2006;8:532–538.
8. Esteller M. The necessity of a human epigenome project. *Carcinogenesis.* 2006;27:1121–1125.
9. Surani MA, Barton SC, Norris ML. Development of reconstituted mouse eggs suggests imprinting of the genome during gametogenesis. *Nature.* 1984;308:548–550.
10. DeChiara TM, Robertson EJ, Efstratiadis A. Parental imprinting of the mouse insulin-like growth factor II gene. *Cell.* 1991;64:849–859.
11. Morison IM, Ramsay JP, Spencer HG. A census of mammalian imprinting. *Trends Genet.* 2005;21:457–465.
12. Dolinoy DC, Weidman JR, Jirtle RL. Epigenetic gene regulation: Linking early developmental environment to adult disease. *Reprod Toxicol.* 2007;23:297–307.
13. Murrell A, Ito Y, Verde G, et al. Distinct methylation changes at the IGF2-H19 locus in congenital growth disorders and cancer. *PLoS ONE.* 2008;3:e1849.

14. Feil R, Berger F. Convergent evolution of genomic imprinting in plants and mammals. *Trends Genet.* 2007;23:192–199.
15. Ubeda F, Wilkins JF. Imprinted genes and human disease: An evolutionary perspective. *Adv Exp Med Biol.* 2008;626:101–115.
16. Kou YC, Shao L, Peng HH, et al. A recurrent intragenic genomic duplication, other novel mutations in NLRP7 and imprinting defects in recurrent biparental hydatidiform moles. *Mol Hum Reprod.* 2008;14:33–40.
17. Oosterhuis JW, Looijenga LH. Testicular germ-cell tumours in a broader perspective. *Nat Rev Cancer.* 2005;5:210–222.
18. Morison IM, Reeve AE. A catalogue of imprinted genes and parent-of-origin effects in humans and animals. *Hum Mol Genet.* 1998;7:1599–1609.
19. Lefebvre L, Viville S, Barton SC, et al. Abnormal maternal behaviour and growth retardation associated with loss of the imprinted gene Mest. *Nat Genet.* 1998;20:163–169.
20. Falls JG, Pulford DJ, Wylie AA, et al. Genomic imprinting: Implications for human disease. *Am J Pathol.* 1999;154:635–647.
21. Chen C, Visootsak J, Dills S, et al. Prader-Willi syndrome: An update and review for the primary pediatrician. *Clin Pediatr (Phila).* 2007;46:580–591.
22. Camprubi C, Coll MD, Villatoro S, et al. Imprinting center analysis in Prader-Willi and Angelman syndrome patients with typical and atypical phenotypes. *Eur J Med Genet.* 2007;50:11–20.
23. Kantor B, Shemer R, Razin A. The Prader-Willi/Angelman imprinted domain and its control center. *Cytogenet Genome Res.* 2006;113:300–305.
24. Maina EN, Webb T, Soni S, et al. Analysis of candidate imprinted genes in PWS subjects with atypical genetics: A possible inactivating mutation in the SNURF/SNRPN minimal promoter. *J Hum Genet.* 2007;52:297–307.
25. Watrin F, Le Meur E, Roeckel N, et al. The Prader-Willi syndrome murine imprinting center is not involved in the spatio-temporal transcriptional regulation of the Necdin gene. *BMC Genet.* 2005;6:1.
26. Yang T, Adamson TE, Resnick JL, et al. A mouse model for Prader-Willi syndrome imprinting-centre mutations. *Nat Genet.* 1998;19:25–31.
27. Kishino T, Lalande M, Wagstaff J. UBE3A/E6-AP mutations cause Angelman syndrome. *Nat Genet.* 1997;15:70–73.
28. Lalande M, Calciano MA. Molecular epigenetics of Angelman syndrome. *Cell Mol Life Sci.* 2007;64:947–960.
29. Fang P, Lev-Lehman E, Tsai TF, et al. The spectrum of mutations in UBE3A causing Angelman syndrome. *Hum Mol Genet.* 1999;8:129–135.
30. Meguro M, Kashiwagi A, Mitsuya K, et al. A novel maternally expressed gene, ATP10C, encodes a putative aminophospholipid translocase associated with Angelman syndrome. *Nat Genet.* 2001;28:19–20.
31. Herzing LB, Kim SJ, Cook EH Jr, et al. The human aminophospholipid-transporting ATPase gene ATP10C maps adjacent to UBE3A and exhibits similar imprinted expression. *Am J Hum Genet.* 2001;68:1501–1505.
32. Weksberg R, Shuman C, Smith AC. Beckwith-Wiedemann syndrome. *Am J Med Genet C Semin Med Genet.* 2005;137C:12–23.
33. Scott RH, Stiller CA, Walker L, et al. Syndromes and constitutional chromosomal abnormalities associated with Wilms tumour. *J Med Genet.* 2006;43:705–715.
34. Cohen Jr MM. Beckwith-Wiedemann syndrome: Historical, clinicopathological, and etiopathogenetic perspectives. *Pediatr Dev Pathol.* 2005;8:287–304.
35. Cooper WN, Luharia A, Evans GA, et al. Molecular subtypes and phenotypic expression of Beckwith-Wiedemann syndrome. *Eur J Hum Genet.* 2005;13:1025–1032.
36. Joyce JA, Lam WK, Catchpoole DJ, et al. Imprinting of IGF2 and H19: Lack of reciprocity in sporadic Beckwith-Wiedemann syndrome. *Hum Mol Genet.* 1997;6:1543–1548.
37. Higashimoto K, Soejima H, Saito T, et al. Imprinting disruption of the CDKN1C/KCNQ1OT1 domain: The molecular mechanisms causing Beckwith-Wiedemann syndrome and cancer. *Cytogenet Genome Res.* 2006;113:306–312.
38. DeBaun MR, Niemitz EL, McNeil DE, et al. Epigenetic alterations of H19 and LIT1 distinguish patients with Beckwith-Wiedemann syndrome with cancer and birth defects. *Am J Hum Genet.* 2002;70:604–611.
39. Cerrato F, Sparago A, Verde G, et al. Different mechanisms cause imprinting defects at the IGF2/H19 locus in Beckwith-Wiedemann syndrome and Wilms' tumour. *Hum Mol Genet.* 2008;17:1427–1435.
40. Diaz-Meyer N, Yang Y, Sait SN, et al. Alternative mechanisms associated with silencing of CDKN1C in Beckwith-Wiedemann syndrome. *J Med Genet.* 2005;42:648–655.
41. Enklaar T, Zabel BU, Prawitt D. Beckwith-Wiedemann syndrome: Multiple molecular mechanisms. *Expert Rev Mol Med.* 2006;8:1–19.
42. Feinberg AP. Phenotypic plasticity and the epigenetics of human disease. *Nature.* 2007;447:433–440.
43. Baujat G, Rio M, Rossignol S, et al. Paradoxical NSD1 mutations in Beckwith-Wiedemann syndrome and 11p15 anomalies in Sotos syndrome. *Am J Hum Genet.* 2004;74:715–720.
44. Feinberg AP, Vogelstein B. Hypomethylation distinguishes genes of some human cancers from their normal counterparts. *Nature.* 1983;301:89–92.
45. Esteller M. Cancer epigenomics: DNA methylomes and histone-modification maps. *Nat Rev Genet.* 2007;8:286–298.
46. Fraga MF, Ballestar E, Villar-Garea A, et al. Loss of acetylation at Lys16 and trimethylation at Lys20 of histone H4 is a common hallmark of human cancer. *Nat Genet.* 2005;37:391–400.
47. Eden A, Gaudet F, Waghmare A, et al. Chromosomal instability and tumors promoted by DNA hypomethylation. *Science.* 2003; 300:455.
48. Karpf AR, Matsui S. Genetic disruption of cytosine DNA methyltransferase enzymes induces chromosomal instability in human cancer cells. *Cancer Res.* 2005;65:8635–8639.
49. Schulz WA. L1 retrotransposons in human cancers. *J Biomed Biotechnol.* 2006;2006:83672.
50. Bjornsson HT, Brown LJ, Fallin MD, et al. Epigenetic specificity of loss of imprinting of the IGF2 gene in Wilms tumors. *J Natl Cancer Inst.* 2007;99:1270–1273.
51. Li Y, Meng G, Guo QN. Changes in genomic imprinting and gene expression associated with transformation in a model of human osteosarcoma. *Exp Mol Pathol.* 2008;84:234–239.
52. Wu J, Qin Y, Li B, et al. Hypomethylated and hypermethylated profiles of H19DMR are associated with the aberrant imprinting of IGF2 and H19 in human hepatocellular carcinoma. *Genomics.* 2008; 91:443–450.
53. Byun HM, Wong HL, Birnstein EA, et al. Examination of IGF2 and H19 loss of imprinting in bladder cancer. *Cancer Res.* 2007;67:10753–10758.
54. Fan T, Schmidtmann A, Xi S, et al. DNA hypomethylation caused by Lsh deletion promotes erythroleukemia development. *Epigenetics.* 2008;3:134–142.
55. Wu H, Chen Y, Liang J, et al. Hypomethylation-linked activation of PAX2 mediates tamoxifen-stimulated endometrial carcinogenesis. *Nature.* 2005;438:981–987.
56. Brueckner B, Stresemann C, Kuner R, et al. The human let-7a-3 locus contains an epigenetically regulated microRNA gene with oncogenic function. *Cancer Res.* 2007;67:1419–1423.
57. Sakai T, Toguchida J, Ohtani N, et al. Allele-specific hypermethylation of the retinoblastoma tumor-suppressor gene. *Am J Hum Genet.* 1991;48:880–888.
58. Herman JG, Baylin SB. Gene silencing in cancer in association with promoter hypermethylation. *N Engl J Med.* 2003;349: 2042–2054.
59. Esteller M, Corn PG, Baylin SB, et al. A gene hypermethylation profile of human cancer. *Cancer Res.* 2001;61:3225–3229.
60. Paz MF, Fraga MF, Avila S, et al. A systematic profile of DNA methylation in human cancer cell lines. *Cancer Res.* 2003;63: 1114–1121.
61. Esteller M, Fraga MF, Guo M, et al. DNA methylation patterns in hereditary human cancers mimic sporadic tumorigenesis. *Hum Mol Genet.* 2001;10:3001–3007.
62. Agrelo R, Cheng WH, Setien F, et al. Epigenetic inactivation of the premature aging Werner syndrome gene in human cancer. *Proc Natl Acad Sci USA.* 2006;103:8822–8827.
63. Jones PA, Baylin SB. The epigenomics of cancer. *Cell.* 2007;128: 683–692.
64. Majid S, Kikuno N, Nelles J, et al. Genistein induces the p21WAF1/CIP1 and p16INK4a tumor suppressor genes in prostate cancer cells by epigenetic mechanisms involving active chromatin modification. *Cancer Res.* 2008;68:2736–2744.

65. Fraga MF, Esteller M. Towards the human cancer epigenome: A first draft of histone modifications. *Cell Cycle.* 2005;4:1377–1381.
66. Seligson DB, Horvath S, Shi T, et al. Global histone modification patterns predict risk of prostate cancer recurrence. *Nature.* 2005;435:1262–1266.
67. He L, Hannon GJ. MicroRNAs: small RNAs with a big role in gene regulation. *Nat Rev Genet.* 2004;5:522–531.
68. Calin GA, Croce CM. MicroRNA signatures in human cancers. *Nat Rev Cancer.* 2006;6:857–866.
69. Lujambio A, Esteller M. CpG island hypermethylation of tumor suppressor microRNAs in human cancer. *Cell Cycle.* 2007;6: 1455–1459.
70. Saito Y, Liang G, Egger G, et al. Specific activation of microRNA-127 with downregulation of the proto-oncogene BCL6 by chromatin-modifying drugs in human cancer cells. *Cancer Cell.* 2006;9:435–443.
71. Lujambio A, Ropero S, Ballestar E, et al. Genetic unmasking of an epigenetically silenced microRNA in human cancer cells. *Cancer Res.* 2007;67:1424–1429.
72. Feinberg AP, Tycko B. The history of cancer epigenetics. *Nat Rev Cancer.* 2004;4:143–153.
73. Lopez-Serra L, Ballestar E, Fraga MF, et al. A profile of methyl-CpG binding domain protein occupancy of hypermethylated promoter CpG islands of tumor suppressor genes in human cancer. *Cancer Res.* 2006;66:8342–8346.
74. Yang XJ. The diverse superfamily of lysine acetyltransferases and their roles in leukemia and other diseases. *Nucleic Acids Res.* 2004;32:959–976.
75. Intini D, Fabris S, Storlazzi T, et al. Identification of a novel IGH-MMSET fusion transcript in a human myeloma cell line with the t(4;14)(p16.3;q32) chromosomal translocation. *Br J Haematol.* 2004;126:437–439.
76. Jaju RJ, Fidler C, Haas OA, et al. A novel gene, NSD1, is fused to NUP98 in the t(5;11)(q35;p15.5) in de novo childhood acute myeloid leukemia. *Blood.* 2001;98:1264–1267.
77. Rosati R, La Starza R, Veronese A, et al. NUP98 is fused to the NSD3 gene in acute myeloid leukemia associated with t(8;11)(p11.2;p15). *Blood.* 2002;99:3857–3860.
78. Bracken AP, Pasini D, Capra M, et al. EZH2 is downstream of the pRB-E2F pathway, essential for proliferation and amplified in cancer. *Embo J.* 2003;22:5323–5335.
79. Huntsman DG, Chin SF, Muleris M, et al. MLL2, the second human homolog of the Drosophila trithorax gene, maps to 19q13.1 and is amplified in solid tumor cell lines. *Oncogene.* 1999;18:7975–7984.
80. Angrand PO, Apiou F, Stewart AF, et al. NSD3, a new SET domain-containing gene, maps to 8p12 and is amplified in human breast cancer cell lines. *Genomics.* 2001;74:79–88.
81. Cloos PA, Christensen J, Agger K, et al. The putative oncogene GASC1 demethylates tri- and dimethylated lysine 9 on histone H3. *Nature.* 2006;442:307–311.
82. Ropero S, Fraga MF, Ballestar E, et al. A truncating mutation of HDAC2 in human cancers confers resistance to histone deacetylase inhibition. *Nat Genet.* 2006;38:566–569.
83. Moinova HR, Chen WD, Shen L, et al. HLTF gene silencing in human colon cancer. *Proc Natl Acad Sci USA.* 2002;99: 4562–4567.
84. Gunduz E, Gunduz M, Ouchida M, et al. Genetic and epigenetic alterations of BRG1 promote oral cancer development. *Int J Oncol.* 2005;26:201–210.
85. Ballestar E, Esteller M, Richardson BC. The epigenetic face of systemic lupus erythematosus. *J Immunol.* 2006;176:7143–7147.
86. Corvetta A, Della Bitta R, Luchetti MM, et al. 5-Methylcytosine content of DNA in blood, synovial mononuclear cells and synovial tissue from patients affected by autoimmune rheumatic diseases. *J Chromatogr.* 1991;566:481–491.
87. Kominato Y, Hata Y, Takizawa H, et al. Expression of human histo-blood group ABO genes is dependent upon DNA methylation of the promoter region. *J Biol Chem.* 1999;274:37240–37250.
88. Menendez L, Walker LD, Matyunina LV, et al. Epigenetic changes within the promoter region of the HLA-G gene in ovarian tumors. *Mol Cancer.* 2008;7:43.
89. Honda T, Tamura G, Waki T, et al. Demethylation of MAGE promoters during gastric cancer progression. *Br J Cancer.* 2004;90:838–843.
90. Nagarajan RP, Hogart AR, Gwye Y, et al. Reduced MeCP2 expression is frequent in autism frontal cortex and correlates with aberrant MECP2 promoter methylation. *Epigenetics.* 2006; 1:e1–e11.
91. Lund G, Andersson L, Lauria M, et al. DNA methylation polymorphisms precede any histological sign of atherosclerosis in mice lacking apolipoprotein E. *J Biol Chem.* 2004;279: 29147–29154.
92. Abel T, Zukin RS. Epigenetic targets of HDAC inhibition in neurodegenerative and psychiatric disorders. *Curr Opin Pharmacol.* 2008;8:57–64.
93. Hu N, Qiu X, Luo Y, et al. Abnormal histone modification patterns in lupus CD4+ T cells. *J Rheumatol.* 2008;35:804–810.
94. Gibbons R. Alpha thalassaemia-mental retardation, X linked. *Orphanet J Rare Dis.* 2006;1:15.
95. De La Fuente R, Viveiros MM, Wigglesworth K, et al. ATRX, a member of the SNF2 family of helicase/ATPases, is required for chromosome alignment and meiotic spindle organization in metaphase II stage mouse oocytes. *Dev Biol.* 2004;272:1–14.
96. Sanlaville D, Verloes A. CHARGE syndrome: An update. *Eur J Hum Genet.* 2007;15:389–399.
97. Jongmans MC, Hoefsloot LH, van der Donk KP, et al. Familial CHARGE syndrome and the CHD7 gene: A recurrent missense mutation, intrafamilial recurrence and variability. *Am J Med Genet A.* 2008;146A:43–50.
98. Spivak G. The many faces of Cockayne syndrome. *Proc Natl Acad Sci USA.* 2004;101:15273–15274.
99. Fousteri M, Vermeulen W, van Zeeland AA, et al. Cockayne syndrome A and B proteins differentially regulate recruitment of chromatin remodeling and repair factors to stalled RNA polymerase II in vivo. *Mol Cell.* 2006;23:471–482.
100. Boerkoel CF, O'Neill S, Andre JL, et al. Manifestations and treatment of Schimke immuno-osseous dysplasia: 14 new cases and a review of the literature. *Eur J Pediatr.* 2000;159:1–7.
101. Boerkoel CF, Takashima H, John J, et al. Mutant chromatin remodeling protein SMARCAL1 causes Schimke immuno-osseous dysplasia. *Nat Genet.* 2002;30:215–220.
102. Brown DC, Grace E, Sumner AT, et al. ICF syndrome (immunodeficiency, centromeric instability and facial anomalies): Investigation of heterochromatin abnormalities and review of clinical outcome. *Hum Genet.* 1995;96: 411–416.
103. Ehrlich M. The ICF syndrome, a DNA methyltransferase 3B deficiency and immunodeficiency disease. *Clin Immunol.* 2003;109:17–28.
104. Xu GL, Bestor TH, Bourc'his D, et al. Chromosome instability and immunodeficiency syndrome caused by mutations in a DNA methyltransferase gene. *Nature.* 1999;402:187–191.
105. Moog U, Smeets EE, van Roozendaal KE, et al. Neurodevelopmental disorders in males related to the gene causing Rett syndrome in females (MECP2). *Eur J Paediatr Neurol.* 2003;7:5–12.
106. Van den Veyver IB, Zoghbi HY. Methyl-CpG-binding protein 2 mutations in Rett syndrome. *Curr Opin Genet Dev.* 2000; 10:275–279.
107. Laccone F, Junemann I, Whatley S, et al. Large deletions of the MECP2 gene detected by gene dosage analysis in patients with Rett syndrome. *Hum Mutat.* 2004;23:234–244.
108. Ballestar E, Ropero S, Alaminos M, et al. The impact of MECP2 mutations in the expression patterns of Rett syndrome patients. *Hum Genet.* 2005;116:91–104.
109. Li MR, Pan H, Bao XH, et al. MECP2 and CDKL5 gene mutation analysis in Chinese patients with Rett syndrome. *J Hum Genet.* 2007;52:38–47.
110. Jones PL, Veenstra GJ, Wade PA, et al. Methylated DNA and MeCP2 recruit histone deacetylase to repress transcription. *Nat Genet.* 1998;19:187–191.
111. Esteller M. Rett syndrome: The first forty years: 1966–2006. *Epigenetics.* 2007;2:1.
112. Abuhatzira L, Makedonski K, Kaufman Y, et al. MeCP2 deficiency in the brain decreases BDNF levels by REST/CoREST-mediated repression and increases TRKB production. *Epigenetics.* 2007;2:214–222.
113. Kesler SR, Simensen RJ, Voeller K, et al. Altered neurodevelopment associated with mutations of RSK2: A morphometric MRI study of Coffin-Lowry syndrome. *Neurogenetics.* 2007; 8:143–147.

114. Field M, Tarpey P, Boyle J, et al. Mutations in the RSK2 (RPS6KA3) gene cause Coffin-Lowry syndrome and nonsyndromic X-linked mental retardation. *Clin Genet.* 2006;70: 509–515.
115. Kang S, Dong S, Guo A, et al. Epidermal growth factor stimulates RSK2 activation through activation of the MEK/ERK pathway and src-dependent tyrosine phosphorylation of RSK2 at Tyr-529. *J Biol Chem.* 2008;283:4652–4657.
116. Merienne K, Jacquot S, Trivier E, et al. Rapid immunoblot and kinase assay tests for a syndromal form of X linked mental retardation: Coffin-Lowry syndrome. *J Med Genet.* 1998;35: 890–894.
117. Roelfsema JH, Peters DJ. Rubinstein-Taybi syndrome: Clinical and molecular overview. *Expert Rev Mol Med.* 2007;9:1–16.
118. Bartsch O, Schmidt S, Richter M, et al. DNA sequencing of CREBBP demonstrates mutations in 56% of patients with Rubinstein-Taybi syndrome (RSTS) and in another patient with incomplete RSTS. *Hum Genet.* 2005;117:485–493.
119. Zimmermann N, Acosta AM, Kohlhase J, et al. Confirmation of EP300 gene mutations as a rare cause of Rubinstein-Taybi syndrome. *Eur J Hum Genet.* 2007;15:837–842.
120. Roelfsema JH, White SJ, Ariyurek Y, et al. Genetic heterogeneity in Rubinstein-Taybi syndrome: Mutations in both the CBP and EP300 genes cause disease. *Am J Hum Genet.* 2005;76:572–580.
121. Tatton-Brown K, Rahman N. Sotos syndrome. *Eur J Hum Genet.* 2007;15:264–271.
122. Rahman N. Mechanisms predisposing to childhood overgrowth and cancer. *Curr Opin Genet Dev.* 2005;15:227–233.
123. Martinez-Glez V, Lapunzina P. Sotos syndrome is associated with leukemia/lymphoma. *Am J Med Genet A.* 2007;143A:1244–1245.
124. Jaenisch R, Bird A. Epigenetic regulation of gene expression: How the genome integrates intrinsic and environmental signals. *Nat Genet.* 2003;33(Suppl):245–254.
125. Waterland RA, Lin JR, Smith CA, et al. Post-weaning diet affects genomic imprinting at the insulin-like growth factor 2 (IGF2) locus. *Hum Mol Genet.* 2006;15:705–716.
126. Kirke PN, Mills JL, Molloy AM, et al. Impact of the MTHFR C677T polymorphism on risk of neural tube defects: Case-control study. *BMJ.* 2004;328:1535–1536.
127. Perez AB, D'Almeida V, Vergani N, et al. Methylenetetrahydrofolate reductase (MTHFR): Incidence of mutations C677T and A1298C in Brazilian population and its correlation with plasma homocysteine levels in spina bifida. *Am J Med Genet A.* 2003;119A:20–25.
128. Giovannucci E. Alcohol, one-carbon metabolism, and colorectal cancer: Recent insights from molecular studies. *J Nutr.* 2004;134:2475S–2481S.
129. Labrune P, Zittoun J, Duvaltier I, et al. Haemolytic uraemic syndrome and pulmonary hypertension in a patient with methionine synthase deficiency. *Eur J Pediatr.* 1999;158:734–739.
130. Benbrahim-Tallaa L, Waterland RA, Styblo M, et al. Molecular events associated with arsenic-induced malignant transformation of human prostatic epithelial cells: Aberrant genomic DNA methylation and K-ras oncogene activation. *Toxicol Appl Pharmacol.* 2005;206: 288–298.
131. Poirier LA, Vlasova TI. The prospective role of abnormal methyl metabolism in cadmium toxicity. *Environ Health Perspect.* 2002;110(Suppl 5):793–795.
132. Silbergeld EK, Waalkes M, Rice JM. Lead as a carcinogen: Experimental evidence and mechanisms of action. *Am J Ind Med.* 2000;38:316–323.
133. Salnikow K, Costa M. Epigenetic mechanisms of nickel carcinogenesis. *J Environ Pathol Toxicol Oncol.* 2000;19:307–318.
134. Wei YD, Tepperman K, Huang MY, et al. Chromium inhibits transcription from polycyclic aromatic hydrocarbon-inducible promoters by blocking the release of histone deacetylase and preventing the binding of p300 to chromatin. *J Biol Chem.* 2004;279:4110–4119.
135. Anway MD, Cupp AS, Uzumcu M, et al. Epigenetic transgenerational actions of endocrine disruptors and male fertility. *Science.* 2005;308:1466–1469.
136. Weaver IC, Cervoni N, Champagne FA, et al. Epigenetic programming by maternal behavior. *Nat Neurosci.* 2004;7:847–854.
137. Niemitz EL, Feinberg AP. Epigenetics and assisted reproductive technology: A call for investigation. *Am J Hum Genet.* 2004;74: 599–609.
138. Rideout WM, 3rd, Eggan K, Jaenisch R. Nuclear cloning and epigenetic reprogramming of the genome. *Science.* 2001;293: 1093–1098.
139. Fraga MF, Ballestar E, Paz MF, et al. Epigenetic differences arise during the lifetime of monozygotic twins. *Proc Natl Acad Sci USA.* 2005;102:10604–10609.

Chapter

9

Clinical Proteomics and Molecular Pathology

Lance A. Liotta . Virginia Espina . Claudia Fredolini . Weidong Zhou . Emanuel Petricoin

UNDERSTANDING CANCER AT THE MOLECULAR LEVEL: AN EVOLVING FRONTIER

Genomic and proteomic research is launching the next era of cancer molecular medicine. Molecular expression profiles can uncover clues to functionally important molecules in the development of human disease and generate information to subclassify human tumors and tailor a treatment to the individual patient. The next revolution is the synthesis of proteomic information into functional pathways and circuits in cells and tissues. Such synthesis must take into account the dynamic state of protein post-translational modifications and protein–protein or protein–DNA/RNA interactions, cross-talk between signal pathways [1], feedback regulation within cells, between cells, and between tissues. This full set of information may be required before we can fully dissect the specific dysregulated pathways driving tumorigenesis. This higher level of functional understanding will be the basis for true rational therapeutic design that specifically targets the molecular lesions underlying human disease.

The rapid progress in molecular medicine is largely due to new insights emanating from data generated by emerging technologies. This chapter summarizes new technologies in the exploding field of proteomics. These technologies hold promise for the early cancer diagnostics from the drop of a patient's blood, to the molecular dissection of a patient's individual tumor cells, to the development of individualized molecularly targeted therapies (Table 9.1).

MICRODISSECTION TECHNOLOGY BRINGS MOLECULAR ANALYSIS TO THE TISSUE LEVEL

Molecular analysis of pure cell populations in their native tissue environment is necessary to understand the microecology of the disease process. Accomplishing this goal is much more difficult than just grinding up a piece of tissue and applying the extracted molecules to a panel of assays. The reason is that tissues are complicated three-dimensional structures composed of large numbers of different types of interacting cell populations. The cell subpopulation of interest may constitute a tiny fraction of the total tissue volume. For example, a biopsy of breast tissue harboring a malignant tumor usually contains the following types of cell populations: (i) fat cells in the abundant adipose tissue surrounding the ducts, (ii) normal epithelium and myoepithelium in the branching ducts, (iii) fibroblasts and endothelial cells in the stroma and blood vessels, (iv) premalignant carcinoma cells in the *in situ* lesions, and (v) clusters of invasive carcinoma. If the goal is to analyze the genetic changes in the premalignant cells or the malignant cells, these subpopulations are frequently located in microscopic regions occupying less than 5% of the tissue volume. After the computer adage "garbage in, garbage out," if the extract of a complex tissue is analyzed using a sophisticated technology, the output will be severely compromised if the input material is contaminated by the wrong cells. Culturing cell populations from fresh tissue is one approach to reducing contamination. However, cultured cells may not accurately represent the molecular events taking place in the actual

Molecular Pathology © 2009, Elsevier, Inc. All Rights Reserved.

Table 9.1 Opportunities, Challenges, and Potential Solutions for Use of Proteomics for Routine Clinical Practice and Patient Care

Opportunities	Challenges	Solutions
Proteomic multiplex Cellular circuit analysis of clinical biopsy specimens	Platform sensitivity precision and accuracy Heterogeneity of tissue populations Perishability: requirement for immediate freezing or preservation of protein analytes Formalin fixation unsuitable for protein or (rna)extraction	Sensitive protein microarrays Microdissection New protocols Surrogate markers for tissue and blood molecular preservation Extraction from formalin fixed state New room temperature tissue and blood preservation technology
Patient stratification individualized therapy based on molecular profiling	Complex trial design and data analysis Patient consent for serial biopsies	New classes of adaptive trial design Dialogues with patient advocates and IRB
Tailored combination therapy	Low number of approved candidate targeted agents Lack of preclinical data Unknown toxicity of combinations Therapies from competing pharmas	Accelerate discovery of novel agents New indications for existing drugs New classes of *ex vivo* treatment models
Rational redesign of therapy following recurrence or outgrowth of metastasis	Safety and justification for rebiopsy Molecular profile of metastasis is different from the primary	Restrict rebiopsy to accessible sites Metastasis-specific tailored therapy

tissue from which they were derived. Assuming methods are successful to isolate and grow the tissue cells of interest, the gene expression pattern of the cultured cells is influenced by the culture environment and can be quite different from the genes expressed in the native tissue state. The reason is that the cultured cells are separated from the tissue elements that regulate gene expression, such as soluble factors, extracellular matrix molecules, and cell-cell communication. Thus, the problem of cellular heterogeneity has been a significant barrier to the molecular analysis of normal and diseased tissue. This problem can now be overcome by new developments in the field of laser tissue microdissection [1–5] (Figure 9.1).

Analysis of critical gene expression and protein patterns in normal developing and diseased tissue progression requires the microdissection and extraction of a microscopic homogeneous cellular subpopulation from its complex tissue milieu [2,3,5]. This subpopulation can then be compared to adjacent, interacting, but distinct, subpopulations of cells in the same tissue. The method of procurement of pure cell populations from heterogeneous tissue should fully preserve the state of the cell molecules if it is to allow quantitative analysis, particularly in sensitive amplification methods based on PCR, reverse transcriptase-PCR, or enzymatic function. Laser capture microdissection (LCM) has been developed to provide scientists with a fast and dependable method of capturing and preserving specific cells from tissue, under direct microscopic visualization. With the ease of procuring a homogeneous population of cells from a complex tissue using the LCM, the approaches to molecular analysis of pathologic processes are significantly enhanced [2,6,7]. The mRNA from microdissected cancer lesions has been used as the starting material to produce cDNA libraries, microchip microarrays, differential display, and other techniques to find new genes or mutations.

The development of LCM allows investigators to determine specific protein expression patterns from tissues of individual patients. Using multiplex analysis, investigators can correlate the pattern of expressed genes and post-translationally modified proteins with the histopathology and response to treatment. Microdissection can be used to study the interactions between cellular subtypes in the organ or tissue microenvironment. Efficient coupling of LCM of serial tissue sections with multiplex molecular analysis techniques is leading to sensitive and quantitative methods to visualize three-dimensional interactions between morphologic elements of the tissue. For example, it will be possible to trace the gene expression pattern and quantitate protein signaling activation state along the length of a prostate gland or breast duct to examine the progression of neoplastic development. The end goal is the integration of molecular biology with tissue morphogenesis and pathology [8].

Beyond Functional Genomics to Cancer Proteomics

Whereas DNA is an information archive, proteins do all the work of the cell. The existence of a given DNA sequence does not guarantee the synthesis of a corresponding protein. The DNA sequence is also not sufficient to describe protein structure, protein-protein, protein-DNA interaction, and cellular location [8,11]. Protein complexity and versatility stem from context-dependent post-translational processes, such as phosphorylation, sulfation, or glycosylation. Nucleic acid profiling (including microRNA) does not provide information about how proteins link together into networks and functional machines in the cell. In fact, the activation of a protein signal pathway, causing a cell to migrate, die, or initiate division,

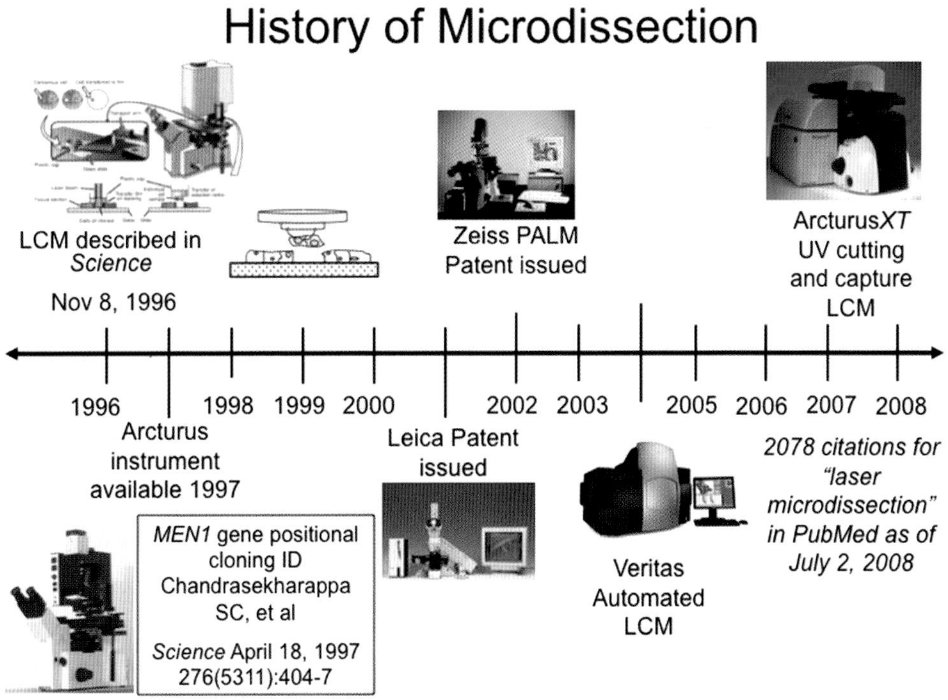

Figure 9.1 **Laser capture microdissection (LCM).** LCM is a technology for procuring pure cell populations from a stained tissue section under direct microscopic visualization. Tissues contain heterogeneous cellular populations (e.g., epithelium, cancer cells, fibroblasts, endothelium, and immune cells). The diseased cellular population of interest usually comprises only a small percentage of the tissue volume. LCM directly procures the subpopulation of cells selected for study, while leaving behind all of the contaminating cells. A stained section of the heterogeneous tissue is mounted on a glass microscope slide and viewed under high magnification. The experimenter selects the individual cell(s) to be studied using a joystick or via a computer screen. The chosen cells are lifted out of the tissue by the action of a laser pulse. The infrared laser, mounted in the optical axis of the microscope, locally expands a thermoplastic polymer to reach down and capture the cell beneath the laser pulse (insert). When the film is lifted from the tissue section, only the pure cells for study are excised from the heterogeneous cellular population. The DNA, RNA, and proteins of the captured cells remain intact and unperturbed. Using LCM, one to several thousand tissue cells can be captured in less than 5 minutes. Using appropriate buffers, the cellular constituents are solubilized and subjected to microanalysis methods. Proteins from all compartments of the cell can be readily procured. Protein conformation and enzymatic activity are retained if the tissues are frozen or fixed in ethanol before sectioning. The extracted proteins can be analyzed by any method that has sufficient sensitivity.

can immediately take place before any changes occur in DNA/RNA gene expression. Consequently, the technology to drive the molecular medicine revolution from the correlation to the causality phase is emerging from protein analytic methods [12].

The term *proteome*, which denotes all the proteins expressed by a genome, was first coined in late 1994 at the Sienna two-dimensional (2D) gel electrophoresis meeting. Proteomics is the next step after genomics. A major goal of investigators in this exciting field is to assemble a complete library of all proteins. To date, only a small percentage of the proteome has been cataloged. Because PCR for proteins does not exist, sequencing the order of 20 possible amino acids in a given protein remains relatively slow and labor intensive compared to nucleotide sequencing. Although a number of new technologies are being introduced for high-throughput protein characterization and discovery [4,7,10,12–18], the traditional method of protein identification continues to be 2D gel electrophoresis [13,17]. Two-dimensional electrophoresis can separate proteins by molecular weight in one dimension and charge in a second dimension. When a mixture of proteins is applied to the 2D gel, individual proteins in the mixture are separated out into signature locations on the display, depending on their individual size and charge. The protein spot can be procured from the gel, and a partial amino acid sequence can be read using mass spectrometry. An experimental 2D gel image can be captured and overlaid digitally with known archived 2D gels. In this way it is possible to immediately highlight proteins that are differentially abundant in one state versus another (for instance, tumor versus normal or before and after hormone treatment).

Two-dimensional gels have traditionally required large amounts of protein-starting material equivalent to millions of cells. Thus, their application has been limited to cultured cells or ground-up heterogeneous tissue. Not unexpectedly, this approach does not provide an accurate picture of the proteins that are in use by cells in real tissue. Tissues are complicated structures composed of hundreds of interacting cell populations in specialized spatial configurations. The fluctuating proteins expressed by cells in tissues may bear little resemblance to the proteins made by cultured cells that are torn from their tissue context and reacting to a new culture environment.

Two classes of technologies were developed with the express purpose to address the tissue-context problem. The first technology is LCM, used to procure specific tissue cell subpopulations under direct microscopic visualization of a standard stained frozen or fixed tissue section on a glass microscope slide. The second class of technology is protein microarrays [4,18–36]. This multiplex analysis tool is sensitive enough to accurately measure the small concentration of molecules in microdissected tissue samples.

While individualized treatments have been used in medicine for years, advances in cancer treatment have now generated a need to more precisely define and identify patients that will derive the most benefit from new-targeted agents. Molecular profiling using gene expression arrays has shown considerable potential for the classification of patient populations in all of these respects [38–41]. Nevertheless, transcript profiling, by itself, provides an incomplete picture of the ongoing molecular network for a number of clinically important reasons. First, gene transcript levels have not been found to correlate significantly with protein expression or the functional (often phosphorylated) forms of the encoded proteins. RNA transcripts also provide little information about protein–protein interactions and the state of the cellular signaling pathways. Finally, most current therapeutics are directed at protein targets, and these targets are often protein kinases and/or their substrates. The human kinome, or full complement of kinases encoded by the human genome, comprises the molecular networks and signaling pathways of the cell. The activation state of these proteins and these networks fluctuate constantly depending on the cellular microenvironment. Consequently, the source material for molecular profiling studies needs to shift from *in vitro* models to the use of actual diseased human tissue. Technologies which can broadly profile and assess the activity of the human kinome in a real biological context will provide a rich source of new molecular informations critical for the realization of patient-tailored therapy (Figure 9.2).

Protein Microarray Tools to Guide Patient-Tailored Therapy

Theoretically, the most efficient way to identify patients who will respond to a given therapy is to determine, prior to treatment initiation, which potential signaling pathways are truly activated in each patient. Ideally, this would come from analysis of tissue material taken from the patient through biopsy procurement. In general, previous traditional proteomic technologies such as 2D gel electrophoresis have significant limitations when they are applied to very small tissue samples, such as biopsy specimens where only a few thousand cells may be procured. Protein microarrays represent an emerging technology that can address the limitations of previous measurement platforms and are quickly becoming powerful tools for drug discovery, biomarker identification, and signal transduction profiling of cellular material. The advantage of protein microarrays lies in their ability to provide a map of known cellular signaling proteins that can reflect, in general, the state of information flow through protein networks in individual specimens. Identification of critical nodes, or interactions, within the network is a potential starting point for drug development and/or the design of individual therapy regimens. Protein microarrays that examine protein-protein recognition events (phosphorylation) in a global, high-throughput manner can be used to profile the working state of cellular signal pathways in a manner not possible with gene arrays. Protein microarrays may be used to monitor changes in protein phosphorylation over time, before and after treatment, between disease and nondisease states and responders versus nonresponders, allowing one to infer the activity levels of the proteins in a particular pathway in real-time to tailor treatment to each patient's cellular circuitry [20–36].

The application of this technology to clinical molecular diagnostics will be greatly enhanced by increasing numbers of high-quality antibodies that are specific for the modification or activation state of target proteins within key pathways. Antibody specificity is particularly critical, given the complex array of biologic proteins at vastly different concentrations contained in cell lysates. A cubic centimeter of biopsy tissue may contain approximately 10^9 cells, whereas a needle biopsy or cell aspirate may contain fewer than 100,000 cells. If the cell population of the specimen is heterogeneous, the final number of actual tumor cells microdissected or procured for analysis may be as low as a few thousand. Assuming that the proteins of interest, and their phosphorylated counterparts, exist in low abundance, the total concentration of analyte proteins in the sample will be very low. Newer generations of protein microarrays with highly sensitive and specific antibodies are now able to achieve adequate levels of sensitivity for analysis of clinical specimens containing fewer than a few thousand cells.

At a basic level, protein microarrays are composed of a series of immobilized spots. Each spot contains a homogeneous or heterogeneous bait molecule. A spot on the array may display an antibody, a cell or phage lysate, a recombinant protein or peptide, or a nucleic acid [20–37]. The array is queried with (i) a probe (labeled antibody or ligand) or (ii) an unknown biologic sample (for instance, cell lysate or serum sample) containing analytes of interest. When the query molecules are tagged directly or indirectly with a signal-generating moiety, a pattern of positive and negative spots is generated. For each spot, the

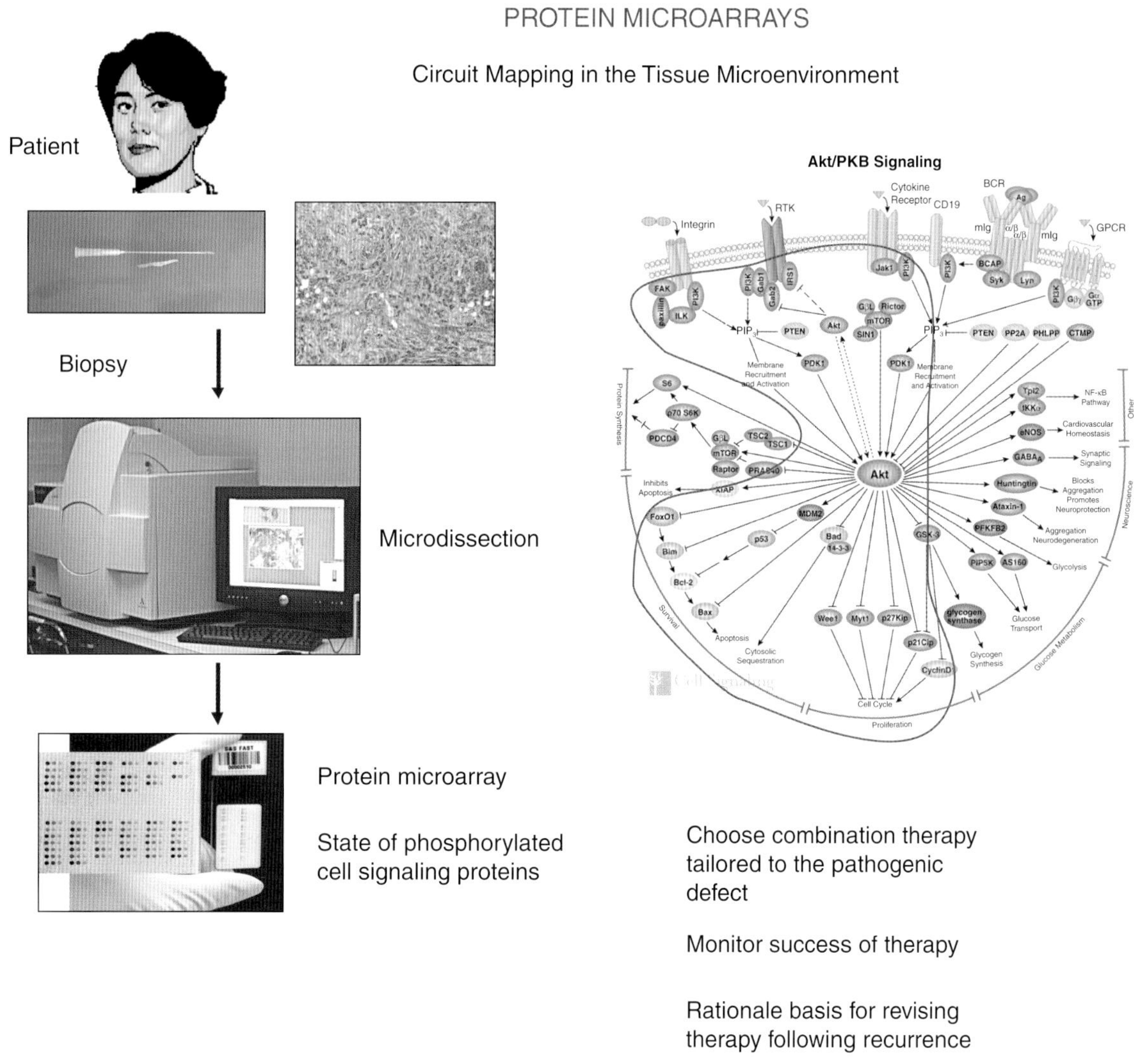

Figure 9.2 **A roadmap for individualized cancer therapy.** Following biopsy or needle aspiration, and laser microdissection, signal pathway analysis is performed using protein microarrays for phosphoproteomic analysis, and RNA transcript arrays.The specific signaling portrait becomes the basis of a patient-tailored therapeutic regime. Therapeutic assesment is obtained by follow-up biopsy, and the molecular portrait of signaling events is reassessed to determine if therapeutic selection should be modified futher.

intensity of the signal is proportional to the quantity of applied query molecules bound to the bait molecules. An image of the spot pattern is captured, analyzed, and interpreted.

Protein microarray formats fall into two major classes, forward phase arrays (FPAs) and reverse phase arrays (RPAs), depending on whether the analyte(s) of interest is captured from solution phase or bound to the solid phase [18]. In FPAs, capture molecules are immobilized onto the substratum and act as the bait molecule. Each spot contains one type of known immobilized protein, fractionated lysate, or other type of bait molecule. In the FPA format, each array is incubated with one test sample (for instance, a cellular lysate from one treatment condition or serum sample from disease/control patients), and multiple analytes are measured at once. A number of excellent reviews summarize recent applications, obstacles, and new advances in FPA technology [19–31]. For example, arrays of human, microbial, or viral recombinant proteins can be used to screen individual serum samples from afflicted and control patients to characterize the immune response and identify potential diagnostic markers, therapeutic, or vaccine targets [42]. Antibody arrays represent another branch of FPAs that have broad applications in both commercial and research settings. Examples of their use in cancer research include the identification of changes in protein levels following treatment of colon cancer cells with ionizing radiation [43], identification of serum protein biomarkers for bladder cancer diagnosis and outcome stratification [44], and prostate cancer diagnosis [31]. Despite their great potential, antibody array use is limited currently by the availability of well-characterized antibodies. A second obstacle to routine use of antibody arrays surrounds detection methods for bound analyte on the array. Current options include the use of specific antibodies recognizing distinct analyte epitopes from the capture antibodies (similar to a traditional sandwich-type ELISA), or the direct labeling of the analytes used for probing the array, both of which present distinct technical challenges.

In contrast to the FPA format, the RPA format immobilizes an individual test sample in each array spot, such that an array is composed of hundreds of different patient samples or cellular lysates. Though not limited to clinical applications, the RPA format provides the opportunity to screen clinical samples that are available in very limited quantities, such as biopsy specimens. Because human tissues are composed of hundreds of interacting cell populations, RPAs coupled with LCM provide a unique opportunity for discovering changes in the cellular proteome that reflect the cellular microenvironment. The RPA format is capable of extremely sensitive analyte detection with detection levels approaching attogram amounts of a given protein and variances of less than 10% [34–36]. The sensitivity of detection for the RPAs is such that low abundance phosphorylated protein isoforms can be measured from a spotted lysate representing fewer than 10 cell equivalents [34–36]. This level of sensitivity combined with analytical robustness is critical if the starting input material is only a few hundred cells from a biopsy specimen. Since the reverse phase array technology requires only one antibody for each analyte, it provides a facile way for broad profiling of pathways where hundreds of phospho-specific analytes can be measured concomitantly. Most importantly, the reverse phase array has significantly higher sensitivity than bead arrays or ELISA [36] such that broad screening of molecular networks can be achieved from tissue specimens routinely procured in the physician's private office or hospital radiology center (such as a needle biopsy specimen).

Molecular Network Analysis of Human Cancer Tissues

A number of studies illustrate the utility of reverse-phase protein microarrays for the analysis of human tissues and demonstrate the potential for the technology to contribute valuable information that can be used in therapeutic decision making [4,18–21,41]. Reverse-phase array technology was first described when it was utilized to demonstrate that prosurvival proteins and pathways are activated during prostate cancer progression [18]. Zha et al. examined the differences in prosurvival signaling between Bcl-2 positive and Bcl-2 negative follicular lymphomas [45]. Comparison of various prosurvival proteins by reverse-phase protein microarrays in Bcl-2 positive and negative samples suggested that there are prosurvival signals independent of Bcl-2 [45]. Pathway mapping of a clinical study set of childhood rhabdomyosarcoma tumors using reverse phase protein arrays revealed that mTOR pathway activation correlated with response to therapy. Moreover the functional significance of suppressing this pathway was tested in xenograft models and shown to profoundly suppress tumor growth [46].

Reverse-phase protein microarrays are also well suited to the analysis of clinical trial material in that they can provide signaling network information that complements standard histological analysis of patient specimens collected before, during, and after treatment. This technology is being applied to several ongoing clinical trials in a variety of cancers [35,39,47–49].

Combination Therapies

There is an increasing evidence demonstrating the promise and potential of combination therapies combining conventional treatments such as chemotherapy or radiotherapy and molecular-targeted therapeutics such as erlotinib (Tarceva®, Roche) and trastuzumab (Herceptin®, Roche) that interfere with kinase activity and protein-protein interactions in specific deregulated pathways [50–53]. However, strategies that target multiple interconnected proteins within a signaling pathway have not been explored to the same extent [50,51]. The view of individual therapeutic targets can be expanded to that of rational targeting of the entire deregulated molecular network, extending both inside and outside the cancer cell. Mathematical modeling of network-targeted therapeutic strategies has revealed that attenuation of downstream signals can be enhanced significantly when multiple upstream nodes or processes are inhibited with small molecule inhibitors compared with inhibition of a single upstream node. Also, inhibition of multiple nodes within a signaling cascade allows reduction of downstream signaling to desired levels with smaller doses of the necessary targeted drugs. While therapeutic strategies incorporating these lower dosages could lead to reduced toxicities and a broadened spectrum of available drugs, it must be recognized that testing these interacting drug modalities will necessitate clinical trials of complex design [52].

Ultimately, targeting, response assessment, and therapeutic monitoring will be individualized, and will reflect the subtle pre-therapy and post-therapy changes at the proteomic level, as well as the protein signaling cascade systems between individuals. The ability to visualize these interconnections both inside and outside a cell could have a profound effect on how we view biology, and can enable the realization of the recent emphasis on personalized combinatorial molecular medicine.

For lung cancer, targeting of the EGFR tyrosine kinase with small molecule inhibitors has received significant attention. Gefitinib (Iressa®, AstraZeneca) and erlotinib have shown significant clinical benefit in specific subsets of patients [54–56]. Promising phase I trial results with gefitinib demonstrated that the drug decreased levels of activated EGFR and mitogen-activated protein kinase in post-treatment skin biopsies indicating that the intended target was being inhibited [57,58]. Subsequent international phase II trials in patients with progressive non-small-cell lung cancer (IDEAL I and II) showed 18.4% and 10% response rates, respectively, based on radiographic assessment [57,58]. These observations led to exploration of clinical factors that distinguish the small minority of patients that respond durably to these EGFR tyrosine kinase inhibitors. Further analysis of this small subset led to the sequencing and identification of *EGFR* mutations that are associated with sensitivity to these drugs [55,56]. The presence of *EGFR* mutations correlates remarkably well with the identified correlative clinical parameters.

To date, the differences in signaling between wild-type and mutant EGFRs are poorly understood. It is not clear if the various *EGFR* mutations possess distinct signaling and/or altered sensitivities to EGFR inhibitors, and there is also a subset of patients who do not have *EGFR* mutations (mutations mainly in exons 19 and 21) that respond to EGFR inhibitors [56]. The EGFR signaling network can be activated in a number of ways: mutation of the receptor, overexpression of the receptor, and mutation of downstream kinases (such as phosphatidylinositol 3-kinase) are just a few examples. This suggests that mutation analysis alone should not serve as the sole criteria for identification and treatment selection, and that further studies incorporating proteomic profiling of tissue may be beneficial in identifying additional patients who will benefit from EGFR tyrosine kinase inhibitor treatment. Proteomic profiling of EGFR-related signaling activity in preclinical and *in vitro* experiments as well as in clinical specimens could provide useful information for characterizing drug responses [35]. Reverse-phase array technology is well suited to assess the signaling differences between mutant and wild-type cells, and these studies are currently underway. Instead of single measurements of selected analytes such as EGFR and ErbB2 levels, future pathology reports can be envisioned to include a phosphoproteomic portrait of the functional state of many relevant specific downstream endpoints, and entire classes of signaling pathways as a guide for therapeutic decision making and prognosis [35,60].

Molecular profiling of the proteins and signaling pathways produced by the tumor microenvironment, host, and peripheral circulation hold great promise in effective selection of therapeutic targets and patient stratification. For many of the more common sporadic cancers, there is significant heterogeneity in cell signaling, tissue behavior, and susceptibility to chemotherapy. Proteomic analysis is particularly useful in this area given the ability to study multiple pathways simultaneously. Cataloguing of abnormal signaling pathways for large numbers of specimens will provide the data necessary for a rationally based formulation of combination therapy that presumably would be more effective than monotherapy, and help to minimize the issues of tumor heterogeneity. The promise of proteomic-based profiling, different from gene transcript profiling alone, is that the resulting prognostic signatures are derived from drug targets (such as activated kinases), not genes, so the pathway analysis provides a direction for therapeutic mitigation. Thus, phosphoproteomic pathway analysis becomes both a diagnostic/prognostic signature as well as a guide to therapeutic intervention.

Protein biomarker stability in tissue: a critical unmet need

The promise of tissue protein biomarkers to provide revolutionary diagnostic and therapeutic information will never be realized unless the problem of tissue protein biomarker instability is recognized, studied, and solved (Figure 9.3). There is a critical need to develop standardized protocols and novel technologies that can be used in the routine clinical setting for seamless collection and immediate preservation of tissue biomarker proteins, particularly those that have been post-translationaly modified, such as phosphoproteins. This critical need transcends the large research hospital environment and extends most acutely to the private practice, where most patients receive therapy. While molecular profiling offers tremendous promise to change the practice of oncology, the fidelity of the data obtained from a diagnostic assay applied to tissue must be monitored and ensured; otherwise, a clinical decision may be based on incorrect molecular data. To date, clinical preservation practices routinely rely on protocols that are decades old, such as formalin fixation, and are designed to preserve specimens for histologic examination. Under the current standard of care, tissue is procured for pathologic examination in three main settings: (i) surgery in a hospital-based operating room, (ii) biopsy conducted in an outpatient clinic, and (iii) image-directed needle biopsies or needle aspirates conducted in a radiologic suite. Based on current reseach standards, tissue must be snap-frozen in order to perform proteomic studies. In the real world of a busy clinical setting, it will be impossible to immediately preserve procured tissue in liquid nitrogen. Moreover, the time delay from patient excision to pathologic examination and molecular analysis is often not recorded and may vary from 30 minutes to many hours depending on the time of day, the length of the procedure, and the number of concurrent cases.

Figure 9.3 depicts the two categories of variable time periods that define the stability intervals for human tissue procurement. Time point A is defined as the moment that tissue is excised from the patient and becomes available *ex vivo* for analysis and processing. The post excision delay time (or EDT) is the time from time point A to the time that the specimen is placed in a stabilized state (immersed in fixative or snap-frozen in liquid nitrogen), herein called time point B. Given the complexity of patient-care settings, during the EDT the tissue may reside at room temperature in the operating room or on the pathologist's cutting board, or it may be refrigerated in a specimen container. The second variable time period is the

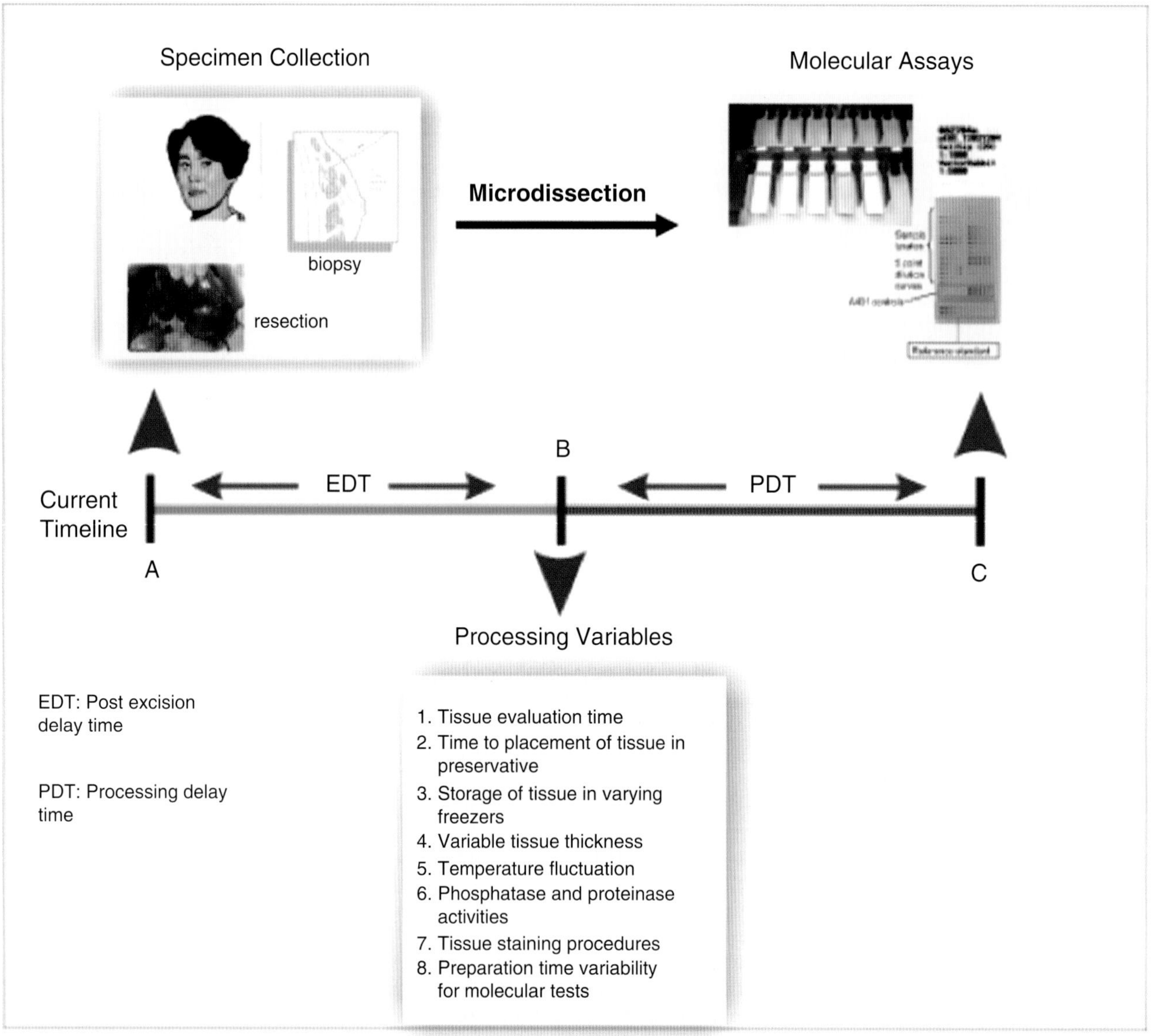

Figure 9.3 Pre-analytical variables during tissue acquisition. Excision delay time, processing delay time, and the type of preservation chemistry selected will affect the quality of molecular analyses.

processing delay time, or PDT. At the beginning of this interval, the tissue is immersed in a preservative solution or stored in a freezer. At the end of this interval, time point C, the tissue is subject to processing for molecular analysis. In addition to the uncertainty about the length of these two time intervals, a host of known and unknown variables can influence the stability of tissue molecules during these time periods. These include (i) temperature fluctuations prior to fixation or freezing; (ii) preservative chemistry and rate of tissue penetration; (iii) size of the tissue specimen; (iv) extent of handling, cutting, and crushing of the tissue; (v) fixation and staining prior to microdissection; (vi) tissue hydration and dehydration; and (vii) the introduction of phosphatases or proteases from the environment at any time. Given current practices, in the face of these uncertainties it would appear virtually impossible to develop a standardized procedure for routine clinical profiling. Even if a strict protocol is followed, there is no ultimate assurance that processing variables are free from compromise up to the time that the molecular profile data are collected.

Protein stability is unrelated to RNA transcript stability

Several studies have been conducted concerning the stability of RNA in tissue *ex vivo* [61,62]. These studies indicate that refrigeration is superior to room temperature and the addition of RNAase inhibitors may be useful as RNA preservatives. While this information is applicable to gene array profiling, it has no bearing on protein stability in general or phosphoprotein stability specifically. Chemical conditions favoring protein stability may be completely different from those for RNA stability. This is true for the following reasons.

a. Gene transcript array data cannot reflect the post-translational state of a protein.
b. Gene transcript array data cannot accurately reflect the activated state of a protein signal pathway or the phosphorylated state of a kinase substrate.
c. Gene array data often do not quantitatively correlate with protein expression for signal proteins and cytokines.

Recognition that the tissue is alive and reactive following procurement

While investigators have worried about the effects of vascular clamping and anesthesia, prior to excision, a much more significant and underappreciated issue is the fact that excised tissue is alive and reacting to *ex vivo* stresses (Figure 9.4). The instant a tissue biopsy is removed from a patient, the cells within the tissue react and adapt to the absence of vascular perfusion, ischemia, hypoxia, acidosis, accumulation of cellular waste, absence of electrolytes, and temperature changes. In as little as 30 minutes post-excision, drastic changes can occur in the protein signaling pathways of the biopsy tissue as the tissue remains in the operating room suite or on the pathologist's cutting board. In response to wounding cytokines, vascular hypotensive stress, hypoxia, and metabolic acidosis, it would be expected that a large surge of stress-related, hypoxia-related, and wound repair-related protein signal pathway proteins and transcription factors will be induced in the tissue immediately following procurement. Over time the levels of candidate proteomic markers (or RNA species) would be expected to widely fluctuate upward and downward. This will significantly distort the molecular signature of the tissue compared to the state of the markers *in vivo* [34]. Moreover, the degree of *ex vivo* fluctuation could be quite different between tissue types and influenced by the pathologic microenvironment. This physiologic fact must be taken into consideration as we plan to implement tissue protein biomarkers in the real world of the clinic, where the living, reacting tissue may remain in the collection basin or on the cutting board for hours.

Formalin Fixation May Be Unsuitable for Quantitative Protein Biomarker Analysis in Tissue

Although it is now possible to extract proteins from formalin-fixed tissue, because of the long period required to formalin tissue fixation, the procedure it may be not optimal for phosphoprotein analysis [63]. For tissue placed directly in formalin, the standard procedure for the past 100 years, the formalin penetration rate is 0.1 mm/hr, so the cellular molecules in the depth of the tissue will have significantly degraded by the time formalin permeates the tissue. Formalin cross-linking, the formation of methylene bridges between amide groups of protein, blocks analyte epitopes, as well as

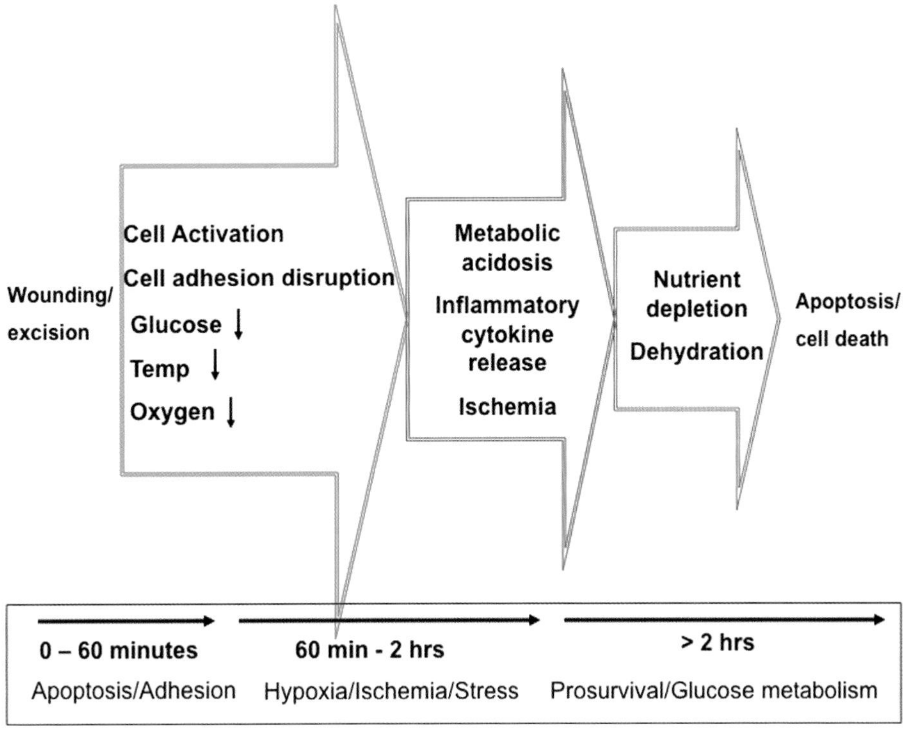

Figure 9.4 **Molecular stages and timeline of tissue cell death.** Post tissue excision, cascades of cellular kinases are activated and deactivated as tissue reacts to wounding, ischemia, inflammation, environmental stresses, hypoxia, and nutrient depletion (Adapted from Espina et al. *Molecular and Cellular Proteomics,* October 2008).

decreases the yield of proteins extracted from the tissue. Since the dimensions of the tissue and the depth of the block that is samples are unknown variables, formalin fixation would be expected to cause significant variability in protein and phosphoprotein stability for molecular diagnostics.

Phosphoprotein stability: The balance between kinases and phosphatases. Phosphoproteins offer a unique minute by minute record of ongoing signal pathway events of high functional relevance to therapeutic target selection and the prediction of toxicity. Phosphorylation and dephosphorylation of structural and regulatory proteins are major intracellular control mechanisms [64]. Protein kinases transfer a phosphate from ATP to a specific protein, typically via serine, threonine, or tyrosine residues. Phosphatases remove the phosphoryl group and restore the protein to its original dephosphorylated state. Hence, the phosphorylation-dephosphorylation cycle can be regarded as a molecular on-off switch. At any point in time within the cellular microenvironment, the phosphorylated state of a protein is a function of the local stoichiometry of associated kinases and phosphatases specific for the phosphorylated residue. During the *ex vivo* time period, if the cell remains alive, it is conceivable that phosphorylation of certain kinase substrates may transiently increase due to the persistence of functional signaling, activation by hypoxia, or some other stress-response signal. On the other hand, the availability of ubiquitous cellular phosphatases would be expected to ultimately destroy phosphorylation sites, given enough time.

Phosphatases determine phosphoprotein stability. Protein phosphatases (PPs) have been classified into three distinct categories: (i) serine/threonine (Ser/Thr)-specific [65], (ii) tyrosine-specific [31,32], and (iii) dual-specificity phosphatases. Based on biochemical parameters, substrate specificity, and sensitivity to various inhibitors, Ser/Thr protein phosphatases are divided into two major classes. Type I phosphatases, which include PP1, can be inhibited by two heat-stable proteins known as Inhibitor-1 (I-1) and Inhibitor-2 (I-2). They preferentially dephosphorylate the β-subunit of phosphorylase kinase. Type II phosphatases are subdivided into spontaneously active (PP2A), Ca2+-dependent (PP2B), and Mg2+-dependent (PP2C) classes of phosphatases. They are insensitive to heat-stable inhibitors and preferentially dephosphorylate the α-subunit of phosphorylase kinase. Protein tyrosine phosphatases (PTPs) remove phosphate groups from phosphorylated tyrosine residues of proteins. PTPs display diverse structural features and play important roles in the regulation of cell proliferation, differentiation, cell adhesion and motility, and cytoskeletal function [66]. They are either transmembrane receptor-like PTPs or cytosolic enzymes. The dual-specificity phosphatases (DSP) play a key role in the dephosphorylation of MAP kinases (MKPs) [34]. MKPs have been divided into three subgroups. All the DSPs share strong amino-acid sequence homology in their catalytic domains. The catalytic domain contains a highly conserved consensus sequence DX26 (V/L)X(V/I)HCXAG(I/V) SRSXT(I/V)XXAY(L/I)M, where X could be any amino acid. The cysteine is required for the nucleophilic attack on the phosphorus of the substrate and the formation of the thiol-phosphate intermediate. The conserved arginine binds the phosphate group of phospho-tyrosine or phospho-threonine, enabling transition-state stabilization, and the aspartate enhances catalysis by protonating oxygen on the departing phosphate group. A variety of chemical-based and protein-based inhibitors of phosphatases exist [67]. Thus, there is adequate chemistry knowlege to design rational stabilizers for the preservation of phosphoprotein stability without freezing.

SERUM PROTEOMICS: AN EMERGING LANDSCAPE FOR EARLY STAGE CANCER DETECTION

The recognition that cancer is a product of the proteomic tissue microenvironment and involves communication networks has important implications. First, it shifts the emphasis away from therapeutic targets being directed solely against individual molecules within pathways and focuses the effort on targeting nodes in multiple pathways inside and outside the cancer cell that cooperate to orchestrate the malignant phenotype. Second, the tumor-host communication system may involve unique enzymatic events and sharing of growth factors. Consequently, the microenvironment of the tumor-host interaction could be a source for biomarkers that could ultimately be shed into the serum proteome.

Application of Serum Proteomics to Early Diagnosis

Cancer is too often diagnosed and treated too late, when the tumor cells have already invaded and metastasized. More than 60% of patients with breast, lung, colon, and ovarian cancer already have hidden or overt metastatic colonies. At this stage, therapeutic modalities are limited in their success. Detecting cancers at their earliest stages, even in the premalignant state, means that current or future treatment modalities might have a higher likelihood of a true cure. Ovarian cancer is a prime example of this clinical dilemma. More than two-thirds of cases of ovarian cancer are detected at an advanced stage, when the ovarian cancer cells have spread away from the ovary surface and have disseminated throughout the peritoneal cavity. Although the disease at this stage is advanced, it rarely produces specific or diagnostic symptoms. Consequently, ovarian cancer is usually treated when it is at an advanced stage. The resulting 5-year survival rate is 35% to 40% for patients with late-stage disease who receive the best possible surgical and chemotherapeutic intervention. By contrast, if ovarian cancer is detected when it is still confined to the ovary (stage I), conventional therapy produces a high rate (95%) of 5-year survival. Thus, early detection of ovarian cancer, by itself, could have a profound effect on the successful treatment of this disease. Unfortunately, early stage ovarian cancer lacks a specific symptom or a specific biomarker and accurate and reliable diagnostic, noninvasive modalities. Because of such profound clinical need, a principal focus of protein marker discovery has been ovarian cancer [68]. A clinically useful biomarker should be measurable in a readily accessible body fluid, such as serum, urine, or saliva. Clinical proteomic methods are especially well suited to discovering such biomarkers. Serum or plasma has been the preferred medium for discovery because this fluid is a protein-rich information reservoir that contains the traces of what has been encountered by the blood during its constant perfusion and percolation throughout the body. Until recently the search for cancer-related biomarkers for early disease detection has been a one-at-a-time approach. However, biomarker discovery is moving away from the idealized single cancer specific biomarker. Despite decades of effort, single biomarkers have not been found that can reach an acceptable level of specificity and sensitivity required for routine clinical use for the detection or monitoring of the most common cancers. Most investigators believe that this is due to the patient-to-patient molecular heterogeneity of tumors. A second level of population heterogeneity exists for tumor location, size, histology, grade, and stage. Moreover, an individual patient's organ may harbor co-existence of multiple stages in the same tissue (such as *in situ* and invasive cancer). Epidemiologic heterogeneity, including differences in age, sex, and genetic background, is a third level of patient-to-patient variability that can reduce cancer biomarker specificity. Taking a cue from gene arrays, the hope is that panels of tens to hundreds of protein and peptide markers may transcend the heterogeneity to generate a higher level of diagnostic specificity. While an individual biomarker candidate may be specific and sensitive only for a certain stage or molecular etiology, combinations of many markers, screened for sensitivity and specificity concomitantly, may be able to bracket across the heterogeneity to reach a higher level of specificity and sensitivity in the aggregate. Thus, while marker A may work for 50% of the population, marker B for 30%, and marker C for 20%, combining A, B, and C has the potential to cover the entire population. The overall specificity and sensitivity required for clinical use depends entirely on the intended use of the marker(s). Markers for general population screening for rare diseases may have to approach 100% specificity to be accepted. On the other hand, markers that are used for high-risk screening or for relapse monitoring can have much lower specificity but require high sensitivity.

The low molecular weight (LMW) range of the serum proteome (generally defined here as peptides less than 50,000 Daltons) is called the peptidome due to the abundance of protein peptides and fragments. While some dismissed the peptidome as noise, biological trash, or too small and unstable to be biologically relevant [69], others have proposed that just the opposite is the case—it may contain a rich, untapped source of disease-specific diagnostic information [68–77]. Tissue proteins that are normally too large to passively diffuse through the endothelium into the circulation can still be represented as fragments of the parent molecule. The information in the peptidome resides in multiple dimensions: (i) the identity of the parent protein [71]; (ii) the peptide fragment isoform identity peptide (such as fragment size and cleavage ends, post-translational glycosylation/phosphorylation sites, and others); (iii) the specific size and cleavage ends of the peptide; (iv) the quantity of the peptide itself; and (v) the nature of the carrier protein to which it is bound.

In 2002, investigators used mass spectrometry to interrogate the circulatory peptidome of patients with ovarian cancer [68], and then later other cancers and non-neoplastic diseases, for disease-related information [70,75,77]. These early studies revealed an apparent abundance of disease-specific information. More recently, investigators have been sequencing and identifying the LMW ions that comprised the underlying key signatures described in the early profiling work [72–77]. The disease-specific ions appear to be fragments of large molecules—either from endogenous high-abundant proteins such as transthyretin or low-abundance cellular and tissue proteins such as BRCA2 [73]. Concomitantly, investigators were exploring the peptide content within the circulation itself and found evidence that disease-specific information may exist. In fact, initial skeptical reaction to some early reports whereby identification of some of the ions that underpinned MS profiles revealed fragment isoforms of common abundant proteins [69,75]. The appreciation of a fragment peptide as a new analyte isoform in and of itself has now given way to recent general optimism about the diagnostic potential of the peptidome [78,79].

The Peptidome: A Recording of the Tissue Microenvironment

Cancer is a product of the tissue microenvironment [80,81]. While normal cellular processes (and the peptide content generated by these processes) are also a manifestation of the tissue microenvironment, the tumor microenvironment, through the process of aberrant cell growth, cellular invasion, and altered immune system function, represents a unique constellation of enzymatic (e.g., kinases, phosphatases) and protease activity (e.g., matrix metalloproteases), resulting in changed stoichiometry of molecules within the peptidome itself compared to the normal milieu. Interactions between the precancerous cells, the surrounding epithelial and stromal cells, vascular channels, the extracellular matrix, and the immune system are mediated by various enzymes, cytokines, extracellular matrix molecules, and growth factors. An added benefit for cancer biomarkers is the leaky nature of newly formed blood vessels and the increased

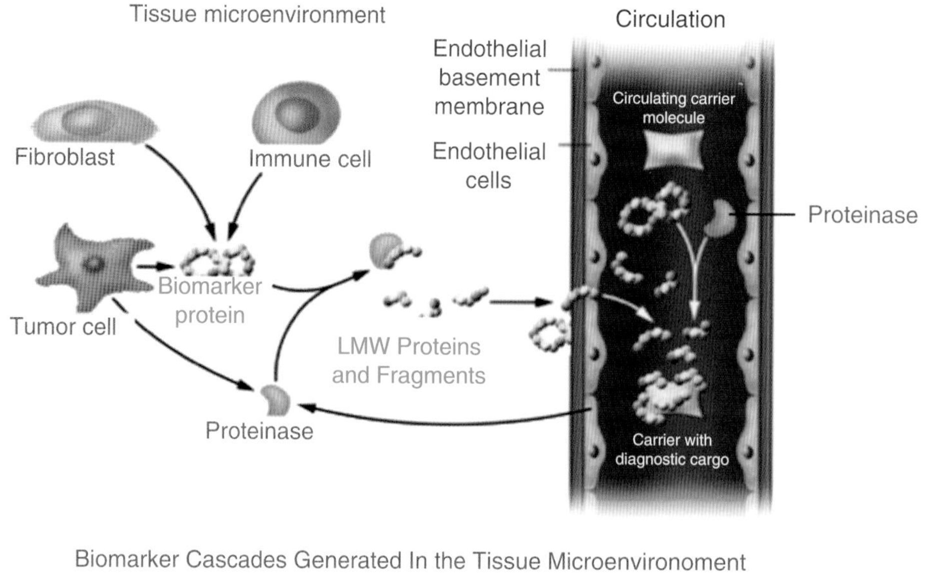

Figure 9.5 **The peptidome hypothesis: Circulating peptides and protein fragments are shed from all cell types in the tissue microenvironment.** Proteolytic cascades within the tissue generate fragments that diffuse into the circulation. The identity and cleavage pattern of the peptides provide two dimensions of diagnostic information.

hydrostatic pressure within tumors [82]. This pathologic physiology would tend to push molecules from the tumor interstitium into the circulation (Figure 9.5). As cells die within the microenvironment, they will shed the degraded products. The mode of death, apoptosis versus necrosis, would be expected to generate different classes of degraded cellular constituents. As a consequence, the blood peptidome may reflect ongoing recordings of the molecular cascade of communication taking place in the tissue microenvironment (Figure 9.5). Combinations of peptidome markers representing the specific interactions of the tumor tissue microenvironment at the enzymatic level can achieve a higher specificity and a higher sensitivity for early stage cancers. This optimism is in part based on the concept that the biomarkers are derived from a population of cells that comprise a volume that is greater that just the small precancerous lesion itself. In this way, the peptidome can potentially supersede individual single biomarkers and transcend the issues of tumor and population heterogeneity.

Physiologic Roadblocks to Biomarker Discovery

Candidate biomarkers are expected to exist in very low concentration, have the potential to be rapidly excreted, and must be separated from high abundance blood proteins such as albumin which exist in a billion-fold excess [86,88]. Early stage disease lesions such as premalignant cancer may arise within a tissue volume less than 0.10 mL. Assuming all the putative biomarkers emanating from this volume are uniformly dispersed within the entire blood volume of 5000 mL, then the dilution factor will be 50,000. We can also reasonably hypothesize that the most physiologically relevant proteins specific for the disease constitute a minor subpopulation of the cellular proteome. Consequently, the greatest challenge to biomarker discovery is the isolation of very rare candidate proteins within a highly concentrated complex mixture of blood proteins massively dominated by seemingly nonrelevant proteins. Because of the low abundance of the biomarkers, analytical sensitivity is the first challenge for biomarker discovery and also routine measurement. During the discovery phase, it is likely that large plasma or serum volumes, including pooled samples, can be available for analysis [86–91]. In contrast, once a candidate marker is taken forward to clinical testing, the volume of blood available for an individual patient's assay may be less than one mL. When one takes all of these factors into consideration, the analytical platform used to measure the candidate marker must have a detection sensitivity sufficient to reliably detect marker concentrations in the subfemtomolar or attomolar concentration.

Requirement for New Classes of Diagnostic Technology

Cancer biomarkers of the future are predicted to emerge from multiplexed measurement of patterns of specific size fragments of known parent molecules. This will require new generations of immunoassay-based technology that can determine both the identity and exact size of the biomarker. Conventional immunoassay platforms such as antibody arrays and bead capture arrays cannot effectively measure panels of fragment analytes [92]. The reason is that immunoassays, by their very definition, rely on antibody-based capture and detection. An antibody-based assay cannot distinguish the parent molecule from its clipped fragments since the antibody recognizes its cognate epitope in both the parent and fragment molecule. Immuno-MS is one possibility: using this two-dimensional immunoassay technology, based on the amino acid sequence of the peptide fragment, a miniaturized affinity antibody column, perhaps in a multiplexed microwell format, is first used to capture all species of molecules that contain the antibody recognition site. Next, the captured fragments are eluted off the antibody column directly into a MS (such as a MALDI-TOF), which can provide an extremely accurate mass determination of the entire population of captured peptides. Thus, in only two automated steps, a panel of peptide fragments derived from a known parent molecule can be rapidly sorted and tabulated. The result is an immediate read-out of identities and specific fragment sizes of a given biomarker candidate, and both dimensions of information are captured concomitantly.

Reduction of Bias in the Discovery Phase of Peptide Biomarkers

The tremendous complexity of the peptidome is entirely predictable from a systems biology standpoint where the composition of the circulation is a true mirror of ongoing cellular and organ system function. While this means that subsets of the blood peptidome can potentially reflect subtle disease events in small tissue volume, it also means that the peptidome is constantly fluctuating due to ongoing daily physiologic events. Epidemiologists and clinical chemists fear that the level of individual blood-borne biomarkers can be greatly influenced by a variety of non-disease-related epidemiologic factors and normal physiologic conditions. This includes stability of the peptidome in the collected blood sample. Thus, the promise of the specificity of the peptidome is counterbalanced by the sensitivity to interfering factors. Consequently, as with any biomarker discovery (single, or as a panel), great care is needed to reduce sample bias during the discovery and validation phase of peptidome biomarker translational research [69,93].

Methods for Discovering and Validating Candidate Protein Biomarkers

Researchers can choose from a series of separation, chromatography, electrophoresis, and mass spectrometry-based methodologies useful for discovering the low molecular weight peptidome. Methods are available for profiling, harvesting, purifying, enriching, and sequencing the peptidome (Table 9.1) [86,98]. Each methodology has advantages and disadvantages. Mass spectrometry profiling technology is useful for rapidly obtaining an ion fingerprint of a test fluid sample. SELDI-TOF is a MALDI-TOF-related research system that is relatively sensitive compared to other MS, and can rapidly read out peptide ion signatures derived from a small sample volume. The sample to be tested is placed on a chemically treated chip surface, peptides of interest stick to the surface, and the unbound proteins are washed away. The bound peptides are ionized by laser-induced desorption and time of flight (TOF) analysis. The disadvantages of SELDI are that the resolution of the instrumentation is usually quite low (many peptides can exist within each ion peak) and the ion peaks of interest cannot be directly identified by mass spectrometry-based sequence determination. However, SELDI-TOF, as with other more common MALDI and ESI instruments, cannot routinely detect proteins in the low abundance range (<4 pg/ml range), which almost all immunoassays can reach with ease. MS profiling can also be conducted in a MS electrospray mode using inline liquid chromatography followed by ESI-MS. This can be done before or after trypsinization of the sample. If the sample is trypsinized, then the identity of the trypsin fragments can be scored, but the original peptide fragment size information may be lost. An intermediate system is represented by a solid particle or bead capture systems. These methods are used to harvest peptides by fairly nonspecific hydrophobic binding. The harvested peptides eluted from the beads can be profiled by MALDI or ES MS, followed by MS sequencing and identification of abundant peaks. Native carrier proteins such as albumin constitute an endogenous resident *in vivo* affinity chromatography system for harvesting peptides. Carrier protein harvesting is a facile method for obtaining high-resolution ion profiles. Furthermore, the captured peptides can be eluted and sequenced with excellent yield. The disadvantages of the bead and carrier protein capture systems are low throughput and minimal prefractionation.

Various methods exist to prefractionate a complex LMW peptidome mixture prior to sequencing. However, given the new finding that a great deal of LMW information appears to exist prebound to albumin and other high-abundant proteins, and that many investigations for biomarker discovery begin with depleting the blood of these high-abundance proteins, a great deal of caution is warranted in using this approach for LMW candidate discovery. An alternate approach would be to denature the input material and dissociate the LMW information from the carrier molecules, and then separate, isolate, and enrich for the LMW archive. One- and two-dimensional preparative gel electrophoresis (with or without labeling methods such as ICAT) can be employed. The disadvantages of 2D gel electrophoresis are an overall lack of reliability of resolving low molecular weight range molecules (mostly below 15 kDa), large sample volume and starting material requirements, and slow throughput. One-dimensional gel preparative electrophoresis exhibits good yield in the low mass range <20 kDa, but is also hampered by slow throughput. Slow throughput and large volume requirements are also a drawback of tagging systems such as ICAT, and these tagging systems inherently ignore those molecules that are truly off-on in disease states, and have the highest predictive value. Thus, a researcher planning an LMW peptidome biomarker discovery project must carefully select the proper sample preparation method tailored to the

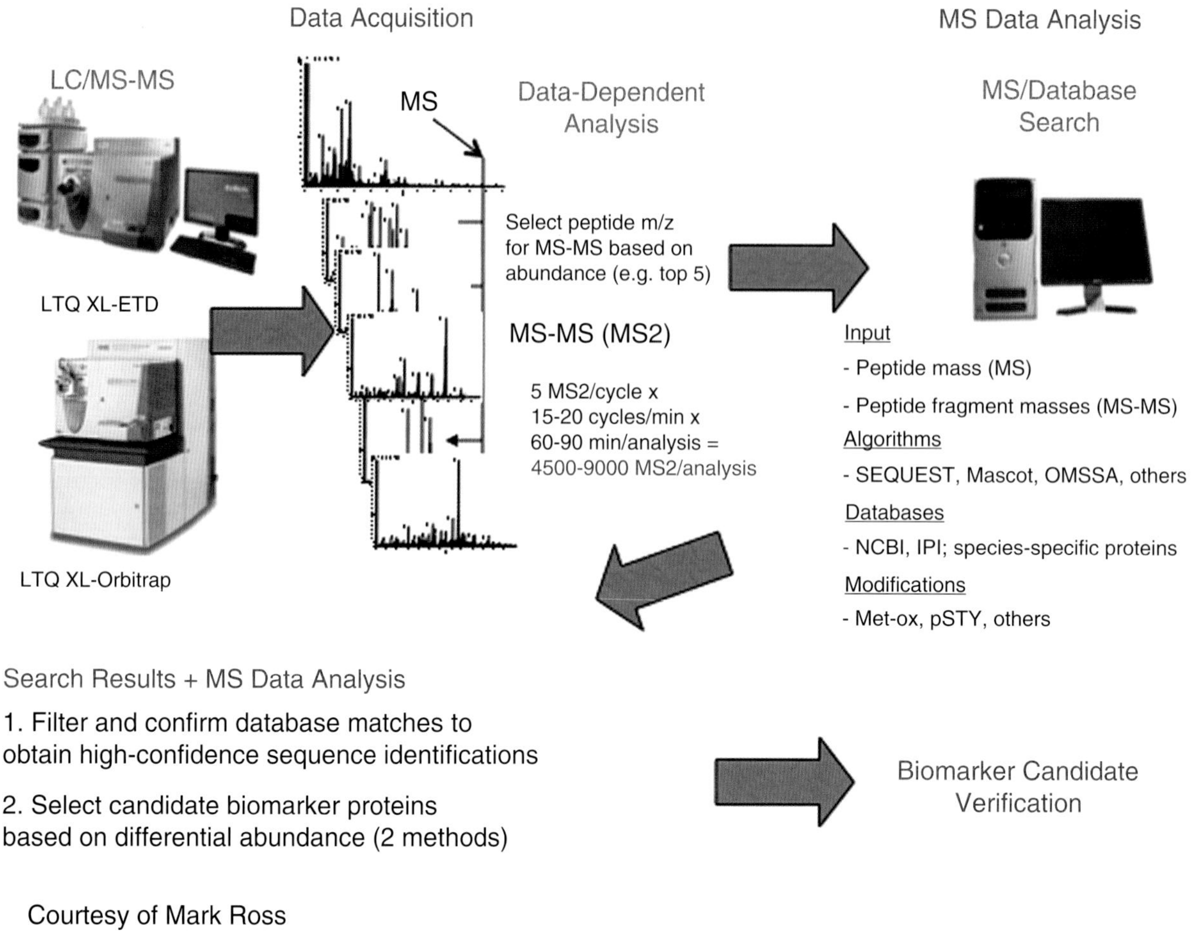

Figure 9.6 Proteomics mass spectrometry analytical workflow.

volume of the available samples, the resolution desired, the required throughput, and the need for identification of the candidate biomarker peptides.

A recommended staging process for biomarker discovery and validation is depicted in Figure 9.6. The starting point for rigorous analysis is the development of a discovery study set consisting of a population of serum or plasma samples from patients who have (i) histologically verified cancer; (ii) benign or inflammatory non-neoplastic disease; and (iii) unaffected, apparently healthy controls or hospital controls, depending on the intended use. The issue of including specimens for initial discovery from patients with no evidence of cancer but with inflammatory conditions, reactive disease, and benign disorders is of critical importance to ensure that specific markers are enriched for from the outset. This issue is critical for cancer research, especially since the disease almost always occurs in the background of inflammatory processes that are part of the disease pathogenesis itself. The peptidome, as a mirror of the ongoing physiology of the entire individual, may be especially sensitive to these processes, which is why care must be taken to at least minimize the chance that nonspecific markers are selected. As an additional level of rigor to reduce or minimize hard-wired bias, the blood collection standard operating protocol for all patients should be exactly identical in methodology, handling, and storage. Prior to the initial peptidome analysis, the biomarker researcher should work with an epidemiologist to develop a discovery subset that is matched in every possible epidemiologic and physiologic parameter. This includes age, sex, hormonal status, treatment and hospitalization status, clinic location, and any other variable that can be known. Reduction of the potential upfront bias is critical prior to undertaking the discovery phase of the research. Because of the overall lability of the proteome and the resultant peptidome, it is critical that samples are collected and handled in a routine and rapid fashion. A rule of thumb should be that any specimen is handled with identical standard operating procedures, processed and stored in identical fashion, and frozen/stored as rapidly as possible after removal from the body.

The next stage is LMW peptidome fractionation, separation, isolation and enrichment, concentration, and mass spectrometry-based identification (Figure 9.6). At this stage it is essential that iterative and repetitive MS-based analysis and MS/MS sequencing be conducted on each sample. Candidate peptides identified repetitively over many iterations within a sample and within a study set have a higher likelihood of being correct. Seven or more iterations has been previously shown to minimize false peptide scores [73,89,90]. The researcher ends up with a list of candidate diagnostic markers that are judged to be differentially abundant in the cancer versus the control populations. The next step is to find or make specific antibodies or other ligands for each candidate peptide marker. After each antibody is validated for specificity using a reference analyte, the antibody can then be used to validate the existence of the predicted peptide marker in the disease and nondiseased discovery set samples. Multiple Reaction Monitoring (MRM) is an emerging generation of ms technology which offers the potential to quantify and identify peptides with such high confidence that antibody validation may not be required (Figure 9.7). The goal is to develop a panel of candidate peptide biomarkers along with measurement reagents that are independent of the analytical technology that will ultimately be used in the clinical lab.

Clinical validation of the candidate biomarkers starts with ensuring the sensitivity and precision of the measurement platform. The antibodies developed in discovery phase can now be applied as capture or detection reagents in the analytical platform. While some immunoassays can utilize antibodies that recognize variant analytes whereby a neo-epitope is formed by post-translational medications, or even recognize fragment-specific termini (such as cleaved caspases), most size-specific discriminations require alternate technologies to be employed, such as immuno-MS. Once the measurement platform is proven to be reliable and reproducible, then the clinical validation can proceed.

The final and most critical stage of research clinical validation is blinded testing of the biomarker panel using independent (not used in discovery), large clinical study sets that are ideally drawn from at least three geographically separate locations. The required size of these test sets for adequate statistical powering depends on both the performance of the peptide analyte panel in the platform validation phase and the intended use of the analyte in the clinic. For example, markers for general population screening require more patients than those for high-risk screening, which in turn require more patients than markers for recurrence/therapeutic monitoring. Employing previously verified controls and calibrators, under standard clinical chemistry guidelines (NCCLS) for immunoassays, the sensitivity and specificity can be determined for a test population. It is important to emphasize that

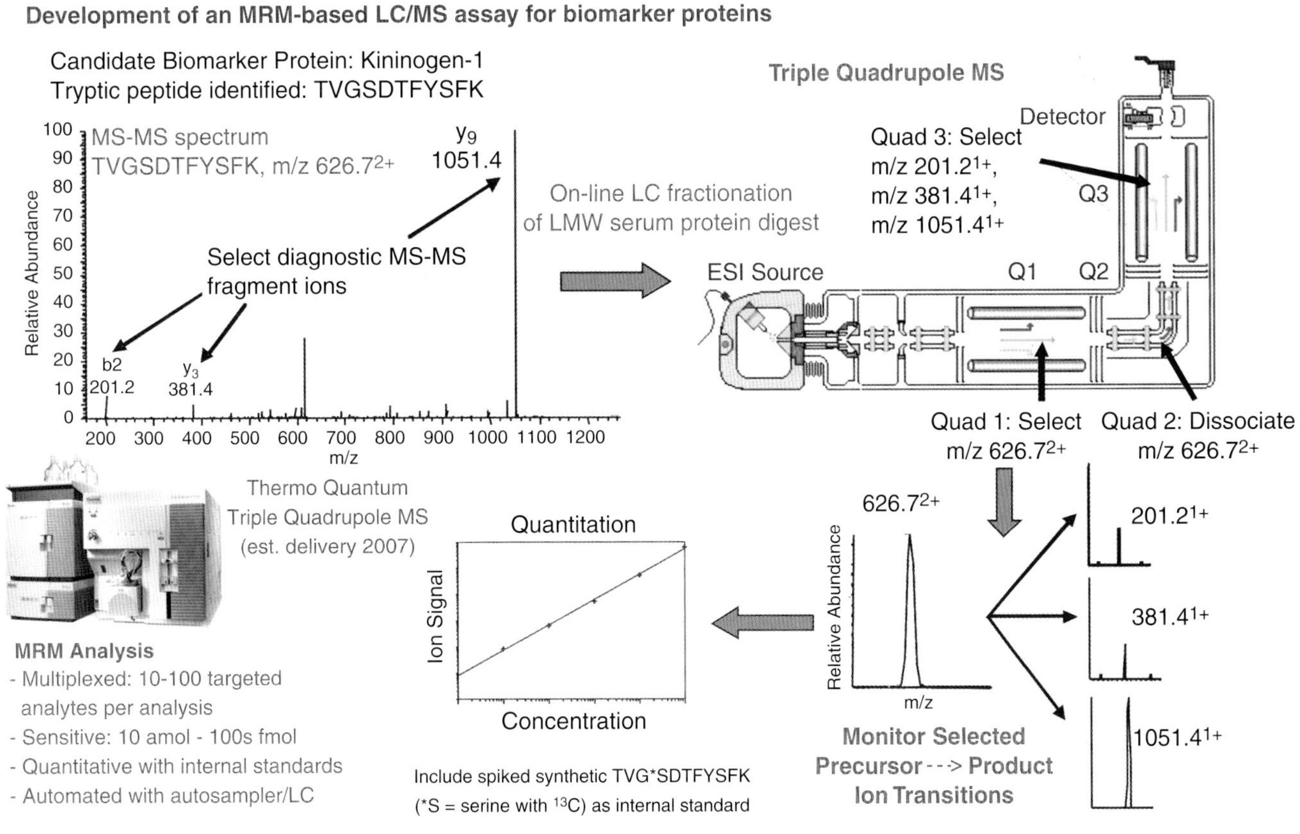

Figure 9.7 Mass spectrometry multiple reaction monitoring (MRM) is an emerging technology for quantifying and specifically identifying protein analytes based on highly accurate mass profiling of a protease peptide fragment. This technology has the potential to validate discovered biomarker candidates without the requirement for a specific antibody.

sensitivity and specificity in an experimental test population does not translate to the positive predictive value that would be seen if the putative test is used routinely in the clinic. The true positive predictive value is a function of the indicated use and the prevalence of the cancer (or other disease condition within the target population). The percentage of expected cancer cases in a population of patients at high genetic risk for cancer is higher than the general population. Consequently, the probability of false positives in the latter population would be much higher. For this reason the ultimate adoption of a peptidome- or proteome-based test will be strongly dependent on the clinical context of its use.

Finally, the information content of the peptidome will never be fully realized unless blood collection protocols and reference sets are standardized, new instrumentation for measuring panels of specific fragments are proven to be reproducible and sensitive, and extensive clinical trial validation is conducted under full CAP/CLIA regulatory guidelines.

Frontiers of Nanotechnology and Medicine

Nanotechnology will have a significant impact on early diagnosis and targeted drug delivery. The development of inorganic nanoparticles that bind specific tumor markers that exist at very low concentrations in serum may be able to be used as serum harvesting agents. In the future, patients may be injected with such nanoparticles that seek out and bind tumor or disease markers of interest. Once the nanoparticles have bound their targets, they can be "harvested" from the serum to enable diagnosis or to monitor disease progression [94–98].

Luchini et al. have created core shell hydrogel nanoparticles to address these fundamental road-blocks to biomarker purification and preservation [98]. The nanoparticles simultaneously conduct molecular sieve chromatography and affinity chromatography, in one step, in solution (Figure 9.8). The molecules captured and bound within the affinity matrix of the particles are protected from degradation

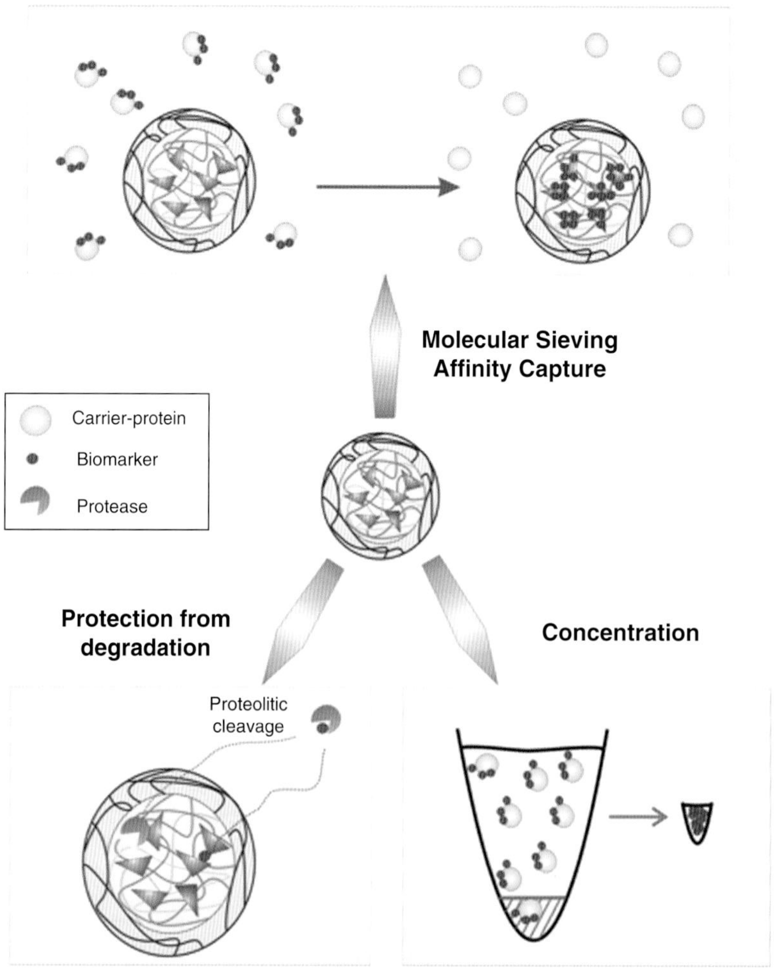

Figure 9.8 Schematic representation of particle structure and function. Particles are constituted by a bait containing core, surrounded by a sieving shell. When introduced into a complex solution, such as serum, core-shell particles remove low molecular weight proteins from carrier albumin with affinity capture and perform molecular weight sieving with total exclusion of high molecular weight proteins. Depending on the starting volume of the fluid sample, the particles can effectively amplify low abundance biomarkers 100- to 1000-fold.

by exogenous or endogenous proteases. These smart nanoparticles conduct enrichment and encapsulation of selected classes of proteins and peptides from complex mixtures of biomolecules such as plasma, purify them away from endogenous high-abundance proteins such as albumin, and protect them from degradation during subsequent sample handling. The particles have a molecular sieving shell surrounding a specific bait core. Each class of smart particle contains a core bait molecule specific for a general category of analyte. Thus, each class of particle will sequester and concentrate a class of all analytes below a molecular weight cutoff that selectively recognize the bait.

When added to a complex molecular solution, the nanoparticles rapidly imbibe and entrap the target molecules with complete exclusion of high molecular weight and abundant proteins such as albumin. Moreover, it is possible to concentrate the biomarkers captured by hydrogel particles into a small volume that is a small fraction of the starting volume. Particles incubated with serum trap the target analytes and are isolated by centrifugation. Candidate biomarkers are then released from particles by means of elution buffers. The ratio of the volume of elution buffer to the original starting solution establishes the concentration amplification factor. This concentration step is a fundamental point for biomarker discovery and measurement because it provides a means to effectively raise the concentration of rare biomarkers that become the input for a measurement system such as an immunoassay platform or mass spectrometry.

Evidence suggests that each organ and tumor has its unique molecular address encoded within the vasculature [99]. Peptides that recognize organ-specific vascular beds have been identified via the screening of phage libraries *in vivo* [100]. With the synthesis of peptide carriers designed to transport drug therapies to specific vascular beds, treatments can be designed to treat specific tissues or tumor sites, leaving nonmalignant tissues unaffected. The use of homing nanoparticles, or semiconductor quantum dots [101], may be an important part of imaging diagnostics, in addition to drug delivery, in the health care of the future.

Future of Cancer Clinical Proteomics

The pathologist of the future will detect early manifestations of disease using proteomic patterns of body-fluid samples and will provide the primary physician a diagnosis based on proteomic signal pathway network signatures as a complement to histopathology. He or she will be able to dissect a patient's individual tumor molecularly, identifying the specific regulatory pathways that are deranged in the cell cycle, differentiation, apoptosis, and invasion and metastasis. Based on this knowledge, recommendations will be made for an individualized selection of therapeutic combinations of molecularly targeted agents that best strike the entire disease-specific protein network of the tumor. The pathologist and the diagnostic imaging physician will assist the clinical team to perform real-time assessment of therapeutic efficacy and toxicity. Proteomic and genomic analysis of recurrent tumor lesions could be the basis for rational redirection of therapy because it could reveal changes in the diseased protein network that are associated with drug resistance. The paradigm shift will directly affect clinical practice, as it has an impact on all of the crucial elements of patient care and management.

REFERENCES

1. Hunter T. Signaling–2000 and beyond. *Cell.* 2000;100:113–127.
2. Sgroi DC, Teng S, Robinson G, et al. In vivo gene expression profile analysis of human breast cancer progression. *Cancer Res.* 1999;59:5656–5661.
3. Emmert-Buck MR, Bonner RF, Smith PD, et al. Laser capture microdissection. *Science.* 1996;274:998–1001.
4. Banks RE, Dunn MJ, Forbes MA, et al. The potential use of laser capture microdissection to selectively obtain distinct populations of cells for proteomic analysis—Preliminary findings. *Electrophoresis.* 1999;20:689–700.
5. Espina V, Wulfkuhle JD, Calvert VS, et al. Laser capture microdissection. *Nature Protocols.* 2006;1:586–603.
6. Paweletz CP, Liotta LA, Petricoin 3rd EF. New technologies for biomarker analysis of prostate cancer progression: Laser capture microdissection and tissue proteomics. *Urology.* 2001;57: 160–163.
7. Page MJ, Amess B, Townsend RR, et al. Proteomic definition of normal human luminal and myoepithelial breast cells purified from reduction mammoplasties. *Proc Natl Acad Sci USA.* 1999; 96:12589–12594.
8. Hancock W, Apffel A, Chakel J, et al. Integrated genomic/proteomic analysis. *Anal Chem.* 1999;71:742A–748A.
9. Mann M, Hendrickson RC, Pandey A. Analysis of proteins and proteomes by mass spectrometry. *Annu Rev Biochem.* 2001;70: 437–473.
10. Baak JP, Path FR, Hermsen MA, et al. Genomics and proteomics in cancer. *Eur J Cancer.* 2003;39: 1199–1215.
11. Ma XJ, Salunga R, Tuggle JT, et al. Gene expression profiles of human breast cancer progression. *Proc Natl Acad Sci USA.* 2003; 100:5974–5979.
12. Celis JE, Gromov P. Proteomics in translational cancer research: Toward an integrated approach. *Cancer Cell.* 2003;3:9–15.
13. Gorg A, Weiss W, Dunn MJ. Current two-dimensional electrophoresis technology for proteomics. *Proteomics.* 2004;4:3665–3685.
14. Gygi SP, Rist B, Gerber SA, et al. Quantitative analysis of complex protein mixtures using isotope-coded affinity tags. *Nat Biotechnol.* 1999;17:994–999.
15. Krutchinsky AN, Kalkum M, Chait BT. Automatic identification of proteins with a MALDI-quadrupole ion trap mass spectrometer. *Analytical Chem.* 2001;73:5066–5077.
16. Washburn MP, Wolters D, Yates JR. Large scale analysis of the yeast proteome by multidimensional protein identification technology. *Nat Biotechnol.* 2001;19:242–247.
17. Zhou G, Li H, DeCamp D, et al. 2D differential in-gel electrophoresis for the identification of esophageal scans cell cancer-specific protein markers. *Mol Cell Proteomics.* 2002;1:117–124.
18. Paweletz CP, Charboneau L, Bichsel VE, et al. Reverse phase protein microarrays which capture disease progression show activation of pro-survival pathways at the cancer invasion front. *Oncogene.* 2001;20:1981–1989.
19. Grubb RL, Calvert VS, Wulkuhle JD, et al. Signal pathway profiling of prostate cancer using reverse phase protein microarrays. *Proteomics.* 2003;3:2142–2146.
20. Gulmann C, Espina V, Petricoin 3rd E, et al. Proteomic analysis of apoptotic pathways reveals prognostic factors in follicular lymphoma. *Clin Cancer Res.* 2005;11:5847–5855.
21. Sheehan KM, Calvert VS, Kay EW, et al. Use of reverse-phase protein microarrays and reference standard development for molecular network analysis of metastatic ovarian carcinoma. *Mol Cell Proteomics.* 2005;4:346–355.

22. Eckel-Passow JE, Hoering A, Therneau TM, et al. Experimental design and analysis of antibody microarrays: Applying methods from cDNA arrays. *Cancer Res.* 2005;65:2985–2989.
23. Haab BB. Antibody arrays in cancer research. *Mol Cell Proteomics.* 2005;4:377–383.
24. Humphery-Smith I, Wischerhoff E, Hashimoto R. Protein arrays for assessment of target selectivity. *Drug Discov World.* 2002;4:17–27.
25. MacBeath G, Schreiber SL. Printing proteins as microarrays for high-throughput function determination. *Science.* 2000;289: 1760–1763.
26. Petach H, Gold L. Dimensionality is the issue: Use of photoaptamers in protein microarrays. *Curr Opin Biotechnol.* 2002;13: 309–314.
27. Weng S, Gu K, Hammond PW, et al. Generating addressable protein microarrays with PROfusion covalent mRNA-protein fusion technology. *Proteomics.* 2002;2:48–57.
28. Zhu H, Snyder M. Protein chip technology. *Curr Opin Chem Biol.* 2003;7:55–63.
29. Leuking A, Cahill DJ, Mullner S. Protein biochips: A new and versatile platform technology for molecular medicine. *Drug Disc Today.* 2005;10:789–794.
30. MacBeath G. Protein microarrays and proteomics. *Nature Genet.* 2002;32:526–532.
31. Miller JC, Zhou H, Kwekel J, et al. Antibody microarray profiling of human prostate cancer sera: Antibody screening and identification of potential biomarkers. *Proteomics.* 2003;3:56–63.
32. Espina V, Wulfkuhle J, Liotta LA, et al. Basic techniques for the use of reverse phase protein microarrays for signal pathway profiling. In: Jorde L, Dunn MJ, and Subramaniam S, ed. *Encyclopedia of Genetics, Genomics, Proteomics and Bioinformatics. Part 3 Proteomics.* West Sussex, UK: John Wiley & Sons; 2005;1979–1986.
33. LaBaer J, Ramachandran N. Protein microarrays as tools for functional proteomics. *Curr Opin Chem Biol.* 2005;9:14–19.
34. Espina VA, Edmiston KH, Heiby M, et al. A portrait of tissue phosphoprotein stability in the clinical tissue procurement process. *Mol Cell Proteomics.* 2008; [In Press].
35. Vanmeter AJ, Rodriguez AS, Bowman ED, et al. LCM and protein microarray analysis of human NSCLC: Differential EGFR phosphorylation events associated with mutated EGFR compared to wild type. *Mol Cell Proteomics.* 2008; [In Press].
36. Grote T, Siwak DR, Fritsche HA, et al. Validation of reverse phase protein array for practical screening of potential biomarkers in serum and plasma: Accurate detection of CA19-9 levels in pancreatic cancer. *Proteomics.* 2008;8:3051–3060.
37. Nishizuka S, Charboneau L, Young L, et al. Proteomic profiling of the NCI-60 cancer cell lines using new high-density reverse-phase lysate microarrays. *Proc Natl Acad Sci USA.* 2003;100: 14229–14234.
38. Liotta LA, Kohn EC, Petricoin EF. Clinical proteomics. Personalized molecular medicine. *JAMA.* 2001;286:2211–2214.
39. Petricoin EF, Zoon KC, Kohn EC, et al. Clinical proteomics: Translating benchside promise into bedside reality. *Nat Rev Drug Discov.* 2002;1:683–695.
40. Liotta L, Petricoin E. Molecular profiling of human cancer. *Nat Rev Genet.* 2000;1:48–56.
41. Petricoin EF 3rd, Bichsel VE, Calvert VS, et al. Mapping molecular networks using proteomics: A vision for patient-tailored combination therapy. *J Clin Oncol.* 2005;23:3614–3621.
42. Wulfkuhle JD, Aquino JA, Calvert VS, et al. Signal pathway profiling of ovarian cancer from human tissue specimens using reverse-phase protein microarrays. *Proteomics.* 2003;3:2085–2090.
43. Sreekumar A, Nyati MK, Varambally S, et al. Profiling of cancer cells using protein microarrays: Discovery of novel radiation-regulated proteins. *Cancer Res.* 2001;61:7585–7593.
44. Sanchez-Carbayo M, Socci ND, Lozano JJ, et al. Profiling bladder cancer using targeted antibody arrays. *Am J Pathol.* 2006;168: 93–103.
45. Zha H, Raffeld M, Charboneau L, et al. Similarities of prosurvival signals in Bcl-2-positive and Bcl-2-negative follicular lymphomas identified by reverse phase protein microarray. *Lab Invest.* 2004;84:235–244.
46. Petricoin 3rd EF, Espina V, Araujo RP, et al. Phosphoprotein pathway mapping: Akt/mammalian target of rapamycin activation is negatively associated with childhood rhabdomyosarcoma survival. *Cancer Res.* 2007;67:3431–3440.
47. Espina V, Dettloff KA, Cowherd S, et al. Use of proteomic analysis to monitor responses to biological therapies. *Expert Opin Biol Ther.* 2004;4:83–93.
48. Posadas EM, Davidson B, Kohn EC. Proteomics and ovarian cancer: Implications for diagnosis and treatment: A critical review of the recent literature. *Curr Opin Oncol.* 2004;16:478–484.
49. Krause DS, Van Etten RA. Tyrosine kinases as targets for cancer therapy. *N Engl J Med.* 2005;353:172–187.
50. Araujo RP, Petricoin EF, Liotta LA. A mathematical model of combination therapy using the EGFR signaling network. *Biosystems.* 2005;80:57–69.
51. Araujo RP, Doran C, Liotta LA, et al. Network-targeted combination therapy: A new concept in cancer treatment. *Drug Disc Today.* 2004;1:425–433.
52. Arteaga CL, Baselga J. Clinical trial design and end points for epidermal growth factor receptor-targeted therapies: Implications for drug development and practice. *Clin Cancer Res.* 2003; 9:1579–1589.
53. Gasparini G, Gion M. Molecular-targeted anticancer therapy: Challenges related to study design and choice of proper endpoints. *Cancer J Sci Am.* 2000;6:117–131.
54. Giaccone G. Epidermal growth factor receptor inhibitors in the treatment of non-small-cell lung cancer. *J Clin Oncol.* 2005;23: 3235–3242.
55. Lynch TJ, Bell DW, Sordella R, et al. Activating mutations in the epidermal growth factor receptor underlying responsiveness of non-small-cell lung cancer to gefitinib. *N Engl J Med.* 2004;350: 2129–2139.
56. Paez JG, Jänne PA, Lee JC, et al. EGFR mutations in lung cancer: Correlation with clinical response to gefitinib therapy. *Science.* 2004;304:1497–1500.
57. Fukuoka M, Yano S, Giaccone G, et al. Multi-institutional randomized phase II trial of gefitinib for previously treated patients with advanced non-small-cell lung cancer. *J Clin Oncol.* 2003;21: 2237–2246.
58. Kris MG, Natale RB, Herbst RS, et al. Efficacy of gefitinib, an inhibitor of the epidermal growth factor receptor tyrosine kinase, in symptomatic patients with non-small cell lung cancer. A randomized trial. *JAMA.* 2003;290:2149–2158.
59. Bailey R, Kris M, Wolf M, et al. Gefitinib, ("Iressa", ZD1839) monotherapy for pretreated advance non-small cell lung cancer in IDEAL 1 and 2: Tumor response is not clinically relevantly predictable from tumor EGFR membrane staining alone. *Lung Cancer* 2003;41:S71.
60. Cappuzzo F, Magrini E, Ceresoli GL, et al. Akt phosphorylation and gefitinib efficacy in patients with advance non-small-cell lung cancer. *J Natl Cancer Inst.* 2004;96:1133–1141.
61. Gillespie JW, Best CJ, Bichsel VE, et al. Evaluation of non-formalin tissue fixation for molecular profiling studies. *Am J Pathol* 2002;160:449–457.
62. Mutter GL, Zahrieh D, Liu C, et al. Comparison of frozen and RNALater solid tissue storage methods for use in RNA expression microarrays. *BMC Genomics.* 2004;5:88.
63. Becker KF, Schott C, Hipp S, et al. Quantitative protein analysis from formalin-fixed tissues: Implications for translational clinical research and nanoscale molecular diagnosis. *J Pathol.* 2007;211: 370–378.
64. Espina V, Wulfkuhle J, Calvert VS, et al. Reverse phase protein microarrays for monitoring biological responses. In: Fisher P, ed. *Cancer Genomics and Proteomics: Methods and Protocols.* Totowa, NJ: Humana Press; 2008 [in press].
65. Khan IH, Mendoza S, Rhyne P, et al. Multiplex analysis of intracellular signaling pathways in lymphoid cells by microbead suspension arrays. *Mol Cell Proteomics.* 2006;5:758–768.
66. Stone RL, Dixon JE. Protein-tyrosine phosphatases. *J Biol Chem.* 1994;269:31323–31326.
67. Neel BG, Tonks NK. Protein tyrosine phosphatases in signal transduction. *Curr Opin Cell Biol.* 1997;9:193–204.
68. Petricoin EF, Ardekani AM, Hitt BA, et al. Use of proteomic patterns in serum to identify ovarian cancer. *Lancet.* 2002;359: 572–577.
69. Diamandis EP. Point: Proteomic patterns in biological fluids: Do they represent the future of cancer diagnostics? *Clin Chem.* 2003;49:1272–1275.

70. Liotta LA, Petricoin EF. Putting the "bio" back into biomarkers: Orienting proteomic discovery toward biology and away from the measurement platform. *Clin Chem*. 2008;54:3–5.
71. Liotta LA, Ferrari M, Petricoin E. Clinical proteomics: Written in blood. *Nature*. 2003;425:905.
72. Tirumalai RS, Chan KC, Prieto DA, et al. Characterization of the low molecular weight human serum proteome. *Mol Cell Proteomics*. 2003;2:1096–1103.
73. Lowenthal MS, Mehta AI, Frogale K, et al. Analysis of albumin-associated peptides and proteins from ovarian cancer patients. *Clin Chem*. 2005;51:1933–1945.
74. Zhou M, Lucas DA, Chan KC, et al. An investigation into the human serum "interactome." *Electrophoresis*. 2004;25:1289–1298.
75. Lopez MF, Mikulskis A, Kuzdzal S, et al. High-resolution serum proteomic profiling of Alzheimer disease samples reveals disease-specific, carrier-protein-bound mass signatures. *Clin Chem*. 2005;51:1946–1954.
76. Brouwers FM, Petricoin 3rd EF, Ksinantova L, et al. Low molecular weight proteomic information distinguishes metastatic from benign pheochromocytoma. *Endocr Relat Cancer*. 2005;12: 263–272.
77. Villanueva J, Shaffer DR, Philip J, et al. Differential exoprotease activities confer tumor-specific serum peptidome patterns. *J Clin Invest*. 2006;116:271–284.
78. Schulz-Knappe P, Schrader M, Zucht HD. The peptidomics concept. *Comb Chem High Throughput Screen*. 2005;8:697–704.
79. Culp WD, Neal R, Massey R, et al. Proteomic analysis of tumor establishment and growth in the B16–F10 mouse melanoma model. *J Proteome Res*. 2006;5:1332–1343.
80. Liotta LA, Kohn ED. The microenvironment of the tumour-host interface. *Nature*. 2001;411:375–379.
81. Jodele S, Blavier L, Yoon JM, et al. Modifying the soil to affect the seed: Role of stromal-derived matrix metalloproteinases in cancer progression. *Cancer Metastasis Rev*. 2006;25:35–43.
82. Hagendoorn J, Tong R, Fukumura D, et al. Onset of abnormal blood and lymphatic vessel function and interstitial hypertension in early stages of carcinogenesis. *Cancer Res*. 2006;66:3360–3364.
83. Zhang Z, Bast Jr RC, Yu Y, et al. Three biomarkers identified from serum proteomic analysis for the detection of early stage ovarian cancer. *Cancer Res*. 2004;64:5882–5890.
84. Traub F, Jost M, Hess R, et al. Peptidomic analysis of breast cancer reveals a putative surrogate marker for estrogen receptor-negative carcinomas. *Lab Invest*. 2006;86:246–253.
85. Skates SJ, Horick N, Yu Y, et al. Preoperative sensitivity and specificity for early-stage ovarian cancer when combining cancer antigen CA-125II, CA 15–3, CA 72–4, and macrophage colony-stimulating factor using mixtures of multivariate normal distributions. *J Clin Oncol*. 2004;22:4059–4066.
86. Anderson NL, Anderson NG. The human plasma proteome: History, character, and diagnostic prospects. *Mol Cell Proteomics*. 2002;1:8458–8467.
87. Deutsch EW, Eng JK, Zhang H, et al. Human plasma peptide-atlas. *Proteomics*. 2005;5:3497–3500.
88. Rai AJ, Stemmer PM, Zhang Z, et al. Analysis of Human Proteome Organization Plasma Proteome Project (HUPO PPP) reference specimens using surface enhanced laser desorption/ionization-time of flight (SELDI-TOF) mass spectrometry: Multi-institution correlation of spectra and identification of biomarkers. *Proteomics*. 2005;5:3467–3474.
89. Hortin GL. The MALDI TOF mass spectrometric view of the plasma proteome and peptidome. *Clin Chem*. 2006;52:1223–1237.
90. Omenn GS, States DJ, Adamski M, et al. Overview of the HUPO Plasma Proteome Project: Results from the pilot phase with 35 collaborating laboratories and multiple analytical groups, generating a core dataset of 3020 proteins and a publicly-available database. *Proteomics*. 2005;5:3226–3245.
91. Shen Y, Kim J, Strittmatter EF, et al. Characterization of the human blood plasma proteome. *Proteomics*. 2005;5:4034–4045.
92. Mehta AI, Ross S, Lowenthal MS, et al. Biomarker amplification by serum carrier protein binding. *Dis Markers*. 2003–2004;19: 1–10.
93. Govorukhina NI, Reijmers TH, Nyangoma SO, et al. Analysis of human serum by liquid chromatography-mass spectrometry: Improved sample preparation and data analysis. *J Chromatogr A*. 2006;1120:142–150.
94. Drake RR, Schwegler EE, Malik G, et al. Lectin capture strategies combined with mass spectrometry for the discovery of serum glycoprotein biomarkers. *Mol Cell Proteomics*. 2006;5: 1957–1967.
95. Gaspari M, Ming-Cheng Cheng M, Terracciano R, et al. Nanoporous surfaces as harvesting agents for mass spectrometric analysis of peptides in human plasma. *J Proteome Res*. 2006;5: 1261–1266.
96. Terracciano R, Gaspari M, Testa F, et al. Selective binding and enrichment for low-molecular weight biomarker molecules in human plasma after exposure to nanoporous silica particles. *Proteomics*. 2006;6:3243–3250.
97. Yocum AK, Yu K, Oe T, et al. Effect of immunoaffinity depletion of human serum during proteomic investigations. *J Proteome Res*. 2005;4:1722–1731.
98. Luchini A, Geho DH, Bishop B, et al. Smart hydrogel particles: Biomarker harvesting: One-step affinity purification, size exclusion, and protection against degradation. *NanoLetters*. 2008;8: 350–361.
99. Ruoslahti E. Specialization of tumour vasculature. *Nat Rev Cancer*. 2002;2:83–90.
100. Ruoslahti E. Targeting tumor vasculature with homing peptides from phage display. *Semin Cancer Biol*. 2000;10:435–442.
101. Akerman ME, Chan WC, Laakkonen P, et al. Nanocrystal targeting in vivo. *Proc Natl Acad Sci USA*. 2002;99: 12617–12621.

Chapter

10

Integrative Systems Biology: Implications for the Understanding of Human Disease

M. Michael Barmada ▪ David C. Whitcomb

"What is now proved was once only imagined."
William Blake

"All models are wrong, but some are useful."
George Box [1]

INTRODUCTION

From the smallest collections of elementary particles to the largest collections of galaxies, all matter is constructed of interacting collections of elements, or networks. Interactions within and between these networks give rise to the observable properties of matter. Biological networks are no exception, from the level of interacting enzymes and substrates which form a pathway, up to the level of the biosphere, incorporating all living things and their relationships. Networks have recently generated considerable interest in biological sciences, not because of any change in our basic knowledge about them, but because of key technological changes that have enhanced our ability to interrogate them and thereby develop descriptive models allowing prediction of their behavior. This shift in ability has fueled new perspectives on how to apply the scientific method to biomedical science, emphasizing integration instead of reduction. Systems (or network) biology focuses on the investigation of complex interactions in biological systems. Much of the promise of systems biology in biomedicine revolves around its potential to explain disease as a perturbation of a normal system. Many disease syndromes include elements from multiple tissue-specific and systemic systems. For example, chronic pancreatitis is defined by variable amounts of maldigestion (loss of pancreatic acinar cell function), diminished bicarbonate secretion (loss of duct cell function), diabetes mellitus (loss of islet cell function), inflammation (immune system), fibrosis (regenerative systems), pain (nervous system), and cancer risk (DNA repair and cell cycle regulation systems). Since genetic and environmental factors affect each of these systems in different ways, it is not surprising that dysfunction of these systems in response to stress or injury will be variable between patients. However, it is also clear that each of these systems interacts with each other. Furthermore, for the systemic systems the commonly dysfunctional components of effector or regulatory pathways will also be dysfunctional in parallel disorders of other tissues or organs. Careful modeling of these different relationships within the phenome (the set of all phenotypes of an organism) or within the interactome (the set of all interactions of an organism) can greatly increase the power to identify functional genomic variation important in disease phenotypes and lead to a better understanding of how to readjust the system to return to a normal state.

At its most basic, systems biology comprises three challenges: (i) how to generate a sufficient quantity of data to analyze variability in networks, (ii) how to properly integrate data from multiple disparate sources into a usable corpus of knowledge, and (iii) how to use that corpus of knowledge to model a component system and optimize it. Model construction allows predictions of the behavior of a system and, ultimately, an understanding of how to change that behavior in predictable ways. The promise of this approach

Molecular Pathology © 2009, Elsevier, Inc. All Rights Reserved.

is vast because the exact definition of a system is not fixed—anything from a particular biochemical pathway up to the level of an entire organism can be considered a system in biomedical applications. In this regard, population-level biosciences, such as epidemiology and population genetics, have long practiced a form of systems biology, but their focus has always been on a collection of individuals as the system of interest. Current systems biology approaches model systems of interest to clinical outcomes, such as metabolic pathways, individual cells, and organs. Modeling at this level has the appeal of allowing translation of results from decades of population-based laboratory research to be applied to individualized clinical medicine.

The field of systems biology borrows from control theory, which deals with the behavior of dynamic systems (systems that fluctuate in a dependent fashion). Control theory proposes the idea that systems can best be modeled as a cycle of controllers which modify inputs to produce the desired outputs, and sensors which provide feedback to the system, producing a steady-state system in which equilibrium is achieved by balancing the input and output concentrations and the feedback signals. In a similar fashion, systems biology includes the definition and measurement of the components of a system, formulation of a model, and the systematic perturbation (either genetically or environmentally) of and remeasurement of the system. The experimentally observed responses are then compared with those predicted by the model, and new perturbation experiments are designed and performed to distinguish between multiple or competing models. This cycle of test-model-retest is repeated until the final model predicts the reaction of the system under a broad range of perturbations.

Systems Biology as a Paradigm Shift

Although systems biology is itself a new discipline, the study of systems in biology is not new. Early studies of enzyme kinetics employed a similar cycle of testing and modeling, followed by retesting of new hypotheses. In a similar fashion, modeling of neurophysiological processes like the propagation of action potentials along a neuron are illustrative of early forays into mathematical modeling of cellular processes. However, like all next-generation methodological shifts, systems biology involves a rethinking of basic principles—a return to a more basic understanding or a more simplistic framework in which to generate hypotheses. Traditional science follows a bottom-up paradigm—that is, break a problem down into its individual components (reductionism), learn everything there is to know about those individual components, and then integrate the information together to get information about a system. Systems biology can work within this paradigm, creating models for individual components, and then integrating the models (creating, in essence, models of models, to explain the behavior of the original system). However, this approach requires that knowledge of the individual components is sufficient to explain the behavior of the system. Apart from a few exceptions, current reductionist research models have not been successful at explaining the large-scale behavior of biologic systems. To this end, systems biology endorses a more top-down approach, allowing modeling to occur at a higher level of organization (not at the level of the components, but at the level of the organ or the individual). In this point of view, a better understanding of the system as a whole can create a better understanding of the components. "Systems biology ... is about putting together rather than taking apart, integration rather than reduction. It requires that we develop ways of thinking about integration that are as rigorous as our reductionist programmes, but different" [2]. Or, put more simply [3]: "You [can] study each part of a Boeing 777 aircraft and describe how it functions, but that still wouldn't tell you how the airplane flies." This has led to the recognition that new paradigms (integration rather than reduction) may be necessary to understand and model systems with multiple interacting quantitative components.

DATA GENERATION

Central to the success of any modeling endeavor is the generation of large quantities of data. Because of the increased interest in systems biology, the theme of the late 1990s and early 21st century reflected exactly this necessity, exemplified by the fact that several papers were published on methods of dealing with the so-called "data deluge" [4–6]. Genomic technologies were the first to enter the high-throughput realm and subsequently the first to produce large quantities of high-quality data. Coupled with a concurrent explosion in computing power, high-throughput data generation made realistic modeling feasible. New technologies produced during this time vastly increased the ability to query an entire system at once and led to the advent of the major international efforts of the early 21st century, such as the Human Genome Project (the effort to sequence the entire complement of human chromosomes) [27]. The HapMap Project (the effort to catalog common genetic variation across multiple human ethnic populations) [7], the ENCODE Project (basically a merger of the Human Genome and HapMap projects—an effort to resequence particular portions of the human genome in multiple individuals from different ethnic populations to explore common and rare genetic variation at a higher resolution) [8], and the 1000 Genomes Project (an effort to completely sequence 1000 human genomes) [28]. Specifically, microarray technology coupled with advances in mass spectroscopy, combinatorial chemistry, and robotics created an explosion of data and unique visualizations of cellular processes. Capitalizing on this explosion of data, the first quantitative model of the metabolism of a whole (hypothetical) cell was published in 1997 [9].

Microarrays

Microarray technology grew out of a confluence of trends and technologies. On the experimental level, microarrays are most similar to Southern blotting, where fragmented DNA is attached to a substrate and then probed (hybridized) with a known gene or fragment to identify complementary sequences. The trend through the 1990s was to increase capacity in individual experiments, allowing the analysis of more and more samples (or more and more markers) in a single experiment. With the advent of paper-blotting techniques (allowing DNA molecules to be immobilized or spotted on special paper, then hybridized), and subsequently glass-blotting techniques, coupled with advances in robotic pipetting (allowing smaller and smaller volumes of liquid to be spotted, in greater and greater densities), microarrays became feasible. Initial microarray experiments [10] began with small numbers of immobilized probes, owing mostly to limited knowledge about the genes that make up a genome. This was quickly followed by arrays with hundreds, thousands, tens of thousands, and now (currently) millions of probes as our understanding of the components of genomes grew (owing, in large part, to early efforts to identify and catalog all expressed genes in organisms and to large-scale efforts like the Human Genome and ENCODE projects).

Microarrays have now found use in multiple experimental venues. Most commonly, the molecules being immobilized on the array are DNA molecules, allowing measurement of expression levels or detection of polymorphisms (such as single-nucleotide polymorphisms or SNPs). A DNA microarray consists of thousands of microscopic spots (or features), each containing a specific DNA sequence. Each feature is typically labeled with a fluorescent tag and then used as probes in hybridization experiments with a DNA or RNA sample (the target), which is labeled with a complementary fluorescent label. Hybridization is quantified by fluorescence-based scanning of the array, allowing determination of the relative abundance of nucleic acid sequences in the target by characterization of the relative abundance of each fluorophore (Figure 10.1).

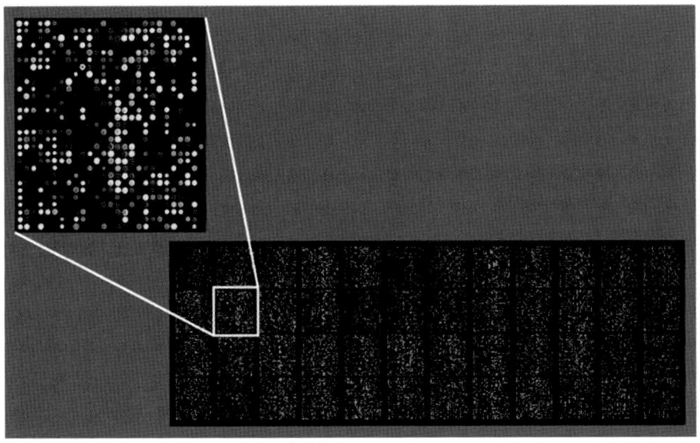

Figure 10.1 Example of an approximately 40,000 probe spotted oligonucleotide microarray with enlarged inset to show detail (Image from Wikimedia Commons).

Transcriptomics

Transcriptomics is the study of relative RNA transcript abundances, using microarray technologies. Chips that are specialized for this purpose are known as RNA microarrays and are typically prepared with a library of transcripts of known origin (representative tags for a known complement of genes, for example—the current human RNA arrays contain approximately 60,000 probes, representative of a majority of known RNA species from the 20,000 or so human genes). These are then interrogated with RNA (typically reverse transcribed into cDNA) from two different samples, labeled with different dyes (commonly green and red dyes). This allows the relative abundance (in one sample versus the other) of each RNA transcript to be assessed. Experiments of this type are routinely used to determine which RNAs are upregulated or downregulated in a disease sample versus a normal sample. However, note that the utility of the information generated from transcriptomics studies is highly variable, as transcription levels are potentially influenced by a wide range of factors, including disease phenotypes. To be of general use, transcriptomic data must be generated under a wide variety of conditions and compared, to eliminate the trivial sources of variation.

Genotyping

One of the earliest explosions of data on a whole-genome scale came from early genetic linkage studies. The first whole-genome genetic linkage scans were performed in the early 1990s using panels of ~350 genetic markers scattered throughout the autosomal chromosomes, as well as the X-chromosome [11]. Subsequent advances in mapping techniques and marker discovery increased the number of markers in the genetic maps, and included both sex chromosomes, as well as the mitochondrial genome. Markers for whole-genome genotyping have followed a rather amusing pattern of oscillation between the use of simple polymorphisms (with two alleles) and complex polymorphisms (with multiple alleles). The earliest genetic markers to be used (easily observable phenotypes such as eye color, sex, handedness, and others) were generally treated as simple binary traits. The first protein biomarkers (blood group protein polymorphisms and HLA polymorphisms) were complex, having many alleles with complicated ethnic and regional variations in frequency. Restriction fragment length polymorphisms (or RFLPs) having only two alleles indicating the presence or absence of a recognition site for restriction enzymes were the next marker type to be in vogue, and were the first to be suggested as usable for whole-genome analysis because of their frequency throughout the genome. This was followed by the discovery of DNA-length polymorphisms, in the form of tandem repeated regions (called Variable Numbers of Tandem Repeats or VNTRs), which are complex polymorphisms. Recognition of the existence of tandem repeats led to the discovery of microsatellite markers—basically short tandem repeats (STRs), on the order of 2–5 base pairs repeated several times. The frequency of STRs in the genome enabled generation of the first whole-genome maps. This was followed in quick succession by the discovery of single nucleotide polymorphisms (or SNPs), which initially were described as having only two alleles, and so continue in use as simple markers (though in fact many SNPs have >2 alleles). Due to their abundance (estimated at 1 SNP per 100 base pairs of DNA) and the relative ease of genotyping these markers, the development of DNA microarray-based genotyping technologies for SNPs occurred quickly. SNPs are now the current standard for genome-wide genetic studies. When SNP content on genotyping arrays became sufficiently dense (and also due to findings from other genomic technologies like array-based comparative genomic hybridization (arrayCGH) techniques), the widespread occurrence of copy number variations (deletions and duplications) was detected, heralding a return to more complex (multiallelic) markers again. Current genotyping arrays now contain a mixture of simple polymorphic markers (SNPs) and so-called copy-number probes, allowing the assessment of copy-number changes across the genome, at a density which provides excellent coverage for most ethnic groups (~1,000,000 SNPs and nearly that many copy number probes, meaning each array carries almost 2 million features). It is expected that chip densities will continue to grow, allowing for higher-throughput genotyping. However, at the same time rapid low-cost sequencing technologies are becoming available that may well allow for affordable whole-genome sequence-based assessment of genomic variation, replacing any need for increased array densities.

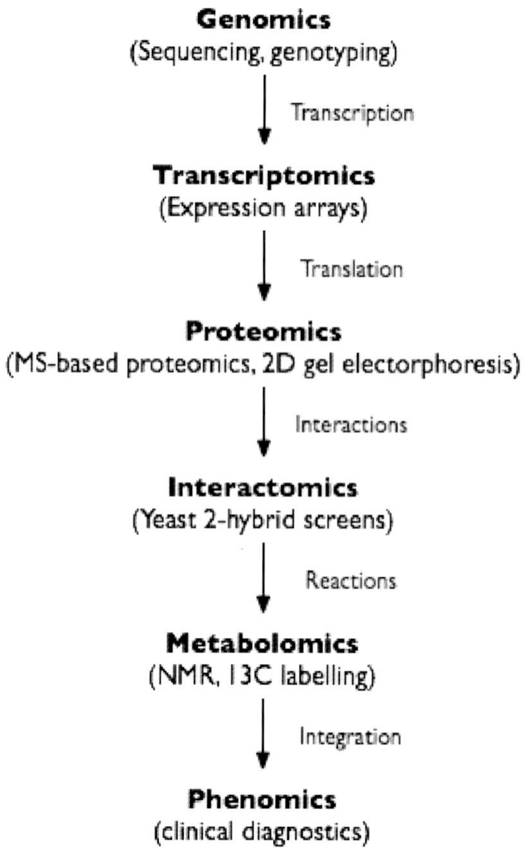

Figure 10.2 Omics technologies gather data on numerous levels. Various "omics" fields are displayed here along with the laboratory techniques used to generate the data, as well as the relationship of data from one level to another. Adapted from [25].

Other Omic Disciplines

Recognizing that many properties of biological systems are emergent, or a result of complex interactions between components that are not understandable at the level of individual system components, scientists in the 21st century turned to analysis of different levels of organization in an organism. Naturally, given the penchant of science for fancy names, each level of organization had to be given an appropriate name. For example, since the genome is the entire set of genes of an organism, genomics is the name given to the study of the whole set of genes of a biological system. Various methods, such as genotyping, sequencing (examining the linear sequence of DNA) and transcriptomics (examining the expression of all genes in an organism) can be used to interrogate the genome. Likewise, given that the proteome is the collection of proteins expressed in a system, proteomics is the name given to the study of the proteome. Proteomic methods include traditional techniques such as mass spectroscopy—identifying compounds based on the mass-charge ratio of ionized particles. But just as microarrays are a modification of traditional genomic techniques for high-throughput experiments, proteomics modifies the techniques of mass spectroscopy in a similar fashion—creating experimental and informatics pipelines that allow a sample (like a blood draw) to be partitioned into various fractions, have those fractions examined via mass spectroscopy, and have the results of the mass spectroscopy experiments compared to other mass spectroscopy profiles to allow for identification of individual components.

Other omic disciplines are similarly named—the general rule being that the respective discipline arises from the study of the associated "ome" (or set). This gives rise to disciplines such as (i) metabolomics—the study of the entire range of metabolites taking part in a biological process; (ii) interactomics—the study of the complete set of interactions between proteins or between these and other molecules; (iii) localizomics—the study of the localization of transcripts, proteins, and other molecules; and (iv) phenomics—the study of the complete set of phenotypes of a given organism. These various "omes" can be organized hierarchically, as demonstrated in Figure 10.2, based on the relationships recognized from years of biological experimentation. As with genomics and proteomics, experimental procedures for these other "omics" disciplines are adaptations of traditional methods (such as microscopy for localizomics, or single metabolite measurements for metabolomics) for high-throughput protocols.

DATA INTEGRATION

While data generation is certainly of paramount importance for systems biology, because the technologies that generate the data have their own unique (and often proprietary) formats for storing the data, systems biologists must also concern themselves with integrating data that span multiple resources. Since 1998, the number of recognized online databases related to biological information has increased 10-fold (from just under 100 in 1998 to over 1000 in 2008) [12]. Resources like the Bioinformatics Links Directory (http://bioinformatics.ca/links_directory/) extend this further by collecting links to molecular resources and tools as well as databases, and currently list 2300 links. However, the problem is more than just quantity. Issues of data quality, lack of standards, lack of interfaces allowing for integration, and longevity (or lack thereof) continue to plague online biological data resources, and make cross-resource querying or integration difficult [12,13]. Nonetheless, tools continue to appear to assist in the integration of data from disparate sources [14]. Using technology adopted from the burgeoning Semantic Web (or Web 2.0) projects [15], biologists and bioinformaticists are able to capture instances of data sufficient for systems biology scale efforts.

Semantic Web Technologies

Many of the problems inherent to integration of biological data resources are similar to those being faced by the larger community of World Wide Web users. The Semantic Web is a vision for how to have computers infer information relating one web page (or element) to another [15]. It is an extension of the current web protocols (primarily the HyperText Transport Protocol or HTTP), which allows for meaning to be imbedded together with content, such that automated agents can make associations between data without needing user input. In essence, creating the Semantic Web involves recasting the information in the World Wide Web (which currently stores relationships in the form of hyperlinks, which can link anything to anything, and so do not represent information content or meaning) into a format which allows relationships to be represented. The eXtensible Markup Language (or XML) is an early example of this type of recasting—allowing content to be stored with representative tags that describe the content. Resource Description Framework (or RDF) builds on XML by using triples (subject-predicate-object) to represent the information in XML tags (or in hyperlinks), and defining a standard set (or schema) of RDF triples to describe a particular object. Each part of a triple names a resource using either a Uniform Resource Identifier (URI) or Uniform Resource Locator (URL), or a literal. The advantage of this format is that RDF schemas are predefined so that meaning can be imbedded in the definition, or by using a hierarchy or ontology, describing the relationship within and between schemas. Also, since the RDF triple can contain a location as well as attribute-value pair, the components of a schema do not need to be located in the same place. Thus, if we have a schema representing a web page in which the sections (header, footer, left panel, right panel, title, and others) are defined by RDF triples, these web pages can easily integrate data from multiple sources.

The representation of information by RDF allows the location of data and data resources to be independent of the location of user interfaces or analytic resources. In essence, this allows the integration of data from multiple repositories and the querying of the integrated data. Coupled with the concept of RDF schemas to accurately describe knowledge about objects and ontologies to organize the relationship of objects to one another, the use of RDF can solve several of the problems mentioned for data integration (namely, the lack of standards and the lack of interfaces for integration). Another current trend—the use of wikis (a website that allows collaborative editing of its content by its users)—addresses the problem of data quality by allowing users to mark data as reliable or not, or by allowing users to update out-of-date or incorrect data.

MODELING SYSTEMS

Once the appropriate types of data have been generated, and the various sources of data collected and integrated, the resulting information is turned into knowledge by interpreting what the data actually mean, and how they address questions that need to be answered. Data modeling is used to understand the relationships important in defining the system. As noted previously, systems biology draws heavily from control theory, which itself is derived from mathematical modeling of physical systems. Although the mathematical derivation of control problems is complex, the fundamental concepts governing the formulation of control models are more intuitive and relatively limited. Three basic concepts can be thought of as central to forming control models: (i) the need for control (regulation or feedback), (ii) the need for fluctuation, and (iii) the need for optimization. A simple control model is presented in Figure 10.3.

Key characteristics of this model include a controller (responsible for the conversion between input and output) and sensor (responsible for determining the degree and direction of the feedback)—each of which can be the result of a single or multiple elements. An initial model of this form can often be derived from an initial data generation step, which can measure a baseline set of conditions for the system and also provide an idea of the components of the system. For example, an expression network (groups of genes that are co-regulated in some fashion) that regulates a biochemical pathway can be thought of as a control model. A single transcriptomics experiment can give you information on genes that are potentially co-regulated (sets of genes that are overexpressed compared to control, and so which might be providing the function of a controller), as well as information about potential sensors (genes that differ in response—one being overexpressed, the other under-expressed). However, the problem with a single experiment (or a single snapshot of the transcriptome) is that the information derived is not sufficient to disentangle the true positives (elements that truly should be components of the model) from the false positives (random variation which transiently mimics the behavior

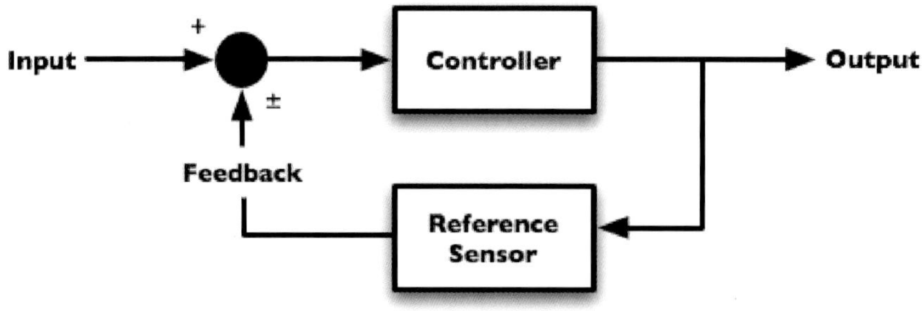

Figure 10.3 **Example of a control module.** This module represents a simple feedback loop, with output from the sensor either upregulating or downregulating the process that converts the input into output.

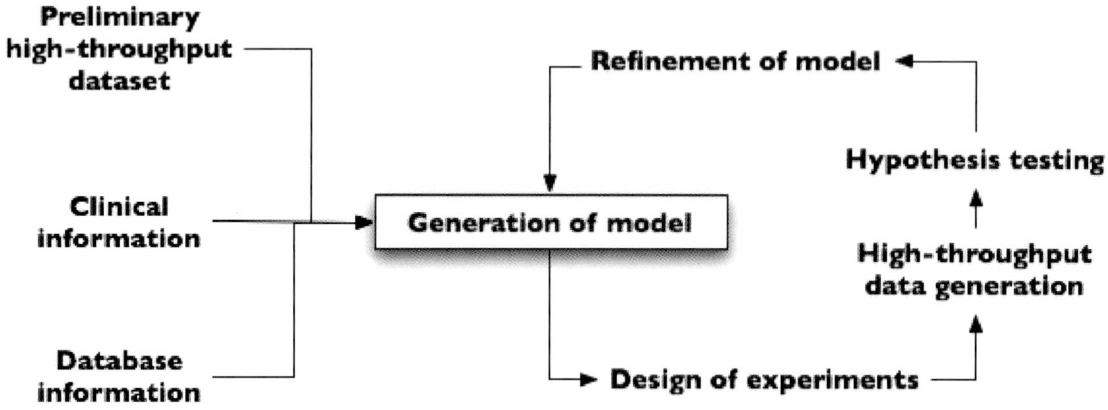

Figure 10.4 **The iterative nature of systems biology research.** Note that every refinement of the model needs additional data generation for retesting of specific hypotheses. Adopted from [26].

of the model). This reflects the need for fluctuation. By perturbing (changing an aspect of the system to produce a predictable outcome) and remeasuring the system, additional confidence regarding the true positives can be achieved assuming the new measurements reflect the predictions of the model. If not, another model which explains the previous data measurements is selected, and another test is devised. This cycle continues until the final model accurately represents the data measured in all tests. In this way, the most optimized model is selected (Figure 10.4).

Lastly, although not immediately obvious from the preceding discussions, data modeling also requires large amounts of computational power. Due to the size of the high-throughput data sets currently in use (expression data on 30,000 genes for multiple time points; SNP genotypes for millions of markers; occurrence of all known protein-protein interactions; computational prediction and annotation of all known protein-DNA binding sites; and others), the advanced mathematical and visualization frameworks currently in use, and the iterative nature of analysis, large amounts of computing power are necessary, often pushing the limits of even parallelized or grid-enabled computational clusters. Continued growth in computing power will be necessary to support the ongoing use of systems biology as the models in use become more and more elaborate and incorporate information from greater numbers of high-throughput methodologies.

IMPLICATIONS FOR UNDERSTANDING DISEASE

A new revolution in medicine was made possible at the end of the 20th century by the sequencing of the human genome. This remarkable feat led to the revelation that millions of variations exist in the DNA sequence of individual humans, with an infinite number of possible variations. Furthermore, genetic studies revealed that many chronic inflammatory diseases, such as Crohn's disease, ulcerative colitis, chronic pancreatitis, rheumatologic disease, heart diseases, and others do not have a single genetic factor causing the disease, but rather several dozen common genetic variations that, in the context of many possible interacting environmental factors, cause a disease that is defined by the location of inflammation and associated signs and symptoms. Unfortunately, the simplicity and early success of the germ theory of disease, in which a single pathological agent was responsible for a specific disease with characteristic signs and symptoms, lulled physicians into thinking that all disease would follow a similar pattern. This reductionist-like approach in medicine allowed physicians and scientists from different disciplines to triangulate on the same target using different methods and perspectives. It has identified numerous components of individual systems—many of which are reused over and over, leading to a modular view of system components—and has provided numerous insights into human disease. Despite these advances, the reductionist approach in medicine has been less successful in identifying the many complex interactions between disease components, and in explaining how system properties (like disease phenotypes) emerge—a concept which becomes important when considering manipulation of systems to produce predictable and desirable outcomes, as required for preventative medicine. As such, new paradigms are required, likewise requiring new approaches and new methods.

One such new paradigm is the advent of personalized medicine, which can be thought of as an application of systems biology-type thinking to medicine. Personalized medicine represents a transition from population-based thinking (describing risks on a population level) to an individual-based approach (describing risks for an individual based on his or her personal genomic/proteomic/metabolomic/phenomic profiles). Reductionist approaches have successfully identified many of the biomarkers required to define disease states in complex disorders on a population level (that is attributing population risks to various changes), but have not been able to describe how individual exposures or biomarkers (or combinations of these components) interact to create the disease state in individuals. To achieve this transition, systemic models which explain the function of systems (cells, organs, etc.) are needed, from which the implications of changes in particular exposures or biomarkers (genetic changes, proteomic changes, and others) can be understood on the system-level.

Redefining Human Diseases

The first major hurdle in transitioning from allopathic medicine to personalized medicine is to redefine common human diseases. In allopathic medicine, diseases are typically defined by characteristic signs and symptoms that occur together and meet accepted criteria. Chronic inflammatory diseases are defined by the location of the inflammation, the persistence of inflammation, and the presence of tissue destruction or scarring. These definitions indicate that the specific mechanism causing a specific organ to develop and to sustain an inflammatory reaction is not known. Furthermore, it has been impossible to identify the cause when using the traditional scientific methods used to identifying a single, causative infectious agent. Instead, the possibility that disease affecting one or more organs can occur through any one of multiple pathways, and that each pathway has multiple steps and regulatory components, and that multiple effects on multiple systems and multiple environmental exposures may be required before disease is manifest must be embraced to transition to personalized medicine. Indeed, physicians recognized that diseases located in specific organs share some common systemic features such as the type of inflammatory response (autoimmune or fibrosing), or chronic pain since they use a limited number of anti-inflammatory or pain medications for a variety of diseases. Additionally, studies which have looked at the clustering of disease phenotypes based on their representative genomics (associated gene/SNP polymorphisms or expression profiles) [16] have demonstrated that even disease phenotypes we once thought were roughly homogeneous (for instance, disease subtypes like Crohn's disease with ileal involvement in Caucasian-only populations) are in fact associated with different genetic loci (heterogeneous) [17]. Personalized medicine must deconvolute systemic and tissue-specific pathways and reconstruct them in a way that leads to precise, patient-specific treatments.

The Transition to Personalized Medicine

The effect of approaching complex inflammatory disorders as a single disease rather than as a complex process is illustrated through the comparison of multiple small studies by meta-analysis. While some of the variance in estimated effect sizes from small genetic studies can be attributed to random chance, the possibility also exists that populations from which the samples were taken were not equivalent, and that different etiologies will lead to the same end-stage signs and symptoms through different, parallel pathways. In cases where a candidate gene is critical to some pathologic pathway, but not others, the wide variance between small genetic studies may reflect the fraction of subjects that progress to disease through that specific gene-associated pathway.

Evidence of this effect was recently demonstrated in an evaluation of reports on the effect of the pancreatic secretory trypsin inhibitor gene (SPINK1) N34S polymorphism in chronic pancreatitis [18]. A model was developed to test the hypothesis that alcohol, which is known to be associated with chronic pancreatitis and fibrosis via the collagen-producing pancreatic stellate cell (PSC), drives pancreatic fibrosis through a recurrent trypsin activation pathway as seen in hereditary pancreatitis [19,20], or through a trypsin-independent pathway, as illustrated in Figure 10.5 [18].

We identified 24 separate genetic association studies of the effect of the SPINK N34S mutation on chronic pancreatitis with effect sizes (odds ratio, OR) reported to be between nonsignificant to ~80. Using meta-analysis, we determined an overall effect of the SPINK1 pN34S on risk of chronic pancreatitis to be high (OR 11.00; 95% CI: 7.59–15.93), but with significant heterogeneity. Subdividing the patients into four groups based in proximal etiological factors (alcohol, tropical region, family history, and idiopathic), we found that the effect of SPINK1 pN34S on alcohol (OR 4.98, 95% CI: 3.16–7.85) was significantly smaller than idiopathic chronic pancreatitis (OR 14.97, 95% CI: 9.09–24.67) or tropical chronic pancreatitis (OR 19.15, 95% CI: 8.83–41.56). Thus, we conclude that alcohol acts through Factor A-type pathway, while tropical chronic pancreatitis and idiopathic chronic pancreatitis act through trypsin-associated pathways. The fact that a higher percentage of alcoholic patients than the control populations had SPINK1 mutations may

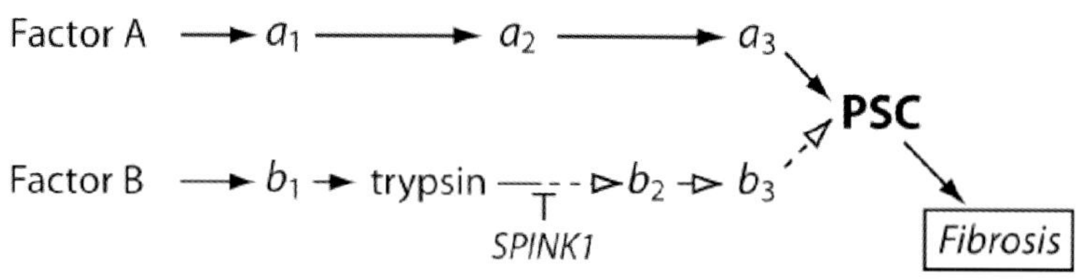

Figure 10.5 Hypothesis of etiology-defined pathways to pancreatic fibrosis. Hypothetical influence diagram illustrating pathologic pathways linking proximal factor (Factor A and B) to PSC (pancreatic stellate cell) and fibrosis through multiple steps (a1, a2, a3). Etiological factors of type B activate trypsinogen to trypsin, and therefore their pathologic pathway to the PSC can be interrupted by SPINK1. Etiological factors of type A are independent of trypsin, and therefore will not be influenced by variations in SPINK1 expression or function.

also mean that some idiopathic patients were misdiagnosed as being alcoholics since the threshold for alcohol-associated risk was previously unknown [21]. It could also mean that alcohol accelerates the progression from recurrent acute pancreatitis to chronic pancreatitis as demonstrated in animal models [22]. Thus, these results are important for understanding the etiology and progression of alcoholic chronic pancreatitis, but have broader implications for understanding the presence and effects of heterogeneity of pathways in complex disorders.

The chronic pancreatitis model also illustrates a major problem for transitioning to personalized medicine. Very large association studies require the recruitment of subjects from many large medical centers in different regions and different countries. This approach will tend to obscure mechanism-based heterogeneity and converge on only the most common features of the disease (population-level effects), even though other factors may have a stronger biological effect in a limited number of patients, while being irrelevant in others. This argues for more detailed phenotyping of study populations (phenomics) and careful analysis of context-dependent effects (interactomics). The application of systems biology to this data would allow for a better understanding of the various pathways leading to disease states, and so a better understanding of the possible interactions (or context-dependent effects) that accurately predict disease in a single individual.

Applications of Systems Biology to Medicine

The realization of personalized medicine requires not only a more systems-like perspective regarding risk factors, it also requires a rethinking of existing classifications, which are largely derived from observation and reductionist approach (common observable phenotypes should be the result of common underlying factors). As an example of this rethinking, a recent study undertook the reclassification of the disease phenome (the space of all associations between disease phenotypes) using the explosion of current information on genetic disease associations [23,24]. Disease-associated genes were clustered based on information about gene-gene or protein-protein interactions (from transcriptomics or interactomics studies), and superimposed with a representation of the organ system of the associated disease phenotype. The resulting network graph (Figure 10.6) demonstrates the association of different genes together in modules, many of which represent associations of proteins in macromolecular complexes (protein-protein interactions), or association of proteins in metabolic or regulatory pathways.

This view of the network of disease genes can at the same time inform us about novel associations we might not have been aware of (such as PAX6 in the ophthalmological cluster, or PTEN and KIT in the cancer cluster), and also direct us toward pathways (clusters) that might be of most benefit in understanding the network more completely (that is, those that are involved in multiple disease states or that have the most connections to them). This network-centric view of disease can also inform clinical medicine in terms of the choice of medications (indicating that medications used in one disease might also function appropriately in another, based on clustering of associated genetic factors).

DISCUSSION

We have outlined many of the reasons that a systems biology-based approach is needed for understanding complex disorders. Once the key components are organized into a logical series, then formal modeling of the disease process can be applied, tested, additional experimental data added, the system calibrated and retested. The optimal model or models are yet to be determined experimentally for any system, but with the continued rapid increase in high-throughput data generation and the continued increase in computational power, large-scale integrations and models can be achieved. The challenge will be to accurately anticipate the information necessary to be included in the model, as accurate assessment of the context is essential to the appropriate understanding of the effect of individual variation and the integration of that understanding into clinical practice. It is not enough to know that certain changes affect a phenotype (like disease risk) in a population, as these overall effects are typically very small (on the order of an odds ratio of 1.1 to 1.2). However, with a proper understanding of the context in which each variant is important, appropriate medical interventions can be implemented.

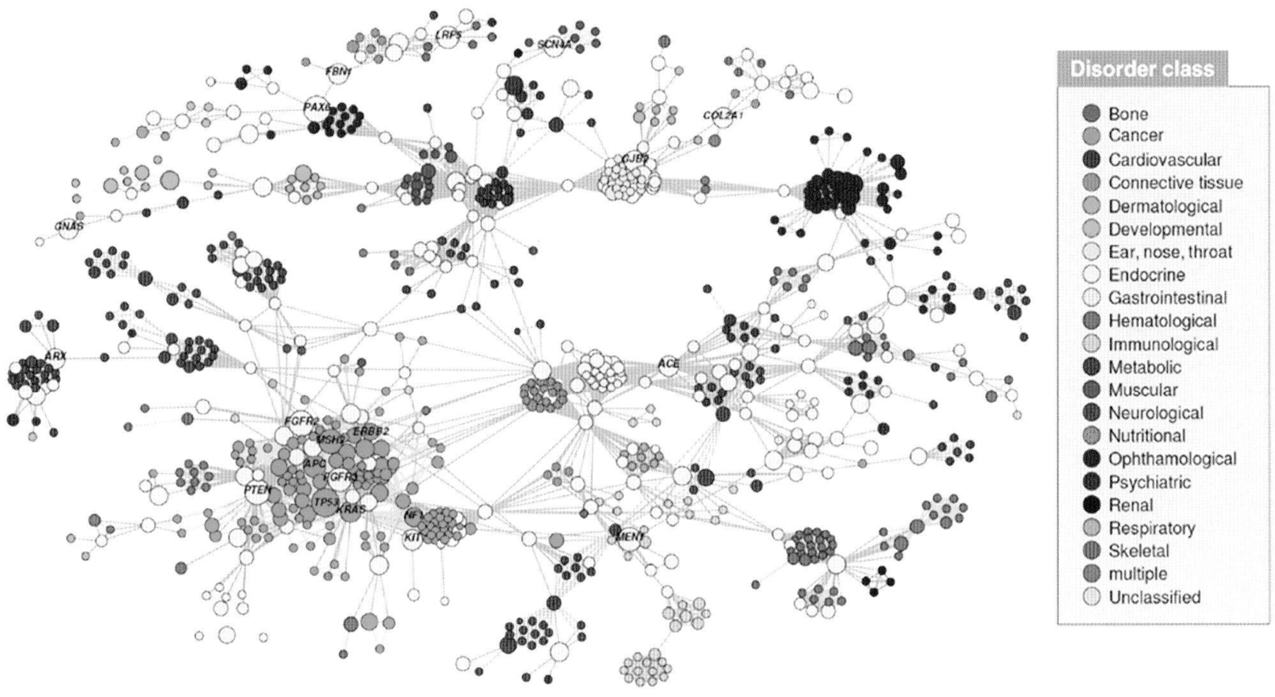

Figure 10.6 **Disease gene network.** Each node is a single gene, and any two genes are connected if implicated in the same disorder. In this network map, the size of each node is proportional to the number of specific disorders in which the gene is implicated. Reproduced with permission from [24].

REFERENCES

1. Box GEP, Draper NR. *Empirical Model-Building and Response Surfaces*. Wiley;1987;424.
2. Noble D. *The Music of Life: Biology Beyond the Genome*. Oxford: Oxford University Press;2006;21
3. UCSF School of Pharmacy—News Archive, *Systems Biology Reshapes Science*, March 9, 2004;http://pharmacy.ucsf.edu/news/2004/03/09/1/
4. Blake JA, Bult CJ. Beyond the data deluge: Data integration and bio-ontologies. *J Biomedical Informatics*. 2006;39:314–320.
5. Lanfear J. Dealing with the data deluge. *Nature Reviews Drug Discovery*. 2002;1:479.
6. Preston JD. Taking advantage of the data deluge. *Int J Prosthodon*. 1996;9:509.
7. The International HapMap Consortium. A haplotype map of the human genome. *Nature*. 2005;437:1299–1320.
8. ENCODE Project Consortium, Birney E, Stamatoyannopoulos JA, et al. Identification and analysis of functional elements in 1% of the human genome by the ENCODE pilot project. *Nature*. 2007;447:799–816.
9. Tomita M, Hashimoto K, Takahashi K, et al. E-CELL: Software environment for whole cell simulation. *Genome Informatics Workshop Genome Informatics*. 1997;8:147–155.
10. Schena M, Shalon D, Davis RW, et al. Quantitative monitoring of gene expression patterns with a complementary DNA microarray. *Science*. 1995;270:467–470.
11. Peltomaki P, Aaltonen LA, Sistonen P, et al. Genetic mapping of a locus predisposing to human colorectal cancer. *Science*. 1993;260:810–812.
12. Galperin MY. The Molecular Biology Database Collection: 2008 update. *Nucleic Acids Research*. 2007;36:D2–D4.
13. Merali Z, Giles J. Databases in peril. *Nature*. 2005;435:1010–1011.
14. Cohen-Boulakia S, Biton O, Davidson S, et al. BioGuideSRS: Querying multiple sources with a user-centric perspective. *Bioinformatics*. 2007;23:1301–1303.
15. Berners-Lee T, Hendler J, Lassila O. The Semantic Web: A new form of web content that is meaningful to computers will unleash a revolution of new possibilities. *Scientific American*. May 17, 2001;29–37.
16. Oti M, Brunner HG. The modular nature of genetic diseases. *Clinical Genetics*. 2007;71:1–11.
17. Rioux JD, Xavier RJ, Taylor KD, et al. Genome-wide association study identifies new susceptibility loci for Crohn disease and implicates autophagy in disease pathogenesis. *Nature Genet*. 2007;39:596–604.
18. Aoun E, Chang CC, Greer JB, et al. Pathways to injury in chronic pancreatitis: Decoding the role of the high-risk SPINK1 N34S haplotype using meta-analysis. *PLoS ONE*. 2008;3:e2003.
19. Whitcomb DC, Gorry MC, Preston RA, et al. Hereditary pancreatitis is caused by a mutation in the cationic trypsinogen gene. *Nature Genet*. 1996;14:141–145.
20. Gorry MC, Gabbaizedeh D, Furey W, et al. Mutations in the cationic trypsinogen gene are associated with recurrent acute and chronic pancreatitis. *Gastroenterology*. 1997;113:1063–1068.
21. Yadav D, Hawes RH, Brand RE. Alcohol Consumption, Cigarette Smoking and the Risk of Recurrent Acute and Chronic Pancreatitis. *Pancreatology* [in press].
22. Deng X, Wang L, Elm MS, et al. Chronic alcohol consumption accelerates fibrosis in response to cerulein-induced pancreatitis in rats. *Am J Pathol*. 2005;166:93–106.
23. Goh KI, Cusick ME, Valle D, et al. The human disease network. *Proceed Natl Acad Sci USA*. 2007;104: 8685–8690.
24. Loscalzo J, Kohane I, Barabási AL. Human disease classification in the postgenomic era: A complex systems approach to human pathobiology. *Mol Syst Biol*. 2007;3:124.
25. Fischer HP. Towards quantitative biology: Integration of biological information to elucidate disease pathways and drug discovery. *Biotechnol Annu Rev*. 2005;11:1–68.
26. Studer SM, Kaminski N. Towards systems biology of human pulmonary fibrosis. *Proceed Am Thorac Soc*. 2007;4:85–91.
27. Human Genome Project: http://www.ornl.gov/sci/techresources/Human_Genome/home.shtml
28. 1000 Genomes Project: http://www.1000genomes.org/page.php

Part

III

Principles and Practice of Molecular Pathology

Chapter

11

Pathology: The Clinical Description of Human Disease

William K. Funkhouser

"...*Future discoveries will not likely be made by morphologists ignorant of molecular biologic findings, or by biologists unaware or scornful of morphologic data, but by those willing and capable of integrating them through a team approach....*"

Rosai [1]

INTRODUCTION

This chapter will discuss the fundamental concepts, terminology, and practice of pathology as the discipline dedicated to the understanding of causes, mechanisms, and effects of diseases. A section on key terms, definitions, and concepts is followed by sections on historical human approaches to diseases, an overview of current diagnostic practice, and a vision for new interface with applied molecular biology.

TERMS, DEFINITIONS, AND CONCEPTS

Pathology (from the Greek word *pathología*, meaning the study of suffering) refers to the specialty of medical science concerned with the cause, development, structural/functional changes, and natural history associated with diseases. Disease refers to a definable deviation from a normal phenotype (observable characteristics due to genome and environment), evident via patient complaints (symptoms), and/or the measurements of a careful observer (signs). The cause of the disease is referred to as its etiology (from the Greek word meaning the study of cause). One disease entity can have more than one etiology, and one etiology can lead to more than one disease. Each disease entity develops through a series of mechanistic chemical and cellular steps. This stepwise process of disease development is referred to as its pathogenesis (from the Greek word meaning generation of suffering). Pathogenesis can refer to the changes in the structure or function of an organism at the gross/clinical level, and it can refer to the stepwise molecular abnormalities leading to changes in cellular and tissue function.

The presentation of a disease to a clinician is in the form of a human patient with variably specific complaints (symptoms), to which the examining physicians can add diagnostic sensitivity and specificity by making observations (screening for signs of diseases). These phenotypic (measurable characteristic) abnormalities reflect the interaction of the genotype (cytogenetic and nucleic acid sequence/expression) of the patient and his/her environment. Patient workup uses present illness history with reference to past medical history, review of other organ systems for other abnormalities, review of family history, physical examination, radiographic studies, clinical laboratory studies (for example, peripheral blood or CSF specimens), and anatomic pathology laboratory studies (for example, tissue biopsy or pleural fluid cytology specimens). As you will see from other chapters in this book, the ability to rapidly and inexpensively screen for chromosomal translocations, copy number variation, genetic variation, and abundance of mRNA and miRNA is adding substantial molecular correlative information to the workup of diseases.

The differential diagnosis represents the set of possible diagnoses that could account for symptoms and signs associated with the condition of the patient. The conclusion of the workup generally results in a specific diagnosis which meets a set of diagnostic criteria, and which explains the patient's symptoms and phenotypic abnormalities. Obviously, arrival at the correct diagnosis is a function of the examining physician and pathologist (fund of knowledge, experience, alertness), the prevalence of the disease in question in the particular patient (age, race, sex, site), and the sensitivity/specificity of the screening tests used (physical exam, vital signs, blood solutes, tissue

Molecular Pathology © 2009, Elsevier, Inc. All Rights Reserved.

stains, genetic assays). The pathologic diagnosis represents the best estimate currently possible of the disease entity affecting the patient, and is the basis for downstream follow-up and treatment decisions. The diagnosis implies a natural history (course of disease, including chronicity, functional impairment, survival) that most patients with this disease are expected to follow. Be aware that not all patients with a given disease will naturally follow the same disease course, so differences in patient outcome do not necessarily correspond to incorrect diagnosis. Variables that independently correlate with clinical outcome differences are called independent prognostic variables, and are routinely assessed in an effort to predict the natural history of the disease in the patient. It is also important to note that medical therapies for specific diseases do not always work. Variables that independently correlate with (predict) responses to therapy are called independent predictive variables.

Diagnosis of a disease and development of an effective therapy for that disease do not require knowledge of the underlying etiology or pathogenesis. For example, Wegener's granulomatosis was understood by morphology and outcomes to be a lethal disease without treatment, yet responsive to cyclophosphamide and corticosteroids, before it was found to be an autoimmune disease targeting neutrophil cytoplasmic protein PR3 (Figure 11.1). However, understanding the molecular and cellular pathogenesis of a disease allows development of screening methods to determine risk for clinically unaffected individuals, as well as mechanistic approaches to specific therapy.

The pathologist is that physician or clinical scientist who specializes in the art and science of medical risk estimation and disease diagnosis, using observations at the clinical, gross, body fluid, light microscopic, immunophenotypic, ultrastructural, cytogenetic, and molecular levels. Clearly, the pathologist has a duty to master any new concepts, factual knowledge, and technology that can aid in the estimation of risk for unaffected individuals, the statement of accurate and timely diagnosis, accurate prognosis, and accurate prediction of response to therapy for affected individuals.

A BRIEF HISTORY OF APPROACHES TO DISEASE

The ability of *H. sapiens* to adapt and thrive has been due in part to the ability of humans to remember the past, respect tradition, recognize the value of new observations, develop tools/symbols, manipulate the environment, anticipate the future, and role-specialize in a social structure. The history of human understanding of diseases has progressed at variable rates, depending on the good and bad aspects of these human characteristics.

Concepts and Practices Before the Scientific Revolution

Our understanding of ancient attitudes toward diseases is limited by the historical written record. Thus, the start point for written medical history corresponds to around 1700 BC for Mesopotamian rules in the code of Hammurabi, and around 1550 BC for the analogous Egyptian rules in the Ebers papyrus. By definition, these philosophers, theologians, and physicians had access and assets to allow a written record, and materials and storage sufficient for the written records to survive. The Mesopotamian records indicate a deity-driven and demon-driven theory and empirical practice by recognized professional physicians. In this context, the prevailing thought was that "Disease was caused by spirit invasion, sorcery, malice, or the breaking of taboos; sickness was both judgment and punishment" [2].

The Greek medical community evolved a theory of disease related to natural causes and effects, with less emphasis on deity-driven theory. The Hippocratic Corpus includes "On the Sacred Disease" (circa 400 BC), which rejected a divine origin for diseases, and postulated a natural rather than supernatural basis for disease etiology ("... nowise more divine nor more sacred than other diseases, but has a natural cause ... like other affections."). Aristotle (384–322 BC) wrote broadly on

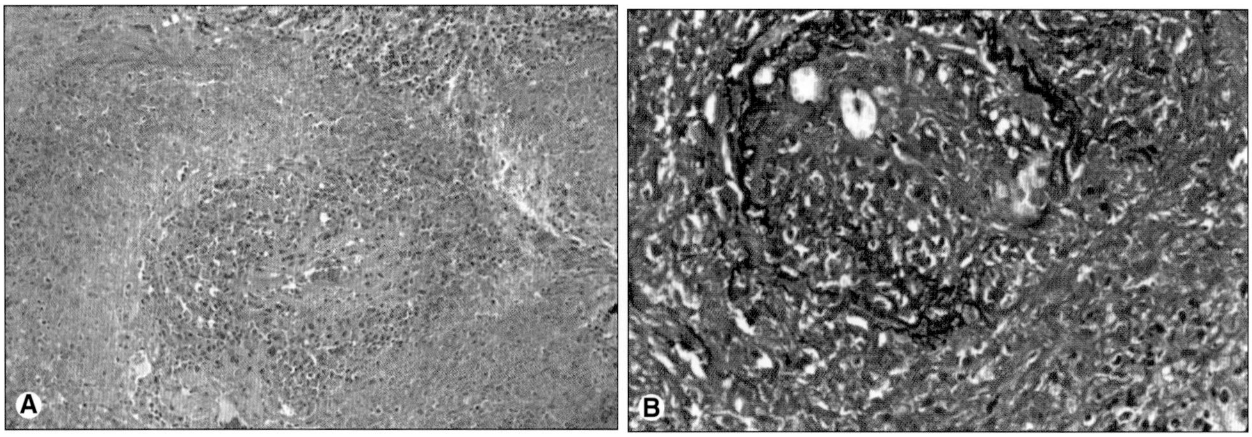

Figure 11.1 Wegener's granulomatosis of the lung. (A) H&E of Wegener's granulomatosis of lung. Necrosis, granulomatous inflammation, and vasculitis are identified. (B) Elastin stain of Wegener's granulomatosis of lung. Elastica disruption of the arterial wall supports a diagnosis of vasculitis.

topics including logic, biology, physics, metaphysics, and psychology. To Aristotle, observations led to a description of causes, or first principles, which in turn could be used logically in syllogisms to predict future observations. We would agree with these basic notions of induction and deduction. However, there was a different background philosophical construct regarding the nature of matter and causality (four elements, four humors, and four causes, including a final or teleologic purpose). We would recognize Aristotle's "efficient" cause of a disease as its etiology. Alexander the Great's conquest of Egypt in the 4th century BC led to Greek (Ptolemaic) leadership of Egypt from 305 BC to 30 BC, with development of the Alexandrian library and University. Faculty such as Euclid developed geometric models of vision (*Optica*), and Herophilus described human anatomy by direct dissection and observation (Greek medicine apparently allowed dissection in Alexandria, including vivisection of the condemned). During the Roman imperial era, Galen (129–207 AD) used dissection and observation of other animals such as the macaque (human dissection was illegal) to extrapolate to human anatomy and physiology. Like Aristotle's approach, Galen's approach to patients, diseases, and treatments was guided by philosophical constructs of four humors (blood, phlegm, yellow bile, and black bile) and the resulting temperaments (buoyant, sluggish, quick-tempered, and melancholic) due to humoral imbalances. It is thought that many of Galen's texts were destroyed with the Alexandrian library before the 7th century AD, but a subset was preserved and translated by Middle Eastern scholars. These ancient classic texts were then retranslated into Latin and Greek when printing houses developed in the 15th century (for instance, Hippocrates' *De Natura Hominis*, circa 1480 AD, and Galen's *Therapeutica*, circa 1500 AD).

The historical picture of the Greco-Roman understanding of disease is one of empirical approaches to diseases based on inaccurate understanding of anatomy, physiology, and organ/cellular pathology. Greek medicine became less superstitious and more natural cause-and-effect oriented, yet philosophy still trumped direct observation, such that evidence was constrained to fit the classic philosophical constructs. Some of the concepts sound familiar; for example, normal represents equilibrium and disease represents disequilibrium. However, we would differ on what variables are in disequilibrium (the historical humors, numbers, and opposites versus contemporary chemical and kinetic equilibria).

Following the collapse of the Western Roman Empire in 476 AD, the classic texts of Aristotle, Hippocrates, and Galen were protected, translated, and built upon in the Byzantine and Arab societies of the near East, and in Spain during the Muslim/Moorish period through the 11th century. During these middle ages for Western Europe outside Spain, there was apparently a retreat to pre-Hellenistic beliefs in supernatural forces that intervened in human affairs, with protecting saints and relics for disease prevention and therapy. Centers of medical learning following the Spanish Muslim model developed in Montpellier, France, and in Salerno, Italy, beginning in the 11th century.

The Scientific Revolution

Aristotle's concept of induction from particulars to general first principles, then use of syllogistic logic to predict particulars, evolved into the scientific method during the Renaissance. Ibn Alhazen's (al-Haitham's) work on the physics of optics in the 11th century challenged Euclid's concepts of vision from the Alexandrian era. (Euclid thought that the eye generated the image, rather than light reflected from the object being received by the eye). In the 13th century, Roger Bacon reinforced this use of observation, hypothesis, and experimentation. The printing press (Gutenberg, 1440 AD) allowed document standardization and reproduction, such that multiple parallel university and city libraries could afford to have similar collections of critical texts, facilitating scholarly publications in journals. Access to publications in libraries and universities led to a system of review, demonstration, discussion, and consensus regarding new scientific findings.

The concept of the body as an elegant machine was captured not only by 15th and 16th centuries Renaissance artists like da Vinci and Michelangelo, but also by anatomists and pathologists interested in the structure/function of health and disease. The ancient models of Aristotle and Galen had become sacrosanct, and newer, evidence-based models were considered heretical to some degree. So it was somewhat revolutionary when Vesalius dissected corpses, compared them with Galenic descriptions, and published *De humani corporis fabrica libri septem* (1543 AD; "Seven Books on the Structure of the Human Body"; the *Fabrica*), challenging and correcting 16th century understanding of normal human anatomy. Vesalius's successors (Colombo, Fallopius, and Eustachius) further improved the accuracy of human anatomic detail. Thus, correction of Galen's anatomical inaccuracies (including the rete mirabilis at the base of the brain, the five lobed liver, and curved humerus) required at least 13 centuries for challenge, scientific disproof, and eventual medical community acceptance.

The Scientific Revolution describes the progressive change in attitude of scientists and physicians toward understanding of the natural world, health, and disease. This revolution began circa 1543 AD, when Copernicus published arguments for a heliocentric universe and Vesalius published the *Fabrica* series on human anatomy. By the 17th century, Galileo, Kepler, Newton, Harvey, and others had used this observation-based, matter-based, and mathematical law-based perspective to develop a scientific approach similar to our own modern approach of testing hypotheses with experimental data and statistics. In human biology, the investigation of structure led to studies of function, initially of human cardiovascular physiology, for example, Harvey's *Anatomical Exercise Concerning the Motion of the Heart and Blood in Animals* (1628). Whereas Galen conceived of parallel but unconnected arteries and veins, with continuous blood production in the liver and continuous blood consumption in the periphery, Harvey demonstrated that blood was pumped by the heart through arteries, through tissue capillaries, to veins, and then back to the heart in a circle (circulation).

Correction of these and other Galenic physiological inaccuracies (such as nasal secretions representing the filtrate of cerebral ventricle fluid) thus required at least 14 centuries before challenge, scientific disproof, and eventual medical community acceptance.

The scientific method facilitates empirical, rational, and skeptical approaches to observational data, and minimizes human dependence on non-evidence-based traditional models. In spite of the scientific method, physicians are still human, and the medical community still shows an inertial reluctance to adapt to new information when it disrupts traditional paradigms. Recent examples would include reluctance to accept an etiologic role for the *H. pylori* bacterium in peptic ulcer disease [3], and reluctance to offer less than radical mastectomy for primary breast carcinoma.

Discovery of the Microscopic World

Before the use of lenses to magnify objects, it was physically impossible to make observations on objects smaller than the resolving limit of the human eye (about 0.1–0.25 mm). Thus, prokaryotic and eukaryotic cells, tissue architecture, and comparisons of normal and disease microanatomy were philosophical speculation until description of the mathematics of optics and lens design. In a real sense, optical technology was rate-limiting for the development of the fields of tissue anatomy, cellular biology, and microbiology. Concepts of optics were written as early as 300 BC in Alexandria (*Optica*, Euclid). Clear glass (crystallo) was developed in Venice in the 15th century. A compound microscope was invented by Janssen in 1590 AD. Microanatomy and structural terminology was begun by Malpighi (1661 AD), who examined capillaries in frog lung, trachea tubes for airflow in silkworms, and stomata in plant leaves. Robert Hooke used a compound microscope to describe common objects in *Micrographia* (1665 AD). Antony Van Leeuwenhoek used self-made simple magnifying lenses to count the threads in cloth in a Dutch dry-goods store, then later published descriptions of bacteria (termed *animalcules*), yeast, and algae, beginning in 1673 AD. Yet the relevance of these observations in microanatomy and microbiology to human diseases required changes in conceptual understanding of the etiology and pathogenesis of diseases. For example, it was 200 years after Van Leeuwenhoek that the common bacterium *Streptococcus* was recognized as the etiologic agent of puerperal fever in post-partum women.

Critical Developments During the 19th Century

Cellular Pathology, Germ Theory, and Infectious Etiologic Agents

The relevance of microanatomy and microbiology to human disease required expansion of conceptual understanding to include morphologic changes in diseased cells and tissues, as well as recognition of an etiologic role for microorganisms. Rokitansky's gross correlates with clinical disease (*A Manual of Pathological Anatomy*, 1846, discussed in [4]), Paget's surgical perspective on gross pathology [5], and Virchow's morphologic correlates with clinical disease [6] were critical to the development of clinico-pathologic correlation, and served to create a role for pathologists to specialize in autopsy and tissue diagnosis. Virchow's description of necrotizing granulomatous inflammation, the morphologic correlate of infections caused by mycobacteria such as TB and leprosy, preceded the discovery of the etiologic agents years later by Hansen (*M. leprae*, 1873 [7]) and Koch (*M. tuberculosis*, 1882 [8]), reviewed in [9].

The causal relationship between microorganisms and clinical disease required scientific demonstration and logical proof before medical community acceptance. For example, it took two centuries for the common bacterium *Streptococcus pyogenes* to be recognized as the etiologic agent of puerperal fever in post-partum women. Identification of this cause-and-effect relationship required an initial recognition of unusual clinical outcomes (clusters of post-partum deaths), then correlation of puerperal fever clusters with obstetrician habits [10,11], then Semmelweiss's experimental demonstration in 1847 that hand-washing reduced the incidence of puerperal fever [12], then the demonstration that particular bacteria (*Streptococci*) are regularly associated with the clinical disease (Koch, circa 1870), and finally by culture of the organism from the blood of patients with the disease (Pasteur, 1879).

During the 19th century, critical causal associations between microorganisms and infectious diseases were proven and accepted. Technical improvements in microscopes (Abbe condenser, apochromatic lenses, oil immersion lenses), the development of culture media, and the development of histochemical stains no doubt made it possible for Koch to identify *M. tuberculosis* in 1882 [8]. To be emphasized is the process of recognition, first of all of the variables associated with the clinical disease, then the scientific demonstration of a causal relationship between one or more of these variables with the clinical disease. This latter step was enunciated as Koch's postulates (1890): (i) the bacteria must be present in every case of the disease, (ii) the bacteria must be isolated from a diseased individual and grown in pure culture, (iii) the specific disease must be reproduced when a pure culture is inoculated into a healthy susceptible host, and (iv) the same bacteria must be recoverable from the experimentally infected host.

Organic Chemistry

Prior to 1828, organic (carbon-containing) compounds were thought to derive from living organisms, and it was thought that they could never be synthesized from nonliving (inorganic) material. This concept (termed vitalism) was disproven by the *in vitro* synthesis of urea by F. Wohler in 1828 [13]. Work from this era initiated the field of organic chemistry. Predictable rules for *in vitro* and *in vivo* organic reactions, structural theory, modeling, separation technologies, and accurate measurement subsequently allowed the chemical description of natural products, and the chemical synthesis of both natural products and synthetic compounds. In addition to setting the stage for a systematic understanding of cellular biochemistry and physiology, organic chemistry set the stage for laboratory synthesis of natural products, such as dyes, vitamins, hormones, proteins, and nucleic acids. In that era, the textile industry was the main consumer of dyes. Access to imported natural product dyes from plants was predictable for seafaring nations, but not for landlocked nations. In parallel with natural product extraction and purification was the recognition that aniline from coal (Perkin, 1856) could be modified to generate a spectrum of colors, catalyzing the development of the German dye industry in the last half of the 19th century. Some of these synthetic dyes were found useful for histochemical staining.

Histotechnology

Morphologic diagnosis requires thin (3–5 microns), contrast-rich (requiring dyes) sections of chemically fixed (cross-linked or precipitated) tissue. Thin sections allow the passage of light through the tissue, but reduce overlapping of cells in the light path. Thus, technologies had to develop for cutting and staining of thin fixed tissue sections (reviewed in [14]). Work leading to our current technique for tissue solidification in paraffin wax was first described by Klebs in 1869. Prototypes of our current mechanical microtome for making thin (~5 micron thick) tissue sections was developed by Minot in 1885. Precursor work leading to our current technique for tissue fixation with diluted formalin was first described by Blum in 1893.

Histochemical stains developed in parallel with dye technology for the textile industry. Botanists used carmine as a stain in 1849, and subsequently Gerlack applied carmine to stain brain tissue in 1858. The current hematoxylin dye used in tissue histochemistry was originally extracted from logwood trees from Central America for the dye industry (to compete with indigo). Metallic ion mordants made oxidized hematoxylin (hematein) colorfast in textiles, and a protocol for tissue staining was published by Boehmer in 1865. Similar to the hematoxylin story, semisynthetic analine dye technology was adapted by histochemists from 1850–1900. Many of these dyes are still routinely used for recognition of tissue structure, peripheral blood cells, and microorganisms, including hematoxylin, eosin, methylene blue, Ziehl-Neelsen, Gram, van Gieson, Mallory trichrome, and Congo red [14].

The most commonly used stains for general tissue diagnosis are the hematoxylin and eosin (H&E) stains, which provide a wealth of nuclear and cytoplasmic detail not visible in an unstained section. Supplemental histochemical stains demonstrate specific structures and organisms: collagen and muscle (trichome), elastin (Verhoff-von Giessen), glycogen/mucin (periodic acid-Schiff, PAS), mucin (PAS diastase, mucicarmine, alcian blue), fungi (Gemori methenamine silver, GMS), mycobacteria (Ziehl-Neelsen, Fite), and bacteria (Gram, Warthin-Starry). Each of these stains is inexpensive ($10–$50), fast (minutes to hours), and automatable, making them extremely valuable for service diagnostic pathology use.

Light microscopy lens technology matured during the last half of the 19th century. Critical were Abbe's introductions of the apochromatic lens system to eliminate chromatic aberration (different focal lengths for different wavelengths of visible light) in 1868, a novel condenser for compound microscopes (to provide better illumination at high magnification) in 1870, and an oil immersion lens (for 10 × objective magnification with visible light) in 1878.

By 1900, maturation of tissue fixation chemistry, histochemical stain protocols, and light microscope technology had evolved into the workhorse technique for evaluation of morphologic abnormalities in tissue examination in anatomic pathology labs, and for the evaluation of morphologic features of microorganisms in microbiology labs. The scope of this chapter is limited to pathology, but it should be clear to the reader that the momentous discoveries of deep general anesthetics (Long, 1841; Morton, 1846), commercial electricity (Edison, 1882), and radiography (Roentgen, 1895) also contributed to the development of the modern medical specialties of diagnostic pathology and laboratory medicine.

Developments During the 20th Century

Humoral and Cellular Immunology

The development of antisera in the 20th century for therapeutic purposes (for instance, treatment of diptheria) led to progressive understanding of the antibody, the efferent arm of the humoral immune response. Similarly, tissue transplantation experiments led to the recognition of cellular rejection [15] due to thymus-derived T-cells. Antibodies and T-cells cooperate to react to foreign (non-self) molecules, common examples being allergic responses, viral infections, and organ transplants. Antibodies were found to be made by B-cells and plasma cells, and were found to be exquisitely specific for binding to their particular antigens (ligands) either in solid phase or in solution. The analogous T-cell receptor (TCR) recognizes a ligand made up of a 15–20mer peptide presented by self MHC (HLA in human) molecules on the surface of antigen-presenting cells (macrophages, dendritic cells, activated B-cells). B- and T-cells activate and proliferate when exposed to non-self proteins, but not to self proteins, attesting to tolerance to self. When B- and T-cell self-tolerance breaks down, autoimmune diseases can result (including myasthenia gravis, Grave's disease, and lupus erythematosus). Antibodies (immunoglobulins) were found to be heterodimers of 50 kD heavy chains and 25 kD light chains, folded together so that a highly variable portion defines the antigen-binding site, and a constant portion defines the isotype (IgM, IgD, IgG, IgE, or IgA). Similarly, T-cell receptors were found to be heterodimers of immunoglobulin-like molecules with a highly variable portion for ligand binding, and a constant portion, but without different isotypes. The range of antigen-binding specificities in a normal mammal is extensive, perhaps infinite, subtracting out only those self proteins to which the animal is tolerant. The genes encoding immunoglobulins and T-cell receptors were sequenced and, surprisingly, the extensive variation of specificities was due to a unique system of gene rearrangements of polymorphic V, D, and J gene segments with random nucleotide addition at the junctions [16,17]. This system allows generation of literally billions of different Ig and TCR binding specificities.

Polyclonal antibodies raised in other species (goat, mouse, rabbit) against an antigen can be used to detect the antigen in diffusion gels (Ochterlony, western blot), in solution (ELISA), and in tissue sections. Use of fluorescent-tagged antibodies for frozen section immunohistochemistry was developed first [18], and immunofluorescence (IF) is still routinely used in renal pathology and dermatopathology. Peroxidase-tagged secondary antibodies and DAB chemistry were developed to generate a stable chromogen in the tissue, and this is now the primary method for detecting antigens in formalin-fixed tissue sections. Improved antibody binding specificity and industrial production required monoclonal antibodies, which required the development of mouse plasmacytomas/myelomas and cell fusion protocols [19]. The net result is that commercial antibodies are now available for antigens of both clinical and research interest, including antibodies specific enough to distinguish minimally modified variant antigens (for instance, phosphorylated versus dephosphorylated proteins).

Natural Product Chemistry and the Rise of Clinical Laboratories

Diseases due to dietary deficiencies (like scurvy) and hormonal imbalances (like diabetes mellitus) were described clinically long before they were understood pathologically. Dietary deficiency diseases prompted searches for the critical metabolic cofactors, so-called vital amines or vitamins. Xerophthalmia was linked to retinol (vitamin A) deficiency in 1917 (McCollum). Rickets was linked to calcitriol (vitamin D) deficiency in 1926. Beri-beri was linked to a deficiency of thiamine (vitamin B1) in 1926. Scurvy was linked to ascorbic acid (vitamin C) deficiency in 1927. Pellagra in the United States was linked to niacin deficiency in 1937. Pernicious anemia was linked to cobalamin (vitamin B12) deficiency in 1948. It is currently unusual to see morphologic features of these diseases in this country. Diseases due to nondietary physiologic imbalances prompted isolation of hormones. For example, hyperthyroidism and hypothyroidism were linked to thyroxine imbalances in 1915, and table salt was iodized starting in 1917. Diabetes mellitus was linked to insulin deficiency in 1921, nonhuman insulin was industrially purified and marketed soon thereafter, and recombinant human insulin was marketed in 1982.

These examples highlight the ability of 20th century chemists to fractionate, purify, synthesize, measure bioactivity, and manufacture these compounds for safe use by humans. Study of diseases due to deficiencies and excesses of single molecule function led to a mechanistic understanding of biochemistry and physiology, with resultant interconnected reaction pathways of byzantine complexity (now referred to as systems biology). Clinical demand for body fluid levels of ions (such as sodium, potassium chloride, and bicarbonate), glucose, creatinine, hormones (such as thyroxine and parathyroid hormone), albumin, enzymes (related to liver and cardiac function), and antibodies (reactive to ASO, Rh, ABO, and HLA antigens) led to the development of clinical laboratories in chemistry, endocrinology, immunopathology, and blood banking. Functional assays for coagulation cascade status were developed, as were methods for estimating blood cell concentration and differential, leading to coagulation and hematology laboratories. Serologic and cell activation assays to define HLA haplotype led to HLA laboratories screening donors and recipients in anticipation of bone marrow and solid organ transplants. Culture medium-based screening for infectious agents led to dedicated clinical microbiology laboratories, which are beginning to incorporate nucleic acid screening technologies for speciation and prediction of treatment response. The clinical laboratories now play a critical, specialized role in inpatient and outpatient management, and their high test volumes (a 700-bed hospital may perform 5 million tests per year) have catalyzed computer databases for central record-keeping of results.

Natural Product Chemistry: Nucleic Acids

The previous vignettes indicate a scientific approach to natural products of the steroid and protein types, but do not indicate how proteins are encoded, what accounts for variation in the same protein in the population, or how inherited diseases are inherited. It turns out that the instruction set for protein sequence is defined by DNA sequence. The role of nucleoproteins as a genetic substance was alluded to by Miescher in 1871, and was shown by Avery to be the pneumococcal transforming principle in 1944. The discovery of X-ray crystallography in 1912 made it possible for Franklin, Wilkins, and Gosling to study DNA crystal structure [20,21], and led to the description of the antiparallel double helix of DNA by Watson and Crick in 1953 [22]. This seminal event in history facilitated dissection of the instruction set for an organism, with recognition that 3-base codons specified amino acids in 1961 [23], and description of the particular codons encoding each amino acid in 1966 [24]. Demonstration of *in situ* hybridization in 1969 [25,26] made it possible to localize specific DNA or RNA sequences within the cells of interest. Recognition and purification of restriction endonucleases and DNA ligases, and the development of cloning vectors, made it possible to clone individual sequences, leading to methods for synthesis of natural products (such as recombinant human insulin in 1978 [27]). Chemical methods were developed for sequencing DNA [28,29], initially with radioisotope-tagged nucleotide detection in plate gels, then with fluorescent nucleotide detection in 1986 [30]. Subsequent conversion to capillary electrophoresis and computer scoring of sequence output allowed high throughput protocols which generated the human genome sequence by 2001 [31,32]. The development of polymerase chain reaction (PCR) [33] made it possible to quickly screen for length polymorphisms (identity testing, donor:recipient ratio after bone marrow transplant, and microsatellite instability) and sequence abnormalities (translocations, rearrangements, insertions, deletions, substitutions) in a targeted fashion. Quantitative PCR methods using fluorescent detection of amplicons made it possible to study DNA and RNA copy number, and to mimic Northern blots/oligo microarrays in estimating RNA transcript abundance for cluster analysis.

CURRENT PRACTICE OF PATHOLOGY

Diseases can be distinguished from each other based on differences at the molecular, cellular, tissue, fluid chemistry, and/or individual organism level. One hundred and fifty years of attention to the morphologic and clinical correlates of diseases has led to sets of diagnostic criteria for the recognized diseases, as well as a reproducible nomenclature for rapid description of the changes associated with newly discovered diseases. Sets of genotypic and phenotypic abnormalities in the patient are used to determine a diagnosis, which then implies a predictable natural history and can be used to optimize therapy by comparison of outcomes among similarly afflicted individuals. The disease diagnosis becomes the management variable in clinical medicine, and management of the clinical manifestations of diseases is the basis for day-to-day activities in clinics and hospitals nationwide. The pathologist is responsible for integration of the data obtained at the clinical, gross, morphologic, and molecular levels, and for issuing a clear and logical statement of diagnosis.

Clinically, diseases present to front-line physicians as patients with sets of signs and symptoms. Symptoms are the patient's complaints of perceived abnormalities. Signs are detected by examination of the patient. The clinical team, including the pathologist, will work up the patient based on the possible causes of the signs and symptoms (the differential diagnosis). Depending on the differential diagnosis, the workup typically involves history-taking, physical examination, radiographic examination, fluid tests (blood, urine, sputum, stool), and possibly tissue biopsy.

Radiographically, abnormalities in abundance, density or chemical microenvironment of tissues allows distinction from surrounding normal tissues. Traditionally, the absorption of electromagnetic waves by tissues led to summation differences in exposure of silver salt photographic film. Tomographic approaches such as CT (1972) and NMR (1973) complemented summation radiology, allowing finely detailed visualization of internal anatomy in any plane of section. In the same era, ultrasound allowed visualization of tissue with density differences, such as a developing fetus or gallbladder stones. More recently, physiology of neoplasms can be screened with positron emission tomography (PET, 1977) for decay of short half-life isotopes such as fluorodeoxyglucose. Neoplasms with high metabolism can be distinguished physiologically from adjacent low-metabolism tissues, and can be localized with respect to normal tissues by pairing PET with standard CT. The result is an astonishingly useful means of identifying and localizing new space-occupying masses, assigning a risk for malignant behavior and, if malignant, screening for metastases in distant sites. This technique is revolutionizing the preoperative decision making of clinical teams, and improves the likelihood that patients undergo resections of new mass lesions only when at risk for morbidity from malignant behavior or interference with normal function.

Pathologically, disease is diagnosed by determining whether the morphologic features match the set of diagnostic criteria previously described for each disease. Multivolume texts are devoted to the gross and microscopic diagnostic criteria used for diagnosis, prognosis, and prediction of response to therapy [1,34,35]. Pathologists diagnose disease by generating a differential diagnosis, then finding the best fit for the clinical presentation, the radiographic appearance, and the pathologic (both clinical lab and morphologic) findings. Logically, the Venn diagram of the clinical, radiologic, and pathologic differential diagnoses should overlap. Unexpected features expand the differential diagnosis and may raise the possibility of previously undescribed diseases. For example, Legionnaire's disease, human immunodeficiency virus (HIV), Hantavirus pneumonia, and severe acute respiratory syndrome (SARS) are examples of newly described diagnoses during the last 30 years. The mental construct of etiology (cause), pathogenesis (progression), natural history (clinical outcome), and response to therapy is the standard approach for pathologists thinking about a disease. A disease may have one or more etiologies (initial causes, including agents, toxins, mutagens, drugs, allergens, trauma, or genetic mutations). A disease is expected to follow a particular series of events in its development (pathogenesis), and to follow a particular clinical course (natural history). Disease can result in a temporary or lasting change in normal function, including patient death. Multiple diseases of different etiologies can affect a single organ, for example, infectious and neoplastic diseases involving the lung. Different diseases can derive from a single etiology, for example, emphysema, chronic bronchitis, and small cell lung carcinoma in long-term smokers. The same disease (for instance, emphysema of the lung) can derive from different etiologies (emphysema from α-1-antitrypsin deficiency or cigarette smoke).

Modern surgical pathology practice hinges on morphologic diagnosis, supplemented by special stains, immunohistochemical stains, cytogenetics, and clinical laboratory findings, as well as the clinical and radiographic findings. Sections that meet all of these criteria are diagnostic for the disease. If some, but not all, of the criteria are present to make a definitive diagnosis, the pathologist must either equivocate or make an alternate diagnosis. Thus, a firm grasp of the diagnostic criteria and the instincts to rapidly create and sort through the differential diagnosis must be possessed by the service pathologist.

The tissue diagnosis has to make sense, not only from the morphologic perspective, but from the clinical and radiographic vantage points as well. It is both legally risky and professionally erosive to make a clinically and pathologically impossible diagnosis. In the recent past, limited computer networking meant numerous phone calls to gather the relevant clinical and radiographic information to make an informed morphologic diagnosis. For example, certain diseases such as squamous and small cell carcinomas of the lung are extremely rare in nonsmokers. Thus, a small cell carcinoma in the lung of a nonsmoker merits screening for a nonpulmonary primary site. Fortunately for pathologists, computing and networking technologies now allow us access to preoperative clinical workups, radiographs/reports, clinical laboratory data, and prior pathology reports. All of these data protect pathologists by providing them with the relevant clinical and radiographic information, and protect patients by improving diagnostic accuracy. Just as research scientists "...ignore the literature at their peril...", diagnostic pathologists "... ignore the presentation, past history, workup, prior biopsies, and radiographs at their peril. ..."

There are limitations to morphologic diagnosis by H&E stains. First, lineage of certain classes of neoplasms (including small round blue cell tumors, clear cell neoplasms, spindle cell neoplasms, and undifferentiated malignant neoplasms) is usually clarified by immunohistochemistry, frequently by cytogenetics (when performed), and sometimes by electron microscopy. Second, there are limitations inherent in a snapshot biopsy or resection. Thus, the etiology and pathogenesis can be obscure or indeterminate, and rates of growth, invasion, or timing of metastasis cannot be inferred. Third, the morphologic changes may not be specific for the underlying molecular abnormalities, particularly the rate-limiting (therapeutic target) step in the pathogenesis of a neoplasm. For example, *Ret* gain of function mutations in a medullary thyroid carcinoma will require DNA level screening to determine germline involvement, familial risk, and presence or absence of a therapeutic target. Fourth, the same morphologic appearance may be identical for two different diseases, each of which would be treated differently. For example, there is no morphologic evidence by H&E stain alone to distinguish host lymphoid response to Hepatitis C viral (HCV) antigens from host lymphoid response to allo-HLA antigens in a liver allograft. This is obviously a major diagnostic challenge when the transplant was done for HCV-related cirrhosis, and the probability of recurrent HCV infection in the liver allograft is high.

Paraffin section immunohistochemistry has proven invaluable in neoplasm diagnosis for clarifying lineage, improving diagnostic accuracy, and guiding customized therapy. If neoplasms are poorly differentiated or undifferentiated, the lineage of the neoplasm may not be clear. For example, sheets of undifferentiated malignant neoplasm with prominent nucleoli could represent carcinoma, lymphoma, or melanoma. To clarify lineage, a panel of immunostains is performed for proteins that are expressed in some of the neoplasms, but not in others. Relative probabilities are then used to lend support (rule in) or exclude (rule out) particular diagnoses in the differential diagnosis of these several morphologically similar undifferentiated neoplasms. The second role is to make critical distinctions in diagnosis that cannot be accurately made by H&E alone. Examples of this would include demonstration of myoepithelial cell loss in invasive breast carcinoma but not in its mimic, sclerosing adenosis (Figure 11.2), or loss of basal cells in invasive prostate carcinoma (Figure 11.3). The third role of immunohistochemistry is to identify particular proteins, such as nuclear estrogen receptor (ER) (Figure 11.4) or the plasma membrane HER2 proteins (Figure 11.5), both of which can be targeted with inhibitors rather than generalized systemic chemotherapy. Morphology remains the gold standard in this diagnostic process, such that immunohistochemical data support or fail to support the H&E findings, not vice versa.

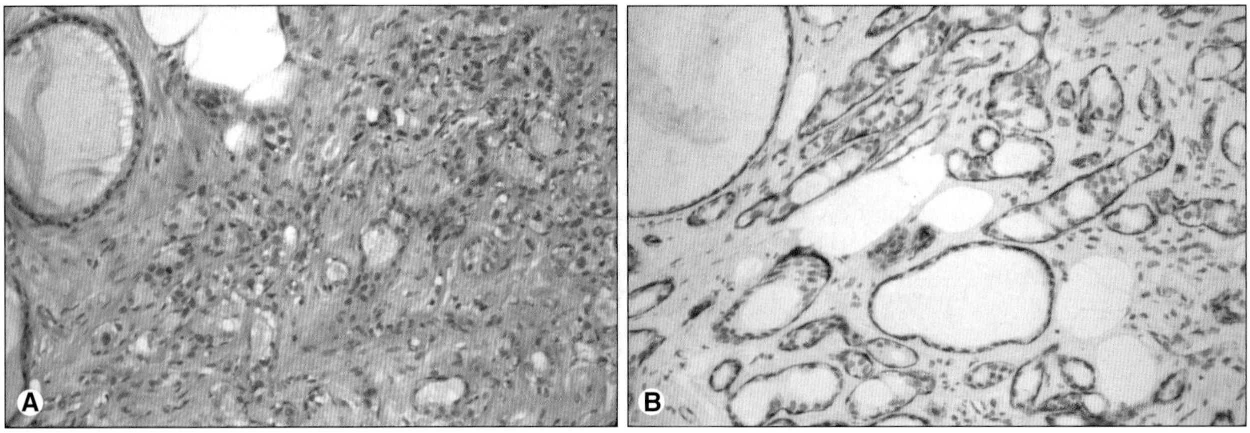

Figure 11.2 Sclerosing adenosis of breast. (A) H&E of sclerosing adenosis of breast. By H&E alone, the differential diagnosis includes infiltrating ductal carcinoma and sclerosing adenosis. (B) Actin immunostain of sclerosing adenosis of breast. Actin immunoreactivity around the tubules of interest supports a diagnosis of sclerosing adenosis and serves to exclude infiltrating carcinoma.

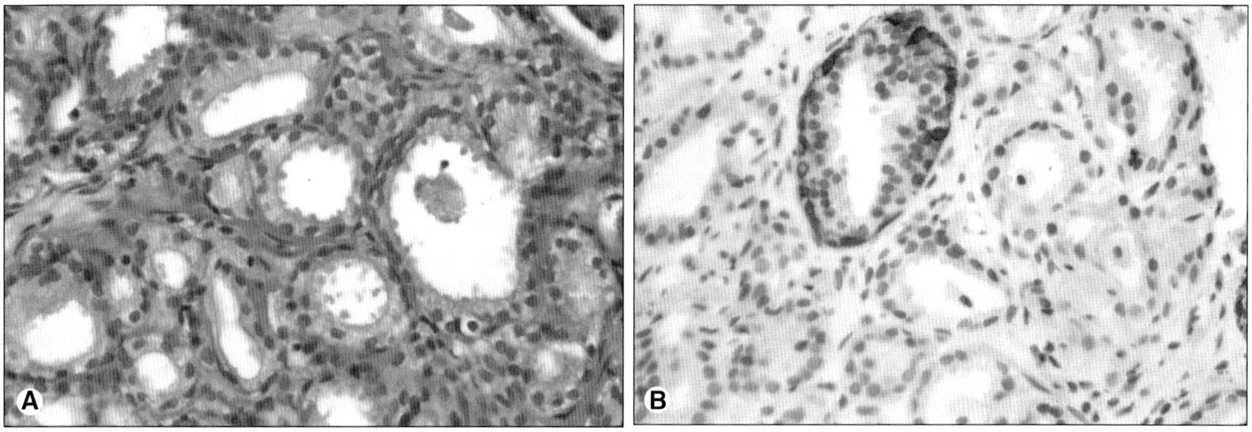

Figure 11.3 **Invasive adenocarcinoma of prostate.** (A) H&E of invasive adenocarcinoma of prostate. By H&E alone, the differential diagnosis includes invasive adenocarcinoma and adenosis. (B) High molecular weight cytokeratin immunostain of invasive adenocarcinoma of prostate. Loss of high molecular weight cytokeratin (34βE12) immunoreactivity around the glands of interest supports a diagnosis of invasive adenocarcinoma.

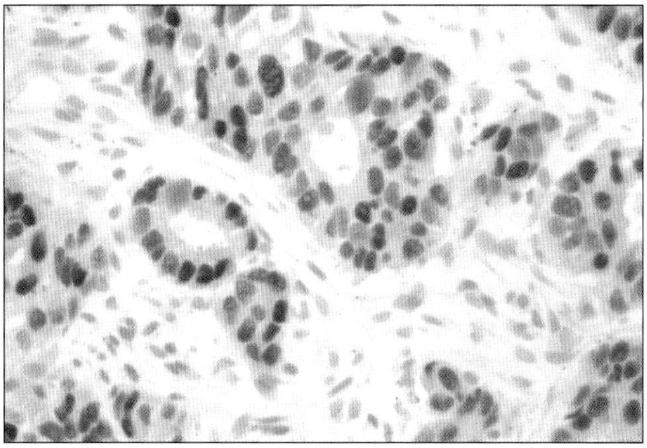

Figure 11.4 **Estrogen receptor immunostain of breast carcinoma.** Strong nuclear immunoreactivity for ER is noted, guiding use of ER inhibitor therapy.

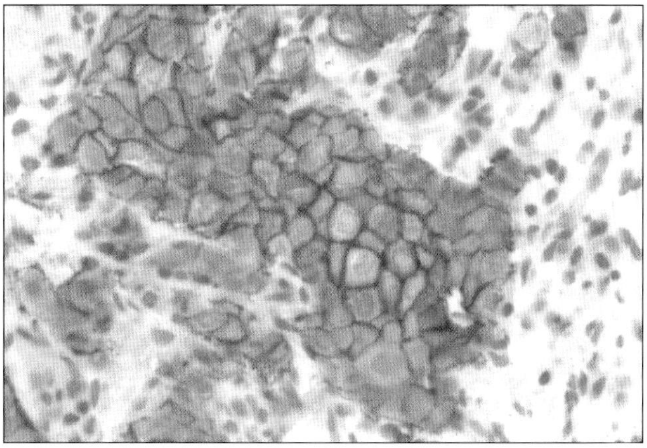

Figure 11.5 **HER2/c-erbB2 immunostain of breast carcinoma.** Strong plasma membrane immunoreactivity for c-erbB2/HER2 is noted, guiding use of either anti-HER2 antibody or HER2 kinase inhibitor therapy.

Probability and statistics are regular considerations in immunohistochemical interpretation, since very few antigens are tissue-specific or lineage-specific. Cytokeratin is positive in carcinomas, but also in synovial and epithelioid sarcomas. This example may imply aspects of the lineage of these two sarcomas that may be helpful in our categorization of these neoplasms. Another example would be the diagnosis of small cell carcinoma in the lung of a non-smoker. Because lung primary small cell carcinoma is extremely uncommon, in non-smokers, this diagnosis would prompt the pathologist to inquire about screening results for other, nonpulmonary, sites. Likewise, immunohistochemistry results are always put into the context of the morphologic, clinical, and radiographic findings. For example, an undifferentiated CD30(+) neoplasm of the testis supports embryonal carcinoma primary in the testis, whereas a lymph node effaced by sclerotic bands with admixed CD30(+) Reed-Sternberg cells supports nodular sclerosing Hodgkin's disease.

Demand for both diagnostic accuracy and report promptness has increased as hospitals come under increasing financial pressure to minimize length of patient stay. Hospitals now manage all but the sickest patients as outpatients. Minimally invasive approaches for the acquisition of tissue samples for diagnosis use flexible endoscopic biotomes or hollow needles that sample 1–2 mm diameter tissue specimens. Multidisciplinary conferences function almost real-time with respect to the initial biopsies. Together, these changes have forced modern pathologists to make critical diagnoses on progressively smaller biopsy specimens, sometimes bordering on the amounts seen in cytopathology aspirates, and to do this in a timely fashion. This requires a clear understanding of the limitations to development of an accurate diagnosis and a willingness on the part of the pathologist to request repeat biopsy for additional tissue when it is necessary for accurate diagnosis.

Diagnostic criteria involving electron microscopic ultrastructure found relevance for the evaluation of neoplasms described as small round blue cell tumors, spindle cell tumors, melanocytic tumors, and neuroendocrine/neuroblastic tumors, as well as delineation of ciliary ultrastructural abnormalities in primary ciliary dyskinesia. Current approaches to these neoplasms are now generally approached using paraffin section immunohistochemistry. Electron microscopy is currently used mainly for nephropathology, for evaluation of ciliary axonemes, for rare cases where immunohistochemistry is not diagnostic, and where demonstration of premelanosomes, neuroendocrine granules, or amyloid is diagnostic.

Adequate sampling of a lesion is critical to making an accurate diagnosis. Undercall diagnostic discrepancies are frequently due to sampling of a small portion of a large lesion that is unrepresentative of the most abnormal portion of the lesion. Insufficient sampling can result in an equivocal diagnosis or, worse, an inaccurate diagnosis. Empirical rules have been adopted over the decades to ensure statistically adequate sampling of masses and organs such as transurethral resections of prostate, soft tissue sarcomas, and heart allograft biopsies.

In spite of the limitations and statistical uncertainties relating to morphologic diagnosis, a wealth of information is conveyed to a service pathologist in a tried-and-true H&E section [36]. Analogous to the fact that a plain chest X-ray is the sum total of all densities in the beam path, the morphologic changes in diseased cells and tissues are the morphologic sum total of all of the disequilibria in the abnormal cells. For most neoplastic diseases, morphologic criteria are sufficient to predict the risk of invasion and metastasis (the malignant potential), the pattern of metastases, and the likely clinical outcomes. For example, the etiology and pathogenesis in small cell lung carcinoma can be inferred (cigarette smoking, with carcinogen-induced genetic mutations) and the outcome predicted (early metastasis to regional nodes and distant organs, with high probability of death within 5 years of diagnosis). New molecular data for both neoplastic and non-neoplastic diseases will most likely benefit unaffected individuals by estimating disease risk, and will most likely benefit patients by defining the molecular subset for morphologically defined diagnostic entities, thus guiding individualized therapy.

THE FUTURE OF DIAGNOSTIC PATHOLOGY

Diagnostic pathology will continue to use morphology and complementary data from protein (immunohistochemical) and nucleic acid (cytogenetics, *in situ* hybridization, DNA sequence, and RNA abundance) screening assays. New data will be integrated into the diagnostic process by reducing the cost and turnaround time of current technologies, and by development of new technologies, some of which are described.

Individual Identity

For transplant candidates, major histocompatability complex (MHC, HLA in human) screening is evolving from cellular assays and serology toward sequencing of the alleles of the class I and II HLA loci. Rapid sequencing of these alleles in newborn cord blood would allow databasing of the population's haplotypes, facilitating perfect matches for required bone marrow or solid organ transplants.

Rapid Cytogenetics

Current uses of *in situ* hybridization to screen for viruses (such as EBV), light chain restriction (in B lymphomas), and copy number variation (for instance, HER2 gene amplification) demonstrate the benefit of *in situ* nucleic acid hybridization assays. It is possible that interphase FISH/CISH will become rapid enough to be used in the initial diagnostic workup of certain patients, including for sarcoma-specific translocations, ploidy analysis in hydatidiform moles, and gene amplification of receptor tyrosine kinase genes.

Rapid Nucleic Acid Sequence and RNA Abundance Screening

Current uses of nucleic acid screening for *bcr-abl* translocation, donor:recipient ratios after bone marrow transplant, microsatellite instability, quantitative viral load (for EBV, BK, CMV, and others), and single gene mutations (for *CFTR*, *Factor 2*, *α-1-antitrypsin*) demonstrate the benefit of nucleic acid screening in diagnosis and management. It is possible that each new neoplasm will be promptly defined as to ploidy, translocations, gene copy number differences, DNA mutations, and RNA expression cluster subset, allowing residual disease screening as well as individualized therapy.

Computer-Based Prognosis and Prediction

Current uses of morphology, immunohistochemistry, and molecular pathology demonstrate their benefit through improved diagnostic accuracy. However, diagnosis, extent of disease, and molecular subsets are currently imperfect estimators of prognosis and response to therapy. Relational databases which correlate an individual's demographic data, family history, concurrent diseases, morphologic features, immunophenotype, and molecular subset, and which integrate disease prevalence by age, sex, and ethnicity using Bayesian probabilities, should improve accuracy of prognosis and prediction of response to therapy. As risk correlates are developed, it is possible that healthy individuals will be screened and given risk estimates for development of different diseases.

Normal Ranges and Disease Risks by Ethnic Group

Current uses of normal ranges for serum chemistry assumes a similar bell-curve distribution across ages, sexes, and races. This may be true for most but not all analytes. Computer reference databases will likely generate normal ranges specific for the particular age/sex/

ethnicity of individual patients. Similarly, familial risk for an inherited disease may vary by ethnic group, and this variation should be used in Bayesian calculations to define risk for unaffected at-risk family members.

Individual Metabolic Differences Relevant to Drug Metabolism

Current uses of liver and renal impairment to guide drug dosage demonstrate the benefit of using patient physiology to customize therapy. It is likely that individual differences in enzymatic metabolism of particular drugs (for instance, warfarin or tamoxifen) will be defined at the enzyme sequence level, and that gene haplotype data will be determined for new patients prior to receipt of these drugs.

Serum Biomarkers

Current uses of prostate specific antigen (PSA) to screen for prostate carcinoma and its recurrence demonstrates the benefit of serum biomarkers in common neoplasms. It is likely that high-sensitivity screening of single and clustered serum analytes will lead to improved methods for early detection and persistence of neoplasms, autoimmune diseases, and infections.

CONCLUSION

Pathologists consider each disease to have a natural, mechanical, physicochemical basis. Each disease has an etiology (initial cause), a pathogenesis (stepwise progression), and a natural history with effects on normal function (clinical outcome). Pathologists collect the data needed to answer patients' and clinicians' questions, simply phrased as "what is it?" (diagnosis), "how it going to behave?" (prognosis), and "how do I treat it?" (prediction of response to therapy). Instincts and diagnostic criteria, as well as the optical, mechanical, chemical, and computing technologies described previously, are the basis for modern service pathology. As the human genome is deciphered, and as the complex interactions of cellular biochemistry are refined, risk of disease in unaffected individuals will be calculable, disease diagnosis will be increasingly accurate and prognostic, and molecular subsets of morphologically defined disease entities will be used to guide customized therapy for individual patients. It is a great time in history to be a pathologist.

REFERENCES

1. Rosai J. *Rosai and Ackerman's Surgical Pathology*. Mosby; 2004.
2. Porter R. *The Greatest Benefit to Mankind: A Medical History of Humanity*. 1st ed. New York, NY: W. W. Norton & Co.; 1998.
3. Marshall BJ, Warren JR. Unidentified curved bacilli in the stomach of patients with gastritis and peptic ulceration. *Lancet*. 1984;1:1311–1315.
4. Jay V. The legacy of Karl Rokitansky. *Arch Pathol Lab Med*. 2000;124:345–346.
5. Paget J. *Lectures on Surgical Pathology*. London: Brown, Green, and Longmans; 1853.
6. Virchow R. *Cellular Pathology*. Berlin: August Hirschwald; 1858.
7. Hansen G. Investigations concerning the etiology of leprosy. *Norsk Mag. Laegervidenskaben*. 1874;4:1–88.
8. Koch R. The etiology of tuberculosis. *Berliner Klinische Wochenschrift*. 1882;15:221–230.
9. Turk JL. Rudolf Virchow—Father of cellular pathology. *J R Soc Med*. 1993;86:688–689.
10. Holmes O. Contagiousness of puerperal fever. *New England Quarterly Journal of Medicine*. 1843;1:503–530.
11. Dunn PM. Oliver Wendell Holmes (1809–1894) and his essay on puerperal fever. *Arch Dis Child Fetal Neonatal Ed*. 2007;92: F325–F327.
12. Raju TN. Ignac Semmelweis and the etiology of fetal and neonatal sepsis. *J Perinatol*. 1999;19:307–310.
13. Wohler F. *Annalen der Physik und Chemie* 1828; 88.
14. Gal AA. In search of the origins of modern surgical pathology. *Adv Anat Pathol*. 2001;8:1–13.
15. Medawar PB. The immunology of transplantation. *Harvey Lect*. 1956;144–176.
16. Early P, Huang H, Davis M, et al. An immunoglobulin heavy chain variable region gene is generated from three segments of DNA: VH, D and JH. *Cell*. 1980;19:981–992.
17. Leder P, Max EE, Seidman JG, et al. Recombination events that activate, diversify, and delete immunoglobulin genes. *Cold Spring Harb Symp Quant Biol*. 1981;45(Pt 2):859–865.
18. Coons AH, Kaplan MH. Localization of antigen in tissue cells; improvements in a method for the detection of antigen by means of fluorescent antibody. *J Exp Med*. 1950;91:1–13.
19. Kohler G, Milstein C. Continuous cultures of fused cells secreting antibody of predefined specificity. *Nature*. 1975;256: 495–497.
20. Wilkins MH, Stokes AR, Wilson HR. Molecular structure of deoxypentose nucleic acids. *Nature*. 1953;171:738–740.
21. Franklin RE, Gosling RG. Molecular configuration in sodium thymonucleate. *Nature*. 1953;171:740–741.
22. Watson JD, Crick FH. Molecular structure of nucleic acids; a structure for deoxyribose nucleic acid. *Nature*. 1953;171: 737–738.
23. Crick FH, Barnett L, Brenner S, et al. General nature of the genetic code for proteins. *Nature*. 1961;192:1227–1232.
24. Nirenberg M, Caskey T, Marshall R, et al. The RNA code and protein synthesis. *Cold Spring Harb Symp Quant Biol*. 1966; 31:11–24.
25. Gall JG, Pardue ML. Formation and detection of RNA-DNA hybrid molecules in cytological preparations. *Proc Natl Acad Sci USA*. 1969;63:378–383.
26. Pardue ML, Gall JG. Molecular hybridization of radioactive DNA to the DNA of cytological preparations. *Proc Natl Acad Sci USA*. 1969;64:600–604.
27. Goeddel DV, Kleid DG, Bolivar F, et al. Expression in Escherichia coli of chemically synthesized genes for human insulin. *Proc Natl Acad Sci USA*. 1979;76:106–110.
28. Sanger F, Nicklen S, Coulson AR. DNA sequencing with chain-terminating inhibitors. *Proc Natl Acad Sci USA*. 1977;74: 5463–5467.
29. Maxam AM, Gilbert W. A new method for sequencing DNA. *Proc Natl Acad Sci USA*. 1977;74:560–564.
30. Smith LM, Sanders JZ, Kaiser RJ, et al. Fluorescence detection in automated DNA sequence analysis. *Nature*. 1986;321:674–679.
31. Lander ES, Linton LM, Birren B, et al. Initial sequencing and analysis of the human genome. *Nature*. 2001;409:860–921.
32. Venter JC, Adams MD, Myers EW, et al. The sequence of the human genome. *Science*. 2001;291:1304–1351.
33. Mullis K, Faloona F, Scharf S, et al. Specific enzymatic amplification of DNA in vitro: the polymerase chain reaction. *Cold Spring Harb Symp Quant Biol*. 1986;51(Pt 1): 263–273.
34. Mills S. *Sternberg's Diagnostic Surgical Pathology*. Lippincott, Philadelphia; 2004.
35. Fletcher C. *Diagnostic Histopathology of Tumors*. Churchill Livingstone; 2007.
36. Rosai J. The continuing role of morphology in the molecular age. *Mod Pathol*. 2001;14:258–260.

Chapter

12

Understanding Molecular Pathogenesis: The Biological Basis of Human Disease and Implications for Improved Treatment of Human Disease

William B. Coleman . Gregory J. Tsongalis

INTRODUCTION

Disease has been a feature of the human existence since the beginning of time. Over time, our knowledge of science and medicine has expanded and with it our understanding of the biological basis of disease. In this regard, the biological basis of disease implies that more is understood about the disease than merely its clinical description or presentation. In the last several decades, we have moved from causative factors in disease to studies of molecular pathogenesis. Molecular pathogenesis takes into account the molecular alterations that occur in response to environmental insults and other contributing factors, to produce pathology. By developing a deep understanding of molecular pathogenesis, we will uncover the pathways that contribute to disease, either through loss of function or gain of function. By understanding the involvement of specific genes, proteins, and pathways, we will be better equipped to develop targeted therapies for specific diseases. Continued growth in our knowledge base with respect to underlying mechanisms of disease has resulted in unprecedented patient management strategies. Identification of genetic variants in genes once associated with the diagnosis of a disease process are now being re-evaluated as they may impact new therapeutic options.

In this chapter we describe three disease entities (Hepatitis C virus infection, acute myeloid leukemia, and cystic fibrosis) as examples of our increased understanding of the pathology represented by these diseases and how novel therapeutics are being introduced into clinical practice.

HEPATITIS C VIRUS INFECTION

Hepatitis C virus (HCV) infection represents the most common chronic viral infection in North America and Europe, and a common viral infection worldwide. In the United States' third National Health and Nutrition Examination Survey, it was estimated that 3.9 million people had detectable antibodies to HCV, indicating a prior exposure to the virus, and 75% of these individuals were positive for HCV RNA, suggesting an active infection [1]. HCV infection has been found to be more common in certain populations, including prison inmates and homeless people, where the prevalence of infection may be as high as 40% [2]. Worldwide, it is estimated that 340 million individuals are chronically infected with HCV [3].

Identification of the Hepatitis C Virus

HCV was first recognized in 1989 using recombinant technology to create peptides from an infectious serum that were then tested against serum from individuals with non-A, non-B hepatitis [4,5]. This approach

Molecular Pathology © 2009, Elsevier, Inc. All Rights Reserved.

resulted in the isolation of a section of the HCV genome [4,5]. Subsequently, the entire HCV genome was sequenced [6]. HCV is a member of the family of flaviviridae. Flaviviruses are positive, single-stranded RNA viruses. The HCV genome encodes a gene for production of a single polypeptide chain of approximately 3000 amino acids. This polypeptide gives rise to a number of specific proteins. The Env proteins are among the most variable parts of the peptide chain and are associated with multiple molecular forms in a single infected person [7,8]. The mutations affecting this portion of the HCV genome (and the endcoded Env protein) seem to be critical for escape of the virus from the host immune response [9,10]. The HCV protein NS5a contains an interferon-response element. Evidence from several studies suggest that mutational variation in the HCV genome encoding this protein are associated with resistance to interferon, the main antiviral agent used in treatment of HCV [11,12]. Other proteins encoded by the HCV genome include the NS3 region that codes for a protease and the NS5b region that codes for an RNA polymerase. Drugs that target the HCV protease or polymerase are now undergoing trials as therapeutic agents to treat HCV infection [13].

There are several HCV strains that differ significantly from each other [14]. The nomenclature adopted to describe these HCV strains is based on division of the HCV RNA into three major levels: (i) genotypes, (ii) subtypes, and (iii) quasispecies [15,16]. There are 6 recognized genotypes of HCV which are numbered from 1 to 6. Among these HCV genotypes there is <70% homology in the nucleotide sequence. HCV subtypes typically display 77%–80% homology in nucleotide sequence, while quasispecies have >90% nucleotide sequence homology [15,16]. Infection of an individual with HCV involves a single genotype and subtype (except in rare instances). However, infected individuals will carry many quasispecies of HCV because these RNA viruses do not contain a proofreading mechanism and acquired mutations in the HCV genome over time are common.

Risk Factors for Hepatitis C Virus Infection

There are a number of recognized risk factors for HCV. Among the most common risk factors for HCV infection are (i) the use of injection drugs and (ii) blood transfusion or organ transplant recipient before 1992 [17]. A significant percentage of people who used recreational injection drugs in the 1960s and 1970s became infected with HCV [18]. Less commonly, HCV infection can be transmitted by dialysis [19–21], by needlestick injury [22], and through vertical transmission from an infected mother to her child [23–25]. The likelihood of infection from needlestick injury or vertical transmission is estimated to be approximately 3%–5%.

Hepatitis C Infection

The primary target cell type for HCV infection is the mature hepatocyte [26], although there is some evidence that infection can also occur in other cell types, particularly circulating mononuclear cells [27]. Following the initial HCV infection, there is a latency period of 2–4 weeks before viral replication is detectable [28]. In most cases, there is no clinical evidence of the infection even after viremia develops. In fact, only 10%–30% of individuals with HCV infection will develop the clinical symptomology of acute hepatitis [29]. When acute HCV hepatitis develops, patients display symptoms of fever, loss of appetite, nausea, diarrhea, and specific liver symptoms, including discomfort and tenderness in the right upper abdomen, jaundice, dark urine, and pale-colored stools. Typically, these symptoms occur 2–3 months after the initial HCV infection and then gradually resolve over a period of several weeks [30]. During this time, liver enzymes such as alanine aminotransferase (ALT) and aspartate aminotransferases (AST) are found at elevated levels in the blood, reflecting hepatocyte injury and death. In acute HCV hepatitis, these enzymes are typically increased from 10-fold to 40-fold the upper reference limit of normal [31]. In the majority of individuals infected with HCV, there are no signs or symptoms that accompany the initial infection. In most of these cases, a chronic HCV infection develops, resulting in chronic hepatitis (ongoing inflammation in the liver). In general, chronic HCV infections can be clinically silent for many years without obvious symptomology associated with the infection or liver injury, or produce only mild, nonspecific symptoms such as fatigue, loss of energy, and difficulty performing tasks that require concentration. The major end-stage diseases that result from chronic hepatitis include cirrhosis and hepatocellular carcinoma. It has been estimated that 20%–30% of individuals with chronic HCV infection will progress to cirrhosis after 20 years of infection, although the fibrotic changes in the liver progress at different rates in different individuals [32,33]. Cirrhosis due to HCV infection has now become the most common indication for liver transplantation in the United States [2].

Testing for Hepatitis C Virus Infection

Most clinical testing for HCV infection begins with detection of antibodies against HCV proteins. The sensitivity of the anti-HCV assay is reported to be in the range of 97%–99% for detecting HCV infection. Most false negative results of the anti-HCV assay occur in the setting of immunosuppression, such as with human immunodeficiency virus (HIV) infection, or in renal failure [19,34]. Anti-HCV antibodies are detectable after 10–11 weeks of infection (on average) using the second generation anti-HCV assays, but the third generation anti-HCV assays show improved sensitivity with positive detection of anti-HCV antibodies by 7–8 weeks after the initial infection [35–37]. At the time of clinical presentation with acute HCV hepatitis, >40% of patients lack detectable anti-HCV [38]. In the current clinical laboratory setting, the major method employed to determine the presence of active HCV infection is HCV RNA measurement. With acute HCV infection, HCV RNA becomes detectable 2–4 weeks after infection, and viral loads climb rapidly [29,33]. Average HCV viral loads are approximately 2–3 million copies per mL. Qualitative assays are designed to determine the presence or absence of HCV RNA, without consideration of actual viral load. Two primary methodologies are used in this type of assay: (i) reverse-transcriptase polymerase chain reaction (RT-PCR) and (ii) transcription-mediated amplification (TMA). The detection limit for assays of this type is <50 IU/mL. The approaches employed for qualitative determination of HCV RNA utilize a known amount of a synthetic standard to enable quantitative measurement of HCV RNA through comparison of the amounts of HCV amplified and the amount of standard amplified using a calibration curve. Determination of HCV viral load has become standard of care in evaluating patients before and during treatment for chronic HCV infection. Real-time PCR allows for reduced carryover amplification, more rapid detection of amplification, increased low-end sensitivity, and a wider dynamic range for detection and quantification [39].

Several techniques have been developed to determine the particular genotype and subtype of HCV causing infection in an individual infected patient. These techniques typically target the 5′-untranslated and/or core regions of the HCV genome which represent the most highly conserved regions. Because most amplification methods for HCV RNA also target the 5′-untranslated region, qualitative PCR methods can provide amplified RNA for use in determination of HCV genotype. The most widely used technique is a commercial line probe assay [40]. In this assay, a large number of oligonucleotide sequences are immobilized on a membrane, incubated with amplified RNA, and then detected using a colorimetric reagent that detects areas of hybridization. The line probe assay enables recognition and identification of most HCV types and subtypes accurately, although there are several subtypes that cannot be distinguished from one another.

Clinical Course of Hepatitis C Virus Infection

Although anyone who is infected with HCV will experience an initial infection incident, in most cases this phase of the infection will be clinically silent without obvious symptoms. Acute infection with HCV is most likely to be detected when it occurs following a needlestick exposure from a person with known HCV, or when the infection arises under other circumstances but produces symptomatic infection and jaundice (estimated to occur in less than one-third of all cases) [29,30,33]. There is some evidence to suggest that patients who develop clinical jaundice are actually more likely to clear the infection and not progress to chronic HCV hepatitis [41]. During the initial incubation period approximately 2 weeks following infection, HCV RNA is either undetectable or can be detected only intermittently. Subsequently, there is a period of rapid increase in the amount of circulating HCV, with an estimated doubling time of <24 hours [28]. HCV viral loads reach very high levels during this period of time, typically reaching values of 10^7 IU/mL and occasionally higher [28]. Evidence of liver injury appears after an additional 1–2 months of HCV infection. This liver injury can be detected secondary to increased serum levels of ALT and AST. Approximately 40%–50% of individuals with acute HCV infection that are clinically diagnosed are detected during this stage of infection, prior to the development of anti-HCV [38]. By 7–8 weeks following infection, anti-HCV becomes detectable using the third generation immunoassays [37]. However, detection of anti-HCV using the second generation immunoassay cannot be accomplished until 10–12 weeks following infection [35]. At the time of this seroconversion, the HCV viral loads decrease, sometimes to undetectable levels. In most individuals who will progress to chronic hepatitis (and in some that eventually clear the infection) the HCV viral load remains detectable, but at reduced levels. In a person suspected of having acute HCV infection, the most reliable test for proving exposure is HCV RNA. Because of the high viral loads seen, either qualitative or quantitative assays would be acceptable for this purpose. Detectable HCV RNA in the absence of anti-HCV is strong evidence of recent HCV infection [42].

Treatment of Hepatitis C Infection

In contrast to other chronic viral infections such as those associated with hepatitis B virus or HIV, treatment has been successful in eradicating replicating HCV and halting progression of liver damage. Interferon alpha-2 is the agent of choice for treatment of chronic HCV infection. There are currently two potential approaches to treatment of chronic HCV: (i) interferon alone or (ii) a combination of interferon plus ribavirin. While ribavirin is ineffective as a single agent for treating HCV [43], it increases the effectiveness of interferon. Application of ribavirin in combination with interferon increases the number of patients who respond to therapy by 2-fold to 3-fold [44]. For many years, the only form of interferon available was standard dose interferon. Using standard dose interferon, large doses (typically 3 million units) were delivered to patients infected with HCV several times each week. The short half-life of this interferon produced widely fluctuating interferon levels in these patients, diminishing its therapeutic effectiveness. In 2001, a longer-acting form of interferon was approved for use in treating HCV infection. The longer-acting interferon was modified by attachment of polyethylene glycol (pegylated interferon), which resulted in increased half-life for the administered drug. Use of pegylated interferon results in sustained high levels of interferon in the patient, reducing the number of required administrations to a single injection each week. There was also an improvement in response rates among patients treated with the pegylated interferon. Currently, the preferred treatment for chronic HCV infection is the combination of pegylated interferon plus ribavirin [45].

Guided Treatment of Hepatitis C Virus

The appropriate duration of treatment for HCV infection varies depending on the HCV strain (genotype) that infects the patient. HCV genotypes 2 and 3 respond much better to standard treatment regimens. Thus, only 24 weeks of therapy are needed to achieve maximum benefit, compared to 48 weeks in persons infected with other HCV genotypes [45]. In current clinical practice, treatment is offered to all patients with HCV infection except those with decompensated cirrhosis, where treatment may lead to worsening of the patient's condition [46,47]. Once treatment is initiated, the most reliable means to determine efficacy is to evaluate the response by measuring HCV RNA. Successful treatment is associated with at least two different phases of viral clearance [48]. The first phase, which occurs rapidly over the course of days, is thought to reflect HCV RNA clearance from a circulating pool through the antiviral effect of interferon. In the second phase of clearance, infected liver cells (the major site of viral replication) undergo cell turnover and are replaced by uninfected cells. The second phase of clearance is more variable in duration. First phase clearance is less specific for detecting success of antiviral treatment; therefore, it is necessary to evaluate whether second phase clearance has occurred.

Summary

HCV infection represents a relatively recently identified infectious agent that has a varied natural history from patient to patient. Intensive research efforts have characterized the phases of HCV infection and the clinical symptomology of acute and chronic HCV infection. Through improved understanding of the biology of the HCV virus and its life cycle in the infected host, effective and sensitive diagnostic tests have been developed. Unlike some other chronic viral infections, HCV infection can be effectively treated using interferon in combination with ribavirin. However, it is now recognized that effective therapy of the patient depends on knowing the genotype of the HCV causing the infection. With continued advances in the understanding of the pathogenesis of HCV infection, new treatments and/or new modes of administration of known anti-HCV drugs will emerge that provide effective control of the viral infection with minimal adverse effects for the patient.

ACUTE MYELOID LEUKEMIA

The human leukemias have been classified as a distinct group of clinically and biologically heterogeneous disorders that are a result of genetic abnormalities that affect specific chromosomes and genes. The acute myeloid leukemias (AML) represent a major form of leukemia. AML is characterized by accumulation of neoplastic immature myeloid cells, consisting of $\geq$30% myeloblasts in the blood or bone marrow and classified on the basis of their morphological and immunocytochemical features. AML can arise (i) *de novo,* (ii) in a setting of a preexisting myelodysplasia, or (iii) secondary to chemotherapy for another disorder.

Chromosomal Abnormalities in Acute Myelogenous Leukemia

Various cytogenetic and/or molecular abnormalities have been associated with various types of AML. Chromosomal translocations are the most common form of genetic abnormality identified in acute leukemias [49–51]. Typically, these translocations involve genes that encode proteins that function in transcription and differentiation pathways [52]. As a result of chromosomal translocation, the genes proximal to the chromosome breakpoints are disrupted, and the $5'$-segment of one gene is joined to the $3'$-end of a second gene to form a novel fusion (chimeric) gene. When the chimeric gene is expressed, a novel protein product is produced from the chimeric mRNA. Other genetic alterations such as point mutations, gene amplifications and numerical gains or losses of chromosomes can also be identified in the acute leukemias. The clinical heterogeneity seen in AML may be due in part to differences in the number and nature of genetic abnormalities that occur in these cancers. However, these same molecular differences define various prognostic and therapeutic characteristics associated with the specific disorder in a given patient.

A major chromosomal translocation in AML involves chromosomes 15 and 17. This genetic abnormality, t(15;17)(q21;q21), occurs exclusively in acute promyelocytic leukemia (APL). APL represents approximately 5%–13% of all *de novo* AMLs [51,53]. The presence of the t(15;17) translocation consistently predicts responsiveness to a specific treatment utilizing all-*trans*-retinoic acid (ATRA). Retinoic acid is a ligand for the retinoic acid receptor (RAR), which is involved in the t(15;17). ATRA is thought to overcome the block in myeloid cell maturation, allowing the neoplastic cells to mature (differentiate) and be eliminated [54,55]. Approximately 75% of patients with APL present with a bleeding diathesis, usually the result of one or more processes including disseminated intravascular coagulation, increased fibrinolysis, and thrombocytopenia, and secondary to the release of procoagulants or tissue plasminogen activator from the granules of neoplastic promyelocytes [55,56]. This bleeding diathesis may be exacerbated by standard cytoreductive chemotherapy. Two morphologic variants of APL have been described, typical (hypergranular) and microgranular, both of which carry the t(15;17) translocation [55–57]. In the typical or hypergranular variant, the promyelocytes have numerous azurophilic cytoplasmic granules that often obscure the border between the cell nucleus and the cytoplasm. Cells with numerous Auer rods in bundles are common. In the microgranular type the promyelocytes contain numerous small cytoplasmic granules that are difficult to discern with the light microscope but are easily seen by electron microscopy.

Consequence of the t(15;17) Translocation in Acute Myelogenous Leukemia

The t(15;17) is a balanced and reciprocal translocation in which the *PML* (for promyelocytic leukemia) gene on chromosome 15 and the *RARα* gene on chromosome 17 are disrupted and fused to form a hybrid gene [58,59]. The *PML-RARα* fusion gene, located on chromosome 15, encodes a chimeric mRNA and a novel protein. On the derivative chromosome 15, both the *PML* and *RARα* genes are oriented in a head-to-tail orientation. The function of the normal *PML* gene is poorly understood. However, the gene is ubiquitously expressed and encodes a protein that contains a dimerization domain and is characterized by an N-terminal region with two zinc-finger-like motifs (known as a ring and a B-box). Given its structural features, the PML protein is thought to be involved in DNA binding [55,58,59]. Furthermore, the normal PML protein appears to have an essential role in cell proliferation. The *RARα* gene encodes a transcription factor that binds to DNA sequences in *cis*-acting retinoic acid-responsive elements. High-affinity DNA binding also requires heterodimerization with another family of proteins, the retinoic acid X receptors. The RARα protein contains transactivation, DNA binding, heterodimerization, and ligand binding domains. The normal RARα protein plays an important role in myeloid differentiation.

There are three major forms of the *PML-RARα* fusion gene, corresponding to different breakpoints in the *PML* gene [60–62]. The breakpoint in the *RARα* gene occurs in the same general location in all cases, involving the sequences within intron 2. Approximately 40%–50% of cases have a *PML* breakpoint in exon 6 (the so-called long form, termed *bcr1*), 40%–50% of cases have the *PML* breakpoint in exon 3 (the so-called short form, termed *bcr3*), and 5%–10% of cases have a breakpoint in *PML* exon 6 that is variable (the so-called variable form, termed *bcr2*). In each form of the translocation, the PML-RARα fusion protein retains the 5′-DNA binding and dimerization domains of PML and the 3′-DNA binding, heterodimerization, and ligand (retinoic acid) binding domains of RARα. Recent studies indicate that the different forms of *PML-RARα* fusion mRNA correlate with clinical presentation or prognosis. In particular, the *bcr3* type of *PML-RARα* correlates with higher leukocyte counts at time of presentation [61,62]. Both higher leukocyte counts and variant morphology are adverse prognostic findings, and the *bcr3* type of *PML-RARα* does not independently predict poorer disease-free survival [62].

Detection of the t(15;17) Translocation in Acute Myelogenous Leukemia

A number of methods may be used to detect the t(15;17) translocation. Conventional cytogenetic methods detect the t(15;17) in approximately 80%–90% of APL cases at time of initial diagnosis. Suboptimal clinical specimens and poor quality metaphases explain a large subset of the negative results. Fluorescence *in situ* hybridization (FISH) is another useful method for detecting the t(15;17) in APL [63]. Different methods employ probes specific for either chromosome 15 or chromosome 17 (or both), and commercial kits are available. Southern blot hybridization is another method to detect gene rearrangements that result from the t(15;17) [64]. The chromosomal breakpoints consistently involve the second intron of the *RARα* gene, and therefore, probes derived from this region are the most often utilized. Virtually all cases of APL can be detected by Southern blot analysis using two or three genomic *RARα* probes. RT-PCR is a very convenient method for detecting the *PML-RARα* fusion transcripts [60]. Primers have been designed to amplify the potential transcripts, and each type of transcript can be recognized. Results using this method are equivalent to or better than other methods at time of initial diagnosis.

Polyclonal and monoclonal antibodies reactive with the PML and RARα proteins have been generated, and immunohistochemical studies to assess the pattern of staining appear to be useful for diagnosis [65,66]. Dyck et al. [65] have studied a number of APLs and have shown that the pattern of PML or RARα immunostaining correlates with the presence of the t(15;17). APL cells immunostaining for either PML or RARα reveals a microgranular pattern. The fusion protein may prevent PML from forming normal oncogenic domains, since treatment with ATRA allows PML reorganization into these domains. For the diagnosis of residual disease or early relapse after therapy, conventional cytogenetic studies, Southern blot analysis, and immunohistochemical methods are limited by low sensitivity. Quantitative RT-PCR and FISH methods are very useful. The sensitivity and rapid turnaround time of RT-PCR makes this method very useful for monitoring residual disease after therapy [67,68].

Summary

Acute promyelocytic leukemia is a distinct subtype of acute myeloid leukemia that is cytogenetically characterized by a balanced reciprocal translocation between chromosomes 15 and 17 [t(15;17)(q21;q21)], which results in a gene fusion involving *PML* and *RARα*. This disease is the most malignant form of acute leukemia with a severe bleeding tendency and a fatal course of only weeks in affected individuals. In the past, cytotoxic chemotherapy was the primary modality for treatment of APL, producing complete remission rates of 75%–80% in newly diagnosed patients, a median duration of remission from 11–25 months, and only 35%–45% of the patients were cured [69]. However, with the introduction of all-trans retinoic acid (ATRA) in the treatment and optimization of the ATRA-based regimens, the complete remission rate increased to 90%–95% and 5-year disease-free survival improved to 74% [69].

CYSTIC FIBROSIS

Cystic fibrosis (CF) is a clinically heterogeneous disease that exemplifies the many challenges of complex genetic diseases and the causative underlying mechanisms [70]. CF is the most common lethal autosomal-recessive disease in individuals of European decent with a prevalence of 1:2500 to 1:3300 live births. While CF occurs most commonly in the Caucasian population, members of other racial and ethnic backgrounds are also at risk for this disease. In the United States, approximately 850 individuals are newly diagnosed on an annual basis, and 30,000 children and adults are affected. The majority of CF diagnoses are made in individuals that are less than one year of age (http://www.genetests.org/).

Cystic Fibrosis Transmembrane Conductance Regulator Gene

The Cystic Fibrosis Transmembrane Conductance Regulator (*CFTR*) gene is responsible for CF. This gene is large, spanning approximately 230 kb on chromosome 7q, and consists of 27 coding exons. The *CFTR* mRNA is 6.5 kb and encodes a CFTR membrane glycoprotein of 1480 amino acids with a mass of ~170,000 daltons [71–73]. CFTR functions as a cAMP- regulated chloride channel in the apical membrane of epithelial cells [74]. To date over 1000 unique mutations in the CFTR gene have been described (Cystic Fibrosis Mutation Data Base, http://www.genet.sickkids.on.ca/cftr/). The most common *CFTR* mutation is the deletion of phenylalanine at position 508 (ΔF508). This mutation affects 70% of patients worldwide. The allelic frequency of *CFTR* mutations varies by ethnic group. For example, the ΔF508 *CFTR* mutation is only present in 30% of the affected Ashkenazi Jewish population.

Diagnosis of Cystic Fibrosis

A diagnosis of CF in a symptomatic or at-risk patient is suggested by clinical presentation and confirmed by a sweat test. In the presence of clinical symptoms (such as recurrent respiratory infections), a sweat chloride above 60 mmol/L is diagnostic for CF. Although the results of this test are valid in a newborn as young as 24 hours, collecting a sufficient sweat sample from a baby younger than 3 or 4 weeks old is difficult. The sweat test can also confirm a diagnosis of CF in older children and adults, but is not useful for carrier detection. Mutations in the *CFTR* gene are grouped into six classes, including (i) Class I, characterized by defective protein synthesis where there is no CFTR protein at the apical membrane; (ii) Class II, characterized by abnormal/defective processing and trafficking where there is no CFTR protein at the apical membrane; (iii) Class III, characterized by defective regulation where there is a normal amount of nonfunctional CFTR at the apical membrane; (iv) Class IV, characterized by decreased conductance where there is a normal amount of CFTR with some residual function at the apical membrane; (v) Class V, characterized by reduced or defective synthesis/trafficking where there is a decreased amount of functional CFTR at the apical membrane; and (vi) Class VI, characterized by decreased stability where there is a functional but unstable CFTR at the apical membrane [75,76]. Of the CFTR mutations, classes I–III are the most common and are associated with pancreatic insufficiency [77]. The ΔF508 CFTR mutation (which is most common worldwide) represents a class II mutation, with varying frequency between ethnic groups [78].

Abnormal Function of CFTR in Cystic Fibrosis

CFTR is a member of an ATP-binding cassette family with diverse functions such as ATP-dependent transmembrane pumping of large molecules, regulation of other membrane transporters, and ion conductance. Mutations in the *CFTR* gene can lead to an abnormal protein with loss or compromised function that results in defective electrolyte transport and faulty chloride ion transport in apical membrane epithelial cells affecting the respiratory tract, pancreas, intestine, male genital tract, hepatobiliary system, and the exocrine system, resulting in complex multisystem disease. The loss of CFTR-mediated anion conductance explains a variety of CF symptoms including elevated sweat chloride, due to a defect in salt absorption by the sweat ducts, and meconium ileus, a defect in fluid secretion by intestinal crypt cells [79]. The malfunction of CFTR as a regulator of amiloride-sensitive epithelia Na+ channel leads to increased Na+ conductance in CF airways, which drives increased absorption of Cl- and water. Most of the symptoms associated with CF, such as meconium ileus, loss of pancreatic function, degeneration of the vas deferens, thickened cervical mucus, and failure of adrenergically mediated sweating are due to the role CFTR plays in Cl-driven fluid secretion.

CFTR is an anion channel that functions in the regulation of ion transport. It plays multiple roles in fluid and electrolyte transport, including salt absorption, fluid absorption, and anion-mediated fluid secretion [79]. Defects in this protein lead to CF, the morbidity of which is initiated by a breach in host defenses and propagated by an inability to clear the resultant infections [80]. Since inflammatory exacerbations precipitate irreversible lung damage, the innate immune system plays an important role in the pathogenesis of CF. Respiratory epithelial cells containing the CFTR also provide a crucial environmental interface for a variety of inhaled insults. The local mucosal mechanism of defense involves mucociliary clearance that relies on the presence and constituents of airway surface liquid (ASL). The high salt in the ASL found in CF patients interferes with the natural antibiotics present in ASL such as defensins and lysozyme [81]. Bals et al. categorized the role of CFTR in the pathogenesis of CF-related lung disease by dividing patients into two groups [80]. The first describes defects in CFTR that result in altered salt and water concentrations of airway secretions. This then affects host defenses and creates a milieu for infection. The second is associated with CFTR deficiency that results in biologically and intrinsically abnormal respiratory epithelia. These abnormal epithelial cells fail as a mechanical barrier and enhance the presence of pathogenic bacteria by providing receptors and binding sites or failing to produce functional antimicrobials.

Much debate exists regarding the relative biologic activity of antibacterial peptides such as beta-defensins and cathelicidins in human ASL and their role in the pathogenesis in CF-related lung disease [79,80,82]. It is possible that the innate immune system provides a first line of host defense against microbial colonization by secreting defensins, small cationic antimicrobial peptides produced by epithelia. The innate antibiotics are thought to possess salt-sensitive bacteriocidal capabilities. Hence, these innate antibiotics demonstrate altered (impaired) function in the lungs of CF patients [82]. Mannose-binding lectin represents another antimicrobial molecule that is present in ASL and is thought to be inactivated by high salt concentrations in the lungs of CF patients. Mannose-binding lectin, an acute phase serum protein produced in the liver, opsonizes bacteria and activates complement. Common variations in the mannose-binding lectin gene (MBL2) are associated with increased disease severity, increased risk of infection with *B. cepacia*, poor prognosis, and early death [83]. The understanding that such naturally occurring peptide antibiotics exist has resulted in the pharmacologic development of these peptides for therapeutics.

Pathophysiology of Cystic Fibrosis

The occurrence of CF leads to clinical, gross, and histologic changes in various organ systems expressing abnormal CFTR, including the pancreas, respiratory, hepatobiliary, intestinal, and reproductive systems. In addition, pathologic changes have been observed in organ systems that do not express the *CFTR* gene (such as the rheumatologic and vascular systems). The current age of individuals affected with CF ranges from 0–74 years, and the predicted survival age for a newly diagnosed child is 33.4 years. The increasing age of survival of CF has led to increased manifestation of pulmonary and extrapulmonary disorders (gastrointestinal, hepatobiliary, vascular, and musculoskeletal) associated with the disease. The extent and severity of disease tends to correlate with the degree of CFTR function. Although all these organ systems are affected, the pulmonary changes are the most pronounced and the major cause of mortality in most cases [84].

Lung infection remains the leading cause of morbidity and mortality in CF patients. It is currently recognized that CF-related lung disease is the consequence of chronic pulmonary consolidation by the well-known opportunistic pathogens *Pseudomonas aeruginosa* (mucoid and nonmucoid), *Burkholderia cepacia*, *Staphylococcus aureus*, and *Haemophilus influenza* [85]. Morbidity and mortality due to persistent lung infection despite therapeutic advances focus attention toward the expanding microbiology of pulmonary colonizers. These increasingly prevalent flora include *Burkholderia cepacia* complex (genomovar I-IX), Methicillin-Resistant *Staphylococcus aureus* (MRSA), *Stenotrophomonas maltophilia*, *Achromobacter xylosoxidans*, *Mycobacterium abscessus*, *Mycobacterium-avium* complex, *Ralstonia* species, and *Pandoraea* species [82,86]. Inflammatory exacerbation precipitates progressive irreversible lung damage, of which bronchiectases are the landmark changes. Bronchial mucous plugging facilitates colonization by microorganisms. Repetitive infections lead to bronchiolitis and bronchiectasis. Other pulmonary changes include interstitial fibrosis and bronchial squamous metaplasia. Often, subpleural bronchiectatic cavities develop and communicate with the subpleural space with resultant spontaneous secondary pneumothorax, the incidence of which increases later in life.

Exocrine pancreas insufficiency is present in the majority of patients with CF. This clinically manifests by failure to thrive, and fatty bulky stools owing to deficiency of pancreatic enzymes. However, pancreatic lesions vary greatly in severity, and the pancreas may be histologically normal in some patients who die in infancy [87]. Early in the postnatal development of the pancreas, patients with CF have a deficiency of normal acinar development. Increased secretory material within the ducts and increased duct volume also contribute to progressive degradation and atrophy of pancreatic acini. These factors result in duct obstruction and progressive pancreatic pathology [88,89]. Exocrine pancreatic disease appears to develop as a result of deficient ductal fluid secretion due to decreased anion secretion. Coupled to normal protein load derived from acinar cell secretion, this then leads to pancreatic protein hyperconcentration within the pancreatic ducts. The protein hyperconcentration increases susceptibility to precipitation and finally obstruction of the duct lumina [90,91]. Hence, the characteristic lesion is cystic ductal dilation, atrophy of pancreatic acini, and severe parenchymal fibrosis.

The manifestation of CF in the hepatobiliary system is directly related to CFTR expression. The liver disease in CF is considered inherited liver disease due to impaired secretory function of the biliary epithelium [92]. While defective CFTR may be expressed, males are more likely to be affected than females and the risk for developing liver disease is between 4% and 17% as assessed by yearly exams and biochemical testing [93,94]. CFTR is expressed in epithelial cells of the biliary tract. Therefore, any or all cells of the biliary tree may be affected. While a variety of liver manifestations exist [95,96], including fatty infiltration (steatosis), common bile duct stenosis, sclerosing cholangitis, and gallbladder disease, the rare but characteristic liver lesion in CF is focal biliary cirrhosis, which develops in a minority of patients and is usually seen in older children and adults [97]. With the increasing life expectancy in patients with CF, liver-related deaths have increased and may become one of the major causes of death in CF [97]. The associated liver disease usually develops before or at puberty, is slowly progressive, and is frequently asymptomatic. There is negligible effect on nutritional status or severity of pulmonary involvement [98]. Only a minority of patients go on to develop a clinically problematic liver disease with rapid progression. Abnormal bile composition and reduced bile flow ultimately lead to intrahepatic bile duct obstruction and focal biliary cirrhosis [97]. Diagnosis of CF-associated liver disease is based on clinical exam findings, biochemical tests, and imaging techniques. Although liver biopsy is the gold standard for the diagnosis of most chronic liver diseases, only rarely is it employed in the diagnostic workup, mainly due to sampling error [97,99].

The gastrointestinal manifestations of cystic fibrosis are seen mainly in the neonatal period and include meconium ileus, distal intestinal obstruction syndrome (DIOS), fibrosing colonopathy, strictures, gastroesophageal reflux, rectal prolapse and constipation in later childhood [93,100–103]. Throughout the intestines CFTR is the determinant of chloride concentration and secondary water loss into the intestinal lumen. Decreased water content results in viscous intestinal contents, with a 10%–15% risk of developing meconium ileus in babies born with cystic fibrosis. This also accounts for DIOS and constipation in older children [104]. DIOS (formerly meconium ileus equivalent) is a recurrent partial or complete obstruction of the intestine in patients with CF and pancreatic insufficiency [103].

Arthritis is a rare but recognized complication of cystic fibrosis that generally occurs in the second decade [105–108]. Three types of joint disease are described in patients with cystic fibrosis: (i) cystic fibrosis arthritis (CFA) or episodic arthritis (EA), (ii) hypertrophic pulmonary osteoarthropathy (HPOA), and (iii) co-existent or treatment-related arthritis [105,106,109,110]. The most common form, episodic arthropathy, is characterized by episodic, self-limited polyarticular arthritis with no evidence of progression to joint damage [105]. Histologic features are minimal with prominent blood vessels and interstitial edema occurring most commonly, or rarely lymphocytic inflammation [111].

Infertility is an inevitable consequence of cystic fibrosis in males occurring in >95% of patients, and is due to congenital bilateral absence or atrophy of the vasa deferentia (CBAVD) and/or dilated or absent seminal vesicles [112]. Spermatogenesis and potency remain normal. Mutations in the *CFTR* gene are present in up to 70% of the patients with CBAVD [113]. Diagnosis of obstructive azoospermia may be diagnosed by semen analysis; however, it must be confirmed by testicular biopsy and no other reason for azoospermia. Fertility in females may be impaired due to dehydrated cervical mucus, but their reproductive function is normal [114]. Advances in techniques such as microscopic epididymal sperm aspiration (MESA) and intracytoplasmic sperm injection have allowed males with cystic fibrosis the ability to reproduce [84].

Summary

Cystic fibrosis is a complex multiorgan system disease that results from mutation in the CFTR gene. Advances in the understanding of the pathogenesis of this disease and related complications (such as recurrent lung infection) have led to improvement in diagnosis and treatment of affected individuals, resulting in improved life expectancy. With continued expansion of our understanding of the molecular pathogenesis of this disease and the variant manifestations of CF-related disorders, it is expected that new treatments will emerge that attempt to counteract or correct the pathologic consequences of *CFTR* mutation.

REFERENCES

1. Alter MJ, Kruszon-Moran D, Nainan OV, et al. The prevalence of hepatitis C virus infection in the United States, 1988 through 1994. *N Engl J Med.* 1999;341:556–562.
2. Kim WR. The burden of hepatitis C in the United States. *Hepatology.* 2002;36:S30–34.
3. Mansell CJ, Locarnini SA. Epidemiology of hepatitis C in the East. *Semin Liver Dis.* 1995;15:15–32.
4. Choo QL, Kuo G, Weiner AJ, et al. Isolation of a cDNA clone derived from a blood-borne non-A, non-B viral hepatitis genome. *Science.* 1989;244:359–362.
5. Kubo Y, Takeuchi K, Boonmar S, et al. A cDNA fragment of hepatitis C virus isolated from an implicated donor of post-transfusion non-A, non-B hepatitis in Japan. *Nucleic Acids Res.* 1989;17:10367–10372.
6. Houghton M, Weiner A, Han J, et al. Molecular biology of the hepatitis C viruses: Implications for diagnosis, development and control of viral disease. *Hepatology.* 1991;14: 381–388.
7. Honda M, Kaneko S, Sakai A, et al. Degree of diversity of hepatitis C virus quasispecies and progression of liver disease. *Hepatology.* 1994;20:1144–1151.
8. Baumert TF, Wellnitz S, Aono S, et al. Antibodies against hepatitis C virus-like particles and viral clearance in acute and chronic hepatitis C. *Hepatology.* 2000;32:610–617.
9. Rehermann B. Interaction between the hepatitis C virus and the immune system. *Semin Liver Dis.* 2000;20:127–141.
10. Rehermann B, Chisari FV. Cell mediated immune response to the hepatitis C virus. *Curr Top Microbiol Immunol.* 2000;242:299–325.
11. Frangeul L, Cresta P, Perrin M, et al. Pattern of HCV antibodies with special reference to NS5A reactivity in HCV-infected patients: Relation to viral genotype, cryoglobulinemia and response to interferon. *J Hepatol.* 1998;28:538–543.
12. Frangeul L, Cresta P, Perrin M, et al. Mutations in NS5A region of hepatitis C virus genome correlate with presence of NS5A antibodies and response to interferon therapy for most common European hepatitis C virus genotypes. *Hepatology.* 1998;28: 1674–1679.
13. Ideo G, Bellobuono A. New therapies for the treatment of chronic hepatitis C. *Curr Pharm Des.* 2002;8:959–966.
14. Bukh J, Miller RH, Purcell RH. Biology and genetic heterogeneity of hepatitis C virus. *Clin Exp Rheumatol.* 1995;13(Suppl 13):S3–7.
15. Bukh J, Miller RH, Purcell RH. Genetic heterogeneity of hepatitis C virus: Quasispecies and genotypes. *Semin Liver Dis.* 1995; 15:41–63.
16. Bukh J, Miller RH, Purcell RH. Genetic heterogeneity of the hepatitis C virus. *Princess Takamatsu Symp.* 1995;25:75–91.
17. Recommendations for prevention and control of hepatitis C virus (HCV) infection and HCV-related chronic disease. Centers for Disease Control and Prevention. *MMWR Recomm Rep.* 1998;47:1–39.
18. McCarthy JJ, Flynn N. Hepatitis C in methadone maintenance patients: Prevalence and public policy implications. *J Addict Dis.* 2001;20:19–31.
19. Bukh J, Wantzin P, Krogsgaard K, et al. High prevalence of hepatitis C virus (HCV) RNA in dialysis patients: Failure of commercially available antibody tests to identify a significant number of patients with HCV infection. Copenhagen Dialysis HCV Study Group. *J Infect Dis.* 1993;168:1343–1348.
20. Courouce AM, Le Marrec N, Girault A, et al. Anti-hepatitis C virus (anti-HCV) seroconversion in patients undergoing hemodialysis: Comparison of second- and third-generation anti-HCV assays. *Transfusion.* 1994;34:790–795.
21. Delarocque-Astagneau E, Baffoy N, Thiers V, et al. Outbreak of hepatitis C virus infection in a hemodialysis unit: Potential transmission by the hemodialysis machine? *Infect Control Hosp Epidemiol.* 2002;23:328–334.
22. Mitsui T, Iwano K, Masuko K, et al. Hepatitis C virus infection in medical personnel after needlestick accident. *Hepatology.* 1992; 16:1109–1114.
23. Mazza C, Ravaggi A, Rodella A, et al. Prospective study of mother-to-infant transmission of hepatitis C virus (HCV) infection. Study Group for Vertical Transmission. *J Med Virol.* 1998; 54:12–19.

24. Okamoto M, Nagata I, Murakami J, et al. Prospective reevaluation of risk factors in mother-to-child transmission of hepatitis C virus: High virus load, vaginal delivery, and negative anti-NS4 antibody. *J Infect Dis.* 2000;182:1511–1514.
25. Roberts EA, Yeung L. Maternal-infant transmission of hepatitis C virus infection. *Hepatology.* 2002;36:S106–113.
26. Mengshol JA, Golden-Mason L, Rosen HR. Mechanisms of disease: HCV-induced liver injury. *Nat Clin Pract Gastroenterol Hepatol.* 2007;4:622–634.
27. El-Awady MK, Ismail SM, El-Sagheer M, et al. Assay for hepatitis C virus in peripheral blood mononuclear cells enhances sensitivity of diagnosis and monitoring of HCV-associated hepatitis. *Clin Chim Acta.* 1999;283:1–14.
28. Naito M, Hayashi N, Hagiwara H, et al. Serial quantitative analysis of serum hepatitis C virus RNA level in patients with acute and chronic hepatitis C. *J Hepatol.* 1994;20:755–759.
29. Seeff LB, Wright EC, Zimmerman HJ, et al. VA cooperative study of post-transfusion hepatitis, 1969–1974: Incidence and characteristics of hepatitis and responsible risk factors. *Am J Med Sci.* 1975;270:355–362.
30. Healey CJ, Sabharwal NK, Daub J, et al. Outbreak of acute hepatitis C following the use of anti-hepatitis C virus—screened intravenous immunoglobulin therapy. *Gastroenterology.* 1996; 110:1120–1126.
31. Hino K, Sainokami S, Shimoda K, et al. Clinical course of acute hepatitis C and changes in HCV markers. *Dig Dis Sci.* 1994;39:19–27.
32. Alberti A, Chemello L, Benvegnu L. Natural history of hepatitis C. *J Hepatol.* 1999;31(Suppl 1):17–24.
33. Seeff LB. Natural history of chronic hepatitis C. *Hepatology.* 2002;36:S35–46.
34. Marcellin P, Martinot-Peignoux M, Elias A, et al. Hepatitis C virus (HCV) viremia in human immunodeficiency virus-seronegative and -seropositive patients with indeterminate HCV recombinant immunoblot assay. *J Infect Dis.* 1994;170:433–435.
35. Gretch DR. Diagnostic tests for hepatitis C. *Hepatology.* 1997;26: 43S–47S.
36. Forns X, Costa J. HCV virological assessment. *J Hepatol.* 2006;44: S35–39.
37. Barrera JM, Francis B, Ercilla G, et al. Improved detection of anti-HCV in post-transfusion hepatitis by a third-generation ELISA. *Vox Sang.* 1995;68:15–18.
38. Barrera JM, Bruguera M, Ercilla MG, et al. Persistent hepatitis C viremia after acute self-limiting posttransfusion hepatitis C. *Hepatology.* 1995;21:639–644.
39. Yang JH, Lai JP, Douglas SD, et al. Real-time RT-PCR for quantitation of hepatitis C virus RNA. *J Virol Methods.* 2002;102:119–128.
40. van Doorn LJ, Kleter B, Stuyver L, et al. Analysis of hepatitis C virus genotypes by a line probe assay and correlation with antibody profiles. *J Hepatol.* 1994;21:122–129.
41. Villano SA, Vlahov D, Nelson KE, et al. Persistence of viremia and the importance of long-term follow-up after acute hepatitis C infection. *Hepatology.* 1999;29:908–914.
42. Alberti A, Boccato S, Vario A, et al. Therapy of acute hepatitis C. *Hepatology.* 2002;36:S195–200.
43. Bodenheimer Jr HC, Lindsay KL, Davis GL, et al. Tolerance and efficacy of oral ribavirin treatment of chronic hepatitis C: A multicenter trial. *Hepatology.* 1997;26: 473–477.
44. Cummings KJ, Lee SM, West ES, et al. Interferon and ribavirin vs interferon alone in the re-treatment of chronic hepatitis C previously nonresponsive to interferon: A meta-analysis of randomized trials. *JAMA.* 2001;285:193–199.
45. Di Bisceglie AM, Hoofnagle JH. Optimal therapy of hepatitis C. *Hepatology.* 2002;36:S121–127.
46. Schalm SW, Fattovich G, Brouwer JT. Therapy of hepatitis C: Patients with cirrhosis. *Hepatology.* 1997;26:128S–132S.
47. Wright TL. Treatment of patients with hepatitis C and cirrhosis. *Hepatology.* 2002;36:S185–194.
48. Layden TJ, Mika B, Wiley TE. Hepatitis C kinetics: Mathematical modeling of viral response to therapy. *Semin Liver Dis.* 2000;20:173–183.
49. Cline MJ. The molecular basis of leukemia. *N Engl J Med.* 1994;330:328–336.
50. Heim S, Mitelman F. Cytogenetic analysis in the diagnosis of acute leukemia. *Cancer.* 1992;70:1701–1709.
51. Mitelman F, Heim S. Quantitative acute leukemia cytogenetics. *Genes Chromosomes Cancer.* 1992;5:57–66.
52. Rabbitts TH. Translocations, master genes, and differences between the origins of acute and chronic leukemias. *Cell.* 1991;67:641–644.
53. Douer D, Preston-Martin S, Chang E, et al. High frequency of acute promyelocytic leukemia among Latinos with acute myeloid leukemia. *Blood.* 1996;87:308–313.
54. Vyas RC, Frankel SR, Agbor P, et al. Probing the pathobiology of response to all-trans retinoic acid in acute promyelocytic leukemia: Premature chromosome condensation/fluorescence in situ hybridization analysis. *Blood.* 1996;87:218–226.
55. Grignani F, Fagioli M, Alcalay M, et al. Acute promyelocytic leukemia: From genetics to treatment. *Blood.* 1994;83:10–25.
56. Warrell Jr RP, de The H, Wang ZY, et al. Acute promyelocytic leukemia. *N Engl J Med.* 1993;329:177–189.
57. Bitter MA, Le Beau MM, Rowley JD, et al. Associations between morphology, karyotype, and clinical features in myeloid leukemias. *Hum Pathol.* 1987;18:211–225.
58. Kakizuka A, Miller Jr WH, Umesono K, et al. Chromosomal translocation t(15;17) in human acute promyelocytic leukemia fuses RAR alpha with a novel putative transcription factor, PML. *Cell.* 1991;66:663–674.
59. de The H, Lavau C, Marchio A, et al. The PML-RAR alpha fusion mRNA generated by the t(15;17) translocation in acute promyelocytic leukemia encodes a functionally altered RAR. *Cell.* 1991;66:675–684.
60. Miller Jr WH, Kakizuka A, Frankel SR, et al. Reverse transcription polymerase chain reaction for the rearranged retinoic acid receptor alpha clarifies diagnosis and detects minimal residual disease in acute promyelocytic leukemia. *Proc Natl Acad Sci USA.* 1992;89:2694–2698.
61. Huang W, Sun GL, Li XS, et al. Acute promyelocytic leukemia: Clinical relevance of two major PML-RAR alpha isoforms and detection of minimal residual disease by retrotranscriptase/ polymerase chain reaction to predict relapse. *Blood.* 1993; 82:1264–1269.
62. Gallagher RE, Willman CL, Slack JL, et al. Association of PML-RAR alpha fusion mRNA type with pretreatment hematologic characteristics but not treatment outcome in acute promyelocytic leukemia: An intergroup molecular study. *Blood.* 1997; 90:1656–1663.
63. Schad CR, Hanson CA, Paietta E, et al. Efficacy of fluorescence in situ hybridization for detecting PML/RARA gene fusion in treated and untreated acute promyelocytic leukemia. *Mayo Clin Proc.* 1994;69:1047–1053.
64. Biondi A, Rambaldi A, Alcalay M, et al. RAR-alpha gene rearrangements as a genetic marker for diagnosis and monitoring in acute promyelocytic leukemia. *Blood.* 1991;77:1418–1422.
65. Dyck JA, Warrell Jr RP, Evans RM, et al. Rapid diagnosis of acute promyelocytic leukemia by immunohistochemical localization of PML/RAR-alpha protein. *Blood.* 1995;86:862–867.
66. Falini B, Flenghi L, Fagioli M, et al. Immunocytochemical diagnosis of acute promyelocytic leukemia (M3) with the monoclonal antibody PG-M3 (anti-PML). *Blood.* 1997;90:4046–4053.
67. Lo Coco F, Diverio D, Pandolfi PP, et al. Molecular evaluation of residual disease as a predictor of relapse in acute promyelocytic leukaemia. *Lancet.* 1992;340:1437–1438.
68. Miller Jr WH, Levine K, DeBlasio A, et al. Detection of minimal residual disease in acute promyelocytic leukemia by a reverse transcription polymerase chain reaction assay for the PML/ RAR-alpha fusion mRNA. *Blood.* 1993;82:1689–1694.
69. Wang ZY, Chen Z. Acute promyelocytic leukemia: From highly fatal to highly curable. *Blood.* 2008;111:2505–2515.
70. Brennan AL, Geddes DM. Cystic fibrosis. *Curr Opin Infect Dis.* 2002;15:175–182.
71. Kerem B, Rommens JM, Buchanan JA, et al. Identification of the cystic fibrosis gene: Genetic analysis. *Science.* 1989;245: 1073–1080.
72. Riordan JR, Rommens JM, Kerem B, et al. Identification of the cystic fibrosis gene: Cloning and characterization of complementary DNA. *Science.* 1989;245:1066–1073.
73. Rommens JM, Iannuzzi MC, Kerem B, et al. Identification of the cystic fibrosis gene: Chromosome walking and jumping. *Science.* 1989;245:1059–1065.

74. Rosenstein BJ, Zeitlin PL. Cystic fibrosis. *Lancet.* 1998;351: 277–282.
75. Vankeerberghen A, Cuppens H, Cassiman JJ. The cystic fibrosis transmembrane conductance regulator: An intriguing protein with pleiotropic functions. *J Cyst Fibros.* 2002;1:13–29.
76. Zielenski J. Genotype and phenotype in cystic fibrosis. *Respiration.* 2000;67:117–133.
77. Correlation between genotype and phenotype in patients with cystic fibrosis. The Cystic Fibrosis Genotype-Phenotype Consortium. *N Engl J Med.* 1993;329:1308–1313.
78. Morral N, Bertranpetit J, Estivill X, et al. The origin of the major cystic fibrosis mutation (delta F508) in European populations. *Nat Genet.* 1994;7:169–175.
79. Wine JJ. The genesis of cystic fibrosis lung disease. *J Clin Invest.* 1999;103:309–312.
80. Bals R, Weiner DJ, Wilson JM. The innate immune system in cystic fibrosis lung disease. *J Clin Invest.* 1999;103:303–307.
81. Goldman MJ, Anderson GM, Stolzenberg ED, et al. Human beta-defensin-1 is a salt-sensitive antibiotic in lung that is inactivated in cystic fibrosis. *Cell.* 1997;88:553–560.
82. Conese M, Assael BM. Bacterial infections and inflammation in the lungs of cystic fibrosis patients. *Pediatr Infect Dis J.* 2001;20:207–213.
83. Chanock SJ, Foster CB. SNPing away at innate immunity. *J Clin Invest.* 1999;104:369–370.
84. Ratjen F, Doring G. Cystic fibrosis. *Lancet.* 2003;361:681–689.
85. Robinson P. Cystic fibrosis. *Thorax.* 2001;56:237–241.
86. LiPuma JJ. Expanding microbiology of pulmonary infection in cystic fibrosis. *Pediatr Infect Dis J.* 2000;19:473–474.
87. King A, Mueller RF, Heeley AF, et al. Diagnosis of cystic fibrosis in premature infants. *Pediatr Res.* 1986;20:536–541.
88. Imrie JR, Fagan DG, Sturgess JM. Quantitative evaluation of the development of the exocrine pancreas in cystic fibrosis and control infants. *Am J Pathol.* 1979;95:697–708.
89. Liu P, Daneman A, Stringer DA, et al. Pancreatic cysts and calcification in cystic fibrosis. *Can Assoc Radiol J.* 1986;37:279–282.
90. Kopelman H, Durie P, Gaskin K, et al. Pancreatic fluid secretion and protein hyperconcentration in cystic fibrosis. *N Engl J Med.* 1985;312:329–334.
91. Kopelman H, Forstner G, Durie P, et al. Origins of chloride and bicarbonate secretory defects in the cystic fibrosis pancreas, as suggested by pancreatic function studies on control and CF subjects with preserved pancreatic function. *Clin Invest Med.* 1989;12:207–211.
92. Tanner MS, Taylor CJ. Liver disease in cystic fibrosis. *Arch Dis Child.* 1995;72:281–284.
93. Modolell I, Guarner L, Malagelada JR. Digestive system involvement in cystic fibrosis. *Pancreatology.* 2002;2:12–16.
94. Scott-Jupp R, Lama M, Tanner MS. Prevalence of liver disease in cystic fibrosis. *Arch Dis Child.* 1991;66:698–701.
95. Colombo C, Battezzati PM, Podda M. Hepatobiliary disease in cystic fibrosis. *Semin Liver Dis.* 1994;14:259–269.
96. Curry MP, Hegarty JE. The gallbladder and biliary tract in cystic fibrosis. *Curr Gastroenterol Rep.* 2005;7:147–153.
97. Colombo C, Crosignani A, Battezzati PM. Liver involvement in cystic fibrosis. *J Hepatol.* 1999;31:946–954.
98. Colombo C, Battezzati PM, Strazzabosco M, et al. Liver and biliary problems in cystic fibrosis. *Semin Liver Dis.* 1998;18:227–235.
99. Ling SC, Wilkinson JD, Hollman AS, et al. The evolution of liver disease in cystic fibrosis. *Arch Dis Child.* 1999;81:129–132.
100. Rubinstein S, Moss R, Lewiston N. Constipation and meconium ileus equivalent in patients with cystic fibrosis. *Pediatrics.* 1986;78:473–479.
101. Pawel BR, de Chadarevian JP, Franco ME. The pathology of fibrosing colonopathy of cystic fibrosis: A study of 12 cases and review of the literature. *Hum Pathol.* 1997;28:395–399.
102. Smyth RL. Fibrosing colonopathy in cystic fibrosis. *Arch Dis Child.* 1996;74:464–468.
103. Lewis TC, Casey SC, Kapur RP. Clinical pathologic correlation: A 3-year-old boy with cystic fibrosis and intestinal obstruction. *J Pediatr.* 1999;134:514–519.
104. Milla PJ. Cystic fibrosis: Present and future. *Digestion.* 1998; 59:579–588.
105. Bourke S, Rooney M, Fitzgerald M, et al. Episodic arthropathy in adult cystic fibrosis. *Q J Med.* 1987;64:651–659.
106. Johnson S, Knox AJ. Arthropathy in cystic fibrosis. *Respir Med.* 1994;88:567–570.
107. Newman AJ, Ansell BM. Episodic arthritis in children with cystic fibrosis. *J Pediatr.* 1979;94:594–596.
108. Schidlow DV, Goldsmith DP, Palmer J, et al. Arthritis in cystic fibrosis. *Arch Dis Child.* 1984;59:377–379.
109. Turner MA, Baildam E, Patel L, et al. Joint disorders in cystic fibrosis. *J R Soc Med.* 1997;90(Suppl 31):13–20.
110. Pertuiset E, Menkes CJ, Lenoir G, et al. Cystic fibrosis arthritis. A report of five cases. *Br J Rheumatol.* 1992;31:535–538.
111. Phillips BM, David TJ. Pathogenesis and management of arthropathy in cystic fibrosis. *J R Soc Med.* 1986;79(Suppl 12):44–50.
112. Dodge JA. Male fertility in cystic fibrosis. *Lancet.* 1995;346: 587–588.
113. Chillon M, Casals T, Mercier B, et al. Mutations in the cystic fibrosis gene in patients with congenital absence of the vas deferens. *N Engl J Med.* 1995;332:1475–1480.
114. Kopito LE, Kosasky HJ, Shwachman H. Water and electrolytes in cervical mucus from patients with cystic fibrosis. *Fertil Steril.* 1973;24:512–516.

Chapter

13

Integration of Molecular and Cellular Pathogenesis: A Bioinformatics Approach

Jason H. Moore ▪ C. Harker Rhodes

INTRODUCTION

Historically, the touchstone of diagnostic pathology has been the histologic appearance of diseased and normal tissues when stained with conventional stains, usually hematoxylin and eosin. The microscopic appearance of these tissues reflects their cellular structure, which in turn is due to the differential expression of approximately 20,000 protein-coding genes in the human genome. The development of immunohistochemical techniques in the 1960s enabled the histologic detection of specific proteins and revolutionized both diagnostic pathology as well as the scientific study of disease. The differential expression of specific proteins is due in large part to changes at the DNA level, changes in transcription and mRNA splicing, and alterations in other aspects of the complex regulation of RNA metabolism and translation. With the sequencing of the human genome and the recent advances in both PCR-based and array-based technologies which allow the easy quantitation of specific RNA levels, the simultaneous detection of tens or hundreds of thousands of different genetic transcripts, and the DNA-based technologies for looking at genetic changes and changes in chromosomal organization, we are now on the verge of another, similar revolution.

The current challenge is to integrate the tremendous wealth of information now available on the molecular changes associated with normal physiology and disease processes with the practice of pathology. In the practice of conventional pathology (based on tissue morphology), the coordinate expression of thousands of genes results in the creation of distinctive microscopic appearances which are recognized by trained pathologists and serve as the basis both for the diagnostic pathology and for the scientific study of disease. With the advent of immunohistochemistry, these subtle and complex images were supplemented with information about the expression of a few specific proteins, one for each stain which is performed. Now, as DNA-based and RNA-based technologies are being introduced into the general practice of pathology, two things are happening. First, information about specific genetic alterations or changes in specific RNA levels is being added to the information available about diseased tissue. The wealth of new insights provided by these techniques rivals that which became available when immunohistochemistry was introduced, but conceptually the integration of this information into the mainstream of pathology is straightforward and similar to what has been done in the past. However, unlike previous technological developments, the nucleic acid-based technologies can be incorporated into arrays that provide thousands or hundreds of thousands of individual pieces of information—genotype information, information about DNA methylation, data about gene copy number, or RNA expression levels. The analysis of this avalanche of data which can be generated from a single biopsy or autopsy specimen requires the techniques of bioinformatics if it is to be reduced to meaningful information.

For example, consider the information currently available from a biopsy of a high-grade glioma, such as a glioblastoma (GBM). The diagnosis is made today as it has been for decades based on the appearance of the lesion in a tissue section stained with hematoxylin and eosin (H&E) based on the presence of endothelial proliferation and tumor necrosis in what is histologically a malignant glial lesion, with the nuclear morphology typical of an astrocytic tumor. For many generations that was all the pathologist could tell the clinician about the tumor, and it was all the clinician needed to know to select a therapy and treat the patient. It had been recognized since the 1940s that these tumors fell into two distinct but overlapping clinical categories, primary GBM and secondary GBM. Primary GBM tended to be found

Molecular Pathology © 2009, Elsevier, Inc. All Rights Reserved.

in older patients, had a shorter clinical course, and appeared to arise as GBM, whereas secondary GBM were found in slightly younger patients and arose by progression from lower grade astrocytic tumors. But the distinction was of no therapeutic importance and there were no pathologic correlates that could be used to distinguish a primary from secondary GBM. The use of immunohistochemical techniques did not change the situation for this particular tumor. Stains for glial fibrillary acidic protein could be used to demonstrate the astrocytic character of the lesion and markers for cycling cells as opposed to those in G0 like the Ki-67 antigen could be used to demonstrate the relatively high proliferative index of the tumor, but they rarely provided clinically important new information. Then in the mid-1990s studies using nucleic acid-based technologies demonstrated that not only did primary and secondary GBM have distinctive clinical histories, but on a molecular basis these histologically identical tumors were completely different with distinctive molecular signatures, including frequent *EGFR* amplification and mutation in the primary GBM, and *p53* gene mutations in secondary GBM. That distinction remained of limited clinical significance, but it helped establish the idea that in spite of the pathologist's inability to separate GBM into subtypes based on H&E histology, they were a molecularly heterogeneous group of tumors. Currently, specific molecularly targeted therapies directed against tumors overexpressing *EGFR* (for example) are entering clinical trials. The molecular differences between primary and secondary GBM and their potential clinical significance are reviewed in Ohgaki and Kleihues [1].

Although these single marker, candidate gene studies of GBM did a great deal to elucidate the molecular pathogenesis of these tumors and define molecularly distinct subsets of GBM, it was clear to most pathologists that the separation of GBM into primary and secondary tumors did not adequately capture the complexity of the situation. Within the last several years, studies using RNA expression arrays to classify these tumors based on unsupervised clustering of thousands of mRNA levels have suggested that GBM are perhaps best thought of as being of three types—those with proneural, proliferative, or mesenchymal molecular signatures [2]. Other studies based on high-resolution copy number analysis using oligonucleotide-based array comparative genomic hybridization have also identified three subsets of GBM—one which seems to correspond to the classically defined primary GBM and two others which represent secondary GBMs [3]. The integration of these studies into a single unified classification system remains to be done, but in the meantime other single-marker studies are providing clinically important information about these tumors. For example, the epigenetic silencing of the gene for the DNA-repair enzyme MGMT has been shown to influence the response of these tumors to conventional therapies, and stratification of GBM patients based on MGMT promoter methylation status is rapidly entering clinical practice [4]. Similarly, the recognition of a subset of tumors which are histologically indistinguishable from other GBM but have loss of the short arm of chromosome 1 and/or the long arm of chromosome 19 (like oligodendrogliomas) and have a better prognosis than the usual GBM is becoming standard clinical practice.

These examples demonstrate the potential for biotechnology to significantly impact our ability to use molecular pathology to understand disease processes. However, our ability to exploit these new technological resources will depend critically on our capability to make sense out of mountains of data collected for a set of pathology samples. The remainder of this chapter will introduce bioinformatics and the resources that are available to pathologists for making full use of genetics, genomics, and proteomics.

OVERVIEW OF BIOINFORMATICS

Bioinformatics is an interdisciplinary field that blends computer science and biostatistics with biomedical sciences such as epidemiology, genetics, genomics, and proteomics. Bioinformatics emerged as an important discipline shortly after the development of high-throughput DNA sequencing technologies in the 1970s [5]. It was the momentum of the Human Genome Project that spurred the rapid rise of bioinformatics as a formal discipline. The word bioinformatics didn't start appearing in the biomedical literature until around 1990 but quickly caught on as the descriptor of this important new field. An important goal of bioinformatics is to facilitate the management, analysis, and interpretation of data from biological experiments and observational studies. Thus, much of bioinformatics can be categorized as database development and implementation, data analysis and data mining, and biological interpretation and inference.

The need to interpret information from whole-genome sequencing projects in the context of biological information acquired in decades of research studies prompted the establishment of the National Center for Biotechnology Information (NCBI) as a division of the National Library of Medicine (NLM) at the National Institutes of Health (NIH) in the United States in November of 1988. When the NCBI was established, it was charged with (i) creating automated systems for storing and analyzing knowledge about molecular biology, biochemistry, and genetics; (ii) performing research into advanced methods of computer-based information processing for analyzing the structure and function of biologically important molecules and compounds; (iii) facilitating the use of databases and software by biotechnology researchers and medical care personnel; and (iv) coordinating efforts to gather biotechnology information worldwide [6]. Since 1988, the NCBI has fulfilled many of these goals and has delivered a set of databases and computational tools that are essential for modern biomedical research in a wide range of different disciplines, including molecular epidemiology. The NCBI and other international efforts such as the European Bioinformatics Institute (EBI) that was established in 1992 [7] have played a very important role in inspiring and motivating the establishment of research groups and centers around the world that are dedicated to providing bioinformatics tools and expertise.

DATABASE RESOURCES

One of the most important prestudy activities is the design and development of one or more databases that can accept, store, and manage molecular pathology data. Haynes and Blach [8] list eight steps for establishing an information management system for genetic studies. These steps are broadly applicable to many different kinds of studies. The first step is to develop the experimental plan for the clinical, demographic, sample, and molecular/laboratory information that will be collected. What are the specific needs for the database? The second step is to establish the information flow. That is, how does the information find its way from the clinic or laboratory to the database? The third step is to create a model for information storage. How are the data related? The fourth step is to determine the hardware and software requirements. How much data needs to be stored? How quickly will investigators need to access the data? What operating system will be used? Will a freely available database such as mySQL (http://www.mysql.com) serve the needs of the project or will a commercial data set solution such as Oracle (http://www.oracle.com) be needed? The fifth step is to implement the database. The important consideration here is to define the database structure so that data integrity is maintained. The sixth step is to choose the user interface to the database. Is a web page portal to the data sufficient? The seventh step is to determine the security requirements. Do HIPAA regulations (http://www.hhs.gov/ocr/hipaa) need to be followed? Most databases need to be password protected at a minimum. The eighth and final step outlined by Haynes and Blach [8] is to select the software tools that will interface with the data for summary and analysis.

Although most investigators choose to develop and manage their own database for security and confidentiality reasons, there is an increasing number of public databases for depositing data so that it is widely available to other investigators. The tradition of making data publicly available soon after it has been analyzed and published can largely be attributed to the community of investigators using gene expression microarrays. Microarrays [9] represent one of the most revolutionary applications that derived from the knowledge of whole genome sequences. The extensive use of this technology has led to the need to store and search expression data for all the genes in the genome acquired in different genetic backgrounds or in different environmental conditions. This has resulted in a number of public databases such as the Stanford Microarray Database [10] (http://genome-www5.stanford.edu), the Gene Expression Omnibus [11,12] (http://www.ncbi.nlm.nih.gov/geo), ArrayExpress [13,14] (http://www.ebi.ac.uk/arrayexpress), and others that anyone can download data from. The nearly universal acceptance of the data sharing culture in this area has yielded a number of useful tools that might not have been developed otherwise. The need for defining standards for the ontology and annotation of microarray experiments has led to proposals such as the Minimum Information About a Microarray Experiment (MIAME) [13,15] (http://www.mged.org/Workgroups/MIAME/miame.html) that provided a standard that greatly facilitates the storage, retrieval, and sharing of data from microarray experiments. The MIAME standards provide an example for other types of data such as SNPs and protein mass spectrometry spectra. See Brazma et al. [16,17] for a comprehensive review of the standards for data sharing in genetics, genomics, proteomics, and systems biology. The success of the different databases depends on the availability of methods for easily depositing data and tools for searching the databases (often after data normalization).

Despite that acceptance of data sharing in the genomics community, the same culture does not yet exist in molecular pathology. One of the few such examples is the Pharmacogenomics Knowledge Base of PharmGKB [18] (http://www.pharmgkb.org). PharmGKB was established with funding from the NIH to store, manage, and make available molecular data in addition to phenotype data from pharmacogenetic and pharmacogenomic experiments and clinical studies [18]. It is anticipated that similar databases for molecular epidemiology will appear and gain acceptance over the next few years as the NIH and various journals start to require data from publicly funded research be made available to the public.

In addition to the need for a database to store and manage molecular pathology data collected from experimental or observational studies, there a number of database resources that can be very helpful for planning a study. A good starting point for database resources are those maintained at the NCBI [19] (http://www.ncbi.nlm.nih.gov). Perhaps the most useful resource when planning a molecular pathology study is the Online Mendelian Inheritance in Man or OMIM database [20,21] (http://www.ncbi.nlm.nih.gov/omim). OMIM is a catalog of human genes and genetic disorders with detailed summaries of the literature. The NCBI also maintains the PubMed literature database with more than 15 million indexed abstracts from published papers in more than 4700 life science journals. The PubMed Central database (http://www.pubmedcentral.nih.gov) is quickly becoming an indispensable tool with more than 400,000 full text papers from over 200 different journals. Rapid and free access to the complete text of published papers significantly enhances the planning, execution, and interpretation phases of any scientific study. The new Books database (http://www.ncbi.nlm.nih.gov/books) provides free access for the first time to electronic versions of many textbooks and other resources such as the NCBI Handbook that serves as a guide to the resources that NCBI has to offer. This is a particularly important resource for students and investigators who need to learn a new discipline such as genomics. One of the oldest databases provided by the NCBI is the GenBank DNA sequence resource [22,23] (http://www.ncbi.nlm.nih.gov/Genbank). DNA sequence data for many different organisms have been deposited in GenBank over the past two decades, now totaling more than 100 gigabases of data. GenBank is a common starting point for the design of PCR primers and other molecular assays that require specific knowledge of gene sequences. Curated information about genes, their chromosomal location, their function, their pathways, and more can be accessed through the Entrez Gene database (http://www.ncbi.nlm.nih.gov/entrez/query.fcgi?db=gene), for example.

Important emerging databases include those that store and summarize DNA sequence variations. NCBI maintains the dbSNP [24,25] (http://www.ncbi.nlm.nih.gov/projects/SNP/) database for single-nucleotide polymorphisms or SNPs. dbSNP provides a wide range of different information about SNPs, including the flanking sequence primers, the position, the validation methods, and the frequency of the alleles in different populations. As with all NCBI databases, it is possible to link to a number of other data sets such as PubMed and OMIM. The recently completed International Haplotype Map (HapMap) project documents genetic similarities and differences among different populations [26]. Understanding the variability of SNPs and the linkage disequilibrium structure plays an important role in determining which SNPs to measure when planning a molecular epidemiology study. The International Hap-Map Consortium maintains an online database with all the data from the HapMap project (http://www.hapmap.org/thehapmap.html). Another useful database is the Allele Frequency Database or ALFRED [27] (http://alfred.med.yale.edu/alfred/index.asp), which currently stores information on more than 3700 polymorphisms across 518 populations.

In addition to databases for storing raw data, a number of databases retrieve and store knowledge in an accessible form. For example, the Kyoto Encyclopedia of Genes and Genomes (KEGG) database stores knowledge on genes and their pathways [28,29] (http://www.genome.jp/kegg). The Pathway component of KEGG currently stores knowledge on 42,937 pathways generated from 307 reference pathways. While the Pathway component documents molecular interaction in pathways, the Brite database stores knowledge on higher-order biological functions. One of the most useful knowledge sources is the Gene Ontology (GO) project, which has created a controlled vocabulary to describe genes and gene products in any organism in terms of their biological processes, cellular components, and molecular functions [30,31] (http://www.geneontology.org). GO descriptions and KEGG pathways are both captured and summarized in the NCBI databases. For example, the description of *p53* in Entrez Gene includes KEGG pathways such as cell cycle and apoptosis. It also includes GO descriptions such as protein binding and cell proliferation.

In general, a good place to start for information about available databases is the annual Database issue and the annual Web Server issue of the journal *Nucleic Acids Research.* These special issues include annual reports from many of the commonly used databases.

DATA ANALYSIS

Once the data are collected and stored in a database, an important goal of molecular pathology is to identify biomarkers or molecular/environmental predictors of disease endpoints. Statistical methods in bioinformatics provide a good starting point for the analysis of molecular pathology data [32]. This can include commonly used methods such as t-tests, analysis of variance, linear regression, and logistic regression [33], or may include more advanced data mining and machine learning methods such as cluster analysis or neural networks [34]. Although many of these methods require special training in mathematics, statistics, or computer science, the good news is that most simple and advanced analysis methods are easily implemented in one or more freely available software packages.

Data Mining Using R

R is perhaps the one software package that everyone should have in his or her bioinformatics arsenal. R is an open source and freely available programming language and data analysis and visualization environment that can be downloaded from http://www.r-project.org. According to the web page, R includes (i) an effective data handling and storage facility; (ii) a suite of operators for calculations on arrays, in particular matrices; (iii) large, coherent, integrated collection of intermediate tools for data analysis; (iv) graphical facilities for data analysis and display either onscreen or on hardcopy; and (v) a well-developed, simple, and effective programming language which includes conditionals, loops, user-defined recursive functions, and input and output facilities. A major strength of R is the enormous community of developers and users that ensure most (many) analysis methods are available. This includes analysis packages such as Rgenetics (http://rgenetics.org) for basic genetic and epidemiologic analysis such as testing for deviations from Hardy-Weinberg equilibrium or haplotype estimation, epitools for basic epidemiology analysis (http://www.epitools.net), geneland for spatial genetic analysis (http://www.inapg.inra.fr/ens_rech/mathinfo/personnel/guillot/Geneland.html), and popgen for population genetics (http://cran.r-project.org/web/packages/popgen/index.html). Perhaps the most useful contribution to R is the Bioconductor project35 (http://www.bioconductor.org). According to the Bioconductor web page, the goals of the project are to (i) provide access to a wide range of powerful statistical and graphical methods for the analysis of genomic data; (ii) facilitate the integration of biological metadata (for instance, using PubMed, GO, and others) in the analysis of experimental data; (iii) allow the rapid development of extensible, scalable, and interoperable software; (iv) promote high-quality and reproducible research; and (v) provide training in computational and statistical methods for the analysis of genomic data.

There are numerous packages for machine learning and data mining that are either part of the base R software or can be easily added. For example, the neural package includes routines for neural network analysis (http://cran.r-project.org/web/packages/neural/index.html). Others include arules for association rule mining (http://cran.r-project.org/web/packages/arules/index.

html), cluster for cluster analysis (http://cran.r-project.org/web/packages/cluster/index.html), genalg for genetic algorithms (http://cran.r-project.org/web/packages/genalg/index.html), som for self-organizing maps (http://cran.r-project.org/web/packages/som/index.html), and tree for classification and regression trees (http://cran.r-project.org/web/packages/tree/index.html). Many others are available. A full list of contributed packages for R can be found at http://cran.r-project.org/web/packages. The primary advantage of using R as your data mining software package is its power. However, the learning curve can be challenging at first. Fortunately, there is plenty of documentation available on the web and in published books. Several essential R books include those by Gentleman et al. [36] and Venables and Ripley [37].

Data Mining Using Weka

One of the most mature open source and feely available data mining software packages is Weka [38] (http://www.cs.waikato.ac.nz/ml/weka). Weka is written in Java and will run in any operating system (including Linux, Mac, Sun, Windows). Weka contains a comprehensive list of tools and methods for data processing, unsupervised and supervised classification, regression, clustering, association rule mining, and data visualization. Machine learning methods include classification trees, k-means cluster analysis, k-nearest neighbors, logistic regression, naïve Bayes, neural networks, self-organizing maps, and support vector machines, for example. Weka includes a number of additional tools such as search algorithms and analysis tools such as cross-validation and bootstrapping. A nice feature of Weka is that it can be run from the command line, making it possible to run the software from Perl or even R (see http://cran.r-project.org/web/packages/RWeka/index.html). Weka includes an experimenter module that facilitates comparison of algorithms. It also includes a knowledge flow environment for visual layout of an analysis pipeline. This is a very powerful analysis package that is relatively easy to use. Further, there is a published book that explains many of the methods and the software [38].

Data Mining Using Orange

Orange is another open source and freely available data mining software package [39] (http://www.ailab.si/orange) that provides a number of data processing, data mining, and data visualization tools. What makes Orange different and in some way preferable to other packages (such as R) is its intuitive visual programming interface. With Orange, methods and tools are represented as icons that are selected and dropped into a window called the canvas. For example, an icon for loading a data set can be selected along with an icon for visualizing the data table. The file load icon is then wired to the data table icon by drawing a line between them. Double-clicking on the file load icon allows the user to select a data file. Once loaded, the file is then automatically transferred by the wire to the data table icon. Double-clicking on the data table icon brings up a visual display of the data. Similarly, a classifier such as a classification tree can be selected and wired to the file icon. Double-clicking on the classification tree icon allows the user to select the settings for the analysis. Wiring the tree viewer icon then allows the user to view a graphical image of the classification tree inferred from the data. Orange facilitates high-level data mining with minimal knowledge of computer programming. A wide range of different data analysis tools are available. A strength of Orange is its visualization tools for multivariate data [40]. Recent additions to Orange include tools for microarray analysis and genomics such as heat maps and GO analysis [41].

Interpreting Data Mining Results

Perhaps the greatest challenge of any statistical analysis or data mining exercise is interpreting the results. How does a high-dimensional statistical pattern derived from population-level data relate to biological processes that occur at the cellular level [42]? This is an important question that is difficult to answer without a close working relationship between pathologists, for example, and statisticians and computer scientists. Fortunately, a number of emerging software packages are designed with this in mind. GenePattern (http://www.broad.mit.edu/cancer/software/genepattern/), for example, provides an integrated set of analysis tools and knowledge sources that facilitate this process [43]. Other tools such as the Exploratory Visual Analysis (EVA) database and software (http://www.exploratoryvisualanalysis.org/) are designed specifically for integrating research results with biological knowledge from public databases in a framework designed for pathologists, for example [44]. These tools and others will facilitate interpretation.

THE FUTURE OF BIOINFORMATICS

We have only scratched the surface of the numerous bioinformatics methods, databases, and software tools that are available to the pathology community. We have tried to highlight some of the important software resources such as Weka and Orange that might not be covered in other reviews that focus on more traditional methods from biostatistics. While there is an enormous number of bioinformatics resources today, the software landscape is changing rapidly as new technologies for high-throughput biology emerge. Over the next few years, we will witness an explosion of novel bioinformatics tools for that analysis of genome-wide association data and, more importantly, the joint analysis of SNP data with other types of data such as gene expression data and proteomics data. Each of these new data types and their associated research questions will require special bioinformatics tools and perhaps special hardware such as faster computers with bigger storage capacity and more memory. Some of these data sets will easily require 1–2 Gb or more of memory or more for analysis and could require as many as 100 processors or more to complete a data mining analysis in a reasonable amount of time. The challenge will be to scale our bioinformatics tools and hardware such that a genome-wide SNP data set can be processed as efficiently as we can process a candidate gene data set with perhaps 20 SNPs today. Only then can molecular pathology truly arrive in the genomics age.

REFERENCES

1. Ohgaki H, Kleihues P. Genetic pathways to primary and secondary glioblastoma. *Am J Pathol.* 2007;170:1445–1453.
2. Phillips HS, Kharbanda S, Chen R, et al. Molecular subclasses of high-grade glioma predict prognosis, delineate a pattern of disease progression, and resemble stages in neurogenesis. *Cancer Cell.* 2006;9:157–173.
3. Maher EA, Brennan C, Wen PY, et al. Marked genomic differences characterize primary and secondary glioblastoma subtypes and identify two distinct molecular and clinical secondary glioblastoma entities. *Cancer Res.* 2006;66:11502–11513.
4. Hau P, Stupp R, Hegi ME. MGMT methylation status: The advent of stratified therapy in glioblastoma? *Dis Markers.* 2007; 23:97–104.
5. Boguski MS. Bioinformatics. *Curr Opin Genet Dev.* 1994;4:383–388.
6. Benson D, Boguski M, Lipman DJ, et al. The National Center for Biotechnology Information. *Genomics.* 1990;6:389–391.
7. Robinson C. The European Bioinformatics Institute (EBI)—Open for business. *Trends Biotechnol.* 1994;12:391–392.
8. Haynes C, Blach C. Information management. In: Haines JL, Pericak-Vance MA, eds. *Genetic Analysis of Complex Disease.* Hoboken, NJ: Wiley; 2006.
9. Schena M, Shalon D, Davis RW, et al. Quantitative monitoring of gene expression patterns with a complementary DNA microarray. *Science.* 1995;270:467–470.
10. Sherlock G, Hernandez-Boussard T, Kasarskis A, et al. The Stanford Microarray Database. *Nucleic Acids Res.* 2001;29:152–155.
11. Barrett T, Suzek TO, Troup DB, et al. NCBI GEO: Mining millions of expression profiles—database and tools. *Nucleic Acids Res.* 2005;33:D562–566.
12. Barrett T, Edgar R. Gene expression omnibus: Microarray data storage, submission, retrieval, and analysis. *Methods Enzymol.* 2006;411:352–369.
13. Brazma A, Hingamp P, Quackenbush J, et al. Minimum information about a microarray experiment (MIAME)—Toward standards for microarray data. *Nature Genetics.* 2001;29:365–371.
14. Brazma A, Parkinson H, Sarkans U, et al. ArrayExpress—A public repository for microarray gene expression data at the EBI. *Nucleic Acids Res.* 2003;31:68–71.
15. Ball CA, Brazma A. MGED standards: Work in progress. *OMICS.* 2006;10:138–144.
16. Brazma A, Kapushesky M, Parkinson H, et al. Data storage and analysis in ArrayExpress. *Methods Enzymol.* 2006;411:370–386.
17. Brazma A, Krestyaninova M, Sarkans U. Standards for systems biology. *Nature Rev Genetics.* 2006;7:593–605.
18. Hewett M, Oliver DE, Rubin DL, et al. PharmGKB: The Pharmacogenetics Knowledge Base. *Nucleic Acids Res.* 2002;30:163–165.
19. Wheeler DL, Barrett T, Benson DA, et al. Database resources of the National Center for Biotechnology Information. *Nucleic Acids Res.* 2006;34:D173–D180.
20. Hamosh A, Scott AF, Amberger J, et al. Online Mendelian Inheritance in Man (OMIM). *Hum Mutat.* 2000; 15:57–61.
21. Hamosh A, Scott AF, Amberger J, et al. Online Mendelian Inheritance in Man (OMIM), a knowledgebase of human genes and genetic disorders. *Nucleic Acids Res.* 2005;33:D514–D517.
22. Benson D, Lipman DJ, Ostell J. GenBank. *Nucleic Acids Res.* 1993;21:2963–2965.
23. Benson DA, Karsch-Mizrachi I, Lipman DJ, et al. GenBank. *Nucleic Acids Res.* 2006;34:D16–D20.
24. Sherry ST, Ward M, Sirotkin K. dbSNP-database for single nucleotide polymorphisms and other classes of minor genetic variation. *Genome Res.* 1999;9:677–679.
25. Sherry ST, Ward MH, Kholodov M, et al. dbSNP: The NCBI database of genetic variation. *Nucleic Acids Res.* 2001;29:308–311.
26. International HapMap Consortium. A haplotype map of the human genome. *Nature.* 2005;437:1299–1320.
27. Cheung KH, Osier MV, Kidd JR, et al. ALFRED: An allele frequency database for diverse populations and DNA polymorphisms. *Nucleic Acids Res.* 2000;28:361–363.
28. Kanehisa M. A database for post-genome analysis. *Trends Genet.* 1997;13:375–376.
29. Ogata H, Goto S, Sato K, et al. KEGG: Kyoto Encyclopedia of Genes and Genomes. *Nucleic Acids Res.* 1999;27:29–34.
30. Ashburner M, Ball CA, Blake JA, et al. Gene ontology: Tool for the unification of biology. The Gene Ontology Consortium. *Nature Genetics.* 2000;25:25–29.
31. Gene Ontology Consortium. The Gene Ontology (GO) project in 2006. *Nucleic Acids Res.* 2006;34:D322–326.
32. Ewens WJ, Grant GR. *Statistical Methods in Bioinformatics.* New York, NY: Springer; 2001.
33. Rosner B. *Fundamentals of Biostatistics.* Pacific Grove, CA: Duxbury; 2000.
34. Hastie T, Tibshirani R, Friedman J. *The Elements of Statistical Learning.* New York, NY: Springer; 2001.
35. Reimers M, Carey VJ. Bioconductor: An open source framework for bioinformatics and computational biology. *Methods Enzymol.* 2006;411:119–134.
36. Gentleman R, Carey VJ, Huber W, et al. *Bioinformatics and Computational Biology Solutions Using R and Bioconductor.* New York, NY: Springer; 2005.
37. Venables WN, Ripley BD. *Modern Applied Statistics with S.* New York, NY: Springer; 2002.
38. Whitten IH, Frank E. *Data Mining.* Boston, MA: Elsevier; 2005.
39. Demsar J, Zupan B, Leban G. Orange: From experimental machine learning to interactive data mining. White Paper (www.ailab.si/orange), Faculty of Computer and Information Science, University of Ljubljana; 2004.
40. Leban G, Bratko I, Petrovic U, et al. VizRank: Finding informative data projections in functional genomics by machine learning. *Bioinformatics.* 2005;21:413–414.
41. Curk T, Demsar J, Xu Q, et al. Microarray data mining with visual programming. *Bioinformatics.* 2005;21:396–398.
42. Moore JH, Williams SW. Traversing the conceptual divide between biological and statistical epistasis: Systems biology and a more modern synthesis. *BioEssays.* 2005;27:637–646.
43. Reich M, Liefeld T, Gould J, et al. GenePattern 2.0. *Nature Genetics.* 2006;38:500–501.
44. Reif DM, Dudek SM, Shaffer CM, et al. Exploratory visual analysis of pharmacogenomic results. *Pacific Symp Biocomput.* 2005; 10:296–307.

Part

IV

Molecular Pathology of Human Disease

Chapter

14

Molecular Basis of Cardiovascular Disease

Amber Chang Liu ▪ Avrum I. Gotlieb

INTRODUCTION

Physicians relied on descriptive studies of the gross and histopathological changes in diseased human cardiovascular organs, discovered meticulously by cardiovascular pathologists, to provide insight into the pathogenesis of infectious, inflammatory, immunological, and degenerative cardiovascular diseases. These studies revealed that the tissue response to disease is characterized by a marked accumulation of cardiovascular cells associated with inflammation, neovascularization, matrix production, eventually fibrosis, and in some cases calcification [1]. Now, mechanistic studies on cardiovascular structure and function are being performed in a new era of innovative heart and blood vessel investigation that is transforming the study of cardiovascular pathobiology from static histopathology research to dynamic mechanistic cell and molecular biology investigation. This transformation was initially advanced with the development of reliable culture methods to study the cells of the cardiovascular system (Table 14.1). The successful field of vascular biology was established in the 1970s once vascular endothelial and smooth muscle cell cultures were developed [2,3]. This was followed by successful characterization of valve interstitial cell (VIC) (Figure 14.1) [4] and valve endothelial cell (VEC) cultures [5]. These cultures allowed for the study of the structure and function of endothelial (EC), smooth muscle (SMC), myocardial [6] and VICs, at the cellular and molecular levels. More recently populations of cells have been identified and isolated in humans that are both renewable and may differentiate along different cardiovascular pathways to participate in cardiovascular repair. These bring hope to be able to repair tissue and restore normal function using cell therapy and tissue engineering approaches.

GENERAL MOLECULAR PRINCIPLES OF CARDIOVASCULAR DISEASES

Several important concepts have emerged as the molecular biology of cardiovascular disease is investigated. (1) Cells of the cardiovascular system have unique properties. (2) Cell function is regulated by the combined actions of specific molecules, some that promote and others that inhibit a cellular process. It is the balance between the bioactivity of all these molecules that dictates the function of the cell at any given moment in time. (3) Depending on conditions, the same molecule may both promote or inhibit a given cellular function, generally by directly or indirectly acting on signaling molecules that regulate different pathways. (4) Signaling pathways interact by sharing downstream molecules. (5) Microenvironments are important in autocrine and paracrine regulation of cardiovascular cell function, including across cell types. (6) Cell-extracellular matrix interactions are critical to normal physiology and pathogenesis of disease. (7) Physical forces regulate functions of cardiovascular cells that are important in both maintaining normal physiology and regulating pathogenesis of disease.

THE CELLS OF CARDIOVASCULAR ORGANS

Vascular Endothelial Cells

Vascular ECs, which are embryologically derived from splanchnopleuric mesoderm, form a thromboresistant barrier on the surface of the vascular tree. The cells are quiescent but have the ability to proliferate once appropriate genes are activated in response to injury and/or disease. These cells are highly metabolically active

Molecular Pathology © 2009, Elsevier, Inc. All Rights Reserved.

Table 14.1 The Cells of the Cardiovascular System

Heart	
	Cardiac Myocytes
	Cardiac Interstitial Fibroblasts
	Valve Interstitial Cells (VICs)
	Valve Endothelial Cells (VECs)
	Endothelial Cells, smooth muscle cells, pericytes of blood vessels
Blood Vessels	
	Endothelial Cells (ECs)
	Smooth Muscle Cells (SMCs)
	Pericytes
	Adventitial Fibroblasts
	Endothelial Cells, smooth muscle cells of vasa vasorum
Stem Cells	
	Endothelial Progenitor Cells (EPCs)
	Mesenchymal stem Cells:
	Bone marrow derived
	Tissue derived
Cells associated with disease	
	Dendritic Cells
	Macrophages/Foam Cells
	Lymphocytes
	Mast Cells
	Giant Cells

and alter their function as their microenvironment changes. These cell functions are balanced between the regulation of physiologic functions that maintain normal homeostasis and the endothelial dysfunction that is associated with pathobiology (Table 14.2). Genetic conditions result in several coagulopathies. A major role of ECs is to transduce hemodynamic shear stress from a physical force to a biochemical signal that regulates gene expression and/or protein secretion of bioactive agents [7]. These shear stress activated molecules include vasoactive compounds, extracellular matrix proteins and degradation enzymes, growth factors, and coagulation and inflammatory factors (Table 14.3).

Vascular Smooth Muscle Cells

Vascular SMCs form the cells of the media and maintain the matrix of the normal vascular wall. Smooth muscle cells are quiescent in this media. However, upon injury, the cells undergo phenotypic transformation to proliferating, secreting, and migrating cells with a capacity to become myofibroblasts and participate in repair. Smooth muscle cells may become foam cells through ingestion of lipids. They participate in autocrine and paracrine pathways, especially in interactions with macrophages and ECs. Smooth muscle cells are important regulators of vascular remodeling.

Valve Endothelial Cells

Valve endothelial cells (VECs) form a single cell layer of adherent cells that cover the surface of the valve. The cells are quiescent, but upon injury, they will proliferate to reconstitute the thromboresistant surface. VECs are heterogenous and show important differences when compared to vascular EC. Using microarray technology, VECs have been shown to differentially express 584 genes on the aortic side versus the ventricular side of normal adult pig aortic valves [8]. Several of these observed differences could help explain the vulnerability of the aortic side of the valve cusp to calcification in diseases such as calcific aortic stenosis (CAS). However, because calcification occurs within the valve tissue, it is likely that VECs may be playing more of a transducing role, regulating VIC function. Valvular ECs also show phenotypic differences in response to shear when compared to vascular ECs [9].

Valve Interstitial Cells

The term valve fibroblasts is still used in the literature. However, it should be abandoned and the term valve interstitial cells (VICs) should be used because these cells do have specific features which are context

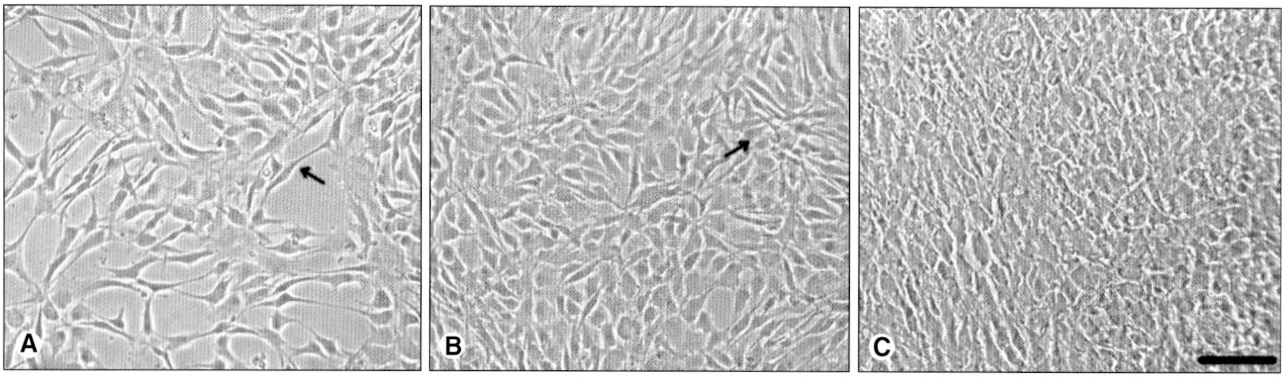

Figure 14.1 Phase contrast photomicrographs of VICs in monolayer culture at moderate (a), confluent (b), and superconfluent (c) densities. Note the elongated morphology in (a) and the overlapping growth pattern in (b) as indicated by arrows. Scale bar represents 20 μm. Magnification 200x. Reprinted from *Am J Pathol.* 2007; **171**:1407–1418 with permission from the *American Society for Investigative Pathology.*

Table 14.2 Endothelial Function

Physiologic Function	Endothelial Dysfunction
Platelet resistant	Platelet adhesion
Anticoagulation	Procoagulation
Fibrinolysis	Antifibrinolysis
Quiescent	Migration/proliferation
Leukocyte resistant	Leukocyte adhesion
Anti-inflammatory	Proinflammatory
Selective impermeability	Enhanced permeability
Quiescent SMC	SMC activation
Provasodilation	Provasoconstriction
Matrix stability	Matrix remodeling
Vessel stability	Angiogenesis

Table 14.3 Shear Stress Regulated Factors in Endothelium

Vasoactive Compounds

- Angiotensin Converting Enzyme (ACE)
- NO — endothelial Nitric Oxide Synthase (eNOS)
- NO — induced Nitric Oxide Synthase (iNOS)
- Prostacyclin
- Endothelin-1

ECM/ECM Degradation Enzymes

- Matrix Metalloproteinase-9 (MMP-9)
- Collagen XII
- Thrombospondin

Growth Factors

- Epidermal Growth Factor (EGF)
- Basic Fibroblast Growth Factor (bFGF)
- Granulocyte Monocyte-Colony Stimulating Factor (GM-CSF)
- Insulin-Like Growth Factor Binding Protein (IGFBP)

Coagulation/Fibrinolysis

- Thrombomodulin
- Tissue Factor
- Tissue Plasminogen Activator (tPA)
- Protease-Activated Receptor-1—thrombin receptor (PAR-1)

Inflammation Factors

- Monocyte chemoattractant protein (MCP-1)
- Vascular cell adhesion molecule (VCAM-1)
- Intercellular adhesion molecule (ICAM-1)
- E-Selectin

Others

- Extracellular superoxide dismutase (ecSOD)
- Sterol regulatory element binding protein (SREBP)
- Platelet/endothelial cell adhesion molecule (PECAM-1)

dependent to the heart valve and show differences when compared to fibroblasts in other tissues and organs.

The valve matrix contains VICs distributed in all three layers of the leaflet: the fibrosa, the spongiosa, and the ventricularis. Compartmentalization occurs at late gestation. However, between 20 and 39 weeks the valves only have a bilaminar structure. It is not known how remodeling of the valve into individual compartments occurs. It is clear that physical forces do play some role because the three layers seen in the adult architecture are not complete until early adulthood [10]. Cell cultures of VICs have been characterized [4] (Figure 14.1) and have provided new information on the cell and molecular biology of these cells [11,12].

Five phenotypes best represent the VIC family of cells (Figure 14.2). Each phenotype exhibits specific sets of cellular functions essential in normal valve physiology and in pathobiologic conditions (Table 14.4) [11]. The phenotypes are referred to as embryonic progenitor endothelial mesenchymal cells, quiescent VICs (qVICs), activated VICs (aVICs), progenitor VICs (pVICs), and osteoblastic VICs (obVICs). The embryonic progenitor endothelial/mesenchymal cells undergo endothelial-to-mesenchymal transformation (EMT) that initiates the process of valve formation in the embryo [13]. The qVICs are at rest in the adult valve and maintain normal valve physiology. The aVICs regulate the pathobiological response of the valve in disease and injury. These cells are a type of myofibroblast cell similar to those found at sites of wound repair in a variety of tissues. The pVICs are poorly defined and consist of a heterogeneous population of progenitor cells that may be important in repair. The obVICs regulate chondrogenesis and osteogenesis. Although these phenotypes may exhibit plasticity and convert from one form to another (Figure 14.2), characterizing VIC function by distinct phenotypes brings clarity to our understanding of the complex VIC biology and pathobiology by focusing investigations on the interaction of each specific VIC phenotype within the valve and the systemic environment in which it resides. Better understanding of the nature of these phenotypes will occur, and new phenotypes will also be discovered.

Leukocytes

There are occasional macrophages and lymphocytes in the normal vessel wall, especially in the intima. Endothelial dysfunction due to injury and/or inflammation promotes monocytes to enter the wall, become activated macrophages, and promote vessel dysfunction and further injury. Macrophages may transform into foam cells. Polymorphonuclear leukocytes are prominent at the interface of necrotic and intact myocardium in a myocardial infarction and in the early stages of several vasculitides.

Vascular and Cardiac Progenitor/Stem Cells

In the last decade, human studies and experimental animal model and cell culture investigations have identified a variety of embryonic and adult-derived cell types that exhibit the potential for vascular or myocardial repair of injured and diseased tissue. Some of these cells are differentiated cells, such as skeletal myoblasts and cardiomyocytes, and others are multipotent embryonic stem cells and multipotent adult stem cells. Some of these cells can adopt both vascular and cardiomyocyte phenotypes such as mesenchymal stromal cells [14] and fetal liver kinase-1 [15]. The stem or progenitor cells are usually rare within a population

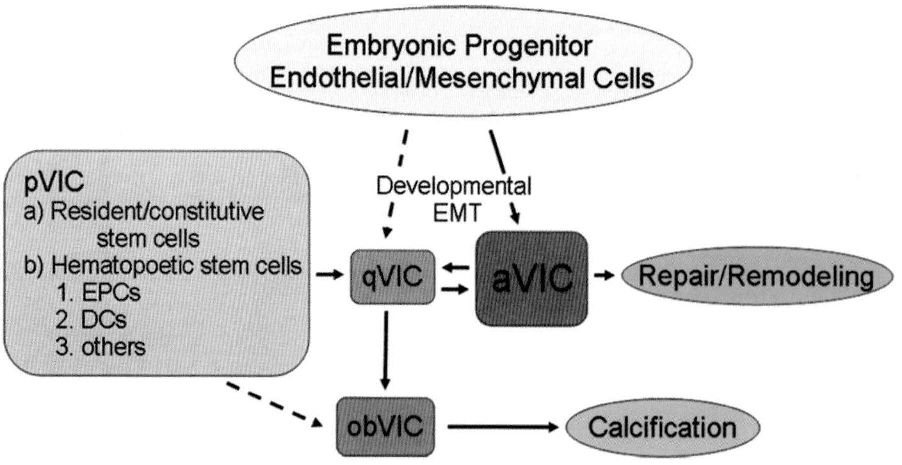

Figure 14.2 **The current literature describes numerous VIC functions which can be conveniently organized into five phenotypes: Embryonic progenitor endothelial/mesenchymal cells, quiescent VICs (qVICs), activated VICs (aVICs), stem cell derived progenitor VICs (pVICs), and osteoblastic VICs (obVICs).** These represent specific sets of VIC functions in normal valve physiology and pathophysiology. Embryonic progenitor endothelial/mesenchymal cells undergo endothelial-mesenchymal transformation in fetal development to give rise to aVIC and/or qVICs resident in the normal heart valve. The VICs undergoing the transformation do have features of aVICs, including migration, proliferation, and matrix synthesis. When the heart valve is subjected to an insult, be it abnormal hemodynamic/mechanical stress or pathological injury, qVICs become activated, giving rise to aVICs which participate in repair and remodeling of the valve. pVICs including bone marrow-derived cells, circulating cells, and resident valvular progenitor cells are another source of aVICs in the adult. The relationship between bone marrow, circulating, and resident pVICs is unknown. Under conditions promoting valve calcification, such as in the presence of osteogenic and chondrogenic factors, qVICs can undergo osteoblastic differentiation into obVICs. It is possible that obVICs are derived from pVICs. obVICs actively participate in the valve calcification process. Compartmentalizing VIC function into distinct phenotypes recognizes as well the transient behavior of VIC phenotypes. The hatched arrows depict possible transitions for which there is no solid evidence currently. Reprinted from *Am J Pathol.* 2007; **171**:1407–1418 with permission from the *American Society for Investigative Pathology.*

Table 14.4 Heart Valve Interstitial Cell Phenotypes

Cell Type	Location	Function
Embryonic progenitor endothelial/mesenchymal cells	Embryonic cardiac cushions	Give rise to resident qVICs, possibly through an activated stage and promoted by FGF2 and TGFβ. EMT can be detected by the loss of endothelial and the gain of mesenchymal markers
qVICs	Valve cusp leaflet	Maintain physiologic valve structure and function and inhibit angiogenesis in the leaflets
aVICs	Valve cusp leaflet	VICs with α-SMA signifying activation of cellular repair processes including proliferation, migration, apoptosis, matrix remodeling, and upregulation of TFGβ[12].
pVICs	Bone marrow, circulation, within valve leaflet cusp	Enter injured valve or are resident in valve to provide aVICs to repair valve, may be CD34–, CD133–, and/or S100-positive
obVICs	Within valve cusp leaflet bone marrow, circulation	Calcification, chondrogenesis, and osteogenesis of valve. Secrete alkaline phosphatase, osteocalcin, osteopontin, bone sialoprotein

Modified from *Am J Pathol.* 2007; 171:1407–1418

of cells, and specific techniques are required to isolate the cells and then expand the population, usually *ex vivo.* The cells are identified by specific markers. The therapeutic potential of these cells following transplantation is considered to be due to paracrine effects since the cells do not readily expand *in vivo* following cell therapy and do not persist for long. The biology and pathobiology of these cells are still poorly understood, and the literature contains numerous controversies, primarily due to the fact that different methods of isolation and different sources of cells have been utilized experimentally and clinically.

Endothelial progenitor cells (EPCs) are a specialized subset of hematopoietic cells found in the adult bone marrow and peripheral circulation arising from hemangioblasts prenatally [16]. They are phenotypically characterized by antigens including CD133, CD34, c-kit, VEGFR2, CD144, and Sca-1. EPCs are immature cells which have the capacity to proliferate, migrate, and differentiate into endothelial lineage cells, but have not yet acquired characteristics of mature ECs, including surface expression of vascular endothelial cadherin and von Willebrand factor and loss of CD133. The discovery of circulating EPCs in the adult changed the view that new blood vessel growth occurs exclusively by angiogenesis postnatally. The process of vasculogenesis in the developing embryo is thought to be re-employed in the adult when EPCs are mobilized and recruited to regions of neovascularization to form new blood vessels. Parallels in the regulatory steps of embryonic and adult vasculogenesis suggest that the underlying initiating stimulus and regulatory pathways may be conserved. EPCs have been the subject of intense experimental and clinical investigations due to their therapeutic potential in cardiovascular regeneration. Since endothelial damage and/or dysfunction may initiate atherosclerosis, bone marrow-derived and circulating EPCs may play an important role in re-establishing a normal endothelium and thus protecting the vessel wall from progression of disease.

The role of EPCs in determining the pathogenesis and prognosis of cardiovascular disease is under study although there are currently no means to precisely track the kinetics of bone marrow-derived EPCs for pathological neovascularization in humans. The cell number and migratory activity of circulating EPCs is decreased in patients with stable coronary artery disease compared to age-matched control subjects and is inversely correlated with the number of coronary risk factors in coronary artery disease patients [17]. Proliferation of EPCs obtained from patients with type II diabetes was decreased by 48% compared with controls [18]. EPCs from subjects at high risk for cardiovascular events had higher rates of *in vitro* senescence than cells from subjects at low risk [19]. These clinical findings indicate that EPCs may be sensitive indicators of heightened risk of cardiovascular diseases.

EPC transplantation to promote collateral circulation is a new therapy offering hope to treat tissue ischemia. When *ex vivo* expanded EPCs obtained from healthy human peripheral blood are intravenously administered into immunodeficient mice with hindlimb ischemia, the result is cell incorporation and *in situ* differentiation into EC lineage, as well as physiological evidence of enhancement of limb blood flow [20]. *Ex vivo* expanded human EPCs intravenously administered into nude rats with acute myocardial infarction contributed to ischemic neovascularization following recruitment into the ischemic area. The EPC therapy also inhibited left ventricular fibrosis and preserved left ventricular function. Human EPC-derived cardiomyocytes and SMC were identified in the rat infarcted myocardium. Whether improvement was due to increased myocardial perfusion and/or increased cardiac muscle mass was not studied [21]. On the basis of the promising outcomes in animal models, clinical applications of EPCs for ischemic diseases are in progress. Intramyocardial injection of autologous bone marrow EPCs into patients with chronic myocardial infarction at the time of coronary artery bypass grafting surgery improved left ventricular global function, improved left ventricular ejection fraction, as well as cardiac perfusion 3–9 months later [22]. Intramyocardial transplantation in intractable angina in chronically ischemia myocardium likely due to induction of neovascularization [23]. However, the life of transplanted EPCs is often short, so many investigators are interpreting the improved function to be due to paracrine effects that the EPCs have on resident cells due to the secretion of bioactive molecules such as cytokines.

Endothelial damage is a well-known trigger of restenosis after percutaneous balloon angioplasty or stenting. EPC-based therapeutic strategies to improve the function and number of ECs following surgery have gained considerable interest. Statins were found to inhibit restenosis in injured murine carotid arteries through enhanced mobilization and incorporation of bone marrow-derived EPCs for re-endothelialization [24]. Implanting VEGF-2 coated stents in rabbit injured iliac arteries also inhibited restenosis through enhanced mobilization of bone marrow-derived EPCs [25]. Thus, EPCs appear to contribute to re-endothelialization in damaged vessels and inhibit restenosis. Intravenous infusion of EPCs in a mouse model of carotid artery injury leads to incorporation of EPCs into the injured vessel wall and re-endothelialization, resulting in inhibition of neointimal hyperplasia. Recently, an innovative approach to EPC-mediated re-endothelialization was carried out. Stainless steel stents coated with anti-CD34 antibody were developed to capture circulating EPCs onto the stent surface to augment re-endothelialization and prevent restenosis and thrombosis [26]. Clinical safety and feasibility of this technology have been demonstrated, and preliminary clinical studies are under way.

Cardiac Stem Cells

The adult heart is considered to be a postmitotic organ comprising fully differentiated cardiomyocytes which survive a lifetime without replenishment. An elevated number of immature cardiomyocytes, capable of mitotic division, have been identified at the infarct border zone of myocardial infarction [27] in some investigations, whereas other studies failed to identify sufficient proliferating cells at sites of myocardial infarction. However, adult heart may contain a population of stem cells either entering from the circulation or resident *in situ* which possess regenerative properties [28]. If the biology of these cells can be characterized, then they may become useful clinically.

The adult myocardium contains a small fraction of stem cells which can give rise to cardiomyocytes that express the cell surface markers c-kit and Sca-1, and are characterized by their ability to exclude Hoechst dye [29]. Human percutaneous endomyocardial biopsy specimens have produced multicellular clusters of cells called cardiospheres which contain cells with the potential for cardiogenic differentiation. Upon injection into the border zone, cardiosphere-derived cells migrate into the infarct zone and improve the proportion of viable myocardium and cardiac function [30]. Another population of endogenous cardiac progenitor cells is the multipotent Isl1+ cardiovascular progenitors that are isolated from neonatal animals and have the potential to form cardiac muscle, smooth muscle, and endothelial cell lineages [31]. More recently, epicardium-derived progenitor cells are thought to undergo epithelial-to-mesenchymal transformation and show extensive migratory capacities to promote coronary vasculogenesis upon stimulation by paracrine factors from the underlying myocardium [32,33]. To date, thymosin β is the only identified factor that stimulates adult epicardium-derived progenitor cells which can differentiate into fibroblasts, endothelial cells, smooth muscle cells, and cardiomyocytes [34]. Transplantation of undifferentiated embryonic stem (ES) cells has resulted in improvement of cardiac function. However, this technology is in its infancy and must overcome problems in control of cardiac differentiation and introduction of undifferentiated cells [35].

These studies suggest that there are many candidate cell types that may promote myocardial repair [36]. Stimulation of exogenous and/or endogenous cardiac stem/progenitor cells to initiate repair *in situ* is a new and exciting area of research with potential for therapeutic cardiac regeneration. Much more investigation is needed to understand this complex area of pathobiology. Clinical success will occur once the basic principles regulating stem/progenitor cell therapy are well understood. Premature clinical studies may delay rather than advance the field.

ATHEROSCLEROSIS

Atherosclerosis is a chronic vascular disease initially developing in the intima of elastic and larger muscular arteries and characterized by the presence of fibro-inflammatory lipid plaques (atheromas) which grow in size to protrude into the vascular lumen and to involve the media of the artery [37]. Focal plaque growth eventually leads to clinical disease characterized by the development of complicated plaques, lumenal stenosis, and focal weakening of vessel walls, especially the aorta. Clinically, atherosclerosis leads to local aneurysm and/or rupture, and to end-stage organ disease including ischemic heart disease, cerebral vascular disease, and peripheral vascular disease. Epidemiologic studies have identified environmental and genetic conditions that increase the risk of developing clinical atherosclerotic disease (Table 14.5). Currently, risk modification has become an important medical approach to prevention and treatment of atherosclerosis. Biomarkers of endothelial dysfunction are being studied including VCAM-1, ICAM-1, ELAM-1, high sensitivity C-reactive protein, IL-1, IL-6, TNFα. It is hoped that these markers will provide a greater predictive capability beyond the traditional risk factors to identify high-risk individuals.

Table 14.5 Risk Factors for Atherosclerosis

LDL receptor mutations	Obesity
Hyperlipidemias	Family History
Diabetes Mellitus	Gender
Hypertension	Advancing Age
Cigarette Smoke	

The interplay between an individual's genetic disposition and the environment adopted by the individual may result in an imbalance between proatherogenic and antiatherogenic factors and processes that then leads to initiation and growth of the atherosclerotic plaque. At present, it is unlikely to identify a single atherogenic gene that explains pathogenesis. Instead, multiple genes (polygenic), including clusters of genes forming networks regulating specific cell functions, interact with the environment and with each other to promote atherogenesis. Recently, the use of high-density genotyping arrays have led to genome-wide association studies showing a strong association of coronary artery disease with a chromosomal locus on chromosome 9p21.3. These studies in subjects of Northern European origin identified four single nucleotide polymorphisms (SNPs), two associated with risk for coronary artery disease and two associated with risk for myocardial infarction. However, cross-race susceptibility has been identified in South Korean [38] and Italian populations [39]. This common variant is located adjacent to the cyclin-dependent kinase inhibitors *CDKN2A* and *CDKN2B* [40]. The pathobiologic role of the 9p21.3 locus is not understood at present. One study has shown that the locus does not have an affect on coronary artery disease through a pathogenetic pathway that affects carotid intima media thickness or brachial artery flow-mediated dilatation, early markers of clinical atherosclerosis.

The development of atherosclerosis may begin as early as the fetal stage, with the formation of intimal cell masses, or perhaps shortly after birth, when fatty streaks begin to evolve. However, the characteristic lesion, which is not initially clinically significant, requires as long as 20 to 30 years to form. Once this lesion is formed, serious acute events may occur, and/or complicated lesions may emerge after several more years of plaque growth.

A convenient way to view pathogenesis is to propose that three stages can be identified over the course of the evolution of the clinical plaque in humans: (i) a plaque initiation and formation stage, (ii) a plaque adaptation stage, and (iii) a clinical stage [37]. Biologically active molecules regulate a number of dynamic cellular functions. As more molecules are identified in the vessel wall and the plaque, their function is studied to determine their antiatherosclerotic (atheroprotective) or atherogenic (atherogenic)potential. In addition, a given molecule may have both atheroprotective and atherogenic effects. Thus, some of the molecules that are considered to regulate the three stages have been identified in a variety of human and experimental studies, especially those associated with lipid metabolism, coagulation and thrombosis, inflammation, cell growth, and matrix regulation. Much effort has gone into identifying genetic variants of specific genes that are described as candidate genes because they have been shown to be important in health and disease of the artery wall [41]. Examples include ACE insertion/deletion, *APOE, APOE2, APOE3, APOE4,* and *MTHFR* C677T. These genetic association studies represent our best understanding at present, but most are not robust and further studies are needed. Caution is required in interpreting genetic association studies [42–44], especially single reports and reports that are difficult to confirm because there is variability in outcomes of several different studies.

Stage I: Plaque Initiation and Formation

The intimal lesions initially occur at vascular sites considered to be predisposed to plaque formation. Endothelial injury is an early event and may be due to several conditions including microorganisms, toxins, hyperlipidemia, hypertension, and immunologic events. Hemodynamic shear stress at branch points and curves may also induce endothelial dysfunction, a predisposition for plaque formation. The accumulation of subendothelial SMCs, as occurs in an intimal cell mass (eccentric intimal thickening, intimal cushion) at branch points and at other sites in certain vessels, particularly the coronary arteries, is considered a predisposing condition for plaque formation since it provides a readily available source of SMCs. It is thought that this intimal thickening is a physiological adaptation to mechanical forces. In humans, atherosclerotic lesions tend to occur at sites where shear stresses are low but fluctuate rapidly, such as at branch points. The fact that hypertension enhances the severity of atherosclerotic lesions in various locations, such as in the pulmonary artery in pulmonary hypertension, further promotes the idea that hemodynamic factors are somehow involved in the development of atherosclerosis. Low shear has been shown to induce cell adhesion molecules on the surface of endothelial cells to promote monocyte attachment. This is regulated by upregulation of vascular cell adhesion molecule (VCAM) on the surface of ECs. The leukocytes first roll along the endothelium mediated by P-selectin and E-selectin and then adhere due to chemokine-induced EC activation and integrin interactions with cell adhesion molecules. The leukocytes penetrate the endothelial barrier at interendothelial sites, regulated by platelet EC adhesion molecule (PECAM, CD31). Low shear also disrupts normal repair following endothelial injury, thus exposing a denuded endothelial surface to blood flow for longer periods of time. Hemodynamic forces induce gene expression of several biologically active molecules in ECs that are likely to promote atherosclerosis, including FGF-2, tissue factor (TF), plasminogen activator, and endothelin (Table 14.3). However, shear stress also induces gene expression of agents that are considered antiatherogenic, including NOS and PAI-1 (Table 14.3).

Lipid accumulation depends on disruption of the integrity of the endothelial barrier through disruption of cell-cell adhesion junctions, cell loss, and/or cell dysfunction. Low-density lipoproteins carry lipids into the intima. Monocyte/macrophages adhere to activated ECs and transmigrate into the intima bringing in lipids. Some macrophages become foam cells, due in part to the uptake of oxidized LDL via scavenger receptors, and undergo necrosis and release lipids. A change in the types of connective tissue and proteoglycans synthesized by the SMCs in the intima also renders these sites prone to lipid accumulation. These proteoglycans have a high binding affinity for lipoproteins such as chondroitin sulfate-rich proteoglycans. Versican and biglycan are thought to promote atherosclerosis, while decorin may be protective [45]. Oxidative stress in ECs and macrophages leads to cellular dysfunction and damage.

Macrophages, in addition to playing a central role by participating in lipid accumulation, secrete several types of cytokines and release growth factors, thereby promoting further accumulation of SMCs. Oxidized lipoproteins induce tissue damage and further macrophage accumulation. Macrophages secrete MCP-1, which promotes macrophage accumulation. Macrophages secrete reactive oxygen species. Monocyte/macrophages synthesize PDGF, FGF, TNF, IL-1, IL-6, interferon-γ (IFN-γ), and TGFβ, each of which can modulate the growth of SMCs and ECs. For example, IFN-γ and TGFβ inhibit cell proliferation and could account for the failure of EC regeneration to maintain an intact surface over the lesion as it protrudes into the lumen. Alternatively, such molecules could inhibit growth-stimulatory peptides. Of particular interest is the discovery that the cytokines IL-1 and TNF stimulate endothelial cells to produce platelet-activating factor (PAF), TF, and plasminogen activator inhibitor (PAI). TF expression is also upregulated by oxidized lipids and by disruption of the fibrous cap. Thus, the normal anticoagulant vascular surface becomes a procoagulant one.

As the lesion progresses, small mural thrombi may develop on the damaged intimal surface which has become prothrombotic. This stimulates the release of numerous molecules from adherent and activated platelets, including PDGF, which accelerates smooth muscle proliferation; TGFβ, which enhances the secretion of matrix components; and thrombin, ADP, and thromboxane, which promote further platelet activation. The thrombus grows as these molecules promote the prothrombotic state. Since thrombosis also initiates fibrinolysis and inhibition of factors in the coagulation pathway, the thrombus may alternatively lyse. Another scenario modulated in part by TGFβ (which regulates secretion of collagen, matrix proteins, and differentiation of SMC into myofibroblasts) is organization of the thrombus and incorporation into the plaque. Further growth of the thrombus is a function of the coagulation cascade, which may continue to be stimulated by cytokines and tissue factor.

The deeper parts of the thickened intima are poorly nourished. Hypoxia promotes HIF-1alpha translocation to the nucleus of SMCs and macrophages which bind to the promoter-specific hypoxia response element,

Table 14.6 Composition of Atherosclerotic Plaque

Cells	- Endothelial - Smooth Muscle - Macrophages - Foam - Lymphocytes (T-cells) - Giant cells
Matrix	- Collagens - Proteoglycans—biglycan, versican, perlecan - Elastin - Glycoproteins
Lipids and lipoproteins, cholesterol crystals	
Serum Proteins	
Platelet and leukocyte products	
Necrotic Debris	
Microvessels	
Hydroxyapatite crystals	

leading to the transcriptional activation of VEGF, and other target genes. The macrophages and SMCs undergo ischemic necrosis, as well as apoptosis. Cell death is also promoted by proteolytic enzymes released by macrophages and by tissue damage caused by oxidized LDL and other reactive oxygen species. VEGF initiates angiogenesis with new vessels forming in the plaque derived from the vasa vasorum. Some regard the presence of neovascularization as a condition that establishes permanency to the plaque and prevents significant regression.

The fibroinflammatory lipid plaque is formed, with a central necrotic core and a fibrous cap which separates the necrotic core from the blood in the lumen (Table 14.6). The plaque is heterogeneous with respect to inflammatory cell infiltration, lipid deposition, and matrix organization. TGFβ is an important regulator of plaque remodeling and extracellular matrix deposition. TGFβ increases several types of collagen, fibronectin, and proteoglycans. It inhibits proteolytic enzymes that promote matrix degradation and enhances expression of protease inhibitors.

The expression of HLA-DR antigens on both ECs and SMCs in plaques implies that these cells have undergone some type of immunological activation, perhaps in response to IFN-γ released by activated T-cells in the plaque. The presence of T-cells reflects an immune response that is important for the progression of atherosclerotic lesions. Possible antigens include oxidized LDL to which antibodies have been identified in the plaque.

Stage II: Adaptation Stage

As the plaque encroaches upon the lumen (in coronary arteries), the wall of the artery undergoes remodeling to maintain the original lumen size likely regulated by TGFβ [46]. Once a plaque encroaches upon half the lumen, compensatory remodeling can no longer maintain normal patency, and the lumen of the artery becomes narrowed (stenosis). Hemodynamic shear stress is an important regulator of vessel wall remodeling acting through the mechanotransduction properties of the ECs. Shear stress activates the expression of a variety of genes that encode for proteins that promote remodeling such as MMPs, collagens, bFGF, TGFβ, and inflammatory factors. Smooth muscle cell turnover-proliferation and apoptosis, and matrix synthesis and degradation modulate remodeling of the vessel and the plaque in the face of atherosclerosis. The molecules that are important in matrix remodeling are metalloproteinases (MMP) and their inhibitors (TIMP). This compensatory remodeling is useful because it maintains patency and blood flow in the lumen. However, it may delay clinical diagnosis of the presence of atherosclerosis since the plaque may be clinically silent without demonstrating any symptoms. At this stage it would be very useful to have a group of biomarkers that can reliably assess the extent of subclinical atherosclerosis present in the unsuspecting person. Even though the plaque is small, plaque rupture with catastrophic results may occur at this stage.

Stage III: Clinical Stage

Plaque growth continues as the plaque encroaches into the lumen. Hemorrhage into a plaque due to leakage from the small fragile vessels of neovascularization may not necessarily result in actual rupture of the plaque but may still increase plaque size. Complications develop in the plaque, including surface erosion, ulceration, fissure formation, calcification, and aneurysm formation. Calcification is driven by chondrogenesis and osteogenesis, regulated in part by TGFβ, osteogenic progenitor cells, and bone-forming proteins. Activated mast cells are found at sites of erosion and may release proinflammatory mediators and cytokines. Continued plaque growth leads to severe stenosis or occlusion of the lumen. Plaque rupture, through the fibrous cap, and ensuing lumen thrombosis and occlusion may precipitate acute catastrophic events in these advanced plaques, such as acute myocardial infarct. Table 14.7 describes risk factors for plaque rupture. Plaques causing less than 50% stenosis may also suddenly rupture. Investigations to discover biomarkers to identify patients with plaques at risk for rupture have not been successful to date.

Table 14.7 Plaque Rupture

- Endothelial erosion, ulceration, fistula
- Thin fibrous cap
- Decreased smooth muscle cells in cap
- Inflammation— Macrophages
- Foam cells
- Hemodynamic shear stress
- Imbalance in matrix synthesis/degradation (metalloproteinases, tissue inhibitors of metalloproteinases)
- Nodular calcification

ISCHEMIC HEART DISEASE

Ischemic heart disease (coronary heart disease) is described clinically as stable angina and acute coronary syndromes including unstable angina, non-ST elevation myocardial infarction, and ST-elevation myocardial infarction. Reliable biomarkers are available to identify cardiac damage, especially sensitive cardiac troponin assays [47,48]. There have been investigations to validate putative risk factors of acute coronary syndromes, but candidate gene variants have not been well characterized. Those in the field are aware that new novel biomarkers will not enter the clinical arena unless they are shown to significantly improve diagnosis and provide clinicians with better tools to monitor and guide treatment than are currently available.

Ischemia of the myocardial cell leads to a series of intracellular structural and biochemical changes which begin almost immediately after onset and evolve over time. Initiation of ATP depletion begins within seconds. Initially the reaction of the cells results in reversible injury; however, by 20–30 minutes the myocardial cells become irreversibly injured and undergo necrosis. In the myocardium, cardiomyocytes in the subendocardium are most at risk for ischemia. Thus, irreversible injury occurs first in the subendocardium and progresses as a wave front toward the epicardium, resulting in a transmural myocardial infarct. Progression of necrosis will involve the full cardiac bed supplied by the occluded coronary artery and usually is complete by 6 hours. The repair of the infarcted myocardium follows a well-characterized sequence of necrosis, inflammation, granulation tissue, remodeling, and scar formation (Table 14.8).

ANEURYSMS

Aneurysms may occur in any vessel. However, the thoracic aorta, the abdominal aorta, and the cerebral arteries are common sites. These aneurysms often result in rupture. Blood dissects along the long axis of the media resulting in a channel filled with blood. There is an intimal tear, the entry point, and often a distal exit tear back into the lumen. Dissection may involve aortic branches. Thoracic dissections are classified based on involvement of the ascending aorta: Type A (ascending aorta involved) and Type B (distal, sparing the ascending aorta). Associated conditions include atherosclerosis, hypertension, bicuspid aortic valve, and idiopathic aortic root dilation.

Medial degeneration characterized by fragmentation and loss of elastic fibers, accumulation of proteoglycans, and depletion of SMCs is a common nonspecific histologic finding in aneurysm or dissection. Degenerative medial change is associated with hyaline and hyperplastic arteriosclerosis of the adventitial vasa vasorum. In all cases of aneurysms and dissection whatever the histologic picture, vascularization is present in the areas of medial destruction in the form of thin-walled, widely patent vessels. Thus, the histopathology is not helpful in establishing the pathogenesis of thoracic aortic aneurysms and dissections. However, genetic information has linked several of these cases to genetic syndromes, including Marafan [49], Ehlers-Danlos (Type IV), and Loeys-Dietz [50] syndromes, and filamin A mutations. In addition, these conditions may also occur as an inherited autosomal dominant condition with decreased penetrance and variable expression without the syndromes. Mutations have been identified in genes for fibrillin-1 (*FBN1*), TGFβ receptor 2 (*TGFβR2*), TGFβ receptor1 (*TGFβR1*), SMC specific β myosin (*MYH11*), and α-actin (*ACTA2*).

Studies suggest that both human and experimental mouse models show that an increase in TGFβ signaling is important in the pathogenesis of Marfan and Loeys-Dietz syndromes (see page 237). Patients with single gene defects suggest that mutations disrupt the contractile functions of vascular SMCs, which leads to activation of the stress and stretch pathways of the SMC [51]. It has been postulated that the stretch pathways promote increased levels of matrix metalloproteins (MMPs; especially MMP2 and MMP9) and proteoglycans, and promote proliferative agents such as IGF-1, TGFβ, and MIP1α and MIP1β. A recent review suggests that the progress in identifying genetic determinants for intracranial aneurysms is very limited [52]. Genome-wide linkage studies have identified two loci on chromosomes 1p34.3-p36.13 and 7q11 with association with positional candidate genes, perlecan gene and elastin and collagen type 1 2A gene, respectively. The authors discuss the difficulties encountered in studies in the genetics of intracranial aneurysms [52].

The pathogenesis of abdominal aortic aneurysms is poorly understood. However, connective tissue degradation, inflammation, and loss of smooth muscle are characteristic features. Rupture is considered to be due to collagen degradation due to increased MMP2, MMP9, and cysteine collagenase [53]. Recent studies identified MMP8; cysteine proteases cathepsin K, L, and S; and osteoclastic proton pump vH+-ATPase as important enzymes that are responsible for proteolysis of the medial and adventitial type I/III fibrillar collagen [54]. Others have focused on the role of cytokines in abdominal aortic aneurysms, including IL1β, TNFα, MCP-1, IL-8, and others. Utilizing expression profiling

Table 14.8 Myocardial Infarction: Stages of Healing

I.	**Ischemic Injury**
	Necrosis
	Hemorrhage
II.	**Inflammation**
	Vascular phase (vasodilatation, edema)
	Polymorphonuclear leukocytes
	Mononuclear leukocytes
III.	**Granulation Tissue**
	Neovasculature
	Fibroblastic activity with myofibroblasts differentiation, proliferation and production of extracellular matrix
IV.	**Scar**
	Matrix remodeling
	Prominent fibrosis
	Blood vessel remodeling and regression

of a 42-cytokine protein array, it was shown that the aneurysm wall exhibits a specific cytokine profile of upregulated proinflammatory cytokines, chemokines, and growth factors. The descriptive analysis of end stage human tissue confirmed previous findings and also identified several new factors, including GCSF, MCSF, IL-13, and others.

VASCULITIS

Vasculitis is inflammation of the blood vessel wall. In general, there are several ways of classifying this condition, including based on the size of the vessel affected, anatomic site, microscopic morphology of the lesions, and/or clinical course. There are numerous overlaps between different vasculitides. Large vessel includes giant cell and Takayasu arteritis; medium includes polyarteritis nodosa and Kawasaki disease; and small vessel includes Wegner granulomatosis, Churg-Strauss syndrome, and microscopic polyarteritis. Although the pathogenesis is still poorly understood for most vasculitides, there is a classification based on putative pathogenesis which includes immunologic and infectious etiology. Infections may be direct or indirect and involve all types of microorganisms. Immunologic pathogenesis may be characterized as (i) immune complex conditions which include those induced by infection, serum sickness, drugs, systemic lupus erythematosis, rheumatoid arthritis, and Henoch-Schönlein purpura; (ii) antineutrophil cytoplasmic antibody conditions (ANCA) including Wegner granulomatosis (autoantigen is proteinase-3), microscopic polyarteritis (autoantigen is myeloperoxidase), and Churg-Strauss syndrome (myeloperoxidase); (iii) antiglomerular basement membrane antibody, as in Goodpasture Syndrome; (iv) antiendothelial cell antibodies, as in Kawasaki disease and systemic lupus erythematosis; (v) cell-mediated response, as in allograft rejection. ANCA titers in patients are associated with a recurrence of active disease, and ANCA has been shown to activate neutrophils, monocytes, and ECs and cause arteritis and glomerulonephritis in experimental animal models.

Polymorphisms of candidate genes have been explored, including immunoregulatory genes, matrix metalloproteinases, nitric oxide synthase, and I-kappaB-like protein [55,56]. These studies are not robust and will need to be larger and shown to be reproducible. The general conceptual approach is that vasculitides are the result of complex interactions between environmental factors and genetically determined host responses. How this works is not known but is being studied.

VALVULAR HEART DISEASE

Mitral Valve Prolapse

Mitral valve prolapse (MVP) is a heart valve condition characterized by progressive thinning of the mitral leaflet tissue, causing leaflets to billow backward during ventricular contraction, prolapsing into the left atrium beyond their normal position of closure at the level of the mitral ring or annulus. The most common etiology of systolic mitral regurgitation in patients with severe mitral insufficiency referred for mitral valve surgery remains myxomatous degeneration. The myxomatous changes are seen as part of connective tissue syndromes or as primary valve disease.

The natural history of asymptomatic MVP is extremely heterogeneous. It can vary from benign, with a normal life expectancy, to adverse, with significant morbidity and mortality attributed to the development of valvular insufficiency. Fortunately, its complications, including heart failure, mitral regurgitation, bacterial endocarditis, thromboembolism, and atrial fibrillation, are extremely uncommon, affecting less than 3% of subjects with MVP. A cause of abrupt clinical deterioration is sudden chordal rupture due to attenuation and thinning of chordal tissue. When associated with systemic disease, like Marfan syndrome, the myxomatous degeneration is more extensive and involves other heart valves.

MVP valves show myxomatous degeneration with greatly increased type III collagen, some increase of type I and V collagens, and an accumulation of dermatan sulfate, a glycosaminoglycan, within the valve matrix. The accompanying loss in elastin and reduction in SMCs is similar to the histological changes in the valve cusps that were described previously in dissecting aneurysm and are often part of syndromes that involve valve and aorta.

Connective Tissue Disorders

MVP is a feature of many patients with Ehlers-Danlos syndrome and Marfan syndrome, which are linked to collagen and *fibrillin-1* mutations, respectively. MVP has also been documented to be more prevalent in patients with osteogenesis imperfecta and other collagen-related disorders. This association of MVP with inherited connective tissue disorders, in addition to the presence of myxomatous degeneration, suggests that abnormalities in matrix proteins of the connective tissues are important in the etiology of MVP.

Ehler-Danlos syndrome is a rare and heterogeneous group of numerous connective tissue heritable disorders characterized by joint hypermobility, skin hyperextensibility, cardiac valvular defects, and tissue fragility. Ehler-Danlos syndrome type IV is associated with mutations in the *COL3A1* gene which encodes type III procollagen.

Marfan syndrome is an autosomal dominant genetic disorder of the connective tissue associated with mutations in *fibrillin-1*, a major component of the microfibrils that form a sheath surrounding amorphous elastin. *Fibrillin-1* is essential for the proper formation of the extracellular matrix including the biogenesis and maintenance of elastic fibers. The extracellular matrix is critical for both the structural integrity of connective tissue but also serves as a reservoir for growth factors which are essential in the normal maintenance of valve structure and function as well as in response to injury and regulation of repair. The interaction of hemodynamic and mechanical forces with the genetically altered extracellular matrix is not well studied. However, it is likely that these physical forces are important determinants of valve dysfunction.

TGFβ Dysregulation

TGFβ, a 25 kDa protein that is a member of the TGFβ superfamily, is a well-known regulator of extracellular matrix deposition and remodeling. It is secreted by numerous cell types including heart valve interstitial cells, the predominant cell type in heart valves, with potent autocrine effects. It is known to promote differentiation of mesenchymal cells into myofibroblasts and to regulate multiple aspects of the myofibroblast phenotype through transcriptional activation of alpha-smooth muscle actin, collagens, matrix metalloproteinases, and other cytokines, such as connective tissue growth factor and basic fibroblast growth factor. TGFβ is secreted in a latent complex containing active TGFβ and latency-associated protein (LAP) (Figure 14.3). This latent complex is tethered through latent TGFβ binding proteins (LTBPs) to matrix proteins to allow cells to tightly regulate TGFβ bioavailability and create special gradients in the cell microenvironment. Fibrillin-1 regulates TGFβ activation. Fibrillin-1 interacts with LTBPs to sequester latent TGFβ at specific locations in the matrix and stabilizes the inactive large latent complex (TGFβ, LAP, and LTBP), rendering it less prone to activation. Reduced or mutated fibrillin-1 leads to increased TGFβ activation and subsequent elevated levels of TGFβ signaling, resulting in cellular responses such as extensive degradation and remodeling of the extracellular matrix. In mouse models of Marfan syndrome with fibrillin-1 mutations, abnormal mitral valves show increased TGFβ activity leading to a MVP-like phenotype. This mitral valve abnormality can be rescued by perinatal administration of neutralizing antibodies to TGFβ.

The importance of the TGFβ pathway in MVP pathogenesis is further highlighted by the discovery of Loeys-Dietz syndrome, which has many similar clinical features to Marfan syndrome [57,58]. Loeys-Dietz syndrome is caused by mutations in the genes encoding TGFβR1 or TGFβR2. Losartan, an angiotensin II receptor antagonist that modulates the interaction of TGFβ with valves and vascular structures leading to blocking of TGFβ activity, can rescue the expression of the lethal aortic aneurysm in the mouse model of Marfan syndrome [59].

The newly discovered filamin A mutation responsible for X-linked valvulopathy with MVP-like phenotypes may also exert its effect through the interaction of *filamin A* with molecules in the TGFβ pathway [60]. *Filamin A* is a ubiquitous phosphoprotein that cross-links actin filaments and links the actin cytoskeleton to the plasma membrane by interacting with transmembrane proteins such as β-integrins. *Filamin A* has also been implicated in regulating many cellular signaling pathways by acting as a scaffold for intracellular proteins involved in signal transduction. *Filamin A* may contribute to the development of myxomatous changes of the heart valves by augmenting TGFβ signaling through its interaction with Smad proteins such as Smad-2 and Smad-5. Thus, defective Smad-mediated TGFβ signaling due to *filamin A* mutations appears to underlie the pathogenesis of X-linked valvulopathy.

Genetic markers of MVP that define specific phenotypes are not available. Characterization of specific genetic markers would be helpful for early clinical detection of MVP, especially in asymptomatic individuals. As only a subset of individuals with MVP

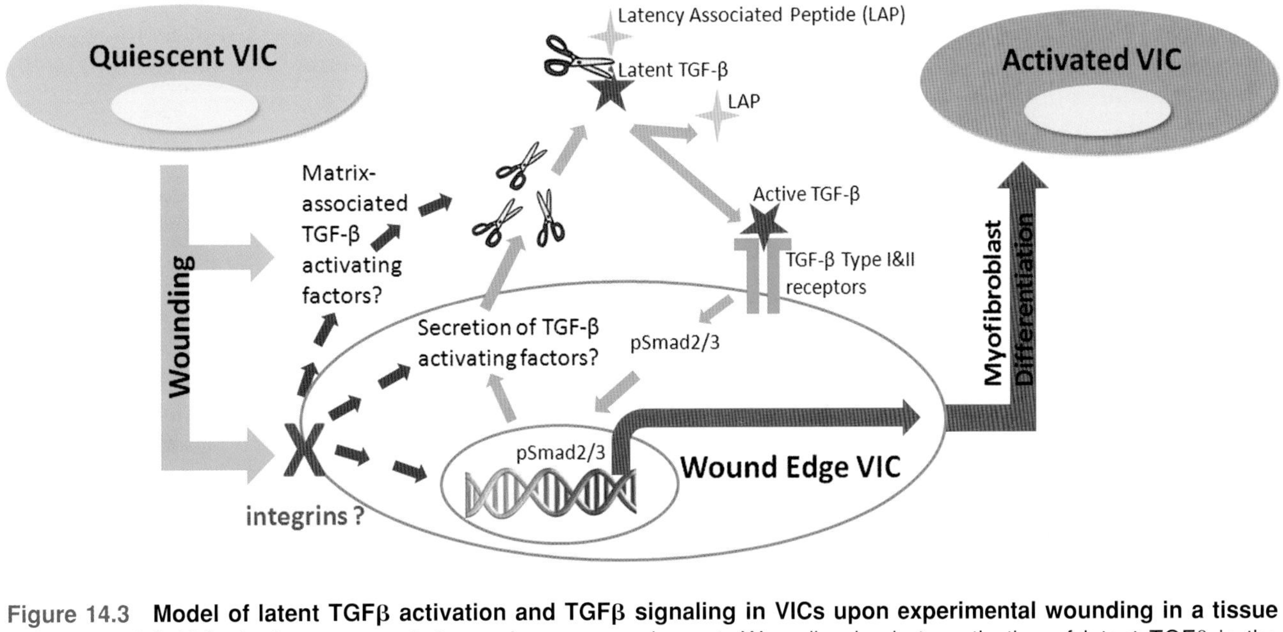

Figure 14.3 Model of latent TGFβ activation and TGFβ signaling in VICs upon experimental wounding in a tissue culture model. VICs in the nonwounded monolayers are quiescent. Wounding leads to activation of latent TGFβ in the extracellular environment of VICs at the wound edge, possibly through changes in integrins or secretion of TGFβ activation proteases. Active TGFβ signals through its cell surface receptors leading to myofibroblast differentiation giving rise to activated VICs at the wound edge.

develops regurgitation, genetics may help predict those patients who are prone to develop valvular insufficiency requiring intervention. Genetic markers may also be used to predict the natural progression of the condition and to identify those cases that might have potentially life-threatening complications. Serum biomarkers for MVP have not been identified to date.

Calcific Aortic Stenosis

Calcific aortic stenosis, or CAS, is an age-related disease which will become much more important as the demographics of our population change. It is commonly attributed to progressive leaflet fibrosis, which in 2%–3% of cases leads to calcification initially in regions of the valve under high mechanical stress. A congenital bicuspid aortic valve is a common congenital abnormality with an incidence of about 1% in the general population that may also become fibrotic and calcified, developing into CAS later in life [61]. The cause of CAS is unknown. However, some bicuspid aortic valves may be heritable [61]. Recently, it has become apparent that CAS shares common morphological changes with atherosclerosis (Table 14.9). Further studies have suggested that there may be a common pathogenic mechanism [62–64], and this may mean that both conditions have similar therapeutic targets and may respond to similar drugs such as statins. Activated VICs appear to regulate several repair processes and may themselves be regulated by TGFβ, an important cytokine in tissue repair, which is present in CAS and appears to be important in repair processes (Figure 14.4).

CARDIOMYOPATHIES

The impact of cell and molecular biology on cardiovascular disease is reflected in the recent reclassification of cardiomyopathies published by the *America Heart Association* [65] and by the *European Society of Cardiology* [66]. Cardiomyopathies are a heterogeneous group of human diseases of the myocardium with primary dysfunction of the cardiomyocytes, the etiology of which is either known or unknown. The former includes ischemia, hypertension, heart valve abnormalities, alcoholism, as well as systemic diseases such as diabetes, hemochromatosis, and amyloidosis. These are usually referred to by their etiology. Those of unknown etiology are classified into dilated, hypertrophic, restrictive, and arrhythmogenic right ventricular cardiomyopathy based on gross and microscopic features and clinical features. Recently, noncompaction cardiomyopathy has been considered either as a distinct cardiomyopathy or as a trait common to several types of cardiomyopathies. As the cell and molecular biology becomes known to us, many of those with unknown etiology turn out to be primary disorders of the myocardium due to one of hundreds of inherited mutations in genes encoding cytoskeletal or sarcomere proteins. Hypertrophic cardiomyopathy (HCM) is considered a disease of the sarcomere; dilated cardiomyopathy, a disease of sarcomere-sarcolemma; and arrythymogenic right ventricular cardiomyopathy, a disease of desmosomes. Although overlaps in mutations in genes encoding proteins of the respective structures occur, the molecular pathogenesis of each cardiomyopathy is unique. However, in most cases it is not known how the specific mutation in the gene results in the dysfunction seen at the cellular level. To date, understanding the genetic causes and the molecular pathogenesis of these distinct cardiomyopathies has allowed improved classification and management of these diseases, as well as the development of better therapies. At the present time the *American Heart Association* classification takes molecular genetics under consideration, and the scope of the classification has been broadened to include ion channelopathies [65]. Clinical genetic testing is available for several monogenic forms of cardiomyopathy. However, at the present time physicians need to discuss the benefits and limitations of such testing with patients and family members.

Table 14.9 Similarities in Calcific Aortic Stenosis and Atherosclerosis

- Histopathology features
- Inflammation
- Reactive oxygen species
- Renin-angiotensin system; angiotensin II
- Calcification, chondrogenesis, osteogenesis, mineralization proteins
- Remodeling of matrix
- Risk factors
- ApoE4 polymorphism risk
- Retrospective clinical studies, statins delay progression of calcific aortic stenosis
- Prospective studies, statins beneficial in atherosclerosis yet to be shown in calcific aortic stenosis
- Hypercholesterolemic rabbit develops calcific aortic stenosis
- LDLr-/-ApoB100/100 mouse shows mineralization of valve

Cardiomyocyte Structure and Function

In order to understand the molecular mechanisms responsible for the development of the clinical phenotypes, one needs an understanding of the cell and molecular biology of the normal myocardium. The myocardium is composed of cardiomyocytes joined in series through intercalated discs containing gap junctions, adherens junctions, and desmosomes. Cardiomyocytes are surrounded by specialized plasma membranes, the sarcolemma, and contain bundles of longitudinally arranged myofibrils. The myofibrils are formed by repeating sarcomeres, the basic contractile units of cardiac muscle, composed of interdigitating thin actin filaments and thick myosin filaments. The thin filaments also contain alpha-tropomyosin and troponins, while the thick filaments also contain myosin binding proteins. Myofibers contain a third filament type formed by the large filamentous protein, titin, which acts as a molecular template for the layout of the sarcomere. The

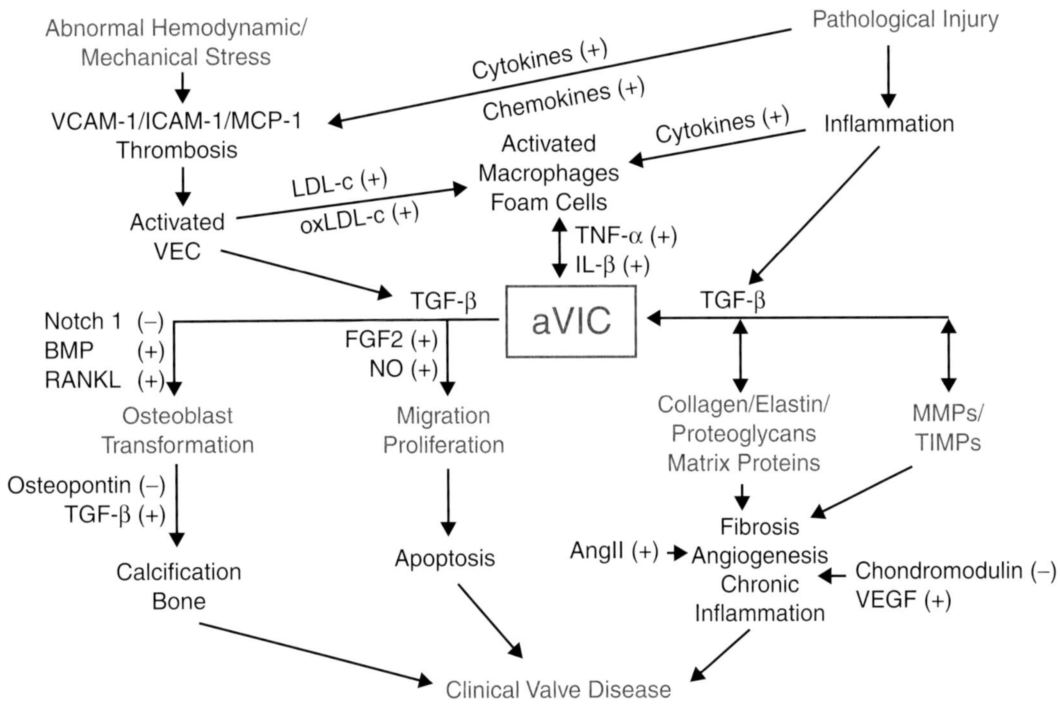

Figure 14.4 **The normal adult heart valve is well adapted to its physiological environment, able to withstand the unique hemodynamic/mechanical stresses under normal conditions.** Under conditions of pathological injury or abnormal hemodynamic/mechanical stresses, VICs become activated through activation of VECs and by inflammation and associated cytokine and chemokine signals. Macrophages will also be activated. aVICs increase matrix synthesis; upregulate expression of matrix remodeling enzymes; migrate, proliferate and undergo apoptosis; as well as undergo osteoblast transformation. These processes are regulated by a variety of factors, several secreted by the aVIC. If the aVICs continue to promote these cellular processes, angiogenesis, chronic inflammation, fibrosis, and calcification result, leading to progressive clinical valve disease. Reprinted from *Am J Pathol.* 2007; **171**:1407–1418 with permission from the *American Society for Investigative Pathology.*

extrasarcomeric cytoskeleton provides structural support for the sarcomere and other subcellular structures and transmits mechanical and chemical signals within and between myocytes. For example, desmin intermediate filaments form a three-dimensional scaffold throughout the extrasarcomeric cytoskeleton, allowing longitudinal connections to adjacent sarcolemma and lateral connections to subsarcolemmal costameres. Costameres are subsarcolemmal interconnections between the various cytoskeletal networks linking the sarcomere and sarcolemma together. They function as anchor sites for stabilization of the sarcolemma and for integration of pathways involved in mechanical force transduction. Costameres contain focal adhesion-type complexes, spectrin-based complexes, and the dystrophin/dystrophin-associated protein complexes (DAPCs). Voltage-gated sodium channels and potassium channels are found to co-localize with dystrophin proteins in DAPCs. Thus, focal structural abnormalities in critical components of these intricate networks within myocytes are likely to lead to clinical phenotypes characterized by functional disruptions of the myocardium, giving rise to specific cardiomyopathies. It should be noted as well that disease genes found in cardiac muscle also cause skeletal muscle disease.

Molecular Genetics and Pathogenesis of Hypertrophic Cardiomyopathy

HCM is a commonly inherited cardiomyopathy characterized by marked thickening of the left ventricular wall in the absence of increased external load and with cardiomyocyte disarray and cardiac fibrosis. Clinical consequences of HCM include heart failure, arrhythmias, and sudden death. HCM is primarily inherited as an autosomal dominant trait, and in rare cases shows mitochondrial inheritance.

Sixty percent of HCM is due to more than 450 different mutations within 8 sarcomeric genes. Most HCM mutations occur in two genes, *MYH7* and *MYBPC3*, encoding the beta-myosin heavy chain and myosin binding protein C, respectively. Mutations in genes that encode cardiac troponin T and cardiac troponin I, essential myosin light chain, regulatory myosin light chain, alpha-tropomyosin, and cardiac actin are less common in HCM patients. Rare mutations in cardiac troponin C and alpha-myosin heavy chain also cause HCM. In addition, mutations in titin, muscle LIM protein, telethonin, and myozenin, which are Z-disc proteins forming a framework connecting sarcomere units to each other, can also cause HCM.

Despite the numerous genetic mutations leading to HCM, it is unclear whether the molecular pathogenesis of HCM is due to impaired sarcomere function or to a gain of new sarcomere function. Findings of mutant myosin fragments and skeletal muscles carrying HCM mutations show decreased actin sliding velocity, supporting the hypothesis that mutations in the myosin components in HCM cause a decrease in motor function of sarcomeres and a compensatory hypertrophic response of the myocardium. Mouse and rabbit models of HCM carrying mutations in myosin heavy chain, myosin binding protein C, and troponin T support the alternate hypothesis that mutant proteins result in abnormal sarcomere function rather than a decrease in motor function [67–69]. These animal models show enhanced actin-activated myosin ATPase activity, increased force generation, and accelerated actin sliding velocity, which may be the molecular mechanisms that result in increased cardiac performance often evident in patients with HCM.

While the functional changes in sarcomeres in HCM remain unclear, the initial step in the pathogenesis of HCM is known to be incorporation of mutant protein with altered mechanical properties to the normal sarcomere. It is likely that the heterogeneity of mutant and wild-type structural proteins within the sarcomere uncouples the normal mechanical coordination between myosin heads and the associated enhanced ATPase activity, resulting in higher levels of energy consumption. This increased energy consumption may trigger hypertrophy, and combined with decreased energy supply as a result of impaired blood flow to the hypertrophied heart, myofibrillar disarray may result, leading to cardiomyocyte death and fibrosis, which are histological features of HCM.

Intracellular Ca2+ is a key regulator of signaling pathways that link sarcomere function and dysfunction to cardiomyocyte biology and pathobiology. Ca2+ homeostasis is disrupted very early in the pathogenesis of HCM. For example, cardiomyocytes from mice with alpha-myosin heavy chain mutations show significant reductions in sarcoplasmic reticulum Ca2+ content and reduced expression levels of proteins responsible for intracellular Ca2+ regulation such as the cardiac ryanodine receptor Ca2+-release channel, the sarcoplasmic reticulum Ca2+ storage protein calsequestrin-2, and the associated anchoring proteins, triadin and junction [70,71]. These changes occur weeks before the onset of HCM, which suggest that intracellular Ca2+ dysregulation is an important early event in the pathogenesis of HCM. A proposed mechanism is that sarcomeric release of Ca2+ at the end of systole is impaired, leading to Ca2+ accumulation within the sarcomere and Ca2+ release from the sarcoplasmic reticulum into the cytoplasm. Elevated cytoplasmic diastolic Ca2+ levels induce hypertrophic responses of cardiomyocytes through the calcineurin-NFAT signaling pathway. Increased intracellular Ca2+ results in calmodulin saturation and calcineurin activation, leading to subsequent dephosphorylation of the nuclear factor of activated T-cells (NFAT) transcription factor in the cytoplasm. Upon dephosphorylation, NFAT translocates to the nucleus, where it induces expression of proliferative and growth-related genes for cardiomyocytes in coordination with other transcription factors [72]. This molecular mechanism of HCM is supported by the evidence that transgenic mice overexpressing activated calcineurin demonstrate a profound hypertrophic response of cardiomyocytes, leading to a 2-fold to 3-fold increase in heart size, which rapidly progresses to dilated heart failure. In the HCM mouse model, administration of the L-type Ca2+ channel inhibitor diltiazem corrects some of the Ca2+-related changes and prevents HCM 50% of the time [70]. This suggests that modulation of cardiomyocyte intracellular Ca2+ concentrations may be beneficial in the treatment of HCM.

Molecular Genetics and Pathogenesis of Dilated Cardiomyopathy

Dilated cardiomyopathy (DCM) is the most common form of cardiomyopathy, involving about 90% of all clinical cardiomyopathies [73]. It occurs more frequently in men than in women. It is characterized by increased ventricular size, reduced ventricular contractility, and left or biventricular dilation. Clinical features include heart failure. Twenty to forty percent of DCM patients have familial forms of the disease, with autosomal dominant inheritance being most common. Autosomal recessive, X-linked, and mitochondrial inheritance of the disease is also found. The majority of mutations are in genes encoding either cytoskeletal or sarcomeric proteins. Mutations in the cytoskeletal proteins desmin, delta-sarcoglycan, and metavinculin, lead to defects of force transmission, while mutations in the sarcomeric proteins lead to defects of force generation in the myocardium. Both of these conditions result in the DCM phenotype. In the majority of instances, the mechanism through which a loss or a defect in these proteins alters function is not well understood. However, these genetic mutations can be used as biomarkers to carry out risk stratification in order to improve management of cardiomyopathies [74]. Further, these proteins may be potential therapeutic targets.

Cytoskeletal Defects

Mutations in 13 genes have been identified to date in autosomal dominant DCM, including desmins, delta-sarcoglycan, metavinculin, alpha-actinin-2, ZASP, actin, troponin T, beta-myosin heavy chain, titin, and myosin binding protein C. The majority of DCM mutations are in genes encoding the cytoskeletal proteins desmin, delta-sarcoglycan, and metavinculin.

Desmin is a cytoskeletal protein which forms intermediate filaments in cardiac, skeletal, and smooth muscle. Desmin is found at the intercalated discs and functions to attach and stabilize adjacent sarcomeres. Furthermore, as previously mentioned, desmin networks also provide a scaffold allowing connections to adjacent sarcolemma and to sarcolemmal costameres. Patients with mutations in desmin also show skeletal myopathy.

Delta-sarcoglycan is a member of the sarcoglycan subcomplex in the DAPC. It is involved in stabilization of the cardiomyocyte sarcolemma as well as signal transduction. In the absence of delta-sarcoglycan, the remaining sarcoglycans (beta, gamma, sigma) cannot assemble properly in the endoplasmic reticulum. Mouse models of delta-sarcoglycan deficiency also sometimes demonstrate HCM instead of DCM. Some patients with delta-sarcoglycan mutations show a form of autosomal recessive limb girdle muscular dystrophy.

Patients with mutations in the cytoskeletal gene metavinculin encoding vinculin and its splice variant metavinculin also present with DCM. Vinculin is ubiquitously expressed and metavinculin is coexpressed with vinculin in cardiac, skeletal, and smooth muscle. It is localized to subsarcolemmal costameres in the heart, where they interact with alpha-actinin, talin, and gamma-actin to form a microfilamentous network linking cytoskeleton and sarcolemma. In addition, vinculin and metavinculin are present in adherens junctions, in intercalated discs and participate in cell-cell adhesion.

Sarcomeric Defects

Mutations in sarcomeric genes may produce DCM or HCM. In the former, sarcomeric mutations giving rise to autosomal dominant inheritance are mainly in genes encoding actin, alpha-tropomyosin, and troponins. Cardiac actin is a sarcomeric protein in sarcomeric thin filaments interacting with tropomyosin and troponin complexes. Mutations in sarcomeric thin filament proteins cardiac actin, alpha-tropomyosin, cardiac troponin T, and troponin I all give rise to DCM. In addition to mutations in thin filament proteins, mutations in the thick filament protein beta-myosin heavy chain also cause DCM. These mutations also perturb the actin-myosin interaction and force generation, as well as alter cross-bridge movement during myocardial contraction.

X-linked and mitochondrial inheritance are the rarer forms of inheritance for DCM. X-linked DCM is caused by mutations in the cytoskeletal protein dystrophin. Multiple mutations in the dystrophin gene identified in DCM patients are affecting the 5′ end of the gene, leading to defective promoter region or N-terminus of the protein. Cardiac actin binds to the N-terminus of dystrophin, linking the sarcomere to the sarcolemma to stabilize the contractile structure.

Complete loss of dystrophin protein or dystrophin with an abnormal N-terminus leads to defective force transmission in the myocardium and DCM. Patients with mitochondrial myopathy due to mitochondrial mutations of the protein MIDNA found in the mitochondrial respiratory chain and responsible for energy generation also present with DCM.

Recent evidence indicates that mutations in lamin A/C in Emery-Dreifuss muscular dystrophy are also associated with DCM [75]. The lamins are proteins in the nucleoplasmic side of the inner nuclear membrane. Lamin A and C are expressed only in heart and skeletal muscle. The mechanisms through which lamin A/C are responsible for the development of DCM are being studied and may be related to their role in maintaining nuclear integrity.

Molecular Genetics and Pathogenesis of Arrhythmogenic Right Ventricular Cardiomyopathy

Arrhythmogenic right ventricular cardiomyopathy (ARVC) is characterized by right ventricular fibro-fatty replacement of myocardial tissue [76]. ARVC presents with palpitations or syncope as a result of ventricular tachyarrhythmias, and is an important cause of sudden death at young ages.

ARVC is familial in about 50% of cases and is mainly autosomal dominant with some autosomal recessive inheritance. Mutations in ARVC have been identified in desmosomal proteins such as plakophilin-2, desmoplakin, plakoglobin, desmoglein-2, and desmocolin-2 [77]. Patients carrying mutations in the *PKP2* gene encoding plakophilin-2, an important desmosomal protein linking cadherins to intermediate filaments, present with ARVC at an earlier age than those without *PKP2* mutations. To date, 50 independent *PKP2* mutations, the majority of which are truncation mutations, are found in 70% of familial cases of ARVC. Mutations in desmoplakin are identified in 6% of ARVC patients in North America, and are associated with the autosomal recessive inherited syndromes—Carvajal syndrome with DCM, wooly hair, and palmoplantar keratoderma, and Naxos-like disorder consisting of ARVC, wooly hair, and an epidermolytic skin disorder. The C-terminus of desmoplakin binds desmin to tightly anchor intermediate filaments, and C-terminal mutations of desmoplakin are associated with an early left ventricular involvement in ARVC, while N-terminus mutations of desmoplakin are associated with a predominantly right ventricular phenotype. Homozygous deletions in the plakoglobin gene are identified in patients with Naxos disease with autosomal recessive inherited ARVC and palmoplantar keratoderma and wooly hair. *In vitro* studies in ECs demonstrate that inhibition of plakoglobin expression induces cell-cell dissociation in response to shear stress. Plakoglobin null mouse embryos die from ventricular rupture when the myocardium undergoes increased mechanical stress in embryonic development [77]. Ten percent of ARVC patients are identified with mutations in *DSG2* encoding desmoglein-2, and 5% of ARVC patients are identified with mutations in *DSC2* encoding desmocollin-2, both major components of desmosomal cadherins. A number of these mutations are splice-site mutations that lead to left ventricular involvement.

The molecular mechanisms through which desmosomal mutations lead to ARVC are currently believed to be related to defects in desmosome composition and function, abnormalities of intercalated discs, and defective Wnt/β-catenin signaling [78,79]. Mutations in genes encoding components of the desmosomal complex lead to insufficient incorporation of the desmosomal proteins into the complex, absence of protein-protein interactions, and incorrect incorporation of these proteins into the complex. These result in disturbed formation or reduced numbers of functional desmosomes. Defects and disruptions in desmosomes also lead to the inability of desmosomes to protect other junctions in the intercalated discs from mechanical stress such as the adhesion junctions between cardiomyocytes leading to loss of cell-cell contacts and cardiomyocyte death. The destabilization of cell adhesion complexes further perturbs the kinetics of gap junctions with smaller and fewer gap junctions present and reduced localization of the gap junction protein connexin 43 at the intercalated discs [78]. This results in heterogeneous conduction, a contributor to the characteristic arrhythmogenesis in ARVC.

The recent discovery that desmosomal components participate in the Wnt/β-catenin pathway involved in a variety of developmental processes including cell proliferation and differentiation help explain the histological findings of fibro-fatty tissue replacement of cardiomyocytes in ARVC [79]. Desmosomal dysfunction results in nuclear translocation of the desmosomal protein plakoglobin, resulting in competition between plakoglobin and β-catenin. This leads to inhibition of Wnt/β-catenin signaling and a shift in cell fate from cardiomyocyte to adipocyte. Plakophilins are also localized to the nucleus, and may be involved in transcriptional regulation or in regulating β-catenin activity in Wnt-signaling. In all cases, defective desmosomes are believed to be unable to maintain tissue integrity under excessive mechanical stress, such as in the thinnest areas of the right ventricle. These include the right ventricular outflow tract, inflow tract, and the apex, which are most vulnerable to ARVC. Thus, it is postulated that patients with desmosomal protein mutations are predisposed to damage in these areas, leading to disruption and subsequent degeneration of cardiomyocytes followed by replacement by fibro-fatty tissue.

Rare mutations in nondesmosomal proteins have also been identified in ARVC. Two families were found to have a mutation in the untranslated regions of the *TGFβ3* gene, leading to altered protein expression [80]. TGFβ3 is a cytokine known to stimulate the production of extracellular matrix components. Aberrant expression or overexpression of TGFβ3 in cardiomyocytes may lead to myocardial fibrosis. Mutations in the *RYR2* gene encoding ryanodine receptor-2, the major Ca2+ release channel of the sarcoplasmic reticulum in cardiomyocytes, have also been found in eight families with highly penetrant ARVC [81,82]. Since their clinical findings show close resemblance to familial catecholaminergic polymorphic ventricular tachycardia, there is controversy as to whether the ARVC in these families should be considered a primary cardiomyopathy.

Molecular Genetics and Pathogenesis of Noncompaction Cardiomyopathy

Noncompaction cardiomyopathy (NCC) is a rare congenital cardiomyopathy [83]. In NCC, myocardial development is hindered during embryogenesis beginning around 8 weeks post-conception. At this stage in development, the myocardium is sponge-like to maximize surface area and to allow perfusion of the myocardium from the left ventricular cavity. However, as the embryo grows, the myocardium compacts and matures. The myocardium of NCC patients fail to fully compact, leaving the myocardium with a spongiform appearance. Symptoms result from poor pumping performance of the heart and tachyarrhythmia, thromboembolisms, and sudden death.

NCC can be inherited in an autosomal dominant manner or through X-linked inheritance [83]. The most common gene responsible for NCC is *TAZ*, an X-linked gene encoding taffazin, which is involved in the biosynthesis of cardiolipin, an essential component of the mitochondrial inner membrane. Mutations in *TAZ* also cause Barth syndrome, a metabolic condition with DCM, with or without noncompaction, neutropenia, skeletal myopathy, and 3-methylglutaconic aciduria. Mutations in cytoskeleton and sarcomere-related genes *DTNA* and *LDB3* can also give rise to NCC. *DTNA* encodes alpha-dystrobrevin, a dystrophin-associated protein involved in maintaining the structural integrity of the sarcolemma. *LDB3* encodes the sarcomeric Z-band protein, LIM domain binding 3 protein. Unfortunately, due to its rare occurrence, the molecular pathogenesis for NCC is not well understood.

Channelopathies

Congenital long QT syndromes, which affect about 1 in 3,000 persons, are a potentially lethal group of cardiac conditions described by delayed repolarization of the myocardium and QT prolongation [84]. This arrhythmogenic disorder is characterized by a significant increased risk of syncope, seizures, and sudden cardiac death. Since several mutations in genes that encode ion channels or their associated proteins account for about 80% of cases, postmortem genetic testing for sudden unexplained death in the young is very useful [83]. In addition new cardiac channel gene mutations and defects in associated proteins will be identified. Recent examples of this include mutant Caveolin 3 [85] and syntrophin mutation [86].

Lymphatic Circulation

The molecular pathogenesis of lymphatic disease, including lymphedema associated with tumor metastasis, infections, and other inflammatory conditions, is currently being addressed through innovative human and experimental studies [87]. This is one of the exciting frontiers of cardiovascular pathology, especially utilizing the techniques developed to culture lymphatic endothelial cells.

The normal physiology of the lymphatic vasculature regulates and maintains tissue fluid balance, is essential for immune surveillance, especially for trafficking of antigen presenting cells from tissues to lymph nodes, and provides a pathway for the absorption of fatty acids in the gut through lacteals present in the intestinal villi. These functions occur as the thin-walled lymphatic vessels regulate the forward flow of lymph with back flow prevented by numerous valves. Specialized junctions that contain adherens and tight junction proteins have been identified in the distal lymphatics [88]. The lymph flows through lymph nodes, which is important in immunosurveillance and leukocyte recirculation, and drains into the internal jugular veins and the venous circulation at the junctions of the left and right subclavian veins.

Over the last few years it has been shown that the lymphatic endothelial cell phenotype differs from blood vessel endothelial cells in that they do not form adherences or tight junctions, are devoid of a basement membrane, and are not surrounded by pericytes. Fox2, the forkhead transcription factor, is required in the latter stages of embryonic development to regulate valve morphogenesis and maintain the lymphatic capillary phenotype. The VEGF-C/VEGFR-3 pathway has been identified and characterized as required in development as it mediates lymphatic endothelial cell migration, proliferation, and survival [89]. Missense mutations in VEGFR-3 result in lymphedema and lymphatic hypoplasia [90]. In addition, PROX-1 is essential for the initial steps in lymphatic formation characterized by budding from the anterior cardinal vein and forming lymph sacs, which then form primary lymphatic plexus under the control of several factors including podoplanin and neuropilin-2 in addition to PROX-1 and VEGFR-3.

Understanding the complex steps of lymphatic development is likely to help in solving the pathogenesis of diseases of the lymphatics [91]. For example, in inflammatory conditions, VEGF-C is upregulated by cytokines, and macrophages express VEGFR-3 and secrete VEGF-C. In mouse models of tumor metastasis, metastasis and lymphangiogenesis can be blocked by inhibiting VEGF-C/VEGFR-3 pathways. An interesting recent finding is that sphingosine-1-phosphate regulates lymphangiogenesis by S1P1/Gi/PLC/Ca2+ signaling pathways [92]. In some patients, primary lymphedema has been shown to be associated with *VEGFR-3* mutations, *FOXC2* mutations, and SOX18 [93].

REFERENCES

1. Butany J, Gotlieb AI. Native valvular heart disease. In: McManus BM, ed. *Atlas of Cardiovascular Pathology for the Clinician, Second Edition*. Philadelphia: Current Medicine Inc., 2001:201–215.
2. Gimbrone MA. Culture of vascular endothelium. *Prog Hemost Thrombo.* 1976;3:1–28.
3. Chamley-Campbell J, Campbell GR, Ross R. The smooth muscle cells in culture. *Physiol Rev.* 1979;59:1–61.
4. Lester W, Rosenthal A, Granton B, et al. Porcine mitral valve interstitial cells in culture. *Lab Invest.* 1988;59:710–719.
5. Butcher JT, Penrod AM, Garcia AJ, et al. Unique morphology and focal adhesion development of valvular endothelial cells in static and fluid flow environments. *Arterioscler Thromb Vasc Biol.* 2004;24:1429–1434.
6. Hollenberg M. Effect of oxygen on growth of cultured myocardial cells. *Circ Res.* 1971;28:148–157.
7. Cunningham KS, Gotlieb AI. The role of shear stress in the pathogenesis of atherosclerosis. *Lab Invest.* 2005;85:9–23.
8. Simmons CA, Grant GR, Manduchi E, et al. Spatial heterogeneity of endothelial phenotypes correlates with side-specific vulnerability to calcification in normal porcine aortic valves. *Circ Res.* 2005;96:792–799.
9. Butcher JT, Tressel S, Johnson T, et al. Transcriptional profiles of valvular and vascular endothelial cell reveal phenotypic differences: Influence of shear stress. *Arterioscler Thrombo Vasc Biol.* 2006;26:69–77.
10. Aikawa E, Whittaker P, Farber M, et al. Human semilunar cardiac valve remodeling by activated cells from fetus to adult: Implications of postnatal adaptation, pathology, and tissue engineering. *Circulation.* 2006;113:1344–1352.
11. Liu AC, Joag VR, Gotlieb AI. The emerging role of valve interstitial cell phenotypes in regulating heart valve pathobiology. *Am J Pathol.* 2007;171:1407–1418.
12. Liu AC, Gotlieb AI. Transforming growth factor-β regulates in vitro heart valve repair by activated valve interstitial cells. *Am J Pathol.* 2008;(in press).
13. Norris RA, Moreno-Rodriguez RA, Sugi Y, et al. Periostin regulates atrioventricular valve maturation. *Dev Biol.* 2008;316:200–213.
14. Psaltis PJ, Zannettino A, Worthley SG, et al. Mesenchymal stromal cells—Potential for cardiovascular repair. *Stem Cells Express.* 2008;26:2201–2210.
15. Kattman SJ, Adler ED, Keller GM. Specification of multipotential cardiovascular progenitor cells during embryonic stem cell differentiation and embryonic development. *Trends Cardiovasc Med.* 2007;17:240–246.
16. Zampetaki A, Kirton JP, Xu Q. Vascular repair by endothelial progenitor cells. *Cardiovasc Res.* 2008;78:413–421.
17. Vasa M, Fichtlschere S, Aicher A, et al. Number and migratory activity of circulating endothelial progenitor cells inversely correlate with risk factors for coronary artery disease. *Circ Res.* 2001;89:E1–E7.
18. Tepper OM, Galiano RD, Capla JM, et al. Human endothelial progenitor cells from type II diabetics exhibit impaired proliferation, adhesion, and incorporation into vascular structures. *Circulation.* 2002;106:2781–2786.
19. Hill JM, Zalos G, Halcox JP, et al. Circulating endothelial progenitor cells, vascular function, and cardiovascular risk. *N Eng J Med.* 2003;348:593–600.
20. Kalka C, Masuda H, Takahashi T, et al. Transplantation of ex vivo expanded endothelial progenitor cells for therapeutic neovascularization. *Proc Natl Acad Sci USA.* 2000;97:3422–3427.
21. Kawamoto A, Tkebuchava T, Yamaguchi J, et al. Intramyocardial transplantation of autologous endothelial progenitor cells for therapeutic neovascularization of myocardial ischemia. *Circulation.* 2003;107:461–468.
22. Stamm C, Westphal B, Kleine HD, et al. Autologous bone-marrow stem-cell transplantation for myocardial regeneration. *Lancet.* 2003;361:45–46.
23. Losordo DW, Schatz RA, White CJ, et al. Intramyocardial transplantation of autologous CD34+ stem cells for intractable angina. A phase I/IIa double-blind, randomized controlled trial. *Circulation.* 2007;115:3165–3172.
24. Walter DH, Rittig K, Bahlmann FH, et al. Statin therapy accelerates reendothelialization: A novel effect involving mobilization and incorporation of bone marrow derived endothelial progenerator cells. *Circulation.* 2002;105:3017–3024.
25. Walter DH, Cejna M, Diaz-Sandoval L, et al. Local gene transfer of phVEGF-2 plasmid by gene-eluting stents: An alternative strategy for inhibition of restenosis. *Circulation.* 2004;110:36–45.
26. Aoki J, Serruys PW, van Beusekom H, et al. Endothelial progenitor cell capture by stents coated with antibody against CD34: The HEALING-FIM (Healthy Endothelial Accelerated Lining Inhibits Neointimal Growth-First In Man) Registry. *J Am Coll Cardiol.* 2005;45:1574–1579.
27. Beltrami AP, Urbanek K, Kajstura J, et al. Evidence that human cardiac myocytes divide after myocardial infarction. *N Engl J Med.* 2001;344:1750–1757.
28. Nakanishi C, Yamagishi M, Yamahara K, et al. Activation of cardiac progenitor cells through paracrine effects of mesenchymal stem cells. *Biochem Biophys Res Communications.* 2008;374:11–16.
29. Martin CM, Meeson AP, Robertson SM, et al. Persistent expression of the ATP-binding cassette transporter, Abcg2, identifies

cardiac SP cells in the developing and adult heart. *Dev Biol.* 2004;265:262–275.
30. Smith RR, Barile L, Cho HC, et al. Regenerative potential of cardiosphere-derived cells expanded from percutaneous endomyocardial biopsy specimens. *Circulation.* 2007;115:896–908.
31. Laugwitz KL, Moretti A, Lam J, et al. Postnatal isl1+ cardioblasts enter fully differentiated cardiomyocyte lineages. *Nature.* 2005; 433:647–653.
32. Smart N, Risebro CA, Melvilles AAD, et al. Thymosin β4 induces adult epicardial progenitor mobilization and neovascularization. *Nature.* 2007;445:177–182.
33. Winter E, Gittenberger-de Groot A. Cardiovascular development: Towards biomedical applicability. Epicardium-derived cells in cardiogenesis and cardiac regeneration. *Cell Mol Life Sci.* 2007;64:692–703.
34. Limana F, Zacheo A, Mocini D, et al. Identification of myocardial and vascular precursor cells in human and mouse epicardium. *Cir Res.* 2007;101:1255–1265.
35. Nussbaum J, Minami E, Laflamme MA, et al. Transplantation of undifferentiated murine embryonic stem cells in the heart: Teratoma formation and immune response. *FASEB J.* 2007;21:1345–1357.
36. Murry CE, Reinecke H, Pabon LM. Regeneration gaps. Observations of stem cells and cardiac repair. *J Am Coll Cardiol.* 2006;47:1777–1785.
37. Gotlieb AI. Blood vessels. In: Rubin T, Strayer DS, eds. *Rubin's Pathology, 5th Edition.* Lippincott, Williams, and Wilkins, Philadelphia; 2007:387–426.
38. Shen GQ, Li L, Rao S, et al. Four SNPs on chromosome 9p21 in a South Korean population implicate a genetic locus that confers high cross-race risk for development of coronary artery disease. *Arterioscler Thromb Vasc Biol.* 2008;28:360–365.
39. Shen GQ, Rao S, Martinelli N, et al. Association between four SNPs on chromosome 9p21 and myocardial infarction is replicated in an Italian population. *J Hum Genet.* 2008; 53:144–150.
40. Schunkert H, Gotz A, Braund P, et al., for the Cardiogenics Consortium. Repeated replication and a prospective meta-analysis of the association between chromosome 9p21.3 and coronary artery disease. *Circulation.* 2008;117:1675–1684.
41. Casas JP, Cooper J, Miller GJ, et al. Review: Investigating the genetic determinants of cardiovascular disease using candidate genes and meta-analysis of association studies. *Annals Human Genetics.* 2006;70:145–169.
42. Hirschhorn JN, Lohmueller K, Byrine E, et al. A comprehensive review of genetic association studies. *Genetic Med.* 2002;4:45–61.
43. Samani NJ, Erdmann J, Hall AS, et al., for the WTCCC and the Cardiogenics Consortium. Genomewide association analysis of coronary artery disease. *N Engl J Med.* 2007;357:443–453.
44. Hamsten A, Eriksson P. Identifying the susceptibility genes for coronary artery disease: From hyperbole through doubt to cautious optimism. *J Intern Med.* 2008;263:538–552.
45. Nakashima Y, Wight TN, Sueishi K. Early atherosclerosis in humans: Role of diffuse intimal thickening and extracellular matrix proteoglycans. *Cardiovasc Res.* 2008;79:14–23.
46. Singh NN, Ramji DP. The role of transforming growth factor-β in atherosclerosis. *Cytokine Growth Factor Reviews.* 2006;17:487–499.
47. Apple FS, Jaffe AS. Cardiovascular disease. In: Brutis CA, Ashwood ER, Bruns DE, eds. *Fundamentals of Clinical Chemistry, Sixth Edition.* St. Louis, Missouri: Elsevier; 2008:614–630.
48. Edwards AVG, White MY, Cordwell SJ. The role of proteomics in clinical cardiovascular biomarker discovery. *Mol Cell Proteomics.* 2008;(in press).
49. Motro M, Fisman EZ, Tenenbaum A. Cardiovascular management of Marfan syndrome. *IMAJ.* 2008;10:182–185.
50. Melenovsky V, Adamira M, Kautznerova D, et al. Aortic dissection in a young man with Loeys-Dietz syndrome. *J Thorac Cardiovasc Surg.* 2008;135:1174–1175.
51. Milewicz DM, Guo D-C, Tran-Fadulu V, et al. Genetic basis of thoracic aortic aneurysms and dissections: Focus on smooth muscle cell contractile dysfunction. *Annu Rev Genomics Hum Genet.* 2008;9:283–302.
52. Ruigork YM, Rinkel GJE. Genetics of intracranial aneurysms. *Stroke.* 2008;39:1049–1055.
53. Thompson RW, Geraghty PJ, Lee JK. Abdominal aortic aneurysms: Basic mechanisms and clinical implications. *Curr Probl Surg.* 2002;39:110–230.
54. Abdul-Hussien H, Soekhoe RGV, Weber E, et al. Collagen degradation in the abdominal aneurysm. A conspiracy of matrix metalloproteinase and cysteine collagenases. *Am J Pathol.* 2007;170: 809–817.
55. Rodriguez-Pla A, Beaty TH, Savino PJ, et al. Association of nonsynonymous single-nucleotide polymorphism of matrix metalloproteinase 9 with giant cell arteritis. *Arthritis Rheumatism.* 2008;58:1849–1853.
56. Slot MC, Sokolowska MG, Savelkouls KG, et al. Immunoregulatory gene polymorphisms are associated with ANCA-related vasculitis. *Clin Immunol.* 2008;128:39–45.
57. Loeys BL, Schwarze U, Holm T, et al. Aneurysm syndromes caused by mutations in the TGF-beta receptor. *N Engl J Med.* 2006;355:788–798.
58. Ng CM, Cheng A, Myers LA, et al. TGF-β-dependent pathogenesis of mitral valve prolapsed in a mouse model of Marfan syndrome. *J Clin Invest.* 2004;114:1543–1592.
59. Habashi JP, Judge DP, Holm TM, et al. Losartan, an AT1 antagonist, prevents aortic aneurysm in a mouse model of Marfan syndrome. *Science.* 2006;312:119–121.
60. Kyndt F, Gueffet JP, Probst V, et al. Mutations in the gene encoding filamin A as a cause for familial cardiac valvular dystrophy. *Circulation.* 2007;115:40–49.
61. Cripe L, Andelfinger G, Martin LJ, et al. Bicuspid aortic valve is heritable. *J Am Coll Cardiol.* 2004;44: 138–143.
62. Helske S, Kupari M, Lindstedt KA, et al. Aortic valve stenosis: An active atheroinflammatory process. *Curr Opin Lipidol.* 2007;18:483–491.
63. Rajamannan NM, Bonow RO, Rahimtoola SH. Calcific aortic stenosis: An update. *Nature Clin Pract Cardiovasc Med.* 2007;4:254–262.
64. Akat K, Borggrefe M, Kaden JJ. Aortic valve calcification—Basic science to clinical practice. *Heart.* 2008;(in press).
65. Maron BJ, Towbin JA, Thiene G, et al. Contemporary definitions and classification of the cardiomyopathies: An American Heart Association Scientific Statement from the Council of Clinical Cardiology, Heart Failure and Transplantation Committee; Quality of Care and Outcomes Research and Functional Genomics and Translational Biology Interdisciplinary Working Groups; and Council on Epidemiology and Prevention. *Circulation.* 2006;113:1807–1816.
66. Elliott P, Andersson B, Arbustini E, et al. Classification of the cardiomyopathies: A position statement from the European Society of Cardiology working group on myocardial and pericardial disease. *Europ Heart J.* 2008;29:270–276.
67. Yang Q, Sanbe A, Osinska H, et al. A mouse model of myosin binding protein C human familial hypertrophic cardiomyopathy. *J Clin Invest.* 1998;102:1292–1300.
68. Marian AJ, Wu Y, Lim DS, et al. A transgenic rabbit model for human hypertrophic cardiomyopathy. *J Clin Invest.* 1999;104: 1683–1692.
69. Tardiff JC, Hewett TE, Palmer BM, et al. Cardiac troponin T mutations result in allele-specific phenotypes in a mouse model for hypertrophic cardiomyopathy. *J Clin Invest.* 1999;104:469–481.
70. Semsarian C, Ahmad I, Giewat M, et al. The L-type calcium channel inhibitor diltiazem prevents cardiomyopathy in a mouse model. *J Clin Invest.* 2002;109:1013–1020.
71. Wilkins BJ, Molkentin JD. Calcium-calcineurin signalling in the regulation of cardiac hypertrophy. *Biochem Biophys Res Commun.* 2004;322:1178–1191.
72. Alcalai R, Seidman JG, Seidman CE. Genetic basis of hypertrophic cardiomyopathy: From bench to clinics. *J Cardiovasc Electrophysiol.* 2008;19:104–110.
73. Towbin JA, Bowles NE. Dilated cardiomyopathy: A tale of cytoskeletal proteins and beyond. *J Cardiovasc Electrophysiol.* 2006;17:919–926.
74. Olivotto I, Girolami F, Ackerman MJ, et al. Myofilament protein gene mutation screening and outcome of patients with hypertrophic cardiomyopathy. *Mayo Clin Proc.* 2008;83:630–638.
75. Bonne G, DiBarletta MR, Varnous S, et al. Mutations in the gene encoding lamin A/C cause autosomal dominant Emery-Dreifuss muscular dystrophy. *Nature Genetics.* 1999;21:285–288.
76. Thiene G, Corrado D, Basso C. Arrhythmogenic right ventricular cardiomyopathy/dysplasia. *Orphanet J Rare Dis.* 2007;2:45.

77. Van Tintelen JP, Hofstra RMW, Wiesfeld ACP, et al. Molecular genetics of arrhythmyogenic right ventricular cardiomyopathy: Emerging horizon? *Curr Opin Cardiol.* 2007;22:185–192.
78. Tandri H, Asimaki A, Dalal D, et al. Gap junction remodeling in a case of arrhythmogenic right ventricular dysplasia due to plakophilin-2 mutation. *J Cardiovasc Electrophysiol.* 2008;(in press).
79. Garcia-Gras E, Lombardi R, Giocondo MJ, et al. Suppression of canonical Wnt/beta-catenin signalling by nuclear plakoglobin recapitulates phenotype of arrhythmogenic right ventricular cardiomyopathy. *J Clin Invest.* 2006;116:2012–2021.
80. Beffagna G, Occhi G, Nava A, et al. Regulatory mutations in transforming growth factor-beta3 gene cause arrhythmogenic right ventricular cardiomyopathy type 1. *Cardiovasc Res.* 2005;65:366–373.
81. Tiso N, Stephan DA, Nava A, et al. Identification of mutations in the cardiac ryanodine receptor gene in families affected with arrhythmogenic right ventricular cardiomyopathy type 2 (ARVD2). *Hum Mol Genet.* 2001;10:189–194.
82. Bauce B, Rampazzo A, Basso C, et al. Screening for ryanodine receptor type 2 mutations in families with effort induced polymorphic ventricular arrhythmias and sudden death: Early diagnosis of asymptomatic carriers. *J Am Coll Cardiol.* 2002;40:341–349.
83. Zaragoza MV, Arbustini E, Narula J. Noncompaction of the left ventricule: Primary cardiomyopathy with an elusive genetic etiology. *Curr Opin Pediatr.* 2007;19:619–627.
84. Tester DJ, Ackerman MJ. Postmortem long QT syndrome genetic testing for sudden unexplained death in the young. *J Am Coll Cardiol.* 2007;49:240–246.
85. Vatta M, Ackerman MJ, Ye B, et al. Mutant caveolin-3 induces persistent late sodium current and is associated with long-QT syndrome. *Circulation.* 2006;114:2104–2112.
86. Ueda K, Valdivia C, Medeiros-Domingo A, et al. Syntrophin mutation associated with long QT syndrome through activation of the nNOS-SCN5A macromolecular complex. *Proc Natl Acad Sci USA.* 2008;105:9355–9360.
87. Bruyere F, Melen-Lamalle L, Blacher S, et al. Modeling lymphangiogenesis in a three-dimensional culture system. *Nature Methods.* 2008;5:431–437.
88. Baluk P, Fuxe J, Hashizume H, et al. Functionally specialized junctions between endothelial cells of lymphatic vessels. *J Exp Med.* 2007;204:2349–2362.
89. Karpanen T, Alitalo K. Molecular biology and pathology of lymphangiogenesis. *Annu Rev Pathol Mech Dis.* 2008;3:367–397.
90. Spiegel R, Ghalamkarpour A, Daniel-Spiegel E, et al. Wide clinical spectrum in a family with hereditary lymphedema type I due to a novel missense mutation in VEGFR3. *J Human Genetics.* 2006;51:846–850.
91. Karpanen T, Alitalo K. Molecular biology and pathology of lymphangiogenesis. *Annu Rev Pathol.* 2008;3:367–397.
92. Yoon CM, Hong BS, Moon HG, et al. Sphingosine-1-phosphate promotes lymphangiogenesis by stimulating S1P1/Gi/PLC/Ca2+ signalling pathways. *Blood.* 2008;112:1129–1138.
93. Ferrell RE, Finegold DN. Biological principles of the lymphatic system: Research perspectives in inherited lymphatic disease—An update. *Ann NY Acad Sci.* 2008;1131:134–139.

Chapter

15

Molecular Basis of Hemostatic and Thrombotic Diseases

Alice D. Ma ▪ Nigel S. Key

INTRODUCTION AND OVERVIEW OF COAGULATION

Blood coagulation is the process whereby cells and soluble protein elements interact to form an intravascular blood clot. When this occurs in response to vessel injury, it is an important protective mechanism that functions to seal vascular bleeds, and thereby prevent excessive hemorrhage. This physiological process is generally referred to as hemostasis. However, in pathological situations, blood coagulation may be triggered by a variety of stimuli and lead to the formation of a maladaptive intravascular clot or thrombus that may obstruct blood flow to or from a critical organ, and/or embolize to a distal site through the circulatory system. This process is known as thrombosis or thromboembolism, and it may affect either the arterial or venous circulations. A comprehensive review of the molecular basis of all of the defects leading to disorders of hemostasis and thrombosis is beyond the scope of this chapter. We will instead focus on broad themes, particularly defects in the soluble coagulation factors, defects in platelet number or function, other defects leading to hemorrhage, and inherited defects predisposing to thrombosis.

The hemostatic system is highly evolutionarily conserved. For example, Horseshoe crabs (*Limulus*), which have existed on earth for 350 million years, have a rudimentary hemostatic system. Their only circulating blood cells are amebocytes (hemocytes) that contain bactericides, and coagulation zymogen proteins, which are released upon activation. The key coagulation system components are serine proteases, clottable protein (coagulogen, factor C), and transglutaminase (a factor XIII-like enzyme). When amebocytes are exposed to endotoxin in seawater following an injury, serine proteases convert coagulogen to coagulin, which is polymerized by transglutaminase. This basic coagulation system enables *Limulus* to locally isolate injuries and invading pathogens, and it is highly analogous to the conversion of soluble fibrinogen (a clottable protein) to insoluble fibrin as the end result of a series of zymogens and their respective derived serine proteases in humans.

Coagulation can be conceptualized as a series of steps occurring in overlapping sequence. Primary hemostasis refers to the interactions between the platelet and the injured vessel wall, culminating in formation of a platelet plug. The humoral phase of clotting (secondary hemostasis) encompasses a series of enzymatic reactions, resulting in a hemostatic fibrin plug. Finally, fibrinolysis and wound repair mechanisms are recruited to restore normal blood flow and vessel integrity. Each of these steps is carefully regulated, and defects in any of the main components or regulatory mechanisms can predispose to either hemorrhage or thrombosis. Depending on the nature of the defect, the hemorrhagic or thrombotic tendency can be either profound or subtle.

Primary hemostasis begins at the site of vascular injury, with platelets adhering to the subendothelium, utilizing interactions between molecules such as collagen and von Willebrand factor (VWF) in the vessel wall with glycoprotein and integrin receptors on the platelet surface. Specifically, the primary platelet receptor for subendothelial VWF under high shear conditions is the glycoprotein (GP) Ib/V/IX complex, while the primary platelet receptors for direct collagen binding are GP VI and the integrin $\alpha_2\beta_1$. Platelet spreading is followed by activation and release of granular components, and exposure to the cocktail of agonists exposed at a wounded vessel amplifies the process of platelet activation. Via a process known as inside-out signaling, the integrin $\alpha_{2b}\beta_3$ (also known as GP

Molecular Pathology © 2009, Elsevier, Inc. All Rights Reserved.

IIbIIIa) undergoes a conformational change in order to be able to bind fibrinogen, which cross-links adjacent platelets and leads to platelet aggregation. Activated GP IIbIIIa can also bind VWF under certain circumstances (such as higher shear rates). Secretion of granular contents is also triggered by external activating signals, further potentiating platelet activation. Lastly, the membrane surface of the platelet changes to serve as a scaffold for the series of enzymatic reactions that result in thrombin generation. This process is primarily dependent on the exposure of negatively charged phospholipids that are normally confined to the inner leaflet of the plasma membrane bilayer.

Our understanding of the process by which fibrin is ultimately generated by thrombin cleavage of soluble fibrinogen has undergone several iterations over the past half century. The waterfall (or cascade) model of coagulation was developed by two groups nearly simultaneously [1,2] and included an extrinsic, intrinsic, and a common pathway leading to fibrin formation. In this model, the intrinsic pathway can be initiated by the action of kaolin, a negatively charged surface activator that interacts with the contact factors (prekallikrein or PK, high molecular weight kininogen or HMWK, and factor XII) to generate factor XIIa. Factor XIIa then cleaves and activates factor XI. Factor XIa activates factor IX to IXa, which cooperates with its cofactor, factor VIIIa, to form the tenase complex, which then proteolytically cleaves factor X to generate factor Xa. Factor Xa, along with its cofactor Va, forms the prothrombinase complex, which generates thrombin from its zymogen (factor II or prothrombin). The thrombin formed by this complex cleaves fibrinogen to allow it to polymerize into insoluble fibrin strands. Factor XIII is also activated by thrombin, and the role of factor XIIIa, a transglutaminase, is to cross-link fibrin strands to provide additional clot strength and stability. The extrinsic pathway requires tissue factor (TF) in complex with factor VIIa to form the tenase complex. Additionally, TF/VIIa can activate factor IX to IXa, serving as an alternate method of activating the intrinsic pathway (Figure 15.1).

While the cascade hypothesis explains the prothrombin time (PT) and the activated partial thromboplastin time (APTT) tests as they are performed *in vitro*, it fails to explain the relative intensity of the bleeding diathesis seen in individuals deficient in factors XI, IX, and VIII, as well as the lack of bleeding in those deficient in factor XII, HMWK, or PK. The first reconciliation of this paradox came with the discovery that the tissue factor/VIIa complex activates both factor X, as well as factor IX [3]. More recently, a cell-based model of hemostasis was proposed to address these deficiencies, and to integrate the role of cell surface-bound coagulation reactions in hemostasis [4]. In this model, a tissue factor-bearing cell such as an activated monocyte or fibroblast serves as the site for generation of a small amount of thrombin and factor IXa. The initial thrombin burst is quickly limited by the interaction of generated factor Xa with tissue factor pathway inhibitor (TFPI), a Kunitz-like inhibitor present in plasma and at cell surfaces. The binary complex of Xa-TFPI then forms an inhibitory

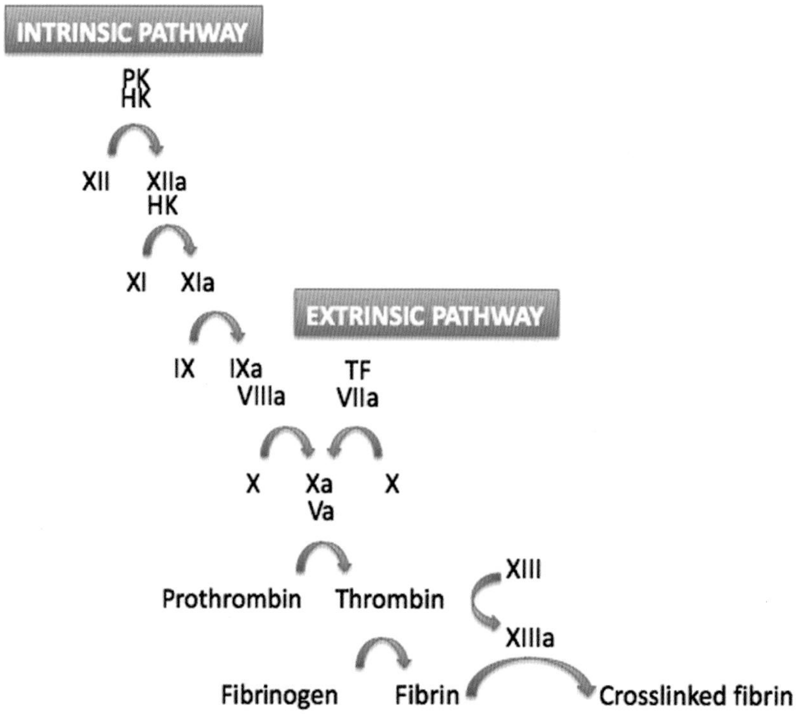

Figure 15.1 Classic view of hemostasis according to the "waterfall" or "cascade" model.

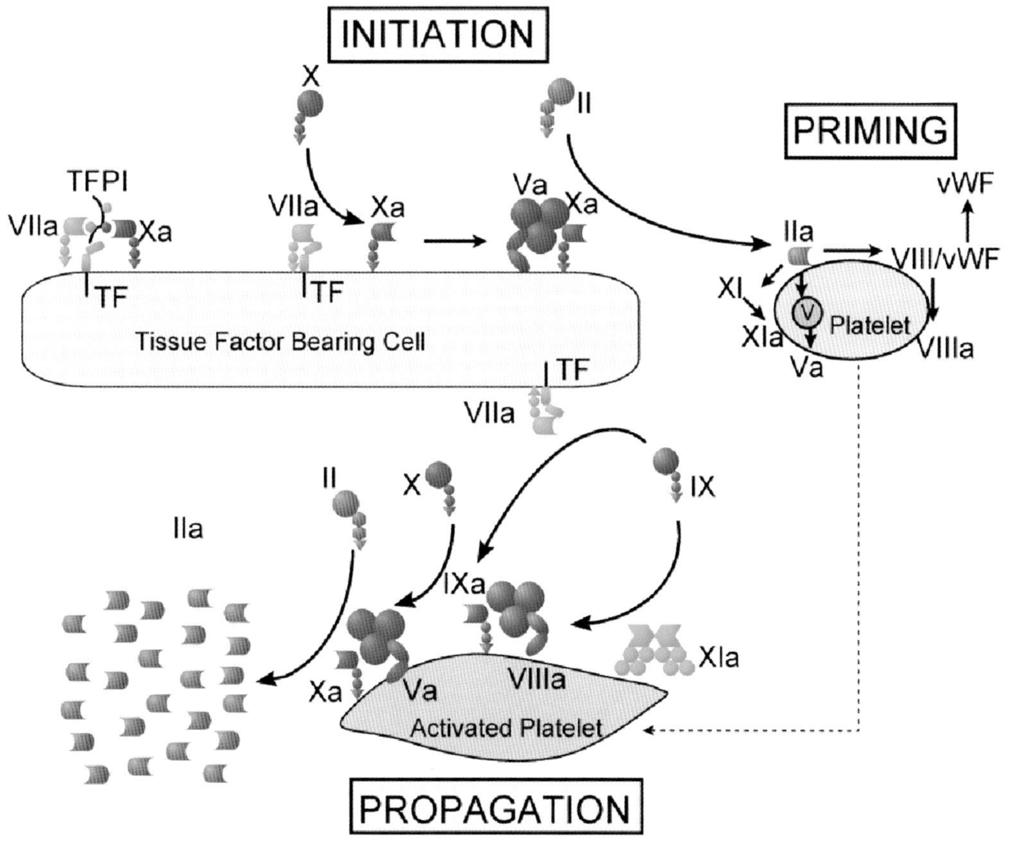

Figure 15.2 The cell-based model of hemostasis. Figure Courtesy of Dr. Dougald Monroe.

quaternary complex with TF-VIIa. The small amount of thrombin generated by this initiation step is insufficient to cleave fibrinogen, but serves to activate platelets and cleave circulating factor VIII from its noncovalently associated VWF, allowing for the formation of VIIIa. The factor IXa formed on the TF-bearing cell cooperates with VIIIa to form the tenase complex on the surface of the activated platelet. The Xa thus formed interacts with the Va generated on the platelet surface to form the prothrombinase complex. This complex generates a large burst of thrombin that is now sufficient to cleave fibrinogen, activate factor XIII, and activate the thrombin activatable fibrinolysis inhibitor (TAFI), thus allowing for formation of a stable fibrin clot (Figure 15.2). In this model, XI binds to the surface of activated platelets where it is activated by thrombin. Factor XIa so formed boosts the activity of the tenase complex, but is not necessary for thrombin generation.

Fibrinolysis leads to clot dissolution, thereby restoring normal blood flow. Plasminogen is activated to plasmin by the action of either tissue plasminogen activator (t-PA) or urokinase plasminogen activator (u-PA). Both plasminogen and the plasminogen activators bind to free lysine residues exposed on fibrin, and in this way fibrin acts as the cofactor for its own destruction. Plasmin is capable of degrading both fibrin and fibrinogen, and can thus dissolve both formed clot as well as its soluble precursor. However, non-fibrin-bound plasmin is inhibited by a number of circulating inhibitors, of which α_2-plasmin inhibitor (α_2-PI) is the most significant. Plasminogen activation is also inhibited by a number of molecules; chief among them is plasminogen activator inhibitor-1 (PAI-1), a serine protease inhibitor (SERPIN) that irreversibly binds to—and inactivates—the plasminogen activators. Lastly, cellular receptors act to localize and potentiate or clear plasmin and plasminogen activators.

DISORDERS OF SOLUBLE CLOTTING FACTORS

While classic hemophilia/hemophilia A (factor VIII deficiency) and Christmas disease/hemophilia B (factor IX deficiency) are the best known examples of the clotting factor deficiencies, the following overview will discuss each deficiency in the numerical order ascribed by the Roman numeral classification system (see Table 15.1). Note that fibrinogen is rarely referred to as factor I even though this is how it is designated using strict nomenclature; similarly, tissue factor, a cell-bound transmembrane glycoprotein, is

Table 15.1 Clinical Features of Inherited Coagulation Factor Deficiency States

Defect		Inheritance Pattern	Bleeding Manifestations	Diagnostic Testing: PT	Diagnostic Testing: PTT	Diagnostic Testing: TCT	Diagnostic Testing: BT	Treatment
Fibrinogen abnormalities	Afibrinogenemia	Autosomal	Severe, but less so than severe Hemophilia A and B	Infinite	Infinite	Infinite	Prolonged	Cryoprecipitate
	Dysfibrinogenemia	Autosomal	Variable bleeding and/or clotting	Prolonged	Prolonged	Prolonged or shortened	Normal	Cryoprecipitate
Prothrombin deficiency		Autosomal	Varies with prothrombin levels	Prolonged	Prolonged	Normal	Normal	PCCs
Factor V deficiency		Autosomal	Mild–moderate	Prolonged	Prolonged	Normal	Prolonged	FFP, potential need for exchange transfusion
Factor VII deficiency		Autosomal	Moderate–severe	Prolonged	Normal	Normal	Normal	Recombinant-activated factor VII
Hemophilia A		X-linked recessive	Variable, depending on Factor VIII level	Normal	Prolonged	Normal	Normal	Factor VIII concentrates, DDAVP in mild cases
Hemophilia B		X-linked recessive	Variable, depending on Factor IX level	Normal	Prolonged	Normal	Normal	Factor IX concentrates
Factor X deficiency		Autosomal	Variable, depending on Factor X level	Prolonged	Prolonged	Normal	Normal	Plasma or PCCs
Factor XI deficiency		Autosomal	Variable, but NOT dependent on Factor XI levels	Normal	Prolonged	Normal	Normal	Plasma or Recombinant-activated factor VII
Deficiency of Factor XII, prekallikrein, or high molecular weight kininogen		Autosomal	None	Normal	Prolonged	Normal	Normal	None needed
Factor XIII deficiency		Autosomal	Severe	Normal	Normal	Normal	Normal	Cryoprecipitate
Deficiency of alpha-2 plasmin inhibitor or plasminogen activator inhibitor-1		Autosomal	Severe	Normal	Normal	Normal	Normal	Antifibrinolytic agents (epsilon aminocaproic acid or tranexamic acid)

Table 15.2 Useful Websites for Documented Mutations in Individual Coagulation Factor Deficiency States

Fibrinogen	www.geht.org/databaseang/fibrinogen/
Prothrombin	www.coagMDB.org/
Factor V	www.lumc.nl/rep/cod/redirect/4010/research/factor_v_gene.html
Factor VII	www.coagMDB.org/
Factor VIII	http://europium.csc.mrc.ac.uk/
Factor IX	www.kcl.ac.uk/ip/petergreen/haemBdatabase.html; www.coagMDB.org/
Factor X	www.coagMDB.org/
Factor XI	www.FactorXI.org/
Factor XIII	www.f13-database.de/(xhgmobrswxgori45zk5jre45)/index.aspx
VWF	www.vwf.group.shef.ac.uk/
Protein S	www.med.unc.edu/isth/ssc/communications/plasma_coagulation/proteins.htm
Protein C	www.coagMDB.org/
Antithrombin	www1.imperial.ac.uk/medicine/about/divisions/is/haemo/coag/antithrombin/

rarely referred to by its designated number (factor III). No tissue factor deficiency states have been described in humans. In addition, factor IV is reagent calcium, and there is no factor VI in the current scheme. While hemophilia A occurs in approximately 1:10,000 of the population, and hemophilia B in about 1:40,000, the remaining factor deficiencies (with the possible exception of factor XI in Ashkenazi Jewish populations) typically occur at a frequency between 1:500,000 and 1:2,000,000 of the general population, and are thus usually known collectively as the rare inherited bleeding disorders.

Rather than attempt to provide an exhaustive list of mutations affecting all coagulation proteins in this text, we have provided a number of useful websites containing described mutations in these proteins in Table 15.2. In addition, a summary of all described mutations in factors II, V, VII, X, XI, XIII, and combined V/VIII deficiency can be found at www.med.unc.edu/isth/mutations-databases/mutations_rare_bleeding_disorders.html. In addition, the Human Gene Mutation database, available at www.hgmd.org, contains detailed information on many genetic mutations in a variety of inherited disorders, including those affecting the coagulation pathways. Table 15.3 summarizes the essential biochemical data on each of the important inherited deficiency states.

Fibrinogen Abnormalities

Fibrinogen abnormalities are inherited in an autosomal pattern and occur in two main patterns: hypo/afibrinogenemia and dysfibrinogenemia [5]. Afibrinogenemia is a very rare disorder that occurs when any one of the three genes coding for the alpha, beta, or gamma chains that make up the fibrinogen molecule is mutated. If the mutation is sufficient to disrupt formation or secretion of any of the three chains, afibrinogenemia results [6]. Afibrinogenemic patients have a severe bleeding disorder manifest by bleeding after trauma into subcutaneous and deeper tissues that may result in dissection. Bleeding from the umbilical stump at birth occurs frequently. Though hemarthroses do occur in these patients, they are less frequent than in the severe forms of hemophilia A and B. Diagnosis is usually apparent with lack of clot formation in screening clotting tests such as the prothrombin time (PT), activated partial thromboplastin time (APTT), or the thrombin clotting time (TCT). The bleeding time (BT) may also be prolonged reflecting the absence of fibrinogen in plasma and in the platelet alpha granules. In the United States, treatment consists of transfusing cryoprecipitate to raise the fibrinogen level to approximately 100 mg/dl. In many other countries, a purified plasma fibrinogen concentrate is available for this purpose. Patients with afibrinogenemia are prone to spontaneous intracranial hemorrhage, which usually mandates prolonged secondary prophylaxis with fibrinogen concentrate or cryoprecipitate. There are a few reports of antifibrinogen antibodies occurring in patients with afibrinogenemia who have been repeatedly exposed to fibrinogen replacement [7,8]. Some patients with a lesser mutation in one of the fibrinogen chains, who have partial deficiency (that is, reduced but measurable circulating levels of fibrinogen), are said to have hypofibrinogenemia. These patients manifest a variable but concordant reduction in fibrinogen antigen and activity levels in plasma.

Dysfibrinogenemia is also rare but is more common than afibrinogenemia, with the majority of patients being heterozygous for the disorder [5,9]. The dysfibrinogens are the result of missense, nonsense, or splice junction mutations. Several hundred mutations have been recorded, many of which result in neither a hemorrhagic nor thrombotic state. However, other dysfibrinogens are associated with bleeding episodes, while a few may be associated with venous or arterial thrombosis. Diagnosis is usually suspected by the observation of a prolonged TCT. In those dysfibrinogens associated with thrombosis, the TCT may occasionally be shortened, but the thrombotic propensity probably originates from the refractory nature of the mutant fibrin to plasmin degradation that can often be demonstrated using an appropriately designed assay system. The reptilase time (reptilase is a thrombin-like enzyme present in the venom of the snake, *Bothrops atrox*) may be more sensitive than the TCT to the presence of a dysfibrinogen, but specific diagnosis requires DNA sequencing of the fibrinogen gene. Bleeding patients should be treated with infusions of cryoprecipitate or fibrinogen concentrate.

Prothrombin (Factor II) Deficiency

Inherited prothrombin deficiency is rare, with fewer than 50 distinct mutations being reported [10]. It is

Table 15.3 Summary of Biochemical Features and Hemostatic Levels of Each Coagulation Factor

	Chromosome	MW (Da)	Number of Chains (Active)	t½ (Half-Life)	Plasma Concentrations	Number of Gla Domains	Number of Described Mutations	Prevalence of Deficiency State	Hemostatic Levels (% ref. Range)
FVII	13	50,000	2	4–6 hr	10 nM	10	>130	1/500,000	15%–20%
FX	13	59,000		40–60 hr		11	~70	$1/10^6$	15%–20%
FII	11	72,000	2	3–4 d	10 μg/ml	10	~40	$1/2\times10^6$	20%–30%
FV	1	330,000	2		20 nM	—	~30	$1/10^6$	15%–25%
FV/FVIII ERGIC 53 MCFD2	18 2	330,000 (FV) (FVIII)	2 (FV) 3 (FVIII)	36 hr (FV) 10–14 hr (FVIII)		—	>15 (ERGIC) >5 (MCFD2)	$1/2\times10^6$	15%–20%
FXI	4	160,000	2	40–72 hr (52 hr)	30 nM (~5 μg/mol)	—	>50	$1/10^6$	15%–20%
FXIII (2 genes)	6 (A chain) 1 (B chain)	320,000	4	11–14d		—	>50	$1/2\times10^6$	2%–5%
Fibrinogen (3 genes)	4 (α, β, and γ chains)	340,000	6	2–4d	150–350 mg/dL	—	>300 (Dysfib) >30 (Hypofib)	$1/10^6$	50 mg/dL

an autosomal recessive disorder, and heterozygotes have no bleeding symptoms. Symptomatic patients may be homozygous or doubly heterozygous for causative mutations. By convention, patients with hypoprothrombinemia manifest a concordant reduction in circulating prothrombin antigen and functional activity, whereas those with hypodysprothrombinemia classically have a reduced concentration of prothrombin antigen with a discordantly low functional activity. Bleeding in affected patients varies from mild to severe, depending on the functional prothrombin activity level. The complete absence of prothrombin probably leads to embryonic lethality. Unlike dysfibrinogenemia, thrombosis is not associated with dysfunctional protein in this or any other of the inherited clotting factor deficiencies. Diagnosis depends on a high index of suspicion in patients with a prolonged PT and APTT (but normal TCT) who do not have other known clotting factor defects. Specific diagnosis requires an assay for factor II activity, followed by a prothrombin antigen assay if reduced. In the neonatal period, levels of prothrombin (as well as the other vitamin K-dependent factors—factors VII, IX, X, protein C, and protein S) may be even lower than true baseline for these patients due to deficiency of vitamin K that is almost invariable. Therefore, repeated testing at 6–12 months of age should be undertaken.

Factor V Deficiency

FV is a large molecule (330 kDa) that shares significant structural homology with FVIII and ceruloplasmin. Factor V deficiency is an autosomal recessive disorder that results from mutations in the factor V gene. Heterozygotes are generally asymptomatic, while homozygotes or combined heterozygotes may have mild to moderately severe bleeding symptoms. While about 40 mutations in the factor V gene have been reported, they seem to occur less frequently than in genes for other clotting factors [10].

Bleeding manifestations are similar to those seen in classic hemophilia, except that they tend to be milder, and hemarthroses are less common. The PT and APTT are prolonged, but the TCT is normal. Factor V deficiency also results in a long bleeding time, presumably because of the lack of platelet factor V. It has been reported that 20% of the circulating factor V mass resides in platelet alpha granules [11]. Platelet FV is both synthesized in megakaryocytes and taken up from plasma.

Treatment consists of replacing factor V with fresh frozen plasma. It can be difficult to raise the factor V level higher than 15%–20% of normal using plasma transfusions alone because of volume overload. Exchange transfusion using fresh frozen plasma may be needed when factor V levels above 15%–20% of normal are required. Inhibitor antibodies against factor V in congenital deficiency are rare, but when present may benefit from the use of platelet transfusions as a source of FV that is inaccessible to humoral inhibitors.

Factor VII Deficiency

Approximately 1% of the total factor VII in plasma circulates in the active serine protease form (FVIIa, 10–100 pmol/L). Controversy still exists as to which enzyme is responsible for basal activation of factor VII *in vivo*, although there is evidence implicating FIXa. In the absence of its cofactor TF, FVIIa is a very weak enzyme.

Factor VII deficiency is an autosomal recessive bleeding disorder that occurs in mild, moderate, and severe forms [12]. It is generally considered to be the most common of the rare bleeding disorders, with a prevalence of 1:300,000 to 1:500,000 in most populations. More than 100 mutations in the gene for factor VII have been reported [13]. Bleeding manifestations vary, but in severely affected patients, bleeding can be as severe as that seen in severe classic hemophilia and may include crippling hemarthroses. Factor VII levels of 10% of normal are probably sufficient to control most bleeding episodes, but in some scenarios, higher levels may be required for hemostasis. For unclear reasons, there exist some patients with almost no measurable factor VII activity who express very few hemorrhagic manifestations. Furthermore, there are reports of thrombotic events occurring in patients with FVII deficiency, such that it has been unclear whether distinct mutations may actually be prothrombotic. However, at present, registry analysis suggests that while factor VII deficiency does not seem to predispose to thrombosis, neither is it protective [12].

Patients with factor VII deficiency have prolongation of the PT, while the APTT, TCT, and BT are all normal. It is recommended that recombinant human tissue factor be used as the thromboplastin source in the PT assay, since tissue factor from other species may give spurious results. The best replacement therapy is to use factor VII concentrates or recombinant factor VIIa. Inhibitor antibodies against factor VII have been reported, usually in those patients whose genetic mutation results in virtual absence of factor VII protein.

Hemophilia A and Hemophilia B (Classic Hemophilia and Christmas Disease)

Hemophilia A and B result from deficiencies of factor VIII and IX, respectively. Factor VIII and IX are necessary for the sustained generation of factor Xa (and ultimately thrombin) to form a normal hemostatic plug-in response to vascular injury. Hemophilia A and B are the only two soluble clotting factor deficiencies that are inherited as X-linked recessive disorders. Several hundred distinct mutations in each gene have been reported [10]. These mutations result in mild, moderate, and severe forms of hemophilia, and the clinical manifestations of hemophilia A and B are, for all practical purposes, indistinguishable. In the severe form (<1% basal activity), both disorders are characterized by recurrent hemarthroses that result in chronic crippling arthropathy unless treated by replacing the deficient factor on a scheduled prophylactic basis. Central nervous system hemorrhage is especially hazardous and remains one of the leading causes of death. Retroperitoneal hemorrhage and bleeding into the pharynx may also be life-threatening.

The diagnosis of hemophilia A or B should be suspected in any male patient with hemarthroses, severe bleeding, or excessive bleeding after trauma or surgery. The APTT is prolonged, and the PT, TCT, and BT are normal. Specific diagnosis requires assays for factors VIII and IX, as the two disorders are indistinguishable on clinical grounds alone.

Safe and effective replacement therapy is available for both hemophilia A and B in the form of highly purified factor VIII and IX concentrates, prepared either from large pools of human plasma or by recombinant DNA technology [14]. The main complication of therapy at the present time is the development of inhibitors to factor VIII, which has a cumulative incidence of up to 30% of treated patients with severe hemophilia A. Factor IX antibody inhibitors occur in about 3% of severely affected hemophilia B patients, usually in those patients whose mutation results in undetectable factor IX antigen. Some patients with factor IX inhibitors develop anaphylaxis and/or the nephrotic syndrome when exposed to factor IX, probably due to the concurrent development of IgE antibodies to factor IX and/or immune complex disease. Hemophilic patients with inhibitors to factor VIII or IX are resistant to treatment and hence subject to more complications with increased morbidity and mortality.

Factor X Deficiency

Like factor VII deficiency, factor X deficiency is inherited in an autosomal recessive fashion and can be mild, moderate, or severe. Numerous mutations have been recorded. Severely affected patients have symptoms similar to severe classic hemophilia, including hemarthroses and chronic crippling hemarthropathy. Since factor X resides in the final common pathway, both the PT and the APTT are prolonged, while the BT and the TCT are normal. The primary substrate for factor Xa is prothrombin, but activation of factors V, VII, and VIII may also be partially dependent on factor Xa generation *in vivo*. About one-third of affected individuals exhibit a type 1 deficiency, while the rest have a type 2 pattern, with factor X antigen levels that are preserved, or at least significantly above the level of factor X activity.

Patients with severe factor X deficiency tend be among the most, if not the most, severely affected of the rare coagulation deficiency states. Treatment of bleeding episodes consists of replacement therapy using either plasma or prothrombin complex concentrates (PCCs) that contain all the vitamin K-dependent factors [15]. The biologic half-life of factor X is about 40 hours, so intermittent plasma therapy is reasonable, though overload of the circulation can still be a problem. Inhibitor antibodies to factor X occur but are not common.

Factor XI Deficiency

Factor XI deficiency is an autosomal recessive disorder that commonly occurs in patients of Ashkenazi Jewish descent. In these communities, homozygotes may be as prevalent as 1:500, compared to a frequency of about 1:1,000,000 in most other populations. Among Jewish factor IX-deficient patients, three mutations account for most cases. One-third of patients homozygous for the type II (Glu117stop) mutation develop factor XI inhibitors following treatment with plasma, whereas inhibitors are rare or nonexistent in patients with other genotypes. Factor XI deficiency generally produces a mild bleeding tendency, but there is a relatively poor relationship between the degree of factor XI deficiency and bleeding manifestations. The lack of clinical severity in some patients may be explained by the fact that factor XI-deficient patients have normal levels of factor VIII and IX to form the tenase complex and normal levels of factors V and X to form the prothrombinase complex [16]. Whether or not factor XI-deficient patients bleed may depend on differences in their ability to generate thrombin, and/or the ability to activate thrombin activatable fibrinolysis inhibitor (TAFI). TAFIa cleaves free lysine residues on fibrin, preventing fibrin-dependent activation of plasminogen by plasminogen activators. Thus, factor XIa has both procoagulant (factor IX activating) and antifibrinolytic activities. In this disorder, the PT and TCT are normal, while the APTT is prolonged. Specific diagnosis depends on an assay for factor XI activity.

Deficiencies of Factor XII, Prekallikrein (PK), and High Molecular Weight Kininogen (HK)

Deficiencies of factors XII, PK, and HK (the so-called contact factors) cause a marked prolongation of the APTT, but other screening tests of coagulation are normal. These defects are inherited in an autosomal recessive fashion. They are not associated with bleeding even after trauma or surgery, although the prolonged APTT may cause a great deal of consternation among those not familiar with these defects. A good history revealing the absence of bleeding in these patients and their family members despite a long APTT is the best indication that one is dealing with one of these defects. A specific assay for each is needed for the exact diagnosis. Factor replacement therapy is not needed. Recent data suggest that although patients with factor XII deficiency do not bleed excessively, they may be relatively protected from arterial thrombosis [17]. The kallikrein-kinin system may be more physiologically important in inflammation, blood pressure regulation, and fibrinolysis than in hemostasis [18].

Factor XIII Deficiency

FXIII is activated by thrombin in the presence of calcium. Factor XIIIa is a plasma transglutaminase that covalently cross-links fibrin alpha and gamma chains through γ-glutamyl-ε-lysine bonds to form an impermeable fibrin clot. Although a clot may form in the absence of factor XIII and be held together by hydrogen bonds, this clot is excessively permeable and is easily dissolved by the fibrinolytic system. The clot formed in the absence of factor XIII does not form a normal framework for wound healing, and abnormal scar formation may occur.

Factor XIII consists of two A chains and two B chains. Factor XIII-A is synthesized in megakaryocytes, monocytes, and macrophages, whereas factor XIII-B is synthesized in hepatocytes. The complete molecule is an A_2B_2 tetramer with the A chains containing the active site and the B chains acting as a carrier for the A subunits. Platelet alpha granules contain A chains but not B chains. Factor XIII deficiency may result from mutations in the genes encoding either the A or B chains, with A chain mutations being more common [19]. Autosomal genes govern hepatic synthesis of the factor, and the disease is expressed as a recessive disorder. Typically, only patients with severe deficiency of factor XIII ($<$3%–5%) are symptomatic.

Bleeding manifestations are generally severe, and hemorrhage can occur into any tissue. Umbilical stump bleeding in the neonatal period is common in factor XIII deficiency. Intracranial hemorrhage is also a relatively common manifestation that may mandate long-term prophylactic factor replacement. All the screening tests of clotting function are normal in this disorder, so the diagnosis requires a high index of suspicion, especially in patients with a striking lifelong history of excessive bleeding and in whom the PT, APTT, and TCT are normal. Screening for the diagnosis is performed by taking the clot from one of the above tests, placing it in a 5M urea solution or a dilute solution of trichloroacetic acid, and measuring the time required for its dissolution. If the clot dissolves in less than 24 hours, then Factor XIII deficiency should be suspected. However, these screening tests are not very sensitive, and specific diagnosis rests on an assay for plasma transglutaminase such as measuring incorporation of putrescine into casein.

The plasma half-life of factor XIII is several days, so weekly prophylactic treatment with concentrate is practical for prophylaxis. Cryoprecipitate is the replacement therapy of choice in the United States, though factor XIII concentrates are available in Europe and are currently in clinical trials in the United States.

Multiple Clotting Factor Deficiencies

The two most common multiple clotting factor deficiencies are a combined deficiency of factors V and VIII and a combined deficiency of the vitamin K-dependent factors (factors II, VII, IX, X, and Protein C and S) [20,21].

A combined deficiency of factors V and VIII is inherited in an autosomal recessive fashion and can be distinguished from a combined inheritance of mild classic hemophilia and mild factor V deficiency by family studies or by genetic analysis. The disorder is due to defects in one of two genes: the *LMAN1* gene and a newly discovered gene called the multiple clotting factor deficiency 2 (*MCFD2*) gene [22]. The products of both genes play a critical role in the transport of factors V and VIII from the endoplasmic reticulum to the Golgi apparatus and are necessary for normal secretion of these factors. The disorder results in a mild to moderate bleeding tendency with factor V and VIII levels ranging from 5%–30% of normal. When both the PT and APTT are prolonged, and levels of either factor V or VIII are found to be decreased, the combined deficiency should be suspected. Factor VIII is easily replaced using factor VIII concentrates, but the only readily available factor V replacement is fresh frozen plasma.

Combined deficiencies of the vitamin K dependent factors can be due to defects in either the gene for vitamin K-dependent carboxylase or the gene for vitamin K epoxide reductase [23]. These are autosomal recessive disorders that may be associated with severe deficiency of prothrombin; factor VII, IX, and X; as well as Protein C and S [23]. In this syndrome both the PT and APTT are prolonged, and assays for the individual factors that influence these tests are necessary. The diagnosis must be distinguished from surreptitious ingestion of coumarin drugs (including rodenticides), which is an acquired disorder with bleeding manifestations of recent onset. Large doses of vitamin K may partially correct the hereditary defect in some but not all cases. Some bleeding episodes will require replacement with prothrombin complex concentrates.

Von Willebrand Disease (VWD)

The most common hereditary bleeding disorder arises from abnormalities in von Willebrand factor (VWF). VWF occurs in plasma as multimers of a 240,000 Dalton subunit, with molecular weights ranging from 1 million to 20 million Daltons. The principal functions of VWF are to act as a carrier for clotting factor VIII and to mediate platelet adhesion to the injured vessel wall. The larger molecular weight multimers are the most effective at mediating platelet adhesion. VWF binds to glycoprotein Ib on the platelet surface and also to collagen in the vessel wall. VWF is particularly important for platelet adhesion under high shear stress vascular beds. It has been elegantly demonstrated that shear stress leads to unfolding of the globular collagen-bound VWF multimers, thereby uncovering the molecular domain responsible for binding to VWF [24]. VWF also cross-links platelets via binding to glycoprotein IIb-IIIa.

There are three major types of von Willebrand disease (VWD): type 1, 2, and 3 [25] (see Table 15.4). Type 1 is autosomal dominant and represents a partial quantitative deficiency of VWF. Generally, it is explained by reduced synthesis of VWF, and analysis of multimers in plasma reveals a global decrease in multimers of all sizes. More recently, it has been appreciated that the pathophysiology of some cases of type 1 VWF is accelerated clearance of VWF from plasma. The prototype of this phenotype is the so-called Vicenza R1205H mutation, which causes a moderate to severe variant of von Willebrand disease. Type 3 is an autosomal recessive severe quantitative deficiency, in which there is an absence of VWF. Type 2 VWD usually occurs as an autosomal dominant disorder with qualitative abnormalities in VWF function. Type 2 occurs in four major forms: 2A, 2B, 2N, and 2M. Types 2A and 2B are characterized by absence of the higher molecular weight multimers of VWF in plasma. Type 2B is also associated with thrombocytopenia as a result of a gain of function mutation resulting in a VWF molecule with higher affinity for the GP Ib receptor, thus enhancing platelet agglutination and accelerated clearance. Type 2M patients show reduced binding of VWF to GPIb, although they have normal VWF multimeric composition in plasma.

Type 2N VWD is a rare autosomal recessive disorder arising from a mutation in the factor VIII binding site

Table 15.4 Von Willebrand Disease Subtypes

				Diagnostic Testing				
		Inheritance Pattern	**Bleeding Manifestations**	**VWF Ag**	**VWF Activity**	**Factor VIII Activity**	**VWF Multimers**	**Treatment**
Type 1		Autosomal dominant	Generally mild	Low	Low	Low	Normal	DDAVP, Factor VIII concentrates rich in VWF
Type 2	2A	Autosomal dominant	Mild–moderate	Low	Lower than antigen	Variable	Absent high molecular weight forms	Factor VIII concentrates rich in VWF
	2B	Autosomal dominant	Mild–moderate	Low	Lower than antigen	Variable	Absent high molecular weight forms	Factor VIII concentrates rich in VWF
	2N	Autosomal recessive	Mild	Normal	Normal	Low	Normal	Factor VIII concentrates rich in VWF
	2M	Autosomal dominant	Mild–moderate	Normal	Lower than antigen	Normal	Normal	DDAVP, Factor VIII concentrates rich in VWF
Type 3		Autosomal recessive	Severe	Near absent	Near absent	Near absent	Absent	Factor VIII concentrates rich in VWF

on the VWF molecule. Without the protection provided by VWF binding, plasma factor VIII levels are reduced because of a markedly decreased half-life. VWF multimers and antigen and activity levels may be normal, while the factor VIII levels are low enough to be confused with mild classic hemophilia. These two disorders can be distinguished by an ELISA-based factor VIII binding assay. Clinically affected patients are either homozygous for one of several gene mutations in the D′ or D3 domain of mature VWF, or are combined heterozygous for a 2N mutation and a type 1 mutation. Specific diagnosis requires either demonstration of the lack of binding of VWF to factor VIII or genetic analysis.

Although VWD is a defect in a soluble clotting factor, bleeding in patients with this disorder is more similar to that produced by a defect in platelet number or function. The bleeding manifestations tend to be more of the "oozing and bruising" variety, with hematoma formation being rare. Bleeding in types 1 and 2 VWD is usually mild to moderate, although severe bleeding may occur with trauma and surgery. Some patients with type 1 VWD may be relatively asymptomatic. Table 15.4 lists the diagnostic features of the various types of VWD.

The diagnosis of VWD should be suspected in any patient with abnormal bruising, bleeding from mucosal surfaces, menorrhagia, or excessive bleeding after surgery, dental work, or trauma [22]. The PT and TCT should be normal. The APTT is variably prolonged, depending on the degree to which the factor VIII level is reduced. The BT is prolonged, except in patients with type 2N VWD. A new diagnostic test, the PFA-100®, is another test of primary hemostasis and has been reported to be more sensitive and specific for diagnosis of VWD [23].

Treatment of type 1 and some cases of type 2A VWD usually consists of administration of desmopressin (DDAVP) either parenterally or by intranasal administration. Desmopressin activates the vasopressin V2 receptors on endothelium, resulting in a release of preformed stores of VWF from Wiebel-Palade bodies. It can be used for 3 to 5 days, but in time, the drug loses its efficacy by depleting endothelial stores of VWF, a process termed tachyphylaxis. The response to DDAVP is usually good in type 1 VWD, although it may be reduced in patients with mutations that result in accelerated clearance of VWF from plasma [26]. DDAVP is theoretically contraindicated in type 2B VWD because thrombocytopenia can be worsened; release of type 2B VWF with enhanced affinity for platelet membrane glycoprotein 1b results in further *in vivo* platelet agglutination. In type 2B and type 3 VWD, factor VIII concentrates rich in VWF should be used for treatment.

DISORDERS OF FIBRINOLYSIS

There are two principal inhibitors of the fibrinolytic system: alpha 2 plasmin inhibitor (α_2-antiplasmin) and plasminogen activator inhibitor-1 (PAI-1). The genes for both proteins have been identified and sequenced, and mutations within each have been described [27,28]. Deficiency of either inhibitor results in excessive fibrinolysis and potentially severe bleeding, including hemarthroses and hematoma formation following trauma or surgery. Both disorders are inherited in an autosomal pattern and, in most cases, bleeding occurs only in homozygotes. The diagnosis is suspected when a patient gives a life-long history of excessive bleeding, often after trauma, and when the usual screening tests of coagulation are normal. In such cases, a euglobulin lysis test should be performed, and if the clot lyses within a few hours (normal lysis times are >24 hours), specific assays for alpha 2 plasmin inhibitor and plasminogen activator inhibitor 1 should be considered. Bleeding episodes respond to administration of antifibrinolytic agents, such as epsilon aminocaproic or tranexamic acid [29].

DISORDERS OF PLATELET NUMBER OR FUNCTION

Disorders of Platelet Production

Inherited disorders causing thrombocytopenia are a heterogeneous group of conditions. Some are associated with a profound thrombocytopathy, some are associated with other somatic changes, while others manifest thrombocytopenia only.

The MYH9-Associated Disorders

The May-Hegglin anomaly is the prototype of a family of disorders now known to be due to a defect in the *MYH9* gene [30]. May-Hegglin anomaly, Sebastian syndrome, Fechtner syndrome, and Epstein syndrome are autosomal dominant macrothrombocytopenias that are distinguished by different combinations of clinical and laboratory signs, such as sensorineural hearing loss, cataract, nephritis, and polymorphonuclear inclusions known as Döhle-like bodies. Mutations in the *MYH9* gene encoding for the nonmuscle myosin heavy chain IIA (NMMHC-IIA) have been identified in all these syndromes. The hallmark of these conditions is macrothrombocytopenia with characteristic Döhle-like inclusions within the leukocytes. The Fechtner syndrome also includes the triad of sensorineural deafness, ocular abnormalities, and nephritis seen in Alport syndrome [31]. Epstein syndrome differs from Fechtner syndrome by the lack of cataracts and the lack of the leukocyte inclusions [32]. Both the May-Hegglin anomaly and the Sebastian syndrome manifest macrothrombocytopenia with leukocyte inclusions. However, these two syndromes may be distinguished by ultrastructural analysis of the inclusions. In the May-Hegglin anomaly, the Döhle-like bodies are composed of cytoplasm surrounding parallel microfilaments with clustered ribosomes, whereas the inclusion bodies in the Sebastian syndrome are composed of highly dispersed microfilaments with few ribosomes [33].

Defects in Transcription Factors

Alterations in megakaryocyte development due to defective transcription factors underlie a large number of the familial thrombocytopenias. Derangements in development of other cell types as well as other somatic mutations can also occur. Mutations in *HOXA11* have been described in two unrelated families with bone marrow failure and skeletal defects [34].

The Paris-Trousseau syndrome is an autosomal dominant condition characterized by macrothrombocytopenia with giant alpha granules. It is caused by hemizygous loss of the *FLI1* gene due to deletion at 11q23. Lack of FLI1 protein leads to lack of platelet production due to arrested megakaryocyte development [35]. The 11q23 deletion is also seen in patients with Jacobsen's syndrome who also have congenital heart disease, trigonocephaly, dysmorphic facies, mental retardation and multiple organ dysfunction as well as macrothrombocytes with abnormal alpha granules [36].

Mutations in *GATA-1* lead to the X-linked congenital dyserythropoietic anemia and thrombocytopenia syndrome [37]. The platelets are large and exhibit defective collagen-induced aggregation. The GATA-1 transcription factor has two zinc fingers. The C-terminal finger binds DNA in a site-specific fashion, while the N-terminal finger stabilizes the DNA binding as well as interacts with FOG-1 (Friend of GATA). Mutations within *GATA-1* may alter DNA-binding, FOG-1 interactions, or both, and phenotypes may differ depending on the site of mutation. X-linked thrombocytopenia without anemia is due to mutations within *GATA-1* that disrupt FOG-1 interactions while leaving DNA-binding intact [38]. By contrast, *GATA-1* mutations which affect binding to DNA while not interrupting FOG interactions lead to a thalassemic phenotype [39]. The acute megakaryoblastic anemia seen in conjunction with Down's syndrome can be associated with mutations in *GATA-1.*

Mutations in *RUNX1* lead to familial thrombocytopenic syndromes with a predisposition to development of acute myelogenous leukemia [40]. *RUNX1* mutations cause an arrest in megakaryocyte development with an expanded population of progenitor cells. The platelets that are produced show defects in aggregation. The development of acute leukemia likely requires a second mutation within *RUNX1* or another gene.

Defects in Platelet Production

Congenital amegakaryocytic thrombocytopenia (CAMT) is due to defects in the *c-MPL* gene encoding the thrombopoietin receptor. Children born with this disorder have severe thrombocytopenia and may go on to develop deficiencies in other cell types [41].

Thrombocytopenia with absent radii (TAR) is a syndrome characterized by severe congenital thrombocytopenia along with absent or shortened radii [42]. The platelets produced show abnormal aggregation. Although thrombopoietin levels are elevated, no defect in *c-MPL* has been identified, and abnormal intracellular signaling pathways are postulated as the cause of this rare disorder.

Perhaps the most common hereditary thrombocytopenia with small platelets is the Wiskott-Aldrich syndrome (WAS), a disorder associated with the triad of immune deficiency, eczema, and thrombocytopenia [43]. This syndrome is X-linked and results from mutations in the gene for Wiskott-Aldrich syndrome protein (WASP). Platelets as well as T-lymphocytes show defective function, and clinical manifestations vary widely. Definitive treatment requires allogeneic stem cell transplantation. As opposed to the macrothrombocytopenic defects, WAS platelets are small and defective in function.

Disorders of Platelet Function

Defects in Platelet Adhesion

Bernard Soulier Syndrome (BSS) is a severe bleeding disorder characterized by macrothrombocytopenia, decreased platelet adhesion to VWF, abnormal prothrombin consumption, and reduced platelet survival. Deficient platelet binding to subendothelial von Willebrand factor is due to abnormalities (either qualitative or quantitative) in the GP-Ib-IX-V complex. Mutations in GPIbα binding sites for P-selectin, thrombospondin-1, factor XI, factor XII, aMb2, and high molecular weight kininogen may mediate variations in the phenotype seen. The product of four separate genes (*GPIBA, GPIBB, GP9,* and *GP5*) assemble within the megakaryocyte to form the GP-Ib-IX-V on the platelet surface. Defects in any of the genes may lead to BSS [44]. Classically, affected platelets from affected individuals aggregate normally in response to all agonists except ristocetin. A database of described causative mutations for BSS may be found at www.bernardsoulier.org/.

Platelet-type von Willebrand disease is due to a gain of function mutation such that plasma VWF binds spontaneously to platelets, and the platelets exhibit agglutination in response to low dose ristocetin. Mutations generally lie within the *GPIBA* gene. High molecular weight multimers of VWF bound to platelets are cleared from the circulation, which may result in bleeding. The phenotype is identical to that seen in type 2B VWD, in which the mutation lies within the VWF rather than its receptor. It can therefore be quite difficult to distinguish between type 2B VWD and platelet-type VWD. Gene sequencing of the VWF gene, the *GPIBA* gene, or both may be required.

Defects in Platelet Aggregation

Glanzmann thrombasthenia is a rare, autosomal recessive disorder characterized by absent platelet aggregation. It is due to absent or defective GP IIbIIIa on the platelet surface. Patients have severe mucocutaneous bleeding, which becomes refractory to platelet transfusions as alloantibodies to transfused platelets form. Recombinant factor VIIa has been used to treat bleeding in this disorder and is postulated to work by enhancing thrombin generation on the platelet surface and allowing fibrin to cross-link platelets via an as yet undetermined receptor [45]. Though demonstration of absent platelet aggregation in response to all agonists (with the exception of ristocetin) will suggest the diagnosis, definitive diagnosis relies on showing absence of functional GPIIbIIIa on the platelet surface, either by flow cytometry or by electron microscopy using immuno-gold labeled fibrinogen imaging. A database of described mutations in the GPIIb and GPIIIa genes that may result in Glanzmann thrombasthenia may be found at http://sinaicentral.mssm.edu/intranet/research/glanzmann/menu. Acquired Glanzmann thrombasthenia has been described in patients who develop autoantibodies against GPIIbIIIa. These patients may have underlying immune thrombocytopenic purpura, but the severity of their bleeding is out of proportion to platelet number. Affected individuals may respond favorably to immunosuppression [46].

Some patients exhibit defects in aggregation in response to specific agonists. Platelets from these patients may show defects in either platelet receptors or in the downstream intracellular signaling pathways leading to activation. Deficiency of GPVI leads to defective collagen-mediated aggregation [47]. Since ADP-mediated platelet secretion is required for generation of the second wave of platelet aggregation needed by weak agonists, mutations within the $P2Y_{12}$ ADP receptor leads to defective aggregation not only to ADP, but also to weak platelet agonists [47]. Defects in the thromboxane A2 (TxA2) receptor lead to defective aggregation in response to arachidonic acid and the thromboxane receptor agonist U46619 [48]. Patients with cyclooxygenase pathway defects, including deficient or dysfunctional cyclooxygenase-1, prostaglandin H synthetase-1, thromboxane synthetase, and lipoxygenase also exhibit defective aggregation in response to arachidonic acid, but their response to U46619 is preserved. Patients deficient in the alpha subunit of the heterotrimeric GTPase protein Gq show defective aggregation in response to agonists which utilize Gq for inside-out signaling [49]. These disorders must be distinguished from the effects of drugs such as aspirin, whose ingestion can produce similar effects on platelet function.

Disorders of Platelet Secretion: The Storage Pool Diseases

Platelets contain two types of intracellular granules: alpha and delta (or dense) granules. Alpha granules contain proteins either synthesized within the megakaryocytes or endocytosed from the plasma, including fibrinogen, factor V, thrombospondin, platelet-derived growth factor, multimerin, fibronectin, factor XIII A chains, high molecular weight kininogen, and VWF among others. Their membrane contains molecules such as P-selectin and CD63 that are translocated to the outer plasma membrane after secretion and membrane fusion. Dense granules contain ATP and ADP as well as calcium and serotonin, and any deficiency of dense granules thus leads to a defective secondary wave of platelet aggregation.

Defects in Alpha Granules

The gray platelet syndrome (GPS) is an autosomal recessive condition that leads to a mild bleeding diathesis. It may be recognized by examination of a Wright-Giemsa stained peripheral blood smear showing platelets that appear gray without the usual red-staining granules. Electron microscopy showing a depletion of alpha granules is a more sensitive method to diagnose the syndrome. GPS is thus classified with the other platelet secretion defects, but may also be classified with the macrothrombocytopenias, since platelets may be slightly larger than usual, albeit not as large as those seen in the giant platelet disorders described previously. Furthermore, the platelet count is only moderately depressed, and bleeding symptoms are mild. Patients with GPS may also develop early onset myelofibrosis, a probable consequence of the impaired storage of growth factors such as PDGF [50].

The Quebec platelet disorder (QPD) is associated with a normal to slightly low platelet count with a mild bleeding disorder. Pathophysiologically, this is due to abnormal proteolysis of alpha granule proteins that appears to be mediated by ectopic intragranular production of urokinase. QPD was first recognized as a specific deficiency of platelet factor V associated with normal concentrations of plasma factor V. The platelets appear normal on peripheral blood smears under the light microscope, and diagnosis depends on showing decreased alpha granule proteins. It is inherited in an autosomal dominant fashion, but the specific genetic defect is unknown [51].

Defects in Dense Granules

The Hermansky-Pudlak syndrome is the association of delta storage pool deficiency with oculocutaneous albinism and increased ceroid in the reticuloendothelial system. There are several subtypes of the Hermansky-Pudlak syndrome resulting from several distinct mutations. The syndrome is inherited in an autosomal recessive pattern. Granulomatous colitis and pulmonary fibrosis are also part of the syndrome. Mutations in at least eight genes (*HPS-1* through *HPS-8*) lead to defects in HPS proteins responsible for organelle biosynthesis and protein trafficking [52]. Described mutations accounting for the Hermansky-Pudlak syndrome are collated in the database available at http://liweilab.genetics.ac.cn/HPSD/.

The Chediak-Higashi syndrome is also associated with storage pool deficiency and is characterized by oculocutaneous albinism, neurologic abnormalities, immune deficiency with a tendency to infections, and giant inclusions in the cytoplasm of platelets and leukocytes. The disorder is rare, and bleeding manifestations are relatively mild. The syndrome is due to mutations in the LYST (lysosomal trafficking regulator) gene; these are listed in the Chediak-Higashi database at http://bioinf.uta.fi/LYSTbase/?content=pin/

IDbases. Affected patients are homozygous, while heterozygotes are phenotypically normal [53].

The Scott Syndrome

In this disorder, platelets, when activated, cannot translocate phosphatidylserine from the inner to the outer platelet membrane when the flip-flop of the membrane leaflet occurs, presumably due to defects in the activity of the scramblase enzyme [54]. Because of this defect, factor Xa and Va are unable to efficiently bind to the membrane to assemble the prothrombinase complex, and thrombin generation on the platelet surface is impaired. Scott syndrome is characterized by a mild bruising and bleeding tendency. It can be detected using flow cytometry with antibodies against annexin V which will show the defective microvesicle formation characteristic of this disorder.

Disorders of Platelet Destruction

Disorders of platelet destruction are too numerous to discuss here.Therefore, the discussion in the section will be limited to those disorders where the molecular pathogenesis is known at a greater level of detail. A more comprehensive description of disorders characterized by platelet destruction can be found in recent reviews [55,56].

Antibody-Mediated Platelet Destruction

Neonatal alloimmune thrombocytopenia (NAIT) is a bleeding disorder caused by transplacental transfer of maternal antibodies directed against fetal platelet antigens inherited from the father. In Caucasians, the antigens most frequently implicated include HPA-1a (PL^{A1}) and HPA-5b (Br^{a}). In Asians, HPA-4a and HPA-3a account for the majority of NAIT cases. NAIT occurs with a lower frequency in Caucasians than is expected by the incidence of HPA-1a negativity in the population, suggesting that other factors influence antibody development. Additionally, NAIT mediated by antibodies against HPA-1a is more clinically severe, perhaps because these antibodies may also block platelet aggregation, since HPA-1a is an antigen expressed on platelet GPIIIa. Mothers who are negative for the antigen in question can develop antiplatelet antibodies that cross the placenta, leading to severe fetal thrombocytopenia. Even the first child may be affected, and intracranial hemorrhage is a feared and devastating complication. Subsequent pregnancies have a near 100% rate of NAIT, and measures including prenatal IVIG with or without corticosteroids given to the mother, or *in utero* transfusions of matched platelets or IVIg have been employed. After birth, affected infants can be treated with IVIg or matched platelets.

Post-transfusion purpura (PTP) is associated with thrombocytopenia resulting from a mismatch between platelet antigens. In this condition, patients previously sensitized against certain platelet antigens (the same ones that lead to NAIT) develop acute, severe thrombocytopenia 5–14 days after transfusion [57]. Though packed red cells are most commonly associated with PTP, transfusion of any blood component may precipitate this disorder. These blood components contain platelet microparticles that express the offensive platelet antigen, leading to an anamnestic production of antibodies. However, paradoxically, these patients develop antibodies directed against their own platelets, either by fusion of the exogenous microparticles with their own platelets, or by a process in which exposure to foreign platelets leads to formation of autoantibodies. Unlike NAIT, transfusion of matched platelets leads to only a transient improvement in this condition. IVIg has been reported to shorten the duration of thrombocytopenia, as has plasma exchange.

Thrombotic Microangiopathies

Thrombotic thrombocytopenic purpura (TTP) is an acute disorder that usually presents in previously healthy subjects. It is highly lethal unless treated promptly, which generally entails plasma exchange with or without additional immunosuppression. In 1982, Moake made the seminal observation that the plasma of patients with TTP contained ultralarge (UL) multimers of VWF, which were absent in normal plasma [58]. He hypothesized that TTP could be due to the absence of a protease or depolymerase responsible for cleaving the UL VWF multimers. The protease was identified in 1996 by the groups of Tsai [59] and Furlan [60], its gene was cloned [61], and the enzyme named ADAMTS-13, when it was found to be a member of the "*a disintegrin-like and metalloprotease with thrombospondin repeats*" family of metalloproteases [62,63]. ADAMTS-13 levels are found to be low in patients with both familial [63] and sporadic TTP, and an IgG autoantibody inhibitor to ADAMTS-13 is found in a majority (but not all) patients with sporadic TTP [64,65].

The hemolytic uremic syndrome (HUS) shares many clinical features with TTP, including microangiopathic hemolytic anemia, thrombocytopenia, and renal insufficiency. Renal findings are more prominent and neurologic findings less so. HUS is divided into diarrhea-associated HUS (D+HUS) and atypical (diarrhea-negative) HUS [66]. Diarrhea-positive HUS is triggered by infection with a Shiga-toxin-producing bacteria, and is much more commonly encountered in children. *Escherichia coli* O157:H7 is implicated in 80% of the cases, but other bacteria including other *E. coli* subtypes and *Shigella dysenteriae* serotype 1 can cause D+HUS. Shiga toxins bind to the glycosphingolipid receptor globotriaosylceramide (Gb3) on the surface of renal mesangial, glomerular, and tubular epithelial cells. Protein synthesis is impaired through inhibition of 60S ribosomes, and cell death occurs [67]. Plasma from patients with HUS demonstrates markers of abnormal thrombin generation. As compared with TTP, ADAMTS-13 levels are typically normal in patients with HUS, and the fibrin microthrombi do not contain VWF strands.

Atypical HUS occurs in patients without a diarrheal prodrome. Underlying conditions, such as organ transplantation or exposure to certain drugs may be present. In as many as 30%–50% of patients, mutations in one of three proteins involved in complement regulation occur [68–70]. Factor H (CFH) and membrane cofactor protein (MCP or CD46) are regulators of complement factor I (CFI), which is a serine protease that cleaves and inactivates surface-bound C3b and C4b. Autoantibodies against these proteins have also been reported, suggesting that unregulated complement activation plays a role in the pathogenesis of HUS [71].

Heparin-Induced Thrombocytopenia

Heparin-induced thrombocytopenia (HIT) is a common iatrogenic thrombocytopenic disorder that can paradoxically lead to thrombosis. It occurs in 1%–5% of patients treated with standard unfractionated heparin for at least 5 days and in <1% of those treated with low molecular weight heparin. Approximately 50% of patients develop venous and/or arterial thromboses. New thromboses develop in 25% of patients, amputations are required in 10%, and reported mortality rates are between 10% and 20% [72]. The pathogenic autoantibodies that cause HIT are directed against neoepitopes on PF4 that are induced by heparin and other anionic glycosaminoglycans (GAGs). PF4 is an abundant protein stored in the alpha granules of platelets in complex with chondroitin sulfate (CS). Upon platelet activation, PF4/CS complexes are released and bind to the platelet surface. Heparin can displace CS, forming PF4/heparin complexes. Binding of IgG anti-PF4/heparin to the platelet leads to Fcγ receptor-mediated clearance of platelets but also leads to platelet activation and generation of procoagulant microparticles via FcγRIIA. PF4/heparin complexes also form on the surface of monocytes and endothelial cells, and antibody binding leads to tissue-factor driven thrombin generation and hence to clot formation. The PF4/heparin complexes are most antigenic when PF4 and heparin are present at equimolar concentrations, where they form ultralarge molecular complexes [73]. Low molecular weight heparin forms these ultralarge complexes less efficiently and at concentrations that tend to be supratherapeutic, perhaps explaining the lower frequency of HIT in patients treated with LMWH as opposed to standard unfractionated heparin.

THROMBOPHILIA

An understanding of how the coagulation system is physiologically regulated is necessary when seeking to determine how it can become deranged. Thrombosis can result from excessive activation of coagulation and/or impaired endogenous regulation. This section will focus on the two major natural anticoagulant pathways that serve to inhibit thrombin generation: the protein C/S pathway and the antithrombin pathway.

The Protein C/S Pathway and Thrombosis

Protein C (PC) is a vitamin K-dependent protein that is activated by thrombin. When bound to the endothelial cell surface protein thrombomodulin, thrombin changes its substrate specificity, losing its ability to cleave fibrinogen and activate platelets. Instead, the thrombin-thrombomodulin complex proteolytically activates zymogen PC to form activated protein C (APC). Activated protein C and its cofactor protein S (another vitamin K-dependent protein) inactivate factors Va and VIIIa, thereby inhibiting thrombin generation. APC also downregulates inflammatory pathways, and inhibits p53-mediated apoptosis of ischemic brain endothelium [74,75]. In fact, the previously shown therapeutic benefit of APC infusions in patients with severe sepsis is believed to be mediated by the anti-apoptotic and anti-inflammatory (rather than the anti-coagulant) activity of APC [74]. The endothelial protein C receptor (EPCR), localized on the surface of endothelial cells, serves to bind PC and thereby enhance its activation by thrombin-thrombomodulin 5-fold. EPCR is also found in a soluble form in plasma, and its levels are enhanced in such multifocal conditions as disseminated intravascular coagulation and systemic lupus erythematosus. EPCR also binds to APC, shifting its substrate specificity to favor activation of the protease activated receptor-1 (PAR-1). This pathway thereby facilitates cross-talk between the coagulation system and inflammatory cell, endothelial, and platelet functions [76].

Heterozygous protein C deficiency is a recognized risk factor for venous thromboembolism, with an odds ratio of 6.5–8 [77–79]. Most of the causative mutations are of the type I variety, with a concordant decrease in activity and antigen. These mutations affect protein folding and lead to unstable molecules that are either poorly secreted or are degraded more rapidly. Type II defects lead to activity levels that are reduced disproportionately to the antigen levels and result in dysfunctional molecules with ineffective protein-protein interactions. Heterozygous protein C deficiency has a prevalence of 0.2%–0.4% in the general population [80,81] and approximately 4%–5% of patients with confirmed deep venous thrombosis [79]. Protein C-deficient individuals with personal and family histories of thrombosis may have a second thrombophilic defect, such as factor V Leiden, to account for the thrombotic tendency. Venous thromboembolic disease (VTE) occurs in 50% of heterozygous individuals in affected families by the age 45, with half of the events being spontaneous in onset [82]. Venous thrombosis at unusual sites (cerebral sinus and intra-abdominal) is a clinical hallmark. Arterial thrombosis is rare, though reported. Homozygous protein C deficiency with levels <1% generally presents with neonatal purpura fulminans and massive thrombosis in affected infants [83]. Individuals with protein C deficiency are predisposed to develop warfarin skin necrosis when anticoagulated with vitamin K antagonists such as Coumadin. Since protein C has a much shorter half-life (8 hours) than the procoagulant vitamin K-dependent factors such as prothrombin and factor X (24–48 hours), a transient hypercoagulable state can occur in patients treated with Coumadin in the absence of an alternate bridging anticoagulant such as heparin. This risk is magnified in patients with underlying deficiency of either protein C or vitamin K [84].

Protein S is a vitamin K-dependent protein that is not a serine protease; rather, it acts as a cofactor for APC. In normal plasma, 60% of protein S is bound to C4b-binding protein (C4BP), and the remainder is present in the free form. Only the free form of protein S can function as the cofactor for APC. Protein S also exhibits anticoagulant activities independent of APC by binding to and inactivating factors Va, VIIIa, and Xa [85]. Most recently, it has been shown that protein S is a cofactor for TFPI-mediated inactivation of tissue factor. Protein S deficiency exists in three forms: (i) type I has equal decrements of antigen and activity; (ii) type II has low activity but normal antigen levels, and (iii) type III shows low free protein S levels, with total protein S levels in the low to normal range. The odds ratio for VTE with protein S deficiency has been variably reported as 1.6 [79], 2.4 [86], 8.5 [78], and 11.5 [87]. More than 50% of VTE events are unprovoked. Arterial thromboses occur at higher frequency, especially among smokers or those with other thrombotic risk factors [88,89]. Laboratory testing needs to be interpreted with caution. Normal levels vary with age and gender, being lower in premenopausal women than men, with further reductions occurring as a result of estrogen therapy or pregnancy. Measured protein S activity can be falsely low in patients with inherited resistance to activated protein C. Acquired protein S deficiency occurs in a variety of conditions, including acute thrombosis, inflammation, liver disease, nephrotic syndrome, vitamin K deficiency, disseminated intravascular coagulation, and in association with the lupus anticoagulant. Antibodies to protein S can be seen in children with varicella or other viral illnesses [90].

Addition of APC to plasma normally causes a prolongation of the APTT. In 1993, Dahlback reported a series of thrombophilic families in which the plasma of the probands and their affected relatives exhibited resistance to APC, with much less prolongation of the APTT than would be expected [91]. Mixing studies showed this defect to be due to a problem with factor V [92], and the genetic defect responsible for APC resistance was shown to be a mutation at the major cleavage site of APC on factor Va from arginine to glutamine (R506Q) [92–95]. This mutation, now known as factor V Leiden, after the Dutch city in which it was first discovered, is the most prevalent inherited mutation leading to thrombophilia. It is found in approximately 5% of Caucasian populations and is felt to be the result of a founder mutation in a single ancestor 21,000 to 34,000 years ago [96]. The mutant factor Va is inactivated by APC 10-fold more slowly, thereby leading to excessive thrombin generation [97]. Factor V Leiden is estimated to account for 20%–25% of inherited thrombophilia. Heterozygosity for this mutation confers a relatively low risk for VTE in younger patients (OR 1.2 in those 40–50), but the risk increases steeply with age (OR 6 for those older than 70) [98]. Approximately 90% of affected individuals do not suffer any venous thromboembolic events during their lifetime. On the other hand, homozygotes have an odds ratio for VTE of 50–100, and half of such individuals will have thromboses during their lives [99]. Coronary artery thrombosis may also occur with greater frequency in young men and women with other risk factors, such as smoking [100–103]. In general, however, factor V Leiden is not considered to be a major risk factor for arterial thrombosis. The risk for venous thrombosis in individuals with factor V Leiden is greatly magnified when other risk factors for thrombosis are present. These risks may be either genetic or acquired, including PC deficiency, PS deficiency, the prothrombin G20210 mutation, elevated levels of factor VIII, antiphospholipid antibodies, hyperhomocysteinemia, prolonged immobility, surgery, malignancy, pregnancy, or use of oral contraceptives. An acquired form of APC resistance may be caused by conditions other than factor V Leiden, including pregnancy, lupus anticoagulants, inflammation, and use of anticoagulants. Testing for APC resistance is best performed using factor V-deficient plasma, which will eliminate the preceding conditions. Genetic testing for factor V Leiden, generally using a PCR-based assay, is also available and is sensitive and specific for the disorder.

A mutation found in 1% of Caucasians is the second most frequent cause of inherited thrombophilia. A mutation in the 3′-untranslated region of the prothrombin gene (G20210A) results in elevated prothrombin synthesis [104]. Thrombotic risk is probably a result of increased thrombin generation and/or decreased fibrinolysis mediated by enhanced activation of TAFI [105,106]. The relative risk for first episode of VTE in heterozygotes is between 2 and 5.5, and 4%–8% of patients presenting with their first VTE will be found to have this mutation [104,107–112]. Homozygosity for the mutation appears to confer a higher risk of VTE [113,114]. Venous clots in odd locations, as well as arterial clots, are found with increased frequency, especially in patients younger than 55, and especially in those with other thrombotic risk factors. PCR amplification of the pertinent region, followed by DNA sequencing is required for the diagnosis. Measurement of factor II levels is neither sensitive nor specific for the disorder.

Antithrombin Deficiency

Antithrombin (AT) is a SERPIN that inactivates thrombin and clotting factors Xa, IXa, and XIa by forming irreversible 1:1 complexes in reactions accelerated by glycosaminoglycans such as heparin or heparan sulfate on the surface of endothelial cells. Deficiency of antithrombin therefore results in potentiation of thrombosis. In type I deficiency, the antigen and activity levels are decreased in parallel, whereas in type II deficiency, a dysfunctional molecule is present. Type IIa mutations affect the active center of the inhibitor, which is responsible for complexing with the active site of the protease. Type IIb mutations target the heparin-binding site, and type IIc mutations are heterogeneous. Severe antithrombin deficiency with levels <5% is rare, resulting from one of several IIb mutations, and leads to severe recurrent arterial and venous thromboses [115–118]. The odds ratio for venous thrombosis in heterozygotes is approximately 10–20 [78,119]. Lower extremity deep vein thrombosis is common, and clots in unusual sites have been reported. Clots tend to occur at a younger age, with 70% presenting before age 35, and 85% before age 50 [120]. Some patients with AT deficiency exhibit resistance to the anticoagulant effects of heparin. Other conditions associated with reduced levels of AT include treatment with heparin, acute thrombosis, disseminated intravascular coagulation, nephrotic syndrome, liver disease, treatment with the chemotherapeutic agent L-asparaginase, and pre-eclampsia [121–126].

REFERENCES

1. Davie EW, Ratnoff OD. Waterfall sequence for intrinsic blood clotting. *Science.* 1964;145:1310–1312.
2. Macfarlane RG. An enzyme cascade in the blood clotting mechanism, and its function as a biochemical amplifier. *Nature.* 1964;202:498–499.
3. Osterud B, Rapaport SI. Activation of factor IX by the reaction product of tissue factor and factor VII: Additional pathway for initiating blood coagulation. *Proc Natl Acad Sci USA.* 1977;74: 5260–5264.
4. Roberts HR, Hoffman M, Monroe DM. A cell-based model of thrombin generation. *Semin Thromb Hemost.* 2006;32(Suppl 1): 32–38.
5. Roberts HR, Stinchcombe TE, Gabriel DA. The dysfibrinogenaemias. *Br J Haematol.* 2001;114:249–257.
6. Anwar M, Iqbal H, Gul M, et al. Congenital afibrinogenemia: Report of three cases. *J Thromb Haemost.* 2005;3:407–409.
7. De Vries A, Rosenberg T, Kochwa S, et al. Precipitating antifibrinogen antibody appearing after fibrinogen infusions in a patient with congenital afibrinogenemia. *Am J Med.* 1961;30: 486–494.
8. Menache D. Abnormal fibrinogens. A review. *Thromb Diath Haemorrh.* 1973;29:525–535.
9. Cunningham MT, Brandt JT, Laposata M, et al. Laboratory diagnosis of dysfibrinogenemia. *Arch Pathol Lab Med.* 2002;126: 499–505.
10. Stenson PD, Ball EV, Mort M, et al. Human Gene Mutation Database (HGMD): 2003 update. *Hum Mutat.* 2003;21:577–581.
11. Chesney CM, Pifer D, Colman RW. Subcellular localization and secretion of factor V from human platelets. *Proc Natl Acad Sci USA.* 1981;78:5180–5184.
12. Mariani G, Herrmann FH, Dolce A, et al. Clinical phenotypes and factor VII genotype in congenital factor VII deficiency. *Thromb Haemost.* 2005;93:481–487.
13. McVey JH, Boswell E, Mumford AD, et al. Factor VII deficiency and the FVII mutation database. *Hum Mutat.* 2001;17:3–17.
14. Key NS, Negrier C. Coagulation factor concentrates: Past, present, and future. *Lancet.* 2007;370:439–448.
15. Leissinger CA, Blatt PM, Hoots WK, et al. Role of prothrombin complex concentrates in reversing warfarin anticoagulation: A review of the literature. *Am J Hematol.* 2008;83:137–143.
16. Oliver JA, Monroe DM, Roberts HR, et al. Thrombin activates factor XI on activated platelets in the absence of factor XII. *Arterioscler Thromb Vasc Biol.* 1999;19:170–177.
17. Gailani D, Renne T. Intrinsic pathway of coagulation and arterial thrombosis. *Arterioscler Thromb Vasc Biol.* 2007;27:2507–2513.
18. Schmaier AH, McCrae KR. The plasma kallikrein-kinin system: Its evolution from contact activation. *J Thromb Haemost.* 2007; 5:2323–2329.
19. Ariens RA, Lai TS, Weisel JW, et al. Role of factor XIII in fibrin clot formation and effects of genetic polymorphisms. *Blood.* 2002;100:743–754.
20. McMillan C, Roberts H. Congenital combined deficiency of coagulation factors II, VII, IX and X. *N Engl J Med.* 1966;274: 1313–1315.
21. Zhang B, Ginsburg D. Familial multiple coagulation factor deficiencies: New biologic insight from rare genetic bleeding disorders. *J Thromb Haemost.* 2004;2:1564–1572.
22. Nichols WL, Hultin MB, James AH, et al. von Willebrand disease (VWD): Evidence-based diagnosis and management guidelines, the National Heart, Lung, and Blood Institute (NHLBI) Expert Panel report (USA). *Haemophilia.* 2008;14:171–232.
23. Favaloro EJ. Appropriate laboratory assessment as a critical facet in the proper diagnosis and classification of von Willebrand disorder. *Best Pract Res Clin Haematol.* 2001;14:299–319.
24. Weiss HJ, Baumgartner HR, Tschopp TB, et al. Correction by factor VIII of the impaired platelet adhesion to subendothelium in von Willebrand disease. *Blood.* 1978;51:267–279.
25. Kessler CM. Diagnosis and treatment of von Willebrand disease: New perspectives and nuances. *Haemophilia.* 2007;13(Suppl 5): 3–14.
26. Castaman G, Lethagen S, Federici AB, et al. Response to desmopressin is influenced by the genotype and phenotype in type 1 von Willebrand disease (VWD): Results from the European Study MCMDM-1VWD. *Blood.* 2008;111:3531–3539.
27. Fay WP, Parker AC, Condrey LR, et al. Human plasminogen activator inhibitor-1 (PAI-1) deficiency: Characterization of a large kindred with a null mutation in the PAI-1 gene. *Blood.* 1997; 90:204–208.
28. Favier R, Aoki N, de Moerloose P. Congenital alpha(2)-plasmin inhibitor deficiencies: A review. *Br J Haematol.* 2001;114:4–10.
29. Morimoto Y, Yoshioka A, Imai Y, et al. Haemostatic management of intraoral bleeding in patients with congenital deficiency of alpha2-plasmin inhibitor or plasminogen activator inhibitor-1. *Haemophilia.* 2004;10:669–674.
30. Seri M, Pecci A, Di Bari F, et al. MYH9-related disease: May-Hegglin anomaly, Sebastian syndrome, Fechtner syndrome, and Epstein syndrome are not distinct entities but represent a variable expression of a single illness. *Medicine (Baltimore).* 2003;82:203–215.
31. Toren A, Amariglio N, Rozenfeld-Granot G, et al. Genetic linkage of autosomal-dominant Alport syndrome with leukocyte inclusions and macrothrombocytopenia (Fechtner syndrome) to chromosome 22q11–13. *Am J Hum Genet.* 1999;65:1711–1717.
32. Epstein CJ, Sahud MA, Piel CF, et al. Hereditary macrothrombocytopathia, nephritis and deafness. *Am J Med.* 1972;52:299–310.
33. Pujol-Moix N, Kelley MJ, Hernandez A, et al. Ultrastructural analysis of granulocyte inclusions in genetically confirmed MYH9-related disorders. *Haematologica.* 2004;89: 330–337.
34. Thompson AA, Nguyen LT. Amegakaryocytic thrombocytopenia and radio-ulnar synostosis are associated with HOXA11 mutation. *Nat Genet.* 2000;26:397–398.
35. Favier R, Jondeau K, Boutard P, et al. Paris-Trousseau syndrome: Clinical, hematological, molecular data of ten new cases. *Thromb Haemost.* 2003;90:893–897.
36. White JG. Platelet storage pool deficiency in Jacobsen syndrome. *Platelets.* 2007;18:522–527.
37. Cantor AB. GATA transcription factors in hematologic disease. *Int J Hematol.* 2005;81:378–384.
38. Freson K, Matthijs G, Thys C, et al. Different substitutions at residue D218 of the X-linked transcription factor GATA1 lead to altered clinical severity of macrothrombocytopenia and anemia and are associated with variable skewed X inactivation. *Hum Mol Genet.* 2002;11:147–152.
39. Balduini CL, Pecci A, Loffredo G, et al. Effects of the R216Q mutation of GATA-1 on erythropoiesis and megakaryocytopoiesis. *Thromb Haemost.* 2004;91:129–140.
40. Michaud J, Wu F, Osato M, et al. In vitro analyses of known and novel RUNX1/AML1 mutations in dominant familial platelet disorder with predisposition to acute myelogenous leukemia: Implications for mechanisms of pathogenesis. *Blood.* 2002;99: 1364–1372.

41. Rose MJ, Nicol KK, Skeens MA, et al. Congenital amegakaryocytic thrombocytopenia: The diagnostic importance of combining pathology with molecular genetics. *Pediatr Blood Cancer.* 2008; 50:1263–1265.
42. Geddis AE. Inherited thrombocytopenia: Congenital amegakaryocytic thrombocytopenia and thrombocytopenia with absent radii. *Semin Hematol.* 2006;43:196–203.
43. Notarangelo LD, Miao CH, Ochs HD. Wiskott-Aldrich syndrome. *Curr Opin Hematol.* 2008;15:30–36.
44. Kunishima S, Kamiya T, Saito H. Genetic abnormalities of Bernard-Soulier syndrome. *Int J Hematol.* 2002;76:319–327.
45. Lisman T, Moschatsis S, Adelmeijer J, et al. Recombinant factor VIIa enhances deposition of platelets with congenital or acquired alpha IIb beta 3 deficiency to endothelial cell matrix and collagen under conditions of flow via tissue factor-independent thrombin generation. *Blood.* 2003; 101:1864–1870.
46. Tholouli E, Hay CR, O'Gorman P, et al. Acquired Glanzmann's thrombasthenia without thrombocytopenia: A severe acquired autoimmune bleeding disorder. *Br J Haematol.* 2004;127: 209–213.
47. Arthur JF, Dunkley S, Andrews RK. Platelet glycoprotein VI-related clinical defects. *Br J Haematol.* 2007;139:363–372.
48. Hirata T, Ushikubi F, Kakizuka A, et al. Two thromboxane A2 receptor isoforms in human platelets. Opposite coupling to adenylyl cyclase with different sensitivity to Arg60 to Leu mutation. *J Clin Invest.* 1996;97:949–956.
49. Rao AK, Jalagadugula G, Sun L. Inherited defects in platelet signaling mechanisms. *Semin Thromb Hemost.* 2004;30:525–535.
50. Nurden AT, Nurden P. The gray platelet syndrome: Clinical spectrum of the disease. *Blood Rev.* 2007;21:21–36.
51. Diamandis M, Veljkovic DK, Maurer-Spurej E, et al. Quebec platelet disorder: Features, pathogenesis and treatment. *Blood Coagul Fibrinolysis.* 2008;19:109–119.
52. Walker M, Payne J, Wagner B, et al. Hermansky-Pudlak syndrome. *Br J Haematol.* 2007;138:671.
53. Kaplan J, De Domenico I, Ward DM. Chediak-Higashi syndrome. *Curr Opin Hematol.* 2008;15:22–29.
54. Zwaal RF, Comfurius P, Bevers EM. Scott syndrome, a bleeding disorder caused by defective scrambling of membrane phospholipids. *Biochim Biophys Acta.* 2004;1636:119–128.
55. McCrae KR, ed. *Thrombocytopenia.* 1st ed. New York: Taylor and Francis; 2006.
56. Michelson AD, ed. *Platelets.* 2nd ed. Amsterdam: Elsevier; 2006.
57. McCrae KR, Herman JH. Posttransfusion purpura: Two unusual cases and a literature review. *Am J Hematol.* 1996;52:205–211.
58. Moake JL, Rudy CK, Troll JH, et al. Unusually large plasma factor VIII: von Willebrand factor multimers in chronic relapsing thrombotic thrombocytopenic purpura. *N Engl J Med.* 1982;307:1432–1435.
59. Tsai HM. Physiologic cleavage of von Willebrand factor by a plasma protease is dependent on its conformation and requires calcium ion. *Blood.* 1996;87:4235–4244.
60. Furlan M, Robles R, Lamie B. Partial purification and characterization of a protease from human plasma cleaving von Willebrand factor to fragments produced by in vivo proteolysis. *Blood.* 1996;87:4223–4234.
61. Zheng X, Chung D, Takayama TK, et al. Structure of von Willebrand factor-cleaving protease (ADAMTS13), a metalloprotease involved in thrombotic thrombocytopenic purpura. *J Biol Chem.* 2001;276:41059–41063.
62. Fujikawa K, Suzuki H, McMullen B, et al. Purification of human von Willebrand factor-cleaving protease and its identification as a new member of the metalloproteinase family. *Blood.* 2001;98:1662–1666.
63. Levy GG, Nichols WC, Lian EC, et al. Mutations in a member of the ADAMTS gene family cause thrombotic thrombocytopenic purpura. *Nature.* 2001;413:488–494.
64. Tsai HM, Lian EC. Antibodies to von Willebrand factor-cleaving protease in acute thrombotic thrombocytopenic purpura. *N Engl J Med.* 1998;339:1585–1594.
65. Furlan M, Robles R, Galbusera M, et al. von Willebrand factor-cleaving protease in thrombotic thrombocytopenic purpura and the hemolytic-uremic syndrome. *N Engl J Med.* 1998;339:1578–1584.
66. Amirlak I, Amirlak B. Haemolytic uraemic syndrome: An overview. *Nephrology (Carlton).* 2006;11:213–218.
67. Tarr PI, Gordon CA, Chandler WL. Shiga-toxin-producing Escherichia coli and haemolytic uraemic syndrome. *Lancet.* 2005;365:1073–1086.
68. Dragon-Durey MA, Fremeaux-Bacchi V. Atypical haemolytic uraemic syndrome and mutations in complement regulator genes. *Springer Semin Immunopathol.* 2005;27:359–374.
69. Fremeaux-Bacchi V, Kemp EJ, Goodship JA, et al. The development of atypical haemolytic-uraemic syndrome is influenced by susceptibility factors in factor H and membrane cofactor protein: Evidence from two independent cohorts. *J Med Genet.* 2005; 42:852–856.
70. Saunders RE, Abarrategui-Garrido C, Fremeaux-Bacchi V, et al. The interactive Factor H-atypical hemolytic uremic syndrome mutation database and website: Update and integration of membrane cofactor protein and Factor I mutations with structural models. *Hum Mutat.* 2007;28:222–234.
71. Dragon-Durey MA, Loirat C, Cloarec S, et al. Anti-Factor H autoantibodies associated with atypical hemolytic uremic syndrome. *J Am Soc Nephrol.* 2005;16:555–563.
72. Levy JH, Hursting MJ. Heparin-induced thrombocytopenia, a prothrombotic disease. *Hematol Oncol Clin North Am.* 2007;21: 65–88.
73. Rauova L, Poncz M, McKenzie SE, et al. Ultralarge complexes of PF4 and heparin are central to the pathogenesis of heparin-induced thrombocytopenia. *Blood.* 2005;105:131–138.
74. Griffin JH, Fernandez JA, Gale AJ, et al. Activated protein C. *J Thromb Haemost.* 2007;5(Suppl 1):73–80.
75. Cheng T, Liu D, Griffin JH, et al. Activated protein C blocks p53-mediated apoptosis in ischemic human brain endothelium and is neuroprotective. *Nat Med.* 2003;9:338–342.
76. Esmon CT. The endothelial protein C receptor. *Curr Opin Hematol.* 2006;13:382–385.
77. Folsom AR, Aleksic N, Wang L, et al. Protein C, antithrombin, and venous thromboembolism incidence: A prospective population-based study. *Arterioscler Thromb Vasc Biol.* 2002;22:1018–1022.
78. Martinelli I, Mannucci PM, De Stefano V, et al. Different risks of thrombosis in four coagulation defects associated with inherited thrombophilia: A study of 150 families. *Blood.* 1998;92:2353–2358.
79. Koster T, Rosendaal FR, Briet E, et al. Protein C deficiency in a controlled series of unselected outpatients: An infrequent but clear risk factor for venous thrombosis (Leiden Thrombophilia Study). *Blood.* 1995;85:2756–2761.
80. Miletich J, Sherman L, Broze G Jr. Absence of thrombosis in subjects with heterozygous protein C deficiency. *N Engl J Med.* 1987;317:991–996.
81. Tait RC, Walker ID, Reitsma PH, et al. Prevalence of protein C deficiency in the healthy population. *Thromb Haemost.* 1995;73: 87–93.
82. Allaart CF, Poort SR, Rosendaal FR, et al. Increased risk of venous thrombosis in carriers of hereditary protein C deficiency defect. *Lancet.* 1993;341:134–138.
83. Seligsohn U, Berger A, Abend M, et al. Homozygous protein C deficiency manifested by massive venous thrombosis in the newborn. *N Engl J Med.* 1984;310:559–562.
84. Eby CS. Warfarin-induced skin necrosis. *Hematol Oncol Clin North Am.* 1993;7:1291–1300.
85. Rezende SM, Simmonds RE, Lane DA. Coagulation, inflammation, and apoptosis: Different roles for protein S and the protein S-C4b binding protein complex. *Blood.* 2004;103:1192–1201.
86. Faioni EM, Valsecchi C, Palla A, et al. Free protein S deficiency is a risk factor for venous thrombosis. *Thromb Haemost.* 1997;78: 1343–1346.
87. Simmonds RE, Ireland H, Lane DA, et al. Clarification of the risk for venous thrombosis associated with hereditary protein S deficiency by investigation of a large kindred with a characterized gene defect. *Ann Intern Med.* 1998;128:8–14.
88. Coller BS, Owen J, Jesty J, et al. Deficiency of plasma protein S, protein C, or antithrombin III and arterial thrombosis. *Arteriosclerosis.* 1987;7:456–462.
89. Allaart CF, Aronson DC, Ruys T, et al. Hereditary protein S deficiency in young adults with arterial occlusive disease. *Thromb Haemost.* 1990;64:206–210.
90. Levin M, Eley BS, Louis J, et al. Postinfectious purpura fulminans caused by an autoantibody directed against protein S. *J Pediatr.* 1995;127:355–363.

91. Dahlback B, Carlsson M, Svensson PJ. Familial thrombophilia due to a previously unrecognized mechanism characterized by poor anticoagulant response to activated protein C: Prediction of a cofactor to activated protein C. *Proc Natl Acad Sci USA.* 1993;90:1004–1008.
92. Bertina RM, Koeleman BP, Koster T, et al. Mutation in blood coagulation factor V associated with resistance to activated protein C. *Nature.* 1994;369:64–67.
93. Dahlback B. Inherited resistance to activated protein C, a major cause of venous thrombosis, is due to a mutation in the factor V gene. *Haemostasis.* 1994;24:139–151.
94. Greengard JS, Sun X, Xu X, et al. Activated protein C resistance caused by Arg506Gln mutation in factor Va. *Lancet.* 1994;343: 1361–1362.
95. Voorberg J, Roelse J, Koopman R, et al. Association of idiopathic venous thromboembolism with single point-mutation at Arg506 of factor V. *Lancet.* 1994;343:1535–1536.
96. Zivelin A, Griffin JH, Xu X, et al. A single genetic origin for a common Caucasian risk factor for venous thrombosis. *Blood.* 1997;89:397–402.
97. Heeb MJ, Kojima Y, Greengard JS, et al. Activated protein C resistance: Molecular mechanisms based on studies using purified Gln506-factor V. *Blood.* 1995;85:3405–3411.
98. Ridker PM, Glynn RJ, Miletich JP, et al. Age-specific incidence rates of venous thromboembolism among heterozygous carriers of factor V Leiden mutation. *Ann Intern Med.* 1997;126:528–531.
99. Rosendaal FR, Koster T, Vandenbroucke JP, et al. High risk of thrombosis in patients homozygous for factor V Leiden (activated protein C resistance). *Blood.* 1995;85:1504–1508.
100. Rosendaal FR, Siscovick DS, Schwartz SM, et al. Factor V Leiden (resistance to activated protein C) increases the risk of myocardial infarction in young women. *Blood.* 1997;89:2817–2821.
101. Inbal A, Freimark D, Modan B, et al. Synergistic effects of prothrombotic polymorphisms and atherogenic factors on the risk of myocardial infarction in young males. *Blood.* 1999; 93:2186–2190.
102. Doggen CJ, Cats VM, Bertina RM, et al. Interaction of coagulation defects and cardiovascular risk factors: Increased risk of myocardial infarction associated with factor V Leiden or prothrombin 20210A. *Circulation.* 1998;97:1037–1041.
103. Atherosclerosis, Thrombosis, and Vascular Biology Italian Study Group. No evidence of association between prothrombotic gene polymorphisms and the development of acute myocardial infarction at a young age. *Circulation.* 2003;107: 1117–1122.
104. Poort SR, Rosendaal FR, Reitsma PH, et al. A common genetic variation in the 3′-untranslated region of the prothrombin gene is associated with elevated plasma prothrombin levels and an increase in venous thrombosis. *Blood.* 1996;88:3698–3703.
105. Kyrle PA, Mannhalter C, Beguin S, et al. Clinical studies and thrombin generation in patients homozygous or heterozygous for the G20210A mutation in the prothrombin gene. *Arterioscler Thromb Vasc Biol.* 1998;18:1287–1291.
106. Colucci M, Binetti BM, Tripodi A, et al. Hyperprothrombinemia associated with prothrombin G20210A mutation inhibits plasma fibrinolysis through a TAFI-mediated mechanism. *Blood.* 2004;103:2157–2161.
107. Leroyer C, Mercier B, Oger E, et al. Prevalence of 20210 A allele of the prothrombin gene in venous thromboembolism patients. *Thromb Haemost.* 1998;80:49–51.
108. Salomon O, Steinberg DM, Zivelin A, et al. Single and combined prothrombotic factors in patients with idiopathic venous thromboembolism: Prevalence and risk assessment. *Arterioscler Thromb Vasc Biol.* 1999;19:511–518.
109. Margaglione M, Brancaccio V, Giuliani N, et al. Increased risk for venous thrombosis in carriers of the prothrombin G->A20210 gene variant. *Ann Intern Med.* 1998;129:89–93.
110. Hillarp A, Zoller B, Svensson PJ, et al. The 20210 A allele of the prothrombin gene is a common risk factor among Swedish outpatients with verified deep venous thrombosis. *Thromb Haemost.* 1997;78:990–992.
111. Cumming AM, Keeney S, Salden A, et al. The prothrombin gene G20210A variant: Prevalence in a U.K. anticoagulant clinic population. *Br J Haematol.* 1997;98: 353–355.
112. Brown K, Luddington R, Williamson D, et al. Risk of venous thromboembolism associated with a G to A transition at position 20210 in the 3′-untranslated region of the prothrombin gene. *Br J Haematol.* 1997;98:907–909.
113. Zawadzki C, Gaveriaux V, Trillot N, et al. Homozygous G20210A transition in the prothrombin gene associated with severe venous thrombotic disease: Two cases in a French family. *Thromb Haemost.* 1998;80:1027–1028.
114. Howard TE, Marusa M, Channell C, et al. A patient homozygous for a mutation in the prothrombin gene 3′-untranslated region associated with massive thrombosis. *Blood Coagul Fibrinolysis.* 1997;8:316–319.
115. Sakuragawa N, Takahashi K, Kondo S, et al. Antithrombin III Toyama: A hereditary abnormal antithrombin III of a patient with recurrent thrombophlebitis. *Thromb Res.* 1983;31:305–317.
116. Fischer AM, Cornu P, Sternberg C, et al. Antithrombin III Alger: A new homozygous AT III variant. *Thromb Haemost.* 1986;55:218–221.
117. Okajima K, Ueyama H, Hashimoto Y, et al. Homozygous variant of antithrombin III that lacks affinity for heparin, AT III Kumamoto. *Thromb Haemost.* 1989;61:20–24.
118. Boyer C, Wolf M, Vedrenne J, et al. Homozygous variant of antithrombin III: AT III Fontainebleau. *Thromb Haemost.* 1986;56: 18–22.
119. van Boven HH, Vandenbroucke JP, Briet E, et al. Gene-gene and gene-environment interactions determine risk of thrombosis in families with inherited antithrombin deficiency. *Blood.* 1999;94:2590–2594.
120. Hirsh J, Piovella F, Pini M. Congenital antithrombin III deficiency. Incidence and clinical features. *Am J Med.* 1989;87: 34S–38S.
121. de Boer AC, van Riel LA, den Ottolander GJ. Measurement of antithrombin III, alpha 2-macroglobulin and alpha 1-antitrypsin in patients with deep venous thrombosis and pulmonary embolism. *Thromb Res.* 1979;15:17–25.
122. Marciniak E, Gockerman JP. Heparin-induced decrease in circulating antithrombin-III. *Lancet.* 1977;2:581–584.
123. Damus PS, Wallace GA. Immunologic measurement of antithrombin III-heparin cofactor and alpha2 macroglobulin in disseminated intravascular coagulation and hepatic failure coagulopathy. *Thromb Res.* 1975;6:27–38.
124. Kauffmann RH, Veltkamp JJ, Van Tilburg NH, et al. Acquired antithrombin III deficiency and thrombosis in the nephrotic syndrome. *Am J Med.* 1978;65:607–613.
125. Buchanan GR, Holtkamp CA. Reduced antithrombin III levels during L-asparaginase therapy. *Med Pediatr Oncol.* 1980;8:7–14.
126. Weenink GH, Treffers PE, Vijn P, et al. Antithrombin III levels in preeclampsia correlate with maternal and fetal morbidity. *Am J Obstet Gynecol.* 1984;148:1092–1097.

Chapter

16

Molecular Basis of Lymphoid and Myeloid Diseases

Joseph R. Biggs . Dong-Er Zhang

DEVELOPMENT OF THE BLOOD AND LYMPHOID ORGANS

Hematopoietic Stem Cells

All hematopoietic cells are derived from hematopoietic stem cells (HSCs) that are capable of both self-renewal and differentiation into all blood cell lineages [1]. In mammals, HSCs are produced at sequential sites beginning with the yolk sac and followed by an area surrounding the dorsal aorta called the aorta-gonad mesonephros (AGM) region, the fetal liver, and finally the bone marrow. HSCs are derived from ventral mesoderm, but the precise nature of the HSC precursor cell has yet to be determined. Yolk sac hematopoiesis is termed primitive because it produces mainly red blood cells and is transient and rapidly replaced by definitive hematopoiesis. Definitive hematopoiesis involves the colonization of the fetal liver, thymus, spleen, and bone marrow by HSCs that have migrated from earlier sites of formation. The AGM has long been viewed as the principal site of HSC production [2], but recent studies have suggested that the yolk sac may also contribute to the adult hematopoietic system [3]. It is not yet clear whether HSCs from the fetal liver circulate to the adult bone marrow and are the source of adult hematopoiesis, or if the fetal liver and bone marrow are seeded with HSCs at the same time during development [4]. All types of mature hematopoietic cells arise from differentiation of the HSCs. Figure 16.1 illustrates normal hematopoietic development, and Figure 16.2 shows the appearance of the different types of normal hematopoietic cells.

HSCs, like all stem cells, depend on their microenvironment (or niche) for normal self-renewal and differentiation [4]. The adult bone marrow is the most widely studied HSC niche, and experimental evidence to date suggests that HSCs may associate with either osteoblasts or with vascular cells, as illustrated in Figure 16.3 [1]. It has been proposed that the precise site of association may regulate HSC function. This regulation is commonly thought to be mediated by cell-cell interactions between the HSC and the osteoblast or vascular cell, and by chemokines secreted by components of the niche [5]. HSCs may be identified by specific marker proteins on the cell surface [6]. Such markers can be tagged with a fluorescently labeled antibody and used to purify HSCs for study and for transplantation by fluorescence-activated cell sorting (FACS). However, no marker has yet been discovered that is absolutely tied to unique stem cell functions or that truly identifies the HSC. Therefore, FACS can be used to isolate cell populations enriched for HSCs, but pure HSCs cannot be obtained in this manner as of yet.

Hematopoietic Differentiation and the Role of Transcription Factors

All types of mature blood cells are produced by lineage-restricted differentiation of HSCs. This process is believed to be regulated by a relatively small group of transcription factors, some required for HSC formation and others for differentiation [7–10]. In the most commonly presented model of hematopoietic differentiation (Figure 16.1), long-term or quiescent HSCs are mobilized from the niche to become proliferating short-term HSCs. The short-term HSCs then differentiate into common myeloid progenitors (CMPs) or common lymphoid progenitors (CLPs). The CMPs give rise to megakaryocyte/erythroid precursors (MEPs) and granulocyte/macrophage precursors (GMPs); the MEPs differentiate into red blood cells and megakaryocytes; and GMPs produce mast cells, eosinophils, neutrophils, and monocyte/macrophages. B- and T-lymphocytes differentiate from the CLPs. There is some evidence for alternative pathways, such as the possibility that MEPs do not originate from CMPs but from an earlier precursor [11]. Each

Molecular Pathology © 2009, Elsevier, Inc. All Rights Reserved.

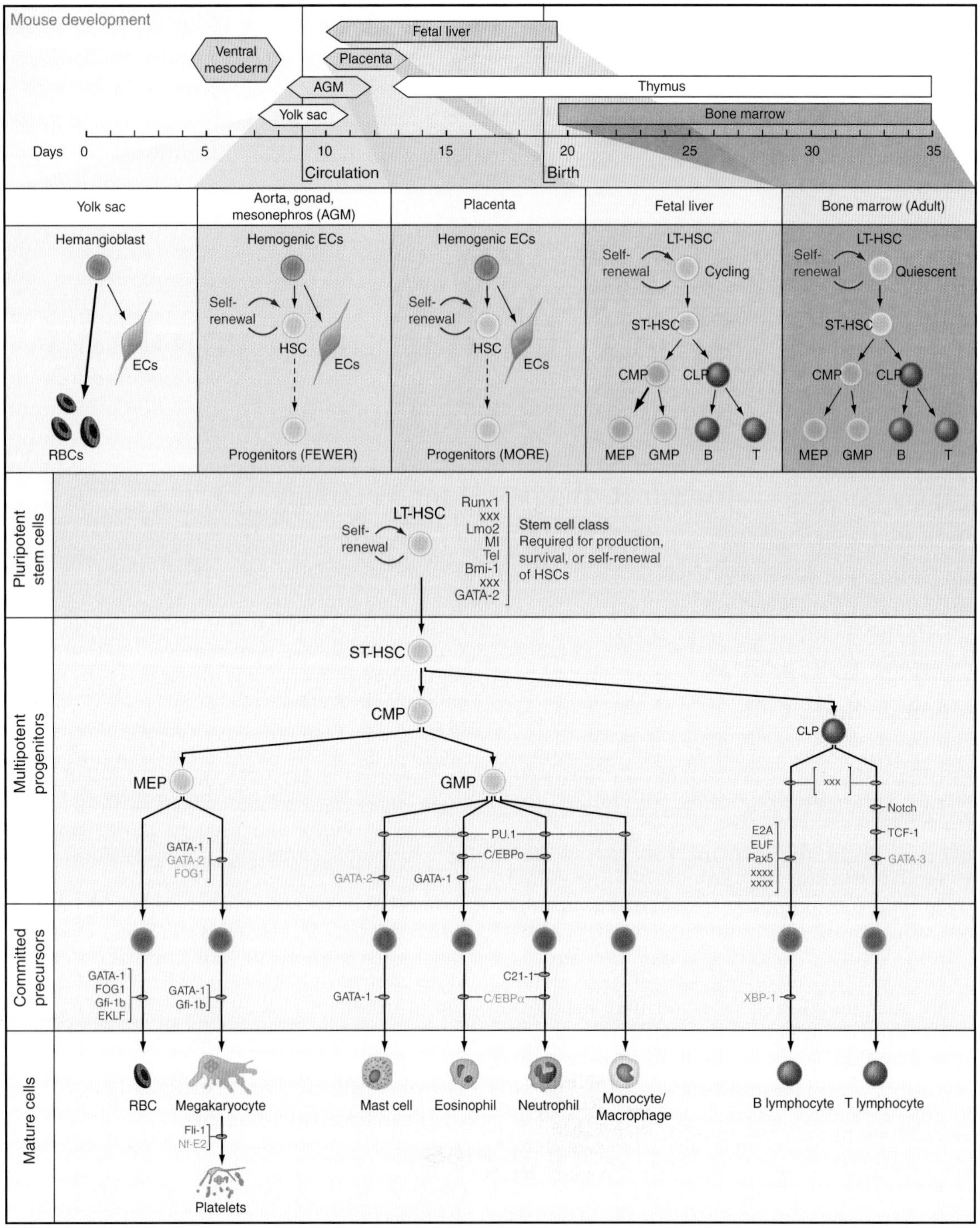

Figure 16.1 Hematopoietic development. The upper panel shows stages of hematopoiesis in the mouse. Hematopoietic stem cells (HSCs) are derived from the ventral mesoderm, and sequential sites of hematopoiesis include the yolk sac, the aorta-gonad-mesonephros (AGM) region, the fetal liver, placenta, and bone marrow. The types of cells produced at each site are illustrated in the middle panel. The main function of primitive hematopoiesis, which occurs in the yolk sac, is to produce red blood cells. The relative contribution of HSCs produced in the AGM region and the placenta to the final pool of adult HSCs remains unknown. Definitive hematopoiesis involves the colonization of the fetal liver, thymus, spleen, and bone marrow. In definitive hematopoiesis, long-term HSCs produce short-term HSCs, which in turn give rise to common myeloid progenitors (CMPs) and common lymphoid progenitors (CLPs). CMPs produce megakaryocyte/erythroid progenitors (MEPs) and granulocyte/macrophage progenitors (GMPs). CLPs produce B and T lymphocytes. The lower panel shows transcription factors that regulate hematopoiesis in mammals. The stages at which hematopoietic development is blocked in the absence of a given factor, as determined through gene knockout, are indicated by red loops. The factors in red are associated with oncogenesis; those in black have not yet been found mutated in hematologic malignancies. Among the genes required for HSC production, survival, or self-renewal are *MLL, Runx1, TEL/EV6, SCL/tal1,* and *LMO2.* These genes account in toto for the majority of known leukemia-associated translocations in patients. From Orkin SH and Zon LI (2008) SnapShot: Hematopoiesis Cell 132: 172.e1.

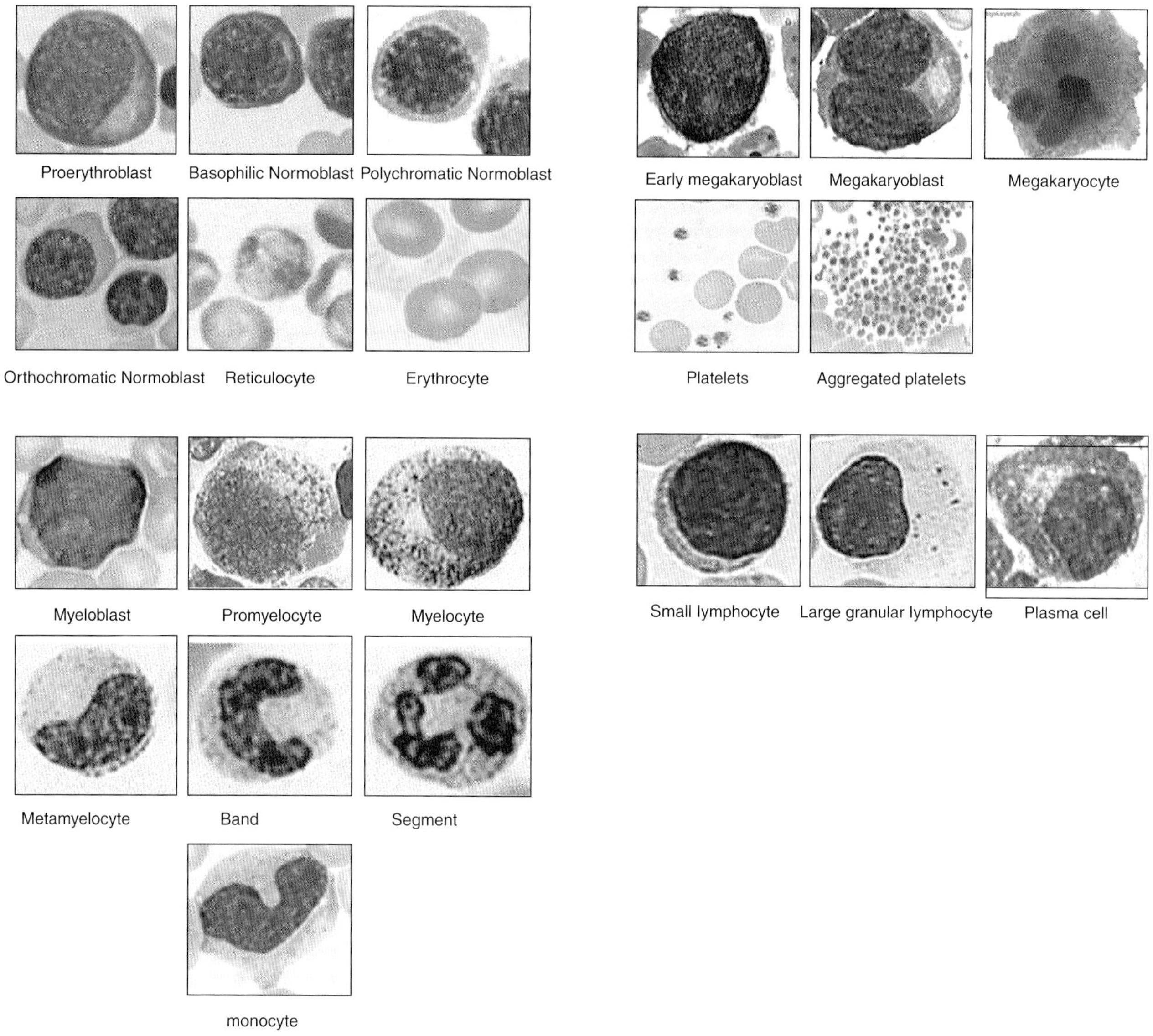

Figure 16.2 **Normal Hematopoietic cells.** Upper left panels show the stages of erythrocyte (red blood cell) development; lower left panels show the stages of granulocyte/monocyte development; upper right panels show differentiation of megakaryocytes and platelets; lower right panels show lymphocytes. It should be noted that not all cells are shown to the same scale. From Amos Cohen M.D. Rabin Medical Center, Israel (inter05@post.tau.ac.il).

stage of this process is controlled by specific transcription factors. The identity of these factors (as with HSC transcription factors) has been determined largely through the study of conventional or conditional gene knock-outs in mice and other model organisms. The fact that most of these factors show lineage and stage-restricted expression also provides information about their function.

Transcription factors essential for the formation of HCSs include SCL/tal-1 and its partner LMO2, as well as Runx1 and its partner CBFβ. The histone methyltransferase MLL, which is necessary to maintain HOX gene expression, also has a vital role in hematopoiesis. In the absence of SCL/tal-1 and LMO2, failure of both primitive and definitive hematopoiesis is observed [8]. In the absence of Runx1 or MLL, HSCs do not appear in the AGM region of the mouse embryo [12–14]. A striking observation is that this set of transcription factors controlling HSC development account for the majority of known leukemia-associated translocations in patients. These translocations either deregulate the expression of the locus or generate chimeric fusion proteins [1].

A second set of transcription factors is required for differentiation of HSCs into specific types of mature blood cells, and the transcription factors involved in the development of HSCs also have roles in later hematopoietic development. Like the factors that control HSC development, these lineage-specific factors have been identified largely through the study of gene knockout models. As examples, loss of the

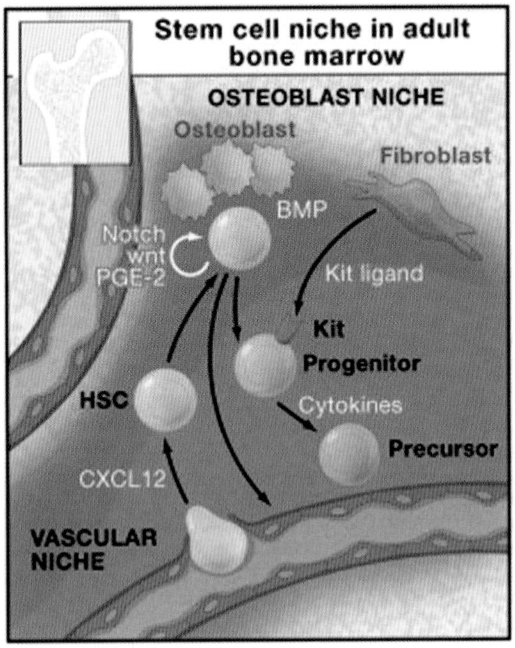

Figure 16.3 Stem cell niche in adult bone marrow. HSCs are found in the osteoblast niche adjacent to osteoblasts that are under the regulation of bone morphogenetic protein (BMP). Pathways involving Notch, wnt, and PGE-2 stimulate HSC self renewal. HSCs are also found adjacent to blood vessels (the vascular niche). The chemokine CXCL12 regulates the migration of HSCs from the circulation to the bone marrow. The osteoblast and vascular niches in vivo lie in close proximity or may be interdigitated. The marrow space also contains stromal cells that support hematopoiesis including the production of cytokines, such as c-Kit ligand, that stimulate stem cells and progenitors. Other cytokines, including interleukins, thrombopoietin, and erythropoietin, also influence progenitor function and survival. From Orkin SH and Zon LI (2008). Cell 132:631–644 (Figure 3).

factor GATA-1 or its cofactor FOG results in failure of erythroid and megakaryocytic differentiation, while mice deficient in the transcription factor C/EBPα lack GMPs and granulocytes [11]. A number of general principles governing regulation of differentiation by transcription factors have been proposed. The first, known as lineage priming, is based on the observation that earlier hematopoietic multipotential progenitors and HSCs express markers of different lineages at low levels [15]. This suggests that early hematopoietic precursors have the potential to differentiate into a number of cell types, and that differentiation consists of the suppression of all but one of the potential pathways. It also suggests the possibility of some degree of plasticity in the differentiation process. Another general observation is that many hematopoietic transcription factors appear to have dual function, simultaneously promoting differentiation toward one lineage while antagonizing factors that promote other lineages [1]. An example of this principle is the interaction between GATA-1, which promotes erythroid differentiation, and PU.1/SPI1, which promotes myeloid differentiation. PU.1/SPI1 physically interacts with GATA-1 to block its pro-erythroid activity, while GATA-1 blocks the formation of a complex between PU.1/SPI1 and its cofactor JUN. Other examples of antagonistic pairs of transcription factors are EKLF and Fli-1 for erythroid versus megakaryocytic choice and Gfi-1 and PU.1/SPI1 for neutrophil versus monocyte differentiation [1]. The model of hematopoietic transcription factors acting as antagonistic complexes suggests a final general principle—that the precise concentration of transcription factors in a cell may determine the choice of differentiation pathway. A substantial body of evidence supports the idea that transcription factor concentrations regulate hematopoietic differentiation [1]. Figure 16.1 illustrates the point of action of many transcription factors involved in hematopoiesis.

Hematopoietic transcription factor levels can be controlled by both transcriptional and post-transcriptional mechanisms. Recent studies on microRNAs (miRNAs) suggest that they provide an additional mechanism for controlling hematopoietic transcription factor levels. MiRNAs bind to the 3′-untranslated region of mRNAs and suppress translation, and several have been shown to affect the levels of transcription factors in hematopoietic cells [16].

Hematopoietic Differentiation and the Role of Signal Transduction

During the processes of proliferation and differentiation, cells respond to external signals such as growth factors or cell-cell contacts. Growth factors act by binding to a specific cell surface receptor and activating intracellular cascades which stimulate or suppress downstream transcription factors. Many of the receptors that regulate normal hematopoiesis are receptor tyrosine kinases (RTKs), such as cFMS (receptor for macrophage colony-stimulating factor/colony-stimulating factor-1), FMS-related tyrosine kinase (FLT3, receptor for FLT3-ligand), c-KIT (receptor for stem cell factor), and platelet-derived growth factor receptor (PDGFR) [17]. Ligand binding activates the tyrosine kinase activity of these RTKs, which then phosphorylate tyrosine residues on associated proteins, thereby triggering cascades in which intracellular kinases are sequentially phosphorylated and activated, until finally the signal is transmitted to nuclear transcription factors. Hematopoietic RTKs usually activate several such cascades, including the Ras/Raf/ERK pathway, the PIK3/Akt pathway, and the JAK/STAT pathway (reviewed in [18]). The activation of these pathways usually favors cell proliferation and survival. Figure 16.4 illustrates several of the pathways commonly activated in response to receptor tyrosine kinases.

Many leukemias are associated with mutations which cause constitutive activation of RTKs (the receptor tyrosine kinase is continuously active, even in the

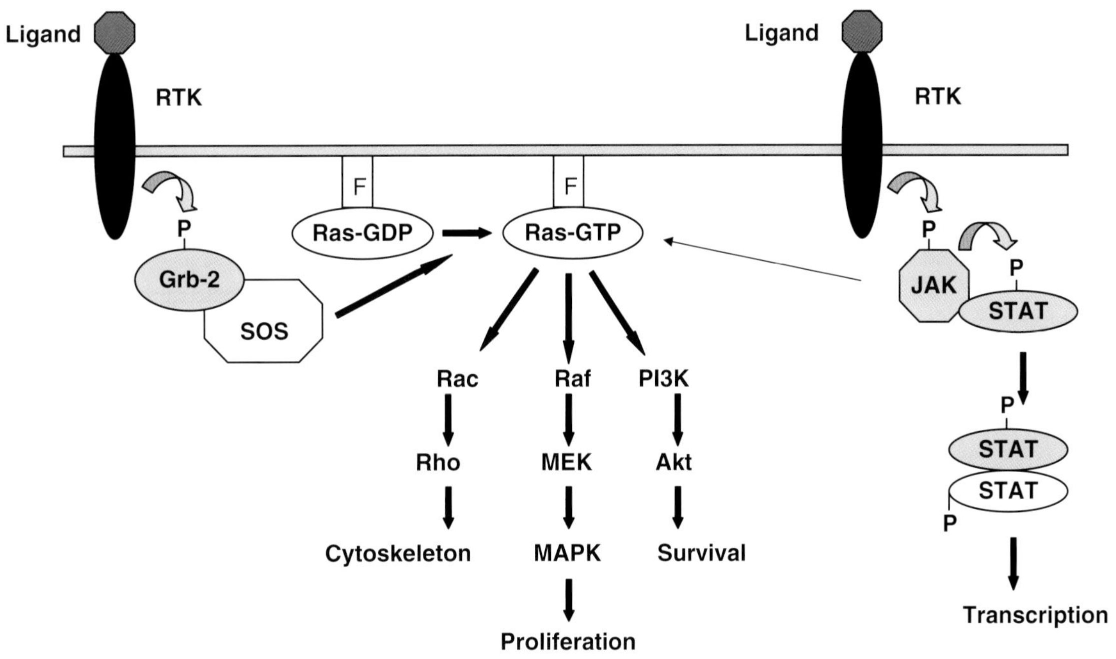

Figure 16.4 **Signal transduction pathways involved in leukemia.** The drawing on the left shows signaling of a receptor tyrosine kinase (RTK) through Ras. Ligand binding causes phosphorylation of Grb-2 by the RTK and formation of a Grb-2/SOS complex. Interaction of Grb-2/SOS with farnesylated (F) Ras-GDP causes conversion to active Ras-GTP, which in turn phosphorylates Rac, Raf, and PI3K leading to stimulation of their respective pathways. The drawing on the right shows the JAK/STAT pathway. Ligand binding causes RTK phosphorylation of JAK, which may then activate the Ras pathway and phosphorylate STATs. The STATs form homodimers or heterodimers with other STATs, and translocate to the nucleus where they activate transcription of specific target genes.

absence of ligand). This results in a continuous signal to the cell favoring growth and survival. Two observations about mutations in leukemia led to the proposal of the two-hit model of leukemogenesis. First, many leukemia patients possess two types of mutation, one affecting a hematopoietic transcription factor, such as *RUNX1/AML1*, and the other affecting a receptor tyrosine kinase or signal-transduction molecule, such as *FLT3*. Second, studies in model systems have shown that many of the leukemia-associated mutations found in patients are unable to induce leukemia by themselves, but can induce leukemia in combination with other mutations. Therefore, the two-hit model proposes that induction of leukemia requires the presence of two types of mutation: (i) a class I mutation in a receptor or signal-transduction molecule which confers a proliferative or survival advantage, and (ii) a class II mutation in a hematopoietic transcription factor which impairs differentiation [19]. This model is almost certainly true for many leukemias, but it is not yet clear if all or most leukemias require mutation of more than one gene.

Spleen

The spleen is a lymphoid organ that also serves as a blood filter. The arteries of the spleen are ensheathed by lymphocytes, which form the white pulp. The white pulp is further subdivided into a T-cell domain and a B-cell domain. The spleen, along with the lymph nodes, is a major repository for lymphocytes [20] and a major site of adaptive immune response to foreign antigens [21]. The remaining internal portion of the spleen is composed of red pulp, which is designed to filter foreign matter from the bloodstream, including damaged blood cells.

Thymus

Mature mammalian T-cells originate in the bone marrow or fetal liver as pluripotent precursors, and these cells then migrate to the thymus, where they proliferate extensively and differentiate into the various mature T-cell lineages [22]. The sole function of the thymus is to serve as the site of T-cell differentiation. Beginning at puberty, the thymus involutes and shrinks, until it eventually consists of groups of epithelial cells depleted of lymphocytes.

Lymph Nodes

Lymph nodes are small glands located in many parts of the body, mainly in the neck, under the arms, and in the groin. Lymph vessels drain fluid from tissue, which then enters the lymph nodes via afferent lymphatic vessels [23]. Lymph nodes are composed of multiple lymphoid lobules surrounded by lymph-filled sinuses and enclosed by a capsule (Figure 16.5) [24]. The smallest nodes may contain only one lobule, while larger nodes may contain a great number. The lobules are divided into regions containing spherical follicles separated by interfollicular cortex and regions containing deep cortical units (DCUs). The DCUs of adjacent lobules in a lymph node often

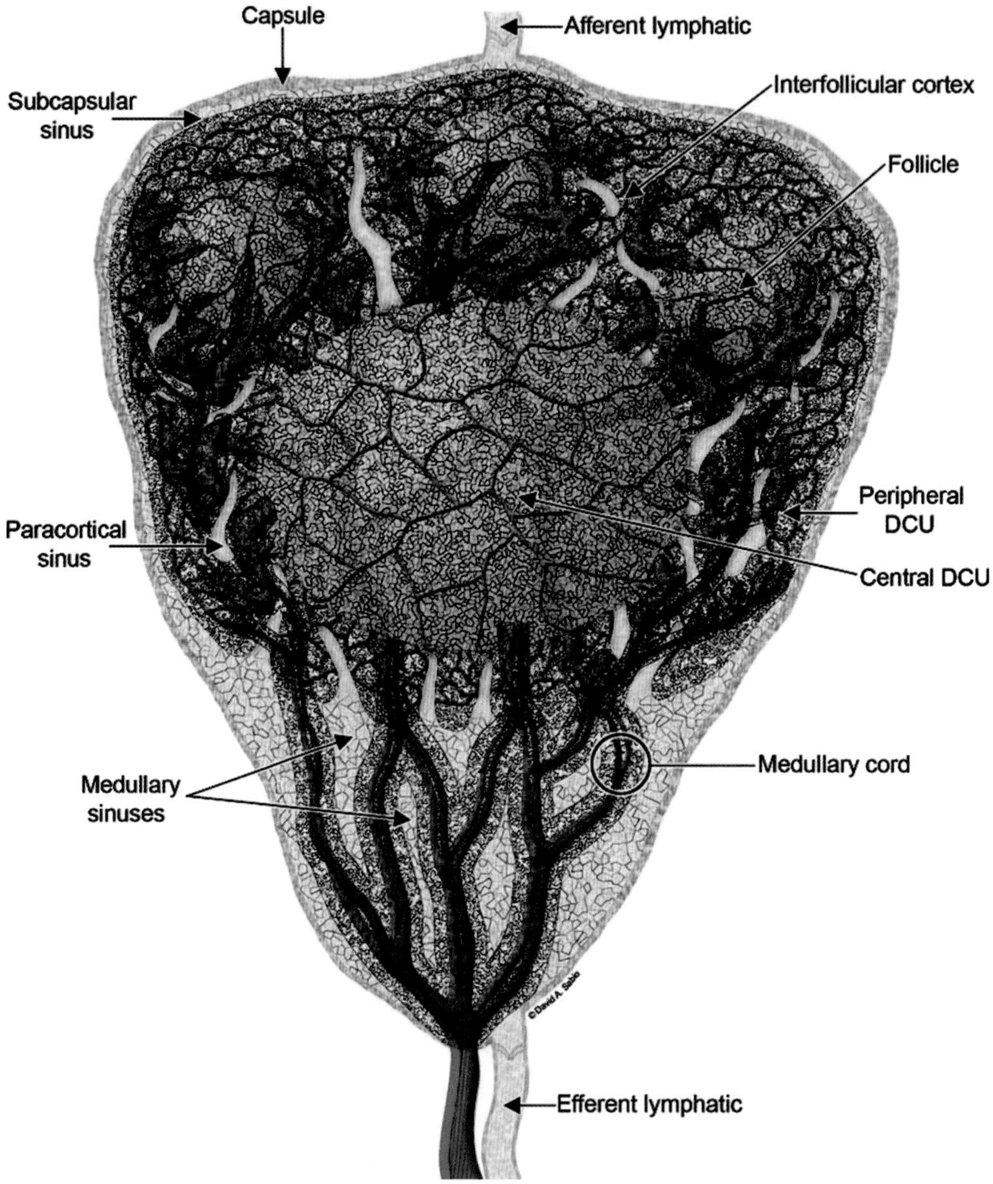

Figure 16.5 Lymph node structure. The left panel shows the simplest possible lymph node containing a single lobule. Lymph from the afferent lymphatic vessel spreads over the apical surface in the subcapsular sinus and then flows through medullary sinuses and exits via the efferent lymphatic vessel. The sinuses are spanned by a reticular meshwork, and the lobule contains a denser meshwork indicated by darker, more condensed background. The meshwork provides a scaffold for lymphocytes, antigen-presenting cells, and macrophages to interact. B cells home to follicles in the superficial cortex, where they interact with dendritic cells. Three follicles are shown as small spheres. Follicles are surrounded and separated by interfollicular cortex. In the deep cortex (paracortex) T cells home to the deep cortical unit (DCU) where they interact with dendritic cells. The right panel shows an idealized section of a small lymph node containing three lymphoid lobules. Taken together, the follicles and interfollicular cortex of these lobules constitute the superficial cortex of the mode, their deep cortical units the paracortex, and their medullary cords and sinuses the medulla. Left lobe shows arterioles (red), venules (blue), and capillary beds (purple). Center lobe as in left panel. Right lobe shows a micrograph from a rat mesenteric lobule as it appears in histological section. From Cynthia L. Willard Mack (2006). Toxicologic Pathology 34:409 (Figures 1 and 2).

Continued

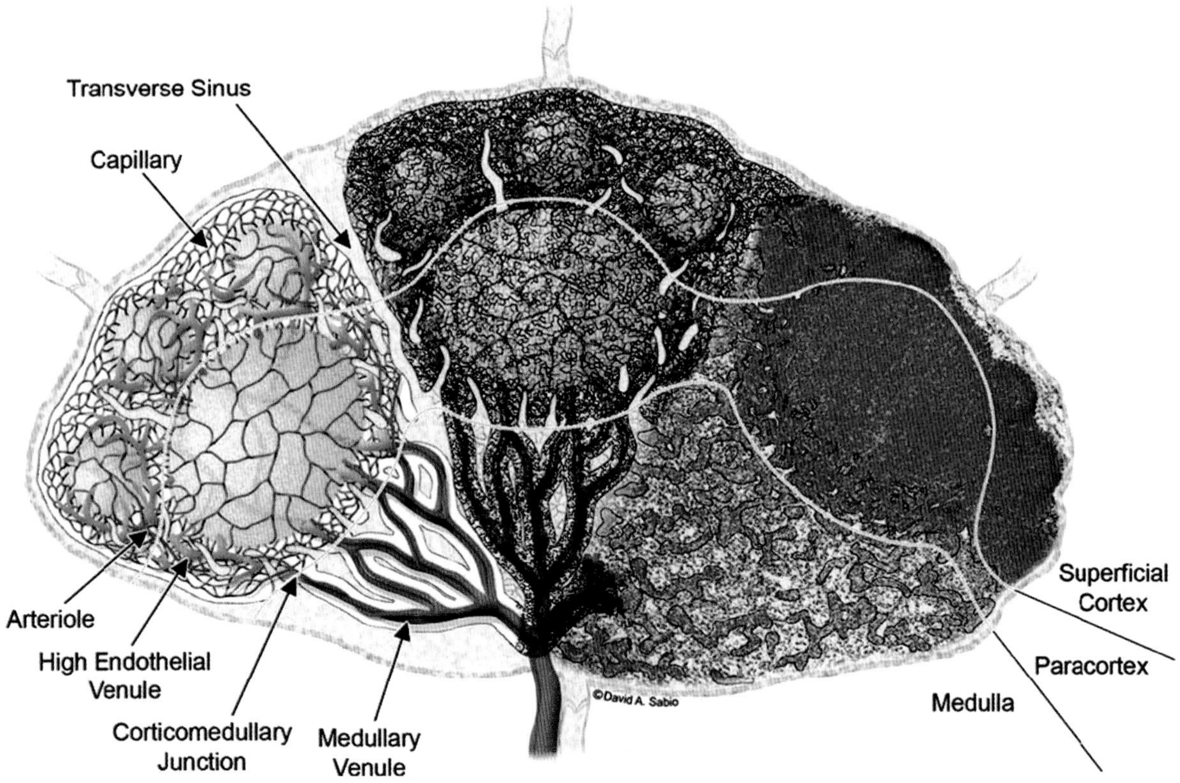

Figure 16.5—Cont'd

fuse into large complexes [24]. Immature B-cells originating in the bone marrow home to follicles where they interact with follicular dendritic cells (FDCs). FDCs trap antigen-antibody complexes that may be collected from lymph carried into the follicle. If a B-cell encounters its antigen displayed on a FDC, it is stimulated to proliferate, and these proliferating B-cells form distinctive germinal centers, which are referred to as secondary follicles. A large number of B-cells undergo apoptosis during this process of proliferation and differentiation. In contrast to B-cells, T-cells migrate to the paracortex and interfollicular cortex and survey dendritic cells. The dendritic cells that interact with T-cells form a separate class from those that interact with B-cells; they collect and process antigens in tissue and then migrate to the lymph nodes. The interfollicular cortex and DCU serve as corridors for the movement of B- and T-cells. Several afferent lymphatic vessels enter the lymph node, but each delivers a stream of lymph to a specific lobule. After passing over the lobules, all the lymph streams exit the lymph node through a single efferent vessel. Thus, individual lobules are exposed to different sets of antigens and cells collected from a specific drainage area by an individual afferent vessel. The constant flow of lymph containing cells and antigens collected from tissue allows the lymph nodes, like the spleen, to serve as a major site of interaction between foreign antigens and lymphocytes [21].

MYELOID DISORDERS

Anemia

Anemia is a condition in which the blood contains a lower than normal number of red blood cells (RBCs), or RBCs that do not contain enough hemoglobin (the normal range is 4.2–6.1 million RBCs per microliter and 12.1–17.2 grams of hemoglobin per deciliter). Anemia may be caused either by lower than normal production of RBCs or higher than normal rates of RBC destruction. Nongenetic causes of anemia include blood loss, iron deficiency, lack of folic acid (vitamin B12), or chronic disease, all of which can impede the production of RBCs. Higher than normal rates of RBC destruction can be caused by inherited disorders such as sickle cell anemia and thalassemia, and certain enzyme deficiencies. Hemolytic anemia occurs when the immune system mistakenly attacks RBCs. Anemia may also be caused by myelodysplastic syndromes, defined as one or more secondary blood cytopenias (cell loss) caused by bone marrow dysfunction.

Neutropenia

Neutropenia may occur as chronic idiopathic neutropenia or severe congenital neutropenia [25]. Chronic idiopathic neutropenia is defined as any unexplained reduction in neutrophil count to below average. The criteria for diagnosis are absolute neutrophil counts below 1.5–1.8 × 10^9/liter blood lasting more than 3 months, absence of evidence of underlying disease associated with the neutropenia, no history of exposure to radiation or chemicals that might cause neutropenia, normal bone marrow karyotype, and serum negative for antineutrophila antibodies. Neutropenia caused by antineutrophil antibodies is known as primary autoimmune neutropenia, and is most common in newborns, where it shows a tendency to resolve spontaneously. Chronic idiopathic neutropenia (CIN) is believed to result from impaired bone marrow granulopoiesis, but the precise molecular mechanism remains unknown. It has been demonstrated that the bone marrow of CIN patients contains activated T-cells producing IFNγ and Fas-ligand, as well as increased local production of TNFα and TGFβ1 and decreased production of IL-10. These changes may lead to increased apoptosis of neutrophil precursors.

Severe chronic neutropenia (SCN) is characterized by life-long neutropenia with an absolute neutrophil count under 0.5 × 10^9/liter, recurrent bacterial infections, and arrest of neutrophil maturation at the promyelocyte stage [25]. Approximately 60% of patients with SCN carry mutations in the neutrophil elastase (*ELA2*) gene. These patients fall into the categories of dominant inheritance of the disease or spontaneous acquisition of the disease. Mutations in the *ELA2* gene are also present in patients with cyclic hematopoiesis, in which the number of neutrophils and other blood cells oscillates in weekly phases. Neutrophil elastase is a protease found in the granules of mature neutrophils. The *ELA2* mutations found in patients with SCN or cyclic hematopoiesis induce the unfolded protein response (UPR) and apoptosis [26,27]. Protein folding occurs in the lumen of the endoplasmic reticulum; misfolded proteins trigger the UPR, which leads to attenuation of translation, expression of ER-resident chaparones, and ER-associated degradation pathways. If this adaptive response is overwhelmed, apoptosis is induced. Specific *ELA2* mutations are associated with either SCN or cyclic hematopoiesis, and it has been hypothesized that cyclic hematopoiesis is caused by *ELA2* mutations, which cause a less drastic activation of the UPR [26]. However, patients with SCN, unlike those with cyclic hematopoiesis, also display a deficiency of the transcription factor LEF-1, leading in turn to reduced levels of the LEF-1 targets C/EBPα, cyclin D1, c-myc, and survivin [28]. The LEF-1 deficiency (not coupled to a mutation in the gene) is present in SCN patients with either mutation of *ELA2* or the gene for HS-1-associated protein X (*HAX-1*). This suggests that LEF-1 deficiency may synergize with the *ELA2* or *HAX-1* mutations to promote neutropenia. In contrast to other SCN patients, those who acquired the disease through recessive inheritance lack mutations in the *ELA2* gene but carry mutations in the *HAX-1* gene [25]. This form of neutropenia was first described by Rolf Kostmann and is also known as Kostmann's disease. HAX-1 is a mitochondria-targeted protein, containing Bcl-2 homology domains, and is critical for maintaining the inner mitochondrial membrane potential. Loss of HAX-1 function causes increased apoptosis in myeloid cells [29]. It therefore appears that mutations in either *ELA2* or *HAX-1* contribute to neutropenia by causing enhanced levels of apoptosis in myeloid precursor cells.

Myelodysplastic Syndromes

Myelodysplastic syndromes (MDS) are diagnosed at a rate of 3.6/100,000 people in the United States. The MDS occur primarily in older patients (>60 years), but occasionally in younger patients. Anemia, bleeding, easy bruising, and fatigue are common, and splenomegaly or hepatosplenomegaly may occasionally be present [30]. The MDS are characterized by abnormal bone marrow and bloodcell morphology. The bone marrow is usually hypercellular, but approximately 15% of patients have hypoplastic bone marrow. Circulating granulocytes are often severely reduced and hypogranular or hypergranular. Early, abnormal myeloid progenitors are identified in the marrow in varying percentages, depending on the type of MDS. Abnormally small megakaryoctyes may be seen in the marrow and hypogranular or giant platelets in the blood.

The MDS are classified according to cellular morphology, etiology, and clinical features. The morphological classification is largely based on percent myeloblasts in the bone marrow and blood, the type of myeloid dysplasia, and the presence of ringed sideroblasts [31]. Beginning in the late 1970s, the MDS were classified according to the French-American-British (FAB) classification scheme under the direction of the FAB Cooperative Group. The FAB classification scheme divides the MDS into refractory anemia, refractory anemia with ringed sideroblasts, refractory anemia with excess blasts, refractory anemia with excess blasts in transformation, and chronic myelomonocytic leukemia. In 1997, a working group of pathologists under the direction of the *World Health Organization* (WHO) agreed to a new classification scheme for hematopoietic and lymphoid malignancies [32]. The WHO scheme attempts to correct weaknesses in the FAB scheme, such as failure to take cytogenetic findings into account. The WHO scheme divides the MDS into refractory anemia, refractory anemia with multilineage dysplasia, refractory anemia with ringed siderblasts, refractory anemia with excess blasts, myelodysplastic syndrome, unclassifiable, and myelodysplastic syndrome associated with del(5q).

Knowledge of the molecular defects underlying the refractory anemias is currently limited. Recurrent deletions of 5q31, 7q22, and 20q12 in MDS suggest that loss of unidentified tumor suppressor genes within these regions contributes to development of MDS [33].

Recently mutations in the *RUNX1* gene have been associated with development of some types of MDS. *RUNX1* point mutations were observed in 23.6% of patients with refractory anemia with excess blasts [34]. This suggests that loss of RUNX1 transcription factor activity, which normally regulates hematopoietic stem cell development and differentiation of some hematopoietic lineages, may contribute to the development of some types of refractory anemia. Approximately 50% of patients with one particular type of refractory anemia, refractory anemia with ringed sideroblasts associated with thrombocytosis (RARS-T), have been found to carry a mutation in the *JAK2* gene, *JAK2* V617F [35,36].

Myelodysplastic/Myeloproliferative Diseases

Myelodysplastic/myeloproliferative diseases have features of both myelodysplastic syndromes and myeloproliferative disorders. A greater than normal number of stem cells develop into one or more types of more mature cells, and the blood cell number increases; but there is also some degree of failure to mature properly. The three main types of myelodysplastic/myeloproliferative diseases are chronic myelomonocytic leukemia (CMML), juvenile myelomonocytic leukemia (JMML), and atypical chronic myelomonocytic leukemia (aCML). A myelodysplastic/myeloproliferative disease that does not match any of the previous types is referred to as myelodysplastic/myeloproliferative disease, unclassifiable (MDS/MPD-UC).

Chronic Myelomonocytic Leukemia

Chronic myelomonocytic leukemia (CMML) is characterized by the overproduction of myelocytes and monocytes, as well as immature blasts. Gradually, these cells replace other cell types, such as red cells and platelets in the bone marrow, leading to anemia or easy bleeding. The specific pathologic features of CMML include persistent monocytosis of greater than 1×10^9/liter in the peripheral blood, no Philadelphia chromosome (*BCR/ABL* fusion gene), fewer than 20% blasts in the blood or bone marrow, and dysplasia involving one or more myeloid lineages [37]. The CMML bone marrow may exhibit hypercellularity (75% of cases), a blast count of less than 20%, granulocytic and monocytic proliferation, micromegakaryocytes or megakaryocytes with lobated nuclei (80% of cases), and fibrosis (30% of cases) [37].

Clonal abnormalities are found in 20%–40% of CMML patients, but none is specific [38]. Point mutations in *RAS* genes are found in up to 40% of patients with CMML, and approximately 3% contain the *JAK2* V617F mutation [39].

Juvenile Myelomonocytic Leukemia

Juvenile myelomonocytic leukemia (JMML) accounts for 2% of all childhood leukemias. The three required criteria for a diagnosis of JMML are no Philadelphia chromosome (*BCR/ABL* fusion gene), peripheral blood monocytosis greater than 1×10^9/liter, and fewer than 20% blasts in the blood and bone marrow [40]. The presence of two or more of the following minor criteria is also required: fetal hemoglobin increased for age, immature granulocytes in the peripheral blood, a white blood cell count greater than 1×10^9/liter, a clonal chromosomal abnormality, and granulocyte-macrophage colony-stimulating factor (GM-CSF) hypersensitivity of myeloid progenitors.

The bone marrow of JMML patients may show hypercellularity with granulocytic proliferation, hypercellularity with erythroid precursors (some patients), monocytes comprising 5%–10% of marrow cells, minimal dysplasia, and reduced numbers of megakaryoctyes.

A distinctive characteristic of JMML leukemic cells is their spontaneous proliferation *in vitro* due to their hypersensitivity to GM-CSF [41]. This hypersensitivity has been attributed to altered Ras pathway signaling as a result of mutually exclusive mutations affecting one of the pathway regulatory molecules including the genes for *RAS*, *PTPN11*, and *NF1* [39,42,43]. Thirty percent of JMML patients display *PTPN11* mutations, while 15%–20% display either *RAS* mutations or *NF1* mutations. Ras is a GTP-dependent protein (G-protein) localized at the inner side of the cell membrane, and transduces signal from growth factor receptors to downstream effectors [18]. Protein-Tyrosine Phosphatase, Nonreceptor-type, 11 (*PTPN11*) encodes the SHP-2 protein, which transmits signals from growth factor receptors to Ras. Neurofibromatosis, type 1 (*NF1*) is a tumor suppressor gene that inactivates Ras through acceleration of Ras-associated GTP hydrolysis [44]. Activating mutations in *RAS* or *PTPN11* or inactivating mutations in *NF1* in JMML cells all result in enhancement of signaling through the Ras pathway and increased stimulus to proliferate (Figure 16.4). This finding has stimulated interest in molecules that inhibit the Ras pathway as possible therapeutic agents [39].

Atypical Chronic Myelogenous Leukemia and Myelodysplastic/Myeloproliferative Disease, Unclassifiable

Atypical chronic myelogenous leukemia (aCML) is characterized pathologically by the following: no Philadelphia chromosome (BCR/ABL fusion gene), peripheral blood leukocytosis with increased numbers of mature and immature neutrophils, prominent dysgranulopoiesis, neutrophil precursors accounting for more than 10% of white blood cells, minimal absolute basophilia (basophils less than 2% of white blood cells), absolute monocytosis (monocytes less than 10% of white blood cells), hypercellular bone marrow with granulocytic proliferation and granulocytic dysplasia, fewer than 20% blasts in blood or bone marrow, and thrombocytopenia [45]. The criteria for MDS/MPD-UC is mixed myeloproliferative and myelodysplastic features that cannot be assigned to any other category of MDS, CMPD, or MDS/MPD. Little is currently known about the molecular pathology of aCML or myelodysplastic/myeloproliferative disease, unclassifiable (MDS/MPD-UC).

Chronic Myeloid Leukemia

Chronic myeloid leukemia (CML) is characterized by less than 10% blasts and promyelocytes in peripheral blood and bone marrow. The transition from chronic phase to the accelerated phase and later blastic phase may occur gradually over a period of 1 year or more, or it may appear suddenly (blast crisis) [46]. Signs of impending progression of CML include progressive leukocytosis, thrombocytosis or thrombocytopenia, anemia, increasing splenomegaly or hepatomegaly. Accelerated phase CML is characterized by 10%–19% blasts in either the peripheral blood or bone marrow, and blast phase CML by 20% or more blast cells in the peripheral blood or bone marrow. Blast crisis is defined as 20% or more blasts plus fever, malaise, and progressive splenomegaly [47]. Examination of bone marrow shows a shift of the myeloid cells to more immature forms that increase in number as the disease progresses. The bone marrow is hypercellular, but the spectrum of mature and immature granulocytes is similar to that in normal marrow. The percentage of lymphocytes is reduced in both blood and marrow, and the ratio of erythroid to myeloid cells is usually reduced [46].

The leukemic cells of almost all CML patients contain a distinctive cytogenetic abnormality, the Philadelphia chromosome. The Philadelphia chromosome is formed by a reciprocal translocation between the long arms of chromosomes 9 and 22, and results in the fusion of the *ABL* gene on chromosome 9 to the *BCR* gene on chromosome 22 [48]. The resulting fusion gene, *BCR-ABL*, produces a fusion protein containing the oligomerization and serine/threonine kinase domains of BCR at the amino terminus and most of the ABL protein at the carboxy-terminus. ABL is a nonreceptor tyrosine kinase, and its activity is normally tightly regulated in cells. The fusion of BCR sequences constitutively activates the ABL tyrosine kinase, transforming ABL into an oncogene [49,50]. BCR-ABL, with the aid of mediator proteins, associates with Ras and stimulates its activation; through stimulation of the Ras-Raf pathway, BCR-ABL increases growth factor-independent cell growth (Figure 16.6) [51]. BCR-ABL is also associated with the Janus kinase and signal transducer and activator of transcription (JAK-STAT) pathway. BCR-ABL activates the phosphatidylinositol-3-kinase (PI3K) pathway, suppressing programmed cell death or apoptosis. BCR-ABL is associated with cytoskeletal proteins leading to a decrease in cell adhesion, and activates pathways that lead to an increase in cell migration [51].

Knowledge of the role of BCR-ABL in the development of CML led to the discovery of imatinib, a small molecule ABL kinase inhibitor, a highly effective therapy for early phase CML [52]. However, some patients develop resistance to imatinib, usually caused by point mutations in the kinase domain of *BCR-ABL* that reduce sensitivity to imantinib [51]. To overcome this problem, second generation of BCR-ABL inhibitors are under development, as well as inhibitors which target oncogenic signaling pathways downstream of BCR-ABL [51].

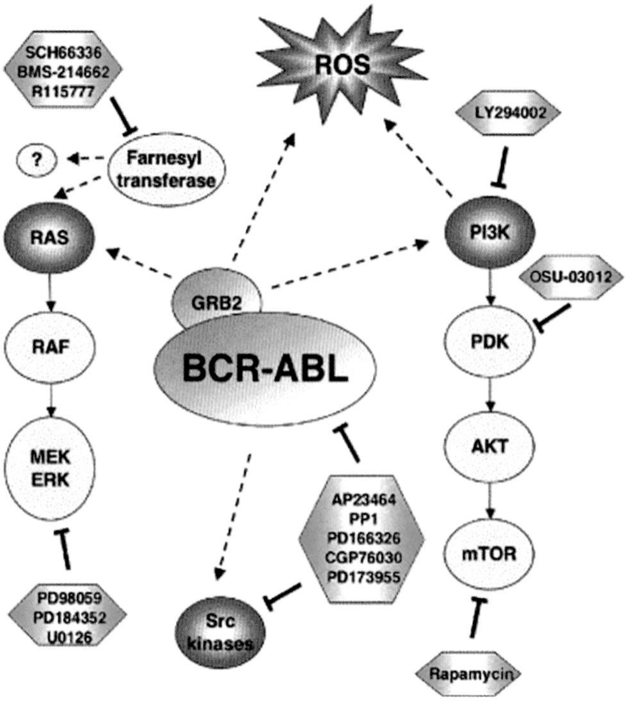

Figure 16.6 Targeting signaling pathways of BCR-ABL. The BCR-ABL onco-protein chronically activates many different downstream signaling pathways to confer malignant transformation in hematopoietic cells. For example, efficient activation of PI3K, Ras and reactive oxygen species (ROS) requires autophosphorylation on Tyr177, a Grb-2 binding site in BCR-ABL. Also, activation of Src family tyrosine kinases have been implicated in the BCR-ABL related disease process. A selection of some inhibitors and pathways discussed in the text are illustrated. From Walz C and Sattler M (2006). Critical Reviews in Oncology/Hematology 57:145–164 (Figure 2).

Chronic Neutrophilic Leukemia

Chronic neutrophilic leukemia (CNL) is a rare disorder characterized by peripheral blood neutrophilia (greater than 25×10^9/liter) and hepatosplenomegaly. The bone marrow is hypercellular, and there is no significant dysplasia in any cell lineage [53]. Cytogenetic studies are normal in 90% of patients.

Approximately 20% of CNL patients are positive for the *JAK2* V617F mutation [39]. JAK2 is a member of the Janus family of tyrosine kinases (JAK1, JAK2, JAK3, and TYK2), which are cytoplasmic kinases that mediate signaling downstream of cytokine receptors

(Figure 16.4) [54]. Activation of a JAK-cytokine receptor complex results in recruitment and phosphorylation of STAT proteins, which then translocate to the nucleus and induce target gene transcription [55].

Polycythemia Vera

In polycythemia vera (PV) too many red blood cells are made in the bone marrow, and the blood becomes thickened with red blood cells. The extra red cells may collect in the spleen, causing it to swell, or may cause bleeding problems and clots [56]. In addition to a blood count, certain specialized tests are used to diagnose polycythemia vera. An arterial blood gas test may be administered to measure the oxygen, carbon dioxide, and pH of the blood. Tests to check serum erythropoietin levels (a hormone that stimulates red blood cell production) and leukocyte alkaline phosphatase levels (an enzyme in leukocytes) may also be administered.

Between 65% and 97% (depending on the study) of polycythemia vera patients have the *JAK2* V617F mutation [57]. As mentioned in the previous section, JAK2 is one of a family of cytoplasmic tyrosine kinases which mediate signaling by growth factor receptors. The V617F mutation changes JAK2 amino acid 617 from valine to phenylalanine, and creates a constitutively active form of JAK2 [58,59]. Valine 617 is located in the pseudokinase domain, which is believed to suppress the activity of the JAK2 catalytic domain; therefore, the V617F mutation could disrupt inhibition of the catalytic domain [60]. JAK2 V617F also escapes negative regulation by the suppressor of cytokine signaling 3 protein [61]. Whatever the precise mechanism, *JAK2* V617F renders cells hypersensitive to the growth-stimulating effects of the erythroid growth factor erythropoietin [58,59] and other growth factors [39]. Constitutive activation of the JAK/STAT, PI3K, ERK, and Akt signal-transduction pathways is also observed in the presence of *JAK2* V617F [39], all of which may promote cell proliferation.

Some of the small number of PV patients who are negative for the *JAK2* V617F mutation have been found to carry one of four mutations in JAK2 exon 12 (F537–K537delinsL, H538QK539L, K539L, or N542–E543del). These mutations have effects similar to those of *JAK2* V617F [62].

The role of *JAK2* V617F in myeloproliferative disorders has stimulated interest in JAK inhibitors as possible therapeutic agents, although no current JAK inhibitors are candidates because of their lack of specificity. Current effort is focused on the development of specific JAK2 inhibitors for use in clinical trials [39].

Essential Thrombocythemia

Essential thrombocythemia (ET) causes an abnormal increase in the number of platelets in the blood and bone marrow. This may inhibit blood flow and lead to problems such as stroke or heart attack [63]. Like PV patients, a high percentage (approximately 50%) of essential thrombocythemia patients carry the *JAK2* V617F mutation [39]. This raises the question of how a single mutation can give rise to several different diseases. Among the theories suggested are transformation of different types of hematopoietic stem or progenitor cells, different genetic backgrounds, or the effects of additional somatic mutations [57]. Transformation of different types of hematopoietic cells is considered unlikely by some because of the belief that the *JAK2* V617F mutation occurs at the level of the hematopoietic stem cell in all cases. It is considered more likely that the activity of *JAK2* V617F is modified by gene dosage, *JAK2* polymorphisms, or cooperating mutations in JAK2-interacting proteins [57].

Chronic Idiopathic Myelofibrosis

Chronic idiopathic myelofibrosis (CIMF) is characterized by the production of too few red blood cells and too many white cells and platelets. An important constant is the production of too many megakaryocytes, which results in overproduction of platelets and cytokine release in the bone marrow. The cytokines stimulate the development of fibrous tissue in the marrow [64]. Megakaryocytes can become so abnormal that platelet production is decreased in some patients.

Approximately 50% of patients with myelofibrosis carry the *JAK2* V617F mutation [39]. As discussed previously, the *JAK2* V617F mutation is also found in patients with PV or ET. Which of these diseases develops as a result of the mutation is thought to depend on genetic background or the presence of secondary mutations. The three diseases are also related to some degree. About 10%–15% of cases of myelofibrosis begin as either PV or ET [64].

A small number of patients with myelofibrosis (about 5%) carry a mutation in the *MPL* gene, *MPL* W55515L/K [39]. Myeloproliferative leukemia virus oncogene (or MPL) encodes the receptor for thrombopoietin, an essential growth factor for megakaryocytes and platelet production [65]. Thrombopoietin acts through MPL and the JAK/STAT signal transduction pathway to stimulate proliferation of megakaryocytes. Mutations in *MPL* have been observed in familial thrombocytosis and congenital omegakaryocytic thrombocytopenia, but not in other myeloproliferative diseases [39]. However, expression of *MPL* is decreased in megakaryocytes from patients with PV and ET [66,67]. In PV, the reduction in MPL function is due to post-translational hypoglycosylation and defective membrane localization [68], while reduced MPL transcription is observed in ET [69].

Chronic Eosinophilic Leukemia/ Hypereosinophilic Syndrome

In chronic eosinophilic leukemia (CEL), a clonal proliferation of eosinophilic precursors results in persistently increased numbers of eosinophils in the blood, bone marrow, and peripheral tissues [70]. Specific characteristics of CEL include an eosinophil count of at least 1.5×10^9/liter, evidence for clonality of eosinophils or an increase in blasts in the blood or bone marrow. Often it is impossible to prove clonality of eosinophils, in which case the diagnosis is hypereosinophilic syndrome.

CEL may be divided into three classes based on the presence or absence of mutations in the *PDGFRA* (platelet-derived growth factor receptor A) or *PDGFRB* genes. The mutations all take the form of fusion genes, and the most widely studied is the *FIP1L1–PDGFRA* fusion [71]. This fusion causes constitutive activation of the PDGFRA tyrosine kinase activity by disruption of the PDGFRA autoinhibitory juxtamembrane motif. *FIP1L1* sequences are not actually required for transforming activity [72]. The active FIP1F1–PDGFRA kinase stimulates the JAK/STAT, PI3K, ERK, and Akt signal-transduction pathways leading to increased survival and proliferation [17]. Other *PDGFRA* fusion mutations are associated with CEL, including *KIF5B-PDGFRA*, *BCR-PDGFRA*, and *CDK5RAP2–PDGFRA*. However, in all cases the *PDGFRA* breakpoint occurs in the region of the juxtamembrane motif [39].

Patients who carry the *FIP1L1–PDGFRA* fusion also develop a particular form of CEL, with symptoms of both CEL and systemic mastocytosis (SM), referred to as CEL-SM. These patients also respond very well to imatinib therapy [73].

PDGFRB fusion mutations can produce CMML with eosinophilia, JMML with eosinophilia, AML with eosinophilia, or UMPD with eosinophilia, which can be classified as CEL-UMPD [39]. In most of these mutations, *PDGFRB* is fused to the N-terminal segment of a protein that encodes for one or more oligomerization domains. As in the case of *PDGRFA* fusion mutations, the *PDBRFB* fusions produce a constitutively active form of *PDGFRB*, and patients who carry *PDGFRB* fusions respond well to treatment with imatinib [74].

Chronic Basophilic Leukemia

Chronic basophilic leukemia (CBL) is an extremely rare disorder, and there is some disagreement about whether it is a separate clinical entity [75]. Four patients with a chronic myeloid disorder but negative for the *BCR-ABL* fusion were identified in the Mayo clinic database from 1975 to 2003. All cases were characterized by prominent basophilia, and had manifestations of basophil mediator-release (ulcerative disease, hyperhistaminemia, coagulopathy, pruritus, and urticaria). The bone marrow was hypercellular with trilineage hyperplasia and basophilia, and bone marrow and peripheral blood showed a lesser degree of eosinophilia. Dysmegakaryopoiesis, with small mononucleated and binucleated forms, was observed. Little is known about the cytogenetics of this disease.

Systemic Mastocytosis

Systemic mastocytosis (SM) is a rare disease in which too many mast cells are found in skin, bones, joints, lymph nodes, liver spleen, and the gastrointestinal tract [76]. The *FIPILI-PDGFRA* fusion causes a disease with symptoms of both chronic eosinophilic leukemia and systemic mastocytosis (CEL-SM). Remaining cases of SM may be divided into those associated with a mutation in the *c-KIT* gene and those without this mutation [77]. The oncogene *c-KIT* encodes a receptor tyrosine kinase. c-KIT and its ligand stem cell factor (SCF) are required for the growth and survival of normal mast cells [78]. SCF ligation to c-Kit activates the Ras/Raf/Erk cascade, the PI3K/Akt pathway, the shp/rac/JNK/c-jun pathway, and the NFkB pathway. Activation of the PI3K/Akt and NFkB pathways have been shown to be necessary for mast cell proliferation [79–81]. Systemic mastocytosis patients carry mutations which cause ligand-independent activation of c-Kit, including D816V, D820G, V560G, D816Y, E839K, D816F, and F522C [39]. Among adult cases of SM, the reported incidence of the D816V mutation ranges from 30% to 100%, and may depend on the study population. The wild-type, V560G, and F522C c-Kit proteins are sensitive to imatinib, but D816V mutant c-Kit is not [77].

Stem Cell Leukemia-Lymphoma Syndrome

Stem cell leukemia-lymphoma syndrome (SCLL) is characterized by concurrent lymphoma, myeloid proliferation/eosinophila often evolving to AML, and cytogenetic abnormalities involving 8p11 [82]. Both myeloid and lymphoid cells exhibit one of a number of 8p11 translocations, all involving the fibroblast growth factor receptor-1 (*FGFR1*) gene [39]. These translocations cause constitutive activation of *FGFR1* by fusing a dimerization domain from the fusion partner to *FGFR1*, a mechanism similar to that observed with the *PDGFRB* fusions associated with CEL-UMPD. *FGFR1* is a receptor tyrosine kinase which when activated stimulates the Ras/Raf/Erk and PI3K/Akt signal transduction cascades [83].

Chronic Myeloproliferative Disease, Unclassifiable

Unclassified myeloproliferative disorder (UMPD) is a category for all cases that do not fit well into the other categories. Examples include patients with CML-like symptoms who are *BCR-ABL* negative. This disorder is sometime referred to as atypical CML or *BCR-ABL*-negative CML. As many as 25% of these patients may carry the *JAK2* V617F mutation [39,77].

Acute Myeloid Leukemia

Normal myeloid stem cells eventually develop into granulocytes, macrophage/monocytes, and megakaryocytes. In acute myeloid leukemia (AML), myeloid stem cells usually develop into a type of immature white blood cell called myeloblasts which are abnormal and do not differentiate. For many years, the different categories of AML were described by the French-American-British (FAB) classification scheme. The eight FAB subtypes are M0 (undifferentiated AML); M1 (myeloblastic, without maturation); M2 (myeloblastic, with maturation); M3 (promyelocytic), or acute promyelocytic leukemia (APL); M4 (myelomonocytic); M4eo (myelomonocytic together with bone marrow eosinophilia); M5 monoblastic leukemia (M5a) or monocytic leukemia (M5b); M6 (erythrocytic), or erythroleukemia; M7 (megakaryoblastic) [84].

Beginning in 1997, the *World Health Organization* (WHO) developed a new classification scheme for acute myeloid leukemias that attempts to incorporate morphology, cytogenetics, and molecular genetics [85]. The WHO scheme also reduced the required blast percentage in the blood or bone marrow for a diagnosis of AML from 30% to 20%. The category of AML with characteristic genetic abnormalities is associated with high rates of remission and favorable prognosis [85].

AML with t(8;21)(q22;q22)

One of the most common genetic abnormalities in AML is (8;21)(q22;q22), which accounts for 5%–12% of all cases of AML [86]. AML with translocation (8;21)(q22;q22) previously fell into FAB classification M2, and is characterized by large blasts often containing azurophilic granules, Auer rods (found in mature neutrophils), smaller blasts in the peripheral blood, dysplasia in the bone marrow (with promyelocytes, myelocytes, and mature neutrophils), abnormal nuclear segmentation, increased eosinophil precursors, reduced monocytes, and normal erythroblasts and megakaryocytes. t(8;21) generates the fusion gene *AML1-ETO*, which fuses sequences coding the amino-terminal portion of the transcription factor *RUNX1* (formerly *AML1*) to almost the entire coding region of *RUNX1T1* (formerly *ETO* or *MTG8*). The resulting fusion protein contains the DNA-binding domain (runt domain) of RUNX1 fused to the RUNX1T1 co-repressor protein [87]. Numerous studies in model systems have demonstrated that expression of the AML1-ETO protein alone is insufficient to induce leukemia but can induce leukemia in cooperation with other mutations [86]. Expression of AML1-ETO does lead to some inhibition of myeloid, lymphoid, and erythroid differentiation, as well as promotion of stem cell self-renewal [88,89]. This is thought to predispose hematopoietic stem cells to leukemia development. It was formerly believed that AML1-ETO changed gene expression patterns by dominant-negative suppression of RUNX1 target genes. However, subsequent gene expression studies found the AML1-ETO activated as many genes as it repressed, suggesting that AML1-ETO promotes leukemogenesis by complex effects on gene expression [88,89]. Model studies suggest that the carboxy-terminus of the ETO/RUNX1T1 protein actually suppresses leukemia development, since mutations or deletions in this region allow AML1-ETO to promote leukemia development without the need for additional mutations [88]. Presumably when full-length AML1-ETO is expressed, such mutations are needed to overcome the antileukemogenic effects of its carboxy-terminal sequences. Among the oncogenic proteins known to promote leukemia development in cooperation with AML1-ETO expression are the TEL-PDGFRb fusion protein and the FLT3 internal duplication (FLT/ITD), both of which stimulate growth-promoting signal-transduction pathways. AML1-ETO will also promote leukemia when expressed in cells lacking the cell cycle inhibitor p21/WAF1/CDKN1A (reviewed in [88]).

AML with inv(16)(p13q22) or t(16;16)(p13q22)

AML with inv(16)(p13q22) or t(16;16)(p13q22) comprises 10%–12% of all cases of AML, and is predominant in younger patients [85]. This type of AML was formerly classified as FAB type M4, and is characterized by monocytic and granulocytic differentiation, abnormal eosinophils with immature granules often with eosinophilia, Auer rods in myeloblasts, and decreased neutrophils in the bone marrow.

Both inv(16)(p13q22) and t(16;16)(p13q22) result in the fusion of the core binding factor-b (CBFb) gene located at 16q22 to the smooth muscle myosin heavy chain (*MYH11*) gene at 16p13. CBFb has no DNA-binding domain, but forms a heterodimer with the AML1/RUNX1 transcription factor and stabilizes AML1/RUNX1 binding to DNA [87]. Since the RUNX1 and CBFb proteins function as a heterodimeric transcription factor, the leukemic fusion protein AML1-ETO and the CBFb/MYH11 fusion protein are predicted to disrupt expression of a similar set of target genes. CBFb/MYH11 binds to RUNX1 with a much higher affinity than CBFb, and two mechanisms have been proposed by which CBFb/MYH11 may disrupt normal RUNX1/CBFb activity [90,91]. CBFb/MYH11 may sequester RUNX1 in the cytoplasm through the interaction of the MYH11 region with the actin cytoskeleton, or the MYH11 sequences may recruit co-repressors when bound with RUNX1 to promoters in the nucleus. It is not yet clear if CBFb/MYH11 utilizes one or both mechanisms.

As was observed with AML1-ETO, expression of CBFb/MYH11 in model systems was not sufficient for leukemogenesis unless secondary mutations are introduced. Secondary mutations that can produce AML in cooperation with CBFb/MYH11 include loss of the cell cycle inhibitors p14ARF, p16INK4a, or p19ARF, or co-expression of FLT3–ITD [92–94]. In patients with inv(16)(p13q22) or t(16;16)(p13q22), 60%–70% have activating mutations in one of the following: FLT3, c-KIT (receptor tyrosine kinases), N-RAS, or K-RAS (signal-transduction proteins) [95]. This suggests that each of these mutations can in fact cooperate with CBFb/MYH11 to induce leukemia.

Acute Promyelocytic Leukemia—AML with t(15;17)(q22q12)

Acute promyeloctyic leukemia (APL) comprises 5%–8% of all cases of AML and is found as typical APL or microgranular APL. Common features of typical APL include promyelocytes with kidney-shaped or bilobed nuclei, cytoplasm densely packed with large granules, bundles of Auer rods in the cytoplasm, larger Auer rods than other types of AML, strongly positive myeloperoxidase reaction in all leukocytes, and only occasional promyelocytes in the blood. Features of microgranular APL include bilobed nuclei, scarce or absent granules, a small number of abnormal promyelocytes with visible granules and bundles of Auer rods, high leukocyte count in the blood, and a strongly positive myeloperoxidase reaction in all promyelocytes [85]. APL was formerly classified as FAB type M3.

In over 98% of cases, the retinoic acid receptor alpha (*RARa*) gene at 17q12 is fused to the *PML* gene at 15q22. In rare cases, *RARa* is fused to another gene, including *PLZF*, *NuMa*, *NPM*, or *STAT5b* [96]. Retinoid signaling is transmitted by two families of nuclear receptors, retinoic acid receptor (RAR) and retinoid X receptor (RXR), which form RAR/RXR heterodimers. In the absence of ligand, the RAR/RXR heterodimer binds to target gene promoters and represses transcription. When a ligand (such as retinoic acid) binds to the complex, it induces a conformational change which transforms the heterodimer into a transcriptional activator. The PML-RARA fusion protein created by t(15;17)(q22q12) binds to RAR/RXR target genes and acts as a potent transcriptional repressor which is not activated by physiological concentrations of ligand. This is due to the fact that all the oncogenic fusion partners of RARa provide a dimerization domain, which results in a dimerized fusion protein with two corepressor binding sites instead of the one found in the RAR/RXR complex. However, recent studies suggest that the PML-RARA fusion protein must have other oncogenic properties, since enforced corepressor binding onto RARA does not initiate APL in model systems [96]. Recent models suggest the PML-RARA leukemogenesis combines enhanced corepressor recruitment and relaxed target specificity to both enhance repression of some genes and target genes not normally bound by RAR/RXR. This disruption of normal gene expression is thought to affect two pathways: myeloid progenitor cell self-renewal and promyelocyte differentiation.

APL is highly sensitive to treatment with all-trans retinoic acid (ATRA), which overcomes the enhanced repression by PML-RARA and induces differentiation of leukemic cells [97]. Although effective, treatment with ATRA alone will cause progressive resistance to the drug, resulting in relapse in 3–6 months. To overcome this problem, treatment of APL now employs a combination of ATRA and other agents, such as the proapoptotic arsenic compound arsenic trioxide (ATO). ATRA and ATO are believed to synergistically enhance differentiation signaling pathways in leukemic cells [97].

AML with 11q23 Abnormalities

AML with 11q23 abnormalities are associated with aberration of the *MLL* gene and comprise 5%–6% of all cases of AML. Two groups of patients show a high frequency of this type of AML: infants and adults with therapy-related AML, usually occurring after treatment with topoisomerase inhibitors. The latter is classified separately and will be discussed later. Common morphologic features include monoblasts and promyelocytes predominant in the bone marrow and showing strong positive nonspecific esterase reactions. AML due to 11q23 abnormalities can be associated with acute myelomonocytic, monoblastic, and monocytic leukemias (FAB M4, M5a, and M5b classifications) and more rarely with leukemias with or without maturation (FAB M2 and M1) [85].

The *MLL* gene encodes a DNA-binding protein that methylates histone H3 lysine 4 (H3K4). *MLL* knockout studies indicate that MLL is necessary for proper regulation of Hox gene expression. Hox genes are a family of transcription factors that regulate many aspects of tissue development. The precise mechanism by which MLL regulates gene expression has not yet been determined. All *MLL* translocations contain the first 8–13 exons of MLL and a variable number of exons from a fusion partner gene. At least 52 *MLL* fusion partner genes have been described, and these fusion partners have diverse functions. Some are nuclear proteins involved in control of transcription and chromatin remodeling; others are cytoplasmic proteins which interact with the cytoskeleton [98]. All MLL fusion proteins have lost the domain necessary for H3K4 methylation. It is believed that leukemogenesis mediated by MLL fusion proteins involves disruption of normal gene expression patterns regulating stem cell differentiation and self-renewal. In some cases the MLL fusion is believed to reactivate the self-renewal program in committed myeloid progenitors [98]. The protein domains contributed by the *MLL* fusion partners are believed to contribute to leukemogenesis through their effects on transcription, chromatin remodeling, and protein-protein interactions.

AML Associated with FLT3 Mutation

Activating mutations in the Fms-like tyrosine kinase (*FLT3*) are present in 20%–30% of all cases of *de novo* AML. Although *FLT3* mutations can be associated with all the major leukemic translocations (*AML1-ETO*, *CBFb-MYH11*, *PML-RARa*, *MLL* fusions), the majority of cases of AML with *FLT3* mutations are cytogenetically normal [19]. Other common clinical features are leukocytosis and monocytic differentiation. Patients with *FLT3* mutations are reported to have an increased rate of relapse and reduced overall survival [99].

Two major types of *FLT3* mutation are found in AML patients. An internal tandem duplication (ITD) of the region of the gene encoding the juxtamembrane domain is found in 25%–35% of adult and 12% of childhood AML [19,100]. The second type of mutation is a missense mutation in the activation loop of the tyrosine kinase domain, commonly affecting codon D835. Both types of mutation result in constitutive phosphorylation of the FLT3 receptor in the absence of ligand and activation of downstream signaling pathways including the PI3K/AKT pathway and the Ras/Raf/ERK pathway.

Studies using model systems have shown that mutated *FLT3* alone is not enough to cause the development of AML. In this respect, FLT3 is similar to fusion proteins such as AML1-ETO or CBFb-MYH11. It is hypothesized that all these mutated proteins require the presence of secondary mutations for AML development.

AML with Multilineage Dysplasia

The characteristics of AML with multilineage dysplasia include 20% or more blasts in the blood or bone marrow and dysplasia in two or more myeloid cell types, generally including megakaryocytes [85]. To qualify as AML with multilineage dysplasia, dysplasia must be present in 50% or more of the cells in at least two lineages and must be present in a pretreatment bone marrow specimen. This classification of AML includes those formerly classified by the FAB system as acute erythroid-myeloid leukemia (M6a) and acute myeloblastic leukemia (M2). In some cases this type of AML arises following a myleodysplastic syndrome (MDS). Diseases that were formerly classified as refractory anemia with excess blasts in transformation (RAEB-t) now fall into the category AML with multilineage dysplasia.

Numerous chromosomal abnormalities involving gain or loss of large segments of certain chromosomes, predominantly chromosomes 5 and 7, are observed in AML with multilineage dysplasia, but little is known of the specific genes affected by these rearrangements [101].

Therapy-Related AML and MDS

This class includes both AML and MDS that arise after chemotherapy or radiation therapy. These diseases are classified according to the mutagenic agents used for treatment, but it can be difficult to attribute a secondary AML to a specific agent because treatment often involves multiple mutagenic agents [102].

Alkylating Agent-Related AML Alkylating agent-related AML usually occurs 5–6 years after exposure to the agent. Typically, this condition is first observed as an MDS with bone marrow failure. Approximately 66% of such cases can be diagnosed as refractory anemia with multilineage dysplasia (RCMD) and another 25% are diagnosed as refractory anemia with excess blasts (RAEB). Some cases evolve into AML, which may correspond to acute myeloid leukemia with maturation (FAB class M2), acute monocytic leukemia (M5b), AMML (M4), erythroleukemia (M6a), or acute megakaryoblastic leukemia (M7) [103].

Cytogenetic abnormalities are observed in more than 90% of cases of therapy-related AML/MDS. Complex abnormalities are the most common finding, often including chromosomes 5 and 7. Recent studies have shown that many patients with therapy-related MDS or AML carry point mutations in *p53* (24% of cases), *RUNX1* (16% of cases), or various other oncogenes [104]. An association between *p53* point mutations and chromosome 5 aberrations, and between *RUNX1* mutations and chromosome 7 aberrations has been observed, suggesting that these sets of mutations may cooperate in the development of therapy-related AML/MDS.

Topoisomerase II Inhibitor-Related AML Topoisomerase II inhibitor-related AML may develop in patients treated with the topoisomerase II inhibitors etoposide, teniposide, doxorubicin, or 4-epi-doxorubicin [103]. Development of AML is observed approximately two years after treatment, and is most commonly diagnosed as acute monoblastic or myelomonocytic leukemia.

AML resulting from treatment with topoisomerase poisons such as etoposide are predominantly associated with translocations of the *MLL* gene at 11q23. Of leukemias that are associated with *MLL*, 5%–10% are therapy-related. Translocations involving other genes associated with leukemogenesis, such as *RUNX1/AML1*, *CBFβ*, and *PML-RARA* have also been observed [105].

Table 16.1 Classifications for AML (not otherwise classified)

1. Acute myeloblastic leukemia, minimally differentiated (FAB M0)
2. Acute myeloblastic leukemia without maturation (FAB M1)
3. Acute myeloblastic leukemia with maturation (FAB M2)
4. Acute promyelocytic leukemia (FAB M3)
5. Acute myelomonocytic leukemia (FAB M4)
6. Acute monoblastic leukemia and acute monocytic leukemia (FAB M5a and M5b)
7. Acute erythroid leukemias (FAB M6a and M6b)
8. Acute megakaryoblastic leukemia (FAB M7)
9. Variant: AML/transient myeloproliferative disorder in Down syndrome
10. Acute basophilic leukemia
11. Acute panmyelosis with myelofibrosis
12. Myeloid sarcoma

AML Not Otherwise Categorized

Cases of AML that do not fall into the other categories are placed in this category. Classification within this category is based on morphology, cytochemistry, and maturation, since these cases lack distinguishing cytogenetic markers [106]. Classifications are shown in Table 16.1.

Acute Leukemias of Ambiguous Lineage

Acute leukemias of ambiguous lineage (also mixed phenotype acute leukemias or hybrid acute leukemias) are leukemias in which the features of the blast population do not allow classification into myeloid or lymphoid categories or share features of both myeloid and lymphoid cells [107].

LYMPHOCYTE DISORDERS

Disorders of lymphocytes include deficiency of lymphocytes (lymphopenia) and overproliferation of lymphocytes. Overproliferation of lymphocytes is due to either reactive proliferation of lymphocytes (lymphocytosis) or to neoplastic problems.

Lymphopenia

Lymphopenia is defined by less than 1,500 lymphocytes/microliter of blood in adults and less than 3,000 lymphocytes/microliter of blood in children. Lymphopenia is relatively rare compared to other leukopenias involving granulocytic cells mentioned previously. Some lymphopenias are due to genetic abnormalities, which are categorized as congenital immunodeficiencies. Most lymphopenias are due to viral infection, chemotherapy, radiation, undernutrition, immunosuppressant drug reaction, and autoimmune diseases.

Lymphocytosis

Lymphocytosis can be divided into relative lymphocytosis and absolute lymphocytosis. In human white blood cells, 20%–40% are lymphocytes. When the percentage exceeds 40%, it is recognized as relative lymphocytosis. When the total lymphocyte count in blood is more than 4,000/microliter in adults, 7,000/microliter in older children, and 9,000/microliter in infants, the patient is diagnosed with absolute lymphocytosis.

The best known lymphocytosis is infectious mononucleosis. This disease is due to an infection of Epstein-Barr Virus (EBV). EBV infection at an early age will not show any specific symptoms. However, infection in adolescents and young adults can cause more severe problems (Kissing Disease), such as fever, sore throat, lymphadenopathy, splenomegaly, hepatomegaly, and increased atypical lymphocytes in blood. EBV is a member of the herpesvirus family. It encodes about 100 proteins. EBV infects B-lymphocytes. In a minority of infected B-cells, EBV infection occurs in the lytic form, which induces cell lysis and virus release. In a majority of cells, EBV infection is nonproductive and the virus is maintained in latent form. The cells with latent viruses are activated and undergo proliferation, and also produce specific antibodies against the virus. The massive expansion of monoclonal or oligoclonal cytotoxic CD8+ T-cells presented as atypical lymphocytes in peripheral blood is the major feature of infectious mononucleosis [108]. Such strong humoral and cellular responses to EBV eventually highly restrict EBV infection.

Lymphadenopathy

Lymphadenopathy refers to the enlargement of lymph nodes. This is a frequent clinical problem. Infection, autoimmune disorders, and neoplasms can induce lymphadenopathy. Lymphadenopathy due to neoplasms (neoplastic lymphadenopathy) will be discussed in more detail in the following sections.

Non-neoplastic lymphadenopathy is the consequence of a variety of infections and inflammatory stimulation triggered proliferation of lymph node cells. Depending on the type of antigen, the major cell types involved in the expansion can be B-lymphocytes, T-lymphocytes, macrophages, or quite frequently a mixture of them. Sometimes, non-neoplastic lymphadenopathy may be hard to distinguish from lymphoma. More careful diagnoses are required.

Neoplastic Problems of Lymphocytes

Lymphocytic leukemia and lymphoma are the two major groups of lymphoid neoplasms. Leukemia and lymphoma do not have a very clear distinction, and the use of the two terms can be confusing. In general, the term lymphocytic leukemia is used for neoplasms involving the general area of the bone marrow and the presence of a large number of tumor cells in the peripheral blood; lymphomas show uncontrolled growth of a tissue mass of lymphoid cells. However, quite often, especially at the late stage of lymphoma, tumor cells originating from the lymphoma mass may spread to peripheral blood and produce a phenotype similar to leukemia.

The second important issue is the classification of lymphocytic neoplasms. According to the *World Health Organization* classification, lymphocytic neoplasms are divided into five major categories: (i) precursor B-cell neoplasms, (ii) peripheral B-cell neoplasms, (iii) precursor T-cell neoplasms, (iv) peripheral T-cell neoplasms, and (v) Hodgkin's lymphoma.

Detailed information on this classification is shown in Table 16.2. It is important to mention that all lymphoid neoplasms develop from a single transformed lymphoid cell. Furthermore, the transformation happens after the rearrangement of antigen receptor genes, including T-cell receptors and immunoglobulin heavy and light chains. Therefore, antigen receptor patterns are generally used to distinguish monoclonal neoplasms from polyclonal reactive lymphadenopathy.

Although there are many different lymphocytic malignancies (as listed in Table 16.2), the majority of adult lymphoid neoplasms are one of four diseases: follicular lymphoma, large B-cell lymphoma, chronic lymphocytic leukemia/small lymphocytic lymphoma, and multiple myeloma; and the majority of childhood lymphoid neoplasms are one of two diseases: acute lymphoblastic leukemia/lymphoma and Burkitt lymphoma.

Acute Lymphoblast Leukemia/Lymphoma

Acute lymphoblast leukemia/lymphoma (ALL) is most common between the age of 2 and 5 years although it affects both adults and children. The majority of ALL is pre-B-cell leukemia. Pre-T-cell leukemia is often reported in adolescent males. Currently, effective treatment of ALL has reached a cure rate of 80% in children, although adult ALL is still a terrible life-threatening problem. Morphologically, it is difficult to separate T and B lineage ALL. Furthermore, patients with T and B ALL also present similar symptoms. Therefore, flow cytometry studies to identify the expression of specific cell surface markers are generally used to distinguish the lineage and differentiation of ALL.

Historically, many cell surface markers are named by their cluster of differentiation (CD) numbers, which were created based on the monoclonal antibodies used to define the specific expression of these surface markers [109]. B- and T-cell malignancies can be classified by observing the appearance of their specific surface markers (Figure 16.7). Most ALL cells are also positive for terminal deoxytransferase (TdT) by immunostaining. TdT is a specialized DNA polymerase and is only expressed in pre-B and pre-T-cells.

Various chromosomal locus translocations are associated with the development of ALL, such as those

Table 16.2 WHO Classifications of Lymphoid Neoplasms

B-CELL NEOPLASMS

Precursor B-cell Lymphoblastic leukemia/ Lymphoma

Peripheral B-Cell Neaplasms

B-cell chronic Lymphocytic leukemia/small Lymphocytic Lymphoma
  Variant: With monoclonal gammopathy! plasmacytoid differentiation
B-cell prolymphocytic leukemia
Immunocytoma/lymphoplasmacytic lymphoma (Waldenstrom's macroglobulinemia)
Mantle cell Lymphoma:
  Variants: Blastic or blastoid
  Pleomorphic
  Small cell
  Monocytoid
Follicular Lymphoma
  Variants: Grade 1 (centroblasts comprise <50% of the follicle area)
  Grade 2 (centroblasts comprise >50% of the follicle area)
Cutaneous follicular Lymphoma
Marginal Zone B-cell Lymphoma of mucosa-associated Lymphoid tissue (MALT) type (monocytoid B cells)
  Variant: Nodal marginal zone Lymphoma monocytoid B-cells
Splenic marginal zone B-cell Lymphoma (villous Lymphocytes)
Hairy cell leukemia
Diffuse large B-cell Lymphoma
  Variants: Burkitt-like Immunoblastic.
  T-cell or histiocyte-rich
  Anaplastic large B-cell
Mediastinal (Thymic) large B-cell Lymphoma
Intravascular large B-cell Lymphoma
Burkitt Lymphoma
  Variant: with plasmacytoid differentiation (AIDS-associated)
Immunosecretory disorders (clinical or pathological variants)
Plasma cell myeloma (multiple myeloma)
Monoclonal gammopathy of undetermined significance (MGUS)
Plasma cell myeloma variants:
  Indolent myeloma
  Smoldering myeloma
  Osteosclerotic myeloma (POEMS syndrome)
  Plasma cell leukemia
  Non-secretory myeloma
Plasmacytomas:
  Solitary plasmacytoma of bone
  Extramedullary plasmacytoma
Waldenstrom's macroglobulinemia (immunocytoma, see above)
Heavy Chain Disease (HCD)
  Gamma HCD
  Alpha HCD
  Mu HCD
Immunoglobulin deposition diseases:
  Systemic light chain disease
  Primary amyloidosis

T-CELL NEOPLASMS

Precursor T-cell lymphoblastic leukemia/lymphoma

Peripheral T-cell and NK-cell neoplasms

T-cell prolymphocytic leukemia
  Variants: Small cell
  Cerebriform cell
T-cell granular Lymphocytic leukemia
Aggressive NK cell leukemia
Nasal and nasal-type NK/T cell Lymphoma
Mycosis fungoides and Sezary syndrome
  Variants: Pagetoid reticulosis
  MF-associated follicular mucinosis
  Granulomatous slack skin disease
Angioimmunoblastic T-cell Lymphoma
Peripheral T-cell Lymphoma unspecified
  Variants: Lymphoepithelioid (Lennert's)
  T-zone
  Pleomorphic, small, mixed, and large
  Immunoblastic
Adult T-cell leukemia/lymphoma (HTLV1)
  Variants: Acute
  Lymphomatous
  Chronic
  Smoldering
  Hodgkin-like
Anaplastic large cell Lymphoma (ALCL) (T and null cell types)
  Variants: Lymphohistiocytic
  Small cell
Primary cutaneous CD-30 positive T-cell Lymphoproliferative disorders
  Variants: Lymphomatoid papulosis (type A and B)
  Primary cutaneous ALCL
  Borderline lesions
Subcutaneous panniculitis-like T-cell Lymphoma
Intestinal T-cell Lymphoma (enteropathy)
Hepatosplenic gamma/delta T-cell Lymphoma

HODGKIN LYMPHOMA (Hodgkin's Disease)

Nodular lymphocyte predominance Hodgkin Lymphoma diffuse areas
Classical Hodgkin Lymphoma
  Nodular sclerosis (Grades I and II)
  Classical Hodgkin lymphoma, lymphocyte-rich
  Mixed cellularity
  Malignant Lymphoma with features of Hodgkin Lymphoma and anaplastic large cell lymphoma (formerly ALCL Hodgkin's-like)

involving the *MLL* gene (chromosome 11q23), the *TCRβ* enhancer (chromosome 7q34), the *TCRα/δ* enhancer (chromosome 14q11), the *E2A* gene (chromosome 19p13), and the *PAX5* gene (chromosome 9p13), t(12;21)(p13;q22), and t(9;22)(q34;q11) [110].

An additional translocation, t(12;21), is identified in approximately 25% of childhood pre-B-cell ALL [111]. The critical fusion protein generated from this translocation is *ETV6-RUNX1* (also known as *TEL-AML1*), which contains 336 amino acids from the N-terminal region of ETV6 and almost the entire RUNX1 protein. Quite frequently, another allele of ETV6 is also lost in t(12;21) ALL patient samples. This finding suggests that TEL is a potential tumor suppressor gene. ETV6-RUNX1 can form dimers via the ETV6 helix-loop-helix domain and contains the RUNX1 DNA binding domain. Therefore, it is believed that ETV6-RUNX1 affects the expression of RUNX1 target genes to promote leukemia development. Interestingly, the ETV6-RUNX1 fusion gene has been identified in neonatal blood spots of children who developed leukemia between 2 and 5 years of age, suggesting that t(12;21) is not sufficient for leukemogenesis without additional malignant promoting factors.

The Philadelphia chromosome caused by t(9;22) is the most frequently identified chromosomal

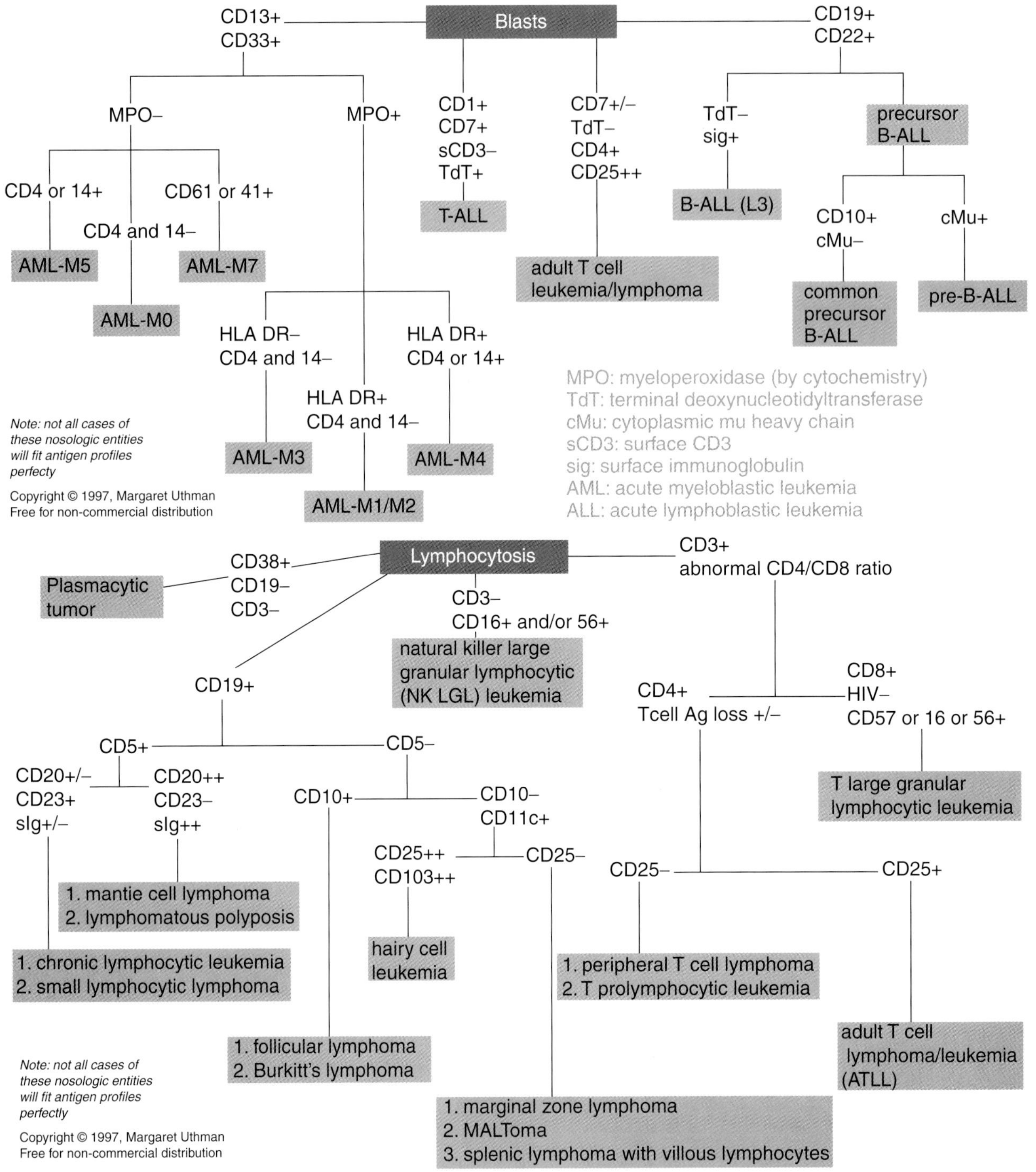

Figure 16.7 Antigen profiles for diagnosis of lymphomas and leukemias. Flow charts for blasts (upper panel) and lymphocytosis (lower panel) are shown indicating CD marker expression patterns used for diagnosis of leukemias and lymphomas. From "Hematopoietic Phenotypes made Mockingly Simple" by Margaret Uthman MD (http://web2.airmail.net/uthman/cdphobia.cdphobia.html).

translocation in adult ALL. The Philadelphia chromosome encodes the fusion protein BCR-ABL. The constitutive activation of the ABL tyrosine kinase and the interaction of this fusion protein with various signaling regulators and proto-oncogene products as indicated in Figure 16.6 also promote B-ALL development.

Through use of high-resolution single-nucleotide polymorphism arrays and genomic DNA sequencing to study 242 B-ALL patient samples, the *PAX5* gene has been identified as the most frequent target of somatic mutation. Approximately 32% of samples present either deletion or point mutation of the *PAX5* gene and result in decreased expression or partial loss

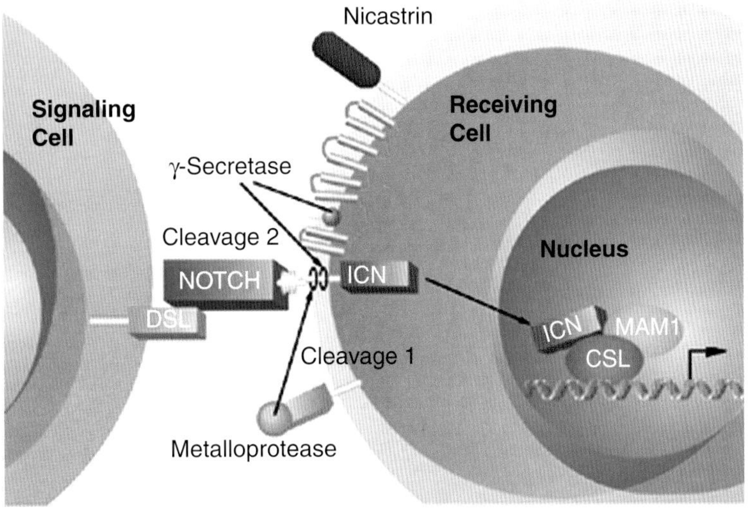

Figure 16.8 NOTCH signaling. Interaction of NOTCH and delta serrate ligand (DSL) stimulates proteolytic cleavage of NOTCH by metalloproteases and γ-secretase. This leads to the release of the intracellular ICN domain, which translocates to the nucleus where it interacts with the DNA binding protein CSL, displaces corepressors and recruits co-activators (MAM1), thereby converting CSL from a repressor to an activator of gene expression. From Armstrong, SA and Look AT (2005) J. Clin. Oncol. 23:6306 (Figure 2).

of its function [112]. *PAX5* is also known as B-cell specific activating protein (BSAP), which plays a crucial role during B lineage commitment and differentiation [113].

The NOTCH signaling pathway plays important roles during hematopoiesis, especially in T-cell lineage development [114]. The interaction of cell surface NOTCH receptors and their ligands of the Delta-Serrate-Lag2 family induces two-step proteolytic cleavage of the NOTCH protein and generates the intercellular domain of NOTCH (ICN) fragment. ICN translocates to the nucleus and activates target gene expression via interaction with the DNA binding transcription factor CSL, displacement of transcription repressors, and recruitment of transcription activators to the DNA binding complexes (Figure 16.8). A NOTCH activating mutation involving somatic alteration of the NOTCH1 gene has been identified in over 50% of T-ALL patients [115]. Furthermore, the FBW7 ubiquitin E3 ligase responsible to the degradation of ICN is also mutated in T-ALL patient samples and cell lines, which increases the cellular concentration of ICN and further enhances NOTCH signaling [110]. The best known NOTCH target gene related to cancer development is the MYC oncogene.

Chronic Lymphocytic Leukemia/Small Lymphocytic Lymphoma

Chronic lymphocytic leukemia (CLL) is characterized by the presence of over 5,000/microliter of mature-appearing lymphocytes in peripheral blood and a specific range of immunophenotypes [116]. The majority of CLL cases are CD19+/CD5+/CD22–/CD79B-low B lineage malignancies. Small lymphocytic lymphoma (SLL) refers to a small percentage of cases in which the tumor cells have a similar immunophenotype to CLL but are restricted to lymph nodes without blood and bone marrow involvement. Due to currently unclear genetic factors, CLL is rare in Asian populations but is the most common form of leukemia in North America and Europe. Furthermore, the incidence of CLL doubles in the Jewish population of North America compared to the non-Jewish North American population and occurs more often in men than women. Hypogammaglobulinemia in CLL patients can cause an increased chance of bacterial infection. Furthermore, some patients may develop autoantibodies against their own red blood cells and platelets, which induce autoimmune hemolytic anemia and thrombocytopenia.

The cytogenetic abnormalities detected by fluorescence *in situ* hybridization (FISH) in CLL are mainly chromosome trisomies and deletions. Analysis of 325 CLL patient samples identified trisomy 12 (18%) and deletions on chromosomes 13q (55%), 11q (16%), and 17p (7%) [117]. Importantly, chromosome 11q and chromosome 17p deletions are related to poor prognosis and an advanced stage of this disease. The well-known tumor suppressor gene *p53* is located in the deleted region of chromosome 17. ATM kinase that regulates p53 activity is located in the deleted region of chromosome 11. Since p53 is a critical inhibitor of cell cycle progression and most chemotherapy drugs target p53-dependent pathways, it is valuable to evaluate cytogenetic conditions before treating CLL patients.

Death associated protein kinase 1 (DAPK1) is a pro-apoptotic protein. Genetic and epigenetic studies have revealed that reduced DAPK1 expression is associated with both familial and sporatic CLL development [118]. The expression of two additional pro-apoptotic proteins,

BAX and BCL-XS, is also downregulated in CLL. Furthermore, several antiapoptosis proteins, such as BCL-XL, BAG1, and MCL1, show increased expression in CML [119]. These results suggest that CLL cells have a decreased apoptosis rate, which may contribute to the accumulation of relatively mature B-cells in patients.

Another interesting phenomenon of CLL is its defect in expression of CD22 and CD79B on the cell surface. CD79B is a subunit of the heterodimer B-cell receptor Igα/Igβ (CD79A/CD79B). Decreased expression of these two proteins on the cell surface is not due to a reduction of their transcription or translation, but is related to abnormal glycosylation and folding of their associated proteins in the endoplasmic reticulum [120], suggesting that dysfunction of these B-cells may be related to abnormal protein processing. However, no precise molecular mechanism for this defect has been identified.

Follicular Lymphoma

The neoplastic cells of follicular lymphoma are derived from germinal center B-cells or cells differentiated toward germinal center B-cells. They are generally CD19+, CD20+, and CD10+, but unlike CLL cells they do not express CD5 on their surface. Furthermore, they present in either a pure follicular pattern or mixed with follicular and diffused areas. The occurrence of this disease is also affected by genetic background. It is one of the most common lymphocytic neoplasms in North America and Europe, but less common in Asia. Follicular lymphoma is a disease of late life with a peak of detection between 60 and 70 years of age.

BCL2 is highly expressed in neoplastic cells in over 90% of follicular lymphoma patients. Therefore, BCL2 immunostaining is used to distinguish normal follicles from follicular lymphoma. This high expression of BCL2 is due to a specific chromosomal translocation [t(14;18)(q32;q21)] that generates a fusion between the immunoglobulin heavy chain enhancer on chromosome 14 and the *BCL2* gene on chromosome 18. This was one of the earliest chromosomal translocations related to cancer development to be discovered [121–123]. The translocation breakpoint on chromosome 14 is at the functional diversity region-joining region (D-J) joint, indicating that mistaken recombination involving the recombination enzymes is the molecular mechanism of generating this translocation [121].

BCL2 is a strong antiapoptotic factor. Normally, most B-cells should be terminated via apoptosis if they are not challenged by specific antigens. With the overexpression of BCL2, follicular lymphoma cells are able to overcome normal apoptotic signals and avoid termination. Therefore, the prolonged life span of follicular cells due to this defect in apoptotic elimination contributes to the development of follicular lymphoma. However, additional cytogenetic lesions besides t(14;18)(q32;q21) are generally observed in most follicular lymphoma cells, which include trisomies, monosomies, deletions, amplifications, and chromosome translocations [124]. These observations suggest that additional mutations besides the overexpression of BCL2 are required for the development of follicular lymphoma.

It is interesting to note that about 10% of patients with follicular lymphoma lack t(14;18)(q32;q21) and do not show increased BCL2 expression by immunostaining. It has been postulated that other genetic defects along a similar antiapoptotic pathway may occur in these patients, which give a similar disease pattern [125]. In follicular lymphoma, 30%–50% will transform into diffuse large B-cell lymphoma after a period of 7 to 9 years.

Diffuse Large B-Cell Lymphoma

The name diffuse large B-cell lymphoma (DLBCL) is based on the morphology and behavior of this group of malignant cells. They typically express B-cell markers CD19, CD20, CD22, and CD79a but lack terminal deoxytransferase (TdT). A few of these tumors also express CD10 and CD5, and are distinguished from mantle cell lymphoma by the lack of cyclin D1 overexpression. DLBCL cells are large and diffusely invade the lymph nodes and extranodal areas. However, this is a highly heterogeneous disease. DLBCL is generally identified in older patients with a median age over 60 and with almost equal distribution between male and female. This is the most common lymphocytic malignancy in adults.

BCL6 is a zinc finger transcription factor and is encoded by a gene located on chromosome 3q27. During normal B-cell development, BCL6 is specifically expressed in germinal center B-cells and plays a critical role during B-cell differentiation and in the formation of the germinal center. Importantly, *BCL6*-involved chromosomal translocations are the most commonly detectable genetic abnormalities in DLBCL, occurring in 35%–40% of cases. One major chromosomal translocation leads to immunoglobulin heavy chain regulatory element directed BCL6 expression. The general feature of these translocations is the constitutive expression of *BCL6* [126]. A significant target of BCL6 related to disease development is *p53*. BCL6 inhibits *p53* expression by directly binding to the *p53* regulatory element and initiating the formation of a histone deacetylase complex which modifies local chromatin structure to generate an inactive condition and represses *p53* transcription [127]. Decreased p53 leads to a reduced rate of apoptosis in response to DNA damage, resulting in the proliferation of malignant clones.

BCL2 is another deregulated gene in DLBCL. *BCL2* is overexpressed in B-cells with t(14;18)(q32;q21), a hallmark of follicular lymphoma. This chromosomal translocation is also observed in 15% of DLBCL, which may come from the transformation of follicular lymphoma to DLBCL. However, abnormally high levels of *BCL2* expression occur in about 50% of DLBCL [128], indicating the involvement of other mechanisms to induce *BCL2* expression in this disease.

Another unique feature of germinal center B-cell maturation is the somatic hypermutation of immunoglobulin variable region (IgV) genes. This process increases antibody diversity and enhances antibody affinity toward antigens. However, somatic hypermutation also increases the opportunity to generate chromosomal translocations and mutations in other genes. *FAS* (CD95), *PAX5*, *PIM1*, *c-Myc*, and *BCL6* are a group of genes reported as targets of abnormal somatic hypermutation and detected in DLBCL [129].

Burkitt Lymphoma

In 1958, Denis Burkitt reported a special type of jaw tumor in African children, and these tumors were later named highly malignant Burkitt lymphoma [130]. Burkitt lymphoma cells are generally monomorphic medium-sized cells (bigger than ALL cells and smaller than DLBCL cells) and with round nuclei and multiple nucleoli. They express CD19, CD20, CD10, and surface IgM. These cells are extremely hyperproliferative and are also highly apoptotic. Close to 100% of Burkitt lymphoma cells are positive for the proliferation marker Ki-67.

According to the *World Health Organization*, Burkitt lymphoma can be further divided into three categories: endemic, sporadic (nonendemic), and immunodeficiency-associated. The common feature of Burkitt lymphoma is the chromosomal translocation-induced overexpression of the *c-Myc* proto-oncogene. The most common form of translocation is t(8;14)(q24;q32), which leads to immunoglobulin heavy chain regulatory element directed expression of the *c-Myc* gene. In rare cases, the immunoglobulin κ or λ chain locus (instead of the heavy chain locus) is involved in the translocation [t(2;8)(p12;q24) or t(8;22)(q24;q11)], each of which leads to the overexpression of *c-Myc*. Interestingly, *c-Myc* was the first gene known to be involved in a chromosome translocation-associated neoplasm via the study of t(8;14)(q24;q32) in Burkitt lymphoma. Since *c-Myc* is also overexpressed in other forms of leukemia and lymphoma, it is believed that other genetic lesions also play critical roles in the development of Burkitt lymphoma.

Endemic Burkitt lymphoma is found mainly among children living in the malaria belt of equatorial Africa. The common sites of endemic Burkitt lymphoma are kidney and jaw. Furthermore, most endemic Burkitt lymphomas are also positive for EBV. Studies suggest that EBV infections occur long before the translocation of *c-Myc*. Several EBV proteins enhance cell proliferation and inhibit apoptosis [131], which can provide premalignant activation conditions for further lymphoma development. Furthermore, the high geographic correlation of endemic Burkitt lymphoma and malaria infection also raises the possible involvement of malaria in the development of Burkitt lymphoma. B-cell proliferation is activated during malaria infection.

Sporadic Burkitt lymphoma has no geographic preference. Furthermore, it also has fewer age restrictions and is detected in adults. Besides *c-Myc* related chromosome translocations, sporadic Burkitt lymphoma patients are generally EBV negative. The lymph nodes and terminal ileum are the common sites of this type of lymphoma.

HIV infection has been related to the development of various forms of lymphoma, including Burkitt lymphoma, diffuse large B-cell lymphoma, low-grade B-cell lymphoma, peripheral T-cell lymphoma, primary effusion lymphoma, and classical Hodgkin lymphoma [132]. Depending on the difference in pathological classification, Burkitt lymphoma is either the most (35%–50%) or the second (after DLBCL) most common lymphoma in HIV patients. Soluble Tat protein encoded by HIV-1 can be released by the infected cells and then taken up by uninfected cells. This protein has been reported to interact with the Rb family member pRb2/p130 and inactivate the normal function of pRb2 [133]. Experimental evidence also suggests that Tat preferentially targets B-cells. pRb2 is one of the three Rb family members that play important roles during cell cycle progression by controlling E2F activity during G1-S phase transition (Figure 16.9). Therefore, Tat may behave as a synergistic factor of *c-Myc* overexpression to promote Burkitt lymphoma development.

Multiple Myeloma

Multiple myeloma is the most important and common plasma cell neoplasm. Plasma cells are mature immunoglobulin-producing cells. Plasma cell neoplasms are a group of neoplastic diseases of terminally differentiated monoclonal immunoglobulin-producing B-cells. They are generally referred to as myeloma. The monoclonal immunoglobulin produced by these cells is considered the M factor of myeloma. In normal plasma cells, the production of immunoglobulin heavy chain and light chain is well balanced. Under neoplastic conditions, normal balance may not be maintained, resulting in the overproduction of either heavy chain or light chain. The free light chains are known as Bence Jones proteins. Multiple myeloma is generally preceded by a premalignant condition called monoclonal gammopathy of undetermined significance (MGUS). MGUS is quite common in older people. About 20% of patients with MGUS will develop myeloma, generally multiple myeloma.

Multiple myeloma is a disease with multiple masses of neoplastic plasma cells in the skeletal system, which is generally associated with pain, bone fracture, and renal failure. It occurs mainly in people of older age (over 45), and affects more men than women. Furthermore, the rate of multiple myeloma in African Americans is about twice that of the rest of the population in the United States.

Like other lymphoid malignancies, multiple myeloma is related to the overexpression of various regulators of cell proliferation and survival due to chromosomal translocations which place these genes under the control of immunoglobulin regulator elements, mainly the immunoglobulin heavy chain locus on chromosome 14 [134]. The five most common chromosomal translocations involving the immunoglobulin heavy chain locus are those involving the *cyclin D1* gene on chromosome 11q13 (16%), the *cyclin D3* gene on chromosome 6p21 (3%), the *MAF* gene on chromosome 16q23, the *MAFB* gene on chromosome 20q12, and the *MMSET* and fibroblastic growth factor receptor 3 (*FGFR3*) on chromosome 4p16 (15%). The well-established MAF targets integrin 7 and cyclin D2 are important in communication with the cellular microenvironment and in regulating cell cycle

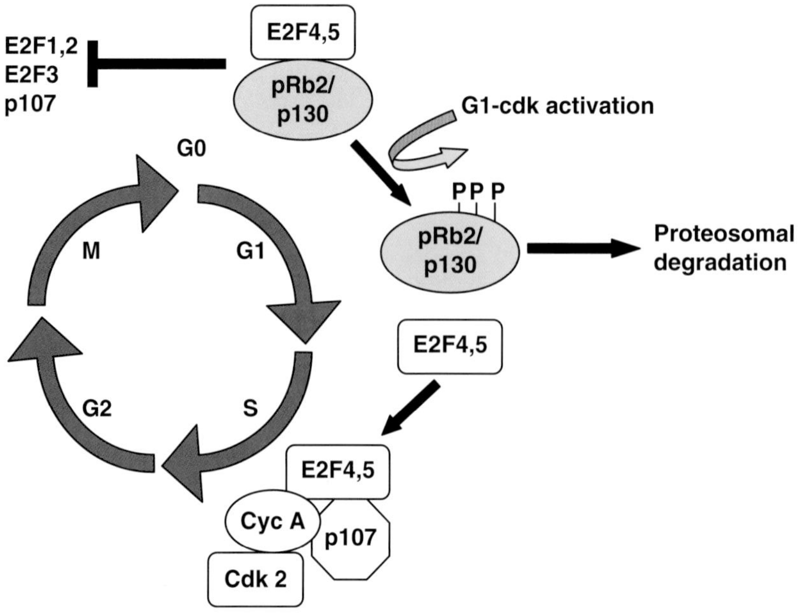

Figure 16.9 **pRb2 function.** In quiescent (G0) cells, the nuclear EsF-pRb2/p130 complex represses several cellular promoters. After activation of the cell cycle, pRb2/p130 is phosphorylated by G1 cyclin-dependent kinases (Cdks) and then degraded by the proteosome. This results in the derepression of various genes, including p107. p107 is then able to interact with E2F4 and E2F5, which have been released from pRb2/p130, and associate with cyclin A and cdk2. Based on Bellan C, Lazzi S, De Falco G, Nyongo A, Giordano A, and Leoncini L. (2003). J. Clin. Pathol. 56:188 (Figure 2).

progression, respectively [135]. MMSET is a histone methyl transferase and is likely involved in the regulation of chromatin structure and protein-protein interactions to regulate gene expression. FGFR3 is a receptor tyrosine kinase and its activation directly promotes cell proliferation and survival. Furthermore, complete chromosome 13 deletion or deletion of chromosome 13q14 was commonly identified in multiple myeloma (50% of patients). It is currently unclear which gene(s) in this region is critical for preventing disease development [135].

Hodgkin's Lymphoma

Thomas Hodgkin, in 1832, was the first to give a detailed report about the macroscopic pathology of the disease currently named Hodgkin's lymphoma. In contrast to non-Hodgkin's lymphoma, Hodgkin's lymphoma starts from a single lymph node or a chain of nodes and spreads in an orderly way from one node to another. Further microscopic analysis revealed that Hodgkin's lymphoma presents a very unique type of malignant cells, called Reed-Sternberg cells (Figure 16.10). The classical morphology of these cells is large size (20–50 micrometers), relative abundance, amphophilic and homogeneous cytoplasm, and two mirror-image nuclei (owl eyes) with one eosinophilic nucleolus in each nucleus. Reed-Sternberg cells occupy only a small portion of the tumor mass. The majority of the cells in the tumors are reactive lymphocytes, macrophages, plasma cells, and eosinophils, which are attracted to the surrounding malignant Reed-Sternberg cells by their secreted cytokines. The World Health Organization classification has divided Hodgkin's lymphoma into several subcategories (Table 16.2).

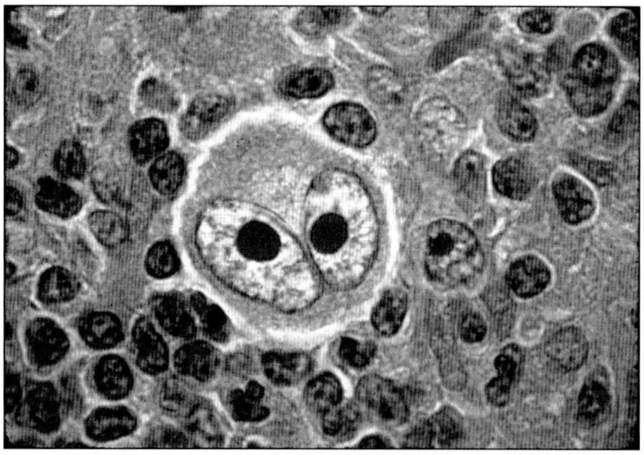

Figure 16.10 **Reed-Sternberg cell.** The figure shows the typical characteristics of the Reed-Sternberg cell: large size (20 – 50 micrometers), amphophilic and homogeneous cytoplasm, and two mirror-image nuclei (owl eyes) with one eosinophilic nucleolus in each nucleus. Reed-Sternberg cells only occupy a small portion of the tumor mass. From a website maintained by the Department of Pathology, Stanford University (http://hematopathology.stanford.edu/).

Hodgkin's lymphoma affects people of relatively young age. It is one of the most common forms of cancer in young adults, with an average age at diagnosis of 32 years. Nowadays, highly developed radiation therapy and chemotherapy treatments have made Hodgkin's lymphoma a curable cancer. However, about 20% of patients still die from this disease. Furthermore, successfully treated patients have a higher risk of dying from late toxicities, such as secondary malignancies and cardiovascular diseases [136].

Studies with highly developed microdissection techniques to isolate Reed-Sternberg cells from their surrounding tumor mass have demonstrated that the majority of Hodgkin's lymphoma cells have a clonal immunoglobulin rearrangement and some show somatic hypermutation of immunoglobulin genes, indicating that Hodgkin's lymphoma is mainly derived from germinal center B-cells [137]. Although there are numerous cytogenetic and genetic lesions reported in Hodgkin's lymphoma [136], the primary cause of Hodgkin's lymphoma is currently unclear.

REFERENCES

1. Orkin SH and Zon LI. Hematopoiesis: An evolving paradigm for stem cell biology. *Cell.* 2008;132:631–644.
2. Dzierzak E and Speck NA. Of lineage and legacy: The development of mammalian hematopoietic stem cells. *Nature Immunol.* 2008; 9:129–136.
3. Samokhvalov IM, Samokhvalova NI, Nishikawa S. Cell tracing shows the contribution of the yolk sac to adult haematopoiesis. *Nature.* 2007;446:1056–1061.
4. Laird DJ, von Andrian UH, Wagers AJ. Stem cell trafficking in tissue development, growth, and disease. *Cell.* 2008;132:612–630.
5. Kiel MJ and Morrison SJ. Uncertainty in the niches that maintain haematopoietic stem cells. *Nature Rev Immunol.* 2008;8:290–301.
6. Ma L, Sun B, Hood L, et al. Molecular profiling of stem cells. *Clin Chim Acta.* 2007;378:24–32.
7. Iwasaki H and Akashi K. Myeloid lineage commitment from the hematopoietic stem cell. *Immunity.* 2007;26:726–740.
8. Kim SI and Bresnick EH. Transcriptional control of erythropoiesis: Emerging mechanisms and principles. *Oncogene.* 2007;26: 6777–6794.
9. Nutt SL and Kee BL. The transcriptional regulation of B-cell lineage commitment. *Immunity.* 2007;26:715–725.
10. Rothenberg EV. Negotiation of the T lineage fate decision by transcription-factor interplay and microenvironmental signals. *Immunity.* 2007;26:690–702.
11. Rosenbauer F and Tenen DG. Transcription factors in myeloid development: Balancing differentiation with transformation. *Nature Rev Immunol.* 2007;7:105–117.
12. Wang Q, Stacy T, Binder M, et al. Disruption of the Cbfa2 gene causes necrosis and hemorrhaging in the central nervous system and blocks definitive hematopoiesis. *Proc Natl Acad Sci USA.* 1996;93:3444–3449.
13. Okuda T, van Deursen J, Hiebert SW, et al. AML1, the target of multiple chromosomal translocations in human leukemia, is essential for normal fetal liver hematopoiesis. *Cell.* 1996;84:321–330.
14. Yu BD, Hess JL, Horning SE, et al. Altered Hox expression and segmental identity in Mll-mutant mice. *Nature.* 1995;378:505–508.
15. Orkin SH. Priming the hematopoietic pump. *Immunity.* 2003; 19:633–634.
16. Fabbri M, Garzon R, Andreeff M, et al. MicroRNAs and noncoding RNAs in hematological malignancies: Molecular, clinical and therapeutic implications. *Leukemia.* 2008;22:1095–1105
17. Matsumura I, Mizuki M, Kanakura Y. Roles for deregulated receptor tyrosine kinases and their downstream signaling molecules in hematologic malignancies. *Cancer Sci.* 2008;99:479–485.
18. Steelman LS, Abrams SL, Whelan J, et al. Contributions of the Raf/MEK/ERK, PI3K/PTEN/Akt/mTOR and Jak/STAT pathways to leukemia. *Leukemia.* 2008;22:686–707.
19. Gilliland DG and Griffin JD. The roles of FLT3 in hematopoiesis and leukemia. *Blood.* 2002;100:1532–1542.
20. Ganusov VV and De Boer RJ. Do most lymphocytes in humans really reside in the gut? *Trends Immunol.* 2007;28:514–518.
21. Millington OR, Zinselmeyer BH, Brewer JM, et al. Lymphocyte tracking and interactions in secondary lymphoid organs. *Inflamm Res.* 2007;56:391–401.
22. Rothenberg EV, Moore JE, Yui MA. Launching the T-cell-lineage developmental programme. *Nature Rev Immunol.* 2008;8:9–21.
23. Blum KS and Pabst R. Keystones in lymph node development. *J Anat.* 2006;209:585–595.
24. Willard-Mack CL. Normal structure, function, and histology of lymph nodes. *Toxicol Pathol.* 2006;34:409–424.
25. Palmblad J and Papadaki HA. Chronic idiopathic neutropenias and severe congenital neutropenia. *Curr Opin Hematol.* 2008; 15:8–14.
26. Grenda DS, Murakami M, Ghatak J, et al. Mutations of the ELA2 gene found in patients with severe congenital neutropenia induce the unfolded protein response and cellular apoptosis. *Blood.* 2007;110:4179–4187.
27. Kollner I, Sodeik B, Schreek S, et al. Mutations in neutrophil elastase causing congenital neutropenia lead to cytoplasmic protein accumulation and induction of the unfolded protein response. *Blood.* 2006;108:493–500.
28. Skokowa J, Cario G, Uenalan M, et al. LEF-1 is crucial for neutrophil granulocytopoiesis and its expression is severely reduced in congenital neutropenia. *Nature Med.* 2006;12:1191–1197.
29. Klein C, Grudzien M, Appaswamy G, et al. HAX1 deficiency causes autosomal recessive severe congenital neutropenia (Kostmann disease). *Nature Genet.* 2007;39:86–92.
30. Ma X, Does M, Raza A, et al. Myelodysplastic syndromes: Incidence and survival in the United States. *Cancer.* 2007;109: 1536–1542.
31. Bennett JM, Catovsky D, Daniel MT, et al. Proposals for the classification of the myelodysplastic syndromes. *Br J Haematol.* 1982;51:189–199.
32. Harris NL, Jaffe ES, Diebold J, et al. The World Health Organization classification of neoplastic diseases of the hematopoietic and lymphoid tissues. Report of the Clinical Advisory Committee meeting, Airlie House, Virginia, November, 1997. *Ann Oncol.* 1999;10:1419–1432.
33. Van Etten RA and Shannon KM. Focus on myeloproliferative diseases and myelodysplastic syndromes. *Cancer Cell.* 2004;6:547–552.
34. Harada H, Harada Y, Niimi H, et al. High incidence of somatic mutations in the AML1/RUNX1 gene in myelodysplastic syndrome and low blast percentage myeloid leukemia with myelodysplasia. *Blood.* 2004;103:2316–2324.
35. Szpurka H, Tiu R, Murugesan G, et al. Refractory anemia with ringed sideroblasts associated with marked thrombocytosis (RARS-T), another myeloproliferative condition characterized by JAK2 V617F mutation. *Blood.* 2006;108:2173–2181.
36. Wang SA, Hasserjian RP, Loew JM, et al. Refractory anemia with ringed sideroblasts associated with marked thrombocytosis harbors JAK2 mutation and shows overlapping myeloproliferative and myelodysplastic features. *Leukemia.* 2006;20:1641–1644.
37. Onida F and Beran M. Chronic myelomonocytic leukemia: Myeloproliferative variant. *Curr Hematol Rep.* 2004;3:218–226.
38. Onida F, Kantarjian HM, Smith TL, et al. Prognostic factors and scoring systems in chronic myelomonocytic leukemia: A retrospective analysis of 213 patients. *Blood.* 2002;99:840–849.
39. Tefferi A and Gilliland DG. Oncogenes in myeloproliferative disorders. *Cell Cycle.* 2007;6:550–566.
40. Niemeyer CM, Kratz CP, Hasle H. Pediatric myelodysplastic syndromes. *Curr Treat Options Oncol.* 2005;6:209–214.
41. Emanuel PD, Snyder RC, Wiley T, et al. Inhibition of juvenile myelomonocytic leukemia cell growth in vitro by farnesyltransferase inhibitors. *Blood.* 2000;95:639–645.
42. Side LE, Emanuel PD, Taylor B, et al. Mutations of the NF1 gene in children with juvenile myelomonocytic leukemia without clinical evidence of neurofibromatosis, type 1. *Blood.* 1998;92: 267–272.
43. Tartaglia M, Niemeyer CM, Fragale A, et al. Somatic mutations in PTPN11 in juvenile myelomonocytic leukemia, myelodysplastic syndromes and acute myeloid leukemia. *Nature Genet.* 2003; 34:148–150.
44. Bollag G, Clapp DW, Shih S, et al. Loss of NF1 results in activation of the Ras signaling pathway and leads to aberrant growth in haematopoietic cells. *Nature Genet.* 1996;12:144–148.

45. Vardiman JW, Imbert M, Pierre R, et al. Atypical chronic myeloid leukemia. In: Jaffe ES, Harris NL, Stein H, and Vardiman JW, eds. *World Health Organization Classification of Tumours: Pathology and Genetics of Tumours of Haematopoietic and Lymphoid Tissues*. Lyon, France: IARC Press; 2001.
46. Sawyers CL. Chronic myeloid leukemia. *N Engl J Med*. 1999;340: 1330–1340.
47. Cortes JE, Talpaz M, O'Brien S, et al. Staging of chronic myeloid leukemia in the imatinib era: An evaluation of the World Health Organization proposal. *Cancer*. 2006;106:1306–1315.
48. Hermans A, Heisterkamp N, von Linden M, et al. Unique fusion of bcr and c-abl genes in Philadelphia chromosome positive acute lymphoblastic leukemia. *Cell*. 1987;51:33–40.
49. Pendergast AM, Muller AJ, Havlik MH, et al. BCR sequences essential for transformation by the BCR-ABL oncogene bind to the ABL SH2 regulatory domain in a non-phosphotyrosine-dependent manner. *Cell*. 1991;66:161–171.
50. McWhirter JR and Wang JY. Activation of tyrosinase kinase and microfilament-binding functions of c-abl by bcr sequences in bcr/abl fusion proteins. *Mol Cell Biol*. 1991;11:1553–1565.
51. Weisberg E, Manley PW, Cowan-Jacob SW, et al. Second generation inhibitors of BCR-ABL for the treatment of imatinib-resistant chronic myeloid leukaemia. *Nat Rev Cancer*. 2007;7:345–356.
52. Kantarjian HM, Talpaz M, O'Brien S, et al. Survival benefit with imatinib mesylate versus interferon-alpha-based regimens in newly diagnosed chronic-phase chronic myelogenous leukemia. *Blood*. 2006;108:1835–1840.
53. Imbert M, Bain B, Pierre R, et al. Chronic neutrophilic leukemia. In: Jaffe ES, Harris NL, Stein H, and Vardiman JW, eds. *World Health Organization Classification of Tumours: Pathology and Genetics of Tumours of Haematopoietic and Lymphoid Tissues*. IARC Press, Lyons, France; 2001:27–28.
54. Rane SG and Reddy EP. JAKs, STATs and Src kinases in hematopoiesis. *Oncogene*. 2002;21:3334–3358.
55. Benekli M, Baer MR, Baumann H, et al. Signal transducer and activator of transcription proteins in leukemias. *Blood*. 2003; 101:2940–2954.
56. Silver RT. Polycythemia vera and other polycythemia syndromes. In: Ansell S, ed. *Rare Hematological Malignancies*. London, Springer; 2008:1–27. Cancer Treatment and Research Series; Vol. 142.
57. Vainchenker W and Constantinescu SN. A unique activating mutation in JAK2 (V617F) is at the origin of polycythemia vera and allows a new classification of myeloproliferative diseases. *Hematology Am Soc Hematol Educ Program*. 2005;195–200.
58. James C, Ugo V, Le Couedic JP, et al. A unique clonal JAK2 mutation leading to constitutive signalling causes polycythaemia vera. *Nature*. 2005;434:1144–1148.
59. Levine RL, Wadleigh M, Cools J, et al. Activating mutation in the tyrosine kinase JAK2 in polycythemia vera, essential thrombocythemia, and myeloid metaplasia with myelofibrosis. *Cancer Cell*. 2005;7:387–397.
60. Lindauer K, Loerting T, Liedl KR, et al. Prediction of the structure of human Janus kinase 2 (JAK2) comprising the two carboxy-terminal domains reveals a mechanism for autoregulation. *Protein Eng*. 2001;14:27–37.
61. Hookham MB, Elliott J, Suessmuth Y, et al. The myeloproliferative disorder-associated JAK2 V617F mutant escapes negative regulation by suppressor of cytokine signaling 3. *Blood*. 2007;109: 4924–4929.
62. Scott LM, Tong W, Levine RL, et al. JAK2 exon 12 mutations in polycythemia vera and idiopathic erythrocytosis. *N Engl J Med*. 2007;356:459–468.
63. Finazzi G. Essential thrombocythemia. In: Ansell S, ed. *Rare Hematological Malignancies*. London: Springer; 2008:51–68. Cancer Treatment and Research; vol. 142.
64. Lichtman MA. Idiopathic myelofibrosis. In: Lichtman MA, Beutler E, and Kipps TJ, eds. *Williams Hematology*, 7th ed. McGraw-Hill Book Co.; 2006:1295–1313.
65. Kaushansky K. Historical review: Megakaryopoiesis and thrombopoiesis. *Blood*. 2008;111:981–986.
66. Matsumura I, Horikawa Y, Kanakura Y. Functional roles of thrombopoietin-c-mpl system in essential thrombocythemia. *Leuk Lymphoma*. 1999;32:351–358.
67. Moliterno AR, Hankins WD, Spivak JL. Impaired expression of the thrombopoietin receptor by platelets from patients with polycythemia vera. *N Engl J Med*. 1998;338:572–580.
68. Moliterno AR and Spivak JL. Posttranslational processing of the thrombopoietin receptor is impaired in polycythemia vera. *Blood*. 1999;94:2555–2561.
69. Li J, Xia Y, Kuter DJ. The platelet thrombopoietin receptor number and function are markedly decreased in patients with essential thrombocythaemia. *Br J Haematol*. 2000;111:943–953.
70. Bain B, Pierre P, Imbert M, et al. Chronic eosinophilic leukemia and the hypersinophillic syndrome. In: Jaffe ES, Harris NL, Stein H, and Vardiman JW, eds. *World Health Organization Classification of Tumours: Pathology and Genetics of Tumours of Haematopoietic and Lymphoid Tissues*. IARC Press, Lyons, France; 2001: 29–31.
71. Cools J, DeAngelo DJ, Gotlib J, et al. A tyrosine kinase created by fusion of the PDGFRA and FIP1L1 genes as a therapeutic target of imatinib in idiopathic hypereosinophilic syndrome. *N Engl J Med*. 2003;348:1201–1214.
72. Stover EH, Chen J, Folens C, et al. Activation of FIP1L1–PDGFRalpha requires disruption of the juxtamembrane domain of PDGFRalpha and is FIP1L1-independent. *Proc Natl Acad Sci USA*. 2006;103:8078–8083.
73. Pardanani A, Brockman SR, Paternoster SF, et al. FIP1L1–PDGFRA fusion: Prevalence and clinicopathologic correlates in 89 consecutive patients with moderate to severe eosinophilia. *Blood*. 2004;104:3038–3045.
74. David M, Cross NC, Burgstaller S, et al. Durable responses to imatinib in patients with PDGFRB fusion gene-positive and BCR-ABL-negative chronic myeloproliferative disorders. *Blood*. 2007;109:61–64.
75. Pardanani AD, Morice WG, Hoyer JD, et al. Chronic basophilic leukemia: A distinct clinico-pathologic entity? *Eur J Haematol*. 2003;71:18–22.
76. Valent P. Systemic mastocytosis. *Cancer Treat Res*. 2008;142:399–419.
77. Tafferi A and Gilliland DG. Classification of chronic myeloid disorders: From Dameshek towards a semi-molecular system. *Best Pract Res Clin Haematol*. 2005;9:365–385.
78. Hu ZQ, Zhao WH, Shimamura T. Regulation of mast cell development by inflammatory factors. *Curr Med Chem*. 2007;14: 3044–3050.
79. Edling CE and Hallberg B. c-Kit—A hematopoietic cell essential receptor tyrosine kinase. *Int J Biochem Cell Biol*. 2007;39: 1995–1998.
80. Shivakrupa R, Bernstein A, Watring N, et al. Phosphatidylinositol 3′-kinase is required for growth of mast cells expressing the kit catalytic domain mutant. *Cancer Res*. 2003;63: 4412–4419.
81. Tanaka A, Konno M, Muto S, et al. A novel NF-kappaB inhibitor, IMD-0354, suppresses neoplastic proliferation of human mast cells with constitutively activated c-kit receptors. *Blood*. 2005; 105:2324–2331.
82. Inhorn RC, Aster JC, Roach SA, et al. A syndrome of lymphoblastic lymphoma, eosinophilia, and myeloid hyperplasia/ malignancy associated with t(8;13)(p11;q11): Description of a distinctive clinicopathologic entity. *Blood*. 1995;85:1881–1887.
83. Schlessinger J. Common and distinct elements in cellular signaling via EGF and FGF receptors. *Science*. 2004;306:1506–1507.
84. Bennett JM, Catovsky D, Daniel MT, et al. Proposed revised criteria for the classification of acute myeloid leukemia. A report of the French-American-British Cooperative Group. *Ann Intern Med*. 1985;103:620–625.
85. Brunning RD, Matute E, Harris NL, et al. Acute myeloid leukemia with multilineage dysplasia. In: Jaffe ES, Harris NL, Stein H, and Vardiman JW, eds. *World Health Organization Classification of Tumours: Pathology and Genetics of Tumours of Haematopoietic and Lymphoid Tissues*, vol. 8. IARC Press, Lyon, France; 2001: 8–89.
86. Peterson LF and Zhang DE. The 8;21 translocation in leukemogenesis. *Oncogene*. 2004;23:4255–4262.
87. Licht JD. AML1 and the AML1-ETO fusion protein in the pathogenesis of t(8;21) AML. *Oncogene*. 2001;20:5660–5679.
88. Peterson LF, Boyapati A, Ahn EY, et al. Acute myeloid leukemia with the 8q22;21q22 translocation: Secondary mutational events and alternative t(8;21) transcripts. *Blood*. 2007;110:799–805.

89. Elagib KE and Goldfarb AN. Oncogenic pathways of AML1-ETO in acute myeloid leukemia: Multifaceted manipulation of marrow maturation. *Cancer Lett.* 2007;251:179–186.
90. Shigesada K, van de SB, Liu PP. Mechanism of leukemogenesis by the inv(16) chimeric gene CBFB/PEBP2B-MHY11. *Oncogene.* 2004;23:4297–4307.
91. Reilly JT. Pathogenesis of acute myeloid leukaemia and inv(16)(p13;q22): A paradigm for understanding leukaemogenesis? *Br J Haematol.* 2005;128:18–34.
92. Yang Y, Wang W, Cleaves R, et al. Acceleration of G(1) cooperates with core binding factor beta-smooth muscle myosin heavy chain to induce acute leukemia in mice. *Cancer Res.* 2002;62:2232–2235.
93. Moreno-Miralles I, Pan L, Keates-Baleeiro J, et al. The inv(16) cooperates with ARF haploinsufficiency to induce acute myeloid leukemia. *J Biol Chem.* 2005;280:40097–40103.
94. Kim HG, Kojima K, Swindle CS, et al. FLT3-ITD cooperates with inv(16) to promote progression to acute myeloid leukemia. *Blood.* 2008;111:1567–1574.
95. Boissel N, Leroy H, Brethon B, et al. Incidence and prognostic impact of c-Kit, FLT3, and Ras gene mutations in core binding factor acute myeloid leukemia (CBF-AML). *Leukemia.* 2006; 20:965–970.
96. Vitoux D, Nasr R, de TH. Acute promyelocytic leukemia: New issues on pathogenesis and treatment response. *Int J Biochem Cell Biol.* 2007;39:1063–1070.
97. Wang ZY and Chen Z. Acute promyelocytic leukemia: From highly fatal to highly curable. *Blood.* 2008;111:2505–2515.
98. Krivtsov AV and Armstrong SA. MLL translocations, histone modifications and leukaemia stem-cell development. *Nat Rev Cancer.* 2007;7:823–833.
99. Yanada M, Matsuo K, Suzuki T, et al. Prognostic significance of FLT3 internal tandem duplication and tyrosine kinase domain mutations for acute myeloid leukemia: A meta-analysis. *Leukemia.* 2005;19:1345–1349.
100. Renneville A, Roumier C, Biggio V, et al. Cooperating gene mutations in acute myeloid leukemia: A review of the literature. *Leukemia.* 2008;22:915–931.
101. Leith CP, Kopecky KJ, Godwin J, et al. Acute myeloid leukemia in the elderly: Assessment of multidrug resistance (MDR1) and cytogenetics distinguishes biologic subgroups with remarkably distinct responses to standard chemotherapy. A Southwest Oncology Group study. *Blood.* 1997;89:3323–3329.
102. Smith SM, Le Beau MM, Huo D, et al. Clinical-cytogenetic associations in 306 patients with therapy-related myelodysplasia and myeloid leukemia: The University of Chicago series. *Blood.* 2003;102:43–52.
103. Brunning RD, Matutes E, Harris NL, et al. Acute myeloid leukemias and myelodysplastic syndromes, therapy related. In: Jaffe ES, Harris NL, Stein H, and Vardiman JW, eds. *World Health Organization Classification of Tumours: Pathology and Genetics of Tumours of Haematopoietic and Lymphoid Tissues.* IARC Press, Lyons, France; 2001:89–91.
104. Pedersen-Bjergaard J, Andersen MK, Andersen MT, et al. Genetics of therapy-related myelodysplasia and acute myeloid leukemia. *Leukemia.* 2008;22:240–248.
105. Felix CA, Kolaris CP, Osheroff N. Topoisomerase II and the etiology of chromosomal translocations. *DNA Repair (Amst).* 2006; 5:1093–1108.
106. Brunning RD, Matutes E, Harris NL, et al. Acute myeloid leukemia not otherwise categorised. In: Jaffe ES, Harris NL, Stein H, and Vardiman JW, eds. *World Health Organization Classification of Tumours: Pathology and Genetics of Tumours of Haematopoietic and Lymphoid Tissues.* IARC Press, Lyons, France; 2001:91–105.
107. Brunning RD, Matutes E, Borowitz M, et al. Acute leukemias of ambiguous lineage. In: Jaffe ES, Harris NL, Stein H, and Vardiman JW, eds. *World Health Organization Classification of Tumours: Pathology and Genetics of Tumours of Haematopoietic and Lymphoid Tissues.* IARC Press, Lyons, France; 2001:106–107.
108. Callan MF, Steven N, Krausa P, et al. Large clonal expansions of CD8+ T-cells in acute infectious mononucleosis. *Nat Med.* 1996;2:906–911.
109. Knapp W, Rieber P, Dorken B, et al. Towards a better definition of human leucocyte surface molecules. *Immunol Today.* 1989; 10:253–258.
110. O'Neil J and Look AT. Mechanisms of transcription factor deregulation in lymphoid cell transformation. *Oncogene.* 2007; 26:6838–6849.
111. Bernard OA, Romana SP, Poirel H, et al. Molecular cytogenetics of t(12;21) (p13;q22). *Leuk Lymphoma.* 1996;23: 459–465.
112. Mullighan CG, Goorha S, Radtke I, et al. Genome-wide analysis of genetic alterations in acute lymphoblastic leukaemia. *Nature.* 2007;446:758–764.
113. Nutt SL, Eberhard D, Horcher M, et al. Pax5 determines the identity of B-cells from the beginning to the end of B-lymphopoiesis. *Int Rev Immunol.* 2001;20:65–82.
114. Allman D, Aster JC, Pear WS. Notch signaling in hematopoiesis and early lymphocyte development. *Immunol Rev.* 2002;187: 75–86.
115. Weng AP, Ferrando AA, Lee W, et al. Activating mutations of NOTCH1 in human T-cell acute lymphoblastic leukemia. *Science.* 2004;306:269–271.
116. Hallek M, Cheson BD, Catovsky D, et al. Guidelines for the diagnosis and treatment of chronic lymphocytic leukemia: A report from the International Workshop on Chronic Lymphocytic Leukemia (IWCLL) updating the National Cancer Institute-Working Group (NCI-WG) 1996 guidelines. *Blood.* 2008; 111:5446–5456.
117. Dohner H, Stilgenbauer S, Benner A, et al. Genomic aberrations and survival in chronic lymphocytic leukemia. *N Engl J Med.* 2000;343:1910–1916.
118. Raval A, Tanner SM, Byrd JC, et al. Downregulation of death-associated protein kinase 1 (DAPK1) in chronic lymphocytic leukemia. *Cell.* 2007;129:879–890.
119. Dighiero G and Hamblin TJ. Chronic lymphocytic leukaemia. *Lancet.* 2008;371:1017–1029.
120. Vuillier F, Dumas G, Magnac C, et al. Lower levels of surface B-cell-receptor expression in chronic lymphocytic leukemia are associated with glycosylation and folding defects of the mu and CD79a chains. *Blood.* 2005;105:2933–2940.
121. Cleary ML and Sklar J. Nucleotide sequence of a t(14;18) chromosomal breakpoint in follicular lymphoma and demonstration of a breakpoint-cluster region near a transcriptionally active locus on chromosome 18. *Proc Natl Acad Sci USA.* 1985; 82:7439–7443.
122. Tsujimoto Y, Finger LR, Yunis J, et al. Cloning of the chromosome breakpoint of neoplastic B-cells with the t(14;18) chromosome translocation. *Science.* 1984;226: 1097–1099.
123. Bakhshi A, Jensen JP, Goldman P, et al. Cloning the chromosomal breakpoint of t(14;18) human lymphomas: Clustering around JH on chromosome 14 and near a transcriptional unit on 18. *Cell.* 1985;41:899–906.
124. Horsman DE, Connors JM, Pantzar T, et al. Analysis of secondary chromosomal alterations in 165 cases of follicular lymphoma with t(14;18). Genes Chromosomes. *Cancer.* 2001; 30:375–382.
125. de JD. Molecular pathogenesis of follicular lymphoma: A cross talk of genetic and immunologic factors. *J Clin Oncol.* 2005; 23:6358–6363.
126. Abramson JS and Shipp MA. Advances in the biology and therapy of diffuse large B-cell lymphoma: Moving toward a molecularly targeted approach. *Blood.* 2005;106:1164–1174.
127. Phan RT and la-Favera R. The BCL6 proto-oncogene suppresses p53 expression in germinal-centre B-cells. *Nature.* 2004;432: 635–639.
128. Gascoyne RD, Adomat SA, Krajewski S, et al. Prognostic significance of Bcl-2 protein expression and Bcl-2 gene rearrangement in diffuse aggressive non-Hodgkin's lymphoma. *Blood.* 1997;90:244–251.
129. Pasqualucci L, Neumeister P, Goossens T, et al. Hypermutation of multiple proto-oncogenes in B-cell diffuse large-cell lymphomas. *Nature.* 2001;412:341–346.
130. Bellan C, Lazzi S, De FG, et al. Burkitt's lymphoma: New insights into molecular pathogenesis. *J Clin Pathol.* 2003;56: 188–192.
131. Brady G, MacArthur GJ, Farrell PJ. Epstein-Barr virus and Burkitt lymphoma. *J Clin Pathol.* 2007;60:1397–1402.
132. Grogg KL, Miller RF, Dogan A. HIV infection and lymphoma. *J Clin Pathol.* 2007;60:1365–1372.

133. De Falco G, Bellan C, Lazzi S, et al. Interaction between HIV-1 Tat and pRb2/p130: A possible mechanism in the pathogenesis of AIDS-related neoplasms. *Oncogene.* 2003;22:6214–6219.
134. Bergsagel PL and Kuehl WM. Molecular pathogenesis and a consequent classification of multiple myeloma. *J Clin Oncol.* 2005;23:6333–6338.
135. Tonon G. Molecular pathogenesis of multiple myeloma. *Hematol Oncol Clin North Am.* 2007;21:985–1006.
136. Re D, Thomas RK, Behringer K, et al. From Hodgkin disease to Hodgkin lymphoma: Biologic insights and therapeutic potential. *Blood.* 2005;105:4553–4560.
137. Kuppers R, Rajewsky K, Zhao M, et al. Hodgkin disease: Hodgkin and Reed-Sternberg cells picked from histological sections show clonal immunoglobulin gene rearrangements and appear to be derived from B-cells at various stages of development. *Proc Natl Acad Sci USA.* 1994;91:10962–10966.

Chapter

17

Molecular Basis of Diseases of Immunity

David O. Beenhouwer

INTRODUCTION

The immune system protects against infection with microorganisms. This system, which has evolved over millions of years, is remarkable in both its complexity and its effectiveness. Without a functioning immune system, humans cannot survive past the first few months of life. However, as extraordinary this system may be, until the most recent century, infections were by far the most common cause of death among humans as microorganisms kept the average life expectancy to about 25–30 years. The increased life expectancy currently seen reflects progress in control of infectious diseases including improved hygiene, vaccines, and antimicrobial drugs. Casanova and Abel have proposed that, by definition, humans who die of infection are immunodeficient and that recent medical progress has masked widespread inherited defects in immunity [1]. A better understanding of the weaknesses and limitations of the immune system will lead to approaches to assist and improve it.

This chapter covers major syndromes of immune dysfunction. These include diseases of deficient immunity, hyperactive immunity (hypersensitivity), and dysregulated immunity (autoimmune diseases). To understand the pathophysiology of these syndromes, one must have an understanding of how the normal immune system functions. This chapter begins with a brief summary of the important cells and molecules of the immune system. However, a detailed description of the immune system and how it functions is beyond the scope of this chapter. For this, the reader is referred to several excellent textbooks in immunology [2–5].

NORMAL IMMUNE SYSTEM

The immune system is made up of several types of cells that carry out specialized functions. Through a remarkable array of specialized cell surface molecules (receptors), immune cells recognize, respond, and adapt to their environment and to foreign invaders. Immune system responses can be thought of as rapid and preprogrammed (innate), or slower but more flexible (adaptive).

Cells

The white blood cells that make up the immune system originate in the bone marrow. These cells then circulate to the peripheral tissues in the bloodstream. They also travel in the lymphatic system, which is a network of ducts connecting lymph nodes throughout the body. All blood cells derive from hematopoietic stem cells, which first differentiate into either lymphoid or myeloid progenitor cells. Lymphoid progenitors give rise to lymphocytes, NK cells, and dendritic cells. Myeloid progenitors give rise to granulocytes (including neutrophils, eosinophils, mast cells and basophils), monocyte/macrophages, and dendritic cells.

B-Cells

B-cells produce antibodies or immunoglobulins. They carry receptors on their cell surface (B-cell receptor, BCR). The BCR is a membrane-bound version of the antibody molecule the B-cell will secrete when activated. This membrane-bound antibody is associated with other molecules capable of transducing signals to the cell upon ligation of the BCR. Antibodies recognize and bind to portions of soluble molecules called antigens, so called because they stimulate antibody generation. Once activated, B-cells may differentiate into plasma cells, which are specialized for producing massive quantities of antibody.

T-Cells

There are two major classes of T-cells: (i) T helper cells and (ii) cytotoxic T-cells. T helper cells express CD4 on the surface membrane and activate other cells such

Molecular Pathology © 2009, Elsevier, Inc. All Rights Reserved.

as B-cells and macrophages. Cytotoxic T-cells (CTLs) express CD8 on their surface membrane and recognize and kill infected cells. T-cells also express an antigen receptor on their cell surface (T-cell receptor or TCR). These antigen receptors are somewhat different from antibodies in that they recognize antigen only when it is bound to a specific cell surface molecule called major histocompatibility complex (MHC).

NK Cells

NK or natural killer cells are part of both innate and adaptive immune responses. They recognize and destroy abnormal cells, such as cancer cells and infected cells. NK cells express receptors for antibodies (Fc receptors) and are capable of recognizing and destroying antibody-coated cells by a process known as antibody-dependent cell-mediated cytotoxicity (ADCC). When stimulated, NK cells release perforin, which forms pores in the cell membrane. Granzymes are also released, which enter through these pores and can then trigger apoptosis in the target cell.

Dendritic Cells

When immature, dendritic cells reside in many host tissues, particularly the skin and mucosa, where they avidly ingest pathogens in a process called phagocytosis. Upon pathogen uptake, these cells mature and travel to lymphoid tissues including the lymph nodes and the spleen. Inside the dendritic cell, pathogens are digested by various enzymes into fragments. Upon maturation, these cells then act as antigen-presenting cells (APCs), displaying pathogen fragments on their cell surface for lymphocytes to recognize. These cells are the primary link between the innate and adaptive immune system.

Macrophages

Macrophages are phagocytic cells that engulf pathogens and cellular debris and play a major role in the innate immune response. Pathogens are taken up into specialized intracellular vesicles called phagosomes, where they are destroyed. Macrophages can also present antigens to lymphocytes. These cells are distributed widely in many different organs and tissues where they may further differentiate and acquire specialized functions. In the liver, they are known as Kupfer cells, and in the central nervous system (CNS), they are called astrocytes.

Granulocytes

Granulocytes are cells that have densely staining granules in their cytoplasm. The granulocytes include neutrophils, eosinophils, mast cells, and basophils. Neutrophils are the most abundant white blood cell and play an essential role in innate protection against bacterial infection. They are phagocytic cells that contain a vast array of microbicidal molecules. Eosinophils carry Fc receptors for a particular form of antibody (IgE). Upon binding multicellular organisms (parasites) coated with IgE, eosinophils discharge their toxic granules, which contain both destructive enzymes and vasoactive substances. Mast cells trigger local inflammatory response to antigen by releasing granules containing histamine and other vasoactive substances. The function of basophils is not well defined.

Molecules

Complement

The complement system consists of a cascade of serum proteins that are involved in both innate and adaptive immunity. There are three pathways through which complement can be activated: classical, alternative, and lectin (Figure 17.1). Activation of complement results in production of several bioactive molecules. The three pathways converge on C3, which when cleaved forms C3b. C3b can bind covalently to pathogens, where it acts as an opsonin, targeting pathogens for uptake and destruction by phagocytes bearing complement receptors. C5a is a potent mediator of vascular permeability and can also recruit neutrophils and monocytes, in a process known as chemotaxis, to areas of inflammation. The membrane attack complex (MAC) is formed by C5–9 and can form lytic membrane pores in cells or pathogens without cell walls, such as gram negative bacteria. Initiation of the classical pathway involves C1q binding to antibody molecules complexed with antigen, either on a pathogen surface or in the form of soluble antibody-antigen complexes. In a similar fashion, initiation of the lectin binding pathway involves mannose binding lectin (MBL) binding to exposed densely spaced mannose residues on pathogen surfaces. On mammalian cells, most mannose residues are masked by other sugars, especially sialic acid. The alternative pathway is initiated by ongoing spontaneous C3 hydrolysis. C3b binds factor B, which then initiates the alternative pathway. Host cells are protected from destruction by this pathway (and others) by several complement regulatory proteins residing on the cell surface. Two of these proteins, protectin (CD59) and decay acceleration factor (DAF), are attached to the cell surface via a glycosylphosphatidyl-inositol (GPI) anchor. A somatic mutation in hematopoietic cells of a gene (*PIG-A*) on the X chromosome that codes for an enzyme involved in GPI synthesis leads to dysfunction in DAF and CD59. This causes the disease paroxysmal nocturnal hemoglobinuria, which is characterized by episodes of hemolysis and thrombosis.

Antibodies

Antibodies (immunoglobulins) are a family of glycoproteins that are produced by B-cells and plasma cells. Antibody molecules are composed of two identical light (L) chain polypeptide chains and two identical heavy (H) polypeptide chains linked together by disulfide bonds (Figure 17.2). The majority of the molecule

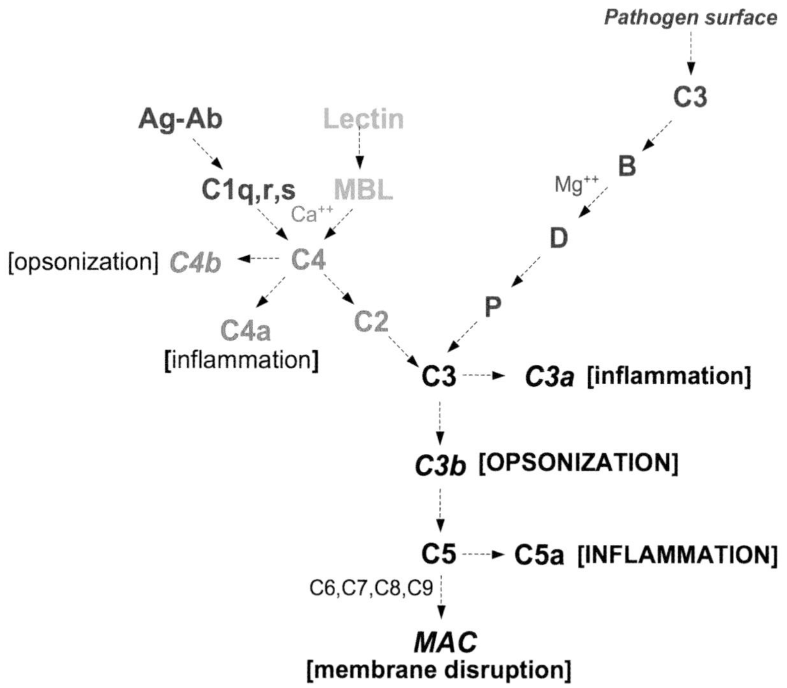

Figure 17.1 **Complement pathways.** The three pathways of complement activation are shown. On the right in red is the alternative pathway, initiated by C3 hydrolysis on pathogen surfaces. In the center in green is the lectin pathway, initiated by MBL binding to exposed mannose residues on pathogen surfaces. On the left, depicted in blue, is the classical pathway, which is initiated upon binding of C1q to bound antibody. Calcium ions are essential for the classical pathway and magnesium ions are essential for the alternative pathway. All three pathways converge at C3. Several mediators of inflammation and opsonization are released upon cleavage of C4, C3, and C5. The membrane attack complex (MAC) forms lytic pores in membranes.

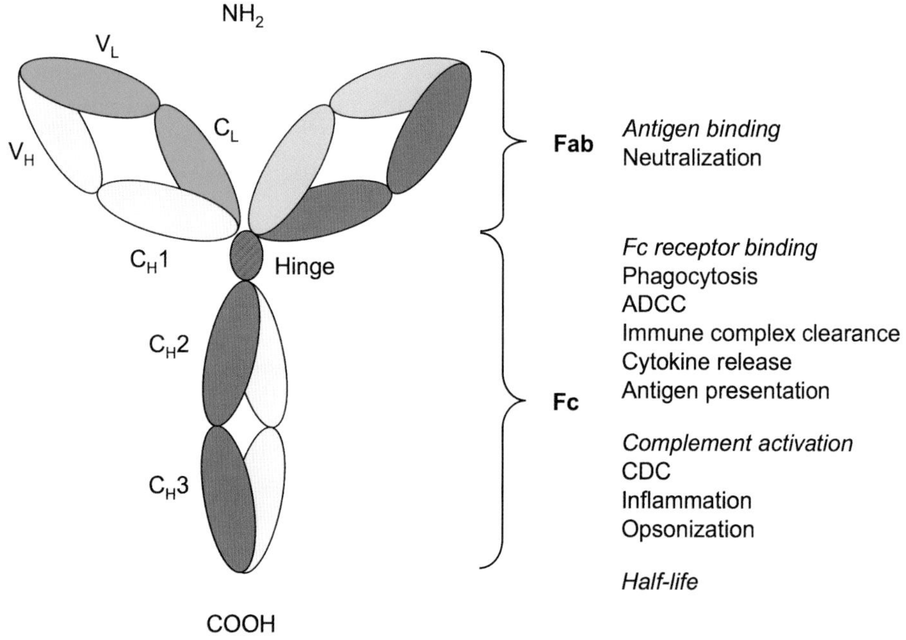

Figure 17.2 **Antibody structure.** Antibody molecules are composed of two light (L) chains (green) and two heavy (H) chains (gray) linked together by disulfide bonds. The L and H chains fold into functional domains, two for the L chain and four or five for the H chain. The amino terminal (NH_2) domain from each chain forms the variable regions, V_L and V_H , which together constitute the antigen binding site. The rest of the molecule has a relatively constant (C) structure. The V regions with the C region of the L chain and the first C region domain of the H chain (C_H1) constitute an Fab region. The remaining C region domains of the H chain (C_H2, C_H3, and in some cases C_H4) constitute the Fc region, which determines the biologic properties of antibodies as indicated. Antibodies with different H chains or isotypes (IgM, IgG, IgA, and IgE) have different effector functions.

consists of one of five protein sequences known as the constant region. However, the amino terminal end is made up of an apparently infinite variety of sequences known as the variable region. The variable (V) region constitutes the antigen binding site. This region is separated from the rest of the molecule by a flexible hinge. The carboxyl terminal portion of the antibody is called the Fc region and can bind to Fc receptors, activate complement, and mediate antibody half-life.

The V region of the heavy chain is composed of three segments: V_H, D_H, J_H. The light chain V region is composed of two segments: V_L and J_L. The extraordinary diversity of the V region is generated during B-cell development when gene segments irreversibly rearrange in a process called VDJ or VJ recombination. In human germline DNA, there are 40 V_H segments, 25 D_H segments, and 6 J_H segments. During VDJ recombination, a single V_H segment first rearranges to a single D_H segment, followed by a second rearrangement to a single J_H segment. A similar process occurs with VJ recombination for the light chain. There are about 6000 different VDJ combinations and 320 different VJ combinations generating a possible 2 million different antibody V regions with different specificities. Further processes, including nucleotide addition/subtraction, enhance this already impressive diversity. Gene recombination is irreversible and, once complete, the BCR/antibody is structurally defined. Upon stimulation with T-cells, B-cells can induce a process called somatic hypermutation, which introduces random mutations in V region genes. This process leads to fine-tuning of the antigen binding site, thereby increasing antigen affinity.

Antibody effector function is determined by the heavy chain or isotype (IgM, IgG, IgA, or IgE). Isotypes have different properties (Table 17.1) including the ability to mediate phagocytosis (IgG), ADCC (IgG and IgE), and complement activation (IgG and IgM). IgA is particularly resistant to proteolysis and is important in host defense at mucosal surfaces. IgE binds to high affinity Fc receptors (FcεRI) on mast cells and eosinophils inducing degranulation. A circulating Fc receptor known as the neonatal Fc receptor (FcRn) binds IgG and contributes to its remarkably long half-life (up to 3 weeks).

T-Cell Receptors

The TCR has a similar structure to the antibody Fab region (Figure 17.1) and consists of two transmembrane glycoprotein chains, TCRα and TCRβ. The generation of diversity of the TCR is very similar to that for antibodies. In contrast to antibodies, TCRs bind to peptides in complex with major histocompatibility complex (MHC) displayed on the surface of APCs. The two different classes of T-cells each bind to different classes of MHC. $CD4^+$ T-cells interact with antigen bound to MHC class II, and $CD8^+$ T-cells interact with antigen presented in the context of MHC class I.

Major Histocompatibility Complex

All nucleated cells express major histocompatibility complex (MHC) class I. Cells infected internally (for instance, in viral infection) will process pathogen peptides and present them on the surface in complex with MHC class I. There they can be recognized by $CD8^+$ T-cells, which are specialized to kill cells recognized in this fashion via apoptosis. MHC class II is expressed by APCs, including dendritic cells, macrophages, and B-cells. $CD4^+$ T-cells recognize peptides processed by APCs and displayed on their surface complexed with MHC class II. Structurally, MHC I is composed of a transmembrane polymorphic α chain that associates noncovalently with a smaller invariant chain called β2-microglobulin. MHC II molecules are composed of two polymorphic transmembrane chains, α and β, that associate noncovalently. MHC class I and II genes are also known as human leukocyte antigen (HLA) genes. There are three class I α-chain genes: HLA-A, -B, and -C. There are three pairs of MHC class II α-chain and β-chain genes: HLA-DR, HLA-DP, and HLA-DQ. Each MHC protein binds a different range of peptides. HLA genes are highly polymorphic. This leads to population-based immunologic diversity. As described in following sections, several autoimmune diseases are linked to specific HLA alleles.

Cytokines

Cytokines are proteins secreted by cells that interact with specific receptors on other cells, thereby affecting function. Macrophages produce a wide array of these molecules, including proinflammatory cytokines, like interleukin (IL) 1β, IL-6, and tumor necrosis factor (TNF), the Th1 cytokine IL-12, and the anti-inflammatory cytokine IL-10. T-cells also produce several cytokines, including those that characterize Th1 (IL-2 and IFN-γ) and Th2 (IL-4, IL-5, and IL-13) responses.

Table 17.1 Properties of Antibody Isotypes

Isotype	Antigen Binding Sites[1]	Complement Activation	Cellular (Fc) Receptors	Half-Life
IgM	10 or 12	yes	1	Medium
IgA	2, 4 or 6	no	1	Short
IgG	2	yes or no	4	Very long
IgE	2	no	2	Very short

[1]IgM and IgA form multimers of 2, 3, 5, or 6 molecules.

Chemokines

Chemokines are similar to cytokines and act specifically to attract cells bearing specific receptors. Chemokines and their receptors have recently undergone a renaming process and frequently have two or more names in current use. The CC chemokines have two adjacent cysteines near the amino terminus, while the CXC chemokines have another amino acid separating the two cysteines. CC chemokines bind to CC receptors (CCRs) and CXC chemokines bind to CXC receptors (CXCRs). Some important chemokines include CXCL-8 (IL-8), which attracts neutrophils bearing CXCR1 and 2; CCL-2 (MCP-1), CCL-3 (MIP-1α), and CCL-5 (RANTES), which attract monocytes and macrophages; and CCL-11 (eotaxin-1), which attracts eosinophils bearing CCR3.

Toll-Like Receptors

Microorganisms produce several molecules that can be immediately recognized as foreign by their general repetitive structural qualities not shared by host molecules, known as pathogen-associated molecular patterns (PAMPs). For example, viruses often express double-stranded RNA (dsRNA), bacteria express unmethylated repeats of the dinucleotide CpG, gram negative bacteria express lipopolysaccharide (LPS) on their surface, and some bacteria possess flagella. The innate immune system recognizes these PAMPs through pattern recognition receptors (PRRs). There are several PRRs expressed in the human immune system, including MBL and toll-like receptors (TLRs). Humans make 10 TLRs, which are expressed on macrophages, dendritic cells, and other cells, and recognize various PAMPs. Some PAMPs recognized by specific TLRs include dsRNA (TLR-3), LPS (TLR-4), flagellin (TLR-5), and CpG DNA (TLR-9). Ligation of TLRs triggers a rapid preprogrammed response (innate immune response). At the same time, the more slowly reactive adaptive immune response is initiated, often in a particular fashion depending on which TLRs were activated. Therefore, while TLRs play a primary role in the innate response, they are also responsible for initiating and shaping adaptive immune responses.

Immune Responses

Innate Immune Responses

The immune system must be capable of responding immediately to a foreign invader. This response, mediated by soluble and cell-attached PRRs, is called an innate immune response. Important cells of the innate immune response include macrophages, NK cells, and granulocytes. Soluble molecules such as complement also play important roles. Recognition of a foreign invader leads to release of cytokines, in particular TNF, as well as chemokines. This leads to infiltration of the infection site with neutrophils, macrophages, and dendritic cells as well as lymphocytes. Most breaches in host defenses are controlled by the innate immune response.

Adaptive Immune Responses

The adaptive immune response primarily involves lymphocytes with their wide diversity of antigen receptors. The adaptive immune response continues to be adjusted during infection and is set in motion following activation of the innate response. APCs, particularly dendritic cells, ingest pathogens and associated antigens and travel from the site of infection to lymph nodes where they encounter dense collections of T-cells and B-cells. Dendritic cells process antigens and present them complexed with MHC class II. T-cells that bind these peptide-MHC complexes are activated to proliferate and produce cytokines that further direct immune responses. This process leads to clonal amplification of pathogen-specific T-cells. Specialized APCs, called follicular dendritic cells (FDCs), do not process antigens into fragments, but carry larger fragments and even entire pathogens on their cell surface, where they can be presented to B-cells, which are then clonally amplified as well. B-cells can also process antigen and present it to T-cells in the context of MHC II. T-cells recognizing this antigen can then provide signals to B-cells leading to maturation of the antibody response including isotype switching and somatic hypermutation.

An important caveat of lymphocyte activation is that it requires two signals: (i) antigen binding, which signals through the antigen receptor, and (ii) a second signal provided by another cell. For T-cells, this second signal is provided by APCs. For B-cells, the second signal is usually provided by activated T-cells. The second signal is delivered through the interaction of certain cell surface molecules, such as CD40–CD40 ligand and CD28–B7, known as co-stimulatory molecules. If lymphocytes receive signaling only through the antigen receptor without co-stimulation, the lymphocyte either becomes anergic (immunologically inactive) or dies via apoptosis. This process ensures that the powerful adaptive immune system is only activated by foreign pathogens initiating an inflammatory response, and not to inert or host antigens.

Another important feature of the adaptive immune response is the development of antigenic memory or recall. The initial adaptive response to a pathogen, known as the primary response, usually takes 7–10 days to reach full effectiveness. In a process that is still rather unclear, during the primary (and subsequent) response, some clonally amplified lymphocytes, known as memory cells, are produced that remain circulating in the host for many years. Upon re-exposure to the same pathogen, these cells are then rapidly stimulated, leading to quick and efficient eradication. This response is also known as a secondary response and requires only a few days to reach maximum intensity and efficacy. For most responses, memory does fade over time. This is why many vaccines require booster immunizations to remain effective.

In the past, there has been a tendency to think about the adaptive immune response as two separate entities—humoral immunity (antibody) and cell-mediated immunity (T-cell and APC). It is now clear that effective responses to most infections require both types of responses and that this dichotomy is largely a didactic one. For example, there is now significant evidence that antibodies play an important role in antigen presentation to T-cells.

MAJOR SYNDROMES

Now that the many important cells and molecules involved in immunity have been introduced, we will turn to a discussion of pathological states of the immune system. One such state is the hypersensitivity reaction, in which innocuous environmental antigens induce an immune response leading to destructive inflammation. Another pathological state is immunodeficiency, in which specific defects in the immune system render it unable to control infection. Finally, autoimmunity arises when the immune system mistakes host tissue as foreign and causes damage.

Hypersensitivity Reactions

In certain cases, harmless environmental antigens may induce an adaptive immune response, and upon re-exposure to the same antigen, an inflammatory state ensues. These antigens are referred to as allergens, and the hypersensitivity responses to them are also known as allergic reactions. While the development of an allergic reaction is divided for instructive purposes into two phases, the sensitization phase and the effector phase, these often occur practically simultaneously. Allergic reactions may range from mildly itchy skin to significant illnesses such as asthma and to life-threatening situations such as anaphylactic shock. Gell and Coombs divided hypersensitivity reactions into four main types for didactic purposes (Table 17.2) [6], although distinguishing these practically may be difficult, as antibody-mediated and cell-mediated immune responses may overlap or occur simultaneously. Also, this classification system tends to maintain the humoral versus cellular dichotomy, which has very little place in modern immunology. More recently, a clinical classification of hypersensitivity has been proposed by the *European Academy of Allergology and Clinical Immunology* (EAACI) [7]. Rajan has further proposed that looking at these phenomena as hypersensitivity reactions tends to overlook the reason they exist in the first place, namely to protect the host against infection [8]. Keeping these limitations in mind, this discussion of the hypersensitivity reactions will utilize the classical scheme proposed by Gell and Coombs [6].

Type I or Immediate Hypersensitivity

Immediate hypersensitivity is classically mediated by IgE and may occur as a local (such as asthma) or systemic (anaphylaxis) reaction. The initial phase occurring within minutes following exposure to allergen is characterized by vasodilation, vascular leakage, smooth muscle spasm, and glandular secretions. A late phase reaction, lasting several days may then occur and is characterized by infiltration of tissues with eosinophils and $CD4^+$ T-cells, as well as tissue destruction.

During the sensitization phase of a type I response, allergen is presented to T-cells by APCs, and signals are generated that cause differentiation of naive $CD4^+$ T-cells into Th2 cells. IL-4 and IL-13 secreted by Th2 cells stimulate B-cells to produce IgE. Allergen-specific IgE then binds to high-affinity FcεRI on mast cells and basophils. During the effector phase, the allergen binds to cytophilic IgE cross-linking FcεRs and leading to activation of mast cells and basophils. This is followed by immediate mast cell degranulation with release of preformed mediators such as histamine and proteolytic enzymes, which cause smooth muscle contraction, increase vascular permeability, and break down tissue matrix proteins. At the same time, there is induction of synthesis of chemokines, cytokines, leukotrienes, and prostaglandins. IL-5 promotes eosinophil production, activation, and chemotaxis. IL-4 and IL-13 promote and amplify Th2 responses. TNF, platelet activating factor (PAF), and macrophage inflammatory protein (MIP-1) induce efflux of effector leukocytes from the bloodstream into tissues. Leukotriene B4 is chemotactic for eosinophils, neutrophils, and monocytes. Leukotrienes C4 and D4 are potent (>1000-fold more active than histamine) mediators of vascular permeability and smooth muscle contraction. Prostaglandin D2 causes increased mucous secretion.

While mast cells initiate the Type I hypersensitivity response, eosinophils play a major role in the Type I

Table 17.2 Hypersensitivity Reactions

Type	Antigen	Mediator	Effector Mechanism	Examples
I Immediate	Soluble antigen, allergen	IgE	FcεRI activation of mast cells	Asthma, anaphylaxis
II Antibody-mediated	Cell-associated antigen	IgM, IgG	FcγR bearing cells, complement	Drug reaction
III Immune-complex	Soluble antigen	IgG	FcγR bearing cells, complement	Serum sickness, Arthus reaction
IV Delayed-type hypersensitivity	Soluble antigen, cell-associated antigen	T-cells	Macrophage activation, direct cytotoxicity	Tuberculin reaction, contact dermatitis

hypersensitivity response, particularly in the late phase. Eosinophils produce major basic protein and eosinophil cationic protein, which are enzymes that can cause extensive tissue destruction. Epithelial cells also produce cytokines (such as TNF and IL-6) and chemokines (like IL-8 and eotaxins) that amplify the inflammatory response.

Susceptibility to Type I hypersensitivity reactions, termed atopy, is familial. However, the genetic basis for this predisposition has not been clearly established and is probably polygenic. Most allergens that trigger a type I reaction are low-molecular weight, highly soluble proteins that enter the body via the mucosa of the respiratory and digestive tracts.

Type II or Antibody-Mediated Hypersensitivity

Type II hypersensitivity is mediated by IgM or IgG targeting membrane-associated antigens. A sensitization phase leads to production of antibodies that recognize substances or metabolites that accumulate in cellular membrane structures. In the effector phase, target cells become coated with antibodies, a process termed opsonization, which leads to cellular destruction by three mechanisms: (i) phagocytosis, (ii) complement-dependent cytotoxicity (CDC), and (iii) ADCC. First, IgG or IgM antibodies coating target cells can bind to Fc receptors present on cells such as macrophages and neutrophils and mediate phagocytosis. IgG or IgM antibodies can also activate complement via the classical pathway. This leads to deposition of C3b, which can mediate phagocytosis. Complement activation also leads to production of the MAC, which forms pores in the cellular membrane resulting in cytolysis (CDC). Finally, IgG antibodies can bind FcγRIII on NK cells and macrophages, thus mediating release of granzymes and perforin and resulting in cell death by apoptosis (ADCC).

The most common cause of type II reactions are medications including penicillins, cephalosporins, hydrochlorothiazide, and methyldopa, which become associated with red blood cells or platelets leading to anemia and thrombocytopenia. The mechanisms involved in type II hypersensitivity also play a role in cellular destruction by autoantibodies.

Type III or Immune Complex Reaction

In this reaction, antibodies bind antigen to form immune complexes, which become deposited in tissues where they elicit inflammation. It is important to note that the formation of immune complexes occurs during many infections and is an effective method of antigen clearance. However, under certain circumstances these complexes sometimes escape clearance by the reticuloendothelial system and are deposited in tissues including the kidney, joints, and blood vessels. The immune complexes fix complement and bind to leukocytes via Fc receptors. Activation of the complement cascade leads to production of vasoactive mediators C5a and C3a, allowing immune complexes to deposit in vessel walls and tissues. Vasoactive mediators, such as PAF, are also released following Fc receptor engagement. The result is tissue edema and deposition of immune complexes in the vessel walls and surrounding tissue. At the same time, chemotactic factors are produced leading to PMN and monocyte recruitment, which upon further activation cause tissue destruction.

Examples of Type III reactions include serum sickness and the Arthus reaction. Serum sickness was originally described in patients suffering from diphtheria who were treated with immune horse serum (antiserum). The condition was characterized by rash, joint pain, lymph node swelling, and fever, with typical onset about 10 days after initial serum exposure. Currently, the most common causes of serum sickness-like illness are antibiotics (such as cephalosporins and sulfonamides) and blood products. The Arthus reaction is a local immune complex mediated vasculitis, usually observed as edema and necrosis in the skin occurring several hours following antigen exposure. As with Type II reactions, the mechanisms involved in Type III hypersensitivity also play a role in certain autoimmune diseases.

Type IV or Delayed-Type Hypersensitivity

While the first three types of hypersensitivity are mediated primarily by antibodies, delayed-type hypersensitivity (DTH) is mediated by T-cells. During the sensitization phase, naïve $CD4^+$ T-cells are exposed to antigens with an induction of an adaptive Th response. In the effector phase, antigen is carried by APCs to lymph nodes where it is presented to memory T-cells, which become activated and then travel back to the site of antigen deposition where they may be stimulated to secrete IFN-γ, thereby initiating a Th1 response and tissue inflammation mediated primarily by macrophages.

The classic example of DTH is the tuberculin skin reaction, which occurs 24–48 hours following intradermal injection of a purified protein derivative (PPD) prepared from *Mycobacterium tuberculosis.* Contact dermatitis from poison ivy represents another common example of DTH.

Immunologic Deficiencies

The first inherited immunodeficiency was described by Bruton over 50 years ago [9]. Many of these diseases have devastating pathological consequences for afflicted individuals. An understanding of the underlying mechanisms responsible for these disease processes has been invaluable in our current understanding of immune system function. There are currently over 200 described inherited immunodeficiencies. A systematic review of these deficiencies is well beyond the scope of this chapter. A few of the more common and important ones will be discussed. In addition, there are acquired immunodeficiencies, which were primarily iatrogenic until the arrival of the Acquired Immune Deficiency Syndrome (AIDS) epidemic in the late 1970s.

Primary Immunodeficiencies

These are inherited disorders that primarily affect immune system cell function. Most are rare with prevalences less than 1 in 50,000 individuals. As would be expected, the less severe tend to be more common. Certain types of infections associate with different classes of immunodeficiency (Table 17.3). Many primary immunodeficiencies lead to significant infections early in life. Deficiencies in multiple lymphocyte lineages tend to be devastating. Breastfeeding neonates with antibody deficiencies usually remain well for a time after birth due to passive transfer of maternal immunoglobulin during the first 6–9 months of life. Then the affected individual develops upper and lower respiratory tract infections including sinusitis, otitis, and pneumonia. Most of these infections are caused by bacteria such as *Streptococcus pneumoniae* and *Haemophilus influenzae* that have polysaccharide capsules, making them resistant to phagocytosis without effective antibody opsonization. If untreated, these recurrent infections cause significant destruction and scarring of lung tissue, resulting in bronchiectasis. Fortunately, treatment with pooled immunoglobulin is effective in treating antibody deficiencies. More recently, it has been recognized that deficiency in one molecule may predispose individuals to only one or a few types of infections. We are only beginning to recognize these syndromes, and it is likely several more will be discovered.

X-Linked Agammaglobulinemia (Bruton's Disease) This condition is diagnosed in a patient with recurrent respiratory tract infections, lack of tonsilar tissue, and a normal white blood cell count with normal lymphocyte percentage. When flow cytometry for surface membrane immunoglobulin (BCR) is performed in these patients, few if any B-cells are found. Serum immunoglobulins are nonexistent. The defect is in Bruton's tyrosine kinase (*Btk*), which is essential for B-cell development and survival. B-cells develop in the bone marrow from pluripotent hematopoietic stem cells, first becoming pro-B cells, then pre-B-cells when the immunoglobulin heavy and light chain V region gene segments undergo rearrangement. During the next stage, the immature B-cell, the BCR is expressed on the cell surface (sIgM), and these cells leave the bone marrow to undergo further maturation steps. These cells eventually develop into mature B-cells and then plasma cells, which, in the presence of antigen and T-cells, undergo antibody class switching and somatic hypermutation. Btk is involved in signal transduction at several stages of B-cell development [10]. At the pre-B-cell stage, Btk signaling appears to be essential for directing light chain gene rearrangement events. Thus, defective Btk leads to arrest of B-cell differentiation at the pre-B-cell stage.

Hyper-IgM Syndrome Hyper-IgM syndrome, which is characterized by the presence of normal or elevated serum levels of IgM and low IgG and IgA, may be caused by one of at least 10 gene defects (Table 17.4). The most common (and the most clinically severe) of these is an X-linked deficiency in CD40 ligand (CD40L). This receptor is transiently expressed on activated T-cells and interacts with CD40 molecules constitutively expressed on the surface of B-cells and APCs. T-cell interaction with B-cells via CD40–CD40L interaction leads to immunoglobulin class switching and differentiation into plasma cells. In the absence of this signal, B-cells only produce IgM. Other T-cell/B-cell interactions can induce class switching, so some small amounts of IgG and IgA may be seen. However, the impaired production of IgG and IgA in CD40L deficiency leads to susceptibility to recurrent bacterial infections, particularly those involving the respiratory tract caused by encapsulated organisms. Importantly, T-cells also interact with APCs via CD40–CD40L. Thus, CD40L deficiency (as well as the more unusual autosomal recessive CD40 deficiency) leads to significant impairment in both antibody production and cell-mediated immunity, resulting in a clinical presentation that may include both infections with encapsulated bacteria and intracellular organisms. Other defects responsible for the hyper-IgM phenotype are either directly or

Table 17.3 Infections in Immunodeficiencies.

Host Defense Affected	Clinical Example	History	Relevant Pathogens
T-cells	AIDS	Disseminated infections Opportunistic infections Persistent viral infections	*Pneumocystis jiroveci* *Cryptococcus neoformans* Herpes viruses
B-cells	X-linked agammaglobulinemia	Recurrent respiratory infections Chronic diarrhea Aseptic meningitis	*Streptococcus pneumoniae* *Haemophilus influenzae* *Giardia lamblia* *Enteroviruses*
Phagocytes	Chronic granulomatous disease	Gingivitis Aphthous ulcers Recurrent pyogenic infections	*Staphylococcus aureus* *Burkholderia cepacia* *Serratia marcescens* *Aspergillus* spp.
Complement	Late complement component (C5, C6, C7, C8, or C9) deficiency	Recurrent bacteremia Recurrent meningitis	*Neisseria* spp.

Table 17.4 Hyper IgM Syndromes [11]

Defective Gene	Genetic Transmission	T-Cell Defect	Somatic Hypermutation
CD40L	X-linked	yes	decreased
CD40	Autosomal recessive	yes	decreased
AID	Autosomal recessive	no	no
AID C terminal	Autosomal dominant	no	yes
UNG	Autosomal recessive	no	yes
AID-targeting	Autosomal recessive	no	yes
DNA repair	Autosomal recessive	no	yes
ATM	Autosomal recessive	yes	yes
MRE-11	Autosomal recessive	yes	?
NB-S1	Autosomal recessive	yes	?

indirectly involved in immunoglobulin class switching [11]. Some do not involve T-cells and therefore do not induce a clinically mixed deficit.

Complement Deficiency Defects have been described for most of the components of complement. There is a propensity for individuals with C3 deficiency and with classical component defects (C1, C2, and C4) to develop systemic lupus erythematosus (SLE). While this is not well understood, classical pathway clearance of apoptotic cells and immune complexes may be important. C2 deficiency is relatively common, occurring in 1 in 20,000 individuals. These individuals have an increased propensity to develop infections, SLE, and myocardial infarctions [12]. Defects in MBP are relatively common, and there is an association with infections in children. Finally, defects in the terminal components (C5–9) lead to a remarkable susceptibility (~5000-fold risk) to recurrent infections caused by pathogenic *Neisseria* spp., especially *N. meningitidis.* Interestingly, mortality due to these infections is 10-fold lower than nondeficient individuals. The reason for observation remains to be determined.

Severe Combined Immunodeficiency (SCID) The severe combined immunodeficiency disorder is characterized by the absence of T-cell differentiation, resulting in severely reduced or absent T-cells. At least 10 autosomal recessive or X-linked genetic defects have been described, which result in several possible phenotypes based on B and NK cell development (Table 17.5). The most common cause of SCID is an X-linked deficiency in the common γ chain shared by receptors for several cytokines including IL-2 and IL-4. This form is characterized by complete absence of T-cells and NK cells. When untreated, SCID results in death within the first year of life from overwhelming opportunistic infections. Improved survival has been reported in infants receiving stem cell transplants. Using retroviral vectors, gene therapy has also been used successfully in several cases [13].

Table 17.5 Genetic Defects Causing SCID

Defective Gene	B-Cell Deficit	NK Cell Deficit
IL-7 receptor α-chain	no	no
CD3 δ-chain	no	no
CD3 ε-chain	no	no
Common γ chain (X-linked)	no	yes
CD45	no	yes
JAK3	no	yes
Artemis	yes	no
RAG1, RAG2	yes	no
Adenosine-deaminase	yes	yes

Common Variable Immunodeficiency (CVID) Common variable immunodeficiency (CVID) is a heterogeneous syndrome, probably comprising several distinct diseases, characterized by recurrent infections and low antibody levels (IgG and IgA and/or IgM). The syndrome is usually diagnosed in the fourth decade of life and with a typical delay of >15 years from first symptoms to diagnosis [14]. The estimated prevalence of CVID is about 1 in 25,000 individuals, making it the most prevalent immunodeficiency requiring medical attention. The main clinical features are recurrent infections of the respiratory tract, chronic diarrhea, autoimmune disease, and malignancy. While most gene defects are unknown, 10–15% are caused by a defect in the transmembrane activator and calcium modulator and cyclophillin interactor (*TACI*) gene [15]. TACI is a member of the TNF receptor family and is expressed on B-cells and activated T-cells. Ligands for TACI include B-cell activating factor of the TNF family (BAFF) and a proliferation inducing ligand (APRIL). Both ligands bind other receptors on B-cells and T-cells. However, in response to ligation by APRIL, TACI mediates isotype switching to IgG and IgA. Similarly, BAFF mediates IgA switching. The penetrance of CVID phenotype in families carrying a mutant TACI gene is quite variable. Recently, a defect in BAFF receptor has also been identified as a cause of CVID.

Chronic Granulomatous Disease (CGD) Phagocytes produce reactive oxygen species (including superoxide and hydrogen peroxide) to kill ingested pathogens. NADPH oxidase is a five-subunit enzyme that catalyzes the production of superoxide from oxygen for this purpose. Over 400 genetic defects in NADPH have been described that result in chronic granulomatous disease (CGD) [16], which is characterized by recurrent, indolent bacterial and fungal infections caused by catalase positive organisms including

Staphylococcus aureus and *Aspergillus spp.* Most mutations occur in the gp91*phox* gene located on the X chromosome. There are also autosomal recessive forms of the disease primarily involving p47*phox*. The lack of microbicidal oxygen species due to NADPH deficiency leads to a severe defect in intracellular killing of phagocytosed organisms. However, organisms that lack catalase produce peroxide as a byproduct of oxidative metabolism, which accumulates and leads to effective intracellular killing of these microbes. Tissue pathology is characterized by granulomas and lipid-filled histiocytes in liver, spleen, lymph nodes, and gut. The disease is treated with antibiotic prophylaxis, and chronic administration of interferon-γ may also be beneficial.

X-Linked Proliferative Disease (Duncan's Syndrome) Caused by a mutation in the *SH2D1A* gene, this rare X-linked proliferative disease is manifested by dysregulated T-cell and NK cell proliferation and responses. *SH2D1A* encodes SLAM-associated protein (SAP), a cytoplasmic adapter protein that binds the transmembrane protein signaling lymphocyte activation molecule (SLAM) and related molecules. SAP inhibits cell activation via SLAM. Defects in SAP lead to T-cell proliferation. Individuals with this disorder have a propensity to develop lymphoma and fulminant Epstein-Barr virus (EBV) infection.

Inherited Susceptibility to Herpes Encephalitis (UNC-93B and TLR-3 Deficiencies) The paradigm of a single gene lesion conferring vulnerability to multiple infections has been challenged by several recent discoveries. While over 80% of young adults are infected with herpes simplex virus type 1 (HSV-1), development of herpes simplex encephalitis (HSE) is quite rare (1/250,000 patient years). Jean Laurent Casanova and colleagues hypothesized that susceptibility to developing HSE was inherited as a monogenic trait [17]. There was evidence that impaired interferon responses might predispose to HSE. Therefore, they screened otherwise healthy children with a history of HSE and found two unrelated children whose leukocytes had defective interferon production in response to HSV-1 antigens. These children had no evidence of increased susceptibility to infections. However, it was determined that they had impaired responses to agonists of TLR-7, TLR-8, TLR-9, and possibly TLR-3. These responses were similar to those recently reported by Bruce Beutler's group in mice lacking UNC-93B, an endoplasmic reticulum protein involved in activation by TLR-3, TLR-7, and TLR-9 [18]. The patients completely lacked mRNA encoding UNC-93B. Each was homozygous for a different mutation in the UNC-93B gene: one had a 4 base deletion and one had a point mutation leading to alternative splicing. Previously, children with a deficiency in interleukin 1 receptor-associated kinase-4 (IRAK-4) had been described. These patients fail to signal through TLR-7, TLR-8, or TLR-9, and show a propensity to developing certain bacterial infections but not HSE. Therefore, Casanova and colleagues proposed that the HSE susceptibility of the UNC-93B deficient children was due to impaired TLR-3 induction of interferon. One year later the same group identified two more otherwise healthy children with HSE and a single point mutation in the *tlr3* gene [19]. TLR-3 is highly expressed in the central nervous system. Together, these data strongly suggest an important role for TLR-3 in protection against HSE and indicate that a single genetic defect may predispose to infection (or severe manifestations of infection) by a specific pathogen. The possibility remains that the TLR-3 pathway may also be important in protection against encephalitis cause by other viruses.

Inherited Susceptibility to Tuberculosis Several recently described genetic defects predispose individuals to severe disease caused by *Mycobacterium tuberculosis* and other less virulent mycobacterial species [20]. The proteins encoded by these genes all play a role in Th1 responses and include IL-12, receptors for interferon-γ and IL-12, and STAT1, which is involved in signaling via the interferon-γ receptor.

Acquired Immunodeficiencies

Exposure to a variety of factors such as infectious agents, immunosuppressive drugs and environmental conditions are much more prevalent causes of immunodeficiency than the genetic defects described above. Malnutrition is the most common cause of immunodeficiency worldwide. Metabolic diseases, such as diabetes, hepatic cirrhosis, and chronic kidney disease also lead to immunosuppression. Many viral and bacterial infections can result in transient immunosuppression. Infection with the human immunodeficiency virus (HIV) can lead to a chronic and severe state of immunosuppression known as the Acquired Immune Deficiency Syndrome (AIDS).

Acquired Immune Deficiency Syndrome It is currently estimated that over 40 million people are infected with HIV, with the majority in sub-Saharan Africa, and South and Southeast Asia. HIV is a retrovirus that contains 9 genes: *gag, pol, env, tat, rev, nef, vif, vpr, vpu*. The *gag* gene product is split by the HIV protease into 5 structural proteins. The *pol* gene product is split into three enzymes: integrase, reverse transcriptase, and protease. The *env* gene product is cleaved to produce two envelope proteins, gp120 and gp41, which together constitute gp160. The other 5 genes encode regulatory proteins. The virus is shed into body fluids and the bloodstream. HIV infection is most typically spread via sexual contact, but transmission via contaminated hypodermic needles and blood products can also occur as can transmission from mother to infant.

HIV enters $CD4^+$ T-cells through interaction of gp160 on the viral envelope and CD4 along with a co-receptor, either CCR5 or CXCR4, on the cell surface. Individuals expressing a *ccr5* mutant gene are protected from HIV infection. Once inside the cell, the virus uncoats and the reverse transcriptase, which is complexed to the RNA genome of the virus, transcribes viral RNA into double-stranded DNA. This is then transported to the cell nucleus where, with the help of virally encoded integrase, it is inserted into the cellular DNA. Thus, the virus establishes lifelong infection of the cell. Replication of the viral genome occurs along with cell replication.

Acute HIV infection is characterized by high viremia, immune activation, and $CD4^+$ T-cell lymphopenia. The acute phase of infection lasts several weeks. Patients often develop nonspecific symptoms of fever, rash, headaches, and myalgias. Occasionally, opportunistic infections may occur in this period. The acute phase is followed by a period of clinical latency generally characterized by absence of significant symptoms. During this period, viral replication continues in the lymphoid tissue, resulting in lymphadenopathy, and there is increased susceptibility to certain infections such as tuberculosis. The latent period lasts several years. Higher viral loads predict shorter clinical latency. While initial host immune responses control viral infection somewhat, they inevitably fail. $CD4^+$ T-cells continue to decline, and eventually the host becomes susceptible to opportunistic infections, including pneumocystis pneumonia, cryptococcal meningitis, disseminated cytomegalovirus (CMV), and mycobacterial infections. If untreated, death occurs on average about 10 years after infection.

While the virus targets $CD4^+$ T-cells, abnormalities in all parts of the immune system occur during HIV infection. There is profound disruption of lymphoid tissue architecture, resulting in an inability to mount responses against new antigens and severely impaired memory responses. $CD4^+$ T-cells that are not killed directly are dysregulated. They have decreased IL-2 production and IL-2 receptor expression, resulting in diminished capacities to proliferate and differentiate. CD28 expression is also reduced. T-cell receptor Vβ repertoire can be significantly reduced in advanced HIV infection. $CD8^+$ T-cells also have decreased IL-2 receptor and CD28 expression. CTL activity is reduced as are production of cytokines and chemokines, including those that block HIV replication. B-cells are hyperactivated by gp41 in HIV infection. gp120 acts as a superantigen for B-cells carrying the V_H3 variable region. This leads to a gap in the B-cell antibody repertoire. B-cell dysregulation explains the propensity to developing infections with encapsulated bacteria. B-cells also express proinflammatory cytokines such as TNF and IL-6, which enhance HIV replication. Macrophages express CD4 and other HIV co-receptors and become infected with HIV. They serve as important reservoirs for HIV. Infection of macrophages also leads to functional abnormalities, including decreased IL-12 secretion, increased IL-10 production, decreased antigen uptake, and impaired chemotaxis. HIV infection of brain tissue macrophages or microglial cells plays a major role in HIV encephalopathy. Other immune cells shown to be dysregulated in HIV infection include NK cells and neutrophils.

Treatment with multiple drugs that target HIV enzymes, known as highly active antiretroviral therapy (HAART), is effective in reducing viremia and restoring normal $CD4^+$ T-cell counts. HAART has drastically reduced mortality rates in HIV infection [21]. After 2–3 weeks, patients treated with HAART may develop a severe inflammatory response to existing opportunistic infections known as the immune reconstitution inflammatory syndrome (IRIS). No treatment regimen has been successful in eradicating infection.

Autoimmune Diseases

T-lymphocytes and B-lymphocytes have incredible diversity in antigen recognition. They also carry a potent armamentarium, capable of destroying cells and causing damaging inflammation. While some pathogen-associated antigens are quite distinct from host molecules, there is often no fundamental difference between host antigens and those of pathogens. Therefore, autoreactive clones will most certainly arise. The host needs to eliminate these self-reactive lymphocytes or suffer self-destruction. Macfarlane Burnet originally formulated the clonal deletion theory, which proposes that all self-reactive lymphocytes (forbidden clones) are destroyed during development of the immune system. In fact, some autoreactive T-cells and B-cells exist in many individuals that do not develop autoimmune disease. Autoimmunity occurs when T-cells or B-cells are activated and cause tissue destruction in the absence of ongoing infection.

Tolerance is the process that neutralizes these autoreactive T-cells and B-cells. Autoimmunity is the failure of tolerance mechanisms. B-cells may produce antibodies that recognize surface proteins, and these may directly cause disease by (i) initiating destruction of host cells or (ii) mimicking receptor ligand, and causing hyperactivation. Antibodies that bind intracellular antigens, such as many of those formed in SLE, are generally believed to be secondary to the autoimmune process itself. Autoreactive B-cells may be deleted in the bone marrow or the lymph nodes and spleen. B-cells must receive a second signal (co-stimulation) following B-cell receptor ligation by antigen. Most often this second signal is delivered by T-cells. Without T-cell help, antigen binding B-cells will die.

Autoreactive T-cells are removed by two separate processes: central and peripheral tolerance. During maturation, T-cells leave the bone marrow and travel to the thymus, where they encounter endogenous peptides complexed with MHC. If the receptors on a given T-cell bind these complexes with significant affinity, it is directed to die by apoptosis in a process called negative selection. Interestingly, moderate binding to self-antigens is necessary for T-cell survival, as lack of binding to antigens presented in the thymus also triggers apoptosis. Not all self-antigens are presented in the thymus, and some autoreactive T-cells escape to the periphery. Similar to B-cells, presentation of antigens to T-cells in the absence of co-stimulation leads to deletion. Activated T-cells express Fas on their cell surface. If they encounter Fas ligand, they will undergo apoptosis. Some tissues such as the eye constitutively express Fas ligand. When activated, T-cells expressing Fas (CD95) enter the anterior chamber of the eye; they encounter Fas ligand and undergo apoptosis without causing tissue damage. Another molecule involved in peripheral tolerance is cytotoxic T-lymphocyte-associated protein (CTLA-4 or CD152) which binds to B7-1 (CD80) on T-cells and

B7-2 (CD86) on B-cells with higher affinity than the co-stimulatory molecule CD28, inhibiting T-cell activation. CTLA-4 polymorphism is associated with an increased predisposition to autoimmune diseases, including SLE, autoimmune thyroiditis, and type 1 diabetes. Another mechanism that plays a role in the development of autoimmunity is called immunological ignorance. Certain tissues, such as the central nervous system, are protected by barriers (such as the blood brain barrier) that do not typically allow entry of peripheral T-cells.

Recently, a new subset of T-cells has been described called regulatory T-cells (Tregs). These cells play an important role in controlling the magnitude and the quality of immune responses. Tregs express CD4 and CD25 on their surface. They also produce a transcription factor, forkhead box P3 (FoxP3), which is critical for the differentiation of thymic T-cells into Tregs. FoxP3 interacts with other transcription factors to control expression of ~700 gene products, and leads to repression of IL-2 production and activation of CTLA-4 and CD25 expression. Tregs mediate immunosuppression by several mechanisms [22]. First, they may compete with specific naïve T-cells for antigen presented by APCs. Second, they may downregulate APCs via CTLA-4-dependent mechanisms. Finally, they may interact with effector T-cells to either kill them or inactivate them with immunosuppressive cytokines such as IL-10. The role of abnormal Tregs in autoimmunity is still being established. Defects in certain genes specifically associated with Tregs may lead to autoimmunity. In addition to CTLA-4, polymorphisms in IL-2 and CD25 are associated with susceptibility to autoimmune diseases. Mutations in *foxP3* lead to an immunodeficiency syndrome, IPEX (immune dysregulation, polyendocrinopathy, enteropathy, X-linked), associated with autoimmune diseases in endocrine organs [23]. Environmental factors may also play a role in Tregs. Tregs have higher metabolic and proliferation rates compared to other T-cells and are therefore more susceptible to ionizing radiation and vitamin deficiencies. Other T-cells may also play a role in immunosuppression and tolerance, including $CD8^+$, $CD4^-CD8^-$, and γ/δ T-cells.

Tolerance can be broken by several mechanisms. Infections are thought to be the main exogenous cause of autoimmunity. Infections can break ignorance by damaging barriers, leading to release of sequestered antigens. Superantigens, which stimulate polyclonal T-cell activation without the need for co-stimulation, are produced by certain microbes. Also, infection can produce significant inflammation with production of inflammatory cytokines and other co-stimulatory molecules, which may activate autoreactive lymphocytes (bystander activation). Another trigger of autoimmunity is a seemingly appropriate immune response to microbial antigens that mimic host antigens. For example, in the demyelinating disease Guillain-Barré syndrome, antibody cross-reactivity has been demonstrated between human gangliosides and *Campylobacter jejuni* lipopolysaccharide. In fact, microbial antigens may evolve to resemble host antigens in order to evade the host immune response. Drugs, such as procainamide, are another significant exogenous trigger of autoimmunity.

Several autoimmune diseases have been described (Table 17.6). These may be either systemic (such as SLE) or organ-specific (for instance, type 1 diabetes). Some of the more common autoimmune diseases are considered in the following sections with the understanding that the underlying mechanisms leading to host immune destruction in this diverse group of diseases are relatively similar.

Systemic Lupus Erythematosus

Systemic lupus erythematosus (SLE) is a diverse systemic autoimmune syndrome with a significant range of symptoms and disease severity [24]. The etiologies of SLE have not been clearly established. An antecedent viral illness often precedes the onset of SLE, and there is a temporal association with EBV infection and SLE. MHC genes (such as HLA-A1, B8, and DR3) are linked to SLE. Deficiencies in the early complement component cascade (C1q, C2, or C4) are strongly associated with SLE. Several other genes may also be linked to SLE. The syndrome probably represents several distinct diseases that result in similar manifestations. There is a major propensity for females to develop this disease (9:1 versus males), implicating a role for female hormones (or male hormones as protective). However, the basis for the sexual preference has not been established.

The hallmark of SLE is the development of autoantibodies, particularly those that bind double-stranded DNA (anti-dsDNA). Many other autoantibodies have been described in SLE and are associated with certain clinical manifestations. Antibodies to dsDNA, Sm antigen, and C1q are correlated with kidney disease. Antibodies to Ro and La antigens are associated with fetal heart problems, while antibodies to phospholipids are associated with thrombotic events and fetal loss in pregnancy. Tissue damage by autoantibodies has been most well studied with anti-dsDNA causing kidney damage. There are two possible mechanisms involved. First, anti-dsDNA may bind to fragments of DNA released by apoptotic cells (nucleosomes) in the bloodstream, forming immune complexes. These complexes are deposited in the glomerular basement membrane in the kidney and cause disease by Type III reactions. Second, anti-dsDNA may cross-react with another antigen expressed on kidney cells. One possible candidate for this antigen is α-actinin, which cross-links actin and is important in maintaining the function of renal podocytes.

Apoptotic cells express intracellular material in the form of blebs on the cell surface and complement may be important in clearing apoptotic debris. An intriguing hypothesis explaining the association of complement deficiencies and SLE is that defects in clearing apoptotic cells leads to continued exposure to self-antigens (such as nucleosomes, Sm, Ro, or La) resulting in SLE [25].

Type 1 Diabetes Mellitus

Pancreatic β-cells produce insulin, which plays an essential role in glucose metabolism. In type 1

Table 17.6 Autoimmune Diseases

Name	Hypersensitivity[1]	Autoantibody[2]
Acute disseminated encephalomyelitis		
Addison's disease		Anti-21-hydroxylase
Ankylosing spondylitis		
Antiphospholipid antibody syndrome		Anticardiolipin
Autoimmune hemolytic anemia	II	
Autoimmune hepatitis		Antismooth muscle actin
Bullous pemphigoid		Antibullous pemphigoid antigen
Celiac disease	IV	Antigliadin
Dermatomyositis		Anti-Jo-1
Diabetes mellitus type 1	IV	Anti-insulin
Goodpasture's syndrome	II	Antibasement membrane collagen type IV
Graves' disease	II	Antithyroid stimulating hormone receptor
Guillain-Barré syndrome	IV	Antiganglioside
Hashimoto's thyroiditis	IV	Antithyroglobulin
Idiopathic thrombocytopenic purpura	II	Antiplatelet membrane glycoprotein IIb-III
Multiple sclerosis	IV	Antimyelin basic protein
Myasthenia gravis	II	Antiacetylcholine receptor
Pemphigus vulgaris	II	Antidesmogein 3
Pernicious anemia	II	Anti-intrinsic factor
Polymyositis		
Primary biliary cirrhosis		Antimitochondrial
Rheumatoid arthritis	III	Rheumatoid factor
Sjögren's syndrome		Anti-SSB/La
Systemic lupus erythematosus	III	Anti-dsDNA
Temporal arteritis	IV	
Wegener's granulomatosis		Antineutrophil cytoplasmic

[1]Hypersensitivity reaction associated with disease pathogenesis.
[2]A single relevant antibody is listed. Many syndromes are associated with more than one autoantibody.

diabetes, cell-mediated destruction of β-cells occurs, leading to complete insulin deficiency. In contrast, in type 2 diabetes there is normal production of insulin, but cells do not respond appropriately. Susceptibility to development of type 1 diabetes is linked to HLA-DR3/4 and DQ8 genes and is also associated with polymorphisms in the gene for CTLA-4. Potential exogenous triggers include viral infections (such as congenital rubella), chemicals (such as nitrosamines), and foods (including early exposure to cow's milk proteins). Interaction with environmental triggers in an individual with genetic predisposition leads to infiltration of the pancreatic islets with $CD4^+$ and $CD8^+$ T-cells, B-cells, and macrophages, and the production of autoantibodies. Autoantibodies to β-cell antigens, insulin, and others (including antiglutamic acid decarboxylase and anti-insulinoma-associated antigen) can precede the onset of type 1 diabetes by years. However, there is no direct evidence that these antibodies play a role in pathogenesis. Th1-activated autoreactive $CD4^+$ T-cells together with cytotoxic $CD8^+$ T-cells induce apoptosis of β-cells mediated via Fas-Fas ligand interaction and release of cytoxic molecules. Clinical expression of disease occurs only after >90% of the β-cells are destroyed.

Multiple Sclerosis

Multiple sclerosis (MS) is a demyelinating disease presumed to be of autoimmune etiology. The disease presents with neurologic deficits and generally has a relapsing course followed by a progressive phase. The disease is more common in females (2:1 versus men), and there appears to be a genetic predisposition. As with SLE and type 1 diabetes, there is an association with certain MHC, including HLA-DR2 and DQ6. However, the inciting factor is believed to be environmental. The disease occurs more commonly in temperate climates, and relapses are often preceded by viral respiratory tract infections. The hallmark of MS is the central nervous system inflammatory plaque containing a perivascular infiltration of myelin-laden macrophages and T-cells (both $CD4^+$ and $CD8^+$). These plaques tend to form in the white matter and involve both the myelin sheath and oligodendrocytes [26]. There is often widespread axonal damage. Myelin-reactive T-cells can be isolated from individuals with and without MS. However, in the former, these T-cells are activated with a Th1 phenotype, whereas in the latter, these cells are naïve. There is significantly increased antibody production in the central nervous system of MS patients. B-cells recovered from cerebrospinal fluid of MS patients also display an activated phenotype.

Celiac Disease

Celiac disease occurs in genetically predisposed individuals following exposure to gluten. The disease results in diarrhea and malabsorption. It is unique among autoimmune diseases, as the environmental precipitant is known. Strict avoidance of the antigen, which is found in wheat, leads to remittance of symptoms in most subjects. Gliadin is the alcohol soluble fraction of gluten and is the primary antigen leading to an inflammatory reaction in the small intestine, characterized by chronic inflammatory infiltrate and villous atrophy. An enzyme, tissue transglutaminase, deamidases gliadin peptides in the lamina propria, increasing their immunogenicity. Gliadin-reactive T-cells recognizing antigen in the context of HLA-DQ2 or DQ8 lead to an inflammatory Th1 phenotype. While these HLA types are necessary for the development of celiac disease, they are not sufficient, and other genetic factors and environmental exposures also play a role in developing disease. IgA antibodies directed against gliadin, tissue transglutaminase, and connective tissue (such as antiendomysial and antireticulin) are found in individuals with celiac disease. These patients are also susceptible to developing several types of cancer, most notably adenocarcinoma of the small intestine and enteropathy-associated T-cell lymphoma.

REFERENCES

1. Casanova JL, Abel L. Inborn errors of immunity to infection: The rule rather than the exception. *J Exp Med.* 2005;202:197–201.
2. Abbas A, Lichtman A, Pillai S. *Cellular and Molecular Immunology.* New York: WB Saunders; 2007.
3. Goldsby R, Kindt T, Osborne B. *Kuby Immunology.* New York: WE Freeman; 2006.
4. Murphy KM, Travers P, Walport M. *Janeway's Immunobiology.* New York: Garland; 2007.
5. Parham P. *The Immune System.* New York: Garland; 2005.
6. Gell PGH, Coombs RRA. *Clinical Aspects of Immunology.* Oxford: Blackwell; 1968.
7. Johansson SG, Hourihane JO, Bousquet J, et al. A revised nomenclature for allergy: An EAACI position statement from the EAACI nomenclature task force. *Allergy.* 2001;56:813–824.
8. Rajan TV. The Gell-Coombs classification of hypersensitivity reactions: A re-interpretation. *Trends Immunol.* 2003;24:376–379.
9. Bruton OC. Agammaglobulinemia. *Pediatrics.* 1952;9:722–727.
10. Maas A, Hendriks RW. Role of Bruton's tyrosine kinase in B cell development. *Dev Immunol.* 2001;8:171–181.
11. Durandy A, Peron S, Fischer A. Hyper-IgM syndromes. *Curr Opin Rheumatol.* 2006;18:369–376.
12. Jonsson G, Truedsson L, Sturfelt G, et al. Hereditary C2 deficiency in Sweden: Frequent occurrence of invasive infection, atherosclerosis, rheumatic disease. *Medicine (Baltimore).* 2005;84:23–34.
13. Cavazzana-Calvo M, Fischer A. Gene therapy for severe combined immunodeficiency: Are we there yet? *J Clin Invest.* 2007;117:1456–1465.
14. Oksenhendler E, Gerard L, Fieschi C, et al. Infections in 252 patients with common variable immunodeficiency. *Clin Infect Dis.* 2008;46:1547–1554.
15. Castigli E, Geha RS. Molecular basis of common variable immunodeficiency. *J Allergy Clin Immunol.* 2006;117:740–746.
16. Heyworth PG, Cross AR, Curnutte JT. Chronic granulomatous disease. *Curr Opin Immunol.* 2003;15:578–584.
17. Casrouge A, Zhang SY, Eidenschenk C, et al. Herpes simplex virus encephalitis in human UNC-93B deficiency. *Science.* 2006; 314:308–312.
18. Tabeta K, Hoebe K, Janssen EM, et al. The Unc93b1 mutation 3d disrupts exogenous antigen presentation and signaling via Toll-like receptors 3, 7 and 9. *Nat Immunol.* 2006;7:156–164.
19. Zhang SY, Jouanguy E, Ugolini S, et al. TLR3 deficiency in patients with herpes simplex encephalitis. *Science.* 2007;317: 1522–1527.
20. Fortin A, Abel L, Casanova JL, et al. Host genetics of mycobacterial diseases in mice and men: Forward genetic studies of BCG-osis and tuberculosis. *Annu Rev Genomics Hum Genet.* 2007;8:163–192.
21. Bhaskaran K, Hamouda O, Sannes M, et al. Changes in the risk of death after HIV seroconversion compared with mortality in the general population. *JAMA.* 2008;300:51–59.
22. Sakaguchi S, Yamaguchi T, Nomura T, et al. Regulatory T cells and immune tolerance. *Cell.* 2008;133:775–787.
23. van der Vliet HJ, Nieuwenhuis EE. IPEX as a result of mutations in FOXP3. *Clin Dev Immunol.* 2007;2007:89017.
24. Rahman A, Isenberg DA. Systemic lupus erythematosus. *N Engl J Med.* 2008;358:929–939.
25. Gaipl US, Munoz LE, Grossmayer G, et al. Clearance deficiency and systemic lupus erythematosus (SLE). *J Autoimmun.* 2007; 28:114–121.
26. Frohman EM, Racke MK, Raine CS. Multiple sclerosis—The plaque and its pathogenesis. *N Engl J Med.* 2006;354:942–955.

Chapter

18

Molecular Basis of Pulmonary Disease

Carol F. Farver ▪ Dani S. Zander

INTRODUCTION

Pulmonary pathology includes a large spectrum of both neoplastic and non-neoplastic diseases that affect the lung. Many of these are a result of the unusual relationship of the lung with the outside world. Every breath that a human takes brings the outside world into the body in the form of infectious agents, organic and inorganic particles, and noxious agents of all types. Although the lung has many defense mechanisms to protect itself from these insults, these are not infallible and so lung pathology arises. Damage to the lung is particularly important given the role of the lung in the survival of the organism. Any impairment of lung function has widespread effects throughout the body, since all organs depend on the lungs for the oxygen they need. Pulmonary pathology catalogs the changes in the lung tissues and the mechanisms through which these occur. What follows is a review of lung pathology and the current state of knowledge about the pathogenesis of each disease. We believe that a clear understanding of both morphology and mechanism is required for the development of new therapies and preventive measures.

NEOPLASTIC LUNG AND PLEURAL DISEASES

Lung cancer is a major cause of morbidity and mortality throughout the world. The most recent estimates available from the Surveillance, Epidemiology, and End Results (SEER) program of the National Cancer Institute are that in 2007 over 213,000 people in the United States were diagnosed with cancer of the lung and bronchus, and over 160,000 will have died due to this disease [1]. However, in the past decade incidence and mortality rates have begun to move in a more positive direction, particularly in men. Overall, men show a decline in lung cancer incidence, while in women, although lung cancer rates grew from 1975 through 1998, they stabilized from 1998 through 2004 [2]. Similarly, cancer death rates due to lung cancer have declined for men and have slowed for women. Although, for women, lung cancer death rates have increased since 1975, the rate of increase has slowed to 0.2% annually from 1995 to 2004 [2]. These trends parallel changes in the prevalence of tobacco smoking, the most important risk factor for development of lung cancer.

Given the tremendous societal and individual impacts of this disease, it is not surprising that the molecular biology of lung cancer is a major focus of investigation. Elucidation of the molecular pathogenesis of these neoplasms has progressed significantly, offering insights into new, targeted therapies, and predictors of prognosis and therapeutic responsiveness. Recognition of precursor lesions for some types of lung cancers has been facilitated by our expanded understanding of early molecular changes involved in carcinogenesis.

The *World Health Organization* (WHO) classification scheme is the most widely used system for classification of these neoplasms (Table 18.1) [3]. Although there are numerous histologic types and subtypes of lung cancers, most of the common malignant epithelial tumors can be grouped into the categories of nonsmall cell lung cancers (NSCLCs) and small cell carcinomas (SCLCs). NSCLCs include adenocarcinomas (ACs), squamous cell carcinomas (SqCCs), large cell carcinomas, adenosquamous carcinomas, and sarcomatoid carcinomas. SCLCs include cases of pure and combined small cell carcinoma. Common pulmonary symptoms associated with these tumors include cough, shortness of breath, chest pain or tightness, and hemoptysis (coughing up blood). Since some tumors cause airway obstruction, they predispose to pneumonia, which can be an important clue to the existence of a tumor in some patients. Constitutional symptoms can include fever, weight loss, and malaise. Some neoplasms will declare themselves with symptoms related

Molecular Pathology © 2009, Elsevier, Inc. All Rights Reserved.

Table 18.1 World Health Organization Histological Classification of Tumors of the Lung

Malignant Epithelial Tumors

- Squamous cell carcinoma
  - Papillary
  - Clear cell
  - Small cell
  - Basaloid
- Small cell carcinoma
  - Combined small cell carcinoma
- Adenocarcinoma
  - Adenocarcinoma, mixed subtype
  - Acinar adenocarcinoma
  - Papillary adenocarcinoma
  - Bronchioloalveolar carcinoma
    - Nonmucinous
    - Mucinous
    - Mixed or indeterminate
  - Solid adenocarcinoma with mucin production
    - Fetal adenocarcinoma
    - Mucinous ("colloid") carcinoma
    - Mucinous cystadenocarcinoma
    - Signet ring adenocarcinoma
    - Clear cell adenocarcinoma
- Large cell carcinoma
  - Large cell neuroendocrine carcinoma
    - Combined large cell neuroendocrine carcinoma
  - Basaloid carcinoma
  - Lymphoepithelioma-like carcinoma
  - Clear cell carcinoma
  - Large cell carcinoma with rhabdoid phenotype
- Adenosquamous carcinoma
- Sarcomatoid carcinoma
  - Pleomorphic carcinoma
  - Spindle cell carcinoma
  - Giant cell carcinoma
  - Carcinosarcoma
  - Pulmonary blastoma
- Carcinoid tumor
  - Typical carcinoid
  - Atypical carcinoid
- Salivary gland tumors
  - Mucoepidermoid carcinoma
  - Adenoid cystic carcinoma
  - Epithelial-myoepithelial carcinoma
- Preinvasive lesions
  - Squamous carcinoma *in situ*
  - Atypical adenomatous hyperplasia
  - Diffuse idiopathic pulmonary neuroendocrine hyperplasia

Mesenchymal Tumors

- Epithelioid hemangioendothelioma
- Angiosarcoma
- Pleuropulmonary blastoma
- Chondroma
- Congenital peribronchial myofibroblastic tumor
- Diffuse pulmonary lymphangiomatosis
- Inflammatory myofibroblastic tumor
- Lymphangioleiomyomatosis
- Synovial sarcoma
  - Monophasic
  - Biphasic
- Pulmonary artery sarcoma
- Pulmonary vein sarcoma

Benign Epithelial Tumors

- Papillomas
  - Squamous cell papilloma
    - Exophytic
    - Inverted
  - Glandular papilloma
  - Mixed squamous cell and glandular papilloma
- Adenomas
  - Alveolar adenoma
  - Papillary adenoma
  - Adenomas of salivary-gland type
    - Mucous gland adenoma
    - Pleomorphic adenoma
    - Others
  - Mucinous cystadenoma

Lymphoproliferative Tumors

- Marginal zone B-cell lymphoma of the MALT type
- Diffuse large B-cell lymphoma
- Lymphomatoid granulomatosis
- Langerhans cell histiocytosis

Miscellaneous Tumors

- Hamartoma
- Sclerosing hemangioma
- Clear cell tumor
- Germ cell tumors
  - Teratoma, mature
  - Immature
  - Other germ cell tumors
- Intrapulmonary thymoma
- Melanoma

Metastatic Tumors

Reprinted with kind permission from Travis, W. D., Brambilla, E., Müller-Hermelink, H. K., and Harris, C. C. (2004). *Pathology and Genetics of Tumours of the Lung, Pleura, Thymus and Heart.* IARC Press, Lyon [290].

to local invasion of adjacent structures such as chest wall, nerves, superior vena cava, esophagus, or heart. SCLCs are known for early and widespread metastasis and are therefore particularly prone to being discovered through presentations as metastases in distant sites. Some tumors are discovered due to pathophysiologic changes triggered by the release of soluble substances from tumor cells. Endocrine syndromes due to elaboration of hormones are well recognized, and include Cushing syndrome, syndrome of inappropriate antidiuretic hormone, hypercalcemia, carcinoid syndrome, gynecomastia, and others. Hypercoagulability commonly occurs with lung cancers, leading to manifestations of venous thrombosis, nonbacterial thrombotic endocarditis, and disseminated intravascular coagulation. Hematologic changes can include anemia, granulocytosis, eosinophilia, and other abnormalities. Other paraneoplastic syndromes such as clubbing of the fingers, myasthenic syndromes, dermatomyositis/polymyositis, and transverse myelitis are noted in subsets of patients.

When lung cancer is suspected, evaluation of the patient includes a thorough clinical, radiologic, and laboratory assessment, with collection of tissue or cytology samples to establish a pathologic diagnosis of malignancy and to classify the tumor type. Fiberoptic bronchoscopy is often performed to collect samples for diagnosis. Sample types can include transbronchial and endobronchial biopsies, bronchial brushings, bronchial washings, bronchoalveolar lavage samples, and transbronchial needle aspirates. Submission of sputum samples for cytologic examination can provide a diagnosis in some cases, particularly for centrally located tumors such as SqCC and SCLC. Tumors arising in a peripheral location can also be sampled, in many cases, by fine needle aspiration or core needle biopsy performed under radiologic guidance. If a pleural effusion is present in combination with a lung parenchymal tumor, analysis of the pleural fluid cytology often allows one to establish a diagnosis. Pleural biopsy, mediastinoscopy with biopsy, and wedge biopsy can also be performed, depending on the clinical and radiologic findings. For tumors with apparent distant metastasis, biopsy of the metastasis focus can both establish a pathologic diagnosis and determine the stage of the tumor.

The prognosis of lung cancers is closely related to tumor stage. For NSCLCs, the *American Joint Commission on Cancer* TNM staging system is widely used (Table 18.2) [4], and for SCLCs, disease is classified as limited (restricted to one hemithorax) or extensive. Overall, for lung cancers, the 5-year survival is 13.4% for men and 17.9% for women [5]. An important factor leading to this relatively poor survival is the late stage at which many lung cancers are diagnosed. Information from the SEER database, from 1996–2003, indicates that 16%, 35%, 42%, and 7% of patients were diagnosed with localized, regional, distant, or unstaged disease, respectively [5]. The corresponding 5-year survival rates are 49.0%, 15.3%, 2.8%, and 8.7%, and 10-year survival rates are 37.8%, 10.3%, 1.6%, and 5.1% [5].

For patients with NSCLCs, treatment depends on stage and comorbid conditions [6]. Surgical resection is the preferred approach to treatment of localized NSCLCs, provided there is no medical contraindication to operative intervention. Lobectomy or more extensive resection (depending on tumor extent) is usually recommended rather than lesser surgeries, unless other comorbid conditions preclude these procedures.

Table 18.2 American Joint Commission on Cancer Lung Cancer Staging

Primary Tumor (T)

TX	Primary tumor cannot be assessed, or tumor proven by presence of malignant cells in sputum or bronchial washings but not visualized by imaging or bronchoscopy
T0	No evidence of primary tumor
Tis	Carcinoma *in situ*
T1	Tumor ≤ 3 cm in greatest dimension, surrounded by lung or visceral pleura, without bronchoscopic evidence of invasion more proximal than the lobar bronchus
T2	Tumor with any of the following features of size or extent: > 3 cm in greatest dimension, involves main bronchus ≥ 2 cm distal to the carina, invades visceral pleura, associated with atelectasis or obstructive pneumonitis that extends to the hilar region but does not involve the entire lung
T3	Tumor of any size that directly invades the chest wall, diaphragm, mediastinal pleura, parietal pericardium; or lies < 2 cm distal to the carina but without involvement of the carina; or is associated with atelectasis or obstructive pneumonitis of the entire lung
T4	Tumor of any size that invades the mediastinum, heart, great vessels, trachea, esophagus, vertebral body, carina; or has separate tumor nodule(s) in same lobe; or is associated with a malignant pleural effusion.

Regional Lymph Nodes (N)

NX	Regional lymph nodes cannot be assessed
N0	No regional lymph node metastasis
N1	Metastasis in ipsilateral peribronchial and/or ipsilateral hilar lymph nodes, including intrapulmonary nodes involved by direct extension of the primary tumor
N2	Metastasis in ipsilateral mediastinal and/or subcarinal lymph node(s)
N3	Metastasis in contralateral mediastinal, contralateral hilar, ipsilateral or contralateral scalene or supraclavicular lymph node(s).

Distant Metastasis (M)

MX	Distant metastasis cannot be assessed
M0	No distant metastasis
M1	Distant metastasis; includes separate tumor nodule(s) in a different lobe.

TNM Stage Groupings

Occult	T0	N0	M0
Stage 0	Tis	N0	M0
Stage IA	T1	N0	M0
Stage IB	T2	N0	M0
Stage IIA	T1	N1	M0
Stage IIB	T2	N1	M0
	T3	N0	M0
Stage IIIA	T1	N2	M0
	T2	N2	M0
	T3	N1	M0
	T3	N2	M0
Stage IIIB	Any T	N3	M0
	T4	Any N	M0
Stage IV	Any T	Any N	M1

Based on Greene, F. L., Page, D. L., Fleming, I. D., Fritz, A., Balch, C. M., Haller, D. G., and Morrow, M. (2002). *AJCC Cancer Staging Manual.* Springer, New York [4].

Intraoperative mediastinal lymph node sampling or dissection is also recommended for accurate pathologic staging and determination of therapy. Subsets of patients also benefit from chemotherapy and radiotherapy. For more advanced NSCLC and for SCLC, chemotherapy and radiotherapy are the primary treatment modalities [6]. Rare patients with limited-stage SCLCs can be considered for surgical resection with curative intent.

Common Molecular Genetic Changes in Lung Cancers

Development of lung cancer occurs with multiple, complex, stepwise genetic and epigenetic changes involving allelic losses, chromosomal instability and imbalance, mutations in tumor suppressor genes (TSGs) and dominant oncogenes, epigenetic gene silencing through promoter hypermethylation, and aberrant expression of genes participating in control of cell proliferation and apoptosis [7]. There are similarities as well as type-specific differences in the molecular alterations between NSCLCs and SCLCs, and between SqCCs and ACs [8–10]. Oncogenes that play a part in the pathogenesis of lung cancer include *MYC, K-RAS* (predominantly ACs), *Cyclin D1, BCL2,* and *ERBB* family genes such as EGFR (epidermal growth factor receptor) (predominantly ACs) and HER2/neu (predominantly ACs) [11,12]. Also, lung cancers often display abnormalities involving TSGs including *TP53, RB,* $p16^{INK4a}$, and new candidate TSGs on the short arm of chromosome 3 *(DUTT1, FHIT, RASFF1A, FUS-1, BAP-1)* [11,13]. As research advances, these lists continue to grow, and as knowledge has expanded about the roles of these genes in carcinogenesis and tumor behavior, new targeted therapeutic agents have been designed to treat this disease (Figure 18.1 and Table 18.3) [14]. Many other agents are under investigation.

In cancers, chromosomal regions harboring TSGs and oncogenes are often deleted or amplified. Allele loss involving loci in 3p14–23 is a consistent feature of lung cancer pathogenesis [15,16]. Wistuba et al. reported allelic losses of 3p, often multiple and discontinuous, in 96% of the lung cancers studied and in 78% of the precursor lesions [15]. Larger segments of allelic loss were noted in most SCLCs (91%) and SqCCs (95%) than in ACs (71%) and preneoplastic/preinvasive lesions [15]. There was allelic loss in the 600-kb 3p21.3 deletion region in 77% of the lung cancers; 70% of the normal or reneoplastic/preinvasive lesions associated with lung cancers; and 49% of the normal, mildly abnormal, or preneoplastic/preinvasive lesions found in smokers without lung cancer, but no loss was seen in the samples from people who had never smoked [15]. 8p21–23 deletions are also frequent and early events in the pathogenesis of lung

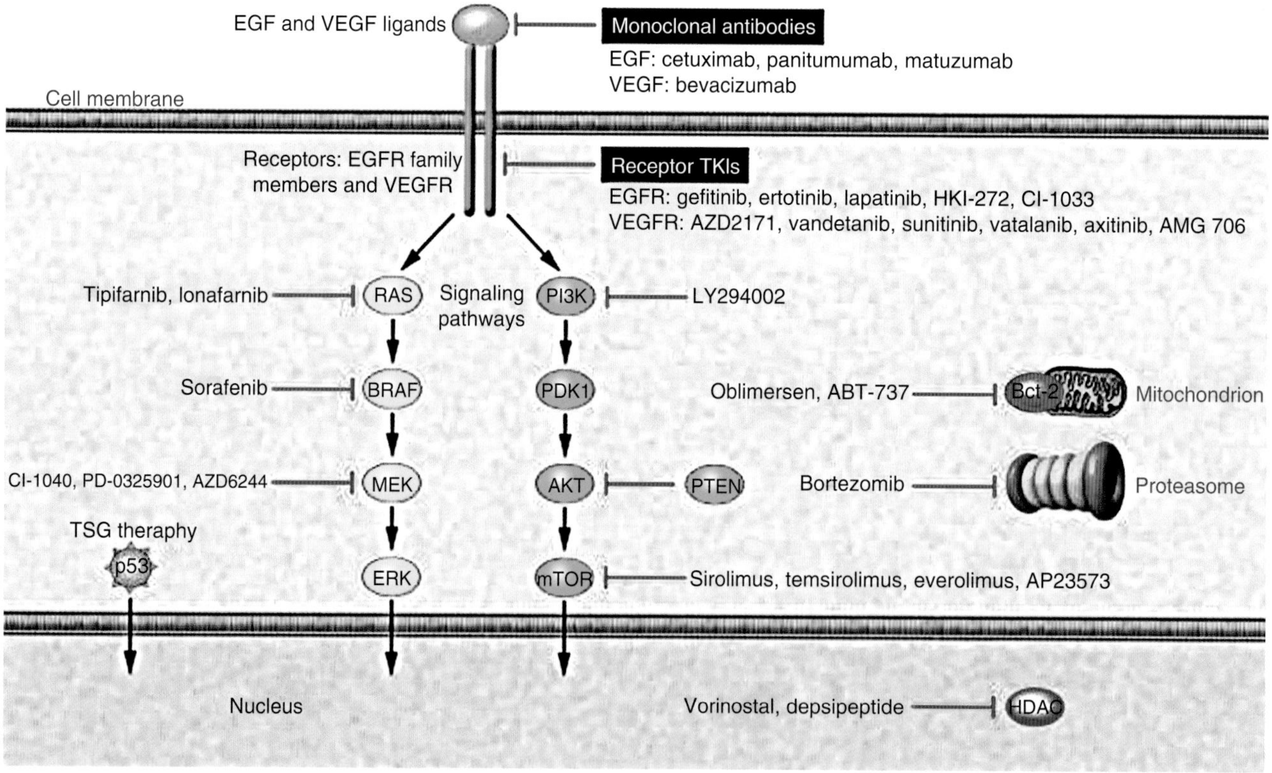

Figure 18.1 Targeted therapies are focused on key oncogenic pathways in lung cancer. These agents are designed to interfere with lung cancer cell proliferation, inhibition of apoptosis, angiogenesis, and invasion. EGFR = epidermal growth factor receptor; VEGFR = vascular endothelial growth factor receptor; TKIs = receptor tyrosine kinase inhibitors; TSG = tumor suppressor gene; PR = proteasome; PDK1 = pyruvate dehydrogenase kinase isoenzyme 1; PTEN = phosphatase and tensin homolog. Reprinted with kind permission from Sun, S., Schiller, J. H., Spinola, M., and Minna, J. D. (2007). New molecularly targeted therapies for lung cancer. *J Clin Invest* **117**, 2740–2750 [14].

Table 18.3 Selected Targeted Agents in Clinical Development for Lung Cancer Treatment

Target	Drug	Trade Name	Stage of Development in Lung Cancer
EGFR pathway inhibitors			
EGFR	Gefitinib	Iressa	Approved for advanced NSCLC
EGFR	Erlotinib	Tarceva	Approved for advanced NSCLC
EGFR	Cetuximab	Erbitux	Phase II/III
EGFR	Matuzumab		Phase I
EGFR	Panitumumab	Vectibix	Phase II
EGFR, HER2	Lapatinib	Tykerb	Phase II
EGFR, HER2	HKI-272		Phase II
EGFR, HER2, ERB4	CI-1033		Phase II
VEGF/VEGFR pathway inhibitors			
VEGF-A	Bevacizumab	Avastin	Approved for advanced NSCLC
VEGFR-2, EGFR	ZD6474; Vandetanib	Zactima	Phase II/III
VEGFR-1–3	AZD2171	Recentin	Phase II/III
VEGFR-1–3, PDGFR, c-KIT, FLT-3	SU11248; Sunitinib	Sutent	Phase II
VEGFR-1–3, PDGFR-β, c-KIT, c-fms	PTK787; Vatalanib		Phase II
VEGFR-1–3, PDGFR, c-KIT	AG-013736; Axitinib	Champix	Phase II
VEGFR-1–3, PDGFR, c-KIT	AMG 706		Phase I
Ras/Raf/MEK pathway inhibitors			
Ras	Tipifarnib (FTI)	Zarnestra	Phase III
Ras	Lonafarnib (FTI)	Sarasar	Phase III
Raf-1, VEGFR-2 and -3, PDGFR, c-KIT	BAY 43–9006; Sorafenib	Nexavar	Phase II
MEK	CI-1040		Phase II
MEK	PD-0325901		Phase I/II
MEK	AZD6244		Phase I
PI3K/Akt/PTEN pathway inhibitors			
PI3K	LY294002		Phase I
mTOR	Rapamycin; Sirolimus	Rapamune	Phase I
mTOR	CCI-779; Temsirolimus		Phase I/II
mTOR	RAD001; Everolimus		Phase I/II
mTOR	AP23573		Phase I
Tumor suppressor gene therapies			
p53	p53 retrovirus		Phase I
p53	p53 adenovirus (Ad5CMV-p53)	Advexin	Phase I
FUS1	FUS1 nanoparticle		Phase I
Proteasome inhibitors			
Proteasomes	Bortezomib	Velcade	Phase II
HDAC inhibitors			
HDAC	SAHA; Vorinostat	Zolinza	Phase II
HDAC	Depsipeptide		Phase I
Telomerase inhibitors			
Telomerase	GRN163L		Phase I

Reprinted with kind permission from Sun, S., Schiller, J. H., Spinola, M., and Minna, J. D. (2007). New molecularly targeted therapies for lung cancer. *J Clin Invest* **117,** 2740–2750 [14].

carcinomas [17], and other common alterations include LOH at 13q, 17q, 18q, and 22p [16].

Allelic losses that are more frequent in SqCCs than ACs include deletions at 17p13 (*TP53*), 13q14 (*RB*), 9p21 ($p16^{INK4a}$), 8p21–23, and several regions of 3p [11,15,17,18]. A recent study utilizing a bacterial artificial chromosome array to perform high-resolution whole genome profiling of SqCC and AC cell lines showed that regions of frequent amplification shared by both types of tumors included 5p; chromosome 7, 8q, 11q13, 19q, and 20q; and common regions of deletion included 3p, 4q, 9p, 10p, 10q; chromosome 18; and chromosome 21 [10]. However, ACs appeared to have higher frequencies of deletion of chromosome 6; 8p, 9q, 15q; and chromosome 16 than SqCCs, and possess small regions of amplification on chromosomes 12 and 14 not seen in SqCCs. Chromosome arms 2q and 13q were frequently deleted in AC but amplified in SqCC cell lines. Both types of tumors showed deletion of chromosome arm 17p, but it was more frequent in the SqCC cell lines, while amplification of chromosome 17p was more frequent in ACs. Amplification of chromosome 3q was common to both types of tumors but showed frequent alteration at 3q23–3q26 in the SqCC lines and at 3q22 in the AC lines.

Inactivation of recessive oncogenes is believed to occur through a two-stage process. It has been suggested that the first allelic inactivation occurs, often via a point mutation, and the second allele is later inactivated by a chromosomal deletion, translocation or other alteration such as methylation of the gene promoter region [19]. Inactivating mutations in the TSG *TP53*, which encodes the p53 protein, are the most frequent mutations in lung cancers. These mutations are found in up to 50% of NSCLCs and over 70% of SCLCs, and are largely attributable to direct DNA damage from cigarette smoke carcinogens [20]. *TP53* mutational patterns show a prevalence of G to T transversions in 30% of smokers' lung cancers versus only 12% of lung cancers in nonsmokers [20]. p53 protein is a transcription factor and a key regulator of cell cycle progression; cellular signals induced by DNA damage, oncogene expression, or other stimuli trigger p53-dependent responses including initiating cell cycle arrest, apoptosis, differentiation, and DNA repair [21]. Loss of p53 function in tumor cells can result in inappropriate progression through the dysregulated cell cycle checkpoints and permits the inappropriate survival of genetically damaged cells [22].

The $p16^{INK4a}$-cyclin D1–CDK4–Rb pathway, which plays a central role in controlling the G1 to S phase transition of the cell cycle, is another important tumor suppressor pathway that is often disrupted in lung cancers. It interfaces with the p53 pathway through $p14^{ARF}$ and $p21^{Waf/Cip1}$. Thirty percent to 70% of NSCLCs contain mutations of $p16^{INK4a}$, including homozygous deletion or point mutations and epigenetic alterations, leading to $p16^{INK4a}$ inactivation [22]. Almost 90% of SCLCs and smaller numbers of NSCLCs, on the other hand, display loss of Rb expression [23], and mutational mechanisms usually responsible include deletion, nonsense mutations, and splicing abnormalities that lead to truncated Rb protein [22]. $p16^{INK4a}$ leads to hypophosphorylation of the Rb protein, which causes arrest of cells in the G1 phase. The active, hypophosphorylated form of Rb regulates other cellular proteins including the transcription factors E2F1, E2F2, and E2F3, which are essential for progression through the G1/S phase transition. Loss of $p16^{INK4a}$ protein or increased complexes of cyclin D-CDK4–6 or cyclin E-CDK2 lead to hyperphosphorylation of Rb with resultant evasion of cell cycle arrest and progression into S phase [21,23]. Cell cycle progression is inhibited by $p21^{Waf/Cip1}$ through its inhibition of the cyclin complexes. The 10%–30% of NSCLCs lacking detectable alterations in $p16^{INK4a}$ and Rb may have abnormalities of cyclin D1 and CDK4, which cause inactivation of the Rb pathway [22]. Figure 18.2 provides an overview of the p53 and retinoblastoma (Rb) pathways, showing the complex interactions between the components [21].

Epigenetic alterations (hypermethylation of the 5′ CpG island) of TSGs are also frequent occurrences during pulmonary carcinogenesis, and methylation profiles of NSCLCs show relationships to smoke exposure, histologic type, and geography. Methylation rates of $p16^{INK4a}$ and *APC* and the mean methylation index (MI) (a reflection of the overall methylation status) in current or former smokers were significantly higher than in never smokers; the mean MI of tumors was highest in current smokers; methylation rates of *APC, CDH13,* and *RARbeta* were significantly higher in ACs than in SqCCs; methylation rates of *MGMT* and *GSTP1* in cases from the United States and Australia significantly exceeded those from Japanese and Taiwanese cases; and no significant gender-related differences in methylation patterns were found [24].

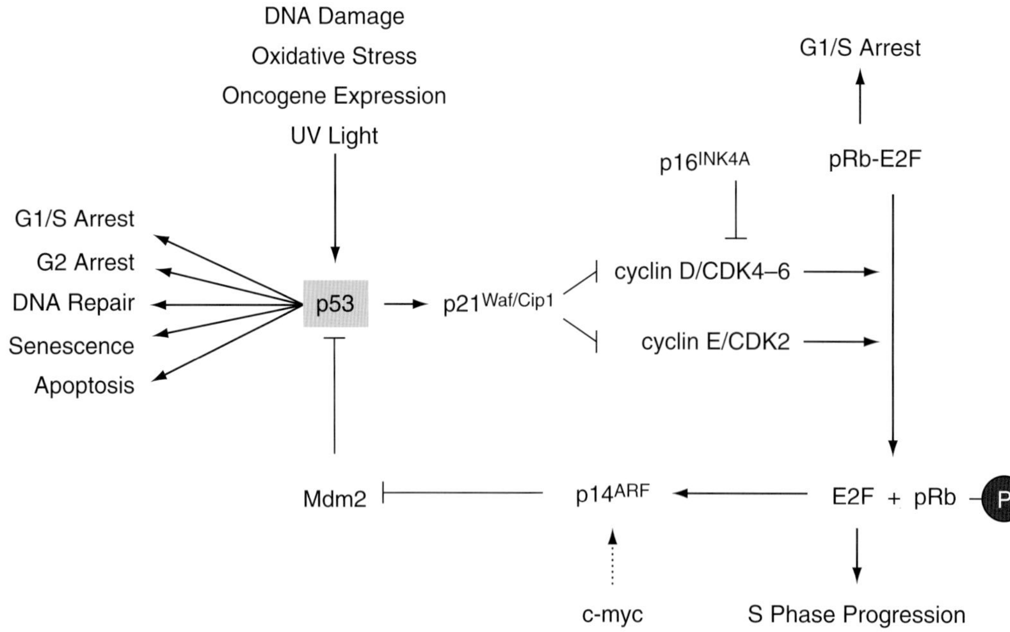

Figure 18.2 The p53 and retinoblastoma (Rb) pathways. A phosphate residue on Rb protein is indicated with a blue P. UV = ultraviolet. Reprinted with kind permission of Springer Science+Business Media, from Stelter, A. A. and Xie, J. (2008). Molecular oncogenesis of lung cancer. In *Molecular Pathology of Lung Diseases* (Zander, D. S., Popper, H. H., Jagirdar, J., Haque, A. K., Cagle, P. T., and Barrios, R., eds.), pp. 169–175, Springer, New York, NY [21].

Proto-oncogene activation and growth factor signaling are important in pulmonary carcinogenesis. The tyrosine kinase epidermal growth factor receptor *(EGFR)* is frequently mutated in NSCLCs, particularly in ACs, and the mutational status is important in determining response to tyrosine kinase inhibitors. A related pathway, the phosphoinositide 3-kinase (PI3K)/Akt/mammalian target of rapamycin (mTOR) pathway, is frequently deregulated in pulmonary carcinogenesis. As reviewed by Marinov et al., this pathway has been reported to mediate the effects of several tyrosine kinase receptors, including EGFR, c-Met, c-Kit, and IGF-IR, on proliferation and survival in NSCLC and SCLC [25]. Clinical trials are ongoing, investigating the efficacy of the mTOR inhibitor rapamycin and its analogues on lung cancer [26]. HER2/neu is another related receptor tyrosine kinase that is upregulated in approximately 20%–30% of NSCLCs [27,28], but unlike the situation with HER2/neu-positive breast cancers, treatment with anti-HER2/neu antibody (trastuzumab) does not seem to yield comparable benefits for NSCLC when used alone or in combination with chemotherapy [28,29]. Point mutations of RAS family proto-oncogenes (most often at *K-RAS* codons 12, 13, or 61) are detected in 20%–30% of lung ACs and 15%–50% of all NSCLCs [22]. Although farnesyl transferase inhibitors prevent Ras signaling, these agents have not shown significant activity as single-agent therapy in untreated NSCLC or relapsed SCLC [30]. MYC family genes *(MYC, MYCN,* and *MYCL),* which play roles in cell cycle regulation, proliferation, and DNA synthesis, are more frequently activated in SCLCs than in NSCLCs, either by gene amplification or by transcriptional dysregulation [22].

Vascular endothelial growth factor (VEGF) is a homodimeric glycoprotein that is overexpressed in many lung cancers and directly stimulates endothelial cell proliferation, promotes endothelial cell survival in newly formed vessels, and induces proteases involved in the degradation of the extracellular matrix needed for endothelial cell migration [31]. Its angiogenic effects are mediated by three receptors: VEGFR-1, VEGFR-2, and VEGFR-3; ligand binding leads to tyrosine kinase activation and activation of the signaling pathways required for angiogenesis [31]. Monoclonal antibodies to VEGF (bevacizumab) and tyrosine kinase inhibitors to VEGFRs have been developed and show promise for treatment of NSCLC. A phase III trial of bevacizumab showed significantly improved overall and progression-free survival when this agent was used in combination with standard first-line chemotherapy in patients with advanced NSCLC, and several small-molecule VEGFR tyrosine kinase inhibitors have yielded favorable results in phase I and II trials in NSCLC [32].

MicroRNAs are a recently discovered class of nonprotein-coding, endogenous, small RNAs which regulate gene expression by translational repression, mRNA cleavage, and mRNA decay initiated by miRNA-guided rapid deadenylation [33]. Some microRNAs such as *let-7* have been suggested to play roles in carcinogenesis by functioning as oncogenes or tumor suppressors, negatively regulating TSGs and/or genes that control cell differentiation or apoptosis [33]. Investigations of the therapeutic potential of microRNAs are also under way.

Adenocarcinoma and Its Precursors

Clinical and Pathologic Features

In the 2004 version of the WHO classification scheme, AC is defined as "a malignant epithelial tumour with glandular differentiation or mucin production, showing acinar, papillary, bronchioloalveolar or solid with mucin growth patterns or a mixture of these patterns" [34]. AC has become the most frequent histologic type of lung cancer in parts of the world. It occurs primarily in smokers, but represents the most common type of lung cancer in people who have never smoked and in women. A small subset of these tumors arise in patients with localized scars or diffuse fibrosing lung diseases such as asbestosis and interstitial pneumonia associated with scleroderma [35]. These neoplasms usually arise in the periphery of the lung, and are more likely to invade the pleura and chest wall than other histologic types of lung cancers. Radiologic studies can show one or more nodules, ground-glass opacities, or mixed solid and ground-glass lesions. On gross examination, the neoplasms are often solitary gray-white nodules or masses, sometimes with necrosis or cavitation, which pucker the overlying pleura. Mucin-producing tumors can have a glistening, gelatinous appearance. Other presentations include a pattern of consolidation resembling pneumonia (usually bronchioloalveolar carcinoma) (Figure 18.3), multiple nodules, diffuse interstitial widening due to lymphangitic spread, endobronchial lesions with submucosal infiltration, and diffuse visceral pleural infiltration and thickening resembling mesothelioma.

Common histologic patterns displayed by ACs include acinar (Figure 18.4), papillary, bronchioloalveolar (Figure 18.5, Figure 18.6), and solid arrangements, and

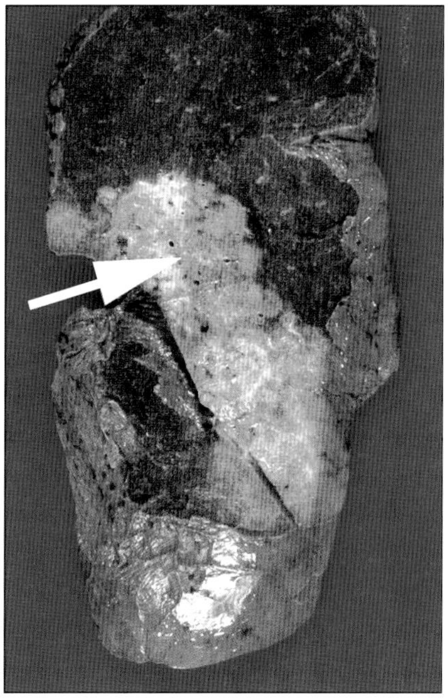

Figure 18.3 Bronchioloalveolar carcinoma. The tan tumor (arrow) replaces a large portion of the normal lung parenchyma.

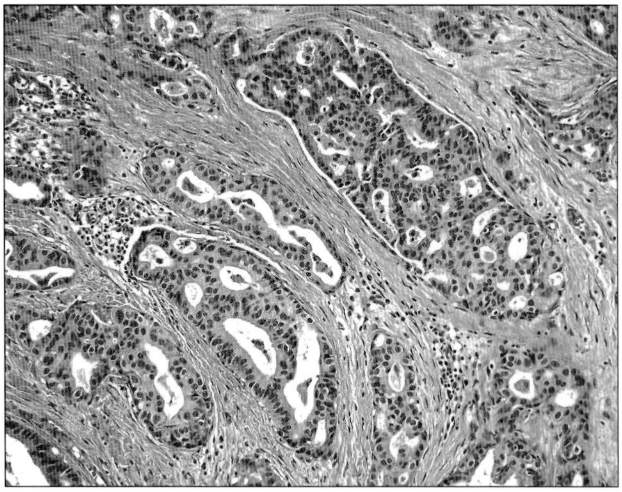

Figure 18.4 Adenocarcinoma with acinar pattern. The tumor consists of abnormal glands, some showing cribriform architecture, in a desmoplastic stromal background. The desmoplastic stroma has a dense collagenized appearance and reflects the presence of invasion. The abnormal glandular structures are lined by columnar tumor cells with abundant cytoplasm and mildly pleomorphic nuclei.

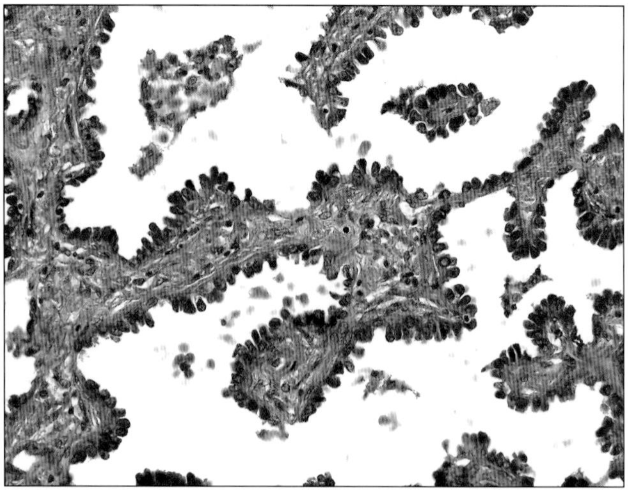

Figure 18.6 Adenocarcinoma with bronchioloalveolar pattern. Columnar tumor cells with a hobnail appearance line thickened alveolar septa. The tumor cells have enlarged, hyperchromatic nuclei. They remain on the surface of the alveolar septa and do not invade the lung tissue. This pattern is considered to represent an *in situ* lesion.

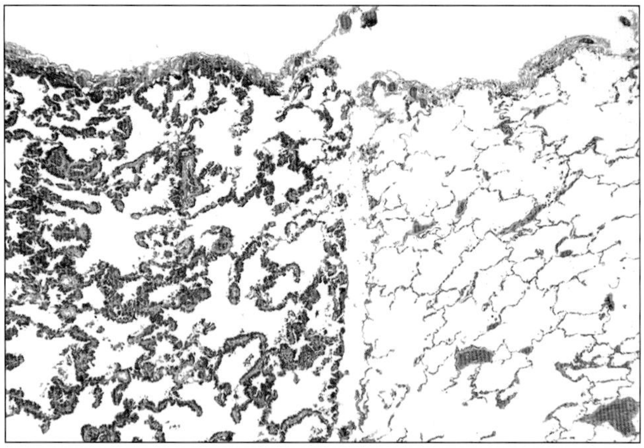

Figure 18.5 Adenocarcinoma with bronchioloalveolar pattern (left) compared with normal lung (right). Bronchioloalveolar carcinoma displays a lepidic growth pattern, in which tumor cells extend along alveolar septa, maintaining the alveolar architecture of the lung. This is illustrated by comparing the section of the figure on the left with that on the right. Notice that although open alveoli are present on both sides, the alveolar septa in the left portion are lined by tumor cells and have a thickened appearance, in contrast to the alveoli on the right, which have thin septa lined by flat pneumocytes.

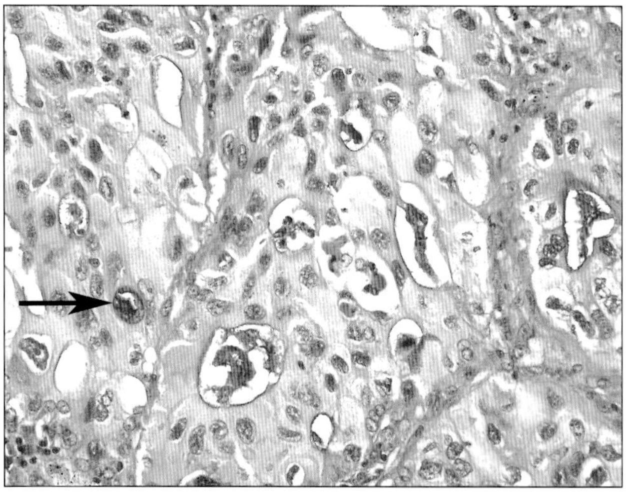

Figure 18.7 Adenocarcinoma (mucicarmine stain). Intracytoplasmic (arrow) and luminal mucin stains dark pink. The production of mucin indicates glandular differentiation.

mixtures of these patterns are very frequent. Less common histologic subtypes include fetal AC, mucinous (colloid) AC, mucinous cystadenocarcinoma, signet ring AC, and clear cell AC [34]. ACs usually exhibit differentiation toward Clara cells or type II pneumocytes or, less often, goblet cells. They manifest a range of differentiation extending from very well-differentiated tumors with extensive gland formation and little cytoatypia, to poorly differentiated, solid tumors that cannot be categorized as ACs unless one orders a mucin stain (Figure 18.7). However, most examples include readily identifiable glands. Invasiveness is reflected by the presence of neoplastic glands that infiltrate through stroma or pleura, stimulating a fibroblastic (desmoplastic) response (Figure 18.4), or by cells in the lumens of blood vessels or lymphatics.

In recent years, atypical adenomatous hyperplasia (AAH) has been recognized as a precursor lesion for peripheral pulmonary ACs. This lesion is defined as "a localized proliferation of mild to moderately atypical cells lining involved alveoli and, sometimes, respiratory bronchioles, resulting in focal lesions in peripheral

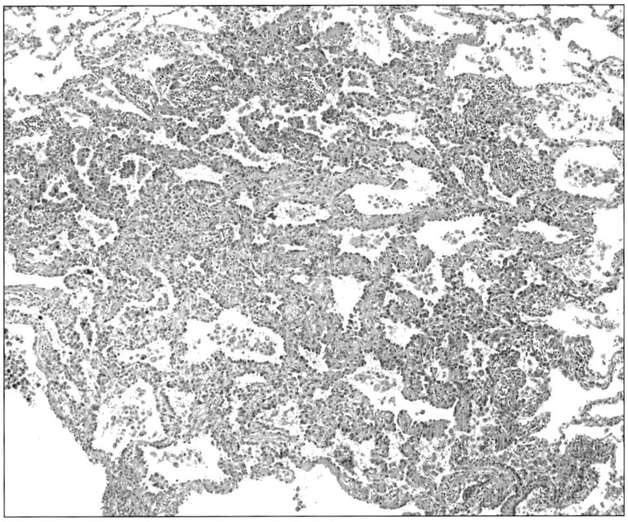

Figure 18.8 Atypical adenomatous hyperplasia. This lesion, which has been defined as a precursor lesion for peripheral pulmonary adenocarcinomas, consists of a well-circumscribed nodule measuring several millimeters in diameter, in which alveolar septa are lined by mildly moderate atypical cells.

alveolated lung, usually less than 5 mm in diameter and generally in the absence of underlying interstitial inflammation and fibrosis" (Figure 18.8) [36]. AAH exists on a histologic continuum with bronchioloalveolar carcinoma (BAC), which is defined as an *in situ* (noninvasive) form of AC, in which the neoplastic cells grow along alveolar septa (lepidic growth) without invasion of stroma or vasculature (Figure 18.5, Figure 18.6) [34]. Most BACs exceed 1 cm in diameter and consist of cells with greater degrees of cytoatypia than AAH. Although AAH is found in approximately 3% of patients without lung cancer at autopsy [37], it has been reported in 9%–21% of lung resection specimens with all types of primary lung cancer and 16%–35% of lung resection specimens with AC [36]. The progenitor cell for BAC and AAH is believed to be an epithelial cell located at the junction between the terminal bronchiole and alveolus, termed the bronchioalveolar stem cell [38].

Molecular Pathogenesis

A recently published large-scale study of primary lung ACs, using dense single nucleotide polymorphism arrays, described 57 significantly recurrent copy-number alterations in these tumors (Table 18.4) [12]. Twenty-six of 39 autosomal chromosome arms showed consistent large-scale copy-number gain or loss, and 31 recurrent focal events, including 24 amplifications and 7 homozygous deletions, were found.

Although some of the alterations involved regions known to harbor a proto-oncogene or TSG, these genes remain to be identified in some of the other regions affected. Amplification of chromosome 14q13.3 was the most common event noted, found in 12% of samples. This region includes *NKX2–1*, which encodes a lineage-specific transcription factor (thyroid transcription factor-1 [TTF-1]) that activates transcription of target genes including the surfactant proteins, and may be an important proto-oncogene involved in a significant fraction of lung ACs. Immunohistochemical staining for TTF-1 can be performed to detect expression of this factor in most lung adenocarcinomas, aiding in the determination of the lung as the site of origin of the tumor (Figure 18.9). Additional work using small interfering RNA (siRNA)-mediated knockdown of this gene in lung cancer cell lines with amplification led to reductions in tumor cell proliferation, through both decreased cell cycle progression and increased apoptosis, suggesting that gene amplification and overexpression contribute to lung cancer cell proliferation rates and survival [39].

EGFR and *K-RAS* mutations are mutually exclusive mutational events in AC of the lung, which suggests the existence of two independent oncogenic pathways [40,41]. EGFR is a receptor tyrosine kinase whose activation by ligand binding leads to activation of cell signaling pathways such as Ras/mitogen-activated protein kinase (MAPK) and phosphatidylinositol-3-kinase, which in turn propagates signals for proliferation, blocking of apoptosis, differentiation, motility, invasion, and adhesion [21]. Tumor-acquired mutations in the tyrosine kinase domain of EGFR, often associated with gene amplification, have been found in approximately 5%–10% of NSCLCs in the United States, and are associated with AC histology, never-smoker status, East Asian ethnicity, and female gender [14,40,42]. *EGFR* mutations are frequently in-frame deletions in exon 19, single missense mutations in exon 21, or in-frame duplications/insertions in exon 20, and occasional missense mutations and double mutations can also be detected [40,43]. *EGFR* mutation has an inverse correlation with methylation of the $p16^{INK4a}$ gene and SPARC (secreted protein acidic and rich in cysteine), an extracellular Ca2+-binding glycoprotein associated with the regulation of cell adhesion and growth [41]. *EGFR* status is an important predictor of response to EGFR kinase inhibitors: patients with *EGFR* mutations are most likely to have a significant response to EGFR tyrosine kinase inhibitor therapy, and EGFR amplification and protein overexpression have been reported to correlate with survival after EGFR tyrosine kinase inhibitor therapy [14,44]. K-Ras is a member of the Ras family of proteins, which function as signal transducers between cell membrane-based growth factor signaling and the MAPK pathways [21]. *K-RAS* mutations are associated with smoking, male gender, and poorly differentiated tumors [43]. *HER2* (also known as *EGFR2* or *ERBB2*), a member of the EGFR family of receptor tyrosine kinases, is mutated in less than 2% of NSCLC, and does not occur in tumors with *EGFR* or *K-RAS* mutation [45]. The *HER2* mutations are in-frame insertions in exon 20 and are significantly more frequent in ACs (2.8%), never smokers (3.2%), Asian ethnicity (3.9%), and women (3.6%), similar to *EGFR* mutations [45].

Alterations in DNA methylation appear to be important epigenetic changes in cancer, contributing to chromosomal instability through global hypomethylation, and aberrant gene expression through alterations in the methylation levels at promoter CpG islands [46].

Table 18.4 Top Focal Regions of Amplification and Deletion

Cytoband*	q Value	Peak Region (Mb)*	Max/Min Inferred Copy No.	Number of Genes* #	Known Proto-Oncogene/Tumor Suppressor Gene in Region*^	New Candidate(s)
Amplifications						
14q13.3	2.26×10^{-29}	35.61–36.09	13.7	2	–	*NKX2–1, MBIP*
12q15	1.78×10^{-15}	67.48–68.02	9.7	3	*MDM2*	–
8q24.21	9.06×10^{-13}	129.18–129.34	10.3	0	*MYC@*	–
7p11.2	9.97×10^{-11}	54.65–55.52	8.7	3	*EGFR*	–
8q21.13	1.13×10^{-7}	80.66–82.55	10.4	8	–	–
12q14.1	1.29×10^{-7}	56.23–56.54	10.4	15	*CDK4*	–
12p12.1	2.83×10^{-7}	24.99–25.78	10.4	6	*KRAS*	–
19q12	1.60×10^{-6}	34.79–35.42	6.7	5	*CCNE1*	–
17q12	2.34×10^{-5}	34.80–35.18	16.1	12	*ERBB2*	–
11q13.3	5.17×10^{-5}	68.52–69.36	6.5	9	*CCND1*	–
5p15.33	0.000279	0.75–1.62	4.2	10	*TERT*	–
22q11.21	0.001461	19.06–20.13	6.6	15	–	–
5p15.31	0.007472	8.88–10.51	5.6	7	–	–
1q21.2	0.028766	143.48–149.41	4.6	86	*ARNT*	–
20q13.32	0.0445	55.52–56.30	4.4	6	–	–
5p14.3	0.064673	19.72–23.09	3.8	2	–	–
6p21.1	0.078061	43.76–44.12	7.7	2	–	*VEGFA*
Deletions						–
9p21.3	3.35×10^{-13}	21.80–22.19	0.7	3	*CDKN2A/ CDKN2B*	–
9p23	0.001149	9.41–10.40	0.4	1	–	*PTPRD*‖
5q11.2	0.005202	58.40–59.06	0.6	1	–	*PDE4D*
7q11.22	0.025552	69.50–69.62	0.7	1	–	*AUTS2*
10q23.31	0.065006	89.67–89.95	0.5	1	*PTEN*	–

*Based on hg17 human genome assembly.
#Ref Seq genes only.
^Known tumor suppressor genes and proto-oncogenes defined as found in either COSMIC30, CGP Census31, or other evidence; if there is more than one known proto-oncogene in the region, only one is listed (priority for listing is, in order: known lung adenocarcinoma mutation; known lung cancer mutation; other known mutation (by COSMIC frequency); listing in CGP Census).
@*MYC* is near, but not within, the peak region.
‖Single gene deletions previously seen, this study provides new mutations as well.
Reprinted with kind permission from Weir, B. A., et al (2007). Characterizing the cancer genome in lung adenocarcinoma. Nature 450, 893–898 [12].

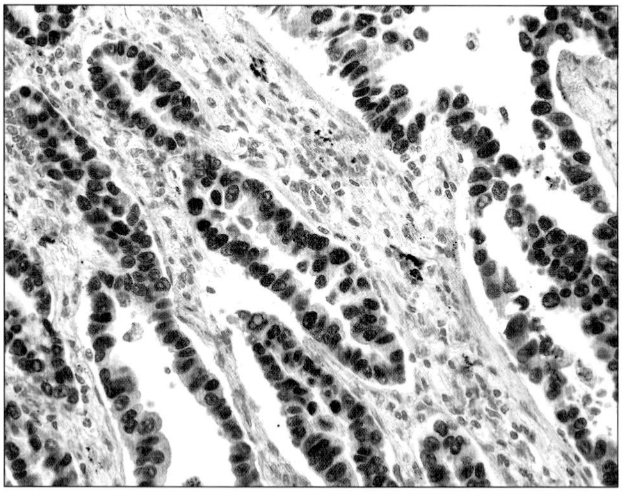

Figure 18.9 Adenocarcinoma (immunohistochemical stain for thyroid transcription factor-1 [TTF-1]). The brown-stained nuclei are positive for TTF-1. TTF-1 is expressed in the majority of pulmonary adenocarcinomas and small cell carcinomas, as well as in the thyroid.

Epigenetic differences exist between *EGFR*-mediated and *K-RAS*-mediated tumorigenesis, and may interact with the genetic changes. A recent study showed that the probability of having *EGFR* mutation was significantly lower among those with $p16^{INK4a}$ and *CDH13* methylation than in those without, and the methylation index was significantly lower in *EGFR* mutant cases than in wild-type. In contrast, *K-RAS* mutation was significantly higher in $p16^{INK4a}$ methylated cases than in unmethylated cases, and the methylation index was higher in *K-RAS* mutant cases than in wild-type [47].

Squamous Cell Carcinoma and Its Precursors

Clinical and Pathologic Features

SqCC is defined as "a malignant epithelial tumour showing keratinization and/or intercellular bridges that arises from bronchial epithelium," in the WHO classification scheme [48]. It is a common histologic type of NSCLC that is closely linked to cigarette smoking. In most patients, this tumor arises in a mainstem, lobar, or segmental bronchus, producing a central mass on imaging studies. Many of these tumors have an endobronchial component that can cause airway obstruction, leading to postobstructive pneumonia, atelectasis, or bronchiectasis. Not infrequently, it is the pneumonia that prompts evaluation of the patient and leads to discovery of the tumor. Less often, SqCCs develop in the periphery of the lung.

Gross examination reveals a tan or gray mass that usually arises in a large bronchus and often includes an endobronchial component (Figure 18.10, Figure 18.11). Partial or complete airway obstruction can be associated with changes of pneumonia, bronchitis, abscess, bronchiectasis, or atelectasis. Necrosis and cavitation are very common in these tumors. Involvement of hilar lymph nodes by tan-gray tumor can be visible in some resected specimens. Microscopically, the key features of this tumor are its keratinization, sometimes with formation of keratin pearls, and intercellular bridges (Figure 18.12). As is true of ACs, the degree of differentiation of this tumor varies from very well differentiated cases, in which there are abundant keratinization and intercellular bridges and little cytoatypia, to very poorly differentiated cases, in which keratinization and intercellular bridges can be quite inconspicuous and the tumor consists of sheets of large atypical cells with marked cytoatypia and frequent mitoses. However, most cases fall more toward the middle of the spectrum. Invasiveness is reflected by the presence of irregular nests and sheets of cells that infiltrate through tissues, stimulating a fibroblastic response, or by cells inside vascular or lymphatic spaces.

Invasive SqCCs are often accompanied by SqCC *in situ* and dysplasia, their precursor lesions. These lesions arise in the bronchi and may be contiguous with the invasive tumor or exist as one or more separate foci. These precursor lesions can also be observed without coexisting invasive carcinoma. Like SqCC, tobacco smoking is the main predisposing factor for SqCC *in situ* and dysplasia. Unlike invasive SqCC, however, these lesions are not invasive—they do not extend through the basement membrane of the bronchial epithelium. Grossly, they may be invisible or appear as flat, tan or red discolorations of the bronchial mucosa, or tan wart-like excrescences. Microscopically, these lesions encompass a

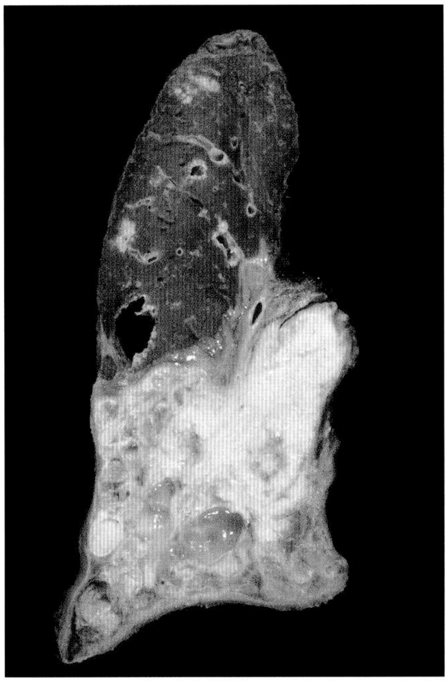

Figure 18.10 **Invasive squamous cell carcinoma with postobstructive pneumonia and abscesses.** This tan tumor lies in the central (perihilar) area of the lung and replaces the normal lung tissue. Distal to the tumor, the lung has extensive cystic changes reflecting abscesses and bronchiectasis, as well as a background of tan consolidation representing pneumonia.

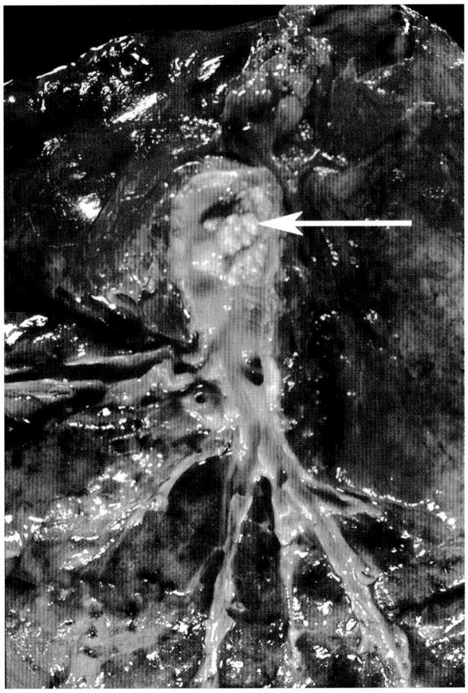

Figure 18.11 **Squamous cell carcinoma.** The tumor has an endobronchial component (arrow) that partially obstructs the airway lumen and has a warty appearance.

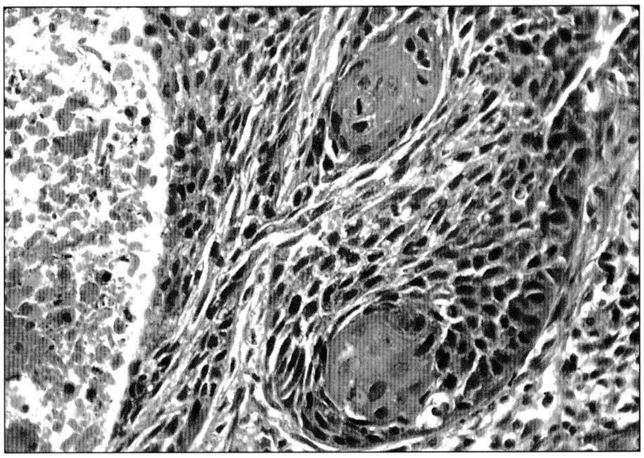

Figure 18.12 **Invasive squamous cell carcinoma.** This tumor consists of cells with hyperchromatic, pleomorphic nuclei and eosinophilic cytoplasm. Two keratin pearls are present (center) and a portion of the tumor is necrotic (left).

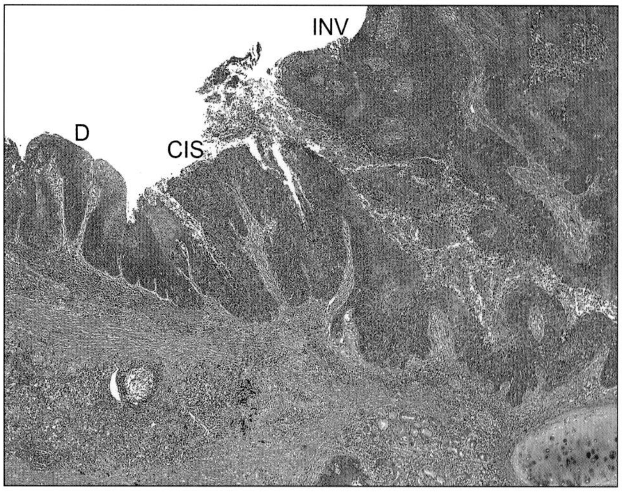

Figure 18.13 Dysplasia, squamous cell carcinoma *in situ*, and invasive squamous cell carcinoma. The dysplastic squamous epithelium (D) demonstrates increased thickness of the basal layer with mild squamous atypia. The atypia is full thickness in the area of carcinoma *in situ* (CIS), and in this area the entire epithelium consists of similar appearing cells with increased nuclear:cytoplasmic ratios. Invasion (INV) into the underlying bronchial tissues is present as well.

range of squamous changes that include alterations in the thickness of the bronchial epithelium, the maturational progress of squamous differentiation, cell size, and nuclear characteristics (Figure 18.13, Figure 18.14) [11,49]. As dysplasia increases from mild to moderate to severe, the epithelium thickens, and maturation is increasingly impaired. The basilar zone expands with epithelial cell crowding, the intermediate zone shrinks, and there is reduced flattening of the superficial squamous cells. Cell size, pleomorphism, and anisocytosis usually increase, and there is coarsening of the chromatin and appearance of nucleoli, nuclear angulations, and folding. In carcinoma *in situ*, although the epithelium may or may not be thickened and the cell size may be small, medium, or large, there is minimal or no maturation from the base to the superficial aspect, and the atypical nuclear features are present throughout the entire thickness of the epithelium. Mitoses appear in the lower third (mild or moderate dysplasia), lower two-thirds (severe dysplasia), or throughout the full thickness of the epithelium (carcinoma *in situ*).

Basal cells in the bronchial epithelium are believed to represent the progenitor cells for invasive SqCC, and the sequence of events leading to SqCC is believed to include basal cell hyperplasia, squamous metaplasia, squamous dysplasia, carcinoma *in situ*, and invasive SqCC (Figure 18.14) [11,49–51]. Regression of lesions preceding invasive SqCC can occur, particularly the earlier lesions [52]. However, severe dysplasia and carcinoma *in situ* are associated with a significantly increased probability of developing invasive SqCC in patients followed over time with surveillance bronchoscopy [53].

Molecular Pathogenesis

Wistuba and colleagues evaluated SqCCs and precursor lesions for loss of heterozygosity (LOH) at 10 chromosomal regions (3p12, 3p14.2, 3p14.1–21.3, 3p21, 3p22–24, 3p25, 5q22, 9p21, 13q14 *RB*, and 17p13 *TP53*)

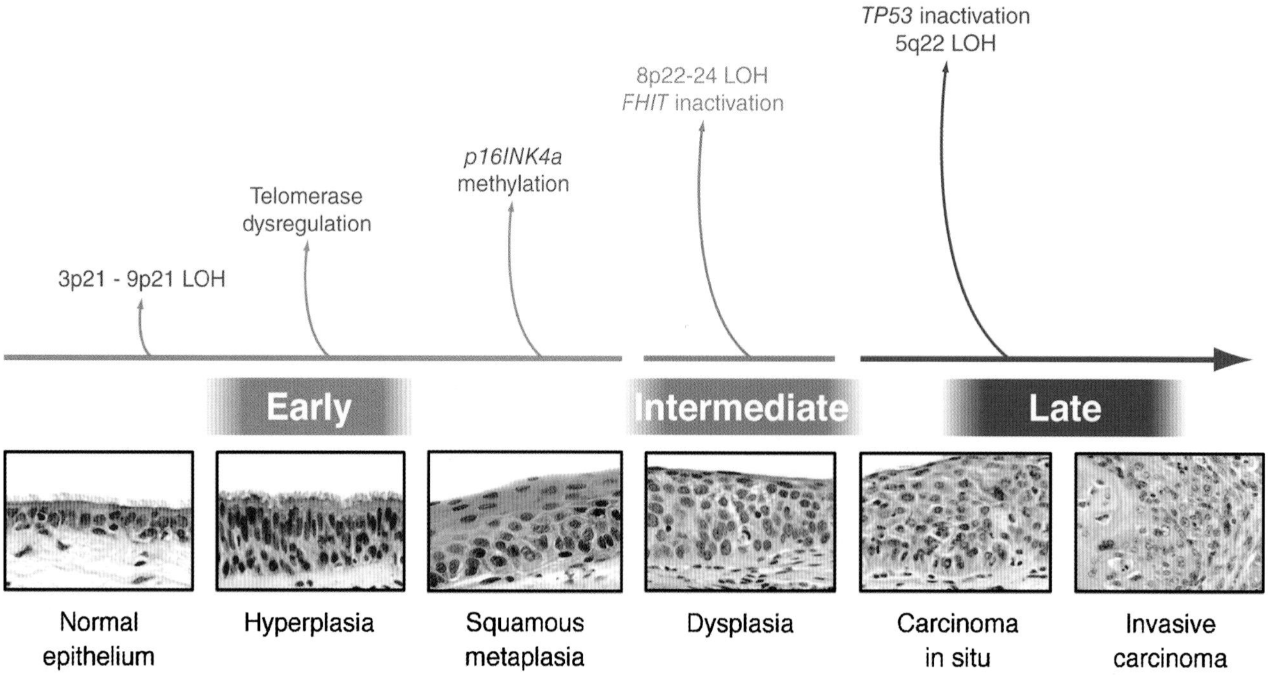

Figure 18.14 Histologic and molecular changes in the development of pulmonary squamous cell carcinoma. These changes occur in a stepwise fashion, beginning in histologically normal epithelium. LOH = loss of heterozygosity. Reprinted with kind permission from Wistuba, II and Gazdar, A. F. (2006). Lung cancer preneoplasia. *Annu Rev Pathol* **1**, 331–348 [11].

frequently deleted in lung cancer and found multiple, sequentially occurring allele-specific molecular changes in separate, apparently clonally independent foci, early in the pathogenesis of SqCCs of the lung, suggesting a field cancerization effect [11,18]. They observed clones of cells with allelic loss at one or more regions in 31% percent of histologically normal epithelium and 42% of specimens with hyperplasia or metaplasia; increasing frequency of LOH within clones with increasing histopathologic lesional severity; the most frequent and earliest regions of allelic loss at 3p21, 3p22–24, 3p25, and 9p21; increasing size of the 3p deletions with progressive histologic changes; and *TP53* allelic loss in many histologically advanced lesions (dysplasia and CIS) [18]. An overview of the sequential molecular events leading to invasive SqCC is shown in Figure 18.14 [11].

Large Cell Carcinoma

Clinical and Pathologic Features

Large cell carcinoma is an undifferentiated NSCLC without light microscopic evidence of squamous or glandular differentiation, although squamous or glandular features may be detectable by ultrastructural examination (Figure 18.15) [54]. Histologic subtypes of large cell carcinoma include large cell neuroendocrine carcinoma (LCNEC), combined LCNEC, basaloid carcinoma, lymphoepithelioma-like carcinoma, clear cell carcinoma, and large cell carcinoma with rhabdoid phenotype [54]. Clinical signs and symptoms resemble those of other types of NSCLC. Most tumors develop as peripheral lung masses, except for basaloid carcinomas, which usually form centrally located masses. Histologically, large cell carcinomas consist of sheets and nests of large cells with vesicular nuclei, prominent nucleoli, and moderate or abundant amounts of cytoplasm. LCNECs demonstrate neuroendocrine architectural features and immunohistochemical or ultrastructural evidence of neuroendocrine differentiation. Basaloid carcinomas display nests of small, monomorphic, rounded or fusiform tumor cells with little cytoplasm, numerous mitoses, comedo-type necrosis, and hyaline or mucoid stromal degeneration. Clear cell carcinoma consists of large tumor cells with clear cytoplasm. Precursor lesions are not currently recognized for any of the subtypes of large cell carcinoma. However, basaloid carcinoma is associated with squamous dysplasia in about one-third of cases [54].

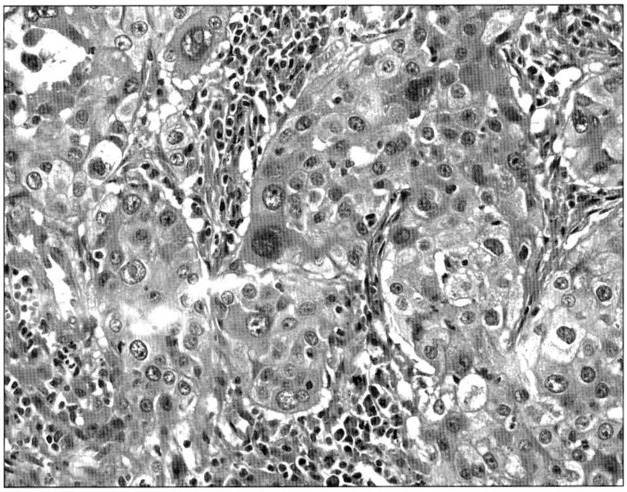

Figure 18.15 Large cell carcinoma. This poorly differentiated tumor displays large cell size with marked nuclear pleomorphism, large nucleoli, nuclear inclusions, and abundant eosinophilic cytoplasm. Evidence of squamous or glandular differentiation is not observed. The intervening stroma is inflamed.

Molecular Pathogenesis

Large cell carcinomas are poorly differentiated carcinomas that can demonstrate features of AC (most frequent), SqCC, or neuroendocrine differentiation when examined by immunohistochemistry, electron microscopy, or molecular methods [55].

These tumors often demonstrate losses of 1p, 1q, 3p, 6q, 7q, and 17p, and gains of 5q and 7p, more closely resembling ACs than other histologic types of lung cancer [56]. Common molecular abnormalities include *TP53* mutation, *C-MYC* amplification, and *p16* promoter hypermethylation, while *K-RAS* mutation is less common [55]. *EGFR* tyrosine kinase domain mutation is not characteristic of large cell carcinomas, and *EGFRvIII* (deletion mutations in the extracellular domain of *EGFR*) is uncommon [57,58].

Neuroendocrine Neoplasms and Their Precursors

Clinical and Pathologic Features

The major categories of pulmonary neuroendocrine (NE) neoplasms include small cell carcinoma (SCLC), large cell neuroendocrine carcinoma (LCNEC), typical carcinoid, and atypical carcinoid. SCLC and LCNEC are high-grade carcinomas, typical carcinoid is a low-grade malignant neoplasm, and atypical carcinoid occupies an intermediate position in the spectrum of biologic aggressiveness. In one large series, the 5-year and 10-year survival rates for typical carcinoid were 87% and 87%, 56% and 35% for atypical carcinoid, 27% and 9% for LCNEC, and 9% and 5% for SCLC, respectively [59]. By light microscopy, these tumors display NE architectural features including organoid nesting, a trabecular arrangement, rosette formation, and palisading. These patterns are more prominent in carcinoids than in LCNECs and may or may not be visible in individual SCLCs. Typical carcinoids contain fewer than 2 mitoses per 2 mm^2 (10 HPF) and lack necrosis (Figure 18.16), while atypical carcinoids show 2–10 mitoses per 2 mm^2 (10 HPF) or necrosis, which is often punctate [60]. SCLC consists of small, undifferentiated tumor cells with scant cytoplasm and finely granular chromatin and absent or inconspicuous nucleoli (Figure 18.17). Nuclear molding is characteristic, necrosis is common, and the mitotic rate is typically high, with a mean of over 60 mitoses per 2 mm^2 [61]. Combined

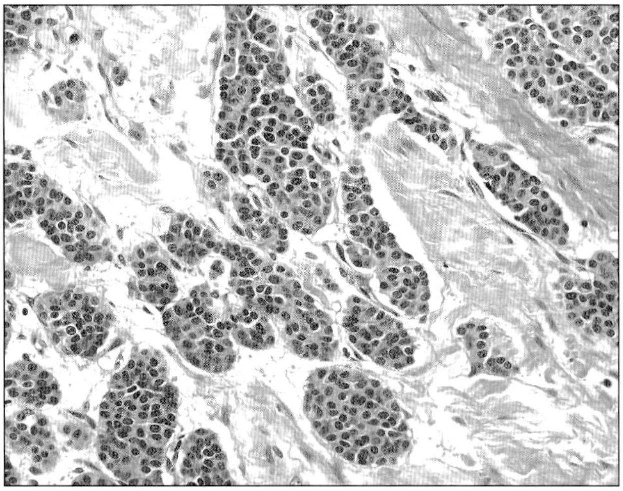

Figure 18.16 **Typical carcinoid.** This tumor consists of nests of uniform tumor cells with round or ovoid nuclei, fine chromatin, and little nuclear cytoatypia. A moderate amount of cytoplasm is present. No necrosis or mitoses are observed. The stromal background is hyalinized.

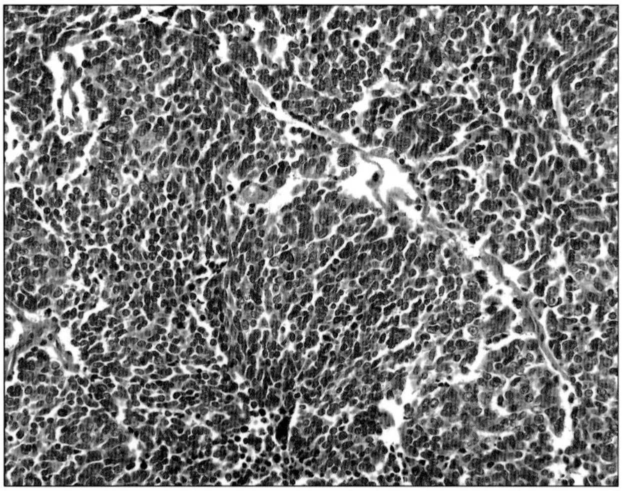

Figure 18.17 **Small cell carcinoma.** Small cell carcinomas typically display sheets of small tumor cells with scant cytoplasm and nuclei demonstrating fine chromatin. Numerous mitoses and apoptotic cells are characteristic.

SCLCs include an SCLC component accompanied by one or more histologic types of NSCLC. LCNECs consist of large tumor cells resembling those of large cell carcinoma, with NE architectural patterns, necrosis, a high mitotic rate (median of 70 per 2 mm^2), and NE differentiation reflected in immunohistochemical staining for one or more NE markers (chromogranin A, synaptophysin, leu-7 [CD57] or N-CAM [neural cell adhesion molecule, or CD56]) or the presence of neurosecretory granules on ultrastructural examination (Figure 18.18) [60].

Differences also exist in the characteristics of patients with carcinoids, as compared to patients with SCLC and LCNEC. Patients with carcinoids are typically younger and less likely to smoke than those with SCLCs and

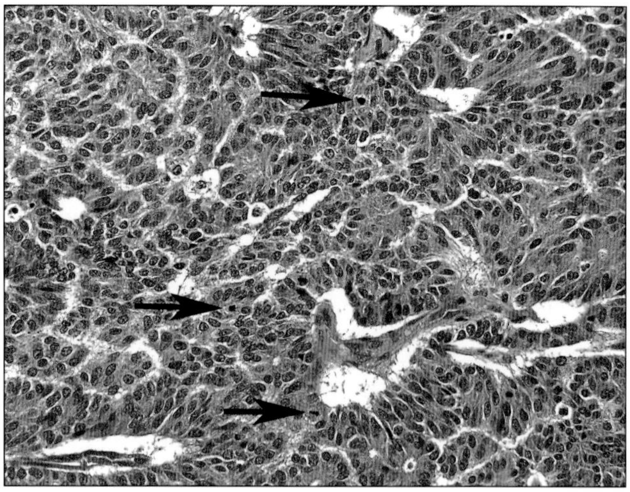

Figure 18.18 **Large cell neuroendocrine carcinoma.** The tumor forms rosettes, a feature that is commonly observed in low-grade neuroendocrine tumors. Although necrosis is not present, mitoses are numerous (several indicated by arrows) and exceed 10 mitoses per 2 mm^2, justifying classification as a large cell neuroendocrine carcinoma.

LCNECs, the vast majority of whom have a current or previous history of tobacco smoking [62,63]. Rare patients with carcinoids have the multiple endocrine neoplasia 1 (MEN1) syndrome, an association that is not seen with SCLCs and LCNECs. In addition, an association with diffuse idiopathic pulmonary neuroendocrine cell hyperplasia (DIPNECH) has been noted for carcinoids but not for SCLCs and LCNECs, leading to classification of DIPNECH as a preinvasive lesion in the most recent version of the WHO classification scheme [64]. DIPNECH is a diffuse proliferation of single cells, small nodules (NE bodies), and linear proliferations of pulmonary NE cells that may reside in the bronchial and/or bronchiolar epithelia (Figure 18.19), and may be accompanied by extraluminal proliferations

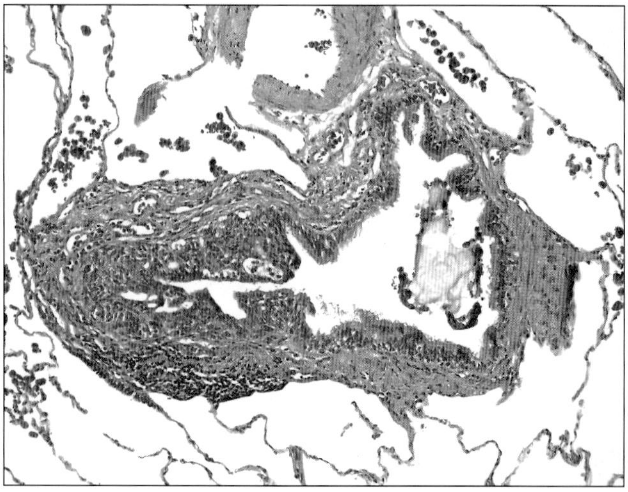

Figure 18.19 **Diffuse idiopathic pulmonary neuroendocrine cell hyperplasia.** A proliferation of neuroendocrine cells expands the epithelium of the left half of this bronchiole.

(tumorlets and carcinoids) [64]. However, morphologically identifiable precursor lesions for SCLC and LCNEC have not been established.

Molecular Pathogenesis

Molecular markers of pulmonary NE tumors include chromogranin A, synaptophysin (Figure 18.20), and N-CAM (CD56). These markers are expressed by all categories of NE tumors, with higher frequencies observed in the carcinoids and atypical carcinoids than in small cell and large cell neuroendocrine carcinomas. Gastrin-releasing peptide, calcitonin, other peptide hormones, the insulinoma-associated 1 *(INSM1)* promotor and the human achaete-scute homolog-1 *(hASH1)* gene have also been reported as overexpressed by these tumors [65,66]. Thyroid transcription factor-1 (TTF-1) is expressed by 80%–90% of SCLCs, 30%–50% of LCNECs, and 0%–70% of carcinoids [67–70].

SCLCs are aneuploid tumors with high frequencies of deletions on chromosomes 3p (including *ROBO1/DUTT1* [3p12.13], *FHIT* [3p14.2], *RASSF1* [3p21.3], *β-catenin* [3p21.3], *Fus1* [3p21.3], *SEMA3B* [3p21.3], *SEMA3F* [3p21.3], *VHL* [3p24.6], and *RARβ* [3p24.6]), 4q (including the proapoptotic gene *MAPK10* [4q21]), 5q, 10q (including the proapoptotic gene *TNFRSF6* [10q23]), 13q (location of the *Rb* gene), and 17p (*TP53*), and gains on 3q, 5p, 6p, 8q, 17q, 19q, and 20q [71–76]. More than 90% of SCLCs and SqCCs demonstrate large, often discontinuous segments of allelic loss on chromosome 3p, in areas encompassing multiple candidate tumor suppressor genes, including some of those listed previously [15,75]. Atypical carcinoids show a higher frequency of LOH at 3p, 13q, 9p21, and 17p than typical carcinoids, but not as high as the high-grade NE tumors [77]. Some typical and atypical carcinoids possess mutations of the *multiple endocrine neoplasia 1 (MEN1)* gene on chromosome 11q13 or LOH at this locus [78], while these abnormalities occur with lower frequencies in SCLCs and LCNECs, supporting separate pathways of tumorigenesis [79]. *MEN1* encodes for the nuclear protein menin, which is believed to play several roles in tumorigenesis by linking transcription factor function to histone-modification pathways, in part through interacting with the activator-protein-1 family transcription factor JunD, modifying it from an oncoprotein into a tumor suppressor protein [80]. Oncogenes frequently amplified in SCLCs include *MYC* (8q24), *MYCN* (2p24), and *MYCL1* (1p34), and additional amplified genes that represent candidate oncogenes include the antiapoptotic genes *TNFRSF4* (1p36), *DAD1* (14q11), *BCL2L1* (20q11), and *BCL2L2* (14q11) [76]. The Myc proteins are transcription factors that are important in cell cycle regulation, proliferation, and DNA synthesis, and can induce $p14^{ARF}$, leading to apoptosis through p53 if cellular conditions do not favor proliferation [21].

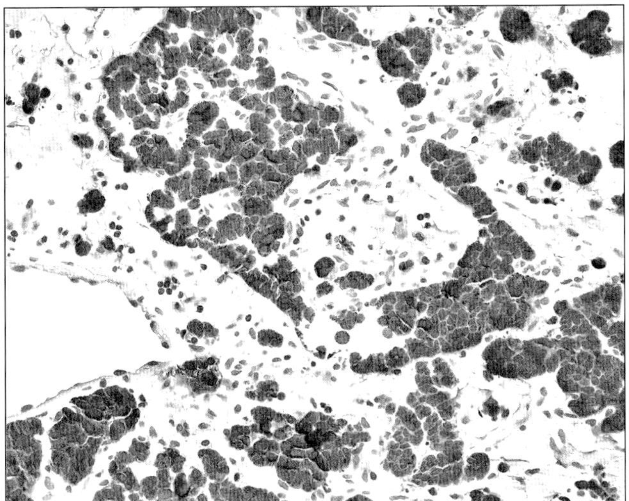

Figure 18.20 Carcinoid (immunohistochemical stain for synaptophysin). The brown-stained cytoplasm contains synaptophysin, a marker of neuroendocrine differentiation.

TSGs are inactivated in the majority of SCLCs. Eighty percent to 90% of SCLCs demonstrate *TP53* mutations, as compared to more than 50% of NSCLCs, fewer atypical carcinoids, and virtually no typically carcinoids [74,81]. Most of the *TP53* mutations in SCLCs are missense point mutations that result in a stabilized p53 mutant protein which can be easily detected by immunohistochemistry [71]. p53 protein overexpression occurs frequently in high-grade NE carcinomas, but is unusual in typical carcinoids and intermediate in atypical carcinoids [82,83]. Dysregulation of p53 produces downstream effects on Bcl-2 and Bax. Antiapoptotic Bcl-2 predominates over proapoptotic Bax in the high-grade NE carcinomas, while the reverse is true for carcinoids [82]. LCNECs resemble SCLCs in their high rates of *TP53* mutation and predominance of Bcl-2 expression over Bax expression [84].

Alterations compromising the $p16^{INK4a}$/cyclin D1/Rb pathway of G1 arrest are consistent in high-grade pulmonary NE carcinomas (92%), primarily through loss of Rb protein, but are less frequent in atypical carcinoids (59%) and are uncommon in typical carcinoids [23]. Mutations in the *RB1* gene exist in many SCLCs, with associated loss of function of the gene product [71,74,85]. In another study, 89% of the NE carcinomas (excluding carcinoids) versus 13% of the non-NE carcinomas exhibited LOH and loss of Rb-protein expression [86]. The hypophosphorylated form of Rb protein functions as a cell cycle regulator for G1 arrest; cyclin D1 overexpression and $P16^{INK4a}$ loss produce persistent hyperphosphorylation of Rb with consequent evasion of cell cycle arrest [23]. Recent data also suggest that in SCLCs, overexpression of *MDM2* (a transcriptional target of p53) or $p14^{ARF}$ loss leads to evasion of cell cycle arrest through the p53 and Rb pathway (Figure 18.2) [71].

The transcription factor E2F-1 appears to play a role in cellular proliferation by activating genes required for S phase entry. E2F-1 product is overexpressed in 92% of SCLCs and 50% of LCNECs, and is significantly associated with a high Ki67 index and Bcl-2:Bax ratio >1 [87]. A mediator of the proteasomal degradation of E2F-1, the S phase kinase-associated protein 2 (Skp2) F-box protein accumulates in high-grade NE carcinomas (86%), and its overexpression has been associated with advanced stage and nodal metastasis in pulmonary NE tumors [88]. In the high-grade NE tumors, Skp2 appears to interact with E2F-1 and stimulate its transcriptional activity toward the cyclin E promoter [87,88].

Telomeres play an important role in the protection of chromosomes against degradation. Telomerases, the enzymes that synthesize telomeric DNA strands, serve to counterbalance losses of DNA during cell divisions. High telomerase activity has been noted in over 80% of SCLCs and LCNECs [89–91] versus 14% or fewer typical carcinoids [91,92]. Expression of human telomerase mRNA component (hTERC) and human telomerase reverse transcriptase (hTERT) mRNA were reported, respectively, in 58% and 74% of typical carcinoids; and in 100% and 100% of atypical carcinoids, LCNECs and SCLCs, and telomere length alterations in LCNECs and SCLCs were greater than in typical carcinoids [92].

Aberrant methylation of cytosine-guanine (CpG) islands in promoter regions of malignant cells is an important mechanism for silencing of TSGs (epigenetic inactivation). Methylation of DNA involves the transfer of a methyl group, by a DNA methyltransferase, to the cytosine of a CpG dinucleotide [93]. *RASSF1A* is a potential TSG that undergoes epigenetic inactivation in virtually all SCLCs and a majority of NSCLCs through hypermethylation of its promoter region [94,95]. NE tumors have lower frequencies of methylation of *p16, APC,* and *CDH13* (H-cadherin) than NSCLCs [95]. SCLCs have higher frequencies of methylation of *RASSF1A, CDH1* (E-cadherin), and *RARβ* than carcinoids [95]. Promoter methylation of *CASP8*, which encodes the apoptosis-inducing cysteine protease caspase 8, was also found in 35% of SCLCs, 18% of carcinoids, and no NSCLCs, suggesting that *CASP8* may function as a TSG in NE lung tumors [96].

Although histologically defined precursors for SCLC are lacking, a higher incidence of genetic abnormalities is found in the normal or hyperplasic airway epithelium of patients with SCLC than NSCLC [97]. By extension, it has been suggested that SCLC may arise directly from histologically normal or mildly abnormal epithelium, rather than evolving through a sequence of recognizable histologic intermediary changes [11]. Relatively little is known about molecular abnormalities in precursors of carcinoids. Although carcinoids have been viewed as arising from tumorlets, 11q13 *(int-2)* allelic imbalance is significantly more common in carcinoids (73%) than in tumorlets (9%), and may represent an early event in carcinoid tumor formation [98]. The *int-2* gene lies in close proximity to *MEN1*, a tumor suppressor gene frequently mutated in NE tumors [98]. The molecular pathology of DIPNECH remains to be elucidated.

Mesenchymal Neoplasms

Mesenchymal neoplasms included in the WHO classification scheme (Table 18.1) encompass a spectrum of malignant and benign proliferations that show differentiation along multiple lineages. Overall, these tumors are much less common in the lung than are epithelial neoplasms. Information about molecular pathogenesis has emerged for some of the mesenchymal neoplasms. Pulmonary inflammatory myofibroblastic tumor (IMT) is a lesion composed of myofibroblastic cells, collagen, and inflammatory cells that primarily occurs in individuals less than 40 years of age, and is the most common endobronchial mesenchymal lesion in childhood (Figure 18.21) [99]. Synovial sarcoma is usually a soft tissue malignancy, but uncommonly arises in the pleura or the lung and often takes an aggressive course [100]. Pulmonary hamartomas are benign neoplasms consisting of mixtures of cartilage, fat, connective tissue, and smooth muscle, which present as coin lesions on chest radiographs and are excised in order to rule out a malignancy (Figure 18.22).

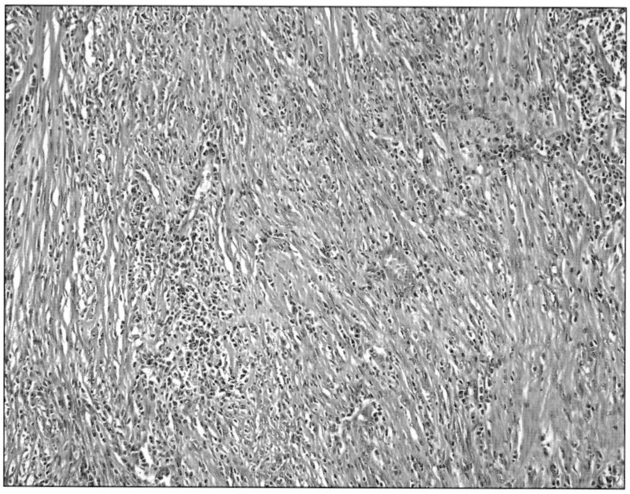

Figure 18.21 Inflammatory myofibroblastic tumor. The tumor consists of a proliferation of cytologically bland spindle cells in a background of collagen, with abundant lymphocytes and plasma cells.

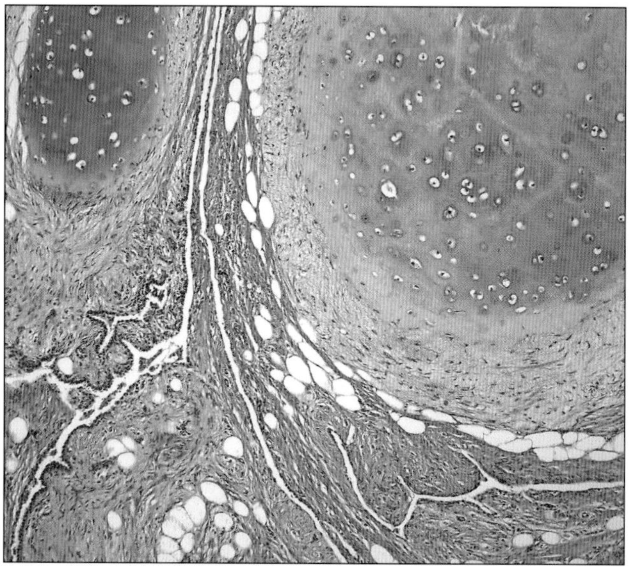

Figure 18.22 Hamartoma. A hamartoma typically includes the components of mature cartilage, adipose tissue, and myxoid or fibrous tissue, all of which are shown here.

Molecular Pathogenesis

Many IMTs demonstrate clonal abnormalities with rearrangements of chromosome 2p23 and the anaplastic lymphoma kinase *(ALK)* gene [101]. The rearrangements involve fusion of tropomyosin (TPM) N-terminal coiled–coil domains to the ALK C-terminal kinase domain, producing two ALK fusion genes, *TPM4–ALK* and *TPM3–ALK*, which encode oncoproteins with constitutive kinase activity [102]. Like their soft tissue counterparts, more than 90% of pulmonary and pleural synovial sarcomas demonstrate a chromosomal translocation t(X;18) *(SYT-SSX)* [103,104]. Detection of this translocation can be very helpful for confirming the diagnosis of synovial sarcoma in this unusual location. Most pulmonary hamartomas show abnormalities of chromosomal bands 6p21, 12q14–15, or other regions [105], corresponding to mutations of high-mobility group (HMG) proteins, a family of nonhistone chromatin-associated proteins that serve an important role in regulating chromatin architecture and gene expression [106].

Pleural Malignant Mesothelioma

Clinical and Pathologic Features

Malignant mesothelioma (MM) is an uncommon, aggressive tumor arising from mesothelial cells on serosal surfaces, primarily the pleura and peritoneum, and less often the pericardium or tunica vaginalis. The most important risk factor for MM is exposure to the subset of asbestos fibers known as amphiboles (crocidolite and amosite) [107]. The incidence of this tumor in the United States peaked in the early to mid-1990s, and appears to be declining, likely related to decreases in the use of amphiboles since their peak period of importation in the 1960s [107]. These tumors are characterized by long latency periods between asbestos exposure and clinical presentation of the tumor, with a mean of 30–40 years [108]. Radiation, a nonasbestos fiber known as erionite, and potentially other processes associated with pleural scarring have also been implicated in the causation of smaller numbers of cases of malignant mesothelioma [108], and a role for Simian virus 40 (SV40) in the genesis of this tumor has been suggested by some, but remains controversial [109,110].

Pleural MM most commonly arises in males over the age of 60. Presenting features typically include a hemorrhagic pleural effusion associated with shortness of breath and chest wall pain. Weight loss and malaise are common. By the time the tumor is discovered, patients usually have extensive involvement of the pleural surfaces. With progression, the tumor typically invades the lung, chest wall, and diaphragm. Lymph node metastasis can cause superior vena caval obstruction, and cardiac tamponade, subcutaneous nodules, and contralateral lung involvement can also occur. From the time of diagnosis, the median survival is 12 months [110]. Treatment may include surgery, chemotherapy, radiotherapy, immunotherapy, or other treatments, often in combination [110]. The intent of surgery is usually palliative. Whether extrapleural pneumonectomy with chemotherapy and radiotherapy can lead to cure is unclear [111]. New agents are currently under investigation for their potential to improve the life expectancy and quality of life in patients with this aggressive malignancy.

Gross pathologic features of MM include pleural nodules which grow and coalesce to fill the pleural cavity and form a thick rind around the lung. A firm tan appearance is common, and occasionally the tumor can have a gelatinous consistency (Figure 18.23). Extension along the interlobar fissures and invasion into the adjacent lung, diaphragm, and chest wall are characteristic. Further spread can occur into the pericardial cavity and around other mediastinal structures, and distant metastases can also develop.

Histologically, MM manifests a wide variety of histologic patterns. The major histologic categories include epithelioid mesothelioma, sarcomatoid mesothelioma, desmoplastic mesothelioma, and biphasic mesothelioma [108]. Epithelioid mesothelioma consists of round, ovoid, or polygonal cells with eosinophilic cytoplasm and nuclei that are usually round with little cytoatypia (Figure 18.24). These cells most often form sheets, tubulopapillary structures, or gland-like arrangements, and some tumors can have a myxoid appearance due to production of large amounts of hyaluronate. Sarcomatoid mesothelioma is composed of malignant-appearing spindle cells occasionally accompanied by mature sarcomatous components (osteosarcoma, chondrosarcoma, others). Desmoplastic mesothelioma can be a diagnostic challenge due to its frequently bland appearance and resemblance to organizing pleuritis. It consists of variably atypical spindle cells in a dense collagenous matrix (Figure 18.25). Helpful features for separating

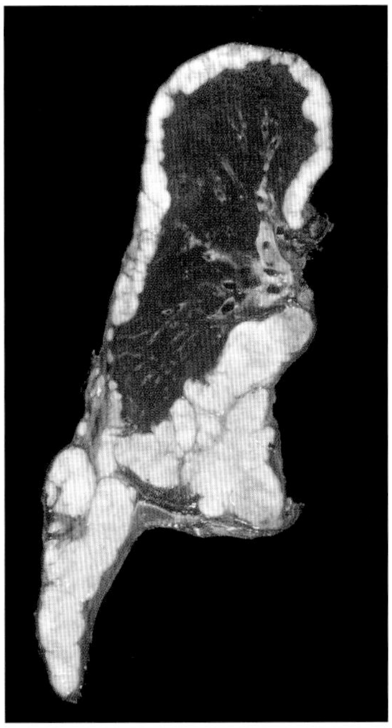

Figure 18.23 Malignant mesothelioma. The tan/white tumor involves the entire pleura surrounding and compressing the underlying parenchyma, which appears congested but relatively unremarkable.

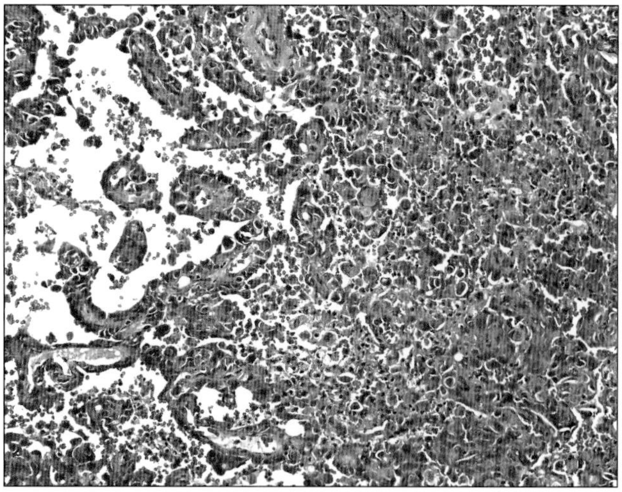

Figure 18.24 **Malignant mesothelioma, epithelioid.** This neoplasm consists of sheets of polygonal cells with pleomorphic nuclei and also forms some papillary structures (left).

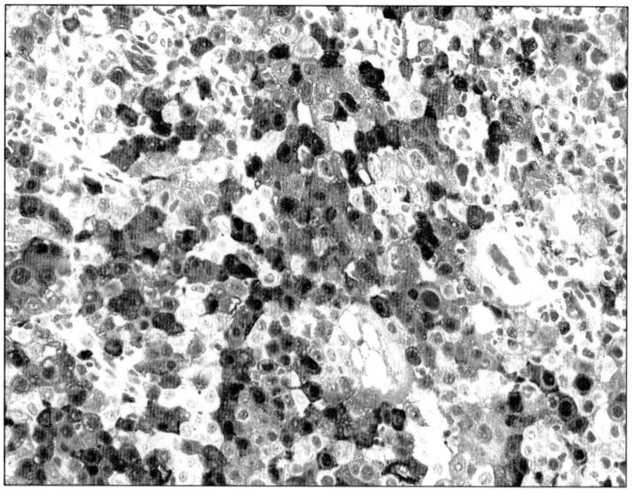

Figure 18.26 **Malignant mesothelioma (immunohistochemical stain for calretinin).** The tumor demonstrates cytoplasmic and nuclear staining (brown) for calretinin, which is expressed by many epithelioid malignant mesotheliomas.

this tumor from organizing pleuritis include invasion of chest wall muscle or adipose tissue and necrosis. Biphasic mesotheliomas include both epithelioid and sarcomatoid elements, each comprising at least 10% of the tumor [108].

Pathologic diagnosis of MM has been greatly assisted by the expanded availability of antibodies for use in immunohistochemistry [112]. Mesothelial differentiation can be supported by immunoreactivity with cytokeratin 5/6, calretinin (Figure 18.26), HBME-1, D2–40, and other antibodies. Histologic distinction of epithelioid mesotheliomas from metastatic ACs is a common need in practice, and a panel approach using calretinin and cytokeratin 5/6, with other antibodies reactive with ACs (CEA, MOC-31, Ber-EP4, leu M1, B72.3, and others) will usually be successful. Electron microscopy can also be helpful in difficult cases by demonstrating long thin microvilli in many MMs with an epithelioid component. Pan-cytokeratin staining is helpful for supporting a diagnosis of sarcomatoid or desmoplastic MM as opposed to sarcoma, since most (but not all) sarcomas will not stain for pan-cytokeratin. Other mesothelial and mesenchymal markers can also be useful for assisting in the differentiation of MM from histologically similar sarcomas.

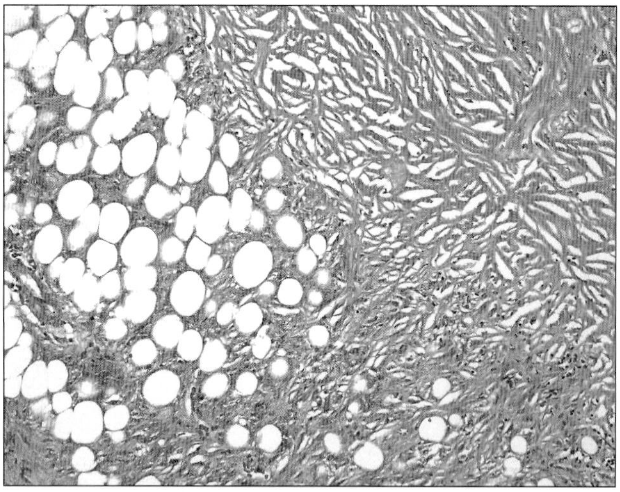

Figure 18.25 **Malignant mesothelioma, desmoplastic.** Abundant dense collagen is characteristic of this tumor, and is shown in the upper right. Tumor cells are spindle shaped and relatively cytologically bland. The slit-like spaces observed in the dense collagen are another frequent feature. The tumor infiltrates adipose tissue, which is helpful in confirming that the tumor is a mesothelioma, as opposed to organizing pleuritis.

Precursor lesions for MM have not been clearly defined from a histologic standpoint, although it is likely that an *in situ* stage exists [108]. The term atypical mesothelial hyperplasia has been recommended for surface (noninvasive) proliferations of mesothelial cells of uncertain malignant potential [108].

Molecular Pathogenesis

Exposure to asbestos fibers is believed to trigger the pathobiological changes leading to the majority of MMs. Currently, it is believed that asbestos may act as an initiator (genetically) and promoter (epigenetically) in the development of MMs [113]. The degree to which tumorigenesis results from direct interactions of the fibers with the mesothelial cells, or through other mechanisms involving oxidative stress (or both), is unresolved [113,114]. Multiple chromosomal alterations are often noted in MMs, and inactivation of TSGs plays an important part in the pathogenesis of MM [113]. A variety of genetic abnormalities have been reported including deletions of 1p21–22, 3p21, 4p, 4q, 6q, 9p21, 13q13–14, 14q, and proximal 15q, monosomy 22, and gains of 1q, 5p, 7p, 8q22–24, and 15q22–

25 [108,115]. The most common genetic abnormality in MM is a deletion in 9p21 encompassing the *CDKN2A* locus encoding the tumor suppressors $p16^{INK4a}$ and $p14^{ARF}$, which participate in the p53 and Rb pathways and inhibit cell cycle progression (Figure 18.2) [113,116]. Recent studies have shown that SV40 large T antigen (present in some MMs) inactivates the TSG products Rb and p53, raising the possibility that asbestos and SV40 could act as co-carcinogens in MM and suggesting that perturbations of Rb- and p53-dependent growth-regulatory pathways may be involved in the pathogenesis of MM [115]. Other common findings include inactivating mutations with allelic loss in the TSG neurofibromin 2 (*NF2)*, found at chromosome 22q12 [117], and inactivation of *CDKN2A/p14*^{ARF} and *GPC3* (another TSG) by promoter methylation [108]. Loss of *CDKN2A/ p14*^{ARF} also results in MDM2-mediated inactivation of p53 [116]. However, in MMs, unlike many other epithelial tumors, mutations in the *TP53, RB,* and *RAS* genes are rare [118].

The Wnt signal transduction pathway is also abnormally activated in MMs and appears to play a role in pathogenesis [119]. Activation of the pathway leads to accumulation of β-catenin in the cytoplasm and its translocation to the nucleus. Interactions with TCF/ LEF transcription factors promote expression of multiple genes including c-*myc* and C*yclin D*. The mechanism of activation does not appear to involve mutations in the *β-catenin* gene, but may instead involve more upstream components of the pathway, such as the disheveled proteins [119]. Recent evidence also suggests that the phosphatidylinositol 3-kinase (PI3–K/AKT) pathway is frequently activated in MMs, and that inhibition of this pathway can increase sensitivity to a chemotherapeutic agent [120]. The *Wilms' tumor gene (WT1)* is also expressed in most MMs, but its role in the pathogenesis of MM is unclear [114]. Finally, EGFR signaling in MMs has recently become a focus of greater attention, and there are some data showing that the EGFR is an early cell membrane target of asbestos fibers and is linked to activation of the MAPK cascade [113]. Unfortunately, a Phase II clinical trial of gefitinib treatment in patients with MMs did not show effectiveness, despite *EGFR* overexpression in over 97% of cases [121]. Another study found that common *EGFR* mutations conferring sensitivity to gefitinib are not prevalent in human malignant mesothelioma [122]. Further investigation continues into new, potentially efficacious agents for the treatment of MM.

NON-NEOPLASTIC LUNG DISEASE

Non-neoplastic pulmonary pathology comprises inflammatory and fibrosing diseases of the conducting airways, alveoli, vessels, and lymphoid tissue. This pathology may be localized or diffuse, may either have an obvious etiology or be idiopathic, and may cause injury that is reparable or irreparable. Most importantly, an understanding of non-neoplastic lung pathology plays a vital role in the clinical management of these diseases. This section covers the major types of obstructive and interstitial diseases, the vascular lesions, the pneumonias, the occupational diseases, the major histiocytic conditions, and the most common developmental anomalies. This list does not include all of the non-neoplastic diseases that can affect the lung, but it represents those that are responsible for the majority of illness. Also, the conditions highlighted within each of these categories are those about which we best understand the molecular biology of the disease mechanisms.

OBSTRUCTIVE LUNG DISEASES

Obstructive lung diseases are characterized by a reduction in airflow due to airway narrowing. This airflow reduction occurs, in general, by two basic mechanisms: (i) inflammation and injury of the airway, resulting in obstruction by mucous and cellular debris within and around the airway lumen; and (ii) destruction of the elastin fibers of the alveolar walls, causing loss of elastic recoil and subsequent premature collapse of the airway during the expiratory phase of respiration. There are four major obstructive lung diseases: asthma, emphysema, chronic bronchitis, and bronchiectasis.

Asthma

Clinical and Pathologic Features

Asthma is a chronic inflammatory disease of the airways that affects more than 150 million people worldwide. The prevalence of disabling asthma has increased over 200% since 1969, ranging from as low as 1% in rural Ethiopia to over 20% among children in parts of Central and South America [123]. In the United States, asthma affects approximately 8%–10% of the population and is the leading cause of hospitalization among children less than 15 years of age [123]. Clinically, the disease is defined as a generalized obstruction of airflow with a reversibility that can occur spontaneously or with therapy. It is characterized by recurrent wheezing, cough, or shortness of breath resulting from airway hyperactivity and mucus hypersecretion. The hyperresponsiveness is a result of acute bronchospasm and can be elicited for diagnostic purposes using histamine or methacholine challenges. The key feature of these symptoms is that they are variable—worse at night or in the early morning, and in some people worse after exercise. It has previously been assumed that these symptoms are separated by intervals of normal physiology. However, evidence is now accumulating that asthma can cause progressive lung impairment due to chronic morphologic changes in the airways. The treatment strategies for this complex disease are myriad. In atopic individuals, allergen avoidance should be the primary therapy. For example, in children, reducing exposure to house dust mites early in life decreases sensitization and the incidence of disease. For those who do develop the disease, avoidance of allergens later in life improves symptom control. Established treatments for asthma flairs include inhaled corticosteroids, and short-acting and long-acting β2-adrenoceptor agonists. Phosphodiesterase (PDE) inhibitors such as theophylline have been used for decades to treat asthmatic bronchoconstriction, but both cardiac and central nervous systems side effects have limited their use. Newer PDE inhibitors without side effects include non-xanthine drugs such as rofumilast.

The pathologic changes to the airways in asthma are very similar to those seen in chronic bronchitis. They consist of a thickened basement membrane with epithelial desquamation, goblet cell hyperplasia, and subepithelial elastin deposition. In the wall of the airway, smooth muscle hypertrophy and submucosal gland hyperplasia are also present (Figure 18.27). In acute asthma exacerbations, a transmural chronic inflammatory infiltrate with variable amounts of eosinophilia may be present, resulting in epithelial injury and desquamation that can become quite pronounced. One sees clumps of degenerating epithelial cells mixed with mucin in the lumen airway. These aggregates of degenerating cells are referred to as Creola bodies and can be seen in expectorated mucin from these patients. Also present in these sputum samples are Charcot-Leyden crystals, rhomboid-shaped structures that represent breakdown products from eosinophil cytoplasmic granules (Figure 18.28). The changes seen in the walls of these airways represent long-term airway remodeling caused by prolonged inflammation. This remodeling may play a role in the pathophysiology of asthma. The amount of airway remodeling is highly variable from patient to patient, but remodeling has been found even in patients with mild asthma. Currently, the effect of the treatment on this chronic pathology is unclear [124].

Molecular Pathogenesis

The pathogenesis of asthma is complex, and most likely involves both genetic and environmental components. Most experts now see it as a disease in which an insult initiates a series of events in a genetically susceptible

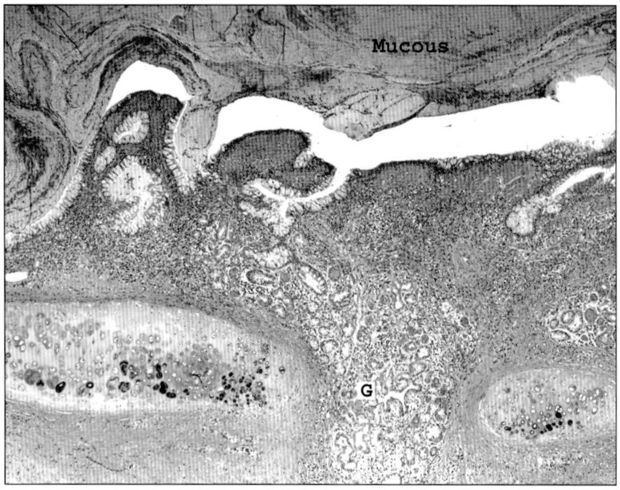

Figure 18.27 **Asthma.** The bronchial wall from a patient with asthma shows marked inflammation with eosinophils, mucosal goblet cell hyperplasia (G), and an increase in the smooth muscle.

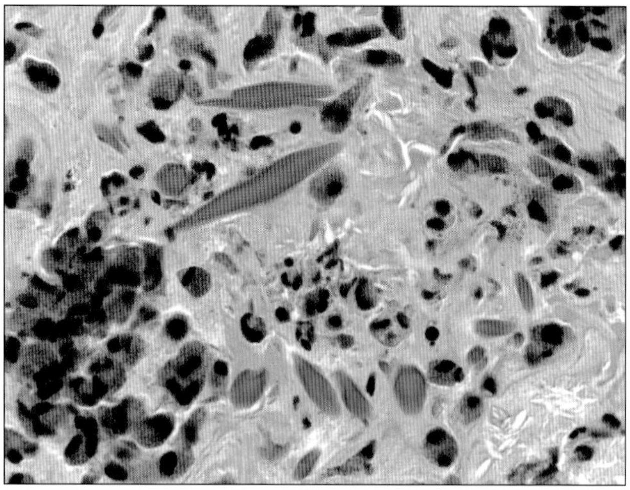

Figure 18.28 **Asthma.** Charcot-Leyden crystals are rhomboid-shaped structures within a mucous plug from an airway of an asthmatic patient. In addition, there are abundant eosinophils. These crystals are made of breakdown products of eosinophils, including major basic protein.

host. No single gene accounts for the familial component of this disease. Genetic analysis of these patients reveals a prevalence of specific HLA alleles, polymorphisms of *FcεRiB*, *IL-4*, and *CD14* [125,126]. Asthma can be classified using a number of different schema. Most commonly, asthma is divided into two categories: atopic (allergic) and nonatopic (nonallergic). Atopic asthma results from an allergic sensitization usually early in life and has its onset in early childhood. Nonatopic asthma is late-onset and, though the immunopathology has not been as well studied, probably has similar mechanisms to atopic asthma. Although this nosology is convenient for purposes of understanding the mechanisms of the disease, most patients manifest a combination of these two categories with overlapping symptoms.

Th0 pathogenetic mechanisms of both types encompass a variety of cells and their products. These include airway epithelium, smooth muscle cells, fibroblasts, mast cells, eosinophils, and T-cells. The asthma response includes two phases: an early response comprising an acute bronchospastic event within 15–30 minutes after exposure, and a late response that peaks approximately 4–6 hours and that can have prolonged effects. If one wants to understand this complex response, it is best to divide it into three components: (i) a type 1 hypersensitivity response, (ii) acute and chronic inflammation, and (iii) bronchial hyperactivity.

Type 1 Hypersensitivity In general, human asthma is associated with a predominance of Type 2 helper cells with a CD4+ phenotype. These Th2-type cells result from the uptake and processing of viral, allergen, and environmental triggers that initiate the episode. The processing includes the presentation of these triggers by the airway dendritic cells to naive T-cells (Th0), resulting in their differentiation into populations of Th1 and Th2. The Th2 differentiation is a result of IL-10 release by the dendritic cells, and the Th2 cells then further propagate the inflammatory reaction in two ways. First, they release a variety of cytokines such as IL-4, IL-5, and IL-13 that mediate a wide variety of responses. IL-4 and IL-13 stimulate B-cells and plasma cells to produce IgE, which, in turn, stimulates mast cell maturation and the release of multiple mediators, including histamine and leukotrienes. Second, these Th2 cells secrete IL-5 that, together with IL-4, also stimulates mast cells to secrete histamine, tryptase, chymase, and the cysteinyl leukotrienes causing the bronchoconstrictor response that occurs rapidly after the exposure to the allergen. IL-5 from these lymphocytes also recruits eosinophils to the airways and stimulates the release of the contents of their granules, including eosinophil cationic protein (ECP), major basic protein (MBP), eosinophil peroxidase, and eosinophil-derived neurotoxin. These compounds not only induce the bronchial wall hyperactivity but are also responsible for the increased vascular permeability that produces the transmural edema in the airways.

The cells can differentiate into Th1 cells as a result of IL-12 produced by dendritic cells. These Th1 cells produce interferon-gamma (IFN-γ), IL-2, and lymphotoxin, which play a role in macrophage activation in delayed-type hypersensitivity reactions as seen in diseases such as rheumatoid arthritis and tuberculosis [123]. These Th1 cells are predominantly responsible for defense against intracellular organisms and are more prominent in normal airways and in airways of patients with emphysema than in asthmatics. However, in severe forms of asthma, Th1 cells are recruited and have the capacity to secrete tumor necrosis factor (TNF)-α and IFN-γ, which may lead to the tissue-damaging immune response one sees in these airways (Figure 18.29) [127,128].

Acute and Chronic Inflammation The role of acute and chronic inflammatory cells, including eosinophils, mast cells, macrophages, and lymphocytes, in asthma is evident in the abundance of these cells in airways, sputum, and bronchoalveolar samples from patients with this disease. The number of eosinophils in the airways correlates with the severity of asthma and the amount of bronchial hyperresponsiveness. Proteins released by these cells including ECP, MCP, and eosinophil-derived neurotoxin cause at least some of the epithelial damage seen in the active form of asthma. Neutrophils are prominent in the more acute exacerbations of asthma and are probably recruited to these airways by IL-8, a potent neutrophil chemoattractant released by airway epithelial cells [123]. These cells also release proteases, reactive oxygen species (ROS), and other proinflammatory mediators that, in addition to the epithelial damage, also contribute to the airway destruction and remodeling that occurs in the more chronic forms of this disease. The susceptibility of the epithelium in asthma to this oxidant injury may be increased due to decreased antioxidants such as superoxide dismutase in these lungs [129]. Finally, mast cells are activated to release an abundance of mediators through the binding of IgE to FcεRI, high-affinity receptors on their surface. Allergens bind to IgE molecules and induce a cross-linking of these molecules, leading to activation of the mast cell and release of a number of mediators, most notably histamine, tryptase, and various leukotrienes, including leukotriene D_4 (LTD_4), and interact with the smooth muscle to induce contraction and the acute bronchospastic response [130].

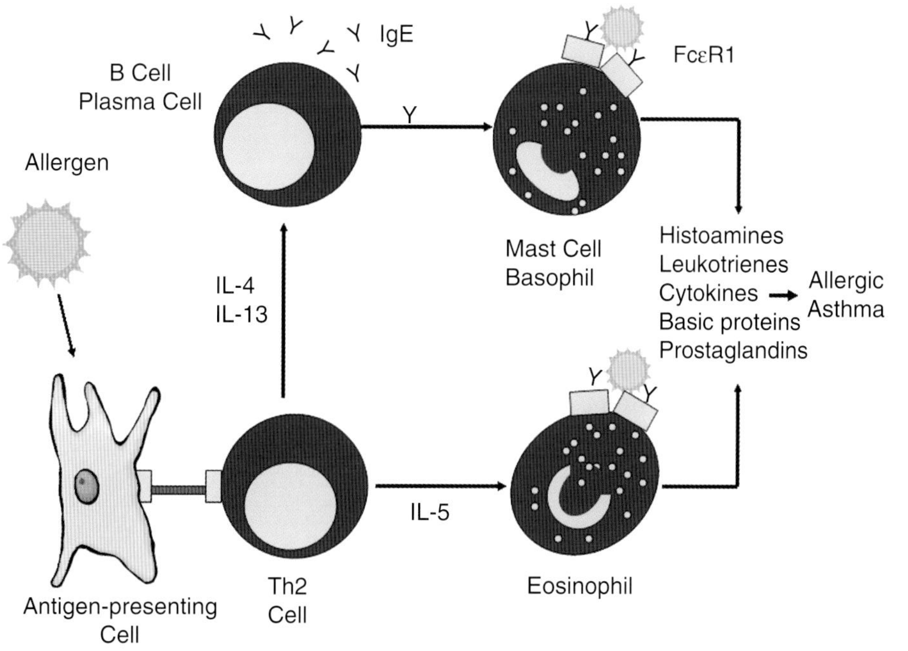

Figure 18.29 Inflammation in asthma. Inflammatory cascade in allergic asthma involves presentation of the allergen by an antigen-presenting cell to Th2 cells. This causes a release of multiple cytokines that lead to the recruitment of eosinophils, macrophages, and basophils [adapted from 128].

Bronchial Hyperactivity The cornerstone of asthma is the hyperactive response of the airway smooth muscle. The mechanism by which this occurs combines neural pathways and inflammatory pathways. As stated, the inflammatory component of this response comes predominantly from the mast cells. The major neural pathway involved is the nonadrenergic noncholinergic (NANC) system. Although cholinergic pathways are responsible for maintaining the airway smooth muscle tone, it is the NANC system that releases bronchoactive tachykinins (substance P and neurokinin A) that bind to NK2 receptors on the smooth muscle and cause the constriction that characterizes the acute asthmatic response [123].

In addition to these acute mechanisms, the airway also undergoes structural alterations to its formed elements. In the mucosa, these changes include goblet cell hyperplasia and basement membrane thickening. Within the submucosa and airway wall, increased deposition of collagen and elastic fibers results in fibrosis and elastosis, and both the smooth muscle cells and the submucosal glands undergo hypertrophy and hyperplasia. These irreversible changes are a consequence of chronic inflammatory insults on the airways through mechanisms that include release of fibrosing mediators such TGFβ and mitogenic mediators such as epidermal and fibroblast growth factors (EGF, FGF). The exact mechanisms by which this occurs are not clearly defined, but the similarity of these factors with those involved in branching morphogenesis of the developing lung has led to a focus on the effect of inflammation on the interaction of the epithelium with the underlying mesenchymal cells [128].

Chronic Obstructive Lung Disease (COPD) – Emphysema/Chronic Bronchitis

Clinical and Pathologic Features

The term chronic obstructive pulmonary disease (COPD) applies to emphysema, chronic bronchitis, and bronchiectasis, those diseases in which airflow limitation is usually progressive, but, unlike asthma, not fully reversible [131]. The prevalence of COPD worldwide is estimated at 9%–10% in adults over the age of 40 [132]. Though there are different forms of COPD with different etiologies, the clinical manifestations of the most common forms of the disease are the same. These include a progressive decline in lung function, usually measured as decreased forced expiratory flow in 1 second (FEV1), a chronic cough, and dyspnea. Emphysema and chronic bronchitis are the most common diseases of COPD and are the result of cigarette smoking. As such, they usually exist together in most smokers. Chronic bronchitis is defined clinically as a persistent cough with sputum production for at least 3 months in at least 2 consecutive years without any other identifiable cause. Patients with chronic bronchitis typically have copious sputum with a prominent cough, more commonly get infections, and typically experience hypercapnia and severe hypoxemia, giving rise to the clinical moniker blue bloater. Emphysema is the destruction and permanent enlargement of the air spaces distal to the terminal bronchioles without obvious fibrosis [133]. These patients have only a slight cough, while the overinflation of the lungs is severe, inspiring the term pink puffers.

The pathologic features of COPD are best understood if one considers the whole of COPD as a spectrum of pathology that consists of emphysematous tissue destruction, airway inflammation, remodeling, and obstruction [134]. The lungs of patients with COPD usually contain all of these features, but in varying proportions. The pathologic features of chronic bronchitis include mucosal pathology that consists of epithelial inflammation, injury, and regenerative epithelial changes of squamous and goblet cell metaplasia. In addition, the submucosa shows changes of remodeling with smooth muscle hypertrophy and submucosal gland hyperplasia. These changes are responsible for the copious secretions characteristic of this clinical disease, although studies have reported no consistent relationship between these pathologic features of the large airways and the airflow obstruction [135].

The pathology definition of emphysema is an abnormal, permanent enlargement of the airspaces distal to the terminal bronchioles accompanied by destruction of the alveolar walls without fibrosis [133]. The four major pathologic patterns of emphysema are defined by the location of this destruction. These include centriacinar, panacinar, paraseptal, and irregular emphysema. The first two of these are responsible for the overwhelming majority of the clinical disease. Centriacinar emphysema (sometimes referred to as centrilobular) represents 95% of the cases and is a result of destruction of alveoli at the proximal and central areas of the pulmonary acinus, including the respiratory bronchioles (Figure 18.30). It predominantly affects the upper lobes

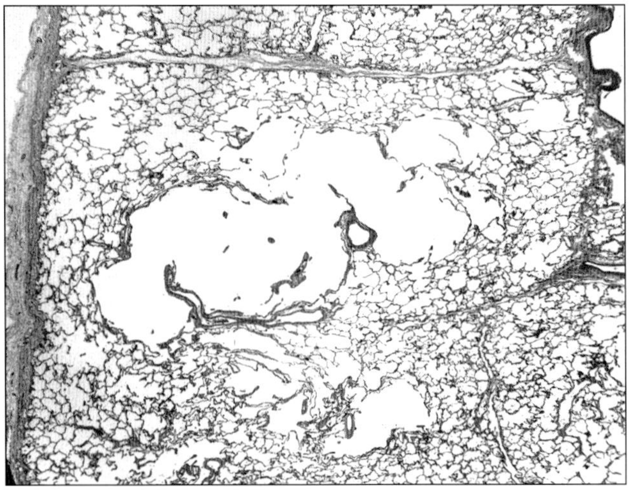

Figure 18.30 Centrilobular emphysema. Tissue destruction in central area of the pulmonary lobule is demonstrated in this lung with a mild centrilobular emphysema. The pattern of tissue destruction is in the area surrounding the small airway where pigmented macrophages release proteases in response to the cigarette smoke.

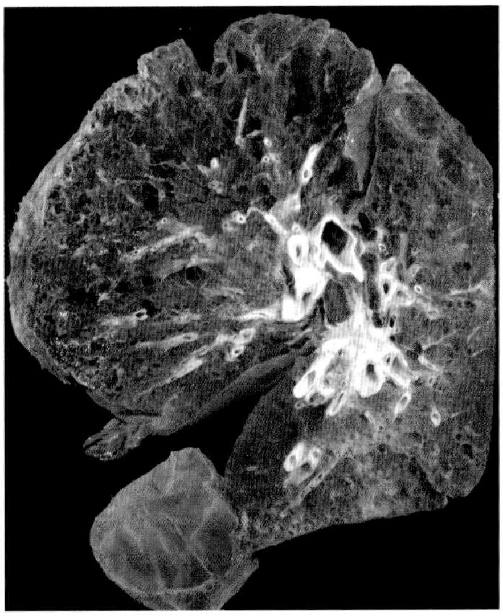

Figure 18.31 **Centrilobular emphysema.** This sagittal cut section of a lung contains severe centrilobular emphysema with significant tissue destruction in the upper lobe and bulla forming in the upper and lower lobes.

(Figure 18.31). Panacinar emphysema, usually associated with α1-antitrypsin (αAT) deficiency, results in a destruction of the entire pulmonary acinus from the proximal respiratory bronchioles to the distal area of the acinus, and affects predominantly the lower lobes (Figure 18.32). The remaining two types of emphysema, paraseptal and irregular, are rarely associated with clinical disease. In paraseptal emphysema, the damage is to the distal acinus, the area that abuts the pleura at the margins of the lobules. Damage in this area may cause spontaneous pneumothoraces, typically in young, thin men [136]. Irregular emphysema is tissue destruction and alveolar enlargement that occurs adjacent to scarring, secondary to the enhanced inflammation in the area. Though this is a common finding in a scarred lung, it is of little if any clinical significance to the patient.

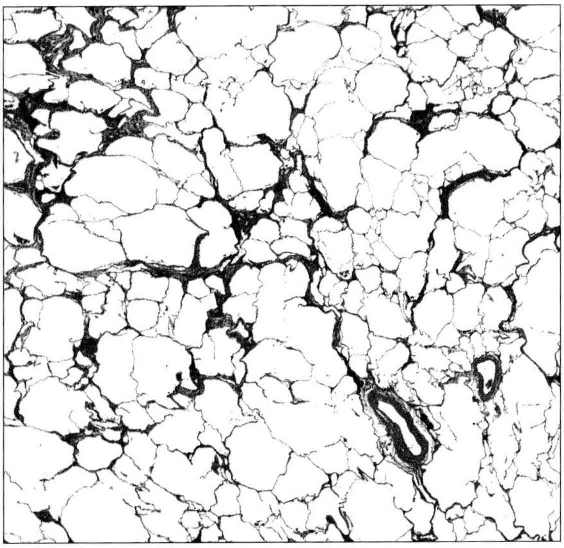

Figure 18.32 **Panacinar emphysema.** Tissue destruction in panacinar emphysema occurs throughout the lobule, producing a diffuse loss of alveolar walls unlike that of centrilobular emphysema with more irregular holes in the tissue.

Though the emphysema in these lungs plays the dominant role in causing the obstruction, small airway pathology is also present. Respiratory bronchiolitis refers to the inflammatory changes found in the distal airways of smokers. These consist of pigmented macrophages filling the lumen and the peribronchiolar airspaces and mild chronic inflammation and fibrosis around the bronchioles (Figure 18.33). The pigment in these macrophages represents the inhaled particulate matter of the cigarette smoke that has been phagocytized by these cells. The macrophages in turn release proteases, which destroy the elastic fibers in the surrounding area, resulting in the loss of elastic recoil and the obstructive symptoms.

Molecular Pathogenesis

In general, COPD is a result of inflammation of the large airways that produces the airway remodeling characteristic of chronic bronchitis as well as inflammation of the smaller airways that results in the destruction of the adjacent tissue and consequent emphysema. The predominant inflammatory cells involved in this process are the alveolar macrophages, neutrophils, and lymphocytes. The main theories of the pathogenesis of COPD support the interaction of airway inflammation with two main systems in the lung: the protease–antiprotease system and the oxidant–antioxidant system. These systems help to protect the lung from the many irritants that enter the lung via the large pulmonary surface area that interfaces with the environment.

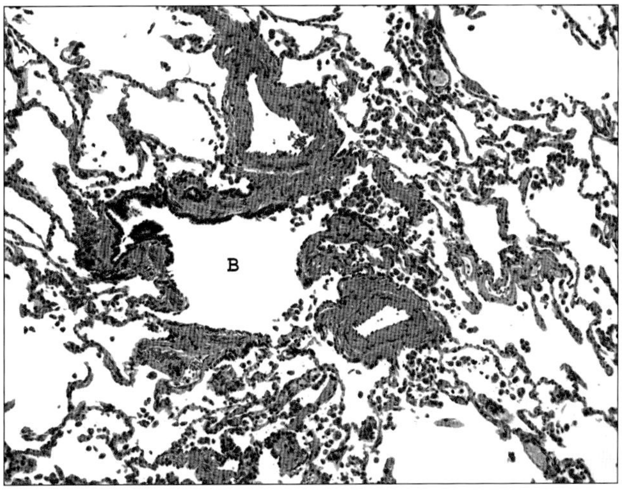

Figure 18.33 **Respiratory bronchiolitis.** Present in the lumen of the small bronchiole (B) and extending into the surrounding alveolar spaces are pigmented macrophages in a lung from a smoker. The pigment in these macrophages represents particulates from the cigarette smoke and stimulates the release of the proteases that are responsible for the tissue destruction in centrilobular emphysema.

In the protease–antiprotease system, proteases are produced by a number of cells, including epithelial cells and inflammatory cells that degrade the underlying lung matrix. The most important proteases in the lung are the neutrophil elastases, part of the serine protease family, and the metalloproteinases (MMPs) produced predominantly by macrophages. These proteases can be secreted in response to invasion by environmental irritants, most notably infectious agents such as bacteria. In this setting, their role is to enzymatically degrade the organism. However, proteases can also be secreted by both inflammatory and epithelial cells in a normal lung to repair and maintain the underlying lung matrix proteins [137]. To protect the lung from unwanted destruction by these enzymes, the liver secretes antiproteases that circulate in the bloodstream to the lung and inhibit the action of the proteases. In addition, macrophages that secrete MMPs also secrete tissue inhibitors of metalloproteinases (TIMPs). A delicate balance of proteases and antiproteases is needed to maintain the integrity of the lung structure. An imbalance that results in a relative excess of proteases (either by overproduction of proteases or underproduction of their inhibitors) leads to tissue destruction and the formation of emphysema.

This imbalance occurs in different ways in the two major types of emphysema: centriacinar and panacinar. In centriacinar emphysema, caused primarily by cigarette smoking, there is an overproduction of proteases primarily due to the stimulatory effect of chemicals within the smoke on the neutrophils and macrophages. Though the exact mechanism is not completely understood, most studies support that nicotine from the cigarette smoke acts as a chemoattractant, and ROS also contained in the smoke, stimulate an increased release of neutrophil elastases and MMPs from activated macrophages, leading to the destruction of the elastin in the alveolar spaces [137]. This inflammatory cell activation may come about through the activation of the transcription factor NFκB that leads to TNFα production [132]. In addition, the elastin peptides themselves may attract additional inflammatory cells to further increase the protease secretion and exacerbate the matrix destruction [137].

Unlike centriacinar emphysema, panacinar emphysema is most commonly caused by a genetic deficiency of antiproteases, usually due to alpha-1 anti-trypsin (αAT) deficiency, a condition that affects approximately 60,000 people in the United States [138]. αAT deficiency is due to a defect in the gene that encodes the protein αAT, a glycoprotein produced by hepatocytes and the main inhibitor of neutrophil elastase. The affected gene is the *SERPINA1* gene (formerly known as P1), located on the long arm of chromosome 14 (14q31–32.3). The genetic mutations that occur have been categorized into four groups: base substitution, in-frame deletions, frame-shift mutations, and exon deletions. These mutations usually result in misfolding, polymerization, and retention of the aberrant protein within the hepatocytes, leading to decreased circulating levels. αAT deficiency is an autosomal codominant disease with over 100 allelic variants, of which the M alleles (M1–M6) are the most common; these alleles produce normal serum levels of a less-active protein [139]. Individuals who manifest the lung disease are usually homozygous for the alleles Z or S (ZZ and SS phenotype) or heterozygous for the 2 M alleles (MZ, or SZ phenotype) [139]. An αAT concentration in plasma of less than 40% of normal confers a risk for emphysema [140]. In individuals with the ZZ genotype, the activity of αAT is approximately one-fifth of normal [141].

The second system in the lung involved in the pathogenesis of emphysema is the oxidant–antioxidant system. As in the protease system, the lung is protected from oxidative stress in the form of ROS by antioxidants produced by cells in the lung. ROS in the lung include oxygen ions, free radicals, and peroxides. The major antioxidants in the airways are enzymes including catalase, superoxide dismutase (SOD), glutathione peroxidase, glutathione S-transferase, xanthine oxidase, and thioredoxin, as well as nonenzymatic antioxidants including glutathione, ascorbate, urate, and bilirubin [142]. The balance of oxidants and antioxidants in the lung prevents damage by ROS. However, cigarette smoke increases the production of ROS by neutrophils, eosinophils, macrophages, and epithelial cells [143]. Evidence that damage to the lung epithelium and matrix is a direct result of ROS includes the presence of exhaled H_2O_2 and 8-isoprostane, decreased plasma antioxidants, and increased plasma and tissue levels of oxidized proteins, including various lipid peroxidation products. In addition to this direct effect, ROS may also induce a proinflammatory response that recruits more inflammatory cells to the lung. In animal models, cigarette smoke induces the expression of proinflammatory cytokines such as IL-6, IL-8, TNFα, and IL-1 from macrophages, epithelial cells, and fibroblasts, perhaps through activation of the transcription factor NFκB [144,145] (Figure 18.34). Finally, there is some evidence that cigarette smoke further disturbs the oxidant–antioxidant balance in the lung by depleting antioxidants such as ascorbate and glutathione [132].

Bronchiectasis

Clinical and Pathologic Features

Bronchiectasis represents the permanent remodeling and dilatation of the large airways of the lung most commonly due to chronic inflammation and recurrent pneumonia. These infections usually occur because airway secretions and entrapped organisms cannot be effectively cleared. This pathology dictates the clinical features of the disease, which include chronic cough with copious secretions and a history of recurrent pneumonia. The five major causes of bronchiectasis are infection, obstruction, impaired mucociliary defenses, impaired systemic immune defenses, and congenital. These may produce either a localized or diffuse form of the disease. Localized bronchiectasis is usually due to obstruction of airways by mass lesions or scars from previous injury or infection. Diffuse bronchiectasis can result from defects in systemic

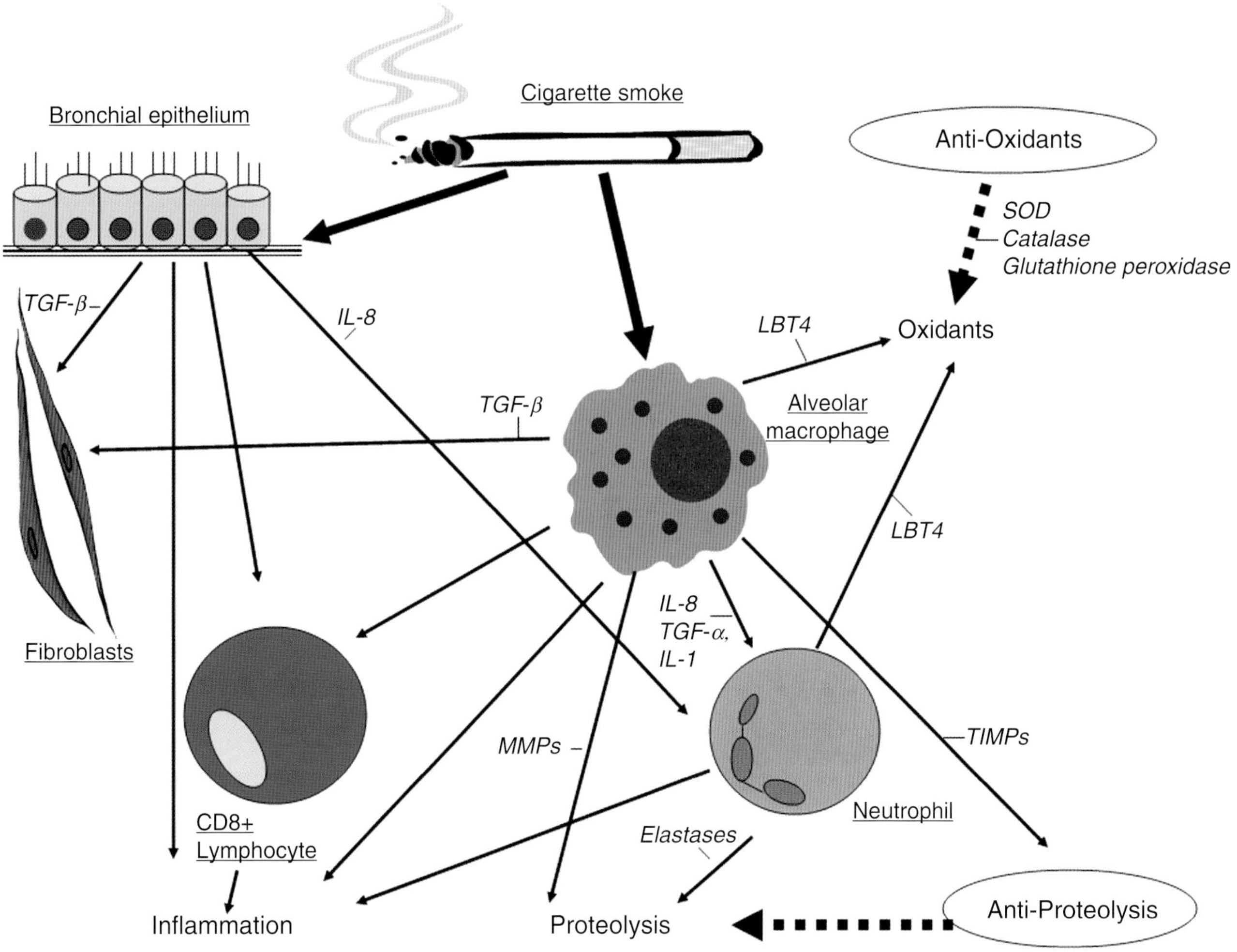

Figure 18.34 Pathogenesis of chronic obstructive pulmonary disease. Inflammatory cells including alveolar macrophages, lymphocytes, and neutrophils are involved in generating inflammation, proteolysis, and oxidative stress in chronic obstructive pulmonary disease caused by cigarette smoke. Antioxidants and antiproteases help to inhibit these effects. Profibrotic mediators stimulate fibroblasts and myofibroblasts to repair and remodel the lung after this injury [adapted from 145].

immune defenses in which either innate or adaptive immunity may be impaired. Diseases due to the former include chronic granulomatous disease (CGD), and diseases due to the latter include agammaglobulinemia/hypogammaglobulinemia and severe combined immune deficiencies. Defects in the mucociliary defense mechanism that is responsible for physically clearing organisms from the lung may also cause diffuse bronchiectasis. These include ciliary dyskinesias that result in cilia with aberrant ultrastructure and cystic fibrosis (CF). Congenital forms of bronchiectasis are rare but do exist. The most common include Mounier-Kuhn's syndrome and Williams-Campbell syndrome, the former causing enlargement of the trachea and major bronchi due to loss of bronchial cartilage, and the latter causing diffuse bronchiectasis of the major airways probably due to a genetic defect in the connective tissue [146,147].

The pathology of bronchiectasis is most dramatically seen at the gross level. One can see dilated airways containing copious amounts of infected secretions and mucous plugs localized either to a segment of the lung or diffusely involving the entire lung as in cystic fibrosis (Figure 18.35). Microscopic features include chronic inflammatory changes similar to those of chronic bronchitis but with ulceration of the mucosa and submucosa leading to destruction of the smooth muscle, and elastic in the airway wall and the characteristic dilatation and fibrosis. These enlarged airways contain mucous plugs comprising mucin and abundant degenerating inflammatory cells, a result of infections that establish themselves in these airways following the loss of the mucociliary defense mechanism. Bacteria may be found in these plugs, most notably *P. aeruginosa.*

Molecular Pathogenesis

The pathogenetic mechanism of bronchiectasis is complex and depends on the underlying etiology. In general, the initial damage to the bronchial epithelium is

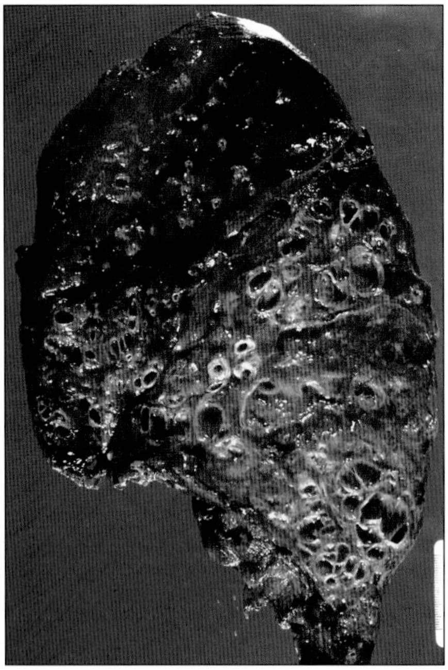

Figure 18.35 **Cystic fibrosis.** This sagittal cut section of a lung from a patient with cystic fibrosis demonstrates a diffuse bronchiectasis illustrated by enlarged, cyst-like airways. This is the typical pathology for cystic fibrosis. The remainder of the lung contains some red areas of congestion.

due to aberrant mucin (cystic fibrosis), dysfunctional cilia (ciliary dyskinesias), and ineffective immune surveillance (defects in innate and antibody-mediated immunity), leading to a cycle of tissue injury, repair, and remodeling that ultimately destroys the normal airway. The initial event in this cycle usually involves dysfunction of the mucociliary mechanism that inhibits the expulsion from the lungs of organisms and other foreign substances that invade the airways. This may be due to defects in the cilia or the mucin. Ciliary defects are found in primary ciliary dyskinesia, a genetically heterogeneous disorder, usually inherited as an autosomal recessive trait that produces immotile cilia with clinical manifestations in the lungs, sinuses, middle ear, male fertility, and organ lateralization [148]. Over 250 proteins make up the axoneme of the cilia, but mutations in 2 genes, *DNAI1* and *DNAH5*, which encode for proteins in the outer dynein arms, most frequently cause this disorder [149]. In CF the main defect affects the mucin. In patients with this autosomal recessive condition, there is a low volume of airway surface liquid (ASL) causing sticky mucin that inhibits normal ciliary motion and effective mucociliary clearance of organisms. This is due to a defect in the cystic fibrosis transmembrane conductance regulator (*CFTR*) gene, located on chromosome 7 that encodes a cAMP-activated channel which regulates the flow of chloride ions in and out of cells and intracellular vacuoles, helping to maintain the osmolality of the mucin. This protein is present predominantly on the apical membrane of the airway epithelial cells, though it is also involved in considerable subapical, intracellular trafficking and recycling during the course of its maturation within these cells. This genetic disease manifests in multiple other organs that depend on chloride ion transport to maintain normal secretions, including the pancreas, intestine, liver, reproductive organs, and sweat glands [150].

The genetic mutations in CF influence the CFTR trafficking in the distal compartments of the protein secretary pathway, and various genetic mutations produce different clinical phenotypes of the disease. Over 1600 mutations of the *CFTR* gene have been found. However, only four of these mutations occur at a frequency of greater than 1%. These mutations are grouped into five classes according to their functional deficit: Group I, CFTR is not synthesized; Group II, CFTR is inadequately processed; Group III, CFTR is not regulated; Group IV, CFTR shows abnormal conductance; Group V, CFTR has partially defective production or processing. Approximately 70% of CF patients are in Group II and have the same mutation, F508Δ CFTR, a deletion of phenylalanine at codon 508 [154]. In these patients, most of the CFTR protein is misfolded and undergoes premature degradation within the endoplasmic reticulum, though a small amount of the CFTR protein is present on the apical membrane and does function normally. CF patients may have a combination of genetic mutations from any of the five groups. However, those patients with the most severe disease involving both the lungs and pancreas usually carry at least two mutations from Group I, II, or III [151].

Systemic immune deficiencies cause bronchiectasis through the establishment of persistent infection and inflammation. There are four major categories of immune deficiencies. The first category consists of a number of genetic diseases that cause either agammaglobulinemia or hypogammaglobulinemia. These include X-linked agammaglobulinemia (XLA) and common variable immunodeficiency (CVI). XLA is caused by a mutation of the Bruton's tyrosine kinase (BTK) gene that results in the virtual absence of all immunoglobulin isotypes and of circulating B lymphocytes. In CVI there is a marked reduction in IgG and IgA and/or IgM, associated with defective antibody response to protein and polysaccharide antigens. As expected, both of these diseases increase susceptibility to infections from encapsulated bacteria. The second category of immune deficiency is hyper-IgE syndrome, a disease with markedly elevated serum IgE levels that is characterized by recurrent staphylococcal infections. The third category is chronic granulomatous disease (CGD), a genetically heterogeneous group of disorders that have a defective phagocytic respiratory burst and superoxide production, inhibiting the ability to kill *Staphylococcus* spp. and fungi such as *Aspergillus* spp. Finally, severe combined immune deficiency (SCID) comprises a group of disorders with abnormal T- cell development and B-cell and/or natural killer cell maturation and function, predisposing these patients to *Pneumocystis jiroveci* and viral infections [152].

After the initial insult, the subsequent steps in the development of bronchiectasis include destruction of the epithelial cells and bronchial wall connective tissue matrix by the proteases and ROS secreted by the neutrophils.

This proinflammatory milieu is produced by multiple factors. First, infections can persist in these lungs due to defective host immune systems and mechanisms certain organisms have developed to evade these immune defenses. For example, *Pseudomonas aeruginosa*, changes from a nonmucoid to a mucoid variant and also releases virulence factors to protect against phagocytosis [153]. Second, in the case of cystic fibrosis, neutrophils are directly recruited by proinflammatory cytokines, such as interleukin-8 (IL-8), released from the bronchial epithelial cells as a result of the defective CGFT protein [154]. Finally, the necrotic cellular debris and other breakdown products act as chemoattractants that recruit more inflammatory cells to the airway wall, further exacerbating the damage.

The final phase of the repair and remodeling begins when macrophages invade and recruit fibroblasts that secrete collagen, leading to the fibrosis seen in the pathology. However, in the absence of effective airway clearance mechanisms, these ectatic airways remain a reservoir of infection that continues the cycle of inflammation and tissue destruction.

INTERSTITIAL LUNG DISEASES

Idiopathic Interstitial Pneumonias – Usual Interstitial Pneumonia

Clinical and Pathology Features

The idiopathic interstitial pneumonias (IIPs) comprise a group of diffuse infiltrative pulmonary diseases with a similar clinical presentation characterized by dyspnea, restrictive physiology, and bilateral interstitial infiltrates on chest radiography [155]. Pathologically, these diseases have characteristic patterns of tissue injury with chronic inflammation and varying amounts of fibrosis. By recognizing these patterns, a pathologist can classify each of these entities and predict prognosis. However, the pathologist cannot establish the etiology, since these pathologic patterns can be seen in multiple clinical settings.

The pathologic classification of these diseases, originally defined by Liebow and Carrington in 1969 [156], has undergone important revisions over the past 35 years with the latest revision by the *American Thoracic Society/European Respiratory Society* in 2003 [157]. The best known and most prevalent entity of the IIPs is idiopathic pulmonary fibrosis (IPF), which is known pathologically as usual interstitial pneumonia (UIP). UIP is a histologic pattern characterized by patchy areas of chronic lymphocytic inflammation with organizing and collagenous type fibrosis. These patients usually present with gradually increasing shortness of breath and a nonproductive cough after having had symptoms for many months or even years. Imaging studies usually reveal bilateral, basilar disease with a reticular pattern [155]. Therapy begins with corticosteroids, advancing to more cytotoxic drugs such as methotrexate and cytoxan, but most current therapies are not effective in stopping the progression of the disease. The current estimates are that 20/100,000 males and 13/100,000 females have the disease, most of whom progress to respiratory failure and death within 5 years [158].

The pathology is characterized by a leading edge of chronic inflammation with fibroblastic foci that begin in different areas of the lung at different times. These processes produce a variegated pattern of fibrosis, usually referred to as a temporally heterogenous pattern of injury [159]. Because it occurs predominantly in the periphery of the lung involving the subpleura and interlobular septae, the gross picture is one of more advanced peripheral and basilar disease (Figure 18.36). The progression from inflammation to fibrosis includes interstitial widening, epithelial injury and sloughing, fibroblastic infiltration, and organizing fibrosis within the characteristic fibroblastic foci. Deposition of collagen by fibroblasts occurs in the latter stages of repair. The presence of the abundant collagen produces stiff lungs that are unable to clear the airway secretions, leading to recurrent inflammation of the bronchiolar epithelium with eventual fibrosis and breakdown of the airway structure. This remodeling produces mucous-filled ectatic spaces giving rise to the gross picture of honeycomb spaces, which is seen in the advanced pathology (Figure 18.36) [160].

Molecular Pathogenesis

Theories of the pathogenesis of IPF have evolved over the past decade. Early theories favored a primary inflammatory process, while current theories favor the concept that the fibrosis of the lung proceeds independently of inflammatory events and develops from aberrant epithelial and epithelial-mesenchymal responses to injury to the alveolar epithelial cells (AECs) [161]. The AECs consist of two populations: the type 1 pneumocytes and the type 2 pneumocytes. In normal lungs, type 1 pneumocytes line 95% of the alveolar wall, and type 2 pneumocytes line the remaining 5%. However, in lung injury, the type 1 cells, which are exquisitely fragile, undergo cell death, and the type 2 pneumocytes serve as progenitor cells to regenerate the alveolar epithelium [162]. Though some studies have suggested that repopulation of the type 2 cells depends on circulating stem cells, this concept remains to be fully proven. According to current concepts, the injury and/or apoptosis of the AECs initiates a cascade of cellular events that produce the scarring in these lungs. Studies of AECs in lungs from patients with IPF have shown ultrastructural evidence of cell injury and apoptosis as well as expression of proapoptotic proteins. Further, inhibition of this apoptosis by blocking a variety of proapoptotic mechanisms such the Fas-Fas ligand pathway, angiotensin, and TNFα production, and caspase activation can stop the progression of this fibrosis [163].

The result of the AEC injury is the migration, proliferation, and activation of the fibroblasts and myofibroblasts that leads to the formation of the characteristic fibroblastic foci of the UIP pathology and the deposition and accumulation of collagen and elastic fibers in the alveoli (Figure 18.37). This unique pathology

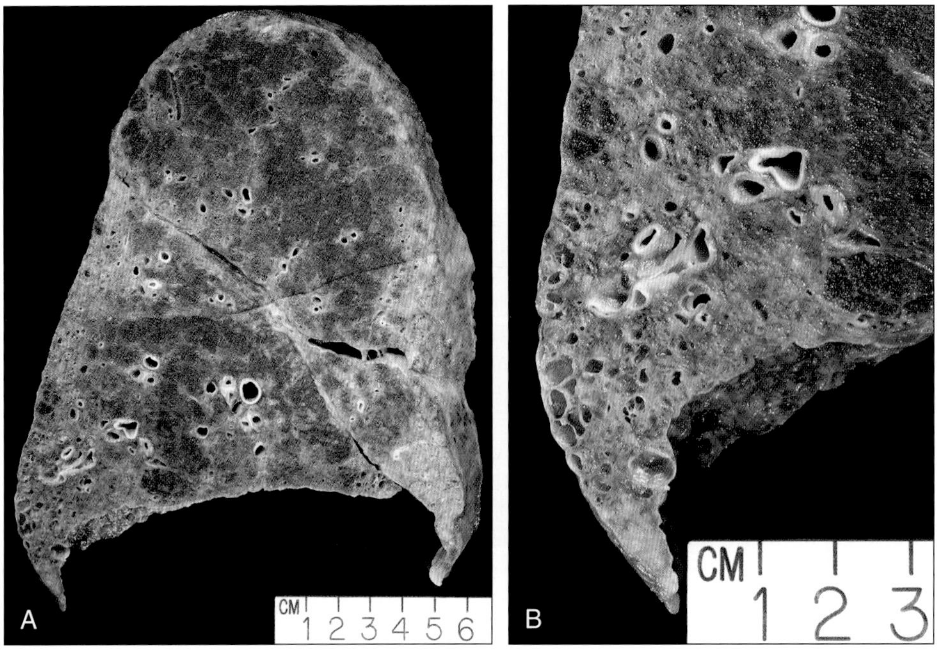

Figure 18.36 **Usual interstitial pneumonia.** A sagittal cut of a lung involved by usual interstitial pneumonia reveals the peripheral and basilar predominance of the dense, white fibrosis (A). A higher power view of the left lower lobe highlights the remodeled honeycomb spaces in the area of the lung with the endstage disease (B).

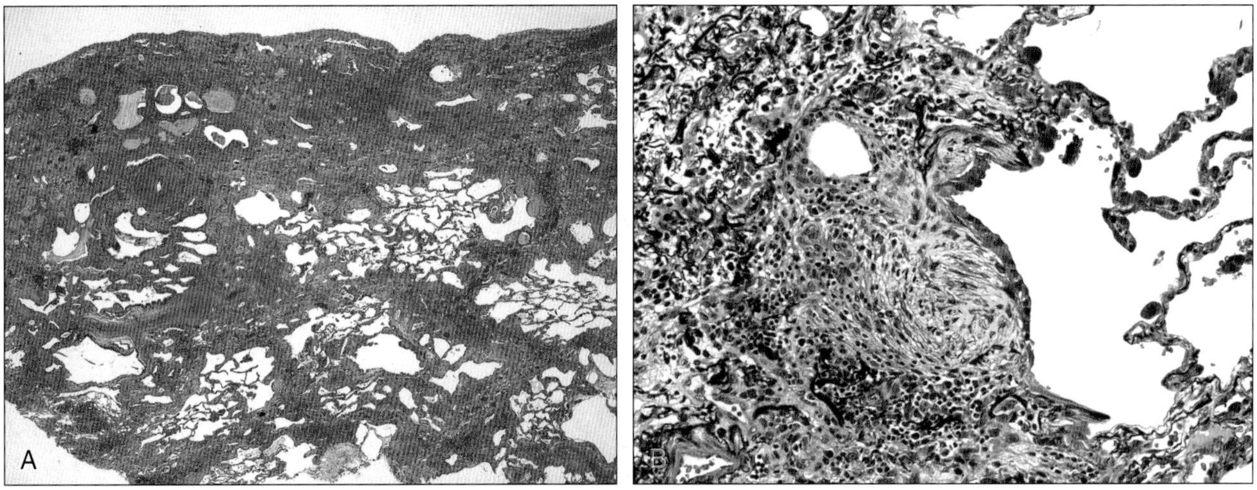

Figure 18.37 **Usual interstitial pneumonia.** The microscopic features of UIP lungs are characterized by inflammation and fibrosis that demonstrate the temporally heterogenous pattern of pathologic injury with normal, inflamed, and fibrotic areas of the lung, all seen at a single lower power view. (A) The leading edge of inflammation is represented by deposition of new collagen in fibroblastic foci. These consist of fibroblasts surrounded by collagen containing mucopolysaccharides highlighted in blue by this connective tissue stain (B).

may be a result of the increased production of profibrotic factors such as transforming growth factor-α (TGFα) and TGFβ, fibroblastic growth factor-2, insulin-like growth factor-1, and platelet-derived growth factor. An alternative pathway might involve overproduction of inhibitors of matrix degradation such as TIMPs (tissue inhibitors of matrix production) [164]. In support of the former mechanism, fibroblasts isolated from the lungs of IPF patients exhibit a profibrotic secretory phenotype [165].

Multiple factors, such as environmental particulates, drug or chemical exposures, and viruses may trigger the initial epithelial injury, but genetic factors also play a role. Approximately 2%–20% of patients with IPF have a family history of the disease with an inheritance pattern of autosomal dominance with variable penetrance. Two genetic mutations have been implicated in this familial form of IPF. One large kindred has been reported with a mutation in the gene encoding surfactant protein C, and six probands have been reported with heterozygous mutations in genes *hTERT* or *hTR*, encoding telomerase reverse transcriptase and telomerase RNA, respectively, resulting in mutant telomerase and short telomeres [166].

Diffuse Alveolar Damage

Clinical and Pathology Features

Adult respiratory distress syndrome (ARDS) represents a constellation of clinical, radiologic, and physiologic features in patients with acute respiratory failure that can occur after a variety of insults. ARDS is defined by clinical criteria that include a rapid onset of severe hypoxemia that is refractory to oxygen therapy, the presence of abnormal chest radiographs with evidence of bilateral alveolar filling and collapse, increased pulmonary artery occlusion pressure, and a resistance to improved oxygenation regardless of mechanical ventilation therapy [167]. Treatment of ARDS includes eliminating the underlying cause, protective ventilation strategies that improve oxygenation, and supportive treatment that may include administration of corticosteroids.

The pathology of ARDS is diffuse alveolar damage (DAD), whose histologic picture is one of inflammation and fibrosis that diffusely involves all of the structures of the alveolus and is similar throughout the affected areas of the lung [168]. DAD is divided into three major phases that follow each other chronologically after the original insult. These are exudative, proliferative, and fibrotic DAD. The initial injury primarily involves the epithelium of the alveolar wall and the endothelium in the capillary, causing the destruction and sloughing of the type 1 pneumocytes into the alveolar space and a breakdown of the tight junctions of the endothelium. In combination, these two events result in the loss of the epithelial-endothelial barrier of the alveolus and leakage of plasma from the capillary into the alveolar space. This flooding of the airspace with fluid markedly decreases oxygen exchange and causes the hypoxia that these patients experience. In addition, acute inflammatory changes of the endothelium also cause thrombi to form in vessels, adding to a decreased amount of blood circulating through the lung and further compromising gas exchange. As air is brought into the alveoli, the positive pressure within the alveolar space forces the plasma against the alveolar wall, producing a membranous morphology referred to as hyalin membranes characteristic of the first phase of DAD, referred to as exudative DAD (Figure 18.38).

This initial injury is followed by a sequence of events that represent the lung's efforts to repair itself. First, type 2 pneumocytes undergo hyperplasia and re-epithelialize the alveolar wall after the loss of the type 1 cells. This re-establishes the epithelial barrier and, because these cells secrete surfactant, results in increased surfactant production, which lowers the surface tension of the alveolus and inhibits its collapse. Because of the increased numbers of type 2 pneumocytes, this is known as the proliferative phase of DAD (Figure 18.38). In the final phase of DAD, fibrotic DAD, fibroblasts migrate in from the adjacent interstitium to the alveolar space and produce organizing and irreversible fibrosis within both the alveolar space and the interstitium. In addition to this mechanism, fibrosis may also occur in those areas where alveolar walls collapse when surfactant is decreased during the initial insult. The histopathologic picture during this fibrotic phase is one of thickened alveolar septa, intra-alveolar granulation tissue, microcyst formation, and areas of irregular alveolar scarring. In rare cases, these microcysts progress to large cysts, an adult equivalent of bronchopulmonary dysplasia.

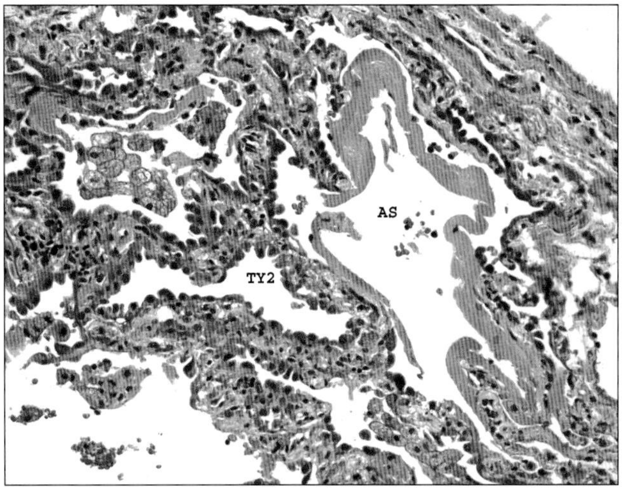

Figure 18.38 Diffuse alveolar damage. This microscopic image reveals both the exudative (right side) and proliferative (left side) phases of diffuse alveolar damage. The eosinophilic hyaline membranes outline the alveolar space (AS), and Type 2 pneumocytes are present on the surface of the adjacent alveolar walls (TY2).

Molecular Pathogenesis

The cellular events of DAD are complex and incompletely understood. In general, the disease can be broken down into two phases. In the first, a large influx of neutrophils and plasma enter the alveolar space. The role the neutrophils play in the initial cellular injury and death is unclear, but it is known that they are necessary for this injury to occur. In addition, clinical studies have shown that within the peripheral blood and bronchoalveolar lavages (BAL) of these patients, neutrophils are present along with a myriad of proinflammatory cytokines, such as IL-8, IL-1, and TGFα, all of which are capable of recruiting them to the lung. Also present in these fluids are mediators that recruit fibroblasts such as TGFβ. All of these mediators are probably the result of upregulation of NFκB, a proinflammatory transcription factor, in alveolar macrophages. The adherence of neutrophils to the capillary endothelium in the lung occurs through adhesion molecules such as selectin, integrin, and immunoglobulins. Neutrophil adherence and subsequent transmigration through the endothelium of the lung capillaries may cause some endothelial damage. However, most speculate that ROS and reactive nitrogen species (RNS) secreted by the neutrophils modulate the majority of this injury [169]. This is supported by the finding that patients with ARDS have products of oxidative damage such as hydrogen peroxide (H_2O_2) in the exhaled breath and myeloperoxidase and oxidized αAT in the BAL.

The cell injury and death of the type 1 pneumocytes most likely occurs via two mechanisms: lipopolysaccharide (LPS)-induced caspase-dependent apoptosis and hyperoxia-induced cell death through apoptosis and non-apoptotic mechanisms [170]. In the former, LPS, an immunogenic component of the outer membrane of gram-negative bacteria, may trigger innate immune and inflammatory responses via toll-like receptors that bind Fas-associated death domain protein and caspase-9, leading to epithelial cell death. In hyperoxia-induced cell death, hyperoxia may induce the expression of angiopoietin 2 (Ang2) in lung epithelial cells. Ang2 is an angiogenic growth factor that can activate caspase pathways and lead to apoptotic cell death [170]. Cell death in ARDS is not limited to these mechanisms, and further study of many of pathways by which this can occur is needed.

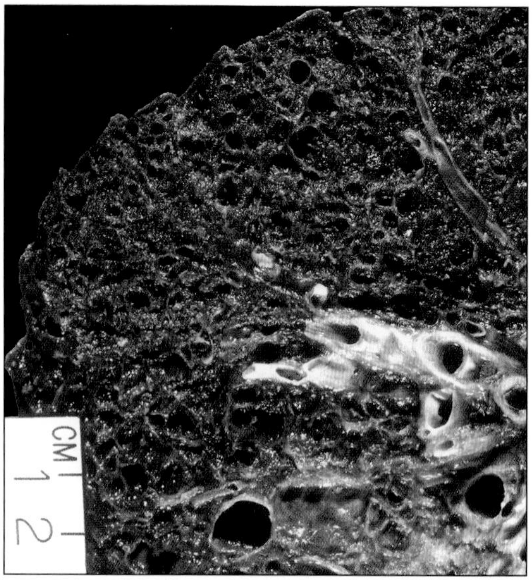

Figure 18.39 Lymphangioleiomyomatosis. The sagittal section of an upper lobe from an explanted lung from a patient with LAM demonstrates cystic features of the red/brown lung parenchyma that are characteristic of this disease.

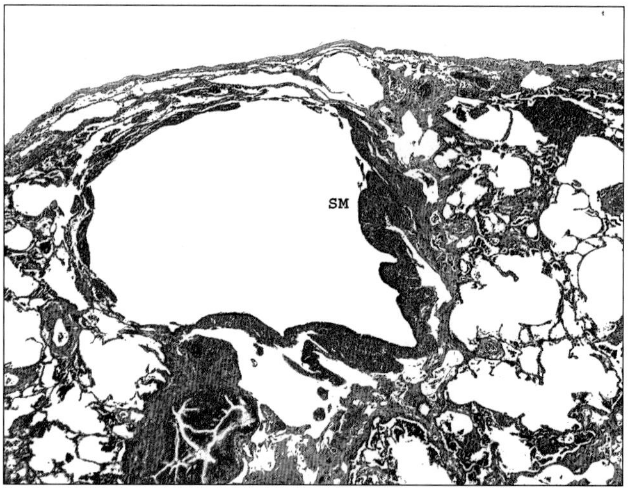

Figure 18.40 Lymphangioleiomyomatosis. The microscopic view of the LAM lung reveals cysts lined by spindled smooth muscle cells (SM). Scattered macrophages surrounding these cysts contain brown hemosiderin pigment.

Lymphangioleiomyomatosis

Clinical and Pathologic Features

Lymphangioleiomyomatosis (LAM) is a rare systemic disease of women, usually in their reproductive years (average age of 35 years), that is characterized by a proliferation of abnormal smooth muscle cells giving rise to cysts in the lungs, abnormalities in the lymphatics, and abdominal tumors, most notably in the kidneys. In addition to sporadic cases (denoted as S-LAM), LAM also affects 30% of women with tuberous sclerosis (denoted as TSC-LAM), a genetic disorder with variable penetrance associated with seizures, brain tumors, and cognitive impairment [171,172]. Global estimates indicate that TSC-LAM may be as much as 5-fold to 10-fold more prevalent than S-LAM, though at least some suggest that TSC-LAM may have a milder clinical course than S-LAM [172]. Clinically, LAM patients usually present with increasing shortness of breath on exertion, obstructive symptoms, spontaneous pneumothoraces, and chylous effusions or with abdominal masses consisting of either angiomyolipomas and/or lymphangiomyomas. Chest imaging studies characteristically reveal hyperinflation with flattened diaphragms and thin-walled cystic changes. Mortality at 10 years from the onset of symptoms is 10%–20% [173].

LAM appears as small, thin-walled cysts (0.5–5.0 cm) randomly throughout both lungs [174] (Figure 18.39). Microscopically, LAM lungs contain a diffuse infiltration of smooth muscle cells, predominantly around lymphatics, veins, and venules. Most notably, one finds smooth muscle cells in the subpleural with hemosiderin-laden macrophages in the adjacent field, and the macrophages are also seen on bronchoalveolar lavage specimens from these patients. The hemosiderin pigment in these lungs is thought to be secondary to microhemorrhages from the obstruction of the veins (Figure 18.40) [175]. The smooth muscle cells in LAM react to antibodies to HMB-45, a premelanosomal protein. Other melanosome-like structures are also found in LAM cells, suggesting that these cells have characteristics of both smooth muscle and melanosomes [176].

Molecular Pathogenesis

The lesional cells in LAM are smooth muscle-like with both spindled and epithelioid morphology [177]. These cells are the same in both S-LAM and TSC-LAM and are a clonal population although they lack other features of malignancy [178]. Molecular studies reveal that the abnormal LAM cell proliferation is caused by mutations in one of two genes linked to tuberous sclerosis: tuberous sclerosis complex 1 or 2 (*TSC1* or *TSC2*). These two genes control cell growth and differentiation through the Akt/mammalian target of rapamycin (mTOR) signaling pathway [172]. In this pathway, a growth factor receptor (such as insulin or PDGF receptors) becomes phosphorylated when an appropriate ligand binds, resulting in activation of downstream effectors and ultimately Akt. The gene products of *TSC1* and *TSC2* are hamartin and tuberin, which act as dimers to maintain Rheb (a member of the Ras family) in a GDP-loaded state via statins, acting as a break to the Akt/mTOR pathway, thereby retarding protein synthesis and cell growth. In LAM cells, loss-of-function mutations in these two genes remove this inhibition, leading to enhanced Rheb activation, mTOR activation (with raptor), and subsequent phosphorylation of downstream molecules which result in uncontrolled cell growth, angiogenesis, and damage to the lung tissue (Figure 18.41) [179].

The abnormal proliferation of LAM cells is thought to damage the lung through overproduction of matrix metalloproteinases (MMPs), which degrade the connective tissue of the lung architecture, destroy the alveolar integrity, and result in cyst formation with air trapping [179]. These destructive capabilities of the LAM cells are enhanced by their secretion of the angiogenic factor VEGF-C, which is thought to cause the proliferation of lymphatic channels throughout the lung [179].

Sarcoidosis

Clinical and Pathologic Features

Sarcoidosis is a multisystemic disease that involves the lung in over 90% of the cases [180]. It is most common in the 20–40-year age group and among females. In the United States, African Americans are more commonly affected than Caucasians [181]. The clinical picture of sarcoidosis is variable, but most patients present with systemic symptoms including fatigue, weight loss, and fever. The most common finding on chest imaging studies is bilateral hilar lymph node enlargement and reticular, reticulonodular, and focal alveolar opacities within the lung parenchyma [182].

Pulmonary sarcoidosis is characterized by granulomas which consist of activated histiocytes, called epithelioid histiocytes that form nodules ranging in size from 15–20 microns (Figure 18.42) [183]. Unlike infectious granulomas that usually contain areas of central necrosis, the granulomas in pulmonary sarcoidosis are predominantly non-necrotizing [184]. Also, the granulomas in sarcoidosis follow a distribution along the lymphatics, which includes the area in the subpleural, along the interlobular septae and around the bronchovascular area containing the bronchiole and branch of the pulmonary artery (Figure 18.42). The granulomas occur much more commonly in the upper lobes, leading to the predominant upper lobe fibrosis and bronchiectasis that can be seen in longstanding sarcoidosis [185].

Molecular Pathogenesis

Despite over 50 years of research on sarcoidosis, the etiology remains unknown. Most agree that the disease is probably a result of environmental triggers acting on a genetically susceptible host [186,187]. A genetic basis of sarcoidosis has been suggested by studies that demonstrate familial clustering and racial variation [188,189]. Further, complex inheritance patterns for the disease suggest that more than one gene may be involved [190]. Several genes of the major histocompatibility complex (MHC) region of the genome have been implicated. Most are clustered on the short arm of chromosome 6 that encompasses the human leukocyte antigen (HLA) domain. The HLA class I MHC molecules associated with sarcoidosis are the *HLA-B7* and *HLA-B8* class I alleles [191,192]. HLA Class II molecules implicated in susceptibility include the *HLA-DR* alleles [193,194]. Genes other than MHC genes thought to regulate the susceptibility to sarcoidosis include those for chemokines such as macrophage inflammatory protein-1α and RANTES (*CCR5* and *CCR3*) [195,196].

Environmental factors that have been implicated are those that are aerosolized. Therefore, these environmental agents have a mode of entry into the lungs and can cause granulomas in the lung, similar to sarcoidosis. These factors can be divided into two major categories, which include infectious and noninfectious agents. The mycobacteria have been the most extensively studied organisms. However, their role in this disease remains controversial due to the difficulty in identifying them by either culture or histochemical stains in sarcoid tissue. Recently, molecular techniques have been able to demonstrate mycobacterial nucleic acid in sarcoid tissue [197,198]. However, even studies using this technology have not produced consistent results, and the role of these organisms in the disease requires further study.

The immune response in sarcoidosis has two major features: (i) the initial event leading to granuloma formation and (ii) the progression of this granulomatous response to either resolution or fibrosis [199]. The formation of the granulomas, triggered by activation of T-cells and antigen-presenting dendritic histiocytes, results in a release of proinflammatory cytokines and chemokines, and recruitment, activation, and proliferation of mononuclear cells, predominantly T-cells. These activated T-cells are predominantly CD4-expressing T-helper (Th) cells, which release IFN-γ and IL-2. Alveolar macrophages at the site release TNFα, IL-12, IL-16, and other growth factors. This results in the granuloma formation and alveolitis, the characteristic morphologic features of the disease [200].

The second phase of this immunologic response that leads to either resolution of the disease or persistence of the granulomas and fibrosis is less well characterized. Ongoing granuloma formation and inflammation may be a result of the persistent presence of antigens, the excessive synthesis of chemotactic factors, or the

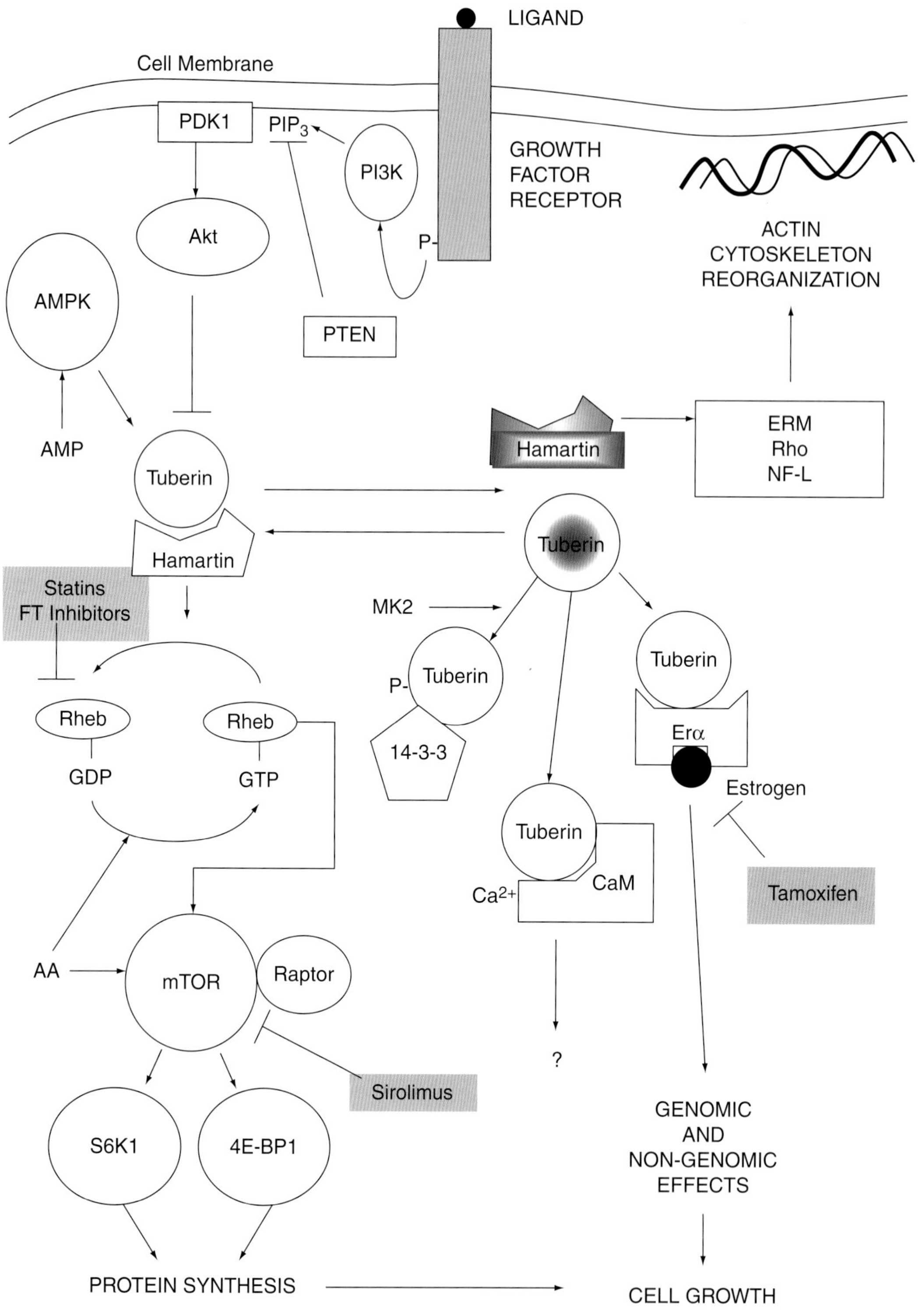

Figure 18.41 **Signal transduction pathways involving the TSC1 and TSC2 gene products, hamartin and tuberin.** *Arrowheads* indicate activating or facilitating influences; *flat-headed lines* indicate inhibitory influences. The harmartin-tuberin dimer maintains Rheb in a GDP-loaded state, thereby preventing activation of mTOR, which requires activated Rheb-GTP. Growth and energy signals tend to inhibit this function of the hamartin-tuberin complex, permitting mTOR activation. The sites of action of several drugs with therapeutic potential in LAM are indicated in gray-shaded boxes. AA, amino acids; FT, farnesyltransferase. (Reprinted with kind permission from Juvet, SC, McCormack FC, Kwiatkowski DJ, Downey GP. (2007). Molecular pathogenesis of lymphangioleiomyomatosis: lesson learned from orphans. *Am J Respir Cell Mol Biol*, 398–408 [179]).

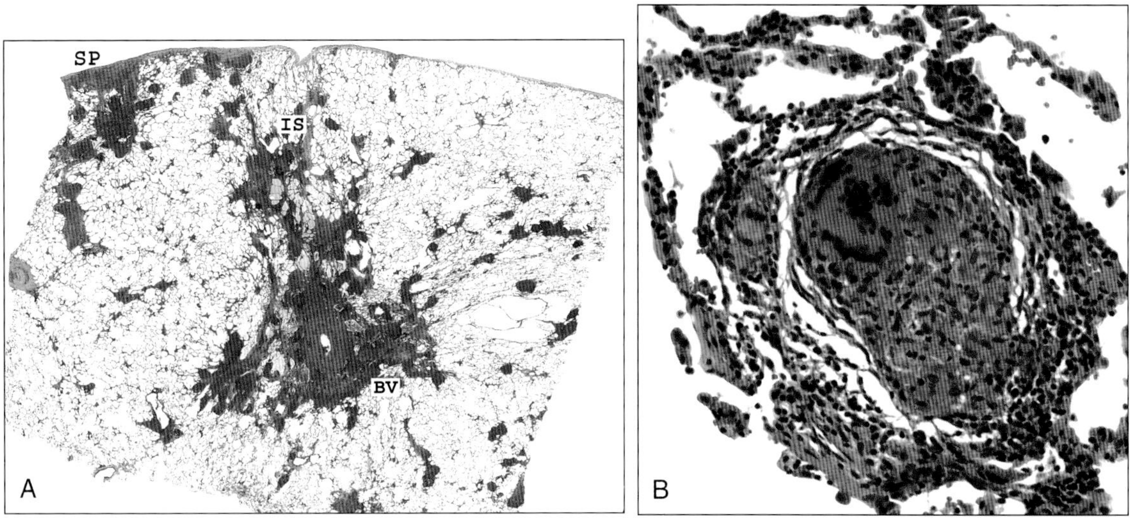

Figure 18.42 **Sarcoidosis.** (A) The distribution of granulomas in sarcoidosis follows the lymphatics in the lung, which includes within the subpleural (SP), along the interlobular septae (IS), and around the bronchovascular areas (BV). (B) The granulomas are composed of activated "epithelioid" histiocytes, usually containing multinucleated giant cells and surrounded by a rim of T-lymphocytes.

persistence of the mononuclear cells within the granulomas. Importantly, the role of the T-cells in these granulomas is to secrete cytokines that attract, stimulate, and ultimately deactivate the fibroblasts that are responsible for the fibrosis that is seen in the chronic disease. The balance between the profibrotic mediators such as TGFβ, insulin-like growth factor-I, platelet-derived growth factor (PDGF), and the antifibrotic mediators, such as IFN-γ, probably dictates the natural history of sarcoidosis in the lung [201]. Genes involved in macrophage-derived cytokines, chemokines, and mediators of fibrosis are all possible candidates for the underlying genetic cause of this complicated disease.

Pulmonary Alveolar Proteinosis

Clinical and Pathologic Features

Pulmonary alveolar proteinosis (PAP) is a rare disease of the lungs characterized by accumulation of surfactant in the alveolar spaces. The names alveolar proteinosis, lipoproteinosis, or perhaps most accurately phospholipoproteinosis, apply equally to this entity. PAP takes three forms clinically: (i) congenital (2%), (ii) secondary (5%–10%), and (iii) idiopathic or primary (88%–93%) [202–204]. PAP arises in previously healthy adults with the median age at diagnosis of approximately 40 years and a male-to-female ratio of 2.7:1. The clinical presentation is variable and usually includes an insidious onset of slowly progressive dyspnea, a dry cough, and other symptoms of respiratory distress, including fatigue and clubbing. However, almost one-third of patients are asymptomatic and are found clinically by abnormal chest X-rays [205,206]. The secondary form of PAP can be found in patients with environmental exposures, including fine silica, aluminum, titanium dioxide, and kaolin dust [206]. Also, secondary PAP may be found in patients with malignancies, most commonly hematologic malignancies such as myelogenous leukemia [207,208].

Chest imaging studies in both the idiopathic and secondary forms most commonly show fine, diffuse, feathery nodular infiltrates, centered in the hilar areas, sparing the peripheral regions [206]. On chest computerized tomographs, the infiltrates may have a geometric-type shape, sometimes referred to as crazy paving [209].

The most prominent microscopic feature of both idiopathic and secondary PAP is the filling of the alveoli with finely granular period acid-Schiff-positive diastase-resistant (PASD) acellular material (Figure 18.43). The

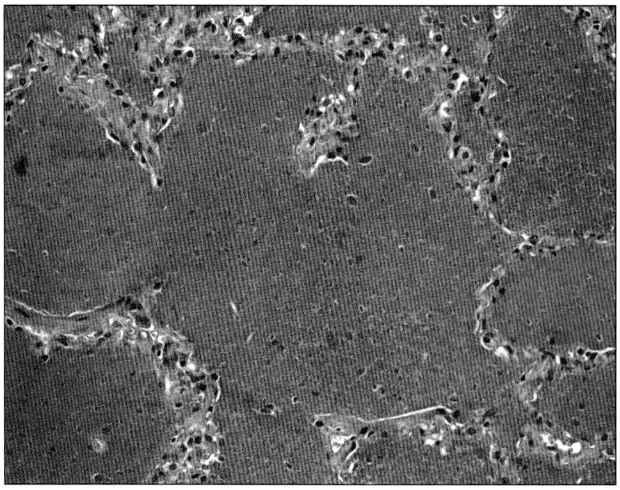

Figure 18.43 **Pulmonary alveolar proteinosis.** The microscopic features of this disease reveal a Periodic acid-Schiff positive surfactant-like substance filling the alveoli that otherwise show only a minimum of inflammatory changes.

material consists of phospholipids (90%); surfactant proteins A, B, C, and D (10%); and carbohydrate (<1%) [210]. Alveolar macrophages (AMs) with prominent foamy cytoplasm are commonly seen, while alveolar septa are remarkably normal in appearance. In some alveolar spaces there are denser, more solid clumps of PAS-D-positive material. Definitive pathologic differences between the idiopathic and secondary forms of PAP have not been well documented [211,212].

Molecular Pathogenesis

The etiologies of the two adult forms of PAP have been well studied with the most known about the idiopathic variant. Theories of the pathogenesis of this form have focused on the abnormal accumulation of the surfactant-like material within the alveolar spaces. Since the regulation of surfactant levels in the alveoli depends on appropriate synthesis, recycling, and catabolism, the two opposing hypotheses have included overproduction versus decreased degradation of this material.

In normal hosts, surfactant is essential to maintaining the low surface tension needed for proper alveolar inflation and gas exchange. The critical role of maintaining the proper composition and amount of surfactant in the alveoli is performed by two cell types: type 2 pneumocytes and alveolar macrophages [213]. The type 2 pneumocytes synthesize surfactant in the endoplasmic reticulum and Golgi, and store it as lamellar bodies [213], which are then delivered to and fuse with the apical plasma membrane, secreting the surfactant into the airways [214]. Catabolism of surfactant is carried out by type 2 pneumocytes and AMs. In PAP, most evidence suggests that the clearance of surfactant by the AM is decreased [203,215].

The first clue as to the underlying mechanism for this defect in AM function came in 1994 when studies revealed that knockout mice deficient in granulocyte-macrophage colony-stimulating factor (GM-CSF) develop lung lesions similar to those in patients with PAP [216]. This rather serendipitous finding prompted explorations centered on the AM and the effect diminished GM-CSF might have on its cellular functions. Subsequent studies from humans with PAP revealed an autoimmune mechanism by which a circulating neutralizing antibody to GM-CSF blocked its binding to the GM-CSF receptor, depressing the effect of GM-CSF on the AMs [217–219]. Neutralizing antibodies to GM-CSF have most often been identified in the idiopathic variant of PAP. However, recently these antibodies have also been reported in patients with secondary PAP [220].

Genes that control many functions in the AM are controlled by signaling pathways initiated by GM-CSF binding to the AM. One pathway is mediated through a transcription factor PU.1 that controls genes involved in surfactant degradation, among other bactericidal functions [221,222]. Another transcription factor, peroxisome-proliferator-activated receptor γ (PPARγ), is also part of a pathway activated by GM-CSF. PPARγ controls the expression of genes involved in intracellular lipid metabolism. AMs from patients with PAP have a deficiency of this transcription factor, which is correctable by GM-CSF therapy [223]. Overall, the lack of GM-CSF-initiated signaling in AMs from patients with PAP leads to inhibition of both PPARγ and PU.1 pathways. This results in decreased surfactant catabolism, intracellular lipid metabolism, and the accumulation of surfactant in the alveoli (Figure 18.44).

PULMONARY VASCULAR DISEASES

Pulmonary Hypertension

Clinical and Pathologic Features

Pulmonary hypertension consists of a group of distinct diseases whose pathology is characterized by abnormal destruction, repair, remodeling, and proliferation of all compartments of the pulmonary vascular tree, including arteries, arterioles, capillaries, and veins. The classification of these diseases has undergone a number of revisions. The most recent revision (in 2003) groups these diseases based on both their pathologic and clinical characteristics [224]. There are five major disease categories in the current classification system: (i) pulmonary arterial hypertension (PAH); (ii) pulmonary hypertension with left heart disease; (iii) pulmonary hypertension associated with lung disease and/or hypoxemia; (iv) pulmonary hypertension due to chronic thrombotic and/or embolic disease; and (v) miscellaneous causes, including sarcoidosis, histiocytosis X, and lymphangioleiomyomatosis. The clinical course of most patients with pulmonary hypertension begins with exertional dyspnea, and progresses through chest pain, syncope, increased mean pulmonary artery pressures and, eventually, right heart failure. The rate of this clinical progression varies among patients, from a few months to many years [225]. Treatment of these diseases focuses on blocking the mediators involved in the pathogenesis of the diseases. However, current therapies rarely prevent progression of the disease, and lung transplantation provides the only hope for long-term survival.

The major group of this classification, PAH, can be subdivided into familial PAH, idiopathic PAH, PAH associated with other conditions (such as connective tissue diseases, HIV, congenital heart disease), and PAH secondary to drugs and toxins (such as anorexigens, cocaine, and amphetamines). In these diseases, the primary pathology is localized predominantly in the small pulmonary arteries and arterioles. However, two other diseases in this group, pulmonary veno-occlusive disease and pulmonary capillary hemangiomatosis, involve predominantly other components of the pulmonary vasculature, the veins, and the capillaries, respectively. The pathologic changes seen in the pulmonary vessels of these patients primarily reflect injury to and repair of the endothelium. Early pathologic changes include medial hypertrophy and intimal fibrosis that narrows and obliterates the vessel lumen. These are followed by remodeling and revascularization, producing a proliferation of abnormal endothelial-lined spaces. These structures are known as plexogenic lesions and are the pathognomonic feature of PAH (Figure 18.45). In the

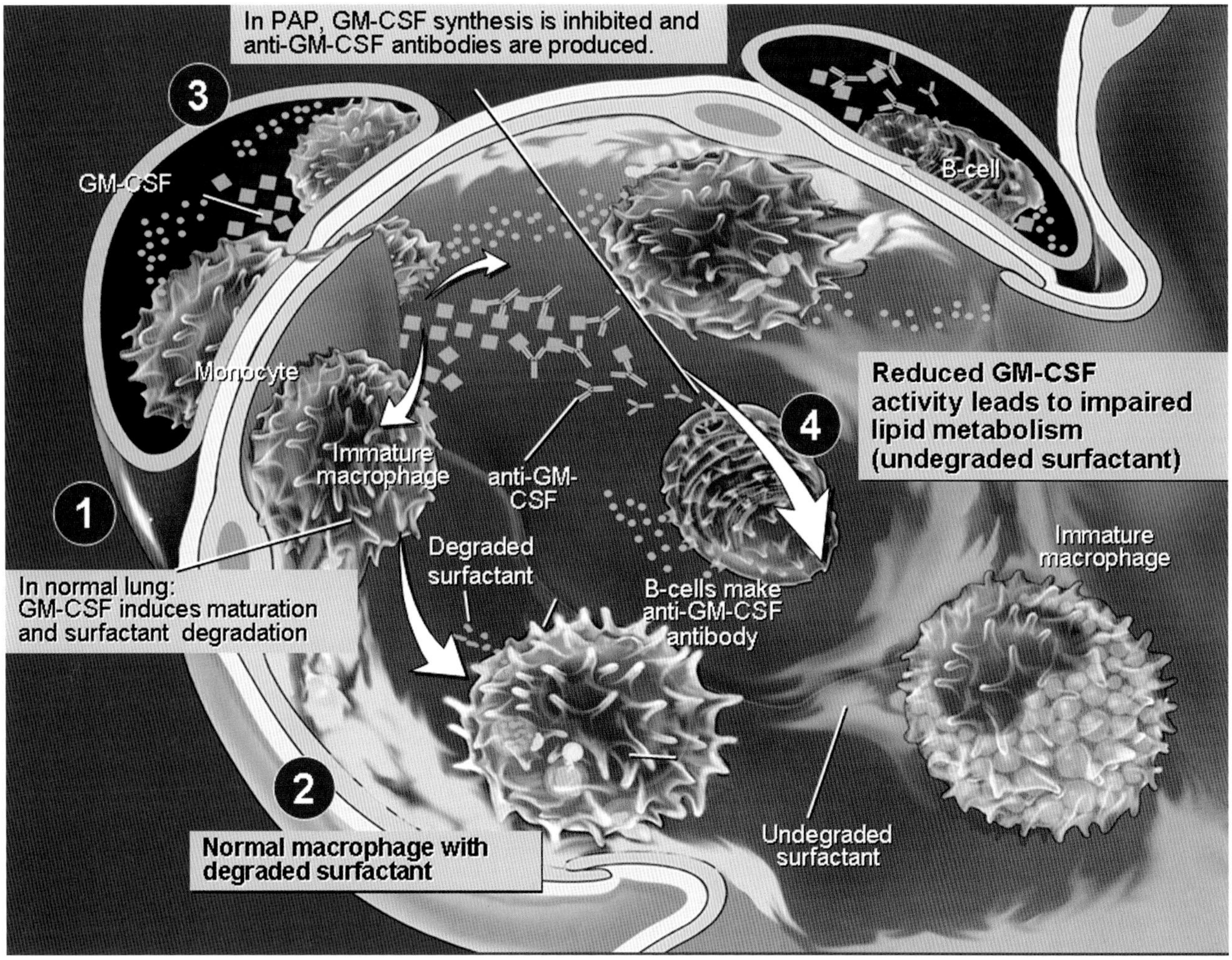

Figure 18.44 Pulmonary alveolar proteinosis. In alveoli of patients with PAP, anti-GM-CSF antibodies produced by B-cells block binding of the GM-CSF to alveolar macrophages (AM), which leads to impaired lipid metabolism and undegraded surfactants.

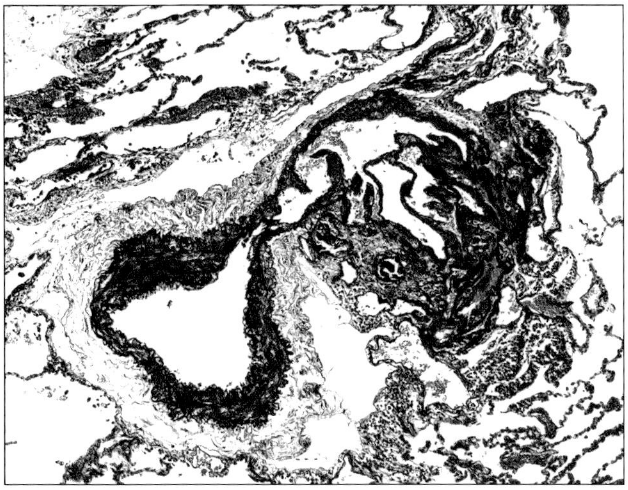

Figure 18.45 Pulmonary hypertension. A plexogenic lesion in a lung from a patient with idiopathic pulmonary hypertension reveals slit-like spaces (upper right corner) emerging from a pulmonary artery. These remodeling vascular spaces represent the irreversible damage done to these vessels in this disease.

most severe pathologic lesions, these abnormal vascular structures become dilated or angiomatoid-like and may develop features of a necrotizing vasculitis with transmural inflammation and fibrinoid necrosis.

Molecular Pathogenesis

Though the exact pathogenetic mechanism of PAH remains unknown, research over the past 10 years has begun to offer some clues. The familial form of PAH, with a 2:1 female-to-male prevalence, has an autosomal dominance inheritance pattern with low penetrance. The genetic basis for this has been found to be germline mutations in the gene encoding the bone morphogenetic protein receptor type 2 (*BMPR2*). These mutations account for approximately 60%–70% of familial PAH and 10%–25% of patients with sporadic PAH [233]. Approximately 140 *BMPR2* mutations have been identified in familial PAH, each resulting in a loss of receptor function, either through alteration in transcription of the gene through missense, nonsense, or frameshift alterations in the codon or by RNA spicing mistakes [226].

The mechanism by which a single mutation to the *BMPR2* gene induces vascular smooth muscle proliferation and decreased apoptosis that is not completely understood, but it most likely involves defects in the BMPR2 signaling pathway. BMPR2 is a receptor for a family cytokines (BMPs) that are members of the TGFβ superfamily of proteins that play a role in the growth and regulation of many cells, including those of the pulmonary vasculature. In the vascular smooth muscle cells of the lung, TGFβ signaling causes a proliferation of smooth muscle in pulmonary arterioles, while BMPR2 signaling causes an inhibition of the proliferation of these cells, favoring an apoptotic environment. The BMPR2 signaling occurs through an activation of a receptor complex (BMPR1 and BMPR2) that leads to phosphorylation and activation of a number cytoplasmic mediators, most notably the Smad proteins (Mothers against decapentaplegic). These Smad proteins, especially the Smad 1, Smad 5, and Smad 8 complex with Smad 4, translocate to the nucleus where they target gene transcription that induces an antiproliferative effect in the cell. In familial PAH, the *BMRPR2* gene mutation may lead to insufficient protein product and subsequent decreased protein function, in this case decreased BMPR2 receptor function, decreased Smad protein activation, and decreased antiproliferative effects in the vascular smooth muscle cells. The imbalance between the proproliferative effects of the TGFβs and the antiproliferative effects of the BMPs results in the formation of the vascular lesions of PAH (Figure 18.46) [227,228].

Despite these advances, questions regarding the pathogenesis of PAH remain. Most notably, why do only 10%–20% of patients with the mutation develop clinical disease? Some speculate that genes confer susceptibility but a second hit is required to develop the clinical disease, such as modifier genes or environmental triggers, perhaps drugs or viral infections [227,229]. In addition, though *BMPR2* mutations have been found in both the familial and the idiopathic form of

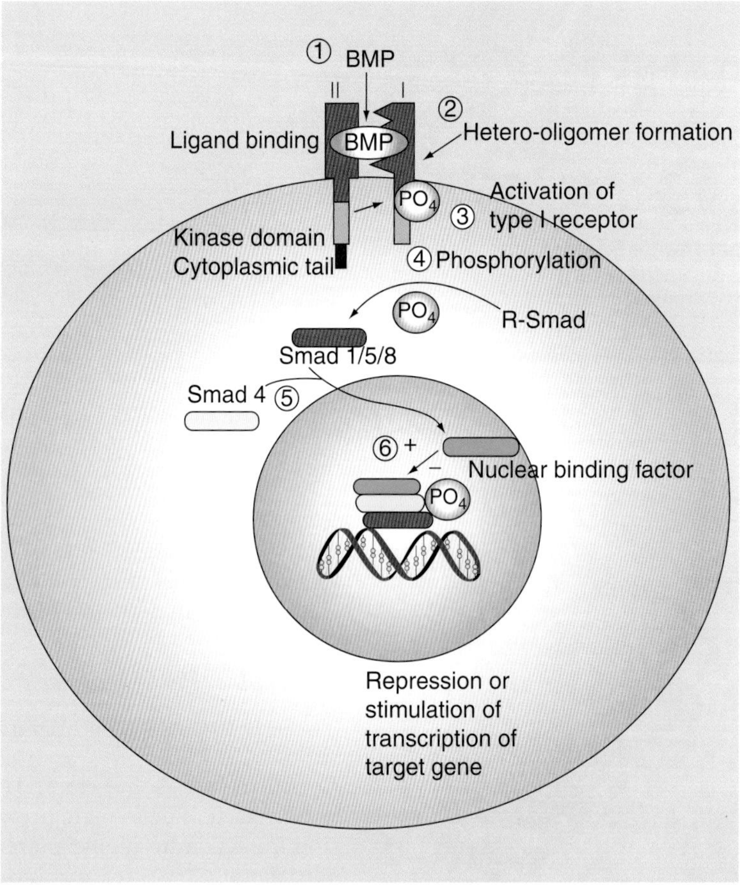

Figure 18.46 Bone-morphogenetic-protein signaling pathway. 1 and 2: BMPR1 and BMPR2 are present on most cell surfaces as homodimers or hetero-oligomers. With ligand (bone morphogenetic proteins; BMP) binding, a complex of ligand, two type I receptors, and two type II receptors is formed. 3: After ligand, two type I receptors and two type II receptors phosphorylate the type I receptor in its juxtamembrane domain. 4: The activated type I receptor then phosphorylates a receptor-regulated Smad (R-Smad); thus, the type I receptors determine the specificity of the signal. 5: Once activated by phosphorylation, the R-Smads interact with the common mediator Smad 4 to form hetero-oligomers that are translocated to the nucleus. 6: In the nucleus, the Smad complex interacts with transcription factors and binds to DNA to induce or suppress transcription of target genes. (Reprinted with kind permission from Runo JR and Loyd JE. (2003). Primary pulmonary hypertension. *Lancet* **316**:1533–1544 [228]).

PAH, they are present in only 30% of all PAH patients, suggesting that further research is needed to uncover additional etiologic agents.

Pulmonary Vasculitides

Pulmonary vasculitides present as diffuse pulmonary hemorrhage and are usually caused by one of three major pulmonary vasculitis syndromes: Wegener's granulomatosis, Churg-Strauss syndrome, and microscopic polyangiitis. All three diseases have similar clinical presentations and considerable overlap in their pathologic features as small vessel systemic vasculitides that affect the lung as well as other organs, most notably the kidney.

Clinical and Pathologic Features

Wegener's granulomatosis (WG) is an unusual disease that affects the upper and lower respiratory tract and the kidneys. It usually presents between 40 and 60 years of age and is slightly more common in men than women. The clinical presentation depends on the affected organ, but when the lung is involved, hemoptysis is the major presenting symptom. Chest imaging studies may show a variety of patterns, most commonly bilateral ground glass opacities with masses, usually in the lower lobes that may cavitate. Immunologic testing of peripheral blood or end organ tissue can be helpful in revealing characteristic immunofluorescent staining patterns for antineutrophilic cytoplasmic antibody (ANCA), an antibody that targets two substances: proteinase 3 (PR3) and myeloperoxidase (MPO). When present in either the blood or the tissue, the pattern of immunofluorescent staining can be cytoplasmic (cANCA) or perinuclear (pANCA). The former pattern is more commonly seen in Wegener's granulomatosis, and the latter is more commonly seen in microscopic polyangiitis and Churg-Strauss syndrome (CSS).

CSS is a systemic disorder defined by the presence of asthma, peripheral blood eosinophilia, and systemic vasculitis. Similar to WG, it usually presents between 40 and 60 years of age, and a clinical diagnosis requires a history of asthma, a peripheral blood eosinophilia, neuropathy, an abnormal chest imaging study, and sinusitis. Other organs involved include the heart, the central nervous system, kidneys (though less commonly than WG), gastrointestinal tract, and skin. Chest imaging usually shows patchy, multifocal infiltrates; masses and cavitation are rare. Laboratory tests reveal positive pANCA tests in 70% of patients.

Microscopic polyangiitis (MPA) is similar to both WG and CSS in that it is a systemic vasculitis that involves the lung and usually presents in the fourth or fifth decade of life. The clinical onset is usually sudden with fever, weight loss, myalgias, and arthralgias. The kidney is the main organ involved, and MPA is the most common cause of pulmonary-renal syndrome. Lung involvement occurs in approximately 50% of the patients, and skin and upper respiratory tract are other common sites. Similar to WG and CSS, ANCA testing is helpful with positive pANCA in 80% of patients. Chest imaging usually shows bilateral infiltrates without masses, similar to CSS. Treatment for all three diseases is immunosuppression with glucocorticoids or cyclophosphamide, and all three usually respond well, although WG has a greater relapse rate after treatment than either CSS or MPA [230].

The pathology of WG, CSS, and MPA have overlapping features of an acute and chronic vasculitis that involves medium- and small-sized vessels in the lung. The inflammatory cell infiltrate that destroys the blood vessels is both lymphocytic and neutrophilic, and areas of fibrinoid necrosis are seen. However, in WG, there are characteristic areas of microabscesses that lead to masses of geographic necrosis with basophilia. Scattered multinucleated giant cells are present, but no well-formed granulomas are seen. This helps to distinguish it from other vasculitides and infection (Figure 18.47). Similarly, the pathology of CSS has distinguishing features, with the early pathology characterized by an eosinophilic pneumonia with areas of loosely formed granulomas with central necrosis containing degenerating eosinophils (Figure 18.48). The infiltrate is predominantly eosinophils, but neutrophils, lymphocytes, and plasma cells are also present. Capillaritis can be seen in WG, CSG, and MPA, and all three have hemosiderin deposition present, both within alveolar macrophages and deposited in the connective tissue of the interstitium and the vessel walls.

Molecular Pathogenesis

The pathogenesis of these three pulmonary hemorrhage syndromes is similar to the mechanisms of these diseases in the kidney. In general, these diseases in the lung and the kidney represent immune-mediated

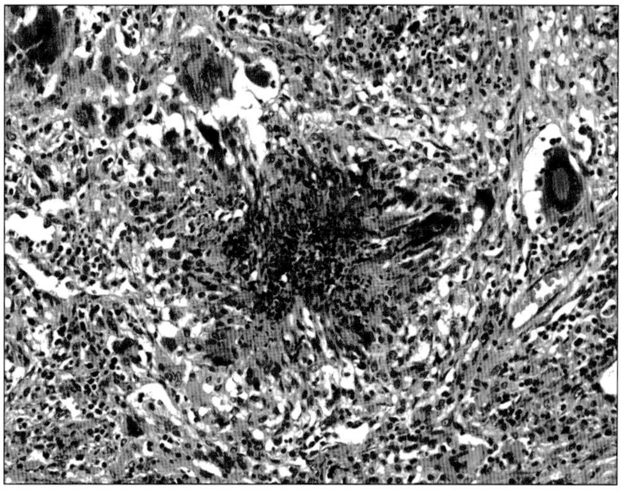

Figure 18.47 **Wegener's granulomatosis.** The inflammation in lungs involved by Wegener's granulomatosis includes neutrophilic microabscesses in a stellate pattern. Giant cells are commonly seen in the surrounding areas. These lesions are thought to be the early form of the larger areas of geographic necrosis that produces the mass-like nodules found in these lungs.

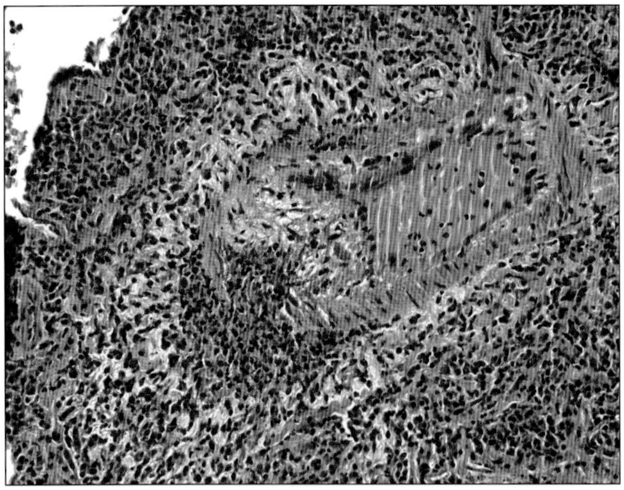

Figure 18.48 **Churg-Strauss Syndrome.** An eosinophilic infiltrate invades the wall of a medium-sized artery in this lung from a patient with CSS. The surrounding lung also contains a dense infiltrate of eosinophils, lymphocytes, and plasma cells.

necrotizing vasculitides that have few or no immune deposits in the vessels but exhibit the presence of ANCA autoantibodies to myeloperoxidase (MPO) and proteinase 3 (PR3), the components of primary granules of neutrophils. MPA and CSS are primarily diseases of MPO antibodies, and WG is primarily a disease of PR3 antibodies. The mechanism by which the ANCAs are induced is not known but may be part of an autoimmune response to environmental exposures early in life. These autoantibodies then inflict damage on the vessels through a mechanism that is not yet completely understood. One theory suggests that circulating ANCAs bind to PR3 and MPO on the surface of neutrophils and initiate a respiratory burst, degranulation, and apoptosis. ROS and proteases are released and inflict endothelial and tissue damage on the adjacent vessel. The ANCA binding may also induce the release of proinflammatory cytokines and chemokines such as IL-1 and TNFα that further contribute to the vascular inflammation. The second theory postulates that circulating immune complexes of excess ANCA antigen (MPO or PR3) and ANCA autoantibodies attach to the vascular endothelium and activate complement that results in the chemotaxis and adhesion of inflammatory cells, causing these cells to undergo a respiratory burst and, as in the first theory, release of ROS and proteases that cause the vascular endothelial damage. In both theories, it is important to remember that MPO and PR3 are also present in monocytes and that ANCA autoantibodies may be involved with monocytes in similar ways to release inflammatory mediators [231].

PULMONARY INFECTIONS

Infectious diseases of the lung are a common cause of pulmonary disease given the constant exposure of the lungs to the environment. Various organisms are capable of causing these infections, including common viruses and bacteria, as well as more uncommon fungi, parasites, and protozoa. The diagnosis of the specific etiologic agent can be challenging given that most have similar clinical features and many are difficult to identify in the lung tissue. This brief overview of the defense mechanisms the lung uses to protect itself will serve to introduce the pathology of these lung infections.

Overview of Pathogenesis of Lower Respiratory Tract Infections

Anatomic Defenses

The lung has multiple anatomic mechanisms by which it defends itself against invasion by various pathogens. First, the upper nasal cavities and respiratory tract serve as anatomic barriers to inhaled organisms. The ciliated epithelium and torturous cavities of the sinuses screen large organisms (typically larger than 10 microns). For those particles that venture further down the respiratory tract, the cough reflex that the upper trachea elicits serves to expel them up and out. Second, the mucociliary tree of the upper respiratory tract captures organisms that evade these two mechanisms. The bronchial epithelium contains cilia of up to 20 microns in length that extend into the air surface liquid (ASL). The ASL is a bilayer of 50–100 microns in thickness consisting of a low-viscosity or watery lower layer that is covered by a high-viscosity or gel upper layer secreted by adjacent goblet cells. This sticky upper layer serves to trap organisms, and the coordinated beating of the cilia moves these entrapped invaders up this mucociliary escalator to the larynx, where they can be expectorated.

Soluble Mediators

Present in the secretions of the large airways and within the surfactant lining the alveolar walls are soluble mediators secreted by various cells. These mediators include lysozyme and lactoferrin, which lyse bacteria and inhibit their growth; the defensins and cathelicidins, small peptides both with microbicidal properties; and surfactant proteins A and D at the alveolar level, which bind to microorganism and enhance phagocytosis and also have direct bactericidal activity [232].

Cells of Innate Immunity

The major cells of the innate immune response of the lung are the alveolar macrophages (AM) and the polymorphonuclear leukocytes (PMN). Neutrophils phagocytize and destroy bacteria such as *S. aureus, S. pneumoniae,* and *H. influenzae* through a respiratory burst that generates NADPH oxidase-dependent ROS. In some instances, AMs may ingest but not kill an organism. This occurs with such organisms as *Mycobacterium* spp., *Nocardia* spp., and *Legionella* spp. Because of the ability of these organisms to continue to replicate within the AM, cell-mediated immunity is required for their complete elimination. Patients with defects in NADPH oxidase are especially prone to respiratory infections by such organisms as *S. aureus*, *Nocardia* spp. and *Aspergillus* spp.

Bronchial epithelial cells are important in innate immunity through secretion of cytokines and molecules including IL-1, IL-5, IL-6, IL-8, and granulocyte-macrophage colony-stimulating factor (GM-CSF). These molecules attract macrophages as well as neutrophils and other inflammatory cells to the area to enhance the inflammatory response to the organism [233]. Bronchial epithelial cells also serve an important role in recognizing pathogens through pattern-recognition receptors (PRRs).

Natural killer (NK) cells are involved in the innate immune response with surface receptors that recognize cells infected with viruses such as RSV, influenza, parainfluenza, and rhinovirus. The NK cells release IFN-γ, which recruit other immune cells to add to the antiviral response.

Dendritic cells are tissue histiocytes positioned around the airways and lymphatics in the lung that recognize pathogens and their antigens and trigger the proliferation and amplification of antigen-specific T-cells. This immune response bridges the innate immune response to the adaptive immune responses and is especially important in fungal infections. This mechanism is mediated through toll-like receptors (TLRs) that are able to distinguish pathogens from self-components by triggering cytokine production through NFκB and AP-1 and expressing co-stimulatory molecules necessary for this T-cell activation [234].

Adaptive Immunity

For those organisms that evade the basic, innate immunity of the lung, there are adaptive immune mechanisms that encompass both humoral and cellular immune mechanisms. Humoral immunity is an important defense against encapsulated bacteria, most notably *S. pneumoniae*, and for other pyogenic bacteria such as *H. influenzae*, and *Staphylococci* spp., and resolution of these infections requires the production of IgG antibodies to the organisms. Cellular immunity is especially important against such respiratory viral infections as influenza, RSV, CMV, varicella, and also against opportunistic infections. These viruses induce a CD4+ and CD8+ T-cell response that clears the lung of these viruses within 8–10 days post infection.

Granulomas

Granulomas are a common inflammatory response to both pathogens and foreign material. The most notable granulomatous infections in the lung are due to mycobacteria and fungal organisms. Activation of CD4+ T-cells by these organisms leads to proliferation and differentiation of these CD4+ T-cells into T-helper-1 cells. The release of IFN-γ by the Th-1 cells activates lung macrophages to form epithelioid macrophages that have an increased ability to kill the microorganisms and express surface molecules that promote cell-to-cell fusion into giant cells. In addition, activation of these macrophages results in the release of numerous cytokines including IFN-γ and TNFα. In patients who are deficient in CD4+ T-cells or IFN-γ, granuloma formation is very poor, altering the pathologic picture of these infections. This effect is most obvious in the nontuberculous mycobacterial infections, which have numerous patterns of injury depending on the immune status of their host.

Clinical Overview

Pneumonias can be broadly categorized into one of five major clinicopathologic categories, including (i) community-acquired pneumonias (acute and atypical), (ii) nosocomial pneumonias, (iii) aspiration pneumonias and lung abscess, (iv) chronic pneumonias, and (v) pneumonias in immunocompromised hosts. Each type presents with a characteristic clinical pattern and may be caused by any of several pathogens so that treatment is many times empiric.

The first category comprises community-acquired pneumonias (CAP). These represent the majority of the lung infections that receive medical treatment, usually on an outpatient basis, with low (<1%) mortality. Patients hospitalized for these infections typically have other comorbidities. The responsible organisms include respiratory syncytial virus (RSV); rhinovirus, parainfluenza, and influenza virus; bacteria, including *Mycoplasma pneumoniae* and rickettsia; and most notably *Chlamydia pneumonia*. Chlamydia causes what is termed atypical pneumonia with a clinical course characterized by a progressive onset of fever without chills, a dry cough, and chest imaging that reveals focal infiltrates. Acute or typical CAP presents abruptly with high fever, chills, productive cough, and radiographs with lobar or segmental consolidation. The most common pathogens are *Streptococcus pneumoniae*, *Haemophilus influenza*, *Staphylococcus aureus*, and *Moraxella catarrhalis*.

The second category, nosocomial pneumonias, consists of infections acquired within the hospital or from healthcare associated facilities. These infections are usually found in patients with predisposing risk factors and are a major source of morbidity and mortality, with some studies reporting a mortality range of 20%–50%. The most common risk factors include respiratory ventilation, artificial airways, nasogastric tubes, supine positioning, and medications that alter gastric emptying. The responsible organisms include *Klebsiella* spp., *Legionella* spp., *Staphylococcus aureus*, and *Pseudomonas aeruginosa*.

The third category includes aspiration pneumonias and lung abscesses. These infections occur in the setting of patients with aberrant swallow or gag reflexes that allow gastric or oral contents into the airways. The organisms where necrosis and cavity formation occurs include *S. aureus*, *K. pneumoniae*, the anaerobic oral flora, and mycobacteria. Clinically, these infections may have an acute course with fever and dyspnea or a more insidious course, many times with patients first presenting with lung cavities, empyemas, or necrotizing pneumonias.

The fourth category, chronic pneumonias, includes indolent infections that cause a localized mass-like lesion in an otherwise healthy host. *Nocardia* and *Actinomyces* spp. are the most common pathogens, but mycobacteria and fungi may also cause these pneumonias. The fifth category includes pneumonias that occur in the setting of an immunocompromised patient. These include a number of organisms that otherwise would not act as pathogens such as the viruses CMV and HSV, the fungi Aspergillosis and Pneumocystis pneumonia, and the bacterium *Mycobacterium avium* complex.

Clinical and Pathologic Features

Bacteria

Streptococcus pneumoniae *Streptococcus pneumoniae*, a gram-positive diplococcus also known as pneumococcus or *Diplococcus pneumonia*, is a common cause of bacterial pneumonia in infants and elderly patients, alcoholics, diabetics, and patients with immunosuppression. This pneumonia usually presents abruptly with chills, a cough with rust-colored sputum and pleuritis, with high fevers, tachycardia, and tachypnea. The characteristic gross pathology is a lobar pneumonia that progresses from a red acute phase to a gray organizing phase. A fibrinous pleuritis is common, which eventually organizes to entrap the lung parenchyma in a fibrous capsule [235]. The microscopic examination reveals abundant fibrin, neutrophils, and extravasated red blood cells within the alveolar space and congested capillaries.

Hemophilus influenzae *Hemophilus influenzae* is a gram-negative bacillus that inhabits the upper respiratory tract and can cause otitis media, epiglottitis, and meningitis, and usually enters the lung through aspiration or hematogenous spread. Six serotypes are defined based on their capsular antigens, with Type B the most common cause of pneumonias. This type of pneumonia is most commonly found in children or in the elderly with underlying chronic lung disease such as emphysema, cystic fibrosis, bronchiectasis, in patients with HIV infection, or in alcoholics. This bacterial pneumonia is usually preceded by a viral or mycoplasma infection that damages the mucociliary elements in the airways and allows for colonization by *H. influenzae*. The symptoms include fever; a productive, purulent cough; and myalgias. The incidence of this pneumonia as a common community-acquired pneumonia in children is quite low due to the advent of effective vaccines. However, it is increasing in incidence as a nosocomial infection [236]. Like pneumococcal pneumonia, the pathology of *H. influenzae* pneumonia is in a lobar distribution with a neutrophilic-rich infiltrate and a pleural effusion. Necrosis and empyema may occur but are uncommon.

Staphylococcus aureus Staphylococcal pneumonia is caused by *Staphylococcus aureus*, gram-positive cocci that usually spread to the lung through the blood from other infected sites, most often the skin. Though a common community pathogen, it is found twice as frequently in pneumonias in hospitalized patients. It often attacks the elderly and patients with CF and arises as a co-infection with influenza viral pneumonia. The clinical course is characterized by high fevers, chills, a cough with purulent bloody sputum, and rapidly progressing dyspnea. The gross pathology commonly reveals an acute bronchopneumonia pattern (Figure 18.49) that may evolve into a necrotizing cavity with congested red/purple lungs and airways that contain a bloody fluid and thick mucoid secretions. The histologic pattern is characterized by a bronchopneumonia that spreads distally from the small airways into to the alveolar spaces (Figure 18.50) to form abscesses that connect with the pleural surface and may result in empyemas. The treatment of this organism has become increasingly problematic due to antibiotic-resistant strains, most notably methicillin-resistant *S. aureus*.

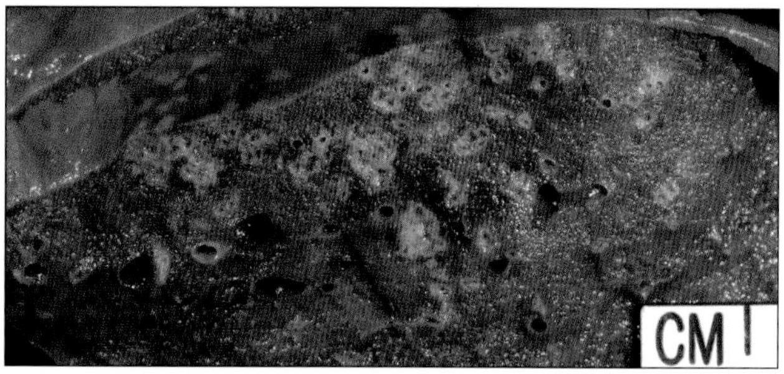

Figure 18.49 **Bronchopneumonia.** The cut surface of an upper lobe of lung reveals scattered tan nodules surrounded by red erythema, features of an early acute bronchopneumonia. The tan areas represent the earliest inflammation surrounding the small airways seen in Figure 18.50.

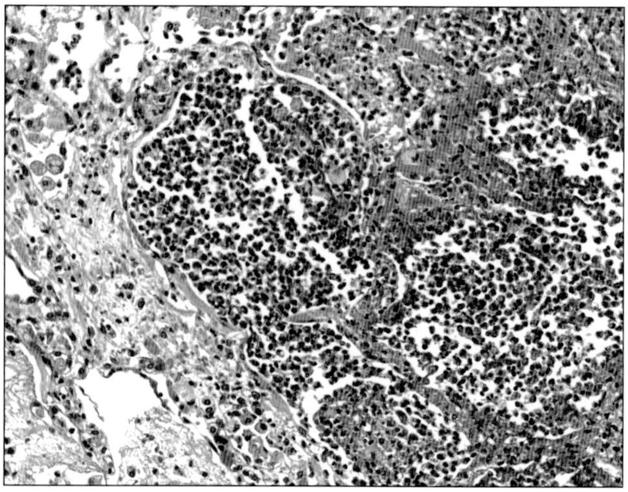

Figure 18.50 **Acute bronchopneumonia.** A microscopic picture of an acute bronchopneumonia caused by *S. aureus* reveals an abundant infiltrate of neutrophils filling a small bronchiole and extending into the adjacent alveoli.

Legionella pneumophila *Legionella* are gram-negative bacilli found predominantly in aquatic habitats such as lakes, rivers, and ponds. Standing pools of water from humidifiers and other water outlets may be other sources. Approximately 50% of air conditioners contain these bacilli. Though 15 serogroups of *Legionella* have been identified, 3 cause the overwhelming majority of human pneumonia. The clinical disease takes two forms: (i) Legionnaires' disease, named after the outbreak of pneumonia at the 1976 American Legion convention in Philadelphia; and (ii) Pontiac fever, a self-limiting flu-like disease with nonspecific symptoms. *Legionella* pneumonia presents as a severe infection of the lung with chills and rigors with a nonproductive cough. It can progress rapidly to systemic symptoms of nausea, vomiting, and diarrhea and can lead to renal failure and death without immediate antibiotic therapy. The infected lungs are remarkably red and congested and appear to be distended with fluid. The microscopic picture reveals fibrinopurulent exudates that fill the alveolar space mixed with a necrotic, cellular infiltrate of degenerating neutrophils and monocytes (Figure 18.51). Hyaline membranes may form in the periphery of the lesions, and pleural effusions consisting of fibrinoserous exudates are common.

Pseudomonas aeruginosa *Pseudomonas aeruginosa* is a gram-negative bacillus that is found throughout the environment and in 50% of the airways of hospitalized patients. It usually enters the body through a disruption of the epithelial surface by cuts, burns, or therapeutic devices such as mechanical ventilators or intravascular catheters. Pneumonias caused by this organism are usually found in intensive care units of hospitals and burn units, in patients with underlying chronic lung diseases including cystic fibrosis, emphysema, and in patients with prolonged hospitalization. The pathology is necrosis with a bronchopneumonia pattern that usually consists of an area of congestion and hemorrhage that is surrounded by a halo of tan/white consolidation (Figure 18.52). A necrotizing vasculitis with abundant organisms in vessel walls can be seen, and cavitation is common (Figure 18.53). In treated lungs, healed cavities or pneumatoceles may appear as smooth-walled fibrous cysts.

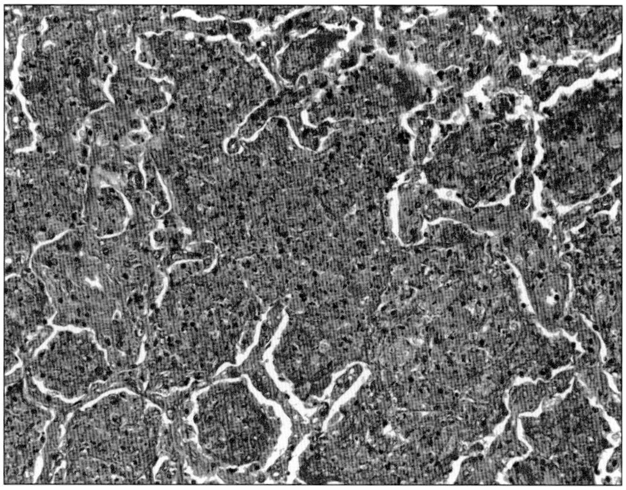

Figure 18.51 ***Legionella pneumonia.*** An eosinophilic exudate containing neutrophilic and histiocyte debris involves the alveolar of a lung infected by *L. pneumoniae.*

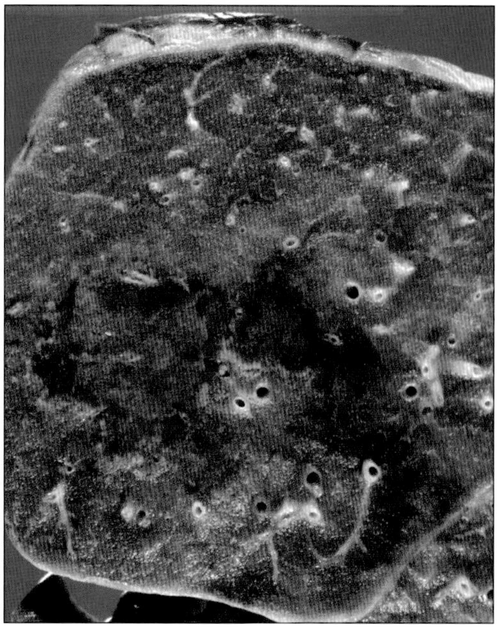

Figure 18.52 ***Pseudomonas aeruginosa.*** The cut surface of this upper lobe of a lung reveals an erythematous patch surrounded by a tan rim, characteristic of an acute pneumonia caused by *P. aeruginosa.* As this infection progresses, it can cavitate and result in abscesses and pneumatoceles that are commonly seen in this pneumonia.

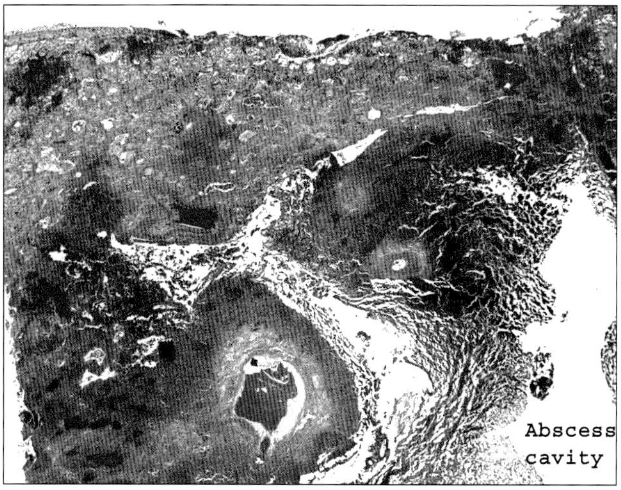

Figure 18.53 ***Pseudomonas aeruginosa.*** A microscopic picture of a more advanced acute infection from *P. aeruginosa* with abscess formation. The blue areas adjacent to the cavity contain neutrophilic debris surrounding large vessels within the lung. The cavity formation in this infection usually results from an ischemic necrosis secondary to a vasculitis caused by the organisms invading into these adjacent vessels.

Other Gram-Negative Bacilli Gram-negative bacilli such as *Klebsiella pneumoniae*, *Acinetobacter*, and various *Enterobacteriaceae* spp. are common nosocomial pathogens. Similar to *P. aeruginosa*, these pathogens colonize the oropharynx and are usually introduced into the lung by inhalation or aspiration of oral contents. The most notable of these is Friedlander's pneumonia caused by *K. pneumoniae*, the most common cause of gram-negative bacterial pneumonia. This typically occurs in men over 40 years of age, usually in the setting of alcoholism, diabetes mellitus, or chronic lung disease. These patients produce large amounts of thick, bloody sputum, a product of the viscous mucopolysaccharide capsule of the organism, and present with severe systemic symptoms of hypotension and generalized weakness. The pathology of these pneumonias is similar to Pseudomonas pneumonia with marked cavitation and abundant organisms on microscopic examination.

Nocardia spp. Nocardiosis of the lung is caused by *Nocardia asteroides*, a gram-positive rod found in the soil or organic matter. This infection is most common in immunocompromised adult patients and can be seen in the setting of pulmonary alveolar proteinosis, chronic lung diseases, and mycobacterial and other granulomatous diseases that affect the lung. Its clinical course is indolent and usually begins 1–2 weeks before the patient presents for medical therapy. Cough is common, often with thick, purulent sputum. In the immunocompromised setting, fever, chills, dyspnea, and hemoptysis are common, and weight loss may occur as the disease progresses. The pathology is remarkable for suppurative abscess formation with multiple cavities filled with green, thick pus. The inflammatory infiltrate consists of neutrophils, macrophages, and abundant necrotic debris with epithelioid histiocytes and giant cells within the wall of the cavity (Figure 18.54). Empyema and pleura involvement occur in the majority of cases.

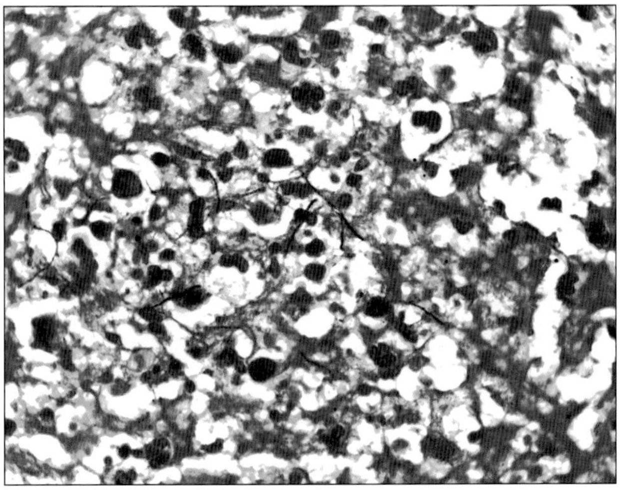

Figure 18.54 ***Nocardia.*** A Fite stain reveals branching filamentous organisms within an abscess caused by *N. asteroides*.

Mycoplasma and Rickettsia Pneumonias *Mycoplasma pneumoniae* pneumonia is among the most common infections of the lower respiratory tract and usually occurs in small epidemics in closed populations. It often presents with atypical features of a progressive onset, fever without chills, a dry cough, diffuse crackles on physical examination, and chest imaging studies that reveal patchy interstitial infiltrates. The pathologic features are a result of the attachment of the organisms to the bronchiolar epithelium where they cause epithelial injury and ulcerations through secretion of peroxide [237]. In cases of severe infection, diffuse alveolar damage may be present.

Chlamydial Pneumonia *Chlamydia* spp. causes pneumonia in a variety of clinical settings. *Chlamydia trachomatis* is an infection found predominantly in the postnatal period, *Chlamydia psittaci* is the result of direct transmission from infected birds, including parakeets, parrots, and pigeons. *Chlamydia pneumoniae* is the most common of the three and is a frequent cause of community-acquired pneumonia. It typically causes a very mild or asymptomatic infection with fever, sore throat, and nonproductive cough. The course of this infection may be severe in the elderly. Chest imaging studies show alveolar infiltrates, and pleural effusions are present in the majority of cases. The pathology has not been well defined since the infection is usually self-limited. However, in experimental animal models there is a neutrophilic response in the early stages, and an interstitial, peribronchiolar, and perivascular infiltrate of lymphocytes, macrophages, and plasma cells in the latter stages of the infection.

Mycobacteria

Mycobacteria, a major cause of lung infections, are nonmotile, aerobic, catalase-producing, acid-fast bacilli. Clinically significant lung infections can be caused by *M. tuberculosis* and by a group of nontuberculous mycobacteria (NTM). The latter group consists of over 100 species, of which three cause the overwhelming majority of pulmonary disease. These are *M. avium-intracellulare* (*M. avium* complex), *M. kansasii*, and *M. fortuitum-chelonae*. Throughout history, tuberculosis (infection with *M. tuberculosis*) was the major disease caused by these organisms and was responsible for worldwide morbidity and mortality. However, over the past two decades lung diseases caused by NTM have become much more common and now represent the majority of the pulmonary mycobacterial disease.

Mycobacterium Tuberculosis Pulmonary tuberculosis is spread by interpersonal contact through aerosolized droplets. Once in the alveoli, the bacteria cause a cell-mediated inflammatory response that is capable of inducing granuloma formation and necrosis. As in all infections, the extent of the disease is a function of the host's immune response. The most susceptible patients are those with certain conditions that include immunosuppression, diabetes, malignancy, renal failure, among others. Clinically, an infected patient has a productive cough, fever, and weight loss, and may develop hemoptysis as the cavitation progresses and erodes into the pulmonary vessels. Extensive involvement of the lung can produce significant dyspnea and pleuritic chest pain.

The pathology of tuberculosis is primarily that of granuloma formation and acute pneumonia. The granulomas are predominantly necrotizing, and the pneumonia usually contains abundant fibrin and neutrophils that fill the alveolar spaces. The gross lesions are referred to as caseous or cheese-like, because of the amount of necrosis present. This caseous material can extend into airways and is commonly coughed up during the active disease. In chronic forms of the disease, the area can undergo fibrosis and involute into a firm, hard scar. There are three major clinicopathologic variants of the disease: (i) primary tuberculosis, (ii) postprimary or reactivation tuberculosis, and (iii) progressive fibrocavitary disease.

Primary tuberculosis. In this form of the disease, the initial site of infection can be anywhere in the lung, but is usually in the lower lobe or anterior segment of the upper lobe, the areas that receive the most ventilation. The lesion usually consists of a dense consolidation with acute pneumonia and necrotizing granulomas. Cavitation may occur, especially in the setting of immunocompromised hosts. From these foci, the organisms may spread through the lymphatics to elsewhere in the lung, the hilar lymph nodes, and the bloodstream, and lay dormant for long periods of time. The combination of the primary site of infection and the involved hilar lymph nodes is known as a Ghon complex [238].

Postprimary tuberculosis. This form of tuberculosis represents reactivation of old, scarred primary lesions long after the initial insult. The lesion can occur anywhere in the lung where the bacteria from the primary lesion have spread, but is usually apical. It consists of a focus or organizing pneumonia and fibrosis with central caseation. In an active lesion, the typical parenchymal pattern is an acute pneumonia with cavitation that expands to include the surrounding lung with aggregates of granulomas. The controversy surrounding this lesion arises as some evidence suggests that these lesions represent exogenous reinfection. The pathology of reactivation or reinfection may be indistinguishable, although reactivation tuberculosis may appear to arise out of a fibrotic, calcified chronic lesion [239].

Progressive fibrocavitary disease. This form of the disease may arise out of either primary or postprimary tuberculosis. However, the latter is the more common scenario. The cavities that develop in this form of the disease begin as a slowly progressive, necrotizing pneumonia with abundant granulomas (Figure 18.55). The active disease may spread through the airways, causing ulceration, necrosis, and fibrosis of the surrounding bronchi and bronchioles. The extension of the disease in this way depends on the host, and patients with depressed immune systems can have large areas of the lung involved with massive pulmonary necrosis. Usually, a fibrous capsule develops in the area of the cavitation, although inspissated necrotic material into the adjacent airways remains a continuous source of inflammation that can lead to reinfection and ongoing scarring [240].

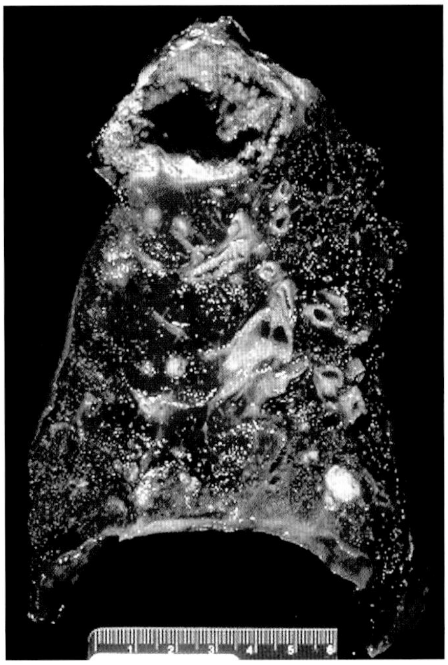

Figure 18.55 **Pulmonary tuberculosis.** This sagittal cut section of lung reveals a remarkable necrotic cavity in the upper lobe of progressive fibrocavitary disease. Within the right lower lobe, there is a tan white nodule, representing infection that has spread through the airways to distal parts of the lung.

Nontuberculous Mycobacteria The nontuberculous mycobacteria (NTM) are ubiquitous inhabitants of our environment, isolated from soil, fresh and brackish water, house dust, birds, animals, and food, and are increasingly important in causing pulmonary disease. There are currently more than 100 NTM species known. Those organisms thought to be pathogenic to the lung include the following: *M. avium* complex, *M. kansasii*, *M. xenopi*, *M. scrofulaceum*, *M. szulgai*, *M. simiae*, *M. fortuitum-chelonei*, and *M. malmonense*. Of these, *M. avium-intracellular*, *M. kansasii*, and *M. fortuitum-chelonei* account for the overwhelming majority of the pulmonary disease caused by NTM. *M. avium* and *M. intracellulare* are usually lumped together as *Mycobacterium avium* complex (MAC) simply because most laboratories are not equipped to distinguish between the two organisms. The clinical presentation of these lung infections can vary from minimally symptomatic small lesions discovered by routine radiography to sudden hemoptysis from advanced disease with severe cavitation (Table 18.5). The two most characteristic lesions are those of diffuse infiltrates in an immunocompromised patient, seen most commonly in the HIV-positive population and an

Table 18.5 Clinical and Pathologic Features of Nontuberculous Mycobacterial Respiratory Infections

Clinical Features	Pathology Features	Organisms
Asymptomatic nodule	Hyalinized, calcified nodule	*M. avium* complex
Immunocompromised	Diffuse, histiocytic infiltrate	*M. avium* complex
Elderly male with chronic lung disease	Apical cavitary pneumonia	*M. avium* complex
Indolent chronic small airway disease in women	Chronic bronchitis and bronchiolitis	*M. avium* complex
Hypersensitivity pneumonitis ('Hot tub lung')	Interstitial pneumonitis with loosely formed granulomas	*M. avium* complex

indolent inflammation of small airways, usually in the right middle lung causing bronchiolectasis, found in middle-aged women.

Viruses Most pulmonary infections are due to viruses from four major groups: influenza, parainfluenza, respiratory syncytial virus (RSV), and adenovirus (Table 18.6) [241]. The clinical presentations of these infections have some common features, including insidious onset, nonproductive cough, fever, and chest pain. Chest imaging studies usually reveal bilateral, multifocal infiltrates, most without evidence of cavitation or pleural involvement. These infections are mild, self-limiting, and require no more than supportive therapy except in immunocompromised hosts, where the clinical course can be much more serious. Also, immunocompromised patients are susceptible to other viruses such as herpesvirus and cytomegalovirus pneumonias, which are not common pathogens in normal hosts [242]. Since the 1980s, a subset of pulmonary viral infections has emerged with a much more aggressive clinical course, most notably SARS, coronavirus, and Hantavirus. These viruses present with systemic symptoms of headache, myalgias, and weakness followed by a deteriorating clinical course with respiratory distress, shock and, in over 50% of the cases, death [243,244]. Therapy for most respiratory viral infections is supportive, although antivirals are available for some viruses, mostly used in the setting of immunocompromised patients. Ribavirin, a guanosine analogue, is the main antiviral used for RSV; M2 inhibitors or adamantanes (amantadine and rimantadine) are used against influenza A and neuraminidase inhibitors (oseltamivir and zanamivir) are used against both influenza A and B [245]. Cytomegalovirus is treated with ganciclovir, foscarnet, or cidofovir, while herpesvirus is treated with acyclovir [241].

The pathologic patterns of injury for most viruses are similar, making morphologic distinctions among them difficult. However, some characteristic patterns emerge, most notably in those viruses that cause cytopathic changes. Influenza, adenovirus, SARS, coronavirus, and Hantavirus all cause an acute lung injury pattern with diffuse alveolar damage, and in the case of the latter two viruses, evidence of hemorrhage and edema. Influenza and adenovirus will also cause a necrotizing bronchiolitis due to their preferential infection of bronchial epithelial cells. Finally, some viral infections can be distinguished by their characteristic cytopathic inclusions. Adenovirus can be identified by characteristic smudge cells that present in advanced stages of the disease and represent adenovirus particles in the nucleus of an infected cell (Figure 18.56). Cytomegalovirus has both nuclear owl's eye inclusions, as well as cytoplasmic inclusions (Figure 18.57). Herpesvirus has glassy intranuclear inclusions and can also have multinucleation (Figure 18.58).

Table 18.6 Common Respiratory Viral Pathogens

Virus	Family	Genome	Respiratory Infection
Influenza A,B	Orthomyxoviridae	Single-stranded RNA	Pneumonia, Pharyngitis, Laryngitis
Parainfluenza 1,2,3	Paramyxoviridae	Single-stranded RNA	Pneumonia, Pharyngitis, Laryngitis
Adenovirus 1,2,3,4,5,7	Adenoviridae	Double-stranded DNA	Pneumonia, Pharyngitis, Laryngitis
Respiratory Syncytial Virus A,B	Paramyxoviridae	Single-stranded RNA	Sinusitis, Pneumonia
CMV	Herpesviridae	Double-stranded DNA	Laryngitis, Pneumonia
HSV	Herpesviridae	Double-stranded DNA	Pharyngitis, Pneumonia
Hantavirus	Bunyaviridae	Single-stranded RNA	Pneumonia
SARS virus	Coronaviridae	Single-stranded RNA	Pneumonia

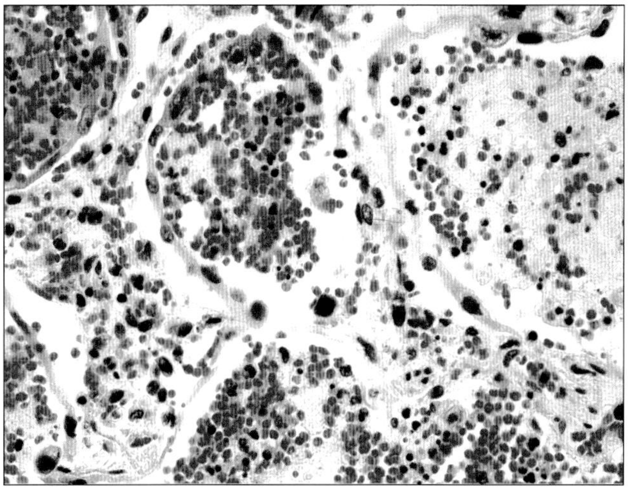

Figure 18.56 Adenovirus. The microscopic features of a severe adenovirus infection in this immunocompromised patient are that of a necrotizing pneumonia with cellular debris within the alveolar space. The viral inclusions are present within the dark nucleus in the center of the field. These cells are called smudge cells due to the obscuring of the nuclear features by these viral inclusions.

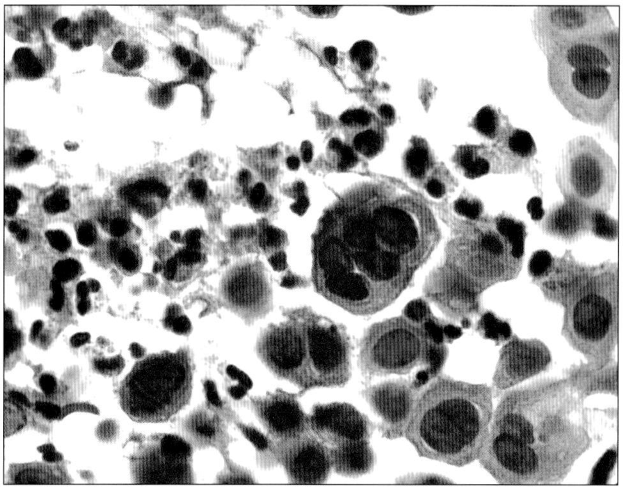

Figure 18.58 Herpesvirus. This microscopic photo is from a lung of an immunocompromised patient with a herpetic simplex pneumonia. In the center of the field is a multinucleated cell representing the cytopathic features of an HSV infection in the lung. These viral inclusions can be found in single cells or syncytia and represent Cowdry A herpetic nuclear inclusions.

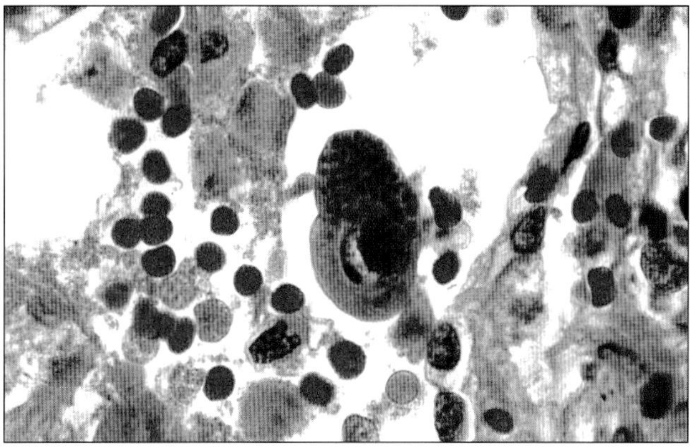

Figure 18.57 Cytomegalovirus. This alveolar macrophage illustrates the microscopic features of a cell infected by the CMV virus. The three features of these cells include (i) cytomegaly, (ii) basophilic cytoplasmic inclusions, and (iii) nuclear amphophilic inclusions.

Fungus

Fungi are larger and more complex than bacteria, and their patterns of injury in the lung are different and in general more destructive. These pathogens are common in our environment and enter the lungs through inhalation. Though many fungi are capable of causing pulmonary disease, most only inhabit the lung as colonizers. Those of most concern for causing clinical disease include the endemic fungi of North America—*Histoplasma capsulatum, Blastomyces dermatitidis,* and *Coccidioides immitis*—and two fungi that are commonly seen in immunocompromised hosts—*Aspergillus fumigatus* and *Pneumocystis jiroveci.*

Histoplasma capsulatum *Histoplasma capsulatum* is a dimorphic fungus most prevalent in the middle portion of the United States from the Great Lakes to Tennessee. The fungus is present in soil that has been contaminated with guano and other debris by nesting birds, most commonly blackbirds and chickens, and by bats. The organism lives in the environment as spores or conidia and germinates to form hyphae. These structures divide to create the yeast forms, which, when inhaled, induce granuloma formation in the lung. Approximately 75% of people have skin tests that are positive for exposure to *H. capsulatum,* but most exposures do not cause clinical disease. Disease typically occurs in people exposed to large amounts of organisms, such as construction workers who move large volumes of dirt or spelunkers who venture into bat-ridden caves. The acute disease has flu-like symptoms which are self-limiting. Healed disease may leave behind calcified granulomas in the lung that appear as buckshot on chest imaging studies. The most chronic forms of this disease may slowly progress, giving rise to cavitating and fibrous lesions. In the immunocompromised host, disseminating histoplasmosis can be seen, although reactivation is uncommon [246].

The pathology reveals characteristic necrotizing granulomas distributed around the airways (Figure 18.59), which contain silver-positive yeast forms of 2–4 microns. These granulomas may resolve into scarred nodules, which can calcify and produce the characteristic chest

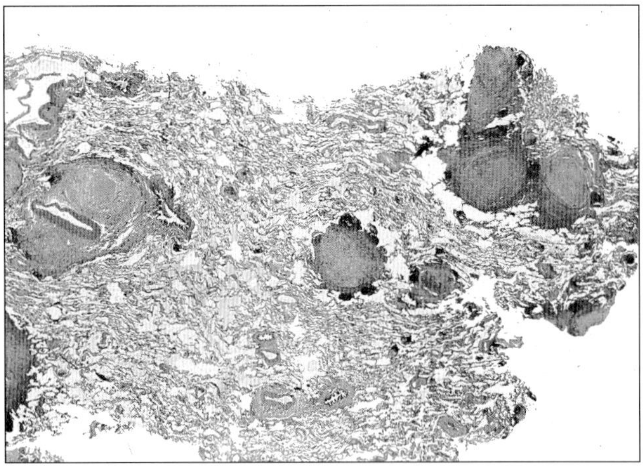

Figure 18.59 **Histoplasmosis.** Necrotizing granulomas centered on the airways within this lung are the characteristic feature of acute histoplasmosis. Within the center of these granulomas, one can find yeast form of *H. capsulatum*.

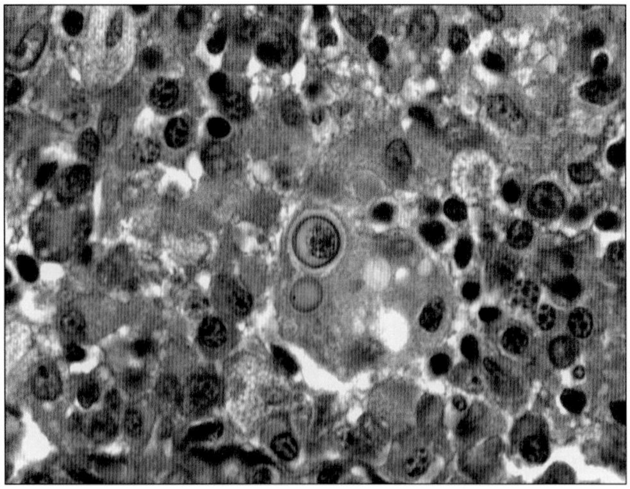

Figure 18.60 **Blastomycosis.** *Blastomyces dermatitidis* is a yeast form found within necrotizing granulomas and characterized by broad-based, single budding with a refractile cell wall that can be seen on the Periodic-acid Schiff stain.

images. Cavities may form in the apices with progression of the disease, and the disseminated form of the disease has an abundance of organisms both within macrophages in the lung and throughout many organs in the body.

Blastomyces dermatitidis *Blastomyces dermatitidis* is also endemic to the middle United States, including the Ohio and Mississippi River valleys. It is found in wooded terrain, usually during the wet seasons, putting campers and outdoorsmen at risk. The clinical disease takes two forms, cutaneous and systemic, the latter beginning in the lungs through inhalation. The acute pulmonary infection takes a nonspecific form with fever, malaise, and chest pain. Imaging studies may show either infiltrates or a mass-like infiltrate. Thus, *Blastomyces* infection may mimic other diseases, and the diagnosis may be delayed. Some patients go on to chronic disease with cavitation or progressive pulmonary blastomycosis, which manifests as acute respiratory distress syndrome, cavitary lesions, and a poor prognosis [247].

The pathology of *Blastomyces* infection is similar to histoplasmosis with necrotizing granulomas. However, the lesions are larger, showing more neutrophilic necrosis. The organisms are also larger (8–15 microns), with prominent broad-based budding, and are apparent on routine hematoxylin and eosin staining (Figure 18.60).

Coccidioides immitis *Coccidioides immitis* is found in the semi-arid desert climate of the southwestern United States. The organisms are inhaled as spores, causing an acute disease characterized by fever, chills, chest pain, dyspnea, and hemoptysis. Chest imaging studies typically show consolidation and cavitation, and hilar lymphadenopathy is common. Reactivation and dissemination are possible in patients with previous infection, whether or not they are immunocompromised patients [248].

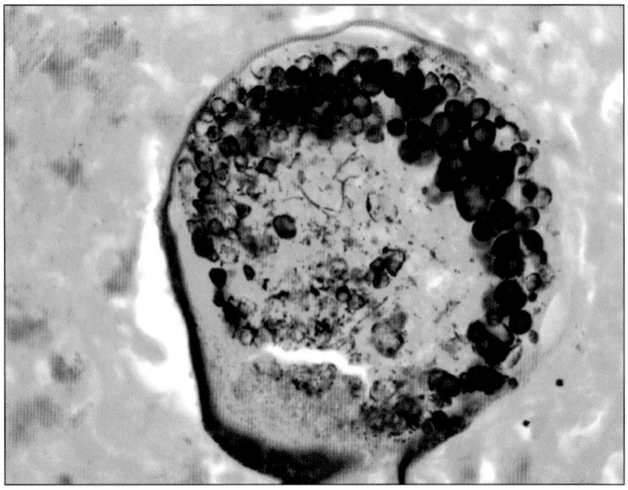

Figure 18.61 **Coccidioidomycosis.** *Coccidioides immitis* can be found in lungs as a large spherule containing endospores using a Grocott methenamine silver stain. These spherules can rupture and endopores can spill into the surrounding lung where they mature into new spherules.

The pathology of pulmonary coccidioidomycosis is neutrophilic, suppurative, and granulomatous. The organisms appear as large spherules containing endospores, visible on silver stains. The spherules are 30–100 microns in diameter and the endospores that are released into the surrounding tissue proceed to mature into new spherules (Figure 18.61). As in histoplasmosis, cavitating lesions may have hyphal forms that begin to germinate.

Aspergillus Fumigatus Aspergilli are asexual mycelial fungi that are ubiquitous in the environment as airborne aspergillus spores. They are weak pathogens

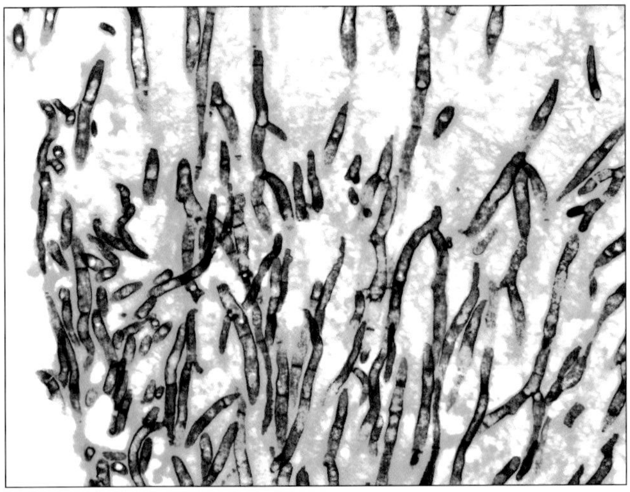

Figure 18.62 **Aspergillosis.** *Aspergillus fumigatus* grows within necrotizing cavities of the lung as branching septated fungal hyphae, as seen on this Grocott methenamine silver stain.

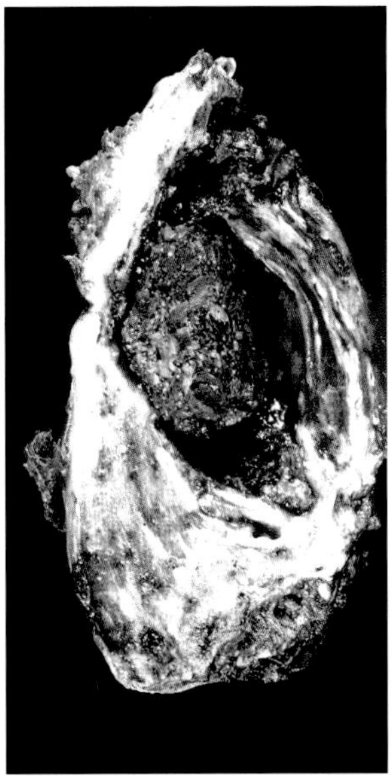

Figure 18.63 **Aspergilloma.** Fungal hyphae from *Aspergillus fumigatus* can colonize chronically inflamed lungs with cavities and may grow to form fungal balls with a dark, green color that are treated by surgical resection, as seen in this case of a lobectomy specimen.

that produce invasive infections predominantly in immunocompromised hosts or in those with significant chronic lung diseases. In tissue, aspergilli form septate hyphae, 3–6 microns in diameter, with characteristic acute-angle, dichotomous branching (Figure 18.62). These organisms affect the lung in three major ways: (i) saprophytic growth in bronchi or pre-existent cavities; (ii) as an allergic or hypersensitivity reaction, predominantly in asthmatics; and (iii) invasive aspergillosis in immunocompromised hosts [249,250]. As a saprophyte, aspergillus produces surface growths or minute masses of hyphae, usually in bronchiectatic cavities, emphysematous bullae, or scars from previous lung diseases such as tuberculosis or sarcoidosis. The pathology is usually that of a fibrous-walled cavity containing degenerating hyphae (Figure 18.63). In this setting, hyphae do not invade into the lung tissue, but surface erosion of a vascularized cavity may cause hemoptysis. Aspergillus causes an immunologic response resulting in mucoid impaction or eosinophilic pneumonia in asthmatics, an entity known as allergic bronchopulmonary aspergillosis (ABPA). Pathologically, one sees mucoid plugs and superficial erosions of the airways with histiocytic inflammation, with only rare hyphal fragments present. The final form of the disease, invasive pulmonary aspergillosis, is found in severely immunocompromised, neutropenic patients. The hyphae, which disseminate through the blood, invade the blood vessels causing thrombosis, hemorrhage, and infarction to form typical targetoid lesions. This form of the disease has a poor prognosis despite aggressive antifungal therapy.

Pneumocystis Jiroveci The taxonomy of *Pneumocystis jiroveci* (formerly *Pneumocystis carinii*) has changed over the past decade. Previously thought to be a protozoan based on the histological characteristics of its trophozoite and cyst life forms, it has recently been placed in the fungal kingdom after ribosomal RNA was found to have sequences compatible with the ascomycetous fungi [251]. The inability to culture *Pneumocystis jiroveci* has slowed the understanding of this organism. Animal models have helped in defining the antigenic and genotypic differences among the various *Pneumocystis* organisms, which has led to the proposal for species-specific strains, with *P. jiroveci* found in human infections [252]. The molecular methods used for the typing these species examine a number of gene loci. Most importantly, sequence analysis of the thymidylate synthase (TS) and superoxide dismutase (SODA) gene loci, the EPSP synthase domain of the multifunctional arom gene, and the mitochondrial small subunit ribosomal RNA (mtSSU rRNA) locus have been used to distinguish the various *Pneumocystis* species that infect different mammalian hosts [253].

Clinically, *P. jiroveci* causes disease predominantly in the immunocompromised setting. *Pneumocystis* pneumonia (PCP) has been found during recent times most commonly in the AIDS population, but prior to this epidemic, it was found in malnourished infants and other severely immunocompromised hosts. Because this organism has not been cultured, the diagnosis of PCP continues to be challenging. The clinical characteristics are nonspecific and vary with the patient's immune status. In the HIV population, patients typically develop a subacute onset of progressive dyspnea, a nonproductive cough, malaise, and a low-grade fever. In the non-HIV population, the presentation is more acute, with fulminant respiratory failure associated with cough and fever, and usually requiring mechanical ventilation [254]. Chest imaging studies typically show bilateral, symmetric, fine reticular interstitial infiltrates involving the perihilar area, which spread to involve the entire lung.

Treatment is usually with trimethoprim/sulfamethoxazole and intravenous pentamidine. Survival is 50%–95% even in severely immunocompromised patients.

The life cycle of *P. jiroveci* consists of three stages: trophozoite, cyst, and sporozoite. The trophozoite form, which adheres to the type 1 epithelium, replicates and enlarges through three precyst stages before maturing into a cyst form that is found in the alveolar space. Sporozoites develop within immature cysts through meiosis and mitosis. The mature cyst contains eight haploid sporozoites. The rupture of the cyst wall releases sporozoites into the surrounding environment where they mature into trophozoites.

The pathology of the infection is predominantly due to the interaction of the organism with the epithelium. The attachment of the organism to the lung epithelium is via glycoprotein A present on the surface of the organism. The binding of the organism to the type 1 cell occurs via surface receptors on the type 1 cell that include macrophage mannose receptors. These interact with glycoprotein A and activate pathways in the organism that induce genes encoding for pathways that induce mating and proliferation responses, and for the formation of pheromone receptors, transcription factors, and heterotrimeric G-protein subunits [263]. In addition to these genetic effects, the cyst wall contains chitins, polymers, and other substances, in particular, 1,3-glucan, that maintain its integrity and induce the inflammatory response of the host. The 1,3-glucan in the wall of the organism stimulates the release by the macrophages of reactive oxidant species and the generation of potent proinflammatory cytokines, such as TNFα, which bind to the organism and exert a toxic effect. Once inside the macrophage, the organism is incorporated into the phagolysosome and degraded. TNFα also directly recruits other inflammatory cells including neutrophils, lymphocytes, and circulating monocytes, and induces the release of IL-8 and IFN-γ that recruit and activate inflammatory cells [255]. In aggregate, the recruitment of these inflammatory cells and the mediators they release is responsible for the damage to the lung epithelium and endothelium that is seen in this disease [255].

The pathology of PCP has typical and atypical variants. Typically, the lung contains a dense interstitial plasma cell pneumonia that expands alveolar walls. The epithelium consists predominantly of Type 2 pneumocytes, and the alveolar spaces contain an eosinophilic, frothy exudate, which contains fine, hemoxylin-stained dots that represent a thickening in the cyst wall (Figure 18.64). In this form of the disease, the organisms are abundant and the diagnosis can usually be made by bronchoalveolar lavage. Atypical pathologic variants include a necrotizing variant that has a pattern similar to the typical form with exudative alveolar infiltrates, but which undergoes necrosis and cavity formation. These cavities heal into fibrous-walled cysts, similar in gross appearance to those found in pseudomonas pneumonia. A third variant has well-formed granulomas involving the airways, a pattern common to histoplasmosis and tuberculosis. In this form, the organisms are rare and very difficult to find, even with tissue organismal stains. In general, the pathologic pattern of injury depends on the host's immune status, with the typical pathology found in severely immunocompromised hosts as the AIDS population and the atypical forms found in hosts with immune systems that are less compromised.

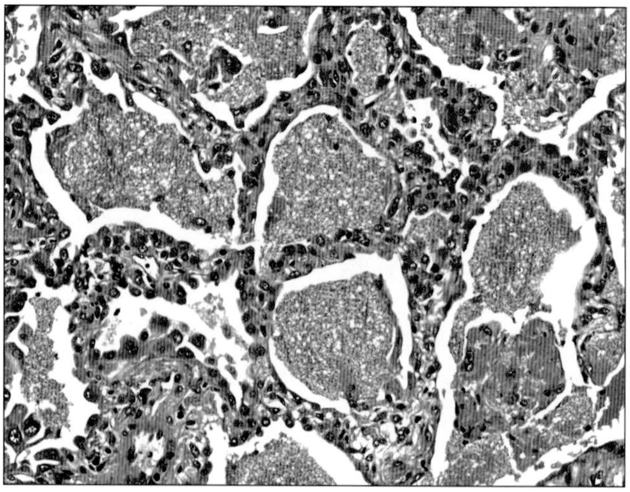

Figure 18.64 ***Pneumocystic jiroveci* pneumonia.** The microscopic features of this pneumonia are characterized by a foamy, eosinophilic infiltrate that fills the alveolar space. Upon close inspection, fine blue dots can be seen that represent the thickening in the cyst walls of the organism.

PULMONARY HISTIOCYTIC DISEASES

Pulmonary Langerhans cell histiocytosis (PLCH) and Erdheim-Chester disease are histiocytic diseases that primarily affect the lung. Other histiocytic diseases may affect the lung, such as Niemann-Pick disease, Gaucher disease, Hermansky-Pudlak and Rosai-Dorfman disease, but these are not considered primarily lung histiocytic diseases.

Pulmonary Langerhans' Cell Histiocytosis

Clinical and Pathologic Features

Pulmonary Langerhans' cell histiocytosis (PLCH) is a disease of the dendritic histiocytes of the lung referred to as Langerhans' cells (LCs). This disease is part of a group of diseases that are characterized by a proliferation of Langerhans cells in organs throughout the body that range from a malignant systemic disease as is seen in children [256] to the pulmonary variant that is seen in adolescents and adults. PLCH is usually the result of inflammatory or neoplastic stimuli in lungs of smokers or in lungs involved by certain neoplasms [257].

Chest radiographs from patients with PLCH usually reveal bilateral nodules, predominantly in the upper lobes, which are worrisome for metastatic disease. Treatment involves smoking cessation and steroid therapy. Typically, the disease undergoes spontaneous regression. Approximately 15%–20% of patients will progress to irreversible end-stage fibrosis [258].

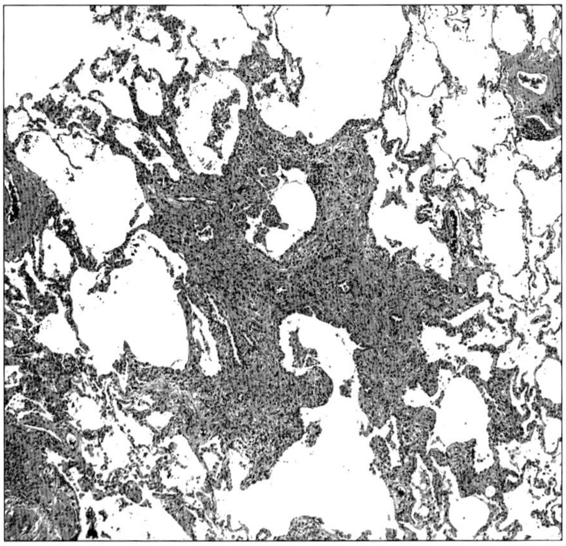

Figure 18.65 Pulmonary langerhans cell histiocytosis. This stellate-shaped scar represents the microscopic features of long-standing PLCH. These lung scars are characteristically found in the upper lung zones and represent chronic injury to the small airways.

The pathology of PLCH consists of airway-based lesions with a proliferation of LCs. The early cellular lesions contain a mixture of cells including Langerhans' cells, lymphocytes, plasma cells, and eosinophils (Figure 18.65). Though it was previously referred to as eosinophilic granuloma, eosinophils are not the major cell type present, and the lesion is, at best, a loosely formed granuloma. Immunohistochemistry reveals the LCs to be diffusely, strongly immunoreactive to S-100 protein and CD1a. Ultrastructural analysis reveals intracytoplasmic organelles called Birbeck granules, a normal constituent of Langerhans' cells, in greater numbers in PLCH [259].

Molecular Pathogenesis

The pathogenetic mechanisms of PLCH focus on defects in the homeostasis of dendritic cells (DCs) in the lungs of smokers and the role tobacco smoke may play in stimulating the proliferation of these cells [260]. Some studies suggest that stimulation of alveolar macrophages by chemicals in smoke results in secretion of such cytokines as GM-CSF, TGFβ, and TNFα [261]. In transgenic mice, accumulation of DCs around airways may be a result of excess GM-CSF [262]. Other theories suggest that cigarette smoke stimulates the secretion of bombesin-like peptide by the neuroendocrine cells in the bronchiolar epithelium and leads to a similar stimulation of alveolar macrophages and a cytokine milieu that promotes the proinflammatory proliferative changes [262].

Not all smokers get PLCH, leading to the suggestion that only smokers with an underlying genetic susceptibility will develop the disease. Studies have established that in some cases the LCs in PLCH are clonal, suggesting that cellular abnormalities must play some part in the pathogenesis of the diseases [263]. To support this, studies have shown genetic mutations and allelic loss of tumor suppressor genes in smokers with PLCH [264].

The mechanisms by which this proliferation of LCs leads to the destruction of the bronchiolar epithelium and the other observed pathology are unclear. LCs in normal lungs have little ability to interact with T-cells or act as effective antigen-presenting cells, but the LCs of PLCH have a mature immunophenotype, expressing B7-1 and B7-2, the co-stimulatory molecules needed for lymphostimulatory activity [265]. Whether this more mature immune phenotype leads to an unregulated immune response and destruction of the bronchial epithelial cells is not known. However, some studies have shown that bronchiolar epithelial cells may induce the expression of this mature phenotype by secreting cytokines in response to environmental stimulants such as cigarette smoke or viral infections, or by the development of hyperplastic or dysplastic lesions that express new foreign antigens [265].

Erdheim-Chester Disease

Clinical and Pathologic Features

Erdheim-Chester disease (ECD) is a systemic non-Langerhans' cell histiocytosis of adults that most commonly involves the long bones. Involvement of other organs, including the lung, has been reported. Lung involvement occurs in approximately 20%–35% of the cases, and the patients usually present with cough, dyspnea, rhonchi, and pleuritic pain. Radiographically, the lungs reveal infiltrates in a lymphatic distribution, predominantly upper lobe, with prominent interstitial septal markings that can mimic sarcoidosis [266–272].

Pulmonary involvement by ECD may have an unfavorable prognosis, and the fibrosis that ensues is one of the most frequently reported causes of death [266,273]. The treatment of ECD is variable with corticosteroids, chemotherapy, surgical resection, and radiation therapy reported [273].

Non-Langerhans' cell histiocytes of dendritic cell phenotype are the main cells present in this disease. This infiltrate contains foamy histiocytes with scattered giant cells, a scant number of lymphocytes or plasma cells, and some fibroblasts. The histiocytes express CD68 (macrophage antigen) and factor XIIIa (dendritic cell antigen), but express S-100 protein weakly or not at all, and do not express CD1a. Ultrastructural analysis reveals phagolysosomes, but no Birbeck granules are present [273]. This infiltrate that involves the lung is usually present in the pleura and subpleura, within the interlobular septa and around the bronchovascular structures. The remainder of the lung parenchyma is unremarkable, though fibrosis and paracicatricial emphysema can appear in the late stages of the disease [266].

Molecular Pathogenesis

The etiology of ECD is not known, but this rare disease has been established as primarily a macrophage disorder [274]. These histiocytes have abundant phagolysosomes and express the antigen CD1a and are consistent with a phagocytic cell, most likely closely related to alveolar macrophages. The peripheral monocytosis and the proinflammatory cytokine profile that is found in these patients might suggest that the histiocytic infiltrate is a result of systemic monocytic activation and invasion of circulating monocytes into the tissues throughout the body [275]. Recently, an ECD patient was successfully treated by an agent toxic to monocytes, supporting the theory that these cells play a part in the disease [275]. Alternatively, end organ cytokine production by local inflammatory cells resulting in proliferation and differentiation of resident immature histiocyte populations may produce a similar picture. Another interesting observation is that Erdheim-Chester has been reported to occur in patients with Langerhans' cell histiocytosis [276], which may suggest that this is a disease where macrophages transition between two different phenotypes along the differentiation spectrum of tissue dendritic cells [276]. Whether this is a benign or malignant proliferation has not been established. Of 5 patients studied, clonality has been demonstrated in 3 by polymerase-chain reaction [277].

PULMONARY OCCUPATIONAL DISEASES

Environmental exposures are a major cause of lung disease and can cause a wide spectrum of both acute and chronic pathology. Many organic and inorganic materials can cause lung damage, and because of their similar patterns of injury and long latent periods, it can be difficult to isolate the exact offending agent without a thorough clinical history. The two occupational lung diseases presented here—asbestosis and silicosis—represent pneumoconiosis, which are defined as diseases which result in diffuse parenchymal lung injury due to inhaled inorganic material. Both have many pathologic patterns of injury that depend on the amount and length of time of exposure, and both can also cause neoplastic diseases of the lung.

Asbestosis

Clinical and Pathologic Features

Asbestos fibers are naturally occurring silicates that are commonly used in construction materials such as cement and insulation and in many textiles. They can be separated into two groups based on their mineralogic characteristics. Serpentine fibers, named as such because they are long and curly, include chrysotile asbestos. Amphibole fibers, more straight and rodlike, include predominantly amosite and crocidolite asbestos. In the United States most of the asbestos is chrysotile. The amphiboles are more pathogenic and are responsible for most of the neoplastic and non-neoplastic pulmonary diseases associated with asbestos exposure.

By definition, asbestosis is bilateral diffuse interstitial fibrosis of the lungs that can be attributed to asbestos exposure. The disease, which mostly affects textile and construction workers, is usually the result of direct exposure over 15–20 years. The latency to clinical disease is inversely proportional to the level of exposure. The symptoms are a gradual onset of shortness of breath, a cough with dry rales at the bases on inspiration, and digital clubbing. In the early disease, the chest x-ray shows basilar disease that begins predominantly as thickening of the subpleural, but progresses as infiltrates and fibrosis that involve the middle zone, eventually leading to thickening of the airways and traction bronchiectasis. The apex of the lung is usually spared. The clinical findings are nonspecific and have considerable overlap with UIP, so the diagnosis is usually made only when a history of significant exposure is discovered.

The gross picture includes a bilateral lower lobe gray/tan fibrosis with honeycomb changes in late disease. Microscopically, asbestosis can cause many patterns of injury in the lung, but the most common is collagenous deposition in the areas of the lymphatics where the fibers are in the highest concentration. These areas include the subpleural, interlobular septae, and around the bronchovascular areas that contain a bronchiole and a branch of the pulmonary artery. Hyalinized pleural plaques are a common manifestation of asbestos exposure but are not specific for asbestos and can be found in the absence of pulmonary parenchymal disease. Eventually, the fibrosis involves the alveoli beyond the bronchioles and causes distortion of the lung architecture to form remodeled, dilated airspaces similar to those seen in UIP. Distinguishing this fibrosis from other forms of fibrosing lung disease can be difficult, but the presence of ferruginous bodies, asbestos fibers coated by iron, proteins, and a mucopolysaccharide coat are indicative of significant asbestos exposures and support this diagnosis (Figure 18.66) [278].

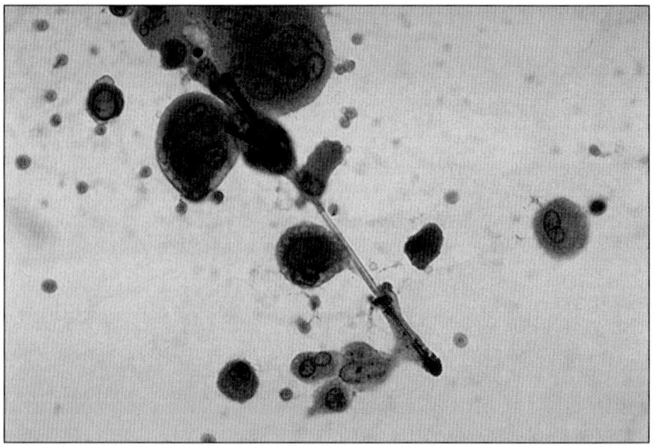

Figure 18.66 Asbestosis. This cytopathologic preparation from a bronchoalveolar lavage specimen illustrates an asbestos fiber coated by an iron-protein-mucopolysaccharide substance and appears as a golden brown, beaded structure known as a ferruginous body.

Silicosis

Clinical and Pathologic Features

Silicosis results from chronic, high-dose exposure to crystalline silica, which consists of silicon and oxygen with trace amounts of other elements, usually iron. The most common silica is quartz, which is present in large amounts in such rocks as granite, shale, and sandstone and is among the more fibrogenic of all silica types. Thus, occupations most at risk for silicosis include sandblasting, quarrying, stone dressing, and foundry work where exposure to quartz is high. The disease takes three major clinical and pathologic forms that have different clinical characteristics.

Simple or nodule silicosis is marked by the presence of fine nodules $\leq$1 cm, on chest imaging studies, usually in the upper lobes. Patients with this condition are typically asymptomatic, with normal respiratory physiology. The pathology in these lungs reveals discrete, hard nodules that have a green/gray color, centered either on the small airways or in the subpleura. Microscopically, these nodules have an early stellate shape that eventually transforms to a more whorled appearance with dust-laden macrophages scattered throughout it. Polarized light examination reveals weakly birefringent material.

Complicated pneumoconiosis represents similar pathologic findings only with larger and more circumscribed nodules, which coalesce into a large upper lobe mass, a condition known as progressive, massive fibrosis (Figure 18.67). These patients are symptomatic with a productive cough and mixed pulmonary function tests with a reduced diffusing capacity as the fibrosis increases. Diffuse interstitial fibrosis may occur; however, unlike asbestosis, this pattern is found in pneumoconiosis. When complicated pneumoconiosis is found with rheumatoid nodules in the setting of a patient with rheumatoid arthritis, this is known as Caplan's syndrome.

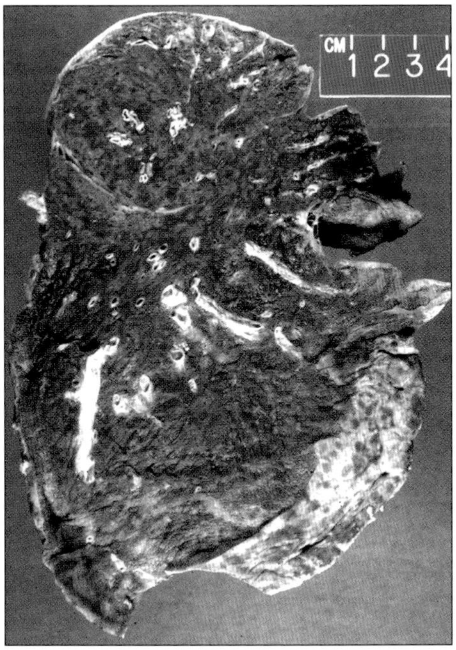

Figure 18.67 Complicated pneumoconiosis/ progressive massive fibrosis. This sagittal cut section of lung reveals a large gray/black mass that extends from the apex to include the majority of the lung. The patient had a long history as a coal mine worker, and the microscopic sections revealed abundant anthracotic pigment and scarring in this area.

Molecular Pathogenesis

The pathogenesis of both asbestosis and silicosis depends upon inflammation and fibrosis caused by the inhaled fibers. In humans, the amount of fiber needed to cause fibrosis varies from person to person. This may be related to a difference in fiber deposition based on the size of the lungs or to the efficacy with which the lung clears these fibers [256]. Some studies have also suggested that fiber length determines the amount of pathology. However, this association has not been confirmed in humans for either asbestosis or silicosis. In both diseases, it is known that other factors increase the risk of developing disease. For example, smokers exhibit worse disease than nonsmokers with similar exposures to asbestosis. The mechanism for this effect is unclear, although speculation centers on the inhibition of fiber clearance in smokers. Also, it is known that smoking enhances the uptake of fibers by pulmonary epithelial cells and in this way may increase the fibrogenic and inflammatory cytokine production by these cells.

The cellular mechanisms by which both asbestos and silica fibers induce the inflammation and fibrosis are mediated predominantly through alveolar macrophages. In the case of silica, it is known that the uptake of these fibers into the alveolar macrophages is by way of a scavenger receptor expressed on the surface of the cell known as MARCO (macrophage receptor with a collagenous structure). Once inside the cells, the fibers activate the release of ROS that can lead to cellular and molecular damage through a number of pathways. First, ROS can directly cause lipid peroxidation, membrane damage, and DNA damage. Second, silica-induced free radicals can trigger phosphorylation of cellular proliferation pathways through mitogen-activated protein kinases (MAPKs), extracellular signal regulated kinases (ERKs), and p38. These pathways are also involved in the proliferation of fibroblasts in asbestosis and of mesothelial and epithelial cells in the neoplastic diseases associated with the inhalation of these fibers [279]. In addition, these fibers can activate proinflammatory pathways controlled by such transcription factors as nuclear NFκB and activator protein 1 (AP-1). These pathways result in the activation of the early response genes *c-fos* and *c-Jun* and the release of proinflammatory cytokines such as IL-1 as well as fibrogenic factors such TNFα [280].

TNFα plays a prominent role in both diseases, and its regulation has been studied in animal models exposed to silica. It is now known that a transcription factor labeled nuclear factor of activated T-cells (NFAT) plays a key role in the regulation of TNFα.

Binding sites for NFAT have been found in the promoter region of the *TNFα* gene. The mediation of silica-induced *TNFα* transcription is probably via O_{2-} but not H_2O_2 [280,281].

DEVELOPMENTAL ABNORMALITIES

Tracheal/Bronchial Atresias and Sequestrations

Clinical and Pathologic Features

Atresia of the lung represents a premature closure of the airway at any level of the bronchial tree including the lobar, segmental, or subsegmental airways. Clinically, these children usually present between 10 and 20 years of age for symptoms of dyspnea, wheezing, recurrent pneumonias, or for incidental findings on a chest imaging study. These lesions are more common in the proximal segmental bronchi, right more often than left. When atresia is associated with anomalies of the vascular supply to the affected airway, the lesion represents a separate, aberrant segment of lung known as a sequestration, either intralobar or extralobar type.

The pathology of bronchial atresias and sequestrations represents sequelae of chronic inflammation due to the accumulation of secretions in these blind-end airways. These features consist of cystically dilated airways with mucus and parenchymal fibrosis with honeycomb changes. In intralobar sequestrations (ILS), the anomalous vessel is a muscular artery that enters through the pleura from an aortic source, usually from the thoracic area. ILS are separate, isolated areas of lung invested with the normal visceral pleura without bronchial or arterial connections (Figure 18.68). Extralobar sequestrations (ELS) are pyramid-shaped accessory pieces of lung that have their own pleura with an artery from the lung but without airway connections.

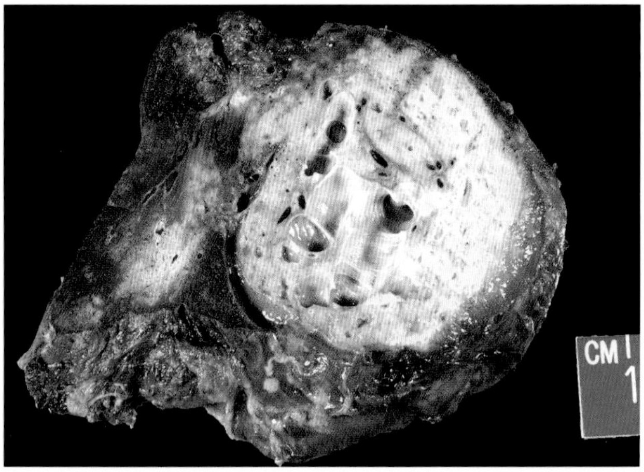

Figure 18.68 Intralobar sequestration. The tan and white mass involving this left lower lobectomy specimen represents chronic pneumonia and fibrosis in the sequestered area of the lung. The dilated airways are features of an endstage fibrosis that is commonly found in this entity.

Congenital Pulmonary Cystic Diseases

Clinical and Pathologic Features

The category of congenital pulmonary cystic diseases represents the majority of congenital pulmonary disease and includes foregut cysts and cystic adenomatoid malformations. Foregut cysts include bronchogenic, esophageal, and thymic cysts that form from defects in the foregut branching. Clinically, these cysts are usually incidental findings on chest imaging studies, but they can present with complications due to infection or hemorrhage.

Pathologic features of these cysts include subtle differences that are usually only apparent after microscopic examination. Grossly, these cysts usually arise proximally either within the mediastinum (over 50%) or in the proximal regions of the lungs, right more commonly than left, along the esophagus, and rarely within the lung parenchyma or below the diaphragm [282]. Microscopically, each cyst contains a simple cuboidal or columnar epithelium, ciliated or nonciliated, that may undergo squamous metaplasia. Distinguishing among the three types of cysts requires the presence of other elements. Bronchogenic cysts have submucosal glands and/or hyaline cartilage within their walls, and thymic cysts may contain residual thymus.

Congenital cystic adenomatoid malformations (CCAM), now more commonly referred to as congenital pulmonary airway malformations (CPAM), are segments of lung with immature airways and alveolar parenchyma. These are usually classified by their predominant cyst size into types 0–4. Type 1 cysts, which contain a main large cyst of up to 10 cm, are the most common. These cysts are distinguished from foregut cysts upon the recognition in the CPAM of immature alveolar duct-like structures connecting to the surrounding lung parenchyma. This type of CPAM is also notable, as it is known to undergo malignant transformation, usually to mucinous bronchioloalveolar cell adenocarcinomas.

Molecular Pathogenesis

These anomalies arise due to defects during the various stages of development and are best considered within these developmental stages. The embryonic stage occurs within the first 3–7 weeks of life when the ventral wall of the foregut separates into the trachea and esophagus and branches to form the left and right lungs. The splanchnic mesenchyme that surrounds this foregut forms the vascular and connective tissues of the lungs. Defects in this phase result in complete lack of lung development known as pulmonary agenesis and incomplete separation of the trachea and esophagus, causing tracheal-esophageal atresias and fistulas. The pseudoglandular stage, between weeks 7–17 of development, is a time of rapid development of the conducting airways including the bronchi and bronchioles and the expansion of the peripheral lung into the acinar buds. The mesenchymal tissue that surrounds these buds begins to thin, becomes vascularized, and forms the cartilage that surrounds the more proximal branching airways. During the canalicular (week 17–24), saccular (weeks 24–38), and alveolar (weeks 36 to maturity) stages of development, the acinar buds continue to expand, and the mesenchyme surrounding this continues to thin. During the canalicular stage, the pulmonary vascular bed begins to organize, the distance between the blood in the vascular spaces and the air in the alveoli narrows, and the respiratory epithelium begins to form. The gas exchange unit of the alveolus becomes functional during the saccular stage with further differentiation of the respiratory epithelium to include Clara cells, ciliated and nonciliated cells, and type 2 cells with the first production of surfactant occurring during this period. This gas exchange unit continues to mature during the alveolar stage with the growth and septation of the alveoli. This process continues postnatally through 6–8 years of age.

The different types of CPAMs arise at different stages of development. CPAMs 0, 1, and 2 are a result of defects during the early embryonic and pseudoglandular stages of development, producing pathology with features of primitive alveolar buds and immature and abnormal airway cartilage structures. CPAMs 3 and 4 result from abnormal formation of the more distal airways and pulmonary parenchyma during the canalicular, saccular, and alveolar phases, causing pathology with immature alveolar, or alveolar simplification with enlarged alveoli [283].

Various genetic defects in the pathways that control lung morphogenesis have been associated with these congenital lung diseases. Two major transcription factors are responsible for the normal branching morphogenesis. The first, thyroid transcription factor-1 (TTF-1), is a member of the Nkx2.1 family of hemeodomain-containing transcription factors. This factor plays a role in the lung epithelial-specific gene expression and proper lung bud development in the embryonic stage, as well as in the maturation of the respiratory epithelium. The second major factor is somatic hedgehog (SHH)/Gli, expressed by endodermally derived cells and required for branching morphogenesis. The development of the lung bud from the foregut endoderm depends on the appropriate expression of these lung-specific genes at the correct time in development. In the presence of genetic defects, aberrant lung development may occur. For example, mutations of various types in the *SHH/Gli* gene have been found to cause tracheoesophageal fistulas, anomalous pulmonary vasculature, and aberrant airway branching. Also, deletions in the *TTF-1* gene are associated with tracheoesophageal fistulas and a variety of forms of lung dysgenesis [284]. Finally, factors present in the surrounding mesenchyme play a role in inducing the proper development of the pulmonary endoderm. A prominent mesenchymal factor in this process is fibroblast growth factor (FGF), which modulates both the proximal and distal lung branching morphogenesis. Deletions in this gene may cause lung agenesis and tracheal malformations [284].

Surfactant Dysfunction Disorders

Clinical and Pathologic Features

Surfactant dysfunction disorders represent a heterogenous group of inherited disorders of surfactant metabolism, found predominantly in infants and children. Pulmonary surfactant includes both phospholipids and surfactant proteins, designated surfactant proteins A, B, C, and D (SP-A, SP-B, SP-C, SP-D), synthesized and secreted by type 2 cells beginning in the canalicular stage of lung development. Damage to type 2 cells during this time period can lead to acquired surfactant deficiencies. However, more commonly these deficiencies are the result of genetic defects of the surfactant proteins themselves.

The major diseases are caused by genetic defects in the surfactant protein B (*SFTPB*, chromosome 2p12-p11.2); surfactant protein C (*SFTPC*, chromosome 8p21); and adenosine triphosphate (APT)-binding cassette transporter subfamily A member 3 (*ABCA3*, chromosome 16p13.3). Defects in *SFTPB* and *ABCA3* have an autosomal recessive inheritance pattern, and defects in *SFTPC* have an autosomal dominant pattern. SP-B deficiency is the most common. It presents at birth with a rapidly progressive respiratory failure and chest imaging studies showing diffuse ground glass infiltrates. The gross pathology in these lungs consists of heavy, red, and congested parenchyma with microscopic features that range from a PAP-like pattern to a chronic pneumonitis of infancy (CPI) pattern. In SP-B deficiency, the PAP pattern predominates with a histologic picture of cuboidal alveolar epithelium and eosinophilic PAS-positive material within the alveolar spaces that appears with disease progression. In the late stages of the disease, the alveolar wall thickens with a chronic inflammatory infiltrate and fibroblasts. This alveolar proteinosis-type pattern of injury can be confirmed with immunohistochemical studies that establish the absence of SP-B within this surfactant-like material. Diseases due to *ABCA3* or *SFTPC* deficiency may present within a week of birth or years later; the former has a poor prognosis, but the latter has a more variable prognosis with some patients surviving into adulthood. Indeed, SP-C mutations have also been recognized in some families as a cause of interstitial pneumonia and pulmonary fibrosis in adults [285]. The pathology of SP-C deficiency has more CPI features and less proteinosis. In contrast, *ABCA* deficiency can have either PAP or CPI features, with the former present early in the disease and the latter present in more chronically affected lungs [286].

Molecular Pathogenesis

The SP-B gene (*SFTPB*) is approximately 10 kb in length and is located on chromosome 2. There are over 30 recessive loss-of-function mutations associated with the *SFTPB* gene. However, the most common mutation is a GAA substitution for C in codon 121, found in about 70% of the cases. The lack of SP-B leads to an abnormal proportion of phosphatidylglycerol and an accumulation of a pro-SP-C peptide, leading to the alveolar proteinosis-like pathology.

SP-C protein deficiency is due to a defect in the *SFTPC* gene localized to human chromosome 8. There are approximately 35 dominantly expressed mutations in *SFTPC* that result in acute and chronic lung disease. Approximately 55% of them arise spontaneously, and the remainder are inherited. The most common mutation is a threonine substitution for isoleucine in codon 73 (I73T), found in 25% of the cases, including both sporadic and inherited disease [287]. This mutation leads to a misfolding of the SP-C protein, which inhibits its progression through the intracellular secretory pathway, usually within the Golgi apparatus or the endoplasmic reticulum [288]. The absence of SP-C within the alveolar space causes severe lung disease in mouse models. Infants with documented mutated proSP-C protein, the larger primary translation product from which SP-C is proteolytically cleaved, can have respiratory distress syndrome (RDS) or CPI. In older individuals, pathologic patterns observed in the lungs with these mutations include nonspecific interstitial pneumonitis (NSIP) and UIP. In this affected adult population, the pathology and age of disease presentation vary even within familial cohorts, suggesting the involvement of a second hit, perhaps an environmental factor [289].

The ABCA3 protein is a member of the family of ATP-dependent transporters, which includes the CFTR, and is expressed in epithelial cells. Mutation in this gene results in severe respiratory failure that is refractory to surfactant replacement. The cellular basis for the lack of surfactant in patients with this genetic mutation is not known. The presence of abnormal lamellar bodies within the type 2 cells by ultrastructural analysis suggests a disruption in the normal surfactant synthesis and packaging in this disease. There is some evidence that this gene contains promoters that share elements consistent with their activation by the transcription factors TTF-1 and Foxa7, and deletions in either or both of these genes may play a role in this disease [289].

REFERENCES

1. http://seer.cancer.gov/csr/1975_2004/results_single/sect_01_table.01.pdf.
2. Espey DK, Wu XC, Swan J, et al. Annual report to the nation on the status of cancer, 1975–2004, featuring cancer in American Indians and Alaska Natives. *Cancer.* 2007;110:2119–2152.
3. Travis WD, Brambilla E, Müller-Hermelink HK, et al. Pathology and Genetics of Tumours of the Lung, Pleura, Thymus and Heart. Lyon: IARC Press; 2004.
4. Greene FL, Page DL, Fleming ID, et al. *AJCC Cancer Staging Manual.* New York: Springer; 2002.
5. Surveillance, Epidemiology, and End Results (SEER) Program (www.seer.cancer.gov) SEER*Stat Database: Incidence - SEER 17 Regs Limited-Use, Nov 2006 Sub (1973–2004 varying), National Cancer Institute, DCCPS, Surveillance Research Program, Cancer Statistics Branch, released April 2007, based on the November 2006 submission.
6. Alberts WM. Diagnosis and management of lung cancer executive summary: ACCP evidence-based clinical practice guidelines (2nd edition). *Chest.* 2007;132:1S–19S.
7. Wistuba II. Genetics of preneoplasia: Lessons from lung cancer. *Curr Mol Med.* 2007;7:3–14.
8. Virmani AK, Fong KM, Kodagoda D, et al. Allelotyping demonstrates common and distinct patterns of chromosomal loss in human lung cancer types. *Genes Chromosomes Cancer.* 1998;21:308–319.
9. Girard L, Zochbauer-Muller S, Virmani AK, et al. Genome-wide allelotyping of lung cancer identifies new regions of allelic loss, differences between small cell lung cancer and non-small cell lung cancer, and loci clustering. *Cancer Res.* 2000; 60:4894–4906.
10. Garnis C, Lockwood WW, Vucic E, et al. High resolution analysis of non-small cell lung cancer cell lines by whole genome tiling path array CGH. *Int J Cancer.* 2006;118:1556–1564.
11. Wistuba II, Gazdar AF. Lung cancer preneoplasia. *Annu Rev Pathol.* 2006;1:331–348.
12. Weir BA, Woo MS, Getz G, et al. Characterizing the cancer genome in lung adenocarcinoma. *Nature.* 2007;450:893–898.
13. Minna JD, Roth JA, Gazdar AF. Focus on lung cancer. *Cancer Cell.* 2002;1:49–52.
14. Sun S, Schiller JH, Spinola M, et al. New molecularly targeted therapies for lung cancer. *J Clin Invest.* 2007;117:2740–2750.
15. Wistuba II, Behrens C, Virmani AK, et al. High resolution chromosome 3p allelotyping of human lung cancer and preneoplastic/preinvasive bronchial epithelium reveals multiple, discontinuous sites of 3p allele loss and three regions of frequent breakpoints. *Cancer Res.* 2000;60:1949–1960.
16. Gazdar A, Franklin WA, Brambilla E, et al. Genetic and molecular alterations. In: Travis WD, Brambilla E, Müller-Hermelink HK, Harris CC, eds. *Pathology and Genetics of Tumours of the Lung, Pleura, Thymus and Heart.* Lyon: IARC Press; 2004:21–25.
17. Wistuba II, Behrens C, Virmani AK, et al. Allelic losses at chromosome 8p21–23 are early and frequent events in the pathogenesis of lung cancer. *Cancer Res.* 1999;59:1973–1979.
18. Wistuba II, Behrens C, Milchgrub S, et al. Sequential molecular abnormalities are involved in the multistage development of squamous cell lung carcinoma. *Oncogene.* 1999;18:643–650.
19. Knudson Jr AG. Hereditary cancers disclose a class of cancer genes. *Cancer.* 1989;63:1888–1891.
20. Pfeifer GP, Denissenko MF, Olivier M, et al. Tobacco smoke carcinogens, DNA damage and p53 mutations in smoking-associated cancers. *Oncogene.* 2002;21: 7435–7451.
21. Stelter AA, Xie J. Molecular oncogenesis of lung cancer. In: Zander DS, Popper HH, Jagirdar J, Haque AK, Cagle PT, Barrios R, eds. *Molecular Pathology of Lung Diseases.* New York, NY: Springer; 2008:169–175.
22. Sekido Y, Fong KM, Minna JD. Molecular genetics of lung cancer. *Annu Rev Med.* 2003;54:73–87.
23. Beasley MB, Lantuejoul S, Abbondanzo S, et al. The P16/cyclin D1/Rb pathway in neuroendocrine tumors of the lung. *Hum Pathol.* 2003;34:136–142.
24. Toyooka S, Maruyama R, Toyooka KO, et al. Smoke exposure, histologic type and geography-related differences in the methylation profiles of non-small cell lung cancer. *Int J Cancer.* 2003;103:153–160.
25. Marinov M, Fischer B, Arcaro A. Targeting mTOR signaling in lung cancer. *Crit Rev Oncol Hematol.* 2007;63:172–182.
26. Gridelli C, Maione P, Rossi A. The potential role of mTOR inhibitors in non-small cell lung cancer. *Oncologist.* 2008;13:139–147.
27. Scheurle D, Jahanzeb M, Aronsohn RS, et al. HER-2/neu expression in archival non-small cell lung carcinomas using FDA-approved Hercep test. *Anticancer Res.* 2000; 20:2091–2096.
28. Hirsch FR, Scagliotti GV, Langer CJ, et al. Epidermal growth factor family of receptors in preneoplasia and lung cancer: Perspectives for targeted therapies. *Lung Cancer.* 2003;41(Suppl 1): S29–42.
29. Clamon G, Herndon J, Kern J, et al. Lack of trastuzumab activity in nonsmall cell lung carcinoma with overexpression of erb-B2: 39810: A phase II trial of Cancer and Leukemia Group B. *Cancer.* 2005;103:1670–1675.
30. Johnson BE, Heymach JV. Farnesyl transferase inhibitors for patients with lung cancer. *Clin Cancer Res.* 2004;10:4254s–4257s.
31. Sun S, Schiller JH. Angiogenesis inhibitors in the treatment of lung cancer. *Crit Rev Oncol Hematol.* 2007;62:93–104.
32. Gutierrez M, Giaccone G. Antiangiogenic therapy in nonsmall cell lung cancer. *Curr Opin Oncol.* 2008;20:176–182.
33. Zhang B, Pan X, Cobb GP, et al. microRNAs as oncogenes and tumor suppressors. *Dev Biol.* 2007;302:1–12.

34. Colby TV, Noguchi M, Henschke C, et al. Adenocarcinoma. In: Travis WD, Brambilla E, Müller-Hermelink HK, Harris CC, eds. *Pathology and Genetics of Tumours of the Lung, Pleura, Thymus and Heart.* Lyon: IARC Press; 2004:35–44.
35. Yang Y, Fujita J, Tokuda M, et al. Lung cancer associated with several connective tissue diseases: With a review of literature. *Rheumatol Int.* 2001;21:106–111.
36. Kerr KM, Fraire AE, Pugatch B, et al. Atypical adenomatous hyperplasia. In: Travis WD, Brambilla E, Müller-Hermelink HK, Harris CC, eds. *Pathology and Genetics of Tumours of the Lung, Pleura, Thymus and Heart.* Lyon: IARC Press; 2004:73–75.
37. Yokose T, Doi M, Tanno K, et al. Atypical adenomatous hyperplasia of the lung in autopsy cases. *Lung Cancer.* 2001; 33:155–161.
38. Kim CF, Jackson EL, Woolfenden AE, et al. Identification of bronchioalveolar stem cells in normal lung and lung cancer. *Cell.* 2005;121:823–835.
39. Kwei KA, Kim YH, Girard L, et al. Genomic profiling identifies TITF1 as a lineage-specific oncogene amplified in lung cancer. *Oncogene.* 2008;27:3635–3640.
40. Shigematsu H, Lin L, Takahashi T, et al. Clinical and biological features associated with epidermal growth factor receptor gene mutations in lung cancers. *J Natl Cancer Inst.* 2005;97:339–346.
41. Suzuki M, Shigematsu H, Iizasa T, et al. Exclusive mutation in epidermal growth factor receptor gene, HER-2, and KRAS, and synchronous methylation of nonsmall cell lung cancer. *Cancer.* 2006;106:2200–2207.
42. Shigematsu H, Gazdar AF. Somatic mutations of epidermal growth factor receptor signaling pathway in lung cancers. *Int J Cancer.* 2006;118:257–262.
43. Tam IY, Chung LP, Suen WS, et al. Distinct epidermal growth factor receptor and KRAS mutation patterns in non-small cell lung cancer patients with different tobacco exposure and clinicopathologic features. *Clin Cancer Res.* 2006;12:1647–1653.
44. Hirsch FR, Varella-Garcia M, Cappuzzo F, et al. Combination of EGFR gene copy number and protein expression predicts outcome for advanced non-small-cell lung cancer patients treated with gefitinib. *Ann Oncol.* 2007;18:752–760.
45. Shigematsu H, Takahashi T, Nomura M, et al. Somatic mutations of the HER2 kinase domain in lung adenocarcinomas. *Cancer Res.* 2005;65:1642–1646.
46. Kerr KM, Galler JS, Hagen JA, et al. The role of DNA methylation in the development and progression of lung adenocarcinoma. *Dis Markers.* 2007;23:5–30.
47. Toyooka S, Tokumo M, Shigematsu H, et al. Mutational and epigenetic evidence for independent pathways for lung adenocarcinomas arising in smokers and never smokers. *Cancer Res.* 2006; 66:1371–1375.
48. Hammar SP, Brambilla C, Pugatch B, et al. Squamous cell carcinoma. In: Travis WD, Brambilla E, Müller-Hermelink HK, Harris CC, eds. *Pathology and Genetics of Tumours of the Lung, Pleura, Thymus and Heart.* Lyon: IARC Press; 2004:26–30.
49. Franklin WA, wistuba II, Geisinger K, et al. Squamous dysplasia and carcinoma in situ. In: Travis WD, Brambilla E, Müller-Hermelink HK, Harris CC, eds. *Pathology and Genetics of Tumours of the Lung, Pleura, Thymus and Heart.* Lyon: IARC Press; 2004:68–72.
50. Giangreco A, Groot KR, Janes SM. Lung cancer and lung stem cells: Strange bedfellows? *Am J Respir Crit Care Med.* 2007;175: 547–553.
51. Bota S, Auliac JB, Paris C, et al. Follow-up of bronchial precancerous lesions and carcinoma in situ using fluorescence endoscopy. *Am J Respir Crit Care Med.* 2001;164:1688–1693.
52. Breuer RH, Pasic A, Smit EF, et al. The natural course of preneoplastic lesions in bronchial epithelium. *Clin Cancer Res.* 2005; 11:537–543.
53. George PJ, Banerjee AK, Read CA, et al. Surveillance for the detection of early lung cancer in patients with bronchial dysplasia. *Thorax.* 2007;62:43–50.
54. Brambilla E, Pugatch B, Geisinger K, et al. Large cell carcinoma. In: Travis WD, Brambilla E, Müller-Hermelink HK, Harris CC, eds. *Pathology and Genetics of Tumours of the Lung, Pleura, Thymus and Heart.* Lyon: IARC Press; 2004:45–50.
55. Eleazar JA, Borczuk AC. Molecular pathology of large cell carcinoma and its precursors. In: Zander DS, Popper HH, Jagirdar J, Haque AK, Cagle PT, Barrios R, eds. *Molecular Pathology of Lung Diseases.* New York: Springer; 2008:279–292.
56. Johansson M, Dietrich C, Mandahl N, et al. Karyotypic characterization of bronchial large cell carcinomas. *Int J Cancer.* 1994; 57:463–467.
57. Ohtsuka K, Ohnishi H, Fujiwara M, et al. Abnormalities of epidermal growth factor receptor in lung squamous-cell carcinomas, adenosquamous carcinomas, and large-cell carcinomas: Tyrosine kinase domain mutations are not rare in tumors with an adenocarcinoma component. *Cancer.* 2007; 109:741–750.
58. Marchetti A, Martella C, Felicioni L, et al. EGFR mutations in non-small-cell lung cancer: Analysis of a large series of cases and development of a rapid and sensitive method for diagnostic screening with potential implications on pharmacologic treatment. *J Clin Oncol.* 2005;23:857–865.
59. Travis WD, Rush W, Flieder DB, et al. Survival analysis of 200 pulmonary neuroendocrine tumors with clarification of criteria for atypical carcinoid and its separation from typical carcinoid. *Am J Surg Pathol.* 1998;22:934–944.
60. Travis WD. The concept of pulmonary neuroendocrine tumours. In: Travis WD, Brambilla E, Müller-Hermelink HK, Harris CC, eds. *Pathology and Genetics of Tumours of the Lung, Pleura, Thymus and Heart.* Lyon: IARC Press; 2004:19–20.
61. Travis WD, Nicholson S, Hirsch FR, et al. Small cell carcinoma. In: Travis WD, Brambilla E, Müller-Hermelink HK and Harris CC, eds. *Pathology and Genetics of Tumours of the Lung, Pleura, Thymus and Heart.* Lyon: IARC Press; 2004: 31–34.
62. Fink G, Krelbaum T, Yellin A, et al. Pulmonary carcinoid: Presentation, diagnosis, and outcome in 142 cases in Israel and review of 640 cases from the literature. *Chest.* 2001;119: 1647–1651.
63. Skuladottir H, Hirsch FR, Hansen HH, et al. Pulmonary neuroendocrine tumors: Incidence and prognosis of histological subtypes. A population-based study in Denmark. *Lung Cancer.* 2002;37:127–135.
64. Gosney JR, Travis WD. Diffuse idiopathic pulmonary neuroendocrine cell hyperplasia. In: Travis WD, Brambilla E, Müller-Hermelink HK, Harris CC, eds. *Pathology and Genetics of Tumours of the Lung, Pleura, Thymus and Heart.* Lyon: IARC Press; 2004: 76–77.
65. Pedersen N, Pedersen MW, Lan MS, et al. The insulinoma-associated 1: A novel promoter for targeted cancer gene therapy for small-cell lung cancer. *Cancer Gene Ther.* 2006;13:375–384.
66. Jiang SX, Kameya T, Asamura H, et al. hASH1 expression is closely correlated with endocrine phenotype and differentiation extent in pulmonary neuroendocrine tumors. *Mod Pathol.* 2004; 17:222–229.
67. Du EZ, Goldstraw P, Zacharias J, et al. TTF-1 expression is specific for lung primary in typical and atypical carcinoids: TTF-1-positive carcinoids are predominantly in peripheral location. *Hum Pathol.* 2004;35:825–831.
68. Saqi A, Alexis D, Remotti F, et al. Usefulness of CDX2 and TTF-1 in differentiating gastrointestinal from pulmonary carcinoids. *Am J Clin Pathol.* 2005;123:394–404.
69. Sturm N, Rossi G, Lantuejoul S, et al. Expression of thyroid transcription factor-1 in the spectrum of neuroendocrine cell lung proliferations with special interest in carcinoids. *Hum Pathol.* 2002;33:175–182.
70. Cai YC, Banner B, Glickman J, et al. Cytokeratin 7 and 20 and thyroid transcription factor 1 can help distinguish pulmonary from gastrointestinal carcinoid and pancreatic endocrine tumors. *Hum Pathol.* 2001;32:1087–1093.
71. Brambilla E. Small cell carcinoma. In: Zander DS, Popper HH, Jagirdar J, Haque AK, Cagle PT, Barrios R, eds. *Molecular Pathology of Lung Diseases.* New York: Springer; 2008: 293–300.
72. Balsara BR, Testa JR. Chromosomal imbalances in human lung cancer. *Oncogene.* 2002;21:6877–6883.
73. Petersen I, Langreck H, Wolf G, et al. Small-cell lung cancer is characterized by a high incidence of deletions on chromosomes 3p, 4q, 5q, 10q, 13q and 17p. *Br J Cancer.* 1997;75:79–86.
74. Righi L, Volante M, Rapa I, et al. Neuro-endocrine tumours of the lung. A review of relevant pathological and molecular data. *Virchows Arch.* 2007;451(Suppl 1):S51–59.

75. Zabarovsky ER, Lerman MI, Minna JD. Tumor suppressor genes on chromosome 3p involved in the pathogenesis of lung and other cancers. *Oncogene.* 2002;21:6915–6935.
76. Kim YH, Girard L, Giacomini CP, et al. Combined microarray analysis of small cell lung cancer reveals altered apoptotic balance and distinct expression signatures of MYC family gene amplification. *Oncogene.* 2006;25:130–138.
77. Onuki N, Wistuba II, Travis WD, et al. Genetic changes in the spectrum of neuroendocrine lung tumors. *Cancer.* 1999;85:600–607.
78. Debelenko LV, Brambilla E, Agarwal SK, et al. Identification of MEN1 gene mutations in sporadic carcinoid tumors of the lung. *Hum Mol Genet.* 1997;6:2285–2290.
79. Debelenko LV, Swalwell JI, Kelley MJ, et al. MEN1 gene mutation analysis of high-grade neuroendocrine lung carcinoma. *Genes Chromosomes Cancer.* 2000;28:58–65.
80. Dreijerink KM, Hoppener JW, Timmers HM, et al. Mechanisms of disease: Multiple endocrine neoplasia type 1-relation to chromatin modifications and transcription regulation. *Nat Clin Pract Endocrinol Metab.* 2006;2:562–570.
81. Wistuba II, Gazdar AF, Minna JD. Molecular genetics of small cell lung carcinoma. *Semin Oncol.* 2001;28:3–13.
82. Brambilla E, Negoescu A, Gazzeri S, et al. Apoptosis-related factors p53, Bcl2, and Bax in neuroendocrine lung tumors. *Am J Pathol.* 1996;149:1941–1952.
83. Przygodzki RM, Finkelstein SD, Langer JC, et al. Analysis of p53, K-ras-2, and C-raf-1 in pulmonary neuroendocrine tumors. Correlation with histological subtype and clinical outcome. *Am J Pathol.* 1996;148:1531–1541.
84. Brambilla E. Neuroendocrine carcinomas and precursors. In: Zander DS, Popper HH, Jagirdar J, Haque AK, Cagle PT and Barrios R, eds. *Molecular Pathology of Lung Diseases.* New York: Springer; 2008:301–306.
85. Kaye FJ. RB and cyclin dependent kinase pathways: Defining a distinction between RB and p16 loss in lung cancer. *Oncogene.* 2002;21:6908–6914.
86. Gouyer V, Gazzeri S, Brambilla E, et al. Loss of heterozygosity at the RB locus correlates with loss of RB protein in primary malignant neuro-endocrine lung carcinomas. *Int J Cancer.* 1994;58:818–824.
87. Eymin B, Gazzeri S, Brambilla C, et al. Distinct pattern of E2F1 expression in human lung tumours: E2F1 is upregulated in small cell lung carcinoma. *Oncogene.* 2001;20:1678–1687.
88. Salon C, Merdzhanova G, Brambilla C, et al. E2F-1, Skp2 and cyclin E oncoproteins are upregulated and directly correlated in high-grade neuroendocrine lung tumors. *Oncogene.* 2007;26:6927–6936.
89. Hiyama K, Hiyama E, Ishioka S, et al. Telomerase activity in small-cell and non-small-cell lung cancers. *J Natl Cancer Inst.* 1995;87:895–902.
90. Lantuejoul S, Soria JC, Moro-Sibilot D, et al. Differential expression of telomerase reverse transcriptase (hTERT) in lung tumours. *Br J Cancer.* 2004;90:1222–1229.
91. Zaffaroni N, Villa R, Pastorino U, et al. Lack of telomerase activity in lung carcinoids is dependent on human telomerase reverse transcriptase transcription and alternative splicing and is associated with long telomeres. *Clin Cancer Res.* 2005;11:2832–2839.
92. Nishio Y, Nakanishi K, Ozeki Y, et al. Telomere length, telomerase activity, and expressions of human telomerase mRNA component (hTERC) and human telomerase reverse transcriptase (hTERT) mRNA in pulmonary neuroendocrine tumors. *Jpn J Clin Oncol.* 2007;37:16–22.
93. Leotlela PD, Jauch A, Holtgreve-Grez H, et al. Genetics of neuroendocrine and carcinoid tumours. *Endocr Relat Cancer.* 2003;10:437–450.
94. Burbee DG, Forgacs E, Zochbauer-Muller S, et al. Epigenetic inactivation of RASSF1A in lung and breast cancers and malignant phenotype suppression. *J Natl Cancer Inst.* 2001;93:691–699.
95. Toyooka S, Toyooka KO, Maruyama R, et al. DNA methylation profiles of lung tumors. *Mol Cancer Ther.* 2001;1:61–67.
96. Shivapurkar N, Toyooka S, Eby MT, et al. Differential inactivation of caspase-8 in lung cancers. *Cancer Biol Ther.* 2002;1:65–69.
97. Wistuba II, Berry J, Behrens C, et al. Molecular changes in the bronchial epithelium of patients with small cell lung cancer. *Clin Cancer Res.* 2000;6:2604–2610.
98. Finkelstein SD, Hasegawa T, Colby T, et al. 11q13 allelic imbalance discriminates pulmonary carcinoids from tumorlets. A microdissection-based genotyping approach useful in clinical practice. *Am J Pathol.* 1999;155:633–640.
99. Yousem SA, Tazelaar HD, Manabe T, et al. Inflammatory myofibroblastic tumour. In: Travis WD, Brambilla E, Müller-Hermelink HK, Harris CC, eds. *Pathology and Genetics of Tumours of the Lung, Pleura, Thymus and Heart.* Lyon: IARC Press; 2004:105–106.
100. Travis WD, Churg A, Aubry MC, et al. Mesenchymal tumours. In: Travis WD, Brambilla E, Müller-Hermelink HK, Harris CC, eds. *Pathology and Genetics of Tumours of the Lung, Pleura, Thymus and Heart.* Lyon: IARC Press; 2004:141–142.
101. Coffin CM, Patel A, Perkins S, et al. ALK1 and p80 expression and chromosomal rearrangements involving 2p23 in inflammatory myofibroblastic tumor. *Mod Pathol.* 2001;14:569–576.
102. Lawrence B, Perez-Atayde A, Hibbard MK, et al. TPM3–ALK and TPM4–ALK oncogenes in inflammatory myofibroblastic tumors. *Am J Pathol.* 2000;157:377–384.
103. Begueret H, Galateau-Salle F, Guillou L, et al. Primary intrathoracic synovial sarcoma: A clinicopathologic study of 40 t(X;18)-positive cases from the French Sarcoma Group and the Mesopath Group. *Am J Surg Pathol.* 2005;29:339–346.
104. Hartel PH, Fanburg-Smith JC, Frazier AA, et al. Primary pulmonary and mediastinal synovial sarcoma: A clinicopathologic study of 60 cases and comparison with five prior series. *Mod Pathol.* 2007;20:760–769.
105. Kazmierczak B, Meyer-Bolte K, Tran KH, et al. A high frequency of tumors with rearrangements of genes of the HMGI(Y) family in a series of 191 pulmonary chondroid hamartomas. *Genes Chromosomes Cancer.* 1999;26:125–133.
106. Hess JL. Chromosomal translocations in benign tumors: The HMGI proteins. *Am J Clin Pathol.* 1998;109:251–261.
107. Weill H, Hughes JM, Churg AM. Changing trends in US mesothelioma incidence. *Occup Environ Med.* 2004;61:438–441.
108. Churg A, Roggli V, Galateau-Salle F, et al. Mesothelioma. In: Travis WD, Brambilla E, Müller-Hermelink HK, Harris CC, eds. *Pathology and Genetics of Tumours of the Lung, Pleura, Thymus and Heart.* Lyon: IARC Press; 2004:128–136.
109. Rizzo P, Bocchetta M, Powers A, et al. SV40 and the pathogenesis of mesothelioma. *Semin Cancer Biol.* 2001;11:63–71.
110. Robinson BW, Lake RA. Advances in malignant mesothelioma. *N Engl J Med.* 2005;353:1591–1603.
111. Maziak DE, Gagliardi A, Haynes AE, et al. Surgical management of malignant pleural mesothelioma: A systematic review and evidence summary. *Lung Cancer.* 2005;48: 157–169.
112. Marchevsky AM. Application of immunohistochemistry to the diagnosis of malignant mesothelioma. *Arch Pathol Lab Med.* 2008;132:397–401.
113. Ramos-Nino ME, Testa JR, Altomare DA, et al. Cellular and molecular parameters of mesothelioma. *J Cell Biochem.* 2006;98:723–734.
114. Galateau-Salle F, Vignaud JM. Diffuse malignant mesothelioma: Genetic pathways and mechanisms of oncogenesis of asbestos and other agents that cause mesotheliomas. In: Zander DS, Popper HH, Jagirdar J, Haque AK, Cagle PT, Barrios R, eds. *Molecular Pathology of Lung Diseases.* New York: Springer; 2008: 347–357.
115. Murthy SS, Testa JR. Asbestos, chromosomal deletions, and tumor suppressor gene alterations in human malignant mesothelioma. *J Cell Physiol.* 1999;180:150–157.
116. Musti M, Kettunen E, Dragonieri S, et al. Cytogenetic and molecular genetic changes in malignant mesothelioma. *Cancer Genet Cytogenet.* 2006;170:9–15.
117. Sekido Y, Pass HI, Bader S, et al. Neurofibromatosis type 2 (NF2) gene is somatically mutated in mesothelioma but not in lung cancer. *Cancer Res.* 1995;55:1227–1231.
118. Toyooka S, Kishimoto T, Date H. Advances in the molecular biology of malignant mesothelioma. *Acta Med Okayama.* 2008;62:1–7.
119. Lee AY, Raz DJ, He B, et al. Update on the molecular biology of malignant mesothelioma. *Cancer.* 2007;109:1454–1461.

120. Ramos-Nino ME, Vianale G, Sabo-Attwood T, et al. Human mesothelioma cells exhibit tumor cell-specific differences in phosphatidylinositol 3-kinase/AKT activity that predict the efficacy of Onconase. *Mol Cancer Ther.* 2005;4:835–842.
121. Govindan R, Kratzke RA, Herndon 2nd JE, et al. Gefitinib in patients with malignant mesothelioma: A phase II study by the Cancer and Leukemia Group B. *Clin Cancer Res.* 2005;11: 2300–2304.
122. Cortese JF, Gowda AL, Wali A, et al. Common EGFR mutations conferring sensitivity to gefitinib in lung adenocarcinoma are not prevalent in human malignant mesothelioma. *Int J Cancer.* 2006;118:521–522.
123. Elias JA, Lee CG, Zheng T, et al. New insights into the pathogenesis of asthma. *J Clin Invest.* 2003;111:291–297.
124. Trigg CJ, Manolitsas ND, Wang J, et al. Placebo-controlled immunopathologic study of four months of inhaled corticosteroids in asthma. *Am J Respir Crit Care Med.* 1994;150:17–22.
125. Tantisira KG, Weiss ST. The pharmacogenetics of asthma: An update. *Curr Opin Mol Ther.* 2005;7:209–217.
126. Israel E. Genetics and the variability of treatment response in asthma. *J Allergy Clin Immunol.* 2005;115:S532–538.
127. Barnes PJ. Immunology of asthma and chronic obstructive pulmonary disease. *Nature Rev Immunol.* 2008;8:183–192.
128. Holgate ST. Pathogensis of asthma. *Clin Exp Allergy.* 2008; 38:872–897.
129. Comhair SA, Xu W, Ghosh S, et al. Superoxide dismutase inactivation in pathophysiology of asthmatic airway remodeling and reactivity. *Am J Pathol.* 2005;166:663–674.
130. Bradding P, Walls AF, Holgate ST. The role of the mast cell in the pathophysiology of asthma. *J Allergy Clin Immunol.* 2006; 117:1277–1284.
131. GOLD. Global Initiative for Chronic Obstructive Lung Disease (GOLD): Global strategy for the diagnosis MOCOPD. NHLBI/WHO Workshop Report. 2003.
132. Churg A, Cosio M, Wright JL. Mechanisms of cigarette smoke-induced COPD: Insights from animal models. *Am J Physiol Lung Cell Mol Physiol.* 2008;294:L612–631.
133. Snider G, Klinerman J, Thurlbeck WM. The definition of emphysema: Report of the National Heart, Lung and Blood Institute Division of Lung Diseases Workshop. *Am Rev Respir Dis.* 1985; 132:182–185.
134. Kim V, Rogers TJ, Criner GJ. Frontiers in emphysema research. *Semin Thorac Cardiovasc Surg.* 2007;19:135–141.
135. Wright JL, Churg A. Advances in the pathology of COPD. *Histopath.* 2006;10:1–9.
136. Thurlbeck WM. Pathology of chronic airflow obstruction. *Chest.* 1990;97:6s–10s.
137. Taraseviciene-Stewart L, Voelkel NF. Molecular pathogenesis of emphysema. *J Clin Invest.* 2008;118:394–402.
138. Lieberman J, Winter B, Sastre A. Alpha 1-antitrypsin Pi-types in 965 COPD patients. *Chest.* 1986;89:370–373.
139. Kohnlein T, Welte T. Alpha-1-antitrypsin deficiency: Pathogenesis, clinical presentation, diagnosis, and treatment. *Am J Med.* 2008;121:3–9.
140. Mulgrew AT, Taggart CC, McElvaney NG. Alpha-1-antitrypsin deficiency: Current concepts. *Lung.* 2007;185:191–201.
141. Ogushi F, Fells GA, Hubbard RC, et al. Z-type alpha 1-antitrypsin is less competent than M1-type alpha 1-antitrypsin as an inhibitor of neutrophil elastase. *J Clin Invest.* 1987;80:1366–1374.
142. Barnes PJ. Mediators of chronic obstructive pulmonary disease. *Pharmacol Rev.* 2004;52:515–548.
143. MacNee W. Oxidative stress and lung inflammation in airways disease. *Eur J Pharmacol.* 2001;429:195–207.
144. Rahman I, Adcock IM. Oxidative stress and redox regulation of lung inflammation of COPD. *Eur Respir J.* 2006;28:219–242.
145. Seifart C, Plagens A. Genetics of chronic obstructive pulmonary disease. *Int J Chron Obstruct Pulmon Dis.* 2007;2:541–550.
146. Mitchell RE, Bury RG. Congenital bronchiectasis due to deficiency of bronchial cartilage (Williams-Campbell syndrome): A case report. *J Pediatr.* 1975;87:230–234.
147. Van Schoor J, Joos G, Pauwels R. Tracheobronchomegaly—The Mounier-Kuhn syndrome: Report of two cases and review of the literature. *Eur Respir J.* 1991;4:1303–1306.
148. Afzelius BA. A human syndrome caused by immotile cilia. *Science.* 1976;193:317–319.
149. Morillas HN, Zariwala M, Knowles MR. Genetic causes of bronchiectasis: Primary ciliary dyskinesia. *Respiration.* 2007;74: 252–263.
150. Davies JC, Alton EWF, Bush A. Cystic fibrosis. *BMJ.* 2007;335: 1255–1259.
151. Riordan JR, Rommens JM, Kerem B, et al. Identification of the cystic fibrosis gene: Cloning and characterization of complementary DNA. *Science.* 1989; 245:1066–1073.
152. Notarangelo LD, Plebani A, Mazzolari E, et al. Genetic causes of bronchiectasis: Primary immune deficiencies and the lung. *Respiration.* 2007;74:264–275.
153. Davies JC, Rubin BK. Emerging and unusual gram-negative infections in cystic fibrosis. *Semin Respir Crit Care Med.* 2007; 28:312–321.
154. Bonfield TL, Konstan MW, Berger M. Altered respiratory epithelial cell cytokine production in cystic fibrosis. *J Allergy Clin Immunol.* 1999;104:72–78.
155. Johkoh T. Imaging of idiopathic interstitial pneumonias. *Clin in Chest Medicine.* 2008;29:133–147.
156. Liebow AA, Carrington CB. The interstitial pneumonias. In: Simon M, Potchen EJ, LeMay M, eds. *Frontiers of Pulmonary Radiology.* New York: Grune & Stratton; 1969:102–141.
157. American Thoracic Society/European Respiratory Society. International multidisciplinary consensus classification of the idiopathic interstitial pneumonias: The Joint Statement of the American Thoracic Society (ATS), and the European Respiratory Society (ERS) was adopted by the ATS Board of Directors, June 2001, and by The ERS Executive Committee, June 2001. *Am J Respir Crit Care Med.* 2002;165:277–304.
158. Lawson WE, Loyd JE. The genetic approach in pulmonary fibrosis. *Proc Am Thorac Soc.* 2006;3:345–349.
159. Katzenstein AL, Myers JL. Idiopathic pulmonary fibrosis: Clinical relevance of pathologic classification. *Am J Respir Crit Care Med.* 1998;157:1301–1315.
160. Farver CF. Pathology of advanced interstitial diseases: Pulmonary fibrosis, sarcoidosis, histiocytosis X, autoimmune pulmonary disease, lymphangioleiomyomatosis. In: Maurer JR, ed. *Non-Neoplastic Advanced Lung Disease* in: *Lung Biology in Health and Disease.* New York: Marcel Dekker, Inc.; 2003:29–58. Chapter 2.
161. Maher TM, Wells AU, Laurent GJ. Idiopathic pulmonary fibrosis: Multiple causes and multiple mechanisms? *Eur Respir J.* 2007;30:835–839.
162. Selman M, Pardo A. Role of epithelial cells in idiopathic pulmonary fibrosis. *Proc Am Thorac Soc.* 2006;3:364–372.
163. Noble PW, Homer RJ. Idiopathic pulmonary fibrosis: New insights into pathogenesis. *Clin Chest Med.* 2004;25:749–758.
164. Chilosi M, Poletti V, Zamò A, et al. Aberrant Wnt/beta-catenin pathway activation in idiopathic pulmonary fibrosis. *Am J Pathol.* 2003;162:1495–1502.
165. Ramos C, Montana M. Fibroblasts from idiopathic pulmonary fibrosis and normal lungs differ. *Am J Resp Cell Mol Biol.* 2001; 24:591–599.
166. Armanios MY, Chen JJ, Cogan JD, et al. Telomerase mutations in families with idiopathic pulmonary fibrosis. *N Engl J Med.* 2007;356:1317–1326.
167. Anderson WR, Thielen K. Correlative study of adult respiratory distress syndrome by light, scanning, and transmission electron microscopy. *Ultrastruct Pathol.* 1992;16:615–628.
168. Tomashefski Jr JF. Pulmonary pathology of the adult respiratory distress syndrome. *Clin Chest Med.* 1990;11:593–619.
169. Tasaka S, Amaya F, Hashimoto S, et al. Roles of oxidants and redox signaling in the pathogenesis of acute respiratory distress syndrome. *Antioxid Redox Signal.* 2008;10:739–753.
170. Tang PS, Mura M, Seth R, et al. Acute lung injury and cell death: How many ways can cells die? *Am J Physiol Lung Cell Mol Physiol.* 2008;294:L632–641.
171. Taylor JR, Ryu J, Colby TV, et al. Lymphangioleiomyomatosis: Clinical course in 32 patients. *N Engl J Med.* 1990;323:1254–1260.
172. McCormack FX. Lymphangioleiomyomatosis: A clinical update. *Chest.* 2008;133:507–516.

173. Johnson SR, Whale CI, Hubbard RB, et al. Survival. *Thorax.* 2004;59:800–803.
174. Muller NL, Chiles C, Kullnig P. Pulmonary lymphangiomyomatosis: Correlation of CT with radiographic and functional findings. *Radiology.* 1990;175:335–339.
175. Steagall WK, Taveira-DaSilva AM, Moss J. Clinical and molecular insights into lymphangioleiomyomatosis. *Sarc Vasc Dif Lung Dis.* 2005;22:S49–S66.
176. Matsumoto Y, Horiba K, Usaki J, et al. Markers of cell proliferation and expression of melanosomal antigen in lymphangioleiomyomatosis. *Am J Respir Cell Mol Biol.* 1999;21:327–336.
177. Krymskaya VP. Smooth muscle-like cells in pulmonary lymphangioleiomyomatosis. *Proc Am Thorac Soc.* 2008;5:119–126.
178. Cornog JL Jr, Enter line HT. Lymphangiomyoma, a benign lesion of chyliferous lymphatics synonymous with lymphangiopericytoma. *Cancer* 1966;19:1909–1930.
179. Juvet SC, McCormack FX, Kwiatkowski DJ, et al. Molecular pathogenesis of lymphangioleiomyomatosis: Lessons learned from orphans. *Am J Respir Cell Mol Biol.* 2007;36:398–408.
180. Thomas PD, Hunninghake GW. Current concepts of the pathogenesis of sarcoidosis. *Am Rev Respir Dis.* 1987;135:747–760.
181. DeRemee RA. Sarcoidoisis. *Mayo Clin Proc.* 1995;70:177–181.
182. Chiles C, Putnam CE. Pulmonary sarcoidosis. *Semin Respir Med.* 1992;13:345–357.
183. Uehlinger ES. The morbid anatomy of sarcoidosis. *Am Rev Respir Dis.* 1961;84:6–13.
184. Ricker W. Sarcoidosis: A clinicopathologic review of three hundred cases, including twenty-two autospies. *Am J Clin Pathol.* 1949;19:725–749.
185. Sheffield EA. Pathology of sarcoidosis. *Clin Chest Med.* 1997;18:741–754.
186. Reich JM. What is sarcoidosis? *Chest.* 2003;124:367–371.
187. Culver DA, Newman LS, Kavuru MS. Gene-environment interactions in sarcoidosis: Challenge and opportunity. *Clin Dermatol.* 2007;25:267–275.
188. Rybicki BA, Maliarik MJ, Major M, et al. Epidemiology, demographics and genetics of sarcoidosis. *Semin Respir Infect.* 1998;13:166–173.
189. Rybicki BA, Iannuzzi MC, Frederick MM, et al. Familial aggregation of sarcoidosis. A case-control etiologic study of sarcoidosis (ACCESS). *Am J Respir Crit Care Med.* 2001;164: 2085–2091.
190. Moller DR, Chen ES. Genetic basis of remitting sarcoidosis: Triumph of the trimolecular complex? *Am J Respir Cell Mol Biol.* 2002;27(4):391–395.
191. McIntyre JA, McKee KT, Loadholt CB, et al. Increased HLA-B7 antigen frequency in South Carolina blacks in association with sarcoidosis. *Transplant Proc.* 1977;9(suppl 1):173–176.
192. Grunewald J, Eklung A, Olerup O. Human leukocyte antigen class I alleles and the disease course in sarcoidosis patients. *Am J Respir Crit Care Med.* 2004;169(6):696–702.
193. Ina Y, Takad K, Yamamoto M, et al. HLA and sarcoidosis in the Japanese. *Chest.* 1989;95(6):1257–1261.
194. Kunikane H, Abe S, Yamaguchi E, et al. Analysis of restriction fragment length polymorphism for the HLA-DR gene in Japanese. *Thorax.* 1994;49(6):573–576.
195. Schurmann M, Lympany PA, Reichet P, et al. Familial sarcoidosis is linked to the major. *Am J Respir Crit Care Med.* 2000; 162:861–864.
196. Schurmann M, Reichel P, Muller-Myhsok B, et al. Results from a genome-wide search for predisposing genes in sarcoidosis. *Am J Respir Crit Care Med.* 2001;164:840–846.
197. Bocart D, Lecossier D, De Lassence A, et al. A search for mycobacterial DNA in granulomatous tissues from patients with sarcoidosis using the polymerase chain reaction. *Am Rev Respir Dis.* 1992;145:1142–1148.
198. Drake WP, Pei Z, Pride DT, et al. Molecular analysis of sarcoidosis and control tissues for Mycobacteria DNA. *Emerg Infect Dis.* 2002;8:1328–1335.
199. Baughman RP, Lynch JP. Difficult treatment issues in sarcoidosis. *J Intern Med.* 2003;253:41–45.
200. Baughman RP, Lower EE, du Bois RM. Sarcoidosis. *Lancet.* 2003;361:1111–1118.
201. Moller DR. Pulmonary fibrosis of sarcoidosis. New approaches, old ideas. *Am J Respir Cell Mol Biol.* 2003;29:S37–41.
202. Presneill JJ, Nakata K, Inoue Y, et al. Pulmonary alveolar proteinosis. *Clin Chest Med.* 2004;25:593–613.
203. Trapnell BC, Whitsett JA, Nakata K. Pulmonary alveolar proteinosis. *N Engl J Med.* 2003;349:2527–2539.
204. Tazawa R, Hamano E, Arai T, et al. Granulocyte-macrophage colony-stimulating factor and lung immunity in pulmonary alveolar proteinosis. *Am J Respir Crit Care Med.* 2005;171:1142–1149.
205. Prakash UB, Barham SS, Carpenter HA, et al. Pulmonary alveolar phospholipoproteinosis: Experience with 34 cases and a review. *Mayo Clin Proc.* 1987;62:499–518.
206. Mazzone PJ, Jane Thomassen M, Kavuru MS. Pulmonary alveolar proteinosis: Recent advances. *Semin Respir Crit Care Med.* 2002;23:115–126.
207. Lakshminarayan S, Schwarz MI, Stanford RE. Unsuspected pulmonary alveolar proteinosis complicating acute myelogenous leukemia. *Chest.* 1976;69:433–435.
208. Bedrossian CW. Alveolar proteinosis as a consequence of immunosuppression. A hypothesis based on clinical and pathologic observations. *Human Pathol.* 1980;11:527–534.
209. Johkoh T, Itoh H, Müller NL, et al. Crazy-paving appearance at thin-section CT: spectrum of disease and pathologic findings. *Radiology.* 1999;211:155–160.
210. Brasch F, Birzele J, Ochs M, et al. Surfactant proteins in pulmonary alveolar proteinosis in adults. *Eur Respir J.* 2004;24:426–435.
211. Singh G, Katyal SL, Bedrossian CW, et al. Pulmonary alveolar proteinosis. Staining for surfactant apoprotein in alveolar proteinosis and in conditions simulating it. *Chest.* 1983;83:82–86.
212. Singh G, Katyal SL. Surfactant apoprotein in nonmalignant pulmonary disorders. *Am J Pathol.* 1980;101:51–61.
213. Ikegami M. Surfactant catabolism. *Respirology.* 2006;11:S24–27.
214. Andreeva AV, Kutuzov MA, Voyno-Yasenetskaya TA. Regulation of surfactant secretion in alveolar type II cells. *Am J Physiol Lung Cell Mol Physiol.* 2007;293:L259–271.
215. Trapnell BC, Whitsett JA. Gm-CSF regulates pulmonary surfactant homeostasis and alveolar macrophage-mediated innate host defense. *Annu Rev Physiol.* 2002;64:775–802.
216. Stanley E, Lieschke GJ, Grail D, et al. Granulocyte/macrophage colony-stimulating factor-deficient mice show no major perturbation of hematopoiesis but develop a characteristic pulmonary pathology. *Proc Natl Acad Sci USA.* 1994;91:5592–5596.
217. Thomassen MJ, Yi T, Raychaudhuri B, et al. Pulmonary alveolar proteinosis is a disease of decreased availability of GM-CSF rather than an intrinsic cellular defect. *Clin Immunol.* 2000;95:85–92.
218. Bonfield TL, Russell D, Burgess S, et al. Autoantibodies against granulocyte-macrophage colony-stimulating factors are diagnostic for pulmonary alveolar proteinosis. *Am J Respir Cell Mol Biol.* 2002;27:481–486.
219. Kitamura T, Tanaka N, Watanabe J, et al. Idiopathic pulmonary alveolar proteinosis as an autoimmune disease with neutralizing antibody against granulocyte/macrophage colony-stimulating factor. *J Exp Med.* 1999;190:875–880.
220. Inoue Y, Trapnell BC, Tazawa R, et al. Characteristics of a large cohort of patients with autoimmune pulmonary alveolar proteinosis in Japan. *Am J Respir Crit Care Med.* 2008;177:752–762.
221. Bonfield TL, Raychaudhuri B, Malur A, et al. Pu.1 regulation of human alveolar macrophage differentiation requires granulocyte-macrophage colony-stimulating factor. *Am J Physiol Lung Cell Mol Physiol.* 2003;285:L1132–L1136.
222. Shibata Y, Berclaz PY, Chroneos ZC, et al. GM-CSF regulates alveolar macrophage differentiation and innate immunity in the lung through pu.1. *Immunity.* 2001;15:557–567.
223. Bonfield TL, Farver CF, Barna BP, et al. Peroxisome proliferator-activated receptor-gamma is deficient in alveolar macrophages from patients with alveolar proteinosis. *Am J Respir Cell Mol Biol.* 2008;29:677–682.
224. Simonneau G, Galie N, Rubin HL, et al. Clinical classification of pulmonary hypertension. *J Am Coll Cardiol.* 2004;43(Suppl 1): 5S–12S.
225. Chin KM, Rubin LH. *J Am Coll Cardiol.* 2008;51:1527–1538.
226. Newman JH, Phillips JA, Loyd JE. Narrative review: The enigma of pulmonary art. *Ann Intern Med.* 2008;148:278–283.
227. Austin ED, Loyd JE. Genetics and mediators in pulmonary arterial hypertension. *Clin Chest Med.* 2007;28:43–57.

228. Runo JR, Loyd JE. Primary pulmonary hypertension. *Lancet.* 2003;361:1533–1544.
229. Humbert M. Update in pulmonary arterial hypertension 2007. *Am J Respir Crit Care Med.* 2008;177:574–579.
230. Churg A. Recent advances in the diagnosis of Churg-Strauss syndrome. *Mod Pathol.* 2001;14:1284–1293.
231. Jennette JC, Falk RJ. New insight into the pathogenesis of vasculitis associated with antineutrophil cytoplasmic autoantibodies. *Curr Opin Rheumatol.* 2008;20:55–60.
232. Ganz T. Defensins: Antimicrobial peptides of innate immunity. *Nat Rev Immunol.* 2003;3:710–720.
233. Bals R, Hiemstra PS. Innate immunity in the lung: How epithelial cells fight against respiratory pathogens. *Eur Respir J.* 2004;23:33.
234. Blasi F, Tarsia P, Aliberti S. Strategic targets of essential host-pathogen interactions. *Respiration.* 2005;72:9–25.
235. Winn WC Jr, Chandler FW. Bacterial infections. In: Dail DH, Hammar SP, eds. *Pulmonary Pathology.* 2nd ed. New York: Springer-Verlag; 1994:255–330.
236. Peltola H. Worldwide Haemophilus influenzae type B disease at the beginning of the 21st century: Global analysis of the disease burden 25 years after the use of the polysaccharide vaccine and a decade after the advent of conjugates. *Clin Microbiol Rev.* 2000;13:302–317.
237. Waites KB, Talkington DF. Mycoplasma pneumoniae and its role as a human pathogen. *Clin Microbiol Rev.* 2004;17:697–728.
238. Stead WW, Kerby GR, Schlueter DP, et al. The clinical spectrum of primary tuberculosis in adults. Confusion with reinfection in the pathogenesis of chronic tuberulosis. *Ann Intern Med.* 1968;68:731–745.
239. van Rie A, Warren R, Richardson M, et al. Exogenous reinfection as a cause of recurrent tuberculosis after curative treatment. *N Engl J Med.* 1999;341:1171–1179.
240. Auerbach O. The natural history of the tuberculous pulmonary lesion. *Med Clin North Am.* 1959;43:239–251.
241. Rabbat A, Huchon GJ. Nonbacterial pneumonia. In: Albert RK, Spiro SG, Jett JR, eds. *Clinical Respiratory Medicine.* 3rd ed. Philadelphia: Mosby Elsevier; 2008:351–364.
242. Stewart S. Pulmonary infections in transplantation pathology. *Arch Pathol Lab Med.* 2007;131:1219–1231.
243. Zaki SR, Khan AS, Goodman RS, et al. Retrospective diagnosis of hanta-virus pulmonary syndrome. *Arch Pathol Lab Med.* 1996;120:134–139.
244. Chong PY, Chui P, Ling AE, et al. Analysis of deaths during the severe acute respiratory syndrome (SARS) epidemic in Singapore: Challenges in determining a SARS diagnosis. *Arch Pathol Lab Med.* 2004;128:195–204.
245. Lee I, Barton TD. Viral respiratory tract infections in transplant patinets. *Drugs.* 2007;67:1411–1427.
246. Bradsher RW. Histoplasmosis and blastomycosis. *Clin Inf Dis.* 1996;22:S102.
247. Davies SF, Sarosi GA. Epidemiological and clinical featues of pulmonary blastomycosis. *Semin Res Inf.* 1997;12:206–218.
248. Stevens DA. Current concepts: Coccidioidomycosis. *N Engl J Med.* 1995;332:1007–1082.
249. Watts JC, Chandler FW. Aspergillosis. In: Connor DH, Chandler FW, et al. eds. *Pathology of Infectious Diseases.* Stamford, CT: Appleton and Lange; 1997:033–941.
250. Yousem S. The histologic spectrum of necrotizing forms of pulmonary aspergillosis. *Hum Pathol.* 1997;28:650–656.
251. Edman JC, Kovacs JA, Masur H, et al. Ribosomal RNA sequence shows *Pneumocystis carinii* to be a member of the fungi. *Nature.* 1988;334:519–522.
252. Stringer JR, Beard CF, Miller RF, et al. A new name (*Pneumocystis jiroveci*) for pneumocystis from humans. *Emerg Infect Dis.* 2002;8:891–896.
253. Beard CB, Roux P, Gilles Nevez, et al. Strain typing methods and molecular epidemiology of *Pneumocystis* pneumonia. *Emerg Infectious Diseases.* 2004;10:1729–1735.
254. Krajicek J, Limper AH, Thomas CF. Advances in the biology, pathogenesis and identification of *Pneumocystis* pneumonia. *Curr Opin Pulm Med.* 2008;14:228–234.
256. Thomas CF Jr., Limper AH. *Pneumocystis* pneumonia. *N Engl J Med.* 2004;350:2487–2498.
256. Mossman BT, Churg A. Mechanisms in the pathogenesis of asbestosis and silicosis. *Am J Respir Crit Care Med.* 1998; 157:1666–1680.
257. Tazi A. Cells of the dendritic cell lineage in human lung carcinomas and pulmonary histiocytosis S. In: Lipscomb M, Russell S, eds. *Lung Macrophages and Dendritic Cells in Health and Disease.* New York: Marcel Dekker; 1997:725–757.
258. Vassallo R, Limper AH. Pulmonary Langerhans' cell histiocytosis. *Semin Respir Crit Care Med.* 2002;23:93–101.
259. Tazi A, Bonay M, Grandsaigne M, et al. Surface phenotype of Langerhans cells and lymphocytes in granulomatous lesions from patients with pulmonary histiocytosis X. *Am Rev Respir Dis.* 1993;147:1531–1536.
260. Vermaelen K, Pauwels R. Pulmonary dendritic cells. *Am J Respir Crit Care Med.* 2005;172:530–551.
261. Annels NE, Da Costa CE, Prins FA, et al. Aberrant chemokine receptor expression and chemokine production by Langerhans cells underlies the pathogenesis of Langerhans cell histiocytosis. *J Exp Med.* 2003;197:1385–1390.
262. Wang J, Snider DP, Hewlett BR, et al. Transgenic expression of granulocyte-macrophage colony-stimulating factor induces the differentiation and activation of a novel dendritic cell population in the lung. *Blood.* 2000;95:2337–2345.
263. Yousem SA, Colby TV, Chen YY, et al. Pulmonary Langerhans cell histiocytosis: Molecular analysis of clonality. *Am J Surg Pathol.* 2001;25:630–636.
264. Dacic S, Trusky C, Baggker A, et al. Genotypic analysis of pulmonary Langerhans cell histiocytosis. *Human Pathol.* 2003;34: 1345–1349.
265. Tazi A, Soler P, Hance AJ. Adult pulmonary Langerhans cell histiocytosis. *Thorax.* 2000;55:405–416.
266. Egan AJ, Boardman LA, Tazelaar HD, et al. Erdheim-Chester disease: Clinical, radiologic, and histopathologic findings in five patients with interstitial lung disease. *Am J Surg Pathol.* 1999;23:17–26.
267. Veyssier-Belot C, Cacoub P, Caparros-Lefebvre D, et al. Erdheim-Chester disease. Clinical and radiologic characteristics of 59 cases. *Medicine (Baltimore).* 1996;75:157–169.
268. Kambouchner M, Colby TV, Domenge C, et al. Erdheim-Chester disease with prominent pulmonary involvement associated with eosinophilic granuloma of mandibular bone. *Histopathology.* 1997;30:353–358.
269. Madroszyk A, Wallaert B, Rémy M, et al. Diffuse interstitial pneumonia revealing Erdheim-Chester's disease. *Rev Mal Respir.* 1994;11:304–307.
270. Murray D, Marshall M, England E, et al. Erdheim-Chester disease. *Clin Radiol.* 2001;56:481–484.
271. Shamburek RD, Brewer Jr HB, Gochuico BR. Erdheim-Chester disease: A rare multisystem histiocytic disorder associated with interstitial lung disease. *Am J Med Sci.* 2001; 321:66–75.
272. Rao RN, Chang CC, Uysal N, et al. Fulminant multisystem non-Langerhans cell histiocytic proliferation with hemophagocytosis: A variant form of Erdheim-Chester disease. *Arch Pathol Lab Med.* 2005;129:e39–43.
273. Rush WL, Andriko JA, Galateau-Salle F, et al. Pulmonary pathology of Erdheim-Chester disease. *Mod Pathol.* 2000;1397:747–754.
274. Devouassoux G, Lantuejoul S, Chatelain P, et al. Erdheim-Chester disease: A primary macrophage cell disorder. *Am J Respir Crit Care Med.* 1998;157:650–653.
275. Myra C, Sloper L, Tighe PJ, et al. Treatment of Erdheim-Chester disease with cladribine: A rational approach. *Br J Ophthalmol.* 2004;88:844–847.
276. Wittenberg KH, Swensen SJ, Myers JL. Pulmonary involvement with Erdheim-Chester disease: Radiographic and CT findings. *Am J Roentgenol.* 2000;174:1327–1331.
277. Chetritt J, Paradis V, Dargere D, et al. Chester-Erdheim disease: A neoplastic disorder. *Hum Pathol.* 1999;30:1093–1096.
278. Craighead JE, Abraham JL, Churg A, et al. The pathology of asbestos-associated disease of the lung and pleural cavities: Diagnostic critera. *Arch Pathol Lab Invest.* 1982;106:544–595.

279. Robledo R, Mossman B. Cellular and molecular mechanisms of asbestos-induced fibrosis. *J Cell Physiol.* 1999;180:158–166.
280. Ke Q, Li J, Ding J, et al. Essential role of ROS-mediated NFAT activation in TNF-alpha induction by crystalline silica exposure. *Am J Physiol Lung Cell Mol Physiol.* 2006;291:L257–264.
281. Tsai EY, Yie J, Thanos D, et al. Cell-type-specific regulation of the human tumor necrosis factor alpha gene in B cells and T cells by NFATp and ATF-2/JUN. *Mol Cell Biol.* 1996;16: 5232–5244.
282. Aktogu S, Yuncu G, Halilcolar H, et al. Bronchogenic cysts: Clinicopathological presentation and treatment. *Eur Respir J.* 1996;9:2017–2021.
283. Stocker JT, Husain AN. Cystic lesions of the lung in children: Classification and controversies. In: Timens W, Popper HH, eds. *Pathology of the Lung.* Sheffield, UK: European Respiratory Society Journals Ltd; 2007:1–20. European Respiratory Monograph 39; Chapter 1.
284. Maeda Y, Dave V, Whisett JA. Transcriptional control of lung morphogenesis. *Physiol Rev.* 2007;87:219–244.
285. Chibbar R, Shih F, Baga M, et al. Nonspecific interstitial pneumonia and usual interstitial pneumonia with mutation in surfactant protein C in familial pulmonary fibrosis. *Mod Pathol.* 2004;17:973–980.
286. Bullard JE, Wert SE, Whitsett JA, et al. ABCA3 mutations associated with pediatric interstitial lung disease. *Am J Respir Crit Care Med.* 2005;172:1026–1031.
287. Comeron HS, Somaschini M, Carrera P, et al. A common mutation in the surfactant protein C gene associated with lung disease. *J Pediatr.* 2005;146:370–375.
288. Hamvas A, Cole FS, Nogee L. Genetic disorders of surfactant proteins. *Neonatology.* 2007;91:311–371.
289. Whitsett JA, Wert SE, Xu Y. Genetic disorders of surfactant homeostasis. *Biol Neonate.* 2005;87:283–287.
290. Travis WD, Brambilla E, Müller-Hermelink HK, et al. *Pathology and Genetics of Tumours of the Lung, Pleura, Thymus and Heart.* Lyon: IARC Press; 2004.

Chapter

19

Molecular Basis of Diseases of the Gastrointestinal Tract

Antonia R. Sepulveda ■ Dara L. Aisner

INTRODUCTION

As we entered the 21st century, much of the progress in the understanding of gastrointestinal disorders has continued to center on the molecular underpinning of gastrointestinal neoplasia. First, the development of cancer in the setting of inflammatory conditions is well represented by the association of *H. pylori* with gastric cancer, and of inflammatory bowel diseases with colorectal cancer. Second, the development of cancer in patients with hereditary predisposition syndromes has shed light not only in the mechanisms of hereditary neoplasia, but has also led to major progress in the understanding of the molecular basis of the more common forms of sporadic cancer. The molecular characterization of the steps of gastrointestinal neoplastic development and progression has led to advances in disease diagnosis and treatment, and has opened the opportunity for development of more targeted approaches to cancer prevention, surveillance, and novel therapeutics.

In this chapter, we will focus on the disease processes that most clearly illustrate the concepts and advances in molecular pathology of the gastrointestinal tract. This includes neoplastic diseases associated with a background of chronic inflammation, well characterized gastrointestinal hereditary cancer syndromes, and the so- called sporadic cancers of the gastrointestinal tract, primarily reviewing gastric and colonic carcinogenesis.

GASTRIC CANCER

Gastric carcinoma is the fourth most frequent cancer worldwide, with highest rates in Asia and countries such as Eastern Europe and areas of Central and South America, while it is less frequent in Western countries [1,2]. Overall, gastric cancer represents the second most common cause of death from cancer (approximately 700,000/year) [2]. About two-thirds of the cases occur in developing countries and 42% in China alone [2]. In the United States, 21,259 new cases of stomach cancer were estimated in 2007 [3].

Most gastric cancers develop as sporadic cancers, without a well-defined hereditary predisposition. A small proportion of gastric cancers arise as a consequence of a hereditary predisposition caused by specific inherited germ line mutations in critical cancer-related genes.

Nonhereditary Gastric Cancer

Risk Factors

Most nonhereditary gastric cancers, also referred to as sporadic cancers, arise in a background of chronic gastritis, which is most commonly caused by *H. pylori* infection of the stomach [4–7]. There may be clustering of gastric cancer within some affected families. Family relatives of patients with gastric carcinoma have an increased risk for gastric carcinoma of about 3-fold [8,9]. Patients with both a positive family history and infection with a CagA positive *H. pylori* strain were reported to have a greater than 8-fold risk of gastric carcinoma as compared to others without these risk factors [10].

The pathogenesis of gastric cancer is multifactorial, resulting from the interactions of host genetic susceptibility factors, environmental exogenous factors that have carcinogenic activity such as dietary elements and smoking [11–13], and the complex damaging effects of chronic gastritis (reviewed in [14]). The diets that have been implicated in increased risk of gastric cancer are predominantly salted, smoked, pickled, and preserved foods (rich in salt, nitrite, and preformed N-nitroso compounds), and diets with reduced vegetables and vitamin intake [11].

Chronic infection of the stomach by *H. pylori* is the most common form of chronic gastritis (reviewed in [15]). Therefore, since most gastric cancers develop in a background of chronic gastritis, *H. pylori* is the most significant known risk factor for the development of gastric cancer. *H. pylori* was first implicated as the causal agent of most cases of chronic gastritis and ulcers in

Molecular Pathology © 2009, Elsevier, Inc. All Rights Reserved.

the seminal studies of Warren and Marshal [16,17]. *H. pylori* was classified as a human carcinogen based on a strong epidemiological association of *H. pylori* gastritis and gastric cancer [4,18–20]. Overall, the risk of gastric cancer in patients with *H. pylori* gastritis is 6-fold higher than that of the population without *H. pylori* gastritis [21]. The risk of gastric cancer increases exponentially with increasing grade of gastric atrophy and intestinal metaplasia, the risk of gastric cancer being about 90 times higher in patients with severe multifocal atrophic gastritis affecting the antrum and corpus of the stomach than in individuals with normal noninfected stomachs [22]. However, even nonatrophic gastritis, as compared to healthy non-*H. pylori* infected individuals, raises the gastric cancer risk to approximately 2-fold [5].

Additional evidence supporting the association of *H. pylori* and gastric cancer includes (i) prospective studies demonstrating gastric cancer development in 2.9% of infected patients over a period of about 8 years, and in 8.4% of patients with extensive atrophic gastritis and intestinal metaplasia during a 10-year surveillance [23–25]; (ii) development of animal models that develop gastric cancer associated with *H. pylori* infection, including Mongolian gerbils and mice [26–30]; and (iii) eradication of *H. pylori* infection in patients with early gastric cancer resulted in the decreased appearance of new cancers [31,32]. *H. pylori* eradication reduced the incidence of gastric cancer in patients without atrophy and intestinal metaplasia at baseline, suggesting that eradication may contribute to preventing progression to gastric cancer [33].

Stepwise Progression of H. pylori *Gastritis to Gastric Carcinoma: Histologic Changes of the Gastric Mucosa*

The chronicity associated with *H. pylori* gastritis is critical to the carcinogenic potential of the infection. *H. pylori* infection is generally acquired during childhood and persists throughout life unless the patient undergoes eradication treatment [34–37]. Gastric cancer develops several decades after acquisition of the infection, associated with the progression of mucosal damage with development of specific histological alterations [38].

H. pylori infection of the stomach activates both humoral and cellular inflammatory responses within the gastric mucosa involving dendritic cells, macrophages, mast cells, recruitment and expansion of T-lymphocytes and B-lymphocytes, and neutrophils [39,40]. Despite a continuous inflammatory response, *H. pylori* organisms are able to evade the host immune mechanisms and persist in the mucosa, causing chronic gastritis.

Histologically, the progression of *H. pylori* associated chronic gastritis to gastric cancer is characterized by a stepwise acquisition of mucosal changes, starting with chronic gastritis, progressive damage of gastric glands resulting in mucosal atrophy, replacement of normal gastric glands by intestinal metaplasia, and development of dysplasia and carcinoma in some patients (Figure 19.1) [4,19,41–43].

The potential role of bone marrow-derived stem cells in chronic gastritis and *H. pylori* associated neoplastic progression has recently been proposed based on studies in animal models [30,44]. The current hypothesis is that *H. pylori* associated inflammation and glandular atrophy create an abnormal microenvironment in the gastric mucosa that favors engraftment of bone marrow-derived stem cells into the inflamed gastric epithelium. It is postulated that engrafted bone marrow-derived stem cells do not follow a normal differentiation pathway and undergo uncontrolled replication, progressive loss of differentiation, and neoplastic behavior [30,44–46]. However, the potential role of bone marrow-derived stem cells in human disease remains unclear.

Stomach cancers are classified according to the World Health Organization classification based on their grade of differentiation into well, moderately, and poorly differentiated adenocarcinomas [47,48]. In addition, gastric adenocarcinomas can be categorized into intestinal and diffuse types (Lauren's classification), based on the morphologic features on H&E stained tumor sections [47–49].

Gastric cancers arising on the inflammatory background of *H. pylori*-associated chronic gastritis are most commonly intestinal type adenocarcinoma, which are predominantly well to moderately differentiated adenocarcinomas, but diffuse type tumors, which are poorly differentiated or are signet ring cell carcinomas, also occur, and may develop in patients with gastric ulcers and chronic active gastritis [20,23,50] (Figure 19.1).

Progression to gastric cancer is higher in patients with extensive forms of atrophic gastritis with intestinal metaplasia involving large areas of the stomach, including the gastric body and fundus. This pattern of gastritis has been described as pangastritis or multifocal atrophic gastritis [6,7,22,51,52]. Extensive gastritis involving the gastric body and fundus results in hypochlorhydria, allowing for bacterial overgrowth and increased carcinogenic activity in the stomach through the conversion of nitrites to carcinogenic nitroso-N compounds [53,54]. *H. pylori*-associated pangastritis is frequently seen in the family relatives of gastric cancer patients, which may contribute to gastric cancer clustering in some families [55].

Molecular Mechanisms Underlying Gastric Epithelial Neoplasia Associated with H. pylori *Infection*

How *H. pylori* gastritis promotes gastric carcinogenesis involves an interplay of mechanisms that include (i) a longstanding inflammation in the mucosa, with increased oxidative damage of gastric epithelium, (ii) a number of alterations of epithelial and inflammatory cells induced by *H. pylori* organisms and by released bacterial products, (iii) inefficient host response to the induced damage, and (iv) mechanisms of response related to host genetic susceptibility, which may mediate the variable levels of damage in different individuals.

H. pylori bacterial products and factors released by activated or injured epithelial and inflammatory cells both contribute to persistent chronic inflammatory

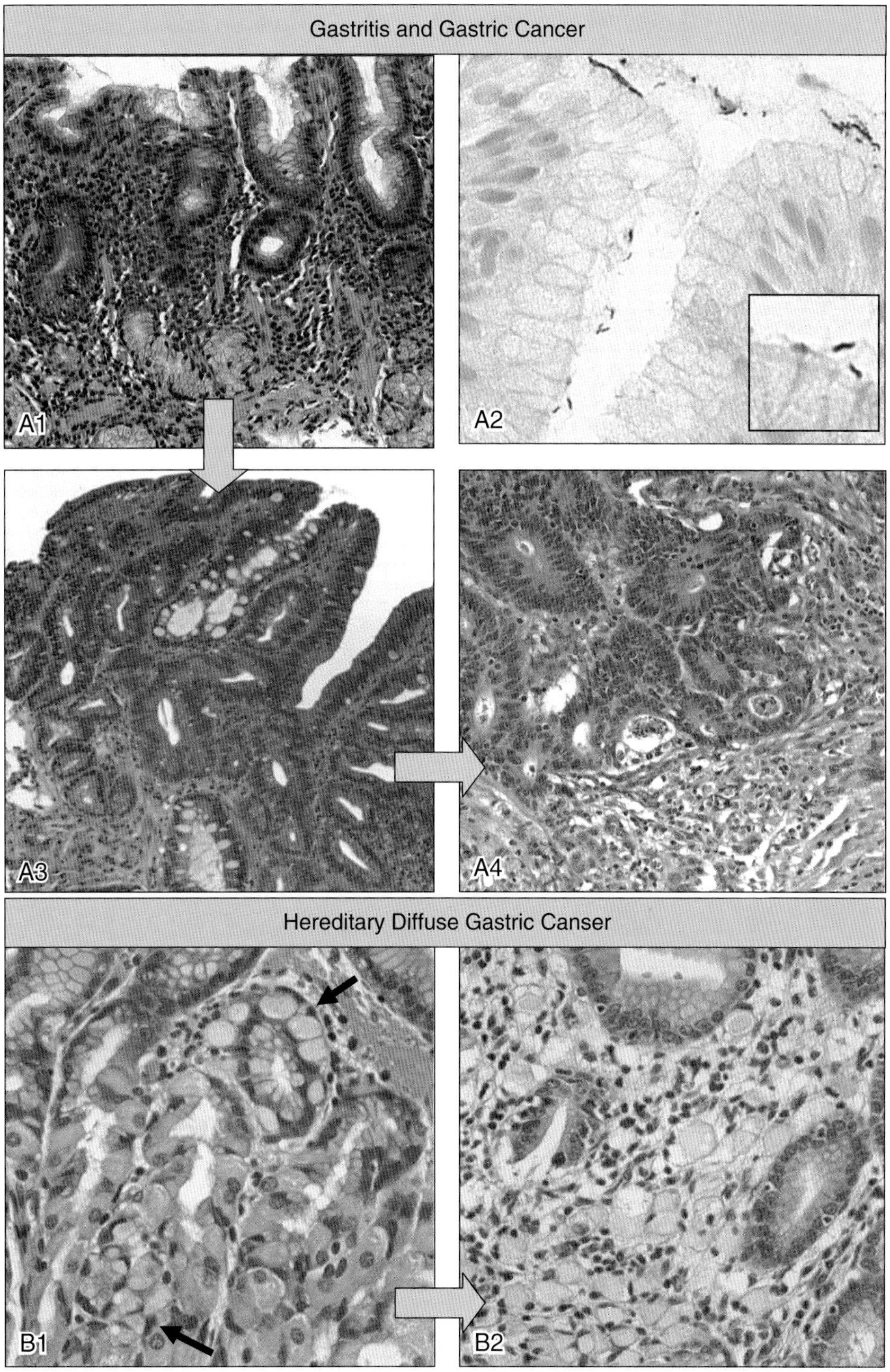

Figure 19.1 **Gastric carcinogenesis: Stepwise Progression of *H. pylori*-Associated Gastric Cancer (Panels A1–A4), and Hereditary Diffuse Gastric Cancer (Panels B1 and B2).** Panel A1. Chronic active gastritis involving the mucosa of the gastric antrum (H&E stain, original magnification 10×); Panel A2. Immunohistochemical stain highlights *H. pylori* organisms with typical S and comma shapes, seen at higher magnification in the inset. *H. pylori* organisms typically appear attached or adjacent to the gastric surface and foveolar epithelium (original magnification 40×); Panel A3. Gastric mucosa with intestinal metaplasia and low-grade dysplasia/adenoma (H&E stain, original magnification 10×); Panel A4. Gastric carcinoma of intestinal type (moderately differentiated adenocarcinoma) (H&E stain, original magnification 10×); Panels B1 and B2 (Courtesy of Dr. Adrian Gologan, Jewish General Hospital, McGill University). Gastric mucosa of patient with hereditary diffuse gastric cancer (HDGC) with *in situ* signet ring cell carcinoma (arrow) (B1) and invasive signet ring cell carcinoma expanding the lamina propria between the gastric glands (E), (H&E stain, original magnification 20×).

response involving the activation of innate and acquired immune responses in the infected gastric mucosa [39,40]. This chronic and continuously active mucosal inflammatory infiltrate is in part responsible for potential damage to the epithelium, through the release of oxygen radicals [56], and the production of chemokines that may alter the normal regulation of molecular signaling in epithelial cells. *H. pylori* organisms and released bacterial products, including the *H. pylori* virulence factors (CagA and VacA), which may directly alter the gastric epithelial cells as well as inflammatory cells, leading to alterations of signaling pathways, gene transcription, and genomic modifications. These changes lead to modifications in cell behavior, such as increased apoptosis and proliferation, as well as increased rates of mutagenesis [57–62]. Once the neoplastic program is activated, then gastric cancer may progress through mechanisms similar to gastric cancers that do not arise in a background of *H. pylori* infection.

A number of genetic susceptibility factors that increase the risk of gastric cancer development in *H. pylori* infected patients have been identified. Interleukin-1 (IL-1) gene polymorphisms, in IL-1beta and IL-1RN (receptor antagonist), have been shown to increase the risk of gastric cancer and gastric atrophy in the presence of *H. pylori*. Individuals with the IL-1B-31*C or IL-1B-511*T, and the IL-1RN*2/*2 genotypes are at increased risk of hypochlorhydria, gastric atrophy, and gastric cancer in response to *H. pylori* infection [63–66]. A gene polymorphism that may affect the function of OGG1, a protein involved in the repair of mutations induced by oxidative stress, was reported frequently in patients with intestinal metaplasia and gastric cancer, suggesting that deficient OGG1 function may contribute to increased mutagenesis during gastric carcinogenesis [67].

Mechanisms and Spectrum of Epigenetic Changes, Mutagenesis, and Gene Expression Changes of the Gastric Epithelium Induced by H. pylori *Infection*

Epigenetic modification and mutagenesis precede tumor development and accompany neoplastic progression during gastric carcinogenesis. Both of these changes have been shown to occur in *H. pylori* gastritis and subsequent preneoplastic and neoplastic mucosal lesions (reviewed in [14]). The combined effects of epigenetic modifications, mutagenesis, and functional gene expression changes in gastric epithelial cells in response to inflammatory mediators and bacterial virulence factors result in abnormal gene expression and function in the various stages of progression to gastric cancer (gastritis, intestinal metaplasia, dysplasia, and cancer). The effects of *H. pylori* organisms on gastric epithelial cells are likely to occur primarily during the phase of gastritis and intestinal metaplasia, while additional molecular events associated with neoplastic progression from dysplasia to invasive cancer may be independent of *H. pylori*. However, the background inflammatory milieu associated with ongoing chronic infection may influence the mechanisms of neoplastic progression.

During gastric carcinogenesis a number of genes are regulated by CpG methylation of the promoter regions at CpG sites, with potential promoter inactivation (reviewed in [14]). For example, Kang et al. described five different classes of methylation behaviors in chronic gastritis, intestinal metaplasia, gastric adenoma, and gastric cancer: (i) genes methylated in gastric cancer only (*GSTP1* and *RASSF1A*); (ii) genes showing low methylation frequency in chronic gastritis, intestinal metaplasia, and gastric adenoma, but significantly higher methylation frequency in gastric cancer (*COX-2*, *MLH1*, and *p16*); (iii) low and similar methylation frequency in all gastric lesions (*MGMT*); (iv) genes with high and similar methylation frequency in all gastric lesions (*APC* and *E-cadherin*); and (v) genes showing an increasing methylation frequency along the oncogenic progression (*DAP-kinase*, *p14*, *THBS1*, and *TIMP3*) [68]. Additional loci demonstrating an association of *H. pylori*-chronic gastritis and CpG island hypermethylation were recently reported [69]. In experimental conditions, gastric epithelial cells co-cultured with *H. pylori* undergo CpG methylation at several gene loci, including the *MLH1* gene promoter, which may contribute to the observed deficiency of DNA mismatch repair associated with *H. pylori* infection [62].

The mechanisms that regulate CpG methylation and gene silencing during *H. pylori*-associated gastritis and resulting gastric mucosal lesions are not currently known. Recent studies showed that proinflammatory interleukin-1-beta polymorphisms were associated with CpG island methylation of target genes [70], and CpG methylation of the *E-cadherin* promoter was induced in cells treated with IL-1beta [71]. These data suggest that components of the inflammatory cascade induced by *H. pylori* may contribute to orchestration of the epigenetic response in *H. pylori*-associated carcinogenesis.

Mutations are likely to accumulate during *H. pylori* chronic gastritis because of increased damaging factors in the mucosa and also because of overall deficiency of some DNA repair functions (reviewed in [14]). DNA damage during *H. pylori* gastritis is caused primarily by reactive oxygen species (ROS), and reactive nitrogen species (RNS). Additionally, when mucosal atrophy develops, the resulting reduced acid levels may allow the overgrowth of other bacteria and activation of environmental carcinogens with mutagenic activity.

ROS are generated by inflammatory cells, as well as by gastric epithelial cells after activation by *H. pylori* bacterial products and cellular released cytokines. The increased ROS levels are associated with increased expression of inducible nitric oxide synthase (iNOS) and increased production of nitric oxide (NO) [72–75]. Increased cyclooxygenase (COX2) has also been reported in *H. pylori*-associated gastritis and may contribute to increased mutagenesis through oxidative stress [72–75]. Further, with reduced level of oxygen radical scavengers, such as glutathione and glutathione-S-transferase, relatively higher levels of oxygen radicals may accumulate in the mucosa of *H. pylori* infected patients [76]. DNA 8-hydroxydeoxyguanosine (8OHdG) can be used as a marker for oxidative DNA damage [61,77]. The gastric mucosa with *H. pylori* gastritis and preneoplastic lesions (intestinal metaplasia and atrophy) contain increased levels of DNA 8-hydroxydeoxyguanosine (8OHdG) and the levels of 8OHdG in the gastric mucosa significantly decrease after eradication of *H. pylori*, supporting the role of active infection in the accumulation of mutations [61,77]. Mutations associated with oxidative damage include point mutations in genes involved in gastric carcinogenesis such as *p53* [78].

Additionally, mutation accumulation during *H. pylori* gastritis may be enhanced because of a relatively deficient DNA repair system, with persistence of ROS-induced mutations and uncorrected DNA sequence replication errors that are transmitted to future epithelial cell generations. Several DNA repair systems are required for correction of DNA damage occurring during *H. pylori* gastritis: (i) the DNA mismatch repair system, which repairs DNA replication-associated sequence errors, and (ii) several other proteins that primarily repair DNA lesions induced by oxidative and nitrosative stress, including MGMT and OGG1 glycosylase. The DNA mismatch repair system functions through the action of MutS proteins (MSH2, MSH3, and MSH6) and MutL proteins (MLH1, PMS1, PMS2, and MLH3) [79–81]. DNA mismatch repair deficiency leads to frameshift mutagenesis, which can generate mutations in the coding region of genes, as well as in repetitive regions known as short tandem repeats or microsatellite regions. The mutations in microsatellite regions result in microsatellite instability (MSI), which can be used as surrogate markers of DNA mismatch repair deficiency [82]. High levels of microsatellite instability (MSI-H), defined as instability in greater than 30% of the microsatellite markers in a panel of 5 microsatellite loci, correlate well with loss of DNA mismatch repair function, and can be detected in tumor tissues with generalized loss of expression of one of the main DNA mismatch repair proteins (most commonly MLH1 or MSH2) [82]. Several studies have reported a role of relative DNA mismatch repair deficiency in the mutation accumulation during *H. pylori* infection [62,83–85]. In experimental conditions, when gastric epithelial cells are co-cultured with *H. pylori* organisms, the levels of DNA mismatch repair proteins, including MSH2 and MLH1, are greatly reduced, and both point mutations and frameshift/microsatellite type mutations accumulate in the *H. pylori* exposed cells [62,83]. *In vivo* microsatellite instability was reported in 13% cases of chronic gastritis, 20% of intestinal metaplasias, 25% of dysplasias, and 38% of gastric cancers, indicating a stepwise accumulation of MSI during gastric carcinogenesis [86]. Microsatellite instability has been detected in intestinal metaplasia from patients with gastric cancer in several studies, indicating that MSI can occur in preneoplastic gastric mucosa [84,87–91]. Several studies reported that patients with MSI-positive tumors showed a significantly higher frequency of active *H. pylori* infection, which supports the notion that *H. pylori* infection can underlie DNA mismatch repair deficiency and MSI in the various steps of gastric carcinogenesis [84,92,93].

Deficient function of other genes involved in the repair of oxidative stress-induced mutations may also contribute to mutagenesis during *H. pylori* gastritis. Repair of 8-OHdG is accomplished by DNA repair proteins including a polymorphic glycosylase (OGG1). This protein may be less efficient in carriers of a gene polymorphism that was reported frequently in patients with intestinal metaplasia and gastric cancer [67]. O6-methylguanine-DNA methyltransferase (MGMT) function includes the repair of O6-alkylG DNA adducts. In the absence of functional MGMT, these adducts mispair with T during DNA replication, resulting in G-to-A mutations. MGMT-promoter methylation has been reported in a subset of cases of *H. pylori* gastritis and in various stages of gastric carcinogenesis, suggesting a possible role for this DNA repair protein in gastric carcinogenesis [94].

Gene expression analysis with microarrays has revealed expression signatures associated with *H. pylori* gastritis and gene expression induced by the effect of *H. pylori* organisms in gastric epithelial cells, confirming the increased expression of some genes by epithelial cells in response to *H. pylori* (such as IL-8) and expanding the knowledge of signaling pathways involved in *H. pylori* pathogenesis [95–97].

Mutational, Epigenetic, Gene Expression and MicroRNA Patterns of Gastric Intestinal Metaplasia, Dysplasia/Adenoma, and Cancer

Intestinal metaplasia, dysplasia, and carcinoma represent cell populations with a clonal origin that manifest epigenetic and genetic alterations incurred by the non-neoplastic epithelium, as well as additional events that occur during neoplastic progression. Figure 19.2 represents the main molecular events that characterize gastric carcinogenesis.

Mutational events during gastric neoplastic development and progression include MSI-type mutations, chromosomal instability manifesting as loss of heterozygosity (LOH), and point mutations of cancer-related genes. High-level of MSI is associated with loss of expression and promoter hypermethylation of *MLH1* in gastric adenomas and cancer [98–101]. MSI was reported in 17%–35% of gastric adenomas [85,102,103], and in 17%–59% of gastric carcinomas [84,85,102–107]. Gastric tumors with MSI-H may also harbor frameshift mutations in the coding regions of cancer-related genes, such as *BAX*, *IGFRII*, *TGFβRII*, *MSH3*, and *MSH6* [89,108–111]. In MSI-H adenomas, frameshift mutations of *TGFβRII* were detected in 38%–88% of the cases, *BAX* in 13%, *MSH3* in 13%, and *E2F-4* in 50% of the cases [112,113].

Point mutations and LOH at multiple gene loci have been detected in intestinal metaplasia, adenomas, and gastric carcinomas [91]. Mutations of *p53* and *APC* genes have been reported in intestinal metaplasia and gastric dysplasia [78,114–117]. *p53* mutations at exons 5 to 8 resulting in G:C to A:T transitions are detected in gastric carcinogenesis [78,116]. *APC* mutations, including stop-codon and frameshift mutations, were reported in 46% of gastric adenomas and 5q allelic loss in 33% of informative cases of gastric adenoma [103] and in 45% of carcinomas [118]. K-*ras* mutations at codon 12 were reported in

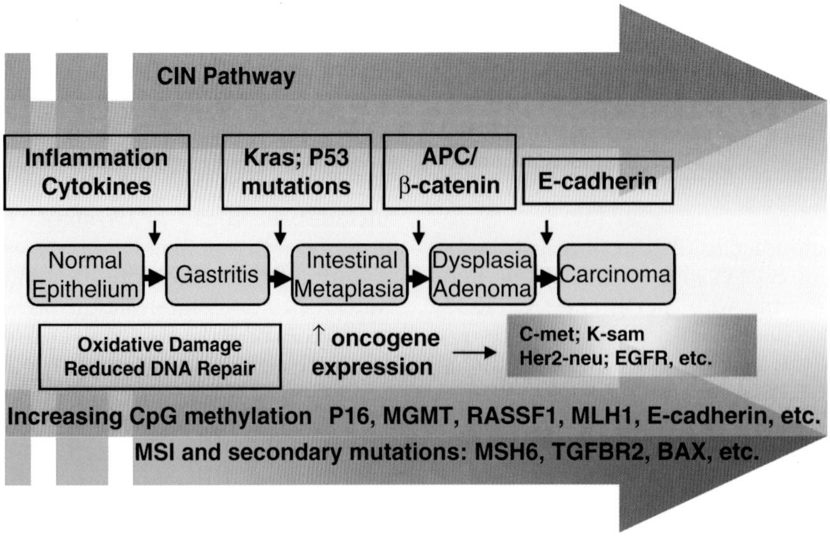

Figure 19.2 Stepwise Progression of Molecular Events During Gastric Carcinogenesis.

14% of biopsies with atrophic gastritis and in less than 10% of adenomas, dysplasia, and carcinomas [104,119].

Gene regulation through epigenetic modification occurs at multiple steps of gastric carcinogenesis. Variations of CpG methylation during the steps of disease progression from *H. pylori* gastritis, intestinal metaplasia, and gastric cancer have been observed for some genes, but not for others (reviewed in [14]). Genes that play a role in cell cycle progression, DNA repair, cell adhesion, and a number of tumor suppressor genes may be regulated by epigenetic mechanisms through promoter methylation. In gastric carcinogenesis CpG island methylation occurs in genes such as *MLH1*, *p14*, *p15*, *p16*, *E-cadherin*, *RUNX3*, thrombospondin-1 (*THBS1*), tissue inhibitor of metalloproteinase 3 (*TIMP-3*), *COX-2*, and *MGMT* [68,120–124]. As pointed out in the previous section, Kang et al. examined the methylation status of 11 genes, demonstrating the association of methylation and chronic inflammation in the gastric mucosa [125]. A number of genes were re-expressed in gastric cancer cell lines after treatment with the demethylating agent 5-aza-2′-deoxycytidine, identifying putative candidate genes involved in gastric carcinogens through epigenetic silencing [126]. Seventeen new DNA methylation markers of gastric cancer, which may serve as useful markers that may identify a distinct subset of gastric cancer, were recently reported [69].

Genome-wide gene expression analysis with microarrays has yielded a huge amount of information on gene expression of the lesions associated with gastric carcinogenesis, with identification of specific profiles that characterize gastritis, intestinal metaplasia, intestinal versus diffuse type gastric adenocarcinoma, and different gastric cancer prognostic groups [127–133]. In several studies, microarrays were used to characterize the gene expression levels in gastric cancer, chronic gastritis, and intestinal metaplasia, and identified unique expression patterns defining preneoplastic lesions and tumor subtypes [96,134,135]. Jinawath et al. reported a signature of diffuse-type cancers which exhibited altered expression of genes related to cell-matrix interaction and extracellular-matrix components, whereas intestinal-type cancers had a pattern of enhancement of cell growth [135]. Meireles et al. reported several combinations of genes that could discriminate between normal and tumor samples, and intestinal metaplasia cases were characterized by a gene expression signature resembling that of adenocarcinoma, supporting the notion that intestinal metaplasia tissue might progress to cancer [96,97]. The lack of reproducibility of studies using microarrays for gene expression has limited the impact of this technology to cancer applications. Further, many of the observed expression signatures await additional studies to confirm their significance in carcinogenesis, and will require integration of expression data using powerful computational methods.

MicroRNAs (miRNAs) are small noncoding RNAs that have been shown to regulate gene expression and may be aberrantly expressed in cancer. There are only a few studies addressing miRNA alterations in gastric carcinogenesis. Available data indicates that most gastric cancers showed overexpression of *miR-21* [136], while *miR-218–2* was consistently downregulated [137]. The functional implications of these findings remain unclear and await further studies.

Familial Gastric Cancer

Genetic and Molecular Basis of Familial and Hereditary Gastric Cancer

Familial gastric cancer may be inherited as an autosomal dominant disease, occurring as the main tumor in hereditary diffuse gastric cancer (HDGC) [138,139] or as one of the types of tumors in a number of cancer predisposing syndromes [140], but it may also occur as family clustering of gastric cancer for which the etiology is likely multifactorial [8,9,141]. Familial gastric cancer represents less than 10% of all stomach cancers [142]. HDGC associated with germline mutations in the

E-cadherin gene account for 30% to 40% of known cases of hereditary diffuse gastric cancer [139,143–146]. In up to 70% of the cases of familial gastric cancer, the underlying genetic defect is unknown [147]. Gastric carcinomas represent one of the types of tumors occurring in the following hereditary cancer syndromes: (i) Germline mutations in the *E-cadherin* gene (CDH1) underlie some but not all HDGC families of note, but germ-line mutation in the *E-cadherin* gene does not occur in hereditary intestinal type gastric cancer families [147]; (ii) Li-Fraumeni syndrome, associated with germline *p53* mutations [148]; (iii) Hereditary nonpolyposis colon cancer (HNPCC) [148], with most HNPCC-associated gastric cancers representing the intestinal type [149,150]; (iv) Gastric cancer may also occur in Peutz-Jeghers syndrome (PJS) [151,152], where hamartomatous polyps in the stomach occur in approximately 24% of patients, but the overall risk of gastric cancer is small [151,152]; (v) Patients with familial adenomatous polyposis (FAP) frequently develop fundic gland polyps in the stomach, and may develop gastric adenomatous polyps in about 10% of individuals with FAP, but the risk of gastric cancer is small [153].

Hereditary Diffuse Gastric Cancer: Genetic Basis

The *CDH1 gene*, which encodes the protein E-cadherin, is the only gene known to be associated with HDGC [139,143–146]. Mutations in other genes may account for susceptibility to HDGC, but the evidence is limited [154]. Gastric cancer occurs in 5.7% of families with the *BRCA2* 6174delT mutation [155], but the type of gastric cancer in these families was not characterized.

The human *CDH1* gene consists of 16 exons that span 100 kb [156]. An excess of 30 germline pathologic mutations has been reported in families with HDGC [143,157]. The identified mutations are scattered throughout the gene and are truncating mutations, caused by frameshift mutations, exon/intron splice site mutations, point mutations, and missense mutations [138,139,143,148,157].

Hereditary Diffuse Gastric Cancer: Molecular Mechanisms, Clinical and Pathologic Features

Natural History and Pathologic Features The average age of diagnosis of HDGC is 38 years, ranging from 14–69 years, with most cases occurring before the age of 40 years [144,158]. The lifetime risk of gastric cancer, by age 80 years is 67% for men and 83% for women. In addition to gastric cancer, women also have a 39% risk for lobular breast cancer [144,158].

Histologically the adenocarcinomas in patients with HDGC are characteristically poorly differentiated carcinomas with signet ring morphology (signet ring cell carcinomas) [138,142,159]. Tumor foci are initially confined mostly in the superficial zone of the gastric mucosa and appear to arise in the lower proliferative zone of the gastric foveolae. Tumor foci may be multifocal. *In situ* signet ring cell carcinoma, characterized by disorderly oriented signet ring cells within glands or foveolar epithelium, may be observed (Figure 19.1). Grossly, the tumors extend through the gastric layers and gastric wall with the development of so-called linitis plastica. These tumors do not arise in a background of intestinal metaplasia and gastric atrophy or dysplastic preneoplastic lesions and are not involved in the progression to cancer. E-cadherin and β-catenin immunohistochemistry shows reduced to absent expression in both the *in situ* and invasive areas of the tumor [142].

Molecular Mechanisms and Pathologic Correlates In hereditary diffuse gastric cancer with known germ line mutations, E-cadherin loss of function caused by pathologic mutations underlie the development of cancer. However, up to 70% of cases of familial gastric cancer of diffuse type do not have a well-defined genetic defect. In HGDC cases, loss of E-cadherin expression is also associated with a transcriptional downregulation of the wild-type *CDH1* allele by promoter hypermethylation [160].

E-cadherin is a member of the cadherin family of transmembrane glycoproteins, characterized by five extracellular domains and a cytoplasmic domain [161]. E-cadherin has been shown to be essential for establishing and maintaining the polarized differentiated organization of epithelia. It plays important roles in signal transduction, gene expression, differentiation, and cell motility. The activity of E-cadherin in cell adhesion is dependent upon its association with the actin cytoskeleton through the interaction with the catenin (α-catenin, β-catenin, and γ-catenin) family of proteins [162,163]. Loss of E-cadherin expression is seen in most diffuse gastric cancers, including sporadic-type gastric carcinomas of diffuse type, and in lobular breast cancer [164,165]. The E-cadherin/catenin complex is important to suppress invasion and metastasis and cell proliferation [166]. Somatic mutation of E-cadherin is associated with increased activation of epidermal growth factor receptor (EGFR) followed by enhanced recruitment of the downstream signal transduction, and activation of *Ras* [167]. The activation of EGFR by E-cadherin mutants in the extracellular domain explains the enhanced motility of cancer cells in the presence of an extracellular mutation of E-cadherin [167].

Hereditary Diffuse Gastric Cancer: Genetic Testing and Clinical Management Criteria for consideration of *CDH1* molecular genetic testing in individuals with diffuse gastric cancer have been recommended [140,147,157]. These criteria are applicable to North America, Northern Europe, and other regions of low gastric cancer incidence, but they may be too broad in regions of high gastric cancer incidence. The criteria are as follows: (i) two or more cases of gastric cancer in a family, with at least one diffuse gastric cancer diagnosed before 50 years of age; (ii) three or more cases of gastric cancer in a family, diagnosed at any age, with at least one documented case of diffuse gastric cancer; (iii) an individual diagnosed with diffuse gastric cancer before 45 years of age; (iv) an individual diagnosed with both diffuse gastric cancer and lobular breast cancer (but no other criteria met); (v) one family member diagnosed with diffuse gastric cancer and another with lobular breast cancer (but no other criteria met); (vi) one family member diagnosed with diffuse gastric cancer and another with signet ring colon cancer (but no other criteria met).

Sequence analysis of the *CDH1* gene is the current test recommended for confirmation of the diagnosis. The clinical management of individuals who have an identified *CDH1* cancer-associated mutation varies from intense surveillance for early detection of gastric cancer or prophylactic gastrectomy.

COLORECTAL CANCER

Colorectal cancer is one of the most frequent types of cancer. Worldwide colorectal cancer ranks fourth in frequency in men and third in women, with approximately one million cases annually [2]. Men and women are similarly affected [2]. In the United States there are approximately 150,000 new cases of colorectal cancer per year [168,169]. Most colorectal cancers do not develop in association with a hereditary cancer syndrome and are known as sporadic colon cancer. Several hereditary colon cancer syndromes have been characterized (Table 19.1). The most frequent is Lynch syndrome or HNPCC, representing 3%–4% of colorectal cancers [170]. Familial adenomatous polyposis (FAP) represents about 1%, and the remaining cancer syndromes (Table 19.1) are responsible for less than 1% of colorectal cancers [171].

Sporadic Colon Cancer

Colon cancers generally arise from precursor lesions of the colonic epithelium described histologically as dysplasia or adenoma, with progression to high-grade dysplasia (previously referred to as carcinoma *in situ*) and invasive adenocarcinomas (Figure 19.3). Adenomas are classified into tubular, tubulovillous, or villous adenomas. The epithelium that constitutes colonic adenomas displays cytologic features of dysplasia. In the initial steps, adenomas consist of epithelium with low-grade dysplasia, which may progress to high-grade dysplasia and invasive adenocarcinomas (Figure 19.3). Recently, another type of neoplastic lesions described as serrated adenomas has been described in the colon [172–174] (Figure 19.4).

The molecular pathways of colon cancer development include a stepwise acquisition of mutations, epigenetic changes, and alterations of gene expression, resulting in uncontrolled cell division and manifestation of invasive neoplastic behavior. The molecular changes underlying colonic neoplasia correlate relatively well with histopathological variants. However, more than one pathway can lead to the development of adenocarcinomas with similar morphology. The

Table 19.1 Gastrointestinal Hereditary Cancer Syndromes and Other Polyposis Syndromes [309,317]

Syndrome	Inheritance	Key Clinical Features	Gene	Gene Product Function
Familial Adenomatous Polyposis and Variants:	Autosomal dominant	>100 colonic adenomas, near 100% lifetime risk for colorectal carcinoma	APC	Growth inhibitory: β-catenin sequestration; targeting of β-catenin for destruction.
- Gardner		FAP plus CHRPE, osteomas, desmoid tumors		
- Turcot*		FAP plus medulloblastoma, glioblastomal		
- Attenuated FAP		>15 but <100 colonic adenomas		
MYH-associated polyposis	Autosomal recessive	FAP-like presentation; no APC mutation identifiable	MYH	DNA damage repair
Hereditary nonpolyposis colorectal cancer (HNPCC; Lynch syndrome) and variants:	Autosomal dominant	70%–85% lifetime risk for colorectal carcinoma; predisposition for extracolonic malignancy including endometrial	MSH2 MLH1 MSH6 PMS2 MLH3	DNA mismatch repair
- Muir-Torre		HNPCC plus sebaceous tumors and extracolonic malignancies		
Peutz-Jeghers	Autosomal dominant	Hamartomatous gastrointestinal polyps; predisposition for multiple extracolonic malignancies	STK11	Serine/threonine kinase
Cowden and **BRR syndrome	Autosomal dominant	Hamartomatous gastrointestinal polyps	PTEN	Protein phosphatase
Juvenile Polyposis	Autosomal dominant	Hamartomatous gastrointestinal polyposis; increased risk for colorectal carcinoma	SMAD4 BMPR1A ENG	TGF beta signaling

CHRPE: Congenital Hypertrophy of the Retinal Pigment Epithelium.
*Two-thirds of Turcot syndrome occur in patients with APC gene mutation and one third in DNA mismatch repair gene mutation.
** BRR: Bannayan-Ruvalcaba-Riley syndrome

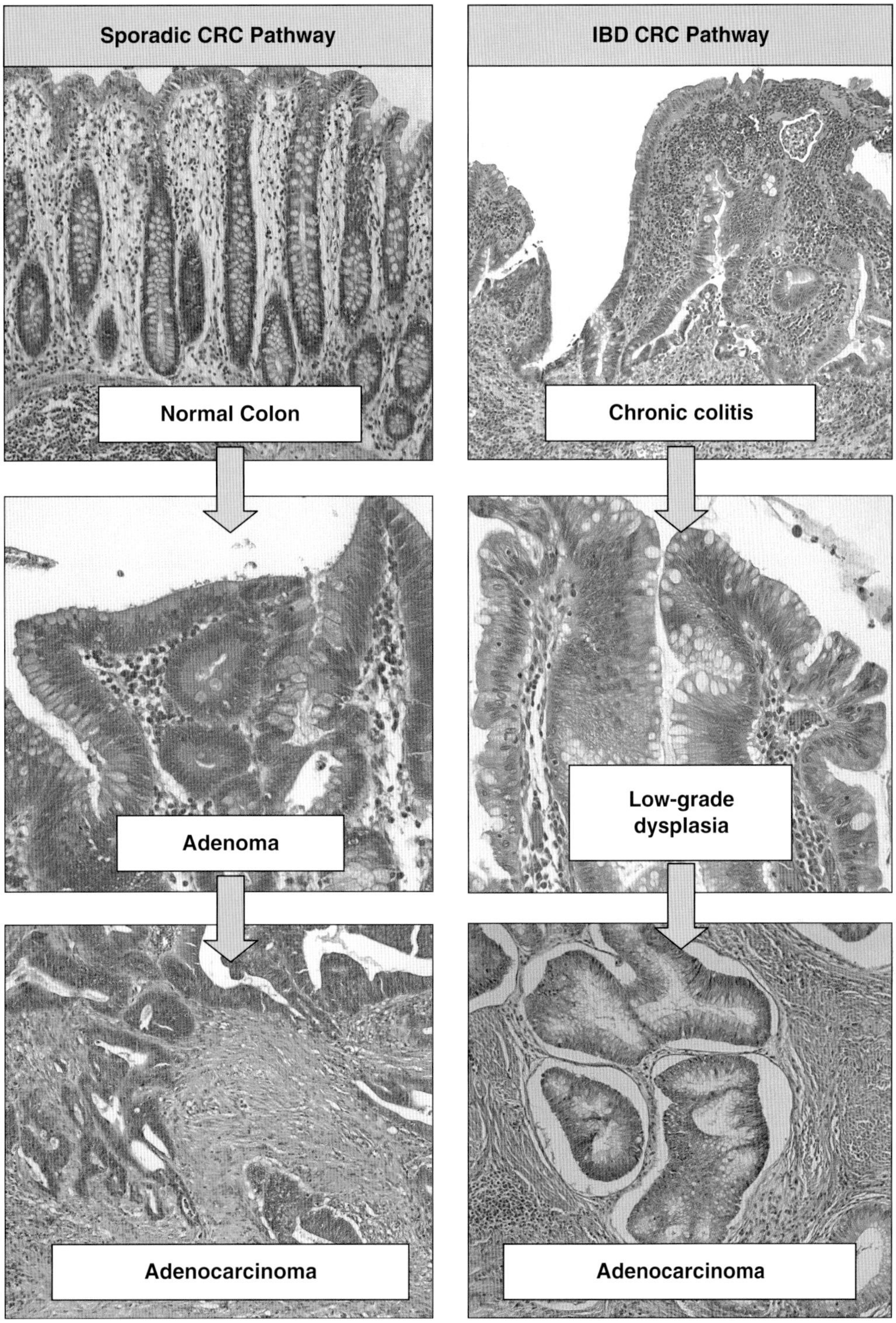

Figure 19.3 **Stepwise Progression of Colorectal Neoplasia: Progression of Neoplasia in Sporadic Colorectal Cancer and in Inflammatory Bowel Disease-Associated Colitis.** In the sporadic colorectal cancer pathway, adenomas characterized in the early stages by low-grade epithelial dysplasia precede the development of high-grade dysplasia, which may then progress to invasive adenocarcinoma. In IBD-associated neoplasia, the background colonic mucosa reveals variable degrees of chronic colitis, and eventually foci of low-grade dysplasia develop, which in turn may progress to high-grade dysplasia and invasive adenocarcinoma. The morphologic features of the neoplastic lesions significantly overlap between sporadic colorectal cancer and IBD-associated neoplasia, but the inflammatory environment that characterizes chronic colitis dictates a number of different molecular mechanisms of neoplastic development and progression.

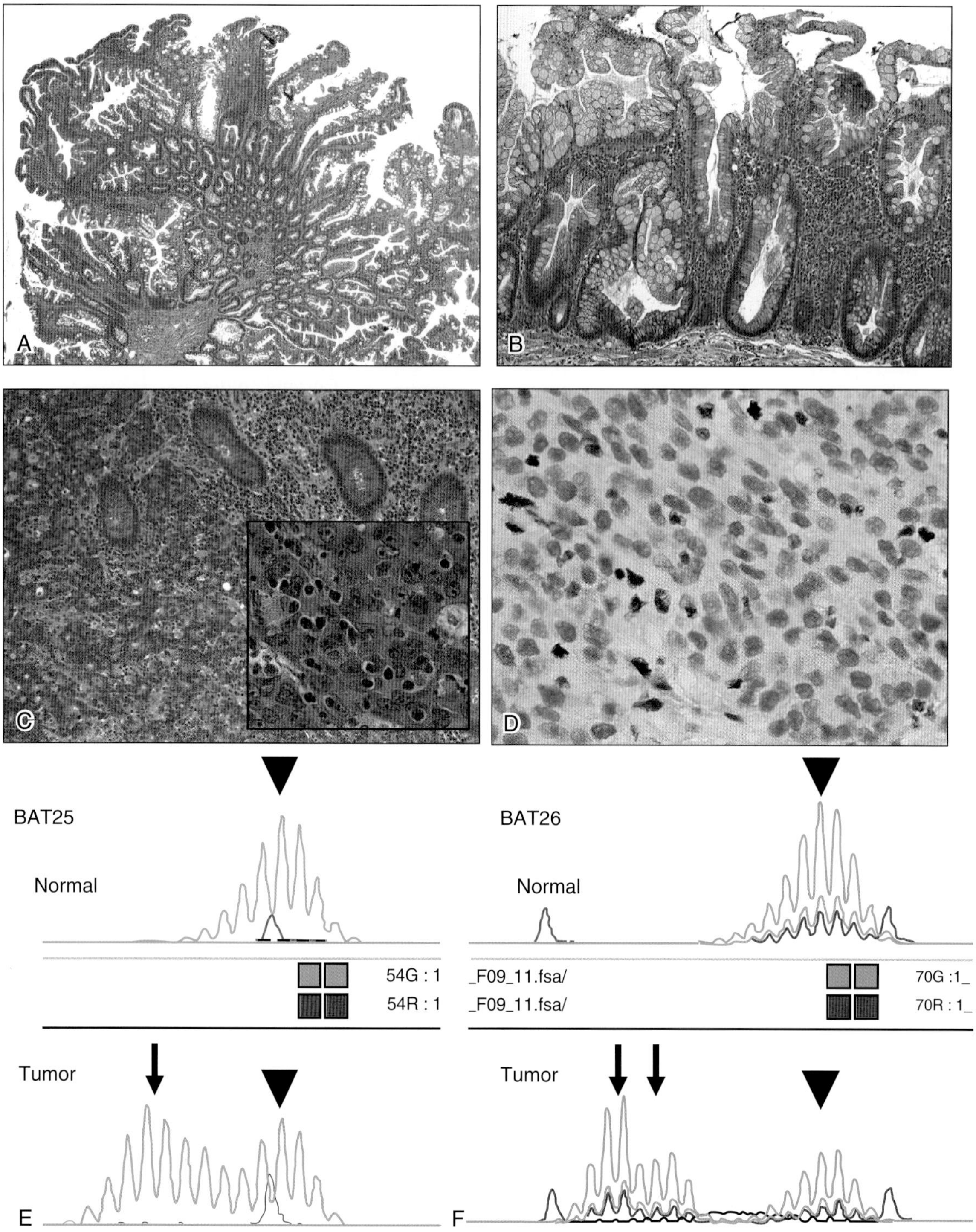

Figure 19.4 Colorectal Cancer Pathways: Histopathology and Molecular Correlates. Panels A and B: The Serrated Pathway. Adenocarcinomas that develop through the serrated pathway arise from serrated polyps that include traditional serrated adenomas; (A) and sessile serrated adenomas (B), (H&E stain, original magnification 5× and 10×, for panels A and B, respectively). Traditional serrated adenomas show both serrated architecture and dysplasia similar to that seen in adenomatous mucosa, whereas sessile serrated adenomas reveal architectural abnormalities but no evidence of classic dysplasia. Panels C and D: The Microsatellite Instability Pathway. Poorly differentiated colonic adenocarcinoma with prominent intratumoral lymphocytes, best seen in the inset (C), (H&E stain, original magnification 10×). By immunohistochemistry, the tumor cells are negative for MSH2, while the surrounding lymphocytes and stromal cells show preserved expression of MSH2 protein in the non-neoplastic cell nuclei (D), (Immunohistochemistry, original magnification 10×). (E) The presence of microsatellite instability in the tumor DNA is demonstrated by microsatellite instability at the microsatellite markers *BAT25* and *BAT26* characterized by the appearance of new PCR amplification peaks of smaller size (tailed arrows) as compared to non-neoplastic DNA from the same patient (arrow tip).

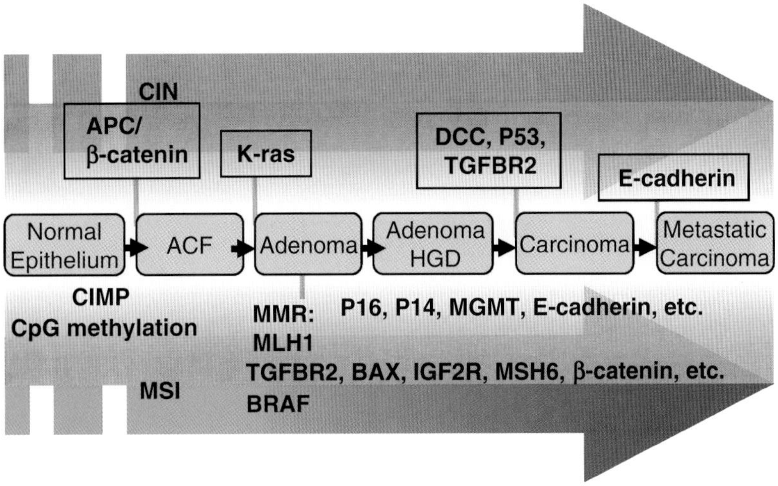

Figure 19.5 Molecular Events in the Stepwise Lesions of Sporadic Colorectal Carcinogenesis.

main molecular pathways of colorectal cancer development characterized to date include (i) the chromosomal instability pathway (CIN) [175–177], (ii) the microsatellite instability pathway (MSI) [177–179], and (iii) the CpG island methylator pathway (CIMP) [180,181] (Figure 19.5).

Most colon cancers progress through a CIN pathway in which large genomic alterations include aneusomy, gains, and losses of chromosomal regions. The tumors that primarily develop through the CIN or tumor suppressor pathway develop frequent cytogenetic abnormalities and allelic losses [182,183] (Figure 19.5). The molecular mechanisms underlying CIN are poorly understood. CIN appears to result from deregulation of the DNA replication checkpoints and mitotic-spindle checkpoints. Mutation of the mitotic checkpoint regulators *BUB1* and *BUBR1*, and amplification of *STK15* are seen in a subset of CIN-colon cancers (reviewed in [184]). In colorectal cancers that follow the CIN or tumor suppressor pathway, the dominant genomic abnormality is inactivation of tumor-suppressor genes, such as *APC* [185] (chromosome 5q), *p53* (chromosome 17p), *DCC* (deleted in colon cancer), *SMAD2*, and *SMAD4* (chromosome 18q) [186,187]. *APC* gene mutations occur early in colonic neoplasia. *APC* mutations are detected in aberrant crypt foci (ACF), the lesion that precedes the development of adenomas, and have been detected in 50% of sporadic adenomas and 80% of sporadic colon cancers [188–190] (Figure 19.5). *DCC* gene loss occurs late in neoplastic progression, with frequent deletion in carcinomas (73%) and in high-grade adenomas (47%) [191,192]. The tumor suppressor gene *p53* is also involved in the later steps of colon carcinogenesis. The p53 protein has DNA binding activity, contains a transcription activation domain, and regulates target genes that mediate cell cycle arrest, apoptosis, and DNA repair. In the sporadic colonic carcinogenesis pathway, *p53* gene mutations occur in adenomas with high grade dysplasia (50%) and carcinomas (75%) [193] (Figure 19.5). The most common mutations are missense point mutations of one allele followed by deletion of the second allele, resulting in LOH of the *p53* gene locus on chromosome 17p [191,194,195].

The activation of oncogenes contributes to both the development and the progression of neoplasia. The *ras* family of oncogenes is frequently activated in cancer. Activation of *ras* leads to activation of several signaling pathways including the Raf/MAPK, PI3K/Akt, and Mekk/JNK pathways. *Ras* mutations in tumors characteristically are point mutations at codons 12, 13, and 61 of K-*ras*, which result in a constitutively activated protein [196]. K-*ras* mutations have been observed in the first stages of colonic neoplasia in aberrant crypt foci [197]. Sporadic, FAP-associated, and CpG island methylator colorectal neoplasms have the highest rates of activating K-*ras* mutations (50%–80%), while K-*ras* mutations are rare in both sporadic MSI-H and HNPCC-associated cancers [198–201]. In sporadic colorectal carcinogenesis, K-*ras* mutation mainly occurs during the formation of ACF [197,202] (Figure 19.5). Of note, in FAP, somatic mutation of *APC* predominantly occurs during ACF formation, followed by K-*ras* mutation [197]. Inactivation of *APC* results in activation of Wnt/beta-catenin signaling, which in turn can induce chromosomal instability in colon cancer [203].

The MSI pathway is seen in approximately 15% of sporadic colorectal cancers [204–206]. The MSI pathway was first characterized in HNPCC or Lynch syndrome [207–210]. Tumors arising through the MSI pathway are characterized by an underlying deficiency of DNA mismatch repair proteins. The main DNA mismatch repair proteins are MLH1, MSH2, MSH6, and PMS2. Loss of function of one of the main DNA mismatch repair proteins results in high levels of mutagenesis, a molecular phenotype described as MSI [211,212]. Through the use of a recommended panel of five microsatellite loci, tumors can be classified by their levels of microsatellite instability. Tumors with high-level MSI (MSI-H) reveal loss of expression of main DNA mismatch repair proteins in tumor cells. In sporadic carcinomas MSI is usually caused by loss of expression of MLH1, secondary to *MLH1* promoter hypermethylation, while in HNPCC the underlying defect is caused by inherited mutations of one of the DNA mismatch repair genes [204–206,213,214]. The deficiency of DNA mismatch repair may facilitate the accumulation of secondary mutations in cancer-related genes. In both sporadic and familial MSI-H colon cancers, mutational inactivation of the *TGFβRII* gene by MSI occurs in 80% of tested cancers and is frequently a late event, occurring at the transition from adenoma to cancer [215]. Studies comparing HNPCC cancers and sporadic MSI colon cancers show that HNPCC tumors are more frequently characterized by aberrant nuclear beta-catenin. Aberrant *p53* expression, 5q loss of heterozygosity, and K-*ras* mutation is infrequent in both MSI-cancer groups [201]. Despite sharing a similar underlying pathway (DNA mismatch repair deficiency), there are differences between sporadic CRC with MSI and HNPCC-associated tumors. Sporadic MSI-H cancers are more frequently poorly differentiated, mucinous, and proximally located than HNPCC tumors [201,216,217]. In sporadic MSI-H cancers, contiguous adenomas are frequently serrated lesions, while traditional adenomas are usually seen in HNPCC. Lymphocytic infiltration is common in both types of tumors [201,218–220] (Figure 19.4).

The CpG island methylator phenotype (CIMP) pathway is characterized by widespread CpG methylation in neoplasms [221,222]. Epigenetic regulation is important in cancer development and progression. The presence of extensive CpG methylation in the CIMP pathway may result in inactivation of important genes involved in tumorigenesis. For example, the DNA repair protein MGMT may be inactivated by promoter hypermethylation in sporadic colorectal cancer (39%–42%) [223], and hypermethylation can be detected in 49% of adenomas [224]. The promoter hypermethylation of *MLH1* as indicated previously triggers the MSI pathway accelerating neoplastic progression.

Recent studies indicate that the CpG island methylator phenotype underlies microsatellite instability in most cases of sporadic colorectal cancers associated with *MLH1* hypermethylation, and is tightly associated with *BRAF* mutation in sporadic colorectal cancer [225]. However, only about one-third of CIMP-positive tumors are MSI-H. Thus, CIMP appears to be independent of microsatellite status [225]. Several studies have confirmed that CIMP-positive tumors are associated with *BRAF* mutations, wild-type *TP53*, inactive WNT/β-catenin, and low-level of genomic instability of the CIN-type [180,181,221,222,226–228].

CIMP phenotype is frequent in lesions of the serrated pathway [229]. *BRAF* mutations are also associated with serrated colorectal lesions, while K-*ras* mutations are associated with hyperplastic polyps and adenomas with tubulovillous morphology [229,230]. Sessile serrated adenomas characteristically have frequent *BRAF* mutation (78%), but have rare K-*ras* mutations (11%) [230]. In contrast, hyperplastic polyps and tubulovillous adenomas show frequent K-*ras* mutation (70% and 55%, respectively), whereas *BRAF* mutation is rare in these lesions (20% and 0%, respectively) [229,230]. *MLH1* promoter methylation is frequent in serrated polyps from patients with cancers showing MSI but not in the lesions from patients with MSS cancers, suggesting that serrated adenomas may give rise to sporadic colorectal carcinomas with MSI [231].

Molecular Mechanisms Of Neoplastic Progression In Inflammatory Bowel Disease

Natural History of Neoplasia in Inflammatory Bowel Disease

Inflammatory bowel diseases (IBD) include both ulcerative colitis and Crohn's disease. Most studies of neoplasia in IBD have been conducted in patients with ulcerative colitis (UC). Patients with inflammatory bowel disease have an increased risk of dysplasia and colorectal cancer, associated with longstanding chronic colitis. Colorectal adenocarcinomas in IBD develop from foci of low-grade dysplasia, which may progress to high-grade dysplasia and ultimately invasive carcinoma (Figure 19.3). The risk of colon cancer for patients with IBD was reported to increase by 0.5%–1.0% every year after 8–10 years of diagnosis [232]. CRC in IBD has an incidence 20-fold higher and is detected on average on patients 20 years younger than colorectal cancer in the non-IBD population [233,234].

Molecular Mechanisms of CRC Development and Progression in IBD-Associated Colitis

In inflammatory bowel disease associated colitis, there is a continued inflammatory environment of the mucosa (Figure 19.3) with damage of the colonic epithelium associated with increased cell proliferation and deregulation of apoptosis. Driving factors in the inflammation-associated cancer pathway include the increased oxidative damage with associated mutagenesis caused by the heightened inflammatory infiltration of the mucosa, resulting in a stepwise progression of neoplastic lesions. The molecular players of IBD-associated carcinogenesis are similar to those seen in sporadic colorectal carcinogenesis, but the timing of occurrence of molecular events is different (Figure 19.6).

During chronic colitis, there is activation of NF-κB in the epithelium [235]. NF-κB activates the expression of COX2; several proinflammatory cytokines including IL-1, TNFα, IL-12p40, and IL-23p19; antiapoptotic factor inhibitor of apoptosis protein (IAP); and B-cell leukemia/lymphoma (Bcl-xL) [236] (Figure 19.5). Prostaglandins and cytokines such as IL-6 are released in the inflammatory milieu and activate intracellular serine-threonine kinase Akt signaling [237,238], with inhibition of proapoptotic factors p53, BAD, and FoxO1, and increased cell survival [239,240].

In colitis-associated cancers, genetic instability includes CIN and MSI, similar to sporadic colon cancers, occurring in 85% and 15% of the cases, respectively [241,242]. In UC-associated colon cancers, *APC* mutations or LOH of the *APC* locus are seen in 14% to 33% of cancers [243–246]. *APC* mutations occur

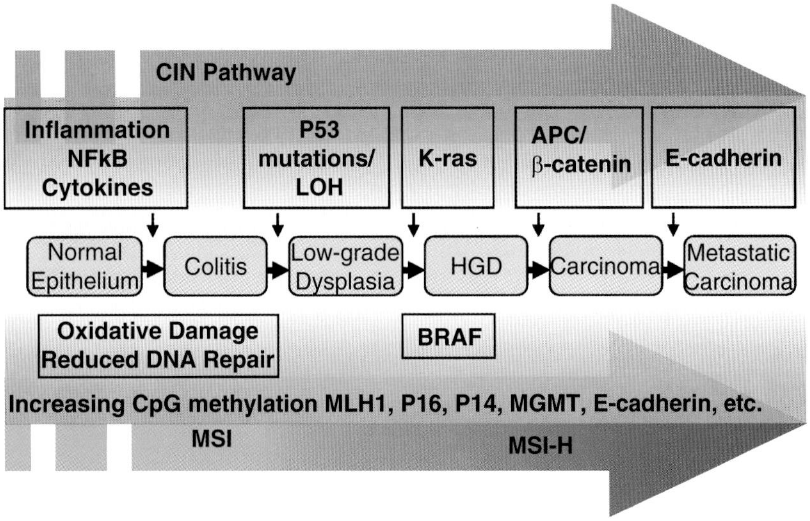

Figure 19.6 Molecular Events in the Stepwise Lesions of Inflammatory Bowel Disease-Associated Neoplasia.

late in UC-neoplasia, in the transition from high-grade dysplasia to carcinoma [241]. This is in contrast with sporadic colon carcinogenesis, where the *APC* gene is mutated in the majority (80%) of tumors and occurs early in the neoplastic process [215]. This significant difference in the timing of genetic events in IBD-associated versus sporadic colon carcinogenesis is likely due to the inflammatory environment in IBD. Since other molecular mechanisms, such as the activation of NF-κB in colonic epithelial cells by inflammatory-cell released cytokines, promote cell proliferation and inhibit apoptosis, APC inactivation may not be necessary to drive the early steps of colitis-associated carcinogenesis [235].

The tumor suppressor gene *p53* is mutated at a high rate in both sporadic and UC colon cancers, but in contrast to sporadic cancers, in UC *p53* mutations occur earlier in the colitis-associated carcinogenesis pathway [247–250]. Mutations in *p53* were detected in 19% of biopsies without dysplasia, with the frequency increasing with higher grades of dysplasia [251]. This is in contrast with sporadic CRC in which *p53* mutations and LOH are associated with the progression from high-grade adenoma to cancer [215]. Early loss of function of p53 in UC-associated carcinogenesis contributes to the rapid progression to colon cancer observed in these patients. Epithelial cells with deficient p53 function have a survival advantage. In the normal mucosa, where there is limited DNA damage, p53 induces *p21* gene expression and delays cell cycle progression in S-phase to allow for repair of the damaged DNA. If DNA damage is more extensive, p53 can induce apoptosis, preventing DNA replication and persistence of significant mutations in daughter cells [252]. In the inflamed mucosa, early mutation of *p53* is associated with inflammation related oxidative damage. The reduced DNA repair, also caused by the inflammatory environment, together with the increased mutation burden and the reduced ability to remove cells with significant mutations, results in clonal selection, expansion of neoplastic cells, and tumor development.

In MSI-positive UC colon cancers, the cause of MSI-H is loss of MLH1 protein expression associated with *MLH1* by promoter hypermethylation [253]. However, MSI in the non-neoplastic mucosa is significantly more frequent in the colitic mucosa as compared to the background mucosa of sporadic colon cancers [235]. This may be explained by reduced expression and function of DNA repair enzymes induced by oxidative stress free-radicals [254,255]. These data point to a role of deficient DNA mismatch repair without complete loss of gene expression underlying mutagenesis in the epithelium before overt neoplasia becomes histologically apparent.

In contrast to sporadic colon cancers, in UC-associated colon cancers with MSI, *TGFβRII* mutations are much less common, observed in only 17% of these cancers (Figure 19.6) [241], typically occurring early in the neoplastic process [256].

Another mechanism that contributes to neoplastic development and progression in UC-associated colorectal cancer is epigenetic gene regulation, in particular CpG island hypermethylation. Methyl binding proteins recognize hypermethylated sites and recruit histone deacetylases, leading to histone deacetylation, chromatin condensation, and inactivation of genes. Several genes have been reported to be hypermethylated in dysplasia and/or cancer in UC, including genes that have been reported as targets of CpG island methylation in sporadic colorectal cancer [257], such as the *MLH1* promoter [253,258], and the *p16INK4a* promoter regions [259]. Issa et al. reported an increased methylation level in high-grade dysplasia (HGD) of patients with UC versus controls without UC, for estrogen receptor (*ESR1*), *MYOD*, *p16* exon 1, and *CSPG2*, and hypermethylation of three of these genes had similar methylation in the colitic mucosa of patients with HGD [260]. The *p16* methylation levels averaged 2% in the mucosa of controls, 3% in UC patients who had mucosa without dysplasia, 8% in the normal appearing epithelium of patients with HGD/CA, and 9% in the dysplastic epithelium (HGD/CA) [260]. Additionally, they found that methylation was present not just in the neoplastic mucosa but also in the non-neoplastic-appearing epithelium from UC patients with HGD/cancer, suggesting that the increased levels of methylation are widespread in the inflammation-afflicted colon and occur early in the process of tumorigenesis, preceding the histological appearance of dysplasia [260].

Methylation of the hyperplastic polyposis gene 1 (*HPP1*) was observed in 50% of UC adenocarcinomas and in 40% of dysplasias. In contrast, no non-neoplastic UC mucosa showed *HPP1* methylation [261]. Methylation of the *CDH1* promoter was detected in 93% of the patients with dysplastic biopsy samples, in contrast to only 6% of the patients without dysplasia and by immunohistochemistry areas of dysplasia displayed reduced E-cadherin expression levels [262]. In IBD, promoter hypermethylation of the gene for DNA repair protein O6-methylguanine-DNA methyltransferase (*MGMT*) was detected in 16.7% adenocarcinomas and in 3.7% of mucosal samples with mild inflammation [223]. Notably, *MGMT* is more frequently methylated in sporadic adenomas and carcinomas than in IBD [223,224]. Extensive methylation characteristic of the CIMP phenotype was observed in 17% of the UC-related cancers, while global DNA methylation measured with a LINE-1 assay was seen in 58% of UC-associated cancers [263].

Activation of the Raf/MEK/ERK (MAPK) kinase pathway-either through *ras* or *BRAF* mutation was detected in 27% of all UC-related cancers. Nondysplastic UC mucosa of patients with UC-cancer did not show *BRAF* mutations, indicating that *BRAF* mutations are not an initiating event in UC-related carcinogenesis, but are associated with mismatch-repair deficiency through *MLH1* promoter hypermethylation in advanced lesions [264]. Conversely, K-*ras* mutations may occur in dysplasia, but also in villous regeneration and active colitis [265]. K-*ras* mutations are inversely correlated with *BRAF* mutations in UC cancers, in that all tested UC cancers with K-*ras* mutations had an intact *BRAF* gene and the cancers with *BRAF* mutations had an intact K-*ras* gene [264].

IBD-Associated Neoplasia: Diagnosis and Clinical Management

The most significant predictor of the risk of malignancy in IBD is the presence of dysplasia on colonic biopsies. Colonoscopy with biopsies to rule out dysplasia is currently used during follow-up of patients with inflammatory bowel disease [266]. The current pathological diagnosis of dysplasia relies on a classification based on morphological criteria established in the early 1980s, and includes the following categories: (i) negative for dysplasia; (ii) indefinite/indeterminate for dysplasia; and (iii) positive for low-grade dysplasia (LGD), high-grade dysplasia (HGD), or invasive cancer [267].

The identification of definitive dysplasia places a patient with chronic IBD in a higher risk group requiring frequent repeat colonoscopies or colonic resection. Through use of currently available histological methods, it is often difficult to determine whether a lesion represents dysplasia or a reactive process, and a diagnosis indicating indefinite/indeterminate for dysplasia is rendered, requiring additional colonoscopy. The identification of molecular markers of dysplasia for a more accurate screening of dysplasia in IBD patients remains a high priority [253].

Hereditary Nonpolyposis Colorectal Cancer

Hereditary Nonpolyposis Colorectal Cancer: Genetic and Molecular Basis

HNPCC is an autosomal dominant cancer predisposition syndrome, characterized by early onset of CRC and tumors from other organs, including endometrium, ovary, urothelium, stomach, brain, and sebaceous glands [82,170,268,269]. HNPCC with underlying DNA mismatch repair mutations and characteristic MSI-positive tumors represent the HNPCC/Lynch syndrome, while a group of patients with clinical features of HNPCC but lacking DNA mismatch repair gene mutations represent a separate group of tumors [270]. HNPCC/Lynch syndrome is estimated to represent 4%–6% of all colorectal cancer cases [170]. At the molecular level, HNPCC/Lynch syndrome patients inherit germ-line mutations in one of the DNA mismatch repair genes, leading to defects in the corresponding DNA mismatch repair proteins [82,268]. Mutations are found most commonly in the *MLH1* and *MSH2* and less frequently in the *MSH6* and *PMS2* genes [216,271–276].

HNPCC/Lynch syndrome is subdivided into Lynch syndrome I, characterized by colon cancer susceptibility, and Lynch syndrome II, which shows all the features of Lynch syndrome I, but patients are also at increased risk for carcinoma of the endometrium, ovary, and other sites [170]. The spectrum of tumors that occur in patients with HNPCC is referred to as HNPCC cancers and includes carcinomas of the colon and rectum, small bowel, stomach, biliary tract, pancreas, endometrium, ovary, urinary bladder, ureter, and renal pelvis [82,277]. Sebaceous gland adenomas and keratoacanthomas are part of the Muir-Torre syndrome, and tumors of the brain, usually glioblastomas, are seen in the Turcot syndrome [82,277] (Table 19.1).

The DNA MMR genes encode the MutS proteins (MSH2, MSH3, and MSH6) and the MutL proteins (MLH1, PMS1, PMS2, and MLH3) [79,81]. Functional DNA MMR requires the formation of MutS and MutL complexes between the different MMR proteins [79,278–281]. The MSH2 protein interacts with either MSH6, forming the MutS-alpha complex, or with MSH3, forming the MutS-beta complex. MutS-alpha and MutS-beta complexes bind to post-DNA replication mismatched sequences. The MutS-alpha complex is required to repair base-base mispairs and small insertion or deletion mispairs, while the MutS-beta complex is primarily involved in the correction of insertion or deletion mispairs [282–284]. MutL heterodimers bind MutS-alpha or MutS-beta [79]. MutL heterodimers include MutL-alpha (MLH1 and PMS2), which appear to be responsible for most of the DNA MMRs. The MutL-beta heterodimer (MLH1 and PMS1) does not appear to be significantly involved in DNA mismatch repair [285,286].

Known germline mutations of DNA MMR genes in HNPCC affect the coding region of *MLH1* (approximately 40%), *MSH2* (approximately 40%), *MSH6* (approximately 10%), and *PMS2* (approximately 5%) [101,210,211,287,288].

Hereditary Nonpolyposis Colorectal Cancer: Natural History, Clinical and Pathologic Features, and Molecular Mechanisms

The average age of presentation of colorectal cancer in HNPCC patients is 45 years of age. Tumors are often multiple or associated with other synchronous or metachronous neoplasms of the HNPCC-cancer spectrum [82,170,277,289]. In addition to colorectal cancers, patients may present with tumors of the Muir-Torre syndrome or with the Turcot syndrome [82,277].

In HNPCC/Lynch syndrome patients, the lifetime risk of colorectal cancer is up to 80%, and it is up to 60% for endometrial carcinoma [82]. Colorectal cancers are located in the proximal colon in two-thirds of cases [201,289], and have detectable microsatellite instability (MSI) in more than 90% of the cases [289].

A few histopathologic features are characteristically associated with MSI-H colorectal adenocarcinomas that may suggest the possibility of HNPCC (Figure 19.4). These features are not specific to HNPCC cancers, in that they are also seen frequently in sporadic MSI-H tumors and in some tumors that are MSS. Three major histopathologic groups of MSI-H cancers can be recognized: (i) poorly differentiated adenocarcinomas, also described as medullary type cancers; (ii) mucinous adenocarcinomas and carcinomas with signet ring cell features; and (iii) well to moderately differentiated adenocarcinomas. The presence of prominent intratumoral-infiltrating lymphocytes (TILs) is the most predictive finding of MSI-H status (Figure 19.4). TILs may be particularly numerous in poorly differentiated cancers, but also occur in the other morphologic types of HNPCC associated cancers. The intratumoral-infiltrating lymphocytes are CD3-positive T-cells and most are CD8-positive cytotoxic T-lymphocytes. Peritumoral lymphocytic inflammation and lymphoid aggregates forming a Crohn's-like reaction are also frequent in MSI-H carcinomas [201,216–220,290].

Patients with HNPCC develop adenomas more frequently and at an earlier age than noncarriers of DNA mismatch repair gene mutations [291]. In HNPCC, patients develop one or few colonic adenomas of traditional type (tubular adenomas and tubulovillous adenomas) [201,218–220]. The progression from adenoma to invasive carcinoma occurs rapidly, in many patients being less than 3 years, contrasting to a mean of 15 years in patients without HNPCC [291,292]. Detection of colonic adenomas in HNPCC mutation carriers occurs at a mean age of 42–43 years (range 24–62 years) [291,293,294]. Compared to sporadic adenomas, HNPCC adenomas are more frequently proximal in the colon, and more frequently show high-grade dysplasia [291,295]. In HNPCC a significant association was found between MSI-H and high-grade dysplasia in adenomas, with associated loss of either MLH1 or MSH2 [296]. Based on these findings it was recommended that immunohistochemical staining/MSI testing of large adenomas with high-grade dysplasia in young patients (younger than 50 years) may be performed to help identify patients with suspected HNPCC [291,296].

The molecular mechanisms that underlie the neoplastic development and progression in HNPCC are those of the MSI pathway [207–210]. The deficient or absent DNA mismatch repair protein(s) in HNPCC/Lynch syndrome is dictated by the gene that carries a germline mutation. A second genetic hit is then responsible for loss of the second allele. Epigenetic silencing through CpG methylation of *MLH1*, which is the underlying mechanism of MSI in sporadic colon cancer, is rare in HNPCC tumors [297]. In addition, the patients with unambiguous germline mutation in DNA mismatch repair genes do not appear to carry *BRAF*-activating mutations in their tumors [298]. As in sporadic MSI carcinomas, loss of DNA mismatch repair may lead to accumulation of mutations in cancer-related genes, such as the *TGFβRII* gene [215]. HNPCC tumors show frequent aberrant nuclear beta-catenin, but aberrant *p53* expression, 5q loss of heterozygosity and K-*ras* mutation is uncommon, similar to sporadic MSI-cancers [201].

Hereditary Nonpolyposis Colorectal Cancer: Molecular Diagnosis, Clinical Management, and Genetic Counseling

To identify patients with HNPCC, criteria known as the Amsterdam criteria were established in the early 1990s, with later revisions [289,299]. Because the Amsterdam criteria do not include some patients with known germline MMR gene mutations, another set of guidelines, known as the Bethesda guidelines, was established to help decide whether or not a patient should undergo further molecular testing to rule out HNPCC [82,101,300–302]. Through use of the Amsterdam II Criteria, patients are diagnosed with HNPCC when the following criteria are present [289]: (i) the family includes three or more relatives with an HNPCC-associated cancer, (ii) one affected patient is a first-degree relative of the other two, (iii) two or more successive generations are affected, (iv) cancer in one or more affected relatives is diagnosed before the age of

50 years, (v) familial adenomatous polyposis is excluded in any cases of colorectal cancer, and (vi) tumors are verified by pathological examination. Alternatively, one of the following criteria (Modified Amsterdam) needs to be met: (i) very small families, which cannot be further expanded, can be considered to have HNPCC with only two colorectal cancers in first-degree relatives if at least two generations have the cancer and at least one case of colorectal cancer was diagnosed by the age of 55 years; (ii) in families with two first-degree relatives affected by colorectal cancer, the presence of a third relative with an unusual early onset neoplasm or endometrial cancer is sufficient; (iii) if an individual is diagnosed before the age of 40 years and does not have a family history that fulfills the preceding criteria (Amsterdam II and Modified Amsterdam criteria), that individual is still considered as having HNPCC.

If an individual has a family history that is suggestive of HNPCC but does not fulfill the Amsterdam, modified Amsterdam or young age at onset criteria, that individual is considered to be HNPCC variant, or familial colorectal cancer X [303,304]. A large group of patients representing 60% of all cases who meet Amsterdam I or Amsterdam II criteria for HNPCC do not have characteristic features of MMR deficiency. Compared to the MSI-HNPCC patients, the MSS HNPCC patient's age at diagnosis is 6 years higher on average, and most colorectal cancers appear on the left side of the colon [270]. The underlying genetic defect for these tumors is not yet known.

The most recently revised Bethesda criteria recommend testing of patients to rule out HNPCC if there is one of the following criteria [82,101,301,302]: (i) patient is diagnosed with colorectal cancer before the age of 50 years; (ii) presence of synchronous or metachronous colorectal or other HNPCC-related tumors (stomach, urinary bladder, ureter and renal pelvis, biliary tract, brain [glioblastoma], sebaceous gland adenomas, keratoacanthomas, and small bowel), regardless of age; (iii) colorectal cancers with a high-microsatellite instability morphology (presence of tumor-infiltrating lymphocytes, Crohn's-like lymphocytic reaction, mucinous or signet ring cell differentiation, or medullary growth pattern) that was diagnosed before the age of 60 years; (iv) colorectal cancer patient with one or more first-degree relatives with colorectal cancer or other HNPCC-related tumors, and one of the cancers must have been diagnosed before the age of 50 years; and (v) colorectal cancer patient with two or more relatives with colorectal cancer or other HNPCC-related tumors, regardless of age.

The application of the revised Bethesda guidelines results in an increased identification of MSI-positive colon cancers as compared to previous guidelines [217]. The effect of setting the age cut-off to less than 50 years and inclusion of histopathologic features in the selection of tumors that should be tested for MSI and DNA mismatch repair protein immunohistochemistry significantly increased the detection rate of potential HNPCC-associated cancers. With the updated Bethesda guidelines, when selected for age (colorectal cancer in patients less than 50 years of age) and histologic features suggestive of MSI-tumor in patients less than 60 years of age, MSI-H tumors were detected in approximately 25% of colorectal cancer cases meeting these revised Bethesda guidelines [217].

Given recent knowledge of the molecular changes underlying MSI-sporadic as compared to HNPCC-associated cancers, algorithms have been established to determine whether a patient has sporadic MSI-positive cancer or has HNPCC/Lynch syndrome [305] (Figure 19.7).

When a patient with HNPCC cancer is identified by the Amsterdam criteria or by the revised Bethesda criteria, the next step is to evaluate MSI with the MSI test and/or immunohistochemical (IHC) analysis of tumors (preferably colorectal cancer tissue if available) for MSH2 and MLH1, MSH6, and PMS2 DNA mismatch repair proteins.

The MSI test is based on the evaluation of instability in small DNA segments that consist of repetitive nucleotides of generally 100 to 200 base pairs in length, called microsatellite regions or short tandem repeats (STRs). The nucleotide sequences within the repetitive elements include mononucleotide repeats of Adenine (A)n or Cytosine-Adenine (CA)n dinucleotide repeats. During DNA replication, these repetitive sequences are susceptible to variations in length because of DNA strand slippage. Such changes in length of the nucleotide repeat are termed microsatellite instability. In cells with deficient DNA mismatch repair, as occurs in patients with HNPCC/Lynch syndrome, these mutations are not repaired and persist in the DNA of future cell generations in the tumor tissue. MSI can be detected in the tumor DNA by PCR approaches. Tissue sections from tumor tissue used for routine pathologic diagnosis embedded in paraffin are adequate for the MSI test. The most used set of microsatellite markers was recommended by an NCI consensus group and consists of two mononucleotide repeat markers (*BAT25* and *BAT26*) and three dinucleotide repeat markers (*D2S123*, *D5S346*, and *D17S250*) [300]. The results of the MSI test using the NCI panel of five microsatellite markers are reported as MSI-High (MSI-H), MSI-Low (MSI-L), and microsatellite stable (MSS). MSI-H tumors show MSI in at least two of the five markers, MSI-Low (MSI-L) tumors show MSI in only one marker, and no instability is detected in any of the five markers in MSS tumors. Figure 19.4 illustrates a colorectal cancer with loss of expression of MSH2 in the tumor cells and associated MSI-H identified by microsatellite instability at both the *BAT25* and *BAT26* markers (Figure 19.4).

If the tumor tissue reveals MSI-H and/or there is loss of expression of one of the DNA repair proteins by immunohistochemistry, germline testing should be performed for the gene encoding the deficient protein, after appropriate genetic counseling of the patient. If tissue testing is not feasible, or if there is sufficient clinical evidence of HNPCC, it is acceptable to proceed directly to germline analysis of the *MSH2*

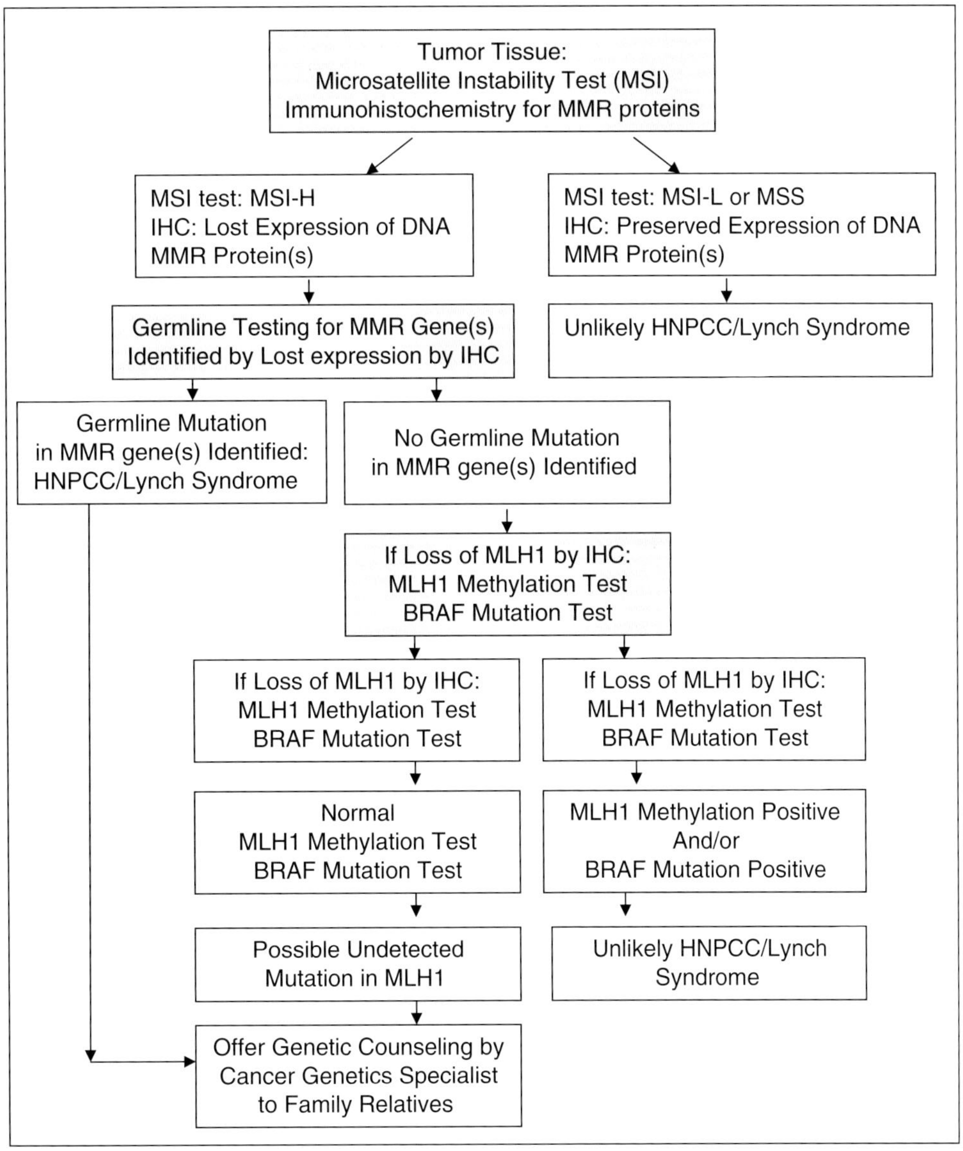

Figure 19.7 Algorithm for Molecular Testing of HNPCC/Lynch Syndrome Colorectal Cancer. Modified from [305].

and/or *MLH1* genes [101,305,306]. Mutation analyses can be performed by several methods, including single-strand conformational polymorphism, denaturing gradient gel-electrophoresis analysis, DNA sequencing, monoallelic expression analysis, Southern analysis, and quantitative PCR [101]. The likelihood of finding a germline mutation in the *MLH1/MSH2* genes of patients with colorectal cancer tumors that are not MSI-H is low [101].

If no loss of expression of *MSH2* or *MLH1* is seen in MSI-H tumors or if the tumor is MSI-L or MSS but there is suspicion of HNPCC, evaluation of other MMR genes, in particular *MSH6* and *PMS2,* should be performed, first by immunohistochemical stains, followed by germline mutational analyses [101,307]. Identification of a germline mutation in index cancer patients is important because it confirms a diagnosis of HNPCC, and the identified mutation may be used to screen at-risk relatives who may be mutation carriers. If the tumor tissue revealed loss of expression of MLH1 by immunohistochemistry, but no mutation in the DNA mismatch repair genes underlying HNPCC/Lynch syndrome are found, two other tests, the *MLH1* methylation and *BRAF* mutation tests, may help discriminate between a sporadic MSI-tumor and HNPCC tumor with undetected *MLH1* mutation (Figure 19.7) [305]. Through use of quantitative methylation analyses, HNPCC patients showed no or low level of *MLH1* promoter methylation, in contrast to high levels of methylation (greater than a cutoff value of 18% methylation) in sporadic MSI cancers [297]. In addition, none of the patients with unambiguous germline mutation in DNA mismatch repair genes demonstrated *BRAF* mutation [298]. Therefore, adding *BRAF* mutation and *MLH1* methylation tests in the algorithm for testing of MSI-H colon tumors with loss of expression of MLH1 protein in the tumor can help determine whether a tumor is likely to be a sporadic or an HNPCC-associated tumor (Figure 19.7).

After a germline mutation is identified or the patient is diagnosed with HNPCC, at-risk relatives should be referred for genetic counseling and tested if they wish. If no mismatch repair gene mutation is found in a proband with an MSI-H tumor and/or a clinical history of HNPCC, the genetic test result is noninformative. The patients and the relatives at-risk should be counseled as if HNPCC was confirmed and high-risk surveillance should be performed [82,101,305].

In families that meet strict clinical criteria for HNPCC, germline mutations in *MSH2* and *MLH1* have been found in 45%–70% of the families, and germline mutations in these two genes account for 95% of HNPCC cases with an identified mutation [287,306]. The reported data show that despite extensive testing there is still a significant number of families without an identified germline mutation that accounts for HNPCC. The germline mutations that occur in *MSH2* and *MLH1* are widely distributed throughout the two genes. Two hundred fifty-nine pathogenic mutations and 45 polymorphisms have been reported in *MLH1*, and 191 pathogenic mutations and 55 polymorphisms have been identified in *MSH2* [308].

As reviewed in Lindor et al. [305] individuals diagnosed as carriers of DNA mismatch repair gene mutations seen in the HNPCC/Lynch syndrome are recommended to have colonoscopic surveillance staring at early age. Colonoscopy is recommended every 1 to 2 years starting at ages 20 to 25 years (age 30 years for those with *MSH6* mutations), or 10 years younger than the youngest age of the person diagnosed in the family. Although there is limited evidence regarding efficacy, the following are also recommended annually: (i) endometrial sampling and transvaginal ultrasound of the uterus and ovaries (ages 30–35 years); (ii) urinalysis with cytology (ages 25–35 years); (iii) history, examination, review of systems, education, and genetic counseling regarding Lynch syndrome (age 21 years). For individuals who will undergo surgical resection of a colon cancer, subtotal colectomy is favored. Further, evidence supports the efficacy of prophylactic hysterectomy and oophorectomy [304,305].

Familial Adenomatous Polyposis (Fap) And Variants

Familial Adenomatous Polyposis: Genetic Basis

FAP is a cancer predisposition syndrome characterized by numerous adenomatous colorectal polyps, with virtually universal progression to colorectal carcinoma at an early age. It accounts for less than 1% of all colorectal carcinoma cases in the United States and affects 1 in 8,000–10,000 individuals [309]. The majority of cases of FAP are caused by germline mutations in the *APC* gene on chromosome 5q, and in its inherited form, FAP is transmitted in an autosomal dominant fashion, although up to a third of cases may present as *de novo* germline mutations [309,310].

It is of historical note that FAP is perhaps one of the best characterized cancer predisposition syndromes, as the elucidation of the molecular basis of the syndrome was paramount in the formulation and advancement of the multistep model of carcinogenesis. The first verified case of adenomatous polyposis was reported in 1881, and familial cases of adenomatous polyposis were the subject of scientific inquiry throughout the 1900s, leading to registries and early intervention with colectomy as early as the 1920s [311]. A series of major breakthroughs both in regard to FAP and as related to the understanding of carcinogenesis started with the association of the syndrome with a deletion on chromosome 5q21–22, and the genetic defect was subsequently identified to reside within the *APC* gene at that locus [312–314].

Familial Adenomatous Polyposis: Natural History and Clinical, Molecular, and Pathologic Features

Patients affected by FAP have a nearly 100% lifetime risk for the development of colorectal carcinoma in the absence of aggressive treatment, which often includes prophylactic colectomy [315]. In addition, patients have a 90% lifetime risk for the development of upper gastrointestinal tract polyps, with a 50% risk of developing advanced duodenal polyposis by the age of 70 years [316]. The lifetime risks for development of gastric adenocarcinoma (<1%) and duodenal and ampullary carcinoma (5–10%) are considerably higher than those of the general population [317]. FAP mutation carriers have increased risk of fundic gland polyposis involving the stomach [318–320]. Among the other associated lesions, congenital hypertrophy of retinal pigmented epithelium (CHRPE) is found in 70%–80% of FAP patients, and desmoid tumors are found in 15%, correlating to a relative risk of ~850 compared to the general population. Thyroid carcinoma, both papillary and follicular type, is estimated to occur in 1%–2% of FAP patients [316]. In young children with a family history of FAP, the relative risk of hepatoblastoma is 847 compared to the general population [316].

Clinically, the common manifestation of FAP and its variants is the presence of numerous (sometimes in excess of 1000) adenomatous polyps distributed throughout the colon and rectum. The characteristic gross findings, combined with the histologic findings and predictable progression to colorectal carcinoma, have historically made FAP a robust model system to better understand carcinogenesis development and progression. There are several variants of FAP, including Gardner Syndrome, Turcot Syndrome, and Attenuated FAP (AFAP), summarized in Table 19.1 [309,317,320,321]. Gardner syndrome is characterized by multiple extra-colonic manifestations including osteomas, desmoid tumors, dental abnormalities, ophthalmologic abnormalities including congenital hypertrophy of retinal pigment epithelium (CHRPE), and cutaneous cysts. Some suggest that some degree of these extra-intestinal manifestations may be identified with close scrutiny in typical FAP patients. Turcot syndrome is the association of the colorectal polyposis and brain tumors, most commonly medulloblastoma. Attenuated FAP demonstrates a reduction in the number of colonic polyps, usually falling short of the 100 polyps necessary for a diagnosis of FAP, but with sufficient colonic polyposis (frequently over 15) to raise suspicion for an underlying polyposis syndrome [317].

FAP and its variants may present clinically in several manners, including through identification of colorectal polyposis or colorectal carcinoma at an unusually early age, the presence of extra-colonic manifestations, or as part of screening in the setting of a known family history [322].

One of the most striking features of FAP and its variants (except AFAP) is the presence of hundreds, perhaps even thousands, of polyps throughout the colon and rectum, leading to a carpet appearance of the colorectal mucosa (Figure 19.8). The polyps are frequently sessile, and appear as early as late childhood/early adolescence, requiring that endoscopic screening in familial cases begin early in life [322]. The histologic features of polyps in FAP are essentially indistinguishable from those seen in sporadic adenomas; however, it is common to identify lesions at various stages in the dysplasia-adenoma-carcinoma sequence, further underscoring the multistep nature of carcinogenesis (Figure 19.8).

The molecular changes in FAP have been extensively characterized and serve as an excellent demonstration of the relationship between the understanding of cellular signaling pathways and their effect on disease process. The detailed examination of the molecular changes of the lesions at various stages in FAP was a cornerstone of the current understanding of carcinogenesis as a multistep process involving gradual increases in tumor size, disorganization, and accumulation of genetic changes [323]. The majority of cases of FAP and its variants are attributed to germline mutation in the *APC* gene, located at the 5q21–22 chromosome locus. The protein product of the *APC* gene serves as key mediator in the Wnt pathway of signal transduction for cellular growth and proliferation (Figure 19.8). Wnt binding to cellular receptors initiates a series of downstream signals which result in increased transcription of cell growth and proliferation-associated genes, through the effect of the protein β-catenin. β-catenin is a transcriptional regulator which must be localized to the nucleus in order to impart its effect on transcription. In the absence of a Wnt-mediated growth signal, APC serves as part of a complex which destabilizes β-catenin through phosphorylation, thereby targeting it for destruction by the proteasome, thus preventing its nuclear localization and transcriptional effects (Figure 19.8A). In the presence of Wnt-mediated growth signal, the APC-β-catenin protein complex is disrupted, and phosphorylation cannot occur, resulting in increased stability and nuclear localization of β-catenin (Figure 19.8B). As is common in many examples of neoplasia, the disease state is one which mimics the activated state, and mutations in APC commonly cause a disruption of the protein complex which destabilizes β-catenin, resulting in constitutive activation of the Wnt pathway (Figure 19.8C) [323–325].

Familial Adenomatous Polyposis and Related Syndromes: Molecular Diagnosis, Clinical Management, and Genetic Counseling

As the majority of cases of FAP have been attributed to mutation in *APC*, there has been extensive investigation of the function of its protein product. In addition to its role in FAP, the understanding of the role of *APC* bears special relevance, as over 70% of nonsyndromic colorectal carcinomas are found to have somatic mutation of *APC* [317]. *APC* codes for a 312 kDa protein that is expressed in a wide range of tissues and is thought to participate in several cellular functions including Wnt-mediated signaling, cell adhesion, cell migration, and chromosomal segregation. Within the APC protein, there are multiple domains which are responsible for these varied functions, including an oligomerization domain; an armadillo domain region, which is thought to be involved in binding of APC to proteins related to cell morphology and motility; β-catenin binding domain, axin binding domain; and a microtubule binding domain [316,326] (Figure 19.9). The majority of germline mutations associated with FAP are either frameshift or nonsense mutations which lead to a truncated protein product, thereby disrupting the interaction with β-catenin, leading to its stabilization and subsequent downstream signaling [326,327]. Two hotspots have been identified for germline mutations in *APC*, at codons 1061 and 1309 and account for 17% and 11% of all germline APC mutations, respectively (Figure 19.9). The region between codons 1286 and 1513 is termed the mutation cluster region (MCR) to reflect the observation that this region encompasses many of the identified *APC* mutations [328].

Both within and beyond the MCR, there is some degree of association between the location of the germline *APC* mutation and the clinical phenotype [329]. Mutations within the MCR are associated with a profuse polyposis, with the development of over 5,000 colorectal polyps. Attenuated polyposis is seen in the settings of mutations in the 5′-end of the *APC* gene (exons 4 and 5), the alternatively spliced form of exon 9, or the 3′-distal end of the gene. An intermediate phenotype is observed with mutations between codon 157 and 1249, and 1465 and 1595. Similarly, genotype-phenotype correlations for extra-intestinal manifestations have been identified. Desmoid tumors and upper gastrointestinal lesions are clinically the most pressing extra-intestinal manifestations, as there is significant morbidity and mortality associated with them. The association of FAP with desmoid tumor formation has been correlated to mutations downstream of codon 1400 [317,329,330]. An unequivocal

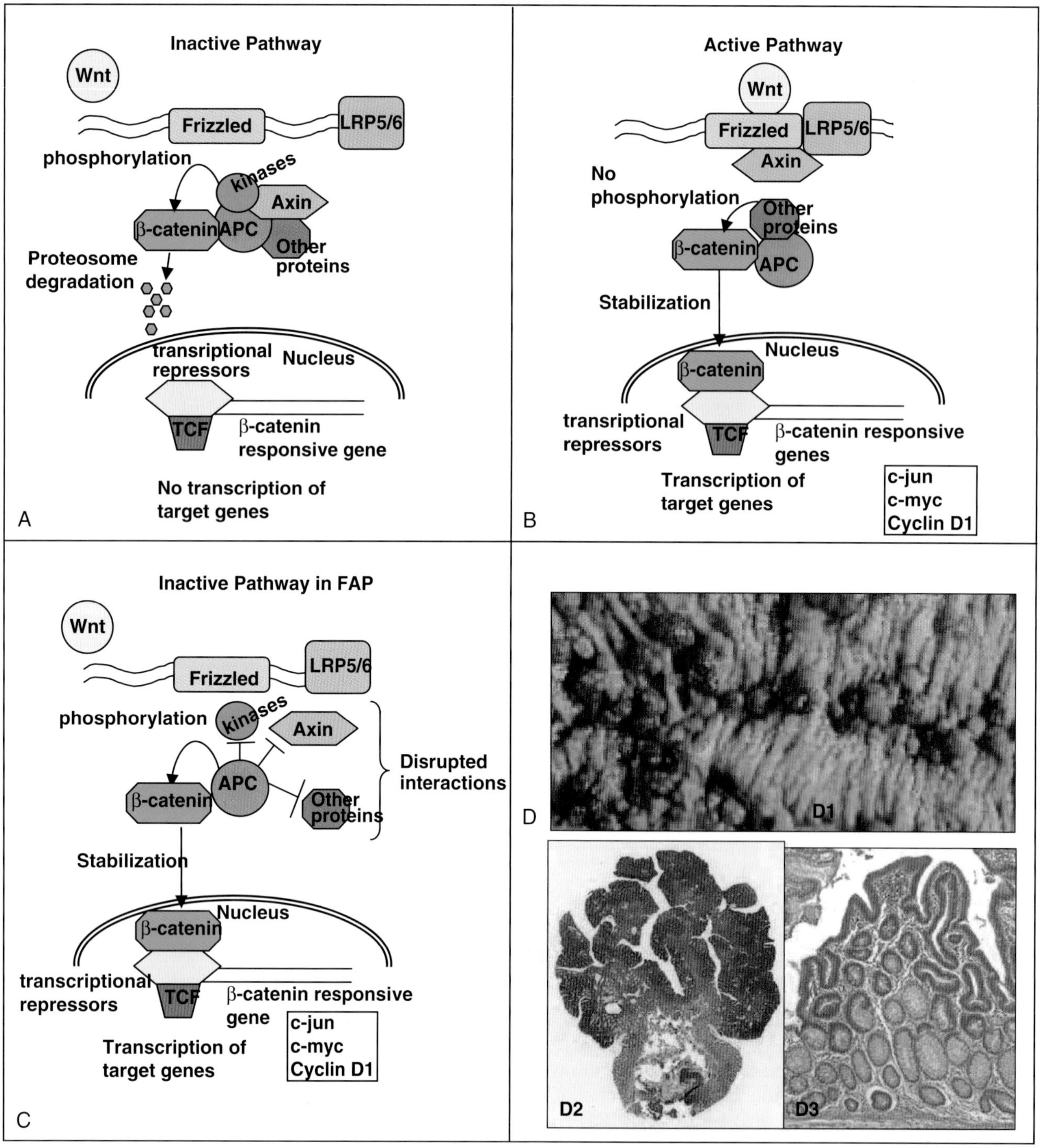

Figure 19.8 The Wnt Pathway and Familial Adenomatous Polyposis. (A) Signaling of the Wnt pathway is mediated through the Frizzled family of receptors, and a co-receptor LRP5 or 6. In the absence of ligand, the pathway is inactive through the negative regulation of the downstream effector β-catenin. When present in sufficient quantity in the nucleus, β-catenin stimulates transcription of target genes. Lack of signaling through the Frizzled receptor results in sequestration of β-catenin in a multiprotein complex including APC, Axin, and other proteins, which exert a negative effect both through the cytoplasmic sequestration of β-catenin and by targeting it for destruction by the proteasome through phosphorylation. (B) In the presence of Wnt ligand, the Frizzled receptor and LRP5/6 form a complex which results in the recruitment and sequestration of Axin at the cell surface, thereby inhibiting the kinase activity of the APC complex. This results in the stabilization and nuclear translocation of β-catenin, thus resulting in transcription of target genes. (C) In the setting of an *APC* gene mutation, the multiprotein APC/Axin/β-catenin complex is disrupted, most commonly due to truncation of the APC protein in domains responsible for protein-protein interaction. This results in stabilization and nuclear translocation of β-catenin, with the net effect of constitutively active transcription of growth-promoting genes otherwise under tight regulation. (D) Typical appearance of polyps in FAP, which are indistinguishable from spontaneous nonsyndromic polyps, but are numerous and may eventually cover most of the surface of the colon with a carpet appearance (D1; From the files of the Department of Pathology, Hospital of the University of Pennsylvania). Tubular adenoma (D2) and adenomatous colonic mucosa (D3) at an early stage before the development of larger adenomas and adenocarcinomas (H&E stain, original magnification 5×).

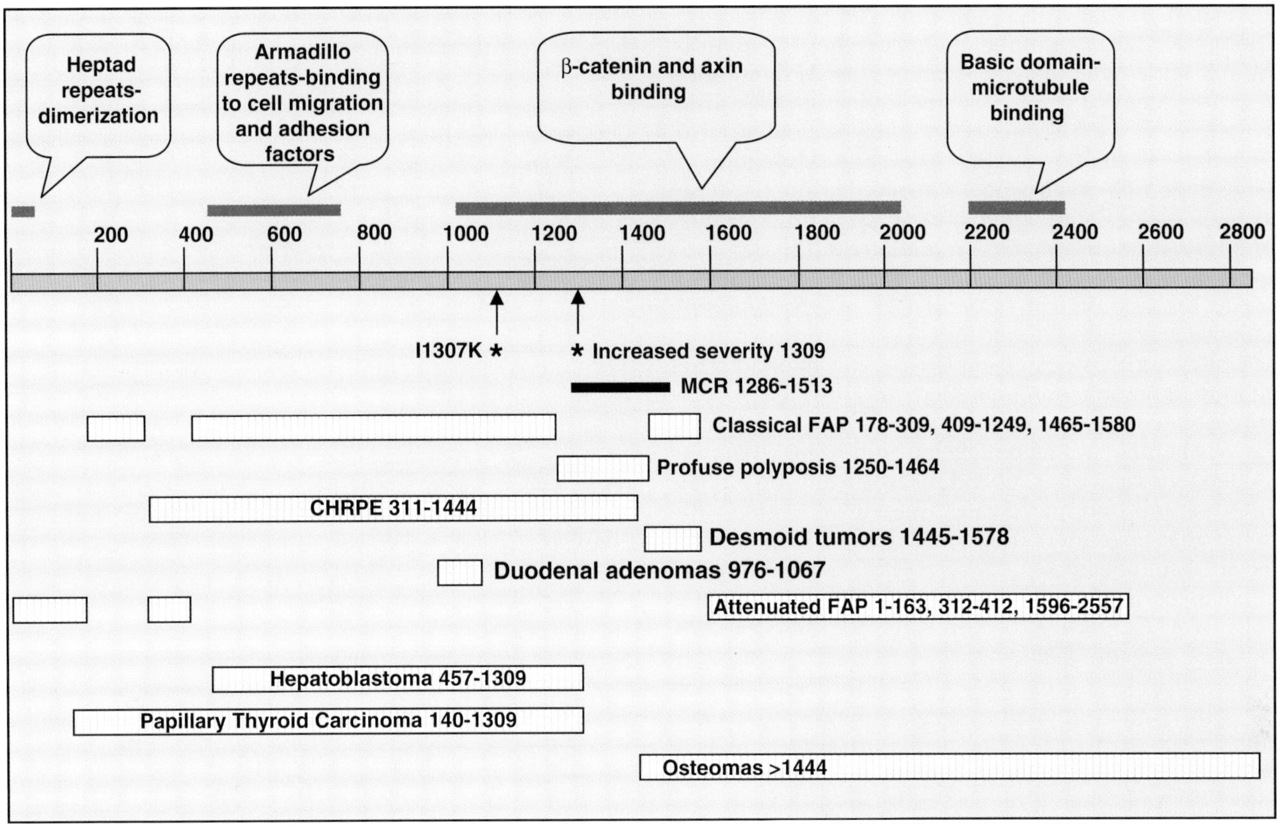

Figure 19.9 **APC Gene Mutation and Phenotype Correlation in FAP.** The figure illustrates the *APC* gene functional domains and mutation-phenotype correlations. The APC protein product consists of 2843 amino acids with multiple functional domains, including a dimerization domain, microtubule domain, and binding sites for β-catenin and axin (pictured as gray boxes). Mutations within certain codon ranges may correlate to a clinical phenotype as depicted in the yellow boxes. Germline mutations generally fall within the entire spectrum of depicted mutational sites, while somatic mutations tend to cluster within the mutation cluster region (MCR). Of note, a profuse polyposis phenotype is seen in patients in whom a germline mutation between codons 1250 and 1464 is seen, with the clinical presentation of >5000 polyps. Mutation at codon 1309 is associated with profuse polyposis and earlier onset of disease. I1307K represents a mutation common in the Ashkenazi Jewish population. Adapted from [316,317,329].

correlation between upper gastrointestinal tumors and specific mutations has not been established, although there are data to suggest that some specific mutations are associated with a higher rate of these lesions, including mutations beyond codon 1395, those beyond codon 934, and within codons 564 to 1465. CHRPE is associated with mutations falling between codons 311 and 1444 (Figure 19.9).

Well-established methodologies are in place for screening patients in whom a diagnosis of FAP is a consideration. Recommendations for indications and approach have been suggested by the American Gastroenterological Association, and the primary indications include clinically high suspicion for FAP (>100 colorectal adenomas), first-degree relatives of FAP patients, >20 cumulative colorectal adenomas (suspected AFAP), and first-degree relatives of patients with AFAP [322]. In *de novo* cases or in cases in which a family mutation is unknown, there are several high-throughput approaches to conduct the mutational analysis [331]. Sequencing of the entire coding region is the gold standard for diagnosis, although other methods such as protein truncation tests and mutation scanning approaches can be used. In some settings, initial targeting of common mutational hotspots is the preferred method for initial screening, followed by more extensive sequence analysis if the common mutations are not identified. Once a mutation is identified, targeted genetic testing can be carried out for potentially affected family members. Additionally, there is a growing use of such targeted genetic testing as part of preimplantation genetic diagnosis [332].

In a subset of patients, a discrete *APC* mutation is not identified using first-line approaches to diagnosis. In some cases, adjusting the approach to screen individual alleles of *APC* yields a diagnostic mutation or, in some cases, large exonic or entire gene deletions [333]. However, in other cases, these approaches do not identify a discrete *APC* mutation. An alternative target for genetic testing has recently been identified as *MUTYH,* in which biallelic mutations are detected in a significant minority of *APC* mutation negative cases of polyposis, and is correlated to cases of attenuated FAP (AFAP) [171]. There have been a number of approaches for identifying the underlying molecular changes in *APC* mutation negative cases not resolved by the evaluation of individual alleles or *MUTYH* gene, including examination of epigenetic regulation of the *APC* gene, evaluation of genes encoding other proteins involved in the β-catenin pathway such as Axin, evaluation of allelic mRNA ratios, and evaluation of somatic *APC* mosaicism [334]. Germline hypermethylation of the *APC* gene has been shown not to be a significant cause of FAP in *APC* mutation negative cases [335]. One study has shown unbalanced *APC* allelic mRNA expression in cases in which no other discrete mutation was identified. Interestingly, in the tumor specimens from these cases, there was loss of the remaining wild-type allele, indicating that the reduced dosage of *APC* and unbalanced *APC* allelic mRNA expression may create a functional haploinsufficiency that engenders the same predisposition to colorectal carcinogenesis [336]. The mechanism of unbalanced allelic mRNA expression is unclear. Mutations in *AXIN2* have also been rarely identified in *APC* mutation negative FAP.

Surveillance of those affected by FAP begins early in life with annual sigmoidoscopy or colonoscopy, beginning at age 10 to 12, followed by prophylactic colectomy, usually by the time the patient reaches his or her early 20s. Additionally, regular endoscopy with full visualization of the stomach, duodenum, and peri-ampullary region has been recommended; however, the optimal timing of these screening evaluations is not well established and is generally managed based on the severity of upper gastrointestinal disease burden. Additional recommendations are based on extra-colonic manifestations which have the propensity for morbidity and mortality. These include annual palpation of the thyroid, with some advocating a low threshold for referral for ultrasound examination and serum AFP levels and abdominal palpation every 6 months in young children in FAP families to detect hepatoblastoma.

REFERENCES

1. World Health Organization. Burden of disease-mortality (as appeared in the 1999 World Health Report): Mortality by sex, cause, and WHO region, estimates for 1998. Available at: http://www-net.who.int/whosis/statistics/burden_whr/burden_whr_mortality_process.cfm/; 1998 Accessed October 3, 2000.
2. Parkin DM, Bray F, Ferlay J, et al. Global cancer statistics, 2002. *CA Cancer J Clin.* 2005;55:74–108.
3. Pickle LW, Hao Y, Jemal A, et al. A new method of estimating United States and state-level cancer incidence counts for the current calendar year. *CA Cancer J Clin.* 2007;57:30–42.
4. Correa P. Helicobacter pylori and gastric carcinogenesis. *Am J Surg Pathol.* 1995;19:S37–S43.
5. Sipponen P, Kekki M, Haapakoski J, et al. Gastric cancer risk in chronic atrophic gastritis: Statistical calculations of cross-sectional data. *Int J Cancer.* 1985;35:173–177.
6. Kimura K. Gastritis and gastric cancer. Asia. *Gastroenterol Clin North Am.* 2000;29:609–621.
7. Dixon MF, Genta RM, Yardley JH, et al. Classification and grading of gastritis. The updated Sydney system. International workshop on the histopathology of gastritis, Houston 1994. *Am J Surg Pathol.* 1996;20:1161–1181.
8. La Vecchia C, Negri E, Franceschi S, et al. Family history and the risk of stomach and colorectal cancer. *Cancer.* 1992;70:50–55.
9. Zanghieri G, Di Gregorio C, Sacchetti C, et al. Familial occurrence of gastric cancer in the 20-year experience of a population-based registry. *Cancer.* 1990;66:2047–2051.
10. Brenner H, Arndt V, Sturmer T, et al. Individual and joint contribution of family history and Helicobacter pylori infection to the risk of gastric carcinoma. *Cancer.* 2000;88:274–279.
11. Palli D. Epidemiology of gastric cancer: An evaluation of available evidence. *J Gastroenterol.* 2000;35(Suppl 12):84–89.
12. Sriamporn S, Setiawan V, Pisani P, et al. Gastric cancer: The roles of diet, alcohol drinking, smoking and Helicobacter pylori in Northeastern Thailand. *Asian Pac J Cancer Prev.* 2002;3:345–352.
13. Brenner H, Arndt V, Bode G, et al. Risk of gastric cancer among smokers infected with Helicobacter pylori. *Int J Cancer.* 2002;98:446–449.
14. Gologan A, Graham DY, Sepulveda AR. Molecular markers in Helicobacter pylori-associated gastric carcinogenesis. *Clin Lab Med.* 2005;25:197–222.
15. Sepulveda AR, Dore MP, Bazzoli F. Chronic gastritis (http://www.emedicine.com/med/topic852.htm). *Emedicine.* 2007;2.
16. Warren JR, Marshall B. Unidentified curved bacilli on gastric epithelium in active chronic gastritis. *Lancet.* 1983;1:1273–1275.
17. Marshall BJ. *Helicobacter pylori*: The etiologic agent for peptic ulcer. *JAMA.* 1995;274:1064–1066.
18. International Agency for Research of Cancer. Shistosomes, liver flukes and Helicobacter pylori. *IARC Monogr Eval Carcinog Risks Hum.* 1994.
19. Parsonnet J, Friedman GD, Vandersteen DP, et al. Helicobacter pylori and the risk of gastric carcinoma. *N Eng J Med.* 1991;325:1127–1131.
20. Huang JQ, Sridhar S, Chen Y, et al. Meta-analysis of the relationship between Helicobacter pylori seropositivity and gastric cancer. *Gastroenterology.* 1998;114:1169–1179.
21. Helicobacter and Cancer Collaborative Group. Gastric cancer and Helicobacter pylori: A combined analysis of 12 case control studies nested within prospective cohorts. *Gut.* 2001;49:347–353.
22. Sipponen P, Marshall BJ. Gastritis and gastric cancer. Western countries. *Gastroenterol Clin North Am.* 2000;29:579–592, v–vi.
23. Uemura N, Okamoto S, Yamamoto S, et al. Helicobacter pylori infection and the development of gastric cancer. *N Engl J Med.* 2001;345:784–789.
24. Whiting JL, Sigurdsson A, Rowlands DC, et al. The long term results of endoscopic surveillance of premalignant gastric lesions. *Gut.* 2002;50:378–381.
25. Sepulveda AR, Coelho LG. Helicobacter pylori and gastric malignancies. *Helicobacter.* 2002;7(Suppl 1):37–42.
26. Watanabe T, Tada M, Nagai H, et al. Helicobacter pylori infection induces gastric cancer in Mongolian gerbils. *Gastroenterology.* 1998;115:642–648.
27. Sugiyama A, Maruta F, Ikeno T, et al. Helicobacter pylori infection enhances N-methyl-N-nitrosourea-induced stomach carcinogenesis in the Mongolian gerbil. *Cancer Res.* 1998;58:2067–2069.
28. Honda S, Fujioka T, Tokieda M, et al. Development of Helicobacter pylori-induced gastric carcinoma in Mongolian gerbils. *Cancer Res.* 1998;58:4255–4259.
29. Shimizu N, Inada K, Nakanishi H, et al. Helicobacter pylori infection enhances glandular stomach carcinogenesis in Mongolian gerbils treated with chemical carcinogens. *Carcinogenesis.* 1999;20:669–676.
30. Houghton J, Stoicov C, Nomura S, et al. Gastric cancer originating from bone marrow-derived cells. *Science.* 2004;306:1568–1571.
31. Uemura N, Mukai T, Okamoto S, et al. Effect of Helicobacter pylori eradication on subsequent development of cancer after endoscopic resection of early gastric cancer. *Cancer Epidemiol Biomarkers Prev.* 1997;6:639–642.
32. Fukase K, Kato M, Kikuchi S, et al. Effect of eradication of Helicobacter pylori on incidence of metachronous gastric carcinoma after endoscopic resection of early gastric cancer: An open-label, randomised controlled trial. *Lancet.* 2008;372:392–397.
33. Wong BC, Lam SK, Wong WM, et al. Helicobacter pylori eradication to prevent gastric cancer in a high-risk region of China: A randomized controlled trial. *JAMA.* 2004;291:187–194.

34. Goodman KJ, Correa P. The transmission of Helicobacter pylori. A critical review of the evidence. *Int J Epidemiol.* 1995;24:875–887.
35. Sipponen P. Helicobacter pylori gastritis—Epidemiology. *J Gastroenterol.* 1997;32:273–277.
36. Malfertheiner P, Megraud F, O'Morain C, et al. Current concepts in the management of Helicobacter pylori infection: The Maastricht III Consensus Report. *Gut.* 2007;56:772–781.
37. Chey WD, Wong BC. American College of Gastroenterology guideline on the management of Helicobacter pylori infection. *Am J Gastroenterol.* 2007;102:1808–1825.
38. Asaka M, Sugiyama T, Nobuta A, et al. Atrophic gastritis and intestinal metaplasia in Japan: Results of a large multicenter study. *Helicobacter.* 2001;6:294–299.
39. Robinson K, Argent RH, Atherton JC. The inflammatory and immune response to Helicobacter pylori infection. *Best Pract Res Clin Gastroenterol.* 2007;21:237–259.
40. Wilson KT, Crabtree JE. Immunology of Helicobacter pylori: Insights into the failure of the immune response and perspectives on vaccine studies. *Gastroenterology.* 2007;133:288–308.
41. Anonymous. Live flukes and *Helicobacter pylori.* IARC Working group on the Evaluation of Carcinogenic Risks to Humans. Lyon. 7–14 June 1994. *IARC Monogr Eval Carcinog Risks Hum.* 1994;61:1–241.
42. Nomura A, Stemmermann GN, Chyou PH, et al. Helicobacter pylori infection and gastric carcinoma among Japanese Americans in Hawaii. *N Engl J Med.* 1991;325: 1132–1136.
43. Dixon MF. Histological responses to Helicobacter pylori infection: Gastritis, atrophy and preneoplasia. *Baillieres Clin Gastroenterol.* 1995;9:467–486.
44. Correa P, Houghton J. Carcinogenesis of Helicobacter pylori. *Gastroenterology.* 2007;133:659–672.
45. Giannakis M, Chen SL, Karam SM, et al. Helicobacter pylori evolution during progression from chronic atrophic gastritis to gastric cancer and its impact on gastric stem cells. *Proc Natl Acad Sci USA.* 2008;105:4358–4363.
46. Katoh M. Dysregulation of stem cell signaling network due to germline mutation, SNP, Helicobacter pylori infection, epigenetic change and genetic alteration in gastric cancer. *Cancer Biol Ther.* 2007;6:832–839.
47. Hamilton RS, Aaltonen LA. Pathology and Genetics of the Digestive System. World Health Organization Classification of Tumors. IARC Press, Lyon, France, 2000:38–52.
48. Greene F, Page D, Fleming I, et al. eds. *AJCC Cancer Staging Manual.* 6th ed. New York: Springer-Verlag; 2002.
49. Lauren P. The two histological main types of gastric carcinoma: Diffuse and so-called intestinal-type carcinoma: An attempt at a histo-clinical classification. *Acta Pathol et Microbiol Scandinav.* 1965;64:31–49.
50. Sepulveda AR, Wu L, Ota H, et al. Molecular identification of main cellular lineages as a tool for the classification of gastric cancer. *Hum Pathol.* 2000;31:566–574.
51. Correa P. Human gastric carcinogenesis: A multistep and multifactorial process—First American Cancer Society Award Lecture on Cancer Epidemiology and Prevention. *Cancer Res.* 1992; 52:6735–6740.
52. Cassaro M, Rugge M, Gutierrez O, et al. Topographic patterns of intestinal metaplasia and gastric cancer. *Am J Gastroenterol.* 2000;95:1431–1438.
53. Recavarren-Arce S, Leon-Barua R, Cok J, et al. Helicobacter pylori and progressive gastric pathology that predisposes to gastric cancer. *Scand J Gastroenterol Suppl.* 1991;181:51–57.
54. Kodama K, Sumii K, Kawano M, et al. Gastric juice nitrite and vitamin C in patients with gastric cancer and atrophic gastritis: Is low acidity solely responsible for cancer risk? *Eur J Gastroenterol Hepatol.* 2003;15:987–993.
55. Sepulveda A, Peterson LE, Shelton J, et al. Histological patterns of gastritis in H. pylori-infected individuals with a family history of gastric cancer. *Am J Gastroenterol.* 2002;97:1365–1370.
56. Shimada T, Watanabe N, Hiraishi H, et al. Redox regulation of interleukin-8 expression in MKN28 cells. *Dig Dis Sci.* 1999; 44:266–273.
57. Handa O, Naito Y, Yoshikawa T. CagA protein of Helicobacter pylori: A hijacker of gastric epithelial cell signaling. *Biochem Pharmacol.* 2007;73:1697–1702.
58. Peek RM Jr, Moss SF, Tham KT, et al. Helicobacter pylori cagA+ strains and dissociation of gastric epithelial cell proliferation from apoptosis [see comments]. *J Natl Cancer Inst.* 1997;89: 863–868.
59. Moss SF, Calam J, Agarwal B, et al. Induction of gastric epithelial apoptosis by Helicobacter pylori. *Gut.* 1996;38:498–501.
60. Shirin H, Sordillo EM, Kolevska TK, et al. Chronic helicobacter pylori infection induces an apoptosis-resistant phenotype associated with decreased expression of p27(kip1). *Infect Immun.* 2000;68:5321–5328.
61. Farinati F, Cardin R, Degan P, et al. Oxidative DNA damage accumulation in gastric carcinogenesis. *Gut.* 1998;42:351–356.
62. Yao Y, Tao H, Park DI, et al. Demonstration and characterization of mutations induced by Helicobacter pylori organisms in gastric epithelial cells. *Helicobacter.* 2006;11:272–286.
63. El-Omar EM, Carrington M, Chow WH, et al. Interleukin-1 polymorphisms associated with increased risk of gastric cancer. *Nature.* 2000;404:398–402.
64. El-Omar EM, Rabkin CS, Gammon MD, et al. Increased risk of noncardia gastric cancer associated with proinflammatory cytokine gene polymorphisms. *Gastroenterology.* 2003;124:1193–1201.
65. El-Omar EM. Role of host genes in sporadic gastric cancer. *Best Pract Res Clin Gastroenterol.* 2006;20:675–686.
66. Machado JC, Figueiredo C, Canedo P, et al. A proinflammatory genetic profile increases the risk for chronic atrophic gastritis and gastric carcinoma. *Gastroenterology.* 2003;125:364–371.
67. Farinati F, Cardin R, Bortolami M, et al. Oxidative DNA damage in gastric cancer: CagA status and OGG1 gene polymorphism. *Int J Cancer.* 2008;123:51–55.
68. Kang GH, Lee S, Kim JS, et al. Profile of aberrant CpG island methylation along the multistep pathway of gastric carcinogenesis. *Lab Invest.* 2003;83:635–641.
69. Kang GH, Lee S, Cho NY, et al. DNA methylation profiles of gastric carcinoma characterized by quantitative DNA methylation analysis. *Lab Invest.* 2008;161–170.
70. Chan AO, Chu KM, Huang C, et al. Association between Helicobacter pylori infection and interleukin 1beta polymorphism predispose to CpG island methylation in gastric cancer. *Gut.* 2007;56:595–597.
71. Qian X, Huang C, Cho CH, et al. E-cadherin promoter hypermethylation induced by interleukin-1beta treatment or H. pylori infection in human gastric cancer cell lines. *Cancer Lett.* 2008;263:107–113.
72. Fu S, Ramanujam KS, Wong A, et al. Increased expression and cellular localization of inducible nitric oxide synthase and cyclooxygenase 2 in Helicobacter pylori gastritis. *Gastroenterology.* 1999;116:1319–1329.
73. Li CQ, Pignatelli B, Ohshima H. Coexpression of interleukin-8 and inducible nitric oxide synthase in gastric mucosa infected with cagA+ Helicobacter pylori. *Dig Dis Sci.* 2000;45:55–62.
74. Grisham MB, Jourd'heuil D, Wink DA. Review article: Chronic inflammation and reactive oxygen and nitrogen metabolism—implications in DNA damage and mutagenesis. *Aliment Pharmacol Ther.* 2000;14(Suppl 1):3–9.
75. Plummer SM, Hall M, Faux SP. Oxidation and genotoxicity of fecapentaene-12 are potentiated by prostaglandin H synthase. *Carcinogenesis.* 1995;16:1023–1028.
76. Verhulst ML, van Oijen AH, Roelofs HM, et al. Antral glutathione concentration and glutathione S-transferase activity in patients with and without Helicobacter pylori. *Dig Dis Sci.* 2000;45:629–632.
77. Hahm KB, Lee KJ, Choi SY, et al. Possibility of chemoprevention by the eradication of Helicobacter pylori: Oxidative DNA damage and apoptosis in H. pylori infection. *Am J Gastroenterol.* 1997;92:1853–1857.
78. Shiao YH, Rugge M, Correa P, et al. p53 alteration in gastric precancerous lesions. *Am J Pathol.* 1994;144:511–517.
79. Kolodner RD, Marsischky GT. Eukaryotic DNA mismatch repair. *Curr Opin Genet Dev.* 1999;9:89–96.
80. Flores-Rozas H, Kolodner RD. The Saccharomyces cerevisiae MLH3 gene functions in MSH3-dependent suppression of frameshift mutations. *Proc Natl Acad Sci USA.* 1998;95: 12404–12409.
81. Lipkin SM, Wang V, Jacoby R, et al. MLH3: A DNA mismatch repair gene associated with mammalian microsatellite instability. *Nat Genet.* 2000;24:27–35.

82. Umar A. Lynch syndrome (HNPCC) and microsatellite instability. *Dis Markers.* 2004;20:179–180.
83. Kim JJ, Tao H, Carloni E, et al. Helicobacter pylori impairs DNA mismatch repair in gastric epithelial cells. *Gastroenterology.* 2002;123:542–553.
84. Leung WK, Kim JJ, Kim JG, et al. Microsatellite instability in gastric intestinal metaplasia in patients with and without gastric cancer. *Am J Pathol.* 2000;156:537–543.
85. Kashiwagi K, Watanabe M, Ezaki T, et al. Clinical usefulness of microsatellite instability for the prediction of gastric adenoma or adenocarcinoma in patients with chronic gastritis. *Br J Cancer.* 2000;82:1814–1818.
86. Ling XL, Fang DC, Wang RQ, et al. Mitochondrial microsatellite instability in gastric cancer and its precancerous lesions. *World J Gastroenterol.* 2004;10:800–803.
87. Semba S, Yokozaki H, Yamamoto S, et al. Microsatellite instability in precancerous lesions and adenocarcinomas of the stomach. *Cancer.* 1996;77:1620–1627.
88. Hamamoto T, Yokozaki H, Semba S, et al. Altered microsatellites in incomplete-type intestinal metaplasia adjacent to primary gastric cancers. *J Clin Pathol.* 1997;50:841–846.
89. Ottini L, Palli D, Falchetti M, et al. Microsatellite instability in gastric cancer is associated with tumor location and family history in a high-risk population from Tuscany. *Cancer Res.* 1997;57:4523–4529.
90. Fang DC, Jass JR, Wang DX, et al. Infrequent loss of heterozygosity of APC/MCC and DCC genes in gastric cancer showing DNA microsatellite instability. *J Clin Pathol.* 1999;52:504–508.
91. Kobayashi K, Okamoto T, Takayama S, et al. Genetic instability in intestinal metaplasia is a frequent event leading to well-differentiated early adenocarcinoma of the stomach. *Eur J Cancer.* 2000;36:1113–1119.
92. Wu MS, Lee CW, Shun CT, et al. Clinicopathological significance of altered loci of replication error and microsatellite instability-associated mutations in gastric cancer. *Cancer Res.* 1998;58:1494–1497.
93. Wu MS, Lee CW, Shun CT, et al. Distinct clinicopathologic and genetic profiles in sporadic gastric cancer with different mutator phenotypes. *Genes Chromosomes Cancer.* 2000;27:403–411.
94. Park TJ, Han SU, Cho YK, et al. Methylation of O(6)-methylguanine-DNA methyltransferase gene is associated significantly with K-ras mutation, lymph node invasion, tumor staging, and disease free survival in patients with gastric carcinoma. *Cancer.* 2001;92:2760–2768.
95. Sepulveda AR, Tao H, Carloni E, et al. Screening of gene expression profiles in gastric epithelial cells induced by Helicobacter pylori using microarray analysis. *Aliment Pharmacol Ther.* 2002;16(Suppl 2):145–157.
96. Meireles SI, Cristo EB, Carvalho AF, et al. Molecular classifiers for gastric cancer and nonmalignant diseases of the gastric mucosa. *Cancer Res.* 2004;64:1255–1265.
97. Meireles SI, Carvalho AF, Hirata R, et al. Differentially expressed genes in gastric tumors identified by cDNA array. *Cancer Lett.* 2003;190:199–211.
98. Baek MJ, Kang H, Kim SE, et al. Expression of hMLH1 is inactivated in the gastric adenomas with enhanced microsatellite instability. *Br J Cancer.* 2001;85:1147–1152.
99. Fleisher AS, Esteller M, Tamura G, et al. Hypermethylation of the hMLH1 gene promoter is associated with microsatellite instability in early human gastric neoplasia. *Oncogene.* 2001;20:329–335.
100. Edmonston TB, Cuesta KH, Burkholder S, et al. Colorectal carcinomas with high microsatellite instability: Defining a distinct immunologic and molecular entity with respect to prognostic markers. *Hum Pathol.* 2000;31:1506–1514.
101. Umar A, Boland CR, Terdiman JP, et al. Revised Bethesda Guidelines for hereditary nonpolyposis colorectal cancer (Lynch syndrome) and microsatellite instability. *J Natl Cancer Inst.* 2004;96:261–268.
102. Kim SS, Bhang CS, Min KO, et al. p53 mutations and microsatellite instabilities in the subtype of intestinal metaplasia of the stomach. *J Korean Med Sci.* 2002;17:490–496.
103. Abraham SC, Park SJ, Lee JH, et al. Genetic alterations in gastric adenomas of intestinal and foveolar phenotypes. *Mod Pathol.* 2003;16:786–795.
104. Lee JH, Abraham SC, Kim HS, et al. Inverse relationship between APC gene mutation in gastric adenomas and development of adenocarcinoma. *Am J Pathol.* 2002;161:611–618.
105. Hayden JD, Martin IG, Cawkwell L, et al. The role of microsatellite instability in gastric carcinoma. *Gut.* 1998;42:300–303.
106. Strickler JG, Zheng J, Shu Q, et al. p53 mutations and microsatellite instability in sporadic gastric cancer: When guardians fail. *Cancer Research.* 1994;54: 4750–4755.
107. Sepulveda AR, Santos AC, Yamaoka Y, et al. Marked differences in the frequency of microsatellite instability in gastric cancer from different countries. *Am J Gastroenterol.* 1999;94:3034–3038.
108. Yamamoto H, Sawai H, Perucho M. Frameshift somatic mutations in gastrointestinal cancer of the microsatellite mutator phenotype. *Cancer Res.* 1997;57:4420–4426.
109. Shinmura K, Tani M, Isogaki J, et al. RER phenotype and its associated mutations in familial gastric cancer. *Carcinogenesis.* 1998;19:247–251.
110. Myeroff LL, Parsons R, Kim SJ, et al. A transforming growth factor beta receptor type II gene mutation common in colon and gastric but rare in endometrial cancers with microsatellite instability. *Cancer Research.* 1995;55:5545–5547.
111. Chung YJ, Park SW, Song JM, et al. Evidence of genetic progression in human gastric carcinomas with microsatellite instability. *Oncogene.* 1997;15:1719–1726.
112. Kim JJ, Baek MJ, Kim L, et al. Accumulated frameshift mutations at coding nucleotide repeats during the progression of gastric carcinoma with microsatellite instability. *Lab Invest.* 1999;79:1113–1120.
113. Kim HS, Woo DK, Bae SI, et al. Microsatellite instability in the adenoma-carcinoma sequence of the stomach. *Lab Invest.* 2000;80:57–64.
114. Correa P, Shiao Y-H. Phenotypic and genotypic events in gastric carcinogenesis. *Cancer Research.* 1994;54(Suppl):1941–1943.
115. Nakatsuru S, Yanagisawa A, Furukawa Y, et al. Somatic mutations of the APC gene in precancerous lesion of the stomach. *Hum Mol Genetics.* 1993;2:1463–1465.
116. Maesawa C, Tamura G, Suzuki Y, et al. The sequential accumulation of genetic alterations characteristic of the colorectal adenoma-carcinoma sequence does not occur between gastric adenoma and adenocarcinoma. *J Pathol.* 1995;176:249–258.
117. Imatani A, Sasano H, Asaki S, et al. Analysis of p53 abnormalities in endoscopic gastric biopsies. *Anticancer Res.* 1996;16:2049–2056.
118. Lee HS, Choi SI, Lee HK, et al. Distinct clinical features and outcomes of gastric cancers with microsatellite instability. *Mod Pathol.* 2002;15:632–640.
119. Hunt JD, Mera R, Strimas A, et al. KRAS mutations are not predictive for progression of preneoplastic gastric lesions. *Cancer Epidemiol Biomarkers Prev.* 2001;10:79–80.
120. Toyota M, Ahuja N, Suzuki H, et al. Aberrant methylation in gastric cancer associated with the CpG island methylator phenotype. *Cancer Res.* 1999;59:5438–5442.
121. Kang GH, Shim YH, Jung HY, et al. CpG island methylation in premalignant stages of gastric carcinoma. *Cancer Res.* 2001;61:2847–2851.
122. To KF, Leung WK, Lee TL, et al. Promoter hypermethylation of tumor-related genes in gastric intestinal metaplasia of patients with and without gastric cancer. *Int J Cancer.* 2002;102:623–628.
123. Waki T, Tamura G, Sato M, et al. Promoter methylation status of DAP-kinase and RUNX3 genes in neoplastic and non-neoplastic gastric epithelia. *Cancer Sci.* 2003;94:360–364.
124. Lee JH, Park SJ, Abraham SC, et al. Frequent CpG island methylation in precursor lesions and early gastric adenocarcinomas. *Oncogene.* 2004;23:4646–4654.
125. Kang GH, Lee HJ, Hwang KS, et al. Aberrant CpG island hypermethylation of chronic gastritis, in relation to aging, gender, intestinal metaplasia, and chronic inflammation. *Am J Pathol.* 2003;163:1551–1556.
126. Mikata R, Yokosuka O, Fukai K, et al. Analysis of genes upregulated by the demethylating agent 5-aza-2′-deoxycytidine in gastric cancer cell lines. *Int J Cancer.* 2006;119:1616–1622.
127. Hippo Y, Taniguchi H, Tsutsumi S, et al. Global gene expression analysis of gastric cancer by oligonucleotide microarrays. *Cancer Res.* 2002;62:233–240.

128. Lee S, Baek M, Yang H, et al. Identification of genes differentially expressed between gastric cancers and normal gastric mucosa with cDNA microarrays. *Cancer Lett.* 2002;184:197–206.
129. Kim B, Bang S, Lee S, et al. Expression profiling and subtype-specific expression of stomach cancer. *Cancer Res.* 2003;63:8248–8255.
130. Wu MS, Lin YS, Chang YT, et al. Gene expression profiling of gastric cancer by microarray combined with laser capture microdissection. *World J Gastroenterol.* 2005;11:7405–7412.
131. Kim SY, Kim JH, Lee HS, et al. Meta- and gene set analysis of stomach cancer gene expression data. *Mol Cells.* 2007;24:200–209.
132. Myllykangas S, Junnila S, Kokkola A, et al. Integrated gene copy number and expression microarray analysis of gastric cancer highlights potential target genes. *Int J Cancer.* 2008;123:817–825.
133. Hasegawa S, Furukawa Y, Li M, et al. Genome-wide analysis of gene expression in intestinal-type gastric cancers using a complementary DNA microarray representing 23,040 genes. *Cancer Res.* 2002;62:7012–7017.
134. Boussioutas A, Li H, Liu J, et al. Distinctive patterns of gene expression in premalignant gastric mucosa and gastric cancer. *Cancer Res.* 2003;63:2569–2577.
135. Jinawath N, Furukawa Y, Hasegawa S, et al. Comparison of gene-expression profiles between diffuse- and intestinal-type gastric cancers using a genome-wide cDNA microarray. *Oncogene.* 2004;23:6830–6844.
136. Chan SH, Wu CW, Li AF, et al. miR-21 microRNA expression in human gastric carcinomas and its clinical association. *Anticancer Res.* 2008;28:907–911.
137. Volinia S, Calin GA, Liu CG, et al. A microRNA expression signature of human solid tumors defines cancer gene targets. *Proc Natl Acad Sci USA.* 2006;103:2257–2261.
138. Guilford P, Hopkins J, Harraway J, et al. E-cadherin germline mutations in familial gastric cancer. *Nature.* 1998;392:402–405.
139. Guilford PJ, Hopkins JB, Grady WM, et al. E-cadherin germline mutations define an inherited cancer syndrome dominated by diffuse gastric cancer. *Hum Mutat.* 1999;14:249–255.
140. Barber M, Fitzgerald RC, Caldas C. Familial gastric cancer—Aetiology and pathogenesis. *Best Pract Res Clin Gastroenterol.* 2006;20:721–734.
141. Lichtenstein P, Holm NV, Verkasalo PK, et al. Environmental and heritable factors in the causation of cancer—Analyses of cohorts of twins from Sweden, Denmark, and Finland. *N Engl J Med.* 2000;343:78–85.
142. Rogers WM, Dobo E, Norton JA, et al. Risk-reducing total gastrectomy for germline mutations in E-cadherin (CDH1): Pathologic findings with clinical implications. *Am J Surg Pathol.* 2008;32:799–809.
143. Gayther SA, Gorringe KL, Ramus SJ, et al. Identification of germ-line E-cadherin mutations in gastric cancer families of European origin. *Cancer Res.* 1998;58:4086–4089.
144. Kaurah P, MacMillan A, Boyd N, et al. Founder and recurrent CDH1 mutations in families with hereditary diffuse gastric cancer. *JAMA.* 2007;297:2360–2372.
145. Keller G. Hereditary aspects of gastric cancer. *Pathologica.* 2002;94:229–233.
146. Yoon KA, Ku JL, Yang HK, et al. Germline mutations of E-cadherin gene in Korean familial gastric cancer patients. *J Hum Genet.* 1999;44:177–180.
147. Caldas C, Carneiro F, Lynch HT, et al. Familial gastric cancer: Overview and guidelines for management. *J Med Genet.* 1999;36:873–880.
148. Shinmura K, Kohno T, Takahashi M, et al. Familial gastric cancer: Clinicopathological characteristics, RER phenotype and germline p53 and E-cadherin mutations. *Carcinogenesis.* 1999;20:1127–1131.
149. Aarnio M, Salovaara R, Aaltonen LA, et al. Features of gastric cancer in hereditary non-polyposis colorectal cancer syndrome. *Int J Cancer.* 1997;74:551–555.
150. Heinimann K, Muller H, Weber W, et al. Disease expression in Swiss hereditary non-polyposis colorectal cancer (HNPCC) kindreds. *Int J Cancer.* 1997;74:281–285.
151. Giardiello FM, Welsh SB, Hamilton SR, et al. Increased risk of cancer in the Peutz-Jeghers syndrome. *N Engl J Med.* 1987;316:1511–1514.
152. Boardman LA, Thibodeau SN, Schaid DJ, et al. Increased risk for cancer in patients with the Peutz-Jeghers syndrome. *Ann Intern Med.* 1998;128:896–899.
153. Wallace MH, Phillips RK. Upper gastrointestinal disease in patients with familial adenomatous polyposis. *Br J Surg.* 1998;85:742–750.
154. Kim IJ, Park JH, Kang HC, et al. A novel germline mutation in the MET extracellular domain in a Korean patient with the diffuse type of familial gastric cancer. *J Med Genet.* 2003;40:e97.
155. Figer A, Irmin L, Geva R, et al. The rate of the 6174delT founder Jewish mutation in BRCA2 in patients with non-colonic gastrointestinal tract tumours in Israel. *Br J Cancer.* 2001;84:478–481.
156. Berx G, Staes K, van Hengel J, et al. Cloning and characterization of the human invasion suppressor gene E-cadherin (CDH1). *Genomics.* 1995;26:281–289.
157. Brooks-Wilson AR, Kaurah P, Suriano G, et al. Germline E-cadherin mutations in hereditary diffuse gastric cancer: Assessment of 42 new families and review of genetic screening criteria. *J Med Genet.* 2004;41:508–517.
158. Pharoah PD, Guilford P, Caldas C. Incidence of gastric cancer and breast cancer in CDH1 (E-cadherin) mutation carriers from hereditary diffuse gastric cancer families. *Gastroenterology.* 2001;121:1348–1353.
159. Oliveira C, Moreira H, Seruca R, et al. Role of pathology in the identification of hereditary diffuse gastric cancer: Report of a Portuguese family. *Virchows Arch.* 2005;446:181–184.
160. Grady WM, Willis J, Guilford PJ, et al. Methylation of the CDH1 promoter as the second genetic hit in hereditary diffuse gastric cancer. *Nat Genet.* 2000;26:16–17.
161. Takeichi M. Cadherin cell adhesion receptors as a morphogenetic regulator. *Science.* 1991;251:1451–1455.
162. Jou TS, Stewart DB, Stappert J, et al. Genetic and biochemical dissection of protein linkages in the cadherin-catenin complex. *Proc Natl Acad Sci USA.* 1995;92: 5067–5071.
163. Kallakury BV, Sheehan CE, Winn-Deen E, et al. Decreased expression of catenins (alpha and beta), p120 CTN, and E-cadherin cell adhesion proteins and E-cadherin gene promoter methylation in prostatic adenocarcinomas. *Cancer.* 2001;92:2786–2795.
164. Caldeira JR, Prando EC, Quevedo FC, et al. CDH1 promoter hypermethylation and E-cadherin protein expression in infiltrating breast cancer. *BMC Cancer.* 2006;6:48.
165. Hirohashi S. Molecular aspects of adhesion-epigenetic mechanisms for inactivation of the E-cadherin-mediated cell adhesion system in cancers. *Verh Dtsch Ges Pathol.* 2000;84:28–32.
166. Birchmeier W. E-cadherin as a tumor (invasion) suppressor gene. *Bioessays.* 1995;17:97–99.
167. Bremm A, Walch A, Fuchs M, et al. Enhanced activation of epidermal growth factor receptor caused by tumor-derived E-cadherin mutations. *Cancer Res.* 2008;68:707–714.
168. American Cancer Society. Estimated New Cancer Cases and Deaths by Sex for All Sites, US. http://www.cancer.org/downloads/MED/Page4.pdf/; 2004 Accessed November 7, 2004.
169. SEER. http://seer.cancer.gov/csr/1975_2005/results_single/sect_01_table.01.pdf/.
170. Lynch HT, Lanspa SJ, Boman BM, et al. Hereditary nonpolyposis colorectal cancer—Lynch syndromes I and II. *Gastroenterol Clin North Am.* 1988;17:679–712.
171. Nielsen M, Hes FJ, Nagengast FM, et al. Germline mutations in APC and MUTYH are responsible for the majority of families with attenuated familial adenomatous polyposis. *Clin Genet.* 2007;71:427–433.
172. Torlakovic E, Snover DC. Sessile serrated adenoma: A brief history and current status. *Crit Rev Oncog.* 2006;12:27–39.
173. Iwabuchi M, Sasano H, Hiwatashi N, et al. Serrated adenoma: A clinicopathological, DNA ploidy, and immunohistochemical study. *Anticancer Res.* 2000;20:1141–1147.
174. East JE, Saunders BP, Jass JR. Sporadic and syndromic hyperplastic polyps and serrated adenomas of the colon: Classification, molecular genetics, natural history, and clinical management. *Gastroenterol Clin North Am.* 2008;37:25–46.
175. Lindblom A. Different mechanisms in the tumorigenesis of proximal and distal colon cancers. *Curr Opin Oncol.* 2001;13:63–69.

176. Tsushimi T, Noshima S, Oga A, et al. DNA amplification and chromosomal translocations are accompanied by chromosomal instability: Analysis of seven human colon cancer cell lines by comparative genomic hybridization and spectral karyotyping. *Cancer Genet Cytogenet.* 2001;126:34–38.
177. Goel A, Arnold CN, Niedzwiecki D, et al. Characterization of sporadic colon cancer by patterns of genomic instability. *Cancer Res.* 2003;63:1608–1614.
178. Charames GS, Bapat B. Genomic instability and cancer. *Curr Mol Med.* 2003;3:589–596.
179. Whitehall VL, Wynter CV, Walsh MD, et al. Morphological and molecular heterogeneity within nonmicrosatellite instability-high colorectal cancer. *Cancer Res.* 2002;62:6011–6014.
180. Grady WM. CIMP and colon cancer gets more complicated. *Gut.* 2007;56:1498–1500.
181. Goel A, Nagasaka T, Arnold CN, et al. The CpG island methylator phenotype and chromosomal instability are inversely correlated in sporadic colorectal cancer. *Gastroenterology.* 2007;132: 127–138.
182. Kern SE, Fearon ER, Tersmette KW, et al. Clinical and pathological associations with allelic loss in colorectal carcinoma [corrected]. *JAMA.* 1989;261:3099–3103.
183. Lengauer C, Kinzler KW, Vogelstein B. Genetic instability in colorectal cancers. *Nature.* 1997;386:623–627.
184. Grady WM. Genomic instability and colon cancer. *Cancer Metastasis Rev.* 2004;23:11–27.
185. Rhyu MG, Park WS, Jung YJ, et al. Allelic deletions of MCC/APC and p53 are frequent late events in human gastric carcinogenesis. *Gastroenterology.* 1994;106:1584–1588.
186. Fearon ER, Cho KR, Nigro JM, et al. Identification of a chromosome 18q gene that is altered in colorectal cancers. *Science.* 1990;247:49–56.
187. Kikuchi-Yanoshita R, Konishi M, Fukunari H, et al. Loss of expression of the DCC gene during progression of colorectal carcinomas in familial adenomatous polyposis and non-familial adenomatous polyposis patients. *Cancer Res.* 1992;52:3801–3803.
188. Suraweera N, Duval A, Reperant M, et al. Evaluation of tumor microsatellite instability using five quasimonomorphic mononucleotide repeats and pentaplex PCR. *Gastroenterology.* 2002;123:1804–1811.
189. Jen J, Powell SM, Papadopoulos N, et al. Molecular determinants of dysplasia in colorectal lesions. *Cancer Res.* 1994;54: 5523–5526.
190. Miyaki M, Konishi M, Kikuchi-Yanoshita R, et al. Characteristics of somatic mutation of the adenomatous polyposis coli gene in colorectal tumors. *Cancer Res.* 1994;54:3011–3020.
191. Vogelstein B, Fearon ER, Hamilton SR, et al. Genetic alterations during colorectal-tumor development. *N Engl J Med.* 1988;319:525–532.
192. Hedrick L, Cho KR, Fearon ER, et al. The DCC gene product in cellular differentiation and colorectal tumorigenesis. *Genes Dev.* 1994;8:1174–1183.
193. Baker SJ, Fearon ER, Nigro JM, et al. Chromosome 17 deletions and p53 gene mutations in colorectal carcinomas. *Science.* 1989;244:217–221.
194. Kikuchi-Yanoshita R, Konishi M, Ito S, et al. Genetic changes of both p53 alleles associated with the conversion from colorectal adenoma to early carcinoma in familial adenomatous polyposis and non-familial adenomatous polyposis patients. *Cancer Res.* 1992;52:3965–3971.
195. Soussi T, Asselain B, Hamroun D, et al. Meta-analysis of the p53 mutation database for mutant p53 biological activity reveals a methodologic bias in mutation detection. *Clin Cancer Res.* 2006;12:62–69.
196. Pruitt K, Der CJ. Ras and Rho regulation of the cell cycle and oncogenesis. *Cancer Lett.* 2001;171:1–10.
197. Takayama T, Ohi M, Hayashi T, et al. Analysis of K-ras, APC, and beta-catenin in aberrant crypt foci in sporadic adenoma, cancer, and familial adenomatous polyposis. *Gastroenterology.* 2001;121:599–611.
198. Losi L, Ponz de Leon M, Jiricny J, et al. K-ras and p53 mutations in hereditary non-polyposis colorectal cancers. *Int J Cancer.* 1997;74:94–96.
199. Toyota M, Ahuja N, Ohe-Toyota M, CpG island methylator phenotype in colorectal cancer. *Proc Natl Acad Sci USA.* 1999;96:8681–8686.
200. Samowitz WS, Holden JA, Curtin K, et al. Inverse relationship between microsatellite instability and K-ras and p53 gene alterations in colon cancer. *Am J Pathol.* 2001;158:1517–1524.
201. Young J, Simms LA, Biden KG, et al. Features of colorectal cancers with high-level microsatellite instability occurring in familial and sporadic settings: Parallel pathways of tumorigenesis. *Am J Pathol.* 2001;159:2107–2116.
202. Yashiro M, Carethers JM, Laghi L, et al. Genetic pathways in the evolution of morphologically distinct colorectal neoplasms. *Cancer Res.* 2001;61:2676–2683.
203. Hadjihannas MV, Bruckner M, Jerchow B, et al. Aberrant Wnt/beta-catenin signaling can induce chromosomal instability in colon cancer. *Proc Natl Acad Sci USA.* 2006;103:10747–10752.
204. Herman JG, Umar A, Polyak K, et al. Incidence and functional consequences of hMLH1 promoter hypermethylation in colorectal carcinoma. *Proc Natl Acad Sci USA.* 1998;95:6870–6875.
205. Kuismanen SA, Holmberg MT, Salovaara R, et al. Genetic and epigenetic modification of MLH1 accounts for a major share of microsatellite-unstable colorectal cancers. *Am J Pathol.* 2000;156:1773–1779.
206. Deng G, Chen A, Hong J, et al. Methylation of CpG in a small region of the hMLH1 promoter invariably correlates with the absence of gene expression. *Cancer Res.* 1999;59:2029–2033.
207. Lynch HT, Boman B, Fitzgibbons Jr RJ, et al. Hereditary nonpolyposis colon cancer: (Lynch syndrome I and II). A challenge for the clinician. *Nebr Med J.* 1989;74:2–7.
208. Lynch HT, Drouhard T, Lanspa S, et al. Mutation of an mutL homologue in a Navajo family with hereditary nonpolyposis colorectal cancer. *J Natl Cancer Inst.* 1994;86:1417–1419.
209. Lynch HT, Lynch JF. 25 years of HNPCC. *Anticancer Res.* 1994;14:1617–1624.
210. Peltomaki P. DNA mismatch repair and cancer. *Mutat Res.* 2001;488:77–85.
211. Liu B, Nicolaides C, Markowitz S, et al. Mismatch repair defects in sporadic colorectal cancers with microsatellite instability. *Nature Genetics.* 1995;9:48–55.
212. Calistri D, Presciuttini S, Buonsanti G, et al. Microsatellite instability in colorectal-cancer patients with suspected genetic predisposition. *Int J Cancer.* 2000;89:87–91.
213. Yamamoto H, Min Y, Itoh F, et al. Differential involvement of the hypermethylator phenotype in hereditary and sporadic colorectal cancers with high-frequency microsatellite instability. *Genes Chromosomes Cancer.* 2002;33:322–325.
214. Wheeler JM, Loukola A, Aaltonen LA, et al. The role of hypermethylation of the hMLH1 promoter region in HNPCC versus MSI+ sporadic colorectal cancers. *J Med Genet.* 2000;37:588–592.
215. Lynch JP, Hoops TC. The genetic pathogenesis of colorectal cancer. *Hematol Oncol Clin North Am.* 2002;16:775–810.
216. Gologan A, Sepulveda AR. Microsatellite instability and DNA mismatch repair deficiency testing in hereditary and sporadic gastrointestinal cancers. *Clin Lab Med.* 2005;25:179–196.
217. Gologan A, Krasinskas A, Hunt J, et al. Performance of the revised Bethesda guidelines for identification of colorectal carcinomas with a high level of microsatellite instability. *Arch Pathol Lab Med.* 2005;129:1390–1397.
218. Alexander J, Watanabe T, Wu TT, et al. Histopathological identification of colon cancer with microsatellite instability. *Am J Pathol.* 2001;158:527–535.
219. Dolcetti R, Viel A, Doglioni C, et al. High prevalence of activated intraepithelial cytotoxic T lymphocytes and increased neoplastic cell apoptosis in colorectal carcinomas with microsatellite instability. *Am J Pathol.* 1999;154:1805–1813.
220. Greenson JK, Bonner JD, Ben-Yzhak O, et al. Phenotype of microsatellite unstable colorectal carcinomas: Well-differentiated and focally mucinous tumors and the absence of dirty necrosis correlate with microsatellite instability. *Am J Surg Pathol.* 2003;27:563–570.
221. Samowitz WS, Albertsen H, Herrick J, et al. Evaluation of a large, population-based sample supports a CpG island methylator phenotype in colon cancer. *Gastroenterology.* 2005;129:837–845.

222. Ogino S, Cantor M, Kawasaki T, et al. CpG island methylator phenotype (CIMP) of colorectal cancer is best characterized by quantitative DNA methylation analysis and prospective cohort studies. *Gut.* 2006;55:1000–1006.
223. Matsumura S, Oue N, Ito R, et al. The promoter methylation status of the DNA repair gene O6-methylguanine-DNA methyltransferase in ulcerative colitis. *Virchows Arch.* 2003;443:518–523.
224. Petko Z, Ghiassi M, Shuber A, et al. Aberrantly methylated CDKN2A, MGMT, and MLH1 in colon polyps and in fecal DNA from patients with colorectal polyps. *Clin Cancer Res.* 2005;11:1203–1209.
225. Weisenberger DJ, Siegmund KD, Campan M, et al. CpG island methylator phenotype underlies sporadic microsatellite instability and is tightly associated with BRAF mutation in colorectal cancer. *Nat Genet.* 2006;38:787–793.
226. Samowitz WS, Slattery ML, Sweeney C, et al. APC mutations and other genetic and epigenetic changes in colon cancer. *Mol Cancer Res.* 2007;5:165–170.
227. Kawasaki T, Nosho K, Ohnishi M, et al. Correlation of beta-catenin localization with cyclooxygenase-2 expression and CpG island methylator phenotype (CIMP) in colorectal cancer. *Neoplasia.* 2007;9:569–577.
228. Ogino S, Kawasaki T, Nosho K, et al. LINE-1 hypomethylation is inversely associated with microsatellite instability and CpG methylator phenotype in colorectal cancer. *Int J Cancer.* 2008;122:2767–2773.
229. O'Brien MJ, Yang S, Mack C, et al. Comparison of microsatellite instability, CpG island methylation phenotype, BRAF and KRAS status in serrated polyps and traditional adenomas indicates separate pathways to distinct colorectal carcinoma end points. *Am J Surg Pathol.* 2006;30:1491–1501.
230. Yang S, Farraye FA, Mack C, et al. BRAF and KRAS Mutations in hyperplastic polyps and serrated adenomas of the colorectum: Relationship to histology and CpG island methylation status. *Am J Surg Pathol.* 2004;28:1452–1459.
231. Hawkins NJ, Ward RL. Sporadic colorectal cancers with microsatellite instability and their possible origin in hyperplastic polyps and serrated adenomas. *J Natl Cancer Inst.* 2001;93:1307–1313.
232. Munkholm P. Review article: The incidence and prevalence of colorectal cancer in inflammatory bowel disease. *Aliment Pharmacol Ther.* 2003;18(Suppl 2):1–5.
233. Harpaz N, Talbot IC. Colorectal cancer in idiopathic inflammatory bowel disease. *Semin Diagn Pathol.* 1996;13:339–357.
234. Itzkowitz SH, Harpaz N. Diagnosis and management of dysplasia in patients with inflammatory bowel diseases. *Gastroenterology.* 2004;126:1634–1648.
235. Boland CR, Luciani MG, Gasche C, et al. Infection, inflammation, and gastrointestinal cancer. *Gut.* 2005;54:1321–1331.
236. Karrasch T, Jobin C. NF-kappaB and the intestine: Friend or foe? *Inflamm Bowel Dis.* 2008;14:114–124.
237. Rose-John S, Scheller J, Elson G, et al. Interleukin-6 biology is coordinated by membrane-bound and soluble receptors: Role in inflammation and cancer. *J Leukoc Biol.* 2006;80:227–236.
238. Rabe B, Chalaris A, May U, et al. Transgenic blockade of interleukin 6 transsignaling abrogates inflammation. *Blood.* 2008;111:1021–1028.
239. Vivanco I, Sawyers CL. The phosphatidylinositol 3-kinase AKT pathway in human cancer. *Nat Rev Cancer.* 2002;2:489–501.
240. Stenson WF. Prostaglandins and epithelial response to injury. *Curr Opin Gastroenterol.* 2007;23:107–110.
241. Itzkowitz SH. Molecular biology of dysplasia and cancer in inflammatory bowel disease. *Gastroenterol Clin North Am.* 2006;35:553–571.
242. Maia L, Dinis J, Cravo M, et al. Who takes the lead in the development of ulcerative colitis-associated colorectal cancers: Mutator, suppressor, or methylator pathway? *Cancer Genet Cytogenet.* 2005;162:68–73.
243. Aust DE, Terdiman JP, Willenbucher RF, et al. The APC/beta-catenin pathway in ulcerative colitis-related colorectal carcinomas: A mutational analysis. *Cancer.* 2002;94:1421–1427.
244. Redston MS, Papadopoulos N, Caldas C, et al. Common occurrence of APC and K-ras gene mutations in the spectrum of colitis-associated neoplasias. *Gastroenterology.* 1995;108:383–392.
245. Tomlinson I, Ilyas M, Johnson V, et al. A comparison of the genetic pathways involved in the pathogenesis of three types of colorectal cancer. *J Pathol.* 1998;184:148–152.
246. Greenwald BD, Harpaz N, Yin J, et al. Loss of heterozygosity affecting the p53, Rb, and mcc/apc tumor suppressor gene loci in dysplastic and cancerous ulcerative colitis. *Cancer Res.* 1992;52:741–745.
247. Burmer GC, Rabinovitch PS, Haggitt RC, et al. Neoplastic progression in ulcerative colitis: Histology, DNA content, and loss of a p53 allele. *Gastroenterology.* 1992;103:1602–1610.
248. Yin J, Harpaz N, Tong Y, et al. p53 point mutations in dysplastic and cancerous ulcerative colitis lesions. *Gastroenterology.* 1993;104:1633–1639.
249. Brentnall TA, Crispin DA, Rabinovitch PS, et al. Mutations in the p53 gene: An early marker of neoplastic progression in ulcerative colitis. *Gastroenterology.* 1994;107:369–378.
250. Hussain SP, Amstad P, Raja K, et al. Increased p53 mutation load in noncancerous colon tissue from ulcerative colitis: A cancer-prone chronic inflammatory disease. *Cancer Res.* 2000;60:3333–3337.
251. Holzmann K, Klump B, Borchard F, et al. Comparative analysis of histology, DNA content, p53 and Ki-ras mutations in colectomy specimens with long-standing ulcerative colitis. *Int J Cancer.* 1998;76:1–6.
252. Meek DW. The p53 response to DNA damage. *DNA Repair (Amst).* 2004;3:1049–1056.
253. Fleisher AS, Esteller M, Harpaz N, et al. Microsatellite instability in inflammatory bowel disease-associated neoplastic lesions is associated with hypermethylation and diminished expression of the DNA mismatch repair gene, hMLH1. *Cancer Res.* 2000;60:4864–4868.
254. Chang DK, Goel A, Ricciardiello L, et al. Effect of H(2)O(2) on cell cycle and survival in DNA mismatch repair-deficient and -proficient cell lines. *Cancer Lett.* 2003;195:243–251.
255. Hofseth LJ, Khan MA, Ambrose M, et al. The adaptive imbalance in base excision-repair enzymes generates microsatellite instability in chronic inflammation. *J Clin Invest.* 2003;112:1887–1894.
256. Souza RF, Lei J, Yin J, et al. A transforming growth factor beta 1 receptor type II mutation in ulcerative colitis-associated neoplasms [see comments]. *Gastroenterology.* 1997;112:40–45.
257. Takahashi T, Shigematsu H, Shivapurkar N, et al. Aberrant promoter methylation of multiple genes during multistep pathogenesis of colorectal cancers. *Int J Cancer.* 2005;
258. Schulmann K, Mori Y, Croog V, et al. Molecular phenotype of inflammatory bowel disease-associated neoplasms with microsatellite instability. *Gastroenterology.* 2005;129:74–85.
259. Hsieh CJ, Klump B, Holzmann K, et al. Hypermethylation of the p16INK4a promoter in colectomy specimens of patients with long-standing and extensive ulcerative colitis. *Cancer Res.* 1998;58:3942–3945.
260. Issa JP, Ahuja N, Toyota M, et al. Accelerated age-related CpG island methylation in ulcerative colitis. *Cancer Res.* 2001;61:3573–3577.
261. Sato F, Shibata D, Harpaz N, et al. Aberrant methylation of the HPP1 gene in ulcerative colitis-associated colorectal carcinoma. *Cancer Res.* 2002;62:6820–6822.
262. Azarschab P, Porschen R, Gregor M, et al. Epigenetic control of the E-cadherin gene (CDH1) by CpG methylation in colectomy samples of patients with ulcerative colitis. *Genes Chromosomes Cancer.* 2002;35:121–126.
263. Konishi K, Shen L, Wang S, et al. Rare CpG island methylator phenotype in ulcerative colitis-associated neoplasias. *Gastroenterology.* 2007;132:1254–1260.
264. Aust DE, Haase M, Dobryden L, et al. Mutations of the BRAF gene in ulcerative colitis-related colorectal carcinoma. *Int J Cancer.* 2005;115:673–677.
265. Benhattar J, Saraga E. Molecular genetics of dysplasia in ulcerative colitis. *Eur J Cancer.* 1995;31A:1171–1173.
266. Itzkowitz SH, Present DH. Consensus conference: Colorectal cancer screening and surveillance in inflammatory bowel disease. *Inflamm Bowel Dis.* 2005;11:314–321.
267. Riddell RH, Goldman H, Ransohoff DF, et al. Dysplasia in inflammatory bowel disease: Standardized classification with provisional clinical applications. *Hum Pathol.* 1983;14:931–968.

268. Peltomaki P. Deficient DNA mismatch repair: A common etiologic factor for colon cancer. *Hum Mol Genet.* 2001;10:735–740.
269. Terdiman JP. HNPCC: An uncommon but important diagnosis. *Gastroenterology.* 2001;121:1005–1008.
270. Llor X, Pons E, Xicola RM, et al. Differential features of colorectal cancers fulfilling Amsterdam criteria without involvement of the mutator pathway. *Clin Cancer Res.* 2005;11: 7304–7310.
271. Leach FS, Nicolaides NC, Papadopoulos N, et al. Mutations of a mutS homolog in hereditary nonpolyposis colorectal cancer. *Cell.* 1993;75:1215–1225.
272. Bronner CE, Baker SM, Morrison PT, et al. Mutation in the DNA mismatch repair gene homologue hMLH1 is associated with hereditary non-polyposis colon cancer. *Nature.* 1994; 368:258–261.
273. Nicolaides NC, Papadopoulos N, Liu B, et al. Mutations of two PMS homologues in hereditary nonpolyposis colon cancer. *Nature.* 1994;371:75–80.
274. Berends MJ, Wu Y, Sijmons RH, et al. Molecular and clinical characteristics of MSH6 variants: An analysis of 25 index carriers of a germline variant. *Am J Hum Genet.* 2002;70:26–37.
275. Kariola R, Raevaara TE, Lonnqvist KE, et al. Functional analysis of MSH6 mutations linked to kindreds with putative hereditary non-polyposis colorectal cancer syndrome. *Hum Mol Genet.* 2002;11:1303–1310.
276. Buttin BM, Powell MA, Mutch DG, et al. Penetrance and expressivity of MSH6 germline mutations in seven kindreds not ascertained by family history. *Am J Hum Genet.* 2004; 74:1262–1269.
277. Lin KM, Shashidharan M, Thorson AG, et al. Cumulative incidence of colorectal and extracolonic cancers in MLH1 and MSH2 mutation carriers of hereditary nonpolyposis colorectal cancer. *J Gastrointest Surg.* 1998;2:67–71.
278. Kolodner R. Biochemistry and genetics of eukaryotic mismatch repair. *Genes & Development.* 1996;10:1433–1442.
279. Genschel J, Littman SJ, Drummond JT, et al. Isolation of MutSbeta from human cells and comparison of the mismatch repair specificities of MutSbeta and MutSalpha. *J Biol Chem.* 1998;273:19895–19901.
280. Umar A, Risinger JI, Glaab WE, et al. Functional overlap in mismatch repair by human MSH3 and MSH6. *Genetics.* 1998;148:1637–1646.
281. Palombo F, Gallinari P, Iaccarino I, et al. GTBP, a 160-kilodalton protein essential for mismatch-binding activity in human cells. *Science.* 1995;268:1912–1914.
282. Marsischky GT, Filosi N, Kane MF, et al. Redundancy of Saccharomyces cerevisiae MSH3 and MSH6 in MSH2-dependent mismatch repair. *Genes Dev.* 1996;10:407–420.
283. Sia EA, Kokoska RJ, Dominska M, et al. Microsatellite instability in yeast: Dependence on repeat unit size and DNA mismatch repair genes. *Mol Cell Biol.* 1997;17:2851–2858.
284. Lu A-L. *Biochemistry of Mammalian DNA Mismatch Repair.* Totowa: Humana Press; 1998.
285. Leung WK, Kim JJ, Wu L, et al. Identification of a second MutL DNA mismatch repair complex (hPMS1 and hMLH1) in human epithelial cells. *J Biol Chem.* 2000;275:15728–15732.
286. Raschle M, Marra G, Nystrom-Lahti M, et al. Identification of hMutLbeta, a heterodimer of hMLH1 and hPMS1. *J Biol Chem.* 1999;274:32368–32375.
287. Liu B, Parsons R, Papadopoulos N, et al. Analysis of mismatch repair genes in hereditary non-polyposis colorectal cancer patients. *Nature Medicine.* 1996;2:169–174.
288. Peltomaki P. Role of DNA mismatch repair defects in the pathogenesis of human cancer. *J Clin Oncol.* 2003;21:1174–1179.
289. Vasen HF, Watson P, Mecklin JP, et al. New clinical criteria for hereditary nonpolyposis colorectal cancer (HNPCC, Lynch syndrome) proposed by the International Collaborative group on HNPCC. *Gastroenterology.* 1999;116:1453–1456.
290. Jass JR. Pathology of hereditary nonpolyposis colorectal cancer. *Ann N Y Acad Sci.* 2000;910:62–73.
291. De Jong AE, Morreau H, Van Puijenbroek M, et al. The role of mismatch repair gene defects in the development of adenomas in patients with HNPCC. *Gastroenterology.* 2004;126:42–48.
292. Vasen HF, den Hartog Jager FC, Menko FH, et al. Screening for hereditary non-polyposis colorectal cancer: A study of 22 kindreds in The Netherlands. *Am J Med.* 1989;86:278–281.
293. Lindgren G, Liljegren A, Jaramillo E, et al. Adenoma prevalence and cancer risk in familial non-polyposis colorectal cancer. *Gut.* 2002;50:228–234.
294. Jarvinen HJ, Aarnio M, Mustonen H, et al. Controlled 15-year trial on screening for colorectal cancer in families with hereditary nonpolyposis colorectal cancer. *Gastroenterology.* 2000;118: 829–834.
295. Rijcken FE, Hollema H, Kleibeuker JH. Proximal adenomas in hereditary non-polyposis colorectal cancer are prone to rapid malignant transformation. *Gut.* 2002;50:382–386.
296. Iino H, Simms L, Young J, et al. DNA microsatellite instability and mismatch repair protein loss in adenomas presenting in hereditary non-polyposis colorectal cancer. *Gut.* 2000;47: 37–42.
297. Bettstetter M, Dechant S, Ruemmele P, et al. Distinction of hereditary nonpolyposis colorectal cancer and sporadic microsatellite-unstable colorectal cancer through quantification of MLH1 methylation by real-time PCR. *Clin Cancer Res.* 2007;13:3221–3228.
298. Bessa X, Balleste B, Andreu M, et al. A prospective, multicenter, population-based study of BRAF mutational analysis for Lynch syndrome screening. *Clin Gastroenterol Hepatol.* 2008;6:206–214.
299. Vasen HF, Mecklin JP, Khan PM, et al. The International Collaborative Group on Hereditary Non-Polyposis Colorectal Cancer (ICG-HNPCC). *Dis Colon Rectum.* 1991;34:424–425.
300. Boland CR, Thibodeau SN, Hamilton SR, et al. A National Cancer Institute Workshop on Microsatellite Instability for cancer detection and familial predisposition: Development of international criteria for the determination of microsatellite instability in colorectal cancer. *Cancer Res.* 1998;58:5248–5257.
301. Syngal S, Fox EA, Eng C, et al. Sensitivity and specificity of clinical criteria for hereditary non-polyposis colorectal cancer associated mutations in MSH2 and MLH1. *J Med Genet.* 2000;37:641–645.
302. Umar A, Risinger JI, Hawk ET, et al. Testing guidelines for hereditary non-polyposis colorectal cancer. *Nat Rev Cancer.* 2004;4:153–158.
303. Lindor NM, Rabe K, Petersen GM, et al. Lower cancer incidence in Amsterdam-I criteria families without mismatch repair deficiency: Familial colorectal cancer type X. *JAMA.* 2005; 293:1979–1985.
304. Guillem JG, Wood WC, Moley JF, et al. ASCO/SSO review of current role of risk-reducing surgery in common hereditary cancer syndromes. *J Clin Oncol.* 2006;24:4642–4660.
305. Lindor NM, Petersen GM, Hadley DW, et al. Recommendations for the care of individuals with an inherited predisposition to Lynch syndrome: A systematic review. *JAMA.* 2006;296: 1507–1517.
306. Grady WM. Genetic testing for high-risk colon cancer patients. *Gastroenterology.* 2003;124:1574–1594.
307. Akiyama Y, Sato H, Yamada T, et al. Germ-line mutation of the hMSH6/GTBP gene in an atypical hereditary nonpolyposis colorectal cancer kindred. *Cancer Res.* 1997;57:3920–3923.
308. MSH2 and MLH1 mutations. http://www.insight-group.org/; Accessed September 2004.
309. Lipton L, Tomlinson I. The genetics of FAP and FAP-like syndromes. *Fam Cancer.* 2006;5:221–226.
310. Schulmann K, Pox C, Tannapfel A, et al. The patient with multiple intestinal polyps. *Best Pract Res Clin Gastroenterol.* 2007;21:409–426.
311. Bulow S, Berk T, Neale K. The history of familial adenomatous polyposis. *Fam Cancer.* 2006;5:213–220.
312. Bodmer WF, Bailey CJ, Bodmer J, et al. Localization of the gene for familial adenomatous polyposis on chromosome 5. *Nature.* 1987;328:614–616.
313. Kinzler KW, Nilbert MC, Su LK, et al. Identification of FAP locus genes from chromosome 5q21. *Science.* 1991;253:661–665.
314. Groden J, Thliveris A, Samowitz W, et al. Identification and characterization of the familial adenomatous polyposis coli gene. *Cell.* 1991;66:589–600.
315. Lynch HT, Boland CR, Rodriguez-Bigas MA, et al. Who should be sent for genetic testing in hereditary colorectal cancer syndromes? *J Clin Oncol.* 2007;25:3534–3542.
316. Galiatsatos P, Foulkes WD. Familial adenomatous polyposis. *Am J Gastroenterol.* 2006;101:385–398.

317. Rustgi AK. The genetics of hereditary colon cancer. *Genes Dev.* 2007;21:2525–2538.
318. Lynch HT, Smyrk T, McGinn T, et al. Attenuated familial adenomatous polyposis (AFAP). A phenotypically and genotypically distinctive variant of FAP [see comments]. *Cancer.* 1995;76:2427–2433.
319. Hofgartner WT, Thorp M, Ramus MW, et al. Gastric adenocarcinoma associated with fundic gland polyps in a patient with attenuated familial adenomatous polyposis. *Am J Gastroenterol.* 1999;94:2275–2281.
320. Ahnen DJ, Axell L. *Clinical Features and Diagnosis of Familial Adenomatous Polyposis.* Waltham, MA; 2008.
321. Ahnen DJ. The genetic basis of colorectal cancer risk. *Adv Intern Med.* 1996;41:531–552.
322. Bonis PA, Ahnen DJ, Axell L. *Screening Strategies in Patients and Families with Familial Colon Cancer Syndromes.* Waltham, MA; 2008.
323. Polakis P. The many ways of Wnt in cancer. *Curr Opin Genet Dev.* 2007;17:45–51.
324. Aoki K, Taketo MM. Adenomatous polyposis coli (APC): A multi-functional tumor suppressor gene. *J Cell Sci.* 2007;120: 3327–3335.
325. Schneikert J, Behrens J. The canonical Wnt signalling pathway and its APC partner in colon cancer development. *Gut.* 2007;56:417–425.
326. Senda T, Iizuka-Kogo A, Onouchi T, et al. Adenomatous polyposis coli (APC) plays multiple roles in the intestinal and colorectal epithelia. *Med Mol Morphol.* 2007;40:68–81.
327. Abdel-Rahman WM, Peltomaki P. Molecular basis and diagnostics of hereditary colorectal cancers. *Ann Med.* 2004;36:379–388.
328. Segditsas S, Tomlinson I. Colorectal cancer and genetic alterations in the Wnt pathway. *Oncogene.* 2006;25:7531–7537.
329. Nieuwenhuis MH, Vasen HF. Correlations between mutation site in APC and phenotype of familial adenomatous polyposis (FAP): A review of the literature. *Crit Rev Oncol Hematol.* 2007;61:153–161.
330. Rozen P, Macrae F. Familial adenomatous polyposis: The practical applications of clinical and molecular screening. *Fam Cancer.* 2006;5:227–235.
331. Hegde MR, Roa BB. Detecting mutations in the APC gene in familial adenomatous polyposis (FAP). *Curr Protoc Hum Genet.* 2006; **Chapter 10**: Unit 10 18.
332. Spits C, De Rycke M, Van Ranst N, et al. Preimplantation genetic diagnosis for cancer predisposition syndromes. *Prenat Diagn.* 2007;27:447–456.
333. Michils G, Tejpar S, Thoelen R, et al. Large deletions of the APC gene in 15% of mutation-negative patients with classical polyposis (FAP): A Belgian study. *Hum Mutat.* 2005;25:125–134.
334. Hes FJ, Nielsen M, Bik EC, et al. Somatic APC mosaicism: An underestimated cause of polyposis coli. *Gut.* 2008;57:71–76.
335. Romero-Gimenez J, Dopeso H, Blanco I, et al. Germline hypermethylation of the APC promoter is not a frequent cause of familial adenomatous polyposis in APC/MUTYH mutation negative families. *Int J Cancer.* 2008;122:1422–1425.
336. Renkonen ET, Nieminen P, Abdel-Rahman WM, et al. Adenomatous polyposis families that screen APC mutation-negative by conventional methods are genetically heterogeneous. *J Clin Oncol.* 2005;23:5651–5659.

Chapter

20

Molecular Basis of Liver Disease

Satdarshan P. Singh Monga ▪ Jaideep Behari

This book chapter is dedicated to Dr. Pramod Behari, a pioneering neurosurgeon and wonderful father and to Dr. Gurdarshan Singh Monga, a loving father and a deeply caring and highly respected family physician.

INTRODUCTION

Liver diseases are a cause of global morbidity and mortality. While the predominance of a specific liver disease varies with geographical location, the breadth of hepatic diseases affecting the underdeveloped, developing, and developed countries is phenomenal, with diseases ranging from infectious diseases of the liver to neoplasia and obesity-related illnesses. This chapter will apprise readers of the progress made toward unraveling the molecular aberrations of several liver diseases that has led to a better understanding of the disease biology with the hope of eventually yielding improved diagnostic, prognostic, and therapeutic tools. Just as in other tissues and organs, identification of molecular basis of normal liver growth and development has been important in identification of aberrations in the pathological states of the liver. We will provide a concise background on the molecular mechanism of liver development and regeneration and briefly summarize additional aspects of hepatic biology. Next, we will discuss the molecular basis of hepatic pathologies with emphasis on alcoholic liver disease, nonalcoholic fatty liver disease, and the benign and malignant tumors of the liver.

MOLECULAR BASIS OF LIVER DEVELOPMENT

Embryonic liver development is characterized by very timely and precise regulatory signals that enable hepatic competence of the foregut endoderm, hepatic specification, and induction followed by hepatic morphogenesis. Clearly, the preceding events are governed by molecular signals that are highly temporal, cell specific, and tightly regulated (Figure 20.1). Liver in mouse begins to arise from the definitive gut endoderm at the embryonic day 8.5 (E8.5) or the 7–8 somite stage [1,2]. At this time the Foxa family of transcription factors specifies the endoderm to express hepatic genes in the process of hepatic competence [2,3]. Next, fibroblast growth factor 1 (FGF1) and FGF2, which originate from the cardiac mesoderm, initiate the expression of liver-specific genes in the endoderm [4]. FGF8, which is important for the morphogenetic outgrowth of the liver, is also expressed during this stage [4]. The hepatic bud next migrates into the septum transversum mesenchyme under the direction of bone morphogenic protein 4 (BMP4) signaling, which is essential for hepatogenesis [5]. One should be reminded that the earliest indicators of successful liver induction are the observed thickening of the hepatic bud endoderm along with *albumin* mRNA expression by E9.0 in the mouse [6].

This stage is followed by the phase of embryonic liver growth characterized by the expansion and proliferation of the resident cells within the hepatic bud. Several transcription factors including Hex, Gata6, and Prox1 are the earliest known mediators of this phase (Figure 20.1) [7–10]. Once the hepatic program is in full swing, the liver growth continues and is now labeled as the stage of hepatic morphogenesis. The epithelial cells at this stage are now considered the hepatoblasts, or the bipotential progenitors, which means that they are capable of giving rise to both major lineages of the liver, the hepatocytes, and the biliary epithelial cells [11]. Hepatoblasts will be undergoing expansion while maintaining their dedifferentiated state during this stage. While distinct from the traditionally known stem cell renewal, this event marks the expansion of a lineage-restricted progenitor population. Several key players at this stage include the HGF/c-Met, β-catenin, TGFβ, embryonic liver fodrin (*Elf*), FGF8, FGF10, Foxm1b, and Hlx, which are regulating the proliferation and survival of resident cells as well as regulating their survival [12–23]. Additional factors at

Molecular Pathology © 2009, Elsevier, Inc. All Rights Reserved.

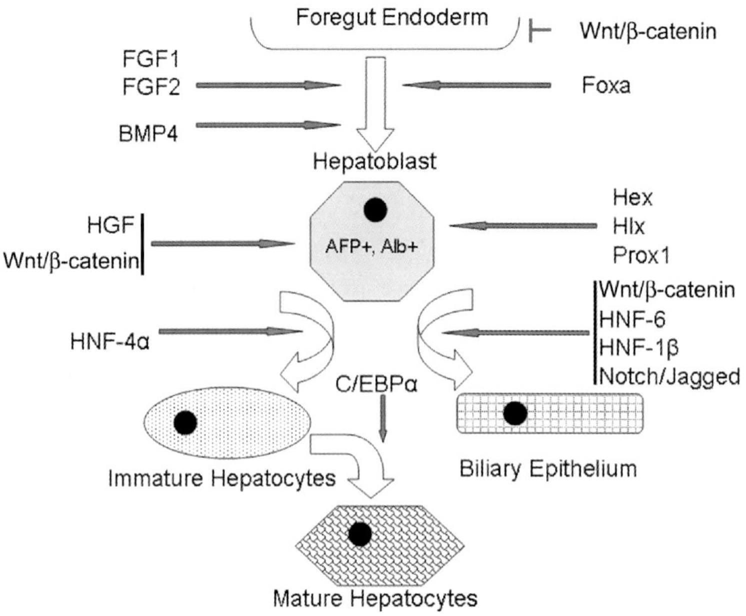

Figure 20.1 Summary of molecular signaling during liver development in mouse. Abbreviations: FGF—Fibroblast growth factor; BMP—Bone morphogenic protein; AFP—α-fetoprotein; Alb—Albumin; HGF—Hepatocyte growth factor; HNF—Hepatocyte nuclear factor; C/EBPα—CCAAT enhancer binding protein-alpha.

this stage have also been identified although the mechanisms are less clear. These include components of NFκB, c-jun, XBP1, K-*ras*, and others [24–28]. At this time the general architecture of the liver is beginning to be established, including the formation of sinusoids and the development of hepatic vasculature [2,29].

The final stage is characterized by the differentiation of hepatoblasts to mature functional cell types: the hepatocytes and the biliary epithelial cells. For maturation into the hepatocytes, a huge change in cell morphology is observed from earlier stages to E17, when the resident epithelial cells acquire a cuboidal morphology with definitive cell polarity and clear cytoplasm after losing their blast characteristics such as the high nuclear to cytoplasmic ratio (Figure 20.2). At the center of the hepatoblast to hepatocyte differentiation process are the liver-enriched transcription factors such as hepatocyte nuclear factor (HNF4α) transcription factors and CCAAT enhancing binding protein-α (C/EBPα) [30,31]. HNF4α is essential for differentiation toward a hepatocyte phenotype, as well as formation of the parenchyma [32]. The liver-enriched transcription factors enable functioning of the fetal hepatocytes by directing the expression of various genes that are classically associated with hepatocyte functions at this stage, including the cytochrome P450s, metabolic, and synthetic enzymes [33]. The hepatocytes at this stage clearly show glycogen accumulation and have been shown to exhibit many functions of adult hepatocyte including xenobiotic metabolism. Again several signaling pathways have been shown to play a role in regulating hepatocyte maturation by regulating the expression of liver-enriched

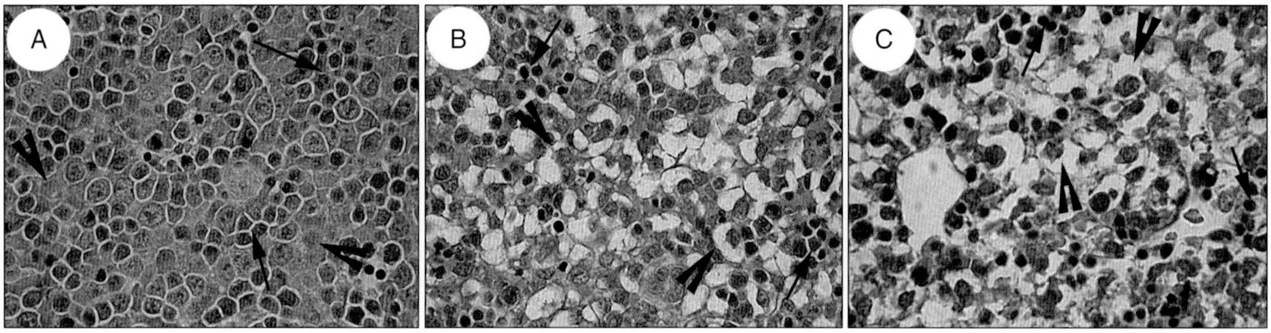

Figure 20.2 H&E of developing mouse livers. (A) Several hematopoietic cells (arrow) are seen interspersed among hepatoblasts (arrowhead), which display large nuclei and scanty cytoplasm, in an E14 liver section. **(B)** Fewer hematopoietic cells (arrow) are observed among the hepatocytes (arrowhead) which begin to show cuboidal morphology and large clear cytoplasm as well as begin to display polarity, in an E17 liver section. **(C)** Similar cuboidal morphology of hepatocytes (arrowhead) is seen in an E19 mouse liver.

transcription factors. Some of these pathways include HGF, EGF, FGFs, Wnt/β-catenin signaling, and others.

While the signals involved in lineage commitment of hepatoblasts to biliary epithelium are not fully understood, there is evidence of the involvement of HNF6 (One-cut-1; OC-1), HNF1β, OC-2, TGFβ, and activin [34–36]. HNF6 promotes hepatocyte over biliary development by activating HNF1β, which in turn attenuates early biliary cell commitment. It has been shown that HNF6 and OC-2 double knockouts display cells with both biliary and hepatocyte programs activated, suggesting their role in segregating the two lineages [34]. They do so by regulating TGFβ/activin signaling gradient in developing livers by controlling expression of TGFβ, TGFβ receptor type II (TBRII), activin B, α2-macroglobulin (TGFβ-antagonist), and follistatin (activin antagonist). At E12.5, TGFβ activity is high in vicinity of portal vein, where hepatoblasts differentiate into biliary cells. While lineage specification begins at E12.5, bile duct differentiation becomes apparent at E15.5, where biliary cells are organized around branches of portal veins and express cytokeratins, show basal lamina (laminin-positive) toward the portal side of the ductal plate, and lack HNF4α. HNF6 is required for bile duct morphogenesis at this stage. Notch signaling pathway is also known to play a role in the development of biliary epithelia [35–37]. Various results support the fact that Jagged/Notch interactions may activate a cascade of events involving HNF6, HNF1β, and *Pkhd1* gene encoding the ciliary protein polyductin or fibrocystin. Additional roles of Foxm1b and Wnt/β-catenin signaling have also been reported in biliary differentiation, although the mechanism is less clear.

This somewhat simplistic outline of embryonic liver development does not take into account the complexity involved in the expression of these growth and transcription factors. Liver development is clearly not a linear process. Rather, there is a significant overlap between gene expression patterns that blur the lines between one stage of liver development and the next. Additionally, activation of one gene may initiate a feedback mechanism that regulates cross-talk between different cell populations. For example, the HNFs are capable of autoregulating their own expression as well as cross-regulating the transcription of other liver-specific genes [11]. Finally, timing remains critical since certain pathways can act at different stages to inhibit or stimulate certain stages or processes of the hepatic development. A classic example is the Wnt/β-catenin pathway, which needs to be initially repressed to induce hepatic program in the foregut endoderm, but immediately following that stage, it becomes indispensable [15,16,20,38,39]. Also, the same pathway plays a role in hepatoblast expansion and survival, but a later stage is indispensable for its maturation into hepatocyte, at the same time playing an early role in biliary differentiation [23,40].

The significance of understanding the molecular basis of hepatic development is really 2-fold. First, this would be critical for the fields of hepatic tissue engineering, stem cell differentiation, and hepatic regenerative medicine. This is highly pertinent as efforts in multiple laboratories across the world are ongoing to identify alternate sources of hepatocytes due to scarcity of organs for orthotopic liver transplantation. For successful, persistent, and reproducible derivation of functioning hepatocytes from sources such as ES cells, mesenchymal stem cells, hematopoietic stem cells, or stem cells derived from skin, adipose tissue, and placenta, it will be critical to identify precise signaling pathways that need to be temporally turned on or off. This mandates a careful study of relevant events as liver development where a perfect and precise control of various signaling molecules results in successful liver formation. Second, many of the molecular signaling pathways that are associated with normal liver development have also been shown to play key roles in hepatocarcinogenesis. Thus, studying regulation of liver development might disclose physiological ways to tame the oncogenic pathways through identification of novel cross-talks and negative regulators of such pathways.

MOLECULAR BASIS OF LIVER REGENERATION

Liver is a unique organ with an innate ability to regenerate, and rightfully so, since it is the gatekeeper to a variety of absorbed materials (both nutrients and toxins) through the intestines [41]. Thus, with an overwhelming ongoing assault on a daily basis, liver must regenerate as and when necessary in order to continue its functions of synthesis, detoxification, and metabolism. This capacity of regeneration is ascertained by activation of multiple signaling pathways as evidenced in many animal models and studies where surgical loss of liver mass triggers the process of regeneration [42]. This ensures proliferation and expansion of all cell types of the liver to enable restoration of lost hepatic mass. It is imperative to identify the various pathways that form the basis of initiation, continuation, and termination of the regeneration process to understand the dysregulation that is often seen in aberrant growth in benign and malignant liver tumors. Additionally, examining the process of liver regeneration for changes in various cytokines and growth factors would allow us to identify novel regulatory loops among several of these pathways, providing novel antitumor therapeutic opportunities to suppress certain oncogenic proteins or booster inhibitory factors. Simultaneously, understanding such a complex mechanism would be critical for regenerative medicine, stem cell transdifferentiation, hepatic tissue engineering, and cell-based therapies. The list of factors independently shown to play important roles in hepatocyte proliferation during regeneration is exhaustive and will be discussed concisely henceforth [43–46].

Partial hepatectomy triggers a sequence of events that proceed in an orderly fashion to restore the lost mass within 7 days in rats, 14 days in mice, and 8–15 days in humans [41]. Following these periods, the liver lobules become larger, and the thickness of hepatocyte plates is doubled as compared to the prehepatectomy livers. However, over several weeks there is gradual lobular and cellular reorganization, leading to an unremarkable and undistinguishable liver histology from a normal liver [47].

Immediately following the hepatectomy, when two-thirds of the hepatic mass is resected, there is a dramatic change in the hemodynamics of the liver, especially since the same volume of portal blood would need to traverse through only one-third of the original tissue mass. Additionally, hypoxic state might be playing a role since portal blood has lower pO_2 when compared to the arterial blood flow that remains relatively unaltered during regeneration. While very few studies conclusively address a role of these changes in initiation of regeneration, there are reports of the impact of such changes on activation of HGF, a major initiator of regeneration, as is discussed in the forthcoming paragraphs [48]. Within minutes, there are specific changes in the gene expression as well as at post-translational levels, which complement each other and lead to a well-orchestrated event of regeneration, during which time the hepatocyte functions are maintained for the functioning of the animal. More than 95% of hepatocytes will undergo cell proliferation during the process of regeneration [41]. The earliest known signals involve both the growth factors and cytokines. While it is difficult to lay out the exact chronology of events, it is important to say that many events are concomitant, ensuring proliferation and maintaining liver functions at the same time. The earliest events observed include activation of uPA enabling activation of plasminogen to plasmin, which induces matrix remodeling that leads to, among other events, activation of HGF from the bound hepatic matrix [49,50]. HGF is a known hepatocyte mitogen that acts through its receptor c-Met, a tyrosine kinase and is a master effector of hepatocyte proliferation and survival both *in vitro* and *in vivo* [51]. Similarly EGF, which is continually present in the portal circulation, is also assisting in hepatocyte proliferation [52]. Other factors activated at this time include the Wnt/β-catenin pathway [53–56], Notch/Jagged pathway [57,58], norepinephrine [59], serotonin [60], and TGFα [61,62]. These factors are working in autocrine and paracrine fashion, and cell sources of various factors include the hepatocytes, Kupffer cells, stellate cells, and sinusoidal endothelial cells. Concurrently, the TNFα [63–65] and IL-6 [66,67] are being released from the Kupffer cells and have been shown to be important in normal liver regeneration through genetic studies. These pathways are known to act through NFκB and Stat3 activation. Bile acids have also been shown to play a vital role in normal liver regeneration through activation of transcription factors such as FoxM1b [68] and c-myc, necessary for cell cycle transition as well, and decreased hepatocyte proliferation after partial hepatectomy was observed in animals depleted for bile acids with the use of cholestyramine or in the *FXR*-null mice [69]. The eventual goal of these changes is to initiate cell cycle in hepatocytes with a successful G1 to S phase transition dependent on the key cyclins such as A, D, and E to ensure DNA synthesis and mitosis [41,44]. Additional growth factors such as FGF, PDGF, and insulin are important in regeneration and might be playing an important role in providing homeostatic support to the regenerating liver (reviewed in [41]).

During the process of regeneration, active matrix remodeling is ongoing as well. Regulation of this process is complex and involves key molecules such as metalloproteinases, TIMPs, and others [70–73]. Additionally, the matrix is a source of matrix-bound growth factors such as HGF, additional glycosylated proteins, hyaluronic acid, decorin, syndecan, and others, which modulate the activity of many mitogenic (HGF, TNFα, EGF) and mitoinhibitory (TGFβ) factors. Thus, active matrix remodeling might be both a cause and an effect of regenerative response after hepatectomy and is again critical for optimal regeneration as demonstrated by genetic models such as *MMP9*-deficient mice [73].

In liver regeneration, while the hepatocytes are the main cell type undergoing proliferation, division of all cells native to the liver is observed at specific times after partial hepatectomy. The peak hepatocyte proliferation in rats is observed at 24 hours after hepatectomy followed by a peak proliferation of the biliary epithelial cells at 48 hours, Kupffer and stellate cells at 72 hours, and sinusoidal endothelial cells at 96 hours [45]. Several growth factors have been shown to be mitogenic for these cell types and work in an autocrine as well as paracrine manner, since several of these factors are produced by one or more cells within the regenerating liver. Factors such as HGF, PDGF, VEGF, FGFs, and angiopoietins are important for Kupffer cells and endothelial cell proliferation and homeostasis during regeneration (reviewed in [41]). Neurotrophins and nerve growth factors have been shown to be important in regeneration and survival of stellate cells during regeneration [74,75].

How does the liver know when to stop? This is complex and incompletely understood. While the liver mass is restored after 14 days in mice and 7 days in rats, it may exceed its original mass, when transient apoptosis occurs and the overall liver mass matches the prehepatectomy liver mass [67]. Based on the mitoinhibitory action of TGFβ on the hepatocytes, it has been suggested to be the terminator of regeneration. However, changes in TGFβ (produced by stellate cells) expression mimics that of many proproliferation factors during regeneration and begins at 2–3 hours after hepatectomy in rats and remains elevated until 72 hours [76]. However, the receptors for TGFβ are downregulated during regeneration, and hence the hepatocytes are resistant to excess TGFβ presence [77,78]. In addition, during regeneration, TGFβ protein is lost first periportally and then gradually toward the central vein [79]. Just behind the leading edge of loss of TGFβ is the wave of hepatocyte mitosis, suggesting that somehow TGFβ is balancing the act of quiescence and proliferation even during regeneration to perhaps keep the growth regulated and to maintain a certain number of hepatocytes in a nonproliferative and differentiated state, to continue their functions necessary for the animal's survival. However, whether TGFβ is the final cytokine that enables termination of the regenerative process has not been shown convincingly and still remains an open-ended question.

As can be appreciated, multiple cytokines and growth factors are activated in response to partial-hepatectomy where two-thirds of the liver is surgically resected [80]. No single signal transduction pathway's suppression has led to the complete abolition of liver regeneration after partial hepatectomy in rodents. Genetic ablation or other modality-mediated suppression of certain growth factors such as EGFR, TGFα, Jagged/Notch, Wnt/β-catenin and of cytokines such as TNFα, IL-6, and of additional modulators such as bile acids, norepinephrine, serotonin, and components of hepatic biomatrix, components of the complement, all show a modest to severe decrease in liver regeneration, but not a complete loss. Interference with the HGF/Met signaling pathway appears to cause the most profound but not complete interference with the regenerative process. Liver regeneration is guided by a significant signaling redundancy, which is paramount to inducing the much-needed cell proliferation within the regenerating liver. In addition, when all else fails, additional cell types such as the oval cells can be called upon to restore hepatocytes [43]. These cells are the facultative hepatic progenitors or transient amplifying progenitor cells that originate from the biliary compartment and appear in the periportal areas when the hepatocytes are unable to proliferate [81–84]. These cells become hepatocytes and rescue the regenerative process. Recent studies have also shown that hepatocytes can transdifferentiate into biliary cells and rescue biliary repair when cholangiocytes are unable to carry it out by themselves [85,86]. Overall, the redundancy in liver regenerative processes, at the level of signaling and at the level of cells, ensures the health of this indispensable Promethean organ.

ADULT LIVER STEM CELLS IN LIVER HEALTH AND DISEASE

Despite the liver's capacity to regenerate and the presence of redundant signaling enabling regeneration on most occasions, there is sometimes a need for stem cell activation in the liver [80]. This term is often used to depict appearance and expansion of the hepatic progenitors in the liver. The basal presence of these facultative stem cells or oval cells or transiently amplifying hepatic progenitors in a normal liver remains debated [41,43,87]. Under certain circumstances in rodents, there is a clear activation of oval cells. This occurs when an injury to the hepatocytes and/or bile ducts in rodents coexists with the presence of an inhibitor of hepatocyte proliferation. This creates a necessity for the regeneration of the injured cells at the same time blocking the proliferation of hepatocytes, allowing for the appearance of the facultative stem cells that stem from the terminal biliary ductular cells in the canal of Herring. These oval cells proliferate and expand and have been shown to be the precursors of both hepatocytes and biliary epithelial cells, based on the type of cell injury [43,88,89]. Based on the ability of hepatocytes and biliary epithelial cells to transdifferentiate into each other, it is quite possible that the oval cells can arise from not just the bile duct cells, but also certain subset of hepatocytes [41,85,86].

Classically, the activation of adult stem cells is observed as atypical ductular proliferation, which can go on to differentiate into polygonal or intermediate hepatocytes and finally mature into hepatocytes. The oval cell response is typically dictated by the kind of injury (biliary versus hepatocytic versus mixed), which in turn determines the proportion of ductular response to intermediate hepatocytes observed [89].

In rodents, various models have been optimized to induce activation of stem cells. The basic premise behind these models is the presence of an injury to the hepatocytes after disabling the proliferative capacity of the hepatocytes. This is classically attained in rats by acetylaminofluorene (AAF), which crosslinks DNA in hepatocytes, followed by two-thirds partial hepatectomy, and leads to appearance of oval cells (Figure 20.3) [90,91]. In mice, the models are more complicated since AAF does not work well. Alternatives used are the administration of diet containing 3,5-diethoxycarbonyl-1,4-dihydrocollidine (DDC) or the choline-deficient, ethionine-supplemented (CDE) diet [84,92]. However, DDC primarily causes a biliary injury that also leads to periportal hepatocyte injury, and this model has been recently characterized as a mouse model of PBC and PSC [93]. Hence, the phenotype observed is a typical and predominant biliary reaction or atypical ductular proliferation (Figure 20.3). While polygonal intermediate hepatocytes are a minor subset observed in this model, it is unclear whether this is differentiation of atypical ductules toward hepatocytes or dedifferentiation of hepatocytes toward biliary cells, since the injury is predominantly biliary. CDE diet has been used successfully in rats and mice to induce oval cell activation [92,94]. It should also be emphasized that remarkable heterogeneity was identified between the oval cells observed in response to specific protocols and at least partly could be explained by the differences in the kind of injury that directs oval cell activation towards hepatocyte or biliary differentiation, for the maintenance of liver function, while on these protocols [95].

Various markers have been applied to detect these oval cells. By histology, these cells are smaller than the hepatocytes and possess high nuclear to cytoplasmic ratio and typically are seen in the periportal region. These cells are concomitantly positive for biliary (CK19, CK7, A6, OV6), hepatocyte (HepPar-1, albumin), and fetal hepatocyte markers (α-fetoprotein) (reviewed in [89]). However, at any given time, only a subset of these cells are positive for all markers. This reflects the different stages of differentiation occupied by the cells or the basic heterogeneity of these cell populations. Reactive ducts are also positive for neural cell adhesion molecules (NCAM) or vascular cell adhesion molecules (VCAM) [93,96,97]. Additional surface markers for the oval cells have been identified in rodent studies, which gives an advantage of cell sorting [98]. The six unique markers include CD133, claudin-7, cadherin 22, mucin-1, Ros1 (oncogene v-ros), and γ-aminobutyrate, type A receptor π (Gabrp).

Using these models, several pathways have been shown to play an important role in the appearance, expansion, and differentiation of stem cells. As can be appreciated, several pathways known to play important roles in liver development and regeneration are also known to be important in oval cell activation and their differentiation. These factors include HGF/c-met, interleukin-6, TGFα, TGFβ, EGF, Wnt/β-catenin pathway, PPARs, IGF, and others [99–110]. These signaling pathways function in an autocrine or paracrine manner to induce oval cell activation.

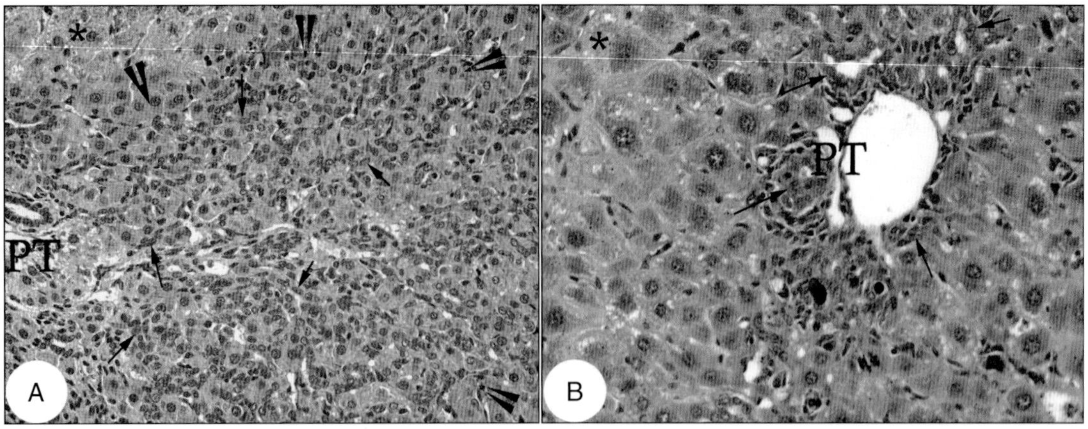

Figure 20.3 H&E demonstrating the activation of oval cells or facultative adult liver stem cells in experimental models. (A) Several oval cells with oval shape, smaller size, and higher cytoplasmic to nuclear ratio (arrows) are observed around the portal triad (PT) in rat livers after AAF/PHx. Several intermediate hepatocytes (arrowheads) are evident as these oval cells undergo progressive differentiation. (*) shows normal hepatocytes adjoining the ongoing oval cell activation, expansion, and differentiation. **(B)** Mouse liver after DDC administration for 2 weeks displays atypical ductular hyperplasia (arrows) equivalent of oval cell activation and adjoining normal hepatocytes (*).

Progenitor cell activation has been seen in patients after various forms of hepatic injury. This is an attempt by the diseased liver to restore the lost cell type in order to maintain hepatic functions. An acute ductular reaction is observed in the setting of submassive necrosis due to hepatitis, drugs, alcohol, or cholestatic disease, which is subtle during the early stages. These cells expand in numbers and show hepatocytic differentiation with passage of time as revealed by limited studies from serial liver biopsies from these patients [97,111]. In some instances, bone marrow-derived stem cells have been observed in these livers, albeit at an extremely low frequency. This has been shown to be due to fusion of a bone marrow-derived stem cell to a hepatocyte, although transdifferentiation of bone marrow-derived stem cells into hepatocytes has not been completely ruled out [112–115]. Noteworthy hepatic progenitor activation has been observed in alcoholic and nonalcoholic fatty liver disease (ALD and NAFLD) patients as well. These scenarios are associated with increased lipid peroxidation, generation of reactive oxygen species, and additional features of elevated oxidative stress, which is a known inhibitor of hepatocyte proliferation [116,117]. The inability of hepatocytes to proliferate in these clinical scenarios might be an impetus for the oval cell activation, especially since high numbers of polygonal and intermediate hepatocytes are observed in both ALD and NAFLD. A positive correlation between the oval cell response and the stage of hepatic fibrosis and the fatty liver disease has been identified [116]. In viral hepatitis also, oval cell activation is observed. This is classically seen at the periportal site and is usually proportional to the extent of inflammatory infiltrate. Moderate to severe inflammation is often associated with higher oval cell response that is composed of intermediate hepatocytes [118]. Based on these findings, a paracrine mechanism of oval cell activation triggered by the inflammatory cells has been proposed. These scenarios also bring into perspective the progenitor or oval cell origin of a subset of hepatocellular cancers (HCC). HCC occurs more frequently in the backgrounds of cirrhosis observed in ALD, NAFLD, and hepatitis than nondiseased liver [119,120]. There are established advantages of stem cells being a target of transformation [121,122]. It is feasible that the oval cell activation in the pathologies as discussed previously, while providing a distinct advantage of maintenance of hepatic function, might also serve as a basis of neoplastic transformation in the right microenvironmental milieu. This Jekyll and Hyde hypothesis is supported by the observation that several early HCC lesions possess hepatic progenitor signatures at genetic and protein levels [123]. However, conclusive studies to this end are still missing.

MOLECULAR BASIS OF HEPATOCYTE DEATH

Hepatocyte death is a common hallmark of much hepatic pathology. This is often seen as diffuse or zonal hepatocyte death and dropout. In many cases the hepatocyte death occurs due to death receptor activation, which leads to hepatocyte apoptosis and ensuing liver injury. These mechanisms of liver injury have been identified in hepatitis, inflammatory hepatitis, alcoholic liver disease, ischemia reperfusion injury, and cholestatic liver disease. Hepatocytes express the typical death receptors that belong to the TNFα receptor superfamily. These include (i) Fas also called Apo 1/CD95 or Tnf receptor superfamily6 (Tnfrsf6), (ii) TNF-receptor 1 (TNF-R1) also called CD120a or Tnfrsf10a, and (iii) TRAIL-R1 or Tnfrsf10A and TRAIL-R2 or Tnfrsf10B. The ligands for these receptors are the Fas ligand (FasL) also called CD178 or Tnfsf6, TNFα or Tnfsf2, and TRAIL or Apo 2L or Tnfsf10, respectively (reviewed in [124]). While the former two subgroups of death receptors and their ligands are known to be associated with the aforementioned liver diseases, TRAIL/TRAIL-R have not yet been identified as a mechanism of hepatic injury pertinent to a liver disease.

Fas-Activation Induced Liver Injury

This mode of death is known to be associated with liver diseases such as viral hepatitis, inflammatory hepatitis, Wilson's disease, cholestasis, and alcoholic liver disease. FasL, present on inflammatory cells or Fas-activating agonistic antibodies such as Jo-2 injection (in experimental models), leads to Fas activation resulting in massive hemorrhagic liver injury with extensive hepatocyte apoptosis and necrosis (Figure 20.4) [125]. Most mice die within 4–6 hrs after Jo-2 injection. Upon activation, homotrimeric association of Fas receptors occurs, which recruits Fas-associated death domain protein (FADD) adapter molecule, and initiator caspase-8 [126]. This complex, or the death-inducing signaling complex (DISC), requires mitochondrial involvement and cytochrome c release, which is inhibited by antideath Bcl-2-family proteins (Bcl-2 and Bcl-x_L) [127–129]. Cytochrome c stimulates the activation of caspase-9 and then caspase-3, inducing cell death. Interestingly, pro-death Bcl-2 family proteins Bid, Bax, and Bak are also needed for hepatocyte death induced by Fas [130–132]. In fact, Bid is cleaved by caspase-8 after Fas activation, and cleaved Bid translocates to mitochondria. This in turn activates Bax or Bak on the mitochondria to stimulate the release of apoptotic factors such as cytochrome c. At the same time Bid also induces mitochondrial release of Smac/DIABLO, which inactivates the inhibitors of apoptosis (IAP). The role of IAPs is at the level of caspase-3 activation, which occurs in two steps. The first step is caspase-8-induced severance of larger subunit of caspase-3. The second step is the removal of the prodomain by its autocatalytic activity, which is essential for caspase-3 activation and inhibited by IAPs [133]. While caspase-8 can directly activate caspase-3, bypassing the mitochondrial involvement, the execution of the entire pathway ensures the process of death in the hepatocytes [134]. Thus, overall it has been suggested that relative expression and activity levels of caspase-8, Bid, and other modulators of this pathway, such as inhibitors of apoptosis (IAPs) and Smac/DIABLO, would finally determine whether Fas-induced hepatocyte apoptosis will or will not utilize mitochondria to induce cell death.

A recent discovery unveils another important regulatory step in the Fas-mediated cell death [135]. Under normal circumstances, Fas receptor was shown to be

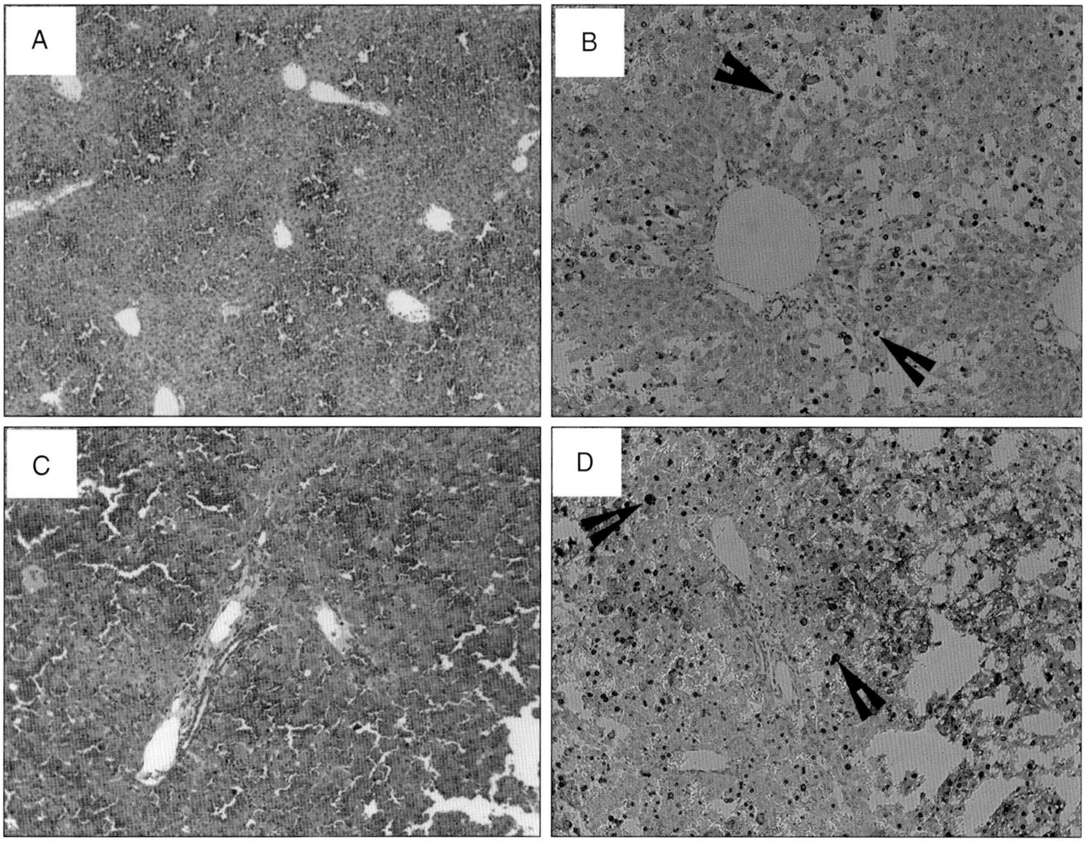

Figure 20.4 H&E and TUNEL immunohistochemistry exhibiting apoptotic cell death after Fas- and TNFα-mediated liver injury. (A) H&E shows massive cell death 6 hours after Jo-2 antibody administration. **(B)** Several TUNEL-positive apoptotic nuclei (arrowheads) are evident in the same liver. **(C)** H&E shows massive cell death 7 hours after D-Galactosamine/LPS administration in mice. **(D)** Several TUNEL-positive apoptotic nuclei (arrowhead) are evident in the same liver.

sequestered with c-Met, the HGF receptor, in hepatocytes. This makes Fas receptor unavailable to the Fas-ligand. Upon HGF stimulation, this complex was destabilized and hepatocytes became more sensitive to Fas-agonistic antibody. On the contrary, transgenic mice overexpressing extracellular domain of c-Met stably sequestered Fas receptor and hence were resistant to anti-Fas-induced liver injury. Recently, lack of this Fas antagonism by Met was identified in fatty liver disease [136]. While this explains how HGF could be pro-death at high doses or in combination with other death signals, a paradoxical effect of HGF on promoting cell survival is also observed, albeit at lower doses. This effect is usually also observed with other growth-promoting factors such as TGFα and are mediated by elevated expression of Bcl-x_L that inhibits mitochondrial release of cytochrome c and inhibits Bid-induced release of Smac/DIABLO [137,138].

TNFα-Induced Liver Injury

This mode of hepatocyte death is commonly observed in ischemia-perfusion liver injury and alcoholic liver disease. In mice, TNFα is induced by bacterial toxin administration such as lipopolysaccharide (25–50 μg/kg LPS) [139]. Since it was identified that LPS alone initiates NFκB-mediated protective mechanisms, an inhibitor of transcription (D-Galactosamine) or translation (Cycloheximide) is used before the LPS injection, for successful execution of cell death [140]. This induced massive hepatocyte death due to apoptosis, as seen by TUNEL immunohistochemistry (Figure 20.4). TNFα binds to the TNFα-R1 on hepatocytes to induce receptor trimerization and DISC formation. DISC is composed of TNFα-R1 and TNFR-associated death domain (TRADD), which can recruit FADD and caspase-8 via an unknown mechanism, to further activate caspase-3 [141,142]. The significance of additional mechanisms is relatively unknown in liver biology (reviewed in [124]). However, one additional mechanism that deserves mention and is relevant to liver injury is that TNFα-R1 engagement also induces Cathepsin B release from lysozyme, which induces cytochrome c release from mitochondria [143]. However, how much effect of TNFα on apoptosis is mediated via the mitochondrial pathway components including Bcl-2, Bcl-x_L, and Bid in TNFα-mediate apoptosis remains debated, but it is believed that some effects are via this intrinsic pathway.

One of the important effects of TNFα stimulation is the unique and concomitant activation of the NF-κB pathway as a protective mechanism. How this occurs is not fully understood, but is relevant as an ongoing protective mode for maintaining hepatic homeostasis. The TNFα-R1 engagement leads to the recruitment of TRADD, which can further induce recruitment of TNF Associated Factor 2 (TRAF2) and Receptor interacting protein (RIP) [144]. This can further recruit and activate IKK complex containing I-κB kinase 1 (IKKα) and IKKβ that lead to the phosphorylation of I-κB and its degradation. This allows for the release and activation of NF-κB p50 and p65 dimers, inducing the phosphorylation of p65, and their nuclear translocation to direct transcriptional activation of cytoprotective genes such as IAP [145,146], iNOS [147], Bcl-xL [148], and others. In addition, it is important to mention that several additional regulators of NFκB exist and also regulate its transcriptional activity. These include glycogen synthase kinase-3 (GSK3) [149] and TRAF2-associated kinase (T2K) [150]. GSK3-knockout embryos die at around E13–E15 stage as a result of massive liver cell death due to TNFα toxicity [27,149,151].

MOLECULAR BASIS OF NONALCOHOLIC FATTY LIVER DISEASE

Accumulation of triglycerides in the liver in the absence of significant alcohol intake is called nonalcoholic fatty liver disease (NAFLD) [152]. The term *NAFLD* encompasses a spectrum of liver disease ranging from simple steatosis to steatosis with inflammation called nonalcoholic steatohepatitis (NASH). The latter condition can progress to fibrosis, cirrhosis, and hepatocellular cancer.

While NAFLD has been associated with many drugs, genetic defects in metabolism, and abnormalities in nutritional states, it is most commonly associated with the metabolic syndrome [153]. The metabolic syndrome is a group of related clinical features linked to visceral obesity, including insulin resistance (IR), dyslipidemia, and hypertension [154]. NAFLD is strongly associated with and considered the hepatic manifestation of the metabolic syndrome. The prevalence of the metabolic syndrome, and therefore also of NAFLD, has been increasing in parallel with the obesity and diabetes epidemic, and NAFLD is now the most common cause of abnormal liver enzyme elevation in the United States [155].

While simple steatosis has a benign course, NASH represents the progressive form of the disease. NASH is characterized by necroinflammatory activity, hepatocellular injury, and progressive fibrosis. The pathogenesis of NAFLD/NASH is incompletely understood, although significant strides have been made in recent years to unravel its underlying molecular processes (Figure 20.5). A two-hit hypothesis has been proposed to explain the pathogenesis of NAFLD and the progression of simple steatosis to NASH that occurs in a subset of patients with steatotic livers [156]. The first hit consists of triglyceride deposition in hepatocytes. A second hit consisting of another cellular event then leads to inflammation and hepatocyte injury and the subsequent manifestations of the disease.

Factors Leading to the Development of Hepatic Steatosis: The First Hit

The accumulation of triglycerides in the liver is the essential characteristic of NAFLD/NASH and can be observed in patients as well as experimental models (Figure 20.6). It results from aberration in metabolic processes in the liver, as well as extrahepatic tissue

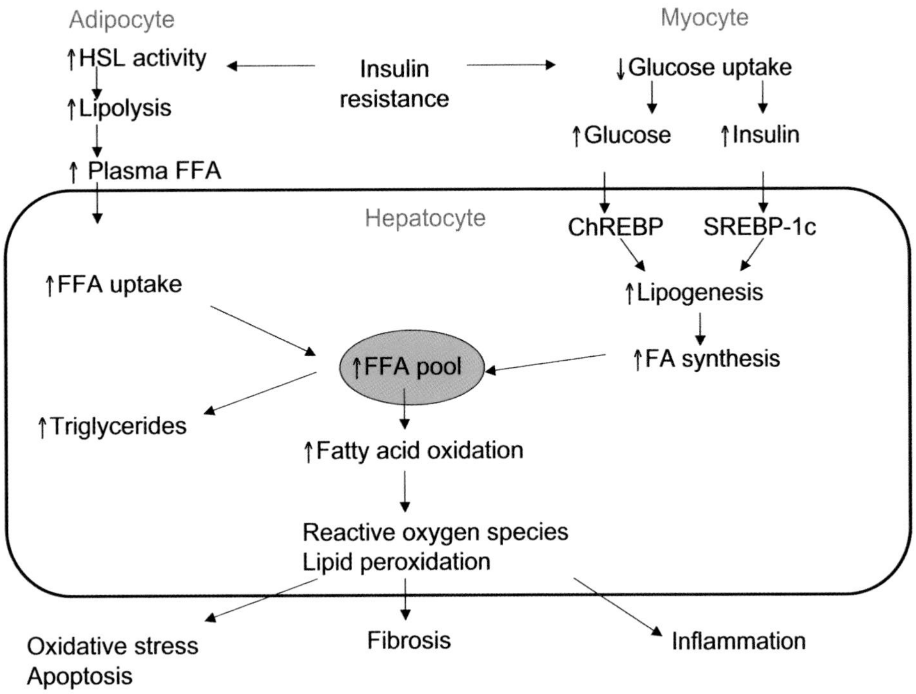

Figure 20.5 **Molecular basis of nonalcoholic steatohepatitis. Insulin resistance results in decreased insulin-induced inhibition of hormone-sensitive lipase (HSL) activity in adipocytes resulting in high rates of lipolysis and increased plasma free fatty acid (FFA) levels.** High plasma FFA levels increase FFA flux to the liver. In myocytes, insulin resistance results in decreased glucose uptake and high plasma glucose and insulin levels. In the liver, hyperinsulinemia activates SREBP-1c, leading to increased transcription of lipogenic genes. High plasma glucose levels simultaneously activate ChREBP, which activates glycolysis and lipogenic genes. SREBP-1c and ChREBP ct synergistically to convert excess glucose to fatty acids. Increased FFA flux and increased lipogenesis increases the total hepatic FFA pool. The increased hepatic FFA pool can either be converted to triglycerides for storage or transport out of the liver as very low density lipoprotein, or undergo oxidation in mitochondria, peroxisomes, or endoplasmic reticulum. Increased reactive oxygen species generated during fatty acid oxidation causes lipid peroxidation, which subsequently increases oxidative stress, inflammation, and fibrosis.

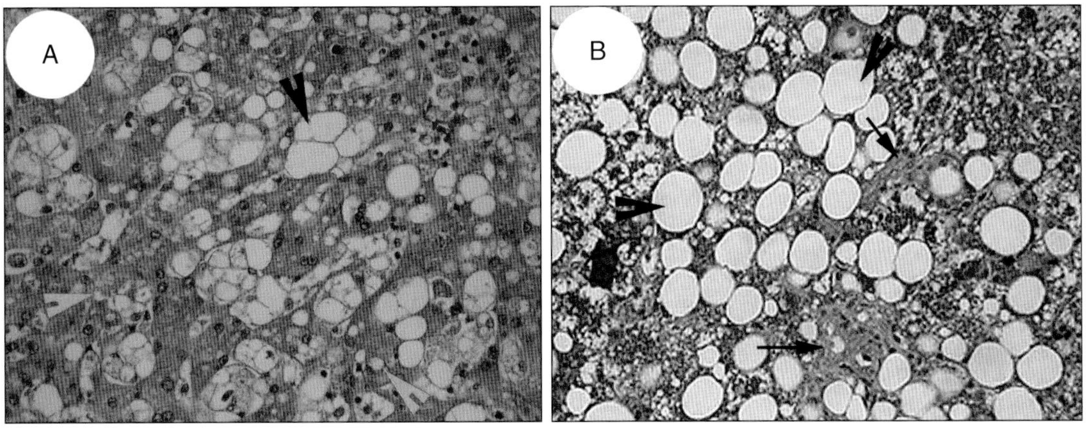

Figure 20.6 **Fatty liver disease along with hepatic fibrosis. (A)** Patient with NAFLD shows significant macrovesicular (arrowhead) and microvesicular steatosis (blue arrowhead) in a representative H&E stain. **(B)** Masson trichrome staining of liver from mouse on methionine and choline-deficient diet for 4 weeks reveals hepatic fibrosis (arrow) along with macrovesicular steatosis (arrowheads).

such as skeletal muscle and adipose tissue that is mediated by insulin resistance (IR). Normally, insulin acts on the insulin receptor of myocytes to tyrosine phosphorylate insulin receptor substrate (IRS). IRS, in turn, activates phosphatidyl inositol 3-kinase and protein kinase B, and results in translocation of the glucose transporter to the plasma membrane with resultant rapid uptake of glucose from the blood into the myocyte. The net effect is decreased blood glucose and therefore decreased insulin secretion from the pancreas [157]. In the setting of obesity, fat-laden myocytes become resistant to the signaling effects of insulin. This inability of skeletal muscle to take up glucose from the circulation on insulin stimulation leads to elevated blood glucose levels and increased insulin secretion from the pancreas, with metabolic consequences in the liver as described in the following text.

Besides skeletal muscle, IR also causes increased blood glucose via decreased insulin action in the liver. Normally, the liver plays an important role in maintaining blood glucose levels regardless of the nutritional state. In the fasting state, gluconeogenesis in the liver results in hepatic glucose production that maintains blood glucose level. However, in the fed state when blood glucose levels are elevated, glucose is converted to pyruvate via glycolysis, which then enters the Krebs cycle to form citrate and ultimately acetyl-CoA, which is utilized for fatty acid biosynthesis. This process is normally activated by insulin, which is also a potent inhibitor of hepatic glucose production. However, in the setting of IR, insulin is unable to suppress hepatic glucose production, which, along with decreased myocyte glucose uptake, leads to high blood glucose levels and high circulating insulin levels [158,159]. The net effect of these changes is increased hepatic lipogenesis and triglyceride accumulation mediated by the synergistic action of two transcription factors: carbohydrate response element binding protein (ChREBP) and sterol regulatory element-binding protein-1c (SREBP-1c). There is evidence from both animal experiments and human studies to support the conclusion that there is increased hepatic fatty acid synthesis in the setting of IR [160,161].

ChREBP is a member of the basic helix-loop-helix-leucine zipper (b-HLH-Zip) transcription factor that after activation by glucose translocates from the cytosol into the nucleus and binds to the E-box motif in the promoter of liver-type pyruvate kinase (L-PK), an important glycolytic enzyme [162,163]. L-PK catalyzes the formation of pyruvate, which enters the Krebs cycle and generates citrate and ultimately acetyl-CoA, which is then utilized for the *de novo* synthesis of fatty acids. Studies with ChREBP knockout mice also revealed that ChREBP also independently stimulates the transcription of fatty acid synthetic enzymes [162,164]. Therefore, in the setting of IR and hyperglycemia, ChREBP mediates the conversion of glucose to fatty acids by upregulating both glycolysis as well as lipogenesis.

Besides glucose, insulin can also regulate *de novo* hepatic fatty acid synthesis via a second transcription factor, SREBP-1c, also a member of the b-HLH-Zip family of transcription factors [165]. There are three SREBP isoforms, but the SREBP-1c isoform is the major isoform in the liver that can activate expression of fatty acid biosynthetic genes and stimulate lipogenesis [166,167]. The important role of SREBP-1c in development of hepatic steatosis was established by the demonstration that transgenic mice overexpressing SREBP-1c have increased lipogenesis and develop fatty liver [168]. Furthermore, in the *ob/ob* mouse model of genetic obesity and insulin resistance, disruption of the *Srebp-1* gene caused a reduction in hepatic triglyceride accumulation [169]. Since SREBP-1c is activated by insulin, it would be expected to be inactive in the setting of IR. Surprisingly, the protein is activated by insulin even when there is IR, resulting in increased fatty acid biosynthesis in the liver [170].

A third process that favors hepatic steatosis results from IR in the adipose tissue. In the fat-laden adipocytes in obese individuals, IR causes defective insulin-mediated inhibition of hormone-sensitive lipase and therefore increased lipolysis and increased free fatty acid (FFA) release into the circulation [171,172]. Increased FFA uptake into the liver from the circulation also favors the development of hepatic steatosis and inflammation. Interestingly, recent studies have shown that high FFA levels in the liver may contribute to lipotoxicity, and triglyceride accumulation in the liver might be a protective response [173]. When diacylglycerol acyltransferase 2, the final step in hepatocyte triglyceride biosynthesis, was inhibited by antisense oligonucleotides in a murine model of steatohepatitis, there was decrease in hepatic steatosis, but increased hepatic FFAs, and increase in makers of inflammation and hepatocyte injury and fibrosis.

Besides SREBP-1c and ChREBP, a transcription factor belonging to the nuclear hormone receptor family, peroxisome-proliferator activated receptor-γ (PPAR-γ) may also contribute to hepatic steatosis. While normal hepatic expression of PPAR-γ is low, increased expression of the transcription factor has been noted in animal models of steatosis [174,175]. Furthermore, in the *ob/ob* mouse model of insulin resistance, deletion of PPAR-γ from the liver markedly decreases hepatic steatosis, suggesting an important role for PPAR-γ in the development of NAFLD.

Once hepatic steatosis is established, outcomes are variable. In some patients with NAFLD there is no further damage to the liver and prognosis in terms of liver-related mortality is good. However, in some patients there is hepatic necrosis, inflammation, hepatocyte apoptosis, and fibrosis. The reason some patients develop NASH is not understood, but several mechanisms have been postulated, as described in the next section.

Progression of Steatosis to NASH: The Second Hit

Absence of noninvasive biomarkers to differentiate NASH from simple steatosis and lack of animal models that completely recapitulates the pathophysiology of human NASH have limited research into the processes that promote this condition. However, several important insights have been gained in recent years on the pathogenesis of NASH and include oxidative stress, lipid peroxidation, mitochondrial dysfunction, inflammatory cytokines and adipokines, and activation of pathways of cell death.

Oxidative Stress and Lipid Peroxidation

Reactive oxygen species (ROS) refers to several short-lived, pro-oxidant chemicals including hydroxyl radical, singlet oxygen molecules, hydrogen peroxide, and superoxide anions. These pro-oxidant molecules lead to oxidative damage to macromolecules in the cell when they overwhelm the protective antioxidant mechanisms of the cell [176].

Functional inactivation of essential cellular biomolecules causes either cell death or production of inflammatory mediators via redox-sensitive transcription factors such as Nrf-1 and NFκB [177,178]. Studies have shown increased serum markers of oxidative stress, and markers of oxidative damage for DNA, and proteins besides lipid peroxides in NASH [179–181]. Additionally, antioxidant factors such as glutathione S-transferase and catalase are reduced [182]. Polyunsaturated fatty acids (PUFAs) in the cell can undergo peroxidation by ROS, resulting in the formation of malondialdehyde (MDA) and trans-4-hydroxy-2-nonenal (4–HNE) byproducts [183].

Lipid peroxides contribute to liver injury by increasing production of TNFα, increasing the influx of inflammatory cells, impairing protein and DNA synthesis, and depleting levels of protective cellular antioxidants like glutathione [184–187].

Mitochondrial Dysfunction

Several lines of evidence suggest that hepatic mitochondrial dysfunction has a role in the pathogenesis of NASH. Ultrastructural studies have shown the presence of hepatocyte megamitochondria with paracrystalline inclusions in patients with NASH [179,188]. Functional differences include impaired ability to synthesize ATP after a fructose challenge, which causes transient liver ATP depletion and lower expression levels of mitochondrial DNA-encoded proteins and lower activity of complexes of the mitochondrial respiratory chain in patients with NASH [189–191].

The mechanisms leading to mitochondrial dysfunction are incompletely understood, but a contributory role may be played by increased mitochondrial oxidation of fat that increases respiratory chain electron delivery. In this setting, mitochondrial respiratory chain dysfunction causes interruption in electron flow and the electrons are then transferred to molecular oxygen, giving rise to hydrogen peroxide and superoxide radicals [192,193]. With progressive decrease in mitochondrial fatty acid oxidation, alternative pathways of fatty acid oxidation, peroxisomal β-oxidation, and microsomal ω-oxidation become active and contribute to the generation of ROS by transfer of electrons to molecular oxygen and a vicious cycle of increasing mitochondrial dysfunction [194–196]. Mitochondrial dysfunction is also worsened by activation of cytochrome P450 2E1, cytochrome P450 4A10, and cytochrome P450 4A14-mediated microsomal ω-oxidation via generation of dicarboxylic acids that uncouple oxidative phosphorylation and add to oxidative stress [197].

Role of Signaling Pathways of Inflammation, Proinflammatory Cytokines, and Adipokines

Inflammation in NASH results from cross-talk between hepatocytes and nonparenchymal cells (activated Kupffer cell, stellate cells, and sinusoidal endothelial cells) mediated by soluble factors, as well as proinflammatory molecules released from visceral adipose tissue. The NFkB pathway has been extensively studied for its role in steatohepatitis and is upregulated in NASH patients and in animal models of the disease [198]. However, its role in pathogenesis of NASH is complex, and while activation of NFkB signaling in liver cells induces inflammation and steatosis, its inactivation also causes steatohepatitis and liver cancer in a mouse model [199,200]. Deletion of the JNK1 isoform of c-Jun N-terminal kinase (JNK), a mediator of TNF-induced apoptosis, is protective in a mouse model of steatohepatitis, suggesting that JNK signaling is important in the pathogenesis of NASH [201].

Proinflammatory cytokines TNFα and IL-6 are increased in NASH patients [202,203]. In animal models of steatohepatitis, decreasing TNFα expression causes decreased hepatic steatosis, cell injury, and inflammation, suggesting a role of this cytokine in the pathogenesis of NASH [204]. However, targeted disruption of TNFα or its receptor TNFR1 does not protect against lipid peroxidation and hepatocyte injury in diet-induced steatohepatitis. Since inactivation of NFκB, of which TNFα is an effector, does protect hepatocytes in this setting, TNFα may represent one of many inflammatory mediators of steatohepatitis.

Adipose tissue, particularly visceral fat, plays an important role in the pathogenesis of IR and development of NASH. Visceral adipocytes overexpress 11 β-hydroxysterol dehydrogenase and more readily mobilize fat and are less mature than subcutaneous adipocytes [205]. In addition, adipocytes are hormonally active and produce several adipokines that affect insulin action and metabolic processes in the liver. Among the important adipokines are leptin and adiponectin [206]. Leptin levels are increased in the plasma of patients with NAFLD [207]. However, leptin has an antisteatotic role and is protective. Therefore, it has been suggested that NAFLD may be associated with leptin resistance and a defective response to leptin.

Adiponectin levels, on the other hand, are low in patients with NASH, and lower levels are associated with more severe liver injury [208,209]. Adiponectin has been shown to increase insulin sensitivity, regulate FFA metabolism, and inhibit gluconeogenesis [210,211]. Adiponectin also has strong anti-inflammatory effects that are mediated by suppressing TNFα levels and activity [212].

Cell Death

Hepatocytes express the membrane receptor Fas, but do not induce apoptosis in their neighboring cells because they do not produce the Fas ligand (FasL). However, patients with NASH have increased hepatocyte FasL expression that can trigger apoptosis [213]. When FasL interacts with Fas, it activates caspase-8, which in turn activates the Bid protein by truncating it. Bid then activates the protein Bax, which forms a complex with Bak and forms a channel in the outer mitochondrial membrane [214]. The resulting increase in permeability of the mitochondrial membrane releases cytochrome c from the inner membrane space and blocks the mitochondrial respiratory chain and increases formation of ROS in the mitochondria. ROS increase permeability and cause rupture of the mitochondrial outer membrane and release of proapoptotic molecules, which activate caspase-9 and caspase-3 and perpetuate a vicious cycle of apoptosis [215].

MOLECULAR BASIS OF ALCOHOLIC LIVER DISEASE

The histological spectrum of alcohol-induced liver disease spans simple steatosis, to steatohepatitis (characterized by inflammatory cell infiltration, hepatocyte ballooning, apoptosis, and necrosis in addition to fat accumulation), fibrosis, and cirrhosis. Steatosis is mainly macrovesicular and most prominent in the centrilobular region. Early stages of alcohol-induced liver injury are reversible, and prevention of steatosis in experimental models has been shown to prevent development of inflammation and fibrosis [216,217].

Pathways of Alcohol Metabolism in the Liver

There are three enzymatic pathways of ethanol metabolism in the liver, and all three contribute to ethanol-induced liver injury [218]. First, oxidation of ethanol occurs through the cytosolic alcohol dehydrogenase (ADH) isoenzymes that produce acetaldehyde, which is then converted into acetate, and these reactions result in the reduction of nicotinamide adenine dinucleotide (NAD) to NADH, its reduced form. Excessive ADH-mediated hepatic NADH generation has multiple important metabolic effects due to inability of hepatocytes to maintain redox homeostasis [219]. This redox imbalance results in inhibition of the Krebs cycle and fatty acid oxidation, which promotes hepatic steatosis. However, with chronic ethanol consumption, the redox state is normalized. Instead, alterations in regulation of lipid metabolism genes appear to be important in this setting. Second, alcohol is metabolized by the microsomal ethanol oxidizing system (MEOS), of which cytochrome P450 2E1 (Cyp2E1) is the key enzyme, although other cytochrome P450 enzymes such as Cyp1A2 and Cyp3A4 also play a role [218–220]. Cyp2E1 is induced by chronic ethanol consumption, and its induction is responsible for the metabolic tolerance to alcohol in alcohol-dependent individuals. Induction of Cyp2E1 plays a role in alcohol-induced liver injury through the production of reactive oxygen species that can overwhelm the cellular defense systems leading to mitochondrial injury, and further exacerbation of oxidative stress [221]. Finally, ethanol is also metabolized via a nonoxidative pathway by fatty acid ethyl ester (FAEE) synthase. The product of this reaction, fatty acid ethyl esters (FAEEs), accumulates in the plasma, lysosomal, and mitochondrial membranes and can interfere with their functioning and with cellular signal-transduction pathways [222].

Changes in Expression of Genes Involved in Lipid Metabolism on Chronic Alcohol Exposure

Recent studies suggest that the development of alcoholic and nonalcoholic steatohepatitis (NASH) share many common pathogenic mechanisms [223]. Insight into the molecular mechanisms of alcoholic-induced steatohepatitis has been obtained through the investigation of pathways and regulatory molecules involved in lipid homeostasis that have previously been shown to play a role in development of NASH. Chronic ethanol consumption affects both oxidation of fatty acids and increases *de novo* lipogenesis, thereby causing accumulation of fat in the liver. Peroxisome proliferators-activated receptor-α (PPARα), a member of the nuclear hormone receptor superfamily, is the master regulator of genes involved in free fatty acid transport and oxidation [224]. PPARα forms a dimer with retinoid X receptor (RXR) and binds to the peroxisome proliferators response element (PPRE) in the promoter regions of its target genes [225]. PPARα targets include mitochondrial and peroxisomal fatty acid oxidation pathway genes, apolipoprotein genes, and the membrane transporter, carnitine palmityl-transferase I (CPT I), which allows long-chain acyl-CoAs to enter mitochondria to initiate β-oxidation. PPARα also regulates CPT I activity and therefore fatty acid oxidation indirectly via regulation of the enzyme malonyl-CoA decarboxylase, which degrades malonyl-CoA, an allosteric regulator of CPT I. Both *in vitro* in primary hepatocyte cultures and hepatoma cells, and *in vivo*, ethanol has been shown to decrease β-oxidation of fatty acids by interfering with PPARα DNA binding and transcription-activation [226,227]. Decreased oxidation of fatty acids by ethanol leads to accumulation of fat in the liver [228].

The second mechanism by which ethanol causes steatosis is by increasing the rate of fat synthesis in the liver by increasing expression levels of lipogenic enzymes that are regulated by the transcription factor SREBP. There are three isoforms of SREBP, called SREBP1a, SREBP1c, and SREBP2 [228,229]. The isoform SREBP1a is mainly expressed in cultured cells, SREBP1c regulates fatty acid synthesis in the liver, and SREBP2 regulates cholesterol synthesis. SREBPs are synthesized as precursor proteins and are attached to the endoplasmic reticulum and nuclear membranes. When the protein is activated, there is proteolytic cleavage of the NH2-terminal fragment (called the mature form of the protein), which then enters the nucleus and binds to target genes via the sterol response elements (SREs) [229]. Recent studies have shown that ethanol-fed mice have increased SREBP1 expression; increase in the amount of mature SREBP1 in the liver; and corresponding upregulation of expression of SREBP-target genes, such as fatty acid synthase, stearoyl-coA desaturase, and ATP citrate lyase, in the fatty acid biosynthetic pathway [166,230,231].

The third critical metabolic regulatory molecule in the liver affected by ethanol is AMP-activated protein kinase (AMPK), which is a metabolic switch regulating pathways of hepatic triglyceride and cholesterol synthesis. AMPK can regulate SREBP1 expression at the transcriptional and post-transcriptional levels [232,233]. Ethanol has been shown to inhibit hepatic AMPK activity, and treatment with AMPK activators partially blocks ethanol-mediated increase in expression of SREBP-dependent genes *in vitro* [234]. Furthermore, adiponectin, a hormone derived from adipocytes that is an activator of AMPK, has been shown to alleviate both alcoholic and nonalcoholic fatty liver disease in mice [235].

Inflammatory Cytokines and Role of Kupffer Cells

Besides altering metabolic pathways, chronic ethanol feeding results in altered expression of several inflammatory mediators, including reactive oxygen species, cytokines, and chemokines [217,236]. An important source of inflammatory mediators from alcohol consumption is hepatic Kupffer cells which produce TNFα, an important mediator of the inflammatory response in mammals. Administration of TNFα to mice causes development of fatty liver, activation of SREBP, and increased fatty acid synthesis [237]. Other studies have shown lipopolysaccharide-mediated increased reactive oxygen species production by Kupffer cells with chronic alcohol feeding and normalization of this response by adiponectin [238,239].

Activation of NFκB by alcohol has been shown in experimental models as well as in livers of patients with alcoholic hepatitis. NFκB activation and its target gene expression play an important role in inflammatory response to bacterial endotoxin and in the activation of the innate immune system in response to necrotic cells [240,241]. In isolated Kupffer cells, chronic ethanol feeding has been shown to increase NFκB binding to the TNFα promoter on LPS stimulation [242].

MOLECULAR BASIS OF HEPATIC FIBROSIS AND CIRRHOSIS

Chronic liver injury is often associated with a wound healing process in the liver that is commonly referred to as liver fibrosis, which in the advanced stage is termed as cirrhosis. Liver cirrhosis can range in its presentation from being asymptomatic to liver failure. Hepatic fibrosis sets in many pathological scenarios including but not limited to alcoholic liver disease, NAFLD, viral hepatitis, autoimmune disorders, Wilson's disease, cholestatic liver disease, and others. The process of fibrosis entails excessive deposition of extracellular matrix in the liver, especially collagen (Figure 20.6). This compromises the hepatic architecture leading eventually to cirrhosis and hepatic failure.

The fundamental cell type involved in the process of hepatic fibrosis is the hepatic stellate cell (HSC). Chronic insult to the liver leads to the activation of hepatic stellate cells. The process of HSC activations entails increased DNA synthesis and proliferation, activation of profibrotic target genes and increased cell contractility (reviewed in [243,244]). The end result of this process is hepatic fibrosis and eventually cirrhosis. It should be mentioned that the overall impact of the fibrosis is not only due to excessive deposition of extracellular matrix and loss of functional hepatocyte compartment, but also indirect effects on circulation secondary to constriction of sinusoids. How do HSCs undergo activation? A multitude of insults converge onto this cell type, which upon activation, bring about the fibrotic phenotype owing to downstream genetic changes. Broadly speaking, the two major insults, including ASH and NASH, both seem to be mediating HSC activation through increased oxidative stress, albeit via unique mechanisms. Alcohol, which is the leading cause of hepatic fibrosis in Western countries, has been shown to elevate oxidative stress after being metabolized by CYP2E1 to generate reactive oxygen species that are known to directly interact with HSC [245,246]. In addition, alcohol is also metabolized to acetaldehyde, which in turn has been shown to be fibrogenic [247,248]. Lastly, Kupffer cells, an additional nonparenchymal cell in the liver, are also involved in the generation of both acetaldehyde and alcohol-induced lipid peroxidation products in alcoholic liver disease [249]. Around 20%–40% of NASH patients also progress to hepatic fibrosis and have been shown to also be secondary to elevated oxidative stress [250]. However, this increase is mediated via production of ROS secondary to PPAR-α mediated increased free fatty acid oxidation that occurs in turn due to excessive high fat intake, experimentally or otherwise [251]. It is worth noting that activation of HSC can very well be secondary to stimuli produced by other cell types present during the injury, including the hepatocytes, Kupffer cells, sinusoidal endothelial cells, and the circulating inflammatory cells.

What signaling mechanisms induce the HSC activation and in turn induce the target gene expression in HSC? Several relevant pathways have been identified, and the most prominent ones include the PDGF axis and TGFβ/Smad signaling pathway. PDGF is a potent mitogen for HSC, and its cognate receptor PDGFR expression goes up during HSC activation [252]. PDGF activation has also been shown to stimulate PI3 kinase, which also induces HSC proliferation [253]. In addition, recruitment of Ras to PDGF receptor also leads to ERK activation via sequential activation of Raf-1, MAPK-1/2, ERK-1, and ERK-2. Nuclear ERK can regulate target genes responsible for proliferation of HSCs. Additionally JNK activation has been shown to positively regulate HSC proliferation [254]. TGFβ is a potent profibrogenic cytokine and is known to be produced by variety of cells including HSC, hepatocytes, and others [255]. TGFβ signaling eventually leads to target gene expression, and several of these targets are the hallmark signatures of HSC activation. TGFβ has been shown to stimulate synthesis of collagens, decorin, elastin, and others. Increased expression of KLF6, a transcription factor and a tumor suppressor, which acts as a chaperone for collagen has also been reported to be increased during HSC activation [256]. Some relevant targets of KLF6 include collagen 1α, TGF-β1 and its receptors, and urokinase plasminogen activator (uPA), which activates latent TGF-β1. Also, connective tissue growth factor (CTGF), which is regulated by TGFβ, is also upregulated during HSC activation, as well as in chronic viral hepatitis, bile duct ligation, and other related scenarios [257,258]. It is also important to mention that decrease in PPAR-γ has been associated with HSC activation [259–262]. In fact PPAR-γ ligands have been shown to inhibit HSC proliferation and activation in cell cultures. Thus, sustained basal expression of PPAR-γ in HSC has been suggested to maintain their quiescence. It is also relevant to mention the phenotype of a quiescent and activated HSC. Quiescent HSCs are typically characterized by the presence of perinuclear lipid droplets containing vitamin A or retinoid [263,264]. These droplets are lost following activation of HSC, when the retinoid is released as retinol. While the causal relationship of this event to HSC activation is unclear, increasing focus is shifting on retinoic acid receptors in the nuclei including RAR and RXR. It is important to point out that decrease in RXR and PPAR-γ receptor is associated with HSC activation, and the converse has been reported in HSC quiescence [265,266].

Once HSC activation occurs, several relevant genes have been shown to be upregulated (reviewed in [244]). The most pertinent include the extracellular matrix genes such as type I collagen (α1 and α2), type III collagen, laminin, and fibronectin, and proteoglycans such as decorin, hyaluronan, heparin sulfate, and chondroitin sulfate. In addition, several genes that are essential for matrix remodeling such as MMP-2, MMP-9, and TIMP-1 have also shown to be elevated. Lastly, genes such as ICAM-1 and α-SMA are upregulated as well. In addition, several studies have reported a genome-wide analysis and either strengthened the existing findings or identified additional aberrations, which would, in the years to come, be exploited for understanding the biology and in turn devising novel therapies for hepatic fibrosis and eventually cirrhosis.

Cirrhosis is often defined as the advanced stage of hepatic fibrosis, which is accompanied by development of regenerative nodules amidst the fibrous bands and follows a chronic liver injury. This pathology leads to vascular distortions, impaired parenchymal flow leading to portal hypertension, and end-stage liver disease. Thus, the molecular pathology of cirrhosis is a continuum of the hepatic fibrosis. An interesting point to remember is that initially during the process of hepatic fibrosis, increased fibrogenesis is being counterbalanced by factors negatively regulating ECM deposition (reviewed in [267]). However, as chronic insult to the liver continues, the process of fibrogenesis, observed as continued activation of myofibroblasts derived from HSC and perivascular fibroblasts, exceeds fibrinolysis. It remains a conundrum to be able to successfully predict the risk of developing cirrhosis in patients with same hepatic pathology. The most promising advances that are being reported include the identification of genetic polymorphisms that are able to predict the risk of progression of hepatic fibrosis. Studies reporting single nucleotide polymorphisms (SNPs) as predictors of progression of fibrosis are beginning to trickle in [268,269]. Recently, cytokine SNPs were identified in successfully predicting disease progression [268]. Similarly SNPs in the DDX5 gene successfully predicted fibrosis progression in Hepatitis C patients as well [269].

Understanding the molecular pathology of hepatic fibrosis and cirrhosis also enables the possibility of prophylactic and therapeutic intervention. To this end, various steps of HSC activation are being investigated, both preclinically and clinically. Broadly speaking, such agents range from general anti-inflammatories to selective biologicals and chemical inhibitors targeting signal transduction components that have been discussed in the preceding sections to play a role in HSC activation [244,267]. In addition, since oxidative stress is implicated in HSC activation, use of antioxidants is also under investigation. Additional drugs being developed include agents regulating ECM deposition, vasoactive modulators, cytokine antagonists, and factors promoting hepatocyte survival and growth.

MOLECULAR BASIS OF HEPATIC TUMORS

Primary liver tumors presenting as hepatic masses are classified based broadly as benign or malignant. The common benign tumors in the liver include hemangiomas, focal nodular hyperplasia (FNH), and liver cell adenoma or hepatic adenoma (HA). Malignant tumors are usually classified based on the lineage of transformed cells and various clinicopathological characteristics. Broadly, the tumors originating from hepatic progenitors or hepatoblasts are referred to as hepatoblastoma (HB) and the tumors originating from more mature hepatocytes are referred to as hepatocellular cancer (HCC).

Benign Liver Tumors

Hemangioma

Cavernous hemangioma tops the list of the benign tumors of the liver [270]. These lesions are usually less than 5 cm and can range from 1–20 cm. These tumors are most common in adult women although they can affect any sex and any age. Most of these tumors occur in the right lobe underneath the capsule and are usually well circumscribed, compressible, and with a darker color. The patients are mostly asymptomatic unless there is an associated complication such as necrosis, thrombosis, infarction, or hemorrhage. The tumor arises from the endothelial cells that line the blood vessels within the liver and is thought to entail ectasia rather than hypertrophic or hyperplastic events. The tumor is composed of large vascular channels lined by endothelial cells and collagen lining and separated by connective tissue septa. These endothelial cells have been shown to possess immunocytochemical properties of vascular rather than sinusoidal endothelial cells [271]. A significant number of the larger tumors are reported to taper into dilated vascular spaces before transitioning to normal hepatic architecture suggesting irregular borders rather than being well circumscribed [272]. The tumors derive their blood supply from the hepatic artery. The pathogenesis of this tumor is poorly understood. Because of its preponderance in adult females, and accelerated growth of tumors observed in high estrogen states such as puberty, pregnancy, and oral contraceptive use, a hormonal mechanism has been suggested, although the reports on the presence of estrogen receptors on a subset of these tumors are conflicting [270,272].

Focal Nodular Hyperplasia

Focal nodular hyperplasia (FNH) are benign hepatic tumors next only to hemangiomas in incidence. These tumors occur at a higher frequency in females and occur between the ages of 20 and 50 [273]. The tumor is usually multinodular, composed of a few hepatocyte-thick plates, and is thought to occur as a hyperplastic response to either a pre-existing developmental arterial malformation or to increased blood flow, thus leading to cellular hyperplasia. There is well-known association of FNH with vascular disorders such as Rendu-Osler-Weber syndrome or hereditary telangiectasia. The molecular basis of FNH remains largely obscure. However, it is clear that FNH is polyclonal in nature in a significant subset of cases with only a minor subset composed of monoclonal tumor cells. Recently, transcriptome analysis of FNH identified activation of Wnt/β-catenin pathway without any mutations in β-catenin gene [274]. While the significance of these findings is unclear, especially since these tumors are thought to be a result of vascular disturbances, these observations might be a result of alternate mechanisms of β-catenin activation such as growth factor-dependent activation [275,276]. Also, levels of genes encoding for angiopoietin 1 and 2 are altered in FNH, with levels of ANGPT1/ANGPT2 ratio being greater in all FNH cases examined [277]. Thus, while the exact molecular basis of FNH remains obscure, aberrations in some pathways are becoming apparent, and additional genetic and proteomic studies would further assist in this endeavor. It should be mentioned that there are no major complications associated with this pathology, and malignant transformation is not known to occur.

Hepatic Adenoma

Hepatic adenomas (HA) are benign liver tumors that occur in greater frequency in females and are observed as benign proliferation of hepatocytes in an otherwise normal liver [273]. These monoclonal tumors have been classically associated with the use of estrogen-containing oral contraceptives or androgen-containing steroid anabolic drugs. Such tumors are usually observed as solitary masses and are asymptomatic. Additionally, glycogen storage diseases, especially Type I (von Gierke) and Type III, are also known risk factors for the development of hepatic adenoma. However, in these circumstances, HAs oftentimes occur as multiple lesions with a greater propensity to undergo malignant transformation. Macroscopically, HAs present as solitary, yellowish masses due to lipid accumulation and give a pseudo-encapsulated appearance because of the compression of adjacent hepatic tissue and can range from 0.5–15 cm in diameter. Some of the histological features include areas of fatty deposits and hemorrhage, cord-like or plate-like arrangements of larger hepatocytes containing excessive glycogen and fat, sinusoidal dilatation (the result of the effects of arterial pressure, as these tumors lack a portal venous supply), absent bile ductules, and presence of few and nonfunctioning Kupffer cells [278] (Figure 20.8). The extensive hypervascularity and lack of a true capsule make this tumor prone to hemorrhage. The tumors can regress in size following discontinuation of inciting factors such as oral contraceptives, while at other times they continue to exhibit a stable or progressive disease. However, the risk of associated hemorrhage and neoplastic transformation is always present. In view of these risks, surgical resection is often recommended. However, performance of surgery is influenced by the multiplicity or size of the tumors and associated symptoms on one hand and surgical risk to the patient on the other.

The molecular basis of the tumors is being increasingly understood [273,279]. Significant subsets of hepatic adenomas display inactivating mutations in *HNF1α* or *TCF1* gene. It should be noted that *HNF1α*-knockout mice develop fatty liver and hepatocyte dysplasia supporting a clear role of this protein in lipid metabolism [280,281]. Biallelic inactivating mutations in *TCF1* genes were identified in around 50% of HA. The tumors in this scenario display marked steatosis and excess glycogen accumulation. These observations were explained by the role of HNF1α in regulating liver fatty acid-binding protein (L-FABP), which plays a role in fatty acid trafficking in hepatocytes, and glucose-6-phosphate increase, respectively [282]. HA with *HNF1α* inactivation display an extremely low risk of malignant transformation. In another subset of HAs, Wnt/β-catenin activation is observed secondary to mutations in *CTNNB1* gene that eventually effect the degradation of β-catenin protein. These adenomas can range from 15% to 46%, but the numbers might be closer to the lower percentage when only *CTNNB1* mutations as a mechanism of β-catenin activation are taken into account [279,283]. These tumors show frequent cytological abnormalities and pseudo-glandular formation and have been shown to occur at abnormally higher frequency in males. Most importantly, these HAs have been shown to possess a higher propensity for malignant transformation. There is also yet another group of HAs that display acute phase inflammation, histologically [273]. The true molecular basis of these HAs remains unknown, but they are seen in patients with high body mass index and excessive alcohol intake. β-catenin activation has also been observed in some of these HAs. Such tumors show sinusoidal dilatation, inflammatory infiltrates, and vessel dystrophy.

Malignant Liver Tumors

Hepatoblastoma

Hepatoblastoma (HB) is a rare primary tumor of the liver (incidence is 1 in 1,000,000 births), but is the

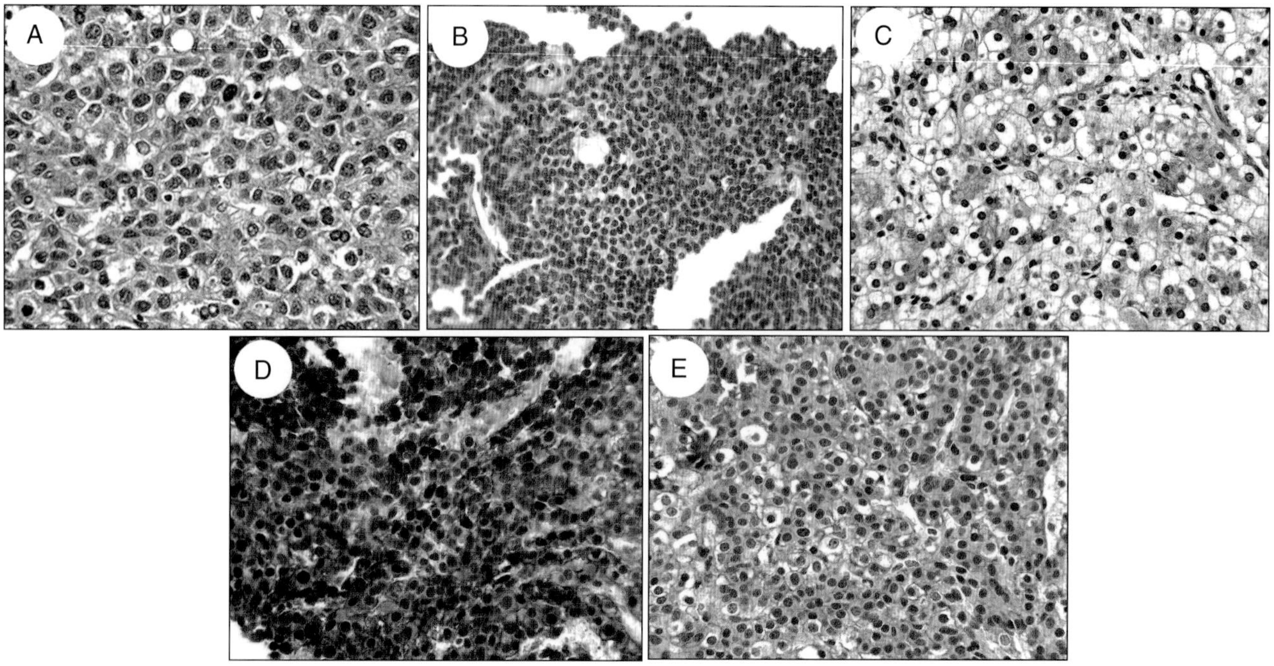

Figure 20.7 Histology and immunohistochemistry for β-catenin in pediatric hepatoblastomas. (A) H&E displaying embryonal pattern of hepatoblastoma. **(B)** H&E displaying small cell embryonal hepatoblastoma. **(C)** H&E displaying fetal hepatoblastoma. **(D)** Nuclear and cytoplasmic localization of β-catenin in an embryonal hepatoblastoma. **(E)** Nuclear localization of β-catenin along with some membranous staining in a fetal hepatoblastoma.

most frequent liver tumor in children under the age of 3 years [284,285]. This tumor is classified based on cellular composition and differentiation. The epithelial group of tumors comprises fetal, embryonal, mixed (embryonal and fetal), macrotrabecular, and small cell undifferentiated subtypes (Figure 20.7). Mixed epithelial-mesenchymal HBs, in addition to containing any of the preceding epithelial cell types, also contain mesenchymal components such as myofibroblastic, chondroid, and osteoid tissues. In addition, teratomatous HBs contain derivatives of all three germ layers. Most HB are sporadic but have been reported as a component of the Beckwith-Wiedemann syndrome (BWS), where its incidence is higher than the general population or in the familial adenomatous polyposis (FAP), which occurs due to germline mutation in the adenomatous polyposis coli (*APC*) gene [286,287].

To discuss the genetic abnormalities in HB, we need to refer to cytogenetic studies that have identified various frequent chromosomal aberrations [288]. Most of these involved chromosomes 2, 20, 1, and 8 and included trisomies and unbalanced translocations. Another important relevant finding is the lack of significant similarity between the molecular cytogenetics by comparative genomic hybridization of HB and HCC [288]. HB typically displayed lower rate of chromosomal imbalances with a small subset showing none, versus HCCs showing multiple chromosomal aberrations ranging from 5–16 per tumor. These studies have suggested tumor genes located in the specific loci to be part of the pathogenetic process, but additional functional studies are lacking. However, by far the most relevant molecular aberration in the HB is anomalous activation of Wnt/β-catenin signaling, also referred to as the canonical Wnt signaling pathway [289]. As discussed previously in the section on liver development, Wnt/β-catenin signaling plays an indispensable role in normal liver development [290,291]. Sequence analysis of β-catenin gene (*CTNNB1*) has revealed missense mutations or interstitial deletions in a significant subset in up to 90% of HBs [292–295]. In addition to mutations in *CTNNB1*, mutations in *AXIN* and *APC* have also been reported in HBs, with the latter being reported in familial cases [294,296]. These events lead to nuclear and/or cytoplasmic accumulation of β-catenin in HB as identified by immunohistochemistry and coincide with upregulation of several targets of the Wnt pathway, such as cyclin-D1 (Figure 20.7). While immunohistochemistry for β-catenin has been attempted to classify HB for prognosis as well as histological subtypes, a more mechanistic insight will be necessary to dissect out the innocent from the inciting β-catenin redistribution. Since β-catenin activation has been implicated in the pathogenesis of a greater subset of HB, it is being discussed as a potential therapeutic target demonstrating efficacy in preclinical setup [297–300].

Hepatocellular Cancer

Hepatocellular cancer (HCC) is the most common primary tumor of the liver, accounting for 85% of all primary malignant tumors. It is the fifth most common malignancy worldwide and third most common cause of death related to cancers [119,301]. Common risk factors of HCC include hepatitis, chronic alcohol abuse, toxins such as aflatoxins, and nonalcoholic fatty liver disease (NAFLD). This disease, which used to be a common malignancy in underdeveloped or developing countries, is now on the rise in developed countries [120,302]. The increasing incidence of HCC in developed countries is attributed to increasing incidence of hepatitis C and hepatitis B, as well as of NAFLD. HCC afflicts three times more men than women and overall incidence increases with age, especially in Western world, although trends are now changing. In fact peak incidence of HCC in a recent study was between the ages of 45 and 60 years [119].

Much of the HCC occurs in the background of cirrhosis. In fact most of the chronic liver insults eventually lead to cirrhosis. During this process, liver function is ascertained by the presence of regenerating nodules, which are a function of the regenerative capacity of surviving hepatocytes [303]. However, some of these nodules evolve into low-grade and high-grade dysplastic nodules, which then lead to HCC [304]. Similarly, a minor subset of HAs proceeds to evolve into HCC (Figure 20.8). Histologically, HCC may be well differentiated to poorly differentiated, depending on degree of nuclear atypia, anaplasia, and nucleolar prominence, and the tumor cells usually arrange in a trabecular pattern composed of uneven layers of hepatocytes, with occasional mitosis (Figure 20.8). Thus, overall, there is a progression of the preneoplastic events into neoplasia, and clearly, understanding the molecular basis of this evolution will have strong clinical implications.

The possibility of a cancer stem cell origin of some of the HCC is also being entertained. Cancer stem cells are formed by mutations in the normal existing stem cells or a progenitor cell within a tissue. It has also been suggested that perfectly mature cells such as hepatocytes can dedifferentiate into progenitor-like cells and also be targets of mutations [305]. Dysplastic lesions within liver have been shown to have preneoplastic precursors. In addition, a subset of HCCs exhibit mixed hepatocholangiocarcinoma phenotype that might represent bipotential stem cell origin of these tumors [306]. In addition, some early HCCs have been shown to harbor smaller proliferating oval cells or stem cells. In fact, the pathways that are known to regulate stem cell renewal such as the Wnt, Hedgehog, and Notch have been shown to play a role in oval cell activation as well as in a subset of HCCs [305].

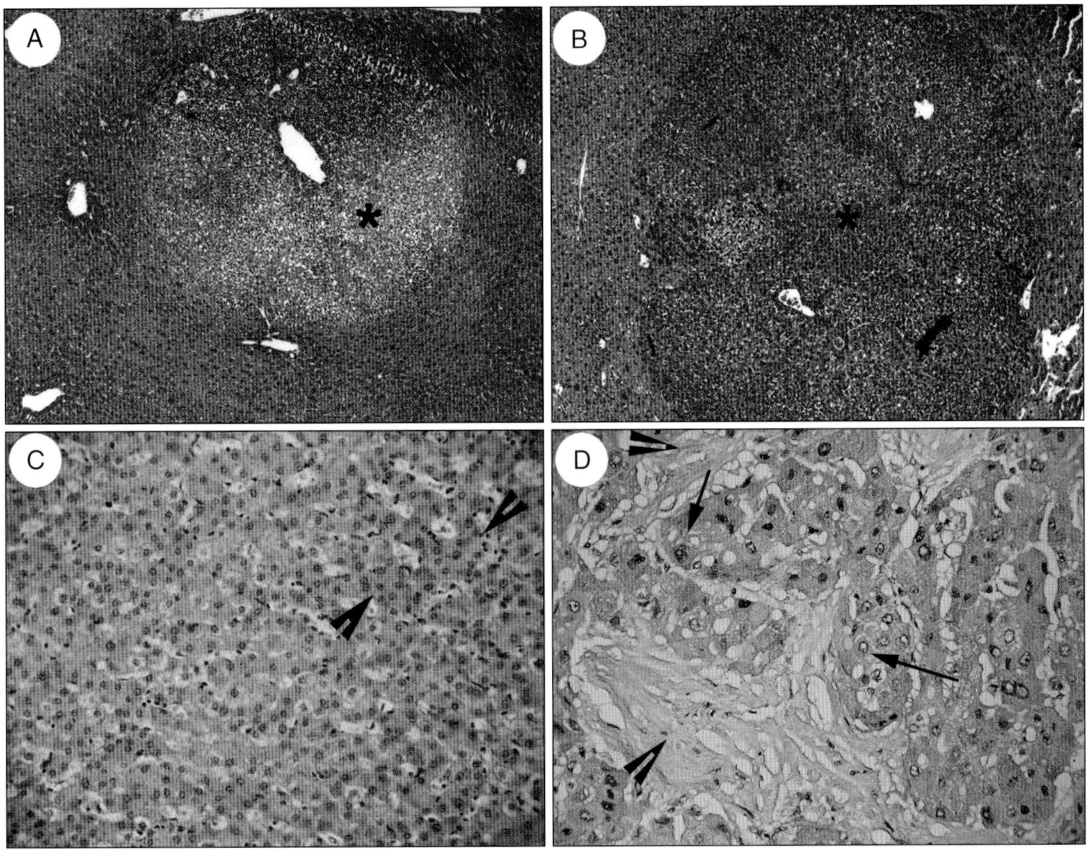

Figure 20.8 **Histology of hepatic adenoma and HCC in mouse model of chemical carcinogenesis and HCC in patients. (A)** A hepatic adenoma (*) is evident on H&E in mice 6 months after diethylnitrosamine (DEN) injection. **(B)** HCC (*) is evident in mouse liver 9 months after DEN injection. **(C)** H&E show abnormal trabecular pattern (arrowheads) in HCC patient. **(D)** H&E showing fibrolamellar variant of HCC in a patient with lamellar fibrosis (arrowhead) pattern surrounding large tumor cells (arrows).

During the process of hepatocarcinogenesis, multiple genetic alterations have been identified to impart a proliferating and cell survival phenotype to the tumor cells. Accumulations of these changes over time are necessary for tumor initiation and progression. While the precise nature of such genetic alterations and chronology of such aberrations, whether stereotypic or heterogeneous from one patient to another, are not known, several alterations ranging from point mutations in individual genes to gain or loss of chromosome arms have been reported [304]. Several candidate genes include *c-myc* (8q), *Cyclin-A2* (4q), *Cyclin-D1* (11q), *Rb1* (13q), *AXIN1* (16p), *p53* (17p), *IGFR-II/M6PR* (6q), *p16* (9p), *E-Cadherin* (16q), *SOCS* (16p) and *PTEN* (10q). The most frequently mutated genes in HCC include *p53, PIK3CA,* and *CTNNB1* (β-catenin gene). All these changes lead to loss of tumor suppressor gene function, gain of oncogene function, or activation of signal transduction pathways that play critical roles in tumor biology, giving the cells stemness, proliferative, and survival advantages, and eventually, invasive and metastatic potential. Various signaling pathways that become aberrantly active in HCC include the Wnt/β-catenin, EGFR, TGFα, VEGFR, IGF, and HGF/Met pathways [120,304]. Various preclinical and clinical studies have shown some advantages of targeting these pathways in HCC, although several studies are preliminary at this stage. Similarly, after any of the preceding factors (excluding Wnt signaling) activate the receptor tyrosine kinases, the signal is transduced via Ras/MAPK, PI3kinase, or Jak/Stat pathways [120,304]. Success of Sorafenib in HCC is attributable to identification of such molecular aberrations in HCC [307,308].

Various agents that are implicated in HCC pathogenesis have shown some preference for the pathways they inflict. Aflatoxins are toxins from the fungi *Aspergillus flavus* and *Aspergillus parasiticus,* which infect foods that are improperly stored in hot and humid conditions and are particularly common in Asia and Africa. Aflatoxins, once ingested, are metabolized by cytochrome P450 to form an active metabolite, which leads to DNA adduct formation that induces hepatocyte transformation [309]. One of the major downstream effects is mutation in a widely studied tumor suppressor gene *p53.* Highest *p53* somatic mutations in fact are noted in HCCs that occur secondary to aflatoxin exposure. Separate studies have shown up to 56% of HCCs due to aflatoxin exposure having *p53* mutations [310]. In Western countries, the *p53* mutations are mostly observed in advanced stages of HCC, suggesting loss of *p53* to be a late event in hepatocarcinogenesis [304]. Additionally, it should be noted that, other than aflatoxins, high rate of *p53* mutations is observed only in hemochromatosis-related HCC [311]. The three leading causes of HCC in the Western world (HCV, HBV, and alcoholic liver disease) mostly employ common molecular and genetic pathways for tumorigenesis. These include the *Rb1, p53,* and *Wnt* pathways. Especially HCC originating in chronic alcoholic liver disease patients display frequent alterations in *Rb1* and *p53,* including loss of *Rb1* due to promoter methylation, *p16* methylation, and *cyclin-D1* amplification [312]. p16 inhibits cyclin-dependent kinases (cdk4/6), which repress G1 cell cycle progression. This occurs secondary to dephosphorylation of retinoblastoma protein, which bind sand inactivates E2F1. Although the mechanism remains unclear, HCV patients demonstrate a high rate of β-catenin gene mutations as well as β-catenin stabilization [313,314]. The mutations typically occur in the exon-3 of *CTNNB1,* which contains serine and threonine sites necessary for GSK3β-mediated and Casein kinase-mediated phosphorylation and eventual proteosomal degradation of β-catenin by ubiquitination. Several known target genes of this pathway such as glutamine synthetase, Cyclin-D1, fibronectin, c-myc, p-450s, and others are upregulated in HCC and might contribute to tumor cell phenotype in HCC.

Fibrolamellar HCC One of the uncommon variants of HCC that deserves a mention is the fibrolamellar HCC (FL-HCC). This usually occurs in younger patients (5–35 years) and in a noncirrhotic hepatic background [315]. The histology is characterized by lamellar pattern of fibrosis surrounding larger tumor cells that contain abundant granular cytoplasm and prominent nucleoli (Figure 20.8). While initially thought to have better prognosis, FL-HCC appears to be equally aggressive with a 45% 5-year survival [316,317]. In fact the improved prognosis appears to be due to absence of cirrhosis in FL-HCC as compared to existing cirrhosis in the conventional HCC cases. The molecular basis of this variant of HCC remains largely obscure. Due to rarity of these tumors, lesser studies are available that have characterized molecular and genetic aberrations in FL-HCC. One study has reported four frequent alterations in FL-HCC than in other tumor types, including gain of chromosome 4q and loss of chromosomes 9p, 16p, and Xq [318]. Another CGH study identified gains of chromosome arm 1q in six out of seven FL-HCCs [319]. This region, which is also preferentially amplified in 58% of conventional HCCs, may contain an essential proto-oncogene commonly implicated in liver carcinogenesis. Recently, increased EGFR protein expression has also been identified in significant numbers of FL-HCC, and this increase appeared to be secondary to gains of chromosome 7 [320]. Thus, EGFR antagonists and other receptor tyrosine kinase inhibitors might have a useful role in treatment of FL-HCC.

REFERENCES

1. Duncan SA. Mechanisms controlling early development of the liver. *Mech Dev.* 2003;120:19–33.
2. Zaret KS. Regulatory phases of early liver development: Paradigms of organogenesis. *Nat Rev Genet.* 2002;3:499–512.
3. Lee CS, Friedman JR, Fulmer JT, et al.The initiation of liver development is dependent on Foxa transcription factors. *Nature.* 2005;435:944–947.
4. Jung J, Zheng M, Goldfarb M, et al. Initiation of mammalian liver development from endoderm by fibroblast growth factors. *Science.* 1999;284:1998–2003.

5. Rossi JM, Dunn NR, Hogan BL, et al. Distinct mesodermal signals, including BMPs from the septum transversum mesenchyme, are required in combination for hepatogenesis from the endoderm. *Genes Dev.* 2001;15:1998–2009.
6. Gualdi R, Bossard P, Zheng M, et al. Hepatic specification of the gut endoderm in vitro: Cell signaling and transcriptional control. *Genes Dev.* 1996;10:1670–1682.
7. Keng VW, Yagi H, Ikawa M, et al. Homeobox gene Hex is essential for onset of mouse embryonic liver development and differentiation of the monocyte lineage. *Biochem Biophys Res Commun.* 2000;276:1155–1161.
8. Martinez Barbera JP, Clements M, Thomas P, et al. The homeobox gene Hex is required in definitive endodermal tissues for normal forebrain, liver and thyroid formation. *Development.* 2000;127:2433–2445.
9. Sosa-Pineda B, Wigle JT, Oliver G. Hepatocyte migration during liver development requires Prox1. *Nat Genet.* 2000;25:254–255.
10. Zhao R, Watt AJ, Li J, et al. GATA6 is essential for embryonic development of the liver but dispensable for early heart formation. *Mol Cell Biol.* 2005;25: 2622–2631.
11. Duncan SA. Transcriptional regulation of liver development. *Dev Dyn.* 2000;219:131–142.
12. Berg T, Rountree CB, Lee L, et al. Fibroblast growth factor 10 is critical for liver growth during embryogenesis and controls hepatoblast survival via beta-catenin activation. *Hepatology.* 2007;46: 1187–1197.
13. Hentsch B, Lyons I, Li R, et al. Hlx homeo box gene is essential for an inductive tissue interaction that drives expansion of embryonic liver and gut. *Genes Dev.* 1996;10:70–79.
14. Krupczak-Hollis K, Wang X, Kalinichenko VV, et al. The mouse Forkhead Box m1 transcription factor is essential for hepatoblast mitosis and development of intrahepatic bile ducts and vessels during liver morphogenesis. *Dev Biol.* 2004;276:74–88.
15. Micsenyi A, Tan X, Sneddon T, et al. Beta-catenin is temporally regulated during normal liver development. *Gastroenterology.* 2004;126:1134–1146.
16. Monga SP, Monga HK, Tan X, et al. Beta-catenin antisense studies in embryonic liver cultures: Role in proliferation, apoptosis, and lineage specification. *Gastroenterology.* 2003;124:202–216.
17. Schmidt C, Bladt F, Goedecke S, et al. Scatter factor/hepatocyte growth factor is essential for liver development. *Nature.* 1995;373: 699–702.
18. Sekhon SS, Tan X, Micsenyi A, et al. Fibroblast growth factor enriches the embryonic liver cultures for hepatic progenitors. *Am J Pathol.* 2004;164:2229–2240.
19. Stenvers KL, Tursky ML, Harder KW, et al. Heart and liver defects and reduced transforming growth factor beta2 sensitivity in transforming growth factor beta type III receptor-deficient embryos. *Mol Cell Biol.* 2003;23:4371–4385.
20. Suksaweang S, Lin CM, Jiang TX, et al. Morphogenesis of chicken liver: Identification of localized growth zones and the role of beta-catenin/Wnt in size regulation. *Dev Biol.* 2004;266:109–122.
21. Tang Y, Katuri V, Dillner A, et al. Disruption of transforming growth factor-beta signaling in ELF beta-spectrin-deficient mice. *Science.* 2003;299:574–577.
22. Weinstein M, Monga SP, Liu Y, et al. Smad proteins and hepatocyte growth factor control parallel regulatory pathways that converge on beta1-integrin to promote normal liver development. *Mol Cell Biol.* 2001;21:5122–5131.
23. Tan X, Yuan Y, Zeng G, et al. beta-Catenin deletion in hepatoblasts disrupts hepatic morphogenesis and survival during mouse development. *Hepatology.* 2008;47:1667–1679.
24. Beg AA, Sha WC, Bronson RT, et al. Embryonic lethality and liver degeneration in mice lacking the RelA component of NF-kappa B. *Nature.* 1995;376:167–170.
25. Hilberg F, Aguzzi A, Howells N, et al. c-jun is essential for normal mouse development and hepatogenesis. *Nature.* 1993; 365: 179–181.
26. Johnson L, Greenbaum D, Cichowski K, et al. K-ras is an essential gene in the mouse with partial functional overlap with N-ras. *Genes Dev.* 1997;11:2468–2481.
27. Li Q, Van Antwerp D, Mercurio F, et al. Severe liver degeneration in mice lacking the IkappaB kinase 2 gene. *Science.* 1999;284:321–325.
28. Reimold AM, Etkin A, Clauss I, et al. An essential role in liver development for transcription factor XBP-1. *Genes Dev.* 2000;14: 152–157.
29. Spear BT, Jin L, Ramasamy S, et al. Transcriptional control in the mammalian liver: Liver development, perinatal repression, and zonal gene regulation. *Cell Mol Life Sci.* 2006;63:2922–2938.
30. Darlington GJ, Wang N, Hanson RW. C/EBP alpha: A critical regulator of genes governing integrative metabolic processes. *Curr Opin Genet Dev.* 1995;5:565–570.
31. Odom DT, Zizlsperger N, Gordon DB, et al. Control of pancreas and liver gene expression by HNF transcription factors. *Science.* 2004;303:1378–1381.
32. Parviz F, Matullo C, Garrison WD, et al. Hepatocyte nuclear factor 4 alpha controls the development of a hepatic epithelium and liver morphogenesis. *Nat Genet.* 2003;34:292–296.
33. Monga SP, Hout MS, Baun MJ, et al. Mouse fetal liver cells in artificial capillary beds in three-dimensional four-compartment bioreactors. *Am J Pathol.* 2005;167:1279–1292.
34. Clotman F, Jacquemin P, Plumb-Rudewiez N, et al. Control of liver cell fate decision by a gradient of TGF beta signaling modulated by Onecut transcription factors. *Genes Dev.* 2005;19:1849–1854.
35. Clotman F, Lannoy VJ, Reber M, et al. The one cut transcription factor HNF6 is required for normal development of the biliary tract. *Development.* 2002;129:1819–1828.
36. Coffinier C, Gresh L, Fiette L, et al. Bile system morphogenesis defects and liver dysfunction upon targeted deletion of HNF1beta. *Development.* 2002;129:1829–1838.
37. Lorent K, Yeo SY, Oda T, et al. Inhibition of Jagged-mediated Notch signaling disrupts zebrafish biliary development and generates multi-organ defects compatible with an Alagille syndrome phenocopy. *Development.* 2004;131:5753–5766.
38. McLin VA, Rankin SA, Zorn AM. Repression of Wnt/beta-catenin signaling in the anterior endoderm is essential for liver and pancreas development. *Development.* 2007;134:2207–2217.
39. Ober EA, Verkade H, Field HA, et al. Mesodermal Wnt2b signalling positively regulates liver specification. *Nature.* 2006;442: 688–691.
40. Decaens T, Godard C, de Reynies A, et al. Stabilization of beta-catenin affects mouse embryonic liver growth and hepatoblast fate. *Hepatology.* 2008;47:247–258.
41. Michalopoulos GK. Liver regeneration. *J Cell Physiol.* 2007;213: 286–300.
42. Higgins GM and Anderson RM. Experimental pathology of the liver. I. Restoration of the white rat following partial surgical removal. *Arch Pathol.* 1931;12:186–202.
43. Fausto N. Liver regeneration and repair: Hepatocytes, progenitor cells, and stem cells. *Hepatology.* 2004;39:1477–1487.
44. Fausto N, Campbell JS, Riehle KJ. Liver regeneration. *Hepatology.* 2006;43:S45–53.
45. Michalopoulos GK, DeFrances MC. Liver regeneration. *Science.* 1997;276:60–66.
46. Taub R. Liver regeneration: From myth to mechanism. *Nat Rev Mol Cell Biol.* 2004;5:836–847.
47. Wagenaar GT, Chamuleau RA, Pool CW, et al. Distribution and activity of glutamine synthase and carbamoylphosphate synthase upon enlargement of the liver lobule by repeated partial hepatectomies. *J Hepatol.* 1993;17:397–407.
48. Marubashi S, Sakon M, Nagano H, et al. Effect of portal hemodynamics on liver regeneration studied in a novel portohepatic shunt rat model. *Surgery.* 2004;136:1028–1037.
49. Mars WM, Liu ML, Kitson RP, et al. Immediate early detection of urokinase receptor after partial hepatectomy and its implications for initiation of liver regeneration. *Hepatology.* 1995;21:1695–1701.
50. Pediaditakis P, Lopez-Talavera JC, Petersen B, et al. The processing and utilization of hepatocyte growth factor/scatter factor following partial hepatectomy in the rat. *Hepatology.* 2001;34:688–693.
51. Stolz DB, Mars WM, Petersen BE, et al. Growth factor signal transduction immediately after two-thirds partial hepatectomy in the rat. *Cancer Res.* 1999;59:3954–3960.
52. Skov Olsen P, Boesby S, Kirkegaard P, et al. Influence of epidermal growth factor on liver regeneration after partial hepatectomy in rats. *Hepatology.* 1988;8:992–996.
53. Monga SP, Pediaditakis P, Mule K, et al. Changes in WNT/beta-catenin pathway during regulated growth in rat liver regeneration. *Hepatology.* 2001;33:1098–1109.

54. Sekine S, Gutierrez PJ, Lan BY, et al. Liver-specific loss of beta-catenin results in delayed hepatocyte proliferation after partial hepatectomy. *Hepatology.* 2007;45:361–368.
55. Sodhi D, Micsenyi A, Bowen WC, et al. Morpholino oligonucleotide-triggered beta-catenin knockdown compromises normal liver regeneration. *J Hepatol.* 2005;43:132–141.
56. Tan X, Behari J, Cieply B, et al. Conditional deletion of beta-catenin reveals its role in liver growth and regeneration. *Gastroenterology.* 2006;131:1561–1572.
57. Croquelois A, Blindenbacher A, Terracciano L, et al. Inducible inactivation of Notch1 causes nodular regenerative hyperplasia in mice. *Hepatology.* 2005;41:487–496.
58. Kohler C, Bell AW, Bowen WC, et al. Expression of Notch-1 and its ligand Jagged-1 in rat liver during liver regeneration. *Hepatology.* 2004;39:1056–1065.
59. Cruise JL, Knechtle SJ, Bollinger RR, et al. Alpha 1-adrenergic effects and liver regeneration. *Hepatology.* 1987;7:1189–1194.
60. Lesurtel M, Graf R, Aleil B, et al. Platelet-derived serotonin mediates liver regeneration. *Science.* 2006;312:104–107.
61. Mead JE, Fausto N. Transforming growth factor alpha may be a physiological regulator of liver regeneration by means of an autocrine mechanism. *Proc Natl Acad Sci USA.* 1989;86: 1558–1562.
62. Russell WE, Dempsey PJ, Sitaric S, et al. Transforming growth factor-alpha (TGF alpha) concentrations increase in regenerating rat liver: Evidence for a delayed accumulation of mature TGF alpha. *Endocrinology.* 1993;133: 1731–1738.
63. Akerman P, Cote P, Yang SQ, et al. Antibodies to tumor necrosis factor-alpha inhibit liver regeneration after partial hepatectomy. *Am J Physiol.* 1992;263:G579–585.
64. Yamada Y, Kirillova I, Peschon JJ, et al. Initiation of liver growth by tumor necrosis factor: Deficient liver regeneration in mice lacking type I tumor necrosis factor receptor. *Proc Natl Acad Sci USA.* 1997;94:1441–1446.
65. Yamada Y, Webber EM, Kirillova I, et al. Analysis of liver regeneration in mice lacking type 1 or type 2 tumor necrosis factor receptor: Requirement for type 1 but not type 2 receptor. *Hepatology.* 1998;28:959–970.
66. Cressman DE, Greenbaum LE, DeAngelis RA, et al. Liver failure and defective hepatocyte regeneration in interleukin-6-deficient mice. *Science.* 1996;274:1379–1383.
67. Sakamoto T, Liu Z, Murase N, et al. Mitosis and apoptosis in the liver of interleukin-6-deficient mice after partial hepatectomy. *Hepatology.* 1999;29:403–411.
68. Wang X, Kiyokawa H, Dennewitz MB, et al. The Forkhead Box m1b transcription factor is essential for hepatocyte DNA replication and mitosis during mouse liver regeneration. *Proc Natl Acad Sci USA.* 2002;99:16881–16886.
69. Huang W, Ma K, Zhang J, et al. Nuclear receptor-dependent bile acid signaling is required for normal liver regeneration. *Science.* 2006;312:233–236.
70. Kim TH, Mars WM, Stolz DB, et al. Expression and activation of pro-MMP-2 and pro-MMP-9 during rat liver regeneration. *Hepatology.* 2000;31:75–82.
71. Mohammed FF, Khokha R. Thinking outside the cell: Proteases regulate hepatocyte division. *Trends Cell Biol.* 2005;15:555–563.
72. Mohammed FF, Pennington CJ, Kassiri Z, et al. Metalloproteinase inhibitor TIMP-1 affects hepatocyte cell cycle via HGF activation in murine liver regeneration. *Hepatology.* 2005;41: 857–867.
73. Olle EW, Ren X, McClintock SD, et al. Matrix metalloproteinase-9 is an important factor in hepatic regeneration after partial hepatectomy in mice. *Hepatology.* 2006;44:540–549.
74. Asai K, Tamakawa S, Yamamoto M, et al. Activated hepatic stellate cells overexpress p75NTR after partial hepatectomy and undergo apoptosis on nerve growth factor stimulation. *Liver Int.* 2006;26:595–603.
75. Passino MA, Adams RA, Sikorski SL, et al. Regulation of hepatic stellate cell differentiation by the neurotrophin receptor p75NTR. *Science.* 2007;315:1853–1856.
76. Jakowlew SB, Mead JE, Danielpour D, et al. Transforming growth factor-beta (TGF-beta) isoforms in rat liver regeneration: Messenger RNA expression and activation of latent TGF-beta. *Cell Regul.* 1991;2:535–548.
77. Chari RS, Price DT, Sue SR, et al. Down-regulation of transforming growth factor beta receptor type I, II, and III during liver regeneration. *Am J Surg.* 1995;169:126–131; discussion 131–122.
78. Houck KA, Michalopoulos GK. Altered responses of regenerating hepatocytes to norepinephrine and transforming growth factor type beta. *J Cell Physiol.* 1989;141:503–509.
79. Jirtle RL, Carr BI, Scott CD. Modulation of insulin-like growth factor-II/mannose 6-phosphate receptors and transforming growth factor-beta 1 during liver regeneration. *J Biol Chem.* 1991;266:22444–22450.
80. Stoick-Cooper CL, Moon RT, Weidinger G. Advances in signaling in vertebrate regeneration as a prelude to regenerative medicine. *Genes Dev.* 2007;21:1292–1315.
81. Alison M, Golding M, Lalani EN, et al. Wholesale hepatocytic differentiation in the rat from ductular oval cells, the progeny of biliary stem cells. *J Hepatol.* 1997;26: 343–352.
82. Evarts RP, Hu Z, Omori N, et al. Precursor-product relationship between oval cells and hepatocytes: Comparison between tritiated thymidine and bromodeoxyuridine as tracers. *Carcinogenesis.* 1996;17:2143–2151.
83. Golding M, Sarraf CE, Lalani EN, et al. Oval cell differentiation into hepatocytes in the acetylaminofluorene-treated regenerating rat liver. *Hepatology.* 1995;22:1243–1253.
84. Preisegger KH, Factor VM, Fuchsbichler A, et al. Atypical ductular proliferation and its inhibition by transforming growth factor beta1 in the 3,5-diethoxycarbonyl-1,4-dihydrocollidine mouse model for chronic alcoholic liver disease. *Lab Invest.* 1999;79:103–109.
85. Michalopoulos GK, Barua L, Bowen WC. Transdifferentiation of rat hepatocytes into biliary cells after bile duct ligation and toxic biliary injury. *Hepatology.* 2005;41:535–544.
86. Michalopoulos GK, Bowen WC, Mule K, et al. Hepatocytes undergo phenotypic transformation to biliary epithelium in organoid cultures. *Hepatology.* 2002;36:278–283.
87. Walkup MH, Gerber DA. Hepatic stem cells: In search of. *Stem Cells.* 2006;24:1833–1840.
88. Forbes S, Vig P, Poulsom R, et al. Hepatic stem cells. *J Pathol.* 2002;197:510–518.
89. Roskams TA, Libbrecht L, Desmet VJ. Progenitor cells in diseased human liver. *Semin Liver Dis.* 2003;23:385–396.
90. Sell S, Leffert HL. An evaluation of cellular lineages in the pathogenesis of experimental hepatocellular carcinoma. *Hepatology.* 1982;2:77–86.
91. Tatematsu M, Ho RH, Kaku T, et al. Studies on the proliferation and fate of oval cells in the liver of rats treated with 2-acetylaminofluorene and partial hepatectomy. *Am J Pathol.* 1984;114:418–430.
92. Akhurst B, Croager EJ, Farley-Roche CA, et al. A modified choline-deficient, ethionine-supplemented diet protocol effectively induces oval cells in mouse liver. *Hepatology.* 2001;34:519–522.
93. Fickert P, Stoger U, Fuchsbichler A, et al. A new xenobiotic-induced mouse model of sclerosing cholangitis and biliary fibrosis. *Am J Pathol.* 2007;171:525–536.
94. Yaswen P, Thompson NL, Fausto N. Oncodevelopmental expression of rat placental alkaline phosphatase. Detection in oval cells during liver carcinogenesis. *Am J Pathol.* 1985;121:505–513.
95. Jelnes P, Santoni-Rugiu E, Rasmussen M, et al. Remarkable heterogeneity displayed by oval cells in rat and mouse models of stem cell-mediated liver regeneration. *Hepatology.* 2007;45:1462–1470.
96. Roskams T, De Vos R, Desmet V. 'Undifferentiated progenitor cells' in focal nodular hyperplasia of the liver. *Histopathology.* 1996;28:291–299.
97. Roskams T, De Vos R, Van Eyken P, et al. Hepatic OV-6 expression in human liver disease and rat experiments: Evidence for hepatic progenitor cells in man. *J Hepatol.* 1998;29:455–463.
98. Yovchev MI, Grozdanov PN, Joseph B, et al. Novel hepatic progenitor cell surface markers in the adult rat liver. *Hepatology.* 2007;45:139–149.
99. Apte U, Thompson MD, Cui S, et al. Wnt/beta-catenin signaling mediates oval cell response in rodents. *Hepatology.* 2008;47:288–295.
100. Fu XX, Su CY, Lee Y, et al. Insulinlike growth factor II expression and oval cell proliferation associated with hepatocarcinogenesis in woodchuck hepatitis virus carriers. *J Virol.* 1988;62: 3422–3430.
101. Hu M, Kurobe M, Jeong YJ, et al. Wnt/beta-catenin signaling in murine hepatic transit amplifying progenitor cells. *Gastroenterology.* 2007;133:1579–1591.
102. Hu Z, Evarts RP, Fujio K, et al. Expression of hepatocyte growth factor and c-met genes during hepatic differentiation and liver development in the rat. *Am J Pathol.* 1993;142:1823–1830.

103. Hu Z, Evarts RP, Fujio K, et al. Expression of transforming growth factor alpha/epidermal growth factor receptor, hepatocyte growth factor/c-met and acidic fibroblast growth factor/fibroblast growth factor receptors during hepatocarcinogenesis. *Carcinogenesis.* 1996;17:931–938.
104. Isfort RJ, Cody DB, Richards WG, et al. Characterization of growth factor responsiveness and alterations in growth factor homeostasis involved in the tumorigenic conversion of mouse oval cells. *Growth Factors.* 1998;15:81–94.
105. Knight B, Yeap BB, Yeoh GC, et al. Inhibition of adult liver progenitor (oval) cell growth and viability by an agonist of the peroxisome proliferator activated receptor (PPAR) family member gamma, but not alpha or delta. *Carcinogenesis.* 2005;26:1782–1792.
106. Matthews VB, Klinken E, Yeoh GC. Direct effects of interleukin-6 on liver progenitor oval cells in culture. *Wound Repair Regen.* 2004;12:650–656.
107. Nagy P, Bisgaard HC, Santoni-Rugiu E, et al. In vivo infusion of growth factors enhances the mitogenic response of rat hepatic ductal (oval) cells after administration of 2-acetylaminofluorene. *Hepatology.* 1996;23:71–79.
108. Park DY, Suh KS. Transforming growth factor-beta 1 protein, proliferation and apoptosis of oval cells in acetylaminofluorene-induced rat liver regeneration. *J Korean Med Sci.* 1999;14: 531–538.
109. Yeoh GC, Ernst M, Rose-John S, et al. Opposing roles of gp130-mediated STAT-3 and ERK-1/2 signaling in liver progenitor cell migration and proliferation. *Hepatology.* 2007;45:486–494.
110. Zhang N, Siegel K, Odenthal M, et al. The role of insulin-like growth factor II in the malignant transformation of rat liver oval cells. *Hepatology.* 1997;25:900–905.
111. Fujita M, Furukawa H, Hattori M, et al. Sequential observation of liver cell regeneration after massive hepatic necrosis in auxiliary partial orthotopic liver transplantation. *Mod Pathol.* 2000;13:152–157.
112. Dorrell C, Grompe M. Liver repair by intra- and extrahepatic progenitors. *Stem Cell Rev.* 2005;1:61–64.
113. Grompe M. Bone marrow-derived hepatocytes. *Novartis Found Symp.* 2005;265:20–27.
114. Krause DS. Engraftment of bone marrow-derived epithelial cells. *Ann NY Acad Sci.* 2005;1044:117–124.
115. Petersen BE, Bowen WC, Patrene KD, et al. Bone marrow as a potential source of hepatic oval cells. *Science.* 1999;284: 1168–1170.
116. Roskams T, Yang SQ, Koteish A, et al. Oxidative stress and oval cell accumulation in mice and humans with alcoholic and nonalcoholic fatty liver disease. *Am J Pathol.* 2003;163: 1301–1311.
117. Sell S. Comparison of liver progenitor cells in human atypical ductular reactions with those seen in experimental models of liver injury. *Hepatology.* 1998;27:317–331.
118. Libbrecht L, Desmet V, Van Damme B, et al. Deep intralobular extension of human hepatic 'progenitor cells' correlates with parenchymal inflammation in chronic viral hepatitis: Can 'progenitor cells' migrate? *J Pathol.* 2000;192:373–378.
119. El-Serag HB. Hepatocellular carcinoma: Recent trends in the United States. *Gastroenterology.* 2004;127:S27–34.
120. McKillop IH, Moran DM, Jin X, et al. Molecular pathogenesis of hepatocellular carcinoma. *J Surg Res.* 2006;136:125–135.
121. Sell S. Stem cell origin of cancer and differentiation therapy. *Crit Rev Oncol Hematol.* 2004;51:1–28.
122. Zhang M, Rosen JM. Stem cells in the etiology and treatment of cancer. *Curr Opin Genet Dev.* 2006;16:60–64.
123. Lee JS, Heo J, Libbrecht L, et al. A novel prognostic subtype of human hepatocellular carcinoma derived from hepatic progenitor cells. *Nat Med.* 2006;12:410–416.
124. Yin XM, Ding WX. Death receptor activation-induced hepatocyte apoptosis and liver injury. *Curr Mol Med.* 2003;3:491–508.
125. Ogasawara J, Watanabe-Fukunaga R, Adachi M, et al. Lethal effect of the anti-Fas antibody in mice. *Nature.* 1993;364:806–809.
126. Ashkenazi A, Dixit VM. Death receptors: Signaling and modulation. *Science.* 1998;281:1305–1308.
127. de la Coste A, Fabre M, McDonell N, et al. Differential protective effects of Bcl-xL and Bcl-2 on apoptotic liver injury in transgenic mice. *Am J Physiol.* 1999;277:G702–708.
128. Lacronique V, Mignon A, Fabre M, et al. Bcl-2 protects from lethal hepatic apoptosis induced by an anti-Fas antibody in mice. *Nat Med.* 1996;2:80–86.
129. Rodriguez I, Matsuura K, Khatib K, et al. A bcl-2 transgene expressed in hepatocytes protects mice from fulminant liver destruction but not from rapid death induced by anti-Fas antibody injection. *J Exp Med.* 1996;183:1031–1036.
130. Eskes R, Antonsson B, Osen-Sand A, et al. Bax-induced cytochrome C release from mitochondria is independent of the permeability transition pore but highly dependent on Mg2+ ions. *J Cell Biol.* 1998;143:217–224.
131. Li H, Zhu H, Xu CJ, et al. Cleavage of BID by caspase 8 mediates the mitochondrial damage in the Fas pathway of apoptosis. *Cell.* 1998;94:491–501.
132. Wei MC, Lindsten T, Mootha VK, et al. tBID, a membrane-targeted death ligand, oligomerizes BAK to release cytochrome c. *Genes Dev.* 2000;14:2060–2071.
133. Deveraux QL, Reed JC. IAP family proteins—Suppressors of apoptosis. *Genes Dev.* 1999;13:239–252.
134. Stennicke HR, Jurgensmeier JM, Shin H, et al. Pro-caspase-3 is a major physiologic target of caspase-8. *J Biol Chem.* 1998;273: 27084–27090.
135. Wang X, DeFrances MC, Dai Y, et al. A mechanism of cell survival: Sequestration of Fas by the HGF receptor Met. *Mol Cell.* 2002;9:411–421.
136. Zou C, Ma J, Wang X, et al. Lack of Fas antagonism by Met in human fatty liver disease. *Nat Med.* 2007;13:1078–1085.
137. Kanda D, Takagi H, Toyoda M, et al. Transforming growth factor alpha protects against Fas-mediated liver apoptosis in mice. *FEBS Lett.* 2002;519:11–15.
138. Kosai K, Matsumoto K, Nagata S, et al. Abrogation of Fas-induced fulminant hepatic failure in mice by hepatocyte growth factor. *Biochem Biophys Res Commun.* 1998;244:683–690.
139. Freudenberg MA, Keppler D, Galanos C. Requirement for lipopolysaccharide-responsive macrophages in galactosamine-induced sensitization to endotoxin. *Infect Immun.* 1986;51: 891–895.
140. Leist M, Gantner F, Bohlinger I, et al. Murine hepatocyte apoptosis induced in vitro and in vivo by TNF-alpha requires transcriptional arrest. *J Immunol.* 1994;153: 1778–1788.
141. Harper N, Hughes M, MacFarlane M, et al. Fas-associated death domain protein and caspase-8 are not recruited to the tumor necrosis factor receptor 1 signaling complex during tumor necrosis factor-induced apoptosis. *J Biol Chem.* 2003;278:25534–25541.
142. Hsu H, Xiong J, Goeddel DV. The TNF receptor 1-associated protein TRADD signals cell death and NF-kappa B activation. *Cell.* 1995;81:495–504.
143. Guicciardi ME, Deussing J, Miyoshi H, et al. Cathepsin B contributes to TNF-alpha-mediated hepatocyte apoptosis by promoting mitochondrial release of cytochrome c. *J Clin Invest.* 2000;106:1127–1137.
144. Wajant H, Pfizenmaier K, Scheurich P. Tumor necrosis factor signaling. *Cell Death Differ.* 2003;10:45–65.
145. Biswas DK, Martin KJ, McAlister C, et al. Apoptosis caused by chemotherapeutic inhibition of nuclear factor-kappaB activation. *Cancer Res.* 2003;63:290–295.
146. LaCasse EC, Baird S, Korneluk RG, et al. The inhibitors of apoptosis (IAPs) and their emerging role in cancer. *Oncogene.* 1998;17:3247–3259.
147. McDonald MC, Mota-Filipe H, Paul A, et al. Calpain inhibitor I reduces the activation of nuclear factor-kappaB and organ injury/dysfunction in hemorrhagic shock. *Faseb J.* 2001;15:171–186.
148. Lee HH, Dadgostar H, Cheng Q, et al. NF-kappaB-mediated up-regulation of Bcl-x and Bfl-1/A1 is required for CD40 survival signaling in B lymphocytes. *Proc Natl Acad Sci USA.* 1999;96:9136–9141.
149. Hoeflich KP, Luo J, Rubie EA, et al. Requirement for glycogen synthase kinase-3 beta in cell survival and NF-kappaB activation. *Nature.* 2000;406:86–90.
150. Bonnard M, Mirtsos C, Suzuki S, et al. Deficiency of T2K leads to apoptotic liver degeneration and impaired NF-kappaB-dependent gene transcription. *Embo J.* 2000;19:4976–4985.
151. Doi TS, Marino MW, Takahashi T, et al. Absence of tumor necrosis factor rescues RelA-deficient mice from embryonic lethality. *Proc Natl Acad Sci USA.* 1999;96:2994–2999.
152. Neuschwander-Tetri BA, Caldwell SH. Nonalcoholic steatohepatitis: Summary of an AASLD Single Topic Conference. *Hepatology.* 2003;37:1202–1219.

153. Angulo P. Nonalcoholic fatty liver disease. *N Engl J Med.* 2002; 346:1221–1231.
154. Marchesini G, Bugianesi E, Forlani G, et al. Nonalcoholic fatty liver, steatohepatitis, and the metabolic syndrome. *Hepatology.* 2003;37:917–923.
155. Clark JM, Brancati FL, Diehl AM. The prevalence and etiology of elevated aminotransferase levels in the United States. *Am J Gastroenterol.* 2003;98:960–967.
156. Day CP, James OF. Steatohepatitis: A tale of two "hits"? *Gastroenterology.* 1998;114:842–845.
157. Shepherd PR, Kahn BB. Glucose transporters and insulin action—Implications for insulin resistance and diabetes mellitus. *N Engl J Med.* 1999;341:248–257.
158. Bugianesi E, Gastaldelli A, Vanni E, et al. Insulin resistance in non-diabetic patients with non-alcoholic fatty liver disease: Sites and mechanisms. *Diabetologia.* 2005;48:634–642.
159. Seppala-Lindroos A, Vehkavaara S, Hakkinen AM, et al. Fat accumulation in the liver is associated with defects in insulin suppression of glucose production and serum free fatty acids independent of obesity in normal men. *J Clin Endocrinol Metab.* 2002;87:3023–3028.
160. Araya J, Rodrigo R, Videla LA, et al. Increase in long-chain polyunsaturated fatty acid n - 6/n - 3 ratio in relation to hepatic steatosis in patients with non-alcoholic fatty liver disease. *Clin Sci (Lond).* 2004;106:635–643.
161. Shimomura I, Shimano H, Korn BS, et al. Nuclear sterol regulatory element-binding proteins activate genes responsible for the entire program of unsaturated fatty acid biosynthesis in transgenic mouse liver. *J Biol Chem.* 1998;273:35299–35306.
162. Kawaguchi T, Osatomi K, Yamashita H, et al. Mechanism for fatty acid "sparing" effect on glucose-induced transcription: Regulation of carbohydrate-responsive element-binding protein by AMP-activated protein kinase. *J Biol Chem.* 2002;277:3829–3835.
163. Yamashita H, Takenoshita M, Sakurai M, et al. A glucose-responsive transcription factor that regulates carbohydrate metabolism in the liver. *Proc Natl Acad Sci USA.* 2001;98:9116–9121.
164. Iizuka K, Bruick RK, Liang G, et al. Deficiency of carbohydrate response element-binding protein (ChREBP) reduces lipogenesis as well as glycolysis. *Proc Natl Acad Sci USA.* 2004;101:7281–7286.
165. Brown MS, Goldstein JL. The SREBP pathway: Regulation of cholesterol metabolism by proteolysis of a membrane-bound transcription factor. *Cell.* 1997;89:331–340.
166. Horton JD, Goldstein JL, Brown MS. SREBPs: Activators of the complete program of cholesterol and fatty acid synthesis in the liver. *J Clin Invest.* 2002;109:1125–1131.
167. Horton JD, Shah NA, Warrington JA, et al. Combined analysis of oligonucleotide microarray data from transgenic and knockout mice identifies direct SREBP target genes. *Proc Natl Acad Sci USA.* 2003;100:12027–12032.
168. Shimano H, Horton JD, Shimomura I, et al. Isoform 1c of sterol regulatory element binding protein is less active than isoform 1a in livers of transgenic mice and in cultured cells. *J Clin Invest.* 1997;99:846–854.
169. Yahagi N, Shimano H, Hasty AH, et al. Absence of sterol regulatory element-binding protein-1 (SREBP-1) ameliorates fatty livers but not obesity or insulin resistance in Lep(ob)/Lep(ob) mice. *J Biol Chem.* 2002;277:19353–19357.
170. Shimomura I, Bashmakov Y, Horton JD. Increased levels of nuclear SREBP-1c associated with fatty livers in two mouse models of diabetes mellitus. *J Biol Chem.* 1999;274:30028–30032.
171. Gastaldelli A, Miyazaki Y, Pettiti M, et al. The effect of rosiglitazone on the liver: Decreased gluconeogenesis in patients with type 2 diabetes. *J Clin Endocrinol Metab.* 2006;91:806–812.
172. Utzschneider KM, Kahn SE. Review: The role of insulin resistance in nonalcoholic fatty liver disease. *J Clin Endocrinol Metab.* 2006;91:4753–4761.
173. Yamaguchi K, Yang L, McCall S, et al. Inhibiting triglyceride synthesis improves hepatic steatosis but exacerbates liver damage and fibrosis in obese mice with nonalcoholic steatohepatitis. *Hepatology.* 2007;45:1366–1374.
174. Chao L, Marcus-Samuels B, Mason MM, et al. Adipose tissue is required for the antidiabetic, but not for the hypolipidemic, effect of thiazolidinediones. *J Clin Invest.* 2000;106:1221–1228.
175. Edvardsson U, Bergstrom M, Alexandersson M, et al. Rosiglitazone (BRL49653), a PPARgamma-selective agonist, causes peroxisome proliferator-like liver effects in obese mice. *J Lipid Res.* 1999;40:1177–1184.
176. Robertson G, Leclercq I, Farrell GC. Nonalcoholic steatosis and steatohepatitis. II. Cytochrome P-450 enzymes and oxidative stress. *Am J Physiol Gastrointest Liver Physiol.* 2001;281:G1135–1139.
177. Schwabe RF, Brenner DA. Nuclear factor-kappaB in the liver: Friend or foe? *Gastroenterology.* 2007;132:2601–2604.
178. Xu Z, Chen L, Leung L, et al. Liver-specific inactivation of the Nrf1 gene in adult mouse leads to nonalcoholic steatohepatitis and hepatic neoplasia. *Proc Natl Acad Sci USA.* 2005;102:4120–4125.
179. Sanyal AJ, Campbell-Sargent C, Mirshahi F, et al. Nonalcoholic steatohepatitis: Association of insulin resistance and mitochondrial abnormalities. *Gastroenterology.* 2001;120:1183–1192.
180. Seki S, Kitada T, Yamada T, et al. In situ detection of lipid peroxidation and oxidative DNA damage in non-alcoholic fatty liver diseases. *J Hepatol.* 2002;37:56–62.
181. Sumida Y, Nakashima T, Yoh T, et al. Serum thioredoxin levels as a predictor of steatohepatitis in patients with nonalcoholic fatty liver disease. *J Hepatol.* 2003;38:32–38.
182. Yesilova Z, Yaman H, Oktenli C, et al. Systemic markers of lipid peroxidation and antioxidants in patients with nonalcoholic fatty liver disease. *Am J Gastroenterol.* 2005;100:850–855.
183. Esterbauer H, Schaur RJ, Zollner H. Chemistry and biochemistry of 4-hydroxynonenal, malonaldehyde and related aldehydes. *Free Radic Biol Med.* 1991;11:81–128.
184. Gardner HW. Oxygen radical chemistry of polyunsaturated fatty acids. *Free Radic Biol Med.* 1989;7:65–86.
185. Infante JP, Huszagh VA. Secondary carnitine deficiency and impaired docosahexaenoic (22:6n-3) acid synthesis: A common denominator in the pathophysiology of diseases of oxidative phosphorylation and beta-oxidation. *FEBS Lett.* 2000;468:1–5.
186. Pan M, Cederbaum AI, Zhang YL, et al. Lipid peroxidation and oxidant stress regulate hepatic apolipoprotein B degradation and VLDL production. *J Clin Invest.* 2004;113:1277–1287.
187. Wagner BA, Buettner GR, Burns CP. Free radical-mediated lipid peroxidation in cells: Oxidizability is a function of cell lipid bis-allylic hydrogen content. *Biochemistry.* 1994;33: 4449–4453.
188. Caldwell SH, Swerdlow RH, Khan EM, et al. Mitochondrial abnormalities in non-alcoholic steatohepatitis. *J Hepatol.* 1999;31: 430–434.
189. Cortez-Pinto H, Chatham J, Chacko VP, et al. Alterations in liver ATP homeostasis in human nonalcoholic steatohepatitis: A pilot study. *JAMA.* 1999;282:1659–1664.
190. Perez-Carreras M, Del Hoyo P, Martin MA, et al. Defective hepatic mitochondrial respiratory chain in patients with nonalcoholic steatohepatitis. *Hepatology.* 2003;38:999–1007.
191. Pessayre D. Role of mitochondria in non-alcoholic fatty liver disease. *J Gastroenterol Hepatol.* 2007;22(Suppl 1):S20–27.
192. Garcia-Ruiz C, Colell A, Morales A, et al. Role of oxidative stress generated from the mitochondrial electron transport chain and mitochondrial glutathione status in loss of mitochondrial function and activation of transcription factor nuclear factor-kappa B: Studies with isolated mitochondria and rat hepatocytes. *Mol Pharmacol.* 1995;48:825–834.
193. Hensley K, Kotake Y, Sang H, et al. Dietary choline restriction causes complex I dysfunction and increased H_2O_2 generation in liver mitochondria. *Carcinogenesis.* 2000;21:983–989.
194. Berson A, De Beco V, Letteron P, et al. Steatohepatitis-inducing drugs cause mitochondrial dysfunction and lipid peroxidation in rat hepatocytes. *Gastroenterology.* 1998;114:764–774.
195. Johnson EF, Palmer CN, Griffin KJ, et al. Role of the peroxisome proliferator-activated receptor in cytochrome P450 4A gene regulation. *Faseb J.* 1996;10:1241–1248.
196. Kersten S, Seydoux J, Peters JM, et al. Peroxisome proliferator-activated receptor alpha mediates the adaptive response to fasting. *J Clin Invest.* 1999; 103:1489–1498.
197. Tonsgard JH, Getz GS. Effect of Reye's syndrome serum on isolated chinchilla liver mitochondria. *J Clin Invest.* 1985; 76:816–825.
198. Dela Pena A, Leclercq I, Field J, et al. NF-kappaB activation, rather than TNF, mediates hepatic inflammation in a murine dietary model of steatohepatitis. *Gastroenterology.* 2005;129:1663–1674.

199. Cai D, Yuan M, Frantz DF, et al. Local and systemic insulin resistance resulting from hepatic activation of IKK-beta and NF-kappaB. *Nat Med.* 2005;11:183–190.
200. Luedde T, Beraza N, Kotsikoris V, et al. Deletion of NEMO/IKK-gamma in liver parenchymal cells causes steatohepatitis and hepatocellular carcinoma. *Cancer Cell.* 2007;11:119–132.
201. Schattenberg JM, Singh R, Wang Y, et al. JNK1 but not JNK2 promotes the development of steatohepatitis in mice. *Hepatology.* 2006;43:163–172.
202. Crespo J, Cayon A, Fernandez-Gil P, et al. Gene expression of tumor necrosis factor alpha and TNF-receptors, p55 and p75, in nonalcoholic steatohepatitis patients. *Hepatology.* 2001;34:1158–1163.
203. Haukeland JW, Damas JK, Konopski Z, et al. Systemic inflammation in nonalcoholic fatty liver disease is characterized by elevated levels of CCL2. *J Hepatol.* 2006;44:1167–1174.
204. Li Z, Yang S, Lin H, et al. Probiotics and antibodies to TNF inhibit inflammatory activity and improve nonalcoholic fatty liver disease. *Hepatology.* 2003;37:343–350.
205. Choudhury J, Sanyal AJ. Insulin resistance in NASH. *Front Biosci.* 2005;10:1520–1533.
206. Diehl AM, Li ZP, Lin HZ, et al. Cytokines and the pathogenesis of non-alcoholic steatohepatitis. *Gut.* 2005;54:303–306.
207. Chitturi S, Farrell G, Frost L, et al. Serum leptin in NASH correlates with hepatic steatosis but not fibrosis: A manifestation of lipotoxicity? *Hepatology.* 2002;36:403–409.
208. Kaser S, Moschen A, Cayon A, et al. Adiponectin and its receptors in non-alcoholic steatohepatitis. *Gut.* 2005;54:117–121.
209. Musso G, Gambino R, Biroli G, et al. Hypoadiponectinemia predicts the severity of hepatic fibrosis and pancreatic Beta-cell dysfunction in nondiabetic nonobese patients with nonalcoholic steatohepatitis. *Am J Gastroenterol.* 2005;100:2438–2446.
210. Bugianesi E, McCullough AJ, Marchesini G. Insulin resistance: A metabolic pathway to chronic liver disease. *Hepatology.* 2005;42:987–1000.
211. Vettor R, Milan G, Rossato M, et al. Review article: Adipocytokines and insulin resistance. *Aliment Pharmacol Ther.* 2005;22 (Suppl 2):3–10.
212. Maeda N, Shimomura I, Kishida K, et al. Diet-induced insulin resistance in mice lacking adiponectin/ACRP30. *Nat Med.* 2002;8:731–737.
213. Feldstein AE, Canbay A, Angulo P, et al. Hepatocyte apoptosis and fas expression are prominent features of human nonalcoholic steatohepatitis. *Gastroenterology.* 2003;125:437–443.
214. Nechushtan A, Smith CL, Lamensdorf I, et al. Bax and Bak coalesce into novel mitochondria-associated clusters during apoptosis. *J Cell Biol.* 2001;153:1265–1276.
215. Feldmann G, Haouzi D, Moreau A, et al. Opening of the mitochondrial permeability transition pore causes matrix expansion and outer membrane rupture in Fas-mediated hepatic apoptosis in mice. *Hepatology.* 2000;31:674–683.
216. Chedid A, Mendenhall CL, Gartside P, et al. Prognostic factors in alcoholic liver disease. VA Cooperative Study Group. *Am J Gastroenterol.* 1991;86:210–216.
217. Sass DA, Shaikh OS. Alcoholic hepatitis. *Clin Liver Dis.* 2006;10:219–237.
218. Lieber CS. Alcohol and the liver: Metabolism of alcohol and its role in hepatic and extrahepatic diseases. *Mt Sinai J Med.* 2000;67:84–94.
219. Lieber CS, Schmid R. The effect of ethanol on fatty acid metabolism; stimulation of hepatic fatty acid synthesis in vitro. *J Clin Invest.* 1961;40:394–399.
220. Thurman RG. II. Alcoholic liver injury involves activation of Kupffer cells by endotoxin. *Am J Physiol.* 1998;275:G605–G611.
221. Dey A, Cederbaum AI. Alcohol and oxidative liver injury. *Hepatology.* 2006;43:S63–74.
222. Lieber CS. Cytochrome P-4502E1: Its physiological and pathological role. *Physiol Rev.* 1997;77:517–544.
223. Tilg H, Diehl AM. Cytokines in alcoholic and nonalcoholic steatohepatitis. *N Engl J Med.* 2000;343:1467–1476.
224. Reddy JK, Rao MS. Lipid metabolism and liver inflammation. II. Fatty liver disease and fatty acid oxidation. *Am J Physiol Gastrointest Liver Physiol.* 2006;290:G852–858.
225. Li J, Hu W, Baldassare JJ, et al. The ethanol metabolite, linolenic acid ethyl ester, stimulates mitogen-activated protein kinase and cyclin signaling in hepatic stellate cells. *Life Sci.* 2003;73:1083–1096.
226. Desvergne B, Michalik L, Wahli W. Transcriptional regulation of metabolism. *Physiol Rev.* 2006;86:465–514.
227. Michalik L, Auwerx J, Berger JP, et al. International Union of Pharmacology. LXI. Peroxisome proliferator-activated receptors. *Pharmacol Rev.* 2006;58:726–741.
228. Rao MS, Reddy JK. Peroxisomal beta-oxidation and steatohepatitis. *Semin Liver Dis.* 2001;21:43–55.
229. Galli A, Pinaire J, Fischer M, et al. The transcriptional and DNA binding activity of peroxisome proliferator-activated receptor alpha is inhibited by ethanol metabolism. A novel mechanism for the development of ethanol-induced fatty liver. *J Biol Chem.* 2001;276:68–75.
230. Horton JD, Shimomura I. Sterol regulatory element-binding proteins: Activators of cholesterol and fatty acid biosynthesis. *Curr Opin Lipidol.* 1999;10:143–150.
231. You M, Fischer M, Deeg MA, et al. Ethanol induces fatty acid synthesis pathways by activation of sterol regulatory element-binding protein (SREBP). *J Biol Chem.* 2002;277: 29342–29347.
232. Ji C, Chan C, Kaplowitz N. Predominant role of sterol response element binding proteins (SREBP) lipogenic pathways in hepatic steatosis in the murine intragastric ethanol feeding model. *J Hepatol.* 2006;45:717–724.
233. Ji C, Kaplowitz N. Betaine decreases hyperhomocysteinemia, endoplasmic reticulum stress, and liver injury in alcohol-fed mice. *Gastroenterology.* 2003;124:1488–1499.
234. Shklyaev S, Aslanidi G, Tennant M, et al. Sustained peripheral expression of transgene adiponectin offsets the development of diet-induced obesity in rats. *Proc Natl Acad Sci USA.* 2003;100:14217–14222.
235. Zhou G, Myers R, Li Y, et al. Role of AMP-activated protein kinase in mechanism of metformin action. *J Clin Invest.* 2001;108:1167–1174.
236. You M, Matsumoto M, Pacold CM, et al. The role of AMP-activated protein kinase in the action of ethanol in the liver. *Gastroenterology.* 2004;127:1798–1808.
237. Xu A, Wang Y, Keshaw H, et al. The fat-derived hormone adiponectin alleviates alcoholic and nonalcoholic fatty liver diseases in mice. *J Clin Invest.* 2003; 112:91–100.
238. Kang L, Chen X, Sebastian BM, et al. Chronic ethanol and triglyceride turnover in white adipose tissue in rats: Inhibition of the anti-lipolytic action of insulin after chronic ethanol contributes to increased triglyceride degradation. *J Biol Chem.* 2007;282:28465–28473.
239. Lawler Jr JF, Yin M, Diehl AM, et al. Tumor necrosis factor-alpha stimulates the maturation of sterol regulatory element binding protein-1 in human hepatocytes through the action of neutral sphingomyelinase. *J Biol Chem.* 1998;273:5053–5059.
240. Thakur V, Pritchard MT, McMullen MR, et al. Adiponectin normalizes LPS-stimulated TNF-alpha production by rat Kupffer cells after chronic ethanol feeding. *Am J Physiol Gastrointest Liver Physiol.* 2006;290:G998–1007.
241. Thakur V, Pritchard MT, McMullen MR, et al. Chronic ethanol feeding increases activation of NADPH oxidase by lipopolysaccharide in rat Kupffer cells: Role of increased reactive oxygen in LPS-stimulated ERK1/2 activation and TNF-alpha production. *J Leukoc Biol.* 2006;79:1348–1356.
242. Li M, Carpio DF, Zheng Y, et al. An essential role of the NF-kappa B/Toll-like receptor pathway in induction of inflammatory and tissue-repair gene expression by necrotic cells. *J Immunol.* 2001;166:7128–7135.
243. Hui AY, Friedman SL. Molecular basis of hepatic fibrosis. *Expert Rev Mol Med.* 2003;5:1–23.
244. Tsukada S, Parsons CJ, Rippe RA. Mechanisms of liver fibrosis. *Clin Chim Acta.* 2006;364:33–60.
245. Albano E, Clot P, Morimoto M, et al. Role of cytochrome P4502E1-dependent formation of hydroxyethyl free radical in the development of liver damage in rats intragastrically fed with ethanol. *Hepatology.* 1996;23:155–163.
246. Knecht KT, Bradford BU, Mason RP, et al. In vivo formation of a free radical metabolite of ethanol. *Mol Pharmacol.* 1990;38:26–30.
247. Anania FA, Potter JJ, Rennie-Tankersley L, et al. Effects of acetaldehyde on nuclear protein binding to the nuclear factor I

consensus sequence in the alpha 2(I) collagen promoter. *Hepatology*. 1995;21:1640–1648.
248. Pares A, Potter JJ, Rennie L, et al. Acetaldehyde activates the promoter of the mouse alpha 2(I) collagen gene. *Hepatology*. 1994;19:498–503.
249. Kono H, Rusyn I, Yin M, et al. NADPH oxidase-derived free radicals are key oxidants in alcohol-induced liver disease. *J Clin Invest*. 2000;106:867–872.
250. Starkel P, Sempoux C, Leclercq I, et al. Oxidative stress, KLF6 and transforming growth factor-beta up-regulation differentiate non-alcoholic steatohepatitis progressing to fibrosis from uncomplicated steatosis in rats. *J Hepatol*. 2003;39:538–546.
251. Svegliati-Baroni G, Candelaresi C, Saccomanno S, et al. A model of insulin resistance and nonalcoholic steatohepatitis in rats: Role of peroxisome proliferator-activated receptor-alpha and n-3 polyunsaturated fatty acid treatment on liver injury. *Am J Pathol*. 2006;169:846–860.
252. Pinzani M, Marra F. Cytokine receptors and signaling in hepatic stellate cells. *Semin Liver Dis*. 2001;21:397–416.
253. Pinzani M. PDGF and signal transduction in hepatic stellate cells. *Front Biosci*. 2002;7:d1720–1726.
254. Schnabl B, Bradham CA, Bennett BL, et al. TAK1/JNK and p38 have opposite effects on rat hepatic stellate cells. *Hepatology*. 2001;34:953–963.
255. Friedman SL. Cytokines and fibrogenesis. *Semin Liver Dis*. 1999;19:129–140.
256. Ratziu V, Lalazar A, Wong L, et al. Zf9, a Kruppel-like transcription factor up-regulated in vivo during early hepatic fibrosis. *Proc Natl Acad Sci USA*. 1998;95:9500–9505.
257. Abou-Shady M, Friess H, Zimmermann A, et al. Connective tissue growth factor in human liver cirrhosis. *Liver*. 2000;20:296–304.
258. Paradis V, Dargere D, Bonvoust F, et al. Effects and regulation of connective tissue growth factor on hepatic stellate cells. *Lab Invest*. 2002;82:767–774.
259. Kon K, Ikejima K, Hirose M, et al. Pioglitazone prevents early-phase hepatic fibrogenesis caused by carbon tetrachloride. *Biochem Biophys Res Commun*. 2002;291:55–61.
260. Sun K, Wang Q, Huang XH. PPAR gamma inhibits growth of rat hepatic stellate cells and TGF beta-induced connective tissue growth factor expression. *Acta Pharmacol Sin*. 2006;27:715–723.
261. Tsukamoto H. Adipogenic phenotype of hepatic stellate cells. *Alcohol Clin Exp Res*. 2005;29:132S–133S.
262. Xu J, Fu Y, Chen A. Activation of peroxisome proliferator-activated receptor-gamma contributes to the inhibitory effects of curcumin on rat hepatic stellate cell growth. *Am J Physiol Gastrointest Liver Physiol*. 2003;285:G20–30.
263. Imai K, Sato M, Kojima N, et al. Storage of lipid droplets in and production of extracellular matrix by hepatic stellate cells (vitamin A-storing cells) in Long-Evans cinnamon-like colored (LEC) rats. *Anat Rec*. 2000;258:338–348.
264. Senoo H, Imai K, Matano Y, et al. Molecular mechanisms in the reversible regulation of morphology, proliferation and collagen metabolism in hepatic stellate cells by the three-dimensional structure of the extracellular matrix. *J Gastroenterol Hepatol*. 1998;13(Suppl):S19–32.
265. Marra F, Efsen E, Romanelli RG, et al. Ligands of peroxisome proliferator-activated receptor gamma modulate profibrogenic and proinflammatory actions in hepatic stellate cells. *Gastroenterology*. 2000;119:466–478.
266. Ohata M, Lin M, Satre M, et al. Diminished retinoic acid signaling in hepatic stellate cells in cholestatic liver fibrosis. *Am J Physiol*. 1997;272:G589–596.
267. Schuppan D, Afdhal NH. Liver cirrhosis. *Lancet*. 2008;371: 838–851.
268. Falleti E, Fabris C, Toniutto P, et al. Genetic polymorphisms of inflammatory cytokines and liver fibrosis progression due to recurrent hepatitis C. *J Interferon Cytokine Res*. 2007;27:239–246.
269. Huang H, Shiffman ML, Cheung RC, et al. Identification of two gene variants associated with risk of advanced fibrosis in patients with chronic hepatitis C. *Gastroenterology*. 2006;130: 1679–1687.
270. Choi BY, Nguyen MH. The diagnosis and management of benign hepatic tumors. *J Clin Gastroenterol*. 2005;39:401–412.
271. Duff B, Weigel JA, Bourne P, et al. Endothelium in hepatic cavernous hemangiomas does not express the hyaluronan receptor for endocytosis. *Hum Pathol*. 2002;33: 265–269.
272. Kim GE, Thung SN, Tsui WM, et al. Hepatic cavernous hemangioma: Underrecognized associated histologic features. *Liver Int*. 2006;26:334–338.
273. Rebouissou S, Bioulac-Sage P, Zucman-Rossi J. Molecular pathogenesis of focal nodular hyperplasia and hepatocellular adenoma. *J Hepatol*. 2008;48:163–170.
274. Rebouissou S, Couchy G, Libbrecht L, et al. The beta-catenin pathway is activated in focal nodular hyperplasia but not in cirrhotic FNH-like nodules. *J Hepatol*. 2008;49:61–71.
275. Chen YW, Jeng YM, Yeh SH, et al. P53 gene and Wnt signaling in benign neoplasms: Beta-catenin mutations in hepatic adenoma but not in focal nodular hyperplasia. *Hepatology*. 2002;36:927–935.
276. Monga SP, Mars WM, Pediaditakis P, et al. Hepatocyte growth factor induces Wnt-independent nuclear translocation of beta-catenin after Met-beta-catenin dissociation in hepatocytes. *Cancer Res*. 2002;62:2064–2071.
277. Bioulac-Sage P, Rebouissou S, Sa Cunha A, et al. Clinical, morphologic, and molecular features defining so-called telangiectatic focal nodular hyperplasias of the liver. *Gastroenterology*. 2005;128:1211–1218.
278. Monga SP. Hepatic adenomas: Presumed innocent until proven to be beta-catenin mutated. *Hepatology*. 2006;43:401–404.
279. Zucman-Rossi J, Jeannot E, Nhieu JT, et al. Genotype-phenotype correlation in hepatocellular adenoma: New classification and relationship with HCC. *Hepatology*. 2006;43:515–524.
280. Lee YH, Sauer B, Gonzalez FJ. Laron dwarfism and non-insulin-dependent diabetes mellitus in the Hnf-1alpha knockout mouse. *Mol Cell Biol*. 1998;18:3059–3068.
281. Shih DQ, Bussen M, Sehayek E, et al. Hepatocyte nuclear factor-1alpha is an essential regulator of bile acid and plasma cholesterol metabolism. *Nat Genet*. 2001;27:375–382.
282. Rebouissou S, Imbeaud S, Balabaud C, et al. HNF1alpha inactivation promotes lipogenesis in human hepatocellular adenoma independently of SREBP-1 and carbohydrate-response element-binding protein (ChREBP) activation. *J Biol Chem*. 2007;282: 14437–14446.
283. Torbenson M, Lee JH, Choti M, et al. Hepatic adenomas: Analysis of sex steroid receptor status and the Wnt signaling pathway. *Mod Pathol*. 2002;15:189–196.
284. Weinberg AG, Finegold MJ. Primary hepatic tumors of childhood. *Hum Pathol*. 1983;14:512–537.
285. Perilongo G, Shafford EA. Liver tumours. *Eur J Cancer*. 1999;35:953–958; discussion 958–959.
286. DeBaun MR, Tucker MA. Risk of cancer during the first four years of life in children from The Beckwith-Wiedemann Syndrome Registry. *J Pediatr*. 1998;132:398–400.
287. Giardiello FM, Offerhaus GJ, Krush AJ, et al. Risk of hepatoblastoma in familial adenomatous polyposis. *J Pediatr*. 1991;119: 766–768.
288. Buendia MA. Genetic alterations in hepatoblastoma and hepatocellular carcinoma: Common and distinctive aspects. *Med Pediatr Oncol*. 2002;39:530–535.
289. Laurent-Puig P, Zucman-Rossi J. Genetics of hepatocellular tumors. *Oncogene*. 2006;25:3778–3786.
290. Nejak-Bowen K, Monga SP. Wnt/beta-catenin signaling in hepatic organogenesis. *Organogenesis*. 2008;4.
291. Thompson MD, Monga SP. WNT/beta-catenin signaling in liver health and disease. *Hepatology*. 2007;45:1298–1305.
292. Blaker H, Hofmann WJ, Rieker RJ, et al. Beta-catenin accumulation and mutation of the CTNNB1 gene in hepatoblastoma. *Genes Chromosomes Cancer*. 1999;25:399–402.
293. Koch A, Denkhaus D, Albrecht S, et al. Childhood hepatoblastomas frequently carry a mutated degradation targeting box of the beta-catenin gene. *Cancer Res*. 1999;59:269–273.
294. Taniguchi K, Roberts LR, Aderca IN, et al. Mutational spectrum of beta-catenin, AXIN1, and AXIN2 in hepatocellular carcinomas and hepatoblastomas. *Oncogene*. 2002;21:4863–4871.
295. Udatsu Y, Kusafuka T, Kuroda S, et al. High frequency of beta-catenin mutations in hepatoblastoma. *Pediatr Surg Int*. 2001;17:508–512.
296. Giardiello FM, Petersen GM, Brensinger JD, et al. Hepatoblastoma and APC gene mutation in familial adenomatous polyposis. *Gut*. 1996;39:867–869.
297. Behari J, Zeng G, Otruba W, et al. R-Etodolac decreases beta-catenin levels along with survival and proliferation of hepatoma cells. *J Hepatol*. 2007;46:849–857.

298. Kuroda T, Rabkin SD, Martuza RL. Effective treatment of tumors with strong beta-catenin/T-cell factor activity by transcriptionally targeted oncolytic herpes simplex virus vector. *Cancer Res.* 2006;66:10127–10135.
299. Sangkhathat S, Kusafuka T, Miao J, et al. In vitro RNA interference against beta-catenin inhibits the proliferation of pediatric hepatic tumors. *Int J Oncol.* 2006;28:715–722.
300. Zeng G, Apte U, Cieply B, et al. siRNA-mediated beta-catenin knockdown in human hepatoma cells results in decreased growth and survival. *Neoplasia.* 2007;9:951–959.
301. Parkin DM, Bray F, Ferlay J, et al. Global cancer statistics, 2002. *CA Cancer J Clin.* 2005;55:74–108.
302. Bergsland EK, Venook AP. Hepatocellular carcinoma. *Curr Opin Oncol.* 2000;12:357–361.
303. Friedman SL. Mechanisms of hepatic fibrogenesis. *Gastroenterology.* 2008;134:1655–1669.
304. Villanueva A, Newell P, Chiang DY, et al. Genomics and signaling pathways in hepatocellular carcinoma. *Semin Liver Dis.* 2007;27:55–76.
305. Sell S, Leffert HL. Liver cancer stem cells. *J Clin Oncol.* 2008;26:2800–2805.
306. Tsuneyama K, Kaizaki Y, Doden K, et al. Combined hepatocellular and cholangiocarcinoma with marked squamous cell carcinoma components arising in non-cirrhotic liver. *Pathol Int.* 2003;53:90–97.
307. Furuse J. Growth factors as therapeutic targets in HCC. *Crit Rev Oncol Hematol.* 2008;67:8–15.
308. Llovet JM, Di Bisceglie AM, Bruix J, et al. Design and endpoints of clinical trials in hepatocellular carcinoma. *J Natl Cancer Inst.* 2008;100:698–711.
309. Smela ME, Currier SS, Bailey EA, et al. The chemistry and biology of aflatoxin B1: From mutational spectrometry to carcinogenesis. *Carcinogenesis.* 2001;22:535–545.
310. Shimizu Y, Zhu JJ, Han F, et al. Different frequencies of p53 codon-249 hot-spot mutations in hepatocellular carcinomas in Jiang-su province of China. *Int J Cancer.* 1999;82:187–190.
311. Vautier G, Bomford AB, Portmann BC, et al. p53 mutations in British patients with hepatocellular carcinoma: Clustering in genetic hemochromatosis. *Gastroenterology.* 1999;117:154–160.
312. Edamoto Y, Hara A, Biernat W, et al. Alterations of RB1, p53 and Wnt pathways in hepatocellular carcinomas associated with hepatitis C, hepatitis B and alcoholic liver cirrhosis. *Int J Cancer.* 2003;106:334–341.
313. Huang H, Fujii H, Sankila A, et al. Beta-catenin mutations are frequent in human hepatocellular carcinomas associated with hepatitis C virus infection. *Am J Pathol.* 1999;155:1795–1801.
314. Street A, Macdonald A, McCormick C, et al. Hepatitis C virus NS5A-mediated activation of phosphoinositide 3-kinase results in stabilization of cellular beta-catenin and stimulation of beta-catenin-responsive transcription. *J Virol.* 2005;79:5006–5016.
315. Craig JR, Peters RL, Edmondson HA, et al. Fibrolamellar carcinoma of the liver: A tumor of adolescents and young adults with distinctive clinico-pathologic features. *Cancer.* 1980;46:372–379.
316. El-Serag HB, Davila JA. Is fibrolamellar carcinoma different from hepatocellular carcinoma? A US population-based study. *Hepatology.* 2004;39:798–803.
317. Katzenstein HM, Krailo MD, Malogolowkin MH, et al. Fibrolamellar hepatocellular carcinoma in children and adolescents. *Cancer.* 2003;97:2006–2012.
318. Terracciano L, Tornillo L. Cytogenetic alterations in liver cell tumors as detected by comparative genomic hybridization. *Pathologica.* 2003;95:71–82.
319. Marchio A, Pineau P, Meddeb M, et al. Distinct chromosomal abnormality pattern in primary liver cancer of non-B, non-C patients. *Oncogene.* 2000;19:3733–3738.
320. Buckley AF, Burgart LJ, Kakar S. Epidermal growth factor receptor expression and gene copy number in fibrolamellar hepatocellular carcinoma. *Hum Pathol.* 2006;37:410–414.

Chapter

21

Molecular Basis of Diseases of the Exocrine Pancreas

Matthias Sendler . Julia Mayerle . Markus M. Lerch

ACUTE PANCREATITIS

Acute pancreatitis presents clinically as a sudden inflammatory disorder of the pancreas, and is caused by premature intracellular activation of pancreatic proteases leading to (i) self-destruction of acinar cells and (ii) autodigestion of the organ. Necrotic cell debris resulting from this process produces a systemic inflammatory reaction, which may lead to multiorgan failure in due course. The incidence of acute pancreatitis differs regionally from 20 to 120 cases per 100,000 population. Acute pancreatitis varies considerably in severity and can be categorized into two forms of the disease. The majority of cases (85%) present with a mild form of disease, classified as edematous pancreatitis, with absent or only transient extrapancreatic organ failure. In the remaining minority of cases (15%), pancreatitis follows a severe course accompanied by sustained multiorgan failure. This latter form of pancreatitis is commonly referred to as severe or necrotizing pancreatitis. Severe necrotizing pancreatitis is associated with high mortality (10%–20%) and may lead to long-term complications such as the formation of pancreatic pseudocysts or impairment of exocrine and endocrine function of the pancreatic gland.

In many patients (approximately 50%) the underlying cause of acute pancreatitis is the migration of a gallstone resulting in obstruction of the pancreatic duct at the papilla of Vater. Development of acute pancreatitis is frequently triggered by alcohol abuse. In 25%–40% of acute pancreatitis patients, increased or excess alcohol consumption is regarded as the cause of the disease. Removal of the underlying disease-causing agent results in complete regeneration of the pancreas and preserved exocrine and endocrine function in the majority of cases. Recurrent attacks of the disease can result from chronic alcohol abuse, repeated gallstone passage, genetic predispositions, sphincter dysfunction, metabolic disorders, or pancreatic duct strictures. All of these factors can contribute to the development of chronic pancreatitis as well. In the remaining cases (10%–20%), no apparent clinical cause or etiology of the disease can be identified. These cases are referred to as idiopathic pancreatitis. It became clear during the last decade that previously unknown genetic factors play a major role in these cases. The initial cellular mechanism causing acute pancreatitis is probably independent of the underlying etiology of the disease [1].

Early Events in Acute Pancreatitis and the Role of Protease Activation

Acute pancreatitis is an inflammatory disorder whose pathogenesis is not well understood. The pancreas is known as the enzyme factory of the human organism, producing and secreting large amounts of potentially hazardous digestive enzymes, many of which are synthesized as pro-enzymes known as zymogens. Under physiological conditions the pancreatic digestive enzymes are secreted in response to hormonal stimulation [2]. Activation of the pro-enzymes (or zymogens) requires hydrolytic cleavage of their activation peptide by protease enzymes. After entering the small intestine, the pancreatic zymogen trypsinogen is first activated to trypsin by the intestinal protease enterokinase (or enteropeptidase). Trypsin then proteolytically processes other pancreatic enzymes to their active forms. Under physiological conditions pancreatic proteases remain inactive during synthesis, intracellular transport, secretion from acinar cells, and transit through the pancreatic duct. They are activated only upon reaching the lumen and brush-border of the small intestine (Figure 21.1).

More than a century ago, Hans Chiari proposed that the underlying pathophysiological mechanism for the development of pancreatitis was autodigestion of the exocrine pancreatic tissue by proteolytic enzymes [3]. Today, this theory is well accepted. Nevertheless, this theory suggests that disease results

Molecular Pathology © 2009, Elsevier, Inc. All Rights Reserved.

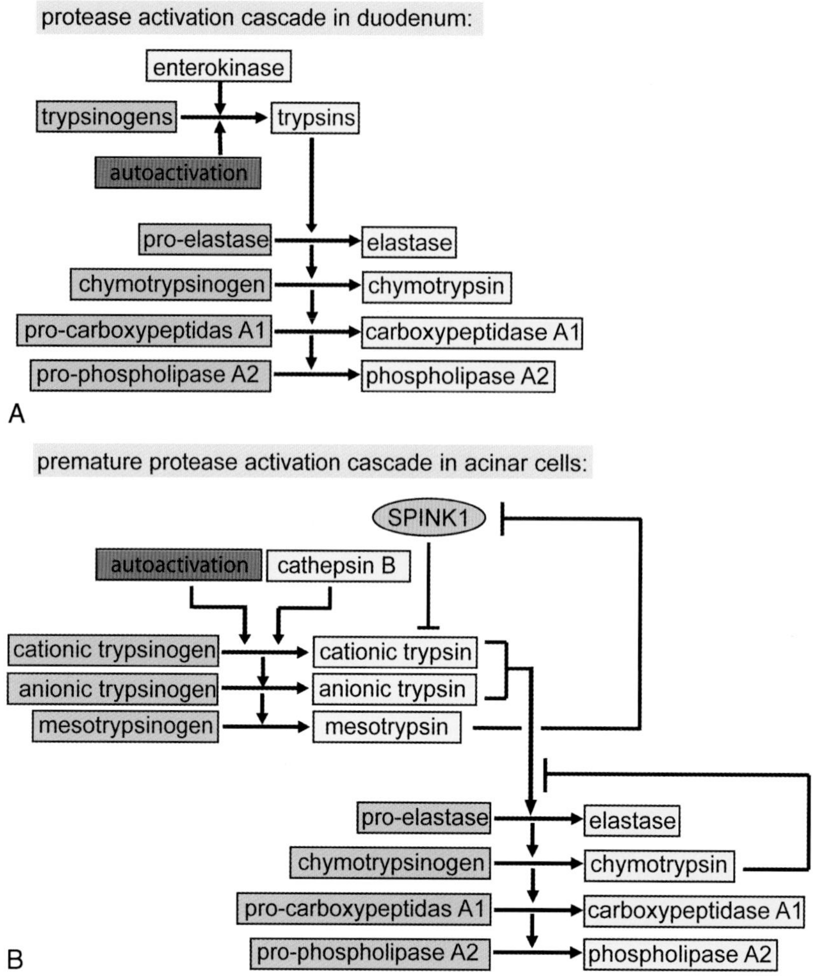

Figure 21.1 **Activation of pancreatic proteases under normal conditions and in disease. (A)** The protease activation cascade in the duodenum. Enterokinase activates trypsinogen by proteolytic cleavage of the trypsin activation peptide (TAP), and autoactivation contributes to this process. Trypsin then activates other digestive pro-enzymes in a cascade-like fashion. **(B)** Within acinar cells, premature activation of trypsinogen by cathepsin B is involved in the setting of the proteolytic cascade. Intracellular trypsin may also activate other digestive pro-enzymes, in spite of presence of the trypsin inhibitor SPINK-1 and the trypsin-degrading enzyme chymotrypsin C.

from the premature intracellular activation of zymogens and that this occurs in the absence of enterokinase. Furthermore, this theory suggests that protease inactivation takes place despite several physiological defense mechanisms, including the synthesis of endogenous protease inhibitors and the storage of proteases in a membrane-confined compartment of zymogen granules.

Much of our current knowledge regarding the onset of pancreatitis was not gained from studies involving the human pancreas or patients with pancreatitis, but from animal or isolated cell models [4]. There are several reasons why these models have been used: (i) the pancreas is a rather inaccessible organ because of its anatomical location in the retroperitoneal space (unlike in the colon or stomach, biopsies of human pancreas are difficult to obtain for ethical and medical reasons), and (ii) patients who are admitted to the hospital with acute pancreatitis have usually progressed beyond the initial stages of the disease where the triggering early events could have been studied. Particularly, the autodigestive process that characterizes this disease has remained a significant impediment for investigations that address initiating pathophysiological events. Therefore, the issue of premature protease activation has mostly been studied in animal models of the disease and before randomized placebo-controlled trials to evaluate antiproteolytic therapeutic concepts in humans could be performed. Experimental models are now irreplaceable tools in studying etiological factors, pathophysiology, new diagnostic tools, and treatment options for acute pancreatitis [5].

Data from various animal models suggest that after the initial insult a variety of pathophysiological factors determine disease onset. These include (i) a block in secretion, (ii) the co-localization of zymogens with lysosomal enzymes, (iii) the activation of trypsinogen and other zymogens, and (iv) acinar cell injury. *In vitro* and *in vivo* studies have demonstrated the importance of premature zymogen activation in the pathogenesis of pancreatitis since the inception of the hypothesis by Chiari [6]. The activation of trypsinogen and other pancreatic zymogens can be demonstrated in pancreatic homogenates from experimental animals and this zymogen activation appears to be an early event. Trypsin activity is detected as early as 10 minutes after supramaximal stimulation with the cholecystokinin-analogue cerulein in rats and increases over time. The activation of trypsinogen requires the hydrolytic cleavage of a 7–10 amino acids propeptide called trypsin activation peptide (TAP) on the N-terminus of trypsinogen. Measuring an increase of immunoactive TAP after cerulein-induced pancreatitis in rats showed trypsinogen activation in the secretory compartment of acinar cells [7]. Furthermore, TAP was detected in serum and urine of patients with pancreatitis [8], and the amount of TAP appears to correlate with the severity of the disease [9]. Not only the activity of trypsin increases in the early phase in acute experimental pancreatitis, but also elastase activity [10]. In addition to the activation peptide of trypsinogen (TAP), the activation peptide of carboxypeptidase A1 (PCA1) can also be identified in serum at an early stage of pancreatitis [11]. Premature activation of these proenzymes leads to the development of necrosis and to autodigestion of pancreatic tissues [12–20]. Recent studies defining the localization of early activation of proenzymes suggest that these processes, which lead to pancreatitis and pancreatic tissue necrosis, originate in acinar cells (Figure 21.1) [6,21,22].

Additional evidence for the association between premature protease activation and severity of pancreatitis resulted from experiments with serine-protease inhibitors, where cell injury could be significantly reduced compared to controls [23,24]. On the other hand, protease inhibitors have a protective effect for the prevention of acute pancreatitis only [25]. Therapeutic application of protease inhibitors after the initial disease phase does not result in a significant beneficial effect on survival or severity of the disease [26].

In conclusion, premature intracellular activation of zymogens to active proteases in the secretory compartment of acinar cells results in acinar necrosis and contributes to the onset of pancreatitis. As a direct result of cellular injury, the acinar cells release chemokines and cytokines which initiate the later events in pancreatitis, including recruitment of inflammatory cells into the tissue. Trypsin seems to be the key enzyme in the process of activating other digestive pro-enzymes prematurely, and one of the crucial questions in understanding the pathophysiology of acute pancreatitis is to identify the mechanism which prematurely activates trypsinogen inside acinar cells. However, it must be noted that the term trypsin, as defined by the cleavage of specific synthetic or protein substrates, comprises a group of enzymes whose individual role in the initial activation cascade may differ considerably.

The Mechanism of Zymogen Activation

One hypothesis for the initiation of the premature activation of trypsinogen suggests that during the early stage of acute pancreatitis, pancreatic digestive zymogens become co-localized with lysosomal hydrolases. Recent data show that the lysosomal cysteine proteinase cathepsin B may play an important role for the activation of trypsinogen. Many years ago *in vitro* data demonstrated activation of trypsinogen by cathepsin B [27–29]. Most lysosomal hydrolases are synthesized as inactive pro-enzymes, but in contrast to digestive zymogens, they are activated by post-translational processing in the cell. During protein sorting in the Golgi system, lysosomal hydrolases are sorted into prelysosomes, whereas zymogens are packaged into condensing vacuoles. The sorting of lysosomal hydrolases depends on a mannose-6-phosphate-dependent pathway [30], which leads to a separation of lysosomal hydrolases from other secretory proteins and to the formation of prelysosomal vacuoles. However, this sorting is incomplete. Under physiological conditions, a significant fraction of hydrolases enter the secretory pathway [31–34]. It has been suggested that these mis-sorted hydrolases play a role in the regulation of zymogen secretion [35]. In acute pancreatitis the separation of digestive zymogens and lysosomal hydrolases is impaired. This leads to further co-localization of lysosomal hydrolases and zymogens within cytoplasmic vacuoles of acinar cells [36]. This co-localization has also been shown in electron microscopy [37], as well as in subcellular fractions isolated by density gradient centrifugation [38]. The redistribution of cathepsin B from the lysosome-enriched fraction was noted within 15 minutes of the start of pancreatitis induction, and trypsinogen activation was observed in parallel [39–41]. There are two main theories trying to explain the colocalization of cysteine and serine proteases: (i) fusion of lysosomes and zymogen granules [42] or (ii) incorrect sorting of zymogens and hydrolases in the process of vacuole maturation [37]. Wortmannin, a phosphoinositide-3-kinase inhibitor, prevents the intracellular mis-sorting of hydrolases and zymogens, and subsequently prevents trypsinogen activation to trypsin during acute pancreatitis [43].

Further experiments focused on cathepsin B as the main enzyme driving the intracellular activation of trypsinogen. Cathepsin B is the most abundant lysosomal hydrolase in acinar cells. Pretreatment of rat pancreatic acini with E64d, a cell-permeable cathepsin B inhibitor, leads to complete inhibition of cathepsin B and completely abrogates trypsinogen activation [44,45]. Final evidence that cathepsin B is involved in activation of trypsinogen during cerulein-induced experimental pancreatitis comes from experiments in cathepsin B knockout mice. In these animals, after induction of experimental pancreatitis, trypsin activity was reduced to less than 20% compared to wild-type animals and the severity of the disease was markedly ameliorated [46]. These data showed unequivocally the importance of cathepsin B for the pathogenesis of acute pancreatitis (Figure 21.1) [47].

The cathepsin B theory implies one further critical point—that trypsinogen is expressed and stored in the presence of different potent intrapancreatic trypsin inhibitors. To activate trypsinogen, cathepsin B needs to override these defensive mechanisms to initiate the premature intracellular activation cascade [18]. Recently, it has become clear that cathepsin B cannot only activate cationic and anionic trypsinogen, but also mesotrypsinogen [48]. Mesotrypsin, the third trypsin isoform expressed in the human pancreas, is resistant against trypsin inhibitors like SPINK-1 or soybean trypsin inhibitor (SBTI) [49]. Moreover, mesotrypsin is able to degrade trypsin inhibitors. Under physiological conditions, mesotrypsin is activated in the duodenum by enterokinase, where it degrades exogenous trypsin inhibitors to ensure normal tryptic digestion. Mesotrypsin rapidly inactivates trypsin inhibitors like SPINK-1 by proteolytic cleavage *in vitro* [48,50]. Therefore, activation of trypsins by cathepsins might not only trigger a proteolytical cascade, but also involve the removal of trypsin inhibitors such as SPINK-1 via the activation of mesotrypsin.

The role of cathepsin B in chronic pancreatitis was recently addressed by an Indian group from Hyderabad. In 140 patients suffering from tropical pancreatitis, they found a significant difference between patients and controls for the C76G polymorphism in the *CTSB* gene [51]. Unfortunately, these data could not be confirmed in a Caucasian cohort [52]. Thus, the role of CTSB in human pancreatitis remains inconclusive. Taken together, these experimental observations represent compelling evidence that cathepsin B can contribute to premature, intracellular zymogen activation and the initiation of acute pancreatitis not only through co-localization with trypsinogen, but also through activation of mesotrypsin, rendering endogenous pancreatic protease inhibitors inactive.

The Degradation of Active Trypsin

During the early phase of pancreatitis, trypsinogen and other zymogens are rapidly activated, while later in the disease course their activity declines to physiological levels, suggesting degradation of the active enzymes. This phenomenon has been termed autolysis or autodegradation. Since this process self-limits autoactivation of trypsinogen, it is regarded as a safety mechanism to counteract premature zymogen activation.

One theory to possibly explain how uncontrolled trypsinogen activation can be antagonized is based on the existence of a serine protease that is capable of trypsin degradation. In 1988 Heinrich Rinderknecht discovered an enzyme which rapidly degrades active cationic and anionic trypsin and named this protease enzyme Y [53]. Recent *in vitro* data suggest that the autodegradation of trypsin is a very slow process and that most of trypsin degradation is not mediated by trypsin itself but by another enzyme. Chymotrypsin C has the capability to proteolytically cleave cationic trypsin at Leu81–Glu82 in the Ca^{2+} binding loop. This leads to rapid autodegradation and catalytic inactivation of trypsin by additional cleavage at the Arg122–Val123 [54]. Thus, chymotrypsin C has the capability to induce trypsin-mediated trypsin autodegradation during pancreatitis and is most likely identical with the enzyme Y that Rinderknecht proposed in 1988. However, chymotrypsin C has also the ability to induce trypsin-mediated trypsinogen autoactivation by proteolytic cleavage at the Phe18–Asp19 position of cationic trypsinogen [55]. The balance between autoactivation and autodegradation of cationic trypsin mediated by chymotrypsin C is regulated via the Ca^{2+} concentration. In the presence of 1 mM Ca^{2+}, degradation of trypsin is blocked and autoactivation of trypsinogen is induced. Under physiological conditions in the duodenum, high Ca^{2+} concentrations facilitate the activation of trypsinogen to promote digestion. In the absence of high Ca^{2+} concentrations, chymotrypsin C degrades active trypsin and protects against premature activation of trypsin (Figure 21.2).

Calcium Signaling

The second messenger calcium plays an important role in multiple different intracellular processes such as metabolism, cellular secretion, cell differentiation, and cell growth. Under physiological conditions, pancreatic acinar cells maintain a Ca^{2+} gradient across the plasma membrane with low intracellular concentration and high extracellular concentration of calcium. In response to hormonal stimulation, Ca^{2+} is released from intracellular stores to regulate signal-secretion coupling. In pancreatic acinar cells, acetylcholine (ACh) and cholecystokinin (CCK) regulate the secretion of digestive enzymes via the generation of repetitive local cytosolic Ca^{2+} signals [56]. In response to secretagogue stimulation with ACh or CCK, Ca^{2+} is initially released from intracellular stores near the apical pole of acinar cells [57]. This induces the fusion of zymogen granules with the apical plasma membrane [58], and activation of Ca^{2+} dependent Cl^- channels in the apical membrane [59]. The pattern of intracellular calcium signal in response to secretagogues stimulation is dependent on the neurotransmitter or hormone concentration. ACh at physiological concentrations elicits repetitive Ca^{2+} spikes and oscillation of Ca^{2+} concentrations, but these oscillations are restricted to the secretory pole of the cell. High concentrations of cholecystokinin lead to short-lasting spikes followed by longer Ca^{2+} transients that spread to the entire cell. Each oscillation is associated with a burst of exocytotic activity and the release of zymogen into the duct lumen [60]. In contrast, supramaximal stimulation of acinar cells induces a completely different pattern of Ca^{2+} signals. Instead of oscillatory activity observed with physiological doses of cholecystokinin, there is a much larger rise followed by a sustained elevation associated to a block of enzyme secretion and premature intracellular protease activation [56,61,62]. Ca^{2+} is released from the endoplasmatic reticulum (ER) in response to stimulation. The ER is located in the basolateral part of the acinar cell with extensions in the apical part enriched with zymogen granules. While the entire ER contains Ca^{2+}, release of Ca^{2+} in response to cholecystokinin or ACh occurs only at the apical pole due to the higher density of Ca^{2+} release channels at the apical pole of

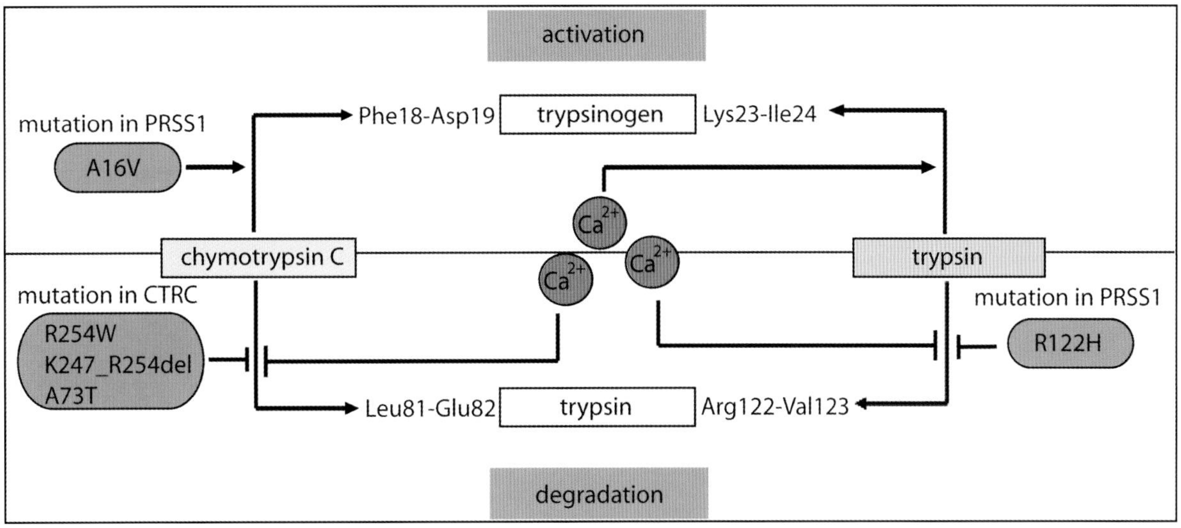

Figure 21.2 Chymotrypsin C has different functions in the processing of trypsin or trypsinogen. The major role in pancreatitis is degrading active trypsin. This function is disturbed by mutations within the *CTRC* gene or by high levels of Ca^{2+}. The activation of trypsinogen to trypsin can also be mediated by chymotrypsin C. Trypsin itself has the capability to autoactivate or to self-degrade. The R122H mutation results in the decreased autolysis of active trypsin.

the ER [63]. Two types of Ca^{2+} channels are expressed in the ER of acinar cells, inositol-triphosphate-receptors (IP_3) and ryanodine receptors. Both of these channels are required for apical Ca^{2+} peaks. ACh activates phospholipase C (PLC) and initiates the Ca^{2+} release via the intracellular messenger IP_3 [64,65], whereas cholecystokinin does not activate PLC but increases the intracellular concentration of nicotinic acid adenine dinucleotide phosphate (NAADP) in a dose-dependent manner [66]. The higher density of Ca^{2+} channels in the apical part of the ER explains the initiation of Ca^{2+} signals in the granule part of the cytoplasm. The apical, zymogen granule-enriched part of the acinar cell is surrounded by a barrier of mitochondria which absorb released calcium and prevent higher Ca^{2+} concentrations from expanding beyond the apical part of acinar cells [67]. The spatially limited release of Ca^{2+} at the apical pole also prevents an unregulated chain reaction across gap junctions, which would affect neighboring cells. The mitochondrial Ca^{2+} uptake leads to increased metabolism and generation of ATP [68,69]. ATP is required for the reuptake of Ca^{2+} in the ER via the sarcoplasmatic endoplasmatic reticulum calcium ATPase (SERCA), and for exocytosis across the apical membrane [70]. Thus, Ca^{2+} homeostasis plays a crucial role for maintaining signal-secretion coupling in pancreatic acinar cells (Figure 21.3).

Elevated Ca^{2+} concentrations in the extracellular compartment or within acinar cells is known to be a risk factor for the development of acute pancreatitis [71–74]. Disturbances in the Ca^{2+} homeostasis of pancreatic acinar cells occur early in the secretagogue-induced model of pancreatitis. An attenuation of Ca^{2+} elevation in acinar cells results from exposure to the cytosolic Ca^{2+} chelator BAPTA-AM, which also prevents zymogen activation, proving that Ca^{2+} is essential for zymogen activation. The sustained elevation that follows the initial intracellular Ca^{2+} spike induced by supramaximal concentrations of cerulein is also attenuated in the complete absence of intracellular calcium and appears to depend on extracellular Ca^{2+}. In the absence of extracellular Ca^{2+}, the activation of trypsinogen induced by supramaximal doses of cerulein is also attenuated, suggesting that the initial and transient rise in Ca^{2+} caused by the release of calcium from the internal stores is not sufficient to permit trypsinogen activation. In contrast, interference with high calcium plateaus by the natural Ca^{2+} antagonist magnesium or a Ca^{2+} chelator *in vivo* abolishes trypsinogen activation as well as pancreatitis [62,75–77].

Acute pancreatitis is characterized by the pathologic activation of zymogens within pancreatic acinar cells. The process requires a rise in cytosolic Ca^{2+} from undefined intracellular stores. Zymogen activation is thereby mediated by ryanodine receptor-regulated Ca^{2+} release, and early zymogen activation takes place in a supranuclear compartment that overlaps in distribution with the ryanodine receptor. Furthermore, *in vivo* inhibition of the ryanodine receptor results in a loss of zymogen activation. Therefore, Ca^{2+} release from the ryanodine receptor mediates zymogen activation but not enzyme secretion [78].

Recent reports have shown that metabolites of alcohol metabolism can have a pathological effect on acinar cell Ca^{2+} homeostasis, suggesting a possible pathogenic mechanism in alcoholic pancreatitis. Nonoxidative metabolites like fatty acid ethyl esters (FAEE) and fatty acids (FA) can cause Ca^{2+}-dependent acinar cell necrosis [79]. It was previously demonstrated that FAEEs are generated in acinar cells incubated with clinically relevant concentrations of ethanol [79]. FAEEs activate the IP_3 receptor after which Ca^{2+} is released from the ER [80]. In contrast, FAs do not activate Ca^{2+} channels but decrease ATP levels in the cytoplasm, which results

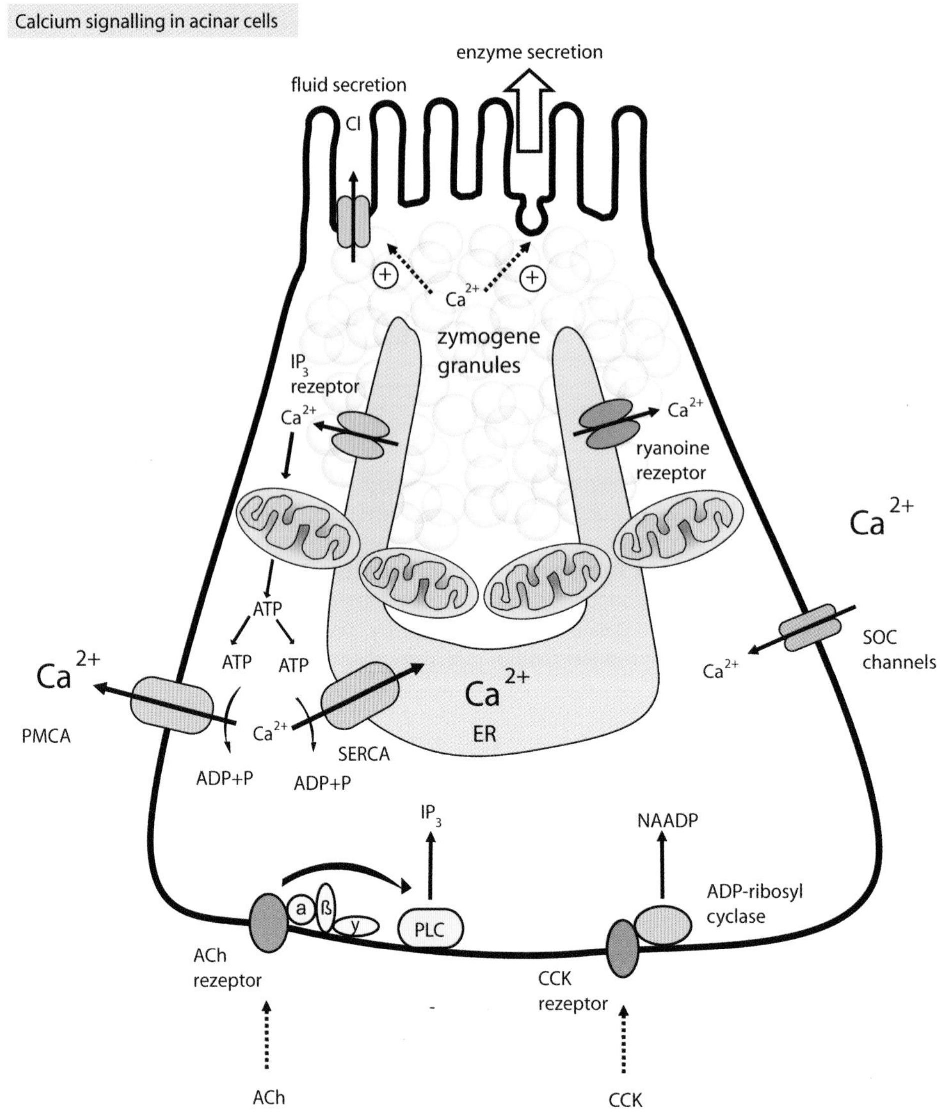

Figure 21.3 Calcium signaling in acinar cells. Intracellular Ca^{2+} homeostasis regulates the secretion of zymogens and fluid. Acetylcholine (ACh) regulates via the second messenger inositol-3-phosphate the apical Ca^{2+} influx from intracellular stores (ER). Choleocystekinin leads to the production of nicotinic acid adenine dinucleotide phosphate (NAADP), which interacts with ryanodine receptor in the ER membrane and also regulates the apical Ca^{2+} influx. The plasma membrane calcium ATPase (PMCA) or the sarcoplasmatic endoriticulum calcium ATPase (SERCA) regulates cytoplasmatic Ca^{2+} decrease. A mitochondrial barrier inhibits a global Ca^{2+} increase by absorbing free Ca^{2+} from the apical zymogen-enriched part of the acinar cell. Interferences in the calcium homeostasis lead to a global Ca^{2+} increase in the cytoplasm, which results in premature activation of zymogens.

in impaired Ca^{2+} reuptake into the ER. Subsequently, increased intracellular Ca^{2+} levels contribute to premature zymogen activation.

Ca^{2+} not only is an important second messenger, but has a direct effect on the activation, activity, and degradation mechanisms of trypsin. Experiments with purified human anionic and cationic trypsinogen in the absence of Ca^{2+} show markedly reduced autoactivation compared to high Ca^{2+} concentration [81]. Moreover, in cytoplasmic vacuoles developing upon supramaximal cholecystokinin stimulation in acinar cells, Ca^{2+} concentrations decreases rapidly to levels which are much lower than optimal for trypsinogen autoactivation [82]. This mechanism could represent a protective mechanism of the endosomes to prevent damage from premature trypsinogen activation.

CHRONIC AND HEREDITARY PANCREATITIS

Chronic pancreatitis is clinically defined as recurrent bouts of a sterile inflammatory disease characterized by persistent and often progressive and irreversible morphological changes, typically causing pain and permanent impairment of pancreatic function. Chronic pancreatitis histologically represents a transformation of focal necrosis into perilobular and intralobular fibrosis of the parenchyma, pancreatic duct obstruction by pancreatic stones and tissue calcification, and the development of pseudocysts. In the course of the disease, progressive loss of endocrine and exocrine function can be monitored. It should be noted that the clinical distinction between acute and chronic pancreatitis is becoming pathophysiologically ever more blurred, and similar or identical onset mechanisms may play a role. These mechanisms include premature and intracellular activation of digestive proteases. Much of our present knowledge about this process has been generated since a genetic basis for pancreatitis was first reported in 1996. Hereditary pancreatitis represents a genetic disorder associated with mutations in the cationic trypsinogen gene and presents with a disease penetrance of up to 80%. Patients with hereditary pancreatitis suffer from recurrent episodes of pancreatitis which do progress in the majority to chronic pancreatitis. The disease usually begins in early childhood, but onset can vary from infancy to the sixth decade of life [83].

Mutations Within the *PRSS1* Gene

Hereditary pancreatitis is associated with genetic mutations in the cationic trypsinogen gene, suggesting that mutations in a digestive protease (such as trypsin) can cause the disease. Hereditary pancreatitis typically follows an autosomal dominant pattern of inheritance with an 80% disease penetrance. The gene coding for cationic trypsinogen (*PRSS1*) is approximately 3.6 kb in size, is located on chromosome 7, and contains 5 exons. The precursor of cationic trypsinogen is a 247 amino acid protein, the first 15 amino acids of which represent the signal sequence; the next 8 amino acids, the activation peptide; and the remaining 224 amino acids form the backbone and catalytic center of the digestive enzyme. Presently, there are two known mutations within the same codon in the *PRSS1* gene: (i) His-122-trypsinogen shows increased autoactivation, and (ii) Cys-122 trypsinogen has reduced activity. The first of these mutations was discovered in 1996, exactly one century after Chiari proposed his theory on autodigestion of the pancreas as a pathogenic mechanism of pancreatitis. Whitcomb et al. reported a mutation in exon 3 of the cationic trypsinogen gene (*PRSS1*) on chromosome 7 (7q36) [84–86] that was strongly associated with hereditary chronic pancreatitis. This single point mutation (CGC to CAC) causes an arginine to histidine (R to H) substitution at position 22 of the cationic trypsinogen gene (R122H). The amino acid exchange is located in the hydrolysis site of trypsin and can prevent the autodegradation of active trypsin [87] (Figure 21.2). Once trypsin has been activated intracellularly, the R122H mutation interferes with the elimination of active trypsin by autodegradation [88]. This conclusion was derived from *in vitro* data using recombinant R122H mutated trypsinogen. Using the same *E. coli* based system, Sahin-Toth and coworkers showed that the R122H mutation leads to an increase in trypsinogen autoactivation [89]. Therefore, the R122H mutation represents a dual gain of function mutation which facilitates intracellular trypsin activity and results in a higher stability of R122H-trypsin [88,90]. The direct pathogenic role of the R122H mutations for the development of pancreatitis was confirmed by the group of Bar-Sagi. These investigators generated a transgenic mouse in which the expression of the murine *PRSS1* mutant R122H (R122H_mPRSS1) was targeted to pancreatic acinar cells by fusion to the elastase promoter. Pancreata from transgenic mice displayed early-onset acinar cell injury and inflammatory cell infiltration. With progressing age, the transgenic mice developed areas of pancreatic fibrosis and displayed acinar cell dedifferentiation [91,92]. Interestingly, no increased trypsin activity was found in these animals under experimental supramaximal stimulation.

Shortly after the identification of the R122H mutation a second mutation was reported in kindreds with hereditary pancreatitis. The R122C mutation is a single amino acid exchange affecting the same codon as the R122H mutation [93,94]. In contrast to the R122H mutation, the R122C mutation causes a decreased trypsinogen autoactivation, and biochemical studies demonstrated a 60%–70% reduced activation in the Cys-122 trypsinogen mutant induced by either enterokinase or cathepsin B activation. The amino acid exchange of the R122C mutation leads to altered cysteine-disulfide bonds and consequently a misfolded protein structure with reduced catalytic activity. In recent years several other attempts have been made to elucidate the pathophysiological role of trypsinogen in the onset of pancreatitis, but the issue is still a matter of intense research [95].

Since the initial discovery, several other mutations (24 to date) in the trypsinogen gene have been reported, but the R122H mutation is still the most common. In addition to the R122H mutation, five other mutations in different regions of the *PRSS1* gene associated with hereditary pancreatitis have been biochemically characterized: A16V [96], D22G [97], K23R [98], E79K [99], and N29I [100]. It has been suggested that these mutations may have different structural effects on the activation and activity of trypsinogen.

The nucleotide substitution from A-T in exon 2 of the *PRSS1* gene (AAC to ATC) in codon 29 of cationic trypsinogen results in an asparagine to isoleucine amino acid change (N29I). This amino acid substitution affects the protein structure of trypsinogen, and seems to stabilize the enzyme [101]. The N29I mutation does not affect the autoactivation of trypsinogen but impairs enzyme degradation *in vivo* [102,103]. The N29I mutation causes a slightly milder course of hereditary pancreatitis compared to the R122H mutation, the onset of disease occurs somewhat later, and the need for in-hospital treatment is lower [83,104].

The A16V mutation results in an amino acid exchange from alanine to valine in the signal peptide of trypsinogen. This mutation is rare [96] and in contrast to R122H and N29I mutations, which are burdened with an 80% penetrance for chronic pancreatitis, develops only in one out of seven carriers of the A16V mutation. Two other mutations were found in the signal peptide of cationic trypsinogen, D22G and K23R. Both mutations result in an increase of autoactivation of trypsinogen [105] but are resistant to cathepsin B activation [106]. Furthermore, in contrast to wild-type trypsinogen, expression of active trypsin and mutated trypsinogens (D22G, K23R) reduced cell viability of AR4–2J cells [107]. This suggests that mutations in the activation peptide of trypsinogen play an important role in premature protease activation, but the biochemical mechanisms remain unsolved.

The EUROPAC-1 study compared genotype to phenotype characteristics of hereditary pancreatitis patients. This study confirmed the importance of *PRSS1* mutations associated with chronic pancreatitis. In a multilevel proportional hazard model employing data obtained from the European Registry of Hereditary Pancreatitis, 112 families in 14 countries (418 affected individuals) were collected [83]: 58 (52%) families carried the R122H mutation, 24 (21%) carried the N29I mutation, and 5 (4%) carried the A16V mutation, while 2 families had rare mutations, and 21 (19%) had no known *PRSS1* mutation. The median time to the start of symptoms for the R122H mutation is 10 years of age (8 to 12 years of age, 95% confidence interval), 14 years of age for the N29I mutation (11 to 18 years of age, 95% confidence interval), and 14.5 years of age for mutation negative patients (10 to 21 years of age, 95% confidence interval; $P = 0.032$). The cumulative risk at 50 years of age for exocrine pancreas failure was 37.2% (28.5%–45.8%, 95% confidence interval), 47.6% for endocrine failure (37.1%–58.1%, 95% confidence interval), and 17.5% for pancreatic resection for pain (12.2%–22.7%, 95% confidence interval). Time to resection was significantly reduced for females ($P < 0.001$) and those with the N29I mutation ($P = 0.014$). Pancreatic cancer was diagnosed in 26 (6%) of 418 affected patients (Table 21.1).

The remaining *Prss1* mutations are very rare and are for the most part detected only in a single family or a single patient. Mutations like P36R, K92N, or G83E were each found in only one patient with idiopathic chronic pancreatitis [108]. The biochemical consequences of these mutations do not result in increased stability or autoactivation of trypsinogen compared to the R122H or N29I mutations [109]. This observation causes difficulties explaining the underlying pathogenic mechanism for these rare mutations.

Mutations Within the *PRSS2* Gene

The fact that mutations in the *PRSS1* gene encoding for cationic trypsinogen are associated with hereditary pancreatitis could suggest that genetic alterations of the anionic trypsinogen gene (*PRSS2*) could also be associated with chronic pancreatitis. The E79K mutation (related to a G to A mutation at codon 237) reduces autoactivation of cationic trypsinogen by 80%–90% but leads to a 2-fold increase in the activation of anionic trypsinogen, suggesting a potential role of *PRSS2* [99]. A direct link between the development of chronic pancreatitis and mutations in the *PRSS2* gene was not established until 2006 [110]. Genetic analysis of the *PRSS2* gene in 2466 chronic pancreatitis patients and 6459 healthy individuals revealed an increased rate of a rare

Table 21.1 Most Common Mutations Associated with Pancreatitis

Gene	Mutation	Comments	Frequency
PRSS1	R122H	increased autoactivation and decreased autolysis of cationic trypsin	most common mutation (>500)
	R122C	decreased autoactivation of trypsinogen, decreased autolysis of trypsin also decreased trypsin activity	5 affected carriers
	N29I	increased autoactivation of trypsinogen	second most common mutation (>160)
	A16V	increased autoactivation of trypsinogen	25 affected carriers
	D22G	increased autoactivation of trypsinogen	rare, 2 carriers
	K23R	increased autoactivation of trypsinogen	rare, 2 carriers
	E79K	increased activation of anionic trypsinogen decreased autoactivation of trypsinogen	8 affected carriers
SPINK1	N34S	no functional defect reported	
	R65Q	60% loss of protein expression	
	G48E	nearly complete loss of protein expression	
	D50E	nearly complete loss of protein expression	
	Y54H	nearly complete loss of protein expression	
	R67C	nearly complete loss of protein expression	
	L14P and L14R	rapid intracellular degradation of SPINK1	
CTRC	R254W	decreased activity of chymotrypsin C	
	K247_R254del	loss of function of chymotrypsin C	
	A37T	decreased trypsin degradation of chymotrypsin C	
CFTR	various >1000	decreased fluid secretion from acinar cells	

mutation in the anionic trypsinogen gene in control subjects. A variant of codon 191 (G191R) was present in 220/6459 (3.4%) control subjects, but in only 32/2466 (1.3%) affected individuals [111]. Biochemical analysis of the recombinantly expressed G191R protein showed a complete loss of tryptic function after enterokinase or trypsin activation, as well as a rapid autolytic proteolysis of the mutant. This was the first report of a loss of trypsin function (in *PRSS2*) that has a protective effect on the onset of pancreatitis.

Mutations in the Chymotrypsin C Gene

Because trypsin degradation is thought to represent a protective mechanism against pancreatitis, Sahin-Toth and coworkers hypothesized that loss-of-function variants in trypsin-degrading enzymes would increase the risk of pancreatitis. Since they knew that chymotrypsin can degrade all human trypsins and trypsinogen isoforms with high specificity, they sequenced the chymotrypsin C gene (*CTRC*) in a German cohort suffering from idiopathic and hereditary pancreatitis. They detected two variants in the *CTRC* gene in association with hereditary and idiopathic chronic pancreatitis. Mutation of codon 760 (C to T) resulted in an R254W variant form of the protein that occurred with a frequency of 2.1% (19/901) in affected individuals compared to 0.6% (18/2804) in healthy controls. In addition, a deletion mutation (c738_761del24) resulting in a K247–R254del variant protein was found in 1.2% of affected individuals compared to 0.1% in controls [112]. In a confirmative cohort of different ethnic background, the authors detected a third mutation in affected patients with a frequency of 5.6% (4/71) compared to 0% (0/84) in control individuals. This mutation leads to an amino acid exchange at the position 73 (A73T) resulting from a G to A mutation at codon 217. The assumed pathogenic mechanism of *CTRC* mutations is based on lowered enzyme activity in the R254W variant and a total loss of function in the deletion mutation (K247–R254del) and the A73T variant (Figure 21.2). Thus, chymotrypsin C is an enzyme that can counteract the disease-causing effect of trypsin, and loss of function mutations impair its protective role in pancreatitis by letting prematurely activated trypsin escape its degradation.

Mutations in Serine Protease Inhibitor Kazal-Type 1

Shortly after the identification of mutations in the trypsinogen gene in hereditary pancreatitis, another important observation was made by Witt et al. [113]. This group found that mutations in the *SPINK-1* gene (the pancreatic secretory trypsin inhibitor or PSTI, OMIM 167790) can be associated with idiopathic chronic pancreatitis in children. *SPINK-1* mutations can be frequently detected in cohorts of patients who do not have a family history but also have none of the classical risk factors for chronic pancreatitis [114,115]. The most common mutation is found in exon 2 of the *SPINK-1* gene (AAT to AGT), which leads to an asparagine to serine amino acid change (N34S) [114]. Homozygote and heterozygote N34S mutations were detected in 10%–20% of patients with pancreatitis compared to 1%–2% of healthy controls, suggesting that *SPINK-1* is a disease-modifying factor [116–118]. Tropical pancreatitis is a common form of pancreatitis in Africa and Asia. This form of pancreatitis is characterized by abdominal pain, intraductal pancreatic calculi, and diabetes mellitus in young nonalcoholic patients. Tropical pancreatitis is associated with an even higher frequency of N34S mutations in the *SPINK-1* gene, representing up to 30% of affected individuals [119]. Structural modeling of SPINK-1 predicted that the N34S region near the lysine41 residue functions as the trypsin-binding pocket of SPINK-1 [120] and that the N34S mutation changes the structure of the trypsin-binding pocket of SPINK-1, resulting in decreased inhibitory capacity of SPINK-1. In contrast to the computer-modeled prediction, *in vitro* experiments using recombinant N34S *SPINK-1* and wild-type *SPINK-1* demonstrated identical trypsin inhibitory activities [116]. To study the role of SPINK-1 *in vivo*, experimenters generated a knockout mouse mode. The murine *SPINK-3* is the functional homolog to human *SPINK-1*. *SPINK-3*-deficient mice show a perturbed embryonic development of the pancreas and die within 2 weeks after birth [121]. No increase in trypsin activity was observed in *SPINK3-/-* mice. This suggests that other yet unknown disease factors are involved in the antiprotease/protease balance, contributing to the disease phenotype, or that trypsin-SPINK-1 interactions differ between murine and human isoforms. Nevertheless, targeted expression of *PSTI* in pancreas of transgenic mice increased endogenous trypsin inhibitor capacity by 190% ($P < 0.01$) compared to control mice. Cerulein administration to transgenic *PSTI* mice produced significantly reduced histologic severity of pancreatitis. There was no difference in trypsinogen activation between cerulein-treated transgenic and wild-type mice. However, trypsin activity was significantly lower in transgenic mice receiving cerulein compared with nontransgenic mice [122]. Recently, two novel *SPINK-1* variants affecting the secretory signal peptide have been reported. Seven missense mutations occurring within the mature peptide of PSTI associated with chronic pancreatitis were analyzed for their expression levels. The N34S and the P55S mutation neither results in a change of PSTI activity nor in a change of expression. The R65Q mutation involves substitution of a positively charged amino acid by a noncharged amino acid, causing ~60% reduction of protein expression. G48E, D50E, Y54H, and R67C mutations, all of which occur in strictly conserved amino acid residues, cause nearly complete loss of PSTI expression. As the authors had excluded the possibility that the reduced protein expression may have resulted from reduced transcription of unstable mRNA, they concluded that these missense mutations probably cause intracellular retention of their respective mutant proteins [123]. In addition, two novel mutants have been described recently. A disease associated codon 41 T to G alteration was found in two European families with an autosomal dominant inheritance pattern, whereas a codon 36 G to C variant was identified as a frequent alteration in subjects of African descent. L14R and L14P mutations resulted in rapid intracellular degradation of the protein and thereby abolished SPINK-1 secretion, whereas the L12F variant showed no such effect [124]. The discovery of *SPINK-1* mutations in humans provides additional support for a role of active trypsin in the development of pancreatitis. SPINK-1 is believed to form the first line of defense in inhibiting prematurely activated trypsinogen in the pancreas.

CFTR Mutations: A New Cause of Chronic Pancreatitis

Cystic fibrosis is an autosomal-recessive disorder with an estimated incidence of 1 in 2500 individuals and is characterized by pancreatic exocrine insufficiency and chronic pulmonary disease. The extent of pancreatic involvement varies between a complete loss of exocrine and endocrine function, to nearly unimpaired pancreatic function. In 1996, Ravnik-Glavac et al. were the first to report mutations in the cystic fibrosis gene in patients with hereditary chronic pancreatitis [125]. Analysis of larger cohorts revealed recurrent episodes of pancreatitis in 1%–2% of patients with cystic fibrosis and normal exocrine function, and rarely in patients with exocrine insufficiency as well. Compared to an unaffected population, 17%–26% of patients who suffer from idiopathic pancreatitis carry mutations in *CFTR*. Chronic pancreatitis now represents, in addition to chronic lung disease and infertility due to vas deferens aplasia, a third disease entity associated with mutations in the *CFTR* gene. It is important to note that pancreatic exocrine insufficiency in patients with cystic fibrosis is a different disease entity and not to be confused with chronic pancreatitis in the presence of *CFTR* mutations [126–130]. CFTR is a chloride channel, regulated by 3′,5′-cAMP and phosphorylation [131,132] and is essential for the control of epithelial ion transport. The level of executable protein function determines the type and the severity of the disease phenotype. *CFTR* knockout mice show a more severe form of experimental pancreatitis induced by supramaximal cerulein stimulation compared with wild-type animals. The underlying hypothesis of the *CFTR*-related pancreatic injury is a disrupted fluid secretion, which leads to impaired secretion of pancreatic digestive enzymes in response to stimulation [133]. Today more than 1000 mutations within the *CFTR* gene are known, and several of them have been reported in direct association with chronic pancreatitis [126,127]. For healthy subjects who are heterozygous carriers of *CFTR* mutations, the risk of developing pancreatitis is about 2-fold [134].

SUMMARY

Recent advances in cell biological and molecular techniques have permitted investigators to address the intracellular pathophysiology and genetics underlying pancreatic diseases in a much more direct manner than was previously considered possible. Studies that have employed these techniques have changed our knowledge about the disease onset. Pancreatitis has long been considered an autodigestive disorder in which the pancreas is destroyed by its own digestive proteases. Under physiological conditions, pancreatic proteases are synthesized as inactive precursor zymogens and stored by acinar cells in zymogen granules. Independent of the pathological stimulus that triggers the disease, the pathophysiological events that eventually lead to tissue destruction begin within the acinar cells and involve premature intracellular activation of proteases. Cell injury subsequently induces a systemic inflammatory response. Much of our present understanding of the underlying pathogenic mechanism comes from genetic studies, which support a crucial role of trypsinogen activation. Different mutations within the *PRSS1* gene (like the R122H mutation), in genes coding for endogenous inhibitors of active trypsin (such as *SPINK-1*), or in trypsin-degrading enzymes (such as *CTRC*), have all been found in association with different varieties of pancreatitis. Nevertheless, the molecular mechanisms that regulate the balance between proteases and antiproteases, as well as the role of individual digestive enzymes in the proteolytic cascades that precede cell injury, still need to be defined by further experimental studies.

REFERENCES

1. Mayerle J, Hlouschek V, Lerch MM. Current management of acute pancreatitis. *Nat Clin Pract Gastroenterol Hepatol.* 2005;2:473–483.
2. Youngs G. Hormonal control of pancreatic endocrine and exocrine secretion. *Gut.* 1972;13:154–161.
3. Chiari H. Über Selbstverdauung des menschlichen Pankreas. *Z Heilkd.* 1896;17:69–96.
4. Saluja AK, Lerch MM, Phillips PA, et al. Why does pancreatic overstimulation cause pancreatitis? *Annu Rev Physiol.* 2007;69: 249–269.
5. Lerch MM, Adler G. Experimental animal models of acute pancreatitis. *Int J Pancreatol.* 1994;15:159–170.
6. Kruger B, Lerch MM, Tessenow W. Direct detection of premature protease activation in living pancreatic acinar cells. *Lab Invest.* 1998;78:763–764.
7. Bialek R, Willemer S, Arnold R, et al. Evidence of intracellular activation of serine proteases in acute cerulein-induced pancreatitis in rats. *Scand J Gastroenterol.* 1991;26:190–196.
8. Hurley PR, Cook A, Jehanli A, et al. Development of radioimmunoassays for free tetra-L-aspartyl-L-lysine trypsinogen activation peptides (TAP). *J Immunol Methods.* 1988;111:195–203.
9. Neoptolemos JP, Kemppainen EA, Mayer JM, et al. Early prediction of severity in acute pancreatitis by urinary trypsinogen activation peptide: A multicentre study. *Lancet.* 2000;355: 1955–1960.
10. Luthen R, Niederau C, Grendell JH. Intrapancreatic zymogen activation and levels of ATP and glutathione during caerulein pancreatitis in rats. *Am J Physiol.* 1995;268:G592–604.
11. Schmidt J, Fernandez-del Castillo C, Rattner DW, et al. Trypsinogen-activation peptides in experimental rat pancreatitis: Prognostic implications and histopathologic correlates. *Gastroenterology.* 1992;103:1009–1016.
12. Lerch MM, Kruger B, Tessenow W, et al. [Role of protease activation in pathophysiology of acute pancreatitis]. *Langenbecks Arch Chir Suppl Kongressbd.* 1998;115:421–426.

13. Scheele G, Bartelt D, Bieger W. Characterization of human exocrine pancreatic proteins by two-dimensional isoelectric focusing/sodium dodecyl sulfate gel electrophoresis. *Gastroenterology.* 1981;80:461–473.
14. Dartsch H, Kleene R, Kern HF. In vitro condensation-sorting of enzyme proteins isolated from rat pancreatic acinar cells. *Eur J Cell Biol.* 1998;75:211–222.
15. Palade G. Intracellular aspects of the process of protein synthesis. *Science.* 1975;189:347–358.
16. Klumperman J, Kuliawat R, Griffith JM, et al. Mannose 6-phosphate receptors are sorted from immature secretory granules via adaptor protein AP-1, clathrin, and syntaxin 6-positive vesicles. *J Cell Biol.* 1998;141:359–371.
17. Rinderknecht H. Activation of pancreatic zymogens. Normal activation, premature intrapancreatic activation, protective mechanisms against inappropriate activation. *Dig Dis Sci.* 1986; 31:314–321.
18. Arias AE, Boldicke T, Bendayan M. Absence of trypsinogen autoactivation and immunolocalization of pancreatic secretory trypsin inhibitor in acinar cells in vitro. *In Vitro Cell Dev Biol.* 1993; 29A:221–227.
19. Lampel M, Kern HF. Acute interstitial pancreatitis in the rat induced by excessive doses of a pancreatic secretagogue. *Virchows Arch A Pathol Anat Histol.* 1977;373:97–117.
20. Kruger B, Weber IA, Albrecht E, et al. Effect of hyperthermia on premature intracellular trypsinogen activation in the exocrine pancreas. *Biochem Biophys Res Commun.* 2001;282:159–165.
21. Lerch MM, Saluja AK, Dawra R, et al. Acute necrotizing pancreatitis in the opossum: Earliest morphological changes involve acinar cells. *Gastroenterology.* 1992;103: 205–213.
22. Grady T, Mah'Moud M, Otani T, et al. Zymogen proteolysis within the pancreatic acinar cell is associated with cellular injury. *Am J Physiol.* 1998;275:G1010–1017.
23. Niederau C, Grendell JH. Intracellular vacuoles in experimental acute pancreatitis in rats and mice are an acidified compartment. *J Clin Invest.* 1988;81:229–236.
24. Keck T, Balcom JH, Antoniu BA, et al. Regional effects of nafamostat, a novel potent protease and complement inhibitor, on severe necrotizing pancreatitis. *Surgery.* 2001;130:175–181.
25. Tsujino T, Kawabe T, Omata M. Antiproteases in preventing post-ERCP acute pancreatitis. *Jop.* 2007;8:509–517.
26. Niederau C, Crass RA, Silver G, et al. Therapeutic regimens in acute experimental hemorrhagic pancreatitis. Effects of hydration, oxygenation, peritoneal lavage, and a potent protease inhibitor. *Gastroenterology.* 1988;95:1648–1657.
27. Greenbaum LM, Hirshkowitz A, Shoichet I. The activation of trypsinogen by cathepsin B. *J Biol Chem.* 1959;234:2885–2890.
28. Lerch MM, Halangk W, Kruger B. The role of cysteine proteases in intracellular pancreatic serine protease activation. *Adv Exp Med Biol.* 2000;477:403–411.
29. Halangk W, Kruger B, Ruthenburger M, et al. Trypsin activity is not involved in premature, intrapancreatic trypsinogen activation. *Am J Physiol Gastrointest Liver Physiol.* 2002;282: G367–374.
30. Brown WJ, Farquhar MG. Accumulation of coated vesicles bearing mannose 6-phosphate receptors for lysosomal enzymes in the Golgi region of I-cell fibroblasts. *Proc Natl Acad Sci USA.* 1984;81:5135–5139.
31. Willemer S, Bialek R, Adler G. Localization of lysosomal and digestive enzymes in cytoplasmic vacuoles in caerulein-pancreatitis. *Histochemistry.* 1990;94:161–170.
32. Hirano T, Saluja A, Ramarao P, et al. Apical secretion of lysosomal enzymes in rabbit pancreas occurs via a secretagogue regulated pathway and is increased after pancreatic duct obstruction. *J Clin Invest.* 1991;87:865–869.
33. Kukor Z, Mayerle J, Kruger B, et al. Presence of cathepsin B in the human pancreatic secretory pathway and its role in trypsinogen activation during hereditary pancreatitis. *J Biol Chem.* 2002;277:21389–21396.
34. Lerch MM, Saluja AK, Runzi M, et al. Luminal endocytosis and intracellular targeting by acinar cells during early biliary pancreatitis in the opossum. *J Clin Invest.* 1995;95: 2222–2231.
35. Rinderknecht H, Renner IG, Koyama HH. Lysosomal enzymes in pure pancreatic juice from normal healthy volunteers and chronic alcoholics. *Dig Dis Sci.* 1979;24:180–186.
36. Steer ML, Meldolesi J, Figarella C. Pancreatitis. The role of lysosomes. *Dig Dis Sci.* 1984;29:934–938.
37. Watanabe O, Baccino FM, Steer ML, et al. Supramaximal caerulein stimulation and ultrastructure of rat pancreatic acinar cell: Early morphological changes during development of experimental pancreatitis. *Am J Physiol.* 1984;246:G457–467.
38. Saluja A, Hashimoto S, Saluja M, et al. Subcellular redistribution of lysosomal enzymes during caerulein-induced pancreatitis. *Am J Physiol.* 1987;253:G508–516.
39. Hofbauer B, Saluja AK, Lerch MM, et al. Intra-acinar cell activation of trypsinogen during caerulein-induced pancreatitis in rats. *Am J Physiol.* 1998;275:G352–362.
40. Grady T, Saluja A, Kaiser A, et al. Edema and intrapancreatic trypsinogen activation precede glutathione depletion during caerulein pancreatitis. *Am J Physiol.* 1996;271:G20–26.
41. Hirano T, Saluja A, Ramarao P, et al. Effects of chloroquine and methylamine on lysosomal enzyme secretion by rat pancreas. *Am J Physiol.* 1992;262:G439–444.
42. Koike H, Steer ML, Meldolesi J. Pancreatic effects of ethionine: Blockade of exocytosis and appearance of crinophagy and autophagy precede cellular necrosis. *Am J Physiol.* 1982;242:G297–307.
43. Singh VP, Saluja AK, Bhagat L, et al. Phosphatidylinositol 3-kinase-dependent activation of trypsinogen modulates the severity of acute pancreatitis. *J Clin Invest.* 2001;108:1387–1395.
44. Saluja AK, Donovan EA, Yamanaka K, et al. Cerulein-induced in vitro activation of trypsinogen in rat pancreatic acini is mediated by cathepsin B. *Gastroenterology.* 1997;113:304–310.
45. Van Acker GJ, Saluja AK, Bhagat L, et al. Cathepsin B inhibition prevents trypsinogen activation and reduces pancreatitis severity. *Am J Physiol Gastrointest Liver Physiol.* 2002;283:G794–800.
46. Halangk W, Lerch MM, Brandt-Nedelev B, et al. Role of cathepsin B in intracellular trypsinogen activation and the onset of acute pancreatitis. *J Clin Invest.* 2000;106:773–781.
47. Lerch MM, Halangk W. Human pancreatitis and the role of cathepsin B. *Gut.* 2006;55:1228–1230.
48. Szmola R, Kukor Z, Sahin-Toth M. Human mesotrypsin is a unique digestive protease specialized for the degradation of trypsin inhibitors. *J Biol Chem.* 2003;278:48580–48589.
49. Rinderknecht H, Renner IG, Abramson SB, et al. Mesotrypsin: A new inhibitor-resistant protease from a zymogen in human pancreatic tissue and fluid. *Gastroenterology.* 1984; 86:681–692.
50. Sahin-Toth M. Human mesotrypsin defies natural trypsin inhibitors: From passive resistance to active destruction. *Protein Pept Lett.* 2005;12:457–464.
51. Mahurkar S, Idris MM, Reddy DN, et al. Association of cathepsin B gene polymorphisms with tropical calcific pancreatitis. *Gut.* 2006;55:1270–1275.
52. Weiss FU, Behn CO, Simon P, et al. Cathepsin B gene polymorphism Val26 is not associated with idiopathic chronic pancreatitis in European patients. *Gut.* 2007;56:1322–1323.
53. Rinderknecht H, Adham NF, Renner IG, et al. A possible zymogen self-destruct mechanism preventing pancreatic autodigestion. *Int J Pancreatol.* 1988;3:33–44.
54. Szmola R, Sahin-Toth M. Chymotrypsin C (caldecrin) promotes degradation of human cationic trypsin: Identity with Rinderknecht's enzyme Y. *Proc Natl Acad Sci USA.* 2007;104:11227–11232.
55. Nemoda Z, Sahin-Toth M. Chymotrypsin C (caldecrin) stimulates autoactivation of human cationic trypsinogen. *J Biol Chem.* 2006;281:11879–11886.
56. Petersen OH. Ca2+ signalling and Ca2+-activated ion channels in exocrine acinar cells. *Cell Calcium.* 2005;38:171–200.
57. Kasai H, Augustine GJ. Cytosolic Ca2+ gradients triggering unidirectional fluid secretion from exocrine pancreas. *Nature.* 1990;348:735–738.
58. Maruyama Y, Inooka G, Li YX, et al. Agonist-induced localized Ca2+ spikes directly triggering exocytotic secretion in exocrine pancreas. *Embo J.* 1993;12:3017–3022.
59. Park MK, Lomax RB, Tepikin AV, et al. OH. Local uncaging of caged Ca(2+) reveals distribution of Ca(2+)-activated Cl(-) channels in pancreatic acinar cells. *Proc Natl Acad Sci USA.* 2001;98:10948–10953.
60. Maruyama Y, Petersen OH. Delay in granular fusion evoked by repetitive cytosolic Ca2+ spikes in mouse pancreatic acinar cells. *Cell Calcium.* 1994;16:419–430.

61. Matozaki T, Goke B, Tsunoda Y, et al. Two functionally distinct cholecystokinin receptors show different modes of action on Ca2+ mobilization and phospholipid hydrolysis in isolated rat pancreatic acini. Studies using a new cholecystokinin analog, JMV-180. *J Biol Chem.* 1990; 265:6247–6254.
62. Kruger B, Albrecht E, Lerch MM. The role of intracellular calcium signaling in premature protease activation and the onset of pancreatitis. *Am J Pathol.* 2000;157:43–50.
63. Lee MG, Xu X, Zeng W, et al. Polarized expression of Ca2+ channels in pancreatic and salivary gland cells. Correlation with initiation and propagation of [Ca2+]i waves. *J Biol Chem.* 1997;272:15765–15770.
64. Ashby MC, Tepikin AV. Polarized calcium and calmodulin signaling in secretory epithelia. *Physiol Rev.* 2002;82:701–734.
65. Saluja AK, Dawra RK, Lerch MM, et al. CCK-JMV-180, an analog of cholecystokinin, releases intracellular calcium from an inositol trisphosphate-independent pool in rat pancreatic acini. *J Biol Chem.* 1992;267:11202–11207.
66. Yamasaki M, Thomas JM, Churchill GC, et al. Role of NAADP and cADPR in the induction and maintenance of agonist-evoked Ca2+ spiking in mouse pancreatic acinar cells. *Curr Biol.* 2005;15:874–878.
67. Park MK, Ashby MC, Erdemli G, et al. Perinuclear, perigranular and sub-plasmalemmal mitochondria have distinct functions in the regulation of cellular calcium transport. *Embo J.* 2001; 20:1863–1874.
68. McCormack JG, Halestrap AP, Denton RM. Role of calcium ions in regulation of mammalian intramitochondrial metabolism. *Physiol Rev.* 1990;70:391–425.
69. Voronina S, Sukhomlin T, Johnson PR, et al. Correlation of NADH and Ca2+ signals in mouse pancreatic acinar cells. *J Physiol.* 2002;539:41–52.
70. Petersen OH, Sutton R, Criddle DN. Failure of calcium microdomain generation and pathological consequences. *Cell Calcium.* 2006;40:593–600.
71. Ward JB, Petersen OH, Jenkins SA, et al. Is an elevated concentration of acinar cytosolic free ionised calcium the trigger for acute pancreatitis? *Lancet.* 1995;346:1016–1019.
72. Mithofer K, Fernandez-del Castillo C, Frick TW, et al. Acute hypercalcemia causes acute pancreatitis and ectopic trypsinogen activation in the rat. *Gastroenterology.* 1995;109:239–246.
73. Klonowski-Stumpe H, Schreiber R, Grolik M, et al. Effect of oxidative stress on cellular functions and cytosolic free calcium of rat pancreatic acinar cells. *Am J Physiol.* 1997;272: G1489–1498.
74. Nicotera P, Bellomo G, Orrenius S. Calcium-mediated mechanisms in chemically induced cell death. *Annu Rev Pharmacol Toxicol.* 1992;32:449–470.
75. Saluja AK, Bhagat L, Lee HS, et al. Secretagogue-induced digestive enzyme activation and cell injury in rat pancreatic acini. *Am J Physiol.* 1999;276:G835–842.
76. Mooren FC, Turi S, Gunzel D, et al. Calcium-magnesium interactions in pancreatic acinar cells. *Faseb J.* 2001;15:659–672.
77. Mooren F, Hlouschek V, Finkes T, et al. Early changes in pancreatic acinar cell calcium signaling after pancreatic duct obstruction. *J Biol Chem.* 2003;278:9361–9369.
78. Husain SZ, Prasad P, Grant WM, et al. The ryanodine receptor mediates early zymogen activation in pancreatitis. *Proc Natl Acad Sci USA.* 2005;102: 14386–14391.
79. Criddle DN, Sutton R, Petersen OH. Role of Ca2+ in pancreatic cell death induced by alcohol metabolites. *J Gastroenterol Hepatol.* 2006;21(Suppl 3):S14–17.
80. Criddle DN, Murphy J, Fistetto G, et al. Fatty acid ethyl esters cause pancreatic calcium toxicity via inositol trisphosphate receptors and loss of ATP synthesis. *Gastroenterology.* 2006;130: 781–793.
81. Kukor Z, Toth M, Sahin-Toth M. Human anionic trypsinogen: Properties of autocatalytic activation and degradation and implications in pancreatic diseases. *Eur J Biochem.* 2003;270:2047–2058.
82. Sherwood MW, Prior IA, Voronina SG, et al. Activation of trypsinogen in large endocytic vacuoles of pancreatic acinar cells. *Proc Natl Acad Sci USA.* 2007;104:5674–5679.
83. Howes N, Lerch MM, Greenhalf W, et al. Clinical and genetic characteristics of hereditary pancreatitis in Europe. *Clin Gastroenterol Hepatol.* 2004;2:252–261.
84. Whitcomb DC, Preston RA, Aston CE, et al. A gene for hereditary pancreatitis maps to chromosome 7q35. *Gastroenterology.* 1996;110:1975–1980.
85. Whitcomb DC, Gorry MC, Preston RA, et al. Hereditary pancreatitis is caused by a mutation in the cationic trypsinogen gene. *Nat Genet.* 1996;14:141–145.
86. Ellis I, Lerch MM, Whitcomb DC. Genetic testing for hereditary pancreatitis: Guidelines for indications, counselling, consent and privacy issues. *Pancreatology.* 2001;1:405–415.
87. Sahin-Toth M. Human cationic trypsinogen. Role of Asn-21 in zymogen activation and implications in hereditary pancreatitis. *J Biol Chem.* 2000;275:22750–22755.
88. Sahin-Toth M, Graf L, Toth M. Trypsinogen stabilization by mutation Arg117—>His: A unifying pathomechanism for hereditary pancreatitis? *Biochem Biophys Res Commun.* 1999;264:505–508.
89. Sahin-Toth M, Toth M. Gain-of-function mutations associated with hereditary pancreatitis enhance autoactivation of human cationic trypsinogen. *Biochem Biophys Res Commun.* 2000;278:286–289.
90. Kukor Z, Toth M, Pal G, et al. Human cationic trypsinogen. Arg (117) is the reactive site of an inhibitory surface loop that controls spontaneous zymogen activation. *J Biol Chem.* 2002; 277:6111–6117.
91. Archer H, Jura N, Keller J, et al. A mouse model of hereditary pancreatitis generated by transgenic expression of R122H trypsinogen. *Gastroenterology.* 2006;131:1844–1855.
92. Simon P, Weiss FU, Zimmer KP, et al. Spontaneous and sporadic trypsinogen mutations in idiopathic pancreatitis. *JAMA.* 2002;288:2122.
93. Simon P, Weiss FU, Sahin-Toth M, et al. Hereditary pancreatitis caused by a novel PRSS1 mutation (Arg-122 —> Cys) that alters autoactivation and autodegradation of cationic trypsinogen. *J Biol Chem.* 2002;277:5404–5410.
94. Pfutzer R, Myers E, Applebaum-Shapiro S, et al. Novel cationic trypsinogen (PRSS1) N29T and R122C mutations cause autosomal dominant hereditary pancreatitis. *Gut.* 2002;50:271–272.
95. Ruthenburger M, Mayerle J, Lerch MM. Cell biology of pancreatic proteases. *Endocrinol Metab Clin North Am.* 2006;35:313–331.
96. Witt H, Luck W, Becker M. A signal peptide cleavage site mutation in the cationic trypsinogen gene is strongly associated with chronic pancreatitis. *Gastroenterology.* 1999;117:7–10.
97. Teich N, Ockenga J, Hoffmeister A, et al. Chronic pancreatitis associated with an activation peptide mutation that facilitates trypsin activation. *Gastroenterology.* 2000;119:461–465.
98. Ferec C, Raguenes O, Salomon R, et al. Mutations in the cationic trypsinogen gene and evidence for genetic heterogeneity in hereditary pancreatitis. *J Med Genet.* 1999;36:228–232.
99. Teich N, Le Marechal C, Kukor Z, et al. Interaction between trypsinogen isoforms in genetically determined pancreatitis: Mutation E79K in cationic trypsin (PRSS1) causes increased transactivation of anionic trypsinogen (PRSS2). *Hum Mutat.* 2004;23:22–31.
100. Gorry MC, Gabbaizedeh D, Furey W, et al. Mutations in the cationic trypsinogen gene are associated with recurrent acute and chronic pancreatitis. *Gastroenterology.* 1997;113:1063–1068.
101. Sahin-Toth M. Hereditary pancreatitis-associated mutation asn (21) —> ile stabilizes rat trypsinogen in vitro. *J Biol Chem.* 1999;274:29699–29704.
102. Gaboriaud C, Serre L, Guy-Crotte O, et al. Crystal structure of human trypsin 1: Unexpected phosphorylation of Tyr151. *J Mol Biol.* 1996;259:995–1010.
103. Chen JM, Ferec C. Molecular basis of hereditary pancreatitis. *Eur J Hum Genet.* 2000;8:473–479.
104. Nishimori I, Kamakura M, Fujikawa-Adachi K, et al. Mutations in exons 2 and 3 of the cationic trypsinogen gene in Japanese families with hereditary pancreatitis. *Gut.* 1999;44:259–263.
105. Chen JM, Kukor Z, Le Marechal C, et al. Evolution of trypsinogen activation peptides. *Mol Biol Evol.* 2003;20:1767–1777.
106. Teich N, Bodeker H, Keim V. Cathepsin B cleavage of the trypsinogen activation peptide. *BMC Gastroenterol.* 2002;2:16.
107. Gaiser S, Ahler A, Gundling F, et al. Expression of mutated cationic trypsinogen reduces cellular viability in AR4–2J cells. *Biochem Biophys Res Commun.* 2005;334:721–728.
108. Chen JM, Piepoli Bis A, Le Bodic L, et al. Mutational screening of the cationic trypsinogen gene in a large cohort of subjects with idiopathic chronic pancreatitis. *Clin Genet.* 2001;59:189–193.

109. Sahin-Toth M. Biochemical models of hereditary pancreatitis. *Endocrinol Metab Clin North Am.* 2006;35:303–312, ix.
110. Idris MM, Bhaskar S, Reddy DN, et al. Mutations in anionic trypsinogen gene are not associated with tropical calcific pancreatitis. *Gut.* 2005;54:728–729.
111. Witt H, Sahin-Toth M, Landt O, et al. A degradation-sensitive anionic trypsinogen (PRSS2) variant protects against chronic pancreatitis. *Nat Genet.* 2006;38:668–673.
112. Rosendahl J, Witt H, Szmola R, et al. Chymotrypsin C (CTRC) variants that diminish activity or secretion are associated with chronic pancreatitis. *Nat Genet.* 2007.
113. Witt H, Luck W, Hennies HC, et al. Mutations in the gene encoding the serine protease inhibitor, Kazal type 1 are associated with chronic pancreatitis. *Nat Genet.* 2000;25:213–216.
114. Weiss FU, Simon P, Witt H, et al. SPINK1 mutations and phenotypic expression in patients with pancreatitis associated with trypsinogen mutations. *J Med Genet.* 2003;40:e40.
115. Witt H, Simon P, Lerch MM. [Genetic aspects of chronic pancreatitis]. *Dtsch Med Wochenschr.* 2001;126:988–993.
116. Hirota M, Kuwata K, Ohmuraya M, et al. From acute to chronic pancreatitis: The role of mutations in the pancreatic secretory trypsin inhibitor gene. *Jop.* 2003;4:83–88.
117. Threadgold J, Greenhalf W, Ellis I, et al. The N34S mutation of SPINK1 (PSTI) is associated with a familial pattern of idiopathic chronic pancreatitis but does not cause the disease. *Gut.* 2002;50:675–681.
118. Bhatia E, Balasubramanium K, Rajeswari J, et al. Absence of association between SPINK1 trypsin inhibitor mutations and Type 1 or 2 diabetes mellitus in India and Germany. *Diabetologia.* 2003;46:1710–1711.
119. Bhatia E, Choudhuri G, Sikora SS, et al. Tropical calcific pancreatitis: Strong association with SPINK1 trypsin inhibitor mutations. *Gastroenterology.* 2002;123:1020–1025.
120. Pfutzer RH, Barmada MM, Brunskill AP, et al. SPINK1/PSTI polymorphisms act as disease modifiers in familial and idiopathic chronic pancreatitis. *Gastroenterology.* 2000;119:615–623.
121. Ohmuraya M, Hirota M, Araki M, et al. Autophagic cell death of pancreatic acinar cells in serine protease inhibitor Kazal type 3-deficient mice. *Gastroenterology.* 2005;129:696–705.
122. Nathan JD, Romac J, Peng RY, et al. Transgenic expression of pancreatic secretory trypsin inhibitor-I ameliorates secretagogue-induced pancreatitis in mice. *Gastroenterology.* 2005;128: 717–727.
123. Boulling A, Le Marechal C, Trouve P, et al. Functional analysis of pancreatitis-associated missense mutations in the pancreatic secretory trypsin inhibitor (SPINK1) gene. *Eur J Hum Genet.* 2007;15:936–942.
124. Kiraly O, Boulling A, Witt H, et al. Signal peptide variants that impair secretion of pancreatic secretory trypsin inhibitor (SPINK1) cause autosomal dominant hereditary pancreatitis. *Hum Mutat.* 2007;28:469–476.
125. Ravnik-Glavac M, Glavac D, di Sant' Agnese P, et al. Cystic fibrosis gene mutations detected in hereditary pancreatitis. *Pflugers Arch.* 1996;431:R191–192.
126. Cohn JA, Friedman KJ, Noone PG, et al. Relation between mutations of the cystic fibrosis gene and idiopathic pancreatitis. *N Engl J Med.* 1998;339:653–658.
127. Sharer N, Schwarz M, Malone G, et al. Mutations of the cystic fibrosis gene in patients with chronic pancreatitis. *N Engl J Med.* 1998;339:645–652.
128. Rich DP, Anderson MP, Gregory RJ, et al. Expression of cystic fibrosis transmembrane conductance regulator corrects defective chloride channel regulation in cystic fibrosis airway epithelial cells. *Nature.* 1990;347:358–363.
129. Kristidis P, Bozon D, Corey M, et al. Genetic determination of exocrine pancreatic function in cystic fibrosis. *Am J Hum Genet.* 1992;50:1178–1184.
130. Durie PR. Pathophysiology of the pancreas in cystic fibrosis. *Neth J Med.* 1992;41:97–100.
131. Berger HA, Anderson MP, Gregory RJ, et al. Identification and regulation of the cystic fibrosis transmembrane conductance regulator-generated chloride channel. *J Clin Invest.* 1991;88: 1422–1431.
132. Picciotto MR, Cohn JA, Bertuzzi G, et al. Phosphorylation of the cystic fibrosis transmembrane conductance regulator. *J Biol Chem.* 1992;267:12742–12752.
133. Dimagno MJ, Lee SH, Hao Y, et al. A proinflammatory, antiapoptotic phenotype underlies the susceptibility to acute pancreatitis in cystic fibrosis transmembrane regulator (-/-) mice. *Gastroenterology.* 2005;129:665–681.
134. Weiss FU, Simon P, Bogdanova N, et al. Complete cystic fibrosis transmembrane conductance regulator gene sequencing in patients with idiopathic chronic pancreatitis and controls. *Gut.* 2005;54:1456–1460.

Chapter

22

Molecular Basis of Diseases of the Endocrine System

Alan Lap-Yin Pang ▪ Malcolm M. Martin ▪ Arline L.A. Martin ▪ Wai-Yee Chan

INTRODUCTION

In the practice of medicine, it has long been recognized that endocrine disorders were caused by too much or too little hormone. However, the etiology of the hormonal excess or deficiency was not known. The advent of molecular biology has made it possible to elucidate many of the steps involved in hormone synthesis, hormone function, and target response, furthering our understanding of some of the causes of endocrine disorders. The endocrine system is extremely complex and affects all human activities. This chapter will limit its attention to several well-established hormonal systems. There are a large number of genes whose mutations are now known to be the cause of or be associated with endocrine disorders, and we will direct our attention to the more common ones. A number of books and reviews devoted to the molecular genetics of endocrine disorders have been published in recent years [1–21]. Our discussion will build on this literature.

THE PITUITARY GLAND

The pituitary gland is located at the base of the brain. Despite its small size, it is one of the most important organs of the body. The pituitary gland functions as a relay between the hypothalamus and target organs by producing, storing, and releasing hormones that affect different target organs in the regulation of basic physiological functions (such as growth, stress response, reproduction, metabolism, and lactation). Anatomically, the pituitary gland is composed of two compartments. The anterior pituitary, the largest part of the gland, is composed of distinct types of hormone-secreting cells: growth hormone (GH) by the somatotrophs, thyroid-stimulating hormone (TSH) by the thyrotrophs, adrenocorticotrophin (ACTH) by the corticotrophs, follicle-stimulating hormone (FSH) and luteinizing hormone (LH) by the gonadotrophs, prolactin (PR) by the lactotrophs, and melanocyte-stimulating hormone (MSH) by the melanotrophs. The posterior pituitary stores and releases hormones (such as antidiuretic hormone and oxytocin) produced by the hypothalamus.

The ontogenesis of the anterior pituitary gland begins during early embryonic development. In humans, it can be identified by the third week of pregnancy [22–24]. The nascent pituitary (Rathke's pouch) is formed under the influence of morphogenetic factors and signaling molecules expressed in the adjacent ventral diencephalon. The expression of early homeodomain transcription factors triggers the expansion of Rathke's pouch by promoting cell proliferation. As Rathke's pouch expands, a signaling gradient is formed that leads to the activation of specific transcription factors for the terminal differentiation of endocrine cells. The function of the pituitary gland is regulated by the hypothalamus. Secretion of anterior pituitary hormones (except prolactin) is stimulated or suppressed by specific hypothalamic releasing or inhibiting factors, respectively. The release of these factors is regulated by feedback mechanisms from hormones produced by the target organs as exemplified by the hypothalamus-pituitary-thyroid axis shown in Figure 22.1. The same feedback mechanism also acts on the pituitary to fine-tune the production of pituitary hormones. At the molecular level, secretion of a specific pituitary hormone is triggered by binding of the respective hypothalamic releasing hormone to the corresponding membrane receptors on specific hormone-producing cells in the anterior pituitary. The pituitary hormone is released into the bloodstream, and its binding to the cell surface receptors in target organs triggers hormone secretion to carry out the relevant physiological functions. Therefore, malformation of the pituitary gland can be caused directly by intrinsic deficits of the

Molecular Pathology © 2009, Elsevier, Inc. All Rights Reserved.

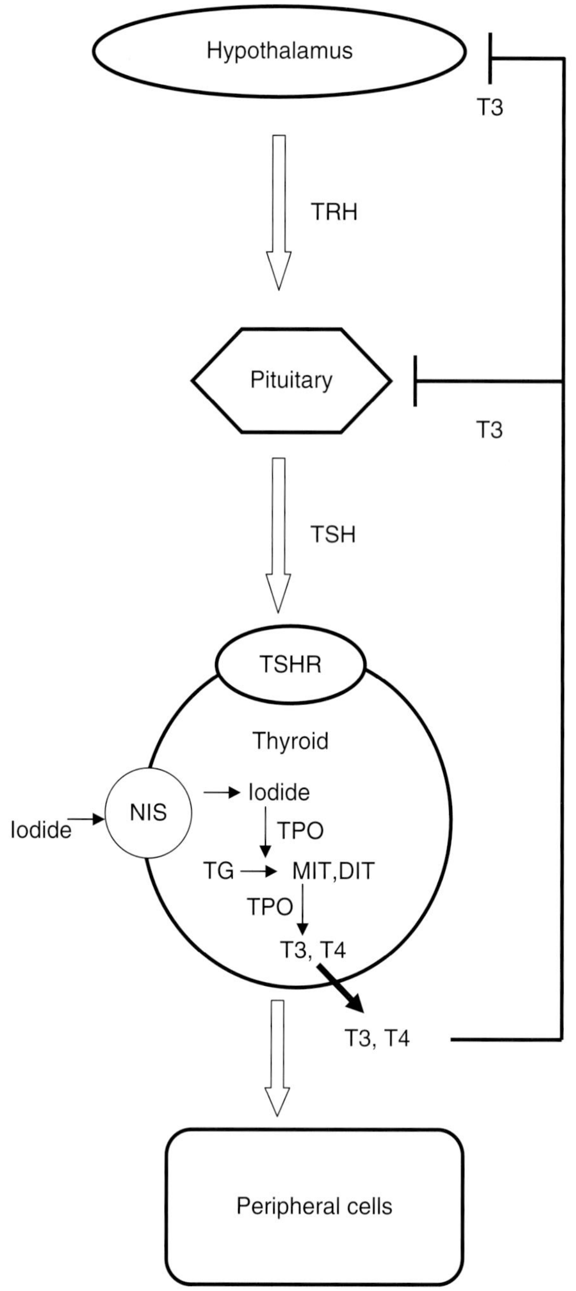

Figure 22.1 The hypothalamus-pituitary-thyroid axis. Hypothalamic thyrotropin-releasing hormone (TRH) stimulates pituitary thyroid-stimulating hormone (TSH) secretion. TSH binds to TSH receptors (TSHR) in the thyroid gland (target organ) to trigger thyroid hormone (T3 and T4) secretion. Thyroid hormones are released into bloodstream and elicit their physiological functions in peripheral cells through receptor-mediated mechanism. Meanwhile, thyroid hormones inhibit further hypothalamic TRH and pituitary TSH secretion through negative feedback. A steady circulating level of thyroid hormones is thus achieved. The example shown reflects the principal mode of regulation of the hypothalamus-pituitary-target organ axis. ⇓ indicates stimulation, ⊢ indicates inhibition. NIS: sodium iodide symporter; TPO: thyroid peroxidase; TG: thyroglobulin; MIT: mono-iodotyrosyl; DIT: di-iodotyrosyl.

gland or indirectly by disturbance in any part of the hypothalamic-pituitary-target organ axis.

Pituitary disorders are caused by either hyposecretion (hypopituitarism) or hypersecretion. Hypopituitarism is the deficiency of a single or multiple pituitary hormones. Target organs usually become atrophic and function abnormally due to the loss of their stimulating factors from the pituitary. Hypopituitarism can be acquired by physical means, including traumatic brain injury, tumors that destroy the pituitary or physically interfere with hormone secretion, vascular lesions, and radiation therapy to the head or neck. In contrast, congenital hypopituitarism is caused mainly by abnormal pituitary development. Mutations in genes encoding transcription factors controlling Rathke's pouch formation, cellular proliferation, and differentiation of cell lineages have been identified. These mutants generally result in combined pituitary hormone deficiency (CPHD), which is characterized by pituitary malformation and a concomitant or sequential loss of multiple anterior pituitary hormones. Isolated pituitary hormone deficiency (IPHD) is caused by mutations in transcription factors controlling the differentiation of a particular anterior pituitary cell line, or mutations affecting the expression or function of individual hormones or their receptors. Pituitary hypersecretion disorders are usually caused by ACTH (Cushing's syndrome) and GH (gigantism or acromegaly).

Combined Pituitary Hormone Deficiency

A number of genes have been shown to cause CPHD. Among these are the homeobox expressed in ES cells 1 (*HESX1*), LIM homeobox protein 3 and 4 (*LHX3/LHX4*), prophet of pit-1 (*PROP1*), POU domain class1, transcription factor 1 (*POU1f1*), and zinc finger protein Gli2 (*GLI2*).

Homeobox Expressed in ES Cells 1

HESX1 is a paired-like homeobox transcriptional repressor essential to the early determination and differentiation of the pituitary. It is one of the earliest markers of the pituitary primordium. *HESX1* is localized on chromosome 3p14.3. Both autosomal recessive and autosomal dominant inheritance have been suggested in HESX1 deficiency. Mutations in *HESX1* are associated with familial cases of septo-optic dysplasia, which is characterized by variable combinations of pituitary abnormalities, midline forebrain defects, and optic nerve hypoplasia. They are also associated with CPHD and IPHD.

Thirteen mutations of *HESX1* have been reported. Five mutations are found in homozygous individuals, including mutations resulting in p.I26T and p.R160C, and an insertional mutation at nucleotide 385 (c.385_386ins315bp), a deletion mutation at nucleotide 449 (c.449_450delAC), and IVS2+2T>C. Eight mutations are found in heterozygotes, including p.Q6H, p.Q117P, p.E149K, p.S170L, p.K176T, p.T181A, g.1684delG, and an insertion at nucleotide 306 (c.306_307insAG). Heterozygotes mostly demonstrate incomplete penetrance and milder phenotypes compared to homozygotes [23–26]. The mutated proteins display diminished DNA binding capacity or impaired recruitment of co-repressor molecules. In two cases (g.1684delG and p.E149K), the ability of HESX1 mutants to interact with PROP1 (an opposing transcription factor to HESX1) is altered. The change in balance between HESX1 on PROP1 is believed to disrupt the normal timing of PROP1-dependent pituitary program, which consequently leads to hypopituitarism.

LIM Homeobox Protein 3 and 4

LHX3 and *LHX4* are LIM-homeodomain transcription factors important to early pituitary development and maintenance of mature anterior pituitary cells. *LHX3* gene is located on chromosome 9q34.3, and *LHX4* is located on 1q25.2. Nine mutations of *LHX3,* including missense mutation, nonsense mutation, small deletion with frameshift, and partial and complete gene deletion, have been identified [27–29]. Inheritance of LHX3 deficiency syndrome follows an autosomal recessive pattern. Accordingly, all individuals with these mutations are homozygous and have CPHD, but in most cases display normal ACTH levels. A hypoplastic pituitary is commonly observed. An enlarged anterior pituitary or microadenoma may be found in some patients. Except for gene deletion, the described mutations cause a change in amino acid residues or removal of the LIM domains and/or homeodomains. When translated, the mutant proteins display compromised DNA binding ability and variable loss of gene transactivation capacity.

Six heterozygous mutations have been reported in *LHX4*. The inheritance pattern of LHX4 deficiency syndrome is autosomal dominant. The first identified mutation is derived from a familial case with complex disease phenotype including CPHD associated with a hypoplastic pituitary. The patients display an intronic G→C transversion at the splice-acceptor site preceding exon 5 [30], which abolishes normal *LHX4* splicing and potentially generates two LHX4 mutant proteins that possess altered DNA binding capacity. The other five mutations are all located in the coding region [31–33]. All mutations are associated with CPHD and/or IPHD, and affected patients show anterior pituitary hypoplasia. The LHX4 mutant proteins show reduced or complete loss of DNA binding and transactivation properties on target gene promoters.

Prophet of Pit-1

PROP1 is a paired-like transcription factor restricted to the developing anterior pituitary. PROP1 is required for the determination of somatotrophs, lactotrophs, and thyrotrophs, and differentiation of gonadotrophs. As reflected by its name, PROP1 expression precedes and is required for *PIT-1* (now called *POU1F1*) expression. In humans, *PROP1* mutations are the leading causes of CPHD, accounting for 30%–50% of familial cases. Nevertheless, the presentation of deficiencies (disease onset and severity) is variable. Some patients also display evolving ACTH/cortisol deficiency. The majority of patients have a hypoplastic or normal anterior pituitary. However, a hyperplastic pituitary with subsequent involution has also been reported.

PROP1 is located on chromosome 5q35.3. Seventeen mutations of *PROP1* have been reported, including 6 missense mutations, 6 nonsense mutations, 4 insertion/deletion frameshift mutations, 1 splicing mutation, and 1 complete gene deletion [23,34–40]. Most mutations were found in homozygous and compound heterozygous patients. However, three of the mutations, namely IVS2–2A>T, c.301_302delAG, and p.R120C, are identified in heterozygotes even though the inheritance of the disorder caused by *PROP1* mutation follows an autosomal recessive mode. Among these mutations, the c.301_302delAG (also known as c.296_297delGA) deletion, which leads to p.S109X, is the most frequently encountered mutation. The single point (IVS2–2A>T) mutation represents the first case of *PROP1* mutation that resides outside the coding region. With the exception of p.W194X mutation, all *PROP1* mutations affect the DNA-binding homeodomain, which lead to a reduced or abolished DNA binding and/or gene transactivation activity of the transcription factor.

POU Domain, Class 1, Transcription Factor 1

POU1F1, also known as *PIT-1,* encodes a POU domain protein essential to the terminal differentiation and expansion of somatotrophs, lactotrophs, and thyrotrophs. It functions as a transcription factor that regulates the transcription of itself and other pituitary hormones and their receptors, including *GH, PRL,* TSH beta subunit (*TSHβ*), TSH receptor (*TSHR*), and growth hormone releasing hormone receptor (*GHRHR*).

POU1F1 is localized on chromosome 3p11. Mutations of *POU1F1* have been shown to be responsible for GH, PRL, and TSH deficiencies. The inheritance and presentation of deficiencies are variable. Both autosomal recessive and autosomal dominant inheritance have been suggested, and both normal and hypoplastic anterior pituitaries have been observed. Twenty-four mutations, including 14 missense mutations, 5 nonsense mutations, 3 insertion/deletion frameshift mutations, 1 splicing mutation, and 1 gene deletion have been found in homozygous or compound heterozygous patients, while 6 distinct mutations (p.P14L, p.P24L, p.K145X, p.Q167K, p.K216E, p.R271W) were found in heterozygotes [3,23,41–44]. Compound heterozygous patients appear to have a more severe disease phenotype. The *POU1F1* mutations are mainly found in the POU-specific and POU-homeodomains, which are important interfaces for high-affinity DNA binding on *GH* and *PRL* genes and protein-protein interaction with other transcription factors. Consequently, DNA-binding and/or gene transactivation capacity is impaired. In some cases the mutant proteins act as dominant inhibitors of gene transcription.

Zinc Finger Protein Gli2

GLI family of transcription factors is implicated as the mediators of Sonic hedgehog (Shh) signals in vertebrates. In humans, Shh signaling is associated with the forebrain defect holoprosencephaly. *GLI2* gene is localized on chromosome 2q14. *GLI2* deficiency phenotype follows an autosomal dominant inheritance pattern. Four heterozygous mutations (c.2274del1, p.W113X, p.R168X, and IVS5+1G>A) were identified in 7 of 390 holoprosencephaly patients who also displayed malformed anterior pituitary and pan-hypopituitarism [45]. All mutations were shown or predicted to lead to premature termination of GLI2 protein.

Growth Hormone

GH1 is a 191 amino acid (~22 kDa) single-chain polypeptide hormone synthesized and secreted by somatotrophs in the anterior pituitary. It is the major player in many physiological processes related to growth and metabolism. For instance, GH1 displays lactogenic, diabetogenic, lipolytic, and protein anabolic effects, and promotes salt and water retention. Thus, its action affects multiple organs. GH1 is considered to affect some tissues directly, but its major effect is mediated by insulin-like growth factor 1 (IGF1). IGF1 is generated predominantly in the liver, and acts through its own receptors to stimulate cell proliferation and maturation in cartilage, bone, muscle, and other tissues.

GH deficiency leads to growth retardation in children and GH insufficiency in adults. The incidence of growth hormone deficiency (isolated or in combination with other pituitary hormone deficiencies) varies from 1 in 4,000 to 1 in 10,000 individuals. Isolated growth hormone deficiency (IGHD) refers to the conditions associated with childhood growth failure owing to the absence of GH action, and not associated with other pituitary hormone deficiencies or organic lesion. Newborns with IGHD are usually of normal length and weight. However, the growth velocity of affected children eventually falls behind progressively with age. Skeletal and dental maturation are delayed in proportion to height, and the appearance is that of a much younger person (dwarfism). Spontaneous puberty is delayed, sometimes well into adulthood, as is epiphyseal fusion. With advancing age, the skin of patients becomes wrinkled and atrophic. While there are many causes of IGHD, familial clustering in some instances has suggested a genetic basis for the disorder.

Based on the presence or absence of GH secretion, clinical characteristics, and inheritance patterns, familial IGHD is classified into four types: IGHD type 1a and 1b, type 2, and type 3. IGHD type 1 is characterized by its autosomal recessive mode of transmission. IGHD type 1a is the most severe form of IGHD, with affected individuals developing severe growth retardation by 6 months of age. Anti-GH antibodies often develop in patients undergoing GH replacement therapy. The phenotype of IGHD type 1b is milder but more variable than type 1a. Patients with IGHD type 1a show no detectable serum GH, whereas low but detectable levels of GH are observed in type 1b patients, and they usually respond well to GH treatment. IGHD type 2 is inherited in an autosomal dominant mode. Affected patients do not always show marked growth retardation. Disease phenotypes vary according to the type of the *GH1* mutation. IGHD type 3 is an X-linked recessively inherited disorder with a highly variable disease phenotype. The precise genetic lesion that causes IGHD type 3 is still unknown.

GHRH-GH-IGF1 Axis

The GHRH-GH-IGF1 axis plays a central role in the regulation of somatic growth. Consequently, genetic lesions affecting any component of the axis usually lead to GHD. The hormones and receptors along the axis represent the hot spots for GHD.

Growth Hormone Releasing Hormone Receptor (GHRHR)

Mutation of GHRH is extremely rare, probably because of its small molecular size (44 amino acids). Mutation of its receptor GHRHR is more prevalent. GHRHR is a G-protein coupled receptor (GPCR) expressed specifically in somatotrophs. It is essential to the synthesis and release of GH in response to GHRH, and the expansion of somatotrophs during the final stage of pituitary development. *GHRHR* is localized on chromosome 7p14. Biallelic mutation in *GHRHR* is a frequent cause of IGHD type 1b. Patients generally display pituitary hypoplasia owing to the lack of somatotroph development. At least 14 different inactivating mutations are found in different parts of the gene, including the promoter, introns, or exons [46–53]. The mutant receptors are unable to bind to their ligands or elicit cyclic AMP (cAMP) responses after GHRH treatment. These mutations are identified in homozygous and compound heterozygous patients in accordance with an autosomal recessive mode of inheritance. However, heterozygous mutant alleles have also been reported. *In vitro* studies show that mutated GHRHR display dominant negative effect on wild-type GHRHR functions. Patients with homozygous mutations tend to display a more severe GHD, but the overall disease phenotype is variable.

Growth Hormone

Human *GH1* resides on chromosome 17q24.2. This chromosomal region contains a cluster of *GH*-like genes arising from gene duplication (namely *GH1*, *CSHP*, *CSH1*, *GH2*, and *CSH2*). The highly homologous nature of the intergenic and coding regions makes the *GH* locus susceptible to gene deletions arising from homologous recombination.

Both autosomal dominant and autosomal recessive inheritance have been suggested for IGHD type 1a. The severity of disease phenotype due to *GH1* mutation varies, with heterozygous mutations giving a milder phenotype than homozygotes. The disorder typically results from deletion or inactivating mutation of *GH1*. At least 5 different large deletions, 37 point mutations and small deletions, and 1 insertion mutation of *GH1* have been reported [1,47,54,55]. A large number of these mutations were found in heterozygous patients. However, the most frequent *GH1* mutations involve the donor splice site of intron 3, which leads to the skipping of exon 3. Mutations resulting in exon 3 skipping were also identified in exon or intron splice enhancer elements. The exon skipping event leads to an internal deletion of amino acid residues 32–71 and the production of a smaller (17.5 kDa) GH. This smaller GH inhibits normal GH secretion in a dominant negative manner. Five missense mutations, namely p.C53S (homozygous mutation), p.R103C, p.D112G, p.D138G, and p.I179M (heterozygous mutations), lead to the production of a biologically inactive GH that can counteract GH action, impair GHR binding and signaling, or prevent productive GHR dimerization [56].

GH Receptor

GHR is localized on chromosome 5p12–13 and encodes a single transmembrane domain glycoprotein belonging to the cytokine receptor superfamily. In mature form, *GHR* encodes 620 amino acid residues, with the first 246 residues (encoded by exons 3 to 7) constituting the extracellular domain for GH binding and receptor dimerization, the middle 24 residues representing the transmembrane domain (exon 8), and the last 350 residues (exon 9 and 10) forming the intracellular domain for GH signaling. GHR is predominantly expressed in liver. The binding of GH triggers receptor dimerization and activation of downstream signaling process leading to IGF-1 production. The absence of GHR activity causes an autosomal recessive disorder called Laron's syndrome. Affected patients display clinical features similar to IGHD but are characterized by a low level of IGF-I and an increased level of GH. Over 50 *GHR* mutations which affect the coding region and splicing of *GHR* transcripts have been identified [47,57–62]. A vast majority of them impact the extracellular domain of GHR. The mutated GHRs display reduced affinity to GH. In other cases, *GHR* mutations lead to defective homodimer formation and thus defective GH binding. The intracellular domain of GHR is lost in certain splice site mutants. The truncated receptor, which carries only the extracellular domain, is unable to anchor to the cell membrane. Although they can bind GH, these GHR mutants fail to activate downstream signaling processes; and they can deplete the binding sites from functional GHRs in a dominant negative manner.

Insulin-Like Growth Factor 1

IGF1, originally called somatomedin C, is a 70-amino acid polypeptide hormone encoded by a gene localized on chromosome 12q22–23. IGF1 is the major mediator of prenatal and postnatal growth. It is produced primarily in liver and serves as an endocrine (as well as paracrine and autocrine) hormone mediating the action of GH in peripheral tissues such as muscle, cartilage, bone, kidney, nerves, skin, lungs, and the liver itself. In the circulation IGF1 exists in a complex with IGF binding protein 3 (IGFBP3) and an acid-labile subunit (ALS).

IGF1 deficiency follows an autosomal recessive mode of transmission. It can be the result of molecular defects that affect any of the upstream components of the GHRH-GH-IGF1 axis and *IGF1* itself. Currently, only four *IGF1* mutations have been described. The first case is a homozygous deletion of 181 nucleotides including exons 4 and 5 of *IGF1*. In patients with this mutation, the translated product of *IGF1* is totally absent. Another case was found in a patient with low circulating IGF1: a homozygous T→A transversion was found at the 3'UTR of *IGF1*. The mutation disrupts the consensus sequence for mRNA polyadenylation so that the E-domain of IGF1 precursor, which is essential to IGF maturation, is altered. In two homozygous missense mutations, the V44M mutant shows a severely reduced affinity to IGF1 receptor. The other mutant, which has a homozygous Arg-to-Gln substitution at the C domain, displays a partial loss of affinity for the receptor [63].

GH Hypersecretion

Excessive production of GH causes IGF1 overproduction leading to gigantism in children and acromegaly in adults. In general, the patients display an abnormally increased somatic growth and distorted body proportions. Affected children show delayed puberty, thickened facial features, muscle weakness, double vision or impaired peripheral vision. In adults, headache and visual impairment are common. Facial features are coarse. Systemic manifestations are seen in a variety of tissues, notably the skin, connective tissue, cartilage, and bone where it causes acral overgrowth. There is frontal bossing, prognathism, large tongue, large hands and feet, as well as thickened fingers and toes. There is generalized organomegaly. Complications are insulin resistance leading to diabetes mellitus and impairment of cardiovascular function that contribute significantly to morbidity and mortality [4].

Almost all GH hypersecretion syndromes are caused by benign pituitary GH-secreting adenomas, either as isolated disorders or associated with other genetic conditions such as multiple endocrine neoplasia (MEN1), McCune Albright's syndrome, neurofibromatosis, or Carney's complex. In rare cases, it can be attributed to a hypothalamic tumor or carcinoid that ectopically secretes GHRH. Activating mutation of the gene encoding G-protein subunit $G_{s\alpha}$ and inactivating mutations of *GHR*, which lead to overactivation of GHRHR and impaired negative feedback on GH production, respectively, are identified in GH-secreting pituitary tumors [64]. A recent report suggests that an inactivating mutation of the gene encoding aryl hydrocarbon receptor-interacting protein may predispose to formation of pituitary adenoma [65].

THE THYROID GLAND

The thyroid, one of the largest endocrine glands, is composed of two types of secretory elements: the follicular and parafollicular cells. The former are responsible for thyroid hormone production, and the latter elaborate calcitonin. The factors that control thyroidal embryogenesis and account for the majority of cases of thyroid agenesis or dysgenesis are presently poorly understood.

Although the thyroid is capable of some independent function, thyroid hormone production is regulated by the hypothalamic-pituitary-thyroid axis (Figure 22.1). Hypothalamic tripeptide thyrotropin-releasing hormone (TRH) stimulates pituitary TSH production. TSH in turn stimulates the thyroid follicular cells to secrete prohormone thyroxine (T4) and its active form triiodothyronine (T3). Since thyroid hormone has a profound influence on metabolic processes, it plays a key role in normal growth and development.

Hypothyroidism

Impaired activity of any component of the hypothalamic-pituitary-thyroid axis will result in decreased thyroid function. Hypothyroidism leads to a slowing of gross metabolic rates. The common causes are faulty embryogenesis, inflammation (most often autoimmune thyroiditis: Hashimoto's thyroiditis), or a reduction in functioning thyroid tissue due to prior surgery, radiation damage, and cancer.

Congenital hypothyroidism is the most common congenital endocrine disorder with an incidence of about 1 in 3500 births. The majority of cases (85%) are nongoitrous and are due to faulty embryogenesis (athyreosis, agenesis, or dysembryogenesis). A small number are goitrous, the result of enzyme defects in hormone biosynthesis (dyshormonogenesis). The molecular causes of athyreosis are not well understood. Clinically congenital hypothyroidism often presents within the first few days of life with constipation, poor feeding, umbilical hernia, decreased activity, or prolonged physiological jaundice.

Childhood onset (juvenile) hypothyroidism is characterized by thyroid enlargement, slow growth, delayed neuromuscular development, delayed tooth eruption, slow speech, dry skin, hoarse voice, coarse facial features, and retarded bone age. Pubertal development may be early or late depending on age of onset and severity of the hormonal deficiency. The majority of cases of juvenile hypothyroidism are due to chronic lymphocytic (Hashimoto's) thyroiditis.

Hypothyroidism in the adult is most often seen in middle age. Females are 8 times more likely to develop this disease than males. It is most common in individuals with a family history of thyroid disease. Hashimoto's thyroiditis is an autoimmune disorder directed against the thyroid gland. The onset of the disease is slow, and it may take months or even years before it is detected. It may occasionally be seen in association with other endocrinopathies and in polyglandular syndromes. Symptoms vary widely depending on the severity and progression of the deficiency. The most common complaints are thyromegaly, increased sensitivity to cold, constipation, pale dry skin, hoarse voice, elevated blood cholesterol, unexplained weight gain, muscle aches and stiffness, weakness, heavier than normal menstrual periods, and frequent depression which worsens with time. Hypothyroidism in the elderly is associated with an increased risk of heart disease and hypertension.

The genes responsible for the development of thyroid follicular cells include thyroid transcription factor 1 (*TIF1*, also known as *TITF1*, *NKX21*, or *T/EBP*), thyroid transcription factor 2 (*TTF2*, also known as *TITF2*, *FOXE1*, or *FKHL15*), paired box transcription factor 8 (*PAX8*), *TSH*, and *TSHR*. Mutation of these genes accounts for about 5% of hypothyroidism patients. *TITF1*, *TITF2*, and *PAX8* mutations give rise to nonsyndromic congenital hypothyroidism, while mutations of *TSH* and *TSHR* cause syndromic congenital hypothyroidism. Most past work had been devoted to the understanding of the molecular genetics of syndromic congenital hypothyroidism. The presence of congenital goiter (which accounts for the remaining 15% of the cases) has been linked to hereditary defects in the enzymatic cascade of thyroid hormone synthesis. Mutations in sodium iodide symporter (*NIS*), pendrin (*PDS*, also known as *SLC264A*), thyroid peroxidase (*TPO*), thyroid oxidase 2 (*THOX2*, also known as dual oxidase DUOX2, *LNOX2*), and thyroglobulin (*TG*) affect organification of iodide and have been linked to congenital hypothyroidism [66].

Etiologically, three types of congenital hypothyroidism can be distinguished: (i) central congenital hypothyroidism including hypothalamic (tertiary) and pituitary (secondary) hypothyroidism, (ii) primary hypothyroidism (impairment of the thyroid gland itself), and (iii) peripheral hypothyroidism (resistance to thyroid hormone).

Central Congenital Hypothyroidism

Central hypothyroidism may be caused by pituitary or hypothalamic diseases leading to deficiency of TSH or TRH. Except for a single case of inactivating mutation of the TSH-releasing hormone (TRH), the majority of cases of central congenital hypothyroidism are caused by *TSH* mutations.

The action of TRH is mediated by its cell surface receptor in the pituitary. TRH receptor mutation is very rare [67]. The majority of cases of congenital isolated TSH deficiency are characterized by low levels of thyroid hormone and low TSH unresponsive to administration of TRH. Prevalence of this condition is estimated to range from 1 in 30,000 to 1 in 50,000 individuals.

Thyroid-Stimulating Hormone. TSH is a heterodimeric glycohormone sharing a common α chain with FSH, LH, and choriogonadotropin (CG). The gene encoding the α chain is on chromosome 6. The gene encoding β chain of TSH is localized on chromosome 1p22. It has 3 exons. Transmission of congenital isolated TSH deficiency follows an autosomal recessive pattern. Patients with homozygous and compound heterozygous mutations in *TSHβ* have been identified in different ethnic populations. Seven different mutations have been identified, with 6 of them located in exon 2 or exon 3 of the gene and the other one affects the donor splice site of intron 2. The most common mutation is a single base deletion in exon 3, resulting in the substitution of cysteine-105 for valine and a frameshift with the appearance of a premature stop codon at position 114 (p.C105fs114X). Cysteine105 is a mutation hot spot in *TSHβ* [5,68,69]. Different mutations have different effects on TSHβ. Nonsense mutations lead to a truncated TSHβ subunit, while other mutations may cause a change in conformation. Thus, mutated TSHβ may be immunologically active but biologically inactive, depending on how significant the mutation is on the structure of the protein. There is also variability in the clinical presentation of patients carrying different mutations.

TSH Receptor–Inactivating Mutations. The effects of TSH on thyroid follicular cells are mediated by TSHR, a cell surface receptor which, together with the receptor for FSH (FSHR) and the receptor for LH and CG (LHR), forms a subfamily of the GPCR family. This subfamily of glycohormone receptors is characterized by the presence of a large amino-terminal extracellular domain (ECD) involved in hormone-binding specificity. The signal of hormone binding is transduced by a transmembrane domain (TMD) containing 7 α-helices. The intracytoplasmic C-terminal tail (ICD) of the receptor is coupled to $G_{s\alpha}$. Binding of hormone to the receptor causes activation of $G_{s\alpha}$ and the adenylyl cyclase cascade resulting in augmented synthesis of cAMP. The receptor is also coupled to G_q and activates the inositol phosphate cascade. *TSHR* is located on chromosome 14q31 and is composed of 10 exons.

A subgroup of patients with familial congenital hypothyroidism due to TSH unresponsiveness has homozygous or compound heterozygous loss-of-function mutations in *TSHR*. A total of 39 mutations of *TSHR* have been identified [67,70–76]. These mutations are distributed throughout the receptor with no obvious mutation hot spot. Monoallelic heterozygous inactivating mutations have been reported. In these cases, the other TSHR allele is probably inactivated by unrecognized mutations in the intronic, 5′-noncoding region, or 3′-noncoding region, because there is no evidence to support a dominant negative effect of an inactivating *TSHR* mutation.

Depending on where the mutation is located, the effects can be defective receptor synthesis due to premature truncation, accelerated mRNA or protein decay, abnormal protein structure, abnormal trafficking, subnormal expression on cell membrane, ineffective ligand binding, and altered signaling. All mutant TSHRs are associated with defective cAMP response to TSH stimulation. There is good correlation between phenotype of resistant patients observed *in vivo* with receptor function measured *in vitro*. These mutated TSHRs result in a wide spectrum of both structural and functional thyroid abnormalities. Thyroid hormone levels in these patients range from hypothyroidism to normal. Variability in the manifestation of TSH abnormalities depends on the degree of TSHR impairment. A good correlation between phenotype and genotype was demonstrated; patients with mutated TSHR with residual activities are less severely affected than those with activity void mutated TSHR.

Thyroid Dyshormonogenesis

Iodide is accrued from the blood into thyroidal cells through sodium/iodide symporter (NIS). It is attached to tyrosyl residue of TG through the action of thyroid peroxidase (TPO) in the presence of hydrogen peroxide (H_2O_2). Two iodinated TG couple with di-iodotyrosyl (DIT) to form T4; one DIT and one TG couple with mono-iodotyrosyl (MIT) to form T3. The coupling reaction is also catalyzed by TPO. TG molecules that contain T4 and T3 upon complete hydrolysis in lysosomes yield the iodothyroxines, T4 and T3. Disorders resulting in congenital primary hypothyroidism have been identified in all major steps in thyroid hormonogenesis.

Sodium Iodide Symporter. A homozygous or compound heterozygous mutation of *NIS* causes an iodide transport defect and an uncommon form of congenital hypothyroidism characterized by a variable degree of hypothyroidism and goiter. It also results in low or no radioiodide uptake by the thyroid and other NIS-expressing organs, and a low saliva-to-plasma iodide ratio. *NIS* is localized on chromosome 19p12–13.2 with 15 exons. Eleven mutations, including missense mutation, nonsense mutation, splicing mutation leading to shifting of reading frame and in-frame deletion, of NIS have been reported. The majority of the patients are homozygous, and about half of them are products of consanguineous marriage. These mutations cause inactivation of NIS resulting in abnormal uptake of iodide by the thyroid [77].

Pendrin. Pendred's syndrome is an autosomal recessive disorder first described in 1896 and characterized by the triad of deafness, goiter, and a partial organification defect. Sensorineural hearing loss is the hallmark of this syndrome. Thyromegaly is the most variable component of the disorder among families and even within the same family, with some individuals developing large goiters, while others present with minimal to no enlargement. While many patients with Pendred's syndrome are euthyroid, others have subclinical or overt hypothyroidism. The variability in clinical expression may be influenced by dietary iodide intake.

In 1996, Pendred's syndrome was mapped to chromosome 7q22–31. The candidate gene, *PDS*, also known as *SLC26A4*, was cloned in the following year and mutations were identified. *PDS* is a member of the solute carrier family 26A. It encodes a 780 amino-acid protein pendrin which is predominantly expressed in thyroid, inner ear, and kidney. More than 100 different mutations of *PDS* have been described (http://www.medicine.uiowa.edu/pendredandbor) in patients with classical Pendred's syndrome. The majority of *PDS* mutations are missense mutations, and some of these mutant proteins appear to be retained in the endoplasmic reticulum. Other forms of mutations include nonsense mutation, splicing mutation, partial duplication, as well as insertion and deletion leading to shifting of reading frame. Individuals with Pendred's syndrome from consanguineous families are usually homozygous for *PDS* mutations, while sporadic cases typically harbor compound heterozygous mutations [6,78–80].

At the thyroid level, the role of pendrin has not been defined. In thyroid follicular cells, pendrin is inserted into the apical membrane and acts as an iodide transporter for apical iodide efflux. Abnormality of pendrin affects iodide transport and may lead to iodide organification defects. This is often mild or absent, leading to the hypothesis that an as-yet-undefined mechanism may compensate for the lack of pendrin.

Thyroid Peroxidase. Thyroid peroxidase (TPO) is a key enzyme in thyroid hormone biosynthesis. It catalyzes both iodination and coupling of iodotyrosine residues in TG. Human *TPO* is located on chromosome 2p25, with 17 exons. The mRNA is about 3 kb encoding a thyroid-specific glycosylated hemoprotein of 110 kDa bound at the apical membrane of thyrocytes. The majority of patients with congenital hypothyroidism have defects in the synthesis or iodination of TG that are attributable to TPO deficiency. Subnormal or absence of TPO activities result in thyroid iodide organification giving rise to goitrous congenital hypothyroidism. Total iodide organification defect (TIOD) as defined by perchlorate discharge test is inherited in an autosomal recessive mode.

The literature of the molecular genetics of TPO is confused by the apparent mixing use of two numbering systems of the *TPO* mRNA, resulting in an ambiguity of the location of identified mutations [81–90]. Some laboratories numbered mutations from the beginning of the coding sequence, while others numbered mutations counting the beginning of the structural gene as nucleotide 1, consequently adding 90 more nucleotides to the former numbering system [91]. Making matters worse, some reports [81] referred to the literature without recognizing the differences between the two numbering systems. Table 22.1 lists all TPO mutations with the first nucleotide of the coding sequence as number 1 and the first amino acid of the encoded polypeptide as residue 1.

TPO mutations are one of the most frequent causes of thyroid dyshormonogenesis. The clinical spectrum of the resulting phenotypes ranges from mild to severe goitrous hypothyroidism. Mutations of *TPO* have also been identified in follicular thyroid carcinoma and adenoma. There are 60 mutations of *TPO* identified in patients with congenital goitrous hypothyroidism (Table 22.1). Many patients with *TPO* mutation are not products of consanguineous marriage. There is no geographical or ethnic prevalence. All forms of mutations have been reported including missense mutations, nonsense mutation, frameshift due to insertion or deletion, and splicing mutations due to deletion of intronic sequences or mutation of nucleotide at intron-exon junctions. Among the 17 exons of *TPO*, mutations have been found in all except the first and the last exon, with exons 8 and 9 sharing the largest number of reported mutations. This is not surprising since exons 8 to 10 encode the catalytic site of the enzyme, and any mutation in these exons will cause enzyme inactivation. Mutations affecting protein folding, such as the c.1429_1449 deletion which removes seven amino acids, will disturb the structure of the enzyme. The fact that TPO is glycosylated suggests that any mutation affecting a potential glycosylation site (N-X-S/T), such as the missense mutation at nucleotide 391 which causes the replacement of Ser131 by Pro and disturbs the glycosylation site at Asn129, may alter TPO activity. Exons 15–17 encode the membrane spanning region and the cytoplasmic tail of TPO. Mutations affecting the membrane spanning region will disturb the insertion of the enzyme into plasma membrane. Frameshift mutations in these exons cause premature truncation of the protein, resulting in improper membrane anchoring. A heterozygous 10 bp deletion at the intron 15-exon 16 boundary identified in a patient presumably causes error in splicing and affects proper membrane insertion of the protein [84]. In another case, a homozygous 10 bp deletion in exon 16 detected in a child of consanguineous parents deletes codon 908 and causes frameshift with the addition of 36 amino acids to the carboxy terminal of the protein. The carboxy terminal of the protein is severely altered. Interestingly, this change appears to affect neither the stability of the protein nor the splicing of the mRNA [86]. Presumably, the observed phenotype is caused by abnormal cellular distribution or cellular trafficking of the protein mutant.

Most cases of congenital hypothyroidism are caused by homozygous or compound heterozygous mutations of *TPO*. Carrier parents of affected individuals have normal thyroid function. There are examples of congenital hypothyroidism due to uniparental disomy for chromosome 2p or monoallelic expression of the mutant allele [90]. In these cases, the genetic defect is confined to the affected patient and is not inherited. TIOD is often associated with inactivating mutations in *TPO*. Mutant *TPO* alleles with residual activities are associated with milder thyroid hormone insufficiency or partial iodide organification defects (PIOD). There are also individuals with PIOD in whom only one mutated allele of *TPO* was identified. In such cases, the other *TPO* allele may have a mutation located in intronic sequences or in a distal region of the gene to affect transcription. Another possibility could be an inactivating mutation in thyroid oxidase 2 (*THOX2*) which has been described in patients with iodide organification defects.

Thyroid Oxidase 2. THOX2 generates H_2O_2 which is used by TPO in the oxidation of iodine prior to iodination of the tyrosyl residues of TG. Human *THOX2* is located on chromosome 15q15.3. There are two highly homologous *THOX* genes, namely *THOX1* and *THOX2*, which are 16kb apart. Even though both *THOX* genes are expressed in thyroid glands, *THOX2* is preferentially expressed in thyroid while *THOX1* is preferentially expressed in airway epithelium and skin. THOX1 is not, but THOX2 is involved in H_2O_2 generation in the thyroid. *THOX2* has 34 exons with the first exon being non-coding. The mRNA encodes a protein of 1548 amino acids, 26 of which comprise a signal peptide that is cleaved off to give the mature protein. The mature protein is glycosylated. The N-terminal segment of THOX2 is homologous to peroxidases, while the C-terminal

Table 22.1 Inactivating Mutations of Thyroid Peroxidase (*TPO*). The First Base of the Coding Sequence Is Assigned Nucleotide Number 1, and the First Amino Acid of the Encoded Polypeptide Is Denoted as Residue Number 1

Exon	Mutation and Nucleotide Position	Amino Acid Position and Effect on Protein
2	c.51dup20bp	fsX92
3	c.157 G > C	p.A53P
3	c.215delA	p.Q72fsX86
4	c.349 G > C	GG/gt to GC/gt
5	c.387delC	p.N129fsX208
5	c.391 T > C	p.S131P
6	c.523 C > T	p.R175X
7	c.718 G > A	p.D240N
8	c.843delC	p.Q252fsX309
8	c.875 C > T	p.S292F
8	c.920 A > C	p.N307T
8	c.976 G > A	p.A326T
8	c.1066 G > A	p.A326T
8	c.1132 G > A	p.E378K
8	c.1152 G > T	p.Q384D
8	c.1159 G > A	p.G387R
8	c.1183_1186insGGCC	p.N396fsX472
8	c.1242 G > T	p.Q384D
8	c.1274 A > G	p.N425S
8	c.1297 G > A	p.V433M
9	c.1335delC	p.A445, fsX472
9	c.1339 A > T	p.I447F
9	c.1357 T > G	p.Y453D
9	c.1373 T > C	p.L458P
9	c.1429_1449del	p.A477_N483del
9	c.1477 G > A	p.G453S
9	c.1496 C > T	p.P499L
9	c.1496delC	p.P499fsX501
9	c.1567 G > A	p.G453S
9	c.1581 G > T	p.W527C
9	c.1597 G > T	p.G533C
9	exon/intron 9 +1G > T	
10	c.1618 C > T	p.R540X
10	c.1690 C > A	p.L564I
10	c.1708 C > T	p.R540X
10	c.1718_1723del	p.D574_L575del
10	c.1768 G > A, AG/gt to AA/gt	
10	intron 10, GA +1, AG/gt to AG/at	
11	c.1955insT	p.F653fsX668
11	c.1978 C > G	p.Q660E
11	c.1993 C > T	p.R655W
11	c.1994 G > A	p.R665Q
11	c.1999 G > A	p.G667S
12	c.2077 C > T	p.R693W
12	c.2153_2154delTT	p.fsF718X
13	c.2243delT	p.fsX831
13	c.2268insT	p.E756fsX756
13	c.2311 G > A	p.G711R
13	c.2386 G > T	p.D796Y
14	c.2395 G > A	p.E799K
14	c.2413delC	p.fsX831
14	c.2415insC	p.fsX879
14	c.2422 T > C	p.C808R
14	c.2422delT	p.C808fsX831
14	c.2512 T > A	p.C838S
15	c.2579 G > A	p.G860R
15	c.2647 C > T	p.P883S
15	Intron 15-exon 16 boundary, g.10delccacaggaca	
16	c.2722delCGGGCCGCAG	p.R908fxX969
16	c.2748 G > A	AG/gt to AA/gt

domain is highly homologous to NADPH oxidases. Thus *THOX2* is also known as dual oxidase 2 (*DUOX2*).

Defects in THOX2 result in a lack or shortage of H_2O_2 and consequently hypothyroidism. A total of 13 mutations in *THOX2* have been identified (Table 22.2) [92–94]. These include missense and nonsense mutations, splicing mutations leading to exon skipping, and small deletions leading to reading frameshift and

Table 22.2 Mutations of Thyroid Oxidase 2 (THOX2) CH Stands for Congenital Hypothyroidism. Numbering System Is the Same as Stated in Table 22.1

Mutation	Genotype	Perchlorate Discharge	CH Phenotype
c.108 G>C, p.Q36H/ c.2895_2898delGTTC, p.S965fsX994	Comp heterozygous	46%, PIOD[b]	Permanent, mild
c.602insG, skip exon 5, p.fsX254	Heterozygous	n.d.[a]	n.d.[a]
c.602insG, skip exon 5, p.fsX254/ c.1516 G>A, p.D506N	Comp heterozygous	n.d.[a]	n.d.[a]
c.1126 C>T, p.R376W/ c.2524 C>T, p.R842X	Comp heterozygous	28%, PIOD[b]	Permanent, mild
c.1253delG, p.G418fsX482/ g.IVS19–2A>C	Comp heterozygous	68%, PIOD[b]	Permanent,
c.1300 C >T, p.R434X	Homozygous	100%, TIOD[c]	Permanent, severe
c.2056 C>T, p.Q686X	Heterozygous	66%, PIOD[b]	Transient, mild
c.2101 C>T, p.R701X	Heterozygous	41%, PIOD[b]	Transient, mild
c.2895_2898delGTTC,p.S965fsX994	Heterozygous	40% PIOD[b]	Transient, mild
		20%–63%	n.d.[a]
c.3329 G>A, p.R1110Q	Homozygous	72.8%, PIOD[b]	Permanent, mild
p.Q1026X	Heterozygous	20%–63%, PIOD[b] n.d.[a]	

[a]n.d.: Not determined

[b]Partial iodide organification defect as defined by the perchlorate discharge test.

[c]Total iodide organification defect as defined by the perchlorate discharge test.

premature truncation of the encoded protein. In one patient, genomic DNA sequencing reveals a G-insertion at position 602 in exon 5. This insertion predicts shifting of the reading frame producing a stop at codon 300. However, sequencing of the cDNA shows that in the mutant allele, exon 5 is skipped. This abnormal splicing leads to a shift of the reading frame beginning at codon 172 and producing a premature stop codon at position 254 [92].

All *THOX2* mutations were identified as congenital hypothyroidism in childhood. The exception is the p.R1110Q mutation which was found in a homozygous adult woman whose thyroid function remained almost normal until the age of 44 [93]. Homozygous and compound heterozygous inactivating mutations in *THOX2* lead to complete disruption of thyroid hormone synthesis and are associated with severe and permanent congenital hypothyroidism. Compound heterozygotes with one or both mutant alleles having residual activity have permanent but mild (subclinical) hypothyroidism, while heterozygotes are associated with milder, transient hypothyroidism, as illustrated in Table 22.2 [93,94]. It is unusual that a genetic defect results in a transient phenotype at a younger age. This can be due to a changing requirement of thyroid hormone with age.

Thyroglobulin. TG is the most abundant protein in the thyroid. It serves the storage of iodine. Some of its tyrosine residues are iodinated by the action of TPO, and the iodinated tyrosines are coupled to form T3 and T4. TG is produced by thyroid follicular cells and secreted into the follicular lumen. Mutations of gene encoding factors responsible for the development and growth of thyroid follicular cells, such as *TTF1, TTF2, PAX8,* and *TSHR,* cause dysgenetic congenital hypothyroidism [5]. Patients present with a congenital goiter, hypothyroidism, high iodide I [131] uptake, normal organification of iodide, elevated serum TSH with simultaneous low or normal serum T4 and T3, and low serum TG concentration in relation to the degree of TSH stimulation [7]. A rarer cause of congenial goitrous hypothyroidism is mutation of *TG* resulting in structural defects of the protein.

TG is encoded by a gene which spans 270 kb on chromosome 8q24.2–24.3, composed of 48 exons, and encodes a protein of 2768 amino acids of which the N-terminal 19-amino acids constitute a signal peptide. So far 38 inactivating mutations in human *TG* have been reported, with 22 being missense mutations, 8 splice site mutations, 5 nonsense mutations, 2 single nucleotide deletions, and 1 single nucleotide insertion resulting in shifting of the reading frame and premature truncation of the protein [7,95]. These mutations are found in exons and introns. Four mutations are the most frequently reported (p.R277X, p.C1058R, p.C1245R, and p.C1977S). Most mutations alter the folding, stability, or intracellular trafficking of the mature protein. Interestingly, some nonsense mutations, such as p.R277X, cause the production of a smaller peptide which is sufficient for the synthesis of T$_4$ in the N-terminal domain. This results in a milder phenotype.

TG deficiency is transmitted in an autosomal recessive mode. Affected individuals are either compound heterozygous or homozygous. The patient phenotypes range from mild to severe goitrous hypothyroidism. No clear phenotype-genotype correlation has been observed.

Resistance to Thyroid Hormone

Resistance to thyroid hormone is characterized by reduced clinical and biochemical manifestations of thyroid hormone action relative to the circulating hormone levels. Patients have high circulating free thyroid hormones, reduced target tissue responsiveness, and an inappropriately normal or elevated TSH [5,8].

The thyroid hormones, T3 and T4, exert their target organ effects via nuclear thyroid hormone receptors (TR). After entering target cells, T4 undergoes deiodination to T3, which enters the nucleus and binds to TR. Upon T3 binding, the conformation of TR changes, causing the release of co-repressors and the recruitment of co-activators resulting in the activation or repression of target gene transcription.

Thyroid Hormone Receptor

There are two TR genes, *TRα* and *TRβ*, located on chromosomes 17 and 3, respectively. Both genes produce two isoforms, *TRα1* and *TRα2* by alternative splicing, and *TRβ1* and *TRβ2* by utilization of different transcription start sites. TR binds as a monomer, or with greater affinity as homodimers or heterodimers, to thyroid hormone response elements (TREs) in target genes to regulate gene expression. The TR protein has a central DNA binding domain; the carboxy-terminal portion of TR contains the T3-ligand binding domain, the transactivation, and the dimerization domains. Unlike the other TRs, TRα2 does not bind thyroid hormone. TRs are expressed at different levels in different tissues and at different stages of development. The biological effects of different TRs are compensatory to some degree. However, some thyroid hormone effects are TR isoform specific.

Thyroid hormone resistance is associated with heterozygous autosomal dominant mutations in the *TRβ*. No *TRα* mutation has been reported, and 10% of families with thyroid hormone hyposensitivity do not have *TRβ* mutation. One hundred twenty-seven different mutations of *TRβ* have been identified in RTH families [8,96–98]. These mutations are located in the ligand binding domain and its adjacent hinge domain that encompass exons 7–10. The vast majority of these mutations cluster within three CpG-rich hot spots, namely amino acids 234–282 (hinge domain), amino acids 310–353, and amino acids 429–460 (ligand-binding domain) [99]. Many of the mutations are found in different families, and the most common of all *TRβ* mutations is p.R338W. Most mutated receptors display reduced T3 binding or abnormal interaction with one of the cofactors involved in thyroid hormone action. Somatic *TRβ* mutations have been identified in TSH-producing pituitary adenomas.

Dimerization of TRs results in dominant negative effects of mutant TRβs. Besides impairing gene transactivation, the mutant receptors also interfere with the function of wild-type receptor by dimerizing with it, resulting in resistance to thyroid hormone. On the other hand, individuals expressing a single wild-type *TRβ* allele are normal despite the fact they lost the other allele due to deletion. Mutations in the hinge domain of TRβ do not interfere with T3 binding and translocation into the nucleus but cause impairment of transactivation function and resistance to thyroid hormone. Differences in the degree of hormonal resistance in different tissues are due to the absolute and relative levels of TRβ and TRα expression.

Thyroid Hormone Cell Transporter

Thyroid hormone action requires transport of the hormone from the extracellular compartments across the cell membrane, which is mediated by the protein encoded by monocarboxylate transporter gene 8 (*MCT8*). Clinical features of patients with *MCT8* mutations are those of Allan-Herndon-Dudley syndrome (AHDS) [9,100,101]. Affected males show a homogeneous neurological phenotype with episodic involuntary movements and absence of development of speech. These patients have unusual combination of high concentration of active thyroid hormone and low level of an inactive metabolite (reverse T3). The severity of the disease is variable among families. In a few families, psychomotor and speech development are less impaired. No clear correlation of phenotype with genotype has been found.

Human *MCT8* is located on chromosome Xq13.2. It has 6 exons. It belongs to the family of genes officially named *SLC16*, the products of which catalyze the proton-linked transport of monocarboxylates, such as lactate, pyruvate, and ketone bodies. *MCT8* is also known as *SLC16A2* (solute carrier family 16, member 2). *MCT8* encodes two putative proteins of 613 and 539 amino acids, respectively, by translation from two in-frame start sites. It is unknown whether there are two human MCT8 proteins expressed *in vivo* [9,101]. MCT8 contains 12 transmembrane domains with both the amino-terminal and carboxyl-terminal peptides located within the cell. It is expressed in a number of tissues, in particular liver and heart. In the pituitary, it is expressed in folliculostellate cells rather than TSH-producing cells.

Twenty-two mutations of *MCT8* have been identified [9,102–104]. They are distributed throughout the coding region of the gene without any obvious mutation hot spot. Mutations of *MCT8* are variable, ranging from missense and nonsense mutations to deletion of a single base-pair, a triplet codon, an entire exon, and several exons. There is also insertion of a triplet codon as well as insertion of a single nucleotide resulting in shifting of reading frame and production of truncated protein. Besides these mutations, there is a family with a missense mutation affecting the putative translational start codon (c.1 A>T, p.M1L). This mutation is present in the patient as well as his healthy brother whose T3 level is normal, indicating that it is not pathogenic. This finding implies that another methionine at amino acid position 75 can serve as the translational start of a functional MCT8 protein [102]. In another family, a single nucleotide deletion in the second to last codon results in a frameshift and bypassing of the natural stop codon until another stop is encountered 195 nucleotide downstream. This mutation leads to an extension of the MCT8 protein by 65 amino acids [104]. In a female patient with full-blown AHDS clinical features, *MCT8* was disrupted at the X-breakpoint of a *de novo* translocation t(X:9)(q13.2;p24). A complete loss of MCT8 expression was observed in cultured fibroblasts from the patient [102].

The majority of the mutations, including a number of missense mutations, result in complete loss of

MCT8 transporter function, which can be due to reduced protein expression, impaired trafficking to plasma membrane, or reduced substrate affinity. Missense *MCT8* mutations (such as p.S194F, p.V235M, p.R271H, p.L434W, and p.L568P) show residual activities when compared with wild-type MCT8 protein. Individuals with these mutations may have a milder phenotype. There is no clearcut phenotype-genotype correlation with *MCT8* mutations. The only consistent feature among mutated MCT8 is elevated level of serum T3 [105].

Familial Nonautoimmune Hyperthyroidism

Patients with a family history of thyrotoxicosis with early disease onset typically have thyromegaly, absence of typical signs of autoimmune hyperthyroidism and recurrence after medical treatment. A number of germline constitutive-activating mutations of the *TSHR* have been identified in patients with nonautoimmune hyperthyroidism (Table 22.3) [10,70,106–109]. The inheritance of the trait follows an autosomal dominant pattern. The majority of these mutations are located in exon 10 of *TSHR*, encoding the transmembrane and intracellular domains. So far only two amino acid residues with 3 different mutations (p.R183K, p.S281N, and p.S281I) which occur outside exon 10 and in exons encoding the extracellular domain have been identified. The majority of the mutations are found in the sixth and seventh transmembrane (TM) helices. In approximately 70% of the activating mutations, a purine is exchanged for a pyrimidine nucleotide, in particular, guanine for cytosine or adenosine for thymidine [10,70]. Aside from germline activating mutations, a number of sporadic mutations have also been identified in patients with nonautoimmune hyperthyroidism (Table 22.3). Germline *TSHR* mutations usually present later in life with milder clinical manifestation, while sporadic *TSHR* mutations cause severe thyrotoxicosis and goiter with neonatal or infancy onset. Missense mutations in one codon resulting in different amino acid exchanges have been observed for several gain-of-function mutations, including codons 281, 505, 568, 597, 631. Mutation hot spots include p.Ala623 and p.Phe631 [10,70,106]. Biological activities of the mutant TSHRs differ in terms of adenylyl cyclase and inositol phosphate pathway activation. Most mutations cause only constitutive activation of the cAMP pathway, whereas a few mutants activate both cAMP and phospholipase C cascades [107].

The expressivity of the activating mutations may be affected by environmental factors such as iodine uptake. Even among cases with the same mutation of *TSHR*, phenotypic expression can be quite variable with respect to the age of onset and severity of hyperthyroidism and thyroid growth [107,108,110]. At least in one report, there is apparent anticipation of disease onset across generations, presumably due to increased iodine supplementation in the area where the patient family resides [110].

Somatic activating mutations of the TSHR have been identified as the cause of solitary toxic adenomas and multinodular goiter and thyroid carcinomas, though less frequently. About 80% of toxic thyroid adenomas (hot nodules) exhibit *TSHR* gain-of-function mutations. An overlap of somatic and germline phenotype has only been observed in 3 (p.M453T, p.L629F, and p.P639S) of 23 known germline activating mutations. This has been ascribed to the fact that hereditary mutations cause a less severely affected phenotype, with a marginal effect on reproductive fitness. Among the 10 sporadic mutations, only 2 occur as somatic mutations in toxic adenomas.

Table 22.3 Activating Mutations of *TSHR* in Familial Nonautoimmune Hyperthyroidism. Numbering System Is the Same as Stated in Table 22.1

Amino Acid Position	Germline	Sporadic	Somatic
p.R183K	X		
p.S281N	X		
p.S281I		X	
P.N372T	X		
p.A428V		X	
p.G431S	X		
p.M453T	X		X
p.M463V	X		
p.A485V	X		
p.S505R	X		
p.S505N	X	X	
p.V509A	X		
p.L512Q	X	X	
p.I568V	X		
p.I568T		X	X
p.V597L		X	
p.V597F	X	X	
p.D617Y	X		
p.A623V	X		
p.M626I	X		
p.L629F	X		X
p.F631L		X	X
p.F631S	X		
p.T632I		X	
p.M637R	X		
p.P639S	X		X
p.N650Y	X		
p.N670S	X		
p.C672Y	X		

THE PARATHYROID GLAND

Histologically, the parathyroid glands comprise two types of cells: chief cells and oxyphil cells; each type may appear in clusters or occasionally mixed. Chief cells have a pale staining cytoplasm with centrally placed nuclei. The cytoplasm contains elongated mitochondria and two sets of granules: secretory granules that take a silver stain and others that stain for glycogen. The oxyphil cells are fewer in number, larger than the chief cells, and their cytoplasm has more mitochondria. The oxyphil cells do not appear until after puberty and their function is unclear.

Parathyroid hormone (PTH) is secreted by the chief cells of the parathyroid glands. It is encoded by a gene located on the short arm of chromosome 11. It has 3 exons encoding a mature peptide of 84 amino acids and a signal peptide of 25 amino acids. The hormone has two main targets: bones and kidneys. In bone, PTH activates lysosomal enzymes in osteoclasts to mobilize calcium; in kidneys it promotes calcium resorption and phosphate and bicarbonate excretion by activating the cAMP signaling pathway. The synthesis and secretion of PTH is regulated by serum ionized calcium concentration.

Calcium Homeostasis

The level of calcium in the blood is kept within a very narrow range between 2.2 and 2.55 mM and is present in three forms: (i) bound, (ii) complexed, and (iii) free. About 30% is bound to protein, mainly albumin; 10% is chelated, mostly to citrate; and the remaining 60% is free or ionized. In humans, the regulation of the ionized fraction of serum calcium depends on the interaction of PTH and the active form of vitamin D-1,25-dihydroxycholecalciferol [1,25$(OH)_2$D].

Vitamin D is a prohormone which requires hydroxylation for its conversion to the active form. Vitamin D from food (ergocalciferol, D_2) is absorbed in the proximal gastrointestinal tract and is also produced in the skin by exposure to ultraviolet light (cholecalciferol, $D_{3.}$). It is carried by a vitamin-D binding protein to the liver where it is hydroxylated to form 25-hydroxyvitamin D. The latter equilibrates with a large stable pool, and the serum concentration of 25-hydroxyvitamin D is the best clinical measure of nutritional vitamin-D status. The 25-hydroxy-Vitamin D is then further hydroxylated in kidneys to 1,25$(OH)_2$D under the influence of PTH to become the physiologically active hormone.

Decreasing serum ionized calcium, acting via a transmembrane G-protein coupled calcium-sensing receptor, stimulates PTH secretion. PTH in turn acts on osteoclasts to release calcium from bones and inhibits calcium excretion from kidneys. PTH, hypocalcemia, and hypophosphatemia furthermore augment the formation of 1,25$(OH)_2$D which stimulates absorption of calcium from the gut and aids PTH in mobilizing calcium from bone, while at the same time inhibiting 24-hydroxylation. The reverse occurs when calcium levels rise, and PTH secretion is inhibited. Whereas 24-hydroxylation is induced, 21-hydroxylation is inhibited, resulting in a shift to the inactive metabolite of vitamin D. Movement of calcium from bone and gut into the extracellular fluid compartment ceases, so does renal reabsorption, thereby restoring serum calcium to its normal physiological concentration. Any disturbance in this regulation may result in hypocalcemia or hypercalcemia. Symptoms and signs may vary from mild, asymptomatic incidental findings to serious life-threatening disorders depending on the duration, severity, and rapidity of onset. Hypocalcemia is the result either of increased loss of calcium from the circulation or insufficient entry of calcium into the circulation; and hypercalcemia, the reverse. Since PTH is the main player in this scenario, its influence is central.

Hypoparathyroidism

Hypoparathyroidism is characterized by levels of PTH insufficient to maintain normal serum calcium concentration. The clinical manifestations of hypoparathyroidism are those of hypocalcemia with increased neuromuscular irritability characterized by numbness, tingling, and paresthesia in the fingertips in mild cases. Facial twitching, tetany, and finally convulsions may occur in more severe cases. Alkalosis, hypokalemia, and hypomagnesemia aggravate the symptoms of hypocalcemia, whereas acidosis ameliorates them.

The most common cause of hypoparathyroidism is injury to the parathyroid glands during head and neck surgery or due to autoimmunity (either isolated or as a polyglandular syndrome). Familial forms of congenital and acquired hypoparathyroidism are relatively rare. Both familial isolated hypoparathyroidism and familial hypoparathyroidism accompanied by abnormalities in multiple organ systems have been described.

Familial Isolated Hypoparathyroidism

Several forms of familial isolated hypoparathyroidism with autosomal dominant, autosomal recessive, and X-linked recessive inheritance have been described [11]. Mutations affecting the signal peptide of pre-pro-PTH have been reported in families with hypoparathyroidism following an autosomal recessive or autosomal dominant inheritance pattern. A patient with a single base substitution resulting in the replacement of Cys18 by Arg, which affects the processing of the mutant pre-pro-PTH, manifested autosomal dominant hypoparathyroidism. In another family with autosomal recessive hypoparathyroidism, a single base substitution gave rise to the replacement of Ser23 by Pro. The mutant pre-pro-PTH cannot be cleaved, resulting in the degradation of mutant peptide in endoplasmic reticulum. A third mutation that occurs at the exon 2-intron 2 boundary results in the skipping of exon 2 and splicing of exon 1 to exon 3. This causes the loss of the initiation codon and signal peptide, and gives rise to autosomal recessive isolated hypoparathyroidism.

Glial cells missing (GCM) belongs to a small family of key regulators of parathyroid gland development. There are two forms: *GCMA* and *GCMB*. The mouse homolog of *GCMB*, Gcm2, is expressed exclusively in parathyroid glands. Five different homozygous mutations of *GCMB* (two missense mutations, p.R47L and p.G63S; an intragenic microdeletion removing exons 1–4; and two single nucleotide deletions leading to frameshift, c.1389delT and c.1399delC) have been identified in patients with isolated hypoparathyroidism. These mutations either disrupt the DNA binding domain or remove the transactivation domain of GCMB, leading to inactivation of the protein. The heterozygous carriers of these mutations are asymptomatic. Thus, this disorder follows an autosomal recessive mode of inheritance [11,111].

Aside from mutations of *PTH* and *GCMB*, X-linked recessive hypoparathyroidism due to mutation of a gene localized to chromosome Xq26-q27 has been reported in two multigenerational kindreds. The candidate gene has not been cloned yet.

Familial Hypoparathyroidism as Part of a Complex Congenital Defect

Hypoparathyroidism may be part of a polyglandular autoimmune disorder or associated with multiple organ abnormalities. Because of the involvement of multiple organs in these disorders, it is likely the abnormalities are not intrinsic to the parathyroid gland but consequential to other abnormalities. Hypoparathyroidism occurs in HAM syndrome (Hypoparathyroidism, Addison's Disease, Mucocutaneous Candidiasis, also known as Polyglandular Failure Type 1 syndrome or Autoimmune Polyendocrinopathy-Candidiasis-Ectodermal Dystrophy (APECED). This syndrome is caused by mutations of *AIRE1* (autoimmune regulator type 1) [11]. Cells destined to become parathyroid glands emerge from the third and fourth pharyngeal pouches. Failure of development of the derivatives of these two pouches results in absence or hypoplasia of the parathyroids and thymus and gives rise to DiGeorge syndrome (DGS). Most cases of DGS are sporadic, but autosomal dominant inheritance of DGS has been observed. The vast majority of cases display imbalanced translocation or microdeletion of 22q11.2 [11,112]. So far, there is no consensus on the gene(s) whose mutation(s) is(are) responsible for DGS. The broad spectrum of abnormalities in DGS patients and the large genomic region in the deletion suggest the possibility of involvement of multiple genes. Similarly, abnormality of *PTH* is excluded in familial Hypoparathyroidism, sensorineural Deafness, and Renal dysplasia (HDR syndrome), which is inherited in an autosomal dominant mode [11]. Deletion analyses of HDR patients indicate the absence of *GATA3* is responsible for HDR. GATA3 is a member of the GATA-binding family of transcription factors, encoded by a gene with 6 exons located on chromosome 10q14. The protein has 44 amino acids with two transactivation domains and two zinc finger domains. GATA3 is expressed in the developing parathyroid gland, inner ear, kidney, thymus and central nervous system. Thus, *GATA3* mutations will affect parathyroid development. So far, 20 mutations of *GATA3* have been identified [113,114], including 5 missense mutations, 1 in-frame deletion of 4 codons, 3 nonsense mutations, 9 insertions or deletions leading to frameshifts generating premature stop codons, 1 donor splice site mutation, and whole gene deletion. These mutations affect DNA-binding, mRNA and protein stability, or reduced protein synthesis. The pathogenic mechanism responsible for HDR syndrome is believed to be GATA3 haploinsufficiency. Other diseases with hypoparathyroidism but with no genetic abnormality of *PTH* include several mitochondrial disorders, Kenney-Caffey and Sanjad-Sakati syndrome, Barakat syndrome, and Blomstrand's disease [11].

Hyperparathyroidism

Primary hyperparathyroidism is a genetically heterogeneous disorder, and usually occurs as a sporadic disorder. It may occur at any age, but is most common in the 60+ age group and more common in females. It results when PTH is secreted in excess due to a benign parathyroid adenoma, parathyroid hyperplasia, multiple endocrine neoplasia (MEN1 or MEN2A), or cancer. An elevated serum calcium together with low serum phosphorus is suggestive of hyperparathyroidism and becomes diagnostic when supported by increased urinary calcium excretion and elevated PTH level.

Familial Isolated Hyperparathyroidism

Familial isolated hyperparathyroidism (FIHP) is a rare disorder characterized by single or multiple glandular parathyroid lesions without other syndromes or tumors. FIHP has been reported in 16 familial hyperparathyroidism patients from 14 families [12]. The mode of transmission is autosomal dominant [115]. Most FIHP kindreds have unknown genetic background. Studies of 36 kindreds reveal that a few families are associated with multiple endocrine neoplasia gene 1 (*MEN1*), with inactivating calcium-sensing receptor (*CaSR*) mutations, or DNA markers near the Hyperparathyroidism-Jaw Tumor (HPT-JT) syndrome locus at 1q25-q32.

Multiple Endocrine Neoplasia Type 1

Multiple Endocrine Neoplasia Type 1 (MEN1) is an autosomal dominant disorder with a high degree of penetrance. Over 95% of patients develop clinical manifestations by the fifth decade. The most common feature of MEN1 is parathyroid tumors which occur in about 95% of patients. Tumors of the pancreas and anterior pituitary also occur in a significant, albeit smaller percentage of patients. MEN1 is caused by germline mutations of *MEN1* which is located on chromosome 11q13. The gene has 10 exons and encodes the 610 amino acid protein menin, a tumor suppressor and a nuclear protein that participates in transcriptional regulations, genome stability, cell division, and proliferation. MEN1 tumors frequently have loss of heterozygosity (LOH) of the *MEN1* locus. Somatic abnormalities of *MEN1* have been reported in MEN1 and non-MEN1 endocrine tumors [11,12,115,116].

There are a total of 565 different *MEN1* mutations among the 1336 mutations reported, with 459 being germline and 167 being somatic (61 mutations occur both as germline and somatic), scattered throughout *MEN1*. Among these mutations, 41% are frameshift deletions or insertions, 23% are nonsense mutations, 20% are missense mutations, 9% are splicing mutations,

6% are in-frame deletions or insertions, and 1% represent whole or partial gene deletion. Somatic missense mutations occur more frequently than germline missense mutations. The majority of these mutations (>70%) are predicted to give rise to truncated or destabilized proteins. Four mutations, namely c.249_252delGTCT (deletion at codons 83–84), c.1546–1547insC (insertion at codon 516), c.1378C>T (R460X), and c.628_631delACAG (deletion at codons 210–211), account for 12.3% of all mutations and are potential mutation hot spots. No phenotype-genotype correlations have been observed in MEN1 [116]. In fact, similar *MEN1* mutations have been observed in FIHP and MEN1 patients. The reason for these altered phenotypes is not known. Between 5% and 10% of MEN1 patients do not harbor mutations in the coding region of *MEN1*.

Multiple Endocrine Neoplasia Type 2

Multiple Endocrine Neoplasia Type 2 (MEN2) describes the association of medullary thyroid carcinoma, pheochromocytomas, and parathyroid tumors. There are three clinical variants of MEN2, the most common of which is MEN2A, which is also known as Sipple's syndrome. MEN2A is inherited in an autosomal dominant manner. All MEN2A patients develop medullary thyroid carcinoma (MTC), 50% develop bilateral or unilateral pheochromocytoma, and 20%–30% develop primary hyperparathyroidism. Primary hyperparathyroidism can occur by the age of 70 in up to 70% of patients. MEN2A is caused by activating point mutations of the *RET* proto-oncogene. *RET* encodes a tyrosine kinase receptor with cadherin-like and cysteine-rich extracellular domains, and a tyrosine kinase intracellular domain. In 95% of patients, MEN2A is associated with mutations of the cysteine-rich extracellular domain, and missense mutation in codon 634 (Cys to Arg) accounts for 85% of MEN2A mutations [11,12,115].

The second clinical variant is MEN2B. These patients exhibit a relative paucity of hyperparathyroidism and development of marfanoid habitus, mucosal neuromas, and intestinal ganglioneuromatosis. Of these patients, 95% present with mutation in codon 918 (Met to Thr) of the intracellular tyrosine kinase domain of *RET* proto-oncogene. The third clinical variant of MEN2 is MTC-only. This variant is also associated with missense mutation of RET in the cysteine-rich extracellular domain, and most mutations are in codon 618. The precise mechanism of the genotype-phenotype relationship in the three clinical variants is unknown [11,12].

Hyperparathyroidism-Jaw Tumor Syndrome

Hyperparathyroidism-jaw tumor syndrome (HPT-JT syndrome) is an autosomal dominant disorder characterized by the development of parathyroid adenomas and carcinomas and fibro-osseous jaw tumors. The gene causing the syndrome, *HRPT2*, is located on chromosome 1q25. It consists of 17 exons and encodes a ubiquitously expressed 531 amino acid protein called parafibromin. *HRPT2* is a tumor-suppressor gene. Its inactivation is directly involved in the predisposition to HPT-TJ syndrome and in the development of some sporadic parathyroid tumors. The function of the gene has not been elucidated. To date, 13 different heterozygous inactivating mutations that predict truncation of parafibromin have been reported [11,12].

Calcium-Sensing Receptor and Related Disorders

Extracellular ionized calcium concentration regulates the synthesis and secretion of PTH as well as parathyroid cell proliferation. The action of calcium is mediated by the calcium-sensing receptor (CaSR). Human *CaSR* is located on chromosome 3q13.3–21. It is composed of 6 exons encoding a protein of 1078 amino acids. CaSR is a member of Family CII of the GPCR superfamily. The protein has a large extracellular domain with 612 amino acids, a transmembrane domain of 250 amino acids containing 7 transmembrane helices, and an intracellular domain of 216 amino acids. The receptor is heavily glycosylated, which is important for normal cell membrane expression. CaSR forms homodimers via intramolecular disulfide linkages within the extracellular domain. The extracellular domain is responsible for calcium binding. However, the stoichiometry between CsSR and calcium ions is unknown. The transmembrane domain is responsible for transducing the signal of calcium binding, while the intracellular domain interacts with $G_{\alpha i}$ or $G_{\alpha o}$ and activates different signal transducing pathways, depending on the cell line. The critical pathway(s) through which CaSR mediates its biological effects has not been defined [11,13,14].

CaSR is expressed in parathyroid cells, kidneys, bones, along the gastrointestinal tract, and in other tissues that are not directly involved in calcium homeostasis. However, the highest cell surface expression of CaSR is found in parathyroid cells, C-cells of the thyroid, and kidneys. Parathyroid cells are capable of recognizing small perturbation in serum calcium and respond by altering the secretion of PTH. Mutations of *CaSR* can result in loss-of-function or gain-of-function of the receptor. Heterozygous mutations of the *CaSR* are the cause of a growing number of disorders of calcium metabolism, which typically manifest as asymptomatic hypercalcemia or hypocalcemia, with relative or absolute hypercalciuria or hypocalciuria. On the other hand, homozygous or compound heterozygous inactivating mutations produce a severe and sometimes lethal disease if left untreated.

Disorders Due to Loss-of-Function Mutations of CaSR

There are two hypercalcemic disorders caused by inactivating mutations in *CaSR:* familial hypocalciuric hypercalcemia and neonatal severe primary hyperparathyroidism.

Familial Hypocalciuric Hypercalcemia. A disorder that must be recognized as distinct from primary hyperparathyroidism is familial hypocalciuric hypercalcemia (FHH), also known as familial benign hypercalcemia (FBH). The hypercalcemia is usually mild and asymptomatic, and is associated with a reduction rather than an increase in urinary calcium excretion. In the majority of cases the cause appears to be an autosomal dominant loss-of-function mutation in *CaSR*. The mutation in FHH reduces the receptor sensitivity to calcium. In the parathyroid glands, this defect means that a higher than normal serum calcium concentration is required to reduce PTH release. In the kidney, this defect leads to an increase in tubular calcium and magnesium resorption. The net effect is hypercalcemia, hypocalciuria, and frequently hypermagnesemia, as well as normal or slightly higher serum PTH concentration.

About two-thirds of the FHH kindreds studied have unique heterozygous mutations [11,13]. The inactivating mutations include 59 missense mutations, 6 nonsense mutations, 7 insertions and/or deletions, including an *Alu* element insertion, and 1 splice mutation. These mutations are described in the *CaSR* mutation database (http://www.casrdb.mcgill.ca). Some of these mutations cause a more severe hypercalcemia by inhibiting the wild-type CaSR in mutant-wild-type heterodimers [13]. There are ~30% of FHH patients without an identifiable mutation. Linkage analyses have linked FHH in some of these patients to the long and short arms of chromosome 19 [11,13,14,117].

Neonatal Severe Primary Hyperparathyroidism. Neonatal severe primary hyperparathyroidism (NSHPT) represents the most severe expression of FHH. Symptoms manifest very early in life with severe hypercalcemia, bone demineralization, and failure to thrive. Most NSHPT patients have either homozygous or compound heterozygous *CaSR* inactivating mutations. There are also three cases of NSHPT in which *de novo* mutation was found in the extracellular domain of CaSR. The patients have only one copy of *CaSR* mutated, and no *CaSR* mutations are found in the parents. There are also asymptomatic patients with homozygous *CaSR* inactivating mutations. Thus, the severity of symptom may be affected by factors other than mutant gene dosage [117].

Disorders Due to Gain-of-Function Mutation of the CaSR

There are two hypocalcemic disorders associated with gain of CaSR function due to activating mutations of the receptor. These mutated receptors are more sensitive to extracellular calcium required for PTH secretion and cause reduced renal calcium resorption. The two disorders are Autosomal Dominant Hypocalcemia (ADH) and Bartter syndrome type V.

Autosomal Dominant Hypocalcemia. ADH is a familial form of isolated hypoparathyroidism characterized by hypocalcemia, hyperphosphatemia, and normal to hypoparathyroidism. Inheritance of the disorder follows an autosomal dominant mode. The patients are generally asymptomatic. A significant fraction of cases of idiopathic hypoparathyroidism may in fact be ADH.

More than 80% of the reported ADH kindreds have CaSR mutations. There are 44 activating mutations of CaSR reported in the literature. These mutations produce a gain of CaSR function when expressed in *in vitro* systems [13,117]. The majority of the ADH mutations are missense mutations within the extracellular domain and transmembrane domain of CaSR. In addition, a deletion in the intracellular domain, p.S895_V1075del, has also been described in an ADH family. The mechanism of CaSR activation by these mutations is not known. Worthwhile noting is that almost every ADH family has its own unique missense heterozygous CaSR mutation [117]. Most ADH patients are heterozygous. The only deletion-activating mutation occurs in a homozygous patient in an ADH family. However, there is no apparent difference in the severity of the phenotype between heterozygous and homozygous patients.

Bartter Syndrome Type V. In addition to hypocalcemia, patients with Bartter syndrome type V have hypercalciuria, hypomagnesemia, potassium wasting, hypokalemia, metabolic alkalosis, elevated renin and aldosterone levels, and low blood pressure. Four different activating mutations of CaSR (p.K29E, p.L125P, p.C131W, and p.A843E) have been identified in patients with Bartter syndrome type V. Functional analyses show that these CaSR mutations result in a more severe receptor activation when compared to the other activating mutations described [13].

THE ADRENAL GLAND

The adrenal glands are two crescent-shaped structures located at the superior pole of each kidney. Each gland is composed of two separate endocrine tissues, the inner medulla and the outer cortex, which have different embryonic origins. The adrenal medulla is responsible for the production of the catecholamines, epinephrine and norepinephrine, which are involved in fight-or-flight responses. The adrenal cortex produces a number of steroid hormones that regulate diverse physiological functions. Despite the difference in origin and physiology, the hormones produced by the medulla and cortex often act in a concerted manner.

The adrenal cortex is composed of three distinct cell layers: (i) the zona glomerulosa, (ii) fasciculata, and (iii) reticularis. The outermost layer (zona glomerulosa) secretes mineralocorticoids, the most important of which is aldosterone. Aldosterone primarily affects the distal renal tubular sodium-potassium exchange mechanism to promote sodium resorption, thereby regulating circulatory volume and having a major impact on cardiac function. The synthesis and release of aldosterone is influenced by angiotensin II, derived from angiotensin I under control of the enzyme renin. Renin is made by the juxtaglomerular cells of the renal cortex in response to variations in renal blood flow. To a lesser extent, the production of aldosterone is also influenced by adrenocorticotropic hormone (ACTH).

The middle layer of the adrenal cortex (zona fasciculata) secretes glucocorticoids. The most important glucocorticoid is cortisol. Its action affects almost every organ and tissue in the body, by maintaining blood glucose level and blood pressure, and modulating stress and inflammatory responses. Secretion of cortisol is regulated by the hypothalamus-pituitary-adrenal axis. The hypothalamic corticotropin-releasing hormone (CRH) stimulates ACTH secretion by the pituitary corticotrophs through ligand-receptor mediated mechanism. ACTH triggers cortisol production in adrenocortical cells by binding to melanocortin receptor 2 (MC2R, which is the ACTH receptor). A series of enzymatic reactions is initiated to mediate the uptake of cholesterol and the biosynthesis of cortisol. The circulating cortisol feeds back to the pituitary and hypothalamus to suppress further ACTH secretion.

The innermost layer of the adrenal cortex (zona reticularis) produces adrenal androgens. The production of adrenal androgens is also regulated by ACTH. In the fetal adrenal cortex (during weeks 7–12 of gestation), androgens are secreted and regulate the differentiation of male external genitalia. However, in adults the contribution of androgens from adrenal glands is quantitatively insignificant.

Congenital Primary Adrenal Insufficiency

As with other endocrine glands, both hypo-function and hyper-function of the adrenal are associated with significant disorders. Adrenal insufficiency is generally referred to an inadequate production of cortisol by the adrenal. It may be primary or secondary. Primary adrenal insufficiency is caused by defects of the adrenals themselves. The most frequent cause is the destruction of adrenal cortex due to autoimmunity. Otherwise, it can result from inflammation, infection, surgical adrenal removal, or neoplasia spreading from other parts of the body. Congenital primary adrenal insufficiency is caused by inactivating mutations of genes responsible for normal adrenal development or cortisol production. Depending on the severity, a deficiency of cortisol alone or with other adrenal steroids would occur.

Adrenal Hypoplasia Congenita

Adrenal hypoplasia congenita (AHC) is the underdevelopment of adrenal glands that typically results in adrenal insufficiency during early infancy. Two forms of AHC are known: (i) an autosomal recessive miniature adult form and (ii) an X-linked cytomegalic form. Patients affected by the latter display low or absent glucocorticoids, mineralocorticoids, and androgens and do not respond to ACTH stimulation. The adrenals are atrophic and structurally disorganized, with no normal zona formation in cortex. Impaired sexual development, owing to hypogonadotropic hypogonadism, manifests in affected males who survive childhood.

The cytomegalic form of AHC is an X-linked disorder due to deletion or inactivating mutation of *NR0B1*, which encodes DAX1 (dosage sensitive sex reversal-AHC critical region on the X chromosome gene 1). As a result, females are unaffected carriers and males only are affected by *NR0B1* mutation. DAX1 is indispensable to the development of adrenal glands, gonads, and kidneys. It encodes an orphan nuclear receptor that interacts with SF1 (Steroidogenic Factor 1; encoded by *NR5A1*). Binding of DAX1 inhibits SF1-mediated transactivation of genes involved in the development of hypothalamus-pituitary-adrenal-gonad axis and biosynthesis of steroid hormones. More than 100 mutations of *NR0B1*, affecting the DNA-binding or ligand-binding domain of DAX1, have been described [118]. The genetic lesions (mostly frameshift, nonsense, and missense mutations) are translated into truncated DAX1 proteins that lose the ability to repress SF1 transactivation activity. Phenotypic heterogeneity, in terms of severity and onset of disorder, is observed in patients with different or even the same *DAX1* mutations [119,120], suggesting that other genetic, epigenetic, or environmental factors may be involved in AHC. Alternatively, the presence of residual activity in *DAX1* mutants (positional effect of mutations) may determine the extent of phenotypic variation. The expression of a novel *NR0B1* splice variant in the adrenal glands is also suspected to influence AHC phenotype.

Similar to DAX1, SF1 is also an orphan nuclear receptor essential to normal adrenogonadal development, besides its pivotal role in steroidogenesis. However, inactivating mutation of *NR5A1* does not always associate with adrenal insufficiency. The cause of such variability is still unknown.

Congenital Adrenal Hyperplasia

Congenital adrenal hyperplasia (CAH) is one of the most common autosomal recessive disorders in humans. It is the result of inborn metabolic errors in adrenal steroid biosynthesis (Figure 22.2). Cholesterol is transported from the cytoplasm into mitochondria in adrenocortical cells by steroidogenic acute regulatory protein (StAR), and is first converted to pregnenolone, the common precursor of all adrenal steroids, by rate-limiting cholesterol side-chain cleavage enzyme P450scc (encoded by *CYP11A1*). Adrenal steroids are produced under the action of various cytochrome P450s, with some of them participating in more than one steroidogenic pathway. Depending on the enzymatic defect, an impaired production of glucocorticoids (and mineralocorticoids or androgens) can occur starting in utero. The feedback to the pituitary is undermined such that ACTH production becomes excessive. Consequently, the adrenals are overstimulated and hyperplastic. The overproduced steroid precursors are shunted to the androgen biosynthetic pathway, leading to androgen overproduction. Genital ambiguity in newborn females (owing to excessive fetal exposure *in utero*) and precocious pseudopuberty in both sexes are commonly observed in CAH patients. In severe cases, salt-wasting CAH occurs with life-threatening vomiting and dehydration within the first few weeks of life.

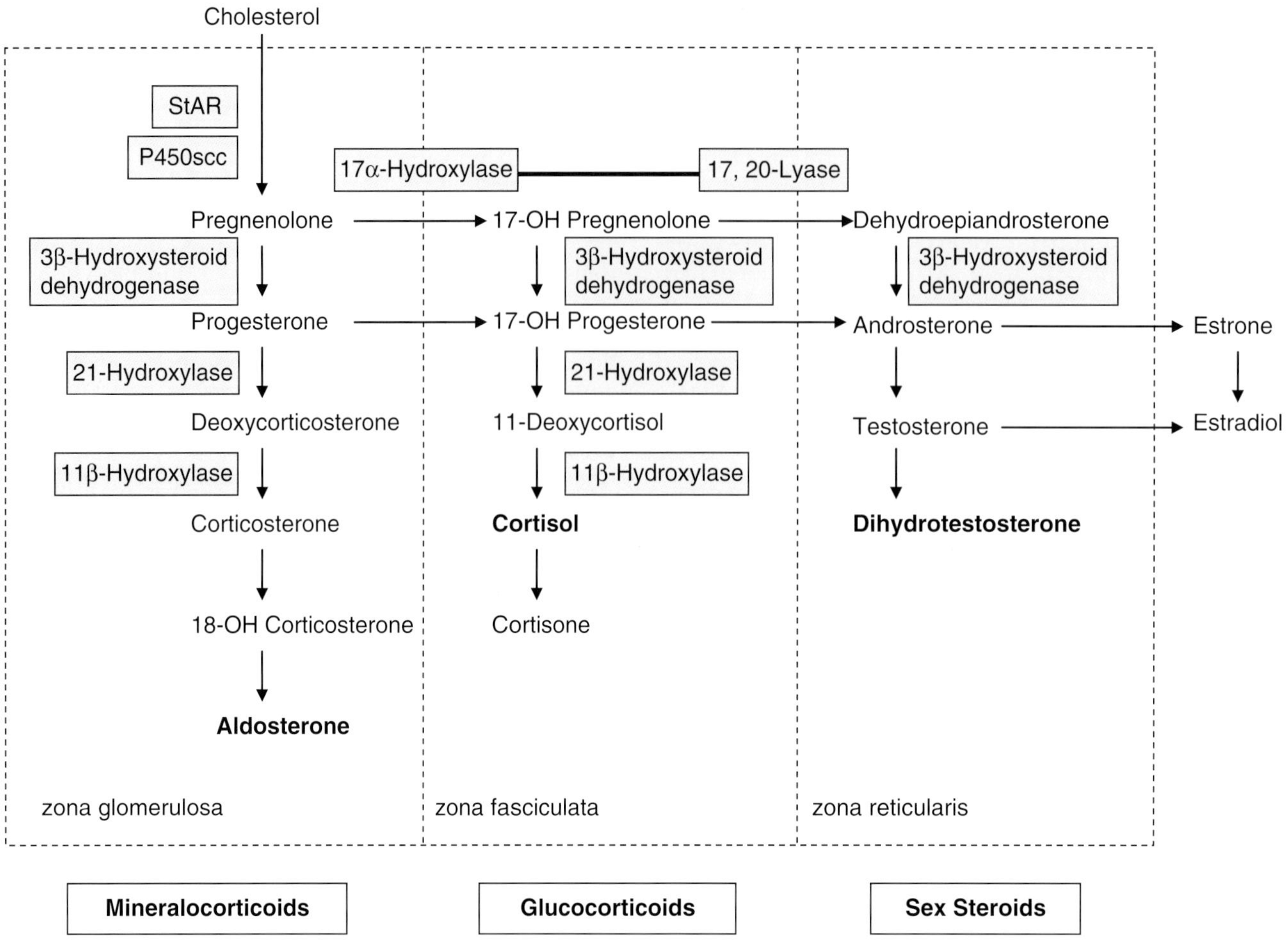

Figure 22.2 **Pathways of adrenal steroidogenesis.** The three classes of adrenal steroids are shown in bold. Enzymes or proteins that cause CAH when defective are shown in gray boxes. The 17α-hydroxylase and 17,20-lyase activities are encoded by the same enzyme (CYP17A1).

21-Hydroxylase Deficiency. The most frequent CAH variant is 21-hydroxylase deficiency (21-OHD), caused by inactivating mutations or deletion of *CYP21A2*, and accounts for ~95% of classical CAH. In many cases the term *CAH* refers to 21-OHD. The failure of conversion of progesterone and 17-hydroxyprogesterone leads to their accumulation and shunting to the androgen biosynthetic pathway. Over 100 mutations, most of which are missense mutations, have been described [121]. Based on the severity of enzyme defect, three clinical forms of 21-OHD CAH are recognized: the most severe classical salt-wasting form (due to aldosterone insufficiency and androgen excess), the moderately severe classical simple-virilizing form (due to androgen excess), and the least severe nonclassical form.

11β-Hydroxylase Deficiency. The second most common cause of CAH is 11β-hydroxylase deficiency (11β-OHD). The enzyme, 11β-hydroxylase (encoded by *CYP11B1*), catalyses the conversion of deoxycorticosterone and 11-deoxycortisol to corticosterone and cortisol, respectively (Figure 22.2). Similar to 21-OHD, the enzymatic defect leads to virilizing CAH due to accumulation of immediate steroid precursors. The accumulation of deoxycorticosterone also causes salt retention and hypertension. More than 50 inactivating *CYB11B1* mutations have been identified; most of them completely abolish the enzymatic activity [122].

A variant form of CAH related to 11β-OHD is caused by chimeric gene formation due to unequal crossing-over. *CYP11B2*, which is >95% identical to and is located ~40 kb upstream of *CYP11B1*, encodes an aldosterone synthase that is exclusively expressed in zona glomerulosa to catalyze the conversion of deoxycorticosterone to corticosterone. The high degree of homology renders the two isoforms susceptible to crossing-over during DNA replication, with regulatory sequence from one fused to the coding region of the other. Use of *CYP11B2* promoter leads to no production of chimeric CYP11B2/B1 protein in zona fasciculata because the promoter is inactive in this region of the adrenal. Thus, carriers of this mutation display 11β-OHD. Conversely, the *CYP11B1* promoter-driven expression of *CYP11B1/B2* chimera, which retains aldosterone synthase activity, leads to overproduction of mineralocorticoids and results in an autosomal dominant disorder called glucocorticoid-suppressible hyperaldosteronism (or familial hyperaldosterone type 1) [123]. A rare case of salt-wasting CAH with 11-OHD due to homozygous internal deletion of the *CYP11B2/B1* chimera has also been reported.

Other Less Common Steroidogenic Enzyme Deficiency in CAH. The key enzyme in steroid hormone biosynthesis is 3β-hydroxysteroid dehydrogenase (3β-HSD). It catalyzes the conversion of progesterone, 17-OH progesterone, and androstenedione from their respective Δ^5 steroid precursors. As a result, inactivation of 3β-HSD leads to incomplete genital development and impaired aldosterone synthesis due to a deficiency of all classes of adrenal steroids. In humans, *HSD3B2* is the isoform expressed specifically in the adrenals and gonads. At least 36 inactivating *HSD3B2* mutations, mostly missense mutations, have been reported. Aside from complete loss of enzymatic activity, some of the mutated enzymes display impaired stability [124,125].

CYP17A1 encodes an enzyme with dual (17α-hydroxylase and 17,20-lyase) activities, mediating the biosynthesis of cortisol and sex steroid precursors (Figure 22.2). Inactivating mutations of the gene frequently produce combined enzyme deficiencies, displaying hypertension, hypokalemia, and sexual infantilism. Despite its rarity in causing CAH, more than 70 inactivating mutations of *CYP17A1* have been identified [126]. Most of the reported mutations alter the enzyme structure or generate truncated products and thus completely abolish both enzymatic activities. Isolated 17,20-lyase deficiency also occurs when lyase activity is selectively impaired by mutations of amino acid residues (such as p.R347 or p.R358) that are crucial to the interaction between CYP17A1 and its redox partners (P450 oxidoreductase and cytochrome b5) in the P450-electron donor complex.

Although uncommon, Lipoid Congenital Adrenal Hyperplasia (LCAH) represents the most severe form of CAH. Life-threatening mineralocorticoid and glucocorticoid deficiencies are common in infants and children. Male infants are undervirilized, and puberty is delayed in both sexes. The enlarged adrenals are filled with lipid globules. Similar damage is observed in the gonads. LCAH results from defective cholesterol to pregnenolone conversion, the first step in steroidogenesis, mediated by P450scc. So far only 5 *CYP11A1* mutations have been described [127]. Most of the genetic lesions actually stem from defective cholesterol transport into the mitochondria by inactivating mutation of *StAR*. The lack of substrate leads to adrenal steroid deficiency and absence of feedback for ACTH suppression. The elevated level of ACTH excessively stimulates cholesterol uptake and adrenal cell growth, resulting in adrenal hyperplasia. At least 48 inactivating mutations of *StAR*, most of which are missense and frameshift mutations, are known [128]. In all cases, the activity of the mutated StARs, in terms of ligand binding and cholesterol-to-pregnenolone conversion, is severely impaired or totally lost.

ACTH Resistance Syndromes

ACTH resistance syndromes represent a group of disorders that lead to unresponsiveness to ACTH in the production of glucocorticoid by adrenal cortex. The disorders can occur in patients with familial glucocorticoid deficiency or Triple A syndrome.

Familial Glucocorticoid Deficiency. Familial Glucocorticoid Deficiency (FGD) is a rare autosomal recessive disease characterized by cortisol deficiency, but without mineralocorticoid deficiency and pituitary structural defects. It is usually diagnosed during neonatal period or early childhood with symptoms like hypoglycemic seizures, failure to thrive, collapse and coma, and recurrent infections. Specifically, FGD patients show an extremely high plasma ACTH level, which indicates a resistance to ACTH action. They almost always develop hyperpigmentation. Adrenarche is absent in children with FGD. Tall stature is observed in some patients.

FGD is caused by defects of MC2R. Based on gene mutations, FGD is subdivided into two types. FGD type 1 is caused by inactivating mutations of MC2R. The mutated receptors are completely inactive or display subnormal cortical response to ACTH stimulation. At least 38 MC2R mutations have been reported [129–131], most of them are missense mutations in the coding region. Nucleotide substitution at the promoter region is also associated with a lower cortisol response to ACTH stimulation and results in FGD in conjunction with a frameshift mutation in the other allele [132]. FGD type 2 is caused by inactivating mutations of the MC2R accessory protein (MRAP). MRAP assists the trafficking and expression of MC2R at cell surface. So far only 9 MRAP mutations, all leading to the production of severely truncated proteins, are identified [133]. Genetic defects of MC2R and MRAP account for 45% of FGD cases, suggesting that more genes may be involved in the etiology of FGD.

Triple A Syndrome. Triple A syndrome is a complex disorder characterized by adrenal failure, alacrima, and achalasia. Not all patients exhibit adrenal failure, but of those with adrenal failure ~80% will have isolated glucocorticoid deficiency. Mutations of the causative gene, AAAS, are identified in patients with the syndrome [134]. However, defects in AAAS do not account for all cases of Triple A syndrome, and the expression pattern of the gene correlates poorly to pathology.

Secondary Adrenal Insufficiency

Secondary adrenal insufficiency can be attributed to a lack of ACTH. The adrenal glands become atrophic and cortisol, but not aldosterone, production is extremely low or undetectable. For acquired cases, ACTH deficiency may result from pituitary tumor or other physical lesions that prevent ACTH secretion. Meanwhile, congenital cases are caused by disorders of the hypothalamus-pituitary axial components that control ACTH production.

ACTH secretion is regulated by the hypothalamic corticotropin-releasing hormone (CRH). The 41 amino acid neuropeptide binds to CRH receptor (CRHR) on pituitary corticotrophs to stimulate ACTH production through enzymatic cleavage of its precursor propiomelanocortin (POMC) under the action of prohormone convertase 1 (PC-1). Malfunctioning of any of these genes can lead to ACTH deficiency. POMC deficiency syndrome, owing to recessive inactivating mutations of *POMC*, was described in 7 patients [135,136], and 2 cases of compound heterozygous mutations of *PC-1* were reported [137]. No *CRH* or *CRHR* mutation is known. The small number of mutations reflects the rarity of these genetic defects in the etiology of ACTH deficiency. In contrast, mutations of *TPIT*, which encodes a T-box transcription factor important to POMC expression and the terminal differentiation of pituitary POMC-expressing cells, cause adrenal insufficiency more frequently [138].

Generalized Glucocorticoid Resistance/Insensitivity

Generalized glucocorticoid resistance is a rare genetic condition characterized by generalized partial target tissue insensitivity to glucocorticoid. Patients show increased levels of ACTH and cortisol, resistance of the hypothalamus-pituitary-adrenal axis to dexamethasone suppression, but no clinical sign of hypercortisolism. Production of mineralocorticoids and androgens is increased as a result of excessive ACTH, which is reflected by symptoms such as hypertension, hypokalemic alkalosis, ambiguous genitalia, and gonadotropin-independent precocious puberty. The disorder is caused by inactivation of the glucocorticoid receptor-α isoform (GRα), which is the classical GR that functions as a ligand-dependent transcription factor. Almost all known mutations are heterozygous, implying that a complete loss of the receptor is incompatible with life. Besides one case of a small deletion at the exon-intron 6 boundary, the remaining 9 *GRα* mutations are missense mutations [139]. All mutations impair one or multiple actions of glucocorticoid, ranging from defective nucleocytoplasmic shuttling, reduction of gene transactivation activity and affinity for ligands, to partial or complete loss of interaction with coactivators. In some cases, the mutated receptors display a dominant negative effect on wild-type receptors.

Hypercortisolism (Cushing's Syndrome)

Excessive hormone production by the adrenal cortex may result from abnormal pituitary ACTH stimulation, pituitary tumors, ectopic ACTH produced by a neoplasm, or pathology within the adrenals themselves. The clinical picture resulting from hypersecretion of cortisol, androgens, and deoxycorticosterone is variable, depending on the etiology. The manifestations, nevertheless, are those of glucocorticoid excess.

Cushing's syndrome is caused by excessive circulating cortisol, which affects 10–15 in every million people at age 20–50. The affected individuals are characterized by moon faces and buffalo hump, with central obesity, severe fatigue, high blood pressure and glucose levels, depression, and gonadal dysfunction commonly observed. The cause of Cushing's syndrome is mostly iatrogenic, resulting from prolonged steroid ingestion (such as from glucocorticoid drug hormone medication). Cushing's syndrome is primarily caused by ACTH-secreting pituitary adenoma (also known as Cushing's disease) which accounts for ~70% of all cases of ACTH-dependent Cushing's syndrome. Other causes include adrenal hyperplasia or neoplasia (the second most common genetic cause) and ectopic ACTH production by other tumors (including small cell lung cancer and carcinoid tumors). All defects result in adrenal gland overgrowth and cortisol overproduction. The elevated level of cortisol promotes protein catabolism, wasting of muscles, thinning of skin, and conversion to fat without weight gain, which leads to the characteristic appearance.

ACTH-independent Cushing's syndrome constitutes the remaining subgroup of the disorder. It is well known to be caused by adrenocortical tumors or inherent endocrine tumor forming-diseases such as MEN1 and primary pigmented nodular adrenocortical disease. Recently, an activating mutation of *MC2R* (p.F278C) was shown to be associated with the disorder and adrenal hyperactivity [140]. Other endocrine factors emerge to play a role in disease etiology. For instance, LHR was shown to be present in adrenocortical cells, and LH/hCG is involved in adrenal hyperfunctions [141].

PUBERTY

Puberty refers to the physical changes by which a child becomes an adult capable of reproduction. The maturation of the reproductive system is very complex, involves almost the entire endocrine system, and occurs in a phasic manner. During fetal development, neuroendocrine cells appear in the rostral forebrain, whence they migrate to an area in the hypothalamus to become the gonadotropin-releasing hormone (GnRH) pulse generator. During infancy, gonadotropin secretion is inhibited centrally by a sensitive negative feedback control, which keeps the reproductive system quiescent.

The onset of puberty is contingent upon maturation of the central pulse generator and disinhibition and reactivation of the hypothalamic-pituitary-gonadal axis. Temporal and developmental changes in GnRH pulse frequency have differential effects on FSH and LH. Slow GnRH pulse frequencies preferentially stimulate FSH synthesis and release, whereas higher frequencies preferentially stimulate LH production. The pulsatile LH secretion promotes growth and maturation of the gonads (ovaries and testes), sex hormone production, and development of secondary sex characteristics. As puberty progresses, there is an increase in the amplitude and frequency of LH pulses, leading eventually to adult steroid production and gametogenesis. The sex hormones stimulate growth as well as function and transformation of the brain, bones, muscle, and skin in addition to the reproductive organs. Growth accelerates at the onset of puberty and stops at its completion.

Before puberty, the body differences between the sexes are mainly confined to the genitalia. During puberty, major differences of size, form, shape, composition, and function develop. Puberty is accompanied by psychological, behavioral, and cultural changes, referred to as adolescence. The time of onset and subsequent course of hormonal and physical changes during puberty are highly variable and influenced by genetic makeup, nutrition, and general health. Breast budding in girls and testicular enlargement in boys are usually the first signs of puberty. Detailed studies of the progression of the physical changes with puberty (Tanner) demonstrate peak growth velocity in girls early in puberty before menarche, whereas in boys it occurs later (in the second half of puberty). The time interval between breast budding and menarche is on average 2½ years, and the average age at menarche is between 2½ and 13 years. During the first 2 years, menstrual cycles are often irregular and nonovulatory.

In addition to the striking changes brought about by activation of the pituitary-gonadal axis, secretion of androgens by the adrenal gland also increases. However, gonadal and adrenal androgen production (adrenarche) is not causally related temporally, and discordance in the two is seen in a number of situations. Adrenarche occurs in response to rising adrenal androgen secretion, usually precedes the pubertal rise in gonadotropins and sex steroids by about 2 years, and is associated with differentiation and growth of the adrenal zona reticularis. The major adrenal androgens are androstenedione and epiandrosterone. They are weak androgens, which in the female, directly or by peripheral conversion, account for about 50% of serum testosterone and are largely responsible for female sexual hair development. Unlike the major gonadal sex steroids, adrenal androgens do not play an important role in the adolescent growth spurt.

Any disturbance in the integrity of the brain, the hypothalamic-pituitary axis, or the peripheral endocrine system may interfere with normal puberty, whether it is developmental, inflammatory, genetic, neoplastic, or functional. Absence of one or more trophic or pituitary hormones interferes with hormonal effectiveness or end organ response, which may be on a genetic or acquired basis. Lack of signs of puberty in a female 13 years of age or a male 14 years of age becomes a concern and deserves evaluation.

Delayed Puberty

The most common cause of failure of pubertal development in the female is gonadal dysgenesis, a disorder of the X chromosome. Short stature and a variety of congenital anomalies associated with elevated serum gonadotropins readily confirm the diagnosis. The exact chromosomal abnormality is established by the karyotype. Primary amenorrhea after some pubertal development may indicate an abnormality of Müllerian development.

The most common cause of delay in the onset of puberty in boys is constitutional. It is often diagnosed in retrospect if boys initiate puberty spontaneously before the age of 18. There is frequently a positive family history. Short stature, a younger appearance, and a delayed bone age are common findings. In the absence of signs of puberty on physical examination, a serum testosterone level >50 mcg/dl foreshadows testicular hormone production and suggests cautious inaction and follow-up. If serum testosterone is <50 mcg/dl, evaluation of the functional integrity of hypothalamic-pituitary-gonadal axis is required to ascertain whether the delay is a temporary or a permanent feature. If hypogonadotropic hypogonadism is present at 18 years of age or older, the diagnosis is Isolated Hypogonadotropic Hypogonadism (IHH) [15]. IHH combined with a history of anosmia/hyposmia suggests Kallmann's syndrome. Otherwise, pituitary deficiency, idiopathic or acquired; pituitary stalk interruption syndrome; hypothyroidism; and a variety of genetic syndromes need to be considered.

Hypogonadotropic Hypogonadism

Isolated hypogonadotropic hypogonadism describes the condition which is the consequence of defects in the pulsatile release of gonadotropins or the deficiency of gonadotropin action. A number of conditions are associated with the Mendelian forms, including developmental defects of the hypothalamus or abnormal pituitary gonadotropin secretion. Structural defects of LH and FSH affect the action of gonadotropins. Mutations in three genes (*KAL1*, *GNRHR*, and *FGF1*) account for most of the known cases of IHH [16].

X-linked Kallmann's Syndrome and KAL1. When associated with anosmia or hyposmia, IHH is referred to as Kallmann's syndrome [15,17]. IHH is caused by GnRH deficiency due to failure of embryonic migration of GnRH-synthesizing neurons, while anosmia is related to hypoplasia or aplasia of the olfactory bulbs. In addition to IHH and anosmia, patients with Kallmann's syndrome may display a variety of nonreproductive anomalies. The phenotype associated with Kallmann's syndrome mutations varies significantly, indicating the potential influence of modifying genes and other factors. There are two forms of Kallmann's syndrome: an X-linked form caused by mutations of *KAL1* and an autosomal-dominant form caused by mutations of *KAL2*.

KAL1 is located on Xp22.3. It is composed of 14 exons encoding an approximately 100 kD extracellular-matrix glycoprotein, anosmin-1, which shares homology with neural cell adhesion molecule. Anosmin-1 has been suggested to provide a scaffold to direct neuronal migration of both GnRH and olfactory neurons to their proper embryonic destination. Mutations in *KAL1* account for approximately 10%–15% of anosmic male patients in sporadic cases but may comprise 30%–60% of familial cases. *KAL1* mutations have not been described in normosmic male patients or in female patients. A variety of mutational changes have been observed with *KAL1*. There are over 40 *KAL1* mutations reported [16,142,143]. Genomic deletions, including microdeletion of chromosomal region containing the *KAL1* locus, deletion of the entire *KAL1* gene, and intragenic deletion of multiple exons, have been reported. A recent study found that 12% of male Kallmann's syndrome patients have intragenic *KAL1* deletions [144]. There are also deletions of 1, 2, 8, or 14 nucleotides, resulting in shifting of the reading frame and occurrence of premature stop codon. Similar frameshift mutations due to insertion of a single nucleotide have been identified. Another mutation shows the replacement of 4 nucleotides and the creation of an in-frame stop codon (c.1651_1654delinsAGCT resulting in p.P551_E552delinsSX). In one patient, a deletion mutation causes a splicing error. The deletion of 9 nucleotides starting at codon 400 (Asn) removes 7 bases from exon 8 and the 2 most 5′ nucleotides of intron 8 (p.N400_IVS8del AACAACAgt). This deletion potentially changes the splicing of intron 8 and creates a different amino acid sequence downstream of residue Asn400. There are also at least 4 splicing mutations reported affecting IVS1, IVS4, IVS6, and IVS12, 2 insertions, resulting in frameshift and early truncation of the encoded protein (c.1166_1167insA, c.570_571insA), and one 4 bp replacement resulting in an in-frame stop codon. In addition to these mutations, there are at least 8 missense mutations and 15 nonsense mutations reported [16,142,143]. These mutations are distributed throughout *KAL1*. The majority of the mutations result in the formation of truncated proteins. Some mutations affect the formation of disulfide bond in the N-terminal cysteine-rich region (WAP domain), while other mutations affect the structure of the four fibronectin type III (FNIII) domains of anosmin-1. The WAP and the first FNIII domain are highly conserved across many species and are believed to be essential for the binding of anosmin-1 with heparin sulfate and its other ligands.

Autosomal Dominant Kallmann's Syndrome and KAL2/FGFR1. The autosomal dominant form of Kallmann's syndrome is caused by an inactivating mutation of *KAL2/FGFR1* which is localized on chromosome 8p11.2-p12. *KAL2/FGFR1* contains 18 exons and encodes fibroblast growth factor receptor 1 (FGFR1). FGFR1 is a single spanning transmembrane receptor. The protein has three immunoglobulin-like domains, a heparin-binding domain, and two tyrosine kinase domains. FGFR1 is expressed in GnRH neurons, and FGF signaling is involved in GnRH neuron specification, migration, and axon targeting. Anosmin-1 is a ligand for FGFR1 and induces neurite outgrowth and cytoskeletal changes through an FGFR1-dependent mechanism [17].

Mutations in *KAL2/FGFR1* occur in approximately 7%–10% of patients (male and female) with autosomal dominant Kallmann's syndrome. Although most *KAL2/FGFR1* mutations have been reported in Kallmann's patients, mutations of this gene in normosmic IHH patients have also been reported. Mutations of *KAL2/FGFR2* show reduced penetrance and variable expressivity.

At least 39 inactivating mutations of *KAL2/FGFR1* leading to IHH with or without anosmia have been described to-date [15–17,142]. About 70% of these mutations are missense mutations, 18% are nonsense mutations, 9% are frameshift deletion/insertions, and two are splicing mutations. A number of the missense mutations cluster in the first immunoglobulin domain of FGFR1, suggesting that this domain is important to FGFR1 function. The nonsense mutations cluster at the C-terminal tyrosine kinase domain and result in truncated receptors lacking the autophosphorylated tyrosine residues. They are likely to impede the signaling activity of the receptor. It is worth mentioning that so far there is only one reported activating mutation of *KAL2/FGFR1* which underlies Pfeiffer syndrome, a form of craniosynostosis.

Normosmic IHH and Gonadotropin-Releasing Hormone Receptor (GnRHR). About 50% of familial cases of normosmic IHH are associated with loss-of-function *GnRHR* mutations. *GnRHR* is located on chromosome 4q21.2 with an open reading frame of 981 nucleotides and encodes a protein of 327 amino acids. It is a member of the GPCR family. GnRHR is expressed on the cell surface of pituitary gonadotrophs. The ligand for GnRHR is gonadotropin-releasing hormone-1 (GnRH), a decapeptide which is derived from a 92 amino-acid pre-pro-protein. It is released in a pulsatile manner in the preoptic area of the hypothalamus and delivered to the anterior pituitary gland. There, it binds and activates GnRHR, resulting in the synthesis and release of gonadotropins (LH and FSH). No GnRH mutation causing IHH has been reported.

Currently, 21 loss-of-function mutations of *GnRHR* have been identified in patients with IHH [16–18,145]. All are missense mutations with the exception of a nonsense mutation and a splicing mutation. These mutations are transmitted in an autosomal recessive mode. Most patients are compound heterozygotes. The two most prevalent mutations are p.Q106R (32%) and p.R262Q (15%). The former mutation reduces GnRH binding affinity, and the latter mutation reduces signal transduction. Slightly more than half of the *GnRHR* mutations interfere with ligand binding, while the remainder affect signal transduction.

There is no report of *GnRHR* mutations in anosmic or hyposmic IHH patients. *GnRHR* mutations are present in 1%–4.6% of all IHH patients, while it is 6%–11% in autosomal recessive IHH families. In most cases the phenotype correlates with the functional alterations of the GnRHR *in vitro*. Patients with complete inactivating GnRHR variants on both alleles present with severe hypogonadism, while patients homozygous for a partially inactivating GnRHR variant present with partial hypogonadism.

More recently, another GPCR, GPR54, has been found to be associated with autosomal recessive IHH [18]. *GPR54* is localized on chromosome 19p13.3 and is composed of 1191 nucleotides encoding a protein of 396 amino acids. At least 7 mutations, 4 missense mutations, 1 nonsense mutation, a 155-bp deletion, and 1 insertion of *GRP54* have been described in homozygous and compound heterozygous IHH patients. Impaired signaling capacity of the mutated receptor has been observed. GPR54 is the receptor for kisspeptins, which are potent stimulators of LH and FSH secretion. Kisspeptin-induced GPR54 signaling is also believed to be a major control point for GnRH release. No kisspeptin gene mutation has been described. Phenotypic expression of *GPR54* inactivating mutation does not differ from that of *GnRHR* mutation.

Isolated Gonadotropin Deficiency. LH and FSH are the main regulators of gonadal steroid secretion, pubertal maturation, and fertility. A small number of hypogonadotropic hypogonadism patients have been found to carry loss-of-function mutations of the hormone-specific β subunit of LH and FSH. Both LH and FSH are heterodimers with an α subunit which is shared among the glycohormones. The β subunit of each hormone gives it its specificity. *LHβ* is located on chromosome 19q13.32, and *FSHβ* is on 11p13. Loss-of-function mutation of the β subunit renders the hormone inactive, giving rise to hypogonadotropic hypogonadism. No mutation of the common α subunit and CGβ subunit is known.

At present there are 7 individuals, 4 women and 3 men, who are hypogonadal due to *FSHβ* mutation. Four different mutations have been identified: p.C51G, p.C82R, p.Y76X, and p.V61delTGfs87X [19,146]. These mutations interfere with dimerization with the α subunit and render the hormone inactive. All four women with homozygous inactivating *FSHβ* mutations have sexual infantilism and infertility because of a lack of follicular maturation, primary amenorrhea, low estrogen production, undetectable serum FSH, and increased LH. The men with homozygous inactivating mutations are all normally masculinized, with normal to delayed puberty, but azoospermic.

Two missense mutations (p.G36D and p.Q54R) and a splicing mutation (IVS2+1G>C) of *LHβ* have been described in 4 men and 1 woman [19,146]. All patients are homozygous carriers of the mutation. The missense mutations either interfere with the formation of active heterodimers or affect the interaction with LH receptor (LHR). The splicing mutation produces a truncated LHβ subunit and abrogates LH secretion. Male homozygotes are normally masculinized at birth but lack postnatal sexual differentiation, are hypogonadal with delayed pubertal development, absence of mature Leydig cells, spermatogenic arrest, and absence of circulating LH and testosterone. The only woman with the homozygous splicing mutation has normal pubertal development, secondary amenorrhea, and infertility. All heterozygotes are fertile and have normal basal gonadotropin and sex-steroid levels.

Hypergonadotropic Hypogonadism

Hypergonadotropic hypogonadism and infertility/subfertility in both sexes are the result of gonadotropin resistance caused by inactivating mutations in receptors of the two gonadotropins, LH and FSH. Both LHR and FSHR are members of GPCR family. The ECDs of the receptors convey ligand specificity and are different among the receptors, while the TMD is highly homologous between species and among the glycohormone receptors.

Leydig Cell Hypoplasia (LCH)—Inactivating Mutations of LHR. LHR is shared between LH and CG, thus the name LH/CG receptor. CG exerts its effect during early embryogenesis to induce Leydig cell maturation. LH also promotes steroidogenesis by Leydig cells, especially around the period of puberty. Inactivating mutations of LHR are recessive in nature. In males with homozygous or compound heterozygous inactivating mutations, a loss of LHR function causes resistance to LH stimulation, resulting in failure of testicular Leydig cell differentiation. This gives rise to Leydig Cell Hypoplasia (LCH), also called Leydig Cell Agenesis. A number of features distinguish LCH from other forms of male pseudohermaphroditism. LCH patients are genetic males with a 46 XY karyotype. The hormonal profile of them shows elevated serum LH level, normal to elevated FSH level, and low testosterone level, which is unresponsive to CG stimulation. Clinical presentation of LCH is variable, ranging from hypergonadotropic hypogonadism with microphallus and hypoplastic male external genitalia to a form of male pseudohermaphroditism with female external genitalia. In between are patients with variable degree of masculinization of the external genitalia. LCH patients show no development of either male or female secondary sexual characteristics at puberty. In females, LHR inactivation causes hypergonadotropic hypogonadism and primary amenorrhea with subnormal follicular development and ovulation, and infertility.

LHR is located on human chromosome 2p21. It has 11 exons. The first 10 exons encode the ECD, while the last exon encodes a small portion of the ECD, the TMD, and the ICD. Currently, 23 inactivating mutations, distributed throughout LHR, have been identified in LCH patients [20,21,147,148]. The majority (17/23) of them are single base substitutions leading to either missense mutations (13/23) or nonsense mutations (4/23). All single base substitutions affect highly conserved amino acids. Both in-frame and out-of-frame loss-of-function insertion mutations have been identified. A 33-bp in-frame insertion has been found as heterozygous as well as homozygous mutation in kindreds with male pseudohermaphrodites. The insertion occurs between amino acid residues 18 and 19, immediately upstream of the signal peptide cleavage site. Partial gene deletions (deletion of exon 8 or exon 10) and minor deletion (6 nucleotides) mutations have also been identified. The majority of the inactivating mutations are found in homozygous patients. In spite of the identification of 23 mutations, there are a number of LCH kindreds in which LHR mutations have not been identified, indicating that inactivating mutations are very heterogeneous or that LCH is caused by mutation of LHR as well as other gene(s).

Effect of the different mutations on LHR activity is variable. Depending on the location of the mutation in the receptor protein, it can cause diminished hormone binding, reduced surface expression, abnormal trafficking, or reduced coupling efficiency, all of which result in reduction or abolition of signal transduction triggered by hormone binding. Clinical presentation of LCH patients can be correlated with the amount of residual activity of the mutated receptor. Mutated LHRs of patients with the most severe phenotype, i.e., male pseudohermaphroditism, have zero or minimal signal transduction activity. On the other hand, patients with male hypogonadism have mutated LHRs with reduced, but not abolished, signal transduction.

Ovarian Dysgenesis—Inactivating Mutation of FSHR. FSHR mediates the action of FSH. In females, FSH function is essential for ovarian follicular maturation. In male, FSH regulates Sertoli cell proliferation before and at puberty, and participates in the regulation of spermatogenesis. Loss-of-function mutations of FSHR are thus expected to be found in connection with hypergonadotropic hypogonadism associated with retarded follicular maturation and anovulatory infertility in women, and small testicles and impaired spermatogenesis in normally masculinized males.

FSHR is located on chromosome 2p, next to LHR. Nine inactivating mutations of FSHR are known, all of which are missense mutations found in the ECD and TMD. The inheritance of these mutations follows an autosomal recessive mode [19,21,149]. The most frequently detected mutation is p.A198V. Female patients homozygous with this mutation present with ovarian dysgenesis that includes hypergonadotropic hypogonadism, primary or early onset secondary amenorrhea, variable pubertal development, hypoplastic ovaries with impaired follicle growth, high gonadotropin and low estrogen levels. Compound heterozygous female patients with totally or partially inactivating FSHR mutations have less prominent phenotypes. The male phenotype of inactivating FSHR mutations is less assuring. The five male subjects with homozygous p.A198V mutation are normally masculinized, with moderately or slightly decreased testicular volume, normal plasma testosterone, normal to elevated LH but high FSH levels, and variable spermatogenic failure. In some cases fertility is maintained, suggesting that FSH action is not compulsory for spermatogenesis.

There is good correlation between the phenotype and the degree of receptor inactivation, as well as the site of mutation and its functional consequences. All mutations in the ECD cause a defect in ligand binding and targeting of FSHR to the cell membrane. In the TMD, the mutations have minimal effect on ligand binding but impair signal transduction to various extents.

Precocious Puberty

Pubertal development is considered precocious if it occurs before 7 or 8 years of age in girls, or 9 years in boys. It is more common in girls and is mostly idiopathic. In contrast, precocious puberty in boys is due to some underlying pathology in half of the cases.

Precocious pubertal development that results from premature activation of the hypothalamic-pituitary-gonadal axis, and hence gonadotropin dependent, is termed central precocious puberty (CPP). Precocious sexual development that is not physiological, not gonadotropin dependent, but the result of abnormal sex hormone secretion, is termed sexual precocity or gonadotropin-independent precocious puberty (also called pseudo-puberty or PPP). CPP is not uncommon in otherwise normal healthy girls.

Clinically, the earliest sign of puberty in girls is breast enlargement (thelarche) in response to rising estrogen secretion, which usually precedes but may accompany or follow the growth of pubic hair. The earliest sign of puberty in boys is an increase in the size of testes and in testosterone production, followed by penile and pubic hair growth. Precocious thelarche and adrenarche may need to be considered but are readily differentiated from true precocious puberty.

Gonadotropin-Dependent Precocious Puberty

A recent study of 156 children with idiopathic central precocious puberty indicates 27.5% of it to be familial. Segregation analysis reveals autosomal dominant transmission with incomplete sex-dependent penetrance [150]. Studies of a girl with CPP show a heterozygous single base substitution leading to the replacement of Arg386 by Pro (c.G1157C) in the carboxy-terminal tail of GPR54. This mutation leads to prolonged activation of intracellular signaling pathways in response to kisspeptin [151]. The kisspeptin-GPR54 signaling complex has been proposed as a gatekeeper of pubertal activation of GnRH neurons and the reproductive axis. Whether mutations affecting the kisspeptin-GPR54 duet are the genetic causes of gonadotropin-dependent precocious puberty need further study.

Gonadotropin-Independent Precocious Puberty

Familial Male-Limited Precocious Puberty (FMPP)—Activating Mutations of LHR. Gain-of-function mutations, resulting in constitutive activation of LHR, cause luteinizing hormone releasing hormone (LHRH)-independent isosexual precocious puberty in boys. Constitutive activity of LHR leads to stimulation of testicular Leydig cells in the fetal and prepubertal period in the absence of the hormone, resulting in autonomous production of testosterone and pubertal development at a very young age. This autosomal dominant condition is termed familial male-limited precocious puberty (FMPP) or testotoxicosis. Signs of puberty usually appear by 2–3 years of age in boys with FMPP. These patients have pubertal to adult levels of testosterone, while the basal and LHRH-stimulated levels of gonadotropins are appropriate for age, i.e., prepubertal. There is also lack of a pubertal pattern of LH pulsatility. Activating mutations of *LHR* have no apparent effect on female carriers.

All activating mutations of the *LHR* identified in FMPP patients are single base substitutions. Sixteen activating mutations in exon 11 of *LHR* have been identified among over 120 kindreds with FMPP [19,20]. These mutations affect 13 amino acids. Half of the mutations are located in transmembrane helix VI (transmembrane VI), which represent 67.5% of all mutations identified in FMPP patients. The most frequently mutated amino acid is Asp578, with >55% of all activating mutations affecting this amino acid. The most common mutation is the c.T1733G transition, which results in the replacement of Asp578 by Gly. This mutation represents >51% of all mutations identified in FMPP patients so far. With the exception of the replacement of Ile542 by Leu in transmembrane V, all mutations result in the substitution of amino acids which are conserved among the glycohormone receptors.

LHR is prone to mutation. Among the kindreds of FMPP with confirmed molecular diagnoses, over 25% are caused by new mutations. There is no difference in clinical manifestation between familial and new mutations. Rare mutations occur more often in patients of non-Caucasian ethnic background. Detection of *LHR* mutations is unsuccessful in about 18% of patients diagnosed to have FMPP.

All FMPP mutations reside in the TMD. Agonist affinity of the mutated receptors is largely unchanged, while cell surface expression is either the same or reduced when compared to the wild-type receptor. All FMPP mutations have been shown to confer constitutive activity to the mutated LHR by *in vitro* expression studies. There is no consensus on the phenotype-genotype correlation in FMPP. However, mutations that give the highest basal level of cAMP in *in vitro* assays are associated with an earlier age of pubertal development.

Besides the germline mutations found in FMPP patients, a somatic activating mutation of *LHR* (p.D578H) has also been identified in tumor tissue of a number of patients with testicular neoplasia. Even though a couple of FMPP patients developed testicular neoplasia, the p.D578H mutation has never been found as a germline mutation in any FMPP patient. Its presence is confined to patients with testicular neoplasia.

Activating Mutation of FSHR. *FSHR* is not prone to mutation. So far, there is only one gain-of-function *FSHR* mutation identified in a single case. The patient had been previously hypophysectomized but maintained normal spermatogenesis in spite of undetectable gonadotropins. He has a heterozygous activating mutation p.D578G in the third intracytoplasmic loop of FSHR. *In vitro* expression of the mutated receptor shows elevated basal activity of the receptor [149].

Ovarian Hyperstimulation Syndrome. Ovarian hyperstimulation syndrome (OHSS) is an iatrogenic complication of ovulation-induction therapy. In its most severe form, this syndrome involves massive ovarian enlargement and the formation of multiple ovarian cysts and can be fatal. OHSS can arise spontaneously during pregnancy owing to a broadening of the specificity of FSHR for hCG at high concentration of the hormone. So far, 6 missense mutations affecting 4 amino acids (p.S128Y, p.T449I, p.T449A, p.I545T, p.D567N, and p.D567G) have been identified in OHSS patients [149,152]. All except one of the mutations are found in the TMD of FSHR. The transmission of these mutations follows an autosomal dominant mode. Patients analyzed so far are heterozygous for these mutations. These mutations relax the ligand specificity of the receptor in such a way that it also binds and becomes activated by hCG. Response of the receptor to TSH was also found for mutations that are located in the TMD. It is possible the promiscuous stimulation of follicles during the first trimester of pregnancy by hCG results in excessive follicular recruitment observed in this disorder.

Acknowledgment

This work was supported in part by the Intramural Research Program of the NIH, Eunice Kennedy Shriver National Institute of Child Health and Human Development.

REFERENCES

1. Mullis PE. Genetics of growth hormone deficiency. *Endocrinol Metab Clin North Am.* 2007;36:17–36.
2. Jameson JL, ed. *Principles of Molecular Medicine.* Humana Press, Inc., New Jersey: Totowa; 1998.
3. Cohen LE, Radovick S. Molecular basis of combined pituitary hormone deficiencies. *Endocr Rev.* 2002;23:431–442.
4. Ayuk J, Sheppard MC. Growth hormone and its disorders. *Postgrad Med J.* 2006;82:24–30.
5. Winter WE, Signorino MR. Review: Molecular thyroidology. *Ann Clin Lab Sci.* 2001;31:221–244.
6. Glaser B. Pendred syndrome. *Pediatr Endocrinol Rev.* 2003;1(suppl 2):199–204; discussion 204.
7. Rivolta CM, Targovnik HM. Molecular advances in thyroglobulin disorders. *Clin Chim Acta.* 2006;374:8–24.
8. Refetoff S, Dumitrescu AM. Syndromes of reduced sensitivity to thyroid hormone: Genetic defects in hormone receptors, cell transporters and deiodination. *Best Pract Res Clin Endocrinol Metab.* 2007;21:277–305.
9. Visser WE, Friesema EC, Jansen J, et al. Thyroid hormone transport in and out of cells. *Trends Endocrinol Metab.* 2008; 19:50–56.
10. Duprez L, Parma J, Van Sande J, et al. TSH receptor mutations and thyroid disease. *Trends Endocrinol Metab.* 1998;9:133–140.
11. Thakker RV. Genetics of endocrine and metabolic disorders: Parathyroid. *Rev Endocr Metab Disord.* 2004;5:37–51.
12. Ferris RL, Simental AA, Jr. Molecular biology of primary hyperparathyroidism. *Otolaryngol Clin North Am.* 2004;37:819–831.
13. Egbuna OI, Brown EM. Hypercalcaemic and hypocalcemic conditions due to calcium-sensing receptor mutations. *Best Pract Res Clin Rheumatol.* 2008;22:129–148.
14. D'Souza-Li L. The calcium-sensing receptor and related diseases. *Arq Bras Endocrinol Metabol.* 2006;50:628–639.
15. Layman LC. Hypogonadotropic hypogonadism. *Endocrinol Metab Clin North Am.* 2007;36:283–296.
16. Bhagavath B, Layman LC. The genetics of hypogonadotropic hypogonadism. *Semin Reprod Med.* 2007;25:272–286.
17. Cadman SM, Kim SH, Hu Y, et al. Molecular pathogenesis of Kallmann's syndrome. *Horm Res.* 2007;67:231–242.
18. de Roux N. GnRH receptor and GPR54 inactivation in isolated gonadotropic deficiency. *Best Pract Res Clin Endocrinol Metab.* 2006;20:515–528.
19. Huhtaniemi IT, Themmen AP. Mutations in human gonadotropin and gonadotropin-receptor genes. *Endocrine.* 2005;26:207–217.
20. Chan WY. Disorders of sexual development caused by luteinizing hormone receptor mutations. *Beijing Da Xue Xue Bao.* 2005;37:32–38.
21. Huhtaniemi I, Alevizaki M. Gonadotrophin resistance. *Best Pract Res Clin Endocrinol Metab.* 2006;20:561–576.
22. Scully KM, Rosenfeld MG. Pituitary development: Regulatory codes in mammalian organogenesis. *Science.* 2002;295:2231–2235.
23. Reynaud R, Saveanu A, Barlier A, et al. Pituitary hormone deficiencies due to transcription factor gene alterations. *Growth Horm IGF Res.* 2004;14:442–448.

24. Zhu X, Lin CR, Prefontaine GG, et al. Genetic control of pituitary development and hypopituitarism. *Curr Opin Genet Dev.* 2005;15:332–340.
25. Coya R, Vela A, Perez de Nanclares G, et al. Panhypopituitarism: Genetic versus acquired etiological factors. *J Pediatr Endocrinol Metab.* 2007;20:27–36.
26. Kelberman D, Dattani MT. Hypothalamic and pituitary development: Novel insights into the aetiology. *Eur J Endocrinol.* 2007;157(suppl 1):S3–S14.
27. Bhangoo AP, Hunter CS, Savage JJ, et al. Clinical case seminar: A novel LHX3 mutation presenting as combined pituitary hormonal deficiency. *J Clin Endocrinol Metab.* 2006;91:747–753.
28. Pfaeffle RW, Savage JJ, Hunter CS, et al. Four novel mutations of the LHX3 gene cause combined pituitary hormone deficiencies with or without limited neck rotation. *J Clin Endocrinol Metab.* 2007;92:1909–1919.
29. Rajab A, Kelberman D, de Castro SC, et al. Novel mutations in LHX3 are associated with hypopituitarism and sensorineural hearing loss. *Hum Mol Genet.* 2008;17:2150–2159.
30. Machinis K, Pantel J, Netchine I, et al. Syndromic short stature in patients with a germline mutation in the LIM homeobox LHX4. *Am J Hum Genet.* 2001;69:961–968.
31. Tajima T, Hattori T, Nakajima T, et al. A novel missense mutation (P366T) of the LHX4 gene causes severe combined pituitary hormone deficiency with pituitary hypoplasia, ectopic posterior lobe and a poorly developed sella turcica. *Endocr J.* 2007;54:637–641.
32. Castinetti F, Saveanu A, Reynaud R, et al. A novel dysfunctional LHX4 mutation with high phenotypical variability in patients with hypopituitarism. *J Clin Endocrinol Metab.* 2008;93:2790–2799.
33. Pfaeffle RW, Hunter CS, Savage JJ, et al. Three novel missense mutations within the LHX4 gene are associated with variable pituitary hormone deficiencies. *J Clin Endocrinol Metab.* 2008;93:1062–1071.
34. Vieira TC, Dias da Silva MR, Cerutti JM, et al. Familial combined pituitary hormone deficiency due to a novel mutation R99Q in the hot spot region of Prophet of Pit-1 presenting as constitutional growth delay. *J Clin Endocrinol Metab.* 2003;88:38–44.
35. Reynaud R, Barlier A, Vallette-Kasic S, et al. An uncommon phenotype with familial central hypogonadism caused by a novel PROP1 gene mutant truncated in the transactivation domain. *J Clin Endocrinol Metab.* 2005;90:4880–4887.
36. Abrao MG, Leite MV, Carvalho LR, et al. Combined pituitary hormone deficiency (CPHD) due to a complete PROP1 deletion. *Clin Endocrinol (Oxf).* 2006;65:294–300.
37. Nose O, Tatsumi K, Nakano Y, et al. Congenital combined pituitary hormone deficiency attributable to a novel PROP1 mutation (467insT). *J Pediatr Endocrinol Metab.* 2006;19:491–498.
38. Fofanova O, Takamura N, Kinoshita E, et al. Compound heterozygous deletion of the PROP-1 gene in children with combined pituitary hormone deficiency. *J Clin Endocrinol Metab.* 1998;83:2601–2604.
39. Parks JS, Brown MR, Hurley DL, et al. Heritable disorders of pituitary development. *J Clin Endocrinol Metab.* 1999;84:4362–4370.
40. Voutetakis A, Maniati-Christidi M, Kanaka-Gantenbein C, et al. Prolonged jaundice and hypothyroidism as the presenting symptoms in a neonate with a novel Prop1 gene mutation (Q83X). *Eur J Endocrinol.* 2004;150:257–264.
41. Rainbow LA, Rees SA, Shaikh MG, et al. Mutation analysis of POUF-1, PROP-1 and HESX-1 show low frequency of mutations in children with sporadic forms of combined pituitary hormone deficiency and septo-optic dysplasia. *Clin Endocrinol (Oxf).* 2005;62:163–168.
42. Miyata I, Vallette-Kasic S, Saveanu A, et al. Identification and functional analysis of the novel S179R POU1F1 mutation associated with combined pituitary hormone deficiency. *J Clin Endocrinol Metab.* 2006;91:4981–4987.
43. Snabboon T, Plengpanich W, Buranasupkajorn P, et al. A novel germline mutation, IVS4+1G>A, of the POU1F1 gene underlying combined pituitary hormone deficiency. *Horm Res.* 2008;69:60–64.
44. McLennan K, Jeske Y, Cotterill A, et al. Combined pituitary hormone deficiency in Australian children: Clinical and genetic correlates. *Clin Endocrinol (Oxf).* 2003;58:785–794.
45. Roessler E, Du YZ, Mullor JL, et al. Loss-of-function mutations in the human GLI2 gene are associated with pituitary anomalies and holoprosencephaly-like features. *Proc Natl Acad Sci USA.* 2003;100:13424–13429.
46. Salvatori R, Fan X, Phillips JA, 3rd, et al. Isolated growth hormone (GH) deficiency due to compound heterozygosity for two new mutations in the GH-releasing hormone receptor gene. *Clin Endocrinol (Oxf).* 2001;54:681–687.
47. Baumann G. Genetic characterization of growth hormone deficiency and resistance: Implications for treatment with recombinant growth hormone. *Am J Pharmacogenomics.* 2002;2:93–111.
48. Salvatori R, Fan X, Mullis PE, et al. Decreased expression of the GHRH receptor gene due to a mutation in a Pit-1 binding site. *Mol Endocrinol.* 2002;16:450–458.
49. Salvatori R, Fan X, Veldhuis JD, et al. Serum GH response to pharmacological stimuli and physical exercise in two siblings with two new inactivating mutations in the GH-releasing hormone receptor gene. *Eur J Endocrinol.* 2002;147:591–596.
50. Carakushansky M, Whatmore AJ, Clayton PE, et al. A new missense mutation in the growth hormone-releasing hormone receptor gene in familial isolated GH deficiency. *Eur J Endocrinol.* 2003;148:25–30.
51. Alba M, Hall CM, Whatmore AJ, et al. Variability in anterior pituitary size within members of a family with GH deficiency due to a new splice mutation in the GHRH receptor gene. *Clin Endocrinol (Oxf).* 2004;60:470–475.
52. Haskin O, Lazar L, Jaber L, et al. A new mutation in the growth hormone-releasing hormone receptor gene in two Israeli Arab families. *J Endocrinol Invest.* 2006;29:122–130.
53. Hilal L, Hajaji Y, Vie-Luton MP, et al. Unusual phenotypic features in a patient with a novel splice mutation in the GHRHR gene. *Mol Med.* 2008;14:286–292.
54. Rojas-Gil AP, Ziros PG, Kanetsis E, et al. Combined effect of mutations of the GH1 gene and its proximal promoter region in a child with growth hormone neurosecretory dysfunction (GHND). *J Mol Med.* 2007;85:1005–1013.
55. Iughetti L, Sobrier ML, Predieri B, et al. Complex disease phenotype revealed by GH deficiency associated with a novel and unusual defect in the GH-1 gene. *Clin Endocrinol (Oxf).* 2008;69:170–172.
56. Hernandez LM, Lee PD, Camacho-Hubner C. Isolated growth hormone deficiency. *Pituitary.* 2007;10:351–357.
57. Rosenfeld RG. Molecular mechanisms of IGF-I deficiency. *Horm Res.* 2006;65(suppl 1):15–20.
58. Fassone L, Corneli G, Bellone S, et al. Growth hormone receptor gene mutations in two Italian patients with Laron Syndrome. *J Endocrinol Invest.* 2007;30:417–420.
59. Gennero I, Edouard T, Rashad M, et al. Identification of a novel mutation in the human growth hormone receptor gene (GHR) in a patient with Laron syndrome. *J Pediatr Endocrinol Metab.* 2007;20:825–831.
60. Ying YQ, Wei H, Cao LZ, et al. Clinical features and growth hormone receptor gene mutations of patients with Laron syndrome from a Chinese family. *Zhongguo Dang Dai Er Ke Za Zhi.* 2007;9:335–338.
61. Fang P, Girgis R, Little BM, et al. Growth hormone (GH) insensitivity and insulin-like growth factor-I deficiency in Inuit subjects and an Ecuadorian cohort: Functional studies of two codon 180 GH receptor gene mutations. *J Clin Endocrinol Metab.* 2008;93:1030–1037.
62. Yamamoto H, Kouhara H, Iida K, et al. A novel growth hormone receptor gene deletion mutation in a patient with primary growth hormone insensitivity syndrome (Laron syndrome). *Growth Horm IGF Res.* 2008;18:136–142.
63. Walenkamp MJ, Wit JM. Genetic disorders in the growth hormone—insulin-like growth factor-I axis. *Horm Res.* 2006;66:221–230.
64. Asa SL, Digiovanni R, Jiang J, et al. A growth hormone receptor mutation impairs growth hormone autofeedback signaling in pituitary tumors. *Cancer Res.* 2007;67:7505–7511.
65. Georgitsi M, Raitila A, Karhu A, et al. Molecular diagnosis of pituitary adenoma predisposition caused by aryl hydrocarbon receptor-interacting protein gene mutations. *Proc Natl Acad Sci USA.* 2007;104:4101–4105.
66. Park SM, Chatterjee VK. Genetics of congenital hypothyroidism. *J Med Genet.* 2005;42:379–389.

67. Beck-Peccoz P, Persani L, Calebiro D, et al. Syndromes of hormone resistance in the hypothalamic-pituitary-thyroid axis. *Best Pract Res Clin Endocrinol Metab.* 2006;20:529–546.
68. Partsch CJ, Riepe FG, Krone N, et al. Initially elevated TSH and congenital central hypothyroidism due to a homozygous mutation of the TSH beta subunit gene: Case report and review of the literature. *Exp Clin Endocrinol Diabetes.* 2006;114:227–234.
69. Miyai K. Congenital thyrotropin deficiency—From discovery to molecular biology, postgenome and preventive medicine. *Endocr J.* 2007;54:191–203.
70. Fuhrer D, Lachmund P, Nebel IT, et al. The thyrotropin receptor mutation database: Update 2003. *Thyroid.* 2003;13: 1123–1126.
71. Kanda K, Mizuno H, Sugiyama Y, et al. Clinical significance of heterozygous carriers associated with compensated hypothyroidism in R450H, a common inactivating mutation of the thyrotropin receptor gene in Japanese. *Endocrine.* 2006;30:383–388.
72. Jeziorowska A, Pniewska-Siark B, Brzezianska E, et al. A novel mutation in the thyrotropin (thyroid-stimulating hormone) receptor gene in a case of congenital hypothyroidism. *Thyroid.* 2006;16:1303–1309.
73. Camilot M, Teofoli F, Gandini A, et al. Thyrotropin receptor gene mutations and TSH resistance: Variable expressivity in the heterozygotes. *Clin Endocrinol (Oxf).* 2005;63:146–151.
74. Tsunekawa K, Onigata K, Morimura T, et al. Identification and functional analysis of novel inactivating thyrotropin receptor mutations in patients with thyrotropin resistance. *Thyroid.* 2006;16:471–479.
75. Tonacchera M, Di Cosmo C, De Marco G, et al. Identification of TSH receptor mutations in three families with resistance to TSH. *Clin Endocrinol (Oxf).* 2007;67:712–718.
76. Grasberger H, Van Sande J, Hag-Dahood Mahameed A, et al. A familial thyrotropin (TSH) receptor mutation provides in vivo evidence that the inositol phosphates/Ca2+ cascade mediates TSH action on thyroid hormone synthesis. *J Clin Endocrinol Metab.* 2007;92:2816–2820.
77. Szinnai G, Kosugi S, Derrien C, et al. Extending the clinical heterogeneity of iodide transport defect (ITD): A novel mutation R124H of the sodium/iodide symporter gene and review of genotype-phenotype correlations in ITD. *J Clin Endocrinol Metab.* 2006;91:1199–1204.
78. Fugazzola L, Cirello V, Dossena S, et al. High phenotypic intrafamilial variability in patients with Pendred syndrome and a novel duplication in the SLC26A4 gene: Clinical characterization and functional studies of the mutated SLC26A4 protein. *Eur J Endocrinol.* 2007;157:331–338.
79. Brownstein ZN, Dror AA, Gilony D, et al. A novel SLC26A4 (PDS) deafness mutation retained in the endoplasmic reticulum. *Arch Otolaryngol Head Neck Surg.* 2008;134:403–407.
80. Palos F, Garcia-Rendueles ME, Araujo-Vilar D, et al. Pendred syndrome in two Galician families: Insights into clinical phenotypes through cellular, genetic, and molecular studies. *J Clin Endocrinol Metab.* 2008;93:267–277.
81. Nascimento AC, Guedes DR, Santos CS, et al. Thyroperoxidase gene mutations in congenital goitrous hypothyroidism with total and partial iodide organification defect. *Thyroid.* 2003;13:1145–1151.
82. Kotani T, Umeki K, Kawano J, et al. Partial iodide organification defect caused by a novel mutation of the thyroid peroxidase gene in three siblings. *Clin Endocrinol (Oxf).* 2003;59:198–206.
83. Rivolta CM, Esperante SA, Gruneiro-Papendieck L, et al. Five novel inactivating mutations in the thyroid peroxidase gene responsible for congenital goiter and iodide organification defect. *Hum Mutat.* 2003;22:259.
84. Tajima T, Tsubaki J, Fujieda K. Two novel mutations in the thyroid peroxidase gene with goitrous hypothyroidism. *Endocr J.* 2005;52:643–645.
85. Rodrigues C, Jorge P, Soares JP, et al. Mutation screening of the thyroid peroxidase gene in a cohort of 55 Portuguese patients with congenital hypothyroidism. *Eur J Endocrinol.* 2005;152:193–198.
86. Pfarr N, Musholt TJ, Musholt PB, et al. Congenital primary hypothyroidism with subsequent adenomatous goiter in a Turkish patient caused by a homozygous 10-bp deletion in the thyroid peroxidase (TPO) gene. *Clin Endocrinol (Oxf).* 2006;64:514–518.
87. Rivolta CM, Louis-Tisserand M, Varela V, et al. Two compound heterozygous mutations (c.215delA/c.2422T–>C and c.387delC/c.1159G–>A) in the thyroid peroxidase gene responsible for congenital goitre and iodide organification defect. *Clin Endocrinol (Oxf).* 2007;67:238–246.
88. Tenenbaum-Rakover Y, Mamanasiri S, Ris-Stalpers C, et al. Clinical and genetic characteristics of congenital hypothyroidism due to mutations in the thyroid peroxidase (TPO) gene in Israelis. *Clin Endocrinol (Oxf).* 2007;66:695–702.
89. Avbelj M, Tahirovic H, Debeljak M, et al. High prevalence of thyroid peroxidase gene mutations in patients with thyroid dyshormonogenesis. *Eur J Endocrinol.* 2007;156:511–519.
90. Deladoey J, Pfarr N, Vuissoz JM, et al. Pseudodominant inheritance of goitrous congenital hypothyroidism caused by TPO mutations: Molecular and in silico studies. *J Clin Endocrinol Metab.* 2008;93:627–633.
91. Kimura S, Hong YS, Kotani T, et al. Structure of the human thyroid peroxidase gene: Comparison and relationship to the human myeloperoxidase gene. *Biochemistry.* 1989;28:4481–4489.
92. Pfarr N, Korsch E, Kaspers S, et al. Congenital hypothyroidism caused by new mutations in the thyroid oxidase 2 (THOX2) gene. *Clin Endocrinol (Oxf).* 2006;65:810–815.
93. Ohye H, Fukata S, Hishinuma A, et al. A novel homozygous missense mutation of the dual oxidase 2 (DUOX2) gene in an adult patient with large goiter. *Thyroid.* 2008;18:561–566.
94. Moreno JC, Visser TJ. New phenotypes in thyroid dyshormonogenesis: Hypothyroidism due to DUOX2 mutations. *Endocr Dev.* 2007;10:99–117.
95. Caputo M, Rivolta CM, Esperante SA, et al. Congenital hypothyroidism with goitre caused by new mutations in the thyroglobulin gene. *Clin Endocrinol (Oxf).* 2007;67:351–357.
96. Kim JH, Park TS, Baek HS, et al. A newly identified insertion mutation in the thyroid hormone receptor-beta gene in a Korean family with generalized thyroid hormone resistance. *J Korean Med Sci.* 2007;22:560–563.
97. Kim JY, Choi ES, Lee JC, et al. Resistance to thyroid hormone with missense mutation (V349M) in the thyroid hormone receptor beta gene. *Korean J Intern Med.* 2008;23:45–48.
98. Sato H, Koike Y, Honma M, et al. Evaluation of thyroid hormone action in a case of generalized resistance to thyroid hormone with chronic thyroiditis: Discovery of a novel heterozygous missense mutation (G347A). *Endocr J.* 2007;54:727–732.
99. Maraninchi M, Bourcigaux N, Dace A, et al. A novel mutation (E333D) in the thyroid hormone beta receptor causing resistance to thyroid hormone syndrome. *Exp Clin Endocrinol Diabetes.* 2006;114:569–576.
100. Refetoff S. Resistance to thyroid hormone: One of several defects causing reduced sensitivity to thyroid hormone. *Nat Clin Pract Endocrinol Metab.* 2008;4:1.
101. Visser WE, Friesema EC, Jansen J, et al. Thyroid hormone transport by monocarboxylate transporters. *Best Pract Res Clin Endocrinol Metab.* 2007;21:223–236.
102. Frints SG, Lenzner S, Bauters M, et al. MCT8 mutation analysis and identification of the first female with Allan-Herndon-Dudley syndrome due to loss of MCT8 expression. *Eur J Hum Genet.* 2008;16:1029–1037.
103. Jansen J, Friesema EC, Kester MH, et al. Functional analysis of monocarboxylate transporter 8 mutations identified in patients with X-linked psychomotor retardation and elevated serum triiodothyronine. *J Clin Endocrinol Metab.* 2007;92:2378–2381.
104. Friesema EC, Jansen J, Heuer H, et al. Mechanisms of disease: Psychomotor retardation and high T3 levels caused by mutations in monocarboxylate transporter 8. *Nat Clin Pract Endocrinol Metab.* 2006;2:512–523.
105. Jansen J, Friesema EC, Kester MH, et al. Genotype-phenotype relationship in patients with mutations in thyroid hormone transporter MCT8. *Endocrinology.* 2008;149:2184–2190.
106. Gozu HI, Mueller S, Bircan R, et al. A new silent germline mutation of the TSH receptor: Coexpression in a hyperthyroid family member with a second activating somatic mutation. *Thyroid.* 2008;18:499–508.
107. Akcurin S, Turkkahraman D, Tysoe C, et al. A family with a novel TSH receptor activating germline mutation (p.Ala485-Val). *Eur J Pediatr.* 2008;167:1231–1237.

108. Nishihara E, Nagayama Y, Amino N, et al. A novel thyrotropin receptor germline mutation (Asp617Tyr) causing hereditary hyperthyroidism. *Endocr J.* 2007;54:927–934.
109. Nishihara E, Fukata S, Hishinuma A, et al. Sporadic congenital hyperthyroidism due to a germline mutation in the thyrotropin receptor gene (Leu 512 Gln) in a Japanese patient. *Endocr J.* 2006;53:735–740.
110. Ferrara AM, Capalbo D, Rossi G, et al. A new case of familial nonautoimmune hyperthyroidism caused by the M463V mutation in the TSH receptor with anticipation of the disease across generations: A possible role of iodine supplementation. *Thyroid.* 2007;17:677–680.
111. Mannstadt M, Bertrand G, Muresan M, et al. Dominant-negative GCMB mutations cause an autosomal dominant form of hypoparathyroidism. *J Clin Endocrinol Metab.* 2008;93:3568–3576.
112. Al-Jenaidi F, Makitie O, Grunebaum E, et al. Parathyroid gland dysfunction in 22q11.2 deletion syndrome. *Horm Res.* 2007;67:117–122.
113. Hernandez AM, Villamar M, Rosello L, et al. Novel mutation in the gene encoding the GATA3 transcription factor in a Spanish familial case of hypoparathyroidism, deafness, and renal dysplasia (HDR) syndrome with female genital tract malformations. *Am J Med Genet A.* 2007;143:757–762.
114. Chiu WY, Chen HW, Chao HW, et al. Identification of three novel mutations in the GATA3 gene responsible for familial hypoparathyroidism and deafness in the Chinese population. *J Clin Endocrinol Metab.* 2006;91:4587–4592.
115. Brandi ML, Falchetti A. Genetics of primary hyperparathyroidism. *Urol Int.* 2004;72(suppl 1):11–16.
116. Lemos MC, Thakker RV. Multiple endocrine neoplasia type 1 (MEN1): Analysis of 1336 mutations reported in the first decade following identification of the gene. *Hum Mutat.* 2008;29:22–32.
117. Raue F, Haag C, Schulze E, et al. The role of the extracellular calcium-sensing receptor in health and disease. *Exp Clin Endocrinol Diabetes.* 2006;114:397–405.
118. Phelan JK, McCabe ER. Mutations in NR0B1 (DAX1) and NR5A1 (SF1) responsible for adrenal hypoplasia congenita. *Hum Mutat.* 2001;18:472–487.
119. Laissue P, Copelli S, Bergada I, et al. Partial defects in transcriptional activity of two novel DAX-1 mutations in childhood-onset adrenal hypoplasia congenita. *Clin Endocrinol (Oxf).* 2006;65:681–686.
120. McCabe ER. DAX1: Increasing complexity in the roles of this novel nuclear receptor. *Mol Cell Endocrinol.* 2007;265–266:179–182.
121. Nimkarn S, New MI. Prenatal diagnosis and treatment of congenital adrenal hyperplasia owing to 21-hydroxylase deficiency. *Nat Clin Pract Endocrinol Metab.* 2007;3:405–413.
122. Bhangoo A, Wilson R, New MI, et al. Donor splice mutation in the 11beta-hydroxylase (CypllB1) gene resulting in sex reversal: A case report and review of the literature. *J Pediatr Endocrinol Metab.* 2006;19:1267–1282.
123. Nimkarn S, New MI. Steroid 11beta- hydroxylase deficiency congenital adrenal hyperplasia. *Trends Endocrinol Metab.* 2008;19:96–99.
124. Simard J, Ricketts ML, Gingras S, et al. Molecular biology of the 3beta-hydroxysteroid dehydrogenase/delta5-delta4 isomerase gene family. *Endocr Rev.* 2005;26:525–582.
125. Welzel M, Wustemann N, Simic-Schleicher G, et al. Carboxyl-terminal mutations in 3-beta-hydroxysteroid dehydrogenase type II cause severe salt-wasting congenital adrenal hyperplasia. *J Clin Endocrinol Metab.* 2008;93:1418–1425.
126. Biason-Lauber A, Kempken B, Werder E, et al. 17-alpha-hydroxylase/17,20-lyase deficiency as a model to study enzymatic activity regulation: Role of phosphorylation. *J Clin Endocrinol Metab.* 2000;85:1226–1231.
127. Kim CJ, Lin L, Huang N, et al. Severe combined adrenal and gonadal deficiency caused by novel mutations in the cholesterol side chain cleavage enzyme, P450scc. *J Clin Endocrinol Metab.* 2008;93:696–702.
128. Fujieda K, Okuhara K, Abe S, et al. Molecular pathogenesis of lipoid adrenal hyperplasia and adrenal hypoplasia congenita. *J Steroid Biochem Mol Biol.* 2003;85:483–489.
129. Artigas RA, Gonzalez A, Riquelme E, et al. A novel adrenocorticotropin receptor mutation alters its structure and function, causing familial glucocorticoid deficiency. *J Clin Endocrinol Metab.* 2008;98:3097–3105.
130. Chan LF, Clark AJ, Metherell LA. Familial glucocorticoid deficiency: Advances in the molecular understanding of ACTH action. *Horm Res.* 2008;69:75–82.
131. Mazur A, Koehler K, Schuelke M, et al. Familial glucocorticoid deficiency type 1 due to a novel compound heterozygous MC2R mutation. *Horm Res.* 2008;69:363–368.
132. Tsiotra PC, Koukourava A, Kaltezioti V, et al. Compound heterozygosity of a frameshift mutation in the coding region and a single base substitution in the promoter of the ACTH receptor gene in a family with isolated glucocorticoid deficiency. *J Pediatr Endocrinol Metab.* 2006;19:1157–1166.
133. Rumie H, Metherell LA, Clark AJ, et al. Clinical and biological phenotype of a patient with familial glucocorticoid deficiency type 2 caused by a mutation of melanocortin 2 receptor accessory protein. *Eur J Endocrinol.* 2007;157:539–542.
134. Collares CV, Antunes-Rodrigues J, Moreira AC, et al. Heterogeneity in the molecular basis of ACTH resistance syndrome. *Eur J Endocrinol.* 2008;159:61–68.
135. Krude H, Biebermann H, Schnabel D, et al. Obesity due to proopiomelanocortin deficiency: Three new cases and treatment trials with thyroid hormone and ACTH4–10. *J Clin Endocrinol Metab.* 2003;88:4633–4640.
136. Farooqi IS, Drop S, Clements A, et al. Heterozygosity for a POMC-null mutation and increased obesity risk in humans. *Diabetes.* 2006;55:2549–2553.
137. Jackson RS, Creemers JW, Farooqi IS, et al. Small-intestinal dysfunction accompanies the complex endocrinopathy of human proprotein convertase 1 deficiency. *J Clin Invest.* 2003;112:1550–1560.
138. Vallette-Kasic S, Couture C, Balsalobre A, et al. The TPIT gene mutation M86R associated with isolated adrenocorticotropin deficiency interferes with protein: Protein interactions. *J Clin Endocrinol Metab.* 2007;92:3991–3999.
139. Charmandari E, Kino T, Ichijo T, et al. Generalized glucocorticoid resistance: Clinical aspects, molecular mechanisms, and implications of a rare genetic disorder. *J Clin Endocrinol Metab.* 2008;93:1563–1572.
140. Swords FM, Baig A, Malchoff DM, et al. Impaired desensitization of a mutant adrenocorticotropin receptor associated with apparent constitutive activity. *Mol Endocrinol.* 2002;16:2746–2753.
141. Carlson HE. Human adrenal cortex hyperfunction due to LH/hCG. *Mol Cell Endocrinol.* 2007;269:46–50.
142. Salenave S, Chanson P, Bry H, et al. Kallmann's syndrome: A comparison of the reproductive phenotypes in men carrying KAL1 and FGFR1/KAL2 mutations. *J Clin Endocrinol Metab.* 2008;93:758–763.
143. Albuisson J, Pecheux C, Carel JC, et al. Kallmann syndrome: 14 novel mutations in KAL1 and FGFR1 (KAL2). *Hum Mutat.* 2005;25:98–99.
144. Pedersen-White JR, Chorich LP, Bick DP, et al. The prevalence of intragenic deletions in patients with idiopathic hypogonadotropic hypogonadism and Kallmann syndrome. *Mol Hum Reprod.* 2008;14:367–370.
145. Bedecarrats GY, Kaiser UB. Mutations in the human gonadotropin-releasing hormone receptor: Insights into receptor biology and function. *Semin Reprod Med.* 2007;25:368–378.
146. Lofrano-Porto A, Barra GB, Giacomini LA, et al. Luteinizing hormone beta mutation and hypogonadism in men and women. *N Engl J Med.* 2007;357:897–904.
147. Leung MY, Steinbach PJ, Bear D, et al. Biological effect of a novel mutation in the third leucine-rich repeat of human luteinizing hormone receptor. *Mol Endocrinol.* 2006;20:2493–2503.
148. Bruysters M, Christin-Maitre S, Verhoef-Post M, et al. A new LH receptor splice mutation responsible for male hypogonadism with subnormal sperm production in the propositus, and infertility with regular cycles in an affected sister. *Hum Reprod.* 2008;23:1917–1923.
149. Meduri G, Bachelot A, Cocca MP, et al. Molecular pathology of the FSH receptor: New insights into FSH physiology. *Mol Cell Endocrinol.* 2008;282:130–142.

150. de Vries L, Kauschansky A, Shohat M, et al. Familial central precocious puberty suggests autosomal dominant inheritance. *J Clin Endocrinol Metab.* 2004;89:1794–1800.
151. Teles MG, Bianco SD, Brito VN, et al. A GPR54-activating mutation in a patient with central precocious puberty. *N Engl J Med.* 2008;358:709–715.
152. De Leener A, Caltabiano G, Erkan S, et al. Identification of the first germline mutation in the extracellular domain of the follitropin receptor responsible for spontaneous ovarian hyperstimulation syndrome. *Hum Mutat.* 2008;29:91–98.

Chapter

23

Molecular Basis of Gynecologic Diseases

Samuel C. Mok ▪ Kwong-kwok Wong ▪ Karen Lu ▪ Karl Munger ▪ Zoltan Nagymanyoki

INTRODUCTION

Gynecologic diseases in general are diseases involving the female reproductive tract. These diseases include benign and malignant tumors, pregnancy-related diseases, infection, and endocrine diseases. Among them, malignant tumor is the most common cause of death. In recent years, the etiology of some of these diseases has been revealed. For example, human papilloma virus (HPV) infection has been shown to be one of the major etiological factors associated with cervical cancer. Inactivation of tumor suppressor gene *BRCA1* has been implicated in hereditary ovarian cancer. In spite of these findings, the molecular bases of most of the diseases remain largely unknown. In this chapter, we will focus on discussing benign and malignant tumors of female reproductive organs as well as pregnancy-related diseases, which have relatively well-understood molecular bases.

BENIGN AND MALIGNANT TUMORS OF THE FEMALE REPRODUCTIVE TRACT

Cervix

Infections of the genital mucosa with human papillomaviruses represent the most common virus-associated sexually transmitted disease, and at age 50 approximately 80% of all females will have acquired a genital HPV infection sometime during their life [1]. At present, approximately 630 million individuals worldwide have a genital HPV infection, with an incidence of approximately 30 million new infections per year [2]. Currently in the United States there are in excess of 20 million people with genital HPV infections, with an estimated annual incidence of 6.2 million new infections [3,4]. Genital HPV infections are particularly prevalent in sexually active younger individuals [5]. Most of these infections are transient and may not cause any overt clinical disease or symptoms. Nonetheless, the total annual cost of clinical care for genital HPV infections exceeds $3 billion in the United States alone [6].

HPVs Associated with Cervical Lesions and Cancer [5]

Human papillomaviruses are members of the Papillomaviridae family. They have a tropism for squamous epithelial cells and cause the formation of generally benign hyperplastic lesions that are commonly referred to as papillomas or warts (reviewed in [7]). Papillomaviruses contain closed circular double-stranded DNA genomes of approximately 8,000 base pairs that are packaged into ~55 nm nonenveloped icosahedral particles. Their genomes consist of three regions; the early (E) region encompasses up to 8 open reading frames (ORFs) designated E followed by a numeral, with the lowest number designating the longest ORF, and the late (L) region encodes the major and minor capsid proteins, L1 and L2, respectively. Only one of the two DNA strands is actively transcribed, and early and late ORFs are encoded on the same DNA strand. A third region, referred to as the long control region (LCR), the upstream regulatory region (URR), or the noncoding region (NCR), does not have significant coding capacity and contains various regulatory DNA sequences that control viral genome replication and transcription (Figure 23.1A) (reviewed in [7]).

In excess of 100 HPV types have been described [6]. HPVs are classified as genotypes based on their nucleotide sequences. A new HPV type is defined when the entire genome has been cloned and the sequence of the L1 open reading frame (ORF), the most conserved ORF among papillomaviruses, is less than 90% identical

Molecular Pathology © 2009, Elsevier, Inc. All Rights Reserved.

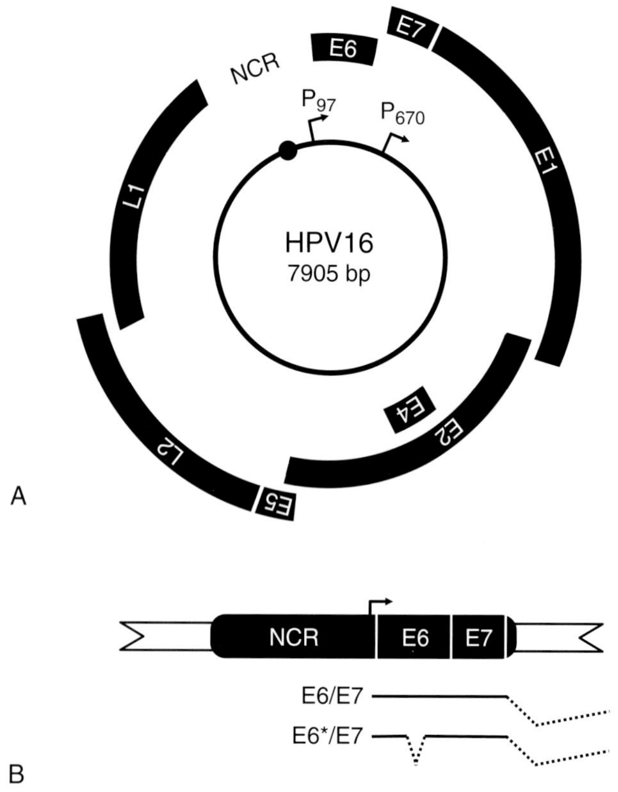

Figure 23.1 The HPV genome. (A) Schematic representation of the HPV16 genome. The double-stranded circular DNA genome is represented by the central circle. Early (E) and late (L) genes are all transcribed from one of the two DNA strands in each of the three possible reading frames. The noncoding region (NCR) does not have extensive coding potential but contains the viral origin of replication (designated by the black circle) as well as the major early promoter, P_{97}, designated by an arrow. The differentiation specific late promoter, P_{670}, is contained within the E7 ORF. See text for details. **(B)** Structure of the minimal HPV16 genome fragment that is consistently retained after integration into a host chromosome. The major E6/E7 transcripts are shown underneath. See text for details.

to a known HPV type. HPVs with higher sequence identity are referred to as subtypes (90% to 98% identity) or variants (>98% identity) [8].

Approximately 30 HPV types infect mucosal epithelia, and these viruses are further classified as low risk or high risk depending on the propensity for malignant progression of the lesions that they cause. Low-risk HPVs, such as HPV6 and HPV11, cause genital warts, whereas high-risk HPVs, such as HPV16 and HPV18, cause intraepithelial neoplasia that can progress to frank carcinoma. Harald zur Hausen's group discovered the association of HPVs with anogenital tract lesions and isolated HPVs from genital warts [9]. Using these sequences as hybridization probes under low stringency conditions, they succeeded in detecting HPV sequences in cervical carcinomas [10]. The most abundant high-risk HPVs are HPV16 and HPV18, which are detected in approximately 50% and 20% of all cervical carcinomas, respectively. HPV18 appears to be frequently associated with adenocarcinomas, whereas HPV16 is mostly detected in squamous cell carcinomas (reviewed in [11]). The following sections are focused on a review of mucosal high-risk HPVs and their contributions to cervical lesions and cancers.

HPV Infection and Life Cycle

The HPV life cycle is closely linked to the differentiation program of the infected squamous epithelial host cell. HPVs infect basal cells, a single layer of actively cycling cells in the squamous epithelium. Basal cells are not readily accessible for viral infection, as they are protected by several layers of differentiated cells that have withdrawn for the cell division cycle. These cell layers are essential for the mechanical stability of the skin and shield the proliferating basal cells from environmental genotoxic insults. HPVs can gain access to basal cells through microabrasions caused by mechanical trauma. Basal-like cells at the cervical squamocolumnar transformation zone in the cervix, however, are particularly accessible and vulnerable to HPV infection. It has been postulated that within the cervical transformation zone, reserve cells, which can give rise to squamous or columnar epithelia, may be physiologically relevant targets for HPV infection [12–14]. The mechanisms of viral entry remain relatively poorly understood but are thought to involve initial binding to heparin sulfate on the cell surface followed by receptor binding and viral uptake, although there is controversy regarding the identity of the virus receptor (reviewed in [15]).

Following infection, HPV genomes are maintained at a low copy number in the nuclei of infected cells and can persist in basal epithelial cells for decades. The productive phase of the viral life cycle, which includes HPV genome amplification, production of capsid proteins, and packaging of newly synthesized genomes, however, occurs exclusively in the terminally differentiated layers of the infected tissue. HPVs are nonlytic, and infectious viral particles are sloughed off with the terminally differentiated, denucleated scales where they remain infectious over extended periods of time (reviewed in [16]).

HPVs encode two proteins, E1 and E2, which directly contribute to viral genome replication. The E1 origin-binding protein is the only virally encoded enzyme and has intrinsic ATPase and helicase activities [17]. E1 forms a complex with the E2 protein, the major HPV-encoded transcriptional regulatory protein. E2 binds with high affinity to specific DNA sequences $ACCN_6GGT$ in the viral regulatory region, whereas E1 binds to the AT-rich replication origin sequences with relatively low affinity. The origin sequence is often flanked by E2 binding sites resulting in high-affinity binding of the E1/E2 complex to the origin of replication [18].

With the exception of the E1/E2 origin-binding complex, HPVs do not encode enzymes that are necessary for viral genome replication and co-opt the host DNA synthesis machinery. Since high-copy number

HPV genome replication and viral progeny synthesis is confined to terminally differentiated cells, which are growth arrested and thus intrinsically incompetent for DNA replication, a major challenge for the viral life cycle is to maintain and/or re-establish a replication-competent milieu in these cells.

The HPV E7 protein contributes to induction and/or maintenance of S-phase competence in differentiating keratinocytes through several mechanisms. Perhaps most importantly, HPV E7 proteins bind to the retinoblastoma tumor suppressor protein pRB and the related p107 and p130 pocket proteins. These proteins have been implicated in regulating G1/S phase transition through members of the E2F family of transcription factors. The G1 specific pRB/E2F complex is a transcriptional repressor that inhibits S-phase entry. In normal cells, pRB is phosphorylated by cyclin/cdk complexes in late G1, the pRB/E2F complex dissociates, and DNA-bound E2Fs act as transcriptional activators. The pRB/E2F complex re-forms when pRB is dephosphorylated at the end of mitosis. This regulatory loop is subverted by HPV E7 proteins, which can associate with pRB and abrogate the inhibitory activity of pRB/E2F complexes. E7 proteins encoded by low-risk HPV associate with pRB with lower affinity than high-risk HPV E7 proteins. Additionally, high-risk HPV E7 proteins induce proteasome-mediated degradation of pRB. Moreover, E7 proteins abrogate the action of cyclin-dependent kinase inhibitors (CKIs) $p21^{CIP1}$ and $p27^{KIP1}$, which regulate cell cycle withdrawal during epithelial cell differentiation, thereby uncoupling epithelial cell differentiation and cell cycle withdrawal. This leads to the formation of hyperplastic lesions, warts, and is necessary for production of progeny virus (reviewed in [19]).

Detection of HPV-Associated Lesions

Papanicolaou tests (also known as Pap tests) are named after their inventor, Georgios Papanicolaou, and serve to detect HPV-associated lesions in the cervix. Upon implementation, this relatively inexpensive test dramatically reduced the incidence and mortality rates of cervical cancer. In the United States the current recommendation is for women to have a Pap test performed at least once every 3 years, and in 2003 approximately 65.6 million Pap tests were performed [20]. The current test involves collecting exfoliated epithelial cells from the outer opening of the cervix and either directly smearing the cells on a slide, or immediately preserving and storing the cells in fixative liquid medium followed by automated processing into a monolayer. In either format, cells are stained and examined for cytological abnormalities. Liquid-based monolayer cytology appears to have a reduced rate of false-positivity presumably due to standardized specimen preparation and immediate fixation of the sampled cells [21,22]. While nuclear features are currently used for diagnoses, attempts are under way to identify new biomarkers for high-risk HPV-associated cervical lesions. The most promising biomarker is $p16^{INK4A}$, an inhibitor of cdk4/cdk6 cyclin D complexes [23], although the molecular basis of $p16^{INK4A}$ overexpression in cervical cancer is currently unknown.

The *American Cancer Society* (ACS) and the *American College of Obstetricians and Gynecologists* (ACOG) also recommend that women over the age of 30 be tested for the presence of HPV DNA [20]. The only currently FDA-approved HPV testing method is based on nucleic acid hybridization and can distinguish between absence and presence of the most frequent low-risk or high-risk HPV types but does not allow identification of individual HPV types. A number of different PCR-based HPV typing methods have been developed for research purposes, but these are currently not FDA approved. While some studies have suggested that HPV testing may be more effective in identifying cervical lesions than Pap smears, this issue clearly requires additional study [24].

Diagnosis and Treatment

Genital warts are a frequent manifestation of low-risk mucosal HPV infections and are diagnosed based on appearance. Such lesions have a very low propensity for malignant progression, and they often regress spontaneously. However, patients generally insist on their removal. No HPV-specific antivirals currently exist for such applications (reviewed in [25]). Standard therapeutic modalities include surgical excision, laser therapy, cryotherapy, topical administration of various caustic chemicals, or immunomodulating agents (reviewed in [7]). In rare cases, low-risk HPV infections of the genital tract can also cause serious disease. One example is the giant condyloma of Buschke-Lowenstein that is caused by infection with low-risk HPVs. In such patients, the immune system is unable to control and/or clear the infection. Although these are slow-growing lesions, they are highly destructive to adjacent normal tissue and eventually can form local and distant metastases (reviewed in [26,27]).

Cytological abnormalities detected by Pap tests are classified according to the Bethesda system as atypical squamous cells (ASC) or squamous intraepithelial lesions (SIL). ASC are further classified as Atypical Squamous Cells of Undetermined Significance (ASCUS) or Atypical Squamous Cells, cannot exclude High-grade squamous intraepithelial lesions (ASC-H), whereas SILs are designated low-grade (LSIL) or high-grade (HSIL) (reviewed in [18]). LSILs are followed up by additional Pap tests, whereas HSILs require analysis by colposcopy. The procedure involves application of an acetic acid-based solution to the cervix whereupon lesions appear as white masses upon evaluation with a colposcope. When lesions are detected, a biopsy is performed and the tissue is examined histologically. Lesions are classified as cervical intraepithelial neoplasias (CIN), carcinoma in situ, or invasive cervical carcinoma. Treatments for CIN include cryotherapy, laser ablation, or loop electrosurgical excision, whereas carcinomas are treated by surgery and/or chemotherapy (reviewed in [11]).

Prevention and Vaccines

Condoms reduce, but do not negate, the risk of infections with HPVs [29,30]. In addition, preclinical studies in a mouse model suggest that the polysaccharide carrageen greatly inhibits HPV transmission, whereas the spermicidal compound nonoxynol-9 appears to increase HPV transmission [31], but these studies await confirmation by clinical studies in humans.

The first generation prophylactic HPV vaccines consist of recombinant HPV L1 proteins that self-assemble into virus-like particles (VLPs). Gardasil® was developed by Merck and has been FDA approved for use in girls and young women of 9 to 26 years of age. It is a quadrivalent formulation that contains VLPs of the most prevalent low-risk (HPV6 and HPV11) and high-risk HPVs (HPV16 and HPV18). It is administered as three doses over the course of 6 months and promises to be highly efficacious in providing type-specific protection from new infections with these HPV types. As such, this vaccine has the potential to reduce the burden of cervical carcinoma and genital warts by up to 70% (reviewed in [32]). Since these prophylactic vaccines lead to the development of humoral immune responses, they are not predicted to affect potential HPV infections at the time of vaccination. Given that cervical cancer generally develops decades after the initial infection, it has been estimated that incidence and mortality rates of cervical cancer will not decrease for 25 to 40 years (reviewed in [33]). Moreover, it is not clear whether other nonvaccine high-risk HPV types will become more prevalent as HPV16 and HPV18 are removed from the biological pool. Hence, recommendations for cervical cytology screening (Pap smears) remain unchanged for vaccinated individuals.

The minor capsid protein L2 contains linear, cross-neutralizing epitopes that may afford more general protection from HPV infections and appears to be an excellent candidate for development of second-generation vaccines (reviewed in [34]).

Molecular Mechanisms of High-Risk HPV-Mediated Cellular Transformation

The transforming activities of high-risk HPVs reflect the necessity of the virus to establish and/or maintain a replication-competent cellular milieu in terminally differentiated keratinocytes. Three high-risk HPV proteins—E5, E6, and E7—have been shown to exhibit transforming activities in cellular and animal systems. Only E6 and E7 are regularly expressed in cervical carcinoma where the HPV genome is frequently integrated into a host chromosome. The integration event is relatively nonspecific with respect to the host chromosome [35]. Integration frequently involves common fragile sites [36], and in some cases HPVs integrate in the vicinity of cellular proto-oncogenes such as *c-myc* [37]. However, integration of high-risk HPV genomes often causes disruption and/or deletion of the E2 ORF, which encodes a transcriptional repressor of E6/E7 expression [38, 39]. Moreover, HPV16 E6/E7 mRNAs are more stable when expressed from integrated copies as compared to episomal HPV16 genomes [40]. Thus, E6/E7 oncoprotein expression is dysregulated in cervical carcinomas (Figure 23.1B). Ectopic expression of HPV E6 and E7 in primary human epithelial cells facilitates cellular immortalization, and when high-risk HPV E6/E7 expressing cells are grown under conditions where they can form multilayered, skin-like structures, they exhibit cellular abnormalities that are reminiscent of high-grade premalignant cervical lesions [41]. Nonetheless, low-passage high-risk HPV immortalized cells are nontumorigenic in immune-deficient mice but can undergo full transformation upon prolonged passaging [42] or when additional oncogenes such as *ras* or *fos* are expressed [43,44]. As such, this mimics the situation *in vivo*, where high-risk HPV-associated cervical lesions progress to cervical cancer after long-term persistent infection (reviewed in [11]). Cervical cancers show hallmarks of chromosomal instability and are generally aneuploid [45]. Cervical cancers that arise upon expression of HPV16 E6/E7 in basal epithelial from a keratin K14 promoter in transgenic animals require chronic treatment with low doses of estrogen, further supporting the notion that additional genomic aberrations are necessary for cancer development [46]. Nonetheless, high-risk HPV E6/E7 oncoproteins directly contribute to malignant progression through induction of genomic instability (reviewed in [47]). Moreover persistent HPV E6/E7 expression is necessary for the maintenance of the transformed phenotype of HPV positive cervical cancer cell lines (reviewed in [48]).

Biological and Biochemical Activities of High-Risk HPV E7 Oncoproteins

Among the high-risk HPV-encoded E7 oncoproteins, HPV16 E7 has been most thoroughly studied. It is a 98 amino acid phosphoprotein that is actively transported to the nucleus [49], but has been detected both in the nucleus and the cytoplasm. HPV16 E7 lacks any known intrinsic enzymatic activities and does not specifically associate with DNA sequences. Rather, it functions by associating with and functionally modifying host cellular regulatory protein complexes. Given its subcellular localization, HPV16 E7 has been shown to associate with nuclear as well as cytoplasmic cellular target proteins. HPV16 E7 shares sequence similarity to a small portion of conserved region 1 (CR1), as well as the entire CR2 of the adenovirus (Ad) E1A oncoprotein. The CR2 homology domain contains the pRB tumor suppressor core-binding site, LXCXE (L, leucine; C, cysteine; E, glutamic acid; X, any amino acid residue), as well as a casein kinase II phosphorylation site. The carboxyl terminus consists of two CXXC motifs separated by a 29 amino acid spacer region that forms a novel zinc binding structure [50,51] (Figure 23.2). Unlike Ad E1A and SV40 TAg, which target pRB through a stoichiometric mechanism, HPV16 E7 inactivates this tumor suppressor protein through ubiquitin-mediated proteasomal degradation, and sequences within the HPV16 E7 CR1 homology domain that do not contribute to pRB binding are necessary for this activity. These HPV16 E7 sequences serve as a binding site for a cullin 2 containing ubiquitin ligase complex, which contributes to pRB degradation [52]. In addition

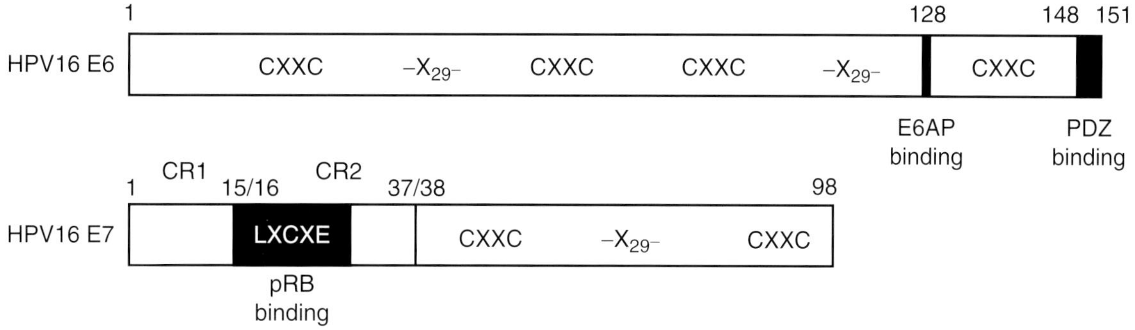

Figure 23.2 **HPV16 E6 (top) and E7 (bottom) oncoproteins.** The positions of the binding sites for E6AP and PDZ proteins on E6 are indicated by black boxes. The amino terminal domains of HPV16 E7 that are similar to a portion of conserved region (CR1) and to CR2 of the Adenovirus E1A protein are indicated. The CR2 domain contains the LXCXE core pRB binding site, which is indicated by a black box that is not to scale. See text for details and references.

the CR1 homology domain also contains a binding site for the N-end rule ubiquitin ligase p600 [53,54], which appears to play an important role in regulating anoikis, a form of apoptosis that is induced upon detachment of cells from a substratum [55–57]. The carboxyl terminus of HPV16 E7 contains additional sequences that are necessary for transformation [48]. While a large number of putative cellular targets, including histone modifying enzymes, that bind to carboxyl terminal sequences have been identified, it is not clear which of these interactions, if any, contribute to cellular transformation (reviewed in [48]).

HPV16 E7 has been shown to associate with pyruvate kinase and modulate the activity of alpha glucosidase and induce a metabolic switch from oxidative phosphorylation to glycolysis, reminiscent of the Warburg effect [59,60]. In addition, HPV16 E7 expression causes aberrant activation of the survival kinase AKT in differentiating keratinocytes, which has been linked to increased cell motility [61,62]. HPV16 E7 has also been shown to cooperate with E6 to induce expression of proteins that modulate angiogenesis [63,64].

Biological and Biochemical Activities of High-Risk HPV E6 Oncoproteins

Like HPV16 E7, HPV16 E6 is a relatively small protein of 151 amino acids that lacks intrinsic enzymatic and specific DNA-binding activities and functions by subverting host cellular protein complexes. HPV16 E6 consists of two zinc-binding motives, each of which consists of 2 CXXC sequences separated by 29 amino acids that each are related to the carboxyl terminal domain of E7 (Figure 23.2). The best known activity of HPV16 E6 is the ability to associate with p53 and the ubiquitin ligase E6-AP, which leads to the ubiquitin-mediated proteasomal degradation of the p53 tumor suppressor. The p53 tumor suppressor senses cellular stress including aberrant S-phase entry induced by expression of the HPV E7 oncoprotein. Activation of p53 results in transcriptional induction of abortive cellular programs such as G1 growth arrest and/or apoptosis, which have been collectively referred to as the trophic sentinel response (Figure 23.3). Inactivation of p53 by the E6 protein presumably serves to abrogate this response, thereby allowing for persistent S-phase competence in differentiated

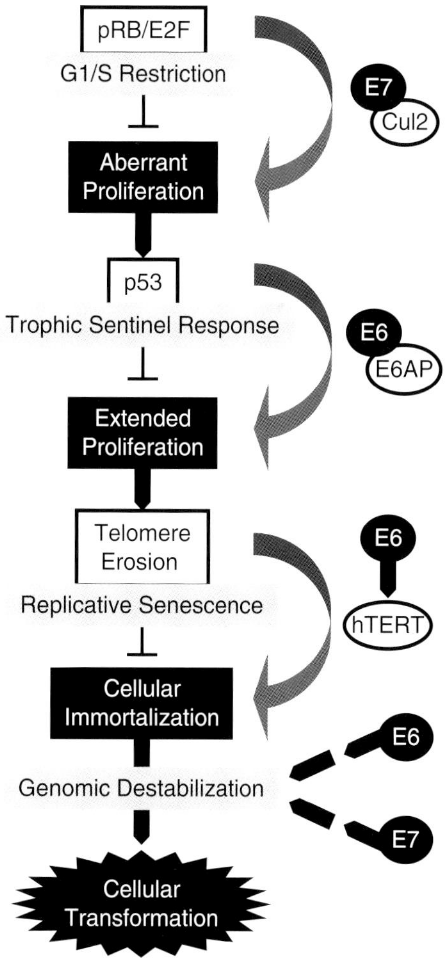

Figure 23.3 **Schematic depiction of some of the major biochemical and biological activities of high-risk HPV oncogenes and how they may cooperate in the development of cervical disease and cancer.**

cells. HPV E6 expression also increases transcription of the catalytic subunit of telomerase, hTERT, which contributes to cellular immortalization of primary human epithelial cells (Figure 23.3).

High-risk HPV E6 proteins contain a short peptide sequence (S/T)-X-V-I-L at their carboxyl termini which mediates association with cellular PDZ proteins [65, 66] (Figure 23.2). The integrity of the PDZ binding sequence on E6 is important for the HPV viral life cycle [39], as well as the transforming activities of E6 [65,68]. Many PDZ proteins act as molecular scaffolds and regulate a number of important processes including cell polarity. It is not clear whether high-risk HPV E6 proteins associate with one single PDZ protein or whether they can target multiple family members, since a number of different candidates have been reported. Since PDZ protein and E6AP binding on E6 are not mutually exclusive, E6-associated PDZ proteins can be targeted for ubiquitination by E6AP (reviewed in [69]).

Many papillomavirus E6 proteins associate with paxillin, and association with E6 leads to disruption of paxillin's association with the focal adhesion proteins vinculin and focal adhesion kinase (FAK). This causes defects in the cellular actin cytoskeleton, as is typically observed in transformed cells [70–73].

Contributions of HPV Oncoproteins to Induction of Genomic Instability

Cervical cancers generally develop years or decades after the initial infection, and these tumors have suffered a multitude of genomic aberrations. The acquisition of some of these genomic alterations appears to define certain stages of disease progression [74,75]. The action of the high-risk HPV E6 and E7 oncoproteins on telomerase activity and the p53 and pRB tumor suppressors are sufficient to lead to extended uncontrolled proliferation and cellular immortalization, but acquisition of additional host genome mutations is necessary for malignant progression (Figure 23.3). A defining biological activity of high-risk HPV E6/E7 proteins is their ability to subvert genomic integrity [76, 77]. Hence, high-risk HPV E6/E7 oncoproteins not only contribute to initiation but also play a key role in malignant progression.

There are two principal mechanisms that lead to genomic instability. Subversion of cell cycle checkpoint mechanisms and DNA repair pathways can lead to perpetuation of mutations induced by environmental triggers such as UV irradiation or exposure to chemical compounds that cause DNA damage. Alternatively, genomic instability can be triggered by active mechanisms that cause genomic destabilization, which have been collectively referred to as a mutator phenotype. Expression of HPV E6/E7 oncoproteins causes genomic instability by both of these mechanisms.

HPV16 E7 has activities of a mitotic mutator and its expression in primary human epithelial cells causes several types of mitotic abnormalities. These include induction of supernumerary centrosomes, lagging chromosomes, and anaphase bridges. Of these, induction of supernumerary centrosomes has been studied in greatest detail. Centrosome-associated multipolar mitoses are a histopathological hallmark of high-risk HPV-associated cervical lesions. HPV16 E7 induces centrosome abnormalities through multiple, cooperating pathways. HPV16 E7 expression causes aberrant activation of cdk2 through several mechanisms, including E2F-mediated transcriptional activation of expression of the cdk2 catalytic subunits cyclins E and A as a consequence of pRB/p107/p130 degradation and inactivation of the cdk2 inhibitors $p21^{CIP1}$ and $p27^{KIP1}$. Induction of supernumerary centrosomes by HPV16 E7 is strictly dependent on cdk2 activity [78,79]. In addition, HPV16 E7 associates with the centrosomal regulatory protein gamma-tubulin and inhibits its loading on centrosomes [80]. As a consequence, the process of centrosome duplication is uncoupled from the cell division cycle [81], resulting in the synthesis of multiple daughter centrioles from a single maternal centriole template [82]. Whereas expression of HPV16 in primary cells effectively induces centrosome overduplication, E6 co-expression is necessary for induction of multipolar mitoses [83].

HPV16 E7 expression also causes a higher incidence of DNA double strand breaks, which can lead to breakage fusion bridge cycles and chromosomal translocations. Specific recurrent chromosomal translocations are well-documented drivers of hematological malignancies and, more recently, similar translocations have also been documented to contribute to the genesis of human solid tumors [84,85]. Chromosomal translocations are regularly detected in cervical cancer specimens [45], and it will be interesting to determine whether some of these directly contribute to cervical carcinogenesis. The mechanistic basis of the formation of mitoses with lagging chromosomal material still awaits full investigation.

Fanconi anemia (FA) patients frequently develop squamous cell carcinomas, and it has been reported that oral cancers arising in FA patients are more frequently HPV-positive than in the general population [86]. The FA pathway, which is normally activated by DNA crosslinking and stalled replication forks (reviewed in [87]), is triggered in response to HPV16 E7 expression, and HPV16 E7 induces DNA double strand breaks more efficiently in FA patient-derived cell lines [88]. Hence, the increased incidence of HPV-associated carcinomas in FA patients may be related to more potent induction of genomic instability by the HPV E7 protein.

The ability of the HPV16 E6 protein to contribute to genomic destabilization is to a large part based on its ability to inactivate the p53 tumor suppressor. As a consequence, the DNA damage-induced G1/S checkpoint is nonfunctional [89], and HPV16 E6-expressing cells also exhibit mitotic checkpoint defects [90,91]. HPV16 E6-mediated p53 degradation also causes subversion of a postmitotic checkpoint that is specifically triggered in cells when they re-enter a G1-like state after a failed mitosis. Such cells have a tetraploid rather than the normal diploid set of chromosomes and contain two centrosomes rather than one. Cells with p53 defects will disregard this checkpoint, re-enter S-phase, and may eventually undergo tetrapolar mitosis, which can lead to generation of aneuploid progeny. Consistent with this model, HPV16 E6 expressing cells show marked nuclear abnormalities [77,92,93].

In addition to passive mechanisms of cell cycle checkpoint subversion, HPV16 E6 may also contribute to genomic destabilization through active mutator mechanisms. HPV16 E6 has been shown to associate with single-strand DNA break repair protein XRCC1 [94] and induce degradation of O6-methylguanine-DNA methyltransferase [95], which is also involved in single-strand DNA break repair. It has also been reported that HPV16 E6 expression decreases the fidelity of DNA end joining [96]. Moreover, while HPV16 E6 expression does not induce centrosome overduplication, it greatly increases the incidence of multipolar mitoses in cells that contain supernumerary centrosomes due to HPV16 E7 expression [83].

Concluding Remarks

High-risk HPVs contribute to the genesis of almost all human cervical carcinomas and have also been associated with a number of other anogenital malignancies including vulvar, anal, and penile carcinomas. In addition approximately 20% of oral carcinomas are also associated with high-risk HPV infections, and oral sex practices have been implicated in transmission of the virus [97,98]. HPV-associated cervical cancers are quite unique in that they represent the only human solid tumor for which the initiating carcinogenic agent has been identified at a molecular level. The fact that the high-risk HPV E6 and E7 oncoproteins contribute to initiation as well as progression and that they are necessary for the maintenance of the transformed phenotype of cervical cancer cells suggests that these proteins and/or the processes that they regulate should provide targets for intervention.

Uterine Corpus

The uterine corpus represents the second common site for malignancy of the female reproductive tract. These neoplasms can be divided into epithelial, mesenchymal tumors, and trophoblastic tumors.

Epithelial Tumors

Uterine cancer is the most common gynecologic cancer in the United States. In 2008, the American Cancer Society estimates that there will be 40,100 new cases of uterine cancer, as compared to 21,650 new cases of ovarian cancer and 11,070 new cases of cervix cancer [99]. Most uterine cancers are adenocarcinomas and develop from the endometrium, the inner lining of the uterus. They are therefore referred to as endometrial cancers. Risk factors for the development of endometrial cancer include obesity, unopposed estrogen use, polycystic ovary syndrome, insulin resistance and diabetes, and estrogen-secreting ovarian tumors [100]. Given the spectrum of risk factors, the development of endometrial cancer has been strongly associated with an excess of systemic estrogen and a lack of progesterone. This hyperestrogenic state is presumed to result in endometrial hyperproliferation and endometrial cancer. Since the early 1990s, there has been increased research into understanding the molecular pathogenesis of endometrial cancer.

Endometrial cancer can be broadly divided into Type I and Type II categories, based on risk factors, natural history, and molecular features. Women with Type I endometrial cancers have the classic risk factors as stated previously, have tumors with low-grade endometrioid histology, frequently have concurrent complex atypical hyperplasia, and typically present with early stage disease. Molecular features important in Type I endometrial cancers include presence of ER and PR receptors, microsatellite instability (MSI), *PTEN* mutations, *KRAS* mutations, and *Beta catenin* mutations. Women with Type II endometrial cancers are typically older, nonobese, have tumors with serous and clear cell histologies as well as high-grade endometrioid histology, and may present with more advanced stage disease. In general, Type II tumors have a poorer clinical prognosis, in part due to their propensity to metastasize even with minimal myometrial invasion. The molecular alterations commonly seen in Type II cancers include *p53* mutations and chromosomal aneuploidy. While not all tumors fall neatly into these categories, the classification is helpful to broadly define endometrial cancers. A number of investigators are working toward developing more specific and rational targeted therapies. This section will focus on defining the molecular changes that have been described for endometrial hyperplasia, as well as for Type I and Type II endometrial cancers, with an emphasis on therapeutic relevance.

Endometrial Hyperplasia . Endometrial hyperplasia is a proliferation of the endometrial glands without evidence of frank invasion. According to the *World Health Organization* (WHO), there are four categories of endometrial hyperplasia: (i) simple hyperplasia, (ii) simple hyperplasia with atypia, (iii) complex hyperplasia, and (iv) complex hyperplasia with atypia. Complex hyperplasia with atypia, or complex atypical hyperplasia (CAH), is considered a precursor lesion to endometrial cancer. Simple hyperplasia is characterized by proliferating glands of irregular size, separated by abundant stroma. Cytologically, there is no atypia. Simple hyperplasia with atypia is characterized by mild architectural changes in the proliferating glands and cytologic atypia, but is rarely found in clinical practice. Complex hyperplasia is characterized by more densely crowded and irregular glands, although intervening stroma is present. Cytologically, there is no atypia. CAH is characterized by densely crowded and irregular glands with intervening stroma, as well as cytologic atypia based on enlarged nuclei, irregular nuclear membranes, and loss of polarity of the cells. From a clinical standpoint, only CAH is associated with a substantial risk of developing endometrial cancer. Kurman et al. estimated that the risk of progression to endometrial cancer is 1% for simple hyperplasia, 3% for complex hyperplasia, and 28.6% for CAH [101]. In a study by Horn et al., complex hyperplasia without atypia had a 2% risk of progression, and CAH had a 51.8% risk of progression [102]. Lacey et al. performed a nested case-control study of 138 patients with endometrial hyperplasia and found that simple hyperplasia and complex hyperplasia without atypia had minimal (relative risk of 2 and 2.8, respectively) risk of progression to cancer. However, CAH had a substantial (relative risk of 14) risk of progression [103]. Recently a *Gynecologic Oncology Group* study (GOG 167) demonstrated that in 289 cases of community diagnosed CAH, 123 (43%) cases had a concurrent Grade I endometrial cancer [104]. Of the 123 cases of endometrial cancer, 38 (31%) had some degree of invasion into the myometrium. In addition, there was substantial discrepancy between the community-diagnosed CAH and subsequent pathologic review by gynecologic pathologists, with both under- and overdiagnosis. Finally, even when a panel of expert gynecologic pathologists reviewed cases of CAH, there was only modest reproducibility with a kappa value of 0.28 [105]. The current clinical management of CAH is simple hysterectomy and bilateral salpingo-oophorectomy. In young women who desire future fertility, progestins (including megestrol acetate) have been used to successfully reverse CAH. An initial dilatation and curettage with careful pathologic evaluation is necessary prior to initiating treatment. In addition, an MRI to rule out an invasive process can also be helpful. Close surveillance with every 3 month endometrial sampling is also recommended.

Risk factors for CAH and Type I endometrial cancers are the same, including obesity, diabetes and insulin resistance, unopposed estrogen use, and polycystic ovarian syndrome. Certain, but not all, molecular features are shared. *PTEN* mutations, *KRAS* mutations, and MSI have all been identified in complex endometrial hyperplasias and likely represent early molecular alterations in the pathogenesis of endometrioid endometrial cancer [106,107]. A novel estrogen-regulated gene, *EIG121*, shows similar increase in expression in CAH and grade I endometrial cancers [108]. *IGFI-R* is also increased and activated in both CAH and grade I endometrial cancers [109].

Endometrioid Endometrial Cancer. Endometrioid endometrial cancer accounts for approximately 70%–80% of newly diagnosed cases of endometrial cancer [110], and these cancers are considered Type I. Risk factors for the development of endometrioid endometrial cancer are associated in general with an excess of estrogen and a lack of progesterone. The most common genetic changes in endometrioid endometrial cancers include mutations in *PTEN*, MSI, mutations in *K-RAS*, and *beta-catenin*. Somatic mutations in the *PTEN* tumor suppressor gene are the most common genetic defect in endometrioid endometrial cancers and have been reported to occur in approximately 40%–50% of cancer cases [111,112]. *PTEN* is a tumor suppressor gene located on chromosome 10. It encodes a lipid phosphatase that acts to negatively regulate AKT. The importance of *PTEN* inactivation in endometrial carcinogenesis has been also demonstrated in a *PTEN* heterozygous mouse model [113]. In this model, 100% of the heterozygote mice will develop endometrial hyperplasia by 26 weeks of age, and approximately 20% will develop endometrial carcinoma. In a study by Vilgelm et al. of this mouse model, phosphorylation of AKT followed by activation of ERα was demonstrated in the mouse endometrium. Reduction of endometrial ERα levels and activity reduced the development of endometrial hyperplasia and cancer [114]. In humans, germline *PTEN* mutations are the underlying genetic defect in individuals with Cowden's syndrome, which is a hereditary syndrome characterized by skin and gastrointestinal hamartomas and increased risk of breast and thyroid cancers. Women with Cowden's syndrome also are at increased risk for endometrial cancer.

In addition to *PTEN* mutations, mutation and inactivation of other components of the PIK3CA/AKT/mTOR pathway have been described. In a study by Hayes et al., mutations in the oncogene phosphatidylinositol-3-kinase *(PIK3CA)* were found in 39% of endometrial carcinomas, but only 7% of CAH cases [115]. Oda et al. reported a 36% rate of mutations in *PIK3CA* and found that there was a high frequency of tumors with both *PIK3CA* and *PTEN* mutations [116]. Our group described loss of TSC2 and LKB1 expression in 13% and 21% of endometrial cancers, with the subsequent activation of mTOR [117]. A heterozygous *LKB1* mouse model has been described recently that develops highly invasive endometrial adenocarcinomas [118]. In humans, germline *LKB1* mutations are responsible for Peutz-Jeghers syndrome. It has not previously been reported that women with Peutz-Jeghers syndrome are at increased risk for endometrial cancer. Clearly, somatic abnormalities in the components of the PTEN/AKT/TSC2/MTOR pathway have been identified in a substantial number of endometrial cancers. Currently, clinical trials are either underway or recently completed examining mTOR inhibitors, including CCI-779 and RAD-001. Further investigation of whether specific alterations in the pathway correlates to response to therapy will be necessary.

Approximately 30% of sporadic endometrioid endometrial cancers demonstrate MSI. MSI identifies tumors that are prone to DNA replication repair errors. Microsatellites are well-defined short segments of repetitive DNA (example: CACACA) scattered throughout the genome. Tumors that demonstrate gain or loss of these repeat elements at specific microsatellite loci, when compared to normal tissue, are considered to have MSI. MSI occurs in approximately 30%–40% of all endometrioid endometrial cancers, but rarely in Type II endometrial cancers [119]. In addition, MSI also occurs in approximately 20% of all colon cancers.

The mechanism of MSI is due to either a somatic hypermethylation or silencing of the *MLH1* promoter or an inherited defect in one of the mismatch repair genes (*MSH1, MSH2, MSH6, PMS2*). An inherited defect in one of the mismatch repair genes is the cause of Lynch syndrome, a hereditary cancer predisposition syndrome. Individuals with Lynch syndrome are at increased risk for colon and endometrial cancer, as well as cancers of the stomach, ovary, small bowel, and ureters. Women with Lynch syndrome have a 40%–60% lifetime risk of endometrial cancer, which equals or exceeds their risk of colon cancer [120,121]. While the risk of endometrial cancer is high in these women, overall Lynch syndrome accounts only for approximately 2%–3% of all endometrial cancers [122]. Endometrial cancers that develop in women with Lynch syndrome almost uniformly demonstrate MSI.

Somatic hypermethylation of *MLH1* and MSI occurs in approximately 30% of all endometrioid endometrial cancers and is an early event in the pathogenesis. MSI has been identified in CAH lesions. It is presumed that MSI specifically targets tumor suppressor genes, resulting in the development of cancer. Reported target genes include *FAS*, *BAX*, *IGF*, the insulin-like growth factor 2 receptor (*IGFIIR*), and transforming growth factor β receptor type 2 (*TGFβRII*) [123]. Studies focusing on the clinical significance of MSI have been mixed. While some investigators have found an association of MSI with a more aggressive clinical course, a more recent study examining only endometrioid endometrial cancers found no difference in clinical outcome between tumors demonstrating MSI and those without [124,125].

The microsatellite instability assay, as well as immunohistochemistry (IHC) for MSH1, MSH2, MSH6, PMS2, can be very useful as a screening tool to identify women with endometrial cancer as having Lynch syndrome [122]. While collecting and interpreting a family history is helpful, these molecular tools can be useful in targeting certain populations that may be at higher risk for Lynch syndrome, such as women under the age of 50 [126,127].

KRAS mutations have been identified in 20%–30% of endometrioid endometrial cancers. There is a higher frequency of *KRAS* mutations in cancers that demonstrate MSI. Mutations in *beta-catenin* have been seen in approximately 20%–30% of endometrioid endometrial cancers [128]. These mutations occur in exon 3 and result in stabilization of the beta catenin protein as well as nuclear accumulation. Nuclear beta-catenin plays an important role in transcriptional activation. One study demonstrated that *PTEN*, MSI, and *KRAS* mutations frequently co-exist. However, beta catenin alterations usually are not seen in these other abnormalities [129]. *Beta catenin* mutations have been identified in CAH, suggesting that this is an early step in the pathogenesis of endometrial cancer.

Type II Endometrial Cancers. Type II endometrial cancers have a more aggressive clinical course and include poorly differentiated endometrioid tumors as well as papillary serous and clear cell endometrial cancers. Patients with Stage I papillary serous endometrial cancer have a 5-year overall survival of 74%, significantly lower than the 90% 5-year overall survival for women with endometrioid endometrial cancer [110]. The average age of diagnosis of patients with papillary serous endometrial cancers is 68 years, and risk factors typically associated with Type I endometrial cancers are not present [130]. In addition, the genetic alterations seen in Type I endometrial cancers, as described previously, are not frequently found in papillary serous endometrial cancers [111,112,131]. Microarray studies examining Type 1 versus Type 2 cancers have shown a distinct set of genes that are up- and downregulated in Type 1 versus Type 2 cancers [111,112]. *TP53* gene mutations are common in uterine papillary serous carcinomas [132]. Her-2/neu overexpression by immunohistochemistry (IHC) was observed in 18% of uterine papillary serous carcinomas [130,133]. Of note, fewer tumors that demonstrated Her-2/neu overexpression by IHC showed *Her-2/neu* gene amplification. Several studies have reported that Her-2/neu overexpression is associated with a poor overall survival for UPSC [134,135]. Interestingly, one study reported that overexpression of Her-2/neu in uterine papillary carcinoma occurs more frequently in African American women.

Mesenchymal Tumors. Endometrial mesenchymal tumors are derived from the mesenchyme of the corpus composed of cells resembling those of proliferative phase endometrial stroma. Among them, uterine fibroids are the most common mesenchymal tumors of the female reproductive tract, which represent benign, smooth muscle tumors of the uterus. Recent studies have shown that the lifetime risk of fibroids in a woman over the age of 45 years is >60%, with incidence higher in African-Americans than in Caucasians [136]. The course of fibroids remains largely unknown and the molecular basis of this disease remains to be determined. Recent molecular and cytogenetic studies have revealed genetic heterogeneity in various histological types of uterine fibroids [137]. Chromosome 7q22 deletions are common in uterine leiomyoma, the most common type of uterine fibroids [67]. Several candidate genes, including *ORC5L* and *LHFPL3*, have been identified, but their roles in the pathogenesis of the disease have not been elucidated [138]. Loss of a portion of chromosome 1 is common in the cellular form of uterine leiomyomata.

Endometrial stromal sarcoma is the most common malignant mesenchymal tumor of the uterus. They often arise from endometriosis [139]. Genetic studies suggested that abnormalities of chromosomes 1, 7, and 11 may play a role in tumor initiation or progression in uterine sarcomas [140]. Fusion of two zinc fingers (*JAZF1* and *JJAZ1*) by translocation t(7;17) has also been described [141].

Ovary and Fallopian Tube

Multiple benign and malignant diseases have been identified in the ovary and the fallopian tube. The most common ones are polycystic ovary syndrome (PCOS), and benign and malignant tumors of the ovary. While molecular studies in all these diseases have revealed changes in multiple genes, the etiology of most of these diseases remains largely unknown.

Polycystic Ovary Syndrome

Polycystic ovary syndrome (PCOS) is a common heterogeneous endocrine disorder associated with amenorrhoea, hyperandrogenism, hirsutism, insulin resistance, obesity, and a 5–10-fold increased risk of type 2 diabetes mellitus [142]. It is a leading cause of female infertility. The inherited basis of this disease was established by epidemiologic studies demonstrating an increased prevalence of PCOS, hyperandrogenemia, insulin resistance, and altered insulin secretion in relatives of women diagnosed with PCOS [143]. From these pathways, several genes have been studied, including genes involved in steroid hormone biosynthesis and metabolism (*StAR, CYP11, CYP17, CYP19 HSD17B1-3, HSD3B1-2*), gonadotropin and gonadal hormone actions (*ACTR1, ACTR2A-B, FS, INHA, INHBA-B, INHC, SHBG, LHCGR, FSHR, MADH4, AR*), obesity and energy regulation (*MC4R, OB, OBR, POMC, UCP2-3*), insulin secretion and action (*IGF1, IGF1R, IGFBPI1-3, INS VNTR, IR, INSL, IRS1-2, PPARG*), and others [34,144–152]. PCOS appears to be associated with the absence of the four-repeat-units allele in a polymorphic region of the *CYP11A* gene, which encodes cytochrome P450scc [153]. Alteration of serine phosphorylation also seems to be involved in the post-translational regulation of 17,20-lyase activity (*CYP17*) [146]. About 50 genes have been demonstrated to have association with PCOS. Linkage and association studies identified a hotspot of candidate genes on chromosome 19p13.3, but to date, no genes are universally accepted as important in the pathogenesis of PCOS [154,155]. Further confirmatory and functional studies are needed to identify key genes in the pathogenesis of PCOS.

Benign Ovarian Cysts

Benign ovarian cysts are a very common condition in premenopausal women. During normal ovulation each month, the follicle (or cyst) created by the ovaries bursts harmlessly. For unknown reasons, this normal physiologic process may sometimes go wrong. The follicle may continue to swell with fluid without releasing its egg, or hormones secreting tissue (corpus luteum) that prepare for pregnancy may persist even though the egg has not been fertilized. Subsequently, a cyst (or fluid-filled sac), which may be as small as a grape or as large as a tennis ball, is formed [156]. Mutation of the *FOXL2* gene has been found in a patient with a large ovarian cyst [157]. Such a cyst is usually benign and disappears within a couple of months. However, if such a cyst persists after several months, it may become a benign semisolid cyst. The most common semisolid cyst is a dermoid cyst, so-called because it is made up of skin-like tissue and can be removed by laparoscopic surgery [158].

Borderline Tumors

Borderline tumors account for 15%–20% of epithelial ovarian tumors, and was first recognized by Howard Taylor in 1929 [159]. He described a group of women of reproductive age with large ovarian tumors whose course was rather indolent [159]. In the early 1970s, the *Federation of Gynecologists and Obstetricians* defined these so-called semi-malignant tumors as borderline ovarian tumors (BOTs) [160]. Later on, at the 2003 *World Health Organization* workshop, the term low malignant potential (LMP) became an accepted synonym for BOTs [161].

Borderline tumors with different cell types (serous, mucinous, endometrioid, clear cell, transitional, and mixed epithelial cells) have been reported. However, the serous and mucinous are the most common types. Mutational analyses have identified several gene mutations (Table 23.1) in borderline tumors. Both *BRAF* and *KRAS* mutations are very common in borderline serous tumor. However, *BRAF* mutation has not been found in borderline mucinous tumors. Moreover, the frequency of *KRAS* mutation is higher in borderline mucinous tumor than in serous tumors. On the other hand, *CTNNB1* and *PTEN* mutations have been found in borderline endometrioid tumors. The difference in mutation spectrum indicates different pathogenic pathways for these histological subtypes.

Whether serous BOT (SBOT) will progress to invasive carcinoma is still controversial [162]. In a recent review of the clinical outcome of 276 patients with SBOTs, approximately 7% of the patients had recurrent disease as invasive low-grade serous carcinoma [163]. In another report, when patients with SBOTs were followed-up for a longer period, over 30% of the patients had recurrent disease as low-grade carcinoma over a period of 3 to 25 years [164]. Studying genetic changes in different types of ovarian tumors provides insight into the pathogenic pathways for ovarian cancer. Mutations in *KRAS* have been found in 63% of mucinous BOTs and 75% of invasive mucinous ovarian cancers. These data suggest that *KRAS* mutations are involved in the development of mucinous BOTs and support the notion that mucinous BOTs may represent a phase of development along the pathologic continuum between benign and malignant mucinous tumors [165]. However, *KRAS* mutation rate in serous BOTs is significantly higher than that in serous invasive cancers. These data suggest that serous BOTs and invasive carcinomas may have different pathogenic pathways, and only a small percentage of serous BOTs may progress to invasive cancers. A mucinous cystadenoma would give rise to a mucinous borderline ovarian tumor (BOT), a subset of which may progress to invasive low-grade and perhaps high-grade mucinous carcinomas. A serous cystadenoma would give rise to a serous BOT, in which benign and borderline features are rare, in contrast to their frequent presence in mucinous BOTs.

Malignant Tumors

Ovarian cancer is a general term that represents a diversity of cancers that are believed to originate in the ovary. Over 20 microscopically distinct types can be identified, which can be classified into three major groups, epithelial cancers, germ cell tumors, and specialized stromal cell cancers. These three groups correspond to the three distinct cell types of different functions in the normal ovary: (i) the epithelial covering may give rise to the epithelial ovarian cancers, (ii) the germ cells may give rise to the germ cell tumors, and (iii) the steroid-producing cells may give rise to the specialized stromal cell cancers.

Epithelial Ovarian Tumors. The majority of malignant ovarian tumors in adult women are epithelial ovarian cancer. Based on the histology of the tumor cells, they are classified into different categories: serous, mucinous, endometrioid, clear cell, transitional, squamous, mixed, and undifferentiated. From the Sanger Center's Catalogues of Somatic Mutations in Cancer (COSMIC) database, we have extracted the most frequently identified mutations for each histological subtype (Table 23.1). In addition to these common mutations, hundreds of rare mutations have also been identified (http://www.sanger.ac.uk/genetics/CGP/cosmic/).

Besides genetic analysis, other high-throughput methods have been exploited to understand the pathogenic pathways in ovarian cancer [166,167]. It is hypothesized that most ovarian tumors develop from ovarian inclusion cysts arising from the OSE. In most cases of serous cancer, the majority of serous BOTs develop directly from ovarian inclusion cysts, without the cystadenoma stage. Alternatively, the epithelial lining of the cyst develops into a mucinous or serous cystadenoma by following one of two distinct pathogenic pathways. Our recent expression profiling analysis of serous BOTs and serous low- and high-grade carcinomas suggested that serous BOTs and low-grade carcinomas may represent developmental stages along a continuum of disease development and progression [167]. Thus, we hypothesize that a majority of high-grade serous carcinomas derive *de novo* from ovarian

Table 23.1 Common Somatic Mutations in Human Sporadic Ovarian Tumors

	Gene Name	Number of Samples Screened	Number of Positive Samples	Percent Mutated
Ovarian adenoma				
	KRAS	152	15	9%
	BRAF	59	3	5%
	BRCA1	13	0	0%
	BRCA2	9	0	0%
	PIK3CA	8	0	0%
Borderline tumors				
Serous	*BRAF*	191	80	41%
	KRAS	141	37	26%
	ERBB2	21	2	9%
	PIK3CA	23	1	4%
	CDKN2A	15	0	0%
Mucinous	*KRAS*	5	4	80%
	CDKN2A	10	3	30%
	SMAD4	5	1	20%
	BRAF	28	0	0%
	BRCA1	5	0	0%
Endometrioid	*CTNNB1*	9	8	88%
	PTEN	8	1	12%
	KRAS	8	0	0%
Carcinomas				
Serous	*KRAS*	447	38	8%
	CDKN2A	227	14	6%
	BRCA1	246	11	4%
	BRAF	250	6	2%
	PIK3CA	230	5	2%
Mucinous	*KRAS*	143	62	43%
	CDKN2A	60	12	20%
	PTEN	36	6	16%
	PIK3CA	30	2	6%
	AATK	2	2	100%
Clear cell	*KRAS*	119	10	8%
	CDKN2A	23	5	21%
	PIK3CA	20	5	25%
	BRCA2	5	2	40%
	BRAF	53	1	1%
Endometrioid	*CTNNB1*	240	65	27%
	PTEN	92	23	25%
	BRAF	107	22	20%
	KRAS	176	12	6%
	PIK3CA	67	10	14%

cysts or ovarian endosalpingiosis. This hypothesis is consistent with the two-tier system recently proposed for the grading serous carcinomas [168,169].

Loss of heterozygosity (LOH) studies have been widely used to identify minimally deleted chromosomal regions where tumor suppressor genes may reside. We previously identified several common loss regions by using 105 microsatellite markers to perform detailed deletion mapping on chromosomes 1, 3, 6, 7, 9, 11, 17, and X in BOTs, invasive ovarian cancers, and serous surface carcinoma of the ovary. Except at the androgen receptor (AR) locus on the X chromosome [170], BOTs showed significantly lower LOH rates (0%–18%) than invasive tumors at all loci screened, suggesting that the LOH rate at autosomes is less important in the development of BOTs than more advanced tumors. Based on these and other results, the importance of LOH at the AR locus in BOTs and invasive cancers remains undetermined. However, several studies have found differences in LOH rates on other chromosomes. LOH at the *p73* locus on 1p36 was found in both high-grade and low-grade ovarian and surface serous carcinomas, but not in borderline ovarian tumors. In one study, LOH rates at 3p25, 6q25.1–26, and 7q31.3 were significantly higher in high-grade serous carcinomas than in low-grade serous carcinomas, mucinous carcinomas, and borderline tumors [171–173]. In another study, LOH rate at a 9-cM region on 6q23-24 were significantly higher in surface serous carcinomas than in serous ovarian tumors [174]. In addition, multiple minimally deleted regions have been identified on chromosomes 11 and 17. LOH at a 4-cM region on chromosome 11p15.1 and an 11-cM region on chromosome 11p15.5 were found only in serous invasive tumors, and the LOH rates in 11p15.1 and 11p15.5 were significantly higher in high-grade than low-grade serous tumors [175]. Similarly, significantly higher LOH rates were identified at the *TP53* locus on 17p13.1 and the *NF1* locus on 17q11.1 in high-grade serous carcinomas than in low-grade and borderline serous tumors and all mucinous tumors [176]. LOH at the region between *THRA1* and D17S1327, including the *BRCA1* locus on 17q21, was found exclusively in high-grade serous tumors [173]. In general, the fact that LOH rates at multiple chromosomal sites were significantly higher in serous than in mucinous tumor subtypes suggests that serous and mucinous tumors may have different pathogenic pathways. Since tumor suppressor genes are implicated to be located in chromosomal regions demonstrating LOH, further analysis of the genes located in the minimally deleted regions may provide insights into genes that may be important for the pathogenesis of serous and mucinous ovarian tumors. Studying genetic changes in endometriosis, endometrioid, and clear cell ovarian cancer also provides insights into pathogenic pathways in the development of these tumor types. Endometriosis is highly associated with endometrioid and clear cell carcinomas (28% and 49%, respectively), in contrast to its very low frequency of association with serous and mucinous carcinomas of the ovary (3% and 4%, respectively) [177]. This fact furnishes strong evidence that endometriosis is a precancerous lesion for both endometrioid and clear cell carcinomas. Obata et al. [178] showed that endometrioid but not serous or mucinous epithelial ovarian tumors had frequent *PTEN/MMAC* mutations. In addition, the loss of PTEN immunoreactivity has been reported in a significantly higher percentage of clear cell and endometrioid ovarian cancers than cancers of other histologic types [179]. Dinulescu et al. [180] demonstrated that the expression of oncogenic *KRAS* or conditional *PTEN* deletion within the OSE induces preneoplastic ovarian lesions with an endometrioid glandular morphology. Furthermore, the combination of the two mutations in the ovary led to the induction of invasive and widely metastatic endometrioid ovarian adenocarcinomas. These data further suggest that tumors with different histologic subtypes may arise through distinct developmental pathways.

Germ Cell Cancer. The ovarian germ cell tumors commonly occur in young women and can be very aggressive. Fortunately, chemotherapy is usually effective in preventing it from recurring. The pathogenesis of these tumors is not well understood. Several subtypes exist, which are dysgerminoma, yolk sac tumor, embryonal carcinoma, polyembryoma, choriocarcinoma, immature teratoma, and mixed GCTs. A recent study has identified *c-KIT* mutation and the presence of Y-chromosome material in dysgerminoma [181]. Furthermore, DNA copy number changes were detected by comparative genomic hybridization in dysgerminomas. The most common changes in DGs were gains from chromosome arms 1p (33%), 6p (33%), 12p (67%), 12q (75%), 15q (42%), 20q (50%), 21q (67%), and 22q (58%) [182]. However, overexpression or mutation of *p53* was not observed in ovarian germ cell tumors [183].

Stromal Cancer. The specialized stromal cell tumors (granulose cell tumors, thecal cell tumors, and Sertoli-Leydig cell tumors) are uncommon [184]. One interesting characteristic of these tumors is that they can produce hormones. Granulosa and thecal cell tumors are frequently mixed and can produce estrogen, which will result in premature sexual development and short stature if the tumors are developed in young girls. Sertoli-Leydig cell tumors produce male hormones, which will cause defeminization. Subsequently, male pattern baldness, deep voice, excessive hair growth, and enlargement of the clitoris will develop in patients with Sertoli-Leydig cell tumors. Overexpression of the *BCL2* gene in Sertoli-Leydig cell tumor of the ovary has been reported [185]. Fortunately, these specialized stromal cell cancers are usually not aggressive cancers and involve only one ovary.

Tumors of the Fallopian Tube. The fallopian tubes are the passageways that connect the ovaries and the uterus. Fallopian tube cancer is very rare. It accounts for less than 1% of all cancers of the female reproductive organs. Only 1,500 to 2,000 cases have been reported worldwide, primarily in postmenopausal women. There is some evidence that women who inherit a mutation in the *BRCA1* gene, a gene already linked to breast and ovarian cancer, seem to have an increased risk of developing fallopian tube cancer. In one recent analysis of several hundred women who were carriers of the *BRCA1* gene mutation, the incidence of fallopian tube cancer was increased more than 100-fold [186]. Likewise, a substantial proportion of women with the diagnosis of fallopian tube cancer test positively for either the *BRCA1* or *BRCA2* gene mutation. Recent recommendations suggest that any woman with a diagnosis of fallopian tube cancer be tested for the *BRCA* mutations. Based on the pathological study of fallopian tubes from *BRCA1* carrier, another hypothesis is being proposed that some serous ovarian tumors may be developed from the secretory fimbrial cells of the tube [187].

Vagina and Vulva

Vaginal intraepithelial neoplasia and squamous cell carcinoma are the most common tumor types affecting the vagina. Persistent infection with high-risk HPV (such as HPV16 and HPV18) is a major etiological factor for both diseases [188,189]. Overexpression of p53 and Ki67 has been identified in 19% and 75% of vaginal cancer cases, respectively [190].

Similar to vaginal malignancies, squamous cell carcinoma is the most common form of neoplasia in the vulva. Precursor lesions include vulvar intraepithelial neoplasia (VIN), the simplex (differentiated) type of VIN, lichen sclerosis, and chromic granulomatous. While vaginal carcinomas and VIN have been shown to be associated with HPV, usually of type 16 and 11, the simplex form of VIN and other precursor lesions are not HPV-associated [191]. Genetic studies identified both *TP53* and *PTEN* mutations in VIN, as well as in carcinomas, suggesting that these molecular changes are early events in the pathogenesis of vulvar cancers [192,193]. Other genetic changes including chromosome losses of 3p, 4p, 5q, 8p, 22q, Xp, 10q, and chromosome gains of 3q, 8p, and 11q have been reported [194–196]. In addition, high frequencies of allelic imbalances have been found in both HPV-positive and HPV-negative lesions at multiple chromosomal loci located on 1q, 2q, 3p, 5q, 8p, 8q, 10p, 10q, 11p, 11q, 15q, 17p, 18q, 21q, and 22q [197].

Other less common epithelial malignancies of the vulva include Paget disease and Bartholin gland carcinoma. Nonepithelial vulvar malignancies include malignant melanomas, which accounts for 2%–10% of vulvar malignancies [198]. The molecular basis of most of these diseases remains largely unknown.

DISORDERS RELATED TO PREGNANCY

Conception, maintenance of the pregnancy, and delivering of a baby are the most important functions of the female reproductive tract. Proper implantation of the embryo is a crucial step in a healthy pregnancy. Failure of this process can compromise the life of the fetus and the mother. During implantation, trophoblast cells basically invade into the endometrium in a controlled fashion and transform the spiral arteries to supply the growing fetus with oxygen and nutrition. Failure of this process results in fetal hypoxia and growth retardation, which might lead to complications like abortion, preeclampsia, and preterm delivery. On the other hand, uncontrolled invasion can lead to deep implantation (placenta accreta, increta, percreta), or in gestational trophoblastic diseases, it can result in persistent disease, metastases, and development of neoplasia.

The trophoblast invasion is regulated by at least three factors: (i) the endometrial environment (including adhesion molecules and vascularization at the implantation site), (ii) the maternal immune system, and (iii) the invasive and proliferative potentials of the trophoblast cells. Complications following normal conceptions are mainly due to the failure of the maternal endometrium or immune system, whereas complications following genetically abnormal conceptions are most often the result of abnormal trophoblastic cell function. Preeclampsia and gestational trophoblastic diseases represent serious conditions that can lead to maternal death. This fact has put these diseases into the focus of clinical and molecular researchers. The molecules discussed here may also play a role in other implantation-related diseases, such as unsuccessful IVF cycles, recurrent spontaneous abortions, preterm deliveries, and others.

In gestational trophoblastic diseases (GTD), the problem is the overly aggressive behavior of the trophoblastic cells. The trophoblast cells in GTD are hyperplastic and have much more aggressive behavior than in normal pregnancy. Trophoblast cells might invade into the myometrium and persist after evacuation, which, without adequate chemotherapy, can compromise the life of the mother. Trophoblast hyperplasia in GTD also can result in true gestational trophoblastic neoplasia (GTN).

In preeclampsia, hypertension and vascular dysregulation are the indirect results of the inadequate placentation due to different combinations of the previously mentioned factors. To date, the primary mechanisms that account for failure of trophoblast cells to invade and transform the spiral arteries are still unknown. However, the mechanisms and factors leading to the classic symptoms after fetal stress seem to be identified, and clinical trials have been started to control these factors to prevent preeclampsia.

Gestational Trophoblastic Diseases

Gestational trophoblastic disease is a broad term covering all pregnancy-related disorders in which the trophoblast cells demonstrate abnormal differentiation and hyperproliferation. Molar pregnancies are the most common gestational trophoblastic diseases, which are characterized by focal (partial mole) or extensive (complete mole) trophoblast cell proliferation on the surface of the chorionic villi. Gestational trophoblastic neoplasms, like gestational choriocarcinoma, can arise from molar pregnancies (5%–20%) and on very rare occasions from normal pregnancies (0.01%) [199].

Partial molar pregnancy is a result of the union of an ovum and two sperms. Therefore, the genetic content of the fetus is imbalanced with overrepresentation of paternal genetic material. Since the maternal chromosomes are present, the embryo can develop to a certain point. The complete mole is the result of a union between an empty ovum and generally two sperms. Rarely, the conception happens with one sperm, and the genetic content of the sperm doubles in the ovum. In this case, the maternal chromosomes are not present. Therefore, no embryo develops [200]. Fisher et al. also have described familial recurrent hydatidiform molar pregnancies where balanced biparental genomic contribution could be found. In these cases, gene mutation is suspected on the long arm of chromosome 19, but to date the responsible gene has not been identified [201].

Epigenetic Changes in Trophoblastic Diseases

Due to unbalanced conception, gene expression in molar pregnancy differs greatly from that of a normal pregnancy. The gene expression differences can partly be explained by the lack or excess of chromosome sets, but more importantly epigenetic parental imprinting has a substantial role in the modification of the gene expression profile. Maternally imprinted genes are normally expressed from one set of paternal chromosomes, while in molar pregnancies these genes are expressed from two or more sets of paternal chromosomes. On the other hand, paternally imprinted genes are normally expressed only from the maternal allele, but in complete moles these genes cannot be expressed at all due to the lack of maternal chromosomes in complete gestational trophoblastic diseases. Complete molar pregnancy, as a unique uniparental tissue, is in the focus of epigenetic studies of placental differentiation [202]. Several imprinted genes (*P57^{kip2}*, *IGF-2, HASH, HYMAI, P19, LIT-1*) have been investigated in complete moles, and paternally imprinted *p57^{kip2}* is now used as a diagnostic tool to differentiate complete mole from partial mole and hydropic pregnancy by detecting maternal gene expression [203]. Table 23.2 shows the known parentally imprinted genes and their functions that are associated with hydatidiform mole.

Immunology and Trophoblastic Diseases

As partial and complete moles are foreign tissues in the maternal uterus, vigorous immune response would be expected from the mother's side [209]. On the other hand, trophoblast cells express several immunosuppressive factors to attenuate the maternal immune response to protect the semiallogeneic fetus.

Trophoblast cells, unlike other human cells, lack class I major histocompatibility complexes (MHC). Instead, they express atypical MHCs like human leukocyte antigen (HLA) G, E, and F [210,211]. The atypical HLAs do not present antigens to the immune cells, but they are still able to inactivate the natural killer (NK) cells. Thus, these atypical HLAs protect the semiallogeneic fetus from the immune cells [212]. The expression of immunosuppressive factors shows significant differences between normal and molar trophoblast cells.

Table 23.3 lists the molecules which may play a role in the development of systematic and local gestational immunosuppression. In gestational trophoblastic diseases, the hyperproliferation of the trophoblast results in an even higher level of immunosuppressive factors in the mother. Furthermore, some of the immunosuppressive factors in GTD and GTN are elevated not only because of the higher number of hyperplastic trophoblast cells, but also due to overexpression of these factors by the trophoblast cells. For example, soluble and membrane bound HLA-G was shown to be overexpressed in molar pregnancies and gestational trophoblastic neoplasms [213]. Besides, molar villous fluid effectively suppressed the cytotoxic T-cells *in vitro* [214]. Shaarawy et al. also found that the soluble IL-2 receptor level was significantly higher in the serum of mothers who subsequently developed persistent gestational trophoblastic disease [215].

Oncogene and Tumor Suppressor Gene Alterations in Trophoblastic Diseases

As molar pregnancies and especially gestational trophoblastic neoplasms demonstrate excessive trophoblast cell proliferation, extensive studies were undertaken to characterize the expression of tumor suppressor and oncogenes. Fulop et al. demonstrated that both complete mole and choriocarcinoma were

Table 23.2 Imprinted Genes in Embryonic Development Associated with Hydatidiform Moles

Gene	Transcribed From	Function
p57^{kip2}	Maternal allele	Negatively regulates cell proliferation by binding G1 cyclin complexes [204]
IGF-2	Paternal allele	Increases the expression of passive amino acid transporters on trophoblast cells; increases trophoblast invasiveness [205]
HYMAI	Maternal allele	Associated with transient neonatal diabetes mellitus [206]
H19	Maternal allele	Necessary for muscular differentiation; associated with Wilms' tumor [205; 207]
LIT-1	Maternal allele	LIT-1 is required for p57kip2 expression and specific deletion results in BWS [208]

IGF-2—Insulin-like growth factor 2; HYMAI—hydatidiform mole-associated imprinted gene; LIT-1—long QT intronic transcript; BWS—Beckwith-Wiedemann syndrome.

Table 23.3 Immunosuppressive Molecules Secreted by Trophoblast Cells (Adapted from [216])

	Increases the Expression	Ligand	Immune Cell or System Affected	Function
HLA molecules (C,G,E,F)	IL-10	KIR	NK	Inhibition
Soluble HLA G	IL-10	KIR	Tc cell	Apoptosis
Fas L	Th2 cytokines and hCG	Fas	Tc cell	Apoptosis
Cytokines (TGFβ, IL-10, IL-4, IL-6)	Unknown	Cytokine receptors	Th cells, Tc cells, NK, Trophoblast cells	Suppression HLA G, E, F overexpression
Indoleamine 2,3-dioxygenase	γ-INF	L-tryptophan	T-cell, Complement system	Inhibition of T-cell maturation by tryptophan depletion, Complement inhibition
DAF, MCP, CD59	Unknown	Multiple	Complement system	MAC inhibition
PI-9	Unknown	Granzyme B	Tc cell, NK cell killing	Granzyme B inhibitor
hCG	Unknown	hCG receptor	Trophoblast Immune cells	Fas L overexpression hCG → Progesterone→PIBF

IL—interleukin; HLA—Human leukocyte antigen; Fas L—Fas ligand; DAF—Decay accelerating factor; MCP— Membrane cofactor protein; INF—interferon; KIR—Killer inhibitor receptor; NK—natural killer; Tc—cytotoxic T cells; Th—Helper T-cells; MAC—membrane attack complex; PIBF—Progesterone-induced blocking factor.

characterized by overexpression of *p53*, *p21*, and *Rb* tumor suppressor genes and *c-myc*, *c-erbB-2*, and *bcl-2* oncogenes, while partial mole and normal placenta generally do not strongly express these molecules [217–219]. *p53* and *Rb* molecules are mainly found in the nuclei of cytotrophoblastic cells, while *p21* could be seen in syncytiotrophoblast cells. *DOC-2/hDab2* tumor suppressor gene expression, on the other hand, was significantly stronger in normal placenta and partial mole compared to complete mole and choriocarcinoma [217].

Kato et al. have compared hydatidiform mole gene expression profile to normal placenta gene expressions by microarray method and identified 508 differentially expressed gene [220]. Most of these genes are members of MAP-RAS kinase, Wnt, and Jak-STAT5 pathways. Some other genes, like versican, might have a role in cell migration or drug interactions [220]. These findings provide important insight into the development of the GTN. However, to date it is not clear which gene has the primary role to develop gestational trophoblastic disease or neoplasia.

Molecular Mechanisms in Trophoblast Invasion

The process of the trophoblast invasion involves the enzymatic degradation of the extracellular matrix of the endometrium. Extracellular proteinases are thought to be important factors in modifying cell-matrix interactions and in the degradation of the extracellular structures [218]. Different types of matrix degrading proteases have been found to be involved in the process of trophoblast invasion, like serine proteinases and matrix metalloproteinases (MMPs). MMPs are regulated by tissue inhibitors of metalloproteinases (TIMPs) [221]. MMP1 and MMP2 have been found to be overexpressed in choriocarcinoma cells compared to molar and normal placental syncytiotrophoblast cells. On the contrary, TIMP-1 has been shown to be downregulated in choriocarcinoma [222]. Therefore, MMPs and their inhibitors may have an effect in the aggressiveness of trophoblastic cells. Table 23.4 and Table 23.5 show the known stimulatory and inhibitory factors regulating trophoblast invasion. Most of these mechanisms have been shown to be altered in gestational trophoblastic diseases.

Detection of Trophoblastic Diseases

Gestational trophoblastic neoplasms are mostly known as chemosensitive malignancies. Human choriogonadotropin (hCG) follow-up is used in every case to diagnose persistent disease or GTN before applying chemotherapy [240]. Several studies have tried to identify a reliable molecular or histologic marker which could predict persistent disease at the time of evacuation, but to present date none of them seems to be sensitive and specific enough to change the current prognostic scoring and hCG follow-up in clinical practice. However, some promising results may help to diagnose GTN and predict prognosis in the future. Cole et al. recently demonstrated that hyperglycosylated hCG (hCG-H) appears to reliably identify active trophoblastic malignancy, but unfortunately hCG-H alone is unable to predict persistence at the time of primary evacuation [241]. Serum IL-2 and soluble IL-2 receptor levels at the time of evacuation have been demonstrated to have association with the clinical outcome and persistence [215]. IL-1, EGFR, c-erbB-3, and antiapoptotic MCL-1 staining on trophoblast cells also significantly correlated with the development of persistent postmolar gestational trophoblastic neoplasia [242–244]. Further studies of potential molecules are needed to find a reliable marker or marker panel to predict persistent gestational trophoblastic diseases.

Table 23.4 Stimulatory Factors Involved in the Regulation of EVT Invasion

Stimulatory Factors	Expression Site: Trophoblast	Expression Site: Decidua	Mechanism
IL-1β*	Syncytiotrophoblasts, Hofbauer cells	NK cells, macrophages	Upregulation of MMP-2, MMP-9, and uPA [223]
IGF-II*	Villous cytotrophoblasts, extravillous trophoblast (EVT)	Not expressed	IGF-II actions mediated by IGF-R II [224]
IGFBP-1*	Not expressed	Decidual cells	Interaction with Arg-Gly-Asp (RGD)-binding sites of EVT surface α5β1 [224]
ET-1	ST, villous trophoblast (VT), and EVT	Blood vessel, endothelium	Binding with ET-1 receptor expressed on EVT [225, 226]
Activin*	Cytotrophoblast, EVT	Decidual cells	Induce cytotrophoblast differentiation, upregulation of MM9, MMP-2, and anti–TGF-β [227, 228]
Heparin binding-EGF, EGF, TGFα*	EGF and receptors expressed on villous cytotrophoblast, EVT	Decidual cells	Inducing cytotrophoblast to EVT differentiation; induce integrin phenotype switching (α6β4 to α1β1) on EVT [229]
MMP-2, MMP-9*	EVT, villous cytotrophoblast	Not expressed	Proteolysis, remodeling, and invasion to type IV collagen [230]
uPA	EVT, cytotrophoblast, syncytiotrophoblast	Decidual cells	Cell-bound proteolysis [231]
OPN*	VT	Decidual cells	Binds to integrins and CD44 [232]
LR1*	Cytotrophoblast, EVT	Decidual cells, macrophages	Changes the conformation of laminin [233]

*Indicates different levels of expression between normal pregnancy and gestational trophoblastic diseases.

Table 23.5 Inhibitory Factors Involved in the Regulation of EVT Invasion.

Inhibitory Factors	Expression Site: Trophoblast	Expression Site: Decidua	Mechanism
TGF-β1–3*	All types of trophoblasts	Secreted by NK and macrophage	Upregulates TIMP and PAI; downregulates MMP-2, MMP-9, uPA [234]
Decorin*	No expression	Decidual products	Interacts with fibronectin with decorin core protein, impedes locomotion of EVTs on their natural substrate [235]
Melanoma cell adhesion molecule (Mel-CAM)*	EVT	Not expressed, but uterine smooth muscle expressed the ligand	Mel-CAM interacts with ligand on uterine smooth muscle to inhibit EVT migration [236]
TIMP-1, TIMP-2*	All trophoblast	Decidual stromal cells	Inhibits the activity of MMP-2 and MMP-9 [230]
PAI-1, PAI-2*	Syncytiotrophoblast, chorionic villi	Endothelial cells, decidual cells	Inhibits the activity of uPA and trophoblast migration [230]
TNFα*	Trophoblast in vitro culture	NK cells, macrophages	Upregulates PAI-1 [237]
IFN-γ*	EVT from cell columns and interstitial trophoblast	Macrophages, vascular endothelial cells	Induces trophoblast apoptosis; upregulates HLA-C and HLA-G on the surface of EVTs; promotes Th1-reaction (antitumor effects) [238, 239]
Hypoxia			Interferes with trophoblast proliferation, fibronectin synthesis, α5 integrin, and MMP-2; increases expression of TNF-α [227].

*Indicates different levels of expression between normal pregnancy and gestational trophoblastic diseases.

Molecular Basis of Preeclampsia

In preeclampsia most of the factors in Table 23.2, Table 23.3, and Table 23.4 might also have a role in the etiology of the disease. To date, the etiology of the disease is not known. In preeclampsia, the failure of the trophoblast implantation indirectly leads to serious consequences. The implantation is sufficient enough to supply the embryo to a certain point, but the capacity of the not completely transformed spiral arteries is limited. Therefore, the fetus and the placenta suffer from hypoxic injury. Preeclampsia is characterized by endothelial dysfunction, which produces altered quantities of vasoactive mediators resulting in general vasoconstriction.

Recent studies demonstrated that soluble fms-like tyrosine kinase-1 (sFlt-1 or sVEGFR-1) might have a crucial role in developing hypertension and proteinuria in pregnant women [245,246]. sFlt-1 is a tyrosine kinase protein which disables proteins that cause blood vessel growth. sFlt-1 prevents the effect of VEGF and placental growth factor (PLGF) on vascular structures. In women developing preeclampsia, the serum level of sFlt-1 is significantly elevated, and the levels of VEGF and PLGF are markedly decreased compared to normotensive women [247]. The sFlt-1 level starts to rise 5 weeks before the onset of preeclampsia [248]. In animal studies, exogenous application of sFlt-1, similar to anti-VEGF bevacizumab, result in hypertension and proteinuria in pregnant rats [249].

Together with sFlt-1, endoglin serum level correlates with the progression and severity of preeclampsia [250]. Endoglin is a soluble transforming growth factor β (TGFβ) coreceptor. Endoglin impairs binding of TGF-β1 to its receptors and downstream signaling, including effects on activation of eNOS and vasodilation, suggesting that endoglin leads to dysregulated TGFβ signaling in the vasculature [251]. These molecules might also serve as diagnostic tools for preeclampsia. Salahuddin et al. demonstrated that sFlt-1 and endoglin are useful tools to differentiate preeclampsia from other types of hypertension in pregnancy [252]. These findings also hold the possibility of developing a new medical therapy for preeclampsia targeting angiogenic factors.

REFERENCES

1. Myers ER, McCrory DC, Nanda K, et al. Mathematical model for the natural history of human papillomavirus infection and cervical carcinogenesis. *Am J Epidemiol.* 2000;151:1158–1171.
2. Scheurer ME, Tortolero-Luna G, Adler-Storthz K. Human papillomavirus infection: biology, epidemiology, and prevention. *Int J Gynecol Cancer.* 2005;15:727–746.
3. Cates W, Jr. Estimates of the incidence and prevalence of sexually transmitted diseases in the United States. American Social Health Association Panel. *Sex Transm Dis.* 1999;26:S2–S7.
4. Koutsky L. Epidemiology of genital human papillomavirus infection. *Am J Med.* 1997;102:3–8.
5. Weinstock H, Berman S, Cates W, Jr. Sexually transmitted diseases among American youth: incidence and prevalence estimates, 2000. *Perspect Sex Reprod Health.* 2004;36:6–10.
6. Chesson HW, Blandford JM, Gift TL, et al. The estimated direct medical cost of sexually transmitted diseases among American youth, 2000. *Perspect Sex Reprod Health.* 2004;36:11–19.
7. Howley PM, Lowy DR. Papillomaviruses. In: Knipe DM, Howley PM, eds. *Fields Virology*, 5th ed., Vol. 2. Philadelphia: Lippincott Williams & Wilkins; 2007:2299–2354. 2 vols.
8. de Villiers EM, Fauquet C, Broker TR, et al. Classification of papillomaviruses. *Virology.* 2004;324:17–27.
9. de Villiers EM, Gissmann L, zur Hausen H. Molecular cloning of viral DNA from human genital warts. *J Virol.* 1981;40: 932–935.
10. Durst M, Gissmann L, Ikenberg H, et al. A papillomavirus DNA from a cervical carcinoma and its prevalence in cancer biopsy samples from different geographic regions. *Proc Natl Acad Sci U S A.* 1983;80:3812–3815.
11. Schiffman M, Castle PE, Jeronimo J, et al. Human papillomavirus and cervical cancer. *Lancet.* 2007;370:890–907.
12. Martens JE, Arends J, Van der Linden PJ, et al. Cytokeratin 17 and p63 are markers of the HPV target cell, the cervical stem cell. *Anticancer Res.* 2004;24:771–775.
13. Smedts F, Ramaekers F, Troyanovsky S, et al. Basal-cell keratins in cervical reserve cells and a comparison to their expression in cervical intraepithelial neoplasia. *Am J Pathol.* 1992;140: 601–612.
14. Tsutsumi K, Sun Q, Yasumoto S, et al. In vitro and in vivo analysis of cellular origin of cervical squamous metaplasia. *Am J Pathol.* 1993;143:1150–1158.
15. Streeck RE, Selinka HC, Sapp M. Viral entry and receptors. In: Garcea RL, DiMaio D, eds. *The Papillomaviruses.* New York: Springer; 2007:89–108.
16. Lee C, Laimins LA. The differentiation-dependent life cycle of human papillomaviruses in keratinocytes. In: Garcea RL, DiMaio D, eds. *The Papillomaviruses.* New York: Springer; 2007:45–68.
17. Hughes FJ, Romanos MA. E1 protein of human papillomavirus is a DNA Helicase/ATPase. *Nucleic Acids Res.* 1993;21:5817–5823.
18. Mohr IJ, Clark R, Sun S, et al. Targeting the E1 replication protein to the papillomavirus origin of replication by complex formation with the E2 transactivator. *Science.* 1990;250: 1694–1699.
19. Munger K, Baldwin A, Edwards KM, et al. Mechanisms of human papillomavirus-induced oncogenesis. *J Virol.* 2004;78: 11451–11460.
20. Solomon D, Breen N, McNeel T. Cervical cancer screening rates in the United States and the potential impact of implementation of screening guidelines. *CA Cancer J Clin.* 2007;57:105–111.
21. Selvaggi SM. Implications of low diagnostic reproducibility of cervical cytologic and histologic diagnoses. *Jama.* 2001;285: 1506–1508.
22. Uyar DS, Eltabbakh GH, Mount SL. Positive predictive value of liquid-based and conventional cervical Papanicolaou smears reported as malignant. *Gynecol Oncol.* 2003;89:227–232.
23. Klaes R, Friedrich T, Spitkovsky D, et al. Overexpression of p16(INK4A) as a specific marker for dysplastic and neoplastic epithelial cells of the cervix uteri. *Int J Cancer.* 2001;92:276–284.
24. Mayrand MH, Duarte-Franco E, Rodrigues I, et al. Human papillomavirus DNA versus Papanicolaou screening tests for cervical cancer. *N Engl J Med.* 2007;357:1579–1588.
25. Phelps WC, Barnes JA, Lobe DC. Molecular targets for human papillomaviruses: prospects for antiviral therapy. *Antivir Chem Chemother.* 1998;9:359–377.
26. Rhea WG, Jr., Bourgeois BM, Sewell DR. Condyloma acuminata: a fatal disease? *Am Surg.* 1998;64:1082–1087.
27. Frega A, Stentella P, Tinari A, et al. Giant condyloma acuminatum or buschke-Lowenstein tumor: review of the literature and report of three cases treated by CO2 laser surgery. A long-term follow-up. *Anticancer Res.* 2002;22:1201–1204.
28. Solomon D, Davey D, Kurman R, et al. The 2001 Bethesda System: terminology for reporting results of cervical cytology. *Jama.* 2002;287:2114–2119.
29. Hippelainen MI, Hippelainen M, Saarikoski S, et al. Clinical course and prognostic factors of human papillomavirus infections in men. *Sex Transm Dis.* 1994;21:272–279.
30. Winer RL, Hughes JP, Feng Q, et al. Condom use and the risk of genital human papillomavirus infection in young women. *N Engl J Med.* 2006;354:2645–2654.
31. Roberts JN, Buck CB, Thompson CD, et al. Genital transmission of HPV in a mouse model is potentiated by nonoxynol-9 and inhibited by carrageenan. *Nat Med.* 2007;13:857–861.
32. Barr E, Tamms G. Quadrivalent human papillomavirus vaccine. *Clin Infect Dis.* 2007;45:609–7.
33. Frazer IH. Prevention of cervical cancer through papillomavirus vaccination. *Nat Rev Immunol.* 2004;4:46–54.
34. Lin YY, Alphs H, Hung CF, et al. Vaccines against human papillomavirus. *Front Biosci.* 2007;12:246–264.

35. Ziegert C, Wentzensen N, Vinokurova S, et al. A comprehensive analysis of HPV integration loci in anogenital lesions combining transcript and genome-based amplification techniques. *Oncogene.* 2003;22:3977–3984.
36. Thorland EC, Myers SL, Persing DH, et al. Human papillomavirus type 16 integrations in cervical tumors frequently occur in common fragile sites. *Cancer Res.* 2000;60:5916–5921.
37. Durst M, Croce CM, Gissmann L, et al. Papillomavirus sequences integrate near cellular oncogenes in some cervical carcinomas. *Proc Natl Acad Sci U S A.* 1987;84:1070–1074.
38. Bernard BA, Bailly C, Lenoir M-C, et al. The HPV18 E2 gene product is a repressor of the HPV18 regulatory region in human keratinocytes. *J Virol.* 1989;63:4317–4324.
39. Thierry F, Yaniv M. The BPV-1 E2 trans-acting protein can be either an activator or repressor of the HPV-18 regulatory region. *EMBO J.* 1987;6:3391–3397.
40. Jeon S, Lambert PF. Integration of human papillomavirus type 16 DNA into the human genome leads to increased stability of E6 and E7 mRNAs: Implications for cervical carcinogenesis. *Proc Natl Acad Sci U S A.* 1995;92:1654–1658.
41. McCance DJ, Kopan R, Fuchs E, et al. Human papillomavirus type 16 alters human epithelial cell differentiation in vitro. *Proceedings of the National Academy of Sciences USA.* 1988;85: 7169–7173.
42. Hurlin PJ, Kaur P, Smith PP, et al. Progression of human papillomavirus type 18-immortalized human keratinocytes to a malignant phenotype. *Proceedings of the National Academy of Sciences USA.* 1991;88:571–574.
43. Pei XF, Meck JM, Greenhalgh D, et al. Cotransfection of HPV-18 and v-fos DNA induces tumorigenicity of primary human keratinocytes. *Virology.* 1993;196:855–860.
44. Durst M, Gallahan D, Jay G, et al. Glucocorticoid-enhanced neoplastic transformation of human keratinocytes by human papillomavirus type 16 and an activated ras oncogene. *Virology.* 1989;173:767–771.
45. Mitelman F, Johansson B, Mertens F. *Mitelman Database of Chromosome Aberrations in Cancer.* http://cgap.nci.nih.gov/Chromosomes/Mitelman; 2007.
46. Arbeit JM, Münger K, Howley PM, et al. Progressive squamous epithelial neoplasia in k14-human papillomavirus type 16 transgenic mice. *J Virol.* 1994;68:4358–4368.
47. Duensing S, Munger K. Mechanisms of genomic instability in human cancer: insights from studies with human papillomavirus oncoproteins. *Int J Cancer.* 2004;109:157–162.
48. Munger K, Howley PM, DiMaio D. Human Papillomavirus E6 and E7 Oncogenes. In: Garcea R, DiMaio D, eds. *The Papillomaviruses.* New York: Springer Verlag; 2007:195–251.
49. Angeline M, Merle E, Moroianu J. The E7 oncoprotein of high-risk human papillomavirus type 16 enters the nucleus via a non-classical Ran-dependent pathway. *Virology.* 2003;317:13–23.
50. Liu X, Clements A, Zhao K, et al. Structure of the human Papillomavirus E7 oncoprotein and its mechanism for inactivation of the retinoblastoma tumor suppressor. *J Biol Chem.* 2006; 281:578–586.
51. Ohlenschlager O, Seiboth T, Zengerling H, et al. 2006;Solution structure of the partially folded high-risk human papilloma virus 45 oncoprotein E7. *Oncogene.*
52. Huh K, Zhou X, Hayakawa H, et al. Human papillomavirus type 16 E7 oncoprotein associates with the cullin 2 ubiquitin ligase complex, which contributes to degradation of the retinoblastoma tumor suppressor. *J Virol.* 2007;81:9737–9747.
53. Nakatani Y, Konishi H, Vassilev A, et al. p600, a unique protein required for membrane morphogenesis and cell survival. *Proc Natl Acad Sci U S A.* 2005;102:15093–15098.
54. Tasaki T, Mulder LC, Iwamatsu A, et al. A family of mammalian E3 ubiquitin ligases that contain the UBR box motif and recognize N-degrons. *Mol Cell Biol.* 2005;25:7120–7136.
55. Huh KW, DeMasi J, Ogawa H, et al. Association of the human papillomavirus type 16 E7 oncoprotein with the 600-kDa retinoblastoma protein-associated factor, p600. *Proc Natl Acad Sci U S A.* 2005;102:11492–11497.
56. DeMasi J, Chao MC, Kumar AS, et al. Bovine papillomavirus E7 oncoprotein inhibits anoikis. *J Virol.* 2007;81:9419–9425.
57. DeMasi J, Huh KW, Nakatani Y, et al. Bovine papillomavirus E7 transformation function correlates with cellular p600 protein binding. *Proc Natl Acad Sci U S A.* 2005;102: 11486–11491.
58. Helt AM, Funk JO, Galloway DA. Inactivation of both the retinoblastoma tumor suppressor and p21 by the human papillomavirus type 16 E7 oncoprotein is necessary to inhibit cell cycle arrest in human epithelial cells. *J Virol.* 2002;76: 10559–10568.
59. Zwerschke W, Mannhardt B, Massimi P, et al. Allosteric activation of acid alpha-glucosidase by the human papillomavirus E7 protein. *J Biol Chem.* 2000;275:9534–9541.
60. Zwerschke W, Mazurek S, Massimi P, et al. Modulation of type M2 pyruvate kinase activity by the human papillomavirus type 16 E7 oncoprotein. *Proc Natl Acad Sci U S A.* 1999;96: 1291–1296.
61. Charette ST, McCance DJ. The E7 protein from human papillomavirus type 16 enhances keratinocyte migration in an Akt-dependent manner. *Oncogene.* 2007;26:7386–7390.
62. Menges CW, Baglia LA, Lapoint R, et al. Human papillomavirus type 16 E7 up-regulates AKT activity through the retinoblastoma protein. *Cancer Res.* 2006;66:5555–5559.
63. Toussaint-Smith E, Donner DB, Roman A. Expression of human papillomavirus type 16 E6 and E7 oncoproteins in primary foreskin keratinocytes is sufficient to alter the expression of angiogenic factors. *Oncogene.* 2004;23:2988–2995.
64. Chen W, Li F, Mead L, et al. Human papillomavirus causes an angiogenic switch in keratinocytes which is sufficient to alter endothelial cell behavior. *Virology.* 2007;367:168–174.
65. Kiyono T, Hiraiwa A, Fujita M, et al. Binding of high-risk human papillomavirus E6 oncoproteins to the human homologue of the Drosophila discs large tumor suppressor protein. *Proc Natl Acad Sci U S A.* 1997;94:11612–11616.
66. Lee SS, Weiss RS, Javier RT. Binding of human virus oncoproteins to hDlg/SAP97, a mammalian homolog of the Drosophila discs large tumor suppressor protein. *Proc Natl Acad Sci U S A.* 1997;94:6670–6675.
67. Lee C, Laimins LA. Role of the PDZ domain-binding motif of the oncoprotein E6 in the pathogenesis of human papillomavirus type 31. *J Virol.* 2004;78:12366–12377.
68. Nguyen ML, Nguyen MM, Lee D, et al. The PDZ ligand domain of the human papillomavirus type 16 E6 protein is required for E6prime;s induction of epithelial hyperplasia in vivo. *J Virol.* 2003;77:6957–6964.
69. Thomas M, Glaunsinger B, Pim D, et al. HPV E6 and MAGUK protein interactions: determination of the molecular basis for specific protein recognition and degradation. *Oncogene.* 2001;20:5431–5439.
70. Chen JJ, Hong Y, Rustamzadeh E, et al. Identification of an alpha helical motif sufficient for association with papillomavirus E6. *J Biol Chem.* 1998;273:13537–13544.
71. Tong X, Howley PM. The bovine papillomavirus E6 oncoprotein interacts with paxillin and disrupts the actin cytoskeleton. *Proc Nat Acad Sci USA.* 1997;94:4412–4417.
72. Tong X, Salgia R, Li JL, et al. The bovine papillomavirus E6 protein binds to the LD motif repeats of paxillin and blocks its interaction with vinculin and the focal adhesion kinase. *J Biol Chem.* 1997;272:33373–33376.
73. Vande Pol SB, Brown MC, Turner CE. Association of Bovine Papillomavirus Type 1 E6 oncoprotein with the focal adhesion protein paxillin through a conserved protein interaction motif. *Oncogene.* 1998;16:43–52.
74. Heselmeyer K, Macville M, Schrock E, et al. Advanced-stage cervical carcinomas are defined by a recurrent pattern of chromosomal aberrations revealing high genetic instability and a consistent gain of chromosome arm 3q. *Genes Chromosomes Cancer.* 1997;19:233–240.
75. Heselmeyer K, Schrock E, du Manoir S, et al. Gain of chromosome 3q defines the transition from severe dysplasia to invasive carcinoma of the uterine cervix. *Proc Natl Acad Sci U S A.* 1996;93:479–484.
76. Hashida T, Yasumoto S. Induction of chromosomal abnormalities in mouse and human epidermal keratinocytes by the

human papillomavirus type 16 E7 oncogene. *Journal of General Virology*. 1991;72:1569–1577.
77. White AE, Livanos EM, Tlsty TD. Differential Disruption of Genomic Integrity and Cell Cycle Regulation in Normal Human Fibroblasts by the HPV Oncoproteins. *Genes Dev.* 1994;8:666–677.
78. Duensing A, Liu Y, Tseng M, et al. Cyclin-dependent kinase 2 is dispensable for normal centrosome duplication but required for oncogene-induced centrosome overduplication. *Oncogene.* 2006;25:2943–2949.
79. Duensing S, Duensing A, Lee DC, et al. Cyclin-dependent kinase inhibitor indirubin-3′-oxime selectively inhibits human papillomavirus type 16 E7-induced numerical centrosome anomalies. *Oncogene.* 2004;23:8206–8215.
80. Nguyen CL, Eichwald C, Nibert ML, et al. Human papillomavirus type 16 E7 oncoprotein associates with the centrosomal component gamma-tubulin. *J Virol.* 2007;81:13533–13543.
81. Duensing S, Duensing A, Crum CP, et al. Human papillomavirus type 16 E7 oncoprotein-induced abnormal centrosome synthesis is an early event in the evolving malignant phenotype. *Cancer Res.* 2001;61:2356–2360.
82. Duensing A, Liu Y, Perdreau SA, et al. Centriole overduplication through the concurrent formation of multiple daughter centrioles at single maternal templates. *Oncogene.* 2007;26:6280–6288.
83. Duensing S, Lee LY, Duensing A, et al. The human papillomavirus type 16 E6 and E7 oncoproteins cooperate to induce mitotic defects and genomic instability by uncoupling centrosome duplication from the cell division cycle. *Proc Natl Acad Sci U S A.* 2000;97:10002–10007.
84. Kim JH, Dhanasekaran SM, Mehra R, et al. Integrative analysis of genomic aberrations associated with prostate cancer progression. *Cancer Res.* 2007;67:8229–8239.
85. Tomlins SA, Laxman B, Dhanasekaran SM, et al. Distinct classes of chromosomal rearrangements create oncogenic ETS gene fusions in prostate cancer. *Nature.* 2007;448:595–599.
86. Kutler DI, Wreesmann VB, Goberdhan A, et al. Human papillomavirus DNA and p53 polymorphisms in squamous cell carcinomas from Fanconi anemia patients. *J Natl Cancer Inst.* 2003;95:1718–1721.
87. Kennedy RD, D'Andrea AD. The Fanconi Anemia/BRCA pathway: new faces in the crowd. *Genes Dev.* 2005;19:2925–2940.
88. Spardy N, Duensing A, Charles D, et al. The human papillomavirus type 16 E7 oncoprotein activates the Fanconi Anemia (FA) pathway and causes accelerated chromosomal instability in FA cells. *J Virol.* 2007;81:in press.
89. Kessis TD, Slebos RJ, Nelson WG, et al. Human papillomavirus 16 E6 expression disrupts the p53-mediated cellular response to DNA damage. *Proceedings of the National Academy of Sciences USA.* 1993;90:3988–3992.
90. Thompson DA, Belinsky G, Chang TH, et al. The human papillomavirus-16 E6 oncoprotein decreases the vigilance of mitotic checkpoints. *Oncogene.* 1997;15:3025–3035.
91. Thomas JT, Laimins LA. Human papillomavirus oncoproteins E6 and E7 independently abrogate the mitotic spindle checkpoint. *J Virol.* 1998;72:1131–1137.
92. Duensing S, Munger K. The human papillomavirus type 16 E6 and E7 oncoproteins independently induce numerical and structural chromosome instability. *Cancer Res.* 2002;62:7075–7082.
93. Havre PA, Yuan J, Hedrick L, et al. p53 inactivation by HPV16 E6 results in increased mutagenesis in human cells. *Cancer Res.* 1995;55:4420–4424.
94. Iftner T, Elbel M, Schopp B, et al. Interference of papillomavirus E6 protein with single-strand break repair by interaction with XRCC1. *EMBO J.* 2002;21:4741–4748.
95. Srivenugopal KS, Ali-Osman F. The DNA repair protein, O(6)-methylguanine-DNA methyltransferase is a proteolytic target for the E6 human papillomavirus oncoprotein. *Oncogene.* 2002;21:5940–5945.
96. Shin KH, Ahn JH, Kang MK, et al. HPV-16 E6 oncoprotein impairs the fidelity of DNA end-joining via p53-dependent and -independent pathways. *Int J Oncol.* 2006;28:209–215.
97. Gillison ML, D'Souza G, Westra W, et al. Distinct risk factor profiles for human papillomavirus type 16-positive and human papillomavirus type 16-negative head and neck cancers. *J Natl Cancer Inst.* 2008;100:407–420.
98. Gillison ML, Koch WM, Capone RB, et al. Evidence for a causal association between human papillomavirus and a subset of head and neck cancers. *J Natl Cancer Inst.* 2000;92: 709–720.
99. (2008). *Cancer Facts and Figures 2008* (Society AC, ed.).
100. Brinton LA, Berman ML, Mortel R, et al. Reproductive, menstrual, and medical risk factors for endometrial cancer: results from a case-control study. *Am J Obstet Gynecol.* 1992;167: 1317–1325.
101. Kurman RJ, Malkasian GD, Jr., Sedlis A, et al. From Papanicolaou to Bethesda: the rationale for a new cervical cytologic classification. *Obstet Gynecol.* 1991;77:779–782.
102. Horn LC, Schnurrbusch U, Bilek K, et al. Risk of progression in complex and atypical endometrial hyperplasia: clinicopathologic analysis in cases with and without progestogen treatment. *Int J Gynecol Cancer.* 2004;14:348–353.
103. Lacey JV, Jr., Ioffe OB, Ronnett BM, et al. Endometrial carcinoma risk among women diagnosed with endometrial hyperplasia: the 34-year experience in a large health plan. *Br J Cancer.* 2008;98:45–53.
104. Trimble CL, Kauderer J, Zaino R, et al. Concurrent endometrial carcinoma in women with a biopsy diagnosis of atypical endometrial hyperplasia: a Gynecologic Oncology Group study. *Cancer.* 2006;106:812–819.
105. Zaino RJ, Kauderer J, Trimble CL, et al. Reproducibility of the diagnosis of atypical endometrial hyperplasia: a Gynecologic Oncology Group study. *Cancer.* 2006;106:804–811.
106. Levine RL, Cargile CB, Blazes MS, et al. PTEN mutations and microsatellite instability in complex atypical hyperplasia, a precursor lesion to uterine endometrioid carcinoma. *Cancer Res.* 1998;58:3254–3258.
107. Mutter GL, Lin MC, Fitzgerald JT, et al. Altered PTEN expression as a diagnostic marker for the earliest endometrial precancers. *J Natl Cancer Inst.* 2000;92:924–930.
108. Deng L, Broaddus RR, McCampbell A, et al. Identification of a novel estrogen-regulated gene, EIG121, induced by hormone replacement therapy and differentially expressed in type I and type II endometrial cancer. *Clin Cancer Res.* 2005;11:8258–8264.
109. McCampbell AS, Broaddus RR, Loose DS, et al. Overexpression of the insulin-like growth factor I receptor and activation of the AKT pathway in hyperplastic endometrium. *Clin Cancer Res.* 2006;12:6373–6378.
110. Kosary CL. Chapter 15: Cancer of the Corpus Uteri. In: Ries LAB YJ, Keel GE, Eisner MP, Lin YD, Horner MJ, ed., *SEER Survival Monograph: Cancer Survival Among Adults:U.S. SEER Program, 1988–2001, Patient and Tumor Characteristics.* Vol. NIH Pub. No. 07-6215. Bethesda, MD: National Cancer Institute, SEER Program; 2007.
111. Risinger JI, Hayes K, Maxwell GL, et al. PTEN mutation in endometrial cancers is associated with favorable clinical and pathologic characteristics. *Clin Cancer Res.* 1998;4: 3005–3010.
112. Risinger JI, Maxwell GL, Chandramouli GV, et al. Microarray analysis reveals distinct gene expression profiles among different histologic types of endometrial cancer. *Cancer Res.* 2003;63:6–11.
113. Stambolic V, Tsao MS, Macpherson D, et al. High incidence of breast and endometrial neoplasia resembling human Cowden syndrome in pten+/- mice. *Cancer Res.* 2000;60:3605–3611.
114. Vilgelm A, Lian Z, Wang H, et al. Akt-mediated phosphorylation and activation of estrogen receptor alpha is required for endometrial neoplastic transformation in Pten+/- mice. *Cancer Res.* 2006;66:3375–3380.
115. Hayes MP, Wang H, Espinal-Witter R, et al. PIK3CA and PTEN mutations in uterine endometrioid carcinoma and complex atypical hyperplasia. *Clin Cancer Res.* 2006;12:5932–5935.
116. Oda K, Stokoe D, Taketani Y, et al. High frequency of coexistent mutations of PIK3CA and PTEN genes in endometrial carcinoma. *Cancer Res.* 2005;65:10669–10673.

117. Lu KH, Wu W, Dave B, et al. Loss of tuberous sclerosis complex-2 function and activation of Mammalian target of rapamycin signaling in endometrial carcinoma. *Clin Cancer Res.* 2008;14:2543–2550.
118. Contreras CM, Gurumurthy S, Haynie JM, et al. Loss of Lkb1 provokes highly invasive endometrial adenocarcinomas. *Cancer Res.* 2008;68:759–766.
119. MacDonald ND, Salvesen HB, Ryan A, et al. Frequency and prognostic impact of microsatellite instability in a large population-based study of endometrial carcinomas. *Cancer Res.* 2000;60:1750–1752.
120. Dunlop MG, Farrington SM, Carothers AD, et al. Cancer risk associated with germline DNA mismatch repair gene mutations. *Hum Mol Genet.* 1997;6:105–110.
121. Aarnio M, Sankila R, Pukkala E, et al. Cancer risk in mutation carriers of DNA-mismatch-repair genes. *Int J Cancer.* 1999;81:214–218.
122. Hampel H, Frankel W, Panescu J, et al. Screening for Lynch syndrome (hereditary nonpolyposis colorectal cancer) among endometrial cancer patients. *Cancer Res.* 2006;66: 7810–7817.
123. Duval A, Reperant M, Compoint A, et al. Target gene mutation profile differs between gastrointestinal and endometrial tumors with mismatch repair deficiency. *Cancer Res.* 2002;62: 1609–1612.
124. Boyd J. Genetic basis of familial endometrial cancer: is there more to learn? *J Clin Oncol.* 2005;23:4570–4573.
125. Zighelboim I, Goodfellow PJ, Gao F, et al. Microsatellite instability and epigenetic inactivation of MLH1 and outcome of patients with endometrial carcinomas of the endometrioid type. *J Clin Oncol.* 2007;25:2042–2048.
126. Lu KH, Schorge JO, Rodabaugh KJ, et al. Prospective determination of prevalence of lynch syndrome in young women with endometrial cancer. *J Clin Oncol.* 2007;25:5158–5164.
127. Berends MJ, Wu Y, Sijmons RH, et al. Toward new strategies to select young endometrial cancer patients for mismatch repair gene mutation analysis. *J Clin Oncol.* 2003;21:4364–4370.
128. Fukuchi T, Sakamoto M, Tsuda H, et al. Beta-catenin mutation in carcinoma of the uterine endometrium. *Cancer Res.* 1998;58:3526–3528.
129. Mirabelli-Primdahl L, Gryfe R, Kim H, et al. Beta-catenin mutations are specific for colorectal carcinomas with microsatellite instability but occur in endometrial carcinomas irrespective of mutator pathway. *Cancer Res.* 1999;59:3346–3351.
130. Slomovitz BM, Burke TW, Eifel PJ, et al. Uterine papillary serous carcinoma (UPSC): a single institution review of 129 cases. *Gynecol Oncol.* 2003;91:463–469.
131. Koul A, Willen R, Bendahl PO, et al. Distinct sets of gene alterations in endometrial carcinoma implicate alternate modes of tumorigenesis. *Cancer.* 2002;94:2369–2379.
132. Sherman ME, Bur ME, Kurman RJ. p53 in endometrial cancer and its putative precursors: evidence for diverse pathways of tumorigenesis. *Hum Pathol.* 1995;26:1268–1274.
133. Slomovitz BM, Broaddus RR, Burke TW, et al. Her-2/neu overexpression and amplification in uterine papillary serous carcinoma. *J Clin Oncol.* 2004;22:3126–3132.
134. Diaz-Montes TP, Ji H, Smith Sehdev AE, et al. Clinical significance of Her-2/neu overexpression in uterine serous carcinoma. *Gynecol Oncol.* 2006;100:139–144.
135. Santin AD, Bellone S, Van Stedum S, et al. Amplification of c-erbB2 oncogene: a major prognostic indicator in uterine serous papillary carcinoma. *Cancer.* 2005;104:1391–1397.
136. Okolo S. Incidence, aetiology and epidemiology of uterine fibroids. *Best Pract Res Clin Obstet Gynaecol.* 2008;22:571–588.
137. Hodge JC, Quade BJ, Rubin MA, et al. Molecular and cytogenetic characterization of plexiform leiomyomata provide further evidence for genetic heterogeneity underlying uterine fibroids. *Am J Pathol.* 2008;172:1403–1410.
138. Ptacek T, Song C, Walker CL, et al. Physical mapping of distinct 7q22 deletions in uterine leiomyoma and analysis of a recently annotated 7q22 candidate gene. *Cancer Genet Cytogenet.* 2007;174:116–120.
139. Bishara M, Scapa E. [Stromal uterine sarcoma arising from intestinal endometriosis after abdominal hysterectomy and salpingo-oophorectomy]. *Harefuah.* 1997;133:353–355, 415.
140. Laxman R, Currie JL, Kurman RJ, et al. Cytogenetic profile of uterine sarcomas. *Cancer.* 1993;71:1283–1288.
141. Hennig Y, Caselitz J, Bartnitzke S, et al. A third case of a low-grade endometrial stromal sarcoma with a t(7;17)(p14 approximately 21;q11.2 approximately 21). *Cancer Genet Cytogenet.* 1997;98:84–86.
142. Pelusi B, Gambineri A, Pasquali R. Type 2 diabetes and the polycystic ovary syndrome. *Minerva Ginecol.* 2004;56:41–51.
143. Givens JR. Familial polycystic ovarian disease. *Endocrinol Metab Clin North Am.* 1988;17:771–783.
144. Fowler DJ, Nicolaides KH, Miell JP. Insulin-like growth factor binding protein-1 (IGFBP-1): a multifunctional role in the human female reproductive tract. *Hum Reprod Update.* 2000;6:495–504.
145. Ibanez L, Ong KK, Mongan N, et al. Androgen receptor gene CAG repeat polymorphism in the development of ovarian hyperandrogenism. *J Clin Endocrinol Metab.* 2003;88: 3333–3338.
146. Nelson-Degrave VL, Wickenheisser JK, Hendricks KL, et al. Alterations in mitogen-activated protein kinase kinase and extracellular regulated kinase signaling in theca cells contribute to excessive androgen production in polycystic ovary syndrome. *Mol Endocrinol.* 2005;19:379–390.
147. Roberts VJ, Barth S, el-Roeiy A, et al. Expression of inhibin/activin system messenger ribonucleic acids and proteins in ovarian follicles from women with polycystic ovarian syndrome. *J Clin Endocrinol Metab.* 1994;79:1434–1439.
148. Sorbara LR, Tang Z, Cama A, et al. Absence of insulin receptor gene mutations in three insulin-resistant women with the polycystic ovary syndrome. *Metabolism.* 1994;43:1568–1574.
149. Villuendas G, Botella-Carretero JI, Roldan B, et al. Polymorphisms in the insulin receptor substrate-1 (IRS-1) gene and the insulin receptor substrate-2 (IRS-2) gene influence glucose homeostasis and body mass index in women with polycystic ovary syndrome and non-hyperandrogenic controls. *Hum Reprod.* 2005;20:3184–3191.
150. Witchel SF, Kahsar-Miller M, Aston CE, et al. Prevalence of CYP21 mutations and IRS1 variant among women with polycystic ovary syndrome and adrenal androgen excess. *Fertil Steril.* 2005;83:371–375.
151. El Mkadem SA, Lautier C, Macari F, et al. Role of allelic variants Gly972Arg of IRS-1 and Gly1057Asp of IRS-2 in moderate-to-severe insulin resistance of women with polycystic ovary syndrome. *Diabetes.* 2001;50:2164–2168.
152. Cisternino M, Dondi E, Martinetti M, et al. Exaggerated 17-hydroxyprogesterone response to short-term adrenal stimulation and evidence for CYP21B gene point mutations in true precocious puberty. *Clin Endocrinol (Oxf).* 1998;48: 555–560.
153. Gaasenbeek M, Powell BL, Sovio U, et al. Large-scale analysis of the relationship between CYP11A promoter variation, polycystic ovarian syndrome, and serum testosterone. *J Clin Endocrinol Metab.* 2004;89:2408–2413.
154. Goodarzi MO. Looking for polycystic ovary syndrome genes: rational and best strategy. *Semin Reprod Med.* 2008;26:5–13.
155. Urbanek M, Woodroffe A, Ewens KG, et al. Candidate gene region for polycystic ovary syndrome on chromosome 19p13.2. *J Clin Endocrinol Metab.* 2005;90:6623–6629.
156. Ovarian disorders. Benign cysts. *Harv Womens Health Watch.* 1999;6:4–5.
157. Raile K, Stobbe H, Trobs RB, et al. A new heterozygous mutation of the FOXL2 gene is associated with a large ovarian cyst and ovarian dysfunction in an adolescent girl with blepharophimosis/ptosis/epicanthus inversus syndrome. *Eur J Endocrinol.* 2005;153:353–358.
158. Yamada T, Okamoto Y, Kasamatsu H, et al. Laparoscopic-assisted surgery for benign ovarian cyst in a young girl. *J Am Assoc Gynecol Laparosc.* 2001;8:295–298.
159. Taylor HC. Malignant and semimalignant tumors of the ovary. *Surgery, Gynecology & Obstetrics.* 1929;48:206–230.
160. FIGO. International Federation of Gynecology and Obstetrics. Classification and staging of malignant tumours in the female pelvis. *Acta Obstet Gynecol Scand.* 1971;50:1–7.

161. Bell DA, Longacre TA, Prat J, et al. Serous borderline (low malignant potential, atypical proliferative) ovarian tumors: workshop perspectives. *Hum Pathol.* 2004;35:934–948.
162. Wong KK, Gershenson D. The continuum of serous tumors of low malignant potential and low-grade serous carcinomas of the ovary. *Dis Markers.* 2007;23:377–387.
163. Longacre TA, McKenney JK, Tazelaar HD, et al. Ovarian serous tumors of low malignant potential (borderline tumors): outcome-based study of 276 patients with long-term (> or =5-year) follow-up. *Am J Surg Pathol.* 2005;29:707–723.
164. Gershenson DM, Sun CC, Lu KH, et al. Clinical behavior of stage II-IV low-grade serous carcinoma of the ovary. *Obstet Gynecol.* 2006;108:361–368.
165. Mok SC, Kwong J, Welch WR, et al. Etiology and pathogenesis of epithelial ovarian cancer. *Dis Markers.* 2007;23:367–376.
166. Thompson ER, Herbert SC, Forrest SM, et al. Whole genome SNP arrays using DNA derived from formalin-fixed, paraffin-embedded ovarian tumor tissue. *Hum Mutat.* 2005;26:384–389.
167. Bonome T, Lee JY, Park DC, et al. Expression profiling of serous low malignant potential, low-grade, and high-grade tumors of the ovary. *Cancer Res.* 2005;65:10602–10612.
168. Shih Ie M, Kurman RJ. Ovarian tumorigenesis: a proposed model based on morphological and molecular genetic analysis. *Am J Pathol.* 2004;164:1511–1518.
169. Malpica A, Deavers MT, Lu K, et al. Grading ovarian serous carcinoma using a two-tier system. *Am J Surg Pathol.* 2004;28:496–504.
170. Edelson MI, Lau CC, Colitti CV, et al. A one centimorgan deletion unit on chromosome Xq12 is commonly lost in borderline and invasive epithelial ovarian tumors. *Oncogene.* 1998;16: 197–202.
171. Edelson MI, Scherer SW, Tsui LC, et al. Identification of a 1300 kilobase deletion unit on chromosome 7q31.3 in invasive epithelial ovarian carcinomas. *Oncogene.* 1997;14:2979–2984.
172. Colitti CV, Rodabaugh KJ, Welch WR, et al. A novel 4 cM minimal deletion unit on chromosome 6q25.1-q25.2 associated with high grade invasive epithelial ovarian carcinomas. *Oncogene.* 1998;16:555–559.
173. Tangir J, Loughridge NS, Berkowitz RS, et al. Frequent microsatellite instability in epithelial borderline ovarian tumors. *Cancer Res.* 1996;56:2501–2505.
174. Huang LW, Garrett AP, Muto MG, et al. Identification of a novel 9 cM deletion unit on chromosome 6q23-24 in papillary serous carcinoma of the peritoneum. *Hum Pathol.* 2000;31:367–373.
175. Lu KH, Weitzel JN, Kodali S, et al. A novel 4-cM minimally deleted region on chromosome 11p15.1 associated with high grade nonmucinous epithelial ovarian carcinomas. *Cancer Res.* 1997;57:387–390.
176. Wertheim I, Tangir J, Muto MG, et al. Loss of heterozygosity of chromosome 17 in human borderline and invasive epithelial ovarian tumors. *Oncogene.* 1996;12:2147–2153.
177. Russell P. The pathological assessment of ovarian neoplasms. III: The malignant "epithelial" tumours. *Pathology.* 1979;11: 493–532.
178. Obata K, Morland SJ, Watson RH, et al. Frequent PTEN/MMAC mutations in endometrioid but not serous or mucinous epithelial ovarian tumors. *Cancer Res.* 1998;58:2095–2097.
179. Hashiguchi Y, Tsuda H, Inoue T, et al. PTEN expression in clear cell adenocarcinoma of the ovary. *Gynecol Oncol.* 2006;101:71–75.
180. Dinulescu DM, Ince TA, Quade BJ, et al. Role of K-ras and Pten in the development of mouse models of endometriosis and endometrioid ovarian cancer. *Nat Med.* 2005;11:63–70.
181. Biermann K, Goke F, Nettersheim D, et al. c-KIT is frequently mutated in bilateral germ cell tumours and down-regulated during progression from intratubular germ cell neoplasia to seminoma. *J Pathol.* 2007;213:311–318.
182. Kraggerud SM, Szymanska J, Abeler VM, et al. DNA copy number changes in malignant ovarian germ cell tumors. *Cancer Res.* 2000;60:3025–3030.
183. Hutson R, Ramsdale J, Wells M. p53 protein expression in putative precursor lesions of epithelial ovarian cancer. *Histopathology.* 1995;27:367–371.
184. Colombo N, Parma G, Zanagnolo V, et al. Management of ovarian stromal cell tumors. *J Clin Oncol.* 2007;25:2944–2951.
185. Truss L, Dobin SM, Rao A, et al. Overexpression of the BCL2 gene in a Sertoli-Leydig cell tumor of the ovary: a pathologic and cytogenetic study. *Cancer Genet Cytogenet.* 2004; 148:118–122.
186. Callahan MJ, Crum CP, Medeiros F, et al. Primary fallopian tube malignancies in BRCA-positive women undergoing surgery for ovarian cancer risk reduction. *J Clin Oncol.* 2007;25: 3985–3990.
187. Jarboe E, Folkins A, Nucci MR, et al. Serous carcinogenesis in the fallopian tube: a descriptive classification. *Int J Gynecol Pathol.* 2008;27:1–9.
188. Bergeron C, Ferenczy A, Shah KV, et al. Multicentric human papillomavirus infections of the female genital tract: correlation of viral types with abnormal mitotic figures, colposcopic presentation, and location. *Obstet Gynecol.* 1987;69:736–742.
189. Frega A, French D, Piazze J, et al. Prediction of persistent vaginal intraepithelial neoplasia in previously hysterectomized women by high-risk HPV DNA detection. *Cancer Lett.* 2007;249:235–241.
190. Koyamatsu Y, Yokoyama M, Nakao Y, et al. A comparative analysis of human papillomavirus types 16 and 18 and expression of p53 gene and Ki-67 in cervical, vaginal, and vulvar carcinomas. *Gynecol Oncol.* 2003;90:547–551.
191. Yang B, Hart WR. Vulvar intraepithelial neoplasia of the simplex (differentiated) type: a clinicopathologic study including analysis of HPV and p53 expression. *Am J Surg Pathol.* 2000;24:429–441.
192. Milde-Langosch K, Albrecht K, Joram S, et al. Presence and persistence of HPV infection and p53 mutation in cancer of the cervix uteri and the vulva. *Int J Cancer.* 1995;63:639–645.
193. Holway AH, Rieger-Christ KM, Miner WR, et al. Somatic mutation of PTEN in vulvar cancer. *Clin Cancer Res.* 2000;6: 3228–3235.
194. Jee KJ, Kim YT, Kim KR, et al. Loss in 3p and 4p and gain of 3q are concomitant aberrations in squamous cell carcinoma of the vulva. *Mod Pathol.* 2001;14:377–381.
195. Worsham MJ, Van Dyke DL, Grenman SE, et al. Consistent chromosome abnormalities in squamous cell carcinoma of the vulva. *Genes Chromosomes Cancer.* 1991;3:420–432.
196. Teixeira MR, Kristensen GB, Abeler VM, et al. Karyotypic findings in tumors of the vulva and vagina. *Cancer Genet Cytogenet.* 1999;111:87–91.
197. Pinto AP, Lin MC, Sheets EE, et al. Allelic imbalance in lichen sclerosus, hyperplasia, and intraepithelial neoplasia of the vulva. *Gynecol Oncol.* 2000;77:171–176.
198. Raber G, Mempel V, Jackisch C, et al. Malignant melanoma of the vulva. Report of 89 patients. *Cancer.* 1996;78:2353–2358.
199. Berkowitz RS, Goldstein DP. Chorionic tumors. *N Engl J Med.* 1996;335:1740–1748.
200. Paradinas F. In: Hancock BW, NE, Berkowitz RS, eds. *Gestational Trophoblastic Disease.* London: Chapman & Hall; 1997.
201. Fisher RA, Hodges MD, Newlands ES. Familial recurrent hydatidiform mole: a review. *J Reprod Med.* 2004;49:595–601.
202. El-Maarri O, Seoud M, Coullin P, et al. Maternal alleles acquiring paternal methylation patterns in biparental complete hydatidiform moles. *Hum Mol Genet.* 2003;12:1405–1413.
203. Genest DR, Dorfman DM, Castrillon DH. Ploidy and imprinting in hydatidiform moles. Complementary use of flow cytometry and immunohistochemistry of the imprinted gene product p57KIP2 to assist molar classification. *J Reprod Med.* 2002;47: 342–346.
204. Matsuoka S, Edwards MC, Bai C, et al. p57KIP2, a structurally distinct member of the p21CIP1 Cdk inhibitor family, is a candidate tumor suppressor gene. *Genes Dev.* 1995;9: 650–662.
205. Wake N, Arima T, Matsuda T. Involvement of IGF2 and H19 imprinting in choriocarcinoma development. *Int J Gynaecol Obstet.* 1998;60(Suppl 1):S1–S8.
206. Arima T, Wake N. Establishment of the primary imprint of the HYMAI/PLAGL1 imprint control region during oogenesis. *Cytogenet Genome Res.* 2006;113:247–252.

207. Rachmilewitz J, Goshen R, Ariel I, et al. Parental imprinting of the human H19 gene. *FEBS Lett.* 1992;309:25–28.
208. Niemitz EL, DeBaun MR, Fallon J, et al. Microdeletion of LIT1 in familial Beckwith-Wiedemann syndrome. *Am J Hum Genet.* 2004;75:844–849.
209. Nagymanyoki Z, Callahan MJ, Parast MM, et al. Immune cell profiling in normal pregnancy, partial and complete molar pregnancy. *Gynecol Oncol.* 2007;107:292–297.
210. King A, Allan DS, Bowen M, et al. HLA-E is expressed on trophoblast and interacts with CD94/NKG2 receptors on decidual NK cells. *Eur J Immunol.* 2000;30:1623–1631.
211. Morales PJ, Pace JL, Platt JS, et al. Placental cell expression of HLA-G2 isoforms is limited to the invasive trophoblast phenotype. *J Immunol.* 2003;171:6215–6224.
212. Ponte M, Cantoni C, Biassoni R, et al. Inhibitory receptors sensing HLA-G1 molecules in pregnancy: decidua-associated natural killer cells express LIR-1 and CD94/NKG2A and acquire p49, an HLA-G1-specific receptor. *Proc Natl Acad Sci U S A.* 1999;96:5674–5679.
213. Rebmann V, Busemann A, Lindemann M, et al. Detection of HLA-G5 secreting cells. *Hum Immunol.* 2003;64: 1017–1024.
214. Fulop V, Feinberg BB, Steller MA, et al. Molar villous fluid suppresses mononuclear cell cytotoxicity. *Gynecol Oncol.* 1992;47: 311–316.
215. Shaarawy M, Darwish NA, Abdel-Aziz O. Serum interleukin-2 and soluble interleukin-2 receptor in gestational trophoblastic diseases. *J Soc Gynecol Investig.* 1996;3:39–46.
216. Wang X, Fu S, Freedman RS, et al. Immunobiology of gestational trophoblastic diseases. *Int J Gynecol Cancer.* 2006;16:1500–1515.
217. Fulop V, Colitti CV, Genest D, et al. DOC-2/hDab2, a candidate tumor suppressor gene involved in the development of gestational trophoblastic diseases. *Oncogene.* 1998;17:419–424.
218. Fulop V, Mok SC, Berkowitz RS. Molecular biology of gestational trophoblastic neoplasia: a review. *J Reprod Med.* 2004;49:415–422.
219. Fulop V, Mok SC, Genest DR, et al. c-myc, c-erbB-2, c-fms and bcl-2 oncoproteins. Expression in normal placenta, partial and complete mole, and choriocarcinoma. *J Reprod Med.* 1998;43:101–110.
220. Kato HD, Terao Y, Ogawa M, et al. Growth-associated gene expression profiles by microarray analysis of trophoblast of molar pregnancies and normal villi. *Int J Gynecol Pathol.* 2002;21:255–260.
221. Vegh GL, Selcuk Tuncer Z, Fulop V, et al. Matrix metalloproteinases and their inhibitors in gestational trophoblastic diseases and normal placenta. *Gynecol Oncol.* 1999;75: 248–253.
222. Okamoto T, Niu R, Yamada S, et al. Reduced expression of tissue inhibitor of metalloproteinase (TIMP)-2 in gestational trophoblastic diseases. *Mol Hum Reprod.* 2002;8: 392–398.
223. Karmakar S, Das C. Regulation of trophoblast invasion by IL-1beta and TGF-beta1. *Am J Reprod Immunol.* 2002;48:210–219.
224. Han VK, Bassett N, Walton J, et al. The expression of insulin-like growth factor (IGF) and IGF-binding protein (IGFBP) genes in the human placenta and membranes: evidence for IGF-IGFBP interactions at the feto-maternal interface. *J Clin Endocrinol Metab.* 1996;81:2680–2693.
225. Hamilton GS, Lysiak JJ, Han VK, et al. Autocrine-paracrine regulation of human trophoblast invasiveness by insulin-like growth factor (IGF)-II and IGF-binding protein (IGFBP)-1. *Exp Cell Res.* 1998;244:147–156.
226. Cervar M, Huppertz B, Barth S, et al. Endothelin A and B receptors change their expression levels during development of human placental villi. *Placenta.* 2000;21:536–546.
227. Caniggia I, Mostachfi H, Winter J, et al. Hypoxia-inducible factor-1 mediates the biological effects of oxygen on human trophoblast differentiation through TGFbeta(3). *J Clin Invest.* 2000;105:577–587.
228. Mohan A, Asselin J, Sargent IL, et al. Effect of cytokines and growth factors on the secretion of inhibin A, activin A and follistatin by term placental villous trophoblasts in culture. *Eur J Endocrinol.* 2001;145:505–511.
229. Leach RE, Kilburn B, Wang J, et al. Heparin-binding EGF-like growth factor regulates human extravillous cytotrophoblast development during conversion to the invasive phenotype. *Dev Biol.* 2004;266:223–237.
230. Isaka K, Usuda S, Ito H, et al. Expression and activity of matrix metalloproteinase 2 and 9 in human trophoblasts. *Placenta.* 2003;24:53–64.
231. Floridon C, Nielsen O, Holund B, et al. Does plasminogen activator inhibitor-1 (PAI-1) control trophoblast invasion? A study of fetal and maternal tissue in intrauterine, tubal and molar pregnancies. *Placenta.* 2000;21:754–762.
232. Batorfi J, Fulop V, Kim JH, et al. Osteopontin is down-regulated in hydatidiform mole. *Gynecol Oncol.* 2003;89:134–139.
233. Nagymanyoki Z, Callahan MJ, Parast MM, et al. Overexpression of laminin receptor 1 on decidual cells in partial and complete mole. *Gynecol Oncol.* 2008;108:121–125.
234. Simpson H, Robson SC, Bulmer JN, et al. Transforming growth factor beta expression in human placenta and placental bed during early pregnancy. *Placenta.* 2002;23:44–58.
235. Xu G, Guimond MJ, Chakraborty C, et al. Control of proliferation, migration, and invasiveness of human extravillous trophoblast by decorin, a decidual product. *Biol Reprod.* 2002;67: 681–689.
236. Shih IM, Kurman RJ. Expression of melanoma cell adhesion molecule in intermediate trophoblast. *Lab Invest.* 1996;75: 377–388.
237. Bauer S, Pollheimer J, Hartmann J, et al. Tumor necrosis factor-alpha inhibits trophoblast migration through elevation of plasminogen activator inhibitor-1 in first-trimester villous explant cultures. *J Clin Endocrinol Metab.* 2004;89:812–822.
238. Yui J, Garcia-Lloret M, Wegmann TG, et al. Cytotoxicity of tumour necrosis factor-alpha and gamma-interferon against primary human placental trophoblasts. *Placenta.* 1994;15:819–835.
239. Liu Z, Chen Y, Yang Y, et al. The effect on MHC class II expression and apoptosis in placenta by IFNgamma administration. *Contraception.* 2002;65:177–184.
240. Feltmate CM, Batorfi J, Fulop V, et al. Human chorionic gonadotropin follow-up in patients with molar pregnancy: a time for reevaluation. *Obstet Gynecol.* 2003;101:732–736.
241. Cole LA, Butler SA, Khanlian SA, et al. Gestational trophoblastic diseases: 2. Hyperglycosylated hCG as a reliable marker of active neoplasia. *Gynecol Oncol.* 2006;102:151–159.
242. Prabha B, Molykutty J, Swapna A, et al. Increased expression of interleukin-1 beta is associated with persistence of the disease and invasion in complete hydatidiform moles (CHM). *Eur J Gynaecol Oncol.* 2001;22:50–56.
243. Fong PY, Xue WC, Ngan HY, et al. Mcl-1 expression in gestational trophoblastic disease correlates with clinical outcome: a differential expression study. *Cancer.* 2005;103:268–276.
244. Tuncer ZS, Vegh GL, Fulop V, et al. Expression of epidermal growth factor receptor-related family products in gestational trophoblastic diseases and normal placenta and its relationship with development of postmolar tumor. *Gynecol Oncol.* 2000;77: 389–393.
245. Davison JM, Homuth V, Jeyabalan A, et al. New aspects in the pathophysiology of preeclampsia. *J Am Soc Nephrol.* 2004;15: 2440–2448.
246. Brockelsby J, Hayman R, Ahmed A, et al. VEGF via VEGF receptor-1 (Flt-1) mimics preeclamptic plasma in inhibiting uterine blood vessel relaxation in pregnancy: implications in the pathogenesis of preeclampsia. *Lab Invest.* 1999;79:1101–1111.
247. Smith GC, Crossley JA, Aitken DA, et al. Circulating angiogenic factors in early pregnancy and the risk of preeclampsia, intrauterine growth restriction, spontaneous preterm birth, and stillbirth. *Obstet Gynecol.* 2007;109:1316–1324.
248. Levine RJ, Maynard SE, Qian C, et al. Circulating angiogenic factors and the risk of preeclampsia. *N Engl J Med.* 2004;350: 672–683.

249. Sane DC, Anton L, Brosnihan KB. Angiogenic growth factors and hypertension. *Angiogenesis*. 2004;7:193–201.
250. Tjoa ML, Levine RJ, Karumanchi SA. Angiogenic factors and preeclampsia. *Front Biosci*. 2007;12:2395–2402.
251. Venkatesha S, Toporsian M, Lam C, et al. Soluble endoglin contributes to the pathogenesis of preeclampsia. *Nat Med*. 2006;12:642–649.
252. Salahuddin S, Lee Y, Vadnais M, et al. Diagnostic utility of soluble fms-like tyrosine kinase 1 and soluble endoglin in hypertensive diseases of pregnancy. *Am J Obstet Gynecol*. 2007;197:28 e1–e6.

Chapter

24

Molecular Pathogenesis of Prostate Cancer: Somatic, Epigenetic, and Genetic Alterations

Carlise R. Bethel ▪ Angelo M. De Marzo ▪ William G. Nelson

INTRODUCTION

Prostate cancer is one of the most common malignancies in Western men, and risk factors include age, race, inherited genes, and environmental factors such as diet [1–4]. Although the precise molecular mechanisms underlying carcinogenesis and progression are currently unknown, histological studies indicate a multistage developmental progression. The existence of premalignant lesions has been demonstrated in the prostate based on shared histological and molecular features with adenocarcinoma, as well as prevalence and severity [5,6]. Ample evidence supports the hypothesis that prostate cancer can progress from intraepithelial neoplasia to invasive carcinoma [7], and ultimately metastasis and androgen-independent lethal disease. Prostatic intraepithelial neoplasia (PIN) is a term currently used to describe the closest precursor lesion to carcinoma [7,8]. Histologically, the transition from normal epithelium to PIN involves nuclear atypia, epithelial cell crowding, and some component of basal cell loss [9,10]. Autopsy studies suggest that PIN lesions precede the appearance of carcinoma [11]. High-grade PIN lesions are often spatially associated with carcinoma and may exhibit molecular alterations similar to those found in prostate tumor cells [8,10] .

Glandular atrophy is frequently observed in aging male prostates [12]. Histologically, luminal spaces appear dilated with flattened epithelial linings. Atrophic lesions in the prostate consist of a number of histological variants, and some of these have also been proposed to be potential precursors to prostate cancer based in part on their frequent occurrence in proximity to carcinoma [5,13–15]. Proliferative inflammatory atrophy (PIA), a term used to describe a range of these morphologies, includes simple atrophy and postatrophic hyperplasia [5]. PIA lesions are highly proliferative and often associated with inflammation [5,16]. Epithelial cells within PIA foci express both luminal and basal markers [17]. The lesions likely result from cellular injury initiated by inflammation and/or carcinogen insult, and at times show evidence of transition to high-grade PIN and, rarely, to carcinoma [14]. Given the intermediate phenotype of these cells and the characteristic expression of stress-associated genes, PIA lesions are considered to be the morphological manifestation of prostate epithelial damage and regeneration [17]. Some of the genetic and epigenetic alterations observed in PIN and adenocarcinoma have also been found in PIA lesions, albeit to a lesser degree [15,18–21]. These findings suggest a potential link between a subset of atrophic lesions and adenocarcinoma.

We have proposed a multistep progression model that leads from normal epithelium to focal prostatic atrophy to PIN and then to carcinoma [2,4,5]. In this model, ongoing injury to the prostate epithelium, as a result of inflammation and/or carcinogen exposures, results in cell damage and cell death. Cell regeneration ensues, and this is manifest morphologically as prostatic atrophy. These atrophic cells may then undergo somatic genome alterations during self-renewal, including methylation of the *GSTP1* promoter, telomere shortening, activation of MYC, leading to neoplastic transformation.

 © 2009, Elsevier, Inc. All Rights Reserved.

Prostate cancer arises from the accumulation of genetic and epigenetic alterations [22]. Cytogenetic analyses have demonstrated the prevalence of chromosomal aberrations associated with prostate tumorigenesis, and many genes that map to deleted or amplified regions have been investigated for their roles in disease progression. In recent years, genome-wide profiling of prostate tumors has led to the identification of a number of biomarkers and pathways that are altered in prostate cancer, some of which may be potential molecular targets for therapy. Despite the advancing knowledge of genes altered in prostate adenocarcinoma, the precise molecular pathways, their combinatorial relation, and the ordering of events utilized in the development of preneoplastic lesions and adenocarcinoma are still being refined. To elucidate mechanisms and to begin to test new therapies and preventative strategies for human prostate cancer, a number of groups have been developing novel animal models of prostate cancer.

These models will help to define many aspects of the molecular and cellular pathogenesis of prostate cancer, such as the role of inflammation, angiogenesis, and stromal-epithelial interactions. In light of previous reviews that extensively address our current knowledge of molecular alterations in prostate tumor progression [2,4,22–28], the primary objectives of this chapter will be to highlight recent advances in our understanding of the molecular pathology of prostatic adenocarcinoma and the latest developments in mouse prostate cancer models.

HEREDITARY COMPONENT OF PROSTATE CANCER RISK

Prostate cancer is known to have a hereditary component, and a large effort has been devoted to uncovering familial prostate cancer genes. A limited set of germline polymorphisms and mutations have been associated with increased prostate cancer risk. Based on linkage analysis, the inherited susceptibility locus Hereditary prostate cancer 1 *(HPC1)*, which encodes Ribonuclease L (RNASEL), was mapped to chromosomal region 1 [26,29,30]. Germline mutations in the Macrophage Scavenger Receptor 1 (*MSR1*) locus have also been linked to prostate cancer risk [31]. The *MSR1* gene is located at chromosomal region 8p22 and is expressed in infiltrating macrophages [4]. Many other loci have been implicated as well, and further characterization of these genes in the etiology of prostate cancer is warranted [32]. Most recently several groups have used Genome Wide Association Studies (GWAS) to show a number of SNPs at novel loci related to prostate cancer risk, several of which map to chromosome 8q24 [33–42].

SOMATIC ALTERATIONS IN GENE EXPRESSION

Acquired somatic gene alterations in prostate cancer cells include cytosine methylation alterations within CpG dinucleotides, point mutations, deletions, amplifications, and telomere shortening [22,43,44]. Inactivation of classical tumor suppressors, for example, *TP53* and *Retinoblastoma* (*RB*), has been found in prostate tumors and cancer cell lines [26]. However, these alterations are far more common in advanced hormone-refractory and/or metastatic cancers. As with other epithelial cancers, cytogenetic studies using fluorescence *in situ* hybridization (FISH) and comparative genomic hybridization (CGH) have identified chromosomal regions frequently gained and lost in prostate cancer. The most common chromosomal abnormalities are losses at 8p, 10q, 13q, and 16q, and gains at 7p, 7q, 8q, and Xq [4,28,45–51].

MYC

MYC protein functions as a nuclear transcription factor that impacts a wide range of cellular processes including cell cycle progression, metabolism, ribosome biogenesis, and protein synthesis [52]. The *MYC* oncogene maps to chromosome 8q24, and this region is amplified in a number of human tumors [52]. The complexity of MYC-regulated transcriptional networks has been under intense study since the late 1990s, yet the precise role of the *MYC* oncogene during neoplastic transformation and its direct molecular targets in prostate tumorigenesis remain largely unknown. Initial reports of increased *MYC* gene expression in PIN lesions and ~30% of tumors [53,54], combined with the correlation of 8q24 amplification with high Gleason grade and metastatic carcinoma, suggested that alterations in MYC were associated with advanced disease [55,56]. This was in contrast to increased levels of *MYC* mRNA detected in most prostate cancers, including low-grade cases [57,58]. However, recent evidence from our group suggests that MYC upregulation at the protein level is an early and common event in primary prostate cancer cases. Using an improved antibody for immunohistochemical analyses on tissue microarrays, nuclear MYC staining was increased in the luminal epithelial cells of PIN, PIA, and carcinoma lesions compared to benign tissues [19]. Interestingly, FISH analysis revealed a positive correlation between gain of 8q24 and Gleason grade, but not overall MYC protein levels. These findings in conjunction with the fact that human MYC overexpression is able to initiate prostate cancer in transgenic mice suggest a key role for MYC upregulation in the initiation of prostate tumorigenesis that is likely independent of gene amplification. Indeed, further investigations of the mechanisms that underlie MYC overexpression in prostate cancer progression are warranted.

NKX3.1

NKX3.1 is one of several candidate tumor suppressor genes located on chromosome 8p. The *NKX3.1* gene encodes a homeodomain transcription factor that is the earliest known marker of prostate epithelium during embryogenesis [59,60]. Its expression persists in the epithelial cells of the adult gland and is required for maintenance of ductal morphology and the regulation of cell proliferation [61–64]. Human *NKX3.1* maps within the minimal deletion interval of chromosomal region 8p21. Loss of heterozygosity at 8p21 has been observed in 63% of high-grade PIN foci and up to ~70% of prostate tumors, although this percentage may be significantly less according to recent findings [18,65,66]. Methylation of CpG dinucleotides upstream of the *NKX3.1* transcriptional start site and the existence of germline variants within the homeodomain have been reported [67]. However, the order of occurrence relative to 8p21 loss and prostate cancer initiation is unknown. In loss of function analyses, *NKX3.1* homozygous mutant mice do not develop invasive carcinoma. However, epithelial hyperplasia and PIN lesions are observed with age in these mice [61,62,64]. These phenotypes are also observed in heterozygous mice, albeit to a lesser extent. When taken together, human and mouse studies support the notion that haploinsufficiency of *NKX3.1* plays a role in prostate cancer development.

Several studies have analyzed NKX3.1 expression by immunohistochemistry in human prostate cancer specimens. In studies of PIN and carcinoma, both decreased intensity and loss of NKX3.1 protein staining compared to benign tissue have been reported [18,67–71]. Decreased expression correlated with high Gleason score, advanced tumor stage, the presence of metastatic disease, and hormone-refractory disease [18,68]. Recently, our immunohistochemical analyses with a novel NKX3.1 antibody revealed a dramatic decrease in the level of NKX3.1 in PIA lesions [18]. As NKX3.1 is thought to regulate cell proliferation [63,72], loss of expression in atrophic epithelial cells may contribute to increased proliferative capacity. This may serve to amplify any genetic changes that may occur in epithelial cells within these regenerative lesions. Consistent with previous reports, NKX3.1 staining intensity was also significantly diminished in PIN and carcinoma lesions. Diminished NKX3.1 protein expression correlated with 8p loss in high-grade tumors, but not PIN or PIA [18,73]. Intriguingly, the levels of *NKX3.1* mRNA and protein were found to be discordant in 7 of 11 carcinoma cases, confirming previous reports and suggesting that multiple mechanisms underlie sporadic loss of NKX3.1 for tumor progression.

Although the precise transcriptional targets of NKX3.1 are largely unknown, microarray analyses suggest that lack of *NKX3.1* results in increased oxidative DNA damage by controlling expression of antioxidant enzymes [74]. A limited set of proteins has been identified to physically interact with NKX3.1, including Serum Response Factor (SRF) and Prostate-derived Ets transcription factor [75,76]. The serine-threonine kinase CK2 and the ubiquitin ligase TOPORS have recently been shown to regulate the stability of *NKX3.1* in prostate cancer cells *in vitro* [77,78]. The functional significance of these interactions in precursor and cancer lesions has yet to be elucidated. It is hoped that with the recent advances in mouse prostate cancer models and the application of high-resolution genomic tools, the mechanisms of *NKX3.1* loss will be uncovered.

PTEN

The Phosphatase and tensin homologue (*PTEN*) gene is a well-characterized tumor suppressor that maps to chromosomal region 10q23 [79] and acts as a negative regulator of the phosphatidylinositol 3-kinase/AKT (PI3K/AKT) signaling, used for cell survival [80–83]. PTEN inhibits growth factor signals sent through PI3 kinase by dephosphorylating the PI3K product, phosphatidylinositol 3,4,5-trisphosphate (PIP3). Loss of PTEN expression results in the downstream activation (phosphorylation) of AKT, an inhibitor of apoptosis and promoter of cell proliferation. Aberrant PTEN expression has been implicated in numerous cancers, including metastatic prostate cancers, which at times exhibit homozygous deletion of *PTEN* [26,43,79]. However, the majority of primary prostate cancer cases with genetic alterations in *PTEN* harbor loss of heterozygosity at the *PTEN* locus without mutations in the remaining allele [84–86]. That PTEN is a haploinsufficient tumor suppressor in the prostate is evident by prostate-specific disruption of *PTEN* in mice, which results in PIN lesions, followed by invasive carcinoma and/or metastasis with age [87,88]. Cooperativity exists between *NKX3.1* and *PTEN* in prostate tumorigenesis, as compound *NKX3.1* and *PTEN* heterozygous mice rapidly develop invasive carcinoma and androgen-independent disease [89–91].

Androgen Receptor

The role of AR is central to prostate pathobiology, as the prostate is dependent on androgens for normal growth and maintenance. AR is a nuclear steroid hormone receptor with high expression in the luminal epithelial cells and little expression in basal epithelial cells. In the absence of ligand, AR is inactive and bound to heat shock chaperone proteins [92]. Upon binding of the active form of testosterone (dihydrotestosterone, DHT), AR is released from heat shock proteins and translocates to the nucleus where it physically associates with cofactors to regulate target gene transcription. AR activity is essential for the development of prostate cancer, and AR expression is evident in high-grade PIN and most adenocarcinoma lesions [93]. Androgen ablation, through the use of antiandrogens, castration, or gonadotropin superagonists is the mainline therapy for advanced, metastatic carcinoma. While the majority of patients respond to this treatment, it eventually fails and the tumors become androgen independent. As tumors progress from androgen dependence to a hormone refractory state, AR can be amplified at the Xq12 region or mutated to respond to a range of ligands for androgen-independent activation and tumor growth [94,95]. In recent years, AR has been reported to engage in crosstalk with other mitogenic signaling pathways such as PI3K/AKT and MAPK as a means of enhancing androgen-independent tumor progression [96,97].

TMPRSS2-ETS Gene Fusions

Ets-related gene-1 (ERG) is an oncogenic transcription factor upregulated in the majority of primary prostate cancer cases [98]. Evidence of recurrent chromosomal rearrangements in prostate cancer was first reported by Tomlins et al. [99]. Through use of a bioinformatic approach to analyze DNA microarray studies, gene fusions between the promoter/enhancer region of the androgen-responsive *TMPRSS2* gene and members of the *ETS* family, including *ERG* and *ETV1*, were found in ~90% of prostate tumors with known overexpression of *ERG*. This discovery has fueled efforts to characterize the functional implications of aberrant chromosomal fusions on tumor progression and clinical outcome. Presently, there are conflicting data regarding the presence of *TMPRSS-Erg* gene fusions and clinical outcome [100–106]. While initial studies suggested these fusions were not present in PIN lesions, it has become clear that some PIN lesions indeed harbor fusions, and this suggests that gene fusion may lead to neoplastic transformation itself and not specifically to the invasive phenotype [107,108]. Additionally, overexpression of *ERG* in the transgenic mouse prostates results in basal cell loss and PIN lesions [109,110], although this latter finding is controversial. These studies demonstrate that upregulation of *ERG* may contribute to transformation of prostate epithelial cells but is insufficient to initiate prostate cancer.

p27

The cyclin-dependent kinase inhibitor $p27^{kip1}$ is a candidate tumor suppressor encoded by the *CDKN1B* gene. To prevent cell cycle progression, $p27^{kip1}$ binds to and inhibits cyclin E/CDK2 and cyclin A/CDK2 complexes [111]. Although mutations are rare, loss of $p27^{kip1}$ expression results in hyperplasia and malignancy in many organs, including the prostate [112–115]. In normal and benign prostate tissues, $p27^{kip1}$ is expressed highly in most luminal epithelial cells, and much more variably in basal cells [116]. However, $p27^{kip1}$ expression is decreased in most high-grade PIN and carcinoma lesions [113,116–119]. Several studies have shown decreased $p27^{kip1}$ to be positively correlated with increased proliferation, PSA relapse, high tumor grade, and advanced stage [113,116,119–122]. The molecular mechanism by which p27 protein is decreased in prostate cancer has not been clarified, although post-transcriptional regulation of $p27^{kip1}$ is evident in carcinoma cases which express high levels of $p27^{kip1}$ mRNA but lack protein expression [117].

Additional support for $p27^{kip1}$ as a tumor suppressor comes from loss of function analyses in mice. Prostates from $p27^{kip1}$-deficient animals exhibit hyperplasia and increased size [117]. Given the cooperativity observed between *NKX3.1* and *PTEN* in tumorigenesis, the relationship between *NKX3.1* and *p27* loss has also been investigated [123]. Both genes have been shown to exhibit haploinsufficiency for tumor suppression [61,91,124]. Anterior prostates of 36-week-old double mutant mice displayed epithelial hyperplasia and dysplasia that was more severe than that observed in the single *p27* mutant mice. However, combined loss of $p27^{kip1}$, *PTEN*, and *NKX3.1* in bigenic and trigenic mouse models results in prostate hyperplasia followed by tumor development [125,126]. These findings suggest that prostate tumor progression is sensitive to *p27* dosage and support the hypothesis that downregulation of $p27^{kip1}$ is an early event in prostatic neoplasia. Expression of these three genes is known to decrease in clinical samples, providing further evidence of their cooperation in tumor progression.

Telomeres

Telomeres are specialized structures composed of repeat DNA sequences at the ends of chromosomes that are complexed with binding proteins and required for maintenance of chromosomal integrity [127]. In cells lacking sufficient levels of the enzyme telomerase, telomeres progressively shorten with cell division as a result of the end replication problem and/or oxidant stress. The enzyme telomerase can add new repeat sequences to the ends of chromosomes, which stabilizes the telomeres and ensures proper telomere length. Excessive shortening can lead to improper segregation of chromosomes during cell division, genomic instability, and the initiation of tumorigenesis. Mice double mutant for telomerase and *p53* display increased epithelial cancer incidence, suggesting a role for short telomeres in cancer initiation [128]. The majority of high-grade PIN and prostate cancer cases have abnormally short telomeres exclusively in the luminal cells. This supports the notion that cells in the luminal compartment may be the target of neoplastic transformation [129]. Oxidative damage can result in telomere shortening, and this may go along with the proposed inflammation-oxidative stress model of prostate cancer progression [2].

MicroRNAs

MicroRNAs (miRNAs) are small noncoding RNA molecules that negatively regulate gene expression by interfering with translation. They are initially generated from primary transcripts (pri-mRNAs) and processed by the RNase III endonucleases Drosha and Dicer to produce the mature miRNA molecule [130,131]. In the cytoplasm, the mature miRNA associates with the RNA-induced silencing complex (RISC) and binds to the 3′ UTR of its target mRNA, leading to degradation or transcriptional silencing [132]. Since their discovery over a decade ago, miRNAs have been shown to play key roles in development, and there is increasing evidence of their widespread dysregulation in cancer [133]. Recently, a limited number of studies have reported a predominant decrease in the levels of miRNAs in prostate carcinoma, including let-7, miR-26a, miR-99, and miR-125-a-b [134–136]. Although a prostate cancer-specific miRNA signature has not emerged from these small-scale studies, greater knowledge of miRNA expression patterns and target genes may reveal the role of miRNAs in the etiology of prostate cancer. MiRNAs that are consistently dysregulated in prostate tumorigenesis and have a strong relationship with disease progression may potentially serve as novel biomarkers for prognosis.

EPIGENETICS

Epigenetic alterations in prostate carcinogenesis include changes in chromosome structure through abnormal deoxycytidine methylation of wild-type DNA sequences and histone modifications (acetylation, methylation). Epigenetic events occur earlier in prostate tumor progression and more consistently than recurring genetic changes. However, the mechanisms by which these changes arise are poorly understood. Silencing of genes through aberrant methylation may occur as a result of altered DNA methyltransferase (DNMT) activity. DNMTs establish and maintain the patterns of methylation in the genome by catalyzing the transfer of methyl groups to deoxycytidine in CpG dinucleotides. Several genes that are silenced by epigenetic alterations have been identified.

GSTP1

Glutathione S-transferases are enzymes responsible for the detoxification of reactive chemical species through conjugation to reduced glutathione. The *GSTP1* gene, encoding the pi class glutathione S-transferase, was the first hypermethylated gene to be characterized in prostate cancer [137]. GSTP1 is thought to protect prostate epithelial cells from carcinogen-associated and/or oxidative stress-induced DNA damage, and loss of GSTP1 may render prostate cells unprotected from such genomic insults. As an example, in LNCaP prostate cancer cells, devoid of GSTP1 as a result of epigenetic gene silencing, restoration of GSTP1 function affords protection against metabolic activation of the dietary heterocyclic amine carcinogen 2-amino-1-methyl-6-phenylimidazo [4,5-β] pyridine (PhIP), known to cause prostate cancer when ingested by rats [138,139]. The loss of enzymatic defenses against reactive chemical species encountered as part of dietary exposures or arising endogenously associated with epigenetic silencing of GSTP1 provides a plausible mechanistic explanation for the marked impact of the diet and of inflammatory processes in the pathogenesis of human prostate cancer.

The most common genomic DNA mark accompanying epigenetic gene silencing in cancer cells is an accumulation of 5-meC bases in CpG dinucleotides clustered into CpG islands encompassing transcriptional regulatory regions. Hypermethylation of these CpG island sequences directs the formation of repressive heterochromatin that prevents loading of RNA polymerase and transcription of hnRNA that can be processed for translation into protein. For *GSTP1*, the 5-meCpG dinucleotides begin to appear in the gene promoter region of rare cells in PIA lesions, with more dense CpG island methylation changes emerging as PIA lesions progress to PIN and carcinoma [21]. The proliferative expansion of cells with hypermethylated *GSTP1* CpG island sequences as PIA lesions progress hints at some sort of selective growth advantage, though how loss of GSTP1 can be selected during prostatic carcinogenesis has not been established. In addition to *GSTP1*, many other critical genes undergo epigenetic silencing during the pathogenesis of prostate cancer [22]. The mechanisms by which epigenetic defects arise in prostate cancer cells, or in other human cancer cells, have not been discerned. However, the consistent appearance of such changes in inflammatory precancerous lesions and conditions, including PIA lesions, inflammatory bowel disease, chronic active hepatitis, and others, supports the contribution of inflammatory processes to some sort of epigenetic or DNA methylation catastrophe [22].

APC

The Adenomatous polyposis coli (APC) protein is a component of the Wnt/β-catenin signaling pathway that negatively regulates cell growth. APC is typically found in a complex with Glycogen synthase kinase-3 (GSK3) and Axin. This complex is responsible for targeting free cytosolic β-catenin for ubiquitin-mediated degradation. The pathway is activated by the binding of the Wnt protein to the Frizzled family of seven transmembrane receptors and LRP5/6, followed by downregulation of GSK3β, which allows accumulation of β-catenin and subsequent translocation of β-catenin to the nucleus. Once inside the nucleus, β-catenin is able to activate transcription of Wnt target genes.

Indirect support of the concept that APC inactivation may be mechanistically tied to prostate cancer progression comes from studies showing frequent hypermethylation of its promoter region and that the extent of methylation correlates with stage, grade, and biochemical recurrence [140–142]. Activating mutations occur in approximately 5% of prostate cancers, and aberrant nuclear localization of β-catenin appears to occur only somewhat more frequently [143]. The latter finding suggests that if APC is functioning as a tumor suppressor in prostate adenocarcinoma, then its primary role in the prostate may not relate to nuclear translocation of β-catenin. More direct support for the notion that APC may be a tumor suppressor in prostate cancer comes from a mouse model in which prostate-specific deletion of *APC* in adult mice resulted in carcinoma induction [144]. Whether APC is involved in prostate cancer formation or progression or not, the methylation of its promoter may become a useful biomarker in prostate cancer diagnosis since this methylation may be detectable in bodily fluids such as urine or blood [145,146].

ADVANCES IN MOUSE MODELS OF PROSTATE CANCER

The human prostate is an encapsulated organ composed of the central, peripheral, and transitional zones. In contrast, the mouse prostate is composed of four individual lobes: designated dorsal, lateral, ventral, and anterior. Despite the organizational differences between the mouse and human prostate glands, rodent prostate cancer models provide the unique opportunity to study prostate cancer development and progression including molecular alterations and histopathology [147]. In recent years, the validity of a subset of mouse models as predictable and valid representations of human disease has been demonstrated.

A number of models of prostate cancer have been developed in mice using gain of function approaches to force expression of known oncogenes in the prostate [147,148]. The most widely used mouse model is the Transgenic Adenocarcinoma Mouse Prostate (TRAMP) model, which utilizes the androgen-regulated rat probasin promoter to target expression of the Simian Virus 40 (SV40) large and small T antigens to the prostate at the onset of sexual maturity [149,150]. This model has been particularly favored since cancers develop quickly and often metastasize to distant sites. At the genetic level, disease progression in the TRAMP model exhibits some similarities to human prostate cancer. It has been demonstrated that somatic AR mutations present in primary human prostate tumors [151] are also found in similar regions of the AR gene in TRAMP mice [152]. Increased expression of IGF-I and Prostate Stem Cell Antigen (PSCA) and downregulation of NKX3.1 and E-cadherin are associated with human cancer progression, and similar, albeit more dramatic, changes are found in TRAMP tumors [149,153–155].

This model and other transgenic lines that express the SV40 early region transforming sequences in the prostate have significant limitations in terms of relevance to human disease. This stems from the increasing realization that the aggressive, poorly differentiated lesions in these models represent small cell, neuroendocrine carcinoma, and not adenocarcinoma as demonstrated by standard histopathology, as well as molecular markers revealed by immunohistochemical staining [156,157]. For example, these lesions often display an absence of, or near absence of, androgen receptor expression, whereas the majority of lethal, metastatic prostate cancers in humans retain high AR expression [158,159], and the poorly differentiated tumors also diffusely express markers of neuroendocrine differentiation such as synaptophysin. Other models of prostate neoplasia have targeted overexpression of growth factors, hormone receptors, and cell cycle regulators to the prostate. However, most of these models fail to progress to invasive carcinoma and result in hyperplasia and mouse PIN lesions [147,160,161].

Recently, several models have been developed through targeted deletion of tumor suppressors or overexpression of oncogenes relevant to prostate cancer progression. These animals recapitulate the early phases of human cancer and provide useful reagents to uncover potential synergism between genes involved in prostate cancer progression. Since excellent reviews exist regarding most of these models [147,148,160,162,163], we will focus on a small number of them that are either most commonly used or are particularly interesting in light of our knowledge of the human disease.

Conditional Loss of Tumor Suppressor Gene Expression in the Prostate

The prostate-specific *PTEN* deletion model was derived using the inducible loxP/Cre recombinase system to cause homozygous loss of PTEN expression in the epithelial cells of the prostate [88,164,165]. Conditional *PTEN* knockout mouse prostates develop hyperplasia by 4 weeks and PIN by the age of 6 weeks. Adenocarcinoma, followed by metastases in the lymph nodes and lungs, was observed in mice one year of age or older [88]. This model mimics the loss of PTEN expression observed in some human prostate cancers, and cDNA microarray analyses performed with *PTEN* null prostates have demonstrated similar alterations as those found in human prostate cancers, including down-regulated *NKX3.1* [88].

Loss of *PTEN*, *NKX3.1*, and *p27* in mice has been shown to cooperate in tumor progression in triple heterozygotes 6 months of age and younger that exhibit increased incidence of high-grade PIN and cancer lesions [126]. These findings suggest that prostate tumor progression is sensitive to *p27* dosage. Expression of these three genes is known to decrease in early prostate tumors, providing further evidence of their cooperation in tumor progression.

PTEN regulates the stability of p53, and p53 can activate *PTEN* transcription [166,167]. Complete loss of *PTEN* in the prostate of knockout mice results in invasive cancer. However, tumor growth is slow, and there is an increase in p53 expression and cellular senescence [168]. The complex relationship between cellular senescence and PTEN loss in the prostate tumorigenesis has been elucidated through prostate-specific conditional inactivation of *Trp53* and *PTEN* in transgenic mice [169]. Animals with loss of PTEN alone displayed high-grade PIN lesions at 10 weeks and eventually developed prostate cancer after 4–6 months. Prostates with homozygous loss of PTEN exhibited diminished proliferative capacity and increased expression of p53 and cell senescence markers, including SA-beta galactosidase staining and p21 expression. Pathological changes were not detected in *Trp53* mutants. However, combined homozygous loss of *PTEN* and *Trp53* resulted in invasive prostate carcinoma in 50% of mice as early as 10 weeks [169]. These results demonstrate that tumor progression in *PTEN* mutant mice is enhanced by loss of *Trp53* and may explain why human prostate tumors select loss of only one *PTEN* allele to escape senescence [169].

To address the role of APC downregulation in prostate cancer progression, Cre-mediated deletion of *APC* in the prostate was carried out using a transgenic line that expressed floxed *APC* alleles [165,170]. Preneoplastic lesions were observed in *Pb-Cre*$^{+}$;*Apc*$^{flox/flox}$ mice as early as 4.5 weeks, and accumulation of β-catenin was evident in hyperplastic regions [144]. Deletion of *APC* resulted in enlargement of all four lobes, and within 7 months of age, all mice developed adenocarcinoma lesions with focal areas of inflammation and reactive stroma. As β-catenin has been implicated in the regulation of AR [171–173], this model provides a valuable resource with which to dissect the role of Wnt signaling in prostate cancer progression.

Loss of RB

Allelic loss or reduced expression of pRB has been reported to occur in 20%–30% of primary human prostate carcinomas [174–176]. However, loss of pRB is more frequent in advanced stage tumors [177]. Conditional knockout of *RB* in the mouse prostate epithelium results in hyperplasia, indicating that loss of *RB* alone cannot initiate PIN or prostate cancer development. To determine if this lack of cancer formation resulted from compensation by the other RB family members, p107 and p130, researchers recently generated transgenic animals that expressed the amino terminus of the SV40 large T antigen in the prostate. The truncated form of T antigen is known to inactivate the RB family of proteins, and these mice develop PIN by 12 weeks and microinvasive carcinoma at 30 weeks of age [178]. Apoptosis was unaffected by loss of *p53;* however, heterozygous loss of *PTEN* reduced apoptosis by 50% and increased the onset of disease [178]. Although metastatic disease is not observed in these animals, the study sheds light on the role of tumor suppressor haploinsufficiency in prostate tumor progression in the context of *RB* deficiency.

Overexpression of Oncogenes in the Prostate

Given that MYC is overexpressed in human prostate cancers, a number of mice that overexpress *Myc* in the prostate have been produced using either the C3 [179] or different versions of the rat probasin promoter to drive expression of the human *MYC* gene in prostate epithelial cells [180]. Ellwood-Yen et al. [180] characterized two different mouse strains. Lo-MYC mice were generated using the rat probasin promoter, and Hi-MYC mice were generated using a modified form of this. Lo-MYC mice were shown to develop PIN lesions by 4 weeks and progress to invasive adenocarcinoma by one year. No metastases were found. Disease progression occurred more rapidly in the Hi-MYC mice, with invasive adenocarcinoma by 3 to 6 months, and some animals eventually developed micrometastatic disease [180]. Additionally, cDNA microarray analyses of aged transgenic animals revealed loss of *NKX3.1* and upregulation of the *Pim-1* kinase, which interacts with Myc in other malignancies [28]. Although metastasis is rare in this model, aspects of its pathobiology are similar to the human disease and demonstrate the usefulness of this model for preclinical applications [162,180].

Overexpression of Mutated AR

Expression of AR is often increased in hormone refractory prostate cancer cases [181]. However, mice that overexpress the wild-type AR in the prostate develop focal PIN lesions with age, suggesting that increased levels of AR may be permissive, but insufficient to drive tumorigenesis [182]. Recently, a mouse prostate cancer model using probasin promoter-driven expression of the AR mutation E231G (AR-E231G) was described [183]. This missense mutation, located in the N-terminal domain of the AR, results in increased responsiveness to coactivators in TRAMP cell lines [152]. Prostates of transgenic mice that overexpressed wild-type AR (AR-WT) or a ligand-binding domain AR mutation (AR-T857A) appeared normal, while AR-E231G mice rapidly developed PIN and invasive adenocarcinoma in the ventral lobe of the prostate. Inflammation accompanied invasive lesions, and metastases were detected in the lungs of AR-E231G transgenic mice after one year [183]. This model is the first to provide *in vivo* evidence of AR as an oncogenic factor. Since increased AR signaling is thought to contribute to tumor progression [23], further biochemical analyses of this model may provide critical insight into the effects of increased AR-coactivator interactions in prostate tumor progression.

Hepsin

According to multiple microarray studies, *hepsin* is one of the most frequently upregulated genes in prostate cancer [184]. Hepsin is a type II transmembrane serine protease of unknown physiological function, and its correlation with disease progression has been investigated. *Hepsin* mRNA and protein expression has been shown to increase with higher tumor grade and metastasis [185–188]. However, one study reported decreased *hepsin* mRNA levels in hormone refractory tumors compared to localized prostate cancer [189]. To assess the functional role of hepsin in prostate cancer progression, Klezovitch and colleagues [190] used the modified probasin promoter (ARR2–PB) to drive expression of hepsin in the ventral prostate of transgenic mice. Increased epithelial expression of hepsin disrupted stromal-epithelial interactions in the basement membrane, but cell proliferation and death remain unaffected [190]. When PB-hepsin mice were mated to the nonmetastatic LPB-Tag prostate cancer model [191], offspring developed invasive carcinoma and metastases of the liver, lung, and bone. These metastatic lesions displayed neuroendocrine differentiation, and the authors concluded that hepsin overexpression enhances cancer progression and the development of neuroendocrine metastases, which may be intrinsic to the SV40 T antigen [190]. Therefore, it would be of great interest to examine the upregulation of hepsin in the context of a non-T-antigen derived prostate cancer model.

CONCLUSION

Mouse models of prostate cancer have shed light on critical molecular events in the development of adenocarcinoma. However, greater relevance to human disease progression requires that these models be further refined to reflect key pathological features of prostatic adenocarcinoma. As tumorigenesis is a multifactorial process, the field of prostate cancer models must continue the use of combinatorial approaches to study the effects of multiple players in prostate cancer development. These models provide the unique opportunity to explore the complex interplay between tumor suppressors and/or oncogenes in the context of an intact immune system and epithelial-stromal interactions.

ACKNOWLEDGMENTS

NIH/NCI Specialized Program in Research Excellence (SPORE) in Prostate Cancer #P50CA58236 (Johns Hopkins), NIH/NCI #R01 CA070196, and The Patrick C. Walsh Foundation.

REFERENCES

1. Chan JM, Giovannucci EL. Vegetables, fruits, associated micronutrients, and risk of prostate cancer. *Epidemiol Rev.* 2001;23:82–86.
2. De Marzo AM, Platz EA, Sutcliffe S, et al. Inflammation in prostate carcinogenesis. *Nat Rev Cancer.* 2007;7:256–269.
3. Kolonel LN. Fat, meat, and prostate cancer. *Epidemiol Rev.* 2001;23:72–81.
4. Nelson WG, De Marzo AM, Isaacs WB. Prostate cancer. *N Engl J Med.* 2003;349:366–381.
5. De Marzo AM, Marchi VL, Epstein JI, et al. Proliferative inflammatory atrophy of the prostate: Implications for prostatic carcinogenesis. *Am J Pathol.* 1999;155:1985–1992.
6. McNeal JE, Bostwick DG. Intraductal dysplasia: A premalignant lesion of the prostate. *Hum Pathol.* 1986;17:64–71.
7. Bostwick DG, Brawer MK. Prostatic intra-epithelial neoplasia and early invasion in prostate cancer. *Cancer.* 1987;59:788–794.
8. McNeal JE, Bostwick DG, Kindrachuk RA, et al. Patterns of progression in prostate cancer. *Lancet.* 1986;1:60–63.
9. Abate-Shen C, Shen MM. Molecular genetics of prostate cancer. *Genes Dev.* 2000;14:2410–2434.
10. Bostwick DG, Qian J. High-grade prostatic intraepithelial neoplasia. *Mod Pathol.* 2004;17:360–379.
11. Sakr WA, Haas GP, Cassin BF, et al. The frequency of carcinoma and intraepithelial neoplasia of the prostate in young male patients. *J Urol.* 1993;150:379–385.
12. McNeal JE. Anatomy of the prostate and morphogenesis of BPH. *Prog Clin Biol Res.* 1984;145:27–53.
13. Franks LM. Atrophy and hyperplasia in the prostate proper. *J Pathol Bacteriol.* 1954;68:617–621.
14. Putzi MJ, De Marzo AM. Morphologic transitions between proliferative inflammatory atrophy and high-grade prostatic intraepithelial neoplasia. *Urology.* 2000;56:828–832.
15. Shah R, Mucci NR, Amin A, et al. Postatrophic hyperplasia of the prostate gland: Neoplastic precursor or innocent bystander? *Am J Pathol.* 2001;158:1767–1773.
16. Ruska KM, Sauvageot J, Epstein JI. Histology and cellular kinetics of prostatic atrophy. *Am J Surg Pathol.* 1998;22:1073–1077.
17. van Leenders GJ, Gage WR, Hicks JL, et al. Intermediate cells in human prostate epithelium are enriched in proliferative inflammatory atrophy. *Am J Pathol.* 2003;162:1529–1537.
18. Bethel CR, Faith D, Li X, et al. Decreased NKX3.1 protein expression in focal prostatic atrophy, prostatic intraepithelial neoplasia, and adenocarcinoma: Association with Gleason score and chromosome 8p deletion. *Cancer Res.* 2006;66:10683–10690.
19. Gurel B, Iwata T, Jenkins RB, et al. Nuclear c-Myc protein overexpression as an early and prevalent marker of prostate carcinogenesis. *Mod Pathol.* 2008;21:1156–1167.
20. Macoska JA, Trybus TM, Wojno KJ. 8p22 loss concurrent with 8c gain is associated with poor outcome in prostate cancer. *Urology.* 2000;55:776–782.
21. Nakayama M, Bennett CJ, Hicks JL, et al. Hypermethylation of the human glutathione S-transferase-pi gene (GSTP1) CpG island is present in a subset of proliferative inflammatory atrophy lesions but not in normal or hyperplastic epithelium of the prostate: A detailed study using laser-capture microdissection. *Am J Pathol.* 2003;163:923–933.
22. Nelson WG, Yegnasubramanian S, Agoston AT, et al. Abnormal DNA methylation, epigenetics, and prostate cancer. *Front Biosci.* 2007;12:4254–4266.
23. Balakumaran BS, Febbo PG. New insights into prostate cancer biology. *Hematol Oncol Clin North Am.* 2006;20:773–796.
24. Bastian PJ, Yegnasubramanian S, Palapattu GS, et al. Molecular biomarker in prostate cancer: The role of CpG island hypermethylation. *Eur Urol.* 2004;46:698–708.
25. De Marzo AM, Nelson WG, Isaacs WB, et al. Pathological and molecular aspects of prostate cancer. *Lancet.* 2003;361:955–964.
26. Karayi MK, Markham AF. Molecular biology of prostate cancer. *Prostate Cancer Prostatic Dis.* 2004;7:6–20.
27. Nelson WG, DeWeese TL, DeMarzo AM. The diet, prostate inflammation, and the development of prostate cancer. *Cancer Metastasis Rev.* 2002;21:3–16.
28. Tomlins SA, Rubin MA, Chinnaiyan AM. Integrative biology of prostate cancer progression. *Annu Rev Pathol.* 2006;1:243–271.
29. Nelson WG, De Marzo AM, DeWeese TL. The molecular pathogenesis of prostate cancer: Implications for prostate cancer prevention. *Urology.* 2001;57:39–45.
30. Palapattu GS, Sutcliffe S, Bastian PJ, et al. Prostate carcinogenesis and inflammation: Emerging insights. *Carcinogenesis.* 2005;26:1170–1181.
31. Xu J, Zheng SL, Komiya A, et al. Germline mutations and sequence variants of the macrophage scavenger receptor 1 gene are associated with prostate cancer risk. *Nat Genet.* 2002;32:321–325.
32. Gillanders EM, Xu J, Chang BL, et al. Combined genome-wide scan for prostate cancer susceptibility genes. *J Natl Cancer Inst.* 2004;96:1240–1247.
33. Croce CM. Oncogenes and cancer. *N Engl J Med.* 2008;358:502–511.
34. Gudmundsson J, Sulem P, Manolescu A, et al. Genome-wide association study identifies a second prostate cancer susceptibility variant at 8q24. *Nat Genet.* 2007;39:631–637.
35. Gudmundsson J, Sulem P, Rafnar T, et al. Common sequence variants on 2p15 and Xp11.22 confer susceptibility to prostate cancer. *Nat Genet.* 2008;40:281–283.
36. Gudmundsson J, Sulem P, Steinthorsdottir V, et al. Two variants on chromosome 17 confer prostate cancer risk, and the one in TCF2 protects against type 2 diabetes. *Nat Genet.* 2007;39:977–983.
37. Ma L, Weinberg RA. MicroRNAs in malignant progression. *Cell Cycle.* 2007;7:570–572.
38. Sun J, Lange EM, Isaacs SD, et al. Chromosome 8q24 risk variants in hereditary and non-hereditary prostate cancer patients. *Prostate.* 2008;68:489–497.
39. Yeager M, Orr N, Hayes RB, et al. Genome-wide association study of prostate cancer identifies a second risk locus at 8q24. *Nat Genet.* 2007;39:645–649.

40. Zanke BW, Greenwood CM, Rangrej J, et al. Genome-wide association scan identifies a colorectal cancer susceptibility locus on chromosome 8q24. *Nat Genet.* 2007;39:989–994.
41. Zheng SL, Sun J, Cheng Y, et al. Association between two unlinked loci at 8q24 and prostate cancer risk among European Americans. *J Natl Cancer Inst.* 2007;99:1525–1533.
42. Zheng SL, Sun J, Wiklund F, et al. Cumulative association of five genetic variants with prostate cancer. *N Engl J Med.* 2008; 358:910–919.
43. Dong JT. Chromosomal deletions and tumor suppressor genes in prostate cancer. *Cancer Metastasis Rev.* 2001;20:173–193.
44. Meeker AK, Hicks JL, Platz EA, et al. Telomere shortening is an early somatic DNA alteration in human prostate tumorigenesis. *Cancer Res.* 2002;6405–6409.
45. Chang BL, Liu W, Sun J, et al. Integration of somatic deletion analysis of prostate cancers and germline linkage analysis of prostate cancer families reveals two small consensus regions for prostate cancer genes at 8p. *Cancer Res.* 2007;67:4098–4103.
46. Gonzalgo ML, Isaacs WB. Molecular pathways to prostate cancer. *J Urol.* 2003;170:2444–2452.
47. Joshua AM, Evans A, Van der Kwast T, et al. Prostatic preneoplasia and beyond. *Biochem Biophys Acta.* 2007;1785:156–181.
48. Liu W, Chang B, Sauvageot J, et al. Comprehensive assessment of DNA copy number alterations in human prostate cancers using Affymetrix 100K SNP mapping array. *Genes Chromosomes Cancer.* 2006;45:1018–1032.
49. Liu W, Chang BL, Cramer S, et al. Deletion of a small consensus region at 6q15, including the MAP3K7 gene, is significantly associated with high-grade prostate cancers. *Clin Cancer Res.* 2007;13:5028–5033.
50. Shand RL, Gelmann EP. Molecular biology of prostate-cancer pathogenesis. *Curr Opin Urol.* 2006;16:123–131.
51. Sun J, Liu W, Adams TS, et al. DNA copy number alterations in prostate cancers: A combined analysis of published CGH studies. *Prostate.* 2007;67:692–700.
52. Dang CV, O'Donnell KA, Zeller KI, et al. The c-Myc target gene network. *Semin Cancer Biol.* 2006;16:253–264.
53. Nesbit CE, Tersak JM, Prochownik EV. MYC oncogenes and human neoplastic disease. *Oncogene.* 1999;18:3004–3016.
54. Qian J, Jenkins RB, Bostwick DG. Detection of chromosomal anomalies and c-Myc gene amplification in the cribriform pattern of prostatic intraepithelial neoplasia and carcinoma by fluorescence in situ hybridization. *Mod Pathol.* 1997; 10:1113–1119.
55. Jenkins RB, Qian J, Lieber MM, et al. Detection of c-Myc oncogene amplification and chromosomal anomalies in metastatic prostatic carcinoma by fluorescence in situ hybridization. *Cancer Res.* 1997;57:524–531.
56. Sato K, Qian J, Slezak JM, et al. Clinical significance of alterations of chromosome 8 in high-grade, advanced, nonmetastatic prostate carcinoma. *J Natl Cancer Inst.* 1999;91:1574–1580.
57. Buttyan R, Sawczuk IS, Benson MC, et al. Enhanced expression of the c-Myc protooncogene in high-grade human prostate cancers. *Prostate.* 1987;11:327–337.
58. Fleming WH, Hamel A, MacDonald R, et al. Expression of the c-Myc protooncogene in human prostatic carcinoma and benign prostatic hyperplasia. *Cancer Res.* 1986;46:1535–1538.
59. Bieberich CJ, Fujita K, He WW, et al. Prostate-specific and androgen-dependent expression of a novel homeobox gene. *J Biol Chem.* 1996;271:31779–31782.
60. He WW, Sciavolino PJ, Wing J, et al. A novel human prostate-specific, androgen-regulated homeobox gene (NKX3.1) that maps to 8p21, a region frequently deleted in prostate cancer. *Genomics.* 1997;43:69–77.
61. Abdulkadir SA, Magee JA, Peters TJ, et al. Conditional loss of NKX3.1 in adult mice induces prostatic intraepithelial neoplasia. *Mol Cell Biol.* 2002;22:1495–1503.
62. Bhatia-Gaur R, Donjacour AA, Sciavolino PJ, et al. Roles for NKX3.1 in prostate development and cancer. *Genes Dev.* 1999;13:966–977.
63. Magee JA, Abdulkadir SA, Milbrandt J. Haploinsufficiency at the NKX3.1 locus. A paradigm for stochastic. dosage-sensitive gene regulation during tumor initiation. *Cancer Cell.* 2003;3:273–283.
64. Schneider A, Brand T, Zweigerdt R, et al. Targeted disruption of the NKX3.1 gene in mice results in morphogenetic defects of minor salivary glands: Parallels to glandular duct morphogenesis in prostate. *Mech Dev.* 2000;95:163–174.
65. Emmert-Buck MR, Vocke CD, Pozzatti RO, et al. Allelic loss on chromosome 8p12–21 in microdissected prostatic intraepithelial neoplasia. *Cancer Res.* 1995;55:2959–2962.
66. Vocke CD, Pozzatti RO, Bostwick DG, et al. Analysis of 99 microdissected prostate carcinomas reveals a high frequency of allelic loss on chromosome 8p12–21. *Cancer Res.* 1996;56:2411–2416.
67. Asatiani E, Huang WX, Wang A, et al. Deletion, methylation, and expression of the NKX3.1 suppressor gene in primary human prostate cancer. *Cancer Res.* 2005;65:1164–1173.
68. Bowen C, Bubendorf L, Voeller HJ, et al. Loss of NKX3.1 expression in human prostate cancers correlates with tumor progression. *Cancer Res.* 2000;60:6111–6115.
69. Korkmaz CG, Korkmaz KS, Manola J, et al. Analysis of androgen regulated homeobox gene NKX3.1 during prostate carcinogenesis. *J Urol.* 2004;172:1134–1139.
70. Ornstein DK, Cinquanta M, Weiler S, et al. Expression studies and mutational analysis of the androgen regulated homeobox gene NKX3.1 in benign and malignant prostate epithelium. *J Urol.* 2001;165:1329–1334.
71. Voeller HJ, Augustus M, Madike V, et al. Coding region of NKX3.1, a prostate-specific homeobox gene on 8p21, is not mutated in human prostate cancers. *Cancer Res.* 1997;57:4455–4459.
72. Lei Q, Jiao J, Xin L, et al. NKX3.1 stabilizes p53, inhibits AKT activation, and blocks prostate cancer initiation caused by PTEN loss. *Cancer Cell.* 2006;9:367–378.
73. Zheng SL, Ju JH, Chang BL, et al. Germ-line mutation of NKX3.1 cosegregates with hereditary prostate cancer and alters the homeodomain structure and function. *Cancer Res.* 2006;66:69–77.
74. Ouyang X, DeWeese TL, Nelson WG, et al. Loss-of-function of NKX3.1 promotes increased oxidative damage in prostate carcinogenesis. *Cancer Res.* 2005;65:6773–6779.
75. Carson JA, Fillmore RA, Schwartz RJ, et al. The smooth muscle gamma-actin gene promoter is a molecular target for the mouse bagpipe homologue, mNkx3–1, and serum response factor. *J Biol Chem.* 2000;275:39061–39072.
76. Chen H, Nandi AK, Li X, et al. NKX-3.1 interacts with prostate-derived Ets factor and regulates the activity of the PSA promoter. *Cancer Res.* 2002;62:338–340.
77. Guan B, Pungaliya P, Li X, et al. Ubiquitination by TOPORS regulates the prostate tumor suppressor NKX3.1. *J Biol Chem.* 2008;283:4834–4840.
78. Li X, Guan B, Maghami S, et al. NKX3.1 is regulated by protein kinase CK2 in prostate tumor cells. *Mol Cell Biol.* 2006;26:3008–3017.
79. Wang SI, Parsons R, Ittmann M. Homozygous deletion of the PTEN tumor suppressor gene in a subset of prostate adenocarcinomas. *Clin Cancer Res.* 1998;4:811–815.
80. Di Cristofano A, Pandolfi PP. The multiple roles of PTEN in tumor suppression. *Cell.* 2000;100:387–390.
81. Maehama T, Dixon JE. The tumor suppressor, PTEN/MMAC1, dephosphorylates the lipid second messenger, phosphatidylinositol 3,4,5-trisphosphate. *J Biol Chem.* 1998;273:13375–13378.
82. Majumder PK, Sellers WR. Akt-regulated pathways in prostate cancer. *Oncogene.* 2005;24:7465–7474.
83. Myers MP, Stolarov JP, Eng C, et al. P-TEN, the tumor suppressor from human chromosome 10q23, is a dual-specificity phosphatase. *Proc Natl Acad Sci USA.* 1997;94:9052–9057.
84. Feilotter HE, Nagai MA, Boag AH, et al. Analysis of PTEN and the 10q23 region in primary prostate carcinomas. *Oncogene.* 1998;16:1743–1748.
85. Hermans KG, van Alewijk DC, Veltman JA, et al. Loss of a small region around the PTEN locus is a major chromosome 10 alteration in prostate cancer xenografts and cell lines. *Genes Chromosomes Cancer.* 2004;39:171–184.
86. Muller M, Rink K, Krause H, et al. PTEN/MMAC1 mutations in prostate cancer. *Prostate Cancer Prostatic Dis.* 2000;3:S32.
87. Ratnacaram CK, Teletin M, Jiang M, et al. Temporally controlled ablation of PTEN in adult mouse prostate epithelium generates a model of invasive prostatic adenocarcinoma. *Proc Natl Acad Sci USA.* 2008;105:2521–2526.

88. Wang S, Gao J, Lei Q, et al. Prostate-specific deletion of the murine PTEN tumor suppressor gene leads to metastatic prostate cancer. *Cancer Cell*. 2003;4:209–221.
89. Abate-Shen C, Banach-Petrosky WA, Sun X, et al. NKX3.1; PTEN mutant mice develop invasive prostate adenocarcinoma and lymph node metastases. *Cancer Res*. 2003;63:3886–3890.
90. Gao H, Ouyang X, Banach-Petrosky WA, et al. Emergence of androgen independence at early stages of prostate cancer progression in NKX3.1; PTEN mice. *Cancer Res*. 2006;66:7929–7933.
91. Kim MJ, Cardiff RD, Desai N, et al. Cooperativity of NKX3.1 and PTEN loss of function in a mouse model of prostate carcinogenesis. *Proc Natl Acad Sci USA*. 2002;99:2884–2889.
92. Prescott J, Coetzee GA. Molecular chaperones throughout the life cycle of the androgen receptor. *Cancer Lett*. 2006;231:12–19.
93. Ruizeveld de Winter JA. Immunocytochemical localization of androgen receptor with polyclonal antibody in paraffin-embedded human tissues. *J Histochem Cytochem*. 1994;42:125–126.
94. Dehm SM, Tindall DJ. Ligand-independent androgen receptor activity is activation function-2-independent and resistant to antiandrogens in androgen refractory prostate cancer cells. *J Biol Chem*. 2006;281:27882–27893.
95. Visakorpi T, Hyytinen E, Koivisto P, et al. In vivo amplification of the androgen receptor gene and progression of human prostate cancer. *Nat Genet*. 1995;9:401–406.
96. Kaarbo M, Klokk TI, Saatcioglu F. Androgen signaling and its interactions with other signaling pathways in prostate cancer. *Bioessays*. 2007;29:1227–1238.
97. Wang Y, Kreisberg JI, Ghosh PM. Cross-talk between the androgen receptor and the phosphatidylinositol 3-kinase/Akt pathway in prostate cancer. *Curr Cancer Drug Targets*. 2007;7:591–604.
98. Petrovics G, Liu A, Shaheduzzaman S, et al. Frequent overexpression of ETS-related gene-1 (ERG1) in prostate cancer transcriptome. *Oncogene*. 2005;24:3847–3852.
99. Tomlins SA, Rhodes DR, Perner S, et al. Recurrent fusion of TMPRSS2 and ETS transcription factor genes in prostate cancer. *Science*. 2005;310:644–648.
100. Attard G, Clark J, Ambroisine L, et al. Duplication of the fusion of TMPRSS2 to ERG sequences identifies fatal human prostate cancer. *Oncogene*. 2008;27:253–263.
101. Demichelis F, Fall K, Perner S, et al. TMPRSS2:ERG gene fusion associated with lethal prostate cancer in a watchful waiting cohort. *Oncogene*. 2007;26:4596–4599.
102. Lapointe J, Kim YH, Miller MA, et al. A variant TMPRSS2 isoform and ERG fusion product in prostate cancer with implications for molecular diagnosis. *Mod Pathol*. 2007;20:467–473.
103. Mehra R, Tomlins SA, Shen R, et al. Comprehensive assessment of TMPRSS2 and ETS family gene aberrations in clinically localized prostate cancer. *Mod Pathol*. 2007;20:538–544.
104. Nam RK, Sugar L, Yang W, et al. Expression of the TMPRSS2: ERG fusion gene predicts cancer recurrence after surgery for localised prostate cancer. *Br J Cancer*. 2007;97:1690–1695.
105. Wang J, Cai Y, Ren C, et al. Expression of variant TMPRSS2/ERG fusion messenger RNAs is associated with aggressive prostate cancer. *Cancer Res*. 2006;66:8347–8351.
106. Winnes M, Lissbrant E, Damber JE, et al. Molecular genetic analyses of the TMPRSS2–ERG and TMPRSS2–ETV1 gene fusions in 50 cases of prostate cancer. *Oncol Rep*. 2007; 17:1033–1036.
107. Mosquera JM, Perner S, Genega EM, et al. Characterization of TMPRSS2–ERG fusion high-grade prostatic intraepithelial neoplasia and potential clinical implications. *Clin Cancer Res*. 2008;14:3380–3385.
108. Perner S, Mosquera JM, Demichelis F, et al. TMPRSS2–ERG fusion prostate cancer: An early molecular event associated with invasion. *Am J Surg Pathol*. 2007;31:882–888.
109. Klezovitch O, Risk M, Coleman I, et al. A causal role for ERG in neoplastic transformation of prostate epithelium. *Proc Natl Acad Sci USA*. 2008;105:2105–2110.
110. Tomlins SA, Laxman B, Varambally S, et al. Role of the TMPRSS2–ERG gene fusion in prostate cancer. *Neoplasia*. 2008;10:177–188.
111. Denicourt C, Dowdy SF. Cip/Kip proteins: More than just CDKs inhibitors. *Genes Dev*. 2004;18:851–855.
112. Ferrando AA, Balbin M, Pendas AM, et al. Mutational analysis of the human cyclin-dependent kinase inhibitor p27kip1 in primary breast carcinomas. *Hum Genet*. 1996;97:91–94.
113. Guo Y, Sklar GN, Borkowski A, et al. Loss of the cyclin-dependent kinase inhibitor p27(Kip1) protein in human prostate cancer correlates with tumor grade. *Clin Cancer Res*. 1997;3:2269–2274.
114. Kawamata N, Seriu T, Koeffler HP, et al. Molecular analysis of the cyclin-dependent kinase inhibitor family: p16(CDKN2/MTS1/INK4A), p18(INK4C) and p27(Kip1) genes in neuroblastomas. *Cancer*. 1996;77:570–575.
115. Ponce-Castaneda MV, Lee MH, Latres E, et al. p27kip1: Chromosomal mapping to 12p12-12p13.1 and absence of mutations in human tumors. *Cancer Res*. 1995;55:1211–1214.
116. De Marzo AM, Meeker AK, Epstein JI, et al. Prostate stem cell compartments: Expression of the cell cycle inhibitor p27kip1 in normal, hyperplastic, and neoplastic cells. *Am J Pathol*. 1998;153:911–919.
117. Cordon-Cardo C, Koff A, Drobnjak M, et al. Distinct altered patterns of p27kip1 gene expression in benign prostatic hyperplasia and prostatic carcinoma. *J Natl Cancer Inst*. 1998;90:1284–1291.
118. Fernandez PL, Arce Y, Farre X, et al. Expression of p27/Kip1 is down-regulated in human prostate carcinoma progression. *J Pathol*. 1999;187:563–566.
119. Yang RM, Naitoh J, Murphy M, et al. Low p27 expression predicts poor disease-free survival in patients with prostate cancer. *J Urol*. 1998;159:941–945.
120. Kato JY, Matsuoka M, Polyak K, et al. Cyclic AMP-induced G1 phase arrest mediated by an inhibitor (p27kip1) of cyclin-dependent kinase 4 activation. *Cell*. 1994;79:487–496.
121. Sanchez-Beato M, Saez AI, Martinez-Montero JC, et al. Cyclin-dependent kinase inhibitor p27kip1 in lymphoid tissue: p27kip1 expression is inversely proportional to the proliferative index. *Am J Pathol*. 1997;151:151–160.
122. Tsihlias J, Kapusta LR, DeBoer G, et al. Loss of cyclin-dependent kinase inhibitor p27kip1 is a novel prognostic factor in localized human prostate adenocarcinoma. *Cancer Res*. 1998;58:542–548.
123. Gary B, Azuero R, Mohanty GS, et al. Interaction of NKX3.1 and p27kip1 in prostate tumor initiation. *Am J Pathol*. 2004; 164:1607–1614.
124. Fero ML, Randel E, Gurley KE, et al. The murine gene p27kip1 is haplo-insufficient for tumour suppression. *Nature*. 1998;396:177–180.
125. Di Cristofano A, De Acetis M, Koff A, et al. PTEN and p27kip1 cooperate in prostate cancer tumor suppression in the mouse. *Nat Genet*. 2001;27:222–224.
126. Gao H, Ouyang X, Banach-Petrosky W, et al. A critical role for p27kip1 gene dosage in a mouse model of prostate carcinogenesis. *Proc Natl Acad Sci USA*. 2004;101:17204–17209.
127. Blackburn EH. Structure and function of telomeres. *Nature*. 1991;350:569–573.
128. Artandi SE, Chang S, Lee SL, et al. Telomere dysfunction promotes non-reciprocal translocations and epithelial cancers in mice. *Nature*. 2000;406:641–645.
129. Meeker AK. Telomeres and telomerase in prostatic intraepithelial neoplasia and prostate cancer biology. *Urol Oncol*. 2006;24:122–130.
130. Bartel DP. MicroRNAs: Genomics, biogenesis, mechanism, and function. *Cell*. 2004;116:281–297.
131. Meister G, Tuschl T. Mechanisms of gene silencing by double-stranded RNA. *Nature*. 2004;431:343–349.
132. Sontheimer EJ. Assembly and function of RNA silencing complexes. *Nat Rev Mol Cell Biol*. 2005;6:127–138.
133. Wiemer EA. The role of microRNAs in cancer: No small matter. *Eur J Cancer*. 2007;43:1529–1544.
134. Lu J, Getz G, Miska EA, et al. MicroRNA expression profiles classify human cancers. *Nature*. 2005;435:834–838.
135. Ozen M, Creighton CJ, Ozdemir M, et al. Widespread deregulation of microRNA expression in human prostate cancer. *Oncogene*. 2008;27:1788–1793.
136. Porkka KP, Pfeiffer MJ, Waltering KK, et al. MicroRNA expression profiling in prostate cancer. *Cancer Res*. 2007;67:6130–6135.

137. Lee WH, Morton RA, Epstein JI, et al. Cytidine methylation of regulatory sequences near the pi-class glutathione S-transferase gene accompanies human prostatic carcinogenesis. *Proc Natl Acad Sci USA.* 1994;91:11733–11737.
138. Nelson CP, Kidd LC, Sauvageot J, et al. Protection against 2-hydroxyamino-1-methyl-6-phenylimidazo[4,5-b]pyridine cytotoxicity and DNA adduct formation in human prostate by glutathione S-transferase P1. *Cancer Res.* 2001;61:103–109.
139. Shirai T, Sano M, Tamano S, et al. The prostate: A target for carcinogenicity of 2-amino-1-methyl-6-phenylimidazo[4,5-b] pyridine (PhIP) derived from cooked foods. *Cancer Res.* 1997;57:195–198.
140. Henrique R, Ribeiro FR, Fonseca D, et al. High promoter methylation levels of APC predict poor prognosis in sextant biopsies from prostate cancer patients. *Clin Cancer Res.* 2007;13:6122–6129.
141. Jeronimo C, Henrique R, Hoque MO, et al. A quantitative promoter methylation profile of prostate cancer. *Clin Cancer Res.* 2004;10:8472–8478.
142. Yegnasubramanian S, Kowalski J, Gonzalgo ML, et al. Hypermethylation of CpG islands in primary and metastatic human prostate cancer. *Cancer Res.* 2004;64:1975–1986.
143. Yardy GW, Brewster SF. Wnt signalling and prostate cancer. *Prostate Cancer Prostatic Dis.* 2005;8:119–126.
144. Bruxvoort KJ, Charbonneau HM, Giambernardi TA, et al. Inactivation of Apc in the mouse prostate causes prostate carcinoma. *Cancer Res.* 2007;67:2490–2496.
145. Roupret M, Hupertan V, Catto JW, et al. Promoter hypermethylation in circulating blood cells identifies prostate cancer progression. *Int J Cancer.* 2008;122:952–956.
146. Roupret M, Hupertan V, Yates DR, et al. Molecular detection of localized prostate cancer using quantitative methylation-specific PCR on urinary cells obtained following prostate massage. *Clin Cancer Res.* 2007;13:1720–1725.
147. Pienta KJ, Abate-Shen C, Agus DB, et al. The current state of preclinical prostate cancer animal models. *Prostate.* 2008;68:629–639.
148. Kasper S. Survey of genetically engineered mouse models for prostate cancer: Analyzing the molecular basis of prostate cancer development, progression, and metastasis. *J Cell Biochem.* 2005;94:279–297.
149. Gingrich JR, Barrios RJ, Morton RA, et al. Metastatic prostate cancer in a transgenic mouse. *Cancer Res.* 1996;56:4096–4102.
150. Greenberg NM, DeMayo F, Finegold MJ, et al. Prostate cancer in a transgenic mouse. *Proc Natl Acad Sci USA.* 1995;92:3439–3443.
151. Tilley WD, Buchanan G, Hickey TE, et al. Mutations in the androgen receptor gene are associated with progression of human prostate cancer to androgen independence. *Clin Cancer Res.* 1996;2:277–285.
152. Han G, Foster BA, Mistry S, et al. Hormone status selects for spontaneous somatic androgen receptor variants that demonstrate specific ligand and cofactor dependent activities in autochthonous prostate cancer. *J Biol Chem.* 2001;276:11204–11213.
153. Bethel CR, Bieberich CJ. Loss of NKX3.1 expression in the transgenic adenocarcinoma of mouse prostate model. *Prostate.* 2007;67:1740–1750.
154. Dubey P, Wu H, Reiter RE, et al. Alternative pathways to prostate carcinoma activate prostate stem cell antigen expression. *Cancer Res.* 2001;61:3256–3261.
155. Kaplan PJ, Mohan S, Cohen P, et al. The insulin-like growth factor axis and prostate cancer: Lessons from the transgenic adenocarcinoma of mouse prostate (TRAMP) model. *Cancer Res.* 1999;59:2203–2209.
156. Chiaverotti T, Couto SS, Donjacour A, et al. Dissociation of epithelial and neuroendocrine carcinoma lineages in the transgenic adenocarcinoma of mouse prostate model of prostate cancer. *Am J Pathol.* 2008;172:236–246.
157. Huss WJ, Gray DR, Tavakoli K, et al. Origin of androgen-insensitive poorly differentiated tumors in the transgenic adenocarcinoma of mouse prostate model. *Neoplasia.* 2007;9:938–950.
158. Roudier MP, True LD, Higano CS, et al. Phenotypic heterogeneity of end-stage prostate carcinoma metastatic to bone. *Hum Pathol.* 2003;34:646–653.
159. Shah RB, Mehra R, Chinnaiyan AM, et al. Androgen-independent prostate cancer is a heterogeneous group of diseases: Lessons from a rapid autopsy program. *Cancer Res.* 2004;64:9209–9216.
160. Kasper S, Smith JA, Jr. Genetically modified mice and their use in developing therapeutic strategies for prostate cancer. *J Urol.* 2004;172:12–19.
161. Park JH, Walls JE, Galvez JJ, et al. Prostatic intraepithelial neoplasia in genetically engineered mice. *Am J Pathol.* 2002; 161:727–735.
162. Roy-Burman P, Wu H, Powell WC, et al. Genetically defined mouse models that mimic natural aspects of human prostate cancer development. *Endocr Relat Cancer.* 2004;11:225–254.
163. Shappell SB, Thomas GV, Roberts RL, et al. Prostate pathology of genetically engineered mice: Definitions and classification. The consensus report from the Bar Harbor meeting of the Mouse Models of Human Cancer Consortium Prostate Pathology Committee. *Cancer Res.* 2004;64:2270–2305.
164. Lesche R, Groszer M, Gao J, et al. Cre/loxP-mediated inactivation of the murine PTEN tumor suppressor gene. *Genesis.* 2002;32:148–149.
165. Wu X, Wu J, Huang J, et al. Generation of a prostate epithelial cell-specific Cre transgenic mouse model for tissue-specific gene ablation. *Mech Dev.* 2001;101:61–69.
166. Freeman DJ, Li AG, Wei G, et al. PTEN tumor suppressor regulates p53 protein levels and activity through phosphatase-dependent and -independent mechanisms. *Cancer Cell.* 2003;3:117–130.
167. Stambolic V, MacPherson D, Sas D, et al. Regulation of PTEN transcription by p53. *Mol Cell.* 2001;8:317–325.
168. Salmena L, Carracedo A, Pandolfi PP. Tenets of PTEN tumor suppression. *Cell.* 2008;133:403–414.
169. Chen Z, Trotman LC, Shaffer D, et al. Crucial role of p53-dependent cellular senescence in suppression of PTEN-deficient tumorigenesis. *Nature.* 2005;436:725–730.
170. Shibata H, Toyama K, Shioya H, et al. Rapid colorectal adenoma formation initiated by conditional targeting of the Apc gene. *Science.* 1997;278:120–123.
171. Terry S, Yang X, Chen MW, et al. Multifaceted interaction between the androgen and Wnt signaling pathways and the implication for prostate cancer. *J Cell Biochem.* 2006;99:402–410.
172. Truica CI, Byers S, Gelmann EP. Beta-catenin affects androgen receptor transcriptional activity and ligand specificity. *Cancer Res.* 2000;60:4709–4713.
173. Willert K, Jones KA. Wnt signaling: Is the party in the nucleus? *Genes Dev.* 2006;20:1394–1404.
174. Brooks JD, Bova GS, Isaacs WB. Allelic loss of the retinoblastoma gene in primary human prostatic adenocarcinomas. *Prostate.* 1995;26:35–39.
175. Ittmann MM, Wieczorek R. Alterations of the retinoblastoma gene in clinically localized, stage B prostate adenocarcinomas. *Hum Pathol.* 1996;27:28–34.
176. Phillips SM, Morton DG, Lee SJ, et al. Loss of heterozygosity of the retinoblastoma and adenomatous polyposis susceptibility gene loci and in chromosomes 10p, 10q and 16q in human prostate cancer. *Br J Urol.* 1994;73:390–395.
177. Tricoli JV, Gumerlock PH, Yao JL, et al. Alterations of the retinoblastoma gene in human prostate adenocarcinoma. *Genes Chromosomes Cancer.* 1996;15:108–114.
178. Hill R, Song Y, Cardiff RD, et al. Heterogeneous tumor evolution initiated by loss of pRb function in a preclinical prostate cancer model. *Cancer Res.* 2005;65:10243–10254.
179. Zhang X, Lee C, Ng PY, et al. Prostatic neoplasia in transgenic mice with prostate-directed overexpression of the c-Myc oncoprotein. *Prostate.* 2000;43:278–285.
180. Ellwood-Yen K, Graeber TG, Wongvipat J, et al. Myc-driven murine prostate cancer shares molecular features with human prostate tumors. *Cancer Cell.* 2003;4:223–238.
181. Stanbrough M, Bubley GJ, Ross K, et al. Increased expression of genes converting adrenal androgens to testosterone in androgen-independent prostate cancer. *Cancer Res.* 2006; 66:2815–2825.
182. Stanbrough M, Leav I, Kwan PW, et al. Prostatic intraepithelial neoplasia in mice expressing an androgen receptor transgene in prostate epithelium. *Proc Natl Acad Sci USA.* 2001;98:10823–10828.

183. Han G, Buchanan G, Ittmann M, et al. Mutation of the androgen receptor causes oncogenic transformation of the prostate. *Proc Natl Acad Sci USA*. 2005;102:1151–1156.
184. Wu Q, Parry G. Hepsin and prostate cancer. *Front Biosci.* 2007;12:5052–5059.
185. Riddick AC, Shukla CJ, Pennington CJ, et al. Identification of degradome components associated with prostate cancer progression by expression analysis of human prostatic tissues. *Br J Cancer.* 2005;92:2171–2180.
186. Stamey TA, Warrington JA, Caldwell MC, et al. Molecular genetic profiling of Gleason grade 4/5 prostate cancers compared to benign prostatic hyperplasia. *J Urol.* 2001;166:2171–2177.
187. Stephan C, Yousef GM, Scorilas A, et al. Hepsin is highly over expressed in and a new candidate for a prognostic indicator in prostate cancer. *J Urol.* 2004;171:187–191.
188. Xuan JA, Schneider D, Toy P, et al. Antibodies neutralizing hepsin protease activity do not impact cell growth but inhibit invasion of prostate and ovarian tumor cells in culture. *Cancer Res.* 2006;66:3611–3619.
189. Fromont G, Chene L, Vidaud M, et al. Differential expression of 37 selected genes in hormone-refractory prostate cancer using quantitative taqman real-time RT-PCR. *Int J Cancer.* 2005;114:174–181.
190. Klezovitch O, Chevillet J, Mirosevich J, et al. Hepsin promotes prostate cancer progression and metastasis. *Cancer Cell.* 2004;6:185–195.
191. Kasper S, Sheppard PC, Yan Y, et al. Development, progression, and androgen-dependence of prostate tumors in probasin-large T antigen transgenic mice: A model for prostate cancer. *Lab Invest.* 1998;78:i–xv.

Chapter

25

Molecular Biology of Breast Cancer

Natasa Snoj . Phuong Dinh . Philippe Bedard . Christos Sotiriou

INTRODUCTION

Breast cancer is the most common cancer among women with a yearly incidence of 109.8 per 100,000 individuals. It is also the second leading cause of cancer-related death [1]. Since the late 1980s, breast cancer-related deaths have significantly declined, partly due to screening and prevention and partly due to improvements in systemic therapy. With the widespread implementation of screening programs, the increasing frequency of screen-detected invasive cancers and noninvasive lesions has shifted the profile of tumor characteristics seen in breast cancer today toward smaller tumors. Furthermore, breast cancer is increasingly being recognized as a biologically heterogeneous disease entity, with distinct subtypes demonstrating different natural history and clinical outcomes. Many of these differences are attributable to the heterogeneity that exists at the molecular level, and as a result, the discovery and development of molecular markers have recently taken center stage. These markers are being investigated for their potential to serve as better predictors of prognosis and response to treatment. Clinically useful biomarkers that can better select patients for tailored clinical trials, according to molecular characteristics, can ultimately lead to individualized treatment for breast cancer patients.

TRADITIONAL BREAST CANCER CLASSIFICATION

Breast cancer has traditionally been classified according to histopathological type, with its anatomical extent is reflected by the TNM classification. For many years, this breast cancer classification was the only tool available to asses the risk of relapse after local therapy and to base decisions on whether or not chemotherapy should be given.

Histopathological Features of Breast Cancer

The histopathological features of breast cancer are the foundation of traditional breast cancer pathology. They consist of (i) histological type, (ii) grade, and (iii) vascular invasion. Histopathological features provide clinicians with essential information for decision making regarding prognosis and treatment.

Histological Type

In the past, breast cancer was divided into lobular and ductal carcinoma according to the common belief that lobular carcinomas emanated from the lobules and ductal carcinomas from the ducts. Today, after demonstration that most breast cancers arise from the same location—the terminal duct lobular unit—these histomorphological differences are regarded as a manifestation of their distinct molecular profiles. The predominant type of breast cancer is ductal carcinoma *in situ* (DCIS). Some breast lesions not classified as breast cancer are also described here, as they are closely related to breast cancer in terms of histopathology, in their expression of molecular markers, and in their clinical presentation.

Lobular Neoplasia. Lobular neoplasia can be classified as (i) lobular intraepithelial neoplasia, (ii) lobular carcinoma *in situ*, and (iii) invasive lobular carcinoma. Lobular intraepithelial neoplasia and lobular carcinoma are commonly named *in situ* lobular neoplasia. Lobular neoplasia (*in situ* lobular neoplasia and invasive lobular carcinoma) is characterized by a population of small aberrant cells with small nuclei, individual private acini, and a lack of cohesion between cells [2]. The distinctive molecular feature of lobular neoplasia (*in situ* and invasive) is the loss of E-cadherin (*CDH1*) which can be demonstrated by immunohistochemistry. Since ductal carcinomas can also be E-cadherin negative, it should

Molecular Pathology © 2009, Elsevier, Inc. All Rights Reserved.

not be considered pathognomic for lobular neoplasia. Other molecular characteristics of lobular carcinomas are epidermal growth factor receptor 1 (EGFR-1) and HER-2 negativity; and positivity for antibody 34b E12 (cytokeratins 1, 5, 10, and 14), ER, and PR. These markers, while helpful, are also not pathognomic, with the pleomorphic variant often described as ER, PR negative, and HER-2 positive. Therefore, such molecular markers, while able to add an extra level of diagnostic information, still need to be correlated with the classical histopathological findings.

The diagnosis of *in situ* lobular neoplasia is often made following surgical excision of a suspicious nodule. It is often bilateral and in 50%–70% of cases is multicentric. There is a high frequency (10%–15% over a 10 year period and lifetime risk of up to 50%) of subsequent invasive ductal carcinoma in patients with a prior *in situ* lobular neoplasia, for reasons that are not well understood. Invasive lobular carcinoma is the second most common invasive breast cancer and represents 10%–20% of all invasive breast cancers.

Ductal Carcinoma *in situ*. DCIS is characterized by the proliferation of malignant cells within the ducts without invasion of the surrounding stromal tissue. The *European Organisation for Research and Treatment of Cancer* (EORTC) grading system, based on cytonuclear pattern, distinguishes well-differentiated from intermediately-differentiated and poorly differentiated DCIS. For the purpose of screening programs, this has been modified into groups based on low, intermediate, and high grade. Compared with normal epithelium, the potential of growth of DCIS is 10 times greater, and its apoptosis rate is increased 15-fold. Comedo type DCIS is typically necrotic in the center of the ducts and has a higher risk of recurrence. Most ductal carcinomas are diffusely positive for luminal cell markers (CK8, CK18, CK19), but negative for basal cell markers (CK5/6 and CK14) [3]. In contrast, benign ductal hyperplasia may show a mosaic staining pattern for any of these markers, indicating a heterogeneous underlying cell population.

DCIS is multifocal in 30% of the cases, often in the same breast. In 2%–6% of cases, axillary lymph node invasion is found at pathological examination, and this is thought to occur as a result of an unidentified invasive component. In the era of widespread mammographic screening, DCIS is more frequently detected and represents 25% of screen-detected breast cancers [4].

The treatment of DCIS is based on the extent of disease within the breast [5]. In patients with extensive or multifocal DCIS, mastectomy with sentinel node biopsy is indicated, with the possibility of reconstruction. Sentinel lymph node biopsy is also indicated when microinvasion is found within DCIS. If breast-conserving surgery is undertaken, additional local irradiation has documented benefit in terms of lowering the incidence of subsequent noninvasive relapse and the progression to frankly invasive disease. Those patients with estrogen receptor positive DCIS may also benefit from adjuvant systemic treatment with tamoxifen in addition to radiotherapy after breast-conserving surgery [6]. However, our understanding of the biology of DCIS remains crude, and further clinical trials are necessary to optimize local and systemic treatment modalities.

Invasive Ductal Carcinoma. Invasive ductal carcinoma is the most common invasive carcinoma of the breast, as it represents roughly 70% of all cases. Invasive lobular carcinoma is the second most common histopathological type representing 10%–20% of all breast malignancies. The other 10%–20% consists mostly of (i) medullary carcinoma (characterized by sharp tumor borders and lymphoid infiltration), (ii) mucoid carcinoma (with large amounts of extracellular mucous), (iii) papillary carcinoma (with well-differentiated papillary structure), and (iv) inflammatory carcinoma which occurs in 1%–2% of all cases. Inflammatory carcinomas typically mimic the clinical presentation of benign inflammatory disease although, histologically, they are characterized by extensive invasion of the lymphatic vessels within the dermis. Other types of rare invasive breast cancer include (i) adenoid cystic carcinoma, (ii) apocrine carcinoma, and (iii) carcinoma with squamous metaplasia.

Another distinct histopathological entity of breast cancer is Paget's disease of the nipple. It is a variant of DCIS, where malignant cells infiltrate the ducts without invasion of the baseline membrane and extend to the major ducts of the nipple and the epidermis of the areola. An invasive component is present in proximally 90% of all cases. There are other malignant diseases of the breast that can resemble breast cancer, including phyllodes tumor, sarcoma, and malignant lymphoma.

Often the distinction between invasive and noninvasive carcinoma is difficult, as carcinoma cells can be found in both lesions and there are no reliable markers that differentiate between invasive and noninvasive cells. E-cadherin is often helpful in distinguishing malignant disease, as it is mostly positive in ductal carcinomas, but negative in lobular carcinomas. Invasive ductal carcinoma may express a variety of markers that have been extensively evaluated.

Other breast lesions, such as benign papillomas, are characterized by the expression of myoepithelial markers (alpha-SMA, myosin, calponin, p63, CD10), although myoepithelial markers may also be present in intraductal papillary carcinomas. Preservation of the myoepithelial layer is the distinguishing characteristic of benign sclerosing lesions, including carcinoma with pseudoinvasive structures. Given their overlapping molecular profiles, comparison with histopathological findings using hematoxylin and eosin is essential to make proper diagnosis.

Vascular Invasion

The term vascular invasion refers to the invasion of lymphatic and blood vessels with tumor cells. It is assessed in hematoxylin-eosin stained slides. Lymphovascular invasion is routinely included in the pathological evaluation and reporting of all breast cancers, although its interpretation is often difficult. In breast cancer, vascular invasion is a poor prognostic factor and predicts for increased local failure and reduced overall survival [7]. Whether lymphatic or hematogenous invasion plays a more dominant role in prognosis determination is still unknown [8]. Lymphovascular invasion by tumor cells is found in approximately 15% of invasive ductal carcinoma of the breast and is present in 10% of tumors without axillary lymph node invasion.

Perineural Invasion

The independent prognostic significance of perineural invasion is not clear, as it has only been assessed in combination with lymphovascular invasion [9].

Histological Grading

The histological grading of malignancy is the classical evaluation method for prognostication in breast cancer patients, and it is the simplest method, requiring only hematoxylin-eosin staining. Histological grading of breast cancer typically consists of three factors: tubule formation, mitotic count, and nuclear atypia. These three factors are all important in identifying patients at high risk of recurrence, although the relative importance of each feature is unclear. In addition, intratumoral heterogeneity and interobserver variation add to the difficulty of accurate prognostication based upon grading. In a review of the prognostic significance of histological grading, only nuclear atypia was correlated with the rate of recurrence [10]. A grading system consisting of mitotic count and nuclear atypia (without tubule formation) was more strongly related to the risk of recurrence than a system using all three factors. Therefore, tubule formation is often excluded from the overall grading system even though it has the lowest heterogeneity within the primary tumor, compared with the other two.

Cells in the mitotic phase can be counted using light microscopy. However, the duration of mitotic phase is often variable, especially in aneuploid tumors, so the number of mitoses is not linearly correlated with cellular proliferation. The recommendation for standardized counting of the mitotic activity index (MAI) includes assessment at ×400 magnification in an area of 1.6 mm^2 in the highest proliferative invasive area at the periphery of the tumor [11]. MAI is not affected by fixation delay, it may impair morphological assessment. Ideally, mitoses should be counted before chemotherapy, but the mitotic index retains prognostic value even after chemotherapy. Mitoses should preferably be counted on excisional biopsies or mastectomies to avoid sampling error. Often underestimated in core biopsies, the MAI has a limited role in the decision of whether neoadjuvant chemotherapy should be given or not. Nonetheless, in several retrospective and prospective studies, the MAI has proven to be a very strong, independent prognostic factor and, subsequently, is regarded as a category I prognostic factor for breast cancer by the *College of American Pathologists.*

TNM

Prognosis estimation has traditionally been based on grouping patients according to the anatomical extent of their disease. Incorporating primary tumor size, lymph node involvement, and distant metastases the TNM system combines these three factors into a stage grouping, with clinical and pathological variants. The TNM system is the most widely used tumor staging system [12], adopted by both the *Union International Contre le Cancer* (UICC) and the *American Joint Committee* (AJC). Although still clinically relevant, this system has limitations because it does not incorporate other important biological features that can influence the overall prognosis of the patient. Other important biological features include steroid hormone receptor content, HER-2 overexpression, tumor grade, indices of tumor proliferation, the presence or absence of vascular or lymphatic vessel invasion [13].

BIOMARKERS

Molecular analyses of cancer have led to the discovery of oncogenes and tumor suppressor genes. In practice, pathologists look for protein products (biomarkers) of the genes mainly through immunohistochemistry or immunocytochemistry. Biomarkers are potentially helpful in distinguishing between various histopathological types of breast cancer, as well as in assessing prognosis and predicting for a response to specific systemic therapy.

Biomarkers typically have continuous values, and cutoff points are arbitrarily established used to simplify the clinical decision-making. An important requirement for a cutoff point in this context is that it should not identify too large or too small a subset of patients for further therapy, as this would lead to overtreatment of one group and undertreatment of the other group.

Estrogen Receptor

Estrogen receptors (ER) are members of a large family of nuclear transcriptional regulators that are activated by steroid hormones, such as estrogen [14]. ERs exist as two isoforms, α and β, that are encoded by two different genes [15]. Although both isoforms are expressed in the normal mammary gland, it appears that only ER-α is critical for normal gland development [16]. However, there is growing evidence that ER-β may antagonize the function of ER-α, and that high levels of ER-β are associated with a more favorable response to tamoxifen treatment [17].

Most research on the biology of ER has focused on its function as a nuclear transcription factor [classical ER-α function, or nuclear-initiated steroid signaling

(NISS)], but there is also mounting evidence that estrogen can bind to ER located in or near the plasma membrane to activate other signaling pathways [nongenomic function, or membrane-initiated steroid signaling (MISS)] [18].

With NISS, ER functions as a hormone-regulated nuclear transcription factor that can induce expression of specific genes in the nucleus [19]. Upon estrogen ligand binding, ER binds to estrogen response elements (ERE) in target genes, recruits a coregulator complex, and regulates specific gene transcription [20]. This ER action is controlled by a number of coregulatory proteins, termed coactivators and corepressors, that recruit enzymes to modulate chromatin structure to facilitate or repress gene transcription [21,22]. It has been postulated that the cellular environment of coregulators may indeed influence whether selective estrogen receptor modulators (SERMS) are agonistic or antagonistic in action [23,24]. Whether levels of coregulators are associated with prognosis and hormone resistance in breast cancer is unknown, but has been the focus of several studies [21,25].

Estrogen can therefore affect many processes, such as cell proliferation, the inhibition of apoptosis, invasion, and angiogenesis. Indeed, c-Myc, vascular endothelial growth factor, bcl-2, insulin-like growth factor (IGF-1), insulin receptor substrate-1, transforming growth factor-α, cyclin-D1, and IGF-2 have all been shown to be regulated by estrogen [26–30].

It has long been recognized that estrogen is also capable of inducing other cellular effects with a much more rapid onset than would be possible via changes imposed by gene transcription through NISS [31]. Such nongenomic effects have been attributed to MISS, involving membrane-bound or cytoplasmic ER. Indeed, ER has been localized outside of the nucleus by biochemical analyses and by direct visualization using immunocytochemistry or more sophisticated microscopy [32,33]. In fact, non-nuclear ER has been shown to exist in complexes with such signaling molecules as the IGF-1 receptor (IGF-1R) [34] and the p85 subunit of phosphatidylinositol-3-OH kinase [35]. While these membrane and non-nuclear ER functions have been well described in experimental model systems, they must still be confirmed in clinical breast cancer.

Targeting the ER—Tamoxifen Therapy

There is a significant body of evidence to suggest that steroid hormones, particularly estrogen, play a major role in the development of breast cancer. Approximately 75% of all breast cancers express estrogen receptors (ER+) [36]. ER expression has a high negative predictive value but a suboptimal positive predictive value for the efficacy of endocrine manipulation. Patients with ER negative (ER−) tumors derive no benefit from endocrine therapy, while for ER+ tumors, estrogen signaling blockade is an important therapeutic strategy that can lead to improved outcomes, including increasing cure rates in early breast cancer, improving response rates and disease control in advanced disease, and reducing breast cancer incidence with prevention.

Tamoxifen is a selective modulator of ER that competitively impedes the binding of estradiol and, in doing so, disrupts a series of mechanisms that regulate cellular replication [37] and proliferation. Tamoxifen has been used in clinical trials since the 1970s, with initial nonselective use across the entire breast cancer population. In these unselected patients, tamoxifen can produce responses of up to 30% [38,39]. In ER+ disease, responses of up to 80% have been observed, while in ER− disease, less than 10% may derive benefit [40].

Testing for ER

The introduction of routine ER testing in the early 1990s changed the way in which endocrine therapy is prescribed, particularly in preventing unnecessary tamoxifen-related toxicity for patients with ER− disease. However, the treatment of individual ER+ patients is less clearcut, largely because of the ambiguity created by varying methodologies and cutoffs employed by different laboratories for ER testing. Indeed, responses have been demonstrated in tumors with as few as 1%–10% of cells positive for ER by immunohistochemistry (IHC) [41]. It is therefore more clinically important to distinguish between ER− absent disease from measurable, but low-level ER expression, variably reported as either ER+ or ER−, depending on which cutoff values are used.

Quantitative ER as a continuous variable has been demonstrated in metastatic studies to be proportionally correlated to the response to hormonal therapy [40]. In the adjuvant setting, a large meta-analysis has shown a similar relationship between tumoral ER expression and tamoxifen efficacy [42], and in the neoadjuvant setting, for tumor response to both tamoxifen and letrozole [43].

Nevertheless, quantitative ER is still an imprecise predictive tool because even tumors in the highest strata of ER-expression can be endocrine-resistant (*de novo* resistance). Furthermore, a significant proportion of ER+ patients who initially respond to endocrine therapy will eventually fail treatment (acquired resistance). Although this resistance is not fully understood, several mechanisms have been hypothesized to be responsible for the development of resistance, including (i) loss of ER in the tumor, (ii) selection clones with ER mutations, (iii) deregulation of cell-cycle components including ER-regulatory proteins, and (iv) cross-talk between ER and other growth factor receptor pathways [44]. Given the limitations of ER assessment, the development of other molecular tools, reflecting the complex biology of these tumors, is needed to improve treatment for patients with ER+ disease.

Progesterone Receptor

The progesterone receptor (*PR*) gene is an estrogen-regulated gene. PR mediates the effect of progesterone in the development of the mammary gland and breast cancer [45]. In the 75% of breast cancers that express ER, more than half of these tumors also express PR [46]. It has been hypothesized that PR levels in breast cancer may be a marker of an intact ER signal transduction pathway and that PR levels may therefore add independent predictive information. Emerging laboratory and clinical data also suggest that among ER+ tumors, PR status may predict differential sensitivity to antiestrogen therapy.

Although the etiology of ER+PR− disease some studies have shown that this phenotype may evolve through the loss of PR, while other studies suggest that they reflect a distinct molecular origin [49], with unique epidemiologic risk factors when compared with ER+PR+ disease [50].

Patients with ER+PR+ tumors appear to derive greater benefit from adjuvant tamoxifen than patients with ER+PR− tumors, as shown in a retrospective analysis [51]. The same analysis also suggested that loss of PR was predictive of outcome independent of quantitative ER levels. These findings have been observed in other smaller studies [52–56].

Apart from being generally predictive for diminished responsiveness to tamoxifen, absence of PR has also been suggested to be predictive of increased responsiveness to aromatase inhibitors. However, this remains controversial, as there have been conflicting results from retrospective analyses of several large adjuvant aromatase inhibitors studies [57–62]. In particular, the initial report of the ATAC [57] trial suggested a greater benefit from aromatase inhibitors compared to tamoxifen for ER+PR− disease based upon local pathology reporting. A subsequent central pathology review of a subset of tumors enrolled in ATAC failed to show a differential benefit for ER+PR− disease.

The HER-2 Receptor

Following the discovery of ER, HER-2 has more recently emerged as an important molecular target in the treatment of breast cancer. Much of the recent success with anti-HER-2 therapy has been a direct result of being able to properly select the right patient subpopulation for targeted treatment.

HER-2 belongs to the human epidermal growth factor receptor family of tyrosine kinases consisting of EGFR (HER-1; erbB1), HER-2 (erbB2, HER-2/*neu*), HER-3 (erbB3), and HER-4 (erbB4). All of these receptors have an extracellular ligand-binding region, a single membrane-spanning region, and a cytoplasmic tyrosine-kinase-containing domain. This intracellular kinase domain is non-function in HER-3. Ligand binding to the extracellular region results in homodimer and heterodimer activation of the cytoplasmic kinase domain and phosphorylation of a specific tyrosine kinase [62]. This leads to the activation of various intracellular signaling pathways, such as the mitogen-activated protein kinase (MAPK) and the phosphatidylinositol 3-kinase (P13K)-AKT pathways [63] involved in cell proliferation and survival.

HER signaling can become dysregulated via a number of mechanisms [64], including overexpression of a ligand; overexpression of the normal HER receptor; overexpression of a constitutively activated, mutation of the HER receptor; and defective HER receptor internalization, recycling, or degradation.

HER-2 was first identified as an oncogene activated by a point mutation in chemically induced rat neuroblastomas [65]. Soon afterward, it was found to be amplified in breast cancer cell lines [66]. Early studies suggested that as many as 30% of breast cancers demonstrate HER-2 overexpression, which is associated with a more aggressive phenotype and a poorer disease-free survival [67–70]. Furthermore, HER-2 positivity appears to be associated with a relative, but not an absolute, resistance to endocrine therapy [71] and resistance to certain chemotherapeutic agents [72–74]. Most importantly, HER-2 status is predictive for benefit from anti-HER-2 therapies, such as trastuzumab.

Identifying the HER-2 Receptor

For the various reasons described above, the accurate determination of HER-2 status is vitally important because it is such a useful marker for therapeutic decision making. Currently, two different methods of determining for HER-2 are routinely available: (i) immunohistochemistry (IHC) and (ii) fluorescence *in situ* hybridization (FISH).

IHC is a semiquantitative method that identifies HER-2 receptor expression on the cell surface using a grading system (0, 1+, 2+, and 3+), where an IHC result of 3+ is regarded as HER-2 overexpression. It is the most widely used technique, performed on paraffin tumor blocks. While fairly easy to perform and relatively low cost, results can vary depending on different fixation protocols, assay methods, scoring systems, as well as the selected antibodies.

FISH is a quantitative method measuring the number of copies of the HER-2 gene present in each tumor cell and is reported as either positive or negative. It is a more reproducible test but is comparatively more time-consuming and expensive.

Concordance between IHC and FISH has been extensively studied. In a study done by Yaziji et al. [75] conducted in 2963 samples using FISH as the standard method, the positive predictive value of an IHC 3+ result was 91.6%, and the negative predictive value of an IHC 0 or 1+ result was 97.2%. They also found that FISH had a significantly higher failure rate (5% versus 0.08%), was more costly, and required more time for testing (36 versus 4 hours) and interpretation (7 minutes versus 45 seconds) than IHC.

Another problem with HER-2 testing is the poor reproducibility between laboratories, even when the same technique is used. When the concordance between local and central evaluation in the NCCTG-N9831 adjuvant trastuzumab trial was examined [76], FISH concordance was 88.1% compared to an IHC concordance of 81.6%. This high rate of discordance was attributed to methodologic factors, such as inadequate quality-control procedures.

In recognizing the importance of accurate HER-2 testing, guidelines from the *American Society of Clinical Oncology* (ASCO)/*College of American Pathologists* (CAP) for HER-2 testing have recently been published [77]. In these guidelines, recommendations for HER-2 evaluation are provided using an algorithm for positive, equivocal, and negative results:

1. HER-2 positive—IHC staining of 3+ (uniform, intense membrane staining of >30% of invasive tumor cells, a FISH result of more than 6.0 HER-2 gene copies per nucleus, or a FISH ratio (HER-2 gene signals to chromosome 17 signals) of more than 2.2;
2. HER-2 negative—IHC staining of 0 or 1+ FISH result of less than 4.0 HER-2 gene copies per nucleus, or FISH ratio of less than 1.8; and
3. HER-2 equivocal—IHC 3+ staining of 30% or less of invasive tumor cells or 2+ staining, a FISH result of 4 to 6 HER-2 gene copies per nucleus, or FISH ratio between 1.8 to 2.2.

Importantly, the Guidelines strongly recommended [77] "validation of laboratory assay or modifications, use of standardized operating procedures, and compliance with new testing criteria to be monitored with the use of stringent laboratory accreditation standards, proficiency testing, and competency assessment."

Targeting HER-2: Trastuzumab

The HER family of receptors is an ideal target for anticancer therapy. To date, two main therapeutic strategies have been developed to target the HER-2 receptor: monoclonal antibodies and small molecule kinase inhibitors.

Trastuzumab (Herceptin; Genentech, South San Francisco) is a recombinant, humanized anti-HER-2 monoclonal antibody and was the first clinically active anti-HER-2 therapy to be developed. Trastuzumab exerts its action through several mechanisms, including (i) induction of receptor downregulation/degradation [78], (ii) prevention of HER-2 ectodomain cleavage [79], (iii) inhibition of HER-2 kinase signal transduction via ADCC [80], and (iv) inhibition of angiogenesis [81].

In metastatic breast cancer (MBC), trastuzumab monotherapy produces response rates ranging from 12% to 34% with a median duration of disease control of 9 months [82,83]. Preclinical studies have also shown additive or synergistic interactions between trastuzumab and multiple cytotoxic agents, including platinum analogues, taxanes, anthracyclines, vinorelbine, gemcitabine, capecitabine, and cyclophosphamide [84]. The use of trastuzumab combined with chemotherapy can further increase response rates (RR), time to progression (TTP), and overall survival (OS) [85,86].

Encouraged by the highly reproducible antitumor activity of trastuzumab in the metastatic setting, four major international studies with enrollment of over 13,000 women were launched in 2000–2001 to investigate the role of trastuzumab in the adjuvant setting: HERA [87,88], the combined North American trials NSABP-B31 and NCCTG/N9831 [89,90], and BCIRG 006 [91,92]. In 2005, the initial results of these four trials, alongside a smaller Finnish trial [93], demonstrated that adjuvant trastuzumab produced significant benefit in reducing recurrence and mortality. Updated analyses for most of these trials have recently been presented, including that of another small trial PACS-04 [94] which, in contrast, failed to show a benefit for adjuvant trastuzumab for reasons that are not well understood (Table 25.1).

Despite differences in patient population and trial design, including chemotherapy regimen, the timing of trastuzumab initiation, and the schedule and duration of trastuzumab administration, remarkly consistent results were reported across these studies: a 33%–58% reduction in the recurrence rate and a 30% reduction in mortality. This degree of benefit in early breast cancer is the largest reported since the introduction of tamoxifen for estrogen receptors positive (ER+) disease.

There are many other novel drugs being developed for use following failure of trastuzumab, either because of *de novo* or acquired resistance. A particular focus of research is in the dual inhibition of EGFR and HER-2, with promising results using drugs such as lapatinib, pertuzumab, and HKI-272.

Proliferative Biomarkers

A high proliferative rate is associated with poor breast cancer survival in untreated patients, but it is also associated with favorable response to chemotherapy [95]. It is still unknown whether some of these proliferative biomarkers also predict better response to specific chemotherapy regimens. Although many proliferative biomarkers are under investigation, none of these are regularly used in clinical practice. The main obstacles lie in the (i) poor standardization of detection methods, (ii) vaguely defined cutoff values, and (iii) requirement for fresh-frozen tissue.

Measurement of Cells in S Phase

Most studies that used fresh or frozen material with sufficient number of patients found an association between S phase fraction and unfavorable prognosis. But several studies done to assess the prognostic value of DNA flow cytometry lacked standardized procedures, sufficient power, and predefined cutoff values. Moreover, the study populations were not controlled for adjuvant treatment. There is also high intratumor heterogeneity of the S phase fraction [95]. Therefore, it cannot be recommended for routine prognostic assessment [96]. Another disadvantage of this method is that it requires a large quantity of tumor material making it inappropriate for smaller tumors identified through mammographic screening.

^{3}H-Thymidine Labeling Index

^{3}H-thymidine labeling index (TLI) was one of the first markers of proliferation used in breast cancer. The number of cells undergoing DNA synthesis is measured by ^{3}H-thymidine uptake using autoradiography for visualization. TLI represents the ratio between the number

Table 25.1 Methods of HER-2 Screening in Different Adjuvant Trastuzumab Trials

	HERA [88, 89]	B-31/N9831 [90, 91]	BCIRG 006 [92, 93]	FinHer [94]
Local Laboratory	IHC/FISH 2+/3+ or FISH+ sent to reference lab	IHC/FISH 3+ or FISH+ sent to reference lab	No role	IHC (any)2+/3+ sent to reference lab
Reference Laboratory	IHC for 3+FISH for 2+Repeat FISH for FISH+	IHC for 3+ FISH for 2+	FISH for all samples	CISH for all samples

of labeled and all counted cells. A similar approach uses IHC technique and halogenated analogue bromodeoxyuridine (BrdU). Limitations of these techniques are the requirement for fresh frozen tissue, the time required to complete the assay, and the use of radioactive tracers [95].

TLI has never been accepted as a standard prognostic marker, although several retrospective studies have shown that it is correlated with poor clinical outcome in terms of relapse-free survival, disease-free survival, distant disease-free survival, and overall survival. Similarly, despite the fact that prospective clinical trials confirmed a benefit for TLI in chemotherapy prediction, it has not been widely adopted in clinical practice [96].

Thymidine Kinase

Thymidine kinase (TK) activity is measured by radioenzymatic assay. TK is an enzyme that catalyses the phosphorylation of deoxythymidine to deoxythymidine monophosphate. Its activity is highest in G1-S translation checkpoint and then declines rapidly in the G2 phase of the cell cycle. In breast cancer, the fetal isoform of TK is present in high levels in the cytoplasm and is cell-cycle regulated [95]. Conflicting data exist for its predictive role, and therefore, TK should not be used for this purpose in clinical practice [96].

Ki67

Ki67 is a nuclear antigen present in mid G1, S, G2, and the entire M phase of the cell cycle, characterized by immunohistochemistry. Although it has been extensively studied, its precise cellular function is still unknown. Most of the studies that have investigated the prognostic significance of Ki67 have shown that overexpression of Ki67 correlates with poor metastases-free survival and overall survival. As with all biomarkers with continuous values, the cutoff value differentiating high Ki67 from low Ki67 tumors is somewhat arbitrary. It has been suggested that the optimal cutoff value for Ki67 is 15% [97]. Nevertheless, routine use of this marker in clinical practice is not recommended because of a lack of adequate reproducibility [96].

MIB1

Similar to Ki67, MIB1 is a nuclear antigen that can be labeled using immunohistochemistry. It can be performed on formalin-fixed and paraffin-embedded blocks. A good correlation between Ki67 and MIB1 assessment has been demonstrated. However, MIB1 is not recommended yet for use in routine clinical practice [95].

Cyclin A

Cyclins are proteins that regulate the cell cycle. Consequently, their concentrations rise and fall throughout the cell cycle, as they are synthesized and degraded at specific stages of the cell cycle. Cyclin A is expressed in the late S, G2, and M phases of the cell cycle and has been associated with poor prognosis in some studies. Other studies have failed to confirm this observation, mainly because of the lack of consensus concerning methodology and cutoff values [97].

Cyclin E

Cyclin E regulates G1 phase progression and entry into S phase. There are two different proteins, cyclin E1 and cyclin E2, that are coded by two different genes with 47% homology. Several splice variants of cyclin E1 not present in normal cells have been identified and these seem to stimulate cells to progress through the cell cycles more than wild-type cyclin E1. The most frequently used determination method for cyclin E1 is IHC, although western blot allows for assessment of total as well as isoform-specific expression. Cyclin E mRNA can also be measured by reverse transcription polymerase chain reaction (RT-PCR).

Elevated levels of both cyclin Es are more frequently found in ER negative tumors. Many retrospective studies have demonstrated an association between high levels of cyclin E and an increased risk of breast cancer-related death, although this finding has not been seen across all the trials. Furthermore, lack of standardization regarding evaluation methods and scoring systems limits its use as a prognostic factor. There is even less evidence for the use of cyclin E as a predictive tool [95].

Cyclin D1

The family of cyclin D consists of at least three different cyclins that regulate progression into G1 phase. The gene *CCND1* that codes for the cyclin D1 protein is located on chromosome 11q13. The function of cyclin D1 is to bind to cyclin-dependent kinases (Cdks) 4 and 6 and phosphorylate downstream proteins such as pRb. These complexes can sequester Cdk inhibitors (p21 and p27). Cyclin D1 acts as a co-factor for ERα in a ligand-independent manner. The concentration of cyclin D1 in the cell is the highest during mid-G1 phase and then gradually declines. Overexpression of cyclin D1 has been found in many human tumors and is believed to promote cell proliferation and differentiation by shortening the G1/S transition. In breast cancer, amplification of the cyclin D1 gene is found in 15% of tumors and overexpression of cyclin D1 at mRNA and protein levels has been observed in up to 50% of tumors, largely those that are ER positive and well differentiated. The most common assessment method of cyclin D1 expression is IHC [95].

The prognostic role of cyclin D1 was evaluated in several retrospective studies. The majority of these studies reported strong correlation between cyclin D1 expression and ER positivity. However, it has not been found to be a strong prognostic marker. One retrospective study suggested that it was predictive for tamoxifen response, as only patients with low and intermediate levels of cyclin D1 benefited from tamoxifen treatment.

p27

p27 is a Cdk inhibitor that acts in the nucleus. It binds to cyclin E-Cdk2 and cyclin A-Cdk2 complexes early in the G1 phase and facilitates the entry of cyclin D1-Cdks complex into the nucleus. It is sequestered through binding to Cdk4 or Cdk6 during periods of cell proliferation. It is mobilized by antiproliferative signals, such as cell-to-cell contact and transforming growth factor β (TGFβ) signaling. It can be assessed using IHC technique with p27-positive cells displaying nuclear staining. The cut-off between low and high p27 expression is set at 50%. In some retrospective analyses of adjuvant clinical trials, low levels of p27 were found to be a predictor for worse clinical outcome, although this was not a universal finding in other trials. Furthermore, in the majority of studies, p27 was positively correlated with ER expression as well as inversely correlated to grade. There is also some evidence that low levels of p27 correlated with HER-2 overexpression and cyclin E expression, whereas high levels correlated with expression of cyclin D1. In *BRCA1/2* mutated tumors, low levels of p27 have also been found. There are preliminary data to suggest that p27 is predictive for greater benefit from chemotherapy, although this requires further validation [95].

Topoisomerase IIα

Topoisomerases II (topo II) are DNA-binding enzymes with nuclease, helicase, and ligase activity. They control and modify topological states of DNA. Two forms of topo II exist: (i) topo IIα is a product of a gene located at 17q21, and (ii) topo IIβ is a product of a gene located at 3p.

While topo IIβ is not cell-cycle dependent and its function remains unknown, the concentration of topo IIα is cell-cycle dependent and is highest in the G2/M transition. Topo IIα produces a double-strand nick on the cleaved DNA that enables the passage of a second DNA through the break and religation of the cleaved one, resulting in reduced DNA supercoiling and twisting. Topo II also regulates chromosome segregation and condensation of newly formed chromosome pairs in dividing cells.

High levels of topo II are found in exponentially growing cells and low levels in quiescent cells. Topo II concentration can be downregulated by the growth of cells in high density and in serum-free conditions. As a result, topo II levels can be regarded as a marker of proliferation [95]. Additionally, different topo II isoforms can differ in their sensitivity to antineoplastic drugs with topo IIα acting as a target for drugs like anthracyclines, epipodophyllotoxins, actinomycin, and mitoxantrone. The topo IIα gene (*TOPO2A*) is located near the *HER-2* oncogene on chromosome 17, although the exact mechanism for the concordance between *TOPO2A* aberration and *HER-2* amplification is not known. Tumors with *HER-2* amplification may have deletion of topo IIα or coamplification of the two genes. Those with *TOPO2A* deletion are associated with low levels of the topo IIα protein, while those with coamplification of *TOPO2A* and *HER-2* appear to be associated with increased sensitivity to anthracyclines. Many studies are currently under way to evaluate the role of topo II in selecting patient subgroups that may or may not benefit from anthracycline therapy. Some evidence also exists to suggest that topo IIα is a prognostic marker for poor disease-free survival and overall survival independent of therapy.

uPA and PAI-1

Central to the process of cancer invasion are matrix-degrading proteases that regulate the tissue scaffold breakdown. Urokinase plasminogen activator (uPA) and plasminogen activator inhibitor (PAI-1), together with receptor for uPA and other inhibitors (PAI-2, PAI-3), are part of the urokinase plasminogen activating system that contributes to this enzymatic degradation system [98]. Apart from invasion, this system has also been shown to be associated with angiogenesis and metastasis.

The IHC assay for detection of uPA and PAI-1 is not specific, and ELISA on a minimum of 300 mg of fresh or frozen breast cancer tissue is recommended [96].

A pooled analysis [99] of uPA/PAI-1 data of breast cancer patients and the first interim report of a prospective trial using uPA and PAI-1 levels to stratify node negative patients [100] confirmed the strong association of uPA and PAI-1 levels with recurrence and survival. Some uncontrolled studies also suggest uPA and PAI-1 may serve as a predictor of sensitivity to hormone therapy or specific types of chemotherapy.

Apart from histological grade, uPA and PAI-1 are the only proliferative markers with level I evidence to support their use as prognostic markers recommended by the *American Society of Clinical Oncology* (ASCO).

GENE EXPRESSION PROFILING

Historically, the classification of breast cancers has been based on histological type, grade, and expression of hormone receptors, but the advent of microarray technology has since demonstrated significant heterogeneity occurring also at the transcriptome level. Through their ability to interrogate thousands of genes simultaneously, microarray studies have allowed for a comprehensive molecular and genetic profiling of tumors. Not only have these studies changed the way in which we have traditionally classified breast cancer, the results of these studies have also yielded molecular signatures with the potential to significantly impact on clinical care by providing a molecular basis for treatment tailoring.

Microarray Technology

Gene expression profiling, using microarray technology, relies on the accurate binding, or hybridization, of DNA strands with their precise complementary copies where one sequence is bound onto a solid-state substrate. These are hybridized to probes of fluorescent cDNAs or genomic sequences from normal or tumor tissue. Through analysis of the intensity of the fluorescence on the microarray chip, a direct comparison of the expression of all genes in normal and tumor cells can be made [101]. At present, there are multiple microarray platforms that use either cDNA-based or oligonucleotide-based microarrays.

cDNA-based microarrays have double-stranded PCR products amplified from expressed sequence tag (EST) clones and then spotted onto glass slides. These have inherent problems with frequent hybridization among homologous genes, alternative splice variants, and antisense RNA [102]. Oligonucleotide-based microarrays are shorter probes with uniform length. Shorter oligonucleotides (25 bases) may be synthesized directly onto a solid matrix using photolithographic technology (Affymetrix), and for longer oligonucleotides (55–70 bases), they may be either deposited by an inkjet process or spotted by a robotic printing process onto glass slides [82].

In the past, there has been much skepticism regarding the reliability and reproducibility of microarray technology. However, the overall reproducibility of the microarray technology has been found to be acceptable, as shown by the MicroArray Quality Control (MAQC) project [103] conducted by the U.S. Food and Drug Administration (FDA). They found that similar changes in gene abundance were detected, despite using different platforms. Furthermore, this reproducibility appears comparable to that of other diagnostic techniques, for example, with immunohistochemical analysis for hormone receptors in breast cancer [104,105].

Molecular Classification of Breast Cancer

One of the most important discoveries stemming directly from microarray studies has been the reclassification of breast cancer into molecular subtypes. This new classification has not only furthered our understanding of tumor biology, but it has also altered the way that physicians and clinical investigators conceptually regard breast cancer—not as one disease, but a collection of several biologically different diseases.

Four main molecular classes of breast cancers have been consistently distinguished by gene expression profiling. Based upon the original classification described by Perou et al., these subtypes are [106] (i) basal-like breast cancers, (ii) HER-2+ breast cancers, (iii) luminal-A breast cancers, and (iv) luminal-B breast cancers.

In the basal-like subtype, there is a high expression of basal cytokeratins 5/6 and 17 and proliferation-related genes, as well as laminin and fatty-acid binding protein 7. In the HER-2+ subtype, there is a high expression of genes in the erbb2 amplicon, such as GRB7. The luminal cancers are ER-positive. Luminal A is characterized by a higher expression of ER, GATA3, and X-box binding protein trefoil factor 3, hepatocyte nuclear factor 3 alpha, and LIV-1. Luminal B cancers are generally characterized by a lower expression of luminal-specific genes.

Beyond differing gene expression profiles, these molecular subtypes appear to have distinct clinical outcomes and responses to therapy that seem reproducible from one study to the next. The basal-like and HER-2+ subtypes are more aggressive with a higher proportion of Major Gene Expression Signatures *TP53* mutations [107,108] and a markedly higher likelihood of being grade III ($P < 0.0001$, and $P = 0.0002$) than luminal A tumors. Despite a poorer prognosis, they tend to respond better to chemotherapy including higher pathologic complete response rates after neo-adjuvant therapy [109].

On the other hand, fewer than 20% of luminal subtypes have mutations in *TP53* and these tumors are often grade I [110]. They tend to be more sensitive to endocrine therapy, less responsive to conventional chemotherapy, and demonstrate overall clinical outcomes.

Despite differences in testing platforms, the collective results of these studies suggest that ER expression is the most important discriminator. Beyond ER expression, distinct expression patterns can also be differentiated, perhaps reflecting distinct cell types of origin [91–93]. However, this molecular classification is not without its inherent limitations, with up to 30% of breast cancers not falling into any of the four molecular categories [111]. Exactly how many true molecular subclasses of breast cancer exist remains uncertain, and it is plausible that the molecular classification will evolve with new technological platforms, with the availability of larger data sets, as well as with improved understanding of tumor biology.

Gene Expression Signatures to Predict Prognosis

Traditional prognostic factors based on clinical and pathological variables are unable to fully capture the heterogeneity of breast cancer. Guidelines like the

National Comprehensive Cancer Network (NCCN) [112] used in the United States and the International St. Gallen Expert Consensus [13] used in Europe to guide treatment decisions take into account relapse risk based traditional anatomical and pathological assesment. These guidelines cannot accommodate for the substantial variability that can exist between patients with similar stages and grades of disease. Using microarray technology, several independent groups have conducted comprehensive gene expression profiling studies with the aim of improving on traditional prognostic markers used in the clinic.

Using the top-down approach, where gene expression data are correlated with clinical outcome without a prior biological assumption, a group of researchers from Amsterdam identified a 70-gene prognostic signature, using the Agilent platform, in a series of 78 systemically untreated node-negative breast cancer patients under 55 years of age [113]. This signature included mainly genes involved in the cell cycle, invasion, metastasis, angiogenesis, and signal transduction. Validated on a larger set of 295 young patients [114], including both node-negative and node-positive disease as well as treated and untreated patients, the 70-gene signature was found to be the strongest predictor for distant metastases-free survival, independent of adjuvant treatment, tumor size, histological grade, and age.

Using the same top-down approach, another group in Rotterdam identified a 76-gene signature [115], using the Affymetrix technology which considered ER-positive patients separately from ER-negative patients. These 76 genes were mainly associated with cell cycle and cell death, DNA replication and repair, and immune response. In a training set of 115 patients and a multicentric validation set of 180 patients [116], they were able to demonstrate comparable discriminative power in predicting the development of distant metastases in untreated patients in all age groups with node-negative breast cancer.

Both the Amsterdam and Rotterdam gene signatures have been independently validated by TRANSBIG, the translational research network of the Breast International Group (BIG) [117,118]. Despite having only three genes in common, both signatures were able to outperform the best validated tools to assess clinical risk. In particular, these signatures were both superior in correctly identifying the low-risk patients but were limited in identifying the high-risk patients, as half of those identified in this category did not, in fact, relapse. This suggests that the highest clinical utility for these molecular signatures may be in potentially reducing overtreatment of low-risk patients.

Also using the top-down approach, Paik et al. [119], in collaboration with Genomic Health Inc., developed a recurrence score (RS) based on 21 genes that appeared to accurately predict the likelihood of distant recurrence in tamoxifen-treated patients with node-negative, ER-positive breast cancer. A final panel of 16 cancer-related genes and 5 reference genes forms the basis for the Oncotype DX^{TM} Breast Cancer Assay.

The RS classifies patients into three risk groups, based on cutoff points from the results of the NSABP trial B-20: high risk of recurrence is assigned if RS >31; intermediate risk if RS is 18–30; and low risk if RS <18. Retrospective validation of this predictor in 675 archival samples of the NSABP trial B-14 [120] showed that the RS was significantly correlated with distant recurrence, relapse-free interval, and overall survival, independent of age and tumor size.

In using a different approach that is hypothesis-driven, otherwise known as the bottom-up approach, Sotiriou et al. [121] looked at whether gene expression patterns associated with histologic grade could improve prognostic capabilities especially within the class of intermediate grade tumor. Accounting for 30%–60% of all breast cancers, these intermediate grade tumors display the most heterogeneity in both phenotype and outcome [122].

Of the unique 97 genes that formed the gene-expression grade index (GGI), most were associated with cell-cycle progression and differentiation. These genes were differentially expressed between low-grade and high-grade breast tumors, without a distinct gene-expression pattern to distinguish the intermediate group. Instead, the intermediate tumors showed expression patterns and clinical outcomes matching those of either low-grade or high-grade cases. The GGI, therefore, could potentially improve treatment decision making for these otherwise problematic patients with intermediate grade by reclassifying them into two distinct and clinically relevant subtypes.

In an examination of genomic grade with ER status, ER-negative tumors with poor clinical outcome were found to be mainly associated with high GGI, although ER-positive tumors were more heterogeneous with a mixture of GGI levels [123]. Thus, these two variables are not entirely independent from each other; with tumor genomic grading capable of providing an extra level of information when stratifying the ER-positive group.

Other prognostic signatures derived from the bottom-up approach include the wound response signature [124], mutant/wild p53 signature [102], invasive gene signature (IGS) [125], and the cancer stem cell signature [103]. These and other prognostic signatures, whether derived with the bottom-up or top-down approach, have only a few genes in common but seem to offer similar prognostic information, with proliferation-related genes being the major driving force. Furthermore, it appears that the prognostic power of many of these signatures is limited to ER-positive patients and is less informative for ER-negative disease (Table 25.2) [126].

Gene Expression Signatures to Predict Site of Recurrence

An interesting theory of breast cancer metastases is that tumors are genetically determined to recur in specific organ sites, and gene expression signatures may be able to serve as site-specific predictors. Massagué and colleagues [127–129], by developing progeny cell

Table 25.2 Major Gene Expression Signatures

	70-Gene Signature. van de Vijver et al. [114]	76-Gene Signature. Wang et al. [115]	21-Gene Signature OncotypeDM™. Paik et al. [119,120]	Wound Response Signature. Chang et al. [124]	Gene Expression Grade Index (GGI). Sotiriou et al. [121]
Tissue	Fresh frozen	Fresh frozen	Paraffin-embedded	Fibroblast cultures	Fresh Frozen
Platform	RNA expression Agilent	RNA expression Affymetrix	RNA expression Genomic Health RT-PCR	RNA expression cDNA "home-made"	RNA expression Affymetrix
No. genes No. tumor samples Training Validation	70 78 295	76 115 171	16 (+5 ref genes) 449 668	512 50 fibroblast cultures 295	97 64 125
ER expression Training Validation	Majority 77%	70% 75%	All	n/a Mixed	all majority
Ax LN involved Training Validation	None Mixed	None None	None	n/a Mixed	None None
Adjuvant Rx H = hormones C = chemotherapyv Training Validation	None H 7%; C 31%; Both 7%	None None	None	n/a H 7%; C 31% Both 7%	H only None
Risk Categories	Good signature 39% Poor signature 61%	Good signature 34% Poor signature 64%	Low risk 51% Intermediate risk 22% High risk 27%	Activated (A) 43% Quiescent (Q) 57%	Histo GGI G2 low / high — n = 216 124 / 92 *(all data sets)*
	HR distant mets (95% CI) Poor vs good 5.1 (2.9–9.0) ($p < 0.0001$) 10y *DMFS Good 85.2% (+/− 4.3%) Poor 50.6% (+/− 4.5%)	HR for 5y *DMFS (95% CI) Good vs poor 5.67 (2.59–12.4) ($p < 0.0001$) HR for 5y **OS (95% CI) Good vs poor 8.62 (2.57–27.9) ($p < 0.0001$)	10y ***DR (95% CI) Low 6.8% (4.0–9.6) Int 14.3% (8.3–20.3) High 30.5% (23.6–37.4) ($p < 0.0001$)	HR for *DMFS (95% CI) 7.25 (1.75–30.0) ($p = 0.006$) 10y *DMFS Activated 51% Quiescent 75%	HR for recurrence in histo grade 2 tumors (95% CI) High vs Low GGI 3.61 (2.25–5.78) ($p < 0.001$)

*DMFS—distant metastases free survival.
**OS—overall survival.
***DR—distant recurrence.

lines of MDA-MB-231 with enhanced ability to metastasize to either bone or lung in immunocompromised mice, identified two different gene sets: the bone gene set and the lung gene set. When applying the bone gene set to a cohort of human breast tumors that eventually developed metastases, they could distinguish tumors that preferentially metastasized to bone from those that preferentially metastasized elsewhere. Similarly, by applying the lung gene set to the same data set used by van't Veer and colleagues [94], they were able to identify a subgroup of patients with worse lung-metastasis-free survival. The site-specific predictive potential of the gene sets was independent from the ER status and the 70-gene signature.

Gene Expression Signatures to Predict Response to Therapy

Improved prognostic tools will help to better identify those patients needing treatment, but there still remains the challenge of knowing which therapy is best to use for the individual breast cancer patient. Only a proportion of patients will respond to any given treatment, but unfortunately, many will experience the adverse side effects. Currently, only ER and HER-2 are used in clinical practice as predictive markers, for the selection of patients likely to respond to specific therapy, such as hormone therapy and trastuzumab, respectively.

Several investigators have recently applied microarray technology to identify gene expression signatures that could predict for drug sensitivity. In predicting for endocrine therapy resistance, several studies have been performed, including the study with 44 genes by Jansen et al. [130] and one using the expression ratio of two genes—*homeobox B13* and *IL17BR*, which could predict for disease survival with 80% accuracy [131–133] when treated with adjuvant tamoxifen. Other endocrine sensitivity tests include the estradiol-induced genes by Oh et al. [134] and the 200-gene signature which predicts for recurrence-free survival after 5 years of tamoxifen therapy [135]. The recurrence score (RS), previously discussed, is also often regarded as a predictive tool for endocrine therapy, rather than a pure prognostic tool, because the NSABP B14 and B20 trials enrolled patients who were treated with tamoxifen [99,100].

In looking at chemotherapy response, fewer studies have been reported so far because these studies ideally require prospective sample collection. Nonetheless, several groups have identified genes associated with response to chemotherapy [130–144], with the majority of the studies using pathologic response rate (pCR) as a surrogate marker for long-term benefit after neoadjuvant chemotherapy. The largest of these studies generated a 30-gene predictor [145] from 82 patients treated with neoadjuvant weekly paclitaxel and 5-fluorouracil, doxorubicin, cyclophosphamide (T/FAC) chemotherapy. In cross-validation of 51 independent cases, this 30-gene predictor had high sensitivity in being able to identify 12/13 (92%) patients with pCR, and high negative predictive value in being able to identify 27/28 (96%) patients with residual disease. Notably, they also showed that their classifier had predictive accuracy similar to that of ER and HER-2 amplification.

Using a hypothesis-driven approach, Bild et al. [146] recently identified several expression patterns associated with the deregulation of a variety of oncogenic pathways that could predict response to different therapeutic agents targeting specific deregulated pathways. By interrogating publicly available drug sensitivity data derived from *in vitro* experiments, they also developed multiple classifiers of response to a variety of chemotherapy drugs and showed that a combination of these classifiers could accurately predict response to preoperative multidrug regimen treatments [147].

While the preliminary results are encouraging, the conclusions that can be derived from these predictive studies are limited by the small sample size as well as heterogeneity in endpoints, treatment regimens, patient populations, and statistical analyses [148]. Furthermore, these predictive tests still need independent validation.

The Future of Gene Expression Profiling

There is significant potential for gene expression profiles to aid treatment tailoring of breast cancer patients. Prognostic signatures can differentiate subpopulations based on risk of relapse, and indeed, there is a suggestion that these signatures may be most useful in identifying low-risk patients who potentially can be spared adjuvant chemotherapy. Predictive signatures, on the other hand, can aid in the decision of which therapy to use for each patient, to maximize individual benefit and minimize individual toxicity.

While the results from these first-generation microarray studies are exciting, there is still a long way to go before these molecular tools can enter routine clinical use. Many of these studies are retrospective, and the gene expression data come from archival material of heterogeneous populations. While level 1 evidence is currently lacking, prospective clinical validation has started for two expression-profiling platforms: OncotypeDx RS with the 21-gene signature in TAILORx, and Mammaprint with the 70-gene signature in MINDACT (Table 25.3).

The TAILORx [Trial Assigning IndividuaLized Options for Treatment (Rx)] trial is a large, randomized prospective study designed to evaluate whether women with node-negative, ER-positive breast cancer need chemotherapy based on the recurrence score (RS). Patients with an RS less than 11 (low risk) will be given only hormonal therapy. An RS more than 25 (high risk) will mean that patients receive chemotherapy in addition to hormone therapy. Patients with RS 11–25 (intermediate risk) will be randomly assigned to receive hormone therapy or chemotherapy followed by hormone therapy. This trial is conducted under the auspices of the U.S. Intergroup and is expected to accrue over 10,000 patients.

The MINDACT (Microarray In Node negative Disease may Avoid ChemoTherapy) Trial is an international prospective, randomized study assessing the potential added

Table 25.3 Features of MINDACT and TAILORx Trials

	MINDACT	TAILORx
Groups/Networks	EORTC, BIG, TRANSBIG	US Intergroup
Population Axillary nodes	Negative	Negative
ER status	ER +/−	ER+
Assay	70 gene Mammaprint^R	21 gene OncotypeDX^{TM}
Tissue	Fresh Frozen	Formalin Fixed Paraffin Embedded
Number	6,000	10,500
Randomized group	Discordant risk	RS 11–25
Randomization	Treatment decision based on clinical vs genomic risk	Treat with hormones +/− chemotherapy
Nonrandomized group	Both low risk (13%): HT or nil Both high risk (55%): CT +/− HT	RS < 11: HT RS > 25 : CT + HT

value of the 70-gene signature classifier to the commonly used clinicopathologic criteria for selecting node-negative breast cancer patients for adjuvant chemotherapy. Through its hypothesis that the 70-gene signature will be able to better select appropriate patients for adjuvant treatment, the benefit would be best seen in patients with good prognostic signatures spared from unnecessary chemotherapy.

Approximately 6000 node-negative patients will have their risk assessment made by using common clinicopathological factors (through a modified version of Adjuvant OnLine) and by the 70-gene signature. Those patients who are classified as high risk by both methods will be offered chemotherapy; those classified as low risk by both methods will not be offered chemotherapy; the discordant group, an estimated 33% (1900 patients) will be randomized between the two methods and will receive or not receive chemotherapy according to the result of the assigned method. This trial is being conducted within the Breast International Group (BIG) network under the sponsorship of the EORTC.

Indeed the technological know-how in this field is rapidly evolving, but the criterion for level 1 evidence must not be uncompromised. As a result, the results of TAILORx and MINDACT are eagerly awaited before routine clinical use. However, it is probable that even with positive validation in large prospective trials, it is unlikely that gene expression profiles will altogether replace existing clinicopathological guidelines, but rather they will become part of an integrative decision-making model based on multiple levels and sources of prognostic data. The best use of these gene-expression signatures may actually be in directing treatment decisions when clinical risk parameters for an individual patient are equivocal. Until there is more certainty regarding the clinical utility of this new molecular technology, treatment guidelines such as those of St. Gallen and the NCCN, having withstood both the test of evidence and time, will continue to be used widely.

CONCLUSION

Breast cancer is a clinically heterogeneous disease, and for many years, this heterogeneity was explained by the differing histopathological characteristics identified mainly by the microscope. Current practice guidelines based on histopathology have been important as risk stratification tools for therapy selection, but these guidelines are limited in the tailoring of treatment for the individual patient. Indeed, it has long been observed that among patients with anatomic and pathological risk profiles, there can be substantial variability in both the natural history and response to treatment.

Novel molecular technologies and a better understanding of the tumor biology of breast cancer have resulted in significant advances in recent years. Previous emphasis on refining the traditional histopathologic criteria, on developing indices of proliferation (S-Phase fraction, Ki67), on understanding genomic instability (e.g., DNA ploidy), and on analyzing single gene expression profiles (*p53* expression) have shifted dramatically to molecular profiling, reflecting the expression of many thousands of genes.

With new knowledge, new challenges emerge. The excitement that comes with each new biomarker must be matched by the scientific rigor of prospective validation with each biomarker being assessed for clinical and economic utility. The challenge also remains how to best integrate multiple new sources and levels of prognostic data with the existing histopathological model. Therefore, it should be with cautious optimism that we move forward in this era of unraveling the mystery behind the many molecules and mutations of this disease.

REFERENCES

1. Pestalozzi BC, Luporsi-Gely E, Jost LM, et al. ESMO Guidelines Task Force. ESMO minimum clinical recommendations for diagnosis, adjuvant treatment and follow-up of primary breast cancer. *Annals of Oncology*. 2005;16(Suppl 1):7–9.
2. Hanby AM, Hughes TA. In situ and invasive lobular neoplasia of the breast. *Histopathology*. 2008;52:58–66.
3. Moriya T, Kasajima A, Ishida K, et al. New trends of immunohistochemistry for making differential diagnosis of breast lesions. *Medical Molecular Morphology*. 2006;39:8–13.
4. Irvine T, Fentiman IS. Biology and treatment of ductal carcinoma in situ. *Expert Review of Anticancer Therapy*. 2007;7:135–145.

5. Patani N, Cutuli B, Mokbel K. Current management of DCIS: a review. *Breast Cancer Research and Treatment.* 2007;(epub ahead of print)
6. Fisher B, Digman J, Wolmark N, et al. Tamoxifen in treatment of intraductal breast cancer: National Surgical Adjuvant Breast and Bowel Project B-24 randomised controlled trial. *Lancet.* 1999;252:1993–2000.
7. Hoda SA, Hoda RS, Merlin S, et al. Issues relating to lymphovascular invasion in breast carcinoma. *Advances in Anatomic Pathology.* 2006;13:308–315.
8. Mohammed RA, Martin SG, Gill MS, et al. Improved methods of detection of lymphovascular invasion demonstrate that it is the predominant method of vascular invasion in breast cancer and has important clinical consequences. *The American Journal of Surgical Pathology.* 2007; 31:1825–1833.
9. McCready DR, Chapman JA, Hanna WM, et al. Factors affecting distant disease-free survival for primary invasive breast cancer: Use of a log-normal survival model. *Ann Surg Oncol.* 2000;7: 416–426.
10. Komaki K, Sano N, Tangoku A. Problems in histological grading of malignancy and its clinical significance in patients with operable breast cancer. *Breast Cancer.* 2006;13:249–253.
11. van Diest PJ, van der Wall E, Baak JPA. Prognostic value of proliferation in invasive breast cancer: A review. *Journal of Clinical Pathology.* 2004;57:675–681.
12. Sobin LH, Wittekinf C. *TNM Classification of Malignant Tumour.* New York: UICC, Wiley-Liss; 2002.
13. Goldhirsch A, Wood WC, Gelber RD, et al. Progress and promise: Highlights of the international expert consensus on the primary therapy of early breast cancer 2007. *Annals of Oncology.* 2007;18: 1133–1144.
14. Parker MG. Steroid and related receptors. *Current Opinion in Cell Biology.* 1993;5:499–504.
15. Mosselman S, Polman J, Kijkema R. Identification and characterization of a novel human estrogen receptor. *FEBS Letters.* 1996;392:49–53.
16. Bocchinfuso WP, Korach KS. Mammary gland development and tumorigenesis in estrogen receptor knockout mice. *Journal of Mammary Gland Biology and Neoplasia.* 1997;2:323–334.
17. Hopp TA, Weiss H, Parra I, et al. Low levels of estrogen receptor beta protein predict resistance to tamoxifen therapy in breast cancer. *Clin Cancer Res.* 2004;10: 7490–7499.
18. Nemere I, Pietras RJ, Blackmore PF. Membrane receptors for steroid hormones: Signal transduction and physiological significance. *Journal of Cellular Biochemistry.* 2003;88:438–445.
19. Horwitz KB, McGuire WL. Estrogen control of progesterone receptor in human breast cancer. *The Journal of Biological Chemistry.* 1978;253:2223–2228.
20. Beato M. Gene regulation by steroid hormones. *Cell.* 1989; 56:335–344.
21. Dobrzycka KM, Townson SM, Jiang S, et al. Estrogen receptor corepressors: A role in human breast cancer? *Endocrine-Related Cancer.* 2003;10:517–536.
22. Smith CL, O'Malley BW. Coregulator function: A key to understanding tissue specificity of selective receptor modulators. *Endocrine Reviews.* 2004;25:45–71.
23. Smith CL, Nawaz Z, O'Malley BW. Coactivator and corepressor regulation of the agonist/antagonist activity of the mixed antiestrogen, 4-hydroxytamoxifen. *Mol Endocrinol.* 1997;11:657–666.
24. Shang Y, Brown M. Molecular determinants for the tissue specificity of SERMs. *Science.* 2002;295:2465–2468.
25. Xu J, Li Q. Review of the in vivo functions of the p160 steroid receptor coactivator family. *Mol Endocrinol.* 2003;17:1681–1692.
26. Schiff R, Massarweh S, Shou J, et al. CK. Breast cancer endocrine resistance: How growth factor signalling and estrogen receptor coregulators modulate response. *Clin Cancer Res.* 2003;9: 447S–454S.
27. Nicholson RI, McClelland RA, Robertson JF, et al. Involvement of steroid hormone and growth factor cross-talk in endocrine response in breast cancer. *Endocrine-Related Cancer.* 1999;6: 373–387.
28. Huynh H, Yang X, Pollak M. Estradiol and antiestrogens regulate a growth inhibitory insulin-like growth factor binding protein 3 autocrine loop in human breast cancer cells. *The Journal of Biological Chemistry.* 1996;271:1016–1021.
29. Lee AV, Cui X, Oesterreich S. Cross-talk among estrogen receptor, epidermal growth factor, and insulin-like growth factor signalling in breast cancer. *Clin Cancer Res.* 2001;7:4429s–4435s.
30. Klinge CM. Estrogen receptor interaction with estrogen response elements. *Nucleic Acids Research.* 2001;29:2905–2919.
31. Levin ER. Cellular functions of the plasma membrane estrogen receptor. *Trends in Endocrinology and Metabolism: TEM.* 1999;10: 374–377.
32. Razandi M, Pedram A, Greene GL, et al. Cell membrane and nuclear estrogen receptors (ERs) originate from a single transcript: studies of ERalpha and ERbeta expressed in Chinese hamster ovary cells. *Mol Endocrinol.* 1999;13:307–319.
33. Pappas TC, Gametchu B, Watson CS. Membrane estrogen receptors identified by multiple antibody labeling and impeded-ligand binding. *FASEB J.* 1995;9:404–410.
34. Kahlert S, Nuedling S, van Eickels H, et al. Estrogen receptor alpha rapidly activates the IGF-1 receptor pathway. *The Journal of Biological Chemistry.* 2000;27:18447–18453.
35. Simoncini T, Hafezi-Moghadam A, Brazil DP, et al. Interaction of oestrogen receptor with the regulatory subunit of phosphatidylinositol-3-OH kinase. *Nature.* 2000;407: 538–541.
36. Nadji M, Gomez-Fernandez P, Ganjei-Azar P, et al. Immunohistochemistry of oestrogen and progesterone receptors reconsidered: Experience with 5,993 breast cancers. *American Journal of Clinical Pathology.* 2005;123:21–27.
37. Jordan VC, Dowse LJ. Tamoxifen as an anti-tumor agent: Effect on oestrogen binding. *J Endocrinol.* 1976;68:297–303.
38. Buzdar AU, Hortobagyi G. Update on endocrine therapy for breast cancer. *Clin Cancer Res.* 1998;4:527–534.
39. Ravdin PM, Green S, Dorr TM, et al. Prognostic significance of progesterone receptor levels in estrogen receptor-positive patients with metastatic breast cancer treated with tamoxifen: Results of a prospective Southwestern Oncology Group study. *J Clin Oncol.* 1992;10:1284–1291.
40. Bezwoda WR, Esser JD, Dansey R, et al. The value of estrogen and progesterone receptor determinations in advanced breast cancer. Estrogen receptor level but not progesterone receptor level correlates with response to tamoxifen. *Cancer.* 1991;68: 867–872.
41. Harvey JM, Clark GM, Osborne CK, et al. Estrogen receptor status by immunohistochemistry is superior to the ligand-binding assay for predicting response to adjuvant endocrine therapy in breast cancer. *J Clin Oncol.* 1999;17:1474–1481.
42. Early Breast Cancer Trialists' Collaborative Group. Tamoxifen for early breast cancer; An overview of the randomized trials. *Lancet.* 1998;351:1451–1467.
43. Ellis MJ, Rosen E, Dressman H, et al. Neoadjuvant comparisons of aromatase inhibitors and tamoxifen: Pretreatment determinants of response and on-treatment effect. *The Journal of Steroid Biochemistry and Molecular Biology.* 2003;86:301–307.
44. Milano A, Dal Lago L, Sotiriou C, et al. What clinicians need to know about antiestrogen resistance in breast cancer therapy? *Eur J Cancer.* 2006;42:2692–2705.
45. Conneely OM, Jericevic BM, Lydon JP. Progesterone receptors in mammary gland development and tumorigenesis. *Journal of Mammary Gland Biology and Neoplasia.* 2003;8:205–214.
46. McGuire WL. Hormone receptors: Their role in predicting prognosis and response to endocrine therapy. *Seminars in Oncology.* 1978;5:428–433.
47. Gross GE, Clark GM, Chamness GC, et al. Multiple progesterone receptor assays in human breast cancer. *Cancer Research.* 1984;44: 836–840.
48. Balleine RL, Earl MJ, Greenberg ML, et al. Absence of progesterone receptor associated with secondary breast cancer in postmenopausal women. *British Journal of Cancer.* 1999;79: 1564–1571.
49. Colditz GA, Rosner BA, Chen WY, et al. Risk factors for breast cancer according to estrogen and progesterone receptor status. *Journal of the National Cancer Institute.* 2004;96:218–228.
50. Potter JD, Cerhan JR, Sellers TA, et al. Progesterone and estrogen receptors and mammary neoplasia in the Iowa Women's Health Study: How many kinds of breast cancer are there? *Cancer Epidemiological Biomarkers and Prevention.* 1995;4:319–326.
51. Bardou VJ, Arpino G, Elledge RM, et al. Progesterone receptor status significantly improves outcome prediction over estrogen

receptor status alone for adjuvant endocrine therapy in two large breast cancer databases. *J Clin Oncol.* 2003;21:1973–1979.

52. Ferno M, Stal O, Baldetorp B, et al. Results of two or five years of adjuvant tamoxifen correlated to steroid receptor and S-phase levels: South Sweden Breast Cancer Group, and South-East Sweden Breast Cancer Group. *Breast Cancer Research and Treatment.* 2000;59:69–76.
53. Lamp PJ, Pujol P, Thezenas S, et al. Progesterone receptor quantification as a strong prognostic determinant in postmenopausal breast cancer women under tamoxifen therapy. *Breast Cancer Research and Treatment.* 2002;76:65–71.
54. Bloom ND, Robin EH, Schreibman B, et al. The role of progesterone receptor in the management of advanced breast cancer. *Cancer.* 1980;45:2992–2997.
55. Osborne CK, Yochmowitz MG, Knight WA, et al. The value of estrogen and progesterone receptors in the treatment of breast cancer. *Cancer.* 1980;46:2884–2888.
56. Ellis MJ, Coop A, Singh B, et al. Letrozole is more effective neoadjuvant endocrine therapy than tamoxifen for ErbB-1 and/or ErbB-2 positive primary breast cancer: Evidence from a phase III randomized trial. *J Clin Oncol.* 2001;19:3808–3816.
57. Dowsett M, Cuzick J, Wale C, et al. Retrospective analysis of time to recurrence in the ATAC trial according to hormone receptor status: A hypothesis-generating study. *J Clin Oncol.* 2005;23:7512–7517.
58. Viale G, Regan MM, Maiorano E, et al. Prognostic and predictive value of centrally reviewed expression of estrogen and progesterone receptors in a randomized trial comparing letrozole and tamoxifen adjuvant therapy for postmenopausal early breast cancer: BIG 1-98. *J Clin Oncol.* 2007;25:3846–3852.
59. Coombes RC, Kilburn LS, Snowdon CF, et al. Intergroup Exemestane Study. Survival and safety of exemestane versus tamoxifen after 2–3 years' tamoxifen treatment (Intergroup Exemestane Study): A randomized controlled trial. *Lancet.* 2007;369:559–570.
60. Dowsett M, Allred DC, on behalf of the TransATAC Investigators. Relationship between quantitative ER and PgR expression and HER status with recurrence in the ATAC trial. *Breast Cancer Research and Treatment.* 2006;100(suppl 1; abstr 48):S21.
61. Goss PE, Ingle JN, Martino S, et al. National Cancer Institute of Canada Clinical Trials Group MA.17. Efficacy of letrozole extended adjuvant therapy according to estrogen receptor and progesterone receptor status on the primary tumor. *J Clin Oncol.* 2007;25:2006–2011.
62. Yarden, Y. The EGFR family and its ligands in human cancer: Signalling mechanisms and therapeutic opportunities. *Eur J Cancer.* 2001;37:S3–S8.
63. Schlessinger J. Common and distinct elements in cellular signalling via EGF and FGF receptors. *Science.* 2004;306:1506–1507.
64. Yarden Y, Sliwkowski M. Untangling the ErbB signalling network. *Nat Rev Mol Cell Biol.* 2001;2:127–137.
65. Schechter AL, Stern DF, Vaidyanathan L, et al. The neu oncogene: An erb-B-related gene encoding a 185,000-M, tumour antigen. *Nature.* 1984;312:513–516.
66. King CR, Kraus MH, Aaronsen SA. Amplification of a novel v-erb-related gene in a human mammary carcinoma. *Science.* 1985;229:974–976.
67. Depowski P, Mulford D, Minot P. Comparative analysis of HER-2/neu protein overexpression in breast cancer using paraffin-embedded tissue and cytologic specimens. *Mod Pathol.* 2002; 15:70A.
68. Joensuu H, Isola J, Lundin M, et al. Amplification of erbB2 and erbB2 expression are superior to estrogen receptor status as risk factors for distant recurrence in pT1N0Mo breast cancer: A nationwide population-based study. *Clin Cancer Res.* 2003;9: 923–930.
69. Press MF, Bernstein PA, Thomas LF, et al. HER-2/neu gene amplification characterized by fluorescence in situ hybridization: Poor prognosis in node-negative breast carcinomas. *J Clin Oncol.* 1997;15:2894–2904.
70. Slamon DJ, Clark GM, Wong SG, et al. Human breast cancer: Correlation of relapse and survival with amplification of the HER-2/neu oncogene. *Science.* 1987;235:177–182.
71. Konecny G, Pauletti G, Pegram M, et al. Quantitative association between HER-2/neu and steroid hormone receptors in hormone receptor-positive primary breast cancer. *Journal of the National Cancer Institute.* 2003;95:142–153.
72. Muss HB, Thor AD, Berry DA, et al. c-erbB-2 expression and response to adjuvant therapy in women with node-positive early breast cancer. *The New England Journal of Medicine.* 1994;330: 1260–1266.
73. Paik S, Bryant J, Park C, et al. erbB-2 and response to doxorubicin in patients with axillary lymph node-positive, hormone receptor-negative breast cancer. *Journal of the National Cancer Institute.* 1998;90:1361–1370.
74. Thor AD, Berry DA, Budman DR, et al. erbB-2, p53, and efficacy of adjuvant therapy in lymph node-positive breast cancer. *Journal of the National Cancer Institute.* 1998;90:1346–1360.
75. Yaziji H, Goldstein L, Barry T, et al. HER-2 testing in breast cancer using parallel tissue-based methods. *JAMA.* 2004;291:1972–1977.
76. Perez EA, Suman VJ, Davidson NE, et al. HER-2 testing by local, central and reference laboratories in specimens from the North Central Cancer Treatment Group N9831 Intergroup Adjuvant Trial. *J Clin Oncol.* 2006;24:3032–3038.
77. Wolff AC, Hammond EH, Schwartz JN, et al. American Society of Clinical Oncology; College of American Pathologists. American Society of Clinical Oncology / College of American Pathologists Guideline Recommendations for Human Epidermal Growth Factor Receptor 2 testing in breast cancer. *J Clin Oncol.* 2007; 25:118–145.
78. Klapper L, Waterman H, Sela M, et al. Tumor-inhibitory antibodies to HER-2/ErbB-2 may act by recruiting c-Cbl and enhancing ubiquitination of HER-2. *Cancer Research.* 2000;60:3384–3388.
79. Molina MA, Codony-Servat J, Albanell J, et al. Trastuzumab (Herceptin), a humanized anti-HER-2 receptor monoclonal antibody, inhibits basal and activated HER-2 ectodomain cleavage in breast cancer cells. *Cancer Research.* 2001;61:4744–4749.
80. Clynes RA, Towers TL, Presta LG, et al. Inhibitory Fc receptors modulate in vivo cytotoxicity against tumor targets. *Nature Medicine.* 2000;6:443–446.
81. Izumi Y, Xu L, di Tomaso E. Tumour biology: Herceptin acts as an anti-angiogenesis cocktail. *Nature.* 2002;416:279–280.
82. Baselga J, Tripathy D, Mendelsohn J, et al. Phase II study of weekly intravenous recombinant humanized anti-p185HER-2 monoclonal antibody in patients with HER-2/neu-overexpressing metastatic breast cancer. *J Clin Oncol.* 1996;14:737–744.
83. Vogel CL, Cobleigh MA, Tripathy D, et al. Efficacy and safety of trastuzumab as a single agent in first-line treatment of HER-2-overexpressing metastatic breast cancer. *J Clin Oncol.* 2002; 20:719–726.
84. Pegram MD, Konecny GE, O'Callaghan C, et al. Rational combinations of trastuzumab with chemotherapeutic drugs used in the treatment of breast cancer. *Journal of the National Cancer Institute.* 2004;96:739–749.
85. Marty M, Cognetti F, Maraninchi D, et al. Randomized phase II trial of the efficacy and safety of trastuzumab combined with docetaxel in patients with human epidermal growth factor receptor 2-positive metastatic breast cancer administered as first-line treatment: The M77001 study group. *J Clin Oncol.* 2005;23: 4265–4274.
86. Slamon DJ, Leyland-Jones B, Shak S, et al. Use of chemotherapy plus a monoclonal antibody against HER2 for metastatic breast cancer that overexpresses HER2. *The New England Journal of Medicine.* 2001;344:783–792.
87. Piccart-Gebhart MJ, Procter M, Leyland-Jones B, et al. Herceptin Adjuvant (HERA) Trial Study Team. Trastuzumab after adjuvant chemotherapy HER2-positive breast cancer. *The New England Journal of Medicine.* 2005;353:1659–1672.
88. Smith I, Procter M, Gelber RD, et al. HERA study team. 2-year follow-up of trastuzumab after adjuvant chemotherapy in HER-2 positive breast cancer: A randomized controlled trial. *Lancet.* 2007;369:29–36.
89. Romond EH, Perez EA, Bryant J, et al. Trastuzumab plus adjuvant chemotherapy for operable HER2-positive breast cancer. *The New England Journal of Medicine.* 2005;353:1673–1684.
90. Perez EA, Romond H, Suman VJ, et al. NCCTG/NSABP. Updated results of the combined analysis of NCCTG N9831 and NSABP B-31 adjuvant chemotherapy with / without trastuzumab in patients with HER2-positive breast cancer. *J Clin Oncol.* 2007;25:512.

91. Slamon D, Eiermann W, Robert N. Phase III randomized trial comparing doxorubicin and cyclophosphamide followed by docetaxel (ACT) with doxorubicin and cyclophosphamide followed by docetaxel and trastuzumab (AC TH) with docetaxel, carboplatin and trastuzumab (TCH) in HER2 positive early breast cancer patients: BCIRG 006 study. *Breast Cancer Res Treat.* 2005; 94(suppl 1):Session 5 abstract 1.
92. Slamon D, Eiermann W, Robert N. BCIRG 006: 2nd interim analysis phase III randomized trial Phase III comparing doxorubicin and cyclophosphamide followed by docetaxel (AC-T) with doxorubicin and cyclophosphamide followed by docetaxel and trastuzumab (AC-TH) with docetaxel, carboplatin and trastuzumab (TCH) in HER2 positive early breast cancer patients. *Breast Cancer Res Treat.* 2006;100:General Session 2, abstract S2.
93. Joensuu H, Kellokumpu-Lehtinen PL, Bono P, et al. FinHer Study Investigators. Adjuvant docetaxel or vinorelbine with or without trastuzumab for breast cancer. *The New England Journal of Medicine.* 2006;354:809–820.
94. Spielmann M, Roche H, Humblet Y. 3-year follow-up of trastuzumab following adjuvant chemotherapy in node positive HER2-positive breast cancer patients: Results of the PACS-04 trial. *San Antonio Breast Cancer Symposium.* 2007;abstract 72.
95. Colozza M, Azambuja E, Cardoso F, et al. Proliferative markers as prognostic and predictive tools in early breast cancer: Where are we now? *Annals of Oncology.* 2005;16:1723–1739.
96. Harris L, Fritsche H, Mennel R, et al. American Society of Clinical Oncology 2007 update of recommendations for the use of tumor markers in breast cancer. *J Clin Oncol.* 2007; 25:5287–5312.
97. Ahlin C, Aaltonen K, Amini RM, et al. Ki67 and cyclin A as prognostic factors in early breast cancer. What are the optimal cut-off values? *Histopathology.* 2007;51:491–498.
98. Stephens RW, Brünner N, Jänicke F, et al. The urokinase plasminogen activator system as a target for prognostic studies in breast cancer. *Breast Cancer Research and Treatment.* 1998; 52:99–111.
99. Look MP, van Putten WLJ, Duffy MJ, et al. Pooled analysis of prognostic impact of urokinase-type plasminogen activator and its inhibitor PAI-1 in 8377 breast cancer patients. *Breast Cancer Research and Treatment.* 1998;52:99–111.
100. Jänicke F, Prechtl A, Thomssen C, et al. German N0 Study Group Randomized adjuvant chemotherapy trial in high-risk, lymph node-negative breast cancer patients identified by urokinase-type plasminogen activator and plasminogen activator inhibitor type 1. *Journal of the National Cancer Institute.* 2001;93:913–920.
101. Barrett JC, Kawasaki K. Microarrays: The use of oligonucleotides and cDNA for the analysis of gene expression. *Drug Discovery Today.* 2003;8:134–141.
102. Cronin M, Pho M, Dutta D, et al. Measurement of gene expression in archival paraffin-embedded tissues: Development and performance of a 92-gene reverse transcriptase-polymerase chain reaction assay. *The American Journal of Pathology.* 2004; 164:35–42.
103. MAQC Consortium. The MicroArray Quality Control (MAQC) project shows inter and intraplatform reproducibility of gene expression measurements. *Nature Biotechnology.* 2006;24:1151–1161.
104. Layfield LJ, Goldstein N, Perkinson KR, et al. Interlaboratory variation in results from immunohistochemical assessment of estrogen receptor status. *The Breast Journal.* 2003;9:257–259.
105. Rhodes A, Jasani B, Barnes DM, et al. Reliability of immunohistochemical demonstration of oestrogen receptors in routine practice: Interlaboratory variance in the sensitivity of detection and evaluation of scoring systems. *Journal of Clinical Pathology.* 2000;53:125–130.
106. Perou CM, Sorlie T, Eisen MB, et al. Molecular portraits of human breast tumors. *Nature.* 200;406:747–752.
107. Sorlie T, Perou CM, Tibshirani R, et al. Gene expression patterns of breast carcinomas distinguish tumor subclasses with clinical implications. *Proc Natl Acad Sci USA.* 2001;98: 10869–10874.
108. Carey LA, Perou CM, Dressler LG. Race and the poor prognosis basal-like breast cancer (BBC) phenotype in the population-based Carolina Breast Cancer Study. *J Clin Oncol.* 2004;suppl: abstract 9510.
109. Rouzier R, Anderson K, Hess KR. Basal and luminal types of breast cancer defined by gene expression patterns respond differently to neoadjuvant chemotherapy. *San Antonio Breast Cancer Symposium.* 2004;abstract 1026.
110. Sotiriou C, Neo SY, McShane LM, et al. Breast cancer classification and prognosis based on gene expression profiles from a population-based study. *Proc Natl Acad Sci USA.* 2003; 100(18):10393–10398.
111. Pusztai L, Mazouni C, Anderson K, et al. Molecular classification of breast cancer: Limitations and potential. *The Oncologist.* 2006;11:868–877.
112. National Comprehensive Cancer Network (NCCN) guidelines. www.nccn.org; 2007.
113. van't Veer LJ, Dai H, van de Vijver MJ, et al. Gene expression profiling predicts clinical outcome of breast cancer. *Nature.* 2002;415:530–536.
114. van de Vijver MJ, He YD, van't Veer LJ, et al. A gene-expression signature as a predictor of survival in breast cancer. *Nature.* 2002;347:1999–2009.
115. Wang Y, Klijn JG, Zhang Y, et al. Gene-expression profiles to predict distant metastases of lymph-node-negative primary breast cancer. *Lancet.* 2005;365:671–679.
116. Foekins JA, Atkins D, Zhang Y, et al. Multicenter validation of a gene expression-based prognostic signature in lymph node-negative primary breast cancer. *J Clin Oncol.* 2006;24:1665–1671.
117. Buyse M, Loi S, van't Veer L, et al. TRANSBIG Consortium. Validation and clinical utility of a 70-gene prognostic signature for women with node-negative breast cancer. *Journal of the National Cancer Institute.* 2006;98:1183–1192.
118. Desmedt C, Piette F, Loi S. Strong time-dependency of the 76-gene prognostic signature for node-negative breast cancer patients in the TRANSBIG multi-centre independent validation series. 2006; Late breaking abstract, Fifth European Breast Cancer Conference.
119. Paik S, Shak S, Tan G, et al. A multigene assay to predict recurrence of tamoxifen-treated, node-negative breast cancer. *The New England Journal of Medicine.* 2004;351:2817–2826.
120. Paik S, Shak S, Tang G. Multi-gene RT-PCR assay for predicting recurrence in node-negative breast cancer patients—NSABP studies B-20 and B-14. *Breast Cancer Research and Treatment.* 2003;82(Suppl 1):S10 abstract 16.
121. Sotiriou C, Wirapati P, Loi S, et al. Gene expression profiling in breast cancer: Understanding the molecular basis of histologic grade to improve prognosis. *Journal of the National Cancer Institute.* 2006;98:262–272.
122. Miller LD, Smeds J, George J, et al. An expression signature for p53 status in human breast cancer predicts mutation status, transcriptional effects, and patient survival. *Proceedings of National Academy of Science USA.* 2005;102:13550–13555.
123. Glinsky GV, Berezovska O, Glinskii AB. Microarray analysis identifies a death-from-cancer signature predicting therapy failure in patients with multiple types of cancer. *J Clin Invest.* 2005;115:1503–1521.
124. Chang HY, Sneddon JB, Alizadeh AA, et al. Gene expression signature of fibroblast serum response predicts human cancer progression: Similarities between tumors and wounds. *PLoS Biology.* 2004;2:E7.
125. Liu R, Wang X, Chen G, et al. The prognostic role of a gene signature from tumorigenic breast-cancer cells. *The New England Journal of Medicine.* 2007;356:217–226.
126. Sotiriou C, Wirapati P, Kunkel S, et al. Integrative meta-analysis of gene-expression profiles in breast cancer: Toward a unified understanding of breast cancer sub-typing and prognosis signatures. *Breast Cancer Research and Treatment.* 2007;106(Suppl 1): S56–57.
127. Kang Y, Siegel PM, Shu W, et al. A multigenic program mediating breast cancer metastasis to bone. *Cancer Cell.* 2003;3:537–549.
128. Minn AJ, Kang Y, Serganova I, et al. Distinct organ-specific metastatic potential of individual breast cancer cells and primary tumour. *J Clin Invest.* 2005;115:44–55.
129. Minn AJ, Gupta GP, Siegel PM, et al. Genes that mediate breast cancer metastasis to lung. *Nature.* 2005;436:518–524.
130. Jansen MP, Foekens J, Dirkzwager-Kiel MM, et al. Molecular classification of tamoxifen-resistant breast carcinomas by gene expression profiling. *J Clin Oncol.* 2005;23:732–740.

131. Ma XJ, Isakoff SJ, Barmettler A, et al. A two-gene expression ratio predicts clinical outcome in breast cancer patients treated with tamoxifen. *Cancer Cell.* 2004;5:607–616.
132. Ma XJ, Ding L, Sgroi DC, et al. The HOXB13:IL17BR expression index is a prognostic factor in early-stage breast cancer. *J Clin Oncol.* 2006;24:4611–4619.
133. Jansen MP, Sieuwerts AM, Look MP, et al. HOXB13-to-IL17BR expression ratio is related with tumor aggressiveness and response to tamoxifen of recurrent breast cancer: A retrospective study. *J Clin Oncol.* 2006;25:662–668.
134. Oh DS, Troester MA, Usary J, et al. Estrogen-regulated genes predict survival in hormone receptor-positive breast cancers. *J Clin Oncol.* 2006;24:1656–1664.
135. Symanns WF, Hatzis C, Sotiriou C, et al. The Microarray-based Endocrine Prediction Collaboration. Ability of a 200-gene endocrine sensitivity index (SET) to predict survival for patients who receive adjuvant endocrine therapy or for untreated patients. *Proceedings of ASCO 2007 Breast Cancer Symposium.* 2007;abstract 25.
136. Folgueira MA, Patrão DF, Barbosa EM, et al. Gene expression profile associated with response to doxorubicin-based therapy in breast cancer. *Clin Cancer Res.* 2005;11:7434–7443.
137. Hannemann J, Oosterkamp HM, Bosch CA, et al. Changes in gene expression associated with response to neoadjuvant chemotherapy in breast cancer. *J Clin Oncol.* 2005;23:3331–3342.
138. Chang JC, Wooten EC, Tsimelzon A, et al. Gene expression profiling for the prediction of therapeutic response to docetaxel in patients with breast cancer. *Lancet.* 2003;362:362–369.
139. Chang JC, Wooten EC, Tsimelzon A, et al. Patterns of resistance and incomplete response to docetaxel by gene expression profiling in breast cancer patients. *J Clin Oncol.* 2005;23:1169–1177.
140. Iwao-Koizumi K, Matoba R, Ueno N, et al. Prediction of docetaxel response in human breast cancer by gene expression profiling. *J Clin Oncol.* 2005;23:422–431.
141. Bertucci F, Finetti P, Rougemont J, et al. Gene expression profiling for molecular characterization of inflammatory breast cancer and prediction of response to chemotherapy. *Cancer Research.* 2004;64:8558–8565.
142. Andre F, Mazouni C, Hortobagyi GN, et al. DNA arrays as predictors of efficacy of adjuvant/neoadjuvant chemotherapy in breast cancer patients: Current data and issues on study design. *Biochimica et Biophysica Acta.* 2006;1766:197–204.
143. Rouzier R, Rajan R, Wagner P, et al. Microtubule-associated protein tau: A marker of paclitaxel sensitivity in breast cancer. *Proc Natl Acad Sci USA.* 2005;102:8315–8320.
144. Gianni L, Zambetti M, Clark K, et al. Gene expression profiles in paraffin-embedded core biopsy tissue predict response to chemotherapy in women with locally advanced breast cancer. *J Clin Oncol.* 2005;23:7265–7277.
145. Hess KR, Anderson K, Symmans WF, et al. Pharmacogenomic predictor of sensitivity to preoperative chemotherapy with paclitaxel and fluorouracil, doxorubicin, and cyclophosphamide in breast cancer. *J Clin Oncol.* 2006;24:4236–4244.
146. Bild A, Yao G, Chang JT, et al. Oncogenic pathway signatures in human cancers as a guide to targeted therapies. *Nature.* 2006;439:353–357.
147. Potti A, Dressman HK, Bild A, et al. Genomic signatures to guide the use of chemotherapeutics. *Nature Medicine.* 2006;12:1294–1300.
148. Sotiriou C, Piccart MJ. Taking gene-expression profiling to the clinic: When will molecular signatures become relevant to patient care? *Nature Reviews. Cancer.* 2007;7:545–553.

Chapter

26

Molecular Basis of Skin Disease

Vesarat Wessagowit ▪ John A. McGrath

SKIN DISEASES AND THEIR IMPACT

Of all the organs of the human body, the skin is the largest—in a 70 kg individual, the skin weighs over 5 kg and covers a surface area approaching 2 m^2. Skin diseases are numerous and common, with the vast majority of work performed by dermatologists in clinical practice being managing common skin conditions, including eczema, psoriasis, acne, tinea, warts, and in some countries skin cancer, and in others, skin infections. Nearly 30% of all consultations with primary care physicians involve skin symptoms or signs and yet fewer than 50% of people seek advice from medical practitioners, with pharmacists being the most common of the other sources of advice.

Skin diseases are not only frequent, but they are very costly [1]. In the United States, spending on skin disease (excluding cosmetic applications) has increased more than 2-fold since the late 1990s, and the medical cost of the top 20 skin conditions exceeds $40 billion US in annual expenditure. The most expensive skin disease categories include skin ulcers, wounds, melanoma, acne, nonmelanoma skin cancer, and atopic dermatitis.

Many skin diseases, although not fatal, can be chronic and cause considerable morbidity; hence, their effects on quality of life can be even more pronounced than other medical conditions. Impact on quality of life measured by willingness to pay for relief from skin conditions is comparable to that of other serious medical conditions. Many symptoms that patients experience are not only burdensome but can be debilitating, both physically and psychologically. Several studies have shown that skin diseases such as psoriasis can reduce physical and mental functioning to levels comparable to those seen in cancer, arthritis, hypertension, heart disease, diabetes, and depression. In children, skin diseases such as atopic dermatitis or urticaria have been shown to have a greater impact on a child's quality of life than conditions such as epilepsy or diabetes mellitus.

The burden of skin disease, given its high prevalence and personal and social cost, is considerable. Prevention of skin disease is consequently an important goal, and in recent years considerable new knowledge has emerged in improving the understanding of molecular and cellular mechanisms relevant to maintaining healthy skin.

MOLECULAR BASIS OF HEALTHY SKIN

A key role of skin is to provide a mechanical barrier against the external environment (Figure 26.1). The cornified cell envelope and the stratum corneum restrict water loss from the skin, while keratinocyte-derived endogenous antibiotics provide innate immune defense against bacteria, viruses, and fungi. Clues to the importance of the skin barrier have recently been highlighted by the findings of very common loss of function mutations in the profilaggrin gene (*FLG*), which encodes a key skin barrier component, filaggrin, expressed in the granular layer of the skin [2;3]. Mutations in *FLG* are the cause of the semidominant condition ichthyosis vulgaris (OMIM146700) and are a major risk factor for the development of atopic dermatitis as well as asthma associated with atopic dermatitis and systemic allergies. These discoveries demonstrate the importance of maintaining an effective skin barrier to minimize and prevent cutaneous inflammation as well as systemic pathology. They also highlight that maintaining an effective functional skin barrier in children may have long-term benefits in reducing atopic asthma and systemic allergies in subsequent years.

Normal skin has also been shown to have a very effective defense system against microbes (Figure 26.2). In the stratum corneum there is an effective chemical barrier maintained by the expression of S100A7 (psoriasin) [4]. This antimicrobial substance is very effective at killing *E. coli*. Subjacent to this in the skin there is another class of antimicrobial peptides such as RNASE7, which

Molecular Pathology © 2009, Elsevier, Inc. All Rights Reserved.

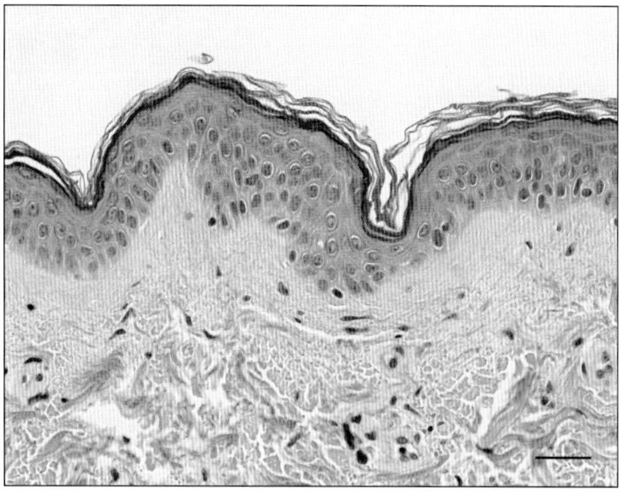

Figure 26.1 **Light microscopic appearances of normal human skin (hematoxylin and eosin; bar = 50 μm).**

is effective against a broad spectrum of micro-organisms, especially *Enterococci*. RNASE7 serves as a protective minefield in the superficial skin layers and helps destroy invading organisms [5]. Below this in the living layers of the skin are other antimicrobial peptides such as a β-defensins [6]. The antimicrobial activity of most peptides occurs as a result of unique structural properties to enable them to destroy the microbial membrane while leaving human cell membranes intact. Some may play a specific role against certain microbes in normal skin, whereas others act only when the skin is injured and the physical barrier is disrupted.

As well as direct protection against micro-organisms, some peptides have additional roles in signaling host responses through chemotactic, angiogenic, growth factor, and immunosuppressive activity. These peptides are known as alarmins [7]. For example, some alarmins not only kill bacteria, but also stimulate expression of factors such as Syndecan 1–4 in dermal fibroblasts, which are critical to the process of wound healing. Alarmins may also stimulate parts of the host defense system, such as barrier repair and recruitment of inflammatory cells. Antimicrobial peptides such as defensins and cathelicidins also greatly increase after infection, inflammation, or injury. Some skin diseases, including atopic dermatitis or rosacea, show altered expression of antimicrobial peptides, partially explaining the pathophysiology of these diseases.

Certain antimicrobial peptides can also influence host cell responses in specific ways. The human cathelicidin peptide LL-37 can activate mitogen-activated protein kinase (MAPK), an extracellular signal-related kinase in epithelial cells, and blocking antibodies to LL-37 hinder wound repair in human skin equivalents [8]. Defensins and cathelicidins have immunostimulatory and immunomodulatory capacities as catalysts for secondary host defense mechanisms, and can be chemotactic for distinct subpopulations of leukocytes as well as other inflammatory cells. Human β-defensins (hBDs) 1–3 are chemotactic for memory T-cells and immature dendritic cells—hBD2 attracts mast cells and activated

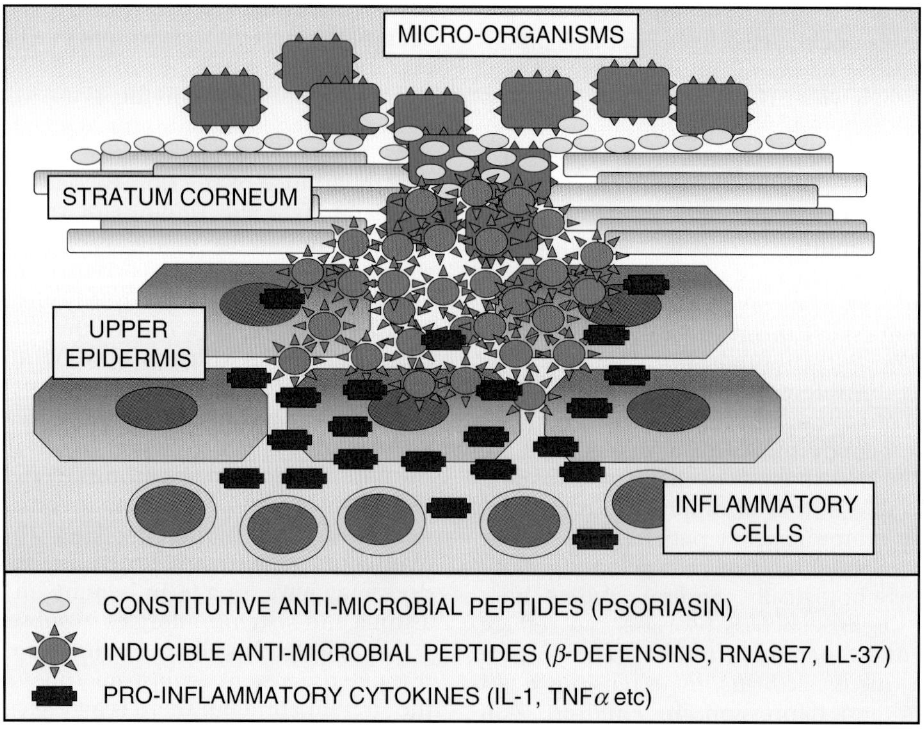

Figure 26.2 **Immune defense system in human skin.** Microorganisms that breach the human epidermis are faced with a constitutive antimicrobial system, for example, psoriasin. Further protection is also provided by inducible antimicrobial peptides, such as the β-defensins RNASE7 and LL-37. Microorganisms may also be targeted by proinflammatory cytokines.

neutrophils, whereas hBD3–4 is also chemotactic for monocytes and macrophages. Cathelicidins are chemotactic for neutrophils, monocytes/macrophages, and CD4 T-lymphocytes. In addition, antimicrobial peptides can also induce keratinocyte proliferation and migration, which involves epidermal growth factor receptor signaling and STAT activation [9].

Skin immunity is also provided by a distinct population of antigen-presenting cells in the epidermis known as Langerhans cells (Figure 26.3). These are dendritic cells of a form similar to melanocytes, but free from pigment and are DOPA-negative. Without stimulation, Langerhans cells can exhibit a unique motion, which has been termed Dendrite Surveillance Extension And Retraction Cycling Habitude (dSEARCH) [10]. This is characterized by rhythmic extension and retraction of dendritic processes between intercellular spaces. When Langerhans cells are exposed to antigen, there is greater dSEARCH motion and also direct cell-to-cell contact between adjacent Langerhans cells. Thus, Langerhans cells function as intraepidermal macrophages, phagocytosing antigens among keratinocytes. Langerhans cells then leave the epidermis and migrate via lymphatics to regional lymph nodes. In the paracortical region of lymph nodes, the Langerhans cell (or interdigitating reticulum cell as it is then known) expresses protein on its surface to present to a T-lymphocyte that can then undergo clonal proliferation. Langerhans cells contribute to several skin pathologies, including infections, inflammation, and cancer; thus, they play a pivotal role in regulating the balance between immunity and peripheral tolerance. Langerhans cells have characteristics that are different from dendritic cells, in that they are more likely to induce Th-2 responses than the Th-1 responses that are usually necessary for cellular immune responses against pathogens. Langerhans cells, or a subset thereof, may also have immunoregulatory properties that counteract the proinflammatory activity of surrounding keratinocytes.

Besides the antigen detection and processing role of epidermal Langerhans cells, cutaneous immune surveillance is also carried out in the dermis by an array of macrophages, T-cells, and dendritic cells (Figure 26.4). These immune sentinel and effector cells can provide rapid and efficient immunologic backup to restore tissue homeostasis if the epidermis is breached. The dermis contains a very large number of resident T-cells. Indeed, there are approximately 2×10^{10} resident T-cells, which is twice the number of T-cells in the circulating blood. Dermal dendritic cells vary in their functionality. Some have potent antigen-presenting capacities, whereas others have potential to develop into CD1a-positive and Langerin-positive cells, while some are proinflammatory. A recent addition to the family of skin immune sentinels is type 1 interferon-producing plasmacytoid predendritic cells, which are rare in normal skin but which can accumulate in inflamed skin [11]. A further component of the dermal immune system is the dermal macrophage. Dermal immune sentinels exhibit flexibility or plasticity in function. Depending on microenvironmental factors and cues, they may acquire an antigen-presenting mode, a migratory mode, or a tissue-resident phagocytic mode.

Inflammation in the skin also has an impact on patients' perception of skin diseases and their overall

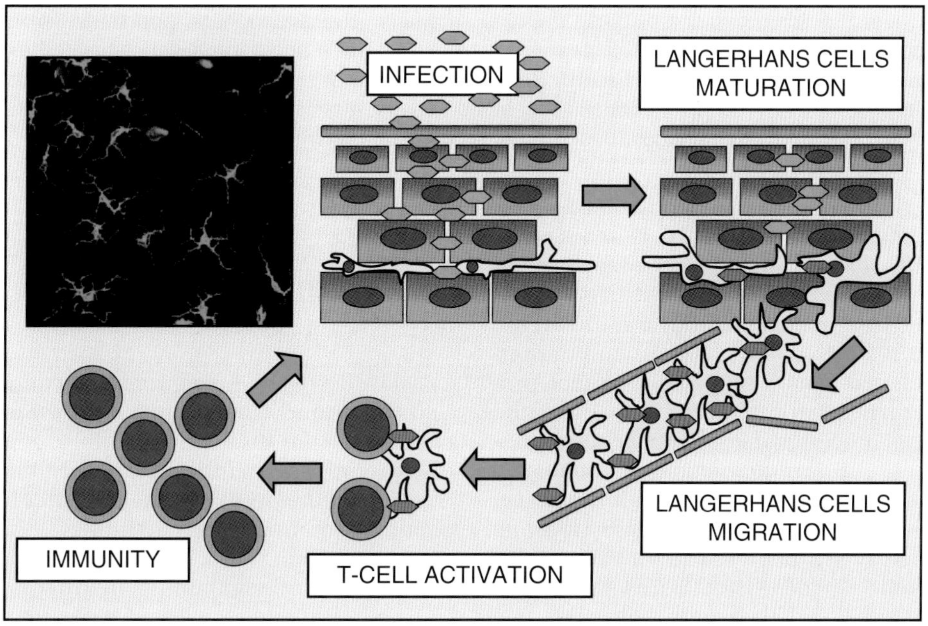

Figure 26.3 **The function of Langerhans cells in human skin.** The photomicrograph (top left) shows the dendritic appearances of Langerhans cells in human epidermis (image kindly supplied by Dr. Rachel Mohr, University of Toledo, TX). Antigenic material from invading peptides or bacteria are phagocytosed and processed by Langerhans cells within the epidermis. These Langerhans cells then mature and migrate to regional lymph nodes. Antigen is presented to T-cells which are then activated, proliferate, and allow for adaptive immune responses.

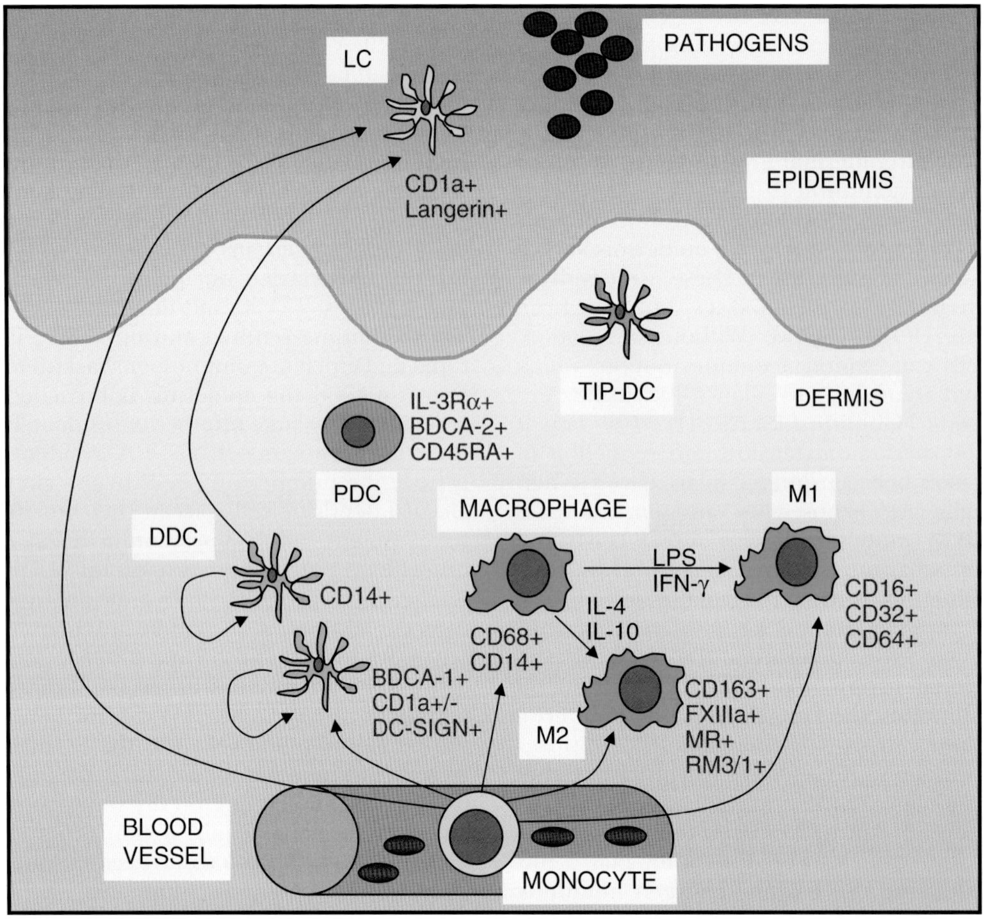

Figure 26.4 **Diversity of immune sentinels in human skin.** These include CD1a+ Langerin+ Langerhans cells located in the epidermis and various subtypes of dendritic cells and macrophages in the dermis. This figure illustrates some of the recent immunophenotypic and functional findings of these immune sentinels. The macrophage population expressing CD68 and CD14 can be further subdivided into classically activated macrophages (M1) and alternatively activated macrophages (M2), which develop under the influence of IL-4 and IL-10. Several cells have self-renewing potential under conditions of tissue homeostasis. Under inflammatory conditions, circulating blood-derived monocytes are potential precursors of Langerhans cells, dermal dendritic cells, and macrophages (based on an original figure by [86]).

well-being. Indeed, inflammatory responses have been shown to have an important role in the pathophysiology of depression [12]. Proinflammatory cytokines, acute phase proteins, chemokines, cellular adhesion molecules, and indeed therapeutic administration of the cytokine interferon-α can all lead to stress and depression. Much of this has a molecular basis, in that proinflammatory cytokines can interact with many of the pathophysiological processes that characterize depression, including neurotransmitter metabolism, neuroendocrine function, synaptic plasticity, and behavior. Stress associated with skin disease can also promote inflammatory responses through effects on sympathetic and parasympathetic nervous system pathways. Targeting proinflammatory cytokines and their signaling pathways, therefore, through greater understanding of the molecular mechanisms involved in skin inflammation may lead to novel treatment strategies to improve skin conditions and the patients' biological response to their disease, including its effects on their personal well-being and quality of life.

SKIN DEVELOPMENT AND MAINTENANCE PROVIDE NEW INSIGHT INTO THE MOLECULAR MECHANISMS OF DISEASE

As skin develops during embryogenesis, a single basal layer of primitive epithelium covers the inner and outer surface of the embryo. As embryogenesis progresses, those epithelia that will be subject to mechanical stress require further multiple layers to offer better protection against the environment. The development of a stratified epithelium such as the skin requires a detailed architecture of maintaining an inner layer of proliferating cells, but that can also give rise to multiple layers of terminally differentiating cells that extend to the body surface and which are subsequently shed. A detailed understanding of this process that generates a self-perpetuating barrier to keep microbes out and essential body fluids in is becoming clearer, and this improved understanding is providing new insights into skin maintenance as well as the pathogenesis and molecular mechanisms underlying certain developmental disorders.

One fundamental issue has been trying to provide an explanation for how epithelial progenitor cells retain a self-renewing capacity. In 1999, it was shown that mice lacking the transcription factor p63 had thin skin and abnormal skin renewal [13]. p63 is an evolutionary predecessor to the p53 protein, part of a family of transcriptional regulators of cell growth differentiation and apoptosis. While p53 is a major player in tumorigenesis, p63 and another family member, p73, appear to have pivotal roles in embryonic development (Figure 26.5) [14]. p73-deficient mice have neurological and inflammatory pathology, whereas p63 knockout mice have major defects in epithelial limb and craniofacial development. These observations suggest that p63 has a crucial role in tissue morphogenesis and maintenance of epithelial stem cell compartments. Lack of p63 compromises skin formation either by creating an absence of lineage commitment and an early block in epithelial differentiation or by causing skin failure through a defect in epithelial stem cell renewal. p63 has been linked to several important signaling pathways such as epidermal growth factor (EGF), fibroblast growth factor (FGF), bone morphogenic protein (BMP), and Notch/Wnt/hedgehog signaling [15]. p63 also directly regulates expression of extracellular matrix adhesion molecules, including the $\alpha_6\beta_4$ basal integrins and desmosomal proteins such as PERP, all of which are essential for epithelial integrity.

The human *p63* gene consists of 16 exons, located on chromosome 3q28. There are 2 different promoter sites and 3 different splicing routes, which create at least 6 different protein isoforms. Several functional domains have been identified (Figure 26.5). These include a central DNA binding domain and an isomerization domain, which are present in all p63 isoforms. The amino-terminal ends are called Transcription Activation (TA) and ΔN. The ΔN isoforms also contain an amino-terminal transactivation domain, denoted TA2. At the carboxy terminal end α, β, and γ termini

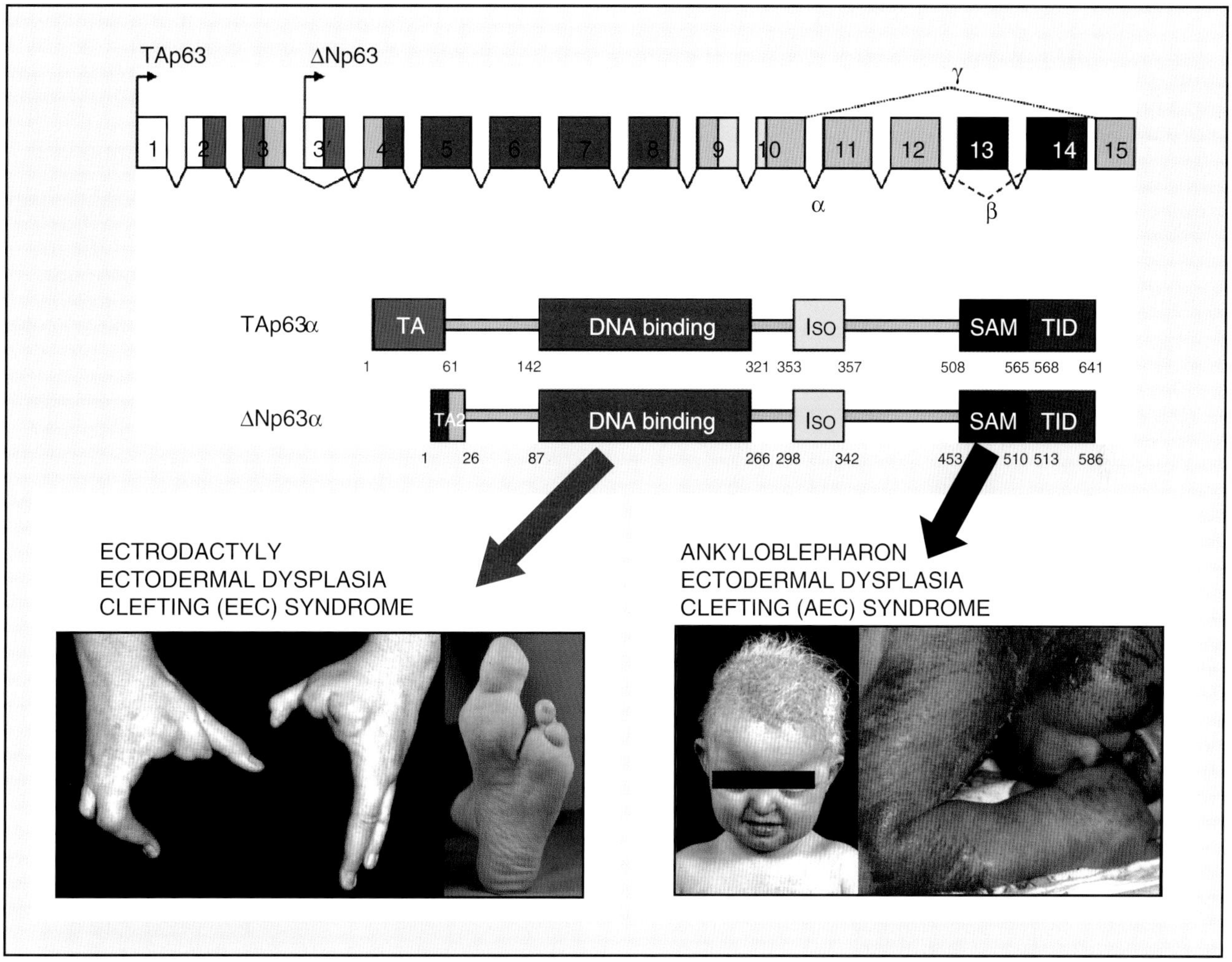

Figure 26.5 **Genomic and functional domain organization of the transcription factor *p63*.** At least 6 different isoforms can be generated by use of alternative translation initiation sites or alternative splicing. The main isoform expressed in human skin is ΔNp63α. Autosomal dominant mutations in the DNA binding domain of the *p63* gene lead to ectrodactyly, ectodermal dysplasia, and clefting (EEC) syndrome. In contrast, autosomal dominant mutations in the SAM domain result in ankyloblepharon, ectodermal dysplasia, and clefting (AEC) syndrome. A number of other ectodermal dysplasia syndromes may also result from mutations in the *p63* gene.

can be synthesized. The carboxy-terminal end has 2 additional domains: (i) the Sterile Alpha Motif (SAM) domain and (ii) a Transactivation Inhibitory (TI) domain. Both of these domains are present in the longest carboxy terminal variant, known as p63α.

A further understanding of the role of p63 in skin development has been gleaned from studies on naturally occurring mutations in this gene. Heterozygous mutations cause developmental disorders, displaying various combinations of ectodermal dysplasia, limb malformations, and oro-facial clefting. Thus far, 7 different disorders have been linked to mutations in the *p63* gene. These conditions may have overlapping genotypic features, but there are some distinct genotype/phenotype correlations [16]. The most common p63-associated ectodermal dysplasia is EEC syndrome (ectrodactyly, ectodermal dysplasia, and cleft/lip palate, OMIM604292). It is characterized by 3 major clinical signs of cleft lip and/or palate, ectodermal dysplasia (abnormal teeth, skin, hair, nails, and sweat glands, or combinations thereof), and limb malformations in the form of split hand/foot (ectrodactyly), and/or fusion of fingers/toes (syndactyly). Another group of *p63*-linked patients are those with Rapp-Hodgkin syndrome (OMIM129400) or AEC/Hay-Wells syndrome (OMIM106260). These syndromes fulfill the criteria of ectodermal dysplasia and oro-facial clefting, but do not have the severe limb malformation(s) seen in EEC syndrome. The features of these syndromes may include eyelid fusion (ankyloblepharon filiform adnatum), severe erosions at birth, and abnormal hair with pili torti or pili canaliculi. The *p63* mutations in EEC syndrome are clustered in the DNA-binding domain and most likely alter the DNA-binding properties of the protein. By contrast, mutations in Rapp-Hodgkin or AEC syndromes are clustered in the SAM and TI domains in the carboxy-terminus of p63α. The SAM domain is involved in protein-protein interactions, whereas the TI domain is combined intra-molecularly to the TA domain, thereby inhibiting transcription activation. All *p63*-associated disorders are inherited in an autosomal dominant manner, and mutations are thought to have either dominant-negative or gain of function effects.

The epidermis contains a population of epidermal stem cells that reside in the basal layer, although it is not clear how many cells within the basal layer have a stem cell capacity (Figure 26.6) [17]. Stem cells are proposed to express elevated levels of β_1 and α_6 integrins and differentiate by delamination and upward movement to form the spinous layer, a granular layer, and the stratum corneum. The proliferation of epidermal stem cells is regulated positively by β_1 integrin and transforming growth factor α, and negatively by transforming growth factor β signaling. Stem cells are also found in sebaceous glands—the latter contain a small number of progenitors that express the transcriptional repressor Blimp1 which reside near or at the base of the sebaceous gland. The proliferative progeny of the cells differentiate into lipid-filled sebocytes. Hair follicle stem cells reside in the bulge compartment below the sebaceous gland. These stem cells are slow cycling and express the cell surface molecules CD34 and VdR as well as the transcription factors TCF3, Sox9, Lhx2, and NFATc1. These bulge area stem cells generate cells of the outer root sheath, which drive the highly proliferative matrix cells next to the mesenchymal papillae. After proliferating, matrix cells differentiate to form the hair channel, the inner root sheath, and the hair shaft.

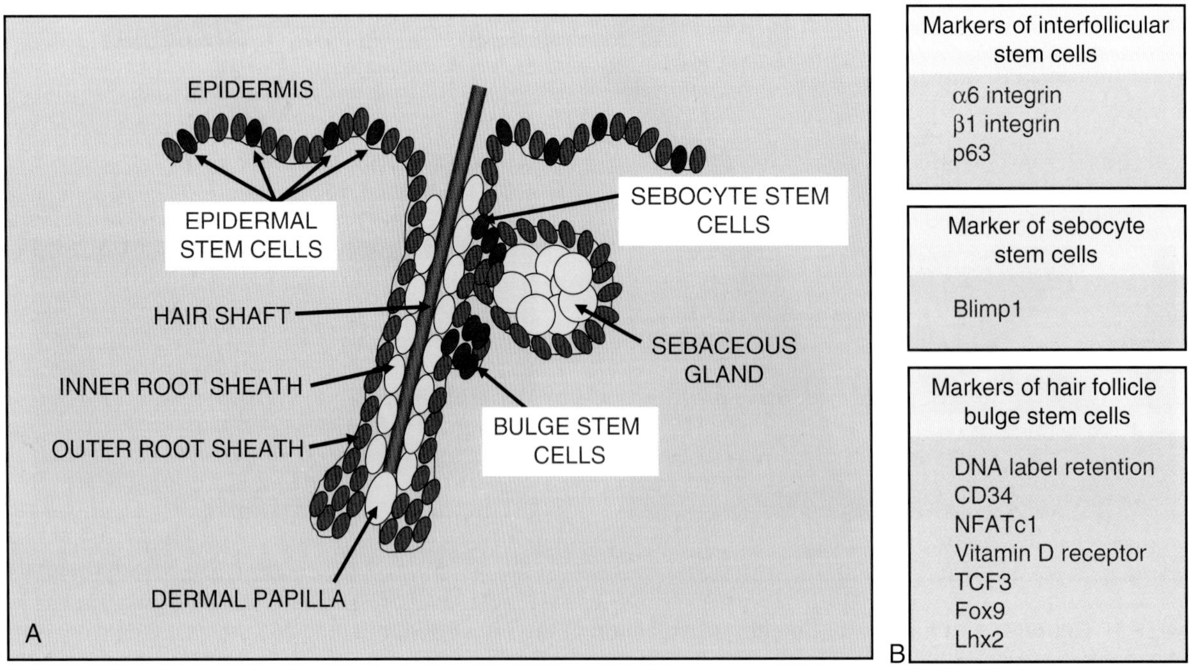

Figure 26.6 **(A) Localization of stem cells in human epidermis. Stem cells are located within the basal layer of interfollicular epidermis, as well as at the base of sebocytes and also in the bulge area of hair follicles.** (B) **These epidermal stem cells are associated with a number of cellular markers.**

The mechanisms that control the stem cell proliferation and differentiation in the skin also provide new insight into skin homeostasis. Epidermal proliferation is controlled by a multitude of transcription factors, including c-Myc and p63 [18]. Differentiation of epidermal cells is controlled by Notch signaling and the transcription factors PPARα, AP2 α/γ, and C/EBPα/β. Proliferation of bulge cells in the hair follicle is positively controlled by Wnt signaling and negatively controlled by BMP signaling, and the transcription factors, NFATc1 and PTEN. Differentiation of the inner root sheath is controlled by Notch and BMP signaling and the transcription factor CDP and GATA3. Hair shaft differentiation is controlled by Wnt signaling and its downstream transcription factor Lef1. Hair matrix cells are controlled by Msx1/2, Ovo1, Foxn1, and Shh. Sebaceous gland stem cells are positively regulated by c-Myc and Hedgehog signaling and negatively regulated by the transcription factor Blimp1 and Wnt signaling. Differentiation of sebocytes is directed by PPARγ expression [19].

MOLECULAR PATHOLOGY OF MENDELIAN GENETIC SKIN DISORDERS

There are approximately 5000 single gene disorders, of which nearly 600 have a distinct skin phenotype. Many of these disorders have also been characterized at a molecular level. Most inherited skin disorders are transmitted either by autosomal dominant, autosomal recessive, X-linked dominant, or X-linked recessive modes of inheritance. However, understanding the precise pattern of inheritance is essential for accurate genetic counseling. For example, the connective tissue disorder pseudoxanthoma elasticum (OMIM264800) was for many years thought to reflect a mixture of autosomal dominant and autosomal recessive genotypes. However, it has been shown that all forms of pseudoxanthoma elasticum are in fact autosomal recessive and that the disease is caused by mutations in the *ABCC6* gene, as the earlier reports of autosomal dominant inheritance were actually pseudo-dominant inheritance in consanguineous pedigrees [20]. Understanding the molecular pathology of pseudoxanthoma elasticum has also identified similar autosomal recessive inherited skin diseases that have an overlapping phenotype, but in which there are subtle clinical differences. For example, a pseudoxanthoma elasticum-like disease with cutis laxa skin changes and coagulopathy (OMIM610842) has been shown to result from mutations in the *GGCX* gene [21]. In both conditions, there is progressive accumulation of calcium phosphate in tissue, resulting in ocular and cardiovascular complications. The knowledge that these conditions are autosomal recessive helps with genetic counseling, prognostication, and management of families with affected members. Moreover, understanding the precise molecular pathology allows for more careful clinical monitoring and patient follow-up.

One of the principal functions of human skin is to provide a mechanical barrier against the external environment. The structural integrity of the skin depends on several proteins. These include intermediate filaments inside keratinocytes, intercellular junctional proteins between keratinocytes, and a network of adhesive macromolecules at the dermal-epidermal junction (Figure 26.7). Since the late 1980s, several Mendelian genetic disorders resulting from autosomal dominant or autosomal recessive mutations in structural proteins in the skin have provided fascinating insights into skin structure and function. In addition, determining the key roles of particular proteins has provided a plethora of new clinically and biologically relevant data.

One of the best characterized groups of disorders is epidermolysis bullosa (EB), a group of skin fragility disorders associated with blister formation of the skin and mucous membranes that occurs following mild trauma. EB simplex is the commonest form of inherited EB and affects approximately 40,000 people in the United States. Transmission is mainly autosomal dominant, but recessive patients (OMIM 601001) have also been reported. Ultrastructurally, the level of split is through the cytoplasm of basal cells, often close to the inner hemidesmosomal plaque. In dominant forms of EB simplex, there may be disruption of keratin tonofilaments or aggregation of keratin filaments into bundles (Figure 26.8). However, sometimes transmission electron microscopy may show only very subtle or indiscernible morphological changes in intermediate filaments. The molecular defects that cause EB simplex lie in either the keratin 5 gene (*KRT5*) or the keratin 14 gene (*KRT14*), or, in cases of the autosomal recessive EB simplex-muscular dystrophy, in the plectin gene (*PLEC1*) [22–24]. The molecular pathology of keratin gene mutations provides some insight into genotype/phenotype correlation. Notably mutations in the helix initiation/helix termination motifs in helices 1A and 2B of the keratin genes result in more severe subtypes. Clinically, EB simplex patients have the mildest skin lesions and scarring is not frequent. Extracutaneous involvement is rare, apart from cases with plectin pathology. The mildest clinical subtype of EB simplex is the Weber-Cockayne variant (OMIM131800) in which blistering occurs mainly on the palms and soles (Figure 26.9). The molecular pathology involves autosomal dominant mutations that occur mainly outside the critical helix boundary motifs. In some forms of EB simplex, such as the Köbner subtype, the disease is more generalized than the Weber-Cockayne variant, although there is considerable overlap. The most severe form of EB simplex is the Dowling-Meara type (OMIM131760). This often presents with generalized blister formation shortly after birth, and it can be fatal in neonates. A characteristic of this condition in later childhood is the grouping of lesions in a herpetiform clustering arrangement (Figure 26.7). This form of EB simplex is associated with the greatest disruption of keratin tonofilaments. Another autosomal dominant form of EB simplex is one associated with mottled skin pigmentation (OMIM131960). Apart from blisters, there is diffuse speckled hyperpigmentation as well as keratoderma of the palms and soles. The hypermelanotic macules are most evident in the axillae, limbs, and lower abdomen. The underlying molecular pathology in all reports is the

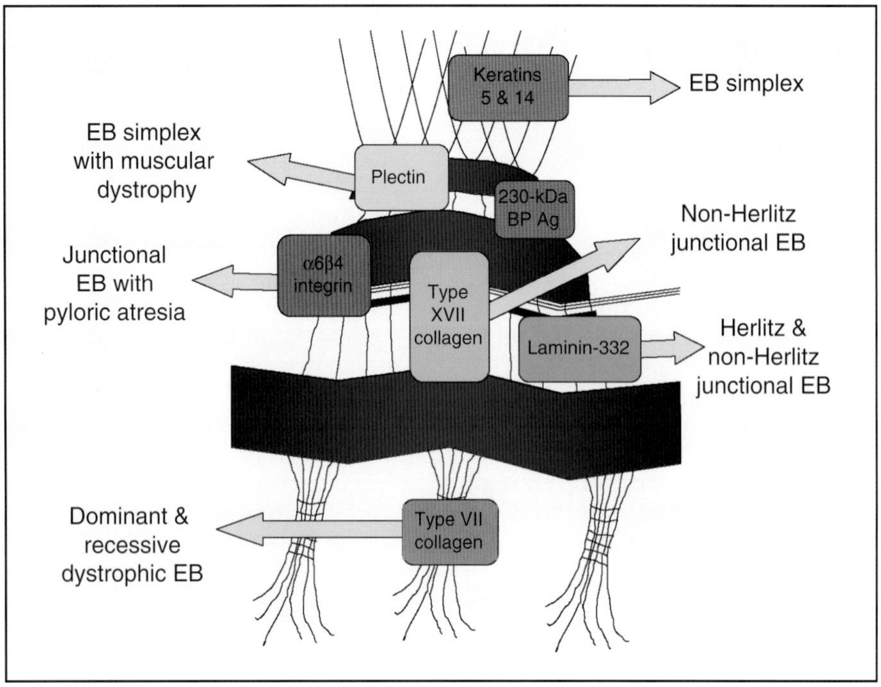

Figure 26.7 Illustration of the integral structural macromolecules present within hemidesmosome-anchoring filament complexes and the associated forms of clinical epidermolysis bullosa that result from autosomal dominant or autosomal recessive mutations in the genes encoding these proteins.

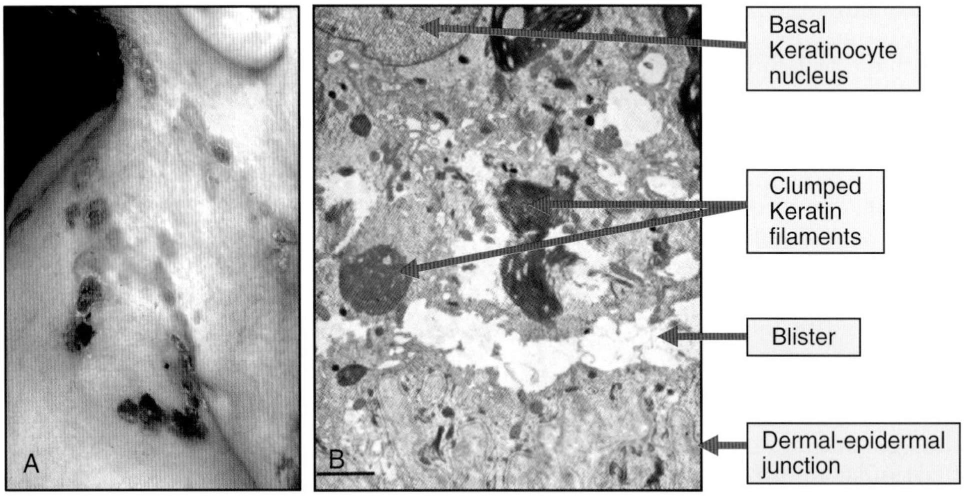

Figure 26.8 Clinicopathological consequences of mutations in the gene encoding keratin 14 (*KRT14*), the major intermediate filament protein in basal keratinocytes. (A) The clinical picture shows autosomal dominant Dowling-Meara epidermolysis bullosa simplex. **(B)** The electron micrograph shows keratin filament clumping and basal keratinocyte cytolysis (bar = 1 μm).

same heterozygous proline to leucine substitution at codon 25 in the nonhelical V1 domain of the *KRT5* gene. This proline residue is expressed on the outer part of polymerized keratin filaments and, when mutated, may result in abnormal interactions with melanosomes or other keratinocyte organelles.

Collectively, these naturally occurring mutations have provided valuable insight into mechanisms of adhesion in human skin as well as the functional and clinical consequences of mutations in individual components. Nevertheless, molecular analysis of other Mendelian disorders has demonstrated that these skin structural proteins may also be mutated in other inherited skin diseases—known as allelic heterogeneity. For example, heterozygous nonsense mutations in the *KRT14* gene have been shown to result in the Naegeli-Franceschetti-Jadassohn form of ectodermal dysplasia (OMIM161000) [25]. In addition, heterozygous loss of function mutations

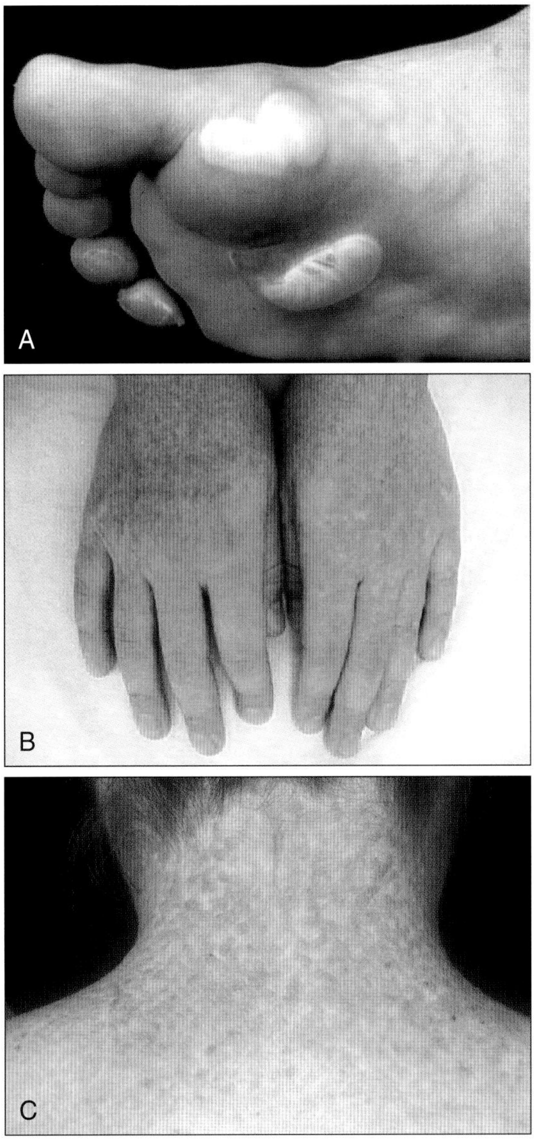

Figure 26.9 Spectrum of clinical abnormalities associated with dominant mutations in keratin 5 (*KRT5*). **(A)** Missense mutations in the nonhelical end domains result in the most common form of EB simplex, which is localized to the hands and feet (Weber-Cockayne variant). **(B)** A specific mutation in keratin 5, p.P25L, is the molecular cause of epidermolysis bullosa simplex associated with mottled pigmentation. **(C)** Heterozygous nonsense or frameshift mutations in the *KRT5* gene leads to Dowling-Degos disease.

in the *KRT5* gene have been shown to underlie the autosomal dominant disorder Dowling-Degos disease (OMIM179850) [26]. This is characterized by clustered skin papules, the histology of which shows seborrheic keratosis-like morphology (Figure 26.9).

Junctional EB is an autosomal recessive condition in which the molecular pathology involves loss of function mutations in any one of at least 6 different genes encoding structural proteins within the hemidesmosome or lamina lucida at the cutaneous basement membrane zone. Clinical features include blistering, atrophic scarring, nail dystrophy, and defective dental enamel as well as other abnormalities affecting the hair, eyes, and genito-urinary tract. Ultrastructurally, the level of split is mainly through the lamina lucida of the basement membrane zone, but blistering may also occur just above the basal keratinocyte plasma membrane as a focal observation in biopsies from patients with junctional EB-associated with pyloric atresia. The most severe type of junctional EB is the Herlitz subtype (OMIM226700). There is typically widespread blistering with mucosal involvement in the mouth and upper respiratory tract. Many affected infants die from overwhelming secondary infection. In later infancy, patients may develop wounds with exuberant granulation tissue, particularly around the mouth, nose, and nails. Dystrophic nails with paronychia and swollen fingertips are frequent findings, and most cases die in early infancy. The underlying molecular pathology involves homozygous or compound heterozygous loss of function mutations in any of the 3 genes that encode the laminin-332 polypeptide: *LAMA3* or *LAMB3* or *LAMC2* (Figure 26.10) [27–29]. Some forms of

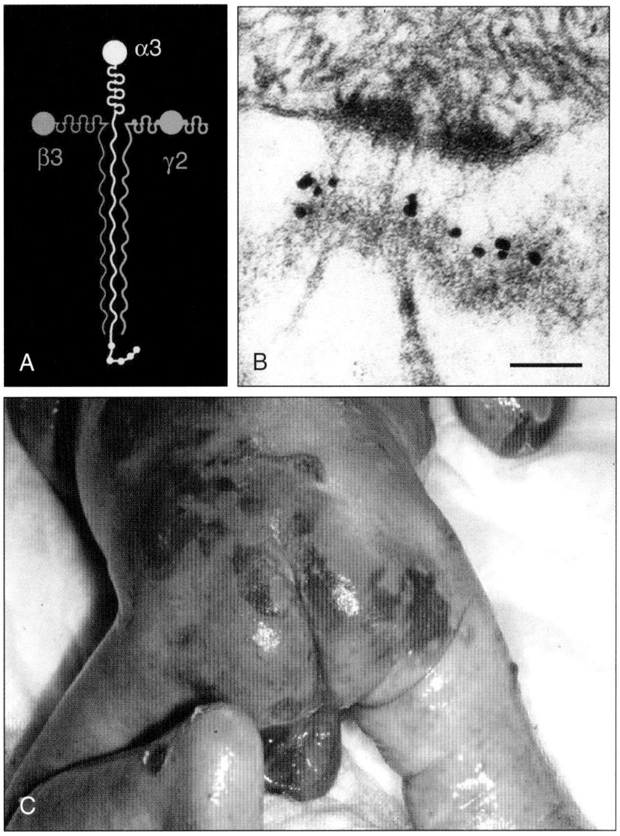

Figure 26.10 Laminin-332 mutations result in junctional epidermolysis bullosa. **(A)** Laminin-332 consists of 3 polypeptide chains: α3, β3, and γ2. **(B)** Immunogold electron microscopy shows laminin-332 staining at the interface between the lamina lucida and lamina densa subjacent to a hemidesmosome (bar = 50 nm). **(C)** Loss of function mutations in any one of these genes encoding the 3 polypeptides chains results in Herlitz junctional epidermolysis bullosa, which is associated with a poor prognosis, usually with death in early infancy.

junctional EB are termed non-Herlitz (OMIM226650). In these cases there may be extensive blisters at birth, but the disease typically lessens in severity with time, although atrophic wounds, abnormal dentition, and nail dystrophy persist, and alopecia is very common. This type of junctional EB is genetically heterogeneous. It may result from mutations in either the *LAMA3, LAMB3,* or *LAMC2* genes (the subcomponents of laminin-332) or alternatively due to loss of function mutations on both alleles of the gene-encoding type XVII collagen, *COL17A1* [30]. The range of *COL17A1* gene mutations includes missense, nonsense, frameshift, or splice-site mutations, but usually there is total ablation of type XVII collagen protein. Another subtype of junctional EB is associated with a further extracutaneous abnormality, namely pyloric atresia. Affected pregnancies are usually complicated by polyhydramnios, as fetuses with pyloric atresia cannot swallow amniotic fluid. Postnatally, feeding results in nonbilious vomiting. The severity of blistering in neonates is variable, but all cases involve molecular pathology in the $\alpha_6\beta_4$ integrin complex [31,32]. Most patients have mutations in the gene encoding β_4 integrin, *ITGB4.* Severe cases usually have nonsense or frameshift mutations, but some missense mutations at conserved amino acid regions or involving critical cysteine residues have been shown to underlie often lethal cases. Mutations in the *ITGA6* gene are also usually associated with a more severe phenotype, but genotype/phenotype correlation can be difficult.

Dystrophic EB represents a third type of inherited skin blistering, which may be inherited by autosomal dominant or autosomal recessive transmission (Figure 26.11). Ultrastructurally, the level of split is below the lamina densa, and the underlying molecular defect in all types of dystrophic EB involves mutations in the type VII collagen gene (*COL7A1*) [33]. Transmission electron microscopy reveals abnormalities in the number and/or morphology of anchoring fibrils, which are principally composed of type VII collagen. Dominant forms of dystrophic EB are usually caused by heterozygous glycine substitution mutations within the type VII collagen triple helix [34]. These result in dominant-negative interference and disruption of anchoring fibril formation. Affected individuals have mild trauma-induced blisters, mainly on skin overlying bony prominences such as the knees, ankles, or fingers. Blistering is followed by scarring and milia formation. Nail dystrophy is common. There are several variants of dominant dystrophic EB that may be distinguishable at a clinical level, including transient bullous dermolysis of the newborn (OMIM131705), pretibial dystrophic EB (OMIM131850), and EB pruriginosa (OMIM604129).

It is clear that the *COL7A1* pathology alone cannot account for the phenotypic variability of dystrophic EB and that other modifying genes or environmental factors may play a role. The most severe form of dystrophic EB is the autosomal recessive Hallopeau-Siemens subtype (OMIM226600). Blister formation in affected individuals starts from birth or early infancy, and the

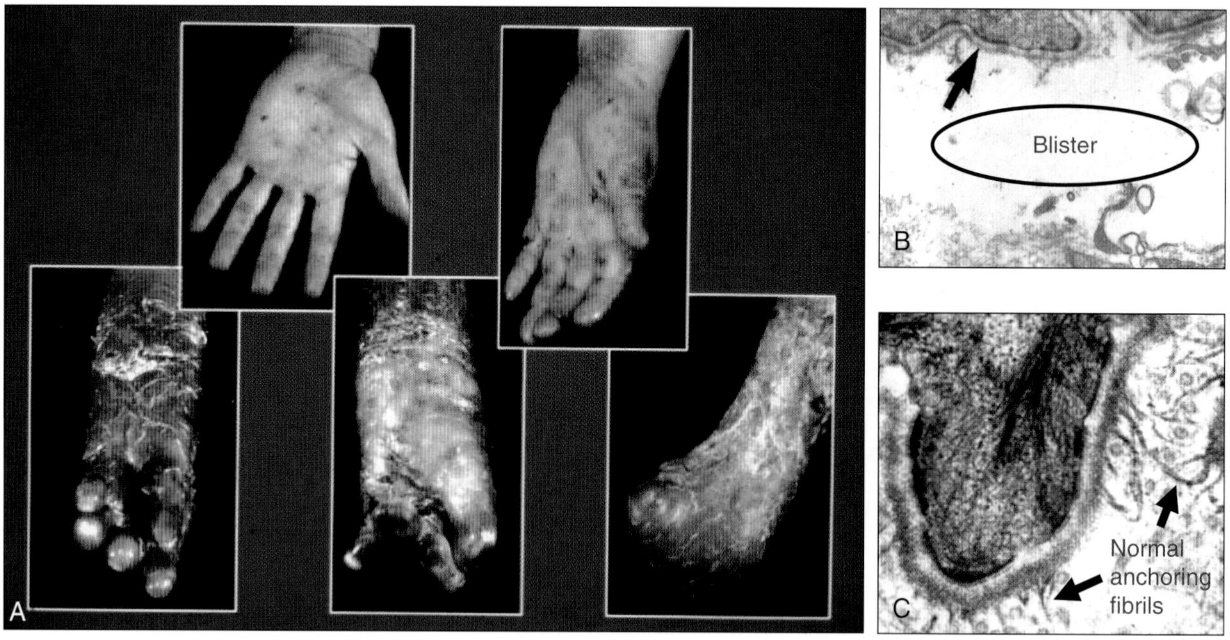

Figure 26.11 Clinicopathological abnormalities in the dystrophic forms of epidermolysis bullosa. (A) This form of epidermolysis bullosa is associated with variable blistering and flexion contraction deformities, here illustrated in the hands. **(B)** The disorder results from mutations in type VII collagen (*COL7A1* gene), the major component of anchoring fibrils at the dermal-epidermal junction. This leads to blister formation below the lamina densa (lamina densa indicated by arrow). **(C)** In contrast, in normal human skin there is no blistering, and the sublamina densa region is characterized by a network of anchoring fibrils.

skin is very fragile. Wound healing is often poor, leading to chronic ulcer formation with exuberant granulation tissue formation, repeated secondary infection, and frequent scar formation. Mucous membranes are extensively affected, and esophageal involvement causes dysphagia and obstruction due to stricture. Affected individuals also have an increased risk of squamous cell carcinoma, particularly in areas of scarring or chronic ulceration (Figure 26.12) [35]. At least 50% of patients with Hallopeau-Siemens subtype die from squamous cell carcinoma by the age of 40 years [36]. The molecular pathology involves loss of function mutations on both alleles of the *COL7A1* gene, leading to markedly reduced or completely absent type VII collagen expression at the dermal-epidermal junction. The reason for the increased risk of malignancy is not clear. In part, it may reflect a response to chronic scarring and protracted tissue injury, but at a molecular level critical domain interactions between type VII collagen and other molecules such as laminin-332 may be relevant to the promotion of malignancy. Less disruptive mutations in *COL7A1* lead to non-Hallopeau-Siemens recessive dystrophic EB or other variants including dystrophic EB inversa (OMIM226450). Overall, genotype/phenotype correlation suggests that in recessive dystrophic EB the amount of type VII collagen that is expressed at the dermal-epidermal junction is inversely proportional to clinical severity in terms of scarring and extent of blistering.

To maintain the structural function of the epidermis, a number of intercellular junctions exist, including desmosomes, tight junctions, gap junctions, and adherens junctions. Mutations in these junctional complexes result in several Mendelian inherited skin diseases. Desmosomes are important cell-cell adhesion junctions found predominantly in the epidermis and the heart. They consist of 3 families of proteins: the armadillo proteins, cadherins, and plakins (Figure 26.13). Mutations in desmosomal proteins result in skin, hair, and heart phenotypes (Figure 26.14). Armadillo proteins contain several 42 amino acid repeat domains and are homologous to the drosophila armadillo protein. They bind to other proteins through their armadillo domains and play a variety of roles in the cell, including signal transduction, regulation of desmosome assembly, and cell adhesion. Mutations in the plakophilin 1 gene (*PKP1*) result in autosomal recessive ectodermal dysplasia-skin fragility syndrome (OMIM604536) [37]. Affected individuals have a combination of skin fragility and inflammation and abnormalities of ectodermal development, such as scanty hair, keratoderma, and nail dystrophy. Pathogenic *PKP1* mutations are typically splice-site or nonsense mutations. Mutations in the plakophilin 2 gene (*PKP2*) result in autosomal dominant familial arrhythmogenic ventricular dysplasia 9 (OMIM609040), but without a skin phenotype. Autosomal dominant and recessive mutations have also been described in the plakoglobin gene (*JUB*). A recessive mutation results in Naxos disease (OMIM601214), a genodermatosis frequently seen on Naxos Island in the Mediterranean where approximately 1 in 1000 individuals is affected with clinical features of arrhythmogenic right ventricular dysplasia, diffuse palmar keratoderma, and woolly hair [38]. Autosomal dominant mutations in plakoglobin can also result in cardiomyopathy.

The cadherins comprise a group of desmogleins and desmocollins, transmembranous glycoproteins that are present between keratinocytes. Heterozygous mutations in the desmoglein 1 gene (*DSG1*) result in autosomal dominant striate palmoplantar keratoderma (OMIM148700) [39]. The molecular pathology in this disorder results from desmoglein 1 haploinsufficiency. Autosomal dominant mutations in the desmoglein 2 gene (*DSG2*) give rise to arrhythmogenic ventricular dysplasia 10 (OMIM609160), but no skin phenotype. Mutations in desmoglein 3 have not been described in humans, but are known to result in the naturally occurring mouse balding phenotype. Mutations in the desmoglein 4 gene (*DSG4*) result in localized autosomal recessive hypotrichosis (OMIM607903) in which affected individuals have hypotrichosis restricted to

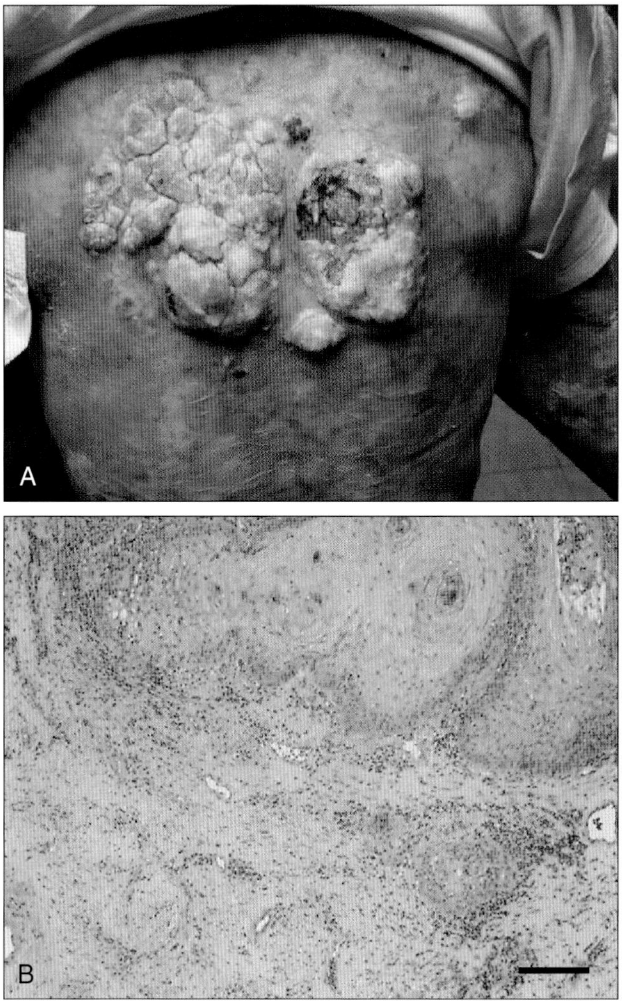

Figure 26.12 **Squamous cell carcinoma (SCC) in recessive dystrophic epidermolysis bullosa. (A)** Affected individuals have a 70-fold increased risk of developing SCC, here illustrated on the mid-back. **(B)** Light microscopy revealsa moderately differentiated SCC.

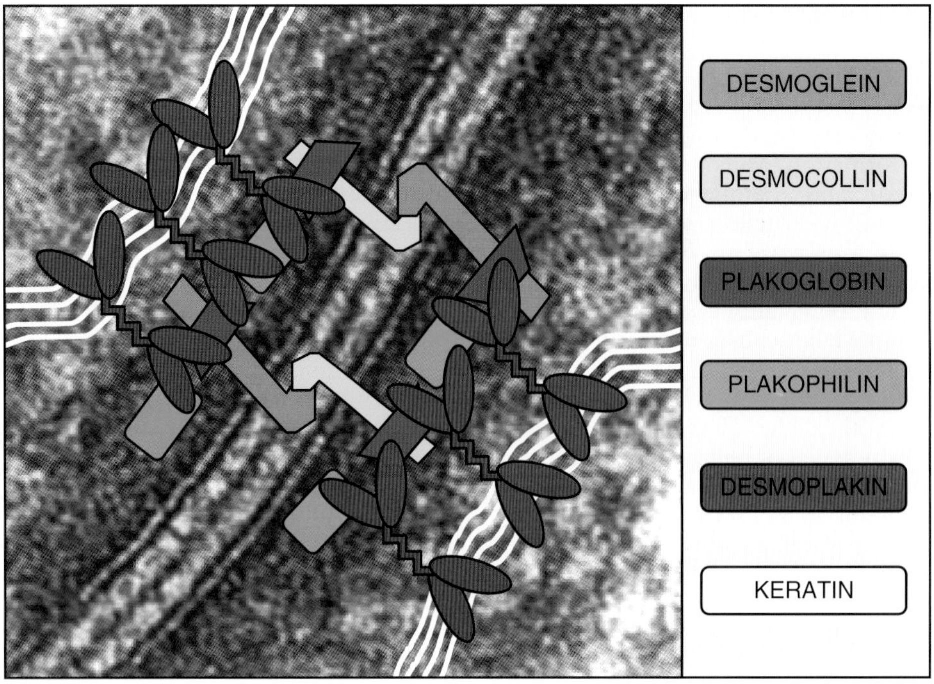

Figure 26.13 **Protein composition of the desmosome linking two adjacent keratinocytes a moderately differentiated SCC.** The major transmembranous proteins are the desmogleins and the desmocollins. Several desmosomal plaque proteins, including desmoplakin, plakophilin, and plakoglobin provide a bridge that links binding between the transmembranous cadherins and the keratin filament network within keratinocytes.

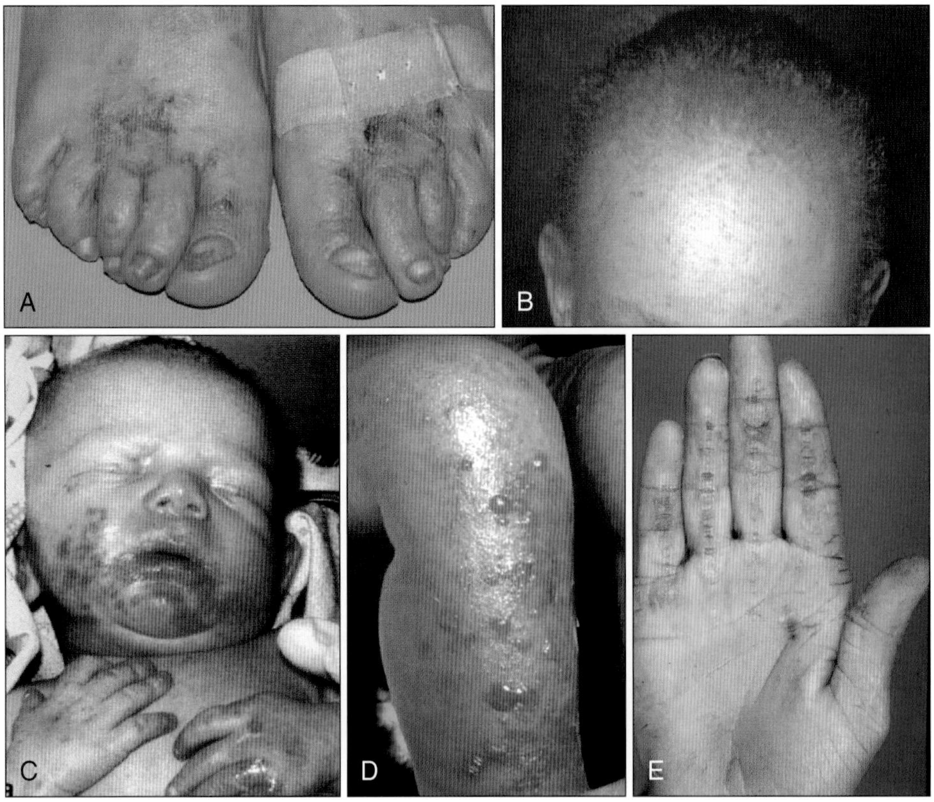

Figure 26.14 **Clinical abnormalities associated with inherited gene mutations in desmosome proteins. (A)** Recessive mutations in plakophilin 1 result in nail dystrophy and skin erosions. **(B)** Woolly hair is associated with several desmosomal gene abnormalities, particularly mutations in desmoplakin. **(C)** Recessive mutations in plakophilin 1 can result in extensive neonatal skin erosions, particularly on the lower face. **(D)** Recessive mutations in desmoplakin can lead to skin blistering. **(E)** Autosomal dominant mutations in desmoplakin do not result in blistering but can lead to striate palmoplantar keratoderma.

the scalp, chest, arms, and legs, but sparing of axillary or pubic hair [40]. Papules on the scalp show atrophic curled-up hair follicles and shafts with marked swelling of the precortical region. Recessive mutations in desmoglein 4 may also underlie some cases of autosomal recessive monilethrix. Mutations in desmocollin 2 (*DSC2*) have been shown to result in autosomal dominant arrhythmogenic ventricular dysplasia 11 (OMIM610476), but there is no skin phenotype. Plakins comprise a family of proteins that cross-link the cytoskeleton to desmosomes. They include desmoplakin, envoplakin, periplakin, plectin, bullous pemphigoid antigen 1, corneodesmosin, and microtubule actin cross-linking factor. Mutations in the desmoplakin gene (*DSP*) result in a variable combination of skin, hair, and cardiac abnormalities [41,42]. These can be autosomal dominant or recessive. The phenotype in autosomal dominant cases ranges from cutaneous striate palmoplantar keratoderma to cardiac arrhythmogenic ventricular dysplasia 8 (OMIM607450). The phenotype in autosomal recessive patients ranges from skin fragility-woolly hair syndrome (OMIM607655) to other syndromes in which the skin and heart may be abnormal, such as dilated cardiomyopathy with woolly hair and keratoderma (OMIM605676). Some recessive *DSP* mutations may just result in a cardiac phenotype. Compound heterozygous mutations that almost completely ablate the desmoplakin tail have recently been demonstrated to produce the most severe clinical subtype known as lethal acantholytic epidermolysis bullosa [43]. Autosomal dominant mutations in the corneodesmosin gene (*CDSN*) result in hypotrichosis of the scalp (OMIM146520) [44].

Collectively, mutations in desmosomal genes give rise to variable abnormalities in skin, hair, and heart. Understanding the molecular pathology of these disorders and the precise phenotypic consequences of specific mutations provides insight into the clinical features in affected individuals and also allows for accurate diagnosis, improved genetic counseling, and more rigorous clinical follow-up to assess potential complications.

MOLECULAR PATHOLOGY OF COMMON INFLAMMATORY SKIN DISEASES

Atopic dermatitis and psoriasis represent two of the most common inflammatory skin dermatoses. Traditional views on their pathologies usually reflect differing and varying contributions from primary abnormalities in keratinocytes or in immunocytes. However, recent molecular insights are now providing new ideas about disease susceptibility and pathogenesis. These discoveries have direct relevance to the classification of inflammatory skin diseases and to the design of future therapies based on specific molecular pathology.

Atopic dermatitis is a chronic itching skin disease that results from a complex interplay between strong genetic and environmental factors [45]. Recent genetic studies suggest that a major abnormality is a generalized dysfunction of the epidermis manifesting as a compromised skin barrier and failure to protect adequately against microbial insults and antigens. The term *atopic dermatitis* encompasses a chronic pruritic inflammatory skin condition that is most common in early childhood and that predominantly affects the skin flexures. The identification and characterization of the pathophysiologic abnormalities in atopic dermatitis are important because of the substantial increase in the prevalence of the disease in industrialized countries where it affects 10%–20% of children and 1%–3% of adults. Atopic dermatitis is also known to be associated with other atopic conditions, including asthma and allergic rhinitis, known as the atopic triad. Of children with severe atopic dermatitis, about 60% have asthma and 35% have allergic rhinitis. There is strong evidence for a genetic contribution to the pathogenesis of atopic dermatitis—with enhanced phenotypic concordance in monozygotic relative to dizygotic twins. The atopic state is typically characterized by positive skin prick test responses to common environmental allergens, the presence of allergen-specific IgE in sera, an increase in total serum IgE levels, or a combination thereof.

There are two forms of atopic dermatitis: the extrinsic and intrinsic forms. The former is associated with IgE-mediated sensitization, whereas the latter is characterized by a normal total serum IgE level and the absence of specific IgE responses to aero-allergens and food-derived allergens. Phenotype-based classifications, however, share several pathological features, including CD4 positive T-cell infiltration and disease histology. Although IgE sensitization is a not a necessary prerequisite for the development of atopic dermatitis, a number of primary immunologic abnormalities can be identified [46]. Compared to extrinsic forms of atopic dermatitis, the intrinsic variants tend to display different pathologic abnormalities, including reduced dermal infiltrate of eosinophils and eosinophil granular proteins, and a more moderate enhancement of lesional cytokine expression, including IL-13, eotaxin, and reduced surface expression of high-affinity receptor for IgE (FcηRI) on epidermal dendritic cells. These observations suggest that the molecular paths of intrinsic disease, therefore, may differ from the extrinsic form, and perhaps genetic influences might contribute in accounting for the dissimilarities. With regards to genetics, parent of origin effects may be relevant. The first evidence for atopy-related maternal inheritance was obtained in the early 1990s when a linkage peak influencing IgE responsiveness was mapped to chromosome 11q through the maternal cell line. Subsequently, mutations in the gene encoding the serine protease inhibitor, Kazal type 5 (*SPINK5*), have also been associated with atopic dermatitis when inherited from maternally derived alleles.

A number of genome-wide linkage screens in atopic dermatitis have been described, and these report linkage to chromosomes 1q, 3q, 3p, and 17q. When phenotypes are combined, additional loci have been matched for atopic dermatitis and asthma to 20p; atopic dermatitis with increased allergen-specific IgE levels to 3p, 4p, and 18q; and atopic dermatitis with total serum IgE level to 16q. Linkage studies have also been complemented by direct candidate gene studies in a bid to explain the molecular pathology of atopic dermatitis. These include the β subunit of the high-affinity (FcηRIβ) on 11q12–13 and the cytokine gene cluster located on chromosome 5q31–33. The latter includes several interleukins, GM-CSF, the CD14 antigen, T-cell immunoglobulin domain, mucin protein 1, and SPINK5. However, considerable recent work has focused on genes within the epidermal differentiation complex.

To dermatologists, the association between atopic dermatitis and the monogenic disorder ichthyosis vulgaris (OMIM146700) has been evident for many years, given that several patients with ichthyosis vulgaris also have atopic dermatitis. Histopathologically, many cases of ichthyosis vulgaris are associated with abnormal (diminished) keratohyalin granules within the granular layer, and there is reduced immunohistochemical labeling for filaggrin in ichthyosis vulgaris skin. Filaggrin is the major component of keratohyalin granules. *Filaggrin* is a composite phrase for filament aggregating proteins, repeat units of complex polypeptides derived from profilaggrin that help aggregate keratin filaments in the formation of the epidermal barrier. Filaggrin expression has also been shown to be reduced in atopic dermatitis skin. Ichthyosis vulgaris has been mapped to the epidermal differentiation complex on chromosome 1q21, one of the loci that atopic dermatitis also maps to. Expression of the filaggrin gene in atopic dermatitis skin has also been shown to be reduced. In 2006, it was shown that ichthyosis vulgaris results from loss of function mutations in the *FLG* gene [2]. Ichthyosis vulgaris is a semi-dominant condition with heterozygotes displaying no phenotype or just mild ichthyosis, whereas homozygotes or compound heterozygotes have a more severe form of ichthyosis vulgaris with skin barrier defects. Filaggrin mutations are very common in the general population, occurring in approximately 10% of Europeans. Subsequently, it has been shown that filaggrin gene mutations are a major primary predisposing risk factor for atopic dermatitis (Figure 26.15) [3]. Atopic dermatitis is present at high frequency in carriers of filaggrin mutations with a relative risk (odds ratio) for atopic dermatitis in heterozygotes of >3. A strong clinical indicator of filaggrin mutations in atopic dermatitis is palmar hyperlinearity. It is evident that approximately 50% of all cases of severe atopic dermatitis harbor mutations in the filaggrin gene. It has also been shown that the presence of filaggrin mutations is also a risk factor for asthma, but only for asthma in combination with atopic dermatitis, and not for asthma alone. This finding suggests that asthma in

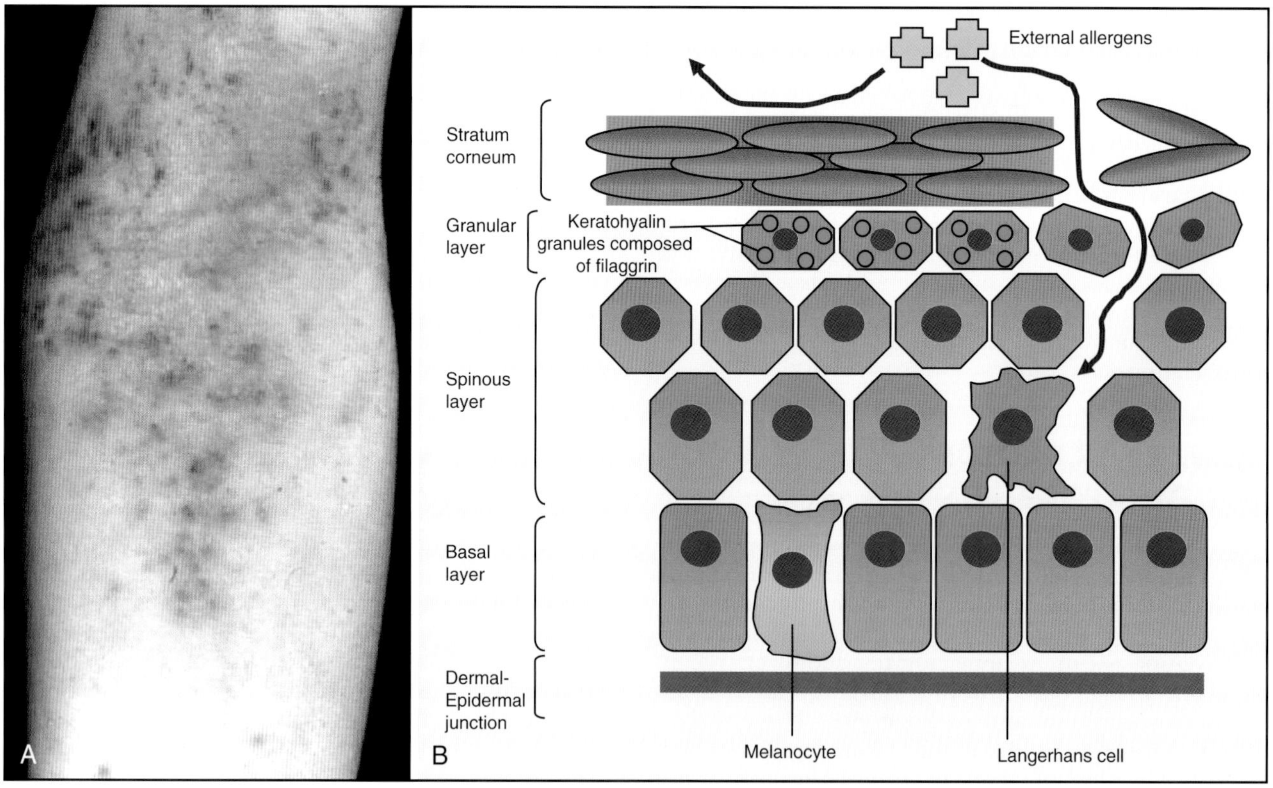

Figure 26.15 Clinicopathological abnormalities in atopic dermatitis. (A) Clinically, there is inflammation in the antecubital fossa with erythema erosions and lichenification. **(B)** Genetic or acquired abnormalities that lead to reduction in filaggrin expression in the granular layer disrupt the skin barrier permeability, which allows penetration of external allergens and presentation to Langerhans cells. Reduced filaggrin in skin may be a major risk factor for atopic dermatitis and increases susceptibility to atopic asthma and systemic allergies.

individuals with atopic dermatitis is secondary to allergic sensitization which develops because of the defective epidermal barrier that allows allergens to penetrate the skin to make contact with antigen-presenting cells. Filaggrin is not expressed in respiratory epithelium, and therefore the new data on filaggrin mutations offer an intriguing new concept that atopic asthma may be initiated as a result of a primary cutaneous (rather than respiratory) abnormality in some individuals. The hypothesis of a defective epidermal barrier underlying asthma, and indeed allergic sensitization, has been verified in several studies which have shown an association between filaggrin gene mutations and extrinsic atopic dermatitis associated with high total serum IgE levels and concomitant allergic sensitizations (Figure 26.16).

Primary defects in filaggrin, however, are not the entire basis of the molecular pathology of atopic dermatitis. It is likely that genes in several other factors, particularly in milder cases of atopic dermatitis, are also important. These factors could include other skin barrier genes, genes affecting protease activity or Langerhans cell function, IgE receptors, mast cell activation, Th-2 cell biology, and cytokines (such as IL-31). It has also been shown that certain cytokines such as IL-4 or IL-12, which may be overexpressed in lesional atopic dermatitis skin, may also lead to secondary reductions in filaggrin protein expression, thereby providing a different route for the same defective skin barrier formation.

The molecular pathology of atopic dermatitis is complex, but the recent focus on abnormalities of the epidermal barrier is providing fascinating new insights into understanding the nature and etiology of atopic dermatitis and perhaps novel treatments. The relevance of these findings, in the near future, is that atopic dermatitis is likely to be subdivided into specific subtypes based on a clearer understanding of the actual pathophysiologic changes. Aside from a structured reclassification of atopic dermatitis, the presence of filaggrin gene mutations is also set to influence and accelerate the design of new treatments that restore filaggrin expression and skin barrier function, given the new evidence that restoration of an intact epidermis may prevent both atopic dermatitis and cases of atopic dermatitis-associated asthma as well as systemic allergies. One of the common mutations in filaggrin is a nonsense mutation (R501X), which may represent an attractive drug target for small molecule approaches that modify post-transcriptional mechanisms designed to increase read-through of nonsense mutations and thereby stabilize mRNA expression. Other approaches that involve drug library screening or *in silico* methods to identify compounds capable of increasing filaggrin expression in the epidermis are also likely to lead to new evidence-based topical preparations suitable for the treatment of atopic dermatitis and ichthyosis vulgaris. Therefore, finding filaggrin mutations in ichthyosis vulgaris and atopic dermatitis represents one of the most significant recent discoveries in dermatologic molecular pathology.

Psoriasis is a common and complex disease affecting approximately 2%–3% of the world's population. It may manifest with inflammation in the skin as well

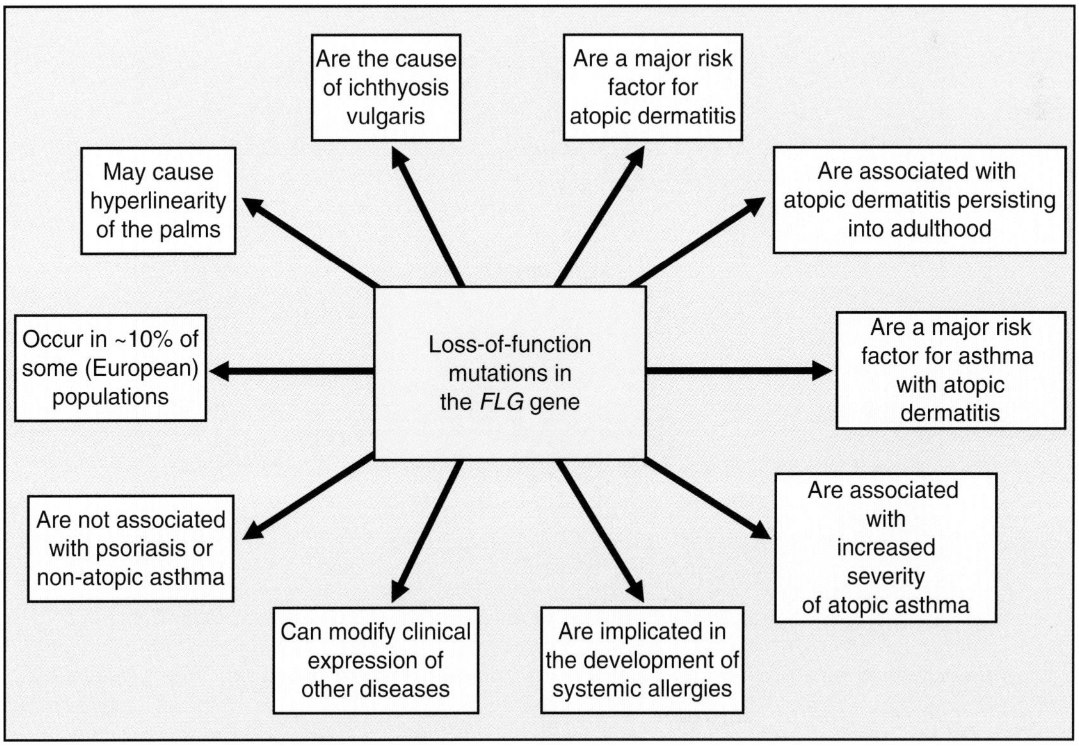

Figure 26.16 Loss of function mutations in the filaggrin gene result in several common disease associations or susceptibilities.

as the nails and joints of patients. Historically, the pathogenesis of psoriasis has been weighted in favor of either primary pathology in keratinocytes or in the immune system. However, in recent years considerable evidence has emerged for specific patterns of immunologic abnormalities and for the critical role of certain cytokines and chemokine networks in psoriasis, findings that may be relevant to the design of new molecular therapies (Figure 26.17).

There is overwhelming evidence that psoriasis has an important genetic component, in that there is a higher incidence among first and second degree relatives of patients than unaffected control subjects and there is a higher concordance in monozygotic compared to dizygotic twins [47]. The risk of a child developing psoriasis has been estimated to be approximately 15% if one parent is affected, approximately 40% if both parents are affected, and approximately 6% if one sibling is affected, compared to 2% when no parent or sibling is affected. The disease has a bimodal distribution of age of onset, with an early peak between 16 and 22 years and a later one between 57 and 60 years. Linkage studies have identified several genetic loci. These include *PSORS1* at 6p21.3, *PSORS2* at 17q, *PSORS3* at 4q, *PSORS4* at 1q, *PSORS5* at 3q, *PSORS6* at 19q, *PSORS5* at 3q, *PSORS6* at 19q, *PSORS7* at 1p, and *PSORS8* at 16q. Early-onset psoriasis has been linked to the *PSORS1* locus, which contains *HLA-Cw6*.

Aside from genetic risk factors, environmental risk factors include trauma (the Köbner phenomenon at sites of injury), infection (including *Streptococcal* bacteria and human immunodeficiency virus infections), drugs (lithium, antimalarials, β-blockers, and angiotensin-converting enzymes), sunlight (in a minority), and metabolic factors (such as high dose estrogen therapy and hypocalcemia). Other factors including psychogenic stress, alcohol, and smoking have also been implicated.

Initial immunologic research in psoriasis focused on the role of prostaglandins and leukotrienes as key mediators of the inflammatory process. Subsequently, emphasis switched from arachidonic acid derivatives to cytokines and chemokines, small proteins that are produced by almost every activated cell in the body and which regulate cell/cell interactions in immune responses and other processes such as wound healing, angiogenesis, hematopoiesis, and tissue remodeling. Cytokines and chemokines may interact with specific receptors and can influence cells in autocrine, juxtacrine, and paracrine fashions. Psoriatic plaques have been shown to contain cytokines such as TNFα, IFN-γ,

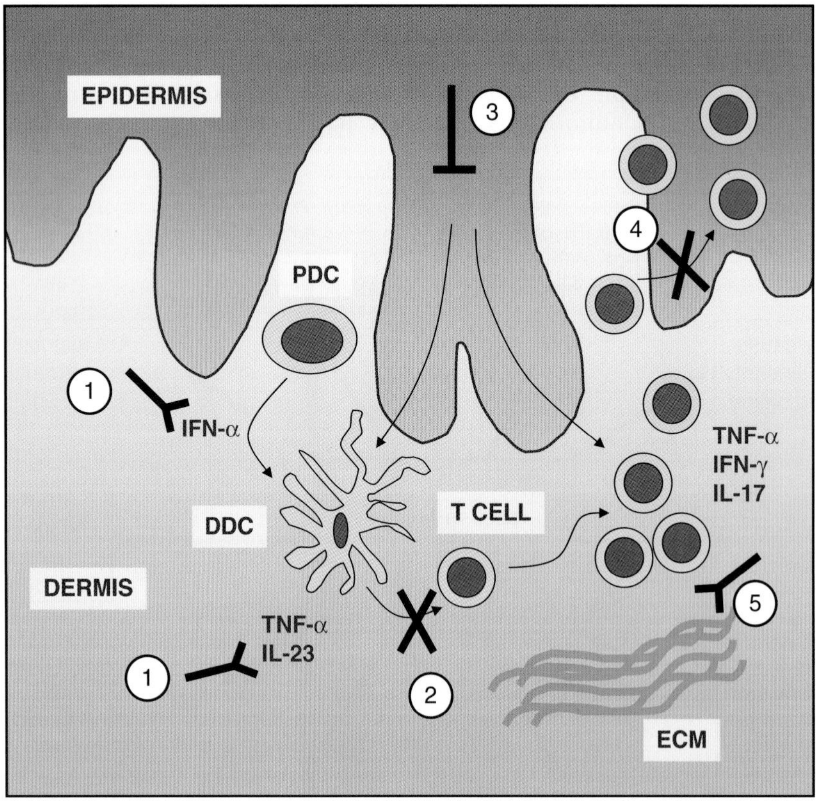

Figure 26.17 **Abnormalities and therapeutic potential for inflamed skin in psoriasis.** There is increasing evidence for a role of tissue-resident immune cells in the immunopathology of psoriasis. New therapies may be developed by **(1)** antagonizing local cytokines and chemokines, such as IFN-α; **(2)** blocking of adhesion molecules (e.g., integrins) and co-stimulatory molecules within the tissue; **(3)** modification of keratinocyte proliferation and differentiation (e.g., use of corticosteroids or vitamin D preparations); **(4)** blocking of entry of dermal T-cells into the epidermis, and **(5)** modification of the microenvironment, including the extracellular matrix (based on original figure published by [87]).

IL-1, IL-2, IL-6, IL-8, IL-15, IL-17, IL-18, IL-19, IL-20, and IL-23, as well as chemokines and chemokine receptors, including CCL2, CCL3, CCL4, CCL5, CCL17, CCL19, CCL20, CCL27, CXCL1, CXCL8, CXCL9, and CXCL10.

A detailed knowledge of cytokine expression has helped define whether psoriasis is a Th-1 or Th-2 disorder. Th-1 diseases are typically associated with production of IL-2, IFN-γ, and TNFα, which promote a T-cell mediated reaction. Conversely in the Th-2 cytokine network response, the immune signals include IL-3, IL-4, IL-5, and IL-10 which contribute to a humoral or B-cell-mediated immune reaction. Psoriasis has traditionally been considered as a Th-1 disease, based on the identification of IFN-γ and TNFα in the lesions with little or no detection of IL-4, IL-5, or IL-10. More recently, other cytokines including IL-18, IL-19, IL-22, and IL-23 have been identified as being upregulated in psoriatic lesions. With relevance to the molecular pathology of psoriasis, one of the first clues came from the observation of clearing of psoriasis in renal transplant patients given the immunosuppressant cyclosporin A. In psoriatic skin that resolved in these individuals, there were reductions in levels of several cytokines, indicating a connection between immunocytes, cytokines, and maintenance of psoriatic plaques. Attempts were then made to therapeutically administer Th-2-type cytokines, such as IL-10, IL-4, or IL-11, to psoriatic patients. However, none of these therapies had any sustained clinical benefits. Nevertheless, the effects of neutralizing a single cytokine can be extremely helpful, particularly with the use of anti-TNFα therapy, a treatment that was initially developed for patients with sepsis. TNFα blockers or receptor blockers are beneficial in patients with psoriasis, but it is unclear if the improvement of psoriatic plaques and arthritis with these agents is due to local or systemic effects.

Intriguing new ideas, however, have recently been gleaned from specific mouse models, in which pre-psoriatic skin is grafted onto a special type of immunodeficient mouse known as an AGR mouse (which lacks type I or IFN-α receptors and lacks type II or IFN-γ receptors, and is RAG deficient). In such mice, psoriatic plaques develop spontaneously. However, if these mice receive injections of anti-TNFα agents, psoriasis does not develop. This demonstrates that resident immunocytes, contained within pre-psoriatic skin, are necessary and sufficient to trigger psoriasis and that the local production of TNFα is critically important in the generation of skin lesions. A pertinent finding from this AGR mouse model is the lack of recirculation or recruitment of cells from the bloodstream necessary to create a psoriatic lesion. Thus, it may not be therapeutically advantageous to try to block trafficking of immunocytes at inflammatory sites by targeting endothelial cell adhesion molecules or chemotactic polypeptides generated in the dermis. For example, clinical trials targeting endothelial cell leukocyte adhesion molecule 1 did not result in clinical benefit in psoriatic patients.

Two recent findings have provided intriguing new insight into the molecular pathology of psoriasis. One observation involves the finding that the innate immune response β cathelicidin molecule LL-37 may be capable of modifying DNA to form an autoantigen. The other discovery is that investigators are now exploring other possible cytokines besides TNFα and IFN-γ and are making promising advances by focusing on the IL-12/IL-23-mediated pathway [48]. A new area of investigation in the field of chronic inflammation centers around a possible inflammatory axis in which IL-12/IL-23 influence levels of a cytokine known as IL-17. This has led to a new paradigm through which Th-17-type T-cells contribute to autoimmunity and chronic inflammation. Thus, besides a Th-1 or Th-2 type immune system, there is also a cytokine network dominated by a Th-17 type response. IL-17 may be important in psoriasis because it can promote accumulation of neutrophils and can affect skin barrier function by inducing release of proinflammatory mediators by keratinocytes. Understanding the molecular role of IL-17 in psoriatic skin may provide new insights for developing novel therapies. However, at present it is unclear whether the primary cellular target within psoriatic plaques involves keratinocytes, macrophages, dendritic cells, T-cells, mast cells, neutrophils, fibroblasts, or endothelial cells, but a clearer understanding of cytokines and chemokine networks in psoriasis is unraveling the molecular pathology of this common inflammatory disease, and this may have potential benefits for patients in the not-too-distant future.

SKIN PROTEINS AS TARGETS FOR INHERITED AND ACQUIRED DISORDERS

The integrity of the skin as a mechanical barrier depends, in part, on adhesive complexes that link cell-to-cell and cell-to-basement membrane. Two key junctional complexes in this task are the hemidesmosomes and the desmosomes. Key insight into the contribution individual components of hemidesmosomes and desmosomes make to skin adhesion has been determined from a variety of *in vitro* knockdown studies and *in vivo* animal models as well as a range of naturally occurring human gene mutations that lead to a diverse group of skin fragility disorders. Defects in many desmosomal or desmosomal proteins also lead to other clinical abnormalities (including extracutaneous involvement) or other cell biologic abnormalities (such as changes in cell proliferation or differentiation). However, in addition to genetic diseases, other clues to the function of hemidesmosomal and desmosomal proteins have been derived from animal models or human diseases, in which the same structural proteins can be targeted and disrupted by autoantibodies (Figure 26.18). Thus, several hemidesmosomal and desmosomal proteins may serve as target antigens for both inherited and acquired disorders (Figure 26.19) [49].

One of the main transmembranous hemidesmosomal proteins is type XVII collagen, also known as the 180 kDa bullous pemphigoid antigen. Autoantibodies against this protein typically result in bullous pemphigoid, a chronic vesiculo-bullous disease that usually

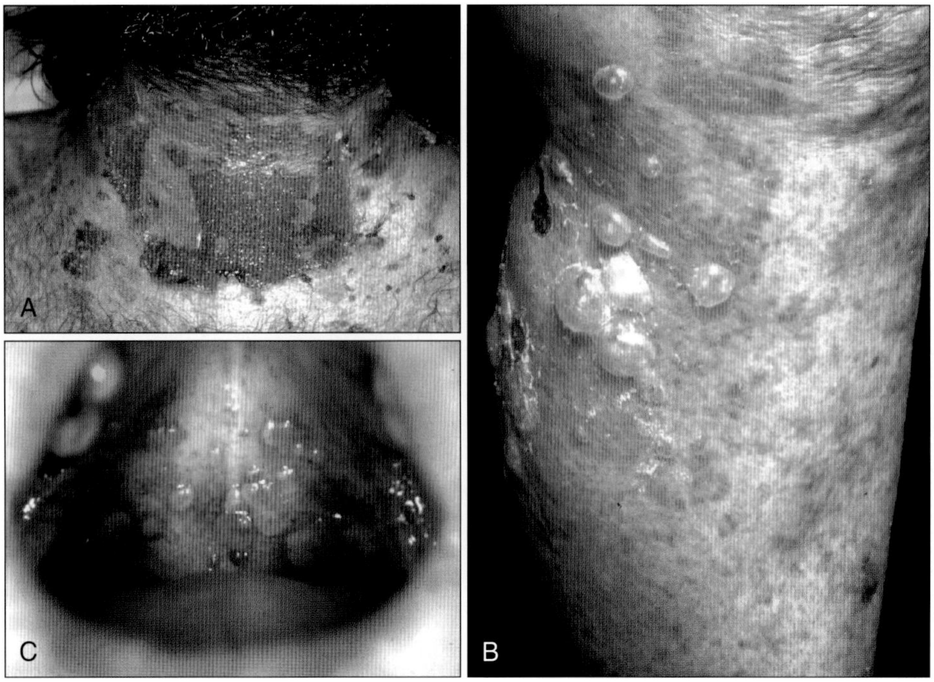

Figure 26.18 Clinical pathology resulting from autoantibodies against desmosomes or hemidesmosomes. **(A)** Pemphigus vulgaris resulting from antibodies against desmoglein 3; **(B)** Bullous pemphigoid associated with antibodies against type XVII collagen; **(C)** Mucous membrane pemphigoid associated with antibodies to laminin-332.

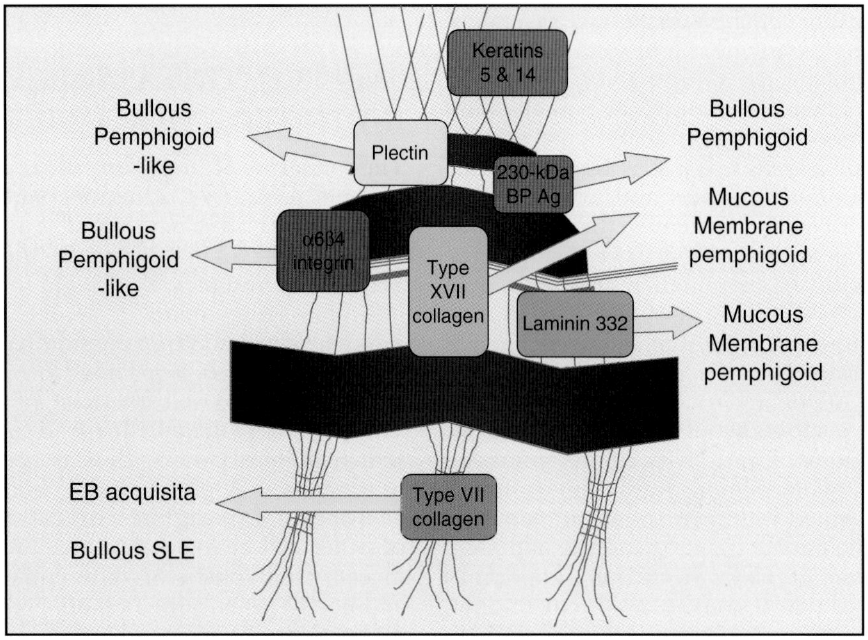

Figure 26.19 Illustration of hemidesmosomal structural proteins and the autoimmune diseases associated with antibodies directed against these individual protein components.

affects the elderly. Histology shows subepidermal blisters with eosinophils, and the pathogenic antibodies are usually directed against a particular epitope (within the NC16A domain) in the first noncollagenous extracellular part of type XVII collagen. IgG1 subclass autoantibodies are found in patients with active skin lesions, whereas IgG4 autoantibodies are found in patients in remission. Besides bullous pemphigoid, other blistering conditions may also have autoantibodies against type XVII collagen. For example, pemphigoid gestationis, which is an acute, pruritic vesiculo-bullous eruption, is seen in pregnant women. Patients produce autoantibodies against the same epitope of type XVII collagen. It has been proposed that this disease is initiated

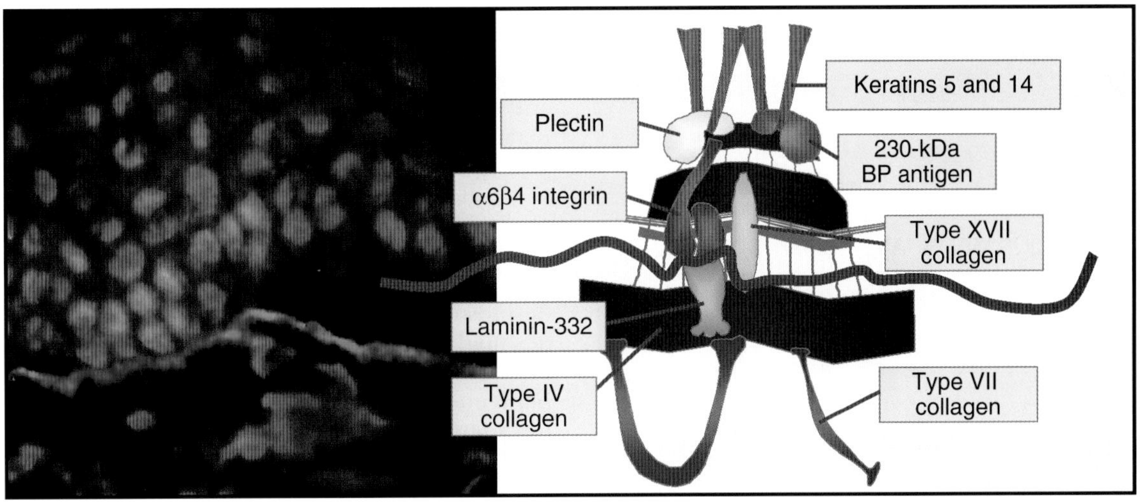

Figure 26.20 **Salt-split skin technique to diagnose immunobullous disease.** Incubation of normal human skin in 1M NaCl overnight at 4 °C results in cleavage through the lamina lucida. This results in separation of some proteins to the roof of the split and some to the base (above and below pink line on the schematic). In the skin labeling shown, immunoglobulin from a patient's serum binds to the base of salt-split skin. Further analysis revealed that the antibodies were directed against type VII collagen. This technique is useful in delineating bullous pemphigoid from epidermolysis bullosa acquisita, both of which are associated with linear IgG at the dermal-epidermal junction in intact skin.

by the aberrant expression of major histocompatibility complex type II antigens of a paternal haplotype within the placenta. IgA autoantibodies against the NC16A domain of type XVII collagen give rise to two different skin diseases: chronic bullous disease of childhood and linear IgA bullous dermatosis. Chronic bullous disease of childhood usually presents in young children with clustered tense bullae, often in the perioral or perineal regions. Clinical manifestations can mimic bullous pemphigoid or dermatitis herpetiformis. In chronic bullous disease of childhood and linear IgA disease, histology shows subepidermal collections of neutrophils with linear IgA deposition at the dermal-epidermal junction. A further disease with autoantibodies against type XVII collagen is lichen planus pemphigoides. Patients typically have lichen planus lesions mixed with tense blisters, either on lichen planus lesions or on normal skin. Autoantibodies react with epitopes within the NC16A domain, but the precise epitope is different from bullous pemphigoid.

Autoantibodies to type VII collagen, the major component of anchoring fibrils, give rise to two different conditions: epidermolysis bullosa acquisita and bullous systemic lupus erythematosus. Patients with epidermolysis bullous acquisita have subepidermal blisters. In some patients, the blistering phenotype can resemble bullous pemphigoid, but in others the inflammation leads to scarring and milia formation at trauma-prone sites, more reminiscent of an inherited form of skin blistering such as dystrophic epidermolysis bullosa. In both epidermolysis bullosa acquisita and bullous systemic lupus erythematosus, there is linear deposition of IgG at the dermal-epidermal junction. However, a useful technique in immunodermatology to distinguish epidermolysis bullosa acquisita from bullous pemphigoid is indirect immunofluorescence using 1 M sodium chloride split-skin (Figure 26.20). Incubation of normal skin in saline results in a split within the lamina lucida. Thus, certain skin antigens such as type XVII collagen map to the roof of the split, whereas others such as type VII collagen map to the base. This means that in bullous pemphigoid and epidermolysis bullosa acquisita, labeling with sera from patients with these conditions on salt-split skin can permit an accurate diagnosis to be made. Specifically, if the IgG maps to the roof of the split, the diagnosis is bullous pemphigoid, whereas if it maps to the base, this is epidermolysis bullous acquisita. The salt-split technique can sometimes be used on skin biopsies taken from patients to see whether *in vivo* bound IgG maps to the roof or base. However, in most laboratories using patient sera on normal salt-split skin is the more widely practiced technique.

Another autoimmune blistering condition that targets hemidesmosomal components is mucous membrane pemphigoid (previously known as cicatricial pemphigoid). This is a chronic progressive autoimmune subepithelial disease characterized by erosive lesions of the skin and mucous membranes that result in scarring. Lesions commonly affect the ocular and oral mucosa. Direct immunofluorescence studies show IgG and/or IgA autoantibodies at the dermal-epidermal junction. Salt-split indirect immunofluorescence can show epidermal, dermal, or roof and base labeling patterns, which reflects the different autoantigens seen in this disease. In most cases, the target epitope is part of the extracellular domain of type XVII collagen, although this differs from the epitope associated with bullous pemphigoid. A minority of patients with mucous membrane pemphigoid display antibodies against laminin-332. This is an important subset of patients to identify since there is association with certain solid tumors, particularly malignancies of the upper aero-digestive tract (Figure 26.21) [50]. Cases of

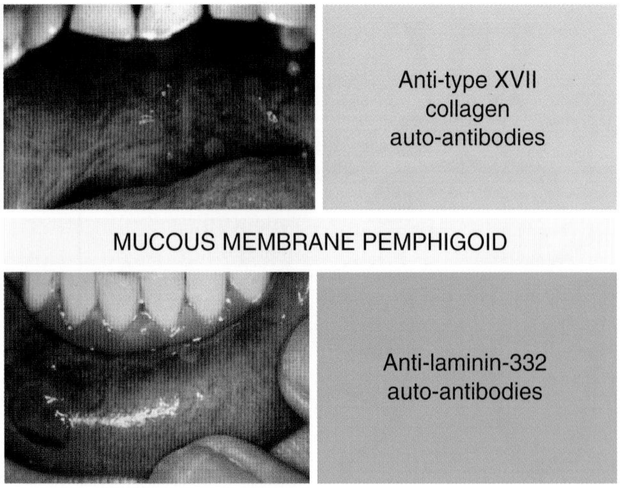

Figure 26.21 **Mucous membrane pemphigoid may be associated with autoantibodies against either type XVII collagen or laminin-332.** Distinction between the two may have clinical relevance since antilaminin-332 antibodies in mucous membrane pemphigoid can be associated with malignancy (especially of the upper aero-digestive tract) in some patients.

mucous membrane pemphigoid with antibodies to β_4 integrin tend to have only ocular involvement and may therefore be referred to as ocular mucous membrane pemphigoid [51].

Although many of the immunobullous diseases that target hemidesmosomal proteins involve immunopathology in which antibodies are initially directed against critical epitopes on individual proteins, it has been shown that in many cases there may be several antigens targeted by humoral immunity. This phenomenon is known as epitope spreading and reflects chronic inflammation that may alter the immunogenicity of other neighboring proteins involved in adhesion [52]. Epitope spreading can involve the generation of antibodies to further epitopes on the same protein or to other epitopes on different structural proteins. It may be an immunologic observation with no clinical sequelae, but in some patients epitope spreading can lead to transition from one autoimmune disease to another. Published examples include transition from mucosal dominant pemphigus vulgaris to mucocutaneous pemphigus vulgaris, pemphigus foliaceus to pemphigus vulgaris, bullous pemphigoid to epidermolysis bullosa acquisita, dermatitis herpetiformis to bullous pemphigoid, pemphigus foliaceus to bullous pemphigoid, and concurrent bullous pemphigoid with pemphigus foliaceus. Antibodies to the plakin protein, desmoplakin, which are characteristically found in paraneoplastic pemphigus, have also been detected in pemphigus vulgaris as well as oral and genital lichenoid reactions. In epidermolysis bullosa acquisita, antibodies to laminin-332 may be found in addition to the typical type VII autoantibodies. In bullous systemic lupus erythematosus, multiple autoantibodies against type VII collagen, bullous pemphigoid antigen 1, laminin-332, and laminin-311 have been reported. The phenomenon of epitope spreading is perhaps best exemplified in paraneoplastic pemphigus, which is characteristically associated with autoantibodies against many desmosomal and hemidesmosomal proteins, including desmogleins, plakins, bullous pemphigoid antigen 1, plectin, and plakoglobin.

In addition to hemidesmosomal proteins, the structural components of desmosomes may also be targets for autoimmune diseases. The skin blistering disease pemphigus is associated with autoantibodies against desmosomal cadherins, principally desmoglein 3 and desmoglein 1. However, the intraepithelial expression patterns of desmoglein 1 and desmoglein 3 differ between the skin and mucous membranes, giving rise to different clinical manifestations when patients have antibodies against different desmogleins. In the skin, desmoglein 1 is expressed throughout the epidermis but more so in superficial parts, as opposed to desmoglein 3, which is predominantly expressed in the basal epidermis. In the oral mucosa, both types of desmogleins are expressed throughout the epithelium, but desmoglein 1 is expressed at a much lower level than desmoglein 3. When there is co-expression of desmoglein 1 and desmoglein 3, these proteins may compensate for each other, but when expressed in isolation, specific pathology can arise when those proteins are targeted by autoantibodies [53]. For example, when there are only desmoglein 1 antibodies in the skin, blisters appear in the superficial epidermis, a site where there is no co-expression of desmoglein 3, whereas in the basal epidermis, the presence of desmoglein 3 can compensate for the loss of function of desmoglein 1. In the oral epithelium, keratinocytes express desmoglein 3 at a much higher level than desmoglein 1, and despite the presence of desmoglein 1 antibodies, no blisters form. Therefore, sera containing only desmoglein 1 antibodies cause superficial blisters in the skin without mucosal involvement, and the clinical consequences are pemphigus foliaceus and its localized form, pemphigus erythematosus. Desmoglein 1 is also the target antigen in endemic pemphigus, known as fogo selvagem. Desmoglein autoantibodies result in skin with very superficial blisters, such that crust and scale are the predominant clinical features. Desmoglein 1 may also be targeted in a different manner—not by autoantibodies, but by bacterial toxins (Figure 26.22). Specifically, exfoliative toxins A-D which are produced by *Staphylococcus aureus*, specifically cleave the extracellular part of desmoglein 1 [54]. This leads to bullous impetigo and Staphylococcal scalded skin syndrome with superficial blistering in the epidermis, findings that are histologically similar to pemphigus foliaceus associated with autoantibodies to the extracellular part of desmoglein 1.

When the sera of a patient contains only desmoglein 3 autoantibodies, co-expressed desmoglein 1 can compensate for the impaired function of desmoglein 3, resulting in no or only limited skin lesions. However, in the mucous membranes, oral erosions predominate, as desmoglein 1 cannot compensate for the impaired desmoglein 3 function because of its low expression. Therefore, the patient typically has painful oral ulcers without much initial skin involvement. This

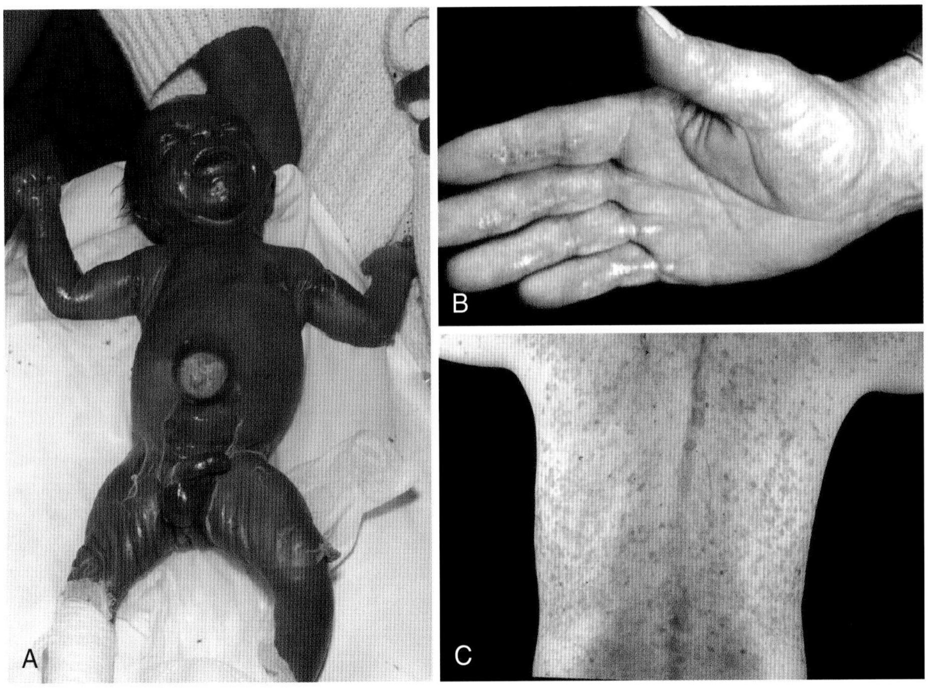

Figure 26.22 **Clinical consequences of disruption of desmoglein 1 in human skin. (A)** Staphylococcal toxins cleave the extracellular part of desmoglein 1 and result in staphylococcal scalded skin syndrome. **(B)** Inherited autosomal dominant mutations in desmoglein 1 can result in striate palmoplantar keratoderma. **(C)** Autoantibodies against desmoglein 1 result in pemphigus foliaceus, which is associated with superficial blistering and crusting in human skin.

accounts for the clinical phenotype in patients with the mucosal dominant type of pemphigus vulgaris. However, when sera contain antibodies to both desmoglein 1 and desmoglein 3, as seen in the mucocutaneous type of pemphigus vulgaris, patients have extensive blisters, mucosal erosions and skin blisters, because the function of both desmogleins is disrupted.

Intercellular autoantibodies in pemphigus are typically composed of IgG isotypes, although some patients with superficial blistering may have intercellular autoantibodies of the IgA subclass. There are two clinical subtypes of IgA pemphigus. One is a subcorneal pustular dermatosis subtype (although this should not be confused with the nonautoimmune Snedden-Wilkinson syndrome), and this is associated with flaccid vesicopustules with clear fluid and pus arranged in an annular/polycyclic configuration. Indirect immunofluorescence studies show IgA autoantibodies reacting against the superficial epidermis. These antibodies usually recognize epitopes within the desmosomal cadherin, desmocollin 1. In contrast, IgA pemphigus patients may have a different phenotype known as the intraepidermal neutrophilic subtype that presents with a sunflower-like configuration and pustules and vesicles with indirect immunofluorescence showing IgA intercellular autoantibodies throughout the entire thickness of the epidermis. As well as spontaneous onset cases of pemphigus, certain drugs such as D-penicillamine can lead to drug-induced pemphigus. Most patients have autoantibodies against the same target epitopes as pemphigus patients. In paraneoplastic pemphigus, the clinical presentation involves progressive blistering and erosions on the upper trunk, head, neck, and proximal limbs with an intractable stomatitis. Erythema multiforme-like lesions on the palms and soles can help distinguish the condition from ordinary pemphigus. Indirect immunofluorescent studies on rat bladder, an organ rich in transitional epithelial cells, demonstrate IgG intercellular autoantibodies consistent with autoimmunity against plakin proteins although numerous antigenic targets are usually present.

Collectively, these autoimmune diseases that target hemidesmosomal or desmosomal proteins illustrate the clinical and biologic consequences of targeting specific proteins through autoimmune assault. There may be some similarities to the inherited mechanobullous diseases in which there are fundamental mutations in the genes that encode these same proteins, but clearly super-added inflammation results in some clinical differences. Nevertheless, the inherited and acquired diseases that target skin proteins do provide new insight into how specific skin proteins contribute to maintaining skin integrity.

MOLECULAR PATHOLOGY OF SKIN CANCER

Skin cancer is very common. Nonmelanoma skin cancers, which include basal cell carcinomas (BCCs) and squamous cell carcinomas (SCCs), are the most common forms of human neoplasia, affecting about 1 million Americans each year [36]. Of these, BCCs alone account for 750,000 new cases per year. It is estimated that 1:3 Caucasians born in the United States after

1994 will develop at least one BCC in their lifetime. Likewise, the incidence of melanoma has been increasing since the mid-1990s at an annual rate of 3%–7% for Caucasian populations. For those Americans born in 2000, there is a lifetime risk of 1:75 of developing melanoma. Currently, approximately 60,000 new melanomas will be diagnosed in the United States each year [55]. However, recent insight into the molecular pathology of both nonmelanoma skin cancer and melanoma is providing new opportunities to understand how and why these tumors occur, as well as providing a basis for molecular target-based chemo-prevention and novel therapeutic management of skin cancer.

The main risk factor for skin neoplasia is environmental exposure to ultraviolet irradiation. The ultraviolet component of sunlight can be divided into 3 energy subtypes: UVC (100–280 nm), UVB (280–315 nm), and UVA (315–400 nm). The action spectra for UVB and UVC closely match the absorption spectrum of DNA, resulting in the formation of pyrimidine dimers, involving nucleotides, thiamine (T), and/or cytosine (C). In contrast, UVA is absorbed by other cellular chromophores, thereby generating oxygen-reactive species such as hydroxyl radicals. These can result in DNA strand breaks and chromosome translocations. Of the ultraviolet light that reaches the earth, UVC is completely absorbed by stratospheric ozone, whereas UVA accounts for more than 90%, and UVB approximately 10%. Most UVB-induced mutations are located almost exclusively at dipyrimidine nucleotide sites (TT, CC, CT, and TC). Approximately 70% of observed mutations are C-to-T transitions and 10% are CC-to-TT, the latter representing an ultraviolet signature mutation [56]. A few other mutagens are known to involve tandem bases. Cellular mechanisms of DNA repair are not always effective, and these ultraviolet B-induced mutations can proceed unchecked, becoming permanent and subsequently inherited by all the progeny of the mutated cell, thereby allowing expression of the aberrant gene/protein function. Aside from ultraviolet radiation inducing DNA point mutations and small deletions, it may also result in gross chromosomal changes. Both UVA and UVB irradiation can induce formation of micronuclei which are cytogenic indicators of chromosomal damage. These round particles of genetic material are believed to result from DNA double-strand breaks, which form as a result of DNA repair of ultraviolet-induced pyrimidine dimers and/or free radical formation.

Collectively, these genetic/chromosomal changes initiate and promote cancer formation as well as the increased genomic instability and loss of heterozygosity (LOH) frequently observed in nonmelanoma skin cancer. Cytogenic analysis has enabled the identification of a number of chromosomal abnormalities associated with nonmelanoma skin cancer and has implicated certain regions containing oncogenes and tumor suppressor genes that may be involved in their development. For example, in BCCs, early LOH studies identified regions on chromosome 9q22 as a common observation specific to these tumors. These regions harbor the Patched tumor suppressor gene (*PTCH*), which is a transmembrane receptor involved in the regulation of hedgehog signaling (Figure 26.23). Subsequently, the discovery of mutations that are known to activate hedgehog signaling pathways, including *PTCH*, Sonic hedgehog *(Shh)*, and Smoothened *(Smo)*, implicates hedgehog signaling as a fundamental transduction pathway in skin tumor development [57].

Shh is responsible for several important functions during embryogenesis, including neural tube patterning and the development of left-right symmetry, limb polarization, and the morphogenesis of various organs including the axial skeleton, limbs, lungs, skin, hair, and teeth. In skin, the Shh pathway is crucial for maintaining the stem cell population and regulating the development of hair follicles and sebaceous glands. Although key embryonic developmental signaling pathways may be switched off during adulthood, aberrant activation of these pathways in adult tissue is often oncogenic. The Shh pathway may be activated in many neoplasms, including BCCs, medulloblastoma, rhabdomyosarcoma, and abnormalities in Shh signaling pathway components—such as Shh, PTCH1, Smo, GLI1, and GLI2—are major contributing factors in the development of BCCs. The function of PTCH1 is to repress Smo signaling. This function is impaired when PTCH1 is mutationally inactivated or when stimulated by Shh binding: both of these lead to uncontrolled Smo signaling. Downstream of Smo are the GLI transcription factors. Overexpression of GLI1 or GLI2 can lead to BCC development, and GLI1 can also activate PDGF-α, the expression of which may be increased in BCCs. The roles of other Smo target molecules such as the suppressor of fused Su(Fu) and protein kinase A (PKA) in the development of BCC are not fully understood. Other components of the pathway include a putative antagonist of Smo signaling, known as hedgehog interacting protein (HIP), and an actin binding protein, missing in metastases (MIM), which is an Shh responsive gene. MIM is a part of the GLI/Su(Fu) complex and potentiates GLI-dependent transcription using domains distinct from those used for monomeric actin binding. Alterations in Shh regulate cell proliferation and associated cell cycle events. Shh overexpression leads to epidermal hyperplasia, accompanied by the proliferation of normally growth-arrested cells. Shh-expressing cells fail to exit the S and G2/M phases in response to calcium-induced differentiation signals and are unable to block the p21CIP1/WAF1/induced growth arrest in skin keratinocytes. Furthermore, PTCH1 protein interacts with phosphorylated cyclin B1 and blocks its translocation to the nucleus. The Shh/GLI pathway also upregulates expression of the phosphatase CDC25b, which is involved in G2/M-transition. Thus, the loss of regulation of cell cyclin control is associated with the development of epithelial cancers, including BCCs.

Patients with the autosomal dominant disorder known as nevoid basal cell carcinoma syndrome or Gorlin syndrome (OMIM109400) have substantially increased susceptibility to BCCs and other tumors, including medulloblastomas, meningiomas, fibromas, rhabdomyomas, and rhabdomyosarcomas. They may also manifest jaw cysts and ectopic calcification, spina

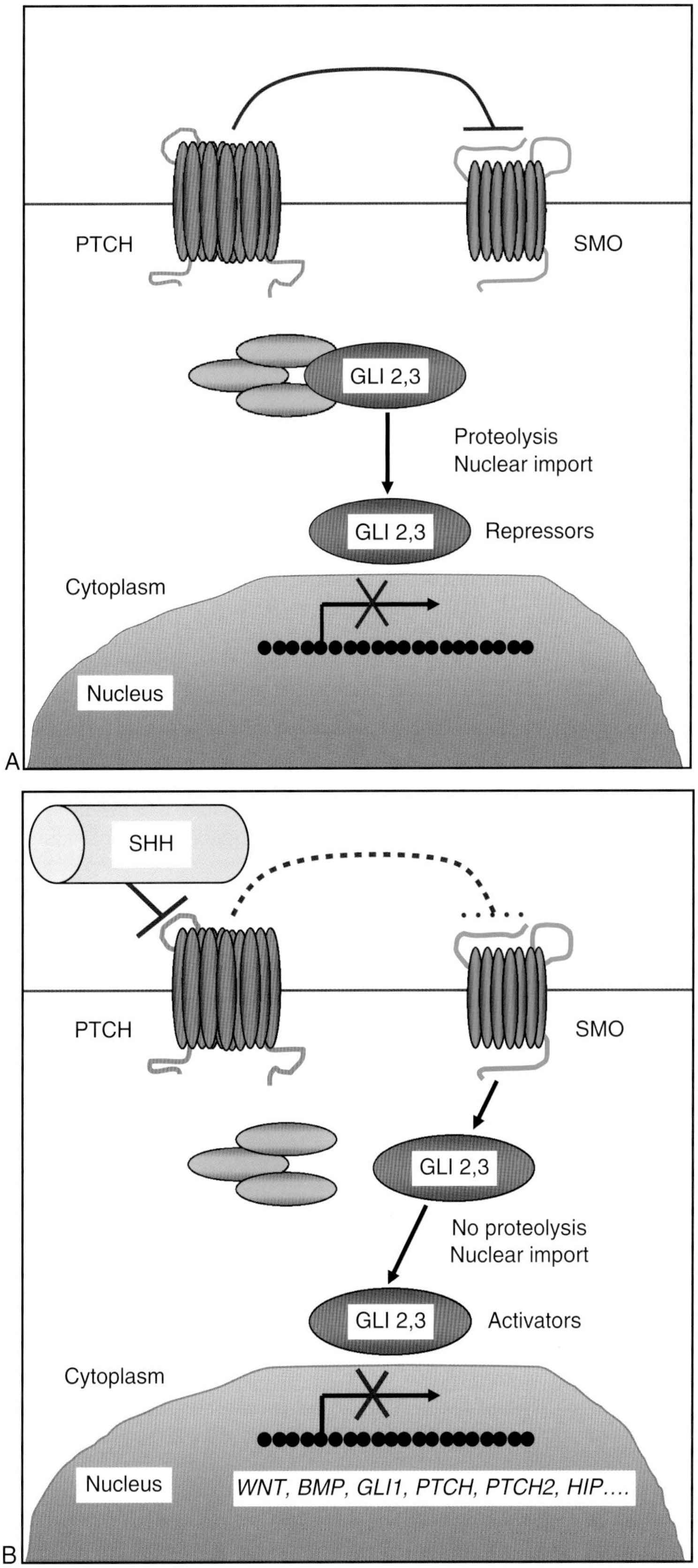

Figure 26.23 **The SHH signaling pathway. (A)** In the absence of SHH, PATCHED constitutively represses smoothened, a transducer of the SHH signal. **(B)** Binding of the ligand SHH to PTCH relieves its inhibition of SMO and transcriptional activation occurs through the GLI family of proteins, resulting in activation of target genes.

(Continued)

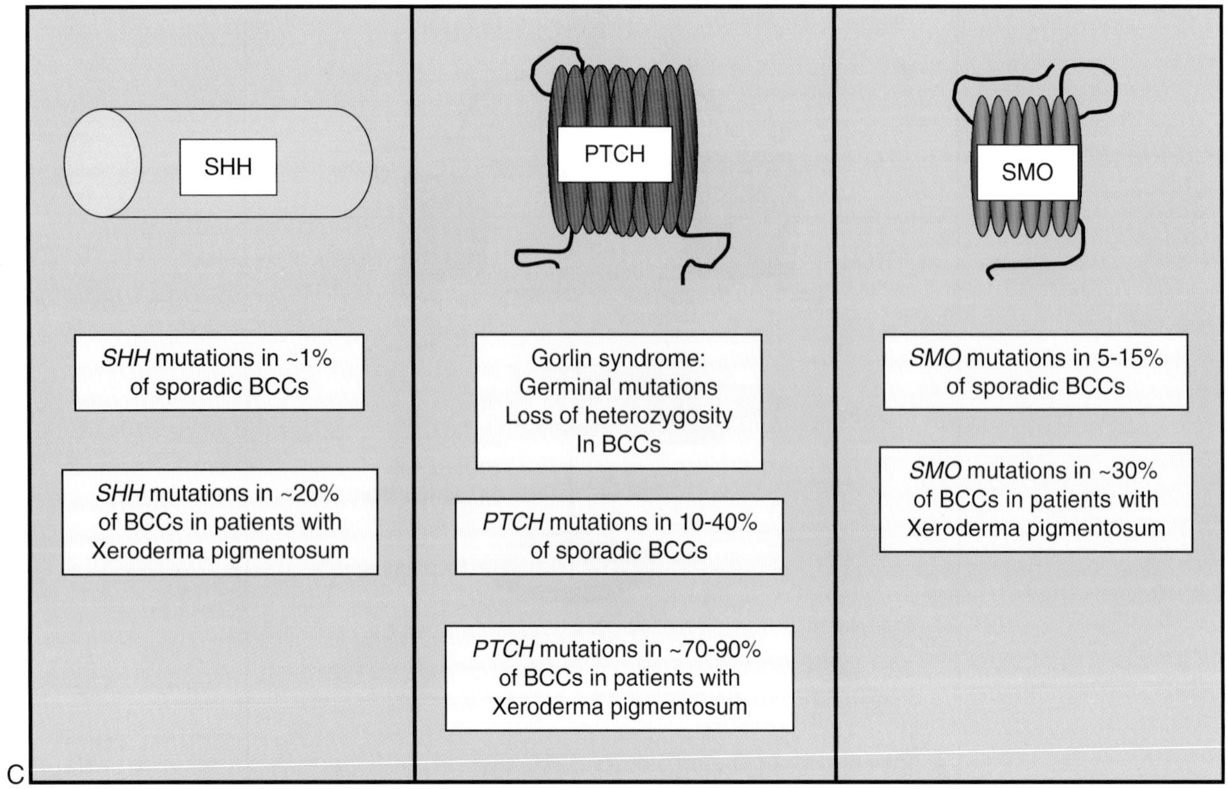

Figure 26.23—Cont'd **(C)** Mutations in *SHH* or *PTCH* or *SMO* may be associated with basal cell carcinomas, both in sporadic tumors as well as in certain genodermatoses, such as xeroderma pigmentosum, that are associated with an increased risk of BCC. Germinal mutations in the *PTCH* gene underlie Gorlin's syndrome. (Based on original figures published by [88]).

bifida, rib defects, palate abnormalities, coarse facies, hypertelorism, microcephaly, and skeletal abnormalities with bony and soft tissue overgrowth. Affected individuals develop BCCs on any part of the body, particularly in sun-exposed skin. These patients have heterozygous germline mutations in the *PTCH* gene [58]. The BCCs in these individuals retain a mutant germline and lose the wild-type allele as a second hit. Mutations in the *PTCH* gene have also been demonstrated in up to 40% of sporadic BCCs, emphasizing that *PTCH* gene mutations are important in the development of BCCs. Other sporadic BCCs which do not have mutations in *PTCH* may carry mutations in the *Smo* gene [59]. An understanding of the mutations that lead to activation of hedgehog signaling has thus expanded our knowledge of the genetic basis of BCCs.

Although most therapy for BCC is straightforward, involving local surgery, cryotherapy, radiotherapy, or topical chemotherapy, insight into the molecular pathology of BCC has led to development of a number of animal models that may be useful in developing chemo-prevention strategies, as well as confirming specific contributions from hedgehog signaling pathway components. For example, drugs such as cyclopamine are known to be a specific inhibitor of Shh signaling. Other possible chemo-prevention strategies might involve small molecule hedgehog signaling inhibitors or immunosuppressive agents such as rapamycin, which is an inhibitor of GLI1.

Overall, the molecular dissection of the Shh pathway has provided novel insight into the molecular pathology of BCCs as well as other cancers. It is likely that this information will lead to the development of various agonists and antagonists of the pathway that can have clinically relevant regulatory or curative roles. Development of small molecules that have specific agonist or antagonist pharmacologic functions might therefore be exploited to develop new therapeutics for the treatment of skin cancers and other tumors that are associated with aberrant activity of this pathway.

Like BCC, melanoma is another tumor that may be entirely curable by early recognition and therapeutic intervention. Nevertheless for patients with advanced metastatic melanoma, the 5-year survival is currently estimated at only 6% with a median survival time of 6 months [60]. It is important, therefore, to try to understand the molecular pathology of melanoma to determine which patients are most at risk and which tumors will exhibit the most aggressive biology.

Dissection of the melanoma genome has revealed a number of driver mutations (those that probably influence the growth of tumors) and passenger mutations (those that do not specifically promote tumor growth). One important pathway is the RAS/Mitogen-activated protein kinase (MAPK) pathway, which regulates cell proliferation and survival in several cell types (Figure 26.24) [61,62]. Activating cell mutations in *NRAS* and *BRAF* have a combined prevalence of approximately

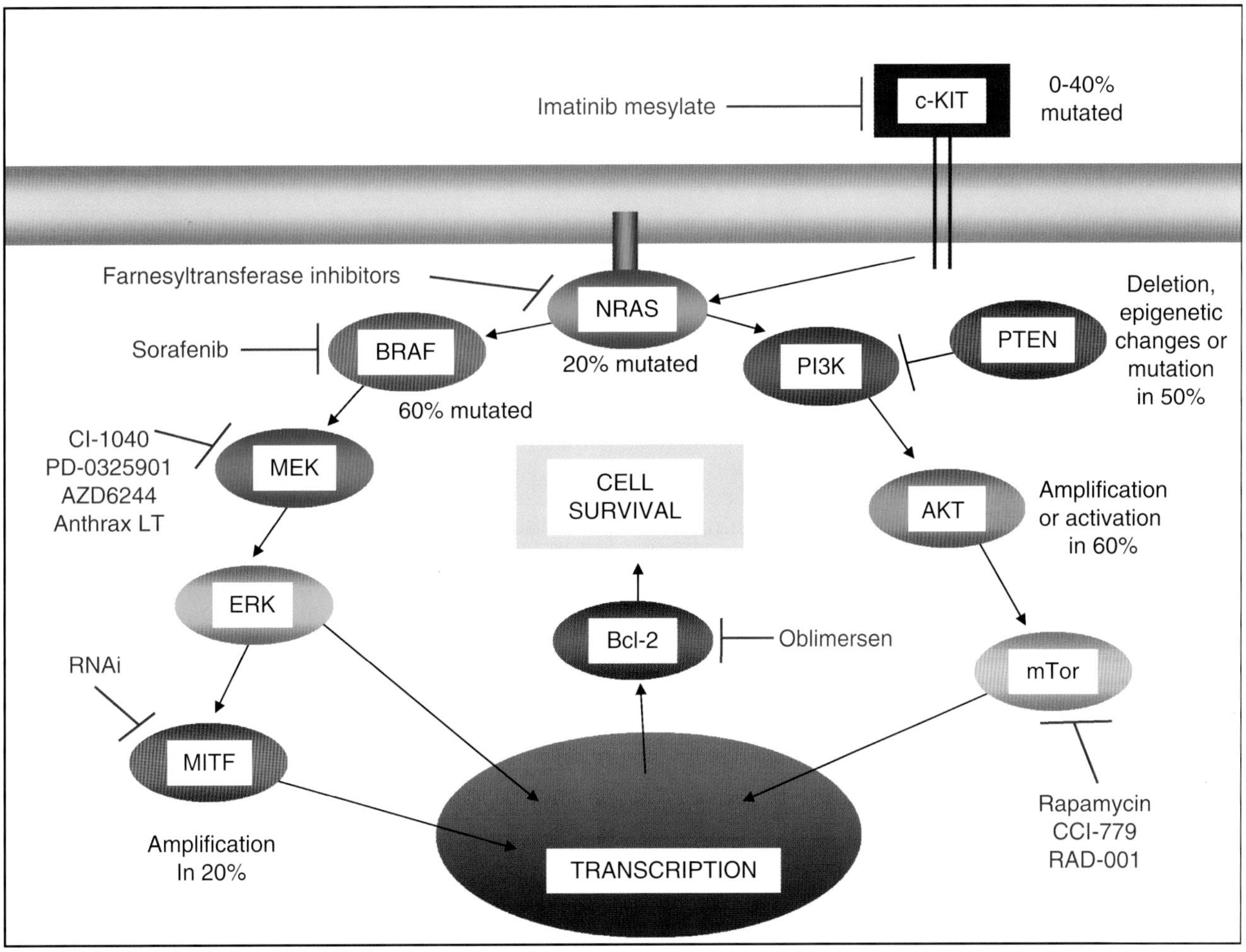

Figure 26.24 **Potential for targeted therapies in melanoma.** Recent improvement in defining the genetics of melanoma has led to the development of targeted therapeutic agents that are directed at specific molecular aberrations involved in tumor proliferation and resistance to chemotherapy. (Based on an original figure published by [89]).

90% in melanoma and in benign melanocytic lesions, suggesting that activation of the MAPK pathway is an early essential step in melanocytic proliferation. Activating somatic *NRAS* mutations occur in 10%–20% of melanomas. Of note, 3 highly recurrent *NRAS* missense changes represent over 80% of all mutations in this gene. *NRAS* mutations are more common on chronically sun-exposed sites and appear to occur early in tumorigenesis as well as being common in congenital nevi. Downstream of NRAS, activating mutations in *BRAF* have been identified, including a common missense mutation at valine 600. This mutation is equally prevalent in benign nevi and in melanoma, suggesting that BRAF activation is necessary for melanocytic proliferation, but not for tumorigenesis. To date, there has been no correlation between *BRAF* mutations, disease progression, and clinical outcome.

A proto-oncogene, *KIT*, encodes the stem cell factor receptor tyrosine-kinase found on numerous cell types, including melanocytes. Oncogenic mutations and increases in copy number of *KIT* appear to be more prevalent in rarer types of melanoma, such as those occurring at acral or mucosal sites. c-KIT signals via the MAPK pathway, and a downstream target is microphthalmia transcription factor (MITF), which is a critical regulator of melanocyte function. MITF regulates the development and differentiation of melanocytes and maintains melanocyte progenitor cells. MITF has a clear role in melanocyte survival, and one of its transcriptional targets is the apoptosis antagonist and proto-oncogene BCL-2. Apart from MAPK signaling, another pathway, the V-RAS/phosphatidylinositol-3 kinase (PIK3) pathway, may also be activated in certain melanomas. Although PIK3 itself is not mutated in melanoma, several downstream components may have a role in melanoma tumorigenesis, including PTEN and Akt. PTEN has a tumor suppressor role in melanomas, and its expression has been shown to be reduced or lost in some primary or metastatic melanomas. Restoration of PTEN in PTEN-deficient melanoma can reduce melanoma tumorigenicity and metastases. Changes in PTEN expression are not usually due to mutations, but involve epigenetic inactivation of both functional alleles. Akt is an important kinase in melanoma survival and progression. Akt3 is the main Akt isoform that is activated in melanomas. Strong expression of Phospho-Akt is found in several melanomas, suggesting a direct role of Akt in tumor progression. Phosphorylation of Akt is associated with increased PIK3 activity. Knowledge of these pathways is important if new treatments are to be developed for advanced melanoma, since it is clear that alterations in survival and growth signaling pathways in melanoma tumor cells can lead to increased tumorigenesis and resistance to chemotherapy.

A detailed knowledge of the molecular pathology of melanomas may also have other therapeutic relevance. For example, it may be possible to render melanoma cells more sensitive to existing forms of chemotherapy, to enhance apoptosis, or to restrict the proliferation of the oncogene-driven cells by targeting these aberrant pathways. New drugs such as Oblimersen (antisense oligonucleotide targeted to the antiapoptotic protein BCL-2) and Sorafenib (a small molecule inhibitor of BRAF that induces apoptosis) are giving some encouraging results, especially when used in combination with traditional chemotherapy [63;64]. Further investigation into the molecular pathology of melanoma is likely to disclose novel therapeutic targets and allows more selective administration of existing therapies to benefit specific patient subgroups.

MOLECULAR DIAGNOSIS OF SKIN DISEASE

Understanding the molecular pathology of skin diseases has the potential to bring about several clinical and translational benefits for patients. Molecular data are helpful in diagnosing both inherited and acquired skin diseases and also contribute to improved disease classification, prognostication, clinical management, and the feasibility for designing and developing new treatments for patients. In many infectious diseases, the gold standard for diagnosis is identification of infectious agents by culture and species identification. However, material yield can be unsatisfactory, and the process can take several weeks, or months, before a pathogen is identified. Indeed, certain diseases can be caused by different micro-organisms, and these may have different responses to antimicrobial agents. For example, subcutaneous mycosis with lymphangiotic spread is typically caused by the mold *Sporothrix schenckii*, but lymphatic sporotrichoid lesions can be caused by diverse pathogens. These include other nontuberculous mycobacteria, most typically *Mycobacterium marinum*, a common pathogen for fish tank granuloma in people who keep tropical fish. Infections by other mycobacteria, other than tuberculosis, that cause the same phenotype include *M. chelonae, M. fortuitum,* and *M. abscessus.* Lymphangiotic sporotrichoid-like lesions can also result from infection by bacteria such as *Nocardia* species, and therefore, precise identification of species is very important in the optimal clinical management of such infections. Mycetoma is another chronic localized skin infection caused by different species of fungi or actinomycetes. Grains from the lesions contain pathogenic organisms, but final organism identification can be protracted. Nevertheless, molecular biology is able to help since sequence-based identification of large ribosomal subunits specific to particular organisms can be used to identify and characterize certain organisms that contribute directly to disease pathogenesis [65]. PCR and sequencing approaches are also useful in identifying subsets of human papilloma virus [66]. For example, epidermodysplasia verruciformis is a rare genodermatosis characterized by profound susceptibility to cutaneous infection with certain human papilloma virus subtypes. Approximately 50% of the patients develop nonmelanoma skin cancers on sun-exposed areas in the 4th or 5th decades of life. The cancer associated viral subtypes are HPV-5, HPV-8, and HPV-14d, and therefore molecular identification of these viruses in susceptible individuals can be helpful in promoting more vigorous surveillance of at-risk patients.

Molecular profiling has also proved to be useful in the identification and classification of cutaneous malignancies. Notably, T-cell receptor gene rearrangements, which arise as a result of variability in the V-J segment, provide important information relevant to the diagnosis and prognosis of cutaneous T-cell lymphoma (Figure 26.25) [67]. The presence of clonality usually (but not always) is associated with the presence of malignancy. Thus, in many cases, a T-cell receptor gene rearrangement can be helpful in distinguishing a malignant from a benign lymphocytic infiltrate,

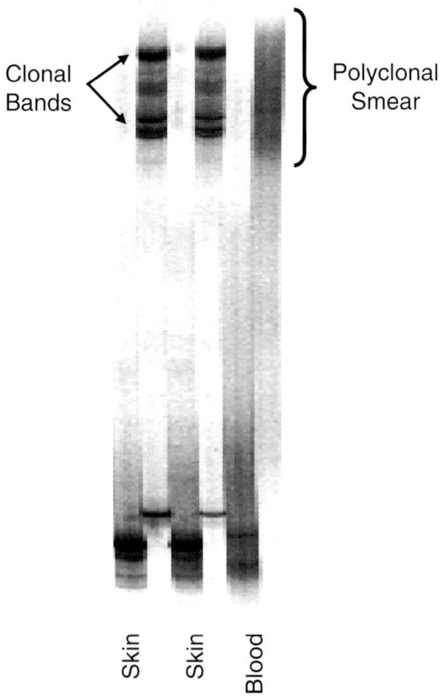

Figure 26.25 Clonal T-cell expansion in a patient with mycosis fungoides (cutaneous T-cell lymphoma). The clinical stage of the patient is Stage 1b. This figure shows single-strand conformational polymorphism (SSCP) analysis and demonstrates an identical clonal T-cell receptor gene rearrangement in two lesional skin biopsies. The matched blood sample is polyclonal.

especially when clinical and histologic features are inconclusive. Clonality can also influence prognosis. For example, Sézary syndrome patients with T-cell clonality are more likely to die from lymphoma/leukemia than their nonclonal counterparts. DNA microarray profiling has also been used in patients with diffuse large B-cell lymphoma. Specific gene signatures have been linked to certain subtypes (such as prominent germinal center B-cell profile or activated B-cell profile), which can be directly associated with different prognoses [68]. Molecular pathologies have also been used to identify specific viral signatures associated with human disease, for example, using sequence and bioinformatics data in the identification of specific polyoma virus sequence in Merkel cell carcinoma [69].

The recent characterization of the molecular basis of the acquired immunobullous diseases has also led to new diagnostic tests. For example, antigen-specific ELISA kits for desmogleins 1 and 3 are now commercially available for the assessment of serum samples from patients with pemphigus [70]. It is known that autoantibodies react against specific desmogleins—type 1 in pemphigus foliaceus or type 3 in mucosal dominant pemphigus vulgaris. These ELISA tests are more sensitive and specific than immunofluorescence microscopy when used in diagnosing pemphigus. More importantly, the titers have been shown to correlate with disease activity, and thus, it is possible to modify a patient's immunosuppressive medication in advance of any clinical change in the status of his or her blistering disease. ELISA tests for the NC16A domain of type XVII collagen are useful in the diagnosis of bullous pemphigoid, although the link to clinical status is less clear than with pemphigus.

For inherited skin blistering diseases, several clinical subtypes can have similar phenotypes, and routine light microscopy is not usually able to assist diagnosis or help determine prognosis. Nevertheless, antibodies to structural components at the dermal-epidermal junction are now commercially available, and these can be helpful in diagnosing recessive forms of epidermolysis bullosa. Testing may involve labeling of skin sections with antibodies to type VII collagen, laminin-332, integrin-α_6 and β_4, plectin, or type XVII collagen. In many recessive forms of epidermolysis bullosa, immunostaining with one of these antibodies is reduced or undetectable [71]. This immunohistochemical approach is therefore a useful prelude to determining the candidate gene that can then be sequenced and used to identify the pathogenic mutations.

Molecular pathology can also provide insight into other genodermatoses that conventional microscopy is unable to provide. For example, in some cases of hereditary leiomyomas, lesions may be multiple, and there can be an autosomal dominant mode of inheritance. In such families, skin leiomyomas usually appear in adolescence or early adulthood (Figure 26.26). Some cases, however, may be complicated by the subsequent development of uterine leiomyomas, usually in the early 20s. The molecular pathology may involve mutations in the fumarase (*FH*) gene [72], and identification of such cases through molecular diagnosis can have important clinical implications. For example, if an *FH* gene mutation is identified in a woman with hereditary multiple leiomyomas who plans to have children at some stage, it may be prudent to advise her to consider having her children at an early age (when she is in her mid to late 20s) before the onset of uterine fibroids (usual onset late 20s or early 30s), which may substantially reduce fertility and the chance of conception. Diagnosis of *FH* gene mutations in hereditary multiple leiomyomas may also have other prognostic significance, in that a subset of patients may be at risk from developing aggressive type II papillary renal cell carcinomas.

Understanding the molecular pathology of inherited skin diseases has also changed clinical practice by allowing the development of newer techniques for prenatal diagnosis for certain disorders (Figure 26.27). Over the past 3 decades the techniques used for prenatal testing have changed from being heavily reliant on the analysis of fetal skin biopsy samples acquired during the second half of the mid-trimester to the examination of fetal DNA from first trimester chorionic villus samples [73]. Advances in fetal medicine have led to further advances in which the molecular pathology of inherited skin diseases can be used to develop new techniques for prenatal testing. These include pre-implantation genetic diagnosis and pre-implantation genetic haplotyping [74]. Further technical advances are likely to lead to the development of less invasive forms of fetal screening, such as the analysis of free fetal DNA in the maternal circulation. Thus, on many levels, understanding the molecular pathology of a wide variety of skin conditions, including infections, malignant, acquired, and inherited diseases, has given rise to a plethora of new diagnostic tests that have a broad and significant impact on clinical practice.

NEW MOLECULAR MECHANISMS AND NOVEL THERAPIES

A detailed understanding of the molecular mechanisms of skin pathology is starting to have significant clinical benefits. New molecular technologies have the potential to increase diagnostic accuracy and to provide important prognostic information. New molecular data also provide a platform to develop novel biological and cytotoxic drugs as well as monoclonal antibodies and other specific forms of immunotherapy. Thus, understanding the molecular pathology of skin diseases can have important implications that benefit patients.

Two groups of dermatological diseases that are benefiting from new molecular-based therapies are the chronic autoinflammatory diseases as well as cutaneous T-cell lymphoma. The autoinflammatory diseases include familial Mediterranean fever, tumor necrosis factor receptor-associated periodic syndrome, hyperimmunoglobulinemia D with periodic fever syndrome, pyogenic arthritis, pyoderma gangrenosum-acne

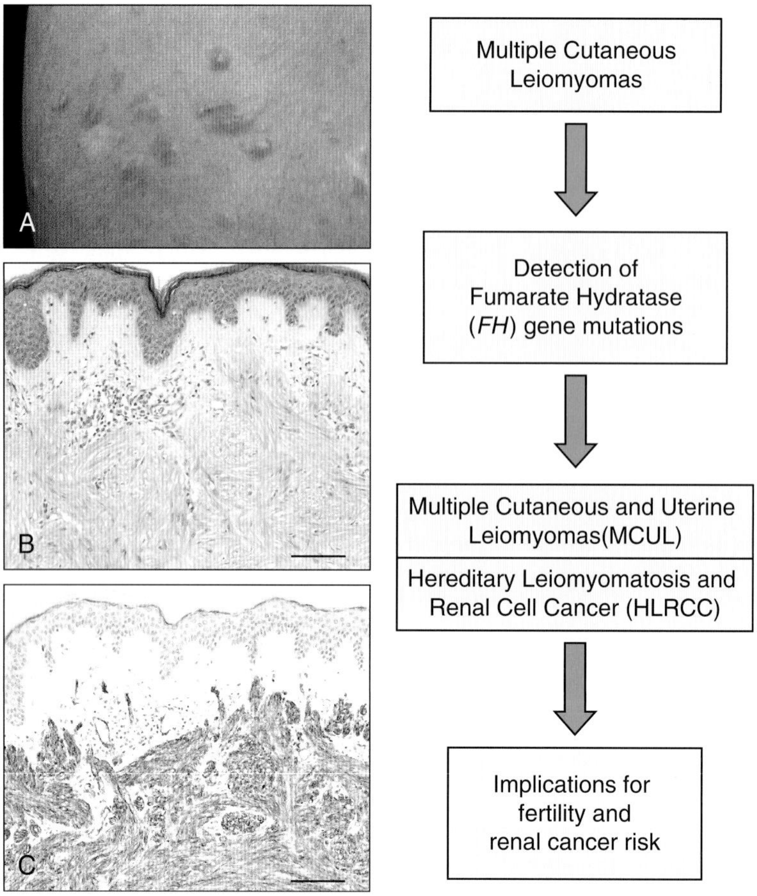

Figure 26.26 **Impact of molecular diagnostics on clinical management. (A)** Clinical appearances of multiple cutaneous leiomyomas. **(B)** Light microscopic appearances show a spindle cell tumor within the dermis (bar = 100 μm). **(C)** Immunostaining with smooth muscle actin identifies the dermal tumor as a leiomyoma (bar = 100 μm). In patients with multiple cutaneous leiomyomas and an autosomal dominant family history, detection of fumarate hydratase (*FH* gene mutations) may indicate a diagnosis of specific syndromes that can have implications for fertility as well as the risk of developing rare forms of renal cancer.

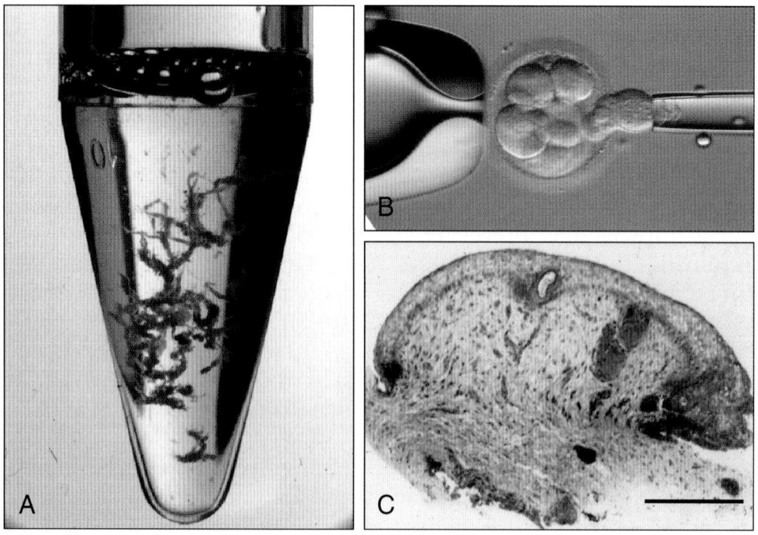

Figure 26.27 **Options for prenatal testing for severe inherited skin diseases. (A)** Chorionic villus samples taken at 10–12 weeks; **(B)** Preimplantation genetic diagnosis, here illustrating single cell extraction from a 72-hour-old embryo; **(C)** Fetal skin biopsy performed at 16–22 weeks gestation, here showing the appearances of normal human fetal skin at 18 weeks (bar = 25 μm).

syndrome, and 3 cryopyrinopathy syndromes, namely neonatal onset multisystem inflammatory disease/chronic infantile neurologic cutaneous and arthropathy syndrome, familial cold autoinflammatory syndrome, and Muckle-Wells syndrome. These conditions are complex and clinically difficult to manage. Nevertheless, new insights into molecular mechanisms are leading to immediate therapeutic benefits. The cryopyrinopathies represent a spectrum of diseases associated with mutations in the cold-induced autoinflammatory syndrome 1 (*CIAS1*) gene that encodes cryopyrin. Cryopyrin and pyrin (the protein implicated in familial Mediterranean fever) belong to the family of pyrin domain containing proteins. Mutations in the gene that encodes for the CD2 binding protein 1 (*CD2BP1*), which binds pyrin, are associated with pyogenic arthritis, pyodermic gangrenosum, and acne syndrome (OMIM604416). Common to all these diseases is activation of the interleukin-1β pathway, and this finding has been translated into direct benefits for patients. Specifically, the recombinant human interleukin-1 receptor antagonist, Anakinra, has proved to be helpful and in some cases results in a quick and dramatic treatment for these chronic autoinflammatory diseases [75]. The *CIAS1* gene mutations result in increased cryopyrin activity. Cryopyrin is expressed in granulocytes, dendritic cells, B- and T-lymphocytes, and also monocytes and epithelial cells lining the oral and genital tracts. It is part of a multiprotein complex termed the NALP3 inflammasone which plays an important role in the regulation of intracellular host defense in response to various signals such as bacterial toxins. Cryopyrin regulates interleukin-1β production, and therefore, targeting the receptor represents a fundamental disease-modifying approach. The clinical response of affected individuals to Anakinra has been nothing short of remarkable, with almost immediate cessation of symptoms and prolonged clinical and biochemical improvements. Although the autoinflammatory diseases are rare disorders, this therapeutic approach represents one of the most selective and specific immunologic interventions and reveals the considerable impact of understanding the detailed molecular pathology to clinical management, particularly in reducing use of other immunosuppressive therapies, such as oral corticosteroids, which lack the target specificity.

Management of the cutaneous T-cell lymphomas is also set to improve following a more detailed understanding of the molecular pathology of these disorders (Figure 26.28) [76]. The most common forms of cutaneous T-cell lymphoma are mycosis fungoides and Sézary syndrome. Mycosis fungoides usually presents in a skin with erythematous patches, plaques, and sometimes tumors. Sézary syndrome is a triad of generalized erythroderma, lymphadenopathy, and the presence of circulating malignant T-cells with cerebriform nuclei (known as Sézary cells). Recent studies have identified a number of changes in various tumor suppressor and apoptosis-related genes in patients with mycosis fungoides and Sézary syndrome. These include Nav3, Fas ligand, Fas, JunB, p16 (INK4a), p15 (INK4b), p14 (ARF), PTEN, p53, and HLA-G [77,78]. Discovery of specific gene abnormalities and patterns of altered gene expression have diagnostic and prognostic value but also offer new insight into developing treatments that transform a malignant disease into a less aggressive chronic illness. The pathogenesis of mycosis fungoides and Sézary syndrome is unclear, but may include chronic antigen stimulation from skin-associated microbes, such as *Staphylococcus aureus* or *Chlamydia* species. Affected patients also have a higher frequency of specific HLA Class II alleles, lending support to the antigen stimulation hypothesis. Viruses such as human T-cell lymphotrophic virus-1, cytomegalovirus, and Epstein-Barr virus have also been implicated. With respect to treatment, there are accumulating data regarding specific immune and genetic abnormalities in these diseases that might lead to new therapies that block the trafficking or proliferation of malignant T-cells. Of note, immunophenotyping suggests that mycosis fungoides and possibly Sézary cells are derived from CLA-positive effector memory cells. The malignant cells demonstrate altered cytokine profiles with IL-7 and IL-18 being upregulated in the plasma and skin of affected individuals. Loss of T-cell diversity is also a feature of these conditions, and antigen-presenting dendritic cells can have an important role in the pathogenesis of the disorders, especially in maintaining the survival of proliferating malignant T-cells. Specific chemokine receptors are associated with mycosis fungoides and Sézary syndrome, and these are directly relevant to skin tropism of malignant T-cells.

Given these abnormalities, new treatments are being developed that go beyond current combinations of nitrogen mustard, corticosteroids, and radiotherapy. The overproduction of Th-2 cytokines suggests that cytokines which promote a Th-1 phenotype might be clinically useful. Indeed, interleukin-12 has shown clinical benefit. Interleukin-12 is a Th-1-promoting cytokine synthesized by phagocytic cells and antigen-presenting cells. It enhances cytolytic T-cell and natural killer cell functions and is necessary for interferon-γ production by activated T-cells. Recombinant IFN-α and IFN-γ can also shift the balance from a Th-2 toward a Th-1 phenotype and may have clinical relevance.

Other approaches using antibodies to CD4 (Zanolimumab) and CD52 (Alemtuzumab) also broadly target T-cells and may help in relieving certain symptoms such as erythroderma and pruritus in Sézary syndrome [79,80] although the immunosuppression may result in complications such as *Mycobacterium* and herpes simplex infections [81]. Vaccine therapy for mycosis fungoides and Sézary syndrome is another intriguing but somewhat preliminary approach [82]. This is due to the scarcity of target antigens, but identifying T-cell receptor sequences expressed by malignant lymphocytes should be technically feasible and may provide targets for immunotherapy. In similar fashion, loading autologous dendritic cells with tumor cells treated with Th-1-priming cytokines is another approach. Vaccination of patients with mimotopes (such as synthetic peptides that stimulate antitumor CDA-positive T-cells) is also a promising clinical option. Specific molecular targeting can also be attempted using

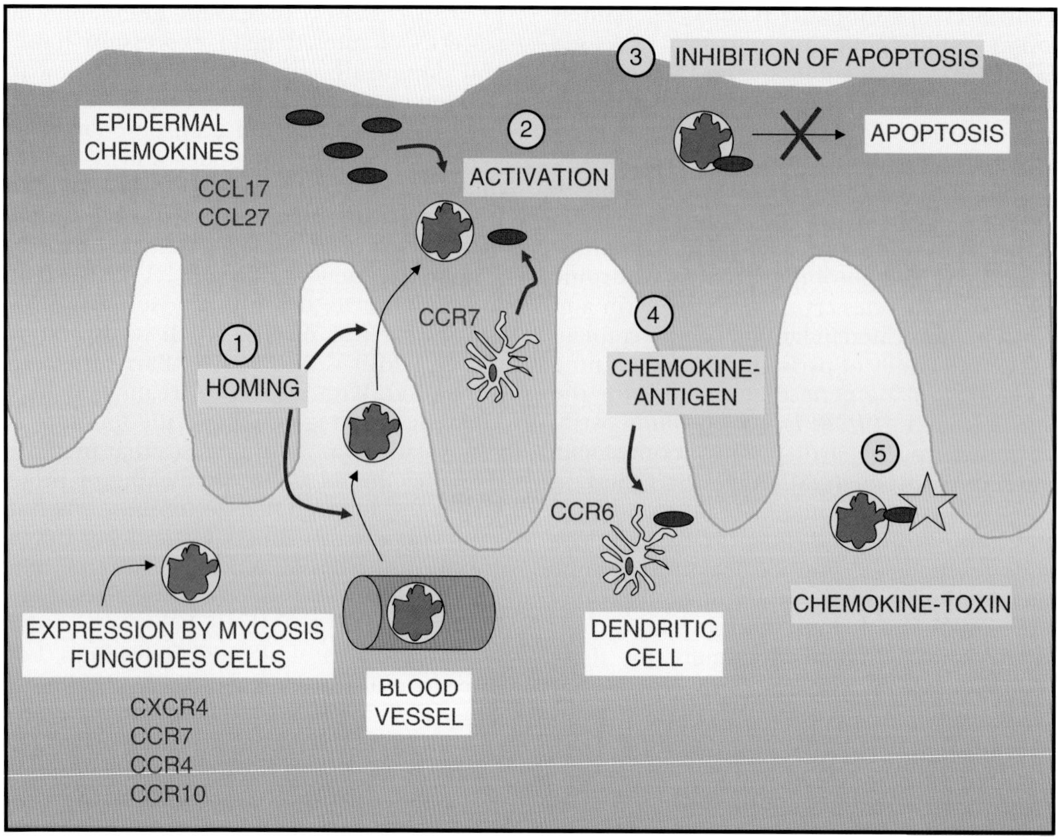

Figure 26.28 **Roles for chemokine receptors and possible therapeutic manipulation in cutaneous T-cell lymphoma.** Chemokine receptors may have important roles in enabling malignant T-cells to enter and survive in the skin. **(1)** Homing: Activation of T-cell integrins permits T-cell adhesion to endothelial cells in the skin and subsequent binding to extracellular matrix proteins. T-cells can then migrate along a gradient of chemokines, e.g., CCL17 and CCL27 to the epidermis. **(2)** Activation: chemokine receptors allow T-cells to interact with dendritic cells such as Langerhans cells, leading to T-cell activation and release of inflammatory cytokines. **(3)** Inhibition of apoptosis: chemokine receptor engagement can lead to upregulation of PI3K and AKT, which are prosurvival kinases. T-cells can therefore survive and proliferate in the skin. **(4)** Chemokine-antigen fusion proteins can be used to target tumor antigens from cutaneous T-cell lymphoma cells to CCR6+ presenting dendritic cells that can stimulate host antitumor immunity. **(5)** Chemokine toxin molecules can also target specific chemokine receptors found on cutaneous T-cell lymphoma cells to mediate direct killing. (Based on an original figure published by [76]).

histone deacetylase inhibitors such as Depsipeptide and Verinostat. These drugs induce growth arrest in conjunction with cell differentiation and apoptosis and have been shown to improve erythroderma and pruritus and to reduce the number of circulating Sézary cells in some patients with Sézary syndrome. Drugs such as Imiquimod, a toll-like receptor agonist, can also benefit cutaneous T-cell lymphoma by inducing IFN-α. Other toll-like receptor agonists may have clinical utility, such as TLR9, which recognizes unmethylated CpG-containing nucleotide motifs that are present in most bacteria and DNA viruses. Treatment with CpG oligodeoxynucleotides leads to plasmacytoid dendritic cell upregulation of co-stimulatory molecules and migratory receptors that subsequently generate type 1 interferons and promote a strong Th-1 immune response and enhance cellular immunity. These drugs may therefore represent a useful adjuvant in immunotherapy or addition to cytotoxic drugs.

Another approved drug for use in cutaneous T-cell lymphoma is Bexarotene [83]. This is a retinoid that modulates gene expression through selective binding to retinoid receptors. These receptors form either homodimers or heterodimers with other nuclear receptors that then act as transcription factors. Bexarotene therapy may also be combined with new treatments such as Denileukin diftitox, which is a fusion molecule containing the IL-2 receptor binding domain and the catalytically active fragment of diphtheria toxin [84]. This targets the high-affinity IL-2 receptor present on activated T- and B-cells. A newer compound which probably works by the same mechanism is anti-Tac (Fv)-PE38 (LMB-2), which is an anti-CD25 recombinant immunotoxin [85]. Another molecular therapy that is currently in clinical practice is extra-corporeal photochemotherapy, a form of light treatment in which leukocytes are treated with psoralen and exposed to UVA light as they pass through a narrow chamber *ex vivo*. The precise mechanism of therapeutic benefit is unclear, but it is thought that psoralen-treated malignant T-cells exposed to light selectively undergo apoptosis and are then engulfed by dendritic cells upon re-infusion in patients. The antigen load of dendritic cells might then trigger a host response against the malignant T-cells, as well as potentially inducing tolerance via the induction of antigen-specific T-regulatory cells. This therapy may therefore trigger vaccine-like effects and induce T-regulatory cells that blunt proliferation of malignant T-cells.

Overall, these new insights into the pathophysiology of cutaneous T-cell lymphoma are providing opportunities to therapeutically target specific aspects of the molecular pathology. These approaches, in which more selective therapy is the goal, are starting to have benefits for patients that result in better survival and fewer side effects.

REFERENCES

1. Bickers DR, Lim HW, Margolis D, et al. The burden of skin diseases: 2004 a joint project of the American Academy of Dermatology Association and the Society for Investigative Dermatology. *J Am Acad Dermatol.* 2006;55:490–500.
2. Smith FJ, Irvine AD, Terron-Kwiatkowski A, et al. Loss-of-function mutations in the gene encoding filaggrin cause ichthyosis vulgaris. *Nat Genet.* 2006;38:337–342.
3. Palmer CN, Irvine AD, Terron-Kwiatkowski A, et al. Common loss-of-function variants of the epidermal barrier protein filaggrin are a major predisposing factor for atopic dermatitis. *Nat Genet.* 2006;38:441–446.
4. Lee KC, Eckert RL. S100A7 (Psoriasin)—Mechanism of antibacterial action in wounds. *J Invest Dermatol.* 2007;127:945–957.
5. Harder J, Schroder JM. RNase 7, a novel innate immune defense antimicrobial protein of healthy human skin. *J. Biol. Chem.,.* 2002;277:46779–46784.
6. Schroder JM, Harder J. Human beta-defensin-2. *Int J Biochem Cell Biol.* 1999;31:645–651.
7. Oppenheim JJ, Tewary P, de la RG, et al. Alarmins initiate host defense. *Adv Exp Med Biol.* 2007;601:185–194.
8. Frohm M, Agerberth B, Ahangari G, et al. The expression of the gene coding for the antibacterial peptide LL-37 is induced in human keratinocytes during inflammatory disorders. *J. Biol. Chem.,.* 1997;272:15258–15263.
9. Schauber J, Gallo RL. Antimicrobial peptides and the skin immune defense system. *J. Allergy Clin. Immunol.* 2008;122:261–266.
10. Nishibu A, Ward BR, Jester JV, et al. Behavioral responses of epidermal Langerhans cells in situ to local pathological stimuli. *J Invest Dermatol.* 2006;126:787–796.
11. Meyer T, Stockfleth E, Christophers E. Immune response profiles in human skin. *Br. J Dermatol.* 2007;157(Suppl 2):1–7.
12. Raison CL, Capuron L, Miller AH. Cytokines sing the blues: Inflammation and the pathogenesis of depression. *Trends Immunol.* 2006;27:24–31.
13. Mills AA, Zheng B, Wang XJ, et al. p63 is a p53 homologue required for limb and epidermal morphogenesis. *Nature.* 1999;398:708–713.
14. Levrero M, De LV, Costanzo A, et al. The p53/p63/p73 family of transcription factors: Overlapping and distinct functions. *J Cell Sci.* 2000;113:1661–1670.
15. Vigano MA, Mantovani R. Hitting the numbers: The emerging network of p63 targets. *Cell Cycle.* 2007;6:233–239.
16. Rinne T, Hamel B, van BH, et al. Pattern of p63 mutations and their phenotypes—update. *Am. J. Med. Genet. A.* 2006;140:1396–1406.
17. Tumbar T, Guasch G, Greco V, et al. Defining the epithelial stem cell niche in skin. *Science.* 2004;303:359–363.
18. Fuchs E, Horsley V. More than one way to skin. *Genes Dev.* 2008;22:976–985.
19. Fuchs E. Skin stem cells: Rising to the surface. *J Cell Biol.* 2008;180:273–284.
20. Ringpfeil F, McGuigan K, Fuchsel L, et al. Pseudoxanthoma elasticum is a recessive disease characterized by compound heterozygosity. *J Invest Dermatol.* 2006;126:782–786.
21. Vanakker OM, Martin L, Gheduzzi D, et al. Pseudoxanthoma elasticum-like phenotype with cutis laxa and multiple coagulation factor deficiency represents a separate genetic entity. *J Invest Dermatol.* 2007;127:581–587.
22. Chen MA, Bonifas JM, Matsumura K, et al. A novel three-nucleotide deletion in the helix 2B region of keratin 14 in epidermolysis bullosa simplex: Delta E375. *Hum Mol Genet.* 1993;2:1971–1972.
23. Chan YM, Yu QC, Fine JD, et al. The genetic basis of Weber-Cockayne epidermolysis bullosa simplex. *Proc. Natl. Acad. Sci. USA.* 1993;90:7414–7418.
24. Smith FJ, Eady RA, Leigh IM, et al. Plectin deficiency results in muscular dystrophy with epidermolysis bullosa. *Nat Genet.* 1996;13:450–457.
25. Lugassy J, Itin P, Ishida-Yamamoto A, et al. Naegeli-Franceschetti-Jadassohn syndrome and dermatopathia pigmentosa reticularis: Two allelic ectodermal dysplasias caused by dominant mutations in KRT14. *Am J Hum Genet.* 2006;79:724–730.
26. Betz RC, Planko L, Eigelshoven S, et al. Loss-of-function mutations in the keratin 5 gene lead to Dowling-Degos disease. *Am J Hum Genet.* 2006;78:510–519.
27. Pulkkinen L, Christiano AM, Gerecke D, et al. A homozygous nonsense mutation in the beta 3 chain gene of laminin 5 (LAMB3) in Herlitz junctional epidermolysis bullosa. *Genomics.* 1994;24:357–360.
28. Pulkkinen L, Christiano AM, Airenne T, et al. Mutations in the gamma 2 chain gene (LAMC2) of kalinin/laminin 5 in the junctional forms of epidermolysis bullosa. *Nat Genet.* 1994;6:293–297.
29. Kivirikko S, McGrath JA, Baudoin C, et al. A homozygous nonsense mutation in the alpha 3 chain gene of laminin 5 (LAMA3) in lethal (Herlitz) junctional epidermolysis bullosa. *Hum. Mol. Genet.* 1995;4:959–962.
30. McGrath JA, Gatalica B, Christiano AM, et al. Mutations in the 180-kD bullous pemphigoid antigen (BPAG2), a hemidesmosomal transmembrane collagen (COL17A1), in generalized atrophic benign epidermolysis bullosa. *Nat Genet.* 1995;11:83–86.
31. Ruzzi L, Gagnoux-Palacios L, Pinola M, et al. A homozygous mutation in the integrin alpha6 gene in junctional epidermolysis bullosa with pyloric atresia. *J Clin Invest.* 1997;99:2826–2831.
32. Vidal F, Aberdam D, Miquel C, et al. Integrin beta 4 mutations associated with junctional epidermolysis bullosa with pyloric atresia. *Nat Genet.* 1995;10:229–234.
33. Christiano AM, Greenspan DS, Hoffman GG, et al. A missense mutation in type VII collagen in two affected siblings with recessive dystrophic epidermolysis bullosa. *Nat Genet.* 1993;4:62–66.
34. Christiano AM, Ryynanen M, Uitto J. Dominant dystrophic epidermolysis bullosa: Identification of a Gly–>Ser substitution in the triple-helical domain of type VII collagen. *Proc. Natl. Acad. Sci. USA.* 1994;91:3549–3553.
35. Wetteland P, Hovding G. Squamous-cell carcinoma in dystrophic epidermolysis bullosa. *Acta Derm Venereol.* 1956;36:27–36.
36. Miller DL, Weinstock MA. Nonmelanoma skin cancer in the United States: Incidence. *J Am Acad Dermatol.* 1994;30:774–778.
37. McGrath JA, McMillan JR, Shemanko CS, et al. Mutations in the plakophilin 1 gene result in ectodermal dysplasia/skin fragility syndrome. *Nat Genet.* 1997;17:240–244.
38. McKoy G, Protonotarios N, Crosby A, et al. Identification of a deletion in plakoglobin in arrhythmogenic right ventricular cardiomyopathy with palmoplantar keratoderma and woolly hair (Naxos disease). *Lancet.* 2000;355:2119–2124.
39. Hunt DM, Rickman L, Whittock NV, et al. Spectrum of dominant mutations in the desmosomal cadherin desmoglein 1, causing the skin disease striate palmoplantar keratoderma. *Eur J Hum Genet.* 2001;9:197–203.
40. Kljuic A, Bazzi H, Sundberg JP, et al. Desmoglein 4 in hair follicle differentiation and epidermal adhesion: Evidence from inherited hypotrichosis and acquired pemphigus vulgaris. *Cell.* 2003;113:249–260.

41. Armstrong DK, McKenna KE, Purkis PE, et al. Haploinsufficiency of desmoplakin causes a striate subtype of palmoplantar keratoderma. *Hum Mol Genet.* 1999;8:143–148.
42. Norgett EE, Hatsell SJ, Carvajal-Huerta L, et al. Recessive mutation in desmoplakin disrupts desmoplakin-intermediate filament interactions and causes dilated cardiomyopathy, woolly hair and keratoderma. *Hum Mol Genet.* 2000;9:2761–2766.
43. Jonkman MF, Pasmooij AM, Pasmans SG, et al. Loss of desmoplakin tail causes lethal acantholytic epidermolysis bullosa. *Am J Hum Genet.* 2005;77:653–660.
44. Levy-Nissenbaum E, Betz RC, Frydman M, et al. Hypotrichosis simplex of the scalp is associated with nonsense mutations in CDSN encoding corneodesmosin. *Nat Genet.* 2003;34:151–153.
45. Morar N, Willis-Owen SA, Moffatt MF, et al. The genetics of atopic dermatitis. *J Allergy Clin Immunol.* 2006;118:24–34.
46. Homey B, Steinhoff M, Ruzicka T, et al. Cytokines and chemokines orchestrate atopic skin inflammation. *J Allergy Clin Immunol.* 2006;118:178–189.
47. Elder JT, Nair RP, Guo SW, et al. The genetics of psoriasis. *Arch Dermatol.* 1994;130: 216–224.
48. Lowes MA, Bowcock AM, Krueger JG. Pathogenesis and therapy of psoriasis. *Nature.* 2007;445:866–873.
49. Hashimoto T. Skin diseases related to abnormality in desmosomes and hemidesmosomes—Editorial review. *J Dermatol Sci.* 1999;20:81–84.
50. Gibson GE, Daoud MS, Pittelkow MR. Anti-epiligrin (laminin 5) cicatricial pemphigoid and lung carcinoma: Coincidence or association? *Br J Dermatol.* 1997;137:780–782.
51. Tyagi S, Bhol K, Natarajan K, et al. Ocular cicatricial pemphigoid antigen: Partial sequence and biochemical characterization. *Proc. Natl. Acad. Sci. USA.* 1996;93:14714–14719.
52. Fairley JA, Woodley DT, Chen M, et al. A patient with both bullous pemphigoid and epidermolysis bullosa acquisita: An example of intermolecular epitope spreading. *J Am Acad Dermatol.* 2004;51:118–122.
53. Hanakawa Y, Matsuyoshi N, Stanley JR. Expression of desmoglein 1 compensates for genetic loss of desmoglein 3 in keratinocyte adhesion. *J Invest Dermatol.* 2002;119:27–31.
54. Hanakawa Y, Schechter NM, Lin C, et al. Molecular mechanisms of blister formation in bullous impetigo and staphylococcal scalded skin syndrome. *J Clin Invest.* 2002;110:53–60.
55. Geller AC, Swetter SM, Brooks K, et al. Screening, early detection, and trends for melanoma: Current status (2000–2006) and future directions. *J Am Acad Dermatol.* 2007;57: 555–572.
56. Young AR, Chadwick CA, Harrison GI, et al. The similarity of action spectra for thymine dimers in human epidermis and erythema suggests that DNA is the chromophore for erythema. *J Invest Dermatol.* 1998;111:982–988.
57. Toftgard R. Hedgehog signalling in cancer. *Cell Mol Life Sci.* 2000;57:1720–1731.
58. Hahn H, Wicking C, Zaphiropoulous PG, et al. Mutations of the human homolog of Drosophila patched in the nevoid basal cell carcinoma syndrome. *Cell.* 1996;85:841–851.
59. Xie J, Murone M, Luoh SM, et al. Activating smoothened mutations in sporadic basal-cell carcinoma. *Nature.* 1998;391:90–92.
60. Barth A, Wanek LA, Morton DL. Prognostic factors in 1,521 melanoma patients with distant metastases. *J Am Coll Surg.* 1995;181:193–201.
61. Fecher LA, Amaravadi RK, Flaherty KT. The MAPK pathway in melanoma. *Curr Opin Oncol.* 2008;20:183–189.
62. Dahl C, Guldberg P. The genome and epigenome of malignant melanoma. *APMIS.* 2007;115:1161–1176.
63. Moreira JN, Santos A, Simoes S. Bcl-2-targeted antisense therapy (Oblimersen sodium): Towards clinical reality. *Rev Recent Clin Trials.* 2006;1:217–235.
64. Pratilas CA, Solit DB. Therapeutic strategies for targeting BRAF in human cancer. *Rev Recent Clin Trials.* 2007;2:121–134.
65. Borman AM, Linton CJ, Miles SJ, et al. Molecular identification of pathogenic fungi. *J Antimicrob Chemother.* 2008;61(Suppl 1): i7–i12.
66. de KM, Struijk L, Feltkamp M, et al. HPV DNA detection and typing in inapparent cutaneous infections and premalignant lesions. *Methods Mol Med.* 2005;119:115–127.
67. Wood GS. T-cell receptor and immunoglobulin gene rearrangements in diagnosing skin disease. *Arch Dermatol.* 2001;137:1503–1506.
68. Alizadeh AA, Eisen MB, Davis RE, et al. Distinct types of diffuse large B-cell lymphoma identified by gene expression profiling. *Nature.* 2000;403:503–511.
69. Feng H, Shuda M, Chang Y, et al. Clonal integration of a polyomavirus in human Merkel cell carcinoma. *Science.* 2008;319:1096–1100.
70. Amagai M, Komai A, Hashimoto T, et al. Usefulness of enzyme-linked immunosorbent assay using recombinant desmogleins 1 and 3 for serodiagnosis of pemphigus. *Br J Dermatol.* 1999;140: 351–357.
71. Fine JD, Eady RA, Bauer EA, et al. Revised classification system for inherited epidermolysis bullosa: Report of the Second International Consensus Meeting on diagnosis and classification of epidermolysis bullosa. *J Am Acad Dermatol.* 2000; 42:1051–1066.
72. Tomlinson IP, Alam NA, Rowan AJ, et al. Germline mutations in FH predispose to dominantly inherited uterine fibroids, skin leiomyomata and papillary renal cell cancer. *Nat Genet.* 2002;30:406–410.
73. McGrath JA, Handyside AH. Preimplantation genetic diagnosis of severe inherited skin diseases. *Exp Dermatol.* 1998;7:65–72.
74. Fassihi H, Renwick PJ, Black C, et al. Single cell PCR amplification of microsatellites flanking the COL7A1 gene and suitability for preimplantation genetic diagnosis of Hallopeau-Siemens recessive dystrophic epidermolysis bullosa. *J Dermatol Sci.* 2006; 42:241–248.
75. Hoffman HM, Rosengren S, Boyle DL, et al. Prevention of cold-associated acute inflammation in familial cold autoinflammatory syndrome by interleukin-1 receptor antagonist. *Lancet.* 2004;364:1779–1785.
76. Hwang ST, Janik JE, Jaffe ES, et al. Mycosis fungoides and Sezary syndrome. *Lancet.* 2008;371:945–957.
77. Karenko L, Hahtola S, Paivinen S, et al. Primary cutaneous T-cell lymphomas show a deletion or translocation affecting NAV3, the human UNC-53 homologue. *Cancer Res.* 2005;65: 8101–8110.
78. Whittaker S. Molecular genetics of cutaneous lymphomas. *Ann NY Acad Sci.* 2001;941:39–45.
79. Kim YH, Duvic M, Obitz E, et al. Clinical efficacy of zanolimumab (HuMax-CD4): Two phase 2 studies in refractory cutaneous T-cell lymphoma. *Blood.* 2007;109:4655–4662.
80. Lundin J, Hagberg H, Repp R, et al. Phase 2 study of alemtuzumab (anti-CD52 monoclonal antibody) in patients with advanced mycosis fungoides/Sézary syndrome. *Blood.* 2003;101: 4267–4272.
81. Nosari A, Tedeschi A, Ricci F, et al. Characteristics and stage of the underlying diseases could determine the risk of opportunistic infections in patients receiving alemtuzumab. *Haematologica.* 2008;93:e30–e31.
82. Seo N, Furukawa F, Tokura Y, et al. Vaccine therapy for cutaneous T-cell lymphoma. *Hematol Oncol Clin North Am.* 2003;17: 1467–1474.
83. Duvic M, Hymes K, Heald P, et al. Bexarotene is effective and safe for treatment of refractory advanced-stage cutaneous T-cell lymphoma: Multinational phase II-III trial results. *J Clin Oncol.* 2001;19:2456–2471.
84. Carretero-Margolis CD, Fivenson DP. A complete and durable response to denileukin diftitox in a patient with mycosis fungoides. *J Am Acad Dermatol.* 2003;48:275–276.
85. Foon KA. Monoclonal antibody therapies for lymphomas. *Cancer J.* 2000;6:273–278.
86. Nestle FO, Nickoloff BJ. Deepening our understanding of immune sentinels in the skin. *J Clin Invest.* 2007;117:2382–2385.
87. Boyman O, Conrad C, Tonel G, et al. The pathogenic role of tissue-resident immune cells in psoriasis. *Trends Immunol.* 2007;28:51–57.
88. Daya-Grosjean L, Couve-Privat S. Sonic hedgehog signaling in basal cell carcinomas. *Cancer Lett.* 2005;225:181–192.
89. Singh M, Lin J, Hocker TL, et al. Genetics of melanoma tumorigenesis. *Br J Dermatol.* 2008;158:15–21.

Chapter

27

Molecular Pathology: Neuropathology

Joshua A. Sonnen ▪ C. Dirk Keene
▪ Robert F. Hevner ▪ Thomas J. Montine

INTRODUCTION

Skin is the most visible organ and one which is readily amenable to molecular dissection. Recent studies have therefore been able to provide fascinating new insight into the development and maintenance of healthy skin and into the range of molecular pathology that underlies a spectrum of inherited and acquired skin diseases. The new information is helpful in establishing better and more accurate diagnoses of skin disorders as well as providing new opportunities to develop more specific therapies that target key molecular events relevant to the pathogenesis of particular diseases. This chapter highlights some of the recent discoveries on the molecular pathology of skin disorders. The focus is on new findings that may have a significant impact on the clinical management of patients. It is evident that an improved understanding of the molecular pathology of the skin is fundamental to making clinical advances in dermatology.

ANATOMY OF THE CENTRAL NERVOUS SYSTEM

The central nervous system (CNS) is composed of cellular components organized in a complex structure that is unlike other organ systems. Broadly, its cellular components can be divided into neuroepithelial and mesenchymal elements. Neuroepithelial elements derive from the primitive neural tube and include neurons and glia. The mesenchymal elements include blood vessels and microglia. Microglia are a bone marrow-derived population of scavenger cells that play a central role in CNS inflammation. At the macroscopic level, the parenchyma of the CNS can be categorized into 2 structurally and functionally unique components: gray and white matter. Gray matter is the location of most neurons and is the site of the integration of neural impulses by neurotransmitters across synapses between neurons. White matter functions to conduct these impulses efficiently and quickly between neurons in different gray matter regions.

Microscopic Anatomy

Gray Matter

Macroscopically, gray matter forms a ribbon of cortex in the human cerebrum and cerebellum as well as the mass of the deep nuclei. Microscopically, it is composed of cells forming and embedded in a finely interdigitating network of cellular processes. This network, referred to as neuropil, is sufficiently dense that individual cellular processes cannot be distinguished. Indeed, under the microscope, neuropil appears as a homogenous matrix surrounding neurons and other cells. Because of this, it was not until 1889 and the work of Santiago Ramón-y-Cajal that the neuron theory was fully applied to the brain, 50 years after cellular theory was proposed by Theodore Schwann [1].

The cellular constituents of gray matter include neurons, glia, endothelium, and perivascular cells of blood vessels and microglia (Figure 27.1). Neurons are the electrically active cells of the brain. A neuron is composed of slender branching dendrites on which other neurons synapse and propagate action potentials, a body or soma where the metabolic and synthetic processes of the neuron are orchestrated, an axonal hillock where electrochemical impulses are integrated, an axon along which the integrated electrochemical impulse is conducted, and an axonal terminal where the electrochemical signal is passed to another neuron's dendrites or an effector cell across a synapse. The soma of neurons is easily identifiable by routine histologic stains. Neurons have generally round nuclei with a prominent nucleolus and open chromatin. The cytoplasm of neurons is remarkable in that it generally contains abundant rough

Molecular Pathology © 2009, Elsevier, Inc. All Rights Reserved.

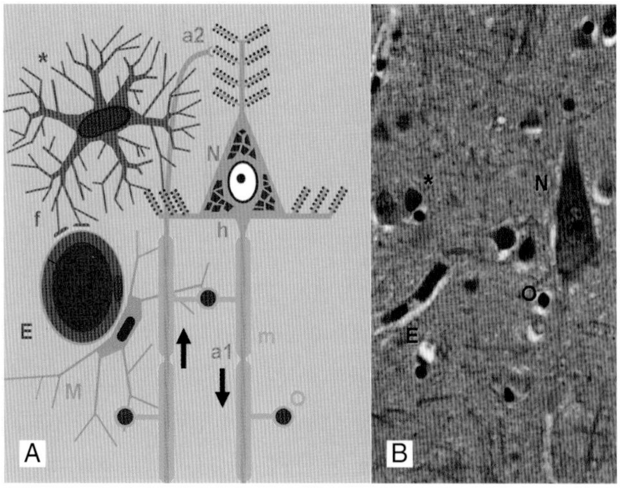

Figure 27.1 **(A) Schematic of the microanatomy of gray matter, (B) Photomicrograph of the microanatomy of gray matter.** (A). A neuron (N) contains a prominent nucleus with open chromatin, a conspicuous nucleolus and cytoplasmic Nissl substance that is composed of abundant rough endoplasmic reticulum. The neuron elaborates numerous apical and lateral dendrites that are decorated with many receptors. Electrochemical impulses are generated at dendrites and integrated across the neuron's body at the axonal hillock (h) and transmitted along the axon (a1). Oligodendrocytes (O) envelop the axon within a segmented myelin (m) sheath allowing more rapid and efficient conduction of impulses. An endothelial cell (E) forms a small capillary space surrounded by a resting microglial cell (M). A nearby astrocyte (*) has numerous cytoplasmic processes, some of which rest foot processes (f) on the vessel, helping to maintain the blood-brain barrier. An axon (a2) from a distant neuron forms an axonal terminal (t) on a dendrite of the pictured neuron, forming a synapse and releasing neurotransmitters to the receptors on the dendrite. (B). Section of frontal cortex stained with hematoxylin and eosin (H&E) and Luxol fast blue (LFB).

endoplasmic reticulum called Nissl substance that is demonstrable by numerous histologic techniques. Loss of Nissl substance is a sign of early neuronal injury and is seen in a variety of conditions including axonal transection and hypoxia/ischemia. The neuron's axons and dendrites are major components of neuropil.

Neurons require a constant supply of oxygen and glucose, and even short interruptions can cause neuronal death. Neurons are the most susceptible to most forms of CNS injury and are the first cells lost to necrosis or apoptosis under stressful conditions. Further, for reasons that are not fully understood, populations of neurons have differential susceptibilities to different types of stress. Cells in one region of the hippocampus may become necrotic in response to hypoxia, while neurons in other regions are spared; this pattern of injury may be reversed in hypoglycemia.

Glia, from Latin for glue, form the bulk of the CNS parenchyma and outnumber neurons on the order of 1000 to 1. The primary glial cell of gray matter is the star-shaped astrocyte. Astrocytes maintain a variety of supportive functions including structure, metabolic support for neurons, management of cellular waste products, uptake and release of neurotransmitters, regulation of extracellular ion concentration, interactions with the vasculature including helping to maintain the blood-brain barrier and responding to injury. Normally, astrocytes have irregular, potato-shaped nuclei and numerous fine cellular processes that are indistinct within the neuropil. In response to noxious stimuli, astrocytes increase production of their characteristic intermediate filament, glial fibrillary acidic protein (GFAP). The astrocyte's soma swells and becomes prominent. This process is known as gliosis and is analogous to scar formation outside the CNS. A second population of glia in gray matter is the oligodendrocytes; these will be discussed in more detail later under white matter.

Microglia are a population of bone marrow-derived scavenger cells. Usually unobtrusive, the quiescent microglia have small rod-shaped nuclei and inapparent cytoplasm. Ramified (quiescent) microglia do not express major histocompatibility complex (MHC) I/II antigens, unlike other scavenger cells. However, in response to injury, microglia replicate and migrate. They produce MHC I/II antigen, activate inflammatory and cytotoxic signaling, and can phagocytize material and process it for antigen presentation to T-cells.

White Matter

White matter is composed of numerous axonal processes (from neurons whose bodies reside in gray matter), glia, blood vessels, and microglia (Figure 27.2). The axons are insulated in segments by layers of myelin,

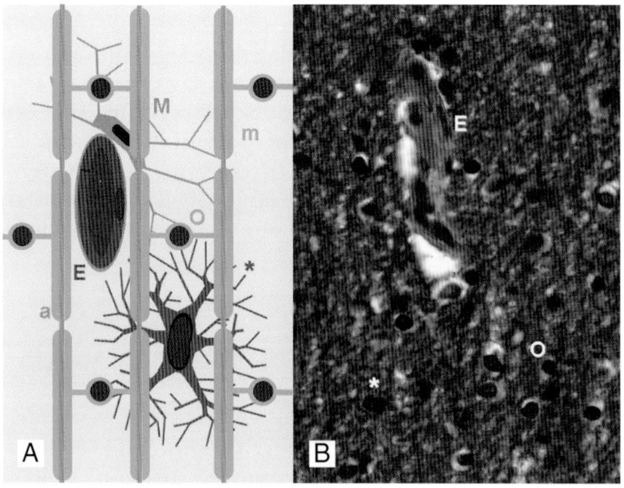

Figure 27.2 **(A) Schematic of the microanatomy of white matter, (B) Photomicrograph of the microanatomy of white matter (H&E/LFB).** Numerous axons (a) from distant neurons pass through white matter conducting nerve impulses. Oligodendroglia (O) are the most common cells and myelinate many adjacent segments of passing axons. The insulating myelin (m) allows rapid and efficient conduction of nerve impulses. Oligodendrocytes have high metabolic needs, and endothelium (E) form numerous capillaries. In the absence of disease, astrocytes (*) and microglia (M) are inconspicuous.

and this insulation allows more rapid and efficient conduction of electrochemical signals along the axon. Myelin is composed of concentric proteolipid membranes which are extensions of cytoplasmic processes of the primary glia of white matter, the oligodendrocyte. Oligodendrocytes have small round nuclei with condensed chromatin and indistinct cytoplasm. A single oligodendrocyte myelinates numerous passing axons. Because oligodendrocytes are responsible for the maintenance of a large amount of myelin, they have relatively high metabolic demands and are consequently relatively sensitive to injury compared to other white matter elements. Injury to oligodendrocytes causes local loss of myelin; this is discussed further in the section on demyelination. Astrocytes and microglia are minority populations within white matter and are less sensitive to injury than oligodendrocytes. Tissue response to noxious stimuli is mediated through astrocytic gliosis and microglial activation as in gray matter.

Cerebrospinal Fluid and the Ventricular System

Within the brain there are interconnected fluid-filled chambers called ventricles. The ventricles are lined by specialized glial cells called ependyma. The ependymal cells form a membrane with tight junctions between cells and microvilli on their apical membrane. The ventricles are filled with cerebrospinal fluid (CSF). Most CSF is produced by a specialized organ within the ventricles called the choroid plexus. The choroid plexus is a network of large capillaries and glia with a specialized epithelial lining. The vessels of the choroid plexus contain fenestrations and produce the CSF filtrate through active ion secretion and passive water flow. The epithelial cells have apical microvilli and cilia and, along with the ependyma, gently push the CSF through the ventricular system. CSF exits the brain through apertures between the base of the cerebellum and brainstem, filling a compartment surrounding the brain known as the arachnoid. CSF surrounding the brain acts as a cushion for the brain and spinal cord. CSF is taken back up into the venous circulation, and the entire volume of CSF (~150 ml) turns over 4 to 5 times a day.

Gross Anatomy

The CNS can be divided into a number of anatomic regions, each with specific neurologic or cognitive functions. Disease or damage to these regions produces neurologic or cognitive deficits that correlate with the anatomic location and extent of disease. There are several ways to divide these structures. The main divisions are the cerebrum, cerebellum, brainstem, and spinal cord. The cerebrum is covered by a folded layer of gray matter called the cortex and is generally believed to be the location of conscious thought. The folding allows greater surface area to fit within the confines of the skull. The folds themselves are called gyri and are characteristic of normal cortical development in humans. Lesions of the cortex cause deficits of cognition and conscious movement or sensation. The cortex can be further divided into lobes which subserve different cognitive domains or neurologic functions (Figure 27.3). The frontal lobes anteriorly are involved in executive function (self-control, planning) and personality. Damage to this region produces personality changes and socially inappropriate behavior. Posteriorly, the frontal lobe houses the primary motor cortex which controls voluntary movement for the opposite side (contralateral) of the body. Damage causes contralateral weakness or paralysis. The parietal lobe contains the primary sensory cortex for

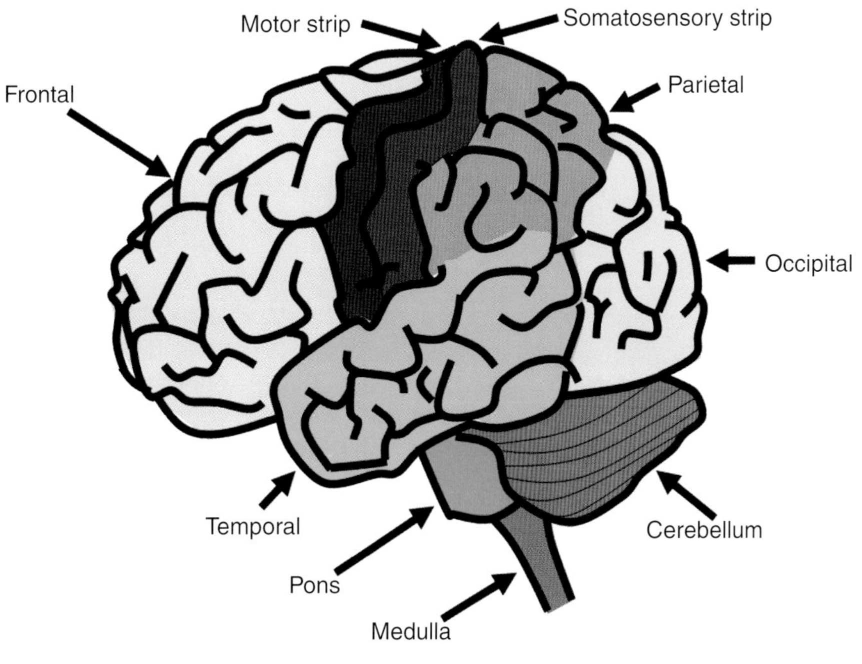

Figure 27.3 Lateral view of the central nervous system.

the contralateral half of the body, and damage causes anesthesia and neglect. The temporal lobe is involved in the conscious processing of sound, but also contains the hippocampus that is integral to the formation of new memories. Damage to the hippocampus produces memory dysfunction, and in cases where both hippocampi are damaged, severe anterograde amnesia wherein the unfortunate individual is incapable of forming new memories. The occipital lobe contains the primary visual cortex for the field of vision on the opposite side of the body, and damage causes partial or complete loss of conscious perception of visual stimuli. The white matter underlying and connecting these cortical regions may also be damaged and produces deficits related to the regions connected. The cerebrum also contains deeply situated gray matter nuclei. The most often implicated of these in disease are the basal ganglia (Figure 27.4). These nuclei form important inhibitory circuits on the cerebrum, and disease causes movement disorders and well as disorders of cognition.

The cerebellum is involved in coordination and balance. The brainstem has 3 primary functional domains. First, it contains white matter tracks that connect the cerebrum to the spinal cord and carries information between these 2 structures. Second, numerous autonomic functions are coordinated by nuclei within the brainstem that serve as integrative centers. Third, the cranial nerve nuclei reside in the brainstem which control motor function of the head and neck and receive sensory input from these same structures. The spinal cord is primarily involved in the transmission of nerve impulses to and from the peripheral nervous system along its abundant white matter tracks, but it also contains central gray matter nuclei that regulate autonomic functions and reflexes.

Summary

The anatomy of the CNS is complex. The correlation between structure and function illuminates an underlying order. Knowledge of the location or cellular target of injury can give great insight into the deficits present and vice versa.

NEURODEVELOPMENTAL DISORDERS

Developmental neuropathology encompasses a broad variety of cerebral malformations and functional impairments caused by disturbances of brain development, manifesting during ages from the embryonic period through adolescence and young adulthood. The neurologic and psychiatric manifestations of neurodevelopmental disorders range widely depending on the affected neural systems and include such diverse manifestations as epilepsy, mental retardation, cerebral palsy, breathing disorders, ataxia, autism, and schizophrenia. In terms of morbidity and mortality, the spectrum is extremely broad: the mildest neurodevelopmental disorders can be asymptomatic, while the worst malformations often lead to intrauterine or neonatal demise. This section will focus on genetic disorders of brain development, caused by mutations of gene loci or chromosomal regions with important neurodevelopmental functions. Excluded from consideration here are other categories of developmental

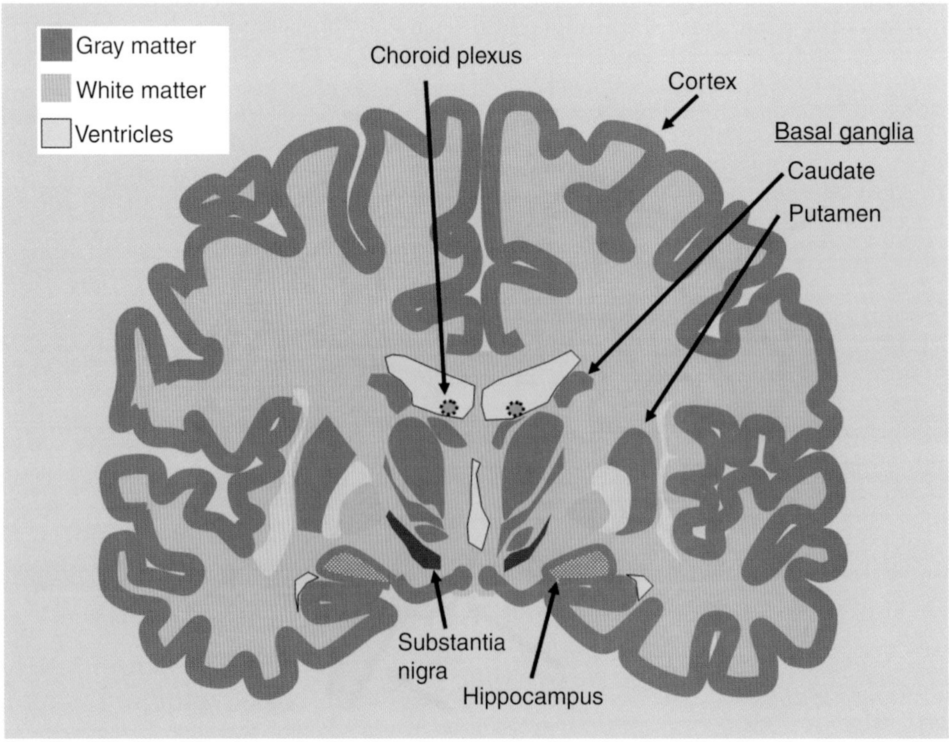

Figure 27.4 **Coronal section of cerebrum through the deep gray matter nuclei.**

neuropathology caused by general metabolic disorders (such as storage diseases and amino acidopathies), disruptions (amniotic web disruption sequence), environmental insults (hypoxia-ischemia or toxic exposure), infection, and neoplasia. More extensive coverage of these categories can be found in other specialized texts, some of which are listed at the end of this chapter.

Current systems for classifying neurodevelopmental malformations emphasize the underlying developmental processes that are perturbed. Recent advances in human genetics and developmental neurobiology have largely substantiated this practical approach, but at the same time have also demonstrated that even single gene mutations (as well as chromosomal alterations) rarely affect one single mechanism in development. Rather, single genes or chromosomal loci often have pleiotropic effects, controlling multiple aspects of brain development. Conversely, each developmental process is controlled by multiple genes. Accordingly, mutations in one gene may cause several different malformations, and the same malformation may be caused by many different genes or other etiologies. Holoprosencephaly, for example, can be caused by mutations in several genes or by toxic exposure [4]. Further complications arise from the modifying effects of genetic background, mosaicism, X chromosome inactivation, and epigenetic effects (imprinting). In practical terms, this means that many neurodevelopmental disorders are genetically heterogeneous and may require additional tests (karyotyping, array hybridization, sequencing) along with family history to gain a clear understanding of the disease.

To understand the pathogenesis of neurodevelopmental disorders, one must first acquire at least a rudimentary knowledge of brain and spinal cord development, including underlying cellular mechanisms and interactions. A brief review of processes highlighting recent progress will be the starting point for subsequent consideration of selected specific brain malformations.

Gastrulation and Neurulation: Neural Tube Defects

Development of the brain and spinal cord begins with gastrulation, when the neural plate forms and differentiates into distinct regional subdivisions along the rostrocaudal and mediolateral axes. As morphogenesis continues, the processes of neurogenesis, gliogenesis, synaptogenesis, and myelination are initiated sequentially, and these processes then continue concurrently with broad overlap. Brain development is mostly complete by the end of adolescence, although neurogenesis continues at decreasing levels in the hippocampus and olfactory system throughout adulthood. In this sense, brain development continues until death.

Organizing the Central Nervous System: Signaling Centers and Regional Patterning

The central nervous system begins as the neural plate, which folds along the midline and closes dorsally to form the neural tube. From its formation, the neural plate is patterned along the rostrocaudal and mediolateral axes in a grid-like fashion by gradients and boundaries of gene expression. *HOX* genes, for example, are differentially expressed along the rostrocaudal axis and play an important role in specifying segmental organization of the spinal cord and hindbrain. Such differences of gene expression are programmed by both the intrinsic developmental history of each region, and by extracellular factors produced in signaling centers that define positional information through interactions with developing neural tissue. Gene expression gradients, compartments, boundaries, and signaling centers continue to be important throughout the embryonic period until each brain subdivision has been generated and acquired its specific identity.

Neural Tube Closure and Wnt-PCP Signaling

Neural tube closure is a key early event in brain and spinal cord development, in which the planar epithelium of the neural plate folds at the midline along the anteroposterior axis, and the lateral edges of the neural plate move dorsally, contact each other, and fuse dorsally to form the neural tube (Figure 27.5). Dorsal fusion first occurs in the region of the cervical spinal cord primordium, followed by separate closure events at midbrain and rostral telencephalic points. The exact location and number of dorsal fusion events varies between and within species. From each of these sites, fusion proceeds rostrally and caudally in a zipper-like mode until closure is complete (primary neurulation) from the rostral end (anterior neuropore) to

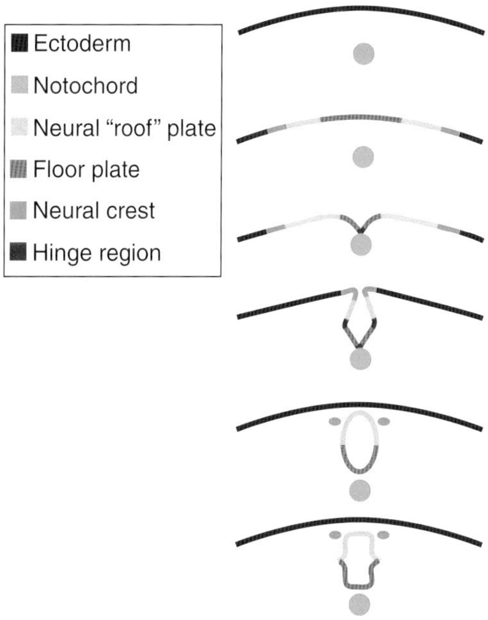

Figure 27.5 Neural tube formation. In cross-section, the neural tube first derives from a planar (two-dimensional) epithelium with mediolateral organization and transforms to a tubular (three-dimensional) with both mediolateral and dorsoventral axes.

the caudal (posterior neuropore). Interestingly, this mode of primary neural tube closure does not quite extend the full caudal length of the spinal cord, but ends around the lumbosacral region. More caudal regions (mainly sacral) appear to develop by a distinct mechanism involving cavitation of the caudal eminence (tail bud) mesenchyme, known as secondary neurulation. A schematic of this process is shown in Figure 27.6.

Primary neurulation depends on deformation of the neural plate due to convergent extension, a process of cell migration within the plane of the neuroepithelium. At the molecular level, convergent extension utilizes the Wnt-planar cell polarity (Wnt-PCP) signaling pathway for cell adhesion and polarity. Many details remain to be elucidated, but this pathway is essential for cells to distinguish mediolateral and anteroposterior directions within the neuroepithelium, and thus migrate in the correct directions. Studies of mutant mouse strains, notably loop-tail (*Lp*), have shown that mutations of several Wnt-PCP genes (such as *Vangl2* in *Lp*) cause craniorachischisis, a severe form of open neural tube defect extending from brain to spinal cord.

The incidence of neural tube defects (NTD) has recently declined in developed countries due to dietary supplementation with folate, but the molecular mechanism of this effect remains unknown. Polymorphisms in enzyme genes involved in folate metabolism are known to confer risk for NTD. One polymorphism of 5,10-methylene tetrahydrofolate reductase (MTHFR) is thermolabile and, if present in fetus or mother, confers increased risk of NTD. Environmental and toxic exposures have also been implicated in the pathogenesis of NTD. A large number of physical agents ranging from X-irradiation to alcohol to folate antagonists have been associated with NTD in epidemiologic studies and can cause NTD in animal models. Maternal health status, including diabetes, obesity, infections, and toxic exposures, also is associated with NTD in humans [6].

Importantly, the process of neurulation transforms the axes of the developing CNS neuroepithelium from a planar to a tubular coordinate system. The planar (two-dimensional) neural plate is defined by rostrocaudal and mediolateral axes, while the tubular (three-dimensional) neural tube is defined by rostrocaudal, mediolateral, and dorsoventral axes. During this reorganization process, the lateral edge of the neural plate becomes the dorsal midline (roof plate) of the neural tube, and the medial edge (midline) of the neural plate becomes the ventral midline (floor plate) of the neural tube (Figure 27.5). The new axes become the substrate for further patterning and remain important throughout later development, although additional axial transformations occur locally in the developing brain.

Finally, neural tube closure is essential for subsequent development of posterior tissues including the vertebral arches and cranial vault, paraspinal muscles, and posterior skin of the head and back. In some cases, neural tube closure proceeds normally, but the overlying skin and mesodermal structures fail to cover the posterior neural tube. It is unknown whether these malformations (such as encephalocele, myelocele, and sacral agenesis) are caused by primary defects of neural tube closure (as frequently assumed in the neuropathology literature) or mesodermal and ectodermal development.

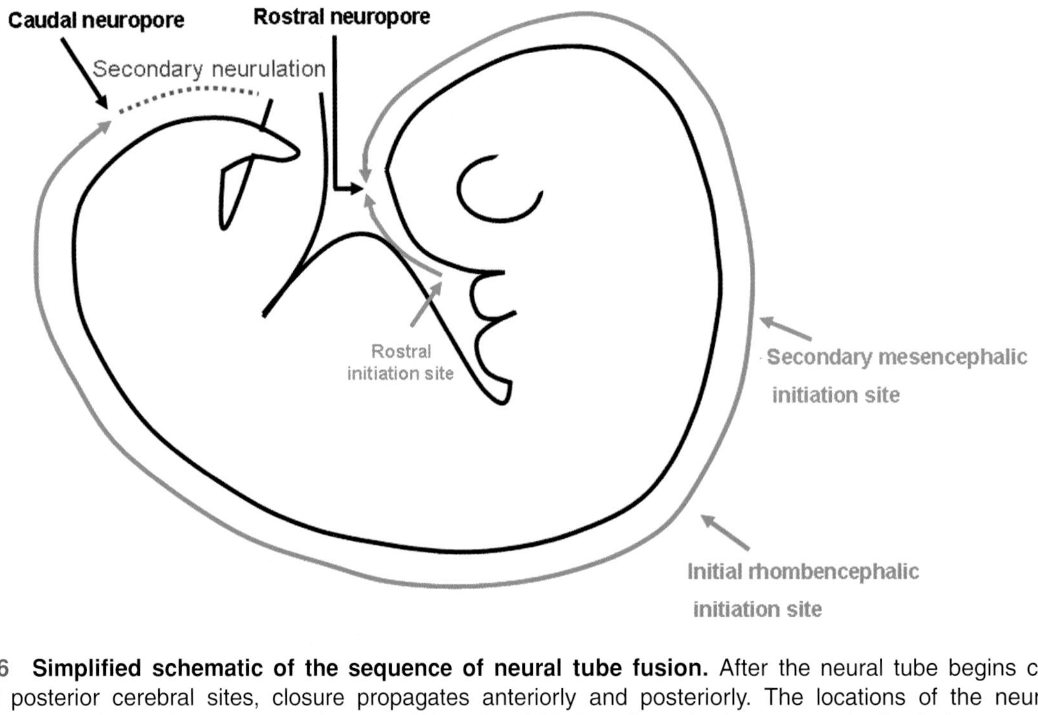

Figure 27.6 **Simplified schematic of the sequence of neural tube fusion.** After the neural tube begins closure from rostral and posterior cerebral sites, closure propagates anteriorly and posteriorly. The locations of the neuropores are potential sites for defects of neural tube closures. Although 2 initiation sites of primary neural tube fusion are shown here, this has not been rigorously proven in humans, and other mammals may have more initiation sites. Adapted from [6].

Human Neural Tube Defects

The most severe neural tube defect, characterized by the complete failure of neural tube closure along the entire craniospinal axis, is called craniorachischisis (Figure 27.7A). This lethal malformation corresponds to the severe form of neural tube defect found in *Lp* mice, described previously, and is probably caused by defects of Wnt-PCP signaling during convergent extension. Closure defects limited to the cranial region result in anencephaly, a severe, lethal defect (Figure 27.7B). In anencephaly, extensive destruction of the brain tissue is secondary to direct exposure to amniotic fluid: initially, the cerebral tissue proliferates and grows outside the surrounding skull base, a transient stage known as exencephaly. The most common human neural tube defects are limited to the lumbosacral region, sites of posterior neuropore closure and secondary neurulation. Typically, the spinal tissue has a ragged interface with mesodermal and ectodermal derivatives and is exposed externally. This appearance is called myelomeningocele (Figure 27.7C), somewhat erroneously because there is usually no cele or closed sac. Less often, malformations with a closed skin surface and sac are also seen. In the great majority of cases, lumbosacral myelomeningocele is accompanied by a malformation of the midbrain, hindbrain, and skull base known as Chiari type II malformation, historically called the Arnold-Chiari malformation (Figure 27.7C).

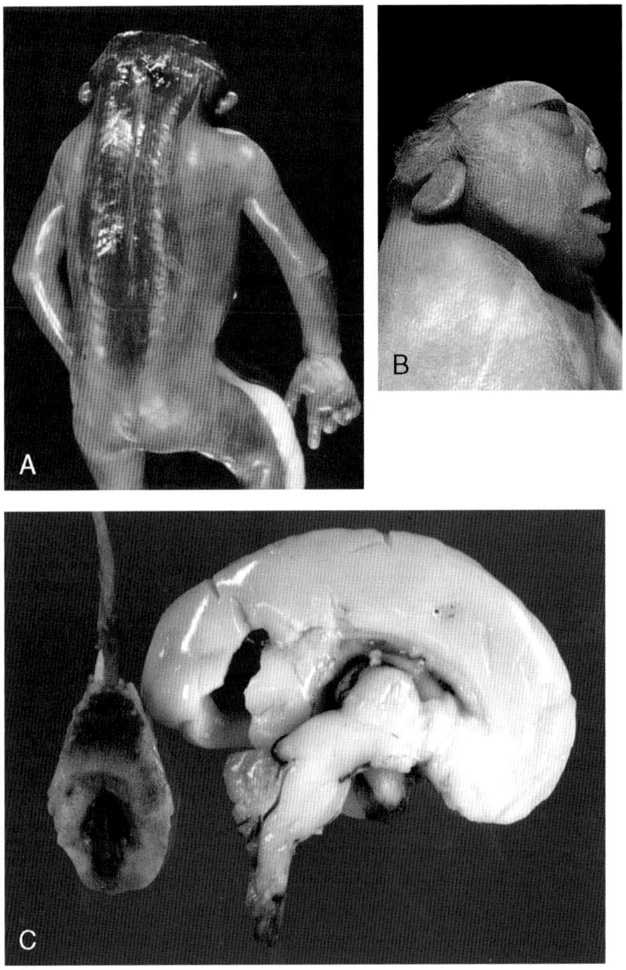

Figure 27.7 **Neural tube defects. (A)** Craniorachischisis. **(B)** Anencephaly. **(C)** Myelomeningocele with Chiari type II malformation. The spinal cord with open lumbosacral myelomeningocele is shown at left; the medial view of brain with hydrocephalus, tectal beaking, and herniation of the medulla over the spinal cord is shown at right.

Rostrocaudal and Dorsoventral Patterning of the Neural Tube: Holoprosencephaly

A program of segmental, compartmentalized gene expression begins to define rostrocaudal subdivisions of the CNS during neural plate stages, and this process accelerates with neurulation. Morphologically, the spinal cord is partitioned into cervical, thoracic, lumbar, and sacral segments associated with mesodermal somites. Morphological development of the brain is more complicated (Figure 27.8), as it is initially divided into 3 vesicles (prosencephalon/forebrain, mesencephalon/midbrain, rhombencephalon/hindbrain), and then further subdivided into 5 vesicles (telencephalon, diencephalon, mesencephalon, metencephalon, myelencephalon). Key molecular players in spinal and hindbrain segmentation are the *HOX* genes, which encode transcription factors expressed in nested patterns. The *HOX* gene family and their functions are highly conserved in evolution: their roles in rostrocaudal segmentation were initially discovered in the fruit fly, *Drosophila melanogaster*, and have been confirmed in other vertebrate and invertebrate species. More rostral segments of the CNS (forebrain and midbrain) also show segment-like subdivisions and nested gene expression patterns, but *HOX* genes are not involved. Instead, these brain areas are patterned by distantly related transcription factors with homologous DNA-binding domains (homeobox), along with other families of transcription factors, expressed in complex tissue patterns that interact with several signaling centers in the embryonic brain. The latter include the isthmus at the midbrain-hindbrain junction and the zona limitans intrathalamica at the junction of rostral and caudal thalamic subdivisions.

Concurrently with rostrocaudal segmentation, the neural tube is patterned along the dorsoventral axis by a different set of molecules and signaling centers. Important ventral signaling centers include the notochord (primordium of vertebral body nucleus pulposus), a mesodermal structure located ventral to the spinal cord and hindbrain; the prechordal plate mesendoderm, located ventral to the forebrain and midbrain; and the floor plate of the neural tube, a specialized neuroepithelial structure in the ventral midline of spinal cord and brain. All three of these ventral signaling centers produce Sonic hedgehog (Shh), a small, post-translationally modified, secreted protein with potent ventralizing activity on neural structures. The key dorsal signaling center is the roof

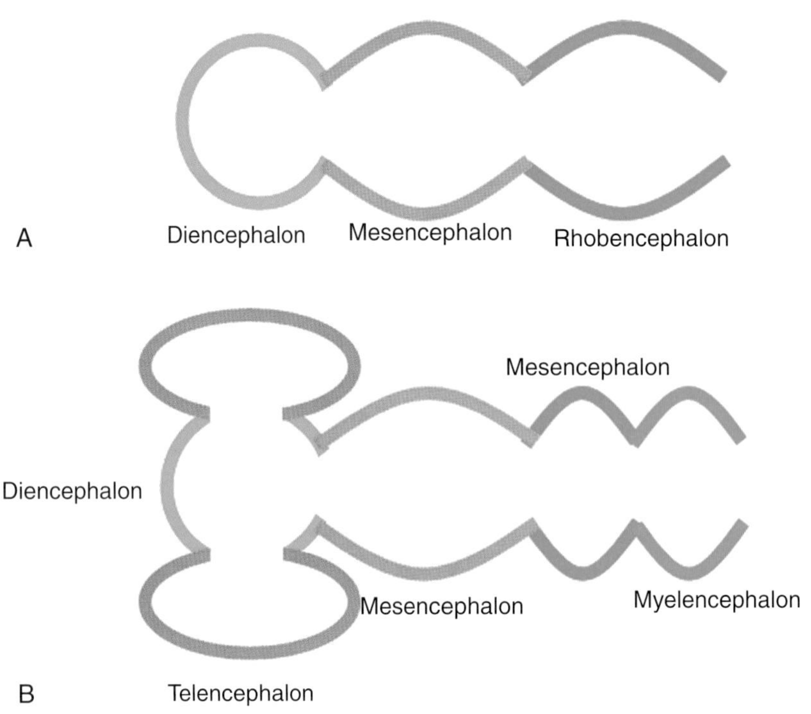

Figure 27.8 **Segmentation and cleavage of the developing brain. (A)** After fusion of the neural tube, the anterior-most portion forms swellings that will become the cerebrum (diencephalon), the midbrain (mesencephalon), and hindbrain (rhombencephalon). **(B)** Following this, proliferation of the cortical neuroepithelium forms telecephalic vesicles and will become the majority of the brain's paired hemispheres, and the rhombencephalon divides into the metencephalon and myelencephalon, which will become the pons and cerebellum and medulla, respectively.

plate, a specialized neuroepithelial structure in the dorsal midline that ultimately develops into choroid plexus. The roof plate, characterized by high-level expression of *ZIC* genes, produces secreted morphogens belonging to the bone morphogenetic protein (BMP) and Wnt families, with potent dorsalizing activity. The balance of antagonistic ventralizing and dorsalizing signals patterns the neural tube into basal and alar subdivisions associated with motor and sensory pathways, respectively. Accordingly, the ventral spinal primordium generates motor neurons, while the dorsal spinal primordium generates sensory relay neurons. Sensory and motor distinctions become more subtle in the midbrain and forebrain. For example, most of the forebrain is composed of alar (sensory) plate neuroepithelium, including such ventral (anatomically speaking) structures as the neural retina and optic pathways. In contrast, the neurohypophysis (ventral hypothalamus) may be considered a motor pathway, inasmuch as hormone-producing neurons generate somatic responses and behaviors.

Holoprosencephaly

The best characterized and most common defect of dorsoventral patterning in humans is holoprosencephaly, defined as complete or partial failure of cerebral hemispheric separation. Holoprosencephaly exhibits a spectrum of severity. The mildest forms (lobar holoprosencephaly) show some development of the interhemispheric fissure with continuity of cortex across the midline. Forms with intermediate severity (semilobar holoprosencephaly) show partial development of the interhemispheric fissure, and severe forms (alobar holoprosencephaly) show complete absence of the interhemispheric fissure, with a single forebrain ventricle (Figure 27.9). The defective midline structures in holoprosencephaly result from abnormalities of ventral or dorsal patterning molecules, including Shh and Zic2. Dorsoventral patterning is essential for differential proliferation and apoptosis of dorsal midline and ventrolateral forebrain structures, the essential underlying mechanisms of hemispheric separation. Overall proliferation is severely reduced in holoprosencephaly, and the brain is invariably small.

Signaling Centers and Transcription Factor Gradients in the Cerebral Cortex: Thanatophoric Dysplasia

Because of its significance for human cognition and awareness, cerebral cortex development has been a subject of great interest in basic research and the findings serve as a paradigm for mechanisms of regional patterning. The cerebral cortex develops as a dorsal outpouching of the alar plate, with midline separation into right and left hemispheres. Early in cortical development, the cleavage of telencephalon into right and

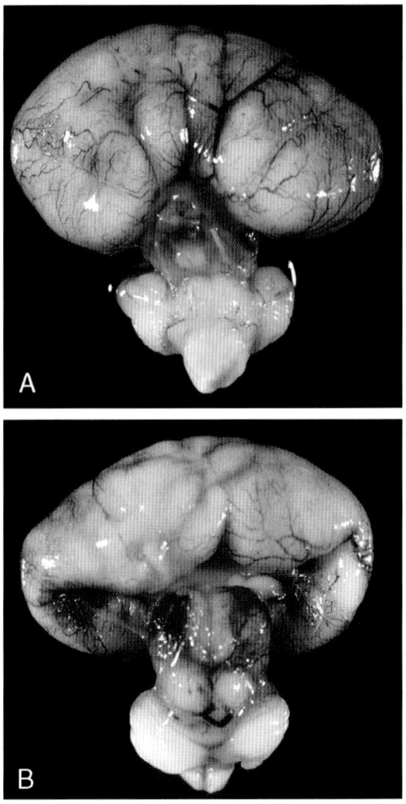

Figure 27.9 **Alobar holoprosencephaly in a 23-week gestational age fetus. (A)** Anterior view: the single "holosphere" exhibits shallow sulci but no deep interhemispheric fissure. **(B)** Posterior view: cortex is continuous across the midline and the posteriorly open (due to disrupted delicate membranes) single forebrain ventricle is visible.

left hemispheres depends on dorsoventral patterning, such that proliferation is initially enhanced in lateral cortical regions (away from the dorsal and ventral midline), while apoptosis is enhanced in the dorsal midline (roof plate). Differential growth of the telencephalic roof plate (slow) and lateral telencephalon (rapid) causes growth of the hemispheric neuroepithelium accompanied by relative reduction of the dorsal midline, leading to formation of the interhemispheric fissure. Accordingly, defective dorsal or ventral patterning (due to mutations in *SHH, ZIC2,* and other genes) results in a failure of separation between the hemispheres.

Even as the cerebral hemispheres begin to separate, patterning of the cortical areas within each hemisphere is already being initiated along the rostrocaudal and mediolateral axes, through the elaboration of additional signaling centers, and their production of secreted morphogens. Principal signaling centers that pattern the cerebral cortex include the commissural plate (rostral signaling center), cortical hem (dorsomedial signaling center), and cortical antihem (ventrolateral signaling center). These signaling centers, located at edges of the cortical neuroepithelium, secrete morphogens that regulate regional growth and identity along diffusion gradients that, by differential activation of receptors, encode positional information in the cortical neuroepithelium. The commissural plate, for example, produces fibroblast growth factor-8 (FGF-8), which specifies frontal lobe identity near the FGF-8 source, and parietal lobe more distally. Remarkably, ectopic expression of FGF-8 in caudal regions of embryonic mouse cortex induces the formation of a duplicate, mirror-image frontoparietal cortex (as indicated by molecular and cytoarchitectonic markers) at the occipital pole. The cortical hem, located at the medial edge of embryonic cortex, produces Wnt and BMP family molecules that specify hippocampus, medial temporal, and medial parietal cortex.

Animal research indicates that FGF-8, Wnt family members, and other signaling molecules specify cortical arealization by inducing the expression of developmental transcription factors in progenitor cells, even before most cortical neurons are produced. Gradients of *EMX1, PAX6, COUP-TF1,* and other transcription factor gene expression have been detected along rostrocaudal and mediolateral axes in rodents and appear to create a molecular coordinate system in which cortical areas are determined by a combinatorial code of transcription factor expression. Some transcription factors (such as PAX6) control both regional identity and growth of cortical areas, thus coordinating these processes. Other transcription factors seem to regulate either identity or growth preferentially. The distinct roles of different signaling molecules and transcription factors provide tremendous flexibility in determining not only the overall size of cerebral cortex, but also the relative size of sensory, motor, and multimodal specific areas.

These findings suggest that the size of human cortical areas may depend on the interplay of signaling centers and transcription factor gradients during embryonic development. This would further imply that aspects of mature human cognitive abilities may be determined by small variations of patterning signals during cortical development. According to this model, the unusual gyral pattern of Albert Einstein's parietal cortex, thought to possibly explain his mathematical gifts, could have arisen by gene interactions or developmental noise during the first trimester of intrauterine development!

Thanatophoric Dysplasia

FGF signaling plays a major role in regulating cortical arealization and proliferation. FGF receptor 3 (FGFR3) is one of four tyrosine kinase receptors for FGF signaling and is highly expressed in the embryonic cerebral cortex in a high caudal–low rostral gradient. Constitutive activating mutations of FGFR3 have been linked to thanatophoric dysplasia, a severe form of achondroplastic dwarfism in which the brain is enlarged and exhibits aberrant, prematurely forming sulci in temporal and occipital lobes (Figure 27.10). The cortical surface area and overall mass are also markedly

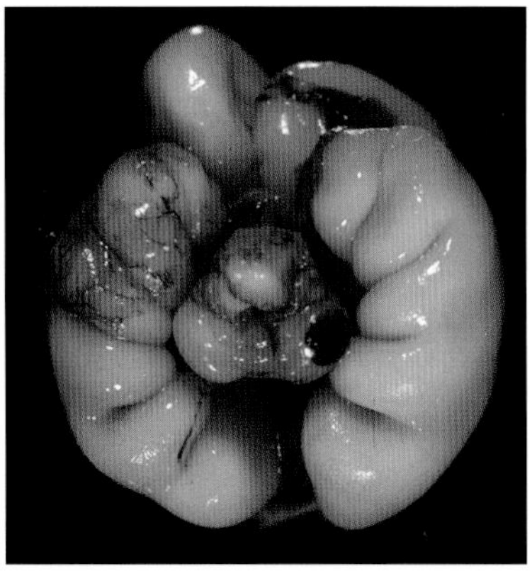

Figure 27.10 Thanatophoric dysplasia in a midgestational fetal brain (inferior view). Aberrant, prematurely formed sulci are present in the medial temporal and occipital lobes, radiating from the hindbrain and cerebellum (with small subarachnoid hemorrhage) in a "bear claw" configuration.

increased in thanatophoric dysplasia, illustrating the coordinate regulation of cortical surface area growth and areal patterning even in a pathological condition.

Compartmentalization of Embryonic Neurogenesis

Mature brain and spinal regions contain a mixture of excitatory and inhibitory cell types that utilize different neurotransmitters. The principal excitatory neurotransmitter at central synapses is L-glutamate, and the principal inhibitory neurotransmitter is γ-aminobutyric acid (GABA). Recently, it has been shown that developmental patterning of the brain and spinal cord plays a fundamental role in defining neurotransmitter types at very early stages, prior to the genesis of postmitotic neurons. Moreover, it has been shown that many distinct regions of the developing brain are patterned to produce either excitatory (glutamatergic) neurons or inhibitory (GABAergic) neurons. Thus, the production of different types of neurons is compartmentalized, and mature circuits are assembled by postmitotic neuronal migrations.

Important examples of compartmentalized neurogenesis have been discovered in development of the cerebral cortex and cerebellum, the two largest brain structures in humans. In the cerebral cortex, it has been shown that glutamatergic (pyramidal) cells, which account for 75%–80% of cortical neurons, are produced locally within the cortical neuroepithelium. In contrast, GABAergic interneurons are produced in subcortical progenitor compartments called the medial, caudal, and lateral ganglionic eminences that have long been known as basal ganglia progenitor regions. The interneurons then migrate long distances tangentially into the developing cortex and integrate into the circuitry along with the locally generated glutamatergic neurons. The compartmentalization of neurogenesis implies that different genes may control the development of pyramidal cells and interneurons, and indeed this is the case. In mice, the cortical and subcortical compartments express different transcription factors, including *EMX* family homeobox genes and *TBR* family T-box genes in the cortex, and *DLX* family homeobox genes in the subcortical regions. Mutations of these transcription factors selectively interfere with the development of the predicted set of neurons in mice, although this remains to be confirmed in humans.

In the cerebellum, glutamatergic and GABAergic neurogenesis are likewise compartmentalized, but in contrast to cortex, glutamatergic neurons migrate tangentially and GABAergic neurons migrate radially. Cerebellar GABAergic neurons (Purkinje cells and inhibitory interneurons) are produced in the cerebellar ventricular neuroepithelium, while glutamatergic neurons (deep nuclei projection neurons, granule cells, and excitatory interneurons) are produced in a specialized compartment known as the rhombic lip, which surrounds the roof of the fourth ventricle, and its derivative, the external granular layer, produced by progenitor migration from the rhombic lip over the entire surface of the cerebellar primordium. Transcription factors play a prominent role in defining the neurogenic regions, such as Ptf1a (also known as *cerebelless*), a basic helix-loop-helix (bHLH) transcription factor, in GABAergic progenitors, and Math1, another bHLH family transcription factor, in rhombic lip glutamatergic progenitors. This view of cerebellar neurogenesis has been established in mice but remains to be confirmed in human brain [5]. No malformations ascribed specifically to one type of neurotransmitter differentiation have been described as yet.

Proliferation, Progenitors, and Apoptosis: Micrencephaly

The overall and relative sizes of the brain and spinal cord, and their component regions, are determined in large part by the balance of proliferation by progenitor cells, and apoptosis by progenitors and postmitotic cells. Cell size also plays a significant role in determining nervous system size. In recent years, there has been tremendous progress in identifying various types of progenitor cells that differ in morphology, proliferative potential, neuronal or glial commitment, and maturation sequences. The analysis of progenitors has been highly stimulated by the discovery of adult neurogenesis and by the explosion of research on stem cells. Likewise, patterns of apoptosis have also been carefully examined throughout CNS development and key molecular regulators identified. Tight control over proliferation and apoptosis is essential not only for determining brain size, but also for preventing neoplasia.

Both neurogenesis and gliogenesis are functions of neural progenitor cells that are heterogeneous and, as a group, undergo progressive maturation and specialization from early to late stages. Timing has great significance, as different types of neurons and glia are produced sequentially according to consistent timetables. In the cortex, for example, deep-layer neurons are produced first, followed by superficial-layer neurons, and then glial cells.

Neuroepithelial Cells and Radial Glia

The nervous system begins as a true epithelium, with apical and basal surfaces that correspond to the ventricular and pial surfaces, respectively. Accordingly, the earliest progenitor cells, which have elongated morphology with processes that span the epithelium, are designated as neuroepithelial cells. Generally, neuroepithelial cells are capable of producing neurons, astrocytes, and oligodendrocytes and have virtually unlimited self-renewal proliferative potential, and thus qualify as a type of neural stem cells. Neuroepithelial cells and other neural stem cells express characteristic markers, such as Nestin (an intermediate filament), although none of these markers are completely specific. With further maturation, the progenitors show evidence of astrocytic differentiation while maintaining their elongated (radial) morphology. These astrocyte-like cells are known as radial glia, and they serve both as progenitors and as a scaffold for the developing nervous system. Radial glia generally have high proliferative potential, but heterogeneous properties of molecular expression and fate commitment (neurogenic or gliogenic). Neuroepithelial cells and radial glia both divide at the apical (ventricular) surface, which may serve to limit their proliferation as well as maintain apical-basal polarity of the pseudostratified neuroepithelium.

Intermediate Neuronal Progenitors

Intermediate neuronal progenitors, also known as transient amplifying cells, are produced from neuroepithelial cells and radial glia, but unlike them, do not contact the ventricular or pial surfaces. Instead, they have short or no processes and divide away from the apical surface; for this reason, they have also been called basal progenitors. Intermediate neuronal progenitors have a very limited proliferative potential, amounting to no more than 1–3 mitotic cycles, perhaps because they are detached from the ventricular surface and thus lack the constraints posed by attachment. Intermediate neuronal progenitors appear to produce only neurons (not glia), and their molecular and morphologic characteristics vary among brain regions. In the cerebral cortex, intermediate progenitors specifically express Tbr2/Eomes, a T-box transcription factor. The role of intermediate neuronal progenitors appears to be expansion of neuronal subsets from limited numbers of neuroepithelial cells or radial glia.

Adult Neurogenesis

Whereas it was long believed that neurogenesis did not occur in mature animals, studies in experimental animals and humans have demonstrated that, in fact, neurogenesis continues at robust levels in two brain regions: the subgranular zone in the dentate gyrus of hippocampus, and the subventricular zone adjacent to striatum. The hippocampal subgranular zone contains Nestin-positive, neural stem-like cells ("like" because they may not have unlimited self-renewal capacity) that produce new glutamatergic neurons that incorporate anatomically and functionally into the dentate gyrus, presumably to subserve new memory formation, although the functional impact of adult hippocampal neurogenesis is still not clear. The striatal subventricular zone likewise contains Nestin-positive, neural stem-like cells that produce new GABAergic (and some glutamatergic) neurons, which migrate along the rostral migratory stream into the olfactory bulb where they incorporate into its circuitry. Interestingly, adult neurogenesis involves stages of progenitor maturation and molecular expression that highly resemble those in embryonic development of the hippocampus and olfactory bulb.

Gliogenesis

Glial cells include astrocytes, oligodendrocytes, microglia, and possible new types such as NG2-positive cells. Generally, gliogenesis follows neurogenesis with relatively little temporal overlap. Astrocytes appear to be produced directly from radial glia, and by continued division of astrocytes and glial progenitors that maintain the potential to proliferate throughout life. In terms of progenitor lineages and molecular expression, oligodendroglia in most regions are more closely related to neurons than to astrocytes. NG2-positive cells are poorly understood glia with some progenitor properties that probably are generated concurrently with other glial types.

Apoptosis

Programmed cell death, or apoptosis, occurs as a normal and essential part of brain development that functions to eliminate unnecessary, excess progenitors and neurons. In fact, apoptosis is most active and important in progenitor compartments. Gene targeting to inactivate (or knockout) genes that are required for apoptosis in mice causes hyperplasia of some neuroepithelial regions, especially the developing cerebral cortex. In mice, the result is a dramatic increase of cortical surface area with aberrant formation of gyri. Unlike humans, mice have smooth-surfaced or lissencephalic brains that normally lack gyri. Apoptosis of postmitotic neurons also does occur, but at lower levels. Here apoptosis may cull excess neurons produced at the end of neurogenesis that fail to integrate into the CNS circuitry.

Micrencephaly

Small brain size (micrencephaly) is caused by defective proliferation of cortical progenitors. Commonly, it is associated with other malformations, such as holoprosencephaly and lissencephaly. Less often, it occurs in the absence of other malformations. This form of micrencephaly has been linked in some cases to mutations in *ASPM*, a gene encoding an essential protein for the mitotic spindle.

Neuronal Migration and Differentiation: Lissencephaly

Postmitotic Events and Integration into Neural Circuits

Much evidence suggests that fundamental aspects of neuronal fate (including positional or laminar fate, neurotransmitter type, axonal connections, and molecular expression) are determined during the last mitotic cycle, when the neuron is produced or born. Thus, newly generated neurons are born with their missions already defined. In order to accomplish its mission, the new neuron must (i) differentiate by expression of general neuronal genes and neuronal subtype-specific genes; (ii) migrate to the correct location; (iii) grow axons and dendrites and make appropriate synaptic connections; (iv) refine connections with other neural elements according to critical environmental stimuli; and (v) acquire glial contacts (myelination and astrocytic contacts) that enhance neuronal function.

Neuronal Differentiation

The vast numbers of distinct neuron types in the brain (numbering in the hundreds) express different combinations of general neuronal genes, as well as neuronal subtype-specific genes. In the cerebral cortex, for example, each cortical layer contains multiple distinct types of glutamatergic (pyramidal) neurons and multiple distinct types of GABAergic (inhibitory) neurons. In general, the specific properties of each neuron type are largely specified by the combinatorial and sequential expression of transcription factors. Initial specification of neuronal fates begins in progenitors by the expression of neurogenic genes, such as *Neurogenin1*, a bHLH transcription factor. Following mitosis, a different set of bHLH transcription factors is activated, belonging to the neuronal differentiation group (such as *Neurod1*). Further differentiation into neuronal subtypes involves other transcription factors belonging to a great variety of families.

Neuronal Migration

Virtually all new neurons migrate from their sites of production in progenitor zones, to sites of integration in postmitotic neural structures. Migrations are remarkably diverse among various brain regions: they may traverse relatively short or long distances, may be mainly radial or tangential or both, and may involve multiple sequential phases (for instance tangential followed by radial migration). One example of a long-range, mainly tangential migration was given previously, in reference to the migration of interneurons from subcortical forebrain into the developing cortex. Indeed, migrations are particularly robust in the cerebral cortex and in the cerebellum, and these regions are hardest hit by mutations that perturb cell migration.

Neuronal migration requires the coordinated activity of numerous molecules belonging to several categories. First, the direction of migration is selected by a leading process with a distal specialization or growth cone where guidance receptors from a variety of families may be expressed (such as Eph family tyrosine kinase receptors). Second, the leading process must grow in the indicated direction, a process that requires actin and microtubule cytoskeletal dynamics, as well as membrane turnover (mediated by endocytic vesicles). Third, the nucleus must be pulled toward the leading process by a network of microtubules oriented from the centrosome. Using these mechanisms, neurons migrate in a phasic rather than continuous manner, with repeated cycles of leading process growth and nuclear translocation. Whereas neuronal migration appears highly sensitive to disturbances in the underlying cellular machinery, it is also clear that the mechanisms of cell migration have a great deal of overlap with axon guidance (likewise mediated by a growth cone), membrane recycling, and cellular proliferation. Accordingly, neuronal migration disorders in humans often display associated abnormalities of axon pathways and brain growth.

Studies of the genetics of human and murine neuronal migration disorders have led to the identification of many key molecules guiding or promoting neuronal migration. In mice, one of the prototypical neuronal migration disorders was found in the spontaneous mutant strain *reeler*, in which the cerebral cortex and cerebellum show severe abnormalities of cellular organization. Further studies revealed that the disorder resulted from deficiency of a large secreted protein encoded by the *reeler* gene. The cognate protein, Reelin, was found to be highly localized in the cortical marginal zone, and in specific locations within the developing cerebellum, suggesting that it might function as a guidance molecule for migrating neurons. Further studies have supported this interpretation, as receptors for Reelin and downstream intracellular signaling molecules have been identified. A human counterpart of the mouse disorder has been identified in patients with lissencephaly and cerebellar hypoplasia, found to have mutations in the human *RELN* gene. Other genetic studies of human patients with lissencephaly or periventricular heterotopia have identified several additional critical genes, notably including *DCX* (Doublecortin), *LIS1* (Lissencephaly-1), *FLNA* (Filamin A), and *ARFGEF2* (a vesicular trafficking molecule). These molecules are important for microtubule dynamics or membrane dynamics, and critical for cellular mechanisms of either nuclear translocation within migrating cells (microtubule dynamics) or addition of membrane to the leading process (membrane trafficking).

Lissencephaly, Periventricular Heterotopia, and Polymicrogyria

Lissencephaly, periventricular heterotopia, and polymicrogyria are the classic disorders of neuronal migration described in human neuropathology. Lissencephaly is an abnormality of cortical development which, in humans, produces a thickened cortical gray matter layer without normal folding (gyri). Lissencephaly may be divided into type I (smooth surface) and type II (cobblestone surface), which have distinct mechanisms. Type I lissencephaly is a primary defect of neuronal proliferation and migration (Figure 27.11A), while type II lissencephaly is a primary defect of basal lamina integrity at the pial surface, in which basal lamina breakdown leads to neuronal migration through the basal lamina defects and, ultimately, disorganization of the cortical structure. Mutations of cell migration molecules such as LIS1 and DCX are the cause of type I lissencephaly, while mutations in glycosylation enzyme genes (essential for basal lamina integrity) are the cause of type II lissencephaly. Periventricular heterotopia, abnormally localized gray matter islands abutting the ventricular surfaces (Figure 27.11B), can be caused by *FLNA* mutations, chromosome abnormalities, and other unknown molecular or tissue abnormalities. Polymicrogyria, abnormally organized cortex with numerous small gyri (Figure 27.11C), is a heterogeneous disorder with variable lobar, unilateral, or bilateral distribution, and genetic as well as nongenetic causes (such as fetal hypoxia-ischemia). Different histological varieties of polymicrogyria have also been described, including four-layered and unlayered types. Whereas lissencephaly and periventricular heterotopia involve defective migration from progenitor compartments (ventricular zone and subventricular zone) to the cortical plate, polymicrogyria involves defective laminar organization of neurons within the cortical plate. Hereditary forms of polymicrogyria have been linked to mutations in *GPR56*, a G-protein coupled receptor of unknown function. On the basis of the genes identified so far, polymicrogyria may be caused by defects in distinct classes of molecules from lissencephaly and periventricular heterotopia.

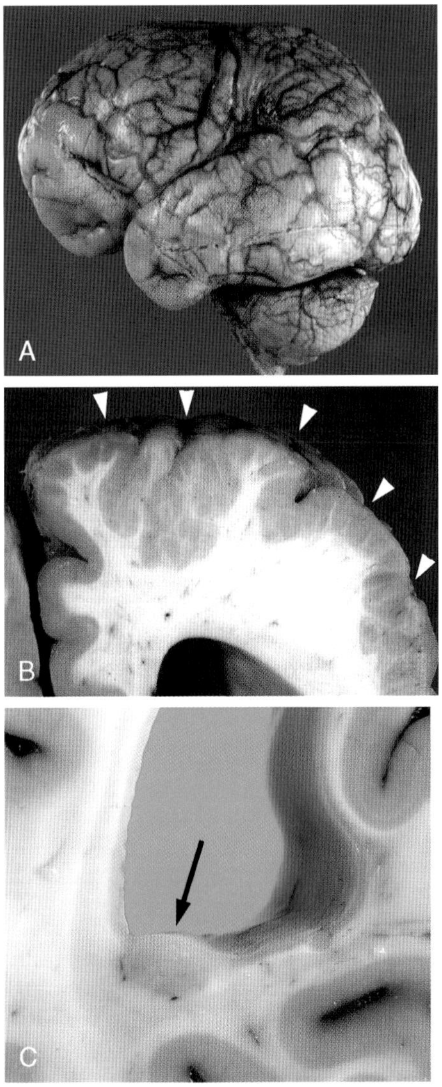

Figure 27.11 Defects of cortical differentiation or migration. (A) Lissencephaly (type I) in Miller-Dieker syndrome (lateral view). The sulci are extraordinarily shallow, lending the brain surface a smooth surface appearance. **(B)** Polymicrogyria affecting lateral frontal cortex. **(C)** Periventricular heterotopia in the occipital horn of the lateral ventricle.

Axon Growth and Guidance: Agenesis of the Corpus Callosum

The cellular complexity of the human brain is exceeded only by its extraordinary connectional complexity. Axonal pathways develop concurrently with neuronal migrations and are mediated by numerous large families of axon guidance molecules, among them the Eph receptors, ephrin ligands, Semaphorins, neuropilins, plexins, Robo receptors, Slit ligands, diverse cell adhesion molecules, secreted morphogens, and extracellular matrix molecules and their receptors. Studies in experimental animals have shown that, in order to effectively navigate long distances, axon growth and guidance typically involve stepwise progression from one intermediate target to another. Hundreds of axon guidance phenotypes have been identified in mice and other experimental animals (fruit flies, worms, zebrafish) with mutations in axon guidance molecules, and this is an area of intense research. Somewhat surprisingly, only a few axon guidance disorders have been found in humans. This likely reflects the difficulty of tracing axon pathways in the human brain, where it is obviously impossible to inject actively transported axon tracers during life.

Agenesis of the Corpus Callosum

Agenesis of the corpus callosum, a relatively common disorder (~1 per 1000), can be asymptomatic, and is believed to be due to a failure of axon guidance.

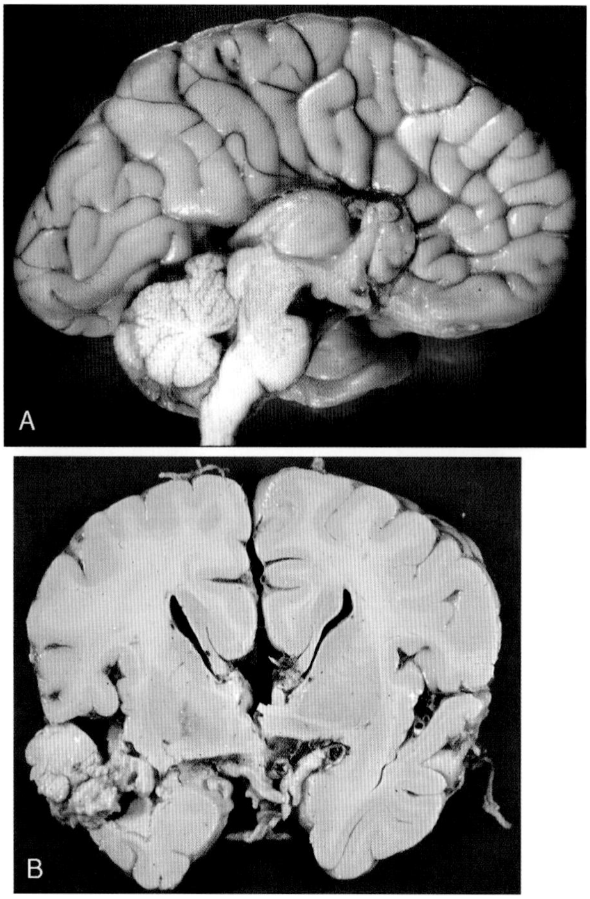

Figure 27.12 Agenesis of the corpus callosum. (A) Medial view: the corpus callosum is absent and the cingulate gyrus and sulcus are malformed. **(B)** Coronal slice: Probst bundles are visible dangling below the malformed cingulate cortex, indicating growth toward the midline but failure to cross.

It has been linked in X-linked pedigrees to mutations of L1 cell adhesion molecule. Often, the callosal axons that fail to cross the midline instead form an aberrant tangle of fibers adjacent to the midline, called a Probst bundle (Figure 27.12).

Synaptogenesis, Refinement, and Plasticity: Autism and Schizophrenia

Synaptogenesis, refinement, and plasticity, relatively late stages of neuronal development, are essential for proper CNS functioning, but not morphogenesis. Thus, perturbations of critical molecules for these processes cause complex cognitive, sensory, or motor disorders without obvious malformations. Human diseases related to these processes are thought to include forms of autism, schizophrenia, and mental retardation. However, these disorders are genetically and phenotypically complex, and remain poorly understood in relation to developmental neurobiology.

Summary

Molecular expression changes due to genetic mutations are a frequent cause of cerebral and spinal malformations. Progress in developmental neurogenetics has transformed developmental neuropathology by providing a mechanistic understanding of malformations, by improving prognostic accuracy, and by providing insights that may one day allow for more effective treatments of these often devastating conditions.

NEUROLOGICAL INJURY: STROKE, NEURODEGENERATION, AND TOXICANTS

Basic Mechanisms of Injury

Before we embark on a discussion of specific types of nervous system injury, it is first worthwhile to consider a few basic mechanisms of damage to the nervous system and the nervous system's response. The following mechanisms are not exclusive to any neurologic disease but are commonly proposed to contribute to varying degrees to the pathogenesis of stroke, neurodegeneration, and neurotoxicant injury.

The adult CNS has limited regenerative ability, as noted previously. The natural history of CNS damage is that cells vulnerable to injury become necrotic or apoptotic and their debris removed by scavenger cells. The overall mass of the effected region is reduced and shrinks in a state called atrophy. The residual neuroepithelial cells, usually astrocytes and microglia, respond by gliosis.

Excitotoxicity

L-glutamate is the most abundant excitatory amino acid (EAA) neurotransmitter in the CNS. It and its cell surface ligand-activated ion channels, such as AMPA and NMDA receptors, participate in a wide array of neurological functions. Figure 27.13A depicts normal activity at an excitatory synapse. The glutamatergic presynaptic terminal releases glutamate upon depolarization into the synapse that then can bind to postsynaptic AMPA receptor to initiate influx of Na^+ into the postsynaptic element that has a potential gradient across its membrane generated primarily by the action of Na^+/K^+-ATPase. Glutamate also binds to the NMDA receptor, but unless the postsynaptic membrane is sufficiently depolarized, flow of cations, including Ca^{2+}, is blocked by Mg^{2+}. Glutamate is removed from the synapse by a number of transporters on neurons and astrocytes; astrocytic glutamate can enter the glutamine cycle to replenish neurotransmitter pools. Excitatory neurotransmission can be subverted to contribute to several neurologic diseases. Indeed, drugs that target excitatory neurotransmission are currently approved for use in patients with Alzheimer's disease and remain a focus of those interested in limiting damage from ischemia or multiple sclerosis. Excessive EAA receptor stimulation sets in motion a

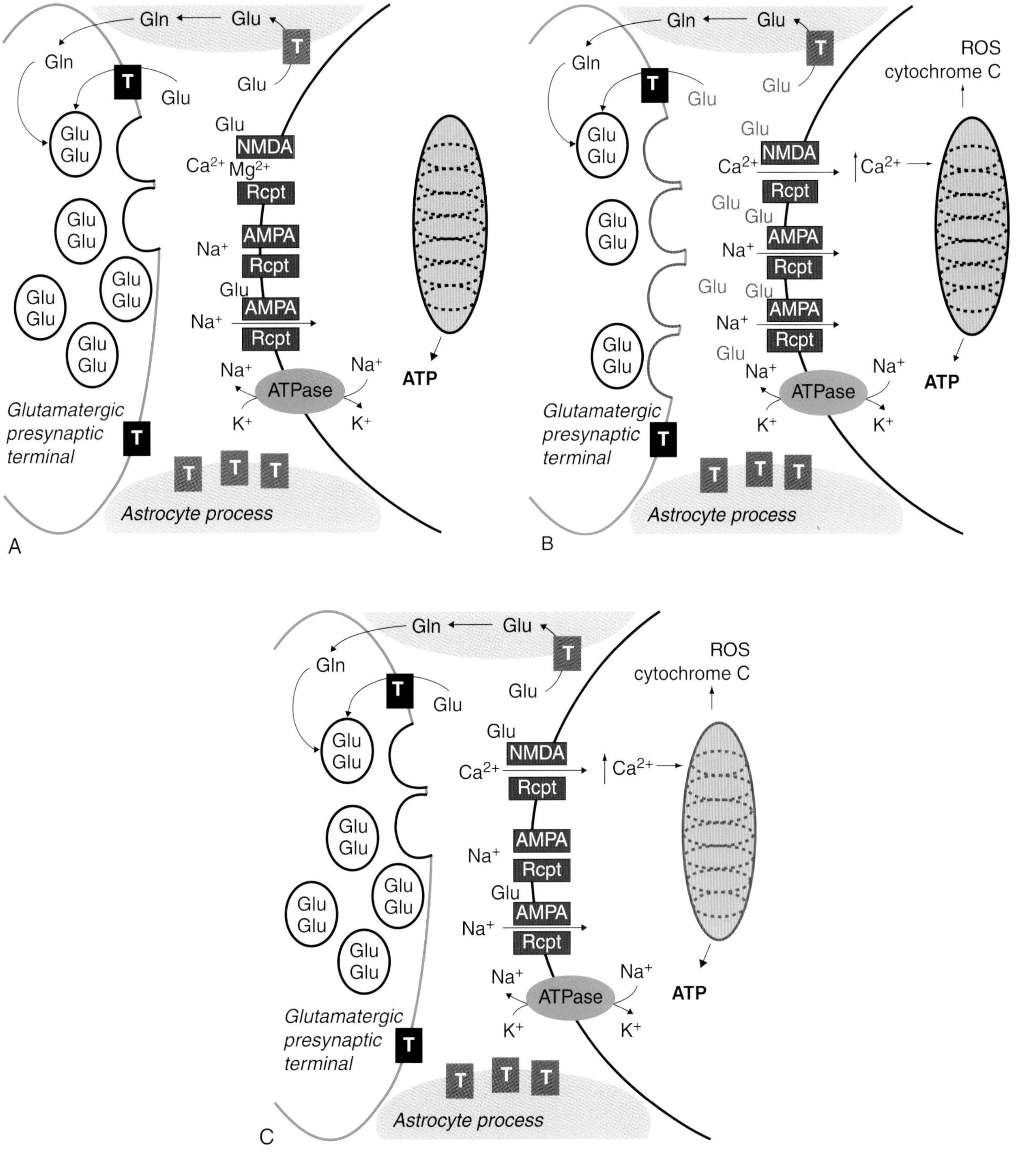

Figure 27.13 Excitotoxicity. Diagrams show excitatory neurotransmission **(A)**, direct excitotoxicity **(B)**, and indirect excitotoxicity (C).

cascade of events that can contribute to neuronal injury through a process called excitotoxicity. Broadly, there are two types of excitotoxicity: direct and indirect.

Figure 27.13B depicts events in direct excitotoxicity. The key initiator in direct excitotoxicity is increased synaptic concentration of EAAs either by release from the presynaptic terminal (shown in red in Figure 27.13), decreased reuptake, or exposure to excitatory neurotoxicants like domoic acid in algal blooms or as occurs in lathyrism. Extensive activation of AMPA receptors sufficiently depolarizes the postsynaptic membrane to relieve the Mg^{2+} block of the NMDA receptor. Rapid injury and lysis can follow this depolarization-dependent increase in ion influx. Delayed toxicity can develop even after EAA has been removed secondary to increased intraneuronal Ca^{2+} levels with subsequent inappropriate activation of calcium-dependent enzymes, mitochondrial damage, increased generation of free radicals, and transcription of proapoptotic genes.

Indirect excitotoxicity is depicted in Figure 27.13C. In distinction from direct excitotoxicity, the key event is impaired mitochondrial function with reduced ATP generation. One consequence is reduced activity of Na^+/K^+-ATPase with partial depolarization of the postsynaptic membrane that, when combined with normal levels of glutamate release, are sufficient to relieve the Mg^{2+} block of NMDA-mediated ion conductance. This again leads to increased intraneuronal Ca^{2+} with consequences similar to those described for direct excitotoxicity.

Mitochondrial Dysfunction

That neuronal mitochondrial dysfunction leads to neuron damage and death is clearly established by toxicants, perhaps the best studied being 1-methyl-4-phenyl-tetrahydropyridine (MPTP). Interrupting electron transport, diminishing the proton gradient, or inhibition of complex V all can result in reduced ATP production and lead to stressors like those described previously. The contribution of mitochondrial dysfunction to neuron injury likely extends far beyond toxicants to include some unusual diseases caused by inherited mutations in the mitochondrial genome or genes encoded in the nucleus whose protein products are incorporated into mitochondria, and perhaps some neurodegenerative diseases like Huntington's disease and Parkinson's disease.

Free Radical Stress

Free radicals are normal components of second messenger signaling pathways. However, excess or uncontrolled free radical production is detrimental to cells either by direct damage to macromolecules or liberation of toxic byproducts. A number of sites for free radical generation exist in the nervous system. A major source appears to be oxidative phosphorylation in mitochondria. Other sources of free radicals may become significant during pathological states. These include excitotoxicity, enzymes such as cyclooxygenase, lipoxygenase, myeloperoxidase, NADPH oxidase, and monoamine oxidase, autoxidation of catechols such as dopamine, and amyloid β peptides.

Research in biological systems has focused on oxygen-, nitrogen-, and carbon-centered free radicals with oxygen-centered free radicals being the most extensively studied. During the sequential four electron reduction of molecular oxygen to water (Figure 27.14, Reaction 1), two free radicals, superoxide anion ($O_2^{\cdot-}$) and hydroxyl radical ($^{\cdot}OH$), are produced along with hydrogen peroxide (H_2O_2). Since H_2O_2 is not a radical, these three molecules are collectively referred to as reduced oxygen species (ROS). Both O_2^- and H_2O_2 are signaling molecules under normal physiological conditions. On the other hand, hydroxyl radical is the most reactive and is capable of oxidizing lipids, carbohydrates, proteins, and nucleic acids at a rate limited by its own diffusion. Hydroxyl radical is formed from O_2^- and H_2O_2 by the Haber-Weiss reaction (Figure 27.14, Reaction 2) or from H_2O_2 by the Fenton reaction (Figure 27.14, Reaction 3). Several metal ions may participate in the Fenton reaction, including Fe(II), Cu(I), Mn(II), Cr(V), and Ni(II).

Nitric oxide ($^{\cdot}NO$), a free radical product of nitric oxide synthase-catalyzed oxidation of L-arginine to citrulline, initiates a cascade of reactive nitrogen species (RNS). $^{\cdot}NO$ has several physiologic functions including vasodilator and neuromodulator. $^{\cdot}NO$ reacts with $O_2^{\cdot-}$ to produce peroxynitrite ($ONOO^-$), a potent oxidant that can modify cellular

$$\underset{\text{Molecular Oxygen}}{O_2} \longrightarrow \underset{\text{Superoxide Anion}}{O_2^-} \longrightarrow \underset{\text{Hydrogen Peroxide}}{H_2O_2} \longrightarrow \underset{\text{Hydroxyl Radical}}{^{\bullet}OH} \longrightarrow \underset{\text{Water}}{H_2O} \qquad (1)$$

$$O_2^{\cdot-} + H_2O_2 \longrightarrow O_2 + OH^- + {^{\bullet}OH} \qquad (2)$$

$$H_2O_2 + Me^+ \longrightarrow OH^- + {^{\bullet}OH} + Me^{++} \qquad (3)$$

$$\underset{\text{Nitric Oxide}}{^{\bullet}NO} + O_2^{\cdot-} \longrightarrow \underset{\text{Peroxynitrite}}{ONOO^-} \qquad (4)$$

$$ONOO^- + CO_2 \longrightarrow \underset{\text{Nitrosoperoxy Carbonate}}{ONOOCO_2^-} \longrightarrow \underset{\text{Nitrogen Dioxide}}{^{\bullet}NO_2} + \underset{\text{Carbonate Anion radical}}{CO_3^{\cdot}} \qquad (5)$$

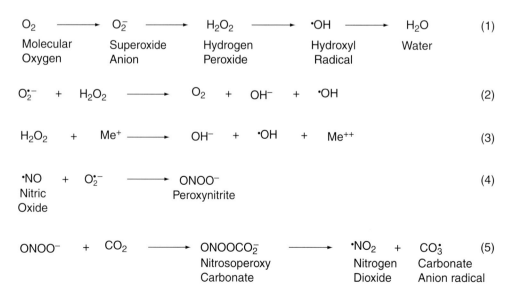

Figure 27.14 **Reactions in the generation of reduced oxygen species (ROS) and reduced nitrogen species (RNS).**

macromolecules (Figure 27.14, Reaction 4). In turn, reaction of $ONOO^-$ with ubiquitous CO_2 forms nitrosoperoxy carbonate that spontaneously cleaves to form a nitrating agent, nitrogen dioxide radical, and an oxidant, carbonate anion radical (Figure 27.14, Reaction 5). Finally, $ONOO^-$ may decompose to nitrous oxide and $^{\cdot}OH$ upon protonation. Carbon-centered free radicals are generated during the metabolism of some toxicants, like carbon tetrachloride, and also are produced on fatty acyl chains during lipid peroxidation.

Lipid Peroxidation

Free radical-mediated damage to polyunsaturated fatty acids, termed lipid peroxidation, differs from other forms of free radical damage to macromolecules because it is a self-propagating process that generates neurotoxic byproducts (Figure 27.15). Lipid peroxidation begins with abstraction of a hydrogen atom from a fatty acyl chain (LH), typically a polyunsaturated fatty acid (PUFA), because the carbon-hydrogen bond is made more acidic by the adjacent carbon-carbon double bond. This leaves a lipid radical ($L^{\cdot}$) that rearranges to a conjugated diene that can accept molecular oxygen to form a peroxyl radical ($LOO^{\cdot}$) that in turn can propagate the reactions by abstracting a hydrogen atom to generate a lipid hydroperoxide (LOOH). Lipid hydroperoxides are reactive molecules that can participate in many reactions, including the Fenton reaction to generate fragmentation products. Thus, there are two deleterious outcomes to lipid peroxidation: (i) structural damage to membranes and (ii) generation of bioactive secondary products. Membrane damage derives from the generation of fragmented fatty acyl chains, lipid-lipid crosslinks, and lipid-protein crosslinks. In addition, lipid peroxyl radicals can undergo endocyclization to produce novel fatty acid esters that may disrupt membranes. Fragmentation of lipid hydroperoxides, in addition to producing abnormal fatty acid esters, liberates a number of diffusable products, some of which are potent electrophiles; the most abundant are chemically reactive aldehydes such as acrolein and 4-hydroxy-2-nonenal. There are now many studies that highlight the potential of these numerous toxins to damage neurons and potentially contribute to neurodegenerative disease.

Carbonyl Stress

Analogous to protein adduction by reactive carbonyls generated by lipid peroxidation, non-enzyme-catalyzed post-translational modification of protein by reducing sugars, a process termed glycation, also is associated with some forms of neurodegeneration. Subsequent irreversible rearrangements, fragmentations, dehydrations, and condensations yield a complex mixture of protein-bound products termed advanced glycation endproducts (AGEs). Like reactive aldehydes from lipid peroxidation, AGEs can modify structural proteins and enzymes and render them inactive or dysfunctional. AGEs also can be redox active and may themselves generate oxidative stress. In addition to these deleterious biochemical reactions, AGE-modified proteins may bind to an AGE receptor (RAGE) that exists on several cell types, including neurons and glia. One role of RAGE is thought to be incorporation of AGE-modified

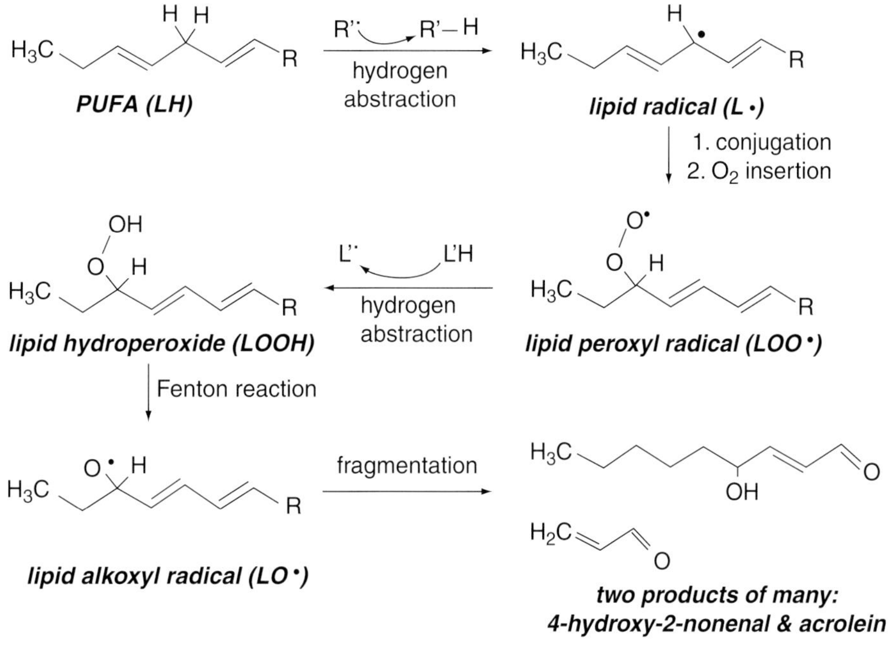

Figure 27.15 **Diagram of reactions in lipid peroxidation.**

proteins for degradation. However, binding of ligands, including Aβ peptides, to RAGE on neurons in culture stimulates production of ROS, and binding to RAGE on microglia in culture leads to their activation.

Innate Immune Activation

Activation of innate immune response in brain, primarily mediated by microglia, is a feature shared by several neurodegenerative diseases. Moreover, microglial-mediated phagocytosis of potentially neurotoxic protein aggregates (*vide infra*) might be an important means of neuroprotection. Indeed, while some aspects of the innate immune activation are an appropriate response to the stressors presented by neurodegenerative diseases, other facets of glial activation, especially when protracted, may contribute to neuronal damage. This balance between beneficial and deleterious actions of immune activation is well appreciated by students of pathology in other organs and is currently an area of very active investigation among neuroscientists with the goal of promoting neurotrophic or neuroprotective actions while suppressing paracrine damage to neurons. Key elements in paracrine neurotoxicity from activated glia appear to include IL-1β, TNFα, and prostaglandin E_2, among others with activation of COX2, iNOS, and NADPH oxidase that interface with excitotoxicity and free radical stress.

Catechol Metabolites

Catechols such as dopamine and norepinephrine are vulnerable to oxidations, rearrangements, and condensations under physiological conditions. Within synaptic vesicles, high concentrations of antioxidants and metal ion chelators stabilize catechols. However, conditions in extracellular fluid and cytosol are more favorable for chemical transformation of the catechol nucleus. Several products from catechols have been proposed to participate in dopaminergic neurodegeneration; these include dopamine quinone, 6-hydroxydopamine, and isoquinolines.

Vascular Disease and Injury

Disease in the cerebrovasculature can lead to compromised perfusion of regions of the brain or spinal cord, a process called vascular brain injury. The most common types of vascular diseases are atherosclerosis and arteriolosclerosis. The most common forms of vascular brain injury are ischemia and hemorrhage.

Ischemia

The clinical consequences of acute ischemic stroke are dramatic and critically dependent on the precise regions of CNS involved. The pathologic consequences of ischemia assume one of three general forms depending on the severity of the insult (Figure 27.16). The first is a complete infarct with necrosis of all parenchymal elements. The second form of injury, an incomplete infarct, occurs when ischemia is less severe, producing necrosis of some cells, but not all tissue elements. Incomplete infarcts demonstrate that neurons and, to a lesser degree, oligodendroglia are more vulnerable than other cells in the CNS to ischemia. The ultimate tissue manifestation of an incomplete infarct resembles the edge of a complete infarct; that is, neuronal and oligodendroglial depopulation, myelin pallor, astrogliosis, and capillary prominence. The final form of ischemic injury results in damage and dysfunction, but without death of parenchymal elements.

Although necrotic brain is essentially irretrievably lost, it is important to realize that the zones with incomplete infarction or damage without necrosis hang in the balance between vulnerability to further injury and salvage or perhaps even regeneration. Indeed, some functional recovery typically occurs in the days, weeks, and even months following an ischemic stroke, in part from resolution of reversible stressors, post-stroke neurogenesis in at least some regions of CNS, and functional reorganization of surviving elements. The balance of deleterious versus beneficial glial and immune responses in these surviving but damaged regions is thought to be key to optimal clinical outcome and is an area of intense investigation.

Regardless of the cause, CNS infarcts share a common evolution of coagulative necrosis, leading to liquefaction that culminates in cavity formation. Although death of cells occurs in CNS tissue within minutes of sufficient ischemia, the earliest structural sign of damage is not apparent until 12 to 24 hours following the ictus, when it takes the form of coagulative necrosis of neurons, or red neuron formation (Figure 27.17). The histologic hallmarks of this

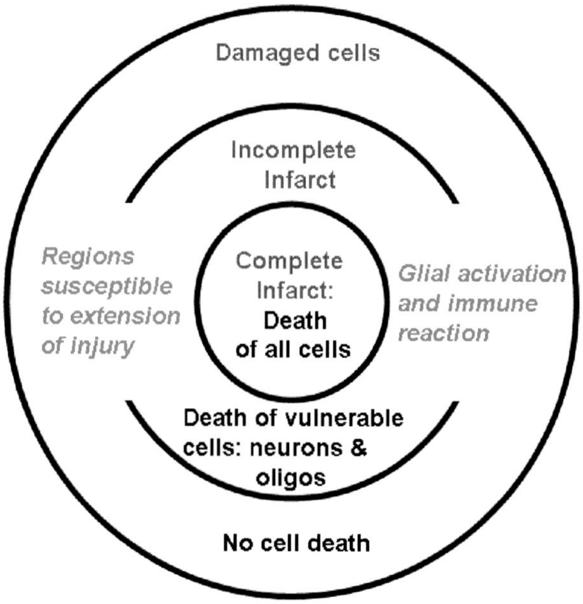

Figure 27.16 Diagram of varying levels of damage and reaction in vascular brain injury.

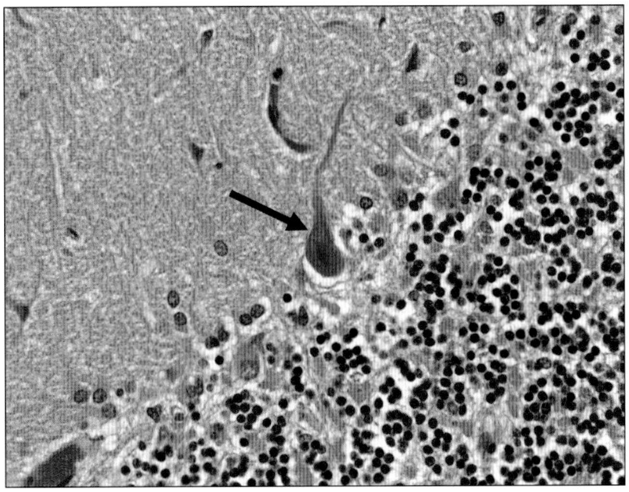

Figure 27.17 **Red neuron.** Photomicrograph of H&E-stained section of cerebellum shows Purkinje neuron with changes of coagulative necrosis, aka "red neuron" (black arrow).

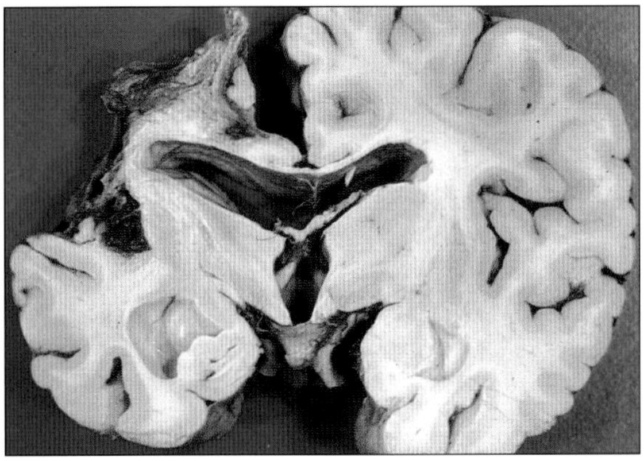

Figure 27.19 **Coronal section of cerebrum shows remote infarct in the territory supplied by the left middle cerebral artery.**

process are neuronal karyolysis and cytoplasmic hypereosinophilia. Infarcts become macroscopically apparent about 1 day after onset as a poorly delineated edematous lesion (Figure 27.18). Subsequent evolution of infarcts is dictated by the inflammatory cell infiltrate. Around 1 to 2 days after infarction, the lesion is characterized by disintegration of necrotic neurons, capillary prominence, endothelial cell hypertrophy, and a short-lived influx of neutrophils that are rapidly replaced by macrophages. Initially, macrophages may be difficult to discern in the inflamed and necrotic tissue, but within several days they accumulate to very high density and become enlarged with phagocytized material; this is the phase of liquefaction. These debris-laden macrophages ultimately return to the bloodstream. When their task is completed, the solid mass of necrotic tissue that characterized the acute infarct has been transformed into a contracted, fluid-filled cavity that is traversed by a fine mesh of atretic vessels but does not develop a collagenous scar that is typical of other organs (Figure 27.19). Usually, the cavitated infarct is surrounded by a narrow zone of astrogliosis and neuron loss that abuts with histologically normal brain parenchyma. Distal degeneration of fiber tracts is also a late manifestation of infarction in the CNS.

The type of vessel occluded leads to characteristic patterns of regional ischemic damage to CNS (Table 27.1). Territorial infarcts describe regions of necrotic tissue secondary to occlusion of an artery, such as the anterior cerebral artery (Figure 27.18). A common mechanism for obstruction of large arteries

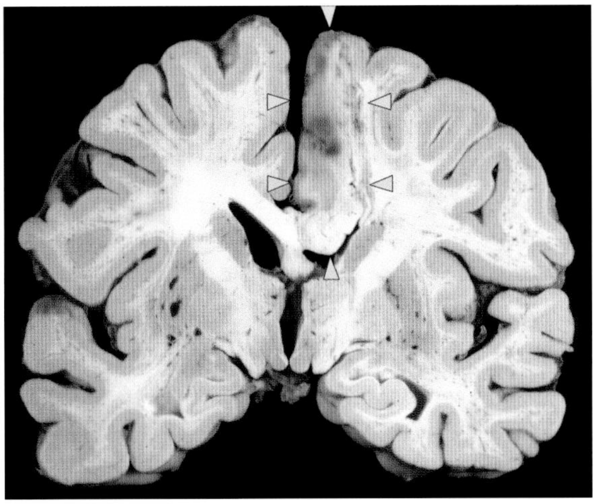

Figure 27.18 **Coronal section of cerebrum shows acute infarct in the territory supplied by the left anterior cerebral artery (blue arrow heads).**

Table 27.1 Characteristic CNS Injuries Related to Size of Vessel Occluded

Vessel	Type of Lesion	Examples of Diseases
Arteries	Territorial infarct, sometimes hemorrhagic	Atherosclerosis, vasculitis, vasospasm, dissection
Large Arterioles	Lacunar infarct Microaneurysm	Arteriolosclerosis, diabetes, hypertension
Microvessels	Widespread injury of varying severity	Arteriolosclerosis, CADASIL, Fat or air emboli
Veins	Hemorrhagic infarct, often bilateral	Thrombosis

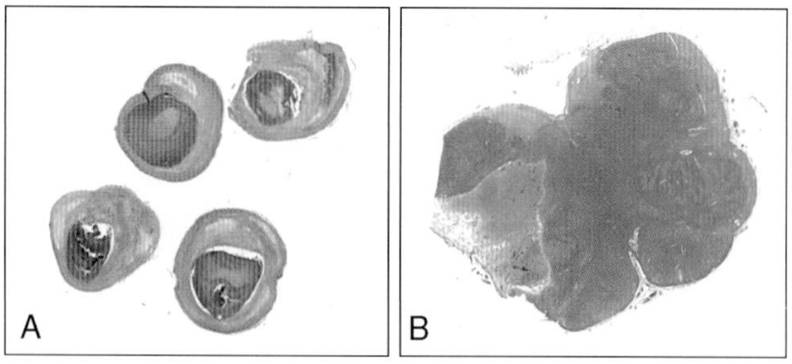

Figure 27.20 Thrombosis. Photomicrographs of H&E/LFB-stained sections show thrombus occluding the basilar artery that also has complicated atherosclerosis **(A)** and the corresponding subacute infarct of the lateral medullary plate **(B)**.

is complicated atherosclerosis with thrombus formation (Figure 27.20A and Figure 27.20B). Some will progress to hemorrhage as the obstruction is lysed and the necrotic regions reperfused. The intraparenchymal large arterioles that arise from arteries at the base of the brain are vulnerable to processes that produce arteriolosclerosis such as hypertension and diabetes mellitus (Figure 27.21). The two major complications of arteriolosclerosis in brain are smaller infarcts (<1 cm), called lacunar infarcts (Figure 27.22), and formation of aneurysms that may rupture and lead to intracerebral hemorrhage.

Rather than focal arteriolar disease leading to discrete lacunar infarcts, widespread disease of small arterioles and capillaries, also known as microvessels, can produce more pervasive ischemic damage to brain that often presents clinically with global neurologic dysfunction. There are several causes of this form of ischemic injury to brain that typically vary in severity

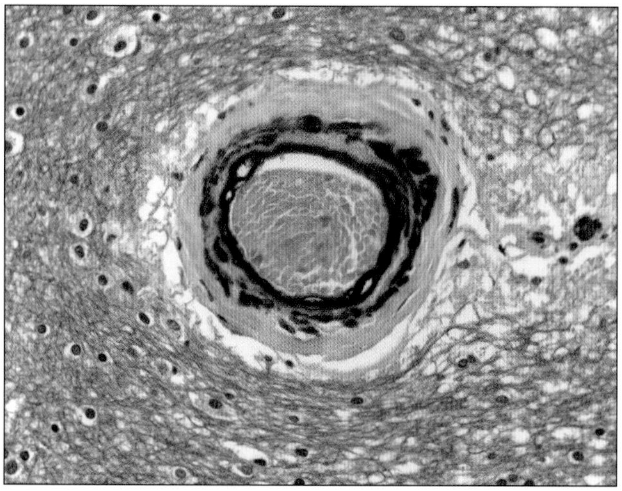

Figure 27.21 Arteriolosclerosis. Section of subcortical white matter demonstrates an arteriole with a thickened vascular wall with immunohistochemical reaction for anti-muscle-specific actin demonstrating continuous hypertrophy Gomori trichrome.

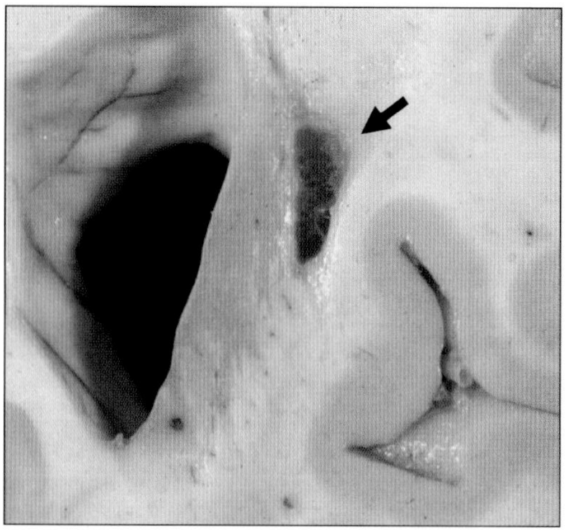

Figure 27.22 Lacunar infarct. Coronal section of right cerebrum shows remote lacunar infarct in the internal capsule adjacent to caudate head (black arrow).

from diffuse degeneration with astrogliosis with or without microinfarcts (discernible only by microscopy) that are sometimes hemorrhagic; this variation in damage likely reflects a gradient of insults from oligemia to ischemia. One form of this type of injury thought to be secondary to widespread arteriolosclerosis is called subcortical arteriosclerotic encephalopathy or Binswanger's disease that is associated with hypertension. Another cause of this type of ischemic injury is Cerebral Autosomal Dominant Arteriopathy with Subcortical Infarcts and Leukoencephalopathy (CADASIL). CADASIL is associated with missense mutations in the *NOTCH3* gene in the vast majority of patients and is characterized by a microangiopathy with degeneration of vascular smooth muscle cells that leads to discontinuous immunoreactivity for smooth muscle actin in arterioles (Figure 27.23). Although CADASIL is relatively rare, knowledge gained from investigation of this disease has spurred the search for other genetic causes and risk factors for ischemic brain injury. Other causes of widespread small arteriole/capillary occlusion are thrombotic thrombocytopenic, rickettsial infections, and fat or air emboli.

Thrombosis of veins leads to stagnation of blood flow in the territories being drained and is observed most frequently as a complication of systemic dehydration, hypercoagulable states, or phlebitis. Edema and extravasation of erythrocytes are observed initially. However, venous thrombosis can lead to infarcts that are commonly bilateral and conspicuously hemorrhagic.

In contrast to regional ischemic damage from diseases that affect blood vessels, global ischemic damage is produced by profound reductions in cerebral perfusion pressure. There are different forms of global ischemic damage that vary with the severity of the insult. In its most extreme form, global ischemia culminates in total cerebral necrosis, or brain death. Although the patient's vital functions may be maintained artificially for some time, cerebral perfusion typically becomes blocked by the massively increased

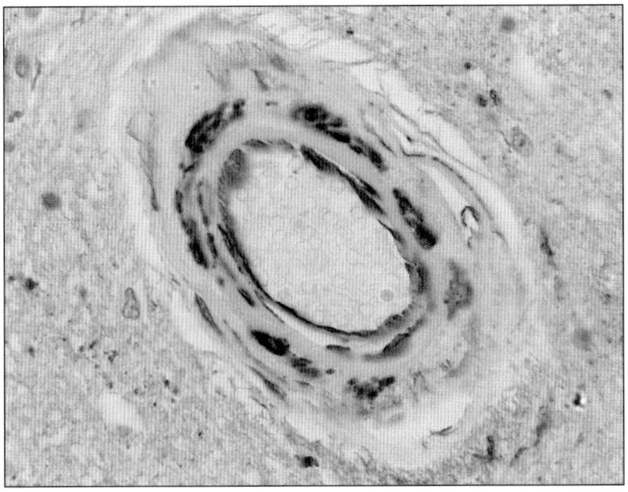

Figure 27.23 **Small vessel in CADASIL.** Photomicrograph of immunohistochemical reaction for antimuscle-specific actin shows discontinuous smooth muscle in a white matter arteriole from a patient with cerebral autosomal dominant arteriopathy with subcortical infarcts and leukoencephalopathy (CADASIL).

intracranial pressure, leading to necrosis of the entire brain. Global ischemia need not be so severe as to produce total cerebral necrosis. These instances are typified by sudden decrease in mean arterial pressure, such as shock or cardiac arrest, from which the patient is resuscitated. The pattern of tissue damage reflects an exaggerated or selective vulnerability of some groups of neurons to ischemic injury. The most susceptible are the pyramidal neurons of the hippocampus and Purkinje cells of the cerebellum. In more severe cases, arterial border zone, or watershed, infarcts occur at the distal extreme of arterial territories, regions of the brain and spinal cord where distal vascular territories overlap. While there are several arterial border zones in adults, incompletely developed anastomoses between the cerebral cortical long penetrating arteries and the basal penetrating arteries produce a periventricular arterial border zone in premature infants that is inordinately sensitive to global ischemia; the resulting lesion is called periventricular leukomalacia.

The pathophysiology of ischemic injury to the CNS is complex. Within seconds to minutes, there is failure of energy production, release of K^+ and glutamate, and massive increase in intraneuronal calcium (direct excitotoxicity). What follows over hours to weeks is an intricate balance of further damage and response to injury that critically involves elements of immune activation. Enormous effort has been spent investigating potential therapeutic targets and experimental interventions focused in the acute phase of ischemic stroke. Sadly, none has translated to general patient care. Indeed, the current therapeutic approach to ischemic stroke is recombinant tissue plasminogen activator to reverse vessel obstruction; however, this can be complicated by hemorrhage into injured brain.

Hemorrhage

Intracranial hemorrhage occurs when a vessel ruptures and releases blood from the intravascular compartment into some other compartment in the cranium. The key to appreciating the clinical significance of intracranial hemorrhages is an understanding of the type and location of the vessel involved (Table 27.2).

Rupture of an artery leads to release of blood under high pressure into the spaces surrounding the brain that can rapidly lead to fatal compression, while rupture of an arteriole releases blood under high pressure into brain parenchyma with equally devastating effects. Rupture of microvessels leads to lesions called microhemorrhages that conspire with microinfarcts to produce apparently progressive widespread neurologic dysfunction. Finally, rupture or tearing of veins that bridge from the subarachnoid space to the dural sinuses releases blood under venous pressure that produces subdural hematomas.

Intraparenchymal hemorrhages from arteriolar disease, which can be large and life-threatening, deserve further discussion. As described in Table 27.1, aneurysm formation in cerebral arterioles is a consequence in at least some individuals with hypertension. Rupture of these weakened regions of vessels is a common cause of intraparenchymal hemorrhage that more commonly occur centrally or deep in the cerebral hemisphere (Figure 27.24). Another common cause of intraparenchymal hemorrhage is rupture of arterioles sufficiently weakened by amyloid deposition, a condition called cerebral amyloid angiopathy (CAA). These more commonly occur in one of the cerebral lobes, such as the occipital lobe, and so are termed lobar hemorrhages (Figure 27.25). Amyloid means starch-like and describes a group of proteins that share common interrelated features such of high β sheet conformation, affinity for certain histochemical dyes, and the capacity to form fibrillar structures of limited solubility. CAA can result from the deposition of amyloid (A) β peptides, cystatin C, prion protein, ABri protein, transthyretin, or gelsolin. CAA with Aβ peptides

Table 27.2 Characteristic CNS Injuries Related to Size of Vessel Ruptured

Vessel	Location of Ruptured Vessel	Outcome
Artery	Epidural Subarachnoid	Medical emergency
Arteriole Microvessel	Intraparenchymal	Cumulative effects can contribute to widespread progressive neurologic deficits
Vein	Subdural	Varies, often slowly progressive

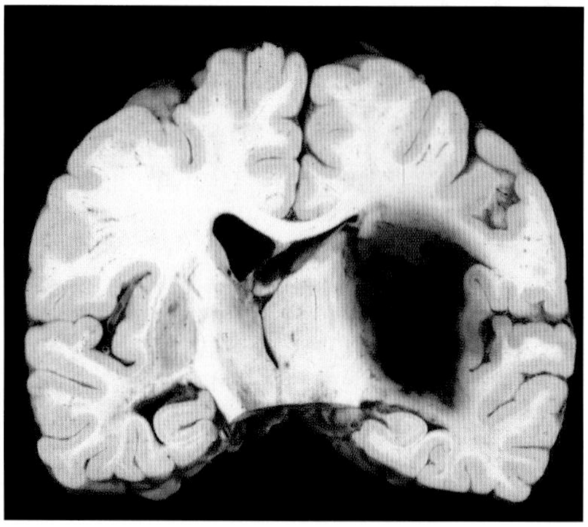

Figure 27.24 **Hemorrhage.** Coronal section of cerebrum shows intracerebral hemorrhage involving deep structures on the right.

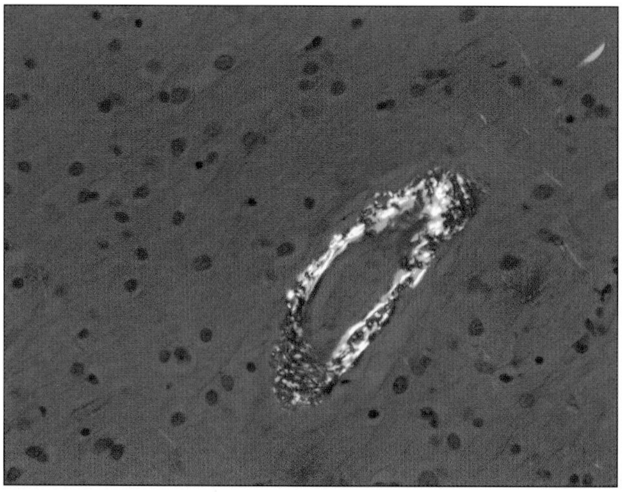

Figure 27.25 **Congophilic amyloid angiopathy.** Polarized light from Congo Red-stained cerebral cortex shows birefringent parenchymal blood vessel. Note also adjacent parenchymal amyloid deposit from a patient with Alzheimer's disease.

can occur apparently sporadically with advancing age and in patients with Alzheimer's disease. Rarely, CAA is caused by autosomal dominant mutations and is called Hereditary Cerebral Hemorrhage with Amyloidosis (HCHWA). These include Dutch, Italian, and Flemish families with mutations in the amyloid precursor protein (*APP*) gene and the Icelandic type that is caused by mutations in the cystatin gene. It is important to note the other mutations in *APP* are rare autosomal dominant causes of Alzheimer's disease. Furthermore, amyloid deposition in blood vessels is not always limited to the cerebrum, for example, HCHWA-Icelandic type typically shows more widespread involvement of vessels throughout the brain.

Degenerative Diseases

Neurodegenerative diseases range from very common illnesses like Alzheimer's disease to rare illnesses like prion diseases. Function is localized within the CNS, so the clinical presentation of neurodegenerative illnesses is dictated by the regions of brain affected. For example, Alzheimer's disease focuses initially in the hippocampus and closely related structures and later involves primarily frontal, temporal, and parietal lobes. Clinically, Alzheimer's disease is characterized by early impairment in declarative memory that is followed by impairments in other cognitive domains. Another example is amyotrophic lateral sclerosis (ALS) in which degeneration of Betz cells in the parietal lobe and anterior horn cells in the spinal cord produces a characteristic combination of weakness and paralysis. A fuller understanding of the human functional neuroanatomy than is possible to discuss here is needed to appreciate the correlations between affected regions and clinical presentation. However, Table 27.3 presents a very broad overview for selected diseases.

Etiology

The cause of each one of these illnesses is known to some extent; however, this mostly concerns forms that have patterns of highly penetrant autosomal dominant inheritance. For example, Huntington's disease is caused by inheritance of an abnormally expanded trinucleotide repeat (CAG) in the *HD* gene that is translated into an expanded glutamine repeat in the Huntintin protein. The situation is more complex for the other four neurodegenerative diseases that we are considering. Alzheimer's disease, Parkinson's disease, ALS, and Creutzfeldt-Jakob disease, the most common type of prion disease, all have an uncommon subset of patients with autosomal dominant forms of the disease, but also much more common sporadic forms that are not caused by inherited mutations, although some have been associated with inherited risk factors (Table 27.4). It is key to understand that identification of genetic causes and risk factors defines relevance, but not mechanism. For example, the mutation that leads to Huntington's disease has been known for 15 years, but the mechanisms of neurodegeneration remain an area of intense investigation. Mechanism of disease is important because it is the foundation for evidence-based therapeutic interventions.

Pathogenesis of Neurodegeneration

While there is evidence for each of the basic mechanisms of neuronal injury contributing to neurodegenerative diseases, a relatively specific hallmark feature of several neurodegenerative diseases, including all of those discussed in this section, is the accumulation of aggregates of misfolded protein that in some instances form amyloid. Although protein misfolding is not limited to neurologic disease, within neurologic disease it seems limited to neurodegenerative conditions; we will discuss this mechanism here. The focus on protein abnormalities in

Table 27.3 Anatomic and Clinical Features of the Neurodegenerative Diseases

Disease	Affected Region	Corresponding Clinical Features
Alzheimer's disease	Hippocampus, regions of cerebral cortex	Dementia
Creutzfeldt-Jakob disease	Cerebral cortex, basal ganglia, cerebellum	Dementia, movement disorders
Parkinson's disease	Midbrain and basal ganglia	Bradykinesia, rigidity, tremor
Huntington's disease	Basal ganglia	Chorea
ALS	Primary motor neurons	Weakness and paralysis

Table 27.4 Autosomal Dominant and Sporadic Forms of the Neurodegenerative Diseases

		Autosomal Dominant		Sporadic	
Disease	Prevalence	Inherited Cause	Frequency	Inherited Risk Factor	Frequency
AD	~20% of people over 65	Mutation in *APP, PSEN1,* or *PSEN2*	Uncommon	APOE ε4 allele	Common
PD	3-5% of people over 65	Mutation in *SNCA, PARK2, UCHL1, DJ1,* or *PINK1*	Uncommon	SNCA polymorphisms	Common
ALS	~4 per million	Mutation in *SOD1*	Uncommon	Not yet identified	Common
CJD	~1 per million	Mutation in *PRNP*	Uncommon	*PRNP* polymorphisms	Common
HD	~3 per 100,000 in Western Europeans	Expanded CAG repeat in *HD*	All	Not applicable	None

neurodegenerative diseases stems from the characteristic amyloid deposits known to neuropathologists for over a century: amyloid plaques in Creutzfeldt-Jakob disease that contain prion protein (PrP) amyloid, senile plaques of Alzheimer's disease that contain amyloid β peptides, and α-synuclein-containing Lewy bodies in Parkinson's disease (Figure 27.26). Our understanding of the role of protein misfolding in neurodegenerative disease has advanced greatly from identification of amyloid. Current ideas about this aspect of neurodegeneration are best demonstrated by prion diseases like Creutzfeldt-Jakob disease (Figure 27.27).

The *PRNP* gene normally is transcribed (step 1) and translated (step 2) to PrP-C (green circle), a glycosyl phosphatidyl inositol- (GPI-) anchored protein that is expressed by many cells but at high levels by neurons. In some forms of prion diseases, PrP-C misfolds from its normal conformation to a pathogenic form (red square), PrP-Sc (step 3), that is high in β-sheet (Sc is an abbreviation for scrapie, a form of prion diseases

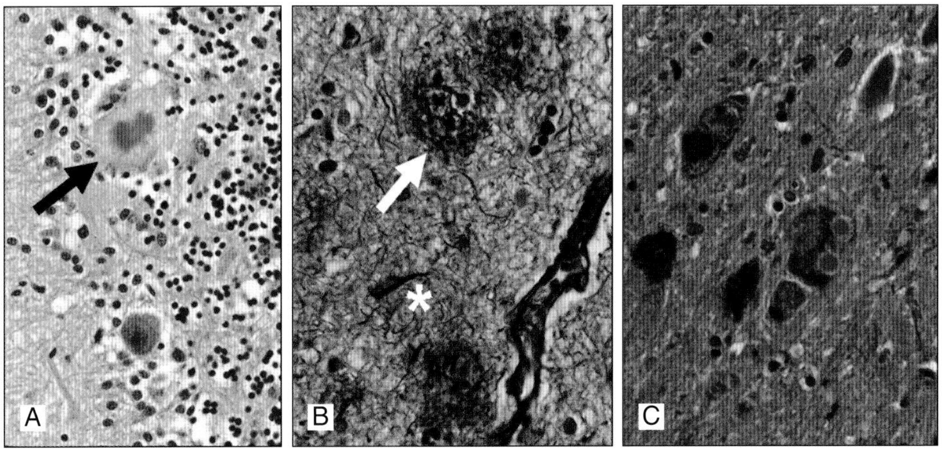

Figure 27.26 Proteinaceous inclusions of neurodegenerative diseases. Photomicrographs show **(A)** plaque (black arrow) in cerebellum from patient with CJD (H&E), **(B)** senile plaque (white arrow) and neurofibrillary tangle (white asterisk) in a patient with AD (modified Bielschowsky), and **(C)** a pair of Lewy bodies, one among several pigmented (dopaminergic) neurons of the substantia nigra from a patient with Parkinson's disease (H&E/LFB).

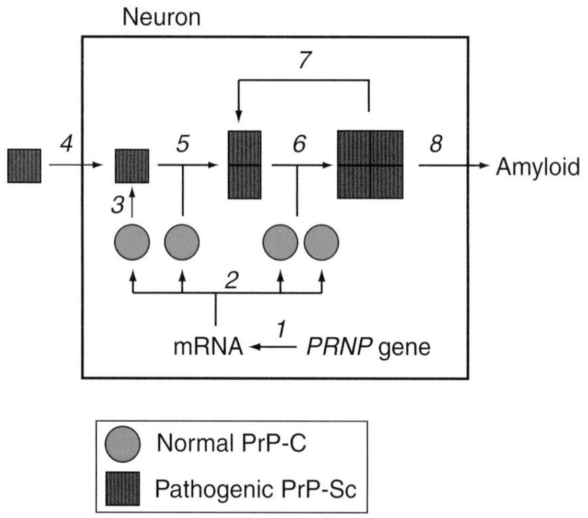

Figure 27.27 Diagram of prion disease pathogenesis. The *PRNP* gene is transcribed (1) and translated (2) to PrP-C (green circle). The pathogenic protein (red square) forms by misfolding of PrP-C (3) or can be transmitted (4). PrP-Sc promotes further recruitment of PrP-C (5) into fibrils (6) that can generate new seeds (7) and form amyloid (8).

in sheep). Some of the prion disease-causing mutations in *PRNP* encode for proteins with increased susceptibility to misfold into these pathogenic forms. These abnormal confomers of PrP-Sc are thought to promote further subversion of PrP-C folding (step 5) leading to apparently self-propagated generation of abnormal confomers that organize progressively into ordered complexes called fibrils (step 6) that can fracture to generate new seeds for further recruitment of PcP-C (step 7), and accumulate as amyloid in brain (step 8). Cellular defenses against protein misfolding, self-aggregation, and accumulation include chaperones, the ubiquitin-proteosome system, and autophagy, among others. Precisely how these abnormal confomers or higher ordered complexes lead to neuron death is not entirely clear but likely involves activation of at least some of the pathogenic processes described previously.

Since similar pathologic features to those described so far for prion diseases are shared by several neurodegenerative diseases, many now propose that similar molecular mechanisms underlie the more common Alzheimer's disease where the misfolded and accumulating proteins are Aβ peptides. The amyloid precursor protein (APP) is the product of the *APP* gene mentioned previously that, when mutated, can cause a highly penetrant form of early onset autosomal dominant Alzheimer's disease. APP is a single membrane-spanning protein that is expressed at high levels by neurons and that undergoes exclusive endoproteolytic cleavages by α-secretase to generate secreted and internalized segments, or by β-secretase and γ-secretase to generate secreted protein, amyloid β peptides, and internalized segments, all of which have biological activity; however, the major research focus has been on the neurotoxic properties of Aβ peptides. The identity of β-secretase is now known and is called BACE1 (β-site APP-cleaving enzyme), while at least part of the multicomponent γ-secretase appear to be the protein products of *PSEN1* and *PSEN2*, mutations of which also cause highly penetrant forms of early onset autosomal dominant Alzheimer's disease. Indeed, the clustering of mutations that cause early onset Alzheimer's disease around the generation of Aβ peptides is the foundation of the amyloid hypothesis for this disease. Promiscuity in the cleavage site by γ-secretase is responsible for generating Aβ peptides of varying lengths. Of these, Aβ that are 40 ($A\beta_{1-40}$) or 42 ($A\beta_{1-42}$) amino acids in length are the most intensely studied with $A\beta_{1-42}$ being more fibrillogenic and neurotoxic in model systems. Key to ultimately understanding Alzheimer's disease will be unraveling the mechanistic connections between accumulation of Aβ peptides, not only in brain parenchyma as shown earlier, but also in cerebral blood vessels (Figure 27.25), and the accumulation of pathologic forms of the microtubule-associated protein tau in structures called neurofibrillary tangles (Figure 27.26B). While this connection may involve some of the processes described previously, it currently remains enigmatic.

It is critically important to emphasize that a unique feature of prion diseases is their transmissibility (step 4 in Figure 27.27). This is a real but rare clinical issue. However, it is achieved routinely in laboratory animals. Indeed, protease-resistant fragments of PrP-Sc are now widely viewed as the transmissible agent in prion diseases. This is not the case for Aβ peptides or aggregated proteins characteristic of other neurodegenerative diseases. Indeed, no other neurodegenerative disease is transmissible. Perhaps this simply reflects varying potency for transmission. Alternatively, this may point to a fundamental difference in mechanism among these diseases.

Before we leave the topic of neurodegenerative diseases, it is important to remember the immense looming public health challenge posed by late-onset Alzheimer's disease (LOAD). LOAD describes those patients with Alzheimer's disease who have onset in later adult life, typically older than 65 years, and who have not inherited a causative mutation; sometimes this is referred to as sporadic Alzheimer's disease. While identification and investigation of autosomal dominant forms have provided invaluable insight into etiology and pathogenesis of Alzheimer's disease, LOAD may present additional facets for investigation. One of these might be the possible intersection of Alzheimer's disease with vascular brain injury in older patients as diagrammed in Figure 27.28.

Neurotoxicants

Neurotoxicology has significance that reaches beyond the identification of xenobiotics (neurotoxicants) or endogenous agents (neurotoxins) that are deleterious to the nervous system. Although considered separate fields of study, neurotoxicology, neurodegeneration, stroke, trauma, and metabolic diseases of the nervous

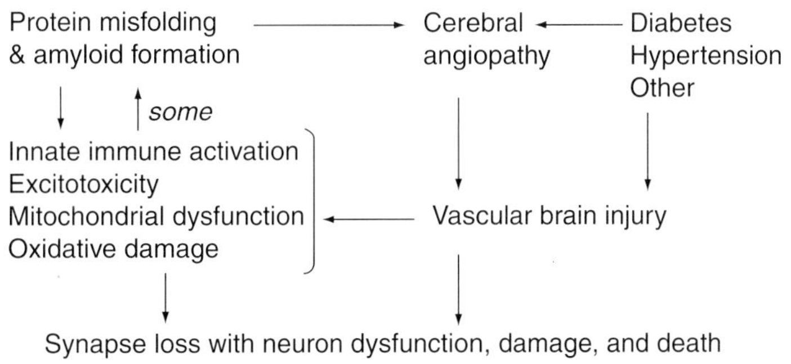

Figure 27.28 **Diagram of potential interactions between Alzheimer's disease and vascular brain injury in dementia.**

system inform each other about the mechanisms of neuronal dysfunction and death, as well as response to injury in the nervous system. Indeed, there are several examples of compounds first identified as neurotoxicants that subsequently came into use as models of human neurodegenerative disease; perhaps the most striking example is 1-methyl-4-phenyl-1,2,3,6-tetrahydropyridine (MPTP). Moreover, many compounds initially identified as neurotoxicants have become fundamental tools used by neuroscientists, such as tetrodotoxin, curare, kainic acid, and 6-hydroxydopamine. Conversely, progress in other fields of neuroscience continually advances understanding of the mechanisms of neurotoxicants.

Neuropathological Changes Caused by Neurotoxicants

For the most part, pathologic changes in brain following neurotoxicant exposure are nonspecific. Edema is the most striking macroscopic finding in acute toxic encephalopathies. The brain can be heavy and swollen, even after fixation, with broadened cortical gyri and obliterated sulci. In severe cases, transtentorial and cerebellar tonsillar herniations may occur. Cerebral edema secondary to vascular damage, as in lead encephalopathy, or direct damage to CNS myelin, as in triethyltin encephalopathy, is largely confined to the white matter. In contrast, edema resulting from diffuse cytotoxic damage, as in thallium intoxication, affects both gray and white matters and has been observed to impart a moth-eaten appearance to the cerebral cortex. The histological manifestations of cerebral edema can be slight: myelin pallor and mild gliosis may be all that is observed by standard histochemical stains.

Neuronal degeneration following exposure to toxicants may be diffuse, for instance, in thallium intoxication. In contrast, a number of neurotoxicants are associated with damage to specific structures, including methylmercury-induced or dimethylmercury-induced degeneration of the calcarine and cerebellar cortices. Whether the primary lesion is focal or generalized, secondary degeneration of fiber pathways may be observed. Intraneuronal inclusions are uncommon in neurotoxicant-induced disease of the CNS; however, they have been reported in neurons following intoxication with heavy metals.

Glial response to injury includes astrocytic gliosis with strong immunoreactive for GFAP. In many instances of systemic exposure to toxicants, GFAP immunoreactivity is most prominent around blood vessels. Another response to injury by astrocytes is formation of Alzheimer type II astrocytes that is associated with encephalopathy from hyperammonemic states. Diffuse microgliomatosis and myelin pallor have been reported following neurotoxicant exposure. Myelin pallor is more commonly due to a combination of edema and gliosis, although examples of demyelination exist, including glue sniffer encephalopathy. Intramyelinic edema follows hexachlorophene or triethyltin intoxication.

Axonal degeneration is a common finding in toxicant-induced peripheral neuropathies. An optimal technique for viewing axons along several internodes is teased fiber preparations showing linear collections of phagocytic cells that are digesting myelin and axonal debris, the appearance of axonal degeneration. Segmental demyelination is another common pathologic lesion in the peripheral nervous system following toxicant exposure. Inappropriately thin myelin sheaths, shortening of internodal distances, and variation of myelin thickness among internodal segments of the same axon are observed during the remyelination that follows demyelination. Recurrent episodes of demyelination with remyelination may lead to Schwann cell hyperplasia. Some neurotoxicant-induced diseases are characterized by giant axonal swellings. These large eosinophilic collections within axons are composed mostly of massive accumulations of neurofilaments. Similar neurofilament-filled axonal swellings also are seen in a rare familial neuropathy termed giant axonal neuropathy. This is in distinction to the morphologically similar axonal spheroids that contain tubulovesicular material and that have been observed in characteristic locations in older individuals as well as in patients with neuraxonal dystrophies.

Biochemical Mechanisms of Selected Neurotoxicants

Domoic Acid. A large number of experiments and trials have provided indirect or pharmacological support for a role for excitotoxicity in neurological injury. Humans accidentally exposed to high doses of EAA receptor

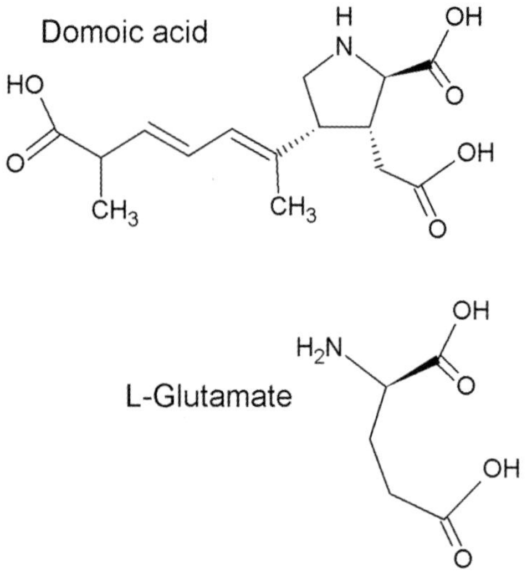

Figure 27.29 **Structures for domoic acid and L-glutamate.**

agonists and who subsequently developed neurologic disease underscore the importance of EAA's in disease. Perhaps the most striking example is the domoic acid intoxications that occurred in the Maritime Provinces of Canada in late 1987 (Figure 27.29). A total of 107 patients were identified who suffered an acute illness that most commonly presented as gastrointestinal disturbance, severe headache, and short-term memory loss within 24 to 48 hours after ingesting mussels. A subset of the more severely afflicted patients was subsequently shown to have chronic memory deficits, motor neuropathy, and decreased medial temporal lobe glucose metabolism by positron emission tomography (PET). Neuropathological investigation of patients who died within 4 months of intoxication disclosed neuronal loss with reactive gliosis that was most prominent in the hippocampus and amygdala, but also affected regions of the thalamus and cerebral cortex. The responsible agent was identified as domoic acid, a potent structural analog of L-glutamate that had been concentrated in cultivated mussels.

MPTP. The history of MPTP-induced parkinsonism in young adults who inadvertently injected themselves with this compound is well known. MPTP is a protoxicant that, after crossing the blood-brain barrier, is metabolized by glial MAO-B to a pyridinium intermediate ($MPDP^+$) that undergoes further two-electron oxidation to yield the toxic metabolite methyl-phenyl-tetrahydropyridinium (MPP^+) that is then selectively transported into nigral neurons via the mesencephalic dopamine transporter (DAT) (Figure 27.30). Once inside these neurons, MPP^+ is thought to act primarily as a mitochondrial toxin by inhibiting complex I activity in the mitochondrial electron transport chain, thereby reducing ATP production and increasing ROS generation. Indeed, MPTP-induced dopaminergic neurodegeneration can be diminished by free

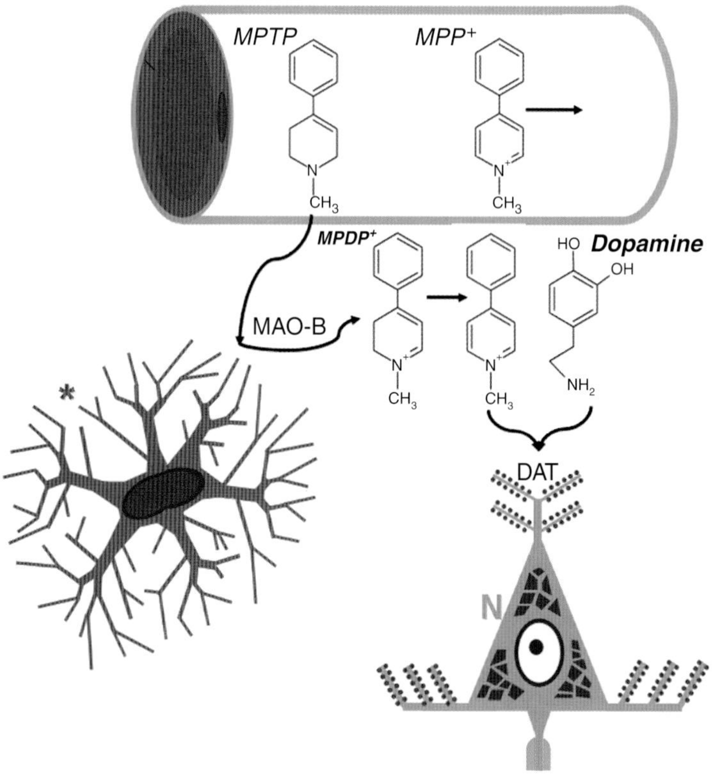

Figure 27.30 **Diagram of steps in dopaminergic toxicity from MPTP exposure.** MAO-B in astrocytes (*) catalyzes oxidation of MPTP to ultimately form MPP^+ which is selectively taken up by dopaminergic neurons (N) expressing the dopamine transporter (DAT).

radical scavengers, inhibitors of the inducible form of nitric oxide synthase, and by EAA receptor antagonists. Alternatively, transgenic mice lacking some elements of antioxidant defenses are significantly more vulnerable to MPTP-induced dopaminergic neurodegeneration. So far, the search for xenobiotics that may act similarly to MPTP and could be potential environmental toxicants that promote PD has not yielded clear candidates.

Axonotoxicants. Distal sensorimotor polyneuropathy is probably the most common clinical manifestation of neurotoxicant exposure in humans. A variety of toxicants, including hexane, methyl n-butylketone (2-hexanone), carbon disulfide (CS2), acrylamide, and organophosphorus esters, results in degeneration of the distal portions of the longest, largest myelinated axons in the peripheral and central nervous systems, an observation encapsulated in the term *central-peripheral distal axonopathy*.

Two of these toxicants, n-hexane and CS_2, are especially interesting because despite their very different chemical structures, chronic exposure to either can produce unusual pathological changes in nerve. The characteristic lesion produced by both n-hexane and CS_2 is multifocal fusiform axonal swellings at the proximal side of nodes of Ranvier at distal but preterminal sites that consist of massive accumulations of disorganized 10 nm neurofilaments, decreased numbers of microtubules, thin myelin, and segregation of axoplasmic organelles and cytoskeletal components. Distal to swellings, axons may become shrunken and then degenerate. With continued exposure, more proximal swellings occur with subsequent degeneration. Investigations of n-hexane and CS_2 have determined that the key to their shared clinical and pathological profiles appears to be the ability of each compound to generate protein-bound electrophilic species that can covalently cross-link proteins.

NEOPLASIA

Neoplasms of the CNS can be divided into primary and metastatic. Metastatic neoplasms arise from cancer cells derived from non-CNS organs, such as lung (carcinoma), skin (melanoma), or blood (lymphoma) that are transported to the CNS via hematogenous or other means. These neoplasms are described briefly at the end of this chapter with their organ systems of origin. Primary CNS neoplasms can be defined as uncontrolled growth of cells derived from normal CNS tissues. The most common histologically resemble glia and are termed gliomas. Although it was previously believed that gliomas were derived from mature glial elements, such as astrocytes or oligodendrocytes, it is more likely that these tumors arise from glio-neuronal progenitor cells, so-called cancer stem cells. Cancer stem cells represent a minority population of tumor cells that derive large numbers of variably differentiated glial elements which make up the bulk of the tumor, and thus represent an attractive therapeutic target.

Gliomagenesis results from a number of known genetic defects associated with control of cell cycle and proliferation, apoptosis pathways, cell motility, and invasive potential. New discoveries continue to be made. Knowledge of the typical genetic alterations is being exploited in animal models of CNS tumors, as potential targets for therapy and, increasingly, in the diagnosis of tumors. Not surprisingly, some mutations underlie the relationship between histologic appearance and biological behavior. Indeed, although assignment of tumor diagnosis and grade (higher grade = more malignant) based on morphologic criteria has been the mainstay of diagnosis in the preceding 100 years, in the future it is likely that molecular diagnosis of CNS neoplasms will be increasingly critical in determining prognosis and guiding therapy.

Unlike other organ systems that use the TNM (local [T]umor growth; regional lymph [N]ode spread; and distant [M]etastasis) staging system for establishing prognosis of tumors, primary CNS tumors are assigned a histologic grade that correlates with their predicted behaviors. The *World Health Organization* (WHO) classification of tumors (2007) identifies tumors as grade I if they are curable by resection alone and have >10 year median survival, grade II if they are not curable by resection alone and have 5–10 year median survival, grade III for 2–3 years median survival, and grade IV if they have <2 years median survival. The histologic features underlying WHO grading include nuclear atypia, mitoses, vascular proliferation, and necrosis. Currently, the WHO recognizes greater than 25 primary CNS tumors (Table 27.5), with up to 3 practical grades each. For the purposes of this section, we will focus on the most common intracranial tumors.

Diffuse Gliomas

A defining characteristic of the diffuse gliomas is their ability to infiltrate widely throughout the CNS parenchyma, causing them to have no clearly recognizable boarder with normal tissue (Figure 27.31). Because of this feature, they are not curable by resection alone. The natural history of diffuse gliomas is a tendency toward progression from low to high grade by the accumulation of additional genetic defects. De novo, grade IV gliomas also occur, but usually by a different set of genetic defects. Broadly, there are 4 types of genetic alterations involved in gliomagenesis: those which affect (i) cell survival, (ii) cell proliferation (cell cycle regulators), (iii) brain invasion, and (iv) neovascularization. Mutations in Rb, p53, receptor tyrosine kinase, integrin, and other signaling pathways critical to cell cycle regulation are important, as are mutations in cell death pathways (TNFR, TRAIL, CD95, Bcl-2, etc.). Transition from low to high grade is heralded by tumor enhancement by neuroimaging which correlates with microvascular proliferation (angiogenesis), in which pathways regulated by VEGF, HIF, and other molecules are critical. Finally, brain invasion, which is probably least understood of all the gliomagenic processes and the most critical to toxicity, is likely

Table 27.5 Abridged List of Neoplasms of the Central Nervous System as Classified by the *World Health Organization* (2007)

Primary Differentiation	Neoplasm	Grade
Astrocytic Tumors	Pilocytic Astrocytoma	I
	Pilomyxoid Astrocytoma	I
	Subependymal Giant Cell Astrocytoma	I
	Pleomorphic Xanthoastrocytoma	I
	Diffuse Astrocytoma	II
	Anaplastic Astrocytoma	III
	Glioblastoma	IV
	Gliosarcoma	IV
	Gliomatosis Cerebri	III
Oligodendroglial Tumors	Oligodendroglioma	II
	Anaplastic Oligodendroglioma	III
Mixed Oligoastrocytic Tumors	Oligoastrocytoma	II
	Anaplastic Oligoastrocytoma	III
Other Neuroepithelial Tumors	Astroblastoma	IV
	Angiocentric Glioma	I
	Chordoid Glioma of the Third Ventricle	I
Neuronal and Mixed Neuronal-Glial Tumors	Dysplastic Gangliocytoma of Cerebellum (Lhermitte-Duclos)	I
	Desmoplastic Infantile Astrocytoma/Ganglioglioma	I
	Dysembryoplastic Neuroepithelial Tumor	I
	Gangliocytoma	I
	Ganglioglioma	I
	Anaplastic Ganglioglioma	III
	Central Neurocytoma	I
	Extraventricular Neurocytoma	I
	Cerebellar Liponeurocytoma	II
	Papillary Glioneuronal Tumor	I
	Rosette-forming Glioneuronal Tumor of the Fourth Ventricle	I
	Paraganglioma	I
Embryonal Tumors	Medulloblastoma	IV
	CNS Primitive Neuroectodermal Tumor	IV
	Atypical Teratoid/Rhabdoid Tumor	IV
Ependymal Tumors	Ependymoma	II
	Anaplastic ependymoma	III
	Myxopapillary ependymoma	I
	Subependymoma	I

Adapted from [12].

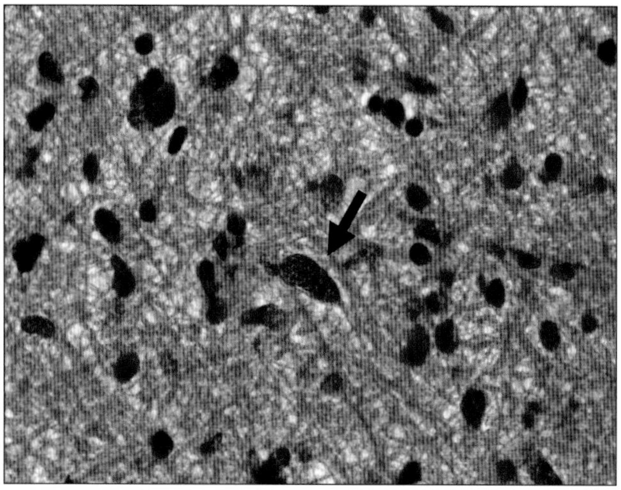

Figure 27.31 Photomicrograph of diffuse astrocytoma. White matter infiltrated by an enlarged, hyperchromatic (darkly stained), pleomorphic (irregularly shaped) neoplastic glial cell (arrow) in a diffuse astrocytoma (H&E 600x magnification).

mediated by metalloproteinases and integrins, among others. For a comprehensive review of astrocytoma genetics, please see [25] and [26].

Diffuse Astrocytoma–Glioblastoma Sequence

Diffuse astrocytomas (WHO grade II–IV) are gliomas in which the neoplastic cells resemble microscopically normal or reactive astrocytes and are often characterized by their cytoplasmic expression of GFAP. Histologically, grade II astrocytomas are typified by mild nuclear atypia and hypercellularity and an absence of mitotic activity, vascular proliferation, or necrosis. Anaplastic astrocytomas (WHO grade III) are histologically characterized by increased hypercellularity with nuclear atypia and mitotic activity in the absence of vascular proliferation and necrosis (Figure 27.32).

Glioblastoma (GBM; WHO grade IV) is the most common primary glioma with an incidence of approximately 3/100,000 population per year. GBMs are characterized by rapid progression and poor survival. They can be classified, based on clinical and molecular

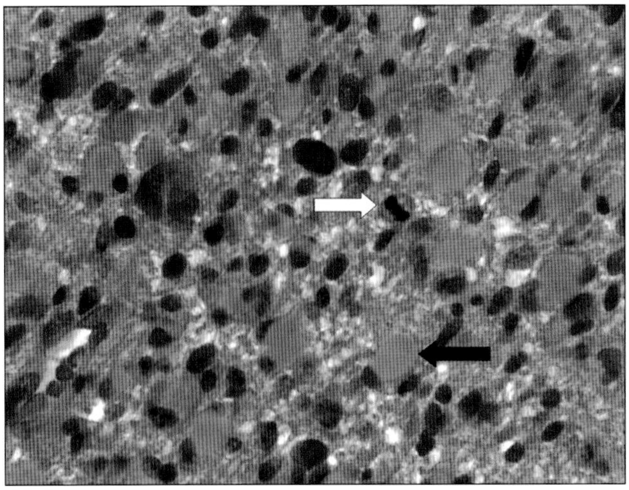

Figure 27.32 **Photomicrograph of an anaplastic astrocytoma.** Numerous enlarged, hyperchromatic (darkly stained) neoplastic glial nuclei have infiltrated brain parenchyma in this anaplastic astrocytoma. A mitotic figure (white arrow) is present. A globule of eosinophilic cytoplasm is seen within neoplastic astrocytes (black arrow) likely representing glial fibrillary acidic protein, the primary intermediate filament of astrocytes (H&E).

characteristics, into those that arise *de novo* and those that progress from lower grade astrocytomas (Figure 27.33).

Primary or *de novo* GBMs account for 90% of cases, generally arise in older patients, and are more prone histologically to be of the small cell subtype. The characteristic abnormalities of signal transduction in primary GBM include overexpression or signal amplification of the epidermal growth factor receptor (*EGFR*) or loss of function mutations of *PTEN* (phosphatase and tensin homology). Deletion of *EGFR* exons 2–7 (EGFRvIII), the most common *EGFR* mutation (found in 20%–30% of GBM and 50%–60% of GBM with *EGFR* amplification), results in constitutive EGFR activation and insensitivity to EGF. EGFRvIII stimulates a different signal transduction pathway than full-length EGFR. EGFR activation operates primarily through the Ras-Raf-MAPK and PI3K-Akt pathways. Constitutive activation of EGFRvIII mutants increases proliferation and survival by preferentially activating (PI3K)/Akt pathway, and possibly through stimulation of a second messenger system not available to nonmutant EGFR. The characteristic genetic alterations involving cell cycle control in primary GBMs include overexpression of *MDM2* (murine double minute 2 protein), which suppresses p53, and deletions of *CDKN2A*, which encodes the tumor suppressor $p16^{INK4A}$, a potent regulator of retinoblastoma (*RB*) tumor suppressor gene, or, through an alternate reading frame (ARF) $p14^{ARF}$, an important accessory to p53 activation. Other common findings include loss of heterozygosity in chromosome 10p and overexpression of Bcl2-like-12 protein, a potent antiapoptotic molecule. The genetic abnormalities occur together in a random distribution and are not progressive. Homozygous deletion of *CDKN2A* is associated with EGFR overexpression, higher proliferative activity, and may account for poorer overall survival of primary GBMs. It is interesting to note that the small cell phenotype common in primary GBMs appears to be most closely associated with EGFR overexpression, suggesting a molecular-histologic link.

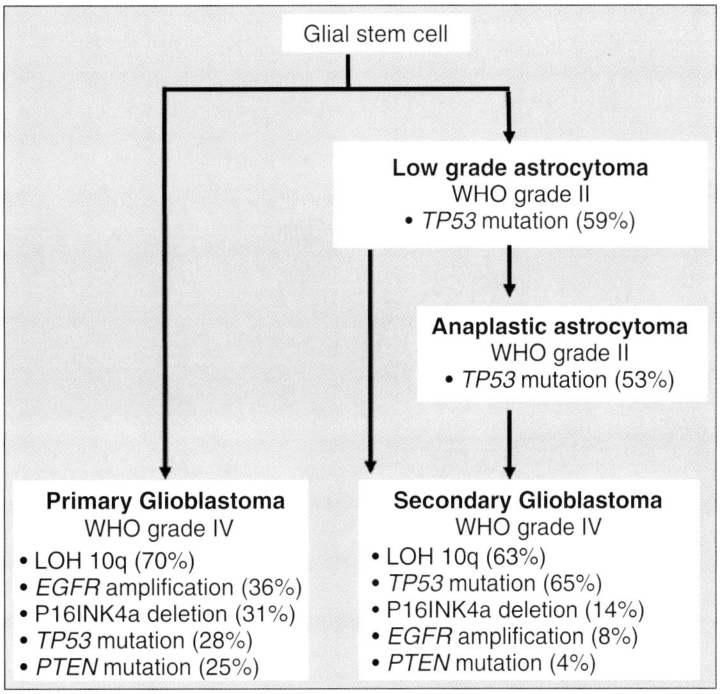

Figure 27.33 **Genetic alterations associated with primary and secondary glioblastoma.** Adapted from [16].

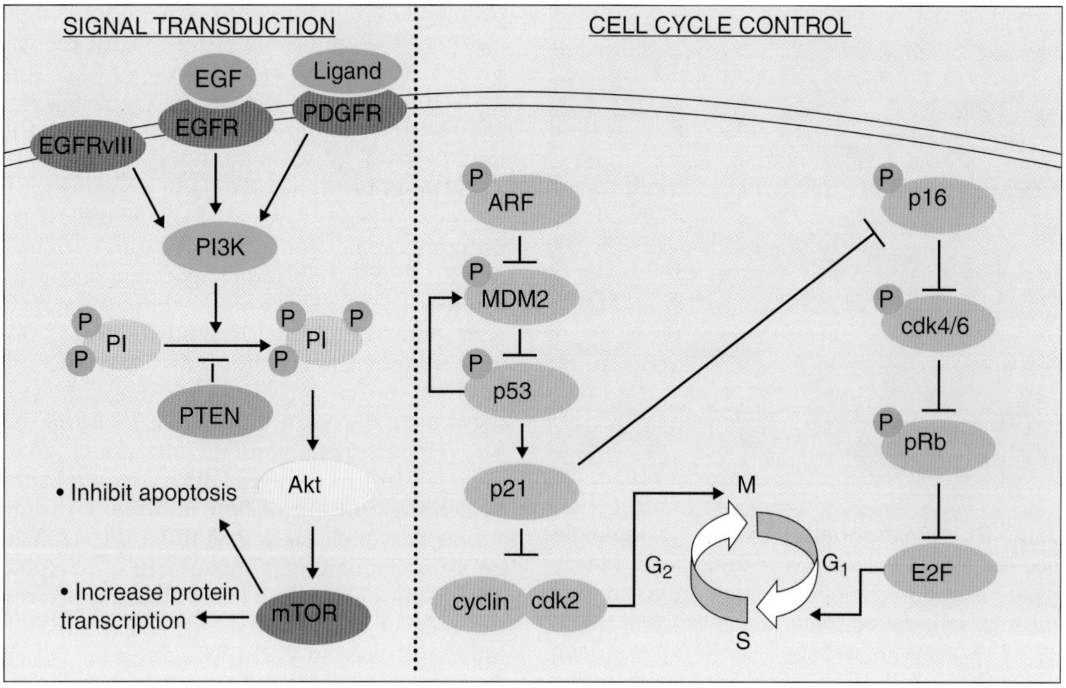

Figure 27.34 Molecular pathways implicated in gliomagenesis. Multiple molecules can be involved in the pathway to glioma. Adapted from [20] and [21].

Secondary GBMs progress from lower grade astrocytomas with stepwise accumulation of additional genetic defects and generally occur in younger individuals. Common early genetic abnormalities include direct mutations of the cell cycle suppressor genes *TP53* and *RB1*, or overexpression of platelet-derived growth factor ligand and/or receptors. Loss of heterozygosity (LOH) of chromosomes 11p and 19q are common in progression of these low-grade astrocytomas to anaplastic astrocytomas. The transition from anaplastic astrocytoma to GBM is less well characterized, but can involve generally LOH of chromosome 10q, mutations indirectly affecting EGFR/PTEN pathways, and alterations to the cell cycle inhibitory $p16^{INK4a}$/RB1 pathway. These pathways are summarized in Figure 27.34.

By definition, all GBMs are prone to spontaneous necrosis and bizarre microvascular proliferation (Figure 27.35). Aberrations in the genetic control of growth and cell cycle give rise to the hypoxia, necrosis, and angiogenesis that underlie these characteristic histologic features. A model of this process (Figure 27.36) begins with genetic alterations in tumor cells which cause a variety of downstream effects which alter the blood-brain barrier, damage endothelium, or directly promote intravascular thrombosis (such as secreting tissue factor). Small vessels in the tumor become occluded, and the resulting damaged tissue releases thrombin as part of the coagulative cascade, which in turn binds to its ligand the protease-activated receptor 1 (PAR-1). Local hypoxia leads to focal necrosis. Tumor cells react to PAR-1 by forming a wave of migration away from the area of hypoxia, the histologic counterpart of which is a pseudopalisade. These migrating cells are severely hypoxic and express high levels of hypoxia-inducible factor (HIF), which activates hypoxia-responsive element domains (HRE), of which VEGF

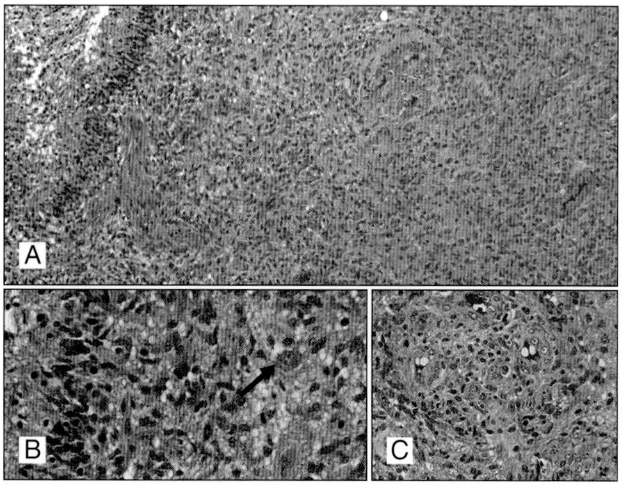

Figure 27.35 Photomicrograph of glioblastoma. (A) Sections from a brain infiltrated by glioblastoma stained with H&E at 100x magnification and **(B)** 600x magnification (below left). A line of parallel nuclei forming a pseudopalisade can be seen extending from the panel above through the left lower panel. The pseudopalisade is adjacent to necrosis. A neuron can be seen surrounded by neoplastic cells (arrow). **(C)** The collagen network of a proliferative vessel is highlighted blue-green by a trichrome stain.

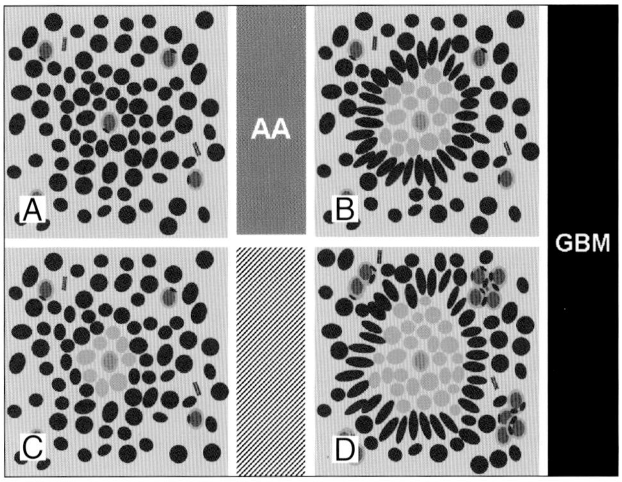

Figure 27.36 **Progression from anaplastic astrocytoma (AA) to glioblastoma (GBM). (A)** AA is mitotically active producing an area of hypercellularity, but no necrosis or vascular cell hyperplasia. **(B)** Neoplastic glia produce thrombophilic compounds such as thrombin and tissue factor causing thrombosis of a microvessel, focal ischemia and hypoxia, and local necrosis. This change is the hallmark of the transition from AA to GBM. **(C)** Neoplastic glia respond to hypoxia by migrating away, forming a "pseudopalisade" of nuclei surrounding the necrosis. **(D)** Migrating neoplastic glia also produce angiogenic factors such as VEGF and HIF, causing vascular cell hyperplasia and abnormal vessel formation. Adapted from [17].

is the most characteristic in GBM. High levels of VEGF cause the vascular proliferation characteristic of this tumor.

An epigenetic feature of GBM that may have therapeutic implications is the methylation status of the DNA repair gene *MGMT* (O^6-methylguanine–DNA methyltransferase). *MGMT* is a DNA-repair protein that removes alkyl groups from the O^6 position of guanine. Epigenetic silencing of this gene is identifiable in a little less than half of GBMs and predicts response to alkylating chemotherapy.

Oligodendroma

Oligodendrogliomas are diffusely infiltrating gliomas that represent the second most common parenchymal tumor of the adult CNS. The neoplastic cells in oligodendrogliomas histologically resemble mature oligodendrocytes. They generally have round nuclei with condensed chromatin and inconspicuous cytoplasm and few cytoplasmic processes. A classical appearance of the neoplastic cells in oligodendroglioma is a round hyperchromatic nucleus with a perinuclear clearing or halo, the so-called fried egg cells. They may express a small amount of cytoplasmic GFAP, but generally less than astrocytic neoplasms. Histologically, they form monotonous proliferations of infiltrating neoplastic cells with occasional hypercellular nodules. They have characteristic chicken-wire vasculature composed of a network of fine branching capillaries

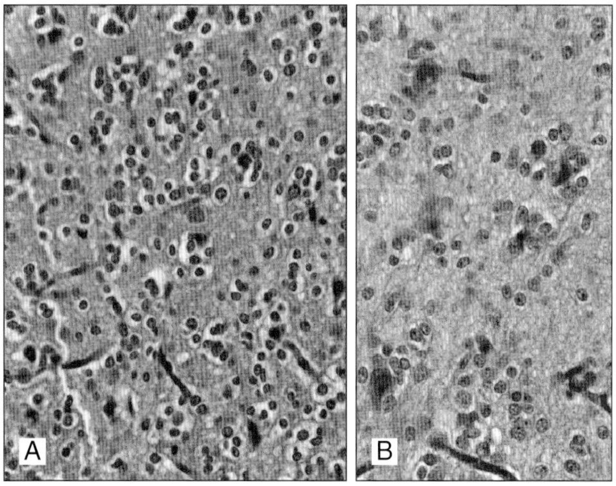

Figure 27.37 **Photomicrograph of oligodendroglioma. (A)** Oligodendrocytes-like neoplastic cells with round nuclei and perinuclear clearing give this oligodendroglioma a fried egg appearance. **(B)** Non-neoplastic neurons are clustered about by neoplastic cells. Fine capillaries are seen in the background of both panels (H&E).

(Figure 27.37A). The neoplastic cells also have a tendency to cluster, or satellite, around infiltrated neurons and vascular structures (Figure 27.37B).

A large subset of oligodendrogliomas contain a chromosomal translocation-mediated loss of the short arm of chromosome 1 (1p) and the long arm of chromosome 19 (19q), commonly detected using *in situ* hybridization or DNA amplification techniques. Loss of 1p/19q has been associated with both classic histomorphology and better clinical behavior. Patients whose tumors have this relative co-deletion demonstrate improved disease-free survival, median survival, and may respond better to alkylating chemotherapeutics. Loss of 1p/19q is inversely correlated with mutations in *TP53*, loss of chromosomal arms 9p and 10q, and amplification of EGFR. Codeletion of 1p/19q is also associated with lower MGMT expression and *MGMT* promoter hypermethylation and may explain the greater chemosensitivity of these tumors. Because of the robust correlation of this specific genetic aberration to both histomorphology and prognosis, oligodendrogliomas exist at the forefront of molecular diagnostics in gliomas. Molecular testing of tumors for loss of 1p/19q is already widely accepted as a prognostic marker and is commonly used by clinicians to guide therapy. Furthermore, although not currently accepted as a diagnostic test for oligodendroglioma, loss of 1p/19q has proven useful in differentiating oligodendrogliomas from histologic mimics and in cases of ambiguous histomorphology. The molecular mechanism(s) by which loss of 1p and 19q mediate their effects has not been elucidated. The chromosomal translocation that underlies most losses of 1p/19q occurs at pericentromeric sites, excluding the possibility of a transgene product driving this neoplasm. Attempts to identify the presumed tumor suppressor genes on

1p and 19q have been unsuccessful. Proteomic analysis of tumors with and without the paired deletion have demonstrated increased expression of proteins associated with malignant behavior in tumors without the deletion, but definitively decreased expression of proteins coded on the lost arms has not been demonstrated.

Oligodendrogliomas are classified in the WHO system into 2 grades. WHO grade II oligodendrogliomas have the same histomorphology as described previously (Figure 27.37). Grade III anaplastic oligodendrogliomas are histologically similar, but display additional nuclear atypia and mitotic activity, and in addition to the classic chicken wire vasculature, they often have vascular proliferation reminiscent of that seen in GBM. Oligodendrogliomas may progress to a histology that is indistinguishable from GBM, although LOH of 1p/19q probably portends a more favorable prognosis than astrocytic GBM.

Circumscribed Gliomas

Circumscribed gliomas are WHO grade I tumors that usually occur in children and young adults. The most common types are pilocytic astrocytoma (PA) and pleomorphic xanthoastrocytoma (PXA). Circumscribed gliomas are different from the diffuse gliomas described previously in that they are generally curable by resection alone. The genetic alterations seen in adult diffuse gliomas are uncommon in the circumscribed gliomas and, when present, portend more malignant behavior. In general, cytogenetic abnormalities are uncommon in both entities, but have been reported, including gains on chromosomes 5 through 9 in PAs and chromosome 7, loss on 8p, and p53 mutations in PXAs. Progression of PAs to higher grade gliomas through the accumulation of additional genetic defects is distinctly uncommon.

Medulloblastoma

Medulloblastoma is the most common CNS malignancy of childhood. They tend to arise in the midline at the cerebellar vermis and are highly malignant, aggressive tumors. They tend to present with either cerebellar signs (truncal ataxia) or with signs of hydrocephalus. Unlike the diffuse gliomas, medulloblastomas do not generally diffusely infiltrate CNS parenchyma, but rather form large necrotic masses which generally displace surrounding brain tissue. Moreover, they have a disturbing tendency to metastasize through the cerebrospinal fluid. Histologically, medulloblastomas form broad sheets of tightly packed cells with large nuclei, finely distributed chromatin and little cytoplasm, an appearance which resembles the primitive neuroglial germinal matrix or external granular layer of cerebellum seen in development. These tumors often have brisk mitotic rates and abundant apoptosis; spontaneous tumor necrosis is also exceptionally common. Occasionally, the tumor cells will form foci with more mature neuroglial elements such as ganglion cells or neuropil.

The molecular mechanisms of tumorigenesis in medulloblastoma have been extensively characterized due to its association with several genetic cancer syndromes (Figure 27.38). The best characterized of these is overactivation of the Sonic hedgehog (Shh) pathway. The Shh pathway was first identified as a pathway in medulloblastoma in Gorlin's syndrome a genetic disorder in which individuals have multiple

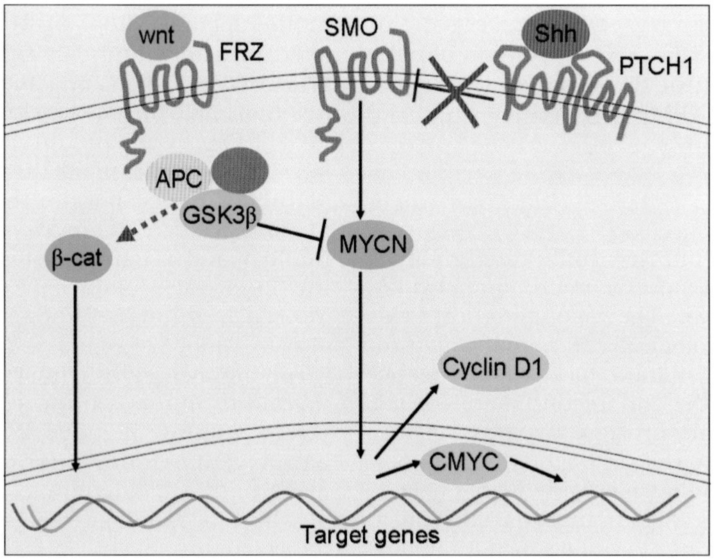

Figure 27.38 **Two pathways implicated in medulloblastoma.** Defects in the Shh and Wnt signaling release and beta catenin (β-cat) and MYCN to activate cell cycle and transcription factors. Abbreviations: FRZ—Frizzled receptor; APC—adenomatous polyposis coli protein; SMO—Smoothened receptor; PTCH1—Patched 1 receptor; GSK3β—glycogen synthase kinase 3 beta; β-cat—beta catenin; MYCN—N-myc oncogene protein; CMYC—Myc transcription factor.

basal cell carcinomas of skin and an increased rate of medulloblastoma. Shh is a glycoprotein which normally activates neuronal precursor proliferation in the primitive external granular layer of the perinatal cerebellum by two primary mechanisms. In the absence of secreted Shh, Patched 1 (PTCH1) inhibits Smoothened (SMO) activation of the Gli family of cell fate and proliferation-related transcription factors. Individuals with Gorlin's syndrome carry an autosomal mutation in *PTCH1*. Shh overactivation also drives proliferation by activation of MYCN, which activates D-type cyclins and inhibits cyclin-dependent kinase inhibitors. As long as MYCN remains active, the proliferating cells cannot exit cell cycle. In normal development, MYCN would be phosphorylated by glycogen synthase kinase-3 beta (GSK3β) and degraded.

A second partially convergent pathway involved in medulloblastoma is the Wnt signaling pathway. This process may be inhibited by the activity of insulin-like growth factor (IGF) in medulloblastoma. Wnt normally binds to its receptor Frizzled and destabilizes a complex containing adenomatous polyposis coli (APC) gene product and the serine-threonine kinase glycogen synthase kinase 3 beta (GSK3β). This action stabilizes β-catenin, which then translocates to the nucleus and positively regulates transcription of genes that ultimately drive cellular proliferation. GSK3β was first implicated in medulloblastoma in a variant of Turcot syndrome, a familial tumor syndrome characterized by colon cancer and malignant brain tumors. Those individuals with Turcot syndrome who are predisposed to develop medulloblastomas carry a loss of function mutation of APC. Less commonly implicated pathways in medulloblastoma include the NOTCH and ErbB signaling pathways or defective DNA repair mechanisms.

Other Neuroepithelial Tumors

There are a number of other tumors that arise in the central nervous system that are beyond the scope of this book. The most common of these are the ependymomas. These tumors have both glial and epithelial differentiation and are most common occurring around the brainstem and spinal cord. For a more in-depth discussion of this and other CNS tumors, see the WHO's *Pathology and Genetics of Tumours of the Nervous System*.

Metastasis and Lymphoma

The most common intracranial tumors are metastases from a distant organ. The most common primary malignancies that metastasize to the brain are, in descending order of frequency, lung, breast, colon, kidney, and skin (melanoma). Unlike diffusely infiltrating gliomas, these tumors tend to form discrete nodules, usually at the gray-white junction in the cortex where arteriolar diameters are smallest.

Primary lymphomas of the CNS are more prevalent in immunosuppressed patients and were once quite rare. However, with the emergence of HIV-AIDS and increasing numbers of individuals taking immune-modulating therapies for organ transplants, lymphomas are increasingly more common. The vast majority of these tumors are diffuse, large B-cell lymphomas and express Epstein-Barr virus-associated genes and proteins, hinting at the oncogenic properties of this virus. Like malignant gliomas, lymphomas can diffusely invade the CNS parenchyma. Further, lymphomas of the CNS have imaging characteristics that are disturbingly similar to high-grade gliomas, including central necrosis and enhancement with contrast agents, and can therefore be challenging radiographic diagnoses.

DISORDERS OF MYELIN

Myelin is composed of layers of a complex proteolipid membrane that allows rapid and efficient conduction of action potentials along the axons of neurons. Loss or dysfunction of the normal myelin sheath causes abnormalities of this normal electrical signaling. There are two broad categories that describe ways myelin can be damaged: dysmyelination and demyelination. Dysmyelination is the loss of abnormally formed myelin or loss of the oligodendrocytes that produce and maintain myelin and is due to biochemically deranged myelin processing. Demyelination is the loss of structurally and biochemically normal myelin either by immune-mediated attack against myelin or oligodendrocytes or due to metabolic derangement of oligodendrocytes.

Leukodystrophies

Leukodystrophies are genetic disorders that cause damage to white matter (*leuko* — Greek: white). The leukodystrophies are autosomal recessive (AR) or X-linked disorders that result from loss of function mutations in enzymes involved in either production of myelin or normal myelin turnover and degradation (Figure 27.39). These enzymatic defects are not necessarily specific only to the CNS, and other organ systems may be affected. Clinically, these diseases typically present in early childhood with progressive loss of motor control, cognitive function, seizures, and eventual death.

Metachromatic leukodystrophy (MLD) is the most common leukodystrophy, is AR, and is caused by deficiency of the arylsulfatase A enzyme, which breaks down galactosyl-3-sulfatide to galactocerebroside in lysosomes. Galactosyl-3-sulfatide accumulates in many tissues, but the symptoms are related to destruction of myelin. MLD is so named because the accumulated material will change the color (metachromasia) of acidified cresyl violet stain from violet to brown in tissue. Krabbe's globoid cell leukodystrophy is another AR leukodystrophy caused by an enzyme deficiency in the sulfatide breakdown pathway. Here, there is a defect in the enzyme galactocerebroside ß-galactocerebrosidase, which breaks down galactocerebroside to galactose and ceramide.

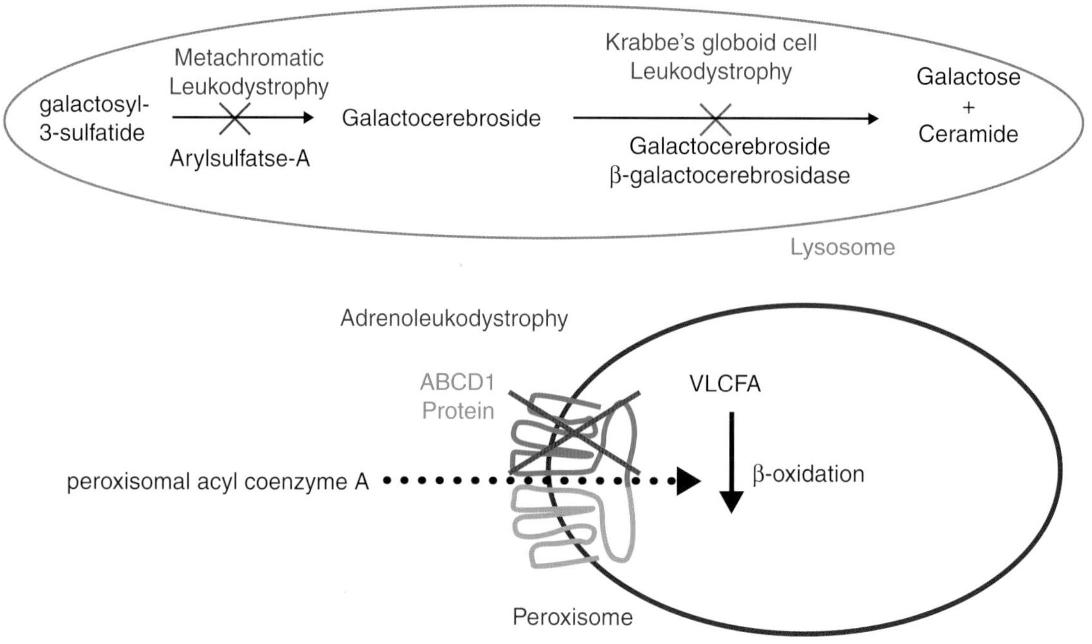

Figure 27.39 **Myelin metabolic pathway and loss of function mutations associated with dysmyelinating disease.**

Adrenoleukodystrophy (ALD) affects both the CNS myelin and adrenal glands. ALD is usually X-linked and is unusual among the leukodystrophies in that it is due to a defect in a peroxisomal transporter protein. ALD is most often caused by a defect in *ABCD1*, a member of the ATP-binding cassette transporter family. *ABCD1* encodes a transmembrane protein which is half of a heterodimeric transporter that transfers the enzyme peroxisomal acyl coenzyme A into peroxisomes, where it ß-oxidizes long chain fatty acids. Deficiencies in this enzyme cause a buildup of very-long chain fatty acids (VLCFA), especially hexacosanoic (C26:0) and tetracosanoic (C24:0) acid. This form of ALD is X-linked. A more severe, autosomal recessive form of the disease is due to absent or reduced peroxisome receptor-1, which causes more widespread peroxisomal dysfunction.

Numerous other leukodystrophies exist, and the molecular defects are known in some. Pelizaeus-Merzbacher leukodystrophy is X-linked and caused by abnormalities in a highly abundant myelin structural protein, proteolipid protein (PLP), which interferes with normal myelin formation. Interestingly, complete loss of the protein causes a relatively mild form of the disease, while missense mutations confer more severe changes, probably because the abnormal protein product interferes conformationally with the compact layering of normal myelin. Alexander's disease is a leukodystrophy in which the pathology is primary to astrocytes, not oligodendrocytes. It is caused by a dominant gain of toxic function mutation in the glial fibrillary acidic protein (GFAP) gene in which the protein forms abnormal aggregates. Large amounts of GFAP and other proteins accumulate in astrocyte processes. The mechanism of damage to oligodendrocytes or to myelin is not known.

Demyelination

Idiopathic Demyelinating Disease. Multiple sclerosis (MS) describes a variety of related disorders characterized by relapsing and remitting, multifocal immune-mediated damage to oligodendrocytes and myelin that are separated by time and anatomic location. These diseases are further clinically subclassified by the rate of progression of disability and the presence or absence of partial recovery between attacks. Histologically, there is loss of myelin with relative preservation of axons (Figure 27.40).

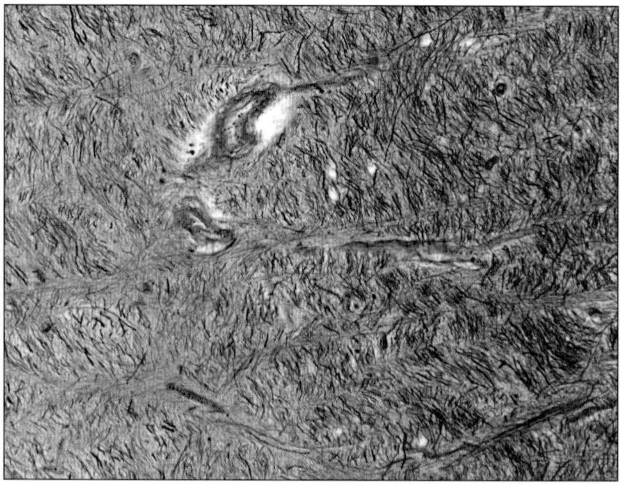

Figure 27.40 **Photomicrograph of old demyelinating lesion in multiple sclerosis.** Sections of cerebral white matter from the edge of demyelinating lesion in multiple sclerosis. A Luxol fast blue demonstrates loss of myelin (blue) in the left half of the figure, while a Holmes silver stain shows preserved axons in black.

Although the patterns and mechanisms of injury in MS are increasingly well characterized, the ultimate cause (s) of MS is unknown. The incidence of the disease is characterized by a markedly uneven geographic distribution generally with higher latitudes having higher risk, perhaps suggesting a role for environmental factors. Genetic factors also clearly have a role, as the incidence of disease is elevated in family members of a proband, with highest risk in monozygotic twins, which reaches a probability of almost 1/3. There is a prevailing multistep hypothesis of MS (Figure 27.41): Inflammatory cells outside the CNS are activated and upregulate adhesion molecules that guide them to the BBB. In this activated state, they are also more reactive to local chemokines and secrete matrix proteases to gain entry into the CNS, where they mediate damage. The disease mechanism of MS is generally understood to be T-cell-mediated autoimmune processes which require some as yet to be characterized trigger. T-cells outside the CNS are activated via the T-cell receptor (TCR) binding with antigen presented on the major histocompatibility complex (MHC) molecule and additional signals by antigen-presenting cells (APC). Additional signaling pathways also play a role in immune activation. One of these pathways that has been characterized involves the B7 signaling molecule on APCs. B7 interacts with signaling molecules on different T-cell subsets (most notably CD28 and CTLA-4) that modify the TCR response which can either cause activation and development of T-cell function, or unresponsiveness and apoptosis [30]. APCs also release chemokines such as interleukin (IL)-12 and IL-23 that generate CD4-positive T-helper type 1 cells (Th_1). Th_1 cells are proinflammatory, interferon γ-producing cells that are part of the normal antiviral response. These and other proinflammatory T-cell subsets have been implicated in demyelinating disease.

Neuromyelitis optica (NMO), also known as Devic's disease, is a variant of MS for which an etiology has been elucidated. Clinically, NMO is characterized by generally monophasic demyelination of the optic nerve, brainstem, and spinal cord without demyelination of the intervening cerebrum. Pathologically, the disease is characterized by a circulating autoantibody called the NMO-immunoglobulin (Ig)G. The NMO-IgG is directed against aquaporin 4, a water transporter in the foot processes of astrocytes where they abut CNS microvessels. The mechanism by which oligodendrocytes and glia are damaged is not fully understood.

Animal Model of Demyelination. Experimental allergic (autoimmune) encephalomyelitis (EAE) is an animal model that was developed to investigate immune-mediated demyelinating disease and, although developed in nonhuman primates and guinea pigs, is now used in inbred mice. Animals are sensitized against brain antigens, myelin basic protein or PLP, that are usually not available to immune surveillance. Commonly, myelin basic protein and PLP are used. Adjuvants designed to open the blood-brain barrier are infused to cause an acute, monophasic T-cell-mediated attack on myelin and demyelination. In its acute, monophasic course, EAE resembles acute disseminated encephalomyelitis (ADEM). ADEM is a human disease

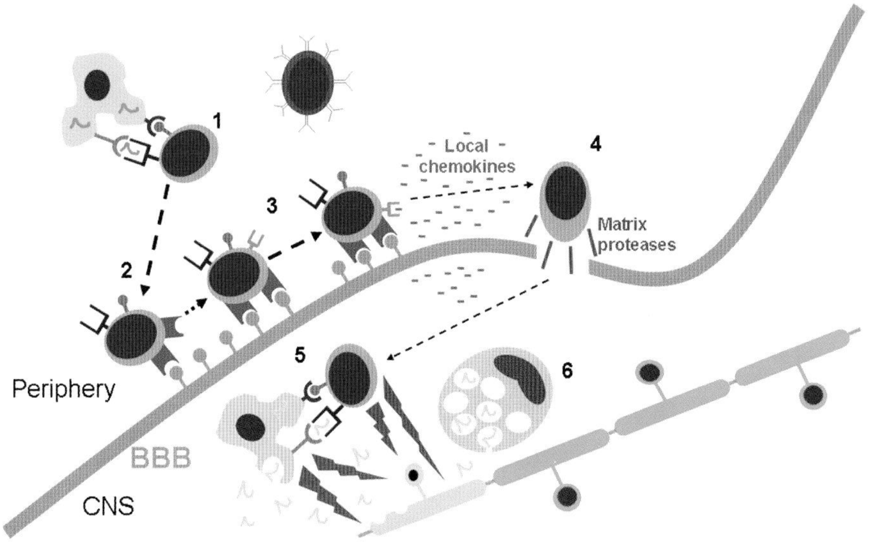

Figure 27.41 **CNS directed autoimmunity in demyelinating disease.** T-cells are activated in the periphery by antigen presenting cells (APCs) through the interaction of the T-cell receptor (TCR) and antigen bound to major histocompatibility complex (MHC) class II along with co-regulatory signals (1). Activated T-cells express molecules surface receptors for adhesion molecules on the luminal surface of the endothelium of the blood-brain barrier (BBB) (2). Activated T-cells also respond to local chemokines (3). The activated T-cells, under the influence of local chemokines, release matrix proteases that allow them to traverse the BBB (4). In the CNS, T-cells interact with microglia-derived or CNS infiltrating APCs through the interaction of the TCR and antigen-MHC and co-stimulatory pathways, thus mediating damage to myelin and oligodendrocytes (5). Macrophages remove debris (6). Adapted from [29].

Table 27.6 Relationship Between Human Demyelinating Disease and Experimental Allergic Encephalomyelitis (EAE)

Time Course	Human	Experimental
Acute	Acute disseminated encephalomyelitis	EAE
	Neuromyelitis optica	
Chronic	Progressive multiple sclerosis	Treated EAE
	Relapsing/remitting multiple sclerosis	

characterized by acute, monophasic demyelination without a chronic phase. It is usually a post viral complication but has been known to follow vaccination, bacterial or parasitic infections, or without known preceding history. Chronic demyelinating diseases like MS can be modeled in EAE with immune-modulating therapy. Table 27.6 shows the relationship between human demyelinating disease and EAE models.

Acquired Metabolic Demyelination. There are 2 primary osmotic demyelination syndromes. Central pontine myelinolysis (CPM) is a dire complication associated with rapid correction of systemic hyponatremia. Rapid replacement of sodium intravenously in persons with low serum sodium concentrations causes an acute loss of myelin that is usually confined to the middle of the base of the pons. The disease usually occurs in the setting of chronic alcoholism, liver or renal disease, post-transplant, or malnourishment. Although the disease mechanism is not well understood, hypotheses include rapid shifts of water in and out of glia, glial dehydration, myelin degradation, or apoptosis of oligodendrocytes.

Marchiafava-Bignami disease is a similar disease entity in which the white matter of the corpus callosum, the major tract that connects the left and right cerebral hemispheres, undergoes a similar demyelination to that seen in CPM, albeit with more necrosis. Initially described in poorly nourished Italian men who consumed large quantities of crude red wine, it is now known to be associated with alcoholism and to have a worldwide distribution.

Infectious Demyelination. Progressive multifocal leukoencephalopathy (PML) is an acute, progressive demyelinating disorder caused by a polyomavirus. JC virus is a small (~50 nm), double-stranded, icosahedral DNA virus lacking a membranous envelope. Approximately 40% of individuals in the developed world are sero-positive for the virus but are asymptomatic. Recrudescence of the JC virus causes disease in immunocompromised individuals, such as in AIDS, of which it is a defining illness. JC virus preferentially infects oligodendrocytes in the CNS. Histologically, PML is characterized by demyelination, lymphocytic inflammation, oligodendrocytes with enlarged nuclei with a ground-glass appearance, and bizarre reactive astrocytes (Figure 27.42). Electron microscopy shows oligodendrocyte nuclei packed with viral particles. The JC virus is oncogenic in rodents producing astrocytomas.

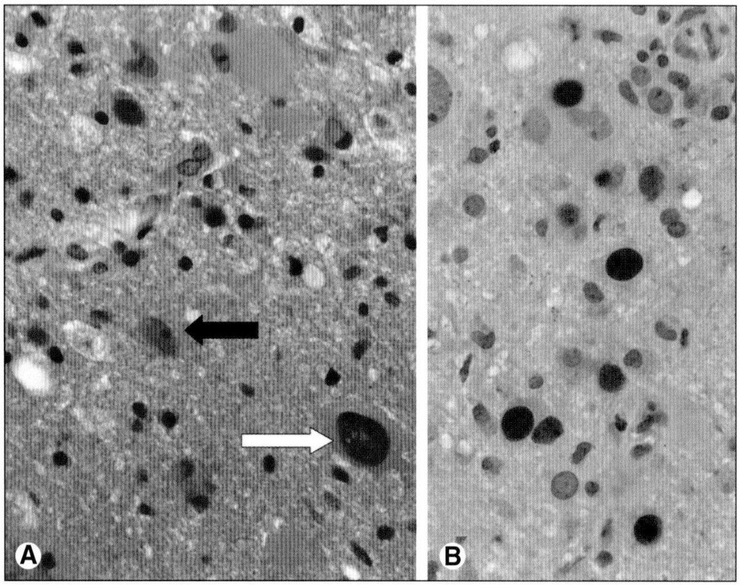

Figure 27.42 **Photomicrograph of Progressive Multifocal Leukoencephalopathy (PML).** Sections of cerebral white matter from a demyelinating lesion of PML. **(A)** H&E stains demonstrate enlarged oligodendrocyte nuclei with homogenous "ground-glass" viral inclusions (black arrow) and bizarre enlarged astrocytes (white arrow) simulating a neoplasm. **(B)** Immunoperoxidase staining for antibodies against Simian virus 40 which cross-react with JC virus nuclear inclusions in oligodendroglia (brown).

REFERENCES

1. López-Muñoz F, Boya J, Alamo C. Neuron theory, the cornerstone of neuroscience, on the centenary of the Nobel Prize award to Santiago Ramón y Cajal. *Brain Research Bulletin.* 2006;70:391–405.
2. Aloisi F. Immune function of microglia. *Glia.* 2001;36:165–179.
3. Baumann, P-D. Biology of oligodendrocyte and myelin in the mammalian central nervous system. *Physiological Reviews.* 2001;18:871–927.
4. Golden JA, Harding BN, eds. *Developmental Neuropathology: Pathology and Genetics.* Basel: ISN Neuropath Press; 2004:42–48.
5. Hoshino M. Molecular machinery governing GABAergic neuron specification in the cerebellum. *Cerebellum.* 2006;5:193–198.
6. Padmanabhan. Etiology, pathogenesis and prevention of neural tube defects. *Congenital Anomalies.* 2006;46:55–67.
7. Sadler TW. Embryology of neural tube development. *Am J Med Genet C Semin Med Genet.* 2005;135C:2–8.
8. Copp AJ. Neurulation in the cranial region—normal and abnormal. *J Anat.* 2005;207:623–635.
9. Endres M, Engelhardt B, Koistinaho J, et al. Improving outcome after stroke: Overcoming the translational roadblock. *Cerebrovasc Dis.* 2008;25:268–278.
10. Prusiner SB. Shattuck lecture—Neurodegenerative diseases and prions. *N Engl J Med.* 2001;344:1516–1526.
11. Langston JW. The etiology of Parkinson's disease with emphasis on the MPTP story. *Neurology.* 1996;47:S153–160.
12. World Health Organization. *Pathology and Genetics of Tumours of the Nervous System.* Geneva: WHO Press; 2007.
13. Holland EC. Progenitor cells and glioma formation. *Curr Opin Neurol.* 2001;14:683–688.
14. Bansal K, Liang ML, Rutka JT. Molecular biology of human gliomas. *Technol Cancer Res Treat.* 2006;5:185–194.
15. Benjamin R, Capparella J, Brown A. Classification of glioblastoma multiforme in adults by molecular genetics. *Cancer J.* 2003;9:82–90.
16. Ohgaki H, Kleihues P. Genetic pathways to primary and secondary glioblastoma. *Am J Pathol.* 2007;170:1445–1453.
17. Rong Y, Durden DL, Van Meir EG, et al. Mechanisms of necrosis in GBM. *J Neuropathol Exp Neurol.* 2006;65:529–539.
18. Hegi ME, Diserens AC, Gorlia T, et al. MGMT gene silencing and benefit from temozolomide in glioblastoma. *N Engl J Med.* 2005;352:997–1003.
19. Aldape K, Burger PC, Perry A. Clinicopathologic aspects of 1p/19q loss and the diagnosis of oligodendroglioma. *Arch Pathol Lab Med.* 2007;131:242–251.
20. Behin A, Hoang-Xuan K, Carpentier AF, et al. Primary brain tumours in adults. *Lancet.* 2003;361:323–331.
21. Fuller CE, Perry A. Molecular diagnostics in central nervous system tumors. *Adv Anat Pathol.* 2005;12:180–194.
22. Polkinghorn WR, Tarbell. Medulloblastoma: Tumorigenesis, current clinical paradigm, and efforts to improve risk stratification. *Nat Clin Pract Oncol.* 2007;4:295–304.
23. McLendon RE, Turner K, Perkinson K, et al. Second messenger systems in human gliomas. *Arch Pathol Lab Med.* 2007;131:1585–1590.
24. Konopka G, Bonni A. Signaling pathways regulating gliomagenesis. *Curr Mol Med.* 2003;3:73–84.
25. Furnari FB, Fenton T, Bachoo RM, et al. Malignant astrocytic glioma: Genetics, biology, and paths to treatment. *Genes Develop.* 2007;21:2683–2710.
26. van den Bent MJ and Kros JM. Predictive and prognostic markers in neuro-oncology. *J Neuropathol Exp Neurol.* 2007;66:1074–1081.
27. Pedersen MW, Meltorn M, Damstrup L, et al. The type III epidermal growth factor receptor mutation. *Ann Oncol.* 2001;12:745–760.
28. Ebers GC. Environmental factors and multiple sclerosis. *Lancet Neurol.* 2008;7:268–277.
29. Bar-Or A. The immunology of multiple sclerosis. *Semin Neurol.* 2008;28:29–45.
30. Green JM. The B7/CD28/CTLA4 T-cell activation pathway implications for inflammatory lung disease. *Am J Respir Cell Mol Biol.* 2000;22:261–264.
31. Kumar S, Fowler M, Gonzalez-Toledo E, et al. Central pontine myelinolysis, an update. *Neurol Res.* 2006;28:360–366.

Part

V

Practice of Molecular Medicine

Chapter

28

Molecular Diagnosis of Human Disease

Lawrence M. Silverman ▪ Grant C. Bullock

INTRODUCTION

From an historical perspective, developments in molecular diagnostics primarily reflect technological breakthroughs which originated in molecular biology, from the description of the Southern blot to the clinical applications of the polymerase chain reaction (PCR) and high throughput sequencing. Specific medical utilization of these molecular biology tools coincided with basic scientific breakthroughs, such as the completion of the Human Genome Project. Superimposed on these areas came changes in the regulatory environment, originated by existing regulatory agencies, professional organizations, and new advisory committees, all of whom strived to describe the good laboratory practices which all clinical laboratories should attempt to attain. In this chapter, we will describe these laboratory and regulatory aspects and the most likely areas for future applications, along with the possible factors which may restrain growth and development.

Initially, we will provide a brief background into the history of molecular diagnostics. No discussion of this evolution would be complete without a description of the impact of regulatory agencies, professional organizations, and ad hoc committees on the analytical and clinical applications, particularly of molecular tests for heritable conditions (notice that this is the preferred term, not genetic or inherited). Next, we will describe examples of molecular testing in infectious diseases, oncology, drug metabolism (or pharmacogenetics), in addition to heritable disorders. For each category we will choose one specific example and describe the preanalytical, analytical, and postanalytical considerations which comprise good laboratory practice. In this section, we will also provide the specific examples to understand new concepts or mechanisms regarding disease pathogenesis and/or treatment. In particular, we will look at these unique applications: (i) clinical diagnosis, (ii) population screening, (iii) selected screening, and (iv) tests associated with disease prediction and risk assessment. Finally, we will focus on future growth areas, including multivariate analyses, personalized medicine, automation, and miniaturization, while acknowledging constraints such as reimbursement and regulatory stagnation.

HISTORY OF MOLECULAR DIAGNOSTICS

Regulatory Agencies and CLIA

The major driving force in clinical molecular diagnostics is the original Clinical Laboratory Improvement Amendments (CLIA) and the impact of subsequent changes. Since the original legislation (1988) and the publication of the final rule (1992), only CLIA-certified laboratories may provide testing of human specimens (other than for forensic and research purposes) which results in clinical decision making. Thus, any changes in CLIA regarding molecular testing will have tremendous impact on the clinical molecular laboratories. Over the years, these changes have been relatively minor, so that the issues of quality of testing and patient safety reflect the general aspects of CLIA rather than a specific area of application. While most of molecular testing is well covered by the existing CLIA mandates, genetic (heritable disorders) testing represents a challenge, particularly regarding preanalytical and postanalytical areas.

It was regarding the preanalytical and postanalytical areas of genetic testing that most of the public and media concern has focused. As early as 1994, the Institute of Medicine of the National Academy of Sciences convened a committee to deal with issues regarding genetic testing. Among the many concerns were the issues surrounding the use of genetic testing to predict future outcomes of individuals and their families. It is this unique aspect of genetic testing that differs from

 © 2009, Elsevier, Inc. All Rights Reserved.

most other clinical laboratory tests and demonstrates the need for special attention in the preanalytical and postanalytical areas.

In 1995, a second committee was commissioned to deal with these issues. The Department of Energy convened a Task Force on Genetic Testing and emphasized the importance of restricting genetic testing to CLIA-certified clinical laboratories. It is important to note that the number of genetic tests and the number of laboratories providing these services was limited (while data are scarce prior to 2000, the number of tests available at this time was probably less than 100). In 2008 over 1500 genetic tests were listed in GeneTests.

Finally, in 1998 the Department of Health and Human Services Secretary Donna Shalala established the Secretary's Advisory Committee on Genetic Testing (SACGT)–which was later renamed SACGHS (Secretary's Advisory Committee on Genetics Health and Society)–to take a more global view of genetic testing and the regulatory agencies who oversaw various aspects described by the previous committees. Over the past 10 years, this committee has been very active in assessing genetic testing and its wider implications, including genetic discrimination, patents, in addition to the roles of various regulatory agencies regarding genetic testing. However, molecular technology and applications continue to develop in previously unanticipated areas.

Because of the nature of molecular genetic testing, many analyses were developed by laboratories for use only in their laboratory. Thus, these laboratory-developed tests (LDTs) presented unusual challenges to traditional regulatory agencies, such as the Food and Drug Administration (FDA). To this day, LDTs are the lifeblood of molecular diagnostics, and differentiate this area from most other areas of the clinical laboratory which depend more on kit and instrument manufacturers. However, regulatory oversight for LDTs is more stringent, by necessity, than those having FDA approval. Thus, we enter the universe of ASR (analyte-specific reagents manufactured for LDTs), RUO (research use only kits), and IVDMIA (*in vitro* diagnostic multivariate index assays). Each of these designations imparts specific FDA guidelines and additional regulatory oversight. Typically, a clinical molecular laboratory uses a mixture of FDA-approved kits and LDTs, which include many ASRs. Occasionally, an RUO kit may be in use for research projects, and perhaps, in the future labs may incorporate IVDMIAs, although currently these are primarily found in commercial laboratories.

In order to maintain CLIA certification, clinical laboratories must undergo inspections, usually every 2 years, by accrediting organizations recognized by the *Centers for Medicare and Medicaid Services* (CMS), such as the *College of American Pathologists* (CAP). A CAP inspection, which is unannounced, uses checklists to monitor laboratory performance and quality, including personnel qualifications, and maintains an ongoing oversight program over molecular testing. However, because of the mixed bag of tests performed and the various clinical applications, the inspection procedure, and the use of multiple checklists which cover the various applications, inspectors are called upon to have extensive inspection experience and training.

Quality Assurance, Quality Control, and External Proficiency Testing

An additional regulatory aspect involves quality assurance (QA) programs consisting of both external proficiency testing (PT) and the use of quality control (QC) materials. While QA is a part of all laboratory testing areas, unique problems surround the dearth of appropriate control materials and external PT programs, which are intimately related for the following reason. Molecular testing is confounded by many factors including (i) the rare nature of many heritable conditions, (ii) the multiplicity of mutations associated with a single disorder (for example, cystic fibrosis), and (iii) the level of tissue heterogeneity, particularly when dealing with malignant conditions. For molecular infectious disease testing, particular care must be given to ensure that lack of an amplified product signifies no measurable target as opposed to amplification failure. In this latter case, use of internal controls added to the amplification reaction assures the laboratory that a negative result is a true negative. Also, since nucleic acid is found in living and dead organisms, the presence of a positive signal may not represent active infectious agents.

Method Validation [25]

Adding to this complexity are the difficulties associated with validating molecular tests. When validation is applied to LDTs, the FDA (via the ASR rules) states "clinical laboratories that develop tests are acting as manufacturers of medical devices and are subject to FDA jurisdiction." To underscore this, CLIA requirements for test validation vary by test complexity and by whether the test is FDA approved, FDA cleared, or laboratory developed. CLIA requires that each lab establish or verify the performance specifications of moderate-complexity and high-complexity test systems that are introduced for clinical use. For an FDA-approved test system without any modifications, the system is validated by the manufacturer; therefore, the laboratory need only verify (confirm) these performance specifications. In contrast, for a modified FDA-approved test or an in-house developed test, the laboratory must establish the performance specifications for a test, including accuracy, precision, reportable range, and reference range. The laboratory must also develop and plan procedures for calibration and control of the test system. In addition, for a modified FDA-approved test or a laboratory-developed test, the laboratory must establish analytic sensitivity and specificity and any other relevant indicators of test performance. How and to what extent performance specifications should be verified or established is ultimately under the purview of medical laboratory professionals and is overseen by the CLIA certification process, including laboratory inspectors.

Clinical Utility

In 1997, the Department of Energy Task Force on Genetic Testing proposed three criteria for the evaluation of genetic tests: (i) analytic validity, (ii) clinical validity, and (iii) clinical utility [1]. By clinical utility, the report referred to "... the balance of benefits to risks. ..." However, more frequently, especially with new molecular genetic tests, clinical utility may not be readily available when a test is put into clinical use. In fact, it may take years to accumulate data on clinical utility. For example, when the molecular test for hereditary hemochromatosis was first introduced, there were limited data on penetrance, or the association of specific phenotype with a specific genotype. In other words, initial data were obtained on individuals demonstrating the phenotype associated with hereditary hemochromatosis. However, clinical utility must be assessed in individuals with some or even no obvious symptoms (through population screening) before penetrance can be estimated. When such studies were performed, penetrance was incomplete, with estimates ranging from 3%–30% of individuals who carried two disease-causing mutations demonstrating the classical phenotype. Thus, at the time the test was introduced, it was assumed that penetrance was close to complete, since almost all patients demonstrating the phenotype had two disease-causing mutations. The clinical utility would be based on positive and negative results defining the disease phenotype. Now we know that homozygous (two disease-causing mutations) individuals frequently do not demonstrate the phenotype, at least at the time of testing.

MOLECULAR LABORATORY SUBSPECIALTIES

In this section, we will describe examples of clinical molecular tests representing the areas of (i) heritable disorders, (ii) infectious disease, (iii) pharmacogenomics, and (iv) oncology. In each case, we will address aspects of (a) method validation, (b) preanalytical and postanalytical concerns, and (c) clinical applications with the benefits and risks associated with both positive and negative results.

Heritable Disorders

Fragile X Syndrome

The most common form of heritable mental retardation is called the Fragile X Syndrome (FRS), so-called because of the cytogenetic appearance of increased fragility at a locus on the long arm of the X chromosome, under specific cell culture conditions [24].

This observation is frequently associated with characteristic clinical features, primarily in males, including mild to moderate mental retardation, a prominent facial appearance with large low-set ears, and enlarged testes in males, postpuberty. Affected females have a less well-defined phenotype, due to the variable affect of the second X chromosome (called lyonization). However, the inherent difficulties of this cytogenetic test led to the development of a molecular test once the gene was identified. The first molecular test was based on Southern blot analysis, coupled with specific restriction enzymes, which could identify the characteristic molecular defect.

The *FMR1* gene contains a CGG repeat region in the 5′ untranslated region (Figure 28.1). The number of CGG repeats is variable in the general population. When increased beyond 200 CGG repeats, the gene is hypermethylated, leading to absence of expression of the gene product, FMR1. In the normal population, the number of repeats varies from 5 to 44 repeats, with a median of 29. In individuals with FRS, the number of repeats usually exceeds 200. Premutation carriers have between 55 and 200 repeats, which has been associated with (i) increased risk of full expansion during female (but not male) meiosis leading to FRS in her offspring, (ii) increased risk of ovarian dysfunction (premature ovarian failure), and (iii) increased risk of FXTAS (fragile x-associated tremor/ataxia syndrome). Clearly, an accurate and precise method for determining the number of CGG repeats, and the associated hypermethylation status, has significant clinical implications. However, the Southern blot technique was able to assess only hypermethylation, but not accurately assess the number of CGG repeats (Figure 28.2).

Validation of Molecular Sizing of FMR1

The Southern blot procedure for FRS is well documented. The focus here is on the validation of a sizing assay for the CGG region of *FMR1* performed on a capillary electrophoresis platform (ABI 3100, Applied Biosystems, Foster City, California). The LDT is based on the report by Fu [26], while an RUO assay is commercially available from Celera Diagnostics (Alameda, California). Inherent in any method validation, well-characterized samples must be available. These can be patient samples analyzed by an independent laboratory, immortalized cell lines (Coriell Cell Repositories), or artificially prepared standards (National Institute of Standards and Technology). Each of these materials has advantages and disadvantages. Ideally, all three sources provide the most complete assessment of method validation. Conditions for both the LDT and the RUO are available from the literature.

PCR and Capillary Electrophoresis

Sizing of the *FMR1* 5′ UTR is performed on an automated capillary electrophoresis platform following PCR amplification of the critical region. Choice of the forward and reverse primers and PCR amplification details are discussed elsewhere, as are the capillary electrophoresis conditions [24].

Statistical Analysis

Validation studies should determine reproducibility (imprecision) and accuracy, allowing the laboratory to report values with confidence intervals (45 ± 2 repeats, for example). These estimates of precision can be

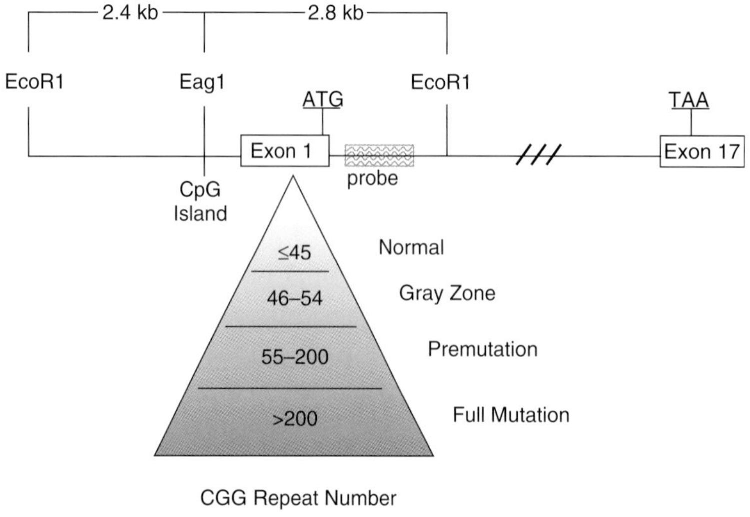

Figure 28.1 Schematic representation of the CGG repeat in exon 1 of *FMR1* and associated alleles. A CGG-repeat number less than or equal to 45 is normal. A CGG-repeat number of 46 to 54 is in the gray zone and has been reported to expand to a full mutation in some families. A CGG-repeat number of 55 to 200 is considered a premutation allele and is prone to expansion to a full mutation during female meiosis. A CGG-repeat number in excess of 200 is considered a full mutation and is diagnostic of fragile X syndrome.

obtained by multiple determinations of the same sample during a single run of samples and by analyzing the same sample on multiple runs, followed by determining the mean and the standard deviation. These terms are referred to as within-run and between-run precision.

Postanalytical Considerations

Interpretation of *FMR1* sizing analysis depends on the clinical considerations for ordering the test. These may be (i) diagnosis, (ii) screening, (iii) predictive within a family with affected members, or (iv) risk assessment. For example, a diagnostic application would include determining whether an individual with the characteristic phenotype has FRS, and depends on the sex of the patient. A male, having a single X chromosome, should have a single band following PCR amplification, with a normal number of CGG repeats (5–44 repeats). One of the limitations of the PCR procedure is the size of the CGG repeat region. Depending on the PCR primers and amplification conditions, the maximum number of CGG repeats that can be amplified is ~100 repeats. Thus, a fully amplified CGG region of >200 repeats would not produce a discernable PCR product. However, a female, with two X chromosomes, with a full expansion would still have an unaffected X chromosome, resulting in a single band following PCR amplification, representing the normal-sized allele. Another limitation with females is that both alleles may be the same size, resulting in a single band following amplification. For females, any analysis which yields a single band following PCR, and with the absence of a band in a male, Southern blot analysis would be necessary, as this analysis is not limited by the constraints of PCR amplification. Thus, Southern blot analysis should detect fully expanded alleles. Screening applications may involve newborn screening in the future. However, methodological constraints are considerable at this time.

If an affected family member has been identified, carrier determination can be performed, exemplifying a predictive use of this testing. However, another level of complexity involves risk assessment of female premutation carriers. These individuals have between 55 and 200 repeats, not large enough for full mutations, but sufficient for the associated conditions mentioned previously (premature ovarian failure, FXTAS, and others), in addition to increased risk of passing on a fully expanded allele to offspring (the risk is proportional to the number of CGG repeats). Note that individual alleles between 45 and 54 repeats are considered intermediate and may carry an increased risk for being permutation carriers, but not for passing on full mutations. No data exist on the risk of these individuals for ovarian dysfunction or FXTAS. Obviously, knowing the precision and accuracy of the assay is paramount in interpreting the number of CGG repeats.

Benefit/Risk Assessment (Clinical Utility)

Since defining clinical utility takes considerable clinical experience with newer assays, it is difficult, if not impossible, to assess clinical utility before setting up an LDT, as advocated by some of the regulatory agencies. For example, *FMR1* sizing assays demonstrate the evolving nature of clinical utility. When Southern blot analysis was the only molecular method available, the clinical utility was based on the contribution of

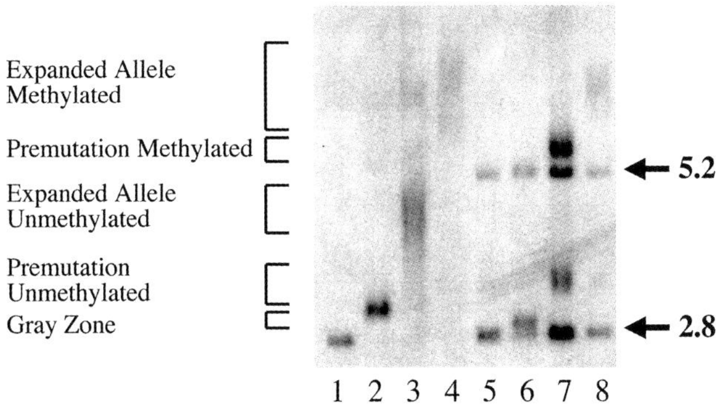

Figure 28.2 Southern blot analysis for the diagnosis of fragile X syndrome. Patient DNA is simultaneously digested with restriction endonucleases *EcoR*1 and *Eag*1, blotted to a nylon membrane, and hybridized with a ^{32}P-labeled probe adjacent to exon 1 of *FMR1* (see Figure 28.1). *Eag*1 is a methylation-sensitive restriction endonuclease that will not cleave the recognition sequence if the cytosine in the sequence is methylated. Normal male control DNA with a CGG-repeat number of 22 on his single X chromosome *(lane 1)* generates a band about 2.8 kb in length corresponding to *Eag*1-*EcoR*1 fragments (see Figure 28.1). Normal female control DNA with a CGG-repeat number of 20 on one X chromosome and a CGG-repeat number of 25 on her second X chromosome *(lane 5)* generates two bands, one at about 2.8 kb and a second at 5.2 kb. *EcoR*1-*EcoR*1 fragments approximately 5.2 kb in length represent methylated DNA sequences characteristic of the lyonized chromosome in each cell that is not digested with restriction endonuclease *Eag*1. DNA in *lane 2* contains an *FMR1* CGG-repeat number of 90 and is characteristic of a normal transmitting male. The banding pattern observed in *lane 3* is representative of a mosaic male with a single X chromosome with a full mutation (>200 repeats). However, the full mutation in some cells is unmethylated; in other cells, the full mutation is fully methylated, hence the term *mosaic*. In those cells in which the full mutation is unmethylated, digestion by both *Eag*1 and *EcoR*1 occurs, and in those cells in which the full mutation is fully methylated, digestion of the DNA by *Eag*1 is inhibited. The banding pattern observed in *lane 4* is diagnostic of a male with fragile X syndrome, illustrating the typical expanded allele fully methylated in all cells. *Lane 6* is characteristic of a female with a normal allele and a CGG-repeat number of 29 and a larger gray zone allele with a CGG-repeat number of 54. *Lane 7* is the banding pattern observed from a premutation carrier female with one normal allele having a CGG-repeat number of 23 (band at about 2.8 kb) and a second premutation allele with CGG repeats of 120 to about 200 (band at about 3.1 kb). In premutation carrier females, in cells in which the X chromosome with the premutation allele is lyonized, the normal 5.2 kb *EcoR*1-*EcoR*1 band is larger because of the increased CGG-repeat number and is about 5.5 kb in length. *Lane 8* is diagnostic of a female with fragile X syndrome with one full expansion mutation allele that is completely methylated and transcriptionally silenced on one X chromosome but with a second normal allele with a CGG-repeat number of 33.

the *FMR1* assay to the diagnosis of FRS. However, by this method, limited information was available for assessing the premutation carrier status. With the development of *FMR1* sizing assays, more accurate assessment of the permutation allele could be made, and the clinical utility for ovarian dysfunction and FXTAS could be assessed. Thus, improvements in methodology can affect the clinical utility of an analyte.

Infectious Diseases

Enteroviral Disease

Enteroviruses (EV) are a general category of single-stranded, positive-sense, RNA viruses of the Picornaviridae family, which includes the subgroups enteroviruses, polioviruses, coxsackieviruses (A and B), and echoviruses [1,2]. Rhinoviruses are closely related members of the Picornaviridae family that are a major cause of the common cold. Most EV infections are asymptomatic. Symptomatic EV infections affect many different organ systems. Diseases of the central nervous system include poliomyelitis, aseptic meningitis, and encephalitis. Diseases of the respiratory system include the common cold, herpangina, pharyngitis, tonsillitis, and pleurodynia. Diseases of the cardiovascular system include myocarditis and pericarditis. EV infections also cause febrile exanthems, conjunctivitis, gastroenteritis, and a sepsis-like syndrome in neonates. EV infections peak during the first month of life and cause the majority of aseptic meningitis worldwide [3,4].

Until recently, a diagnosis of EV infection was based on clinical presentation and the isolation of virus in cell culture from throat, stool, blood, or cerebrospinal fluid (CSF) specimens. EV detection by culture methods is insensitive and takes too long to be useful in clinical management decisions. By contrast, reverse transcription-PCR (RT-PCR) is more rapid and sensitive and has been used to detect EV RNA in CSF, throat swabs, serum, stool, and muscle biopsies [5]. EV RNA detection by molecular methods is now considered the gold standard for the diagnosis of enteroviral meningitis. At the time of validation, several ASRs were available for the detection of enteroviral RNA from CSF specimens.

Validation of LightCycler-Based, RT-PCR Assay for Enterovirus

We developed a real-time PCR assay using the LightCycler platform (LC-PCR) [6] because our molecular diagnostics laboratory has extensive experience with this technology. In addition, others had demonstrated the utility of this approach [7]. For rapid detection of EV RNA in CSF specimens, we established a qualitative semi-nested LC-PCR method based on EV RNA isolation and cDNA synthesis followed by PCR amplification with primers that target the highly conserved 5′-noncoding region (5′-NCR) [9]. To improve sensitivity and specificity, we used a second round of semi-nested PCRs. The semi-nested PCR uses two different viral subgroup-specific forward primers (one for polioviruses and one for coxsackieviruses), and a common reverse primer that generates 85 bp to 87 bp amplicons depending on the viral subgroup. The amplicons bind the EV-TaqMan probe which allows for real-time fluorometric detection and excludes cross-hybridization with rhinoviral sequences. Viral RNA is isolated from infected CSF using either the MagNA Pure total Nucleic Acids Extraction Kit (Roche Diagnostics, Indianapolis, Indiana) or the High Pure Viral RNA Kit (Roche Diagnostics). The extracted nucleic acids are subjected to RT-PCR in the LightCycler using the LC RNA Master Hybridization Probes Kit (Roche Diagnostics). Amplicons are diluted with sterile water, and subsequent semi-nested PCR is carried out in the LightCycler with the EV-TaqMan probe using the LC Fast Start DNA Master Hybridization Probes Kit (Roche Diagnostics).

QC materials include external positive and negative controls and a no template control (water blank). An appropriate internal positive control was not available at the time of assay development. Therefore, a technical failure cannot be ruled out when a patient sample is negative. The crossing-thresholds for known positive samples were determined during the validation set by a limiting dilution approach using negative CSF samples spiked with known amounts of cell culture grown EVs. A minimum cycle threshold (Ct) of 30 cycles was established. Limits of detection for several EVs were determined and compared to results from a reference lab for the same specimen control. In the validation set, the sensitivity of the MagNA Pure RNA extraction was compared to the High Pure Column RNA extraction method. The number and type of controls need to be re-evaluated periodically after the implementation of a new assay. For example, the development of a molecular EV assay that uses internal positive controls to confirm negative results for patient samples is needed and once this is available it should be integrated into the assay.

Preanalytical Considerations

A semi-nested RT-PCR method is especially susceptible to contamination by nucleases or unrelated nucleic acids, which cause false negative or false positive results, respectively. Laboratory organization and equipment should be designed to avoid contamination, and specimens should be handled with gloves in a biosafety cabinet to prevent contamination with fingertip RNases and biohazard exposures, as described [8]. The quality of viral RNA in CSF depends on what happens to the specimen before RNA sample extraction occurs. Therefore, only 0.5 mL or greater volumes of CSF specimens collected in the appropriate sterile plastic containers are accepted. The EV virion is very stable and protects the viral RNA from degradation. Sample stabilization is not required as long as specimens are dispatched to the laboratory within hours of acquisition. If there will be a delay in specimen dispatch, then the specimen should be refrigerated or placed on ice. If the delay will be more than 24 hours, then the specimen should be stored frozen and kept frozen until RNA extraction. Neonatal serum or plasma specimens for suspected EV sepsis require approval of the laboratory director and may be acceptable if handled properly. Plasma should be collected in K_2EDTA or ACD tubes [9]. Heparin should be avoided, since it inhibits RT and PCR reactions, as described in [10]. Heparin inhibition of RT-PCR reactions can be reversed using lithium chloride if necessary [11]. The suitability of this approach was validated during assay development. Samples that have been frozen before dispatch and arrive in the state of thawing out should be rejected.

Postanalytical Considerations

Fluoresence signal detected above background before the 30th cycle of the semi-nested EV LC-PCR is diagnostic of EV infection. Ct values that are equal to or greater than 30 cycles are negative. In a patient with the clinical signs and symptoms of viral meningitis, a positive result is consistent with a diagnosis of EV meningitis. Interpretation of a negative result is more difficult due to the absence of an internal positive control for the RT-PCR. A negative result could indicate a technical failure with the RT or PCR steps in the method. Alternatively, the sample may not contain EV RNA. Thus, a negative result does not exclude EV infection and needs to be communicated in the report to the ordering physician. In addition the pediatricians who will be ordering this test most often need to be made aware of the meaning of a negative result. Finally, because EV is a reportable disease to the state health department, a positive result must be called to the physician and reported to hospital epidemiology.

Benefit/Risk Assessment (Clinical Utility)

The distinction between EV and bacterial infections in neonates can be difficult, and patients are usually treated empirically with antibiotics until an EV infection is confirmed, at which point supportive care is the treatment choice. Therefore, a rapid molecular method for detecting EV RNA has clinical utility if available daily. Several studies have suggested that the use of molecular EV testing in the pediatric patient population can lead to cost savings by reducing the length of hospital stay, reducing antibiotic use, and reducing imaging studies [12]. A positive molecular EV test helps to exclude bacterial or other etiologies that may require longer hospitalization for treatment. A negative EV test is not very useful clinically because it does not rule out EV meningitis.

Hepatitis C Viral Genotype

Approximately 4 million people in the United States are chronically infected with HCV, and many are unaware that they are infected. Recent studies estimate that, in the United States, 30,000 individuals per year are newly infected with HCV, and approximately 10,000 people will die annually from the sequelae of hepatitis C. HCV is also the most frequent indication for liver transplant in adults in the United States. Since the outcome of HCV infection depends on both the viral load and HCV genotype, the determination of HCV genotype provides important clinical information regarding the duration and type of antiviral therapy used for a given HCV-infected patient [13]. In addition, the genotype is an independent predictor of the likelihood of sustained HCV clearance after therapy [14]. There are more than 11 genotypes and more than 50 subtypes of HCV. In the United States, genotypes 1a, 1b, 2a/c, 2b, and 3a are most common. Genotypes 4, 5, 6, and 7 are endemic to Egypt, South Africa, China, and Thailand, respectively, and are rarely found in the United States. Since patients infected with HCV genotype 1 may benefit from a longer course of therapy, and genotypes 2 and 3 are more likely to respond to combination interferon-ribavirin therapy, the clinically relevant distinction among the affected population of the United States is between genotype 1 and nontype 1. Not enough data currently exists to determine the likelihood of therapeutic response for HCV genotypes 4, 5, 6, and 7. Therefore, it has been recommended that infections with genotypes 4, 5, 6, or 7 be treated similar to patients infected with HCV type 1.

In addition to viral load, HCV genotype is important clinically. Line-probe and sequencing-based methods are available for HCV genotyping, but are expensive, time consuming, and provide more information than is currently needed by clinicians for patient management decisions. With this in mind, we developed a rapid HCV genotyping assay for the LightCycler to differentiate HCV type 1 from nontype 1 infections.

Validation of HCV Genotyping

The determination of HCV genotype is based on isolation of HCV viral RNA and cDNA synthesis followed by PCR amplification of the 5′ untranslated region (5′UTR). Viral RNA is isolated from infected serum after HCV viral load determination using the MagNA Pure Total Nucleic Acids Extraction Kit or the Qiagen EZ1 BioRobot. The extracted nucleic acids are subjected to RT-PCR using primers specific for 5′UTR sequence that is conserved among all known HCV genotypes. The RT-PCR step is performed using an Applied Biosystems GeneAmp 9600 PCR System (block cycler). A second PCR amplification using a semi-nested pair of primers is carried out in the LightCycler in the presence of a pair of FRET oligonucleotide probes. The FRET detection probe allows the discrimination of HCV genotypes 1a/b, 2a/c, 2b, 3a, and 4 during melting curve analysis (Figure 28.3), because it hybridizes with differential affinity to a region of

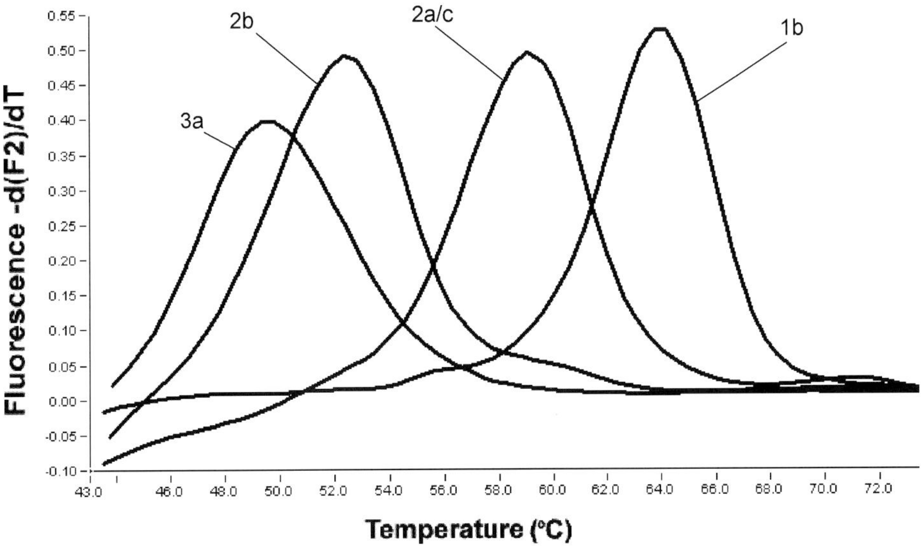

Figure 28.3 **HCV genotyping assay.** Comparison of samples from patients infected with HCV genotypes 1b, 2a/c, 2b, and 3a. Shown are genotype-specific melting transitions for four samples in 2 mM $MgCl_2$. Data were obtained by monitoring the fluorescence of the LCRed640-labeled FRET sensor probe during heating from 40 to 80°C at a temperature transition rate of 0.1°C/s.

the 5′UTR that varies among the different HCV genotypes. Magnesium concentration was critical for optimal genotype discrimination. Primer and probe design and PCR details have been described [15]. For method validation, patient samples analyzed by an independent reference laboratory were used. Positive QC materials were pooled patient serum samples of previously determined genotype (genotypes 1, 2a/c, 2b, and 3a). Negative QC material was pooled patient serum samples previously determined to be HCV negative. A water blank is also always included with each run. A serial limiting dilution of positive patient samples with known viral titers was used to determine the limits of detection. The semi-nested method described will fail when viral titers are less than 12,600 IU per mL. A more sensitive fully nested PCR method was developed and is able to genotype specimens with 130 IU per mL. The fully nested PCR method uses the freshly extracted RNA and the same RT and first PCR steps used for the semi-nested approach. The second PCR step uses a new set of nested primers and is done in the block cycler. The products of this reaction are subjected to melting curve analysis in the LightCycler to determine genotype using the same set of FRET probes described for the semi-nested genotyping method. The fully nested approach is rarely needed because patients usually have high viral loads upon initial presentation, and this is the ideal time to assess the HCV genotype.

Statistical Analysis

In addition to determination of the accuracy of the genotyping assay, melting point precision was determined within run and between runs using QC materials. An analysis of assay performance after 18 months showed no assay failures for 70 runs with 411 patient samples using 3 different LightCyclers operated by 6 different technicians. The genotype-specific melting temperatures remained nonoverlapping with coefficients of variation that were less than 1% for all genotypes [16]. Melting temperatures are continuously monitored to assure assay performance.

Preanalytical Considerations

The semi-nested or a nested RT-PCR method is especially susceptible to contamination by RNases, previously amplified DNA, or unrelated nucleic acids. Unidirectional flow of samples through physically separated preamplification and postamplification areas should be maintained to prevent contamination of reagents or specimens with amplified DNA. Equipment, reagents, plasticware, and specimens should be handled with gloves to prevent contamination with fingertip RNases or cross-contamination of specimens. The use of plastic microfuge tube cap openers instead of gloved fingertips is recommended to prevent cross-contamination. Samples should be added one at a time, first followed by positive QC, negative QC, and finally the water blank. The only known interference to this procedure is heparin, which may inhibit the RT-PCR reaction. Evaluation of isolation of nucleic acid by the MagNA Pure system has shown that heparin is removed during the isolation of nucleic acid by this method. A minimum of 500 microliters of serum is the required sample type for this method. Whole blood samples should be centrifuged, and serum should be separated within 6 hours of collection. HCV is an enveloped RNA virus and is labile to repeated freeze-thaw cycles. Viral titers will drop upon storage at −20°C. Therefore, fresh samples are preferred. Since aqueous solutions of RNA are unstable and susceptible to spontaneous hydrolysis even at −20°C, it is best to perform the RT-PCR step on the same day as RNA extraction. MagNA Pure-extracted nucleic acid samples should be stored at −80°C if the RT-PCR step cannot be performed on the day of isolation.

Postanalytical Considerations

Interpretation of HCV genotype is reported as type 1, type 2, type 3, type 4, or none detected if negative. Both early viral load measurements and viral genotype are used to determine likelihood of clinical response and influence management decisions during combination antiviral therapy. Therefore, these tests should be ordered together during the initial workup of a patient with viral hepatitis. Reflexive genotyping of patients who are positive for HCV with sufficiently high viral loads and no prior genotype should be considered. Once the genotype is determined, there is no need for additional genotyping unless there is clinical suspicion of a second infection with a different HCV genotype. The HCV LC-PCR assay is able to detect two genotypes in one patient sample. One limitation with this assay is the inability to detect some rare genotypes that may occur in the United States, such as genotypes 5 or 6. Thus, the detection of a novel melting temperature could indicate an infection with a non-1, -2, -3, or -4 HCV genotype. This specimen would be sent out for sequencing by a reference laboratory. Similarly a specimen with a high viral load but no detectable genotype indicates a technical failure or a novel genotype and would also be sent to a reference laboratory for sequencing.

Benefit/Risk Assessment (Clinical Utility)

HCV genotyping assays are an essential component of an evolving clinical decision tree used to determine the duration of therapy and the likelihood of sustained virological response (SVR) to this therapy. Absolute viral load, log decline in viral load from baseline, and viral genotype are clinically useful in predicting SVR and nonsustained virological response (NSVR) to treatment [17]. The *NIH Consensus Development Conference Statement, Management of Hepatitis C:2002*, stresses the importance of early prediction of SVR and NSVR in patient management in the first 12 weeks of therapy [18]. Combination interferon plus ribavirin therapy is given for 24 or 48 weeks, depending on viral genotype and viral load. The therapy is expensive and difficult for patients to tolerate due to side effects. HCV genotype and viral loads are used to predict the likelihood of SVR or NSVR. The likelihood of SVR and NSVR is useful for clinicians and patients when making decisions about the duration of therapy. Nontype 1 genotypes have a much greater rate of SVR than genotype 1 infected patients. HCV type 1 with a high viral load requires a longer course of therapy than nontype 1 infections. HCV genotype determination is clinically useful in risk assessment and making therapeutic decisions instead of making a diagnosis.

Oncology

B-Cell and T-Cell Clonality

B-cell and T-cell clonality assays are based on the detection of a predominant antigen receptor gene rearrangement that represents the outgrowth of a predominant lymphocyte clone. In the correct clinical and pathologic context, the presence of a predominant clonal lymphocyte population is consistent with lymphoma or lymphoid leukemia. Southern blots were traditionally the method used to detect a predominant antigen receptor gene rearrangement. Multiplex PCR-based B-cell and T-cell clonality assays are replacing Southern-blot-based assays because they are robust, faster to perform, and easier to interpret. Unlike Southern blots, PCR-based assays are amenable to formalin-fixed, paraffin-embedded (FFPE) tissues. PCR-based assays amplify the DNA between primers that target the conserved framework (FR) and joining (J) sequence regions within the lymphoid antigen receptor genes. These conserved regions flank the V-J gene segments that randomly rearrange during the maturation of B-lymphocytes and T-lymphocytes.

Random genetic rearrangement of the V-J segments occurs in the immunoglobulin heavy chain gene (*IGH*) and the kappa and lambda immunoglobulin light chain genes in prefollicular center B-cells. Random genetic rearrangement of the V-J segments also occurs in the alpha (*TCRA*), beta (*TCRB*), gamma (*TCRG*), and delta (*TCRD*) T-cell receptor genes during thymic T-cell development. Antigen receptor gene rearrangement occurs in a sequential fashion one allele at a time. If the first allele fails to successfully recombine (produces a nonfunctional protein product), then the other allele will undergo rearrangement. It is important to remember that the nonfunctional rearrangement can still be detected by the clonality assay. Because of the sequential manner in which these genes rearrange, a mature lymphocyte may carry one (mono-allelic clone) or two (bi-allelic clone) rearranged antigen receptor genes of a given type. Each maturing B-cell or T-cell has either one or two gene-specific V-J rearrangements unique in both sequence and length. Therefore, when the V-J region is amplified, one or two unique amplicons will be produced per cell. If the population of lymphocytes being examined is a lymphoma, then one or two unique amplicons will dominate the population. In contrast, if the population is a normal polyclonal lymphoid infiltrate, then there will be a Gaussian distribution of many different sized amplicons centered around a statistically favored, average-sized rearrangement (Figure 28.4).

Because the antigen receptors are polymorphic and subject to mutagenesis as part of the normal rearrangement process, multiple primer sets are used to increase the ability of the assay to detect the majority of possible V-J rearrangements. Each set of multiplex primers has a defined valid amplicon size range. After antigen exposure, germinal center B-cells undergo somatic hypermutation of the V regions, and this may affect the binding of the V-region specific (FR) primers and prevent the detection of lymphoma clones arising from the germinal center or postgerminal center B-cell compartments. Primer sets that target *IGH* D-J rearrangements are less susceptible to somatic hypermutation and allow the detection of more terminally differentiated B-cell malignancies.

Validation of B-Cell and T-Cell Clonality Assays

PCR-based clonality assays are routinely used by many laboratories. However, prior to 2004, referral lab PCR-based clonality assays used nonstandardized, lab-specific PCR primer sets and methods, making proficiency testing and lab performance comparisons difficult. In addition, many of these institution-specific B-cell and T-cell clonality assays are not comprehensive and target a limited number of possible V-J rearrangements. To address this concern, we used the optimized InVivoScribe (IVS) multiplex PCR primer master mixes. The IVS master mixes have been standardized and extensively validated by testing more than 400 clinical samples using the WHO Classification scheme. The validation was done at more than 30 prominent independent testing centers throughout Europe in a collaborative study known as the BIOMED-2 Concerted Action [19].

The *TCRG* locus is on chromosome 7 (7q14) and includes 14 V gene segments divided into 4 subgroups. Six of the 14 V segments are functional, 3 are open reading frames, and 5 are pseudogenes. The 200 kilobase pair *TCRG* region also contains 5 J segments and 2 C genes. Although most mature T-cells express the alpha/beta TCR, the gamma/delta TCR genes rearrange first and are carried by almost all alpha/beta T-cells. The IVS *TCRG* assay uses only 2 master mixes and is able to detect 89% of all T-cell lymphomas tested in the BIOMED-2 Concerted Action. Mastermix tube A targets the V gamma 1–8 + 10 and most J gamma gene segments. Mastermix tube B targets V gamma 9 + 11 and most J gamma segments. The included positive and negative kit controls are used to assess each of the master mixes. The specimen control size standard is a separate master mix included with the kit that evaluates each patient DNA sample by targeting six different housekeeping genes potentially producing amplicons of 84, 96, 200, 300, 400, and 600 base pairs. Under ideal conditions and using capillary electrophoresis to analyze the amplicons, the *TCRG* assay can resolve a 1% clonal population in a polyclonal background.

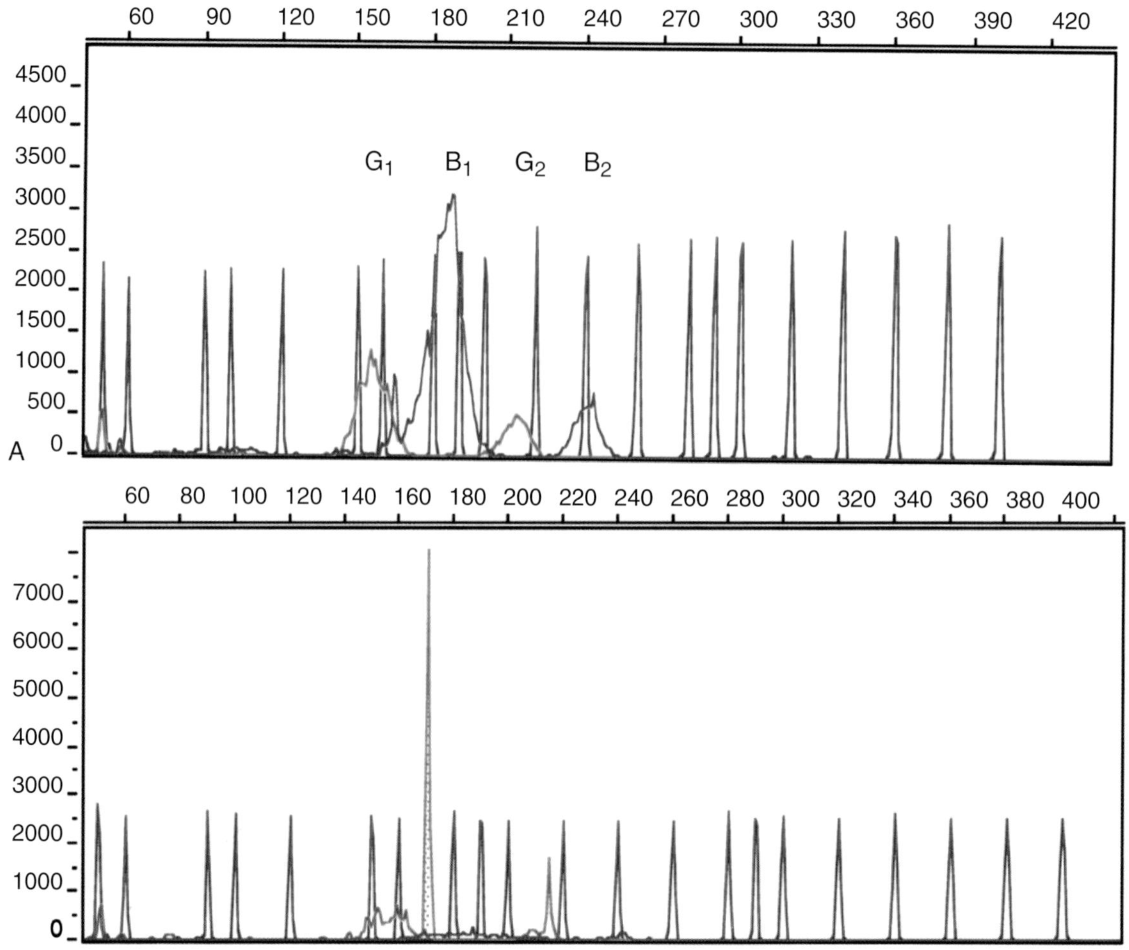

B

Dye/Sample Peak	Minutes	Size	Peak Height	Peak Area	Data Point
G, 1	16.02	170.59	8065	66810	4369

Figure 28.4 T-cell receptor gamma chain PCR assay for clonality. (**A**) polyclonal reactive T-cell proliferation pattern. A polyclonal population of T-cells with randomly rearranged T-cell receptor gamma chain genes produces a normal or Gaussian distribution of fluorescently labeled PCR products from each primer pair in the multiplex reaction. This produces 4 "bell-shaped curves" that represent the valid size range for an individual primer pair. Two of the valid size ranges are green and two are blue; G_1, B_1, G_2, B_2. The red peaks represent size standards. (**B**) clonal T-cell proliferation pattern. A clonal T-cell proliferation results in a relative dominance of a single T-cell receptor gamma chain gene producing a predominant spike of a discrete size on the corresponding electropherogram. Data were obtained using an ABI 3100 capillary electrophoresis system and ABI Prism Software.

The *IGH* locus is on chromosome 14 (14q32.3) and includes more than 66 V, 27 D, and 6 J gene segments. The IVS *IGH* assay uses 5 master mixes and is able to detect 92% of all B-cell lymphomas tested in the BIOMED-2 Concerted Action validation study. Master mixes A, B, and C target the framework (FR) 1, 2, and 3 regions of the V gene segments, respectively, and all J gene segments. Master mixes D and E target the D and J regions. The *IGH* locus rearranges the D and J segments first during maturation, followed by the addition of a random V segment to the previously rearranged D-J fusion product. In addition, the D segments are less likely to be mutated by somatic mutation during postgerminal center development. Because the primers in tubes D and E target D-J regions, these tubes are more likely to detect clonal *IGH* rearrangements in lymphoma cells that arose either very early or late in B-cell ontogeny. Positive, negative, and specimen extraction controls are included and run on each master mix and specimen, respectively. The specimen control size standard is used to assess the integrity of the DNA extracted from the specimen and is identical to the specimen control size standard described previously for the *TCRG* kit. The specimen control master mix is useful in determining the amount of DNA target molecule degradation that has occurred during fixation and DNA extraction. Capillary electrophoresis (CE) using the

ABI 3100 Avant Genetic Analyzer was better at determining the size of predominant amplicons than both agarose and polyacrylamide gel systems during method development. Under ideal conditions and using CE to analyze the amplicons, the IGH assay can resolve a 1% clonal population in a polyclonal background.

Genomic DNA is isolated from whole blood, body fluids, fresh tissues, and formalin-fixed paraffin-embedded tissue blocks, using the appropriate Qiagen DNA isolation protocol. A slightly modified Qiagen DNA isolation protocol for FFPE samples was selected after a comparison to three other DNA isolation protocols during validation. After DNA isolation, multiple dilutions are tested for beta-globin amplification by PCR to assess DNA quality prior to analysis using the IVS system. Samples that fail to amplify beta-globin at any concentration are quantified by UV spectrophotometry. Approximately 100 to 200 ng of input DNA are ideal for *TCRG* or *IGH* PCR reactions. Samples that are undetectable by UV spectrophotometry and fail beta-globin PCR should be reported as quantity insufficient and retested by increasing the amount of starting material by 2- to 3-fold if possible. Samples that fail beta-globin PCR but contain sufficient quantities of DNA are most likely nonamplifiable due to inhibitors and should be reported as nonamplifiable due to PCR inhibitors in the sample. Additional dilutions could be retested by beta-globin PCR if desired. The least dilute DNA that gives a strong beta-globin signal should be used for the InVivoScribe B-cell or T-cell clonality assay. Each of the products of the IVS PCR is combined with 0.5 mL ROX400HD size standards, denatured at 95°C in formamide and electrophoresed through a capillary that contains the POP4 matrix. The amplified fragments are detected by the 6FAM, HEX, or NED fluorescent labels on the conserved, downstream *IGH* or *TCRG* J-region PCR primers. The size standards are detected by the ROX label. After assay validity is confirmed by checking all controls, the results are analyzed and a report is generated.

After the method was established, a variety of clinically and histopathologically diagnostic cases of B-cell and T-cell lymphomas were used to evaluate the performance of the method. This laboratory modified method was able to detect 92% of B-cell lymphomas and 85% of T-cell lymphomas used in the validation set. An ideal QC control material is not available for this assay. The ideal QC material would consist of a high and low positive FFPE QC and a negative FFPE QC. FFPE QC material would control for the DNA extraction in addition to the PCR and interpretation phases of this method. The currently available QC and PT control materials do not assess the DNA extraction phase of this test.

Preanalytical Considerations

Special considerations regarding specimen type and handling are important for this method. One of the most common specimens submitted for *IGH* or *TCRG* gene rearrangements is FFPE tissue blocks. PCR inhibitors are commonly found with this specimen type. Xylene, heme, and excessive divalent cation chelators are also inhibitory and need to be removed during DNA extraction procedures. The fixative used prior to paraffin embedment is critical for optimum results because formalin of inferior quality can introduce PCR inhibitors or prevent the extraction of amplifiable DNA. Fresh, high-quality, 10% neutral-buffered formalin is acceptable, and this should be communicated to the pathologists who will be submitting specimens for gene rearrangement studies. Fixatives that contain heavy metals such as Hg (Zenkers and B5) should not be used for PCR. Decalcified bone marrow biopsy specimens are unacceptable as are specimens that were fixed with strong acid-based fixatives like Bouin's solution. Tissues should be fixed at room temperature using proper histologic standards. Fixation depends on formaldehyde-mediated cross-linking of proteins and nucleic acids. Excessive cross-linking will inhibit DNA extraction and shorten the average size of the recovered DNA molecules; therefore, overfixation should be avoided. Fixation using the approved fixatives at room temperature progresses at a rate of approximately 1 mm/hour. Typical 0.5 mm thick tissue sections should fix for a minimum of 6 hours and a maximum of 48 hours. After 48 hours in fixative, tissues can be stored in 50% ethanol for up to 2 weeks prior to processing and embedding. The volume of fixative used should be at least 10 times the volume of tissue to achieve optimal results. Some laboratories use razor blades to gouge tissue out of the block, but this destroys the block and prevents any additional studies that may be necessary in the future. The best way to obtain FFPE samples for PCR is to cut 4 micron thick sections on the microtome. Histotechnologists should cut the sections for PCR first thing in the morning with a fresh microtome blade and new water. Several initial sections are discarded to remove any possible contaminating DNA on the surface of the block. One to five sections are obtained, depending on the size and type of tissue sample and the relative proportion of the tissue that is involved by the suspicious lymphoid infiltrate. PCR of lymphoid infiltrates in FFPE skin or brain tissue blocks are somewhat refractory to PCR analysis and require more sections for analysis. Skin tissues are more likely to be overfixed and yield more fragmented DNA. Brain tissues could be problematic due to the higher lipid content than most tissues submitted for lymphocyte gene rearrangement studies.

To snap freeze fresh tissues, one should cut them into 1 mm^3 to 3 mm^3 pieces, place them into an appropriately labeled cryovial, snap freeze them in liquid nitrogen, and then store them frozen at −70°C until processed. Specimens snap frozen at a different location should be shipped on dry ice to arrive during normal business hours and be stored at −70°C until processed. Whole blood specimens should be at least 200 μL EDTA or citrated blood, stored at 4°C for up to 3 days before DNA extraction. Bone marrow aspirates should be at least 200 μL EDTA or citrated bone marrow, stored at 4°C for up to 3 days before DNA extraction.

CSF specimens should be at least 200 μL depending on cellularity. All CSF samples should be collected in an appropriate sterile container and stored at 4°C for up to 24 hours or at −20°C for long-term storage. Other sample types, such as stained or unstained tissue on slides, may be accepted with the approval of the laboratory director. Extracted DNA samples should be stored at −20°C for a minimum of 2 months after analysis is complete.

Postanalytical Considerations

The results of this test must be interpreted in the context of the other clinicopathologic data for the specimen. The detection of a clonal lymphoid proliferation in a tissue sample does not necessarily indicate a diagnosis of lymphoma or leukemia. On the other hand, the failure to detect a clonal lymphoid proliferation does not necessarily rule out a lymphoma/leukemia. The test is one puzzle piece that must be integrated with other clinicopathologic data before rendering a diagnosis. The molecular B-cell and T-cell clonality assay is an adjunct ordered by pathologists for cases that are suspicious but not obviously lymphoma/leukemia, and the results of this test do not stand alone.

Although helpful guidelines for the interpretation of results are provided by the manufacturer, this assay is not an out-of-the-box assay. After method development, an important part of validation is the development of interpretive guidelines that are appropriate for the established method, specimen types, and the patient population. An important consideration is how the results will be interpreted by the ordering group of pathologists and oncologists. Specimen handling, DNA extraction procedures, and fragment analysis procedures will affect the final results from this assay, and this will affect the ability to detect a clonal proliferation in a polyclonal background. Our laboratory has established a set of interpretive guidelines that include a required minimum peak height for positive peaks and positive peak to background peak ratio values that are appropriate for our method and reduce the number of false positives. The details of our interpretive guidelines are beyond the scope of this chapter and are described in [20–22].

Results are interpreted and reported to the ordering pathologist in an interpretive report (CoPath) format. The results are interpreted as (i) Positive for a clonal lymphoid proliferation; (ii) Suspicious but not diagnostic for a clonal lymphoid proliferation (with a recommendation to repeat the assay if clinical concerns persist); (iii) Oligoclonal reactive pattern (commonly seen in the peripheral blood of elderly patients with rheumatologic disease or viral infections); (iv) Polyclonal lymphoid population; (v) Insufficient material/Inadequate quality. The reports also contain information on the pertinent clinical and laboratory history, the specimen submitted for PCR, and the relevant surgical pathology case if appropriate. Suspicious results are usually repeated for confirmation, especially *TCRG* studies.

Benefit/Risk Assessment (Clinical Utility)

In many cases of atypical lymphoid proliferations suspicious for lymphoma, the morphologic, immunohistochemical, flow cytometric, or clinical features are equivocal or contradictory. Detection of a predominant antigen receptor gene rearrangement (detection of a T-cell or B-cell clone) can support a diagnosis of lymphoma. Early diagnosis and appropriate treatment of lymphoma can prolong patient survival and in many cases result in a complete remission or cure. Because the morphology of many lymphomas is indistinguishable from benign lymphoid hyperplasia, ancillary tests that demonstrate an atypical clonal lymphocyte proliferation are essential in early diagnosis and subsequent treatment of many lymphoma cases. Multiplex PCR assays for clonality have become the standard of practice in lymphoma diagnosis and may eventually be used on all cases of atypical lymphoid proliferations to either rule in or rule out lymphoma earlier. For those cases that are obviously lymphoma, molecular characterization of the specific lymphoma cell clone may also prove useful in patient follow-up after treatment to screen for residual or recurrent disease. In addition many of the morphologic subtypes of lymphoma have not been extensively characterized from the molecular diagnostics standpoint. For example, for biopsy cases that are suspicious but not diagnostic for lymphoma, detection of a clonal B-cell or T-cell population predicts lymphoma behavior or patient outcome. The diagnostic tools that allow these types of correlative studies have only recently become available, and this assay represents one of the assays that will allow these kinds of questions to be considered. In the future, B-cell and T-cell clonality assays will become a standard part of a lymphoma workup and may alter clinical outcomes by providing a more comprehensive earlier diagnosis of lymphoma and preventing unnecessary treatment of nonlymphomas.

Another important consideration is the improved performance and diagnostic utility of the newer generation of multiplex PCR assays since 2004. Many descriptions on the utility of PCR-based clonality assays refer to multiplex PCR assays that cannot detect all possible *IGH* or *TCRG* gene rearrangements because of suboptimal primer sets. This can increase the number of false negatives because the PCR primers that could recognize the clonal gene rearrangement are missing. In addition, oligoclonal reactive proliferations may appear to be clonal because a limiting number of primers may not detect all of the members of the oligoclonal proliferation. To address this issue, we have initiated an investigation of the diagnostic utility of the IVS assay at our institution over the past 4 years [23].

FUTURE APPLICATIONS

Beyond the predictable advancements in technology, several new areas of molecular applications have appeared in the past year which are indicative of increased emphasis on molecular techniques in clinical medicine: (i) array-based comparative genomic hybridization (aCGH),

(ii) gene expression profiling, and (iii) quantitative reverse-transciptase PCR (qRT-PCR). Additional applications that deserve mention include (a) tissue-based pharmacogenomics, (b) detection of fetal DNA markers in maternal serum, and (c) personalized medicine. All of these developments are dependent on the appropriate balance between regulatory control and reimbursement issues, without falling victim to regulatory stagnation.

Array-Based Comparative Genomic Hybridization

Molecular cytogenetics has evolved since fluoresence *in situ* hybridization (FISH) became a routine clinical procedure. Recently, a research technique, called comparative genomic hybridization (CGH), has been modified using microarray technology to create array-based CGH. Without extensive details, a CGH scans the entire genome for imbalances, including duplications, deletions, and copy-number variants, creating another molecular cytogenetic tool. Traditional karyotypic studies, limited by the resolution of all microscopic procedures, are complemented and sometimes replaced by microarrays which increase resolution by 1000-fold or greater. The impact of this technology on both constitutional and tissue cytogenetics cannot be overestimated.

Gene Expression Profiling

RNA expression can be assessed in tissue by an array-based technique. While the information can be overwhelming, sophisticated software has been applied to arrays that assess ~25,000 genes. For example, breast cancer samples have been profiled and reveal specific gene expression profiles that correlate with prognosis and/or response to specific therapies. Once these specific genes are identified, unique quantitative reverse-transcriptase PCR assays replace the expression arrays and provide useful quantitative data which similarly predict disease progression and therapeutic responses at a significantly reduced cost. Currently, gene expression studies are performed by commercial or clinical research laboratories.

Tissue-Based Pharmacogenomics

Pharmacogenomics refers to using specific genomic data to predict response to therapeutic drugs. When applied to tissue, particularly malignant tissue, similar predictions can be made. For example, inactivating members of the epidermal growth factor receptor (EGFR) family is a new approach to treating epithelial malignancies. One example involves the Her-2 oncogene and response to Herceptin, a monoclonal antibody directed against Her-2 and used to treat breast cancer patients. Patients with amplified Her-2 were found to respond more favorably to Herceptin than patients without amplification. Another example involves Erbitux, a monoclonal antibody to EGFR used to treat gastrointestinal adenocarcinomas. In this case, mutations in the *kRAS* oncogene predict a less favorable response, particularly in patients with colorectal adenocarcinoma. Molecular tests can be used to evaluate these oncogenes and predict therapeutic outcomes.

Fetal DNA in Maternal Serum

Prenatal diagnosis conventionally involves obtaining fetal cells by either amniocentesis or chorionic villus sampling. These invasive techniques are associated with some risk to the developing fetus and discomfort to the mother. Through examination of maternal serum, fetal DNA can be detected that provides similar information, without the fetal risk or maternal discomfort. Analytical procedures have been reported for detection of trisomy 21 (Down syndrome), trisomy 18, and trisomy 13. Commercial laboratories are currently evaluating these procedures.

Personalized Medicine

Many of the new applications discussed in the preceding sections are collectively referred to as *personalized medicine*. In broad terms, personalized medicine uses genomic and/or gene expression data to predict outcome or select specific therapeutic modalities. Of course, this information must be interpreted with patient and family history and other clinical and laboratory data. An extension of personalized medicine is the storage of biological material, including sperm, cord blood, and DNA, where applicable, for future considerations or comparisons between normalcy and disease.

REFERENCES

1. Romero JR. Reverse-transcription polymerase chain reaction detection of the enteroviruses. *Arch Pathol Lab Med.* 1999;123: 1161–1169.
2. Ruckert R. Picornaviridae and their replication. In: Fields BN and Knipe DM, eds. *Fields Virology*, 2nd ed. New York: Raven Press; 1990.
3. Rotbart HA, Sawyer MH, Fast S, et al. Diagnosis of enteroviral meningitis by using PCR with a colorimetric microwell detection assay. *J Clin Microbiol.* 1994;32:2590–2592.
4. Young PP, Buller RS, Storch GA. Evaluation of a commercial DNA enzyme immunoassay for detection of enterovirus reverse transcription-PCR products amplified from cerebrospinal fluid specimens. *J Clin Microbiol.* 2000;38:4260–4261.
5. Pozo F, Casas I, Tenorio A, et al. Evaluation of a commercially available reverse transcription-PCR assay for diagnosis of enteroviral infection in archival and prospectively collected cerebrospinal fluid specimens. *J Clin Microbiol.* 1998;36:1741–1745.
6. Wittwer CT, Ririe KM, Andrew RV, et al. The LightCycler: A microvolume multisample fluorimeter with rapid temperature control. *Biotechniques.* 1997;22:176–181.
7. Watkins-Riedel T, Woegerbauer M, Hollemann D, et al. Rapid diagnosis of enterovirus infections by real-time PCR on the LightCycler using the TaqMan format. *Diagn Microbiol Infect Dis.* 2002; 42:99–105.
8. Neumaier M, Braun A, Wagener C. Fundamentals of quality assessment of molecular amplification methods in clinical diagnostics. International Federation of Clinical Chemistry Scientific Division Committee on Molecular Biology Techniques. *Clin Chem.* 1998;44:12–26.

9. Haverstick DM. Specimen collection and processing. In: Bruns DE, Ashwood ER, Burtis CA, eds. *Fundamentals of Molecular Diagnostics*. St. Louis, MO: Elsevier Saunders; 2007.
10. Beutler E, Gelbart T, Kuhl W. Interference of heparin with the polymerase chain reaction. *Biotechniques*. 1990;9:166.
11. Jung R, Lubcke C, Wagener C, et al. Reversal of RT-PCR inhibition observed in heparinized clinical specimens. *Biotechniques*. 1997;23:24, 26, 28.
12. Bruns DE, Ashwood ER, Burtis CA, eds. *Fundamentals of Molecular Diagnostics*. St. Louis, MO: Elsevier Saunders; 2007.
13. Shindo M, Di Bisceglie AM, Cheung L, et al. Decrease in serum hepatitis C viral RNA during alpha-interferon therapy for chronic hepatitis C. *Ann Intern Med*. 1991;115:700–704.
14. Germer JJ & Zein NN. Advances in the molecular diagnosis of hepatitis C and their clinical implications. *Mayo Clin Proc*. 2001; 76:911–920.
15. Bullock GC, Bruns DE, Haverstick DM. Hepatitis C genotype determination by melting curve analysis with a single set of fluorescence resonance energy transfer probes. *Clin Chem*. 2002; 48:2147–2154.
16. Haverstick DM, Bullock GC, Bruns DE. Genotyping of hepatitis C virus by melting curve analysis: Analytical characteristics and performance. *Clin Chem*. 2004;50:2405–2407.
17. Terrault NA, Pawlotsky JM, McHutchison J, et al. Clinical utility of viral load measurements in individuals with chronic hepatitis C infection on antiviral therapy. *J Viral Hepat*. 2005;12:465–472.
18. NIH. National Institutes of Health Consensus Development Conference Statement: Management of hepatitis C 2002 (June 10–12, 2002). *Gastroenterology*. 2002;123:2082–2099.
19. van Dongen JJ, Langerak AW, Bruggemann M, et al. Design and standardization of PCR primers and protocols for detection of clonal immunoglobulin and T-cell receptor gene recombinations in suspect lymphoproliferations: Report of the BIOMED-2 Concerted Action BMH4–CT98–3936. *Leukemia*. 2003;17: 2257–2317.
20. Bullock GC, SK, Haverstick DM, et al. Technical validation and adaptation of the InVivoScribe TCR-gamma and IgH gene rearrangement assays for detection of clonal lymphoid populations (Abstract). *Mod Pathol*. 2006;19(suppl 1):217A.
21. Chute DJ, Cousar JB, Mahadevan MS, et al. Detection of immunoglobulin heavy chain gene rearrangements in classic Hodgkin lymphoma using commercially available BIOMED-2 primers. *Diagn Mol Pathol*. 2008;17:65–72.
22. Mikesh LM, Crowe SE, Bullock GC, et al. Celiac disease refractory to a gluten-free diet? *Clin Chem*. 2008;54:441–444.
23. Shumaker NR, BG, Silverman LM, Cousar JB. The clinical utility of PCR-based clonality assessment in diagnostic hematopathology. (Abstract). *Modern Pathology*. 2008;21:274A.
24. Wilson JA, et al. Consensus characterization of 16 FMR1 Reference Materials: A consortium study. *Journal of Molecular Diagnostics*. 2008;10:2–12.
25. Jennings L, Van Deerlin VM, Gulley MA. Principles and Practice for Validation of Clinical Molecular Pathology Tests. *Archives of Pathology and Laboratory Medicine*. [in press].
26. Fu YH, Kuhl DP, Pizzuti A, et al. Variation of the CGG repeat at the fragile X site results in genetic instability: Resolution of the Sherman paradox. *Cell* 1991;67:1047–1058.

Chapter

29

Molecular Assessment of Human Disease in the Clinical Laboratory

Joel A. Lefferts . Gregory J. Tsongalis

INTRODUCTION

New technologies that were once labeled *For Research Use Only* have made a rapid transition into the clinical laboratory as user-developed assays (UDAs), analyte-specific reagents (ASRs), or FDA-cleared/approved diagnostic assays. Molecular diagnostics has spanned the entire spectrum of applications that will be performed on a routine clinical basis in most hospital laboratories. While the ability for clinical laboratories to detect human genetic variation has historically been limited to a rather small number of traditional genetic diseases where no clinical laboratory testing was ever available, our current understanding of many disease processes at a molecular level has expanded our testing capabilities to both human and nonhuman applications. The identification of numerous new genes and disease-causing mutations, as well as benign polymorphisms that may influence a specific phenotype, has created a rapid demand for molecular diagnostic testing.

There is a growing need for clinical laboratories to provide high-quality nucleic acid-based tests that have significant clinical relevance, excellent performance characteristics, and shortened turnaround-times. This need was initially driven by the completion of the Human Genome Project, which identified thousands of genes, millions of human single nucleotide polymorphisms (SNPs) and culminated in disease associations. An unprecedented demand has now been placed on the clinical laboratory community to provide increased diagnostic testing for rapid and accurate identification and interrogation of genomic targets.

Molecular technologies first entered the clinical laboratories in the early 1980s as manual, labor-intensive procedures that required a working knowledge of chemistry and molecular biology, as well as an exceptional skill set. Testing capabilities have moved very quickly away from labor intense, highly complex, and specialized procedures to more user-friendly, semiautomated procedures [1]. This transition began with testing for relatively high-volume infectious diseases such as Chlamydia trachomatis, HPV, HIV-1, HCV, and others. Much of this was championed by the availability of FDA-cleared kits and higher throughput, semiautomated instruments such as the Abbott LCX and Roche Cobas Amplicor systems. These two instruments that were based on the ligase chain reaction (LCR) and polymerase chain reaction (PCR), respectively, fully automated the detection steps of the assays with manual specimen processing and amplification reactions.

In the early 1980s, the Southern blot was the routine method used in the few laboratories performing clinical testing for a variety of clinical applications, even though the turnaround time was in excess of 2–3 weeks. The commercial availability of restriction endonucleases and various agarose matrices helped in making this technique routine. The PCR revolutionized blotting technologies and detection limits by offering increased sensitivity and the much-needed shortened turnaround time for clinical result availability. Initial PCR thermal cycling occurred in different temperature waterbaths, until the discovery of Taq polymerase, the first commercially available thermostable polymerase. Programmable thermal cyclers with reaction vessels containing all of the amplification reagents have gone through several generations to current instrumentation. Many modifications to the PCR have been introduced, but none as significant as real-time capability [2–5]. The elimination of post-PCR detection systems and the ability to perform the entire assay in a closed vessel had significant advantages for the clinical laboratory. Various detection chemistries for

Molecular Pathology © 2009, Elsevier, Inc. All Rights Reserved.

real-time PCR were rapidly introduced, and many of the older detection methods (including gel electrophoresis, ASO blots, and others) vanished from the laboratory.

More recently, real-time PCR has become a method of choice for most molecular diagnostics laboratories. This modification of the traditional PCR allows for the simultaneous amplification and detection of amplified nucleic acid targets for both qualitative and quantitative applications. The main advantages of real-time PCR are the speed with which samples can be analyzed, as there are no post-PCR processing steps required, and the closed-tube nature of the technology that helps to prevent contamination. The analysis of results via amplification curve and melt curve analysis is very simple and contributes to its being a much faster method for delivering PCR results.

DNA sequencing technologies, fragment sizing, and loss of heterozygosity studies have benefited from automated capillary electrophoresis instrumentation. Several forms of microarrays are currently available as post-PCR detection mechanisms for multiplexed analysis of SNPs, gene expression, and pathogen detection. Today, molecular diagnostic laboratories are well equipped with various instruments and technologies to meet the challenges set forth by molecular pathology needs (Table 29.1).

This chapter will introduce the new molecular diagnostic paradigm and several new directions for the molecular assessment of human diseases.

THE CURRENT MOLECULAR INFECTIOUS DISEASE PARADIGM

Since early molecular technologies began entering the clinical laboratory, clinical applications were few in number, techniques were labor intense, and turnaround times were in excess of 2 weeks. Nonetheless, there was significant clinical utility for Southern blot transfer assays such as the IgH and TCR gene rearrangements, Fragile X Syndrome, and linkage analysis. These studies were qualitative interrogations of specific genetic sequences to determine the presence or absence of a disease-associated abnormality, directly or indirectly. It was not until molecular applications for infectious disease testing were being developed that the need for quantitative and resistance testing became apparent.

Molecular infectious disease testing posed the first comprehensive testing paradigm using molecular techniques. That is, clinical applications of molecular diagnostics to infectious diseases included the molecular trinity: qualitative testing, quantitative testing, and resistance genotyping [6–10]. This paradigm spanned the spectrum of current molecular capabilities and has provided significant insight to the nuances of molecular testing for other disciplines (Table 29.2). Molecular diagnostic testing for infectious disease applications continues to be the highest volume of testing being performed in clinical laboratories (Table 29.3).

While accurate and timely diagnosis of infectious diseases is essential for proper patient management, traditional testing methods for many pathogens did not allow for rapid turnaround time. Prompt detection of the microbial pathogen would allow providers to institute adequate measures to interrupt transmission to a susceptible hospital or community population and/or to begin proper therapeutic management of the patient. Unlike traditional microbial techniques, molecular techniques offered higher sensitivity and specificity with turnaround times that were unprecedented once amplified technologies were introduced to the clinical laboratory. However, a qualitative test to identify the presence or absence of an organism did not meet all of the necessary clinical needs, especially when trying to monitor therapeutic efficacy. Clinically, it was more useful to identify how much of an organism was present versus if it was or was not present. This would not only allow for the use of the assay as a therapeutic monitoring tool, but also help distinguish a clinically significant infection from a latent or resolving infection. Quantitative assays, initially for viral load testing in HIV-1 infected patients, were developed to monitor the success of newly introduced

Table 29.1 Current Technologies Being Used in the Molecular Pathology Laboratory at the Dartmouth Hitchcock Medical Center

Technology	Platform/Instrument
DNA/RNA extraction	Manual Spin column Qiagen EZ1 robot
Fluorescence *in situ* hybridization (FISH)	Hybrite oven Molecular imaging system
bDNA	Siemens System 340
Hybrid capture	Digene HCII System
Real-time PCR	Cepheid GeneXpert ABI 7500 Cepheid Smartcycler Roche Taq48
Capillary electrophoresis	Beckman CEQ Beckman Vidiera
Traditional PCR	MJ Research DNA Engines
Microarray	Luminex Superarray Nanosphere

Table 29.2 Nuances of Molecular Diagnostic Testing

Preanalytical	Analytical	Postanalytical
1. Specimen type	1. Technology used	1. LIS
2. Specimen collection device	2. Type of assay: qualitative vs quantitative	2. Result reporting and interpretation
3. Specimen transport	3. Controls	3. Consultation
4. DNA vs RNA	4. Quality assurance	4. Standardization

Table 29.3 Molecular Infectious Disease Tests That Are Currently FDA Cleared and Being Performed in Clinical Laboratories

Bacteria	Viruses
Bacillus anthracis	Avian flu
Candida albicans	Cytomegalovirus
Chlamydia trachomatis	Enterovirus
Enterococcous faecalis	HBV (quantitative)
Francisella tularensis	Hepatitis C virus (qualitative and quantitative)
Gardnerella, Trichomonas, and Candida spp.	HIV drug resistance
Group A Streptococci	HIV (quantitative)
Group B Streptococci	HBV/HCV/HIV blood screening assay
Legionella pneumophila	Human Papillomavirus
Methicillin-resistant Staphylococcus aureus	Respiratory viral panel
Mycobacterium tuberculosis	West Nile virus
Mycobacterium spp.	
Staphylococcus aureus	

protease and reverse transcriptase inhibitors. This application for HIV-1 viral load testing also represented the first true need for a companion molecular diagnostic assay, not in the sense of prescribing a therapeutic but rather in monitoring its efficacy effectively. To do so, PCR became quantitative, and other quantitative chemistries such as branched DNA (bDNA), transcription-mediated amplification (TMA), and nucleic acid sequenced-based amplification (NASBA) were born. More recently, real-time PCR capabilities have been further developed and routinely include more accurate methods for quantification of target sequences.

HIV-1 once again became the poster child for a novel molecular-based application when it was realized that the virus often mutated in response to therapy. The need to detect viral genome mutations also represented the first major routine application of sequencing methods in the clinical laboratory. These assays would predict response to regimens of antiretroviral therapy and help guide the practitioner's choices of therapy. In this sense, HIV-1 resistance testing also represented the first routine application of personalized medicine. This application resulted in the need to address the implementation and maintenance of accurate databases as well as deal with specimens co-infected with different mutant viruses. From a single disease moiety, AIDS, a molecular paradigm evolved that would set the stage and expectations for all future molecular diagnostic applications.

A NEW PARADIGM FOR MOLECULAR DIAGNOSTIC APPLICATIONS

Our ability to test individuals for many types of human and nonhuman genetic alterations in a clinical setting is expanding at an enormous rate. This is in part due to our better understanding of human disease processes, advanced technologies, and expansion of the molecular trinity—qualitative, quantitative, and resistance genotyping—for more common diseases such as cancer. Human cancer represents a complex set of abnormal cellular processes that culminates in widespread systemic disease. Add to this the rapid identification of new biomarkers, introduction of novel therapeutics and progressive management strategies, and human cancer now becomes a model for the new molecular diagnostic paradigm demanding the three phases of testing (qualitative, quantitative, and resistance genotyping) capabilities of the clinical laboratory.

Phase I: Qualitative Analysis

Qualitative testing results in the determination of presence or absence of a particular genetic sequence. This target can be associated with an infectious agent, can be some human genomic alteration associated with a clinical phenotype, or some benign polymorphism associated with personalized medicine traits. Using the numerous tools available to the Molecular Pathology Laboratory, performing this testing has become routine for many applications, including such targets as parvovirus B19, mutation screening for cystic fibrosis, and *CYP2D6* genotyping (Table 29.4).

One such oncology target is the *JAK2* V617F mutation. V617F is a somatic point mutation in a conserved region of the autoinhibitory domain of the Janus kinase 2 tyrosine kinase that is associated with chronic myeloproliferative diseases (CMPD) [11–13]. The G>T mutation results in a substitution of phenylalanine for valine at position 617. This single point mutation leads to loss of JAK2 autoinhibition and constitutive activation of the kinase.

CMPD is a group of hematopoietic proliferations characterized by increased circulating red blood cells, extramedullary hematopoiesis, hyperplasia of hematopoietic cell lineages, and bone marrow dysplasia and fibrosis. CMPDs include chronic myelogenous leukemia, polycythemia vera, essential thrombocythemia, and chronic idiopathic myelofibrosis [13]. The diagnosis of a CMPD is made from clinical and morphologic features which can often be confounded by the stage of disease. Numerous molecular methods have been developed for identifying this mutation including PCR-RFLP, allele-specific PCR, DNA sequencing, and real-time PCR (Figure 29.1). Identification of the *JAK2* V617F mutation in CMPD provides a significant molecular biomarker as it is found in almost all cases of polycythemia vera and can be used to rule out other CMPDs [14]. Due to the lack of accurate diagnostic tests for the CMPDs, the utilization of the *JAK2* mutation assay has increased significantly and has in many cases become part of a testing algorithm for the CMPDs.

Phase II: Quantitative Analysis

Quantitative molecular testing refers to our ability to enumerate the target sequence(s) present in any given tissue type. As with infectious diseases, this application

Table 29.4 Qualitative Molecular Diagnostic Tests that are Routinely Performed in the Dartmouth Molecular Pathology Laboratory

Genetics	Hematology	Infectious	Oncology	Personalized Medicine
AAT	*BCL-2*	B. holmesii	*BRAF*	*CYP2C9*
ApoE	*FLT3*	B. parapertussis	*MSI*	*CYP2D6*
CFTR	*IGH*	B. pertussis	*SLN*	*EGFR*
Factor II	*JAK2*	Brucella sp.		*GSTP1*
Factor V	*TCR*	GBS		HCV Genotype
HFE		HHV6		*K-ras*
MTHFR		HPV		*UGT1A1*
SNCA Rep1		MRSA		*VKORC1*
		Parvo B19		

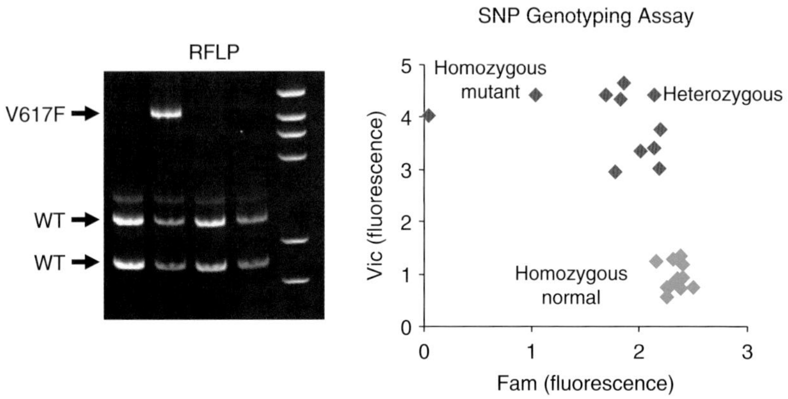

Figure 29.1 Detection of the *JAK2* V617F mutation using PCR-RFLP (left) and real-time PCR allelic discrimination (right).

has resulted mainly from the need to monitor therapy and determine eligibility for therapy. Commonly performed quantitative molecular assays are listed in Table 29.5. Typical of the diagnosis of a human cancer is the identification of numerous protein targets through immunohistochemical techniques that help pathologists in their diagnostic decision making. In addition, molecular markers have become part of this armamentarium not so much for the diagnosis of disease, but more so for the selection and monitoring of therapies. The first example of this type of application was the quantitative detection of ERBB2 (v-erb-b2 erythroblastic leukemia viral oncogene homolog 2) in human breast cancers.

Table 29.5 Quantitative Molecular Diagnostic Tests That are Routinely Performed in the Dartmouth Molecular Pathology Laboratory

Hematology	Infectious	Personalized Medicine
BCR-ABL	BKV	HER2
Chimerism	HCV	
t(15;17)	HIV	

The human epidermal growth factor receptor 2 (*ERBB2*, commonly referred to as *HER2*) gene had been recognized as being amplified in breast cancers for many years [15,16]. In fact, this gene is amplified in up to 25% of all breast cancers. Amplification of the *HER2* gene is the primary mechanism of HER2 overexpression that results in increased receptor tyrosine kinase activity (Figure 29.2). HER2-positive breast cancers have a worse prognosis and are often resistant to hormonal therapies and other chemotherapeutic agents. Trastuzumab (Herceptin), the first humanized monoclonal antibody against the HER2 receptor, was approved by the FDA in 1998. The availability of a therapeutic now made it necessary for laboratories to determine HER2 gene copy number or protein overexpression before patients would be eligible for treatment. Several FDA-cleared tests for immunohistochemical detection of HER2 protein and fluorescence *in situ* hybridization (FISH) detection of gene amplification are commercially available and approved as companion diagnostics for Herceptin (Figure 29.3). Recently, the American Society of Clinical Oncology and the College of American Pathologists published guidelines for performing and interpreting *HER2* testing [17]. These guidelines attempt to standardize *HER2* testing by addressing preanalytical, analytical, and postanalytical variables that could lessen result variability

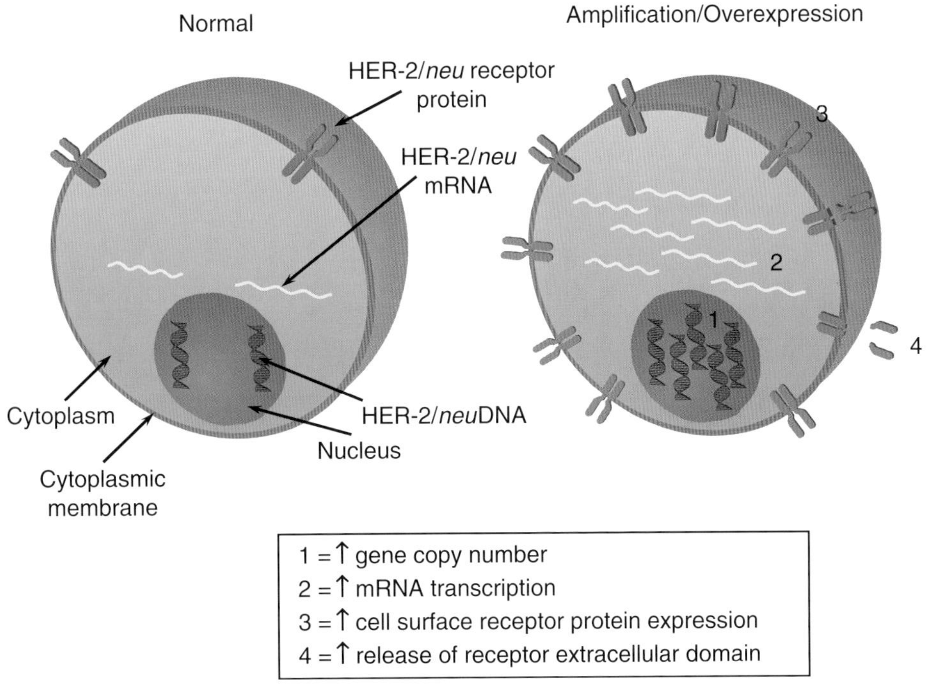

Figure 29.2 **Schematic diagram of the *HER2* amplification process in normal and tumor breast epithelial cells.**

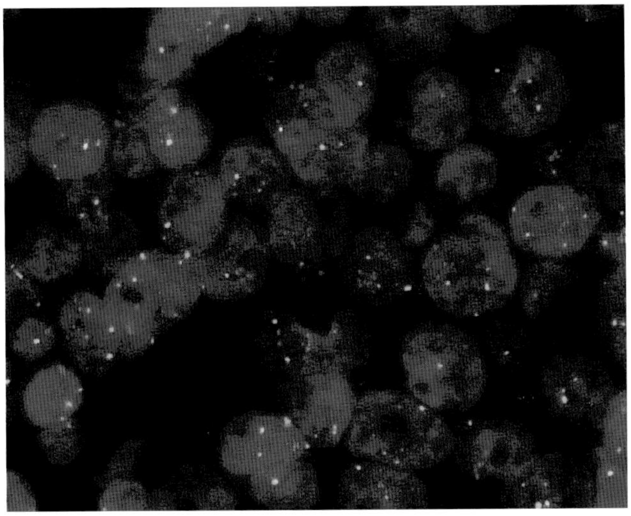

Figure 29.3 ***HER2* amplification as detected by FISH analysis (orange signal, LSI HER2; green signal, CEP17).**

due to technical and interpretative subjectivity. While many laboratories have adopted an algorithm for screening breast cancer cases by immunohistochemical staining and reflexing equivocal results for FISH analysis, others have gone to a primary FISH screen for *HER2* gene amplification.

Phase III: Resistance Testing

With respect to resistance genotyping, our infectious disease experience has taught us that this can be attributed to identification of subtypes of organisms such as in Hepatitis C virus (HCV) or in actual mutations of the viral genome as in the HIV-1 virus. It is well known that tumor cells develop resistance to therapeutic agents via multiple mechanisms and, more importantly, are sensitive to newer therapeutic modalities if certain genetic abnormalities are or are not present in the tumor cells. For example, significant improvements have been made in the response rates, progression-free survival, and overall survival for colorectal cancers, the second leading cause of cancer death in the United States, in the last decade. This has been due to new combinations of chemotherapeutic agents and new small molecule targeted therapies [18–20].

The overall efficacy of traditional chemotherapeutic agents for treating cancer has been extremely low. More modern targeted therapies hope to improve on this. Selecting patients most likely to benefit from molecularly targeted therapies, however, has been challenging. Recently, K-*ras* mutations have been shown to be an independent prognostic factor in patients with advanced colorectal cancer who are treated with Cetuximab [21]. Cetuximab is known to be active in metastatic colorectal cancer patients which are resistant to irinotecan, oxaliplatin, and fluoropyrimidine. In these studies, K-*ras* mutation has also been identified as a marker of tumor cell resistance to Cetuximab. Amado et al. have also associated the presence of wild-type K-*ras* sequences with panitumumab (a humanized monoclonal antibody against the epidermal growth factor receptor) efficacy in patients with metastatic disease [22]. These data demonstrate the need for identifying resistant mutations in other genes, as well as identifying the presence of the target for these novel therapies.

K-*ras* mutations are present in >30% of colorectal cancers and play a critical role in the pathogenesis of this disease. K-*ras* is a member of the guanosine-5′-triphosphate (GTP)-binding protein superfamily. Stimulation of a growth factor receptor by ligand binding results in activated K-*ras* by binding to GTP. This results in downstream activation of pathways such as RAF/MAP kinase, STAT, and PI3K/AKT, that in turn results in cellular proliferation, adhesion, angiogenesis, migration, and survival [23–25]. While the association of K-*ras* mutations and cancer had been known for quite some time, the association of mutated K-*ras* with a response to a novel and expensive therapeutic has made its interrogation a prerequisite to receiving therapy. It is now estimated that all colorectal cancer patients who will receive these therapies will first be tested for K-*ras* mutations.

BCR-ABL: A MODEL FOR THE NEW PARADIGM

Chronic myelogenous leukemia (CML) is considered one of the diseases in the group of myeloproliferative disorders. These are clonal hematopoietic malignancies characterized by the proliferation and survival of one or more of the myeloid cell lineages [26,27]. CML has an estimated 5,000 newly diagnosed cases per year, and it accounts for about 20% of all adult cases of leukemia [28,29]. Typically, the disease progresses through three clinical phases: (i) a chronic phase (which may be clinically asymptomatic), (ii) an accelerated phase, and (iii) a blast phase or blast crisis. Most patients are diagnosed with the disease during the chronic phase by morphologic, cytogenetic, and molecular genetic techniques.

The (9;22)(q34;q11) translocation, which results in the *BCR-ABL* fusion gene, is a defining feature of this disease [30–32]. *BCR-ABL* analysis warrants the three phases of the presented paradigm for qualitative diagnostic testing, quantitative monitoring, and resistance mutation analysis. CML has traditionally been diagnosed and monitored by cytogenetics, FISH, and Southern blot analysis (Figure 29.4). Conventional cytogenetics, such as bone marrow karyotyping that is performed on 30–50 cells, can be used to visualize the translocation by G-banding analysis. FISH, typically performed on 100–1000 cells in interphase or metaphase, allows the visualization of the translocation through the use of fluorescently labeled probes (Figure 29.4). These techniques are limited in sensitivity and cannot detect more than a 2-log reduction in residual CML cells and have increased turnaround times [33].

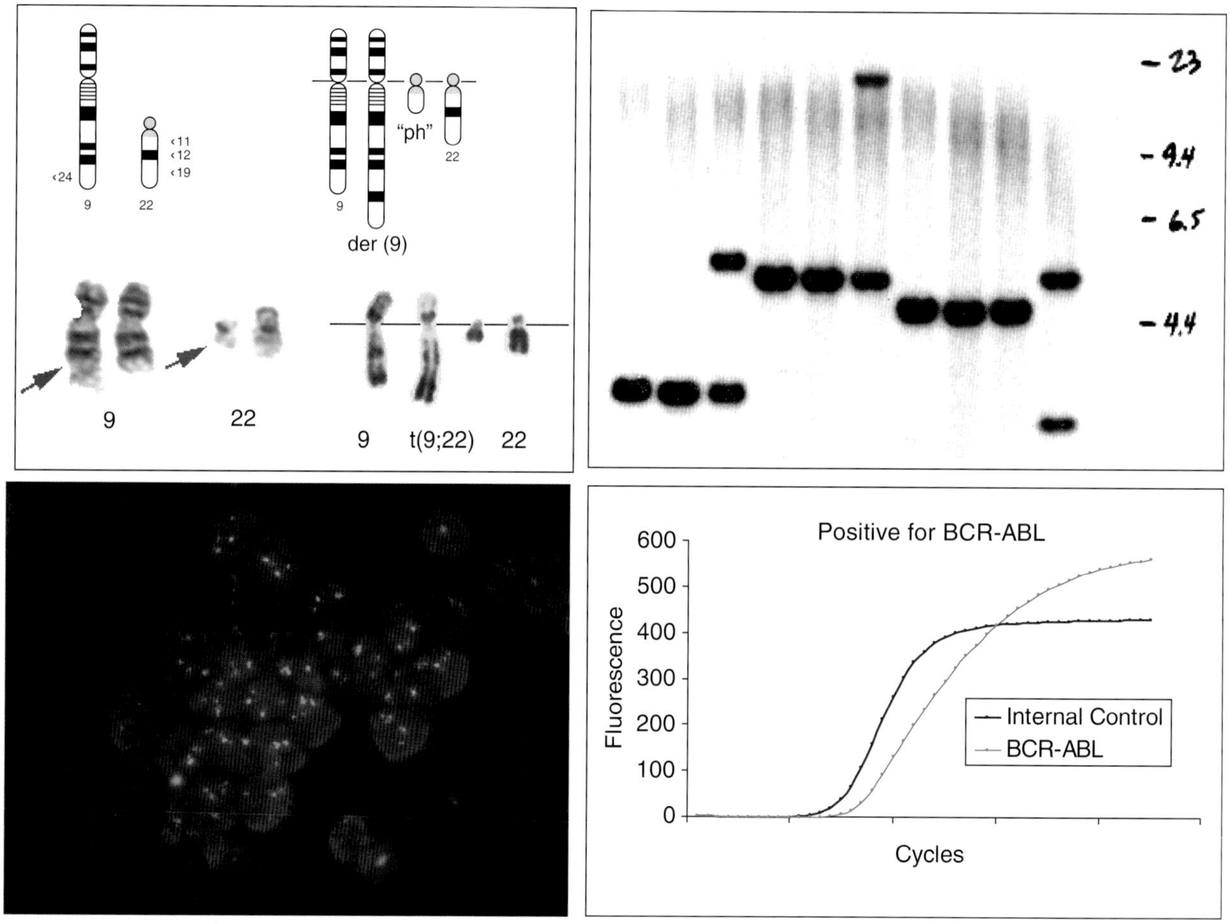

Figure 29.4 Testing methods for the detection of the *BCR-ABL* translocation. Traditional cytogenetic karyotyping, Southern blot transfer analysis, FISH, and real-time PCR.

Real-time reverse transcriptase quantitative PCR (qPCR) has become the preferred method for monitoring the second phase of the paradigm, minimal residual disease in CML patients treated with imatinib (Figure 29.4) [33–39]. The major advantage of qPCR is that it has a sensitivity on the order of one CML cell per 10^5 normal cells, thus achieving a 3-log improvement in sensitivity compared to cytogenetics and FISH. Novel instrumentation now allows clinical laboratories the ability to provide quantitative analysis using real-time PCR in under 2 hours (Figure 29.5) [40,41].

The detection of low levels of disease in CML patients is critical to the proper management of these patients. Studies have shown that a rapid 3-log quantitative reduction in *BCR-ABL* transcript predicts risk of progression in CML patients being treated with imatinib [38,39]. Similarly, a significant increase in *BCR-ABL* transcript in a patient being treated with imatinib may indicate an acquired mutation which confers resistance to treatment. Molecular analysis of the transcript is performed to identify mutations that confer therapeutic resistance, the third phase of the paradigm. It has been recommended that patients who fail to achieve a 3-log reduction in *BCR-ABL* transcript levels after 18 months of imatinib therapy have an increase in their dose in an attempt to achieve a major molecular response [39].

The recognition of the role of BCR-ABL kinase activity in CML has provided an attractive target for the development of therapeutics. Although interferon therapy and stem cell transplantation were previously the first-line treatments of choice, tyrosine kinase inhibitors such as imatinib mesylate (Gleevec®; Novartis, Basel, Switzerland), and dasatinib (SPRYCEL®; Bristol-Myers Squibb, New York, NY) have now become first- and second-line therapies. Imatinib binds to the inactive form of the wild-type Bcr-Abl protein, preventing its activation. However, acquired mutations in the *BCR-ABL* kinase domains prevent imatinib binding. Dasatinib binds to both the inactive and active forms of the protein, and also binds to all imatinib-resistant mutants except the threonine to isoleucine mutation at amino acid residue 315 (T315I). The introduction of more efficacious therapeutics such as imatinib and dasatinib has redefined the therapeutic responses that are achievable in the CML patient (Table 29.6). CML now represents a molecular oncology paradigm similar to HIV-1 whereby qualitative, quantitative, and resistant genotyping can be performed routinely.

Table 29.6 Therapeutic Goals for the CML Patient

Therapeutic Goal	Therapeutic Response
Hematologic response	Normal peripheral blood valves, normal spleen size
Cytogenetic response 1. complete 2. partial 3. minor	Reduction of Ph+ cells in blood or bone marrow 1. 0% Ph+ cells 2. 1%–35% Ph+ cells 3. 36%–95% Ph+ cells
Molecular response	Reduction or elimination of BCR-ABL mRNA in peripheral blood or bone marrow

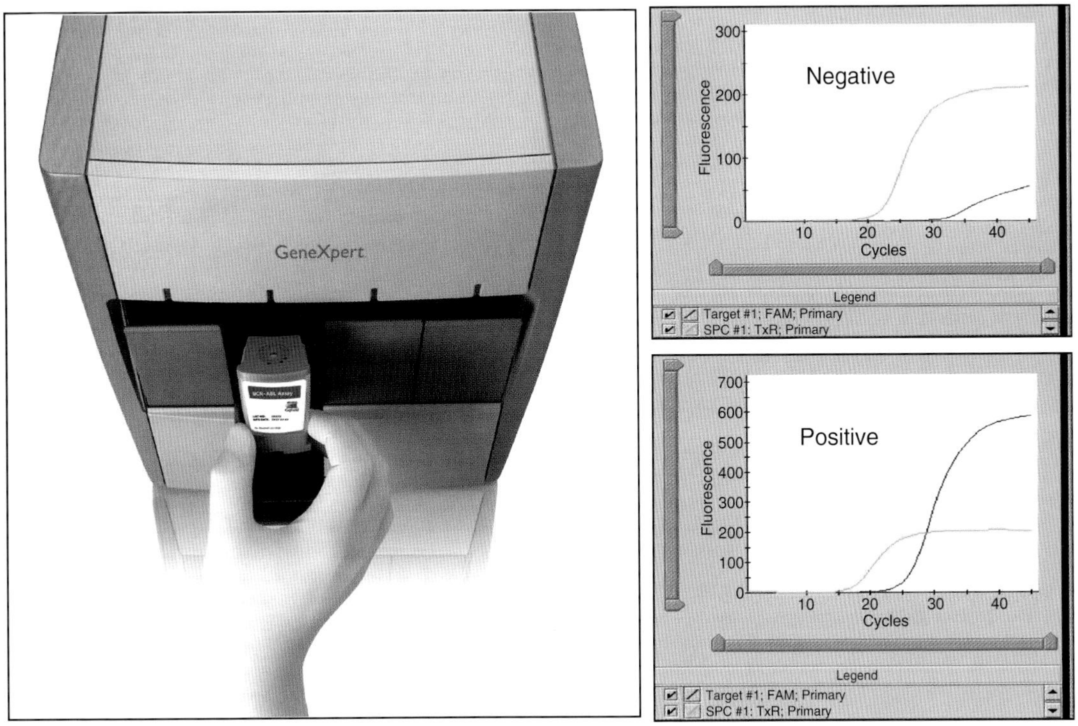

Figure 29.5 **GeneXpert real-time PCR system and cartridge for quantification of the *BCR-ABL* transcript.**

CONCLUSION

Numerous genes and mutations have been identified for many human diseases, including cancer. Yet little of this knowledge has made it into the clinical laboratory as validated diagnostic tests. This is beginning to change with our increased medical and biological knowledge of these diseases. Genomics and proteomics in the clinical context of diagnostic testing are progressing at record speeds. Technology, as is typical in the clinical laboratory, has far outpaced proven clinical utility. Routine applications of information from the "omics-era" are currently being performed for a new paradigm in molecular diagnostic testing that consist of qualitative, quantitative, and resistance genotyping for various clinical conditions.

REFERENCES

1. Tsongalis GJ, Coleman WB. Clinical genotyping: The need for interrogation of single nucleotide polymorphisms and mutations in the clinical laboratory. *Clin Chim Acta*. 2006;363:127–137.
2. Wilhelm J, Pingoud A. Real-time polymerase chain reaction. *Chembiochem*. 2003;4:1120–1128.
3. Wittwer CT, Herrmann MG, Moss AA, et al. Continuous fluorescence monitoring of rapid cycle DNA amplification. *Biotechniques*. 1997;22:130–131, 134–138.
4. Parks SB, Popovich BW, Press RD. Real-time polymerase chain reaction with fluorescent hybridization probes for the detection of prevalent mutations causing common thrombophilic and iron overload phenotypes. *Am J Clin Pathol*. 2001;115:439–447.
5. Wittwer CT, Reed GH, Gundry CN, et al. High-resolution genotyping by amplicon melting analysis using LCGreen. *Clin Chem*. 2003;49:853–860.
6. Ratcliff RM, Chang G, Kok T, et al. Molecular diagnosis of medical viruses. *Curr Issues Mol Biol*. 2007;9:87–102.
7. Tsongalis GJ. Branched DNA technology in molecular diagnostics. *Am J Clin Pathol*. 2006;126:448–453.
8. Yang S, Rothman RE. PCR-based diagnostics for infectious diseases: Uses, limitations, and future applications in acute-care settings. *Lancet Infect Dis*. 2004;4:337–348.
9. Tan TY. Use of molecular techniques for the detection of antibiotic resistance in bacteria. *Expert Rev Mol Diagn*. 2003;3:93–103.
10. Versalovic J, Lupski JR. Molecular detection and genotyping of pathogens: More accurate and rapid answers. *Trends Microbiol*. 2002;10:S15–S21.
11. James C, Ugo V, Le Couédic JP, et al. A unique clonal JAK2 mutation leading to constitutive signalling causes polycythemia vera. *Nature*. 2005;434:1144–1148.
12. Kralovics R, Passamonti F, Buser AS, et al. A gain of function mutation of JAK2 in myeloproliferative disorders. *N Engl J Med*. 2005;352:1779–1790.
13. Levine RL, Wadleigh M, Cools J, et al. Activating mutation in the tyrosine kinaase JAK2 in polycythemia vera, essential thrombocythemia, and myeloid metaplasia with myelofibrosis. *Cancer Cell*. 2005;7:387–397.
14. Steensma DP. Jak2 V617F in myeloid disorders: Molecular diagnostic techniques and their clinical utility. *J Molec Diagn*. 2006;8:397–411.
15. Moasser MM. The oncogene HER2: Its signaling and transforming functions and its role in human cancer pathogenesis. *Oncogene*. 2007;26:6469–6487.
16. Moasser MM. Targeting the function of the HER2 oncogene in human cancer therapeutics. *Oncogene*. 2007;26:6577–6592.
17. Wolff AC, Hammond ME, Schwartz JN, et al. American Society of Clinical Oncology; College of American Pathologists. American Society of Clinical Oncology/College of American Pathologists guideline recommendations for human epidermal growth factor receptor 2 testing in breast cancer. *J Clin Oncol*. 2007;25:118–145.
18. de Gramont A, Figer A, Seymour M, et al. Leucovorin and fluorouracil with or without oxaliplatin as first-line treatment in advanced colorectal cancer. *J Clin Oncol*. 2000;18:2938–2947.
19. Saltz LB, Cox JV, Blanke C, et al. Irinotecan plus fluorouracil and leucovorin for metastatic colorectal cancer. Irinotecan Study Group. *N Engl J Med*. 2000;343:905–914.
20. Cunningham D, Humblet Y, Siena S, et al. Cetuximab monotherapy and cetuximab plus irinotecan in irinotecan-refractory metastatic colorectal cancer. *N Engl J Med*. 2004;351:337–345.
21. Lièvre A, Bachet JB, Boige V, et al. KRAS mutations as an independent prognostic factor in patients with advanced colorectal cancer treated with cetuximab. *J Clin Oncol*. 2008;26:374–379.
22. Amado RG, Wolf M, Peeters M, et al. Wild-type KRAS is required for panitumumab efficacy in patients with metastatic colorectal cancer. *J Clin Oncol*. 2008;26:1626–1634.
23. Krause DS, Van Etten RA. Tyrosine kinases as targets for cancer therapy. *N Engl J Med*. 2005;353:172–187.
24. Nguyen H, Tran A, Lipkin S, et al. Pharmacogenomics of colorectal cancer prevention and treatment. *Cancer Invest*. 2006; 24:630–639.
25. Mendelsohn J, Baselga J. Epidermal growth factor receptor targeting in cancer. *Semin Oncol*. 2006;33:369–385.
26. Tefferi A, Gilliland DG. Oncogenes in myeloproliferative disorders. *Cell Cycle*. 2007;6:550–566.
27. Delhommeau F, Pisani DF, James C, et al. Oncogenic mechanisms in myeloproliferative disorders. *Cellul Molec Life Sci*. 2006; 63:2939–2953.
28. Schiffer CA. BCR-ABL tyrosine kinase inhibitors for chronic myelogenous leukemia. *N Engl J Med*. 2007;357:258–265.
29. Quintas-Cardama A, Cortes JE. Chronic myeloid leukemia: Diagnosis and treatment. *Mayo Clinic Proc*. 2006;81:973–988.
30. Sawyers CL. Chronic myeloid leukemia. *N Engl J Med*. 1999; 340:1330–1340.
31. Ben-Neriah Y, Daley GQ, Mes-Masson AM, et al. The chronic myelogenous leukemia-specific P210 protein is the product of the bcr/abl hybrid gene. *Science*. 1986;233:212–214.
32. Ren R. Mechanisms of BCR-ABL in the pathogenesis of chronic myelogenous leukemia. *Nature Rev Cancer*. 2005;5:172–183.
33. Goldman J. Monitoring minimal residual disease in BCR-ABL-positive chronic myeloid leukemia in the imatinib era. *Curr Opin Hematol*. 2005;12:33–39.
34. Martinelli G, Iacobucci I, Soverini S, et al. Monitoring minimal residual disease and controlling drug resistance in chronic myeloid leukaemia patients in treatment with imatinib as a guide to clinical management. *Hematol Oncol*. 2006;24:196–204.
35. Hughes T, Branford S. Molecular monitoring of BCR-ABL as a guide to clinical management in chronic myeloid leukaemia. *Blood Rev*. 2006;20:29–41.
36. Marin D, Kaeda J, Szydlo R, et al. Monitoring patients in complete cytogenetic remission after treatment of CML in chronic phase with imatinib: Patterns of residual leukaemia and prognostic factors for cytogenetic relapse. *Leukemia*. 2005;19:507–512.
37. van der velden VHJ, Hochhaus A, Cazzaniga G, et al. Detection of minimal residual disease in hematologic malignancies by real time quantitative PCR: Principles, approaches, and laboratory aspects. *Leukemia*. 2003;17:1013–1034.
38. Hochhaus A, La Rosee P. Imatinib therapy in chronic myelogenous leukemia: Strategies to avoid and overcome resistance. *Leukemia*. 2004;18:1321–1331.
39. Baccarani M, Saglio G, Goldman J, et al. Evolving concepts in the management of chronic myeloid leukemia: Recommendations from an expert panel on behalf of the European Leukemia Net. *Blood*. 2006;108:1809–1820.
40. Dufresne SD, Belloni DR, Levy NB, et al. Quantitative assessment of the BCR-ABL transcript using the Cepheid Xpert BCR-ABL Monitor assay. *Arch Pathol Lab Med*. 2007;131:947–950.
41. Winn-Deen ES, Helton B, Van Atta R, et al. Development of an integrated assay for detection of BCR-ABL RNA. *Clin Chem*. 2007;53:1593–1600.

Chapter

30

Pharmacogenomics and Personalized Medicine in the Treatment of Human Diseases

Hong Kee Lee . Gregory J. Tsongalis

INTRODUCTION

In 2001, the first draft of the Human Genome Project was simultaneously published by two groups [1,2]. This was the beginning of an era that promised better diagnostic and prognostic testing that would lead to preventive medicine and more personalized therapy. As part of this promise, the ability to select and dose therapeutic drugs through the use of genetic testing has become a reality. Pharmacogenomics represents the role that genes play (pharmacogenetics) in the processing of drugs by the body (pharmacokinetics) and how these drugs interact with their targets to give the desired response (pharmacodynamics) (Figure 30.1). While other environmental factors also play a role in this selection, an individual's genetic makeup provides new insights into the metabolism and targeting of commonly prescribed as well as novel targeted therapies [3–5]. The importance of implementing this knowledge into clinical practice is highlighted by the more than 2 million annual hospitalizations and greater than 100,000 annual deaths in the United States due to adverse drug reactions [6].

Current medical practices often utilize a trial-and-error approach to select the proper medication and dosage for a given patient (Figure 30.2). Pharmacogenetics (PGx) utilizes our knowledge of genetic variation to assess an individual's response to therapeutic drugs, and when interpreted in the context of the entire being, pharmacogenomics can result in better personalized medicine. This application of human genetics is our attempt to more accurately and efficaciously determine genomic sequence variations and expression patterns involved in response to the absorption, distribution, metabolism, and excretion of therapeutic drugs at a systemic level [3,7–9]. A second component to this application of genomics is in the identification of target genes for novel therapies at the cellular level, as well as acquired mutations in genes of specific pathways that may result in a better or worse response to a novel therapy. The overall aim of PGx testing is to decrease adverse responses to therapy and increase efficacy by ensuring the appropriate selection and dose of therapy.

Numerous genetic variants or polymorphisms in genes that code for drug metabolizing enzymes, drug transporters, and drug targets that alter response to therapeutics have been identified (www.pharmgkb.org/) (Table 30.1). Classification of these enzymes includes specific nomenclature, CYP2D6*1, such that the name of the enzyme (CYP) is followed by the family (for example, 2), subfamily (for example, D), and gene (for example, 6) associated with the biotransformation. Allelic variants are indicated by an asterisk (*) followed by a number (for example, *1). The technologies that are currently available in clinical laboratories and are employed to routinely test for these variants are shown in Table 30.2 [10].

Single nucleotide polymorphisms (SNPs) in these genes may result in no significant phenotypic effect, change drug metabolism by >10,000-fold, or alter protein binding by >20-fold [11]. More than 3.5 million SNPs have been identified in the human genome (www.hapmap.org), making analysis quite challenging. As shown in Table 30.1, different genes can

 © 2009, Elsevier, Inc. All Rights Reserved.

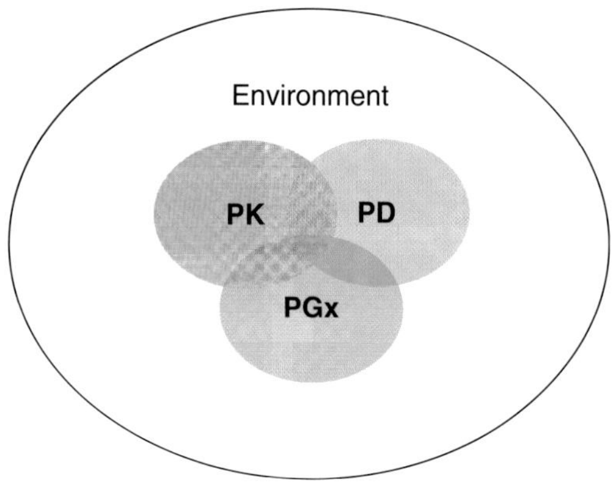

Figure 30.1 **Venn diagram of pharmacogenomics showing the interactions of pharmacogenetics (PGx), pharmacokinetics (PK), and pharmacodynamics (PD), along with other environmental factors that affect drug response.**

contain various numbers of SNPs that alter the function of the coded protein. It is important to clearly differentiate these benign polymorphisms (present in greater than 1% of the general population) from rare disease-causing mutations that are present in less than 1% of the general population. The SNPs associated with pharmacogenetics do not typically result in any form of genetic disease. We do know that these SNPs may be used to evaluate individual risk for adverse drug reactions (ADRs) so that we may decrease ADRs, select optimal therapy, increase patient compliance, develop safer and more effective drugs, revive withdrawn drugs, and reduce time and cost of clinical trials [6,12].

In addition to the genotype predicting a phenotype, drug metabolizing enzyme activity can be induced or inhibited by various drugs. Induction leads to the production of more enzyme within 3 or more days of exposure to inducers [13]. Enzyme inhibition by commonly prescribed drugs can be an issue in polypharmacy and is usually the result of competition between two drugs for metabolism by the same enzyme. A better understanding of drug metabolism has led to a number of amended labels for commonly prescribed drugs (Table 30.3).

Table 30.1 Examples of Enzymes and Their Designated Polymorphic Alleles

Enzyme	Alleles
Cytochrome P450 2D6 (CYP2D6)	*2, *3, *4, *5, *6, *7, *8, *11, *12, *13, *14, *15, *16, *18, *19, *20, *21, *38, *40, *42, *10, *17, *36, *41
Cytochrome P450 2C9 (CYP2C9)	*1, *2, *3, *4, *5, *6
Cytochrome P450 2C19 (CYP2C19)	*2, *3
UGT1A1	*6, *28, *37, *60

Historical Perspective

Friedrich Vogel was the first to use the term pharmacogenetics in 1959 [14]. However, in 510 BC, Pythagoras recognized that some individuals developed hemolytic anemia with fava bean consumption [15]. In 1914, Garrod expanded on these early observations to state enzymes detoxify foreign agents so that they may be excreted harmlessly, but some people lack these enzymes and experience adverse effects [16]. Hemolytic anemia due to fava bean consumption was later determined to occur in glucose-6-phosphate dehydrogenase deficient individuals [17,18].

Through the early 1900s PGx evolved as investigators combined Mendelian genetics with observed phenotypes. In 1932, Snyder performed the first global study of ethnic variation and deduced that taste deficiency was inherited. As such, the phenylthiourea non-taster phenotype was an inherited recessive trait, and the frequency of occurrence differed between races [19]. Similarly, polymorphisms in the N-acetyl transferase enzyme also segregate by ethnicity [20]. Shortly thereafter other genetic differences such as aldehyde dehydrogenase and alcohol dehydrogenase deficiencies were discovered. In 1956, it was recognized that variants of glucose-6-phosphate dehydrogenase caused primaquine-induced hemolysis [21]. It was shown that the metabolism of nortriptyline and desipramine was highly variable among individuals, and two phenotypes were identified in 1967 [22]. Subsequently, genetic deficiencies in other enzyme

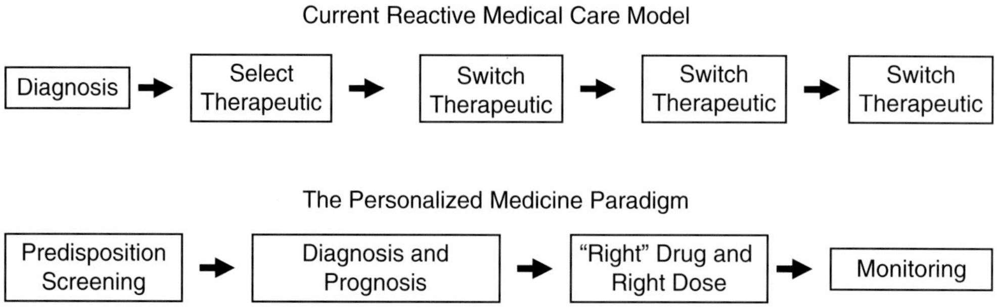

Figure 30.2 **Schematic diagram illustrating current medical practice and the changes that occur in a personalized medicine model.**

Table 30.2 Genotyping Methodologies Currently in Use in the Clinical Laboratory

Technology	Methodology*	Company*
Gel electrophoresis	Southern blot PCR-RFLP*	Generic equipment
Real-time PCR	Allele-specific PCR	Applied Biosystems
	Allelic discrimination	Biorad
		Cepheid
		Roche Molecular
DNA Sequencing	Sanger—fluorescent detection	Applied Biosystems
		Beckman Coulter
		Pyrosequencing
Arrays	Liquid beads	Autogenomics
	Microarray	Luminex
	Microfluidic	Nanosphere
		Roche Molecular

*Representative examples.
**PCR-RFLP, polymerase chain reaction-restriction fragment length polymorphism analysis.

systems were documented and shown that metabolism of drugs played a critical role in a patient's response as well as risk for development of ADRs [23–26]. The central dogma for assessing human diseases at a molecular level, DNA → RNA → protein, became the model for moving the newly discovered knowledge base and technologies forward at the molecular level.

Genotyping Technologies

We currently have the means to successfully and accurately genotype patients for polymorphisms and mutations on a routine basis [10] (Table 30.2). Genotyping involves the identification of defined genetic variants that give rise to the specific drug response phenotypes [27]. Genotyping methods are easier to perform and more cost effective than the more traditional phenotyping methods. Because an individual's genotype is not expected to change over time, most genotyping applications need be performed only once in an individual's lifetime. The exception to this is for the acquired genetic variants that occur in certain disease conditions such as cancer where variants of tumor cells may change during the course of the disease. While a single genotype can determine responses to numerous

Table 30.3 Examples of Therapeutic Drugs That Include PGx Information or Have Had Labels Amended Due to New Knowledge of Metabolism

Atomoxetine	Strattera®
Thioridazine	Mellaril®
Voriconazole	Vfend®
6-mercaptopurine	Purinethol®
Azathioprine	Imuran®
Irinotecan	Camptosar®
Warfarin	Coumadin®

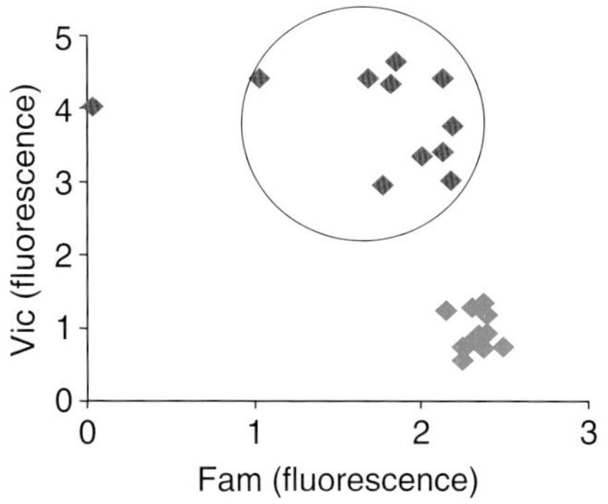

Figure 30.3 Allelic discrimination using real-time PCR and Taqman probes to identify individuals who are homozygous or heterozygous (circle) for a particular SNP.

therapeutic drugs, it is worth noting that patients and providers will need to refer back to genotype information on numerous occasions.

Early genotyping efforts utilized conventional procedures such as Southern blotting and the polymerase chain reaction (PCR) followed by restriction endonuclease digestion to interrogate human gene sequences for polymorphisms [27,28]. The introduction of real-time PCR to clinical laboratories has resulted in assays which can be performed much faster and in a multiplexed fashion so that more variants are tested for at the same time [29] (Figure 30.3). Direct sequencing reactions have been developed on automated capillary electrophoresis instruments to detect specific base changes that result in a variant allele.

More recently, platforms utilizing various array technologies for genotyping have been introduced into the clinical laboratory. Roche Molecular was the first to introduce an FDA-cleared microarray, AmpliChip CYP450 test, for PGx testing on an Affymetrix platform which detects 31 known polymorphisms in the CYP2D6 gene, including gene duplication and deletion, as well as two variations in the CYP2C19 gene [30]. Other array platforms being introduced into the clinical laboratory are the AutoGenomics INFINITI™ [31], Luminex 100 and 200 microbead-array technologies [32], and the Nanosphere Verigene system [10].

PGx and Drug Metabolism

Most of the enzymes involved in drug metabolism are members of the cytochrome P450 (CYP450) superfamily [33–35]. CYP450 enzymes are mainly located in the liver and gastrointestinal tract and include greater than 30 isoforms [33]. The most polymorphic of these enzymes responsible for the majority of biotransformations are the CYP3A, CYP2D6, CYP2C19, and CYP2C9. Benign genetic variants or polymorphisms in these genes can lead to the following phenotypes: poor, intermediate, extensive, and ultrarapid metabolizers. Poor metabolizers (PM) have no detectable enzymatic activity; intermediate metabolizers (IM) have decreased enzymatic activity; extensive metabolizers (EM) are considered normal and have at least one copy of an active gene; ultrarapid metabolizers (UM) contain duplicated or amplified gene copies that result in increased drug metabolism. The following are several examples of polymorphic drug metabolizing enzymes that can affect response to therapy.

CYP2D6

CYP2D6 is an example of one of the most widely studied members of this enzyme family. It is highly polymorphic and contains 497 amino acids. The *CYP2D6* gene is localized on chromosome 22q13.1 with two neighboring pseudogenes, *CYP2D7* and *CYP2D8* [36,37]. More than 50 alleles of *CYP2D6* have been described, of which alleles *3, *4, *5, *6, *7, *8, *11, *12, *13, *14, *15, *16, *18, *19, *20, *21, *38, *40, *42, and *44 were classified as nonfunctioning and alleles *9, *10, *17, *36, and *41 were reported to have substrate-dependent decreased activity [38]. CYP2D6 alone is responsible for the metabolism of 20%–25% of prescribed drugs (Table 30.4). Screening for CYP2D6*3, *4, and *5 alleles identifies at least 95% of poor metabolizers in the Caucasian population. Based on the type of metabolizer, an individual can determine the response to a therapeutic drug.

CYP2C9

A more specific example of CYP450 metabolizing enzyme polymorphisms and drug metabolism is demonstrated by *CYP2C9* and warfarin. In the past six years, there has been tremendous interest in studying the effect of genetics on warfarin dosing [39]. *CYP2C9* polymorphisms (CYP2C9*2 and CYP2C9*3) were first found to reduce the metabolism of S-warfarin, which can lead to dosing differences [40,41]. Soon after, the identification of a polymorphism in the vitamin K epoxide reductase (*VKORC1*) gene proved that drug target polymorphisms are also important in warfarin dosing [42,43]. Since then, several warfarin dosing algorithms that take into consideration patient demographics, *CYP2C9* and *VKORC1* polymorphisms, and concurrent drug use have been developed [44–48]. In August 2007, the FDA announced an update to the warfarin package insert to include information on *CYP2C9* and *VKORC1* testing [49].

Table 30.4 Some Therapeutic Drugs Metabolized by CYP2D6

Cytochrome P450 2D6	
Amitriptyline	Lidocaine
Aripiprazole	Metoclopramide
Atomoxetine	Metoprolol
Carvedilol	Nortriptyline
Chlorpromazine	Oxycodone
Clomipramine	Paroxetine
Codeine	Propafenone
Desipramine	Propranolol
Dextromethorphan	Risperidone
Duloxetine	Tamoxifen
Flecainide	Thioridazine
Fluoxetine	Timolol
Fluvoxamine	Tramadol
Haloperidol	Venlafaxine
Imipramine	Zuclopenthixol

UGT1A1

The uridine diphosphate glucuronosyltransferases (UGT) superfamily of endoplasmic reticulum-bound enzymes is responsible for conjugating a glucuronic acid moiety to a variety of compounds, thus allowing these compounds to be more easily eliminated. It is a member of this family that catalyzes the glucuronidation of bilirubin, allowing it to be excreted in the bile. As irinotecan therapy for advanced colorectal cancers became more widely used, it was observed that patients who had Gilbert syndrome (defects in UGT leading to mild hyperbilirubinemia) suffered severe toxicity [50]. Irinotecan is converted to SN-38 by carboxylesterase-2, and SN-38 inhibits DNA topoismoerase I activity [51,52]. SN-38 is glucuronidated by uridine diphosphate glucuronosyl-transferase (UGT), forming a water-soluble metabolite, SN-38 glucuronide, which can then be eliminated [53] (Figure 30.4). The observation made in Gilbert syndrome patients revealed that SN-38 shares a glucuronidation pathway with bilirubin. The decreased glucuronidation of bilirubin and SN-38 can be attributed to polymorphisms in the *UGT1A1* gene. The wild-type allele of this gene, *UGT1A1*1*, has six tandem TA repeats in the regulatory TATA box of the *UGT1A1* promoter. The most common polymorphism associated with low activity of *UGT1A1* is the *28 variant, which has seven TA repeats. In August 2005, the FDA amended the irinotecan (Camptosar®) package insert to recommend genotyping for the UGT1A1 polymorphism and suggested a dose reduction in patients homozygous for the *28 allele.

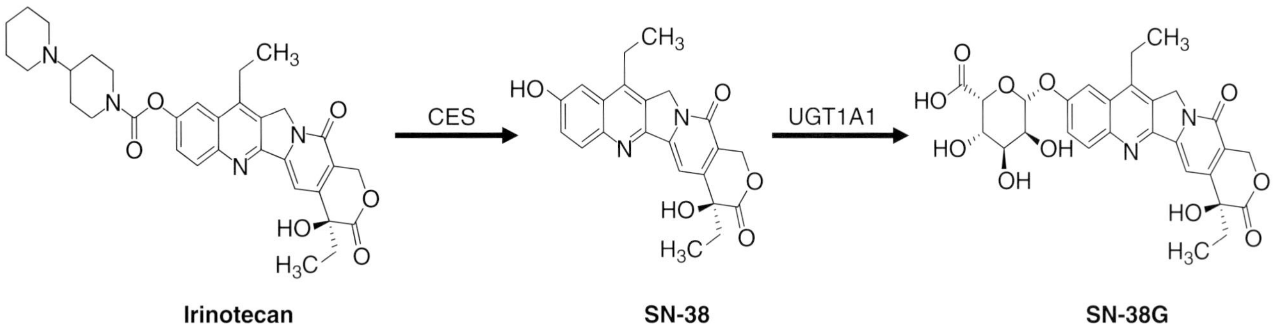

Figure 30.4 Schematic diagram of irinotecan metabolism.

PGx and Drug Transporters

Although the genes that code for drug metabolizing enzymes have received more attention in recent years as markers for PGx testing, the genes that code for proteins used to transport drugs across membranes also need to be considered when discussing PGx. These drug transporter proteins move substrates across cell membranes, bringing them into cells or removing them from cells. These proteins are essential in the absorption, distribution, and elimination of various endogenous and exogenous substances including pharmaceutical agents.

Several groups of drug transporters that may be significant in the field of pharmacogenomics exist, including multidrug resistance proteins (MDRs), multidrug resistance-related proteins (MRPs), organic anion transporters (OATs), organic anion transporting polypeptides (OATPs), organic cation transporters (OCTs), and peptide transporters (PepTs). ABCB1 (MDR1) is a member of the multidrug resistance protein family and is one example of a drug transporter protein important in the field of PGx.

ABCB1

ABCB1 is a member of the ATP-binding cassette (ABC) superfamily of proteins. Also known as P-glycoprotein (P-gp) or MDR1, it is a 170 kDa glycosylated membrane protein expressed in various locations including the liver, intestines, kidney, brain, and testis [54–56]. Generally speaking, ABCB1 is located on the membrane of cells in these locations and serves to eliminate metabolites and a wide range of hydrophobic foreign substances, including drugs, from cells by acting as an efflux transporter [57,58]. Due to the localization of ABCB1 on specific cells, ABCB1 aids in eliminating drugs into the urine or bile and helps maintain the blood-brain barrier [56].

Like other eukaryotic ABC proteins, the ABCB1 protein is composed of two similar halves, each half containing a hydrophobic membrane binding domain and a nucleotide-binding domain. The membrane binding domains are each composed of six hydrophobic transmembrane helices, and the two hydrophilic nucleotide-binding domains are located on the intracellular side of the membrane where they bind ATP [59–61].

ABCB1 was first identified in cancer cells that had developed a resistance to several anticancer drugs because of an overexpression of the transporter. When expressed at normal levels in noncancerous cells, ABCB1 has been shown to transport other classes of drugs out of cells including cardiac drugs (digoxin), antibiotics, steroids, HIV protease inhibitors, and immunosuppressants (cyclosporin A) [61].

Genetic variations in the ABCB1 gene expressed in normal cells have been shown to have a role in interindividual variability in drug response [62,63]. Although many polymorphisms have been detected in the ABCB1 gene, correlations between genotype and either protein expression or function have been described for only a few of the genetic variants. Most notable among these is the 3435 C>T polymorphism, found in exon 26, which has been found to result in decreased expression of ABCB1 in individuals homozygous for the T allele. These results, however, have been found to be slightly controversial [62,64–67]. The possible importance of the 3435 C>T is particularly interesting considering the mRNA levels are not affected by this polymorphism, which is found within a coding exon. Although the correlation between the 3435 C>T polymorphism and ABCB1 protein levels may be due to linkage of this polymorphism to others in the ABCB1 gene, a recent study showed that this polymorphism alone does not affect ABCB1 mRNA or protein levels but does result in ABCB1 protein with an altered configuration. The altered configuration due to 3435 C>T is hypothesized to be caused by the usage of a rare codon which may affect proper folding or insertion of the protein into the membrane, affecting the function, but not the level of ABCB1 [68]. Genetic variations affecting the expression or function of drug transporter proteins, such as ABCB1, could drastically alter the pharmacokinetics and pharmacodynamics of a given drug.

PGx and Drug Targets

Most therapeutic drugs act on targets to elicit the desired effects. These targets include receptors, enzymes, or proteins involved in various cellular events such as signal transduction, cell replication, and others. Investigators have now identified polymorphisms in these targets that render them resistant to the particular therapeutic agent. While these polymorphisms in receptors may not show dramatic increases or decreases in drug activity as the drug metabolizing enzymes, biologically significant effects occur frequently [3,15,33]. The human μ-opioid receptor gene (*OPRM1*) and *FLT3* are two examples of the effects of polymorphisms on drug targets.

OPRM1

The human μ-opioid receptor, coded by the *OPRM1* gene, is the major site of action for endogenous β-endorphin, and most exogenous opioids. Forty-three SNPs have been discovered in the *OPRM1* gene, but the 118A>G SNP is the most well studied [69]. The allele frequency of *OPRM1* 118A>G SNP in Caucasians is 10%–30% [69,70]. This SNP causes an increased affinity of the μ-opioid receptor for β-endorphin, which improves pain tolerance in humans [71,72]. However, the effect of this SNP is controversial in people taking exogenous opioids. Studies showed that 118G allele carriers required an increased dose of morphine in patients with acute and chronic pain [73,74], but a decreased dose of fentanyl in patients requiring epidural or intravenous fentanyl for pain control during labor [75]. The exact mechanism of how this happens is still unknown.

FLT3

FLT3 is a receptor tyrosine kinase expressed and activated in most cases of acute myeloid leukemia (AML), which has a relatively high relapse rate due to acquired resistance to traditional chemotherapies. An internal tandem duplication (ITD) mutation in the FLT3 gene is found in up to 30% of AML patients, while point mutations have been shown to account for approximately 5% of refractory AML. The FLT3-ITD induces activation of this receptor and results in downstream constitutive phosphorylation in STAT5, AKT, and ERK pathways. This mutation in FLT3 is a negative prognostic factor in AML. Recently, a novel multitargeted receptor tyrosine kinase inhibitor, ABT-869, has been developed to suppress signal transduction from constitutively expressed kinases such as a mutated FLT3 [76].

PGx Applied to Oncology

Cancer represents a complex set of deregulated cellular processes that are often the result of underlying molecular mechanisms. While there is recognized interindividual variability with respect to an observed chemotherapeutic response, it is also apparent that there is considerable cellular heterogeneity within a single tumor that may account for this lack of efficacy. It is becoming clear that molecular genetic variants significantly contribute to an individual's response to a particular therapy of which some is due to polymorphisms in genes coding for drug-metabolizing enzymes. However, in the cancer patient the acquired genetics of the tumor cell must also be taken into account. Unlike traditional PGx testing, those tests for the cancer patient must be prepared to identify acquired or somatic genetic alterations that deviate from the underlying genome of the individual.

Cancer patients exhibit a heterogeneous response to chemotherapy with only 25%–30% efficacy. PGx can improve on chemotherapeutic and targeted therapy responses by providing a more informative evaluation of the underlying molecular determinants associated with tumor cell heterogeneity. In the cancer patient, PGx can be used to integrate information on drug responsiveness with alterations in molecular biomarkers. Thus, as with previous examples of PGx testing, therapeutic management of the cancer patient can be tailored to the individual patient or tumor phenotype.

Estrogen Receptor

One of the first and most widely used parameters for a targeted therapy is the evaluation of breast cancers for expression of the estrogen receptor (ER) (Figure 30.5). ER-positive breast cancers are then treated with hormonal therapies that mimic estrogen. One of these estrogen analogues is Tamoxifen (TAM), which itself has now been shown to have altered metabolism due to CYP450 genetic polymorphisms [77,78]. TAM is used to treat all stages of estrogen receptor-positive breast cancers [79]. TAM and its metabolites compete with estradiol for occupancy of the estrogen receptor, and in doing so inhibit estrogen-mediated cellular proliferation. Conversion of TAM to its active metabolites occurs predominantly through the CYP450 system [80] (Figure 30.6). Conversion of TAM to primary and secondary metabolites is important because these metabolites can have a greater affinity for the estrogen receptor than TAM itself. For example, 4-OH *N*-desmethyl-TAM (endoxifen) has approximately 100 times greater affinity for the estrogen receptor than tamoxifen. Activation of tamoxifen to endoxifen is primarily due to the action of CYP2D6 [81]. Therefore, patients with defective *CYP2D6* alleles derive less benefit from tamoxifen therapy than patients with functional copies of *CYP2D6* [82]. The most common null allele among Caucasians is *CYP2D6*4*, a splice site mutation (G1934A) resulting in loss of enzyme activity [83] and, therefore, lack of conversion of TAM to endoxifen. This

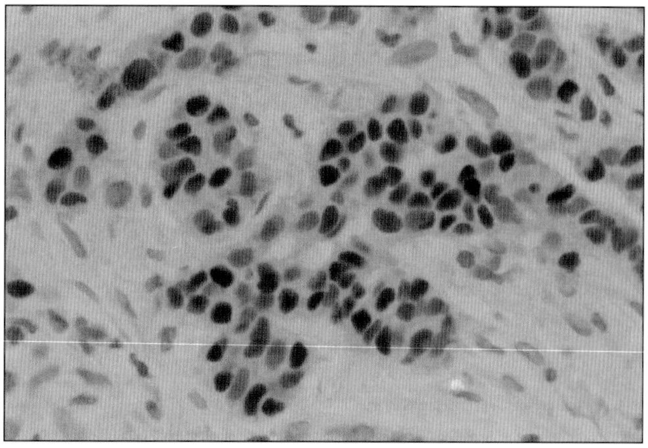

Figure 30.5 Immunohistochemical staining for the estrogen receptor in breast cancer.

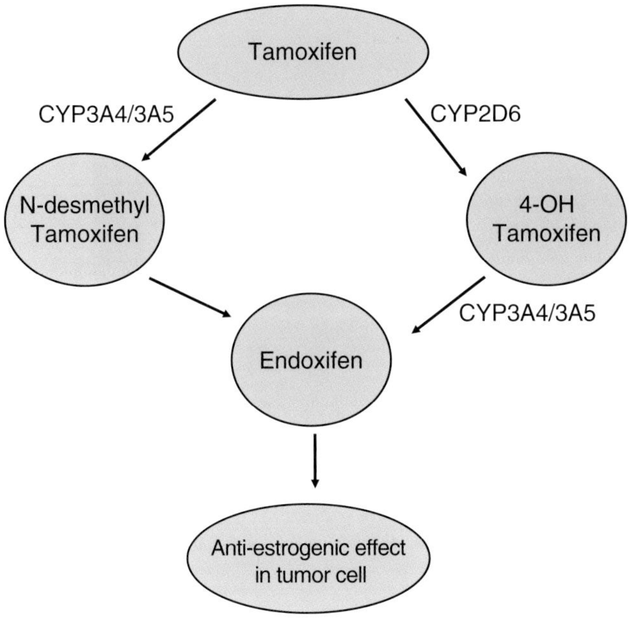

Figure 30.6 **Simplified schematic diagram showing metabolism of tamoxifen by CYP450 enzymes with the resulting production of the active metabolite endoxifen.**

would result in significantly decreased response to this commonly used antihormonal therapy. Common polymorphisms such as this make it feasible to accurately genotype patients so that treatments may be optimized.

HER2

The human epidermal growth factor receptor 2 (*ERBB2* or *HER2*) gene is amplified in up to 25% of all breast cancers. While this gene can be expressed in low levels in a variety of normal epithelia, amplification of the *HER2* gene is the primary mechanism of HER2 overexpression that results in increased receptor tyrosine kinase activity. HER2 status has been implemented as an indicator of both prognosis as well as a predictive marker for response to therapy such that HER2-positive breast cancers have a worse prognosis and are often resistant to hormonal therapies and other chemotherapeutic agents [84–86]. As a predictive marker, HER2 status is utilized to determine sensitivity to anthracycline-based chemotherapy regimens. Determination of HER2 status is also recognized as the first FDA-approved companion diagnostic since Trastuzumab (Herceptin), the first humanized monoclonal antibody against the HER2 receptor, was approved by the FDA in 1998 [87,88]. Introduction of this therapeutic into routine use made it necessary for laboratories to determine the HER2 status in breast cancer cells before patients would be eligible for treatment (Figure 30.7). Several FDA-cleared tests for immunohistochemical detection of HER2 protein and fluorescence *in situ* hybridization (FISH) detection of gene amplification are commercially available and approved as companion diagnostics for Herceptin therapy [89]. Recently, the American Society of Clinical Oncology and the College of American Pathologists published guidelines for performing and interpreting HER2 testing [90]. These guidelines attempt to standardize HER2 testing by addressing preanalytical, analytical, and postanalytical variables that could lessen result variability due to technical and interpretative subjectivity.

TPMT

Thiopurine S-methyltransferase (TPMT) is a cytosolic enzyme that inactivates thiopurine drugs such as 6-mercaptopurine and azathioprine through methylation. Thiopurines are frequently used to treat childhood acute lymphoblastic leukemia [91]. Variability in activity levels of TPMT enzyme function exists between individuals, and it has been found that this variability can be attributed to polymorphisms of the *TPMT* gene [92,93]. The most common variant alleles, *TPMT*2*, *TPMT*3A*, and *TPMT*3C*, account for 95% of TPMT deficiency [94]. Molecular testing is a relatively convenient method for assessment of TPMT enzyme function in patients before treatment with thiopurines. Low TPMT activity levels could put a patient at risk for developing toxicity, since too much drug would be converted to 6-thioguanine nucleotides (6-TGNs), the cytotoxic active metabolite incorporated into DNA. On the other hand, a patient with high TPMT activity levels would need higher than standard doses of a thiopurine drug to respond well to the therapy, since a large amount of the drug is being inactivated before it can be converted to 6-TGNs [95].

EGFR

The epidermal growth factor receptor (*EGFR* or *HER1*) is a member of the ErbB family of tyrosine kinases that also includes HER2. Once the ligand binds to the receptor, the receptor undergoes homodimerization or heterodimerization. This activation via phosphorylation initiates signaling to downstream pathways such as PI3K/AKT and RAS/RAF/MAPK which in turn regulates cell proliferation and apoptosis.

EGFR is expressed in some lung cancers and is the target of newly developed small molecule drugs, including gefitinib and erlotinib. It has been shown that tumors that respond to these tyrosine kinase (TK) inhibitors (TKIs) contain somatic mutations in the EGFR TK domain [96–98]. Up to 90% of mutations in non-small cell lung cancers (NSCLC) can be attributed to two mutations, an inframe deletion of exon 19 and a single point mutation in exon 21 (T2573G). While these mutations are associated with a favorable response to the new TKIs, other mutations are associated with a poor response and potential resistance to these therapeutics (for example, in-frame insertion at exon 20 confers resistance). Mutation analysis in the *EGFR* represents a new application for molecular diagnostics as some mutations confer a favorable response, while others are not so favorable. It becomes critical to the management of the NSCLC patient that the clinical molecular diagnostics laboratory be able to identify such mutations on a routine basis.

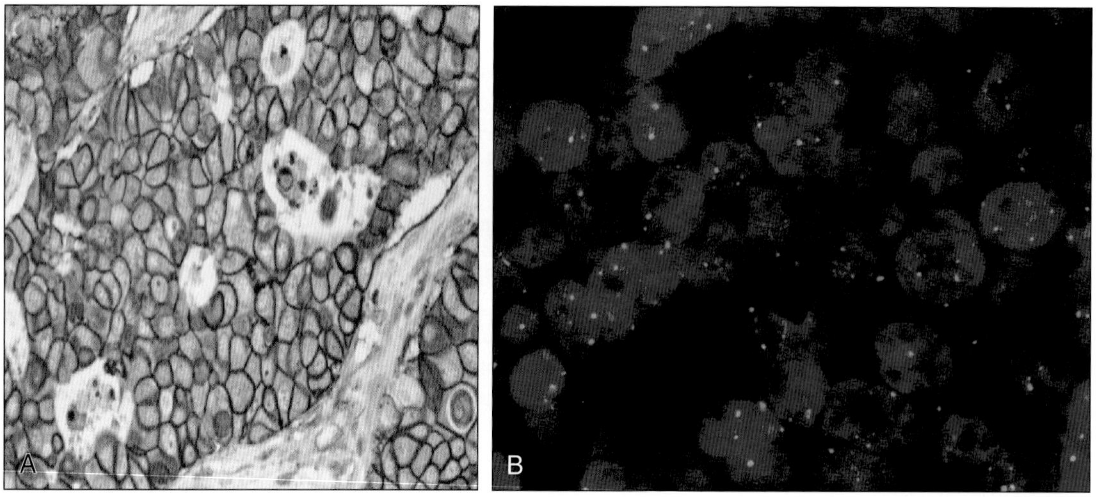

Figure 30.7 (A) Determination of HER2 status in breast cancer cells using immunohistochemistry and (B) fluorescence *in situ* hybridization (FISH).

CONCLUSION

Our knowledge base of PGx and clinical applications of such testing are progressing at record speeds. Technology is allowing us to perform these tests routinely in the clinical laboratory. Currently, PGx testing can be performed to detect polymorphisms in the genes for metabolizing enzymes and some drug targets. Clearly, one growing application in cancer patients is to detect polymorphisms and mutations associated with responses to newly developed small molecule targeted therapies. The ultimate goal of these efforts is to truly provide a "personalized medicine" approach to patient management for the purpose of eliminating ADRs, selecting more efficacious therapeutics, and improving the overall well-being of the patient.

REFERENCES

1. Lander ES, Linton LM, Birren B, et al. Initial sequencing and analysis of the human genome. *Nature.* 2001;409:860–921.
2. Venter JC, Adams MD, Myers EW, et al. The sequence of the human genome. *Science.* 2001;291:1304–1351.
3. Evans WE, McLeod HL. Pharmacogenomics—Drug disposition, drug targets, and side effects. *N Engl J Med.* 2003;348:538–549.
4. Kirchheiner J, Fuhr U, Brockmoller J. Pharmacogenetics-based therapeutic recommendations—Ready for clinical practice. *Nat Rev Drug Discov.* 2005;4(8):639–647.
5. Long RM. Planning for a national effort to enable and accelerate discoveries in pharmacogenetics: The NIH Pharmacogenetics Research Network. *Clin Pharmacol Ther.* 2007;81(3):450–454.
6. Shastry BS. Pharmacogenetics and the concept of individualized medicine. *Pharmacogenomics J.* 2006;6(1):16–21.
7. Bailey DS, Bondar A, Furness LM. Pharmacogenomics—It's not just pharmacogenetics. *Curr Opin Biotechnol.* 1998;9(6):595–601.
8. de Leon J, Armstrong SC, Cozza KL. Clinical guidelines for psychiatrists for the use of pharmacogenetic testing for CYP450 2D6 and CYP450 2C19. *Psychosomatics.* 2006;47(1):75–85.
9. Weinshilboum R, Wang L. Pharmacogenomics: Bench to bedside. *Nat Rev Drug Discov.* 2004;3(9):739–748.
10. Tsongalis GJ, Coleman WB. Clinical genotyping: The need for interrogation of single nucleotide polymorphisms and mutations in the clinical laboratory. *Clin Chim Acta.* 2006;363(1–2):127–137.
11. Nebert DW, Ingelman-Sundberg M, Daly AK. Genetic epidemiology of environmental toxicity and cancer susceptibility: Human allelic polymorphisms in drug-metabolizing enzyme genes, their functional importance, and nomenclature issues. *Drug Metab Rev.* 1999;31(2):467–487.
12. Lazarou J, Pomeranz BH, Corey PN. Incidence of adverse drug reactions in hospitalized patients: A meta-analysis of prospective studies. *JAMA.* 1998;279(15):1200–1205.
13. Lee HK, Lewis LD, Tsongalis GJ, et al. Negative urine opioid screening caused by rifampin-mediated induction of oxycodone hepatic metabolism. *Clin Chim Acta.* 2006;367(1–2):196–200.
14. Vogel F. Moderne probleme der Humangenetik. *Ergeb Inn Med Kinderheilkd.* 1959;12:52–125.
15. Nebert DW. Pharmacogenetics and pharmacogenomics: Why is this relevant to the clinical geneticist. *Clin Genet.* 1999;56(4):247–258.
16. Weber WW. Populations and genetic polymorphisms. *Mol Diagn.* 1999;4(4):299–307.
17. Mager J, Glaser G, Razin A, et al. Metabolic effects of pyrimidines derived from fava bean glycosides on human erythrocytes deficient in glucose-6-phosphate dehydrogenase. *Biochem Biophys Res Commun.* 1965;20(2):235–240.
18. Podda M, Fiorelli G, Ideo G, et al. In-vitro-effect of a fava bean extract and of its fractions on reduced glutathione in glucose-6-phosphate dehydrogenase deficient red cells. *Folia Haematol Int Mag Klin Morphol Blutforsch.* 1969;91(1):51–55.
19. Snyder L. Studies in human inheritance. The inheritance of taste deficiency in man. *Ohio J Sci.* 1932;32:436–468.
20. Evans DA. N-acetyltransferase. *Pharmacol Ther.* 1989;42(2):157–234.
21. Alving AS, Carson PE, Flanagan CL, et al. Enzymatic deficiency in primaquine-sensitive erythrocytes. *Science.* 1956;124(3220):484–485.
22. Hammer W, Sjoqvist F. Plasma levels of monomethylated tricyclic antidepressants during treatment with imipramine-like compounds. *Life Sci.* 1967;6(17):1895–1903.
23. Evans WE, Relling MV. Pharmacogenomics: Translating functional genomics into rational therapeutics. *Science.* 1999;286(5439):487–491.
24. Hughes HB, Biehl JP, Jones AP, et al. Metabolism of isoniazid in man as related to the occurrence of peripheral neuritis. *Am Rev Tuberc.* 1954;70:266–273.
25. Kalow W. Familial incidence of low pseudocholinesterase level. *Lancet.* 1956;211:576–577.

26. Weinshilboum R. Inheritance and drug response. *N Engl J Med.* 2003;348(6):529–537.
27. Linder MW, Prough RA, Valdes R, Jr. Pharmacogenetics: A laboratory tool for optimizing therapeutic efficiency. *Clin Chem.* 1997;43(2):254–266.
28. Schur BC, Bjerke J, Nuwayhid N, et al. Genotyping of cytochrome P450 2D6*3 and *4 mutations using conventional PCR. *Clin Chim Acta.* 2001;308(1–2):25–31.
29. Lucas T, Losert D, Allen M, et al. Combination allele-specific real-time PCR for differentiation of beta 2-adrenergic receptor coding single-nucleotide polymorphisms. *Clin Chem.* 2004;50(4):769–772.
30. Jain KK. Applications of AmpliChip CYP450. *Mol Diagn.* 2005;9(3):119–127.
31. Vairavan R. AutoGenomics, Inc. *Pharmacogenomics.* 2004;5(5):585–588.
32. Bortolin S, Black M, Modi H, et al. Analytical validation of the tag-it high-throughput microsphere-based universal array genotyping platform: Application to the multiplex detection of a panel of thrombophilia-associated single-nucleotide polymorphisms. *Clin Chem.* 2004;50(11):2028–2036.
33. Gardiner SJ, Begg EJ. Pharmacogenetics, drug-metabolizing enzymes, and clinical practice. *Pharmacol Rev.* 2006;58(3):521–590.
34. Ingelman-Sundberg M. Pharmacogenetics of cytochrome P450 and its applications in drug therapy: The past, present and future. *Trends Pharmacol Sci.* 2004;25(4):193–200.
35. Ingelman-Sundberg M. Genetic polymorphisms of cytochrome P450 2D6 (CYP2D6): Clinical consequences, evolutionary aspects and functional diversity. *Pharmacogenomics J.* 2005;5(1):6–13.
36. Heim MH, Meyer UA. Evolution of a highly polymorphic human cytochrome P450 gene cluster: CYP2D6. *Genomics.* 1992;14(1):49–58.
37. Ledesma MC, Agundez JA. Identification of subtypes of CYP2D gene rearrangements among carriers of CYP2D6 gene deletion and duplication. *Clin Chem.* 2005;51(6):939–943.
38. Andersson T, Flockhart DA, Goldstein DB, et al. Drug-metabolizing enzymes: Evidence for clinical utility of pharmacogenomic tests. *Clin Pharmacol Ther.* 2005;78(6):559–581.
39. D'Andrea G, D'Ambrosio R, Margaglione M. Oral anticoagulants: Pharmacogenetics relationship between genetic and nongenetic factors. *Blood Rev.* 2008;22(3):127–140.
40. Higashi MK, Veenstra DL, Kondo LM, et al. Association between CYP2C9 genetic variants and anticoagulation-related outcomes during warfarin therapy. *JAMA.* 2002;287(13):1690–1698.
41. Sanderson S, Emery J, Higgins J. CYP2C9 gene variants, drug dose, and bleeding risk in warfarin-treated patients: A HuGEnet systematic review and meta-analysis. *Genet Med.* 2005;7(2):97–104.
42. Li T, Chang CY, Jin DY, et al. Identification of the gene for vitamin K epoxide reductase. *Nature.* 2004;427(6974):541–544.
43. Rost S, Fregin A, Ivaskevicius V, et al. Mutations in VKORC1 cause warfarin resistance and multiple coagulation factor deficiency type 2. *Nature.* 2004;427(6974):537–541.
44. Gage BF, Eby C, Johnson JA, et al. Use of pharmacogenetic and clinical factors to predict the therapeutic dose of warfarin. *Clin Pharmacol Ther.* 2008;84(3):326–331.
45. Sconce EA, Khan TI, Wynne HA, et al. The impact of CYP2C9 and VKORC1 genetic polymorphism and patient characteristics upon warfarin dose requirements: Proposal for a new dosing regimen. *Blood.* 2005;106(7):2329–2333.
46. Wu AHB. Use of genetic and nongenetic factors in warfarin dosing algorithms. *Pharmacogenomics.* 2007;8(7):851–861.
47. Wu AHB, Wang P, Smith A, et al. Dosing algorithm for warfarin using CYP2C9 and VKORC1 genotyping from a multi-ethnic population: Comparison with other equations. *Pharmacogenomics.* 2008;9(2):169–178.
48. Zhu Y, Shennan M, Reynolds KK, et al. Estimation of warfarin maintenance dose based on VKORC1 (-1639 G>A) and CYP2C9 genotypes. *Clin Chem.* 2007;53(7):1199–1205.
49. Thompson CA. FDA encourages genetics-aided warfarin dosing. *Am J Health Syst Pharm.* 2007;64(19):1994–1996.
50. Wasserman E, Myara A, Lokiec F, et al. Severe CPT-11 toxicity in patients with Gilbert's syndrome: Two case reports. *Ann Oncol.* 1997;8(10):1049–1051.
51. Hartmann JT, Lipp HP. Camptothecin and podophyllotoxin derivatives: Inhibitors of topoisomerase I and II — mechanisms of action, pharmacokinetics and toxicity profile. *Drug Saf.* 2006;29(3):209–230.
52. Humerickhouse R, Lohrbach K, Li L, et al. Characterization of CPT-11 hydrolysis by human liver carboxylesterase isoforms hCE-1 and hCE-2. *Cancer Res.* 2000;60(5): 1189–1192.
53. Maitland ML, Vasisht K, Ratain MJ. TPMT, UGT1A1 and DPYD: Genotyping to ensure safer cancer therapy. *Trends Pharmacol Sci.* 2006;27(8):432–437.
54. Borst P, Schinkel AH, Smit JJ, et al. Classical and novel forms of multidrug resistance and the physiological functions of P-glycoproteins in mammals. *Pharmacol Ther.* 1993;60(2):289–299.
55. Kusuhara H, Suzuki H, Sugiyama Y. The role of P-glycoprotein and canalicular multispecific organic anion transporter in the hepatobiliary excretion of drugs. *J Pharm Sci.* 1998;87(9): 1025–1040.
56. Tanigawara Y. Role of P-glycoprotein in drug disposition. *Ther Drug Monit.* 2000;22(1):137–140.
57. Ambudkar SV, Dey S, Hrycyna CA, et al. Biochemical, cellular, and pharmacological aspects of the multidrug transporter. *Annu Rev Pharmacol Toxicol.* 1999;39:361–398.
58. Urquhart BL, Tirona RG, Kim RB. Nuclear receptors and the regulation of drug-metabolizing enzymes and drug transporters: Implications for interindividual variability in response to drugs. *J Clin Pharmacol.* 2007;47(5):566–578.
59. Higgins CF, Callaghan R, Linton KJ, et al. Structure of the multidrug resistance P-glycoprotein. *Semin Cancer Biol.* 1997;8(3): 135–142.
60. Rosenberg MF, Callaghan R, Ford RC, et al. Structure of the multidrug resistance P-glycoprotein to 2.5 nm resolution determined by electron microscopy and image analysis. *J Biol Chem.* 1997;272(16):10685–10694.
61. Sakaeda T, Nakamura T, Okumura K. MDR1 genotype-related pharmacokinetics and pharmacodynamics. *Biol Pharm Bull.* 2002;25(11):1391–1400.
62. Hoffmeyer S, Burk O, von Richter O, et al. Functional polymorphisms of the human multidrug-resistance gene: Multiple sequence variations and correlation of one allele with P-glycoprotein expression and activity in vivo. *Proc Natl Acad Sci USA.* 2000;97(7):3473–3478.
63. Jamroziak K, Robak T. Pharmacogenomics of MDR1/ABCB1 gene: The influence on risk and clinical outcome of haematological malignancies. *Hematology.* 2004;9(2):91–105.
64. Jamroziak K, Balcerczak E, Cebula B, et al. No influence of 3435C>T ABCB1 (MDR1) gene polymorphism on risk of adult acute myeloid leukemia and P-glycoprotein expression in blast cells. *Ther Drug Monit.* 2006;28(5):707–711.
65. Schaich M, Soucek S, Thiede C, et al. MDR1 and MRP1 gene expression are independent predictors for treatment outcome in adult acute myeloid leukaemia. *Br J Haematol.* 2005;128(3): 324–332.
66. van der Holt B, Van den Heuvel-Eibrink MM, Van Schaik RH, et al. ABCB1 gene polymorphisms are not associated with treatment outcome in elderly acute myeloid leukemia patients. *Clin Pharmacol Ther.* 2006;80(5):427–439.
67. van der Kolk DM, de Vries EG, Noordhoek L, et al. Activity and expression of the multidrug resistance proteins P-glycoprotein, MRP1, MRP2, MRP3 and MRP5 in de novo and relapsed acute myeloid leukemia. *Leukemia.* 2001;15(10):1544–1553.
68. Kimchi-Sarfaty C, Oh JM, Kim IW, et al. A "silent" polymorphism in the MDR1 gene changes substrate specificity. *Science.* 2007;315(5811):525–528.
69. Rollason V, Samer C, Piguet V, et al. Pharmacogenetics of analgesics: Toward the individualization of prescription. *Pharmacogenomics.* 2008;9(7):905–933.
70. Landau R, Cahana A, Smiley RM, et al. Genetic variability of mu-opioid receptor in an obstetric population. *Anesthesiology.* 2004;100(4):1030–1033.
71. Bond C, LaForge KS, Tian M, et al. Single-nucleotide polymorphism in the human mu opioid receptor gene alters beta-endorphin binding and activity: Possible implications for opiate addiction. *Proc Natl Acad Sci USA.* 1998;95(16):9608–9613.
72. Fillingim RB, Kaplan L, Staud R, et al. The A118G single nucleotide polymorphism of the mu-opioid receptor gene (OPRM1) is associated with pressure pain sensitivity in humans. *J Pain.* 2005;6(3):159–167.

73. Janicki PK, Schuler G, Francis D, et al. A genetic association study of the functional A118G polymorphism of the human mu-opioid receptor gene in patients with acute and chronic pain. *Anesth Analg.* 2006;103(4):1011–1017.
74. Oertel B, Lotsch J. Genetic mutations that prevent pain: Implications for future pain medication. *Pharmacogenomics.* 2008;9(2): 179–194.
75. Landau R. Pharmacogenetics and obstetric anesthesia. *Anesthesiol Clin.* 2008;26(1):183–195, 8–9.
76. Shankar DB, Li J, Tapang P, et al. ABT-869, a multitargeted receptor tyrosine kinase inhibitor: Inhibition of FLT3 phosphorylation and signaling in acute myeloid leukemia. *Blood.* 2007;109 (8):3400–3408.
77. Goetz MP, Rae JM, Suman VJ, et al. Pharmacogenetics of tamoxifen biotransformation is associated with clinical outcomes of efficacy and hot flashes. *J Clin Oncol.* 2005;23(36):9312–9318.
78. van Schaik RH. CYP450 pharmacogenetics for personalizing cancer therapy. *Drug Resist Updat.* 2008;11(3):77–98.
79. Group E. Tamoxifen for early breast cancer: An overview of the randomised trials. Early Breast Cancer Trialists' Collaborative Group. *Lancet.* 1998;351(9114):1451–1467.
80. Poon GK, Chui YC, McCague R, et al. Analysis of phase I and phase II metabolites of tamoxifen in breast cancer patients. *Drug Metab Dispos.* 1993;21(6):1119–1124.
81. Klein TE, Chang JT, Cho MK, et al. Integrating genotype and phenotype information: An overview of the PharmGKB project. Pharmacogenetics Research Network and Knowledge Base. *Pharmacogenomics J.* 2001;1(3):167–170.
82. Jin Y, Desta Z, Stearns V, et al. CYP2D6 genotype, antidepressant use, and tamoxifen metabolism during adjuvant breast cancer treatment. *J Natl Cancer Inst.* 2005;97(1):30–39.
83. Sachse C, Brockmoller J, Bauer S, et al. Cytochrome P450 2D6 variants in a Caucasian population: Allele frequencies and phenotypic consequences. *Am J Hum Genet.* 1997;60(2):284–295.
84. Press MF, Bernstein L, Thomas PA, et al. HER-2/neu gene amplification characterized by fluorescence in situ hybridization: Poor prognosis in node-negative breast carcinomas. *J Clin Oncol.* 1997;15(8):2894–2904.
85. Slamon DJ, Godolphin W, Jones LA, et al. Studies of the HER-2/ neu proto-oncogene in human breast and ovarian cancer. *Science.* 1989;244(4905):707–712.
86. Tandon AK, Clark GM, Chamness GC, et al. HER-2/neu oncogene protein and prognosis in breast cancer. *J Clin Oncol.* 1989; 7(8):1120–1128.
87. Jones SE. Metastatic breast cancer: The treatment challenge. *Clin Breast Cancer.* 2008;8(3):224–233.
88. Moasser MM. Targeting the function of the HER2 oncogene in human cancer therapeutics. *Oncogene.* 2007;26(46):6577–6592.
89. Laudadio J, Quigley DI, Tubbs R, et al. HER2 testing: A review of detection methodologies and their clinical performance. *Expert Rev Mol Diagn.* 2007;7(1):53–64.
90. Wolff AC, Hammond ME, Schwartz JN, et al. American Society of Clinical Oncology/College of American Pathologists guideline recommendations for human epidermal growth factor receptor 2 testing in breast cancer. *J Clin Oncol.* 2007;25(1):118–145.
91. Lennard L, Rees CA, Lilleyman JS, et al. Childhood leukaemia: A relationship between intracellular 6-mercaptopurine metabolites and neutropenia. 1983. *Br J Clin Pharmacol.* 2004;58(7):S867–871; discussion S872–874.
92. Salavaggione OE, Wang L, Wiepert M, et al. Thiopurine S-methyltransferase pharmacogenetics: Variant allele functional and comparative genomics. *Pharmacogenet Genomics.* 2005;15(11):801–815.
93. Weinshilboum RM, Sladek SL. Mercaptopurine pharmacogenetics: Monogenic inheritance of erythrocyte thiopurine methyltransferase activity. *Am J Hum Genet.* 1980;32(5):651–662.
94. Yates CR, Krynetski EY, Loennechen T, et al. Molecular diagnosis of thiopurine S-methyltransferase deficiency: Genetic basis for azathioprine and mercaptopurine intolerance. *Ann Intern Med.* 1997;126(8):608–614.
95. Lennard L, Lilleyman JS, Van Loon J, et al. Genetic variation in response to 6-mercaptopurine for childhood acute lymphoblastic leukaemia. *Lancet.* 1990;336(8709): 225–229.
96. Lynch TJ, Bell DW, Sordella R, et al. Activating mutations in the epidermal growth factor receptor underlying responsiveness of non-small-cell lung cancer to gefitinib. *N Engl J Med.* 2004;350 (21):2129–2139.
97. Paez JG, Janne PA, Lee JC, et al. EGFR mutations in lung cancer: Correlation with clinical response to gefitinib therapy. *Science.* 2004;304(5676):1497–1500.
98. Pao W, Miller V, Zakowski M, et al. EGF receptor gene mutations are common in lung cancers from "never smokers" and are associated with sensitivity of tumors to gefitinib and erlotinib. *Proc Natl Acad Sci USA.* 2004;101(36):13306–13311.

Index

A

B

C

D

H

I

M

N

O

P

Q

R

S

T

U

V

W

X

Y